VOLUME FIVE HUNDRED AND SIXTY EIGHT

METHODS IN ENZYMOLOGY

Intermediate Filament Proteins

METHODS IN ENZYMOLOGY

VOLUME FIVE HUNDRED AND SIXTY EIGHT

METHODS IN ENZYMOLOGY

Intermediate Filament Proteins

Edited by

M. BISHR OMARY

Department of Molecular & Integrative Physiology, Department of Medicine, University of Michigan, and VA Ann Arbor Healthcare System, Ann Arbor, Michigan, USA

RONALD K.H. LIEM

Department of Pathology and Cell Biology, Taub Institute for Research on Alzheimer's Disease and the Aging Brain, Columbia University College of Physicians and Surgeons, New York, USA

[redacted]

ELSEVIER

AMSTERDAM • BOSTON • HEIDELBERG • LONDON
NEW YORK • OXFORD • PARIS • SAN DIEGO
SAN FRANCISCO • SINGAPORE • SYDNEY • TOKYO
Academic Press is an imprint of Elsevier

Academic Press is an imprint of Elsevier
50 Hampshire Street, 5th Floor, Cambridge, MA 02139, USA
525 B Street, Suite 1800, San Diego, CA 92101–4495, USA
The Boulevard, Langford Lane, Kidlington, Oxford OX5 1GB, UK
125 London Wall, London, EC2Y 5AS, UK

First edition 2016

Notices

Knowledge and best practice in this field are constantly changing. As new research and experience broaden our understanding, changes in research methods, professional practices, or medical treatment may become necessary.

Practitioners and researchers must always rely on their own experience and knowledge in evaluating and using any information, methods, compounds, or experiments described herein. In using such information or methods they should be mindful of their own safety and the safety of others, including parties for whom they have a professional responsibility.

To the fullest extent of the law, neither the Publisher nor the authors, contributors, or editors, assume any liability for any injury and/or damage to persons or property as a matter of products liability, negligence or otherwise, or from any use or operation of any methods, products, instructions, or ideas contained in the material herein.

ISBN: 978-0-12-803470-5
ISSN: 0076-6879

DEDICATION

We dedicate this book to four major leaders in the intermediate filament field who have essentially established the field and made seminal contributions that gave us the joy of being able to work in this exciting area: Drs. Elaine Fuchs, Werner Franke, Robert Goldman, and Klaus Weber. Dr. Elaine Fuchs' laboratory cloned several of the keratins genes, and by using reverse genetics identified epidermolysis bullosa simplex as the first disease to be caused by an intermediate filaments gene mutation; Drs. Werner Franke's, Robert Goldman's, and Klaus Weber's individual laboratories initiated many of the studies on intermediate filaments and taught us fundamentals of the field. Dr. Fuchs has moved her interests in skin biology to stem cells, but she remains an ardent supporter of the field. Dr. Weber is now Professor Emeritus at the Max Planck Institute, while Dr. Franke is the Helmholtz Professor at the German Cancer Center in Heidelberg. Drs. Franke and Weber remain an inspiration to us and other investigators in the field. Fortunately for us, Dr. Goldman continues to be at the forefront as a very active investigator in the intermediate filament field, and is a coauthor of the chapter on vimentin. Finally, we also acknowledge the late Peter Steinert, who left us much too soon. Dr. Steinert made many seminal contributions to the field, and was the first Chair of the Gordon Conference on Intermediate Filaments (which he initiated) that remains the major meeting for researchers to exchange ideas and present new developments in the exciting intermediate filament field.

CONTENTS

Contributors *xv*
Preface *xxiii*
Acknowledgments *xxv*

Part I
General Methods to Study IF Proteins

1. How to Study Intermediate Filaments in Atomic Detail **3**
Anastasia A. Chernyatina, John F. Hess, Dmytro Guzenko, John C. Voss, and Sergei V. Strelkov

1. Introductory Remarks 4
2. Bioinformatics 5
3. X-ray Crystallography 7
4. Electron Paramagnetic Resonance with Site-Directed Spin Labeling 21
5. Concluding Remarks 27

Acknowledgments 28
References 28

2. Mechanical Properties of Intermediate Filament Proteins **35**
Elisabeth E. Charrier and Paul A. Janmey

1. Introduction 36
2. Viscoelasticity of Purified IFs *In Vitro* 36
3. IFs and the Mechanical Properties of Cells 42
4. Conclusion 53

Acknowledgments 54
References 54

3. Multidimensional Monitoring of Keratin Intermediate Filaments in Cultured Cells and Tissues **59**
Nicole Schwarz, Marcin Moch, Reinhard Windoffer, and Rudolf E. Leube

1. Introduction 60
2. 3D Imaging of Keratin Intermediate Filaments in Cultured Cells 62
3. 3D Imaging of Keratin Intermediate Filaments in Murine Preimplantation Embryos 75
4. Outlook 80

Acknowledgments 81
References 81

4. Phospho-Specific Antibody Probes of Intermediate Filament Proteins **85**

Hidemasa Goto, Hiroki Tanaka, Kousuke Kasahara, and Masaki Inagaki

1. Introduction 86
2. Production of Site- and Phosphorylation State-Specific Antibodies 91
3. Characterization of Site- and Phosphorylation State-Specific Antibodies 97
4. Immunocytochemistry 100
5. Other Applications 105
6. Conclusions 105

Acknowledgments 106

References 106

5. Assays for Posttranslational Modifications of Intermediate Filament Proteins **113**

Natasha T. Snider and M. Bishr Omary

1. Introduction 114
2. Extraction of IF Proteins from Tissues and Cells for Biochemical Analysis of IF PTMs 120
3. Methods for Monitoring Specific PTMs on IF Proteins 126
4. Conclusions 135

Acknowledgments 135

References 135

6. Immunofluorescence and Immunohistochemical Detection of Keratins **139**

Cornelia Stumptner, Margit Gogg-Kamerer, Christian Viertler, Helmut Denk, and Kurt Zatloukal

1. Introduction 140
2. Impact of Molecular Structure on the Detection of Keratins in Frozen Tissue Samples 142
3. Keratin Immunohistochemistry on Paraffin-Embedded Tissue 146
4. Conclusions 159
5. Pearls and Pitfalls 159

Acknowledgments 160

References 160

7. High-Throughput Screening for Drugs that Modulate Intermediate Filament Proteins **163**

Jingyuan Sun, Vincent E. Groppi, Honglian Gui, Lu Chen, Qing Xie, Li Liu, and M. Bishr Omary

1. Overview of Intermediate Filaments and Their Associated Diseases 164
2. Current Targeted Therapeutic Approaches for IF-Pathies 165

3. Unbiased Drug Screening to Target IF Mutations 170
4. High-Throughput Steps to Identify Drugs that Target Intermediate Filaments 171
5. Available Libraries and Vendors for Drug Screening 179
6. Pearls and Pitfalls 182
Acknowledgments 183
References 183

8. The Use of Withaferin A to Study Intermediate Filaments 187

Royce Mohan and Paola Bargagna-Mohan

1. Introduction 188
2. WFA: A Novel Chemical Tool for IF Biology and Pharmacology 194
3. Targeting IFs 197
4. Primary Cell Culture Systems 207
5. Ocular Injury Model of Fibrosis 209
6. Techniques 211
7. Perspectives 213
Acknowledgments 214
References 214

9. Assays to Study Consequences of Cytoplasmic Intermediate Filament Mutations: The Case of Epidermal Keratins 219

Tong San Tan, Yi Zhen Ng, Cedric Badowski, Tram Dang, John E.A. Common, Lukas Lacina, Ildikó Szeverényi, and E. Birgitte Lane

1. Introduction 221
2. Identification of Keratin Mutations 223
3. Developing Experimental Model Systems 227
4. Keratin Mutations in Stress Assays 238
5. Conclusion, Pearls, and Pitfalls 246
Acknowledgments 248
References 248

10. Using Data Mining and Computational Approaches to Study Intermediate Filament Structure and Function 255

David A.D. Parry

1. Introduction 256
2. Methodology 259
3. Summary 270
References 273

Part II
Mammalian IF Proteins

11. Isolation and Analysis of Keratins and Keratin-Associated Proteins from Hair and Wool **279**

Santanu Deb-Choudhury, Jeffrey E. Plowman, and Duane P. Harland

1. Introduction 280
2. Chemical Extraction of Whole Fibers 281
3. Isolation and Digestion of Other Components of the Fiber 286
4. Combination of Sodium Deoxycholate and C18 Empore™ for an Efficient Extraction of Keratin Peptides from Gel-Resolved Keratin Proteins 290
5. Ionic Liquid-Assisted Extraction of Fiber Keratins 294
6. Conclusions 299

References 300

12. Skin Keratins **303**

Fengrong Wang, Abigail Zieman, and Pierre A. Coulombe

1. Introduction 305
2. Collection of Mouse Skin Tissue for Analysis 311
3. Cell Culture Studies 319
4. *In Vitro* Methods to Study Keratin Proteins 330
5. Pearls and Pitfalls 337
6. Conclusions 340

Acknowledgments 340

Appendix 340

References 346

13. Simple Epithelial Keratins **351**

Pavel Strnad, Nurdan Guldiken, Terhi O. Helenius, Julia O. Misiorek, Joel H. Nyström, Iris A.K. Lähdeniemi, Jonas S.G. Silvander, Deniz Kuscuoglu, and Diana M. Toivola

1. Introduction 352
2. Isolation of SEKs 354
3. Studying SEKs in Cell Culture 357
4. Genetic Mouse Models of SEKs 365
5. MDBs—Keratin Aggregates in Liver Disease 372
6. SEK Variant Detection in Humans 378
7. Conclusions 379

Acknowledgments 380

References 380

14. Methods for Determining the Cellular Functions of Vimentin Intermediate Filaments **389**

Karen M. Ridge, Dale Shumaker, Amélie Robert, Caroline Hookway, Vladimir I. Gelfand, Paul A. Janmey, Jason Lowery, Ming Guo, David A. Weitz, Edward Kuczmarski, and Robert D. Goldman

1. Introduction 391
2. Disruption of Vimentin IFs 391
3. Analysis of Vimentin Dynamics Using Photoactivatable and Photoconvertible Protein Tags 398
4. Investigating Vimentin–Protein Interactions 405
5. Investigating the Mechanical Properties of Vimentin IF Networks 411
6. Investigating the Role of Vimentin IFs in Cell Mechanics 415
7. Conclusion 420

Acknowledgments 421

References 421

15. Strategies to Study Desmin in Cardiac Muscle and Culture Systems **427**

Antigoni Diokmetzidou, Mary Tsikitis, Sofia Nikouli, Ismini Kloukina, Elsa Tsoupri, Stamatis Papathanasiou, Stelios Psarras, Manolis Mavroidis, and Yassemi Capetanaki

1. Introduction: Desmin's Scaffold—The Fine-Tuning Machinery of Striated Muscle 428
2. Cell Systems Used for Desmin Studies 430
3. Methods for Desmin Detection (Expression and Localization) 440
4. Methods for Desmin Isolation and Assembly 447
5. Model Animals for Desmin Mutation Studies 451
6. Identification of Desmin-Associated Proteins 452
7. Pearls and Pitfalls 454

References 454

16. Genetic Manipulation of Neurofilament Protein Phosphorylation **461**

Maria R. Jones, Eric Villalón, and Michael L. Garcia

1. Introduction 462
2. Genetic Manipulation of Mice 464
3. Steps for Creating a Gene-Targeted Mutant of a Mouse 465
4. Conclusions and Applications 471

References 473

17. α-Internexin and Peripherin: Expression, Assembly, Functions, and Roles in Disease **477**
Jian Zhao and Ronald K.H. Liem

1. α-Internexin 478
2. Peripherin 489
3. Conclusion 497
References 498

18. Studying Nestin and its Interrelationship with Cdk5 **509**
Julia Lindqvist, Num Wistbacka, and John E. Eriksson

1. Introduction 510
2. Cellular Techniques to Study Nestin Functions in Cells 520
3. *In Vivo* Mouse Models to Study Nestin Function 528
4. Conclusions 530
Acknowledgments 530
References 530

19. Synemin: Molecular Features and the Use of Proximity Ligation Assay to Study Its Interactions **537**
Madhumita Paul and Omar Skalli

1. Synemin 538
2. Reagents and Tools to Study Synemin 545
3. PLA for *In Situ* Detection of Synemin Interaction with Binding Partners 546
4. Conclusions 552
References 552

20. Targeting Mitogen-Activated Protein Kinase Signaling in Mouse Models of Cardiomyopathy Caused by Lamin A/C Gene Mutations **557**
Antoine Muchir and Howard J. Worman

1. Introduction 558
2. Mouse Models of Cardiomyopathy Caused by *LMNA* Mutations 560
3. Altered MAP Kinase Signaling in Hearts of $Lmna^{H222P/H222P}$ Mice 563
4. MAP Kinase Inhibitor Treatment Studies in $Lmna^{H222P/H222P}$ Mice 568
5. Assessment of $Lmna^{H222P/H222P}$ Mice After Treatment with MAP Kinase Inhibitors 570
6. Conclusions 576
Acknowledgments 577
References 577

21. *In vivo, Ex Vivo,* and *In Vitro* Approaches to Study Intermediate Filaments in the Eye Lens **581**

Miguel Jarrin, Laura Young, Weiju Wu, John M. Girkin, and Roy A. Quinlan

1. Introduction 582
2. Models to Study IFs in the Eye Lens 586
3. Methods to Study IFs 594
4. Concluding Remarks 603

References 603

Part III
Non-Mammalian IF Protein Systems

22. Compartment-Specific Phosphorylation of Squid Neurofilaments **615**

Philip Grant and Harish C. Pant

1. Introduction 616
2. Squid Giant Fiber System 617
3. Isolation of Axoplasm 619
4. Squid NF Genes 621
5. Squid NF Antibodies 622
6. Developmental Regulation of NF Expression 622
7. NF-Associated Protein Kinases 627
8. Squid NFs Are Included in Compartment-Specific Multimeric Protein Complexes: P13suc1 Affinity Chromatography 628

References 630

23. Using *Xenopus* Embryos to Study Transcriptional and Posttranscriptional Gene Regulatory Mechanisms of Intermediate Filaments **635**

Chen Wang and Ben G. Szaro

1. Introduction 636
2. Preparation of Expression Plasmids 641
3. Microinjection of *Xenopus* Embryos 644
4. Assays for Gene Expression 649
5. Assays for Effects of *cis*-Regulatory Elements on RNA Processing, Trafficking, and Translation 651
6. Conclusions 654

Acknowledgments 654

References 654

24. Intermediate Filaments in *Caenorhabditis elegans* **661**

Noam Zuela and Yosef Gruenbaum

1. Introduction 662
2. Essential Roles of Cytoplasmic IFs in *C. elegans* 662
3. Essential Roles of Lamin in *C. elegans* 663
4. Assembly of *C. elegans* Lamins 668
5. Assembly of *C. elegans* IFs 670
6. Methods 670
7. Summary 676

Acknowledgments 677
References 677

25. Mechanical Probing of the Intermediate Filament-Rich *Caenorhabditis Elegans* Intestine **681**

Oliver Jahnel, Bernd Hoffmann, Rudolf Merkel, Olaf Bossinger, and Rudolf E. Leube

1. Introduction 682
2. Imaging of Intermediate Filaments in *C. elegans* Intestines by Epifluorescence Microscopy 686
3. Outline of Intestinal Rings in *C. elegans* by a Fluorescent Apical Junction Reporter 688
4. Dissection of Intestines and Vitality Testing 691
5. Experimental Setup for Micropipette Measurements 695
6. Outlook 702

Acknowledgments 703
References 703

26. Using *Drosophila* for Studies of Intermediate Filaments **707**

Jens Bohnekamp, Diane E. Cryderman, Dylan A. Thiemann, Thomas M. Magin, and Lori L. Wallrath

1. Introduction 708
2. Methods 711
3. Conclusions 723

References 723

Author Index *727*
Subject Index *775*

CONTRIBUTORS

Cedric Badowski
Institute of Medical Biology, Singapore

Paola Bargagna-Mohan
Department of Neuroscience, University of Connecticut Health Center, Farmington, Connecticut, USA

Jens Bohnekamp
Institute of Biology and Translational Center for Regenerative Medicine, University of Leipzig, Leipzig, Germany

Olaf Bossinger
Institute of Molecular and Cellular Anatomy, RWTH Aachen University, Aachen, Germany

Yassemi Capetanaki
Center of Basic Research, Biomedical Research Foundation, Academy of Athens, Athens, Greece

Elisabeth E. Charrier
Institute for Medicine and Engineering, University of Pennsylvania, Philadelphia, Pennsylvania, USA

Lu Chen
Department of Infectious Diseases, Ruijin Hospital, Jiaotong University School of Medicine, Shanghai, PR China

Anastasia A. Chernyatina
Department of Pharmaceutical and Pharmacological Sciences, KU Leuven, Leuven, Belgium

John E.A. Common
Institute of Medical Biology, Singapore

Pierre A. Coulombe
Department of Biochemistry and Molecular Biology, Bloomberg School of Public Health; Department of Biological Chemistry; Department of Dermatology, and Department of Oncology, School of Medicine, Johns Hopkins University, Baltimore, Maryland, USA

Diane E. Cryderman
Department of Biochemistry, University of Iowa, Iowa City, Iowa, USA

Tram Dang
Institute of Medical Biology, Singapore

Santanu Deb-Choudhury
Food and Bio-Based Products Group, AgResearch Ltd., Christchurch, New Zealand

Helmut Denk
Institute of Pathology, Medical University of Graz, Graz, Austria

Antigoni Diokmetzidou
Center of Basic Research, Biomedical Research Foundation, Academy of Athens, Athens, Greece

John E. Eriksson
Cell Biology, Biosciences, Faculty of Science and Engineering, and Turku Centre for Biotechnology, University of Turku and Åbo Akademi University, Turku, Finland

Michael L. Garcia
Department of Biological Sciences, and C.S. Bond Life Sciences Center, University of Missouri, Columbia, Missouri, USA

Vladimir I. Gelfand
Department of Cell and Molecular Biology, Northwestern University, Feinberg School of Medicine, Chicago, Illinois, USA

John M. Girkin
Centre for Advanced Instrumentation and Biophysical Sciences Institute, Department of Physics, The University of Durham, Durham, United Kingdom

Margit Gogg-Kamerer
Institute of Pathology, Medical University of Graz, Graz, Austria

Robert D. Goldman
Department of Cell and Molecular Biology, Northwestern University, Feinberg School of Medicine, Chicago, Illinois, USA

Hidemasa Goto
Division of Biochemistry, Aichi Cancer Center Research Institute, and Department of Cellular Oncology, Graduate School of Medicine, Nagoya University, Nagoya, Aichi, Japan

Philip Grant
CPR, NINDS, NIH, Bethesda, MD, USA

Vincent E. Groppi
Department of Pharmacology, The Center for Chemical Genomics, University of Michigan, Ann Arbor, Michigan, USA

Yosef Gruenbaum
Department of Genetics, The Alexander Silberman Institute of Life Sciences, The Hebrew University of Jerusalem, Jerusalem, Israel

Honglian Gui
Department of Molecular & Integrative Physiology; Department of Medicine, University of Michigan; VA Ann Arbor Healthcare System, Ann Arbor, Michigan, USA, and Department of Infectious Diseases, Ruijin Hospital, Jiaotong University School of Medicine, Shanghai, PR China

Nurdan Guldiken
Department of Internal Medicine III and IZKF, University Hospital Aachen, Aachen, Germany

Ming Guo
School of Engineering and Applied Sciences, Harvard University, Cambridge, Massachusetts, USA

Dmytro Guzenko
Department of Pharmaceutical and Pharmacological Sciences, KU Leuven, Leuven, Belgium

Duane P. Harland
Food and Bio-Based Products Group, AgResearch Ltd., Christchurch, New Zealand

Terhi O. Helenius
Faculty of Science and Engineering, Department of Biosciences, Cell Biology, Åbo Akademi University, and Turku Center for Disease Modeling, University of Turku, Turku, Finland

John F. Hess
Department of Cell Biology and Human Anatomy, University of California, Davis, California, USA

Bernd Hoffmann
Institute of Complex Systems, ICS-7: Biomechanics, Jülich, Germany

Caroline Hookway
Department of Cell and Molecular Biology, Northwestern University, Feinberg School of Medicine, Chicago, Illinois, USA

Masaki Inagaki
Division of Biochemistry, Aichi Cancer Center Research Institute, and Department of Cellular Oncology, Graduate School of Medicine, Nagoya University, Nagoya, Aichi, Japan

Oliver Jahnel
Institute of Molecular and Cellular Anatomy, RWTH Aachen University, Aachen, Germany

Paul A. Janmey
Institute for Medicine and Engineering, and Departments of Physiology and Physics & Astronomy, University of Pennsylvania, Philadelphia, Pennsylvania, USA

Miguel Jarrin
Integrative Cell Biology Laboratory, School of Biological and Biomedical Sciences, The University of Durham, Durham, United Kingdom

Maria R. Jones
Department of Biological Sciences, and C.S. Bond Life Sciences Center, University of Missouri, Columbia, Missouri, USA

Kousuke Kasahara
Division of Biochemistry, Aichi Cancer Center Research Institute, and Department of Oncology, Graduate School of Pharmaceutical Sciences, Nagoya City University, Nagoya, Aichi, Japan

Ismini Kloukina
Center of Basic Research, Biomedical Research Foundation, Academy of Athens, Athens, Greece

Edward Kuczmarski
Department of Cell and Molecular Biology, Northwestern University, Feinberg School of Medicine, Chicago, Illinois, USA

Deniz Kuscuoglu
Department of Internal Medicine III and IZKF, University Hospital Aachen, Aachen, Germany

Lukas Lacina
Institute of Medical Biology, Singapore

Iris A.K. Lähdeniemi
Faculty of Science and Engineering, Department of Biosciences, Cell Biology, Åbo Akademi University, and Turku Center for Disease Modeling, University of Turku, Turku, Finland

E. Birgitte Lane
Institute of Medical Biology, Singapore

Rudolf E. Leube
Institute of Molecular and Cellular Anatomy, RWTH Aachen University, Aachen, Germany

Ronald K.H. Liem
Department of Pathology and Cell Biology, Taub Institute for Research on Alzheimer's Disease and the Aging Brain, Columbia University College of Physicians and Surgeons, New York, USA

Julia Lindqvist
Cell Biology, Biosciences, Faculty of Science and Engineering, and Turku Centre for Biotechnology, University of Turku and Åbo Akademi University, Turku, Finland

Li Liu
Hepatology Unit, Department of Infectious Diseases, and Department of Radiation Oncology, Nanfang Hospital, Southern Medical University, Guangzhou, PR China

Jason Lowery
Division of Pulmonary and Critical Care Medicine, Chicago, Illinois, USA

Thomas M. Magin
Institute of Biology and Translational Center for Regenerative Medicine, University of Leipzig, Leipzig, Germany

Manolis Mavroidis
Center of Basic Research, Biomedical Research Foundation, Academy of Athens, Athens, Greece

Rudolf Merkel
Institute of Complex Systems, ICS-7: Biomechanics, Jülich, Germany

Julia O. Misiorek
Faculty of Science and Engineering, Department of Biosciences, Cell Biology, Åbo Akademi University, and Turku Center for Disease Modeling, University of Turku, Turku, Finland

Marcin Moch
Institute of Molecular and Cellular Anatomy, RWTH Aachen University, Aachen, Germany

Royce Mohan
Department of Neuroscience, University of Connecticut Health Center, Farmington, Connecticut, USA

Antoine Muchir
Center of Research in Myology, UPMC-Inserm UMR974, CNRS FRE3617, Institut de Myologie, G.H. Pitie Salpetriere, Paris Cedex, France

Yi Zhen Ng
Institute of Medical Biology, Singapore

Sofia Nikouli
Center of Basic Research, Biomedical Research Foundation, Academy of Athens, Athens, Greece

Joel H. Nyström
Faculty of Science and Engineering, Department of Biosciences, Cell Biology, Åbo Akademi University, and Turku Center for Disease Modeling, University of Turku, Turku, Finland

M. Bishr Omary
Department of Molecular & Integrative Physiology; Department of Medicine, University of Michigan, and VA Ann Arbor Healthcare System, Ann Arbor, Michigan, USA

Harish C. Pant
CPR, NINDS, NIH, Bethesda, MD, USA

Stamatis Papathanasiou
Center of Basic Research, Biomedical Research Foundation, Academy of Athens, Athens, Greece

David A.D. Parry
Institute of Fundamental Sciences and Riddet Institute, Massey University, Palmerston North, New Zealand

Madhumita Paul
Department of Biological Sciences, The University of Memphis, Memphis, Tennessee, USA

Jeffrey E. Plowman
Food and Bio-Based Products Group, AgResearch Ltd., Christchurch, New Zealand

Stelios Psarras
Center of Basic Research, Biomedical Research Foundation, Academy of Athens, Athens, Greece

Roy A. Quinlan
Integrative Cell Biology Laboratory, School of Biological and Biomedical Sciences, The University of Durham, Durham, United Kingdom

Karen M. Ridge
Division of Pulmonary and Critical Care Medicine; Department of Cell and Molecular Biology, Northwestern University, Feinberg School of Medicine, and Veterans Administration, Chicago, Illinois, USA

Amélie Robert
Department of Cell and Molecular Biology, Northwestern University, Feinberg School of Medicine, Chicago, Illinois, USA

Nicole Schwarz
Institute of Molecular and Cellular Anatomy, RWTH Aachen University, Aachen, Germany

Dale Shumaker
Division of Pulmonary and Critical Care Medicine, and Department of Cell and Molecular Biology, Northwestern University, Feinberg School of Medicine, Chicago, Illinois, USA

Jonas S.G. Silvander
Faculty of Science and Engineering, Department of Biosciences, Cell Biology, Åbo Akademi University, and Turku Center for Disease Modeling, University of Turku, Turku, Finland

Omar Skalli
Department of Biological Sciences, The University of Memphis, Memphis, Tennessee, USA

Natasha T. Snider
Department of Cell Biology and Physiology, University of North Carolina, Chapel Hill, North Carolina, USA

Sergei V. Strelkov
Department of Pharmaceutical and Pharmacological Sciences, KU Leuven, Leuven, Belgium

Pavel Strnad
Department of Internal Medicine III and IZKF, University Hospital Aachen, Aachen, Germany

Cornelia Stumptner
Institute of Pathology, Medical University of Graz, Graz, Austria

Jingyuan Sun
Department of Molecular & Integrative Physiology; Department of Medicine, University of Michigan; VA Ann Arbor Healthcare System, Ann Arbor, Michigan, USA; Hepatology Unit, Department of Infectious Diseases, and Department of Radiation Oncology, Nanfang Hospital, Southern Medical University, Guangzhou, PR China

Ben G. Szaro
Department of Biological Sciences, University at Albany, State University of New York, Albany, New York, USA

Ildikó Szeverényi
Institute of Medical Biology, Singapore

Tong San Tan
Institute of Medical Biology, Singapore

Hiroki Tanaka
Division of Biochemistry, Aichi Cancer Center Research Institute, Nagoya, Aichi, Japan

Dylan A. Thiemann
Department of Biochemistry, University of Iowa, Iowa City, Iowa, USA

Diana M. Toivola
Faculty of Science and Engineering, Department of Biosciences, Cell Biology, Åbo Akademi University, and Turku Center for Disease Modeling, University of Turku, Turku, Finland

Mary Tsikitis
Center of Basic Research, Biomedical Research Foundation, Academy of Athens, Athens, Greece

Elsa Tsoupri
Center of Basic Research, Biomedical Research Foundation, Academy of Athens, Athens, Greece

Christian Viertler
Institute of Pathology, Medical University of Graz, Graz, Austria

Eric Villalón
Department of Biological Sciences, and C.S. Bond Life Sciences Center, University of Missouri, Columbia, Missouri, USA

John C. Voss
Department of Cell Biology and Human Anatomy, University of California, Davis, California, USA

Lori L. Wallrath
Department of Biochemistry, University of Iowa, Iowa City, Iowa, USA

Chen Wang
Department of Biological Sciences, University at Albany, State University of New York, Albany, New York, USA

Fengrong Wang
Department of Biochemistry and Molecular Biology, Bloomberg School of Public Health, Johns Hopkins University, Baltimore, Maryland, USA

David A. Weitz
School of Engineering and Applied Sciences, and Department of Physics, Harvard University, Cambridge, Massachusetts, USA

Reinhard Windoffer
Institute of Molecular and Cellular Anatomy, RWTH Aachen University, Aachen, Germany

Num Wistbacka
Cell Biology, Biosciences, Faculty of Science and Engineering, and Turku Centre for Biotechnology, University of Turku and Åbo Akademi University, Turku, Finland

Howard J. Worman
Department of Medicine, and Department of Pathology and Cell Biology, College of Physicians and Surgeons, Columbia University, New York, USA

Weiju Wu
Integrative Cell Biology Laboratory, School of Biological and Biomedical Sciences, The University of Durham, Durham, United Kingdom

Qing Xie
Department of Infectious Diseases, Ruijin Hospital, Jiaotong University School of Medicine, Shanghai, PR China

Laura Young
Centre for Advanced Instrumentation and Biophysical Sciences Institute, Department of Physics, The University of Durham, Durham, United Kingdom

Kurt Zatloukal
Institute of Pathology, Medical University of Graz, Graz, Austria

Jian Zhao
Department of Pathology and Cell Biology, Taub Institute for Research on Alzheimer's Disease and the Aging Brain, Columbia University College of Physicians and Surgeons, New York, USA

Abigail Zieman
Department of Biochemistry and Molecular Biology, Bloomberg School of Public Health, Johns Hopkins University, Baltimore, Maryland, USA

Noam Zuela
Department of Genetics, The Alexander Silberman Institute of Life Sciences, The Hebrew University of Jerusalem, Jerusalem, Israel

PREFACE

We are honored to have the opportunity to coedit this *Methods in Enzymology* volume on Intermediate Filaments (IFs). An exciting aspect of our effort is that this volume comes with a companion, also published in January 2016 (coedited by Kathleen Wilson and Arnoud Sonnenberg) which focuses solely on IF-associated proteins. These two companion volumes are indeed complimentary, with the need for the second related volume simply reflecting the expansion of the IF field as more binding protein partners have become identified and studied.

IFs along with microfilaments and microtubules form the three filamentous systems in the cell. Microfilaments and microtubules are composed of the highly conserved actins and tubulins, respectively. There are six mammalian actin genes with each gene encoding one protein (Perrin & Ervasti, 2010) and six mammalian tubulin gene families (Oakley, 2000). In contrast, IFs vary widely in sequence and in size, and form a large family of proteins. Many researchers may only be familiar with individual subgroups of this large family of IF proteins, or may know specific subgroup members, such as keratins (epithelial cell or hair, which make up types I and II); desmin, glial fibrillary acidic protein, peripherin, and vimentin (type III); α-internexin, nestin, neurofilaments, syncoilin, and synemin (type IV); lamins (type V); and the lens beaded filament structural proteins 1 and 2 (type VI). In all, IFs encompass 70 human genes, with >70 related human diseases (Omary, 2009; Szeverenyi et al., 2008), that include among them 54 functional human keratin genes (Schweizer et al., 2006). Except for those in the IF field, not many may realize that all these proteins belong to a single family that share a prototype structure, with family members having unique cell and tissue distribution. The name *intermediate filaments* comes in part because of their intermediate size (10 nm) as compared to microfilaments (4–5 nm) and microtubules (25 nm), although their initial description compared them to actin and myosin filaments (Ishikawa, Bischoff, & Holtzer, 1968).

In reaching out to the outstanding group of authors who have published extensively in the area that they covered in their respective chapters, we had two primary goals in mind. First, we aimed to assemble a volume that serves the community of investigators at-large who are either working directly in the IF field or whose work is bringing them to the IF field. Second, we wanted to put together not only an update but also a somewhat unique volume, as compared with the most recent IF-related methods book that was published in 2004 (*Methods in Cell Biology*, Volume 78, edited by Bishr Omary and Pierre Coulombe). Although some of the tools have remained unchanged, there have been many refinements in the way by which science is done including a huge expansion of available resources and reagents. Furthermore, the 2004 book had five chapters on IF-associated proteins, but now the field has developed to such an extent that inclusion of a separate volume on IF-associated proteins became clearly warranted.

This current volume includes 26 chapters that make up three major sections. The first section includes 10 chapters that cover general methods related to studying IF proteins. The second section includes 11 chapters that cover different mammalian IF proteins. The third section includes 5 chapters that span nonmammalian IF protein systems. All the chapters provide essential how-to techniques, including suppliers of the reagents being covered, and caveats and pearls that are typically not found in the Materials and Methods sections of manuscripts. Our goal is to help investigators avoid unnecessary efforts of trials and errors, while being able to quickly explore how their experimental system or to-be-tested hypothesis might relate to a given IF protein.

We are very excited about the broad spectrum of methods that have been covered in this book and hope that readers at all levels of their research careers or training will find the chapters in this volume and the companion volume of benefit and interest.

M. Bishr Omary and Ronald K.H. Liem

REFERENCES

Ishikawa, H., Bischoff, R., & Holtzer, H. (1968). Mitosis and intermediate-sized filaments in developing skeletal muscle. *Journal of Cell Biology*, *38*, 538–555.

Oakley, B. R. (2000). An abundance of tubulins. *Trends in Cell Biology*, *10*, 537–542.

Omary, M. B. (2009). "IF-pathies": A broad spectrum of intermediate filament-associated diseases. *Journal of Clinical Investigation*, *119*, 1756–1762.

Perrin, B. J., & Ervasti, J. M. (2010). The actin gene family: Function follows isoform. *Cytoskeleton*, *67*, 630–634.

Schweizer, J., Bowden, P. E., Coulombe, P. A., Langbein, L., Lane, E. B., Magin, T. M., et al. (2006). New consensus nomenclature for mammalian keratins. *Journal of Cell Biology*, *174*, 169–174.

Szeverenyi, I., Cassidy, A. J., Chung, C. W., Lee, B. T., Common, J. E., Ogg, S. C., et al. (2008). The Human Intermediate Filament Database: Comprehensive information on a gene family involved in many human diseases. *Human Mutation*, *29*, 351–360.

ACKNOWLEDGMENTS

We wish to thank all the authors of the chapters for their hard work and for sending us their manuscripts in a timely manner, and for working with us on revising the chapters to stay within the publication schedule. We thank Professor Anna Marie Pyle, Editor of *Methods of Enzymology*, for providing us the opportunity to showcase intermediate filaments and their associated proteins. We especially thank Sarah Lay for providing us with outstanding and prompt editorial support while keeping us on task, and Preethy Simonraj for the superb copy editing of the chapters.

PART I

General Methods to Study IF Proteins

CHAPTER ONE

How to Study Intermediate Filaments in Atomic Detail

Anastasia A. Chernyatina*, John F. Hess†, Dmytro Guzenko*, John C. Voss†, Sergei V. Strelkov*,[1]

*Department of Pharmaceutical and Pharmacological Sciences, KU Leuven, Leuven, Belgium
†Department of Cell Biology and Human Anatomy, University of California, Davis, California, USA
[1]Corresponding author: e-mail address: sergei.strelkov@pharm.kuleuven.be

Contents

1. Introductory Remarks 4
2. Bioinformatics 5
3. X-ray Crystallography 7
   3.1 Design of IF Protein Fragments for Crystallization 7
   3.2 Protein Expression 11
   3.3 Purification 13
   3.4 Crystallization 14
   3.5 X-ray Data Collection 16
   3.6 Phasing by Molecular Replacement 16
   3.7 Experimental Phasing Using Heavy Atoms 17
   3.8 Structure Refinement and Validation 19
   3.9 Limitations and Pitfalls 19
   3.10 Obtained Structures and Impact 20
4. Electron Paramagnetic Resonance with Site-Directed Spin Labeling 21
   4.1 Principle 21
   4.2 Sample Preparation and Measurements 22
   4.3 Data Interpretation and Impact 24
   4.4 Limitations and Outlook 26
5. Concluding Remarks 27
Acknowledgments 28
References 28

Abstract

Studies of the intermediate filament (IF) structure are a prerequisite of understanding their function. In addition, the structural information is indispensable if one wishes to gain a mechanistic view on the disease-related mutations in the IFs. Over the years, considerable progress has been made on the atomic structure of the elementary building block of all IFs, the coiled-coil dimer. Here, we discuss the approaches, methods and practices that have contributed to this advance. With abundant genetic information

Methods in Enzymology, Volume 568
ISSN 0076-6879
http://dx.doi.org/10.1016/bs.mie.2015.09.024

on hand, bioinformatics approaches give important insights into the dimer structure, including the head and tail regions poorly assessable experimentally. At the same time, the most important contribution has been provided by X-ray crystallography. Following the "divide-and-conquer" approach, many fragments from several IF proteins could be crystallized and resolved to atomic resolution. We will systematically cover the main procedures of these crystallographic studies, suggest ways to maximize their efficiency, and also discuss the possible pitfalls and limitations. In addition, electron paramagnetic resonance with site-directed spin labeling was another method providing a major impact toward the understanding of the IF structure. Upon placing the spin labels into specific positions within the full-length protein, one can evaluate the proximity of the labels and their mobility. This makes it possible to make conclusions about the dimer structure in the coiled-coil region and beyond, as well as to explore the dimer–dimer contacts.

ABBREVIATIONS

ASU asymmetric unit
DEER double electron–electron resonance
DTT dithiothreitol
E. coli *Escherichia coli*
EM electron microscopy
EPR electron paramagnetic resonance
HEPES 4-(2-hydroxyethyl)piperazine-1-ethanesulfonic acid
IF intermediate filament
Ig immunoglobulin
IMAC immobilized ion affinity chromatography
MES 2-(*N*-morpholino)ethanesulfonic acid
MR molecular replacement
MTSL (1-oxyl-2,2,5,5-tetramethylpyrroline-3-methyl)methanethiosulfonate
NMR nuclear magnetic resonance
PCR polymerase chain reaction
PDB Protein Data Bank
PEG polyethylene glycol
SAD single-wavelength anomalous diffraction
SDSL site-directed spin labeling
SEC size-exclusion chromatography
SeMet selenomethionine
SUMO small ubiquitin-like modifier
TEV tobacco etch virus
Tris tris(hydroxymethyl)aminomethane

1. INTRODUCTORY REMARKS

The importance of studying the molecular structure of IFs cannot be overestimated. In particular, it is the structure of the elementary dimer that defines IF assembly properties, ultimately leading to the formation of the typical smooth filaments with a diameter of about 10 nm. For certain IF proteins such as vimentin (type III) or keratins (types I and II) this process can be reliably reproduced in a test tube upon adjusting the solution conditions to a lower pH and a higher salt concentration (Herrmann, Hofmann, & Franke, 1992; Herrmann, Kreplak, & Aebi, 2004; Herrmann et al., 1996). This opens the possibility for experimental investigation of the assembly process by a number of biophysical methods. Yet the interpretation of the latter at the molecular level strongly relies on the availability of an atomic model of the elementary dimer. Likewise, the structural studies of IFs have a strong impact on the quickly expanding research on the diseases caused by mutations in the IF proteins. When trying to rationalize the debilitating effects of disease-related mutations (a majority of which are just single amino acid substitutions; Szeverenyi et al., 2008) it is imperative to start from a clear understanding what changes these mutations induce in the atomic structure.

In this chapter, we will briefly cover the approaches, methods, and practices that have contributed to the current knowledge of the atomic structure of IF proteins. In particular, we will highlight the use of X-ray crystallography as well as electron paramagnetic resonance with site-specific spin labeling (SDSL-EPR), the two methods that were most instrumental in the field thus far.

2. BIOINFORMATICS

The large number of available IF protein sequences from various species provides a good basis for the initial assessment of their structural features using bioinformatics. Apart from the standard protein sequence databases, a valuable information source is the Intermediate Filament Database (http://www.interfil.org; Szeverenyi et al., 2008). In particular, the human genome contains a total of 73 IF protein genes. As widely known (see, e.g., Chernyatina, Guzenko, & Strelkov, 2015), the amino acid sequence of all IF proteins shows a considerable conservation of many regions, and, as the result, includes a set of distinct features that were dubbed as the "IF signature." The most important of those is the tripartite organization, including the poorly ordered N- (head) and C-terminal (tail) domains and the predominantly α-helical central "rod" domain. The latter is due to the formation of a so-called coiled-coil structure by the two chains of the dimer, which can be either homomeric or heteromeric depending on the particular IF.

Nowadays, there is an abundant supply of efficient and highly automated sequence analysis tools, typically available as Web-based servers, capable of revealing the main structural features of an IF protein of interest. In particular, standard secondary structure prediction algorithms, such as Jpred (http://www.compbio.dundee.ac.uk/jpred4/index.html; Drozdetskiy, Cole, Procter, & Barton, 2015) and PSIPRED (http://bioinf.cs.ucl.ac.uk/psipred; Buchan, Minneci, Nugent, Bryson, & Jones, 2013), delineate the borders of the central α-helical domain in a reliable fashion. These algorithms also clearly suggest that, for a vast majority of IF proteins, the central domain is composed of three α-helical segments, termed coils 1A, 1B, and 2, interconnected by nonhelical linkers L1 and L12 (Chernyatina et al., 2015). Another useful type of bioinformatics analysis is the prediction of intrinsic disorder, which can be done efficiently using the DisMeta server (http://www-nmr.cabm.rutgers.edu/bioinformatics/disorder; Huang, Acton, & Montelione, 2014), which runs a number of contemporary algorithms and outputs consolidated results. Such analyses for IF proteins clearly suggest that both the head and tail domains are rather poorly ordered, which correlates with only a small amount of predicted α- or β-structure. A prominent exception here is the immunoglobulin (Ig) fold present within the tail domain of lamins (Dhe-Paganon, Werner, Chi, & Shoelson, 2002). This intrinsic disorder of the head domain in particular—notwithstanding the clear-cut role of the latter in the assembly process (Herrmann et al., 1996)—is also appearing as the main reason why very little structural information about this domain could be obtained thus far using any experimental methods including crystallography.

The major part of the α-helical segments of the IF rod contains a characteristic pattern of hydrophobic amino acids along the sequence which has a 7-residue (heptad) periodicity. Such a periodicity, designated $(abcdefg)_n$, is typical for the "classical" left-handed coiled-coil structure (Harbury, Zhang, Kim, & Alber, 1993). Here, the residues situated in the positions *a* and *d* are forming the stabilizing hydrophobic core of the latter. The program Coils (Lupas, Van Dyke, & Stock, 1991) which is available at http://www.ch.embnet.org/software/COILS_form.html readily picks the heptads for the major part of IF rod. In addition, some of the α-helical regions of the rod, in particular, the N-terminal part of coil2, contain a different, 11-residue repeat pattern of hydrophobic residues, which is responsible for a distinct "parallel bundle" structure (Peters, Baumeister, & Lupas, 1996). The reader should be warned that, at present, the standard sequence analysis packages do not include the detection of such periodicities, and will therefore misclassify these regions. To avoid this limitation, a customized analysis of the IF sequence is necessary (Chernyatina, Nicolet, Aebi, Herrmann, & Strelkov, 2012; Chernyatina et al., 2015). Here, the prediction of the residue accessibility along the sequence, such as the one provided by the NetSurfP algorithm (http://www.cbs.dtu.dk/services/NetSurfP; Petersen, Petersen, Andersen, Nielsen, & Lundegaard, 2009), appears suitable to detect the correct hydrophobic periodicity in various regions. Both the left-handed heptad-based regions and the parallel α-helical bundle regions delineated by this approach could also be confirmed experimentally, as will be discussed below.

3. X-RAY CRYSTALLOGRAPHY

X-ray studies of IF proteins necessarily rely on the crystallization of multiple 5–10 kDa fragments thereof, as we will explain shortly. The ample experience obtained by now indicates that bulk of the commonly used crystallographic protocols are perfectly applicable to the IF fragments, which, in this regard, should simply be considered as proteins of modest size. At the same time, most of the IF fragments studied contain a lot of coiled-coil structure, and this causes specific difficulties at various stages, such as crystallization and structure solution in particular.

3.1 Design of IF Protein Fragments for Crystallization

Full-length IF proteins cannot be used for growing 3D crystals, due to their intrinsic capacity to self-assemble into filaments (Herrmann, Bar, Kreplak, Strelkov, & Aebi, 2007). To circumvent this problem, a "divide-and-conquer" approach, i.e., production, crystallization, and structure determination of multiple smaller fragments of the full-length IF protein of interest, has been proposed (Strelkov et al., 2001). Over the years, many dozens of such fragments have been overexpressed, purified, and screened toward crystallization (Aziz et al., 2012; Chernyatina & Strelkov, 2012; Chernyatina et al., 2012; Kapinos, Burkhard, Herrmann, Aebi, & Strelkov, 2011; Kapinos et al., 2010; Lee, Kim, Chung, Leahy, & Coulombe, 2012; Meier et al., 2009; Strelkov, Kreplak, Herrmann, & Aebi, 2004; Strelkov, Schumacher, Burkhard, Aebi, & Herrmann, 2004; Strelkov et al., 2001, 2002). The design of such fragments is crucial. The regions which lack a regular structure, such as the head and tail domains of most IF proteins, should be best excluded. At the same time, the α-helical coiled-coil fragments can often be crystallized, but our experience suggests that the success rate decreases with the fragment length. Indeed, while longer coiled-coil fragments are more likely to fold into the correct native-like conformation, they represent elongated and somewhat flexible structures which may prevent crystallization. As seen from Tables 1 and 2, most solved IF fragments include between 70 and 110 residues.

Table 1 Crystal Structures of Human Vimentin Fragments[a]

PDB Code	Residue	Structural Part and Reference	Crystallization Conditions
1GK7*	102–138	Monomeric coil1A (Strelkov et al., 2002) 	2.0 M NH_4 acetate, 10% dioxane, 0.1 M MES-Na, pH 6.5, 20 °C
3G1E	101–139	Coil1A with a stabilizing mutation Y117L (Meier et al., 2009) 	33% Isopropanol, 20% PEG, 0.1 M Na_3 citrate, pH 5.6, 25 °C
3SSU	99–189	Fragment including coil1A, L1, and part of coil1B (Chernyatina et al., 2012) 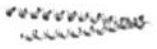	0.04 M Ca acetate, 6% isopropanol, 0.1 M MES, pH 6, 4 °C
3S4R	99–189	Fragment including coil1A, L1, and part of coil1B, with a stabilizing mutation Y117L (Chernyatina et al., 2012) 	0.45 M Mg acetate, 20% MPD, 0.1 M Na cacodylate, pH 6.5, 4 °C
3UF1	144–251	Coil1B (Aziz et al., 2012) 	0.15 M $(NH_4)_2SO_4$, PEG 3350 18%, 0.1 M Tris–HCl, pH 7.5, 20 °C
3SWK	153–238	Major part of coil1B (Chernyatina et al., 2012) 	0.2 M NH_4 acetate, 25% w/v PEG 3350, 0.1 M Bis-Tris, pH 6.5, 20 °C
4YV3*	161–238	Major part of coil1B (Anastasia Chernyatina & Sergei Strelkov, unpublished) forming a trimer 	2.2 M $(NH_4)_2SO_4$, 0.2 M NaSCN, 38 mM NaCl, 10 mM Tris, pH 8, 20 °C
4YPC*	161–243	Major part of coil1B (Anastasia Chernyatina & Sergei Strelkov, unpublished) forming a trimer 	1 M Na_3 citrate, 38 mM NaCl, 10 mM Tris–HCl, pH 8, 0.1 M Na cacodylate, pH 6.5, 20 °C
3KLT	263–334	First half of coil2 (Nicolet, Herrmann, Aebi, & Strelkov, 2010) 	27 mM $CaCl_2$, 7.5% (v/v) glycerol, 27% methoxy PEG 550, 10 mM DTT, 0.1 M Bis–Tris, pH 6.5, 25 °C
3TRT	261–335	First half of coil2 with a dimerizing mutation L265C (Chernyatina & Strelkov, 2012) 	2 M $(NH_4)_2SO_4$, 0.1 M Tris–HCl, pH 8.5, 20 °C

Table 1 Crystal Structures of Human Vimentin Fragments—cont'd

PDB Code	Residue	Structural Part and Reference	Crystallization Conditions
1GK4	328–411	Second half of coil2 (Strelkov et al., 2002) 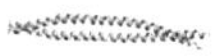	0.17 *M* Na acetate, 25.5% PEG 8000, 0.1 *M* Na cacodylate, pH 6.5, 20 °C
1GK6	385–412	C-terminal section of coil2 fused to GCN4 zipper (Strelkov et al., 2002) 	0.55 *M* $NH_4H_2PO_4$, pH 9, 20 °C

PDB Code	Resol. (Å)	Space Group and Unit Cell a, b, c (Å); β (°)	Phasing Method[b]	R_{work}/R_{free}	Dimers/ ASU[c]	Remark
1GK7*	1.4	$P6_222$ 56.8, 56.8, 58.4	MR	0.198/0.216	Monomer	Fragment does not form a coiled coil
3G1E	1.83	$P4_12_12$ 35.5, 35.5, 108.0	MR	0.253/0.295	1	
3SSU	2.6	$I2_12_12_1$ 45.6, 83.5, 118.6	SeMet	0.240/0.272	1	N-terminus up to residue 143 is disordered
3S4R	2.45	$R32$ 65.3, 65.3, 377.3	SeMet	0.293/0.315	1	Coil1A part forms an antiparallel assembly of four helices
3UF1	2.81	$P2_12_12_1$ 61.0, 86.8, 114.9	SeMet	0.239/0.284	2	Two dimers form a biological tetramer
3SWK	1.7	$P2_1$ 40.2, 54.5, 44.1; 109.8	SeMet	0.234/0.272	1	
4YV3*	2.0	$P2_1$ 39.1, 60.9, 48.9, 104.3	SeMet	0.255/0.296	Trimer	Fragment forms a nonnatural trimer
4YPC*	1.44	$R3$ 35.4, 35.4, 200.1	MR	0.245/0.276	1/3 trimer	Fragment forms a nonnatural trimer
3KLT	2.7	$P2_1$ 63.2, 30.1, 83.0, 102.7	HA ($SmCl_3$)	0.294/0.331	2	N-termini of dimers form a four-helix antiparallel bundle
3TRT	2.3	$I222$ 53.4, 75.5, 86.9	SeMet	0.234/0.298	1	Disulfide bonds between the introduced Cys265
1GK4	2.3	$I222$ 76.6, 84.3, 240.8	HA (Sm acetate)	0.242/0.262	3	
1GK6	1.9	$P3_121$ 98.8, 98.8, 36.5	MR	0.201/0.227	1/2	

[a]Each of the Tables 1 and 2 consists of two parts separated by a double line.
[b]MR, molecular replacement; SeMet, SeMet-SAD; HA, heavy-metal labeling by soaking and SAD.
[c]Number of dimers in the asymmetric unit (ASU) of the crystal.
The atomic models were drawn using PyMOL (DeLano, 2002). Structures deviating from the biologically relevant dimers are marked with asterisks.

Table 2 Crystal Structures of Human Keratin (K) and Lamin A (LA) and B1 (LB1) Fragments

PDB Code	Residue	Structural Part and Reference	Crystallization Conditions
3TNU	K14: 295–422 K5: 350–477	Second half of coil2 of keratin 5 and keratin 14 heterodimer (Lee et al., 2012) 	2.8 *M* NaCl, 0.1 *M* Tris–HCl, pH 8.5, 20 °C
1X8Y	LA: 305–387	Second half of lamin A coil2 (Strelkov, Schumacher, et al., 2004) 	0.7 *M* Na/K tartrate, pH 7.5, 25 °C
3V5B	LA: 313–386	Second half of lamin A coil2 (Bollati et al., 2012) 	0.5 *M* $(NH_4)_2SO_4$, 30% 2-methyl-2,4-pentanediol, 0.1 *M* HEPES, pH 7.5, 20–37 °C
3V4Q	LA: 313–386	Second half of lamin A coil2 (R335W mutant) (Bollati et al., 2012) 	0.2 *M* Li_2SO_4, 15% ethanol, 0.1 *M* Na_3 citrate, pH 5.5, 20–37 °C
3V4W	LA: 313–386	Second half of lamin A coil2 (E347K mutant) (Bollati et al., 2012) 	0.2 *M* NaCl, 30% 2-methyl-2,4-pentanediol, 0.1 *M* Na acetate, pH 4.6, 20–37 °C
3TYY	LB1: 311–388	Second half of lamin B1 coil2 (Ruan et al., 2012) 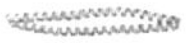	0.2 *M* Li_2SO_4, 0.1 *M* $CrCl_3$, 25% PEG 3350, 0.1 *M* Bis–Tris, pH 6.5, 18 °C
2XV5	LA: 328–398	Second half of lamin A coil2 with a small tail section (Kapinos et al., 2011) 	73% 2-methyl-2,4-pentanediol, 10 m*M* Tris–HCl, pH 8.5, 0.1 *M* NaCl, 20 °C

PDB Code	Resol. (Å)	Space Group and Unit Cell a, b, c (Å)	Phasing Method	R_{work}/R_{free}	Dimers/ASU	Remark
3TNU	3.01	*R32* 151.0, 151.0, 141.6	SeMet	0.210/0.234	2	
1X8Y	2.2	$P6_222$ 89.3, 89.3, 75.6	MR	0.276/0.305	1/2	
3V5B	3.00	$P6_522$ 90.0, 90.0, 74.7	MR	0.286/0.326	1/2	
3V4Q	3.06	$P6_522$ 90.5, 90.5, 75.4	MR	0.276/0.282	1/2	
3V4W	3.7	$P6_522$ 90.2, 90.2, 74.9	MR	0.317/0.339	1/2	
3TYY	2.4	$P3_221$ 46.0, 46.0, 203.2	MR	0.236/0.261	1	Disulfide bond between Cys317 residues (*d* position)
2XV5	2.4	$P2_12_12_1$ 35.1, 50.1, 89.7	SeMet	0.245/0.284	1	Biological significance of the observed antiparallel right-handed coiled-coil dimer is not clear

In our experience, for many produced fragments no conditions yielding even small crystals could be found. In this case, a sensible approach is to prepare additional fragments with varying N- and C-termini, and repeat crystallization screening. We could not establish any rules toward the preferred choice of the terminal residues. For most fragments that were resolved (Tables 1 and 2), the borders do not correspond exactly to the predicted ends of either coil1A, coil1B, or coil2. In addition, there can be complications at later stages, e.g., inability to grow bigger crystals with good diffraction quality, inability to phase the diffraction data, etc. As the result, in our estimate, only about one prepared fragment out of 10 could be eventually resolved crystallographically.

3.2 Protein Expression

Predictably, recombinant overexpression of IF fragments in *Escherichia coli* (*E. coli*) has been the basis of all crystallographic studies. Of various expression vectors used in the past, our favorite is the pPEP-TEV vector, a modification of the pPEP-T vector (Kammerer et al., 1998). The pPEP-TEV vector (3 kbp) provides for the overexpression of a fusion protein including an N-terminal 6xHis-tag, a laminin spacer (5 kDa), a tobacco etch virus (TEV) protease cleavage site, and the fragment of interest (Fig. 1). The

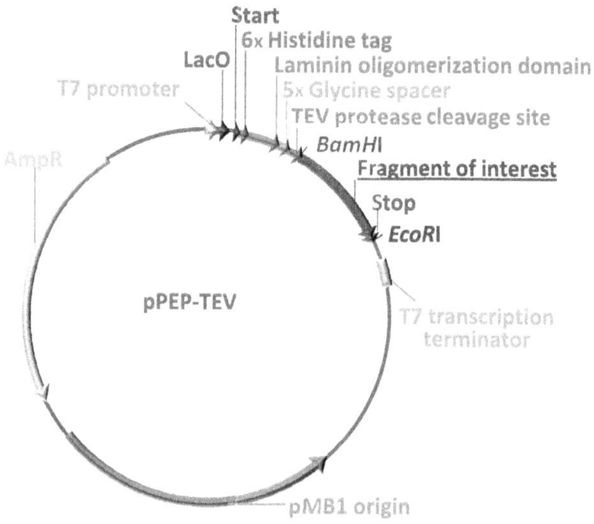

Figure 1 Scheme of the pPEP-TEV expression vector. AmpR, ampicillin resistance gene; pMB1, origin of DNA replication; LacO, binding site for the Lac operator that allows inducible protein expression.

use of this system typically provides for high expression levels, and up to 50 mg of pure protein per 1 l of culture can be obtained. Further advantages include the ease of purification and the removal of the His-tagged part from the final product using subtractive immobilized ion affinity chromatography (IMAC), as will be described below. We also found that the cleavage by TEV protease (as opposed to the cleavage by thrombin when the original pPEP-T vector (Brandenberger, Kammerer, Engel, & Chiquet, 1996; Kammerer, Antonsson, Schulthess, Fauser, & Engel, 1995) was used) is highly efficient and minimizes the chance of undesired cleavage within the IF fragment. As the result of the engineered cleavage sites for TEV protease and the BamH1 endonuclease, the final purified product contains two exogenous residues GlySer at the N-terminus.

A plasmid carrying the cDNA of the IF protein of interest can be obtained through OriGene (http://www.origene.com). For the polymerase chain reaction (PCR) amplification of the designed fragment, the direct primer is designed to include an overhang with the *Bam*HI recognition site, and the reverse primer carries an overhang with an *Eco*RI site followed by a stop codon. The primers are designed using the ApE program (http://biologylabs.utah.edu/jorgensen/wayned/ape) to have an annealing temperature of 52–55 °C. Thereafter both, the PCR product and the vector, are simultaneously digested with the *Bam*HI and *Eco*RI enzymes (New England Biolabs), purified using the Qiagen DNA purification kit and ligated with the T4 DNA ligase (New England Biolabs).

As an alternative, one can use the pETSUK vector which also permits the overexpression of an IF protein fragment as a fusion product—in this case including an N-terminal His-tag, the small ubiquitin-like modifier (SUMO) sequence and the fragment of interest (Weeks, Drinker, & Loll, 2007). The purification again follows the subtractive IMAC principle, whereby the fusion is cleaved by the SUMO hydrolase, leaving no exogenous residues. At the same time, we found that the efficiency of the cleavage by SUMO hydrolase was low in some cases, apparently depending on the sequence at the N-terminus of the IF fragment.

Expression constructs for IF protein fragments are transformed into competent cells using the heat-shock method (Bergmans, van Die, & Hoekstra, 1981). For optimal results, we employ Rosetta (DE3) pLysS *E. coli* cells (Novagen). Typically, the use of autoinduction growth medium ZYP-5052 (Studier, 2005) produces good protein yields. The cells should be cultivated at 24 °C until OD_{600} of 6 and afterwards seven more hours at 18 °C. Alternatively, the culture may be grown on LB medium at 37 °C until OD_{600} reaches 0.6, and thereafter protein expression is induced by adding

1 m*M* IPTG, followed by overnight cultivation at 18 °C. The induction protocol is used if a slowdown of the culture growth is observed, suggesting that the overexpressed protein is toxic for the bacteria, or if a specific protein labeling (see below) is necessary.

3.3 Purification

The course of the protein purification is shown in Fig. 2. Overexpressed IF protein fragments often form inclusion bodies. Correspondingly, the

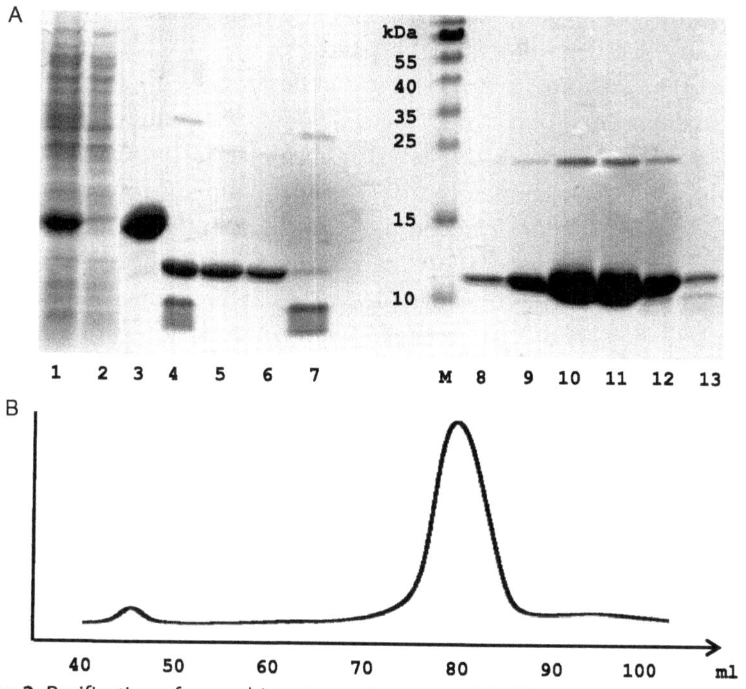

Figure 2 Purification of recombinant proteins, exemplified by a lamin A fragment. The fragment carries an introduced Cys residue in a *d* position near its C-terminus. (A) The main purification steps shown on a reducing SDS-PAGE. Lanes: 1—initial cell lysate; 2—flow-through of the first IMAC column; 3—His-tagged fusion construct (theoretical mass 17 kDa) eluting from the column in 300 m*M* imidazole; 4—sample after digestion by TEV protease including bands for the protease (27 kDa), the lamin fragment (10.8 kDa) and the His-laminin tag (6.2 kDa); 5 and 6—the lamin fragment eluting in both the flow-through and wash (12.5 m*M* imidazole) steps of the second IMAC column; 7—the His-laminin tag and the TEV protease (also with a His-tag) eluting from the column in 300 m*M* imidazole; M—molecular weight marker; 8–13—the main peak fractions (75–85 ml) of the final SEC run (B; the gel reveals some disulfide-crosslinked dimers which persist even in the presence of 10 m*M* dithiothreitol (DTT) used for the SDS-PAGE). (B) The SEC elution profile obtained using a Superdex 200 16/60 column.

pelleted cells are resuspended in buffer A: 8 *M* urea, 40 m*M* Tris buffer (pH 8), 50 m*M* NaCl, 10 m*M* imidazole, 5 m*M* β-mercaptoethanol, and lysed using Branson Digital Sonifier. The cellular dust is then removed by centrifugation (13,000 × *g*, 30 min). The solution containing denatured proteins (50 ml) is loaded on a 10 ml Ni-chelating column (His-select, Sigma) preequilibrated with the same buffer A, and the bound His-tagged protein is eluted with buffer A supplemented with 300 m*M* imidazole. The resulting fraction is dialyzed against the 40 m*M* Tris–HCl (pH 8), 50 m*M* NaCl (cleavage buffer), and then the fusion protein is digested with the appropriate protease. Recombinant TEV protease carrying an N-terminal His-tag (GenScript) can be used in molar ratio 1:100–1:50 for overnight digestion at 4 °C. Thereafter, a second pass through the Ni-chelating column in the cleavage buffer is used to retain the His-tag-containing components, i.e., the tag, the uncleaved fusion and the protease (which carries a 6xHis-tag as well). The untagged protein, found in the flow-through, is concentrated to 10–30 mg/ml using a centrifugal concentration device (Millipore Amicon ultra-centrifugal filter concentrator) and gel-filtered on Superdex 200 16/60 column (GE Healthcare) in the same buffer using the Äkta chromatographic system. Due to the extended shape of the IF fragments they are retained on a size-exclusion chromatography (SEC) column to a lesser extent than globular proteins of the same molecular weight. This justifies the use of a resin with bigger pore size such as Superdex 200.

3.4 Crystallization

As a final step before crystallization, pure IF protein fragments are dialyzed against a neutral low molarity buffer such as 10 m*M* Tris–HCl (pH 7.5) and concentrated. The final concentration should be ~10 mg/ml if possible, even though IF fragments could be successfully crystallized in the past from solutions of only few up to several dozens of mg/ml.

Crystallization experiments generally include two stages: the initial trial-and-error screening for conditions, followed by their optimization toward obtaining bigger crystals suitable for X-ray diffraction collection. Both stages typically utilize the popular "vapor diffusion" setups, in either the hanging drop or the sitting drop variant (Rhodes, 2006). The initial screening is most efficiently performed using kits with precipitant solutions available commercially. In particular, the Index (Hampton Research), PACT (Qiagen) and Wizard I + II (Emerald) screens together provide 288 different conditions which appear reasonable for the initial effort. The screening is conveniently done on 96-well MRC sitting drop plates (Molecular Dimensions). Crystallization robot such as Mosquito (TTP LabTech) allows mixing only ~100 nl of protein solution with an equal amount of reservoir solution, which is about 10 times less than necessary for manual pipetting. If available, the crystallization setups are automatically photographed using an imaging system such as Rock Imager 1000 (Formulatrix) over the course of several weeks. We typically perform the initial screening at 20 °C, but if no crystals are seen then the screening is repeated at 4 °C.

Once initial crystallization hits are found, a systematic sampling of factors such as the precipitant concentration, pH or temperature around the original condition should be carried out; ample advice on the topic is available (Luft et al., 2007; McPherson & Cudney, 2014). If the amount of protein sample is limited, the crystal optimization can be done in the 96-well plate format. However, we recommend to use 1–3 μl hanging drops on 24-well crystallization plates (Molecular Dimensions), as the upscaling often helps obtaining crystals of optimal size (at least 100 μm in each direction). The X-ray data collection is typically done at 100 K (Garman, 2003). Toward this end, the crystals are briefly soaked in the original growth condition supplemented with 30% glycerol or PEG 400, mounted on nylon cryo-loops (Hampton Research) and flash-cooled in liquid nitrogen.

Tables 1 and 2 include the crystallization conditions for all 19 IF rod fragments for which the crystal structure could be determined thus far. Figure 3

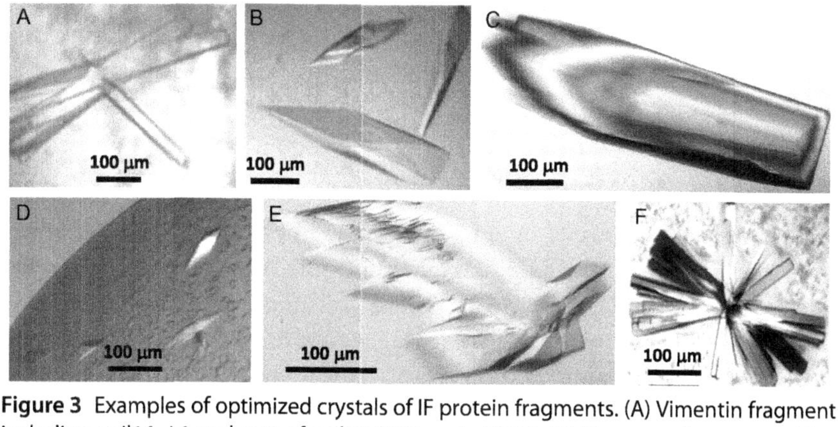

Figure 3 Examples of optimized crystals of IF protein fragments. (A) Vimentin fragment including coil1A, L1 and part of coil1B (PDB code 3SSU). (B) The same fragment with a stabilizing mutation Y117L (PDB code 3S4R). (C)–(E) Three related fragments each corresponding to a major part of vimentin coil1B (PDB codes 3SWK, 4YV3, and 4YPC, respectively). (F) First half of vimentin coil2 with the L265C mutation (PDB code 3TRT). (See the color plate.)

shows some of the obtained crystals. As can be seen, the crystallization conditions are quite variable, including both inorganic salts and organic compounds such as PEGs as the main precipitating agent, and buffers of varying pH. In our experience, no particular "magic" condition, a single preferred precipitant or additive, could be found.

3.5 X-ray Data Collection

X-ray data collection from IF fragment crystals is usually straightforward, given their modest molecular weight, which typically translates to moderate unit cell parameters (see Tables 1 and 2). Standard X-ray cameras and collection protocols, preferably utilizing synchrotron radiation, can be used (Rhodes, 2006). At the same time, diffraction pattern pathologies, in particular anisotropy of diffraction (which seems to be related to the elongated shape of the IF fragments) but also twinning and other forms of disorder, are frequently observed. For the initial indexing and integration of the diffraction images, we typically use the XDS program (Kabsch, 2010) started via the XDSME wrapper (https://code.google.com/p/xdsme) or XDSAPP (Krug, Weiss, Heinemann, & Mueller, 2012). In difficult cases, HKL2000 (Otwinowski & Minor, 1997) or iMosflm (Powell, Johnson, & Leslie, 2013) can also be tried. The bulk of subsequent data processing can be done with the CCP4 (Winn et al., 2011) or PHENIX (Adams et al., 2010) suites.

As the phases of diffracted beams cannot be recorded directly, the collection of the native diffraction pattern is not *per se* sufficient for the crystal structure solution (Taylor, 2010). Several ways to resolve this so-called "phase problem" were used for IF fragment crystals, including molecular replacement (MR) and experimental phasing via the introduction of "heavy" atoms (see Tables 1 and 2).

3.6 Phasing by Molecular Replacement

MR has been successfully used toward phasing of many IF fragment structures, including several vimentin fragments (Protein Data Bank (PDB) codes 1GK7, 3G1E, 4YPC, and 1GK6; Table 1), a lamin A coil2 fragment (PDB code 1X8Y) and its mutants that crystallize isomorphously (3V5B, 3V4Q, 3V4W), as well as a lamin B1 coil2 fragment (3TYY; Table 2). This method is based on the ability to (1) find a model structure that approximates (part of) the crystallized protein and to (2) place such a model correctly into the unit cell of the latter. From the positioned model, an initial estimate of the crystallographic phases can be calculated (Rhodes, 2006). In our hands, for phasing IF fragment structures, two highly popular MR algorithms, Phaser (McCoy et al., 2007) and Molrep (Vagin & Teplyakov, 1997), could be used with similar efficiency, while trying both of them was sometimes useful in difficult cases.

Finding a suitable search model for short IF fragments appears relatively easy. Indeed, coiled coils are periodic and symmetric structures with limited variability in general (Lupas & Gruber, 2005). Moreover, a considerable sequence conservation across various IF proteins suggests that once some IF protein fragment has been resolved, crystallographic data for homologous IF fragments can be phased using MR. This has indeed been the case for e.g., the lamin A coil2 fragment 1X8Y (phased using the corresponding vimentin fragment 1GK4) and a similar fragment 3TYY from lamin B1 (phased using the lamin A fragment structure).

Nevertheless, in practice many IF fragment structures could not be phased using MR even after extensive trials. The likely reason is that a MR procedure with a coiled-coil molecule is inherently challenging even if the search model is appropriate, which is due to both its elongated shape and internal periodicity. Indeed, for a standard heptad-based coiled coil, a screw movement corresponding to a seven-residue shift overlaps the structure onto itself. As the result, it is often impossible to find the correct rotational and translational position among many false positives. For similar reasons, the values obtained using the standard scoring functions of the MR solution (such as the log-likelihood gain and translation function Z-score in Phaser) can often appear misleadingly high even for a false solution. Thus, while phasing of IF fragment structures using MR may be considered the first option to try, the success of this procedure can never be guaranteed. The final proof of the correct MR solution should always be a pronounced decrease of the crystallographic working and free R-factors during subsequent model refinement.

3.7 Experimental Phasing Using Heavy Atoms

For the majority of resolved IF protein fragments, experimental phasing was necessary (see Tables 1 and 2). Toward this end, the crystal must contain well-bound "heavy" atoms, i.e., those that are substantially heavier than carbons. The traditional way of preparing such crystals is by soaking in a solution containing a heavy-metal salt solution, which is a trial-and-error process (Blundell & Johnson, 1976; Drenth, 1999). This procedure was successful for two vimentin fragments, 3KLT and 1GK4. In either case, a

Sm3+ ion could be used, rather by coincidence. The heavy-atom-containing crystals are exposed to synchrotron radiation with appropriate wavelength, allowing recording the anomalous component of the scattering (Drenth, 1999). Typically, phasing from single-wavelength anomalous diffraction (SAD) data is sufficient to resolve the structure.

A more reliable way of introducing the heavy atoms is the overexpression of the target protein in a medium-containing selenomethionine (SeMet), so that the final product incorporates SeMet residues in the place of methionines, and its crystallization. Phasing by SeMet-SAD was highly successful for various IF fragments (Tables 1 and 2). If the native sequence of a particular IF fragment does not contain any methionines, point replacements of selected amino acid residues (usually two) to methionines can be used. In our experience, exchanging Leu residues for Met can work very well. To avoid an impact on the stability of the coiled coil, it is probably safer not to mutate Leu residues which are located in the core positions. For IF fragments of up to ~100 residues, the presence of two well-ordered Se atoms usually provides for a reliable phasing (for more detailed insights, see Boggon & Shapiro, 2000). As an example, readily interpretable electron density maps were obtained from phasing two Se sites per monomer as exemplified by the structure solution of fragments 3TRT and 3S4R.

Recombinant production of SeMet-containing protein can be achieved using a protocol of Cowie and Cohen (1957). The culture is grown on M9 minimal medium (6 g Na_2HPO_4, 3 g KH_2PO_4, 0.5 g NaCl, 1 g NH_4Cl, 2 ml of 1 *M* $MgSO_4$, 20 ml of 20% glucose, 0.1 ml of 1 *M* $CaCl_2$ per 1 l) at 37 °C until $OD_{600}=0.6$. At this point SeMet (50 mg/l, Sigma) complemented by natural amino acids (50–100 mg/l, excluding methionine) is added. After 15 min the protein expression is induced by the addition of 1 m*M* IPTG, and the culture growth is continued overnight at 18 °C. Alternatively, Studier's protocol (Studier, 2005) can be used. Purification procedure for the SeMet proteins is the same as for the unlabeled. It should be noted that the degree of incorporation of SeMet should always be verified using mass spectrometry, as in our experience only partial SeMet incorporation is not uncommon.

The processing of SAD data can be done by a multitude of programs (see http://www.iucr.org/resources/other-directories/software). Some of these are incorporated in the automated Auto-Rickshaw pipeline (http://www.embl-hamburg.de/Auto-Rickshaw; Panjikar, Parthasarathy, Lamzin, Weiss, & Tucker, 2005).

3.8 Structure Refinement and Validation

The starting atomic model can result either from the MR procedure or the interpretation of the experimental electron density map. The latter can be accomplished with the programs SHELXE (Sheldrick, 2010), Buccaneer (Cowtan, 2006) or the Autobuild module of PHENIX (Adams et al., 2010). Manual model completion can be done using the interactive graphics program Coot (Emsley & Cowtan, 2004). For model refinement, Refmac (Vagin et al., 2004) or PHENIX (Adams et al., 2010) can be used. Especially in the cases when high resolution data (above 2 Å) are available, the addition of solvent molecules which are ordered on the surface of the coiled coil can considerably improve the crystallographic statistics. Interactions within the final structure are analyzed with the program PISA (Krissinel & Henrick, 2007). The program Twister (Strelkov & Burkhard, 2002) can be useful toward the analysis of the geometry of the coiled coil as well as the structure of its hydrophobic core.

We noticed that, for IF fragment structures, the final values of working and free R-factors tend to be at the high end of the generally acceptable values, even after extensive model rebuilding and refinement. Out of 19 structures included in Tables 1 and 2, five had the final free R-factor above 30%. While we do not have a full explanation for this observation, it may be related to the elongated shape of the coiled-coil molecule which has a large exposed surface area compared to its mass. Such a molecule may deviate stereochemically from an average globular protein and also involve complex disorder of the solvent-exposed residues, which may make the automated refinement with a standard target less efficient.

3.9 Limitations and Pitfalls

Experience shows that only relatively short IF protein fragments (up to about 100 residues) have a fair chance to be crystallized. At the same time, it cannot be excluded that shorter fragments will fail to assemble into the correct parallel dimeric coiled coil. In the past, there were indeed several examples of such aberrant behavior. In particular, a 39-residue vimentin fragment corresponding to coil1A has initially been crystallized as a monomer (Strelkov et al., 2002). Later, the correct dimerization of this fragment could be induced by an artificial point mutation Y117L (Meier et al., 2009). Moreover, of several available coil1B fragments, the longer ones (residues 144–251 and 153–238, respectively, see Table 1) yielded the correct dimers, but the slightly shorter ones (residues 161–238 and 161–243) were found to form nonnatural trimers. It is therefore advisable to check for the correct oligomerization of each newly obtained fragment, which should be best done using analytical ultracentrifugation. To overcome the wrong oligomerization problem, a mutation to Cys was introduced in a core position near one of the ends of the fragment (Chernyatina & Strelkov, 2012). It was shown that such cysteines form a disulfide, promoting the correct dimeric and parallel structure.

A central question is of course whether the crystal structure of an isolated fragment represents the true conformation within the full-length protein. In stark contrast to globular proteins, the "linear" structure of a coiled coil suggests a positive answer here, provided that the correct dimerization has been secured in the first place. This has indeed been confirmed by a good match observed for the crystal structures of multiple overlapping vimentin fragments (Chernyatina & Strelkov, 2012; Chernyatina et al., 2012). At the same time, it should be expected that some parts of the IF dimer still change conformation at higher assembly stages up to the mature filaments, due to dimer–dimer interactions. Finally, it should be noted that sometimes parts of the crystallized fragment were not resolved in the final electron density, the most prominent example being the structure of a fragment including vimentin residues 99–189 (Chernyatina et al., 2012), suggesting disorder in the crystals and possibly in solution as well. For the full-length dimer and certainly within the assembled filament, a more ordered structure should be expected.

3.10 Obtained Structures and Impact

Above we have outlined the approaches and methods that, until now, have led to the crystal structure determination of the crystal structures of 12 fragments of human vimentin, one heterodimeric fragment of keratins K5 and K14, five (related) fragments of lamin A and one of lamin B1 (Tables 1 and 2). In addition, there is a number of structures (including disease-related mutations) of the Ig fold which is present in the tail domain of nuclear lamins (Krimm et al., 2002; Ruan et al., 2012). This globular fold, highly specific for lamins, is readily amenable to crystallization and structure determination.

The crystal structures obtained for human vimentin jointly cover the nearly complete length of its rod domain (with linker L12 the only unresolved region remaining). Overlapping structures as well as further homologous regions established for lamins and the keratin heterodimer provide a redundant view that adds extra security with respect to the correctness of the overall structure. For a detailed discussion of the latter and its biological implications, please refer to Chernyatina et al. (2012, 2015).

4. ELECTRON PARAMAGNETIC RESONANCE WITH SITE-DIRECTED SPIN LABELING

The site-directed spin labeling (SDSL-EPR) technique has proven to be very useful in describing the structural features of the IF dimer (Budamagunta, Hess, Fitzgerald, & Voss, 2007; Hess, Budamagunta, Aziz, FitzGerald, & Voss, 2013; Pittenger, Hess, Budamagunta, Voss, & Fitzgerald, 2008). Thus far such experiments have always been carried out on human vimentin as a model system. As the result, independent evidence toward the coiled-coil structure, supporting both the bioinformatics predictions and the crystal structures, could be obtained. Moreover, the SDSL-EPR provided interesting insights into the association of dimers into tetramers as well as further aspects of IF organization such as the structure of the head and tail domains.

4.1 Principle

By definition, EPR is a magnetic resonance technique that measures the interaction between the magnetic moments (spins) of electrons in a magnetic field and electromagnetic radiation in the microwave range (Eaton, Eaton, & Salikhov, 1998). In the experiments with proteins, a special "spin label," i.e., a chemical moiety that contains a single unpaired electron, is typically used. It is most convenient to employ a spin-labeled compound that covalently reacts with cysteine residues. The applied magnetic field generates an energy difference between the two possible spin orientations (parallel or antiparallel). As the result, at the resonance conditions transition of the spin to the higher-energy (i.e., antiparallel) orientation and thus absorbance of the microwave energy are observed. In the nitroxide spin label, the unpaired electron is localized near the nitrogen, where the local magnetic field from its nuclear spin generates three discrete resonance conditions. This hyperfine coupling produces the typical three-line EPR spectrum (usually plotted as the first derivative of absorption over the magnetic field, see Fig. 4). The exact shape of the spectrum reflects the extent of the modulation of the nuclear-spin orientation relative to the external magnetic field, which depends on how well the nitroxide label is ordered. In fact, the principal affecters of the EPR signal are (a) the mobility of the spin label at the particular position and (b) the interaction between closely located spins, if any

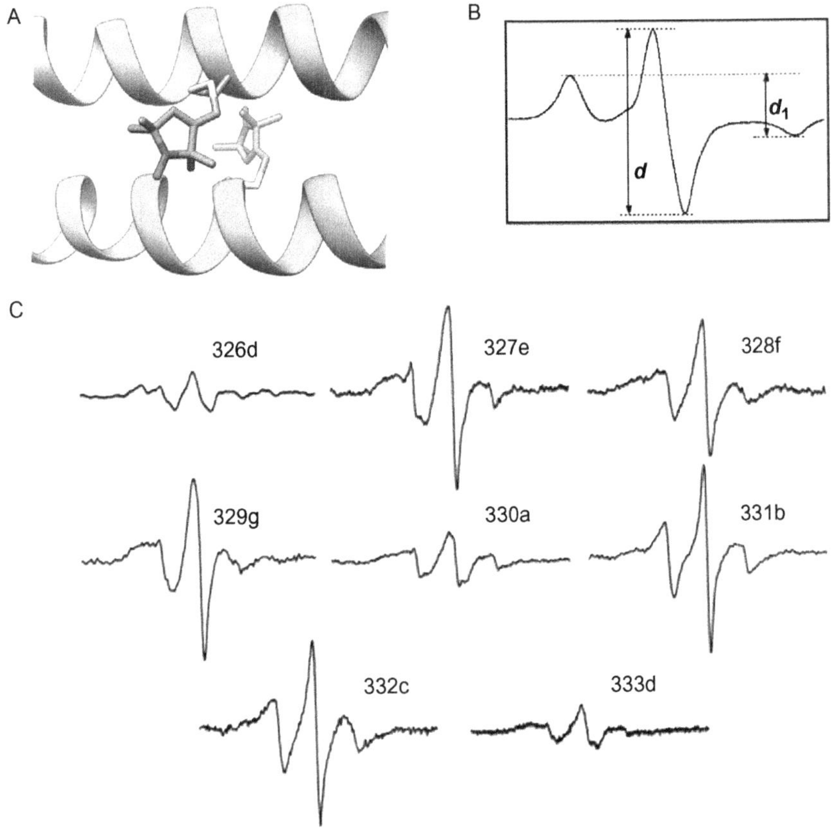

Figure 4 SDSL-EPR experiments on human vimentin. (A) Molecular model of a left-handed dimeric coiled coil (shown as a main-chain ribbon) with spin label MTSL attached to a Cys residue in a core *d* position (sticks). The NO group with unpaired electron is shown in blue (dark gray in the print version) and red (light gray in the print version). (B) Parameters of the recorded EPR spectra. (C) A series of recorded spectra for the label placed in positions 326–333 within coil2 segment of vimentin, with the heptad positions indicated by a letter after the residue number (Hess, Voss, & FitzGerald, 2002).

(Hubbell, Cafiso, & Altenbach, 2000; Langen, Isas, Luecke, Haigler, & Hubbell, 1998). Correspondingly, analysis of the obtained spectra (outlined below) allows extracting the local structural properties of the labeled molecules.

4.2 Sample Preparation and Measurements

Untagged full-length human vimentin can be efficiently overexpressed in *E. coli*. The resulting protein is found in inclusion bodies, which are subsequently solubilized in 8 *M* urea. The protein is isolated to purity using chromatography in denaturing conditions. Several related protocols can be used for this purpose (Herrmann et al., 1992, 2004; Hess et al., 2002). After purification, dialysis into 5 m*M* Tris–HCl buffer (pH 8.4) yields a solution of vimentin tetramers (Mucke et al., 2004).

Experiments with vimentin employed the popular (1-oxyl-2,2,5,5-tetramethylpyrroline-3-methyl)methanethiosulfonate (MTSL) compound which carries a nitroxide spin label (Fig. 4A). When covalently attached to a cysteine residue, it adds 184 Da to the protein's mass, which is comparable to adding one extra bigger amino acid such as tryptophan. For the labeling, purified vimentin solution at 1–5 mg/ml is treated with O.1 m*M* *tris*(2-carboxyethyl)phosphine for 30 min to reduce cysteines, and thereafter an excess (0.5 m*M*) of MTSL (Toronto Research Chemicals) is added. After 30 min of incubation at room temperature, unreacted MTSL is separated from the spin-labeled vimentin by chromatography on Source Q or S columns (GE Healthcare).

The wild-type vimentin contains a single cysteine residue at position 328. In addition, well over 100 different double mutants of vimentin were constructed which had the cysteine residue shifted to various positions in the sequence. To achieve this, the Cys328Ser vimentin mutant was prepared first, and thereafter this mutant was further modified to introduce a mutation to cysteine in a desired position. All point mutations were made using the QuikChange method and kit (Stratagene). To use this method, one first needs direct and reverse oligonucleotide primers that are complementary to the desired region of mutation except for coding for a cysteine in a specific position. Subsequently, a PCR reaction using the WT plasmid template and the primers is performed, resulting in the amplification of the mutated sequence. The mutations must always be confirmed by DNA sequencing.

It is important to verify that the double mutation and/or attachment of the spin label do not affect the properties of the protein. To this end, it is convenient to employ a sensitive *in vitro* assembly test that has been extensively used in the past (Rogers, Herrmann, & Franke, 1996). This electron microscopy (EM) based test shows whether the key feature of the protein, namely, the ability to form 10 nm filaments, is preserved. It should be noted that even minor modifications to IF proteins such as single point mutations can often jeopardize their assembly properties, resulting in a complete loss of the filament-forming capacity (see, e.g., Bar et al., 2005). To perform the assembly test, the purified, spin-labeled protein is diluted to 0.5 mg/ml with 5 m*M* Tris–HCl (pH 8.4) and then dialyzed against the "assembly buffer"

(25 m*M* Tris 7.5, 160 m*M* NaCl). Thereafter, the sample is negatively stained with uranyl acetate on standard EM grids. The long and smooth 10 nm filaments are clearly discernible in EM images (see, e.g., Herrmann et al., 1996), while the appearance of only short, thick or irregular filaments, or no filaments at all can point to assembly problems. Experience shows that most, but not all, spin-labeled vimentin mutants carrying a displaced cysteine are indeed capable of apparently normal *in vitro* filament assembly (Hess, Budamagunta, FitzGerald, & Voss, 2005).

EPR experiments can be carried out with a JEOL X-band spectrometer as described (Chomiki, Voss, & Warden, 2001; Hubbell, Froncisz, & Hyde, 1986). Labeled vimentin can be measured either in the tetrameric form or in the filamentous form. In the latter case, the solution of tetramers is supplemented with 1/9 volume of 10× "assembly buffer" and then immediately pipetted into the EPR capillary. The protein concentration is typically 0.02–0.1 m*M* (1–5 mg/ml). 5 and 50 μl samples are required for a room temperature and a low-temperature experiment, respectively. For the latter, the sample is quickly frozen in a dry ice ethanol bath and thereafter maintained at −100 °C by a cold N_2 gas flow. The obtained derivative spectra are evaluated using the Origin software.

4.3 Data Interpretation and Impact

It can be shown that, for samples in solution, the quickly ($\tau < 1$ ns) moving spin labels produce three sharp spectral lines, while more restrained labels yield flattened, lower-amplitude spectra (for detail, see e.g., Fig. 2 in Budamagunta et al., 2007). This provides information about the degree of order versus flexibility in the corresponding part of the protein structure. Such information, collected for vimentin labeled at various positions, could offer important structural insights. As an example, Fig. 4C shows room-temperature spectra from a series of vimentin samples that were systematically labeled at positions 326–333 within the coil2 region. As can be seen, the variation in the spectra is considerable. The observed periodicity has originally given evidence for a coiled-coil structure in this region (Hess et al., 2002). Indeed, the labeling in positions 326, 330, and 333 resulted in particularly "flat" derivative spectra. Later on, the crystal structure solution has confirmed that the latter positions correspond to the core residues of the coiled coil (heptad positions *d*, *a*, and *d*, respectively; Nicolet et al., 2010). Thus, the SDSL-EPR procedure can readily reveal the core residues that are typically tightly packed and therefore more restrained in motion.

For instance, the complex coiled-coil organization of the vimentin coil2 which contains both hendecad and heptad repeats has been independently established by both EPR and crystallography (Chernyatina & Strelkov, 2012; Hess, Budamagunta, Shipman, FitzGerald, & Voss, 2006; Nicolet et al., 2010).

Besides the motional freedom, the line shape of the SDSL-EPR spectra can also be affected by the interference of multiple spin labels located close to each other within the sample. To evaluate the latter, the measurements should be best done at low temperatures (−100 °C) in order to remove the contribution of the spin label motion. Thereafter, one can observe (calculate) a semiempirical parameter, the d_1/d ratio (Fig. 4B). This ratio directly correlates with the spin–spin interaction, and is practically useful for spins within ~2 nm (Kokorin, 2012; Likhtenshtein, 1993).

It can be easily seen that in a parallel, dimeric coiled coil the pairwise distance between the equivalent residues in the two chains is much shorter for the core positions than for the rest. Correspondingly, the labels placed in the core positions should be close to each other, and therefore exhibit a strong spin–spin interaction and increased d_1/d values. This has indeed been observed (Fig. 5) upon systematic labeling of vimentin dimer within two regions, residues 224–238 corresponding to the C-terminal part of coil1B, and residues 266–291 corresponding to the beginning of coil2 (Hess et al., 2006; John Hess, unpublished). Here again, a perfect correlation has been found between the d_1/d maxima of low-temperature spectra and the core positions of the atomic model based on multiple crystallographic structures.

Beyond the identification of the coiled-coil regions of the IF dimer, there can be other uses of the spin–spin interaction data. In particular, it was established that the labeling of vimentin residue 191 shows strong spin interference, even though this residue is located in a heptad position *c* (Hess, Budamagunta, Voss, & FitzGerald, 2004). This observation has been attributed to the interaction of labels that are located on different dimers, but come in a close proximity in assembled filaments. Specifically, this is realized in the A_{11} type dimer–dimer contact that aligns the respective coils 1B in antiparallel fashion (Hess et al., 2004). Recently, the crystal structure determination of the tetrameric 1B vimentin fragment (PDB code 3UF1) (Aziz et al., 2012) could confirm this interpretation. It turns out that, of the total of four spin labels residing on residues 191, two are likely to locate in close proximity (Fig. 6).

Finally, further SDSL-EPR studies (Aziz, Hess, Budamagunta, FitzGerald, & Voss, 2009; Aziz, Hess, Budamagunta, Voss, & FitzGerald,

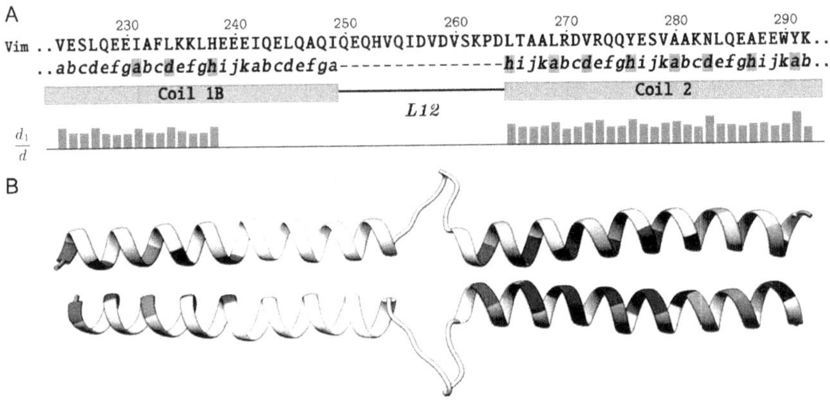

Figure 5 Structure of the vimentin dimer region (residues 224–292) encompassing the C-terminal part of coil1B, linker 12, and the beginning of coil2. (A) The amino acid sequence, with the heptad (yellow) and hendecad (violet) repeats of the coiled coil indicated above. The d_1/d values calculated from the EPR experiments are shown as a bar graph. (B) Ribbon model of the same region, based on crystallographic structures of the coiled-coil segments and modeling of the L12 linker (Chernyatina et al., 2015). The model is colored according to the d_1/d values from light blue (lowest) to dark blue (highest); white = no data. (See the color plate.)

2010; Hess et al., 2013) could provide interesting insights into the structure of vimentin head and tail domains. For instance, it was established that residue 17 of the vimentin head folds back on the coil1A region, coming in proximity with residue 137.

4.4 Limitations and Outlook

The SDSL-EPR technique is highly labor-intensive, since for each single position along the sequence a new mutant has to be cloned, purified, labeled, and measured. A sensible selection of the labeling positions is therefore important. To this end, it is advisable that planning of an SDSL-EPR experiment starts with a careful evaluation of the available hypotheses such as provided by 3D structural modeling in particular. In addition, in certain cases the introduction of a new Cys residue and label attachment can result in a loss of a proper filament formation, suggesting that the structure has been affected. At the same time, an important advantage of this approach is that it utilizes full-length proteins, while the crystallographic procedure necessarily relies on fragments.

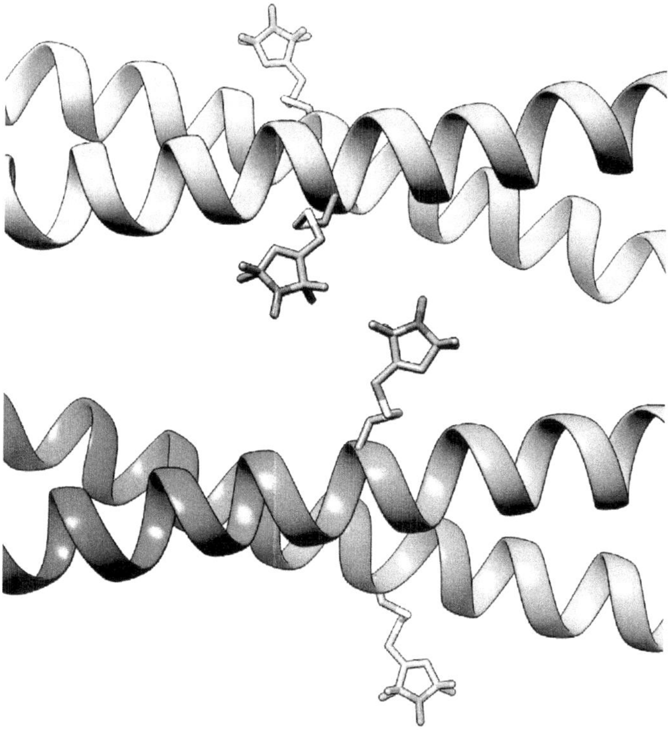

Figure 6 Middle part of the tetrameric arrangement seen in the crystals of the 1B vimentin fragment (PDB code 3UF1). The expected (modeled) conformation of the MTSL labels attached in positions 191 of the four chains is shown.

In addition, the d_1/d analysis of spin–spin interaction is sensitive to distances of ~2 nm or less. Double electron–electron resonance (DEER) (Sale, Song, Liu, Perozo, & Fajer, 2005) is a conceptually similar method that allows to detect the interlabel distances of up to 5 nm. As a proof of principle, a vimentin sample was doubly spin labeled at positions 225 and 240 and mixed 1:5 with unlabeled vimentin. Thereafter, the DEER measurements in 5 m*M* Tris buffer (pH 7.5) yielded a distance of 2.2 nm, consistent with the molecular structure of the solution A_{11} tetramer (Lishan Liu, Robert McCarrick, Gary Lorrigan, & John Hess, unpublished).

5. CONCLUDING REMARKS

To date, X-ray crystallography has been the main experimental method to resolve the IF structure in atomic detail. The considerable experience obtained over the years indicates that the mainstream protein crystallography approaches are applicable to IF fragments. At the same time, the latter, being largely coiled-coil structures, remain a very difficult object for crystallographic studies. As an alternative to X-ray crystallography, the structure determination of IF fragments might be possible using the nuclear magnetic resonance (NMR) technique, although little has been done this thus far in this direction. At present, NMR structures only exist for one very specific domain of the IF dimer, namely the globular Ig fold present within the lamin tail (Krimm et al., 2002). For the latter, a number of X-ray structures is available as well (Dhe-Paganon et al., 2002; Ruan et al., 2012).

It should be noted that the published series of SDSL-EPR experiments on vimentin have covered the majority of its coiled-coil rod as well as its head and tail domains. Importantly, with respect to the core structure of the rod domain, there was a nearly perfect match between the crystallographic and EPR data, as we have recently discussed in detail for each individual coiled-coil segment (Chernyatina et al., 2015). The EPR measurements are therefore supportive of the "divide-and-conquer" crystallographic approach. Moreover, the SDSL-EPR is able to provide complementary evidence on the structure of the terminal domains of vimentin, which are not suitable for crystallization. Overall, the excellent convergence of the spin labeling, crystallographic and bioinformatics data is very encouraging.

ACKNOWLEDGMENTS

We thank Prof. Harald Herrmann (German Cancer Research Center, Heidelberg) for continuous support of our structural investigations and Dr. Gábor Bunkóczi for advice on molecular replacement. J.F.H. and J.C.V. were supported by the NEI grant RO1 EY08747, NEI Core Facilities Grant P30EY012576, and US Army grant DAMD 17-02-1-0664. S.V.S. was supported by the KU Leuven grant OT13/097 and the Research Foundation Flanders (FWO) grant G070912N.

REFERENCES

Adams, P. D., Afonine, P. V., Bunkoczi, G., Chen, V. B., Davis, I. W., & Echols, N. (2010). PHENIX: A comprehensive python-based system for macromolecular structure solution. *Acta Crystallographica. Section D: Biological Crystallography*, *66*(Pt. 2), 213–221.

Aziz, A., Hess, J. F., Budamagunta, M. S., FitzGerald, P. G., & Voss, J. C. (2009). Head and rod 1 interactions in vimentin: Identification of contact sites, structure, and changes with phosphorylation using site-directed spin labeling and electron paramagnetic resonance. *Journal of Biological Chemistry*, *284*(11), 7330–7338.

Aziz, A., Hess, J. F., Budamagunta, M. S., Voss, J. C., & FitzGerald, P. G. (2010). Site-directed spin labeling and electron paramagnetic resonance determination of vimentin head domain structure. *Journal of Biological Chemistry*, *285*(20), 15278–15285.

Aziz, A., Hess, J. F., Budamagunta, M. S., Voss, J. C., Kuzin, A. P., & Huang, Y. J. (2012). The structure of vimentin linker 1 and rod 1B domains characterized by site-directed spin-labeling electron paramagnetic resonance (SDSL-EPR) and x-ray crystallography. *Journal of Biological Chemistry*, *287*(34), 28349–28361.

Bar, H., Mucke, N., Kostareva, A., Sjoberg, G., Aebi, U., & Herrmann, H. (2005). Severe muscle disease-causing desmin mutations interfere with in vitro filament assembly at distinct stages. *Proceedings of the National Academy of Sciences of the United States of America*, *102*(42), 15099–15104.

Bergmans, H. E., van Die, I. M., & Hoekstra, W. P. (1981). Transformation in Escherichia coli: Stages in the process. *Journal of Bacteriology*, *146*(2), 564–570.

Blundell, T. L., & Johnson, L. N. (1976). *Protein crystallography*. London: Academic press.

Boggon, T. J., & Shapiro, L. (2000). Screening for phasing atoms in protein crystallography. *Structure*, *8*(7), R143–R149.

Bollati, M., Barbiroli, A., Favalli, V., Arbustini, E., Charron, P., & Bolognesi, M. (2012). Structures of the lamin A/C R335W and E347K mutants: Implications for dilated cardiolaminopathies. *Biochemical and Biophysical Research Communications*, *418*(2), 217–221.

Brandenberger, R., Kammerer, R. A., Engel, J., & Chiquet, M. (1996). Native chick laminin-4 containing the beta 2 chain (s-laminin) promotes motor axon growth. *Journal of Cell Biology*, *135*(6 Pt. 1), 1583–1592.

Buchan, D. W., Minneci, F., Nugent, T. C., Bryson, K., & Jones, D. T. (2013). Scalable web services for the PSIPRED Protein Analysis Workbench. *Nucleic Acids Research*, *41*(Web Server issue), W349–W357.

Budamagunta, M., Hess, J., Fitzgerald, P., & Voss, J. (2007). Describing the structure and assembly of protein filaments by EPR spectroscopy of spin-labeled side chains. *Cell Biochemistry and Biophysics*, *48*(1), 45–53.

Chernyatina, A. A., Guzenko, D., & Strelkov, S. V. (2015). Intermediate filament structure: The bottom-up approach. *Current Opinion in Cell Biology*, *32C*, 65–72.

Chernyatina, A. A., Nicolet, S., Aebi, U., Herrmann, H., & Strelkov, S. V. (2012). Atomic structure of the vimentin central alpha-helical domain and its implications for intermediate filament assembly. *Proceedings of the National Academy of Sciences of the United States of America*, *109*(34), 13620–13625.

Chernyatina, A. A., & Strelkov, S. V. (2012). Stabilization of vimentin coil2 fragment via an engineered disulfide. *Journal of Structural Biology*, *177*(1), 46–53.

Chomiki, N., Voss, J. C., & Warden, C. H. (2001). Structure-function relationships in UCP1, UCP2 and chimeras: EPR analysis and retinoic acid activation of UCP2. *European Journal of Biochemistry*, *268*(4), 903–913.

Cowie, D. B., & Cohen, G. N. (1957). Biosynthesis by Escherichia coli of active altered proteins containing selenium instead of sulfur. *Biochimica et Biophysica Acta*, *26*(2), 252–261.

Cowtan, K. (2006). The Buccaneer software for automated model building. 1. Tracing protein chains. *Acta Crystallographica. Section D: Biological Crystallography*, *62*(Pt. 9), 1002–1011.

DeLano, W. L. (2002). *The PyMOL molecular graphics system*. Retrieved from, http://www.pymol.org.

Dhe-Paganon, S., Werner, E. D., Chi, Y. I., & Shoelson, S. E. (2002). Structure of the globular tail of nuclear lamin. *Journal of Biological Chemistry*, *277*(20), 17381–17384.

Drenth, J. (1999). *Principles of protein X-ray crystallography*. New York: Springer.

Drozdetskiy, A., Cole, C., Procter, J., & Barton, G. J. (2015). JPred4: A protein secondary structure prediction server. *Nucleic Acids Research*, *43*(W1), W389–W394.

Eaton, G. R., Eaton, S. S., & Salikhov, K. M. (1998). *Foundations of modern EPR*. Singapore: World Scientific.

Emsley, P., & Cowtan, K. (2004). Coot: Model-building tools for molecular graphics. *Acta Crystallographica. Section D: Biological Crystallography*, *60*(Pt. 12 Pt. 1), 2126–2132.

Garman, E. (2003). 'Cool' crystals: Macromolecular cryocrystallography and radiation damage. *Current Opinion in Structural Biology*, *13*(5), 545–551.

Harbury, P. B., Zhang, T., Kim, P. S., & Alber, T. (1993). A switch between two-, three-, and four-stranded coiled coils in GCN4 leucine zipper mutants. *Science*, *262*(5138), 1401–1407.

Herrmann, H., Bar, H., Kreplak, L., Strelkov, S. V., & Aebi, U. (2007). Intermediate filaments: From cell architecture to nanomechanics. *Nature Reviews Molecular Cell Biology*, *8*(7), 562–573.

Herrmann, H., Haner, M., Brettel, M., Muller, S. A., Goldie, K. N., & Fedtke, B. (1996). Structure and assembly properties of the intermediate filament protein vimentin: The role of its head, rod and tail domains. *Journal of Molecular Biology*, *264*(5), 933–953.

Herrmann, H., Hofmann, I., & Franke, W. W. (1992). Identification of a nonapeptide motif in the vimentin head domain involved in intermediate filament assembly. *Journal of Molecular Biology*, *223*(3), 637–650.

Herrmann, H., Kreplak, L., & Aebi, U. (2004). Isolation, characterization, and in vitro assembly of intermediate filaments. *Methods in Cell Biology*, *78*, 3–24.

Hess, J. F., Budamagunta, M. S., Aziz, A., FitzGerald, P. G., & Voss, J. C. (2013). Electron paramagnetic resonance analysis of the vimentin tail domain reveals points of order in a largely disordered region and conformational adaptation upon filament assembly. *Protein Science*, *22*(1), 47–55.

Hess, J. F., Budamagunta, M. S., FitzGerald, P. G., & Voss, J. C. (2005). Characterization of structural changes in vimentin bearing an epidermolysis bullosa simplex-like mutation using site-directed spin labeling and electron paramagnetic resonance. *Journal of Biological Chemistry*, *280*(3), 2141–2146.

Hess, J. F., Budamagunta, M. S., Shipman, R. L., FitzGerald, P. G., & Voss, J. C. (2006). Characterization of the linker 2 region in human vimentin using site-directed spin labeling and electron paramagnetic resonance. *Biochemistry*, *45*(39), 11737–11743.

Hess, J. F., Budamagunta, M. S., Voss, J. C., & FitzGerald, P. G. (2004). Structural characterization of human vimentin rod 1 and the sequencing of assembly steps in intermediate filament formation in vitro using site-directed spin labeling and electron paramagnetic resonance. *Journal of Biological Chemistry*, *279*(43), 44841–44846.

Hess, J. F., Voss, J. C., & FitzGerald, P. G. (2002). Real time observation of coiled-coil domains and subunit assembly in intermediate filaments. *Journal of Biological Chemistry*, *16*, 16.

Huang, Y. J., Acton, T. B., & Montelione, G. T. (2014). DisMeta: A meta server for construct design and optimization. *Methods in Molecular Biology*, *1091*, 3–16.

Hubbell, W. L., Cafiso, D. S., & Altenbach, C. (2000). Identifying conformational changes with site-directed spin labeling. *Nature Structural Biology*, 7(9), 735–739. http://dx.doi.org/10.1038/78956.

Hubbell, W. L., Froncisz, W., & Hyde, J. S. (1986). Continuous and stopped flow EPR spectrometer based on a loop gap resonator. *Review of Scientific Instruments*, *58*(1879).

Kabsch, W. (2010). Xds. *Acta Crystallographica. Section D: Biological Crystallography*, *66*(Pt. 2), 125–132.

Kammerer, R. A., Antonsson, P., Schulthess, T., Fauser, C., & Engel, J. (1995). Selective chain recognition in the C-terminal alpha-helical coiled-coil region of laminin. *Journal of Molecular Biology*, *250*(1), 64–73.

Kammerer, R. A., Schulthess, T., Landwehr, R., Lustig, A., Engel, J., & Aebi, U. (1998). An autonomous folding unit mediates the assembly of two-stranded coiled coils. *Proceedings of the National Academy of Sciences of the United States of America*, *95*(23), 13419–13424.

Kapinos, L. E., Burkhard, P., Herrmann, H., Aebi, U., & Strelkov, S. V. (2011). Simultaneous formation of right- and left-handed anti-parallel coiled-coil interfaces by a coil2 fragment of human lamin A. *Journal of Molecular Biology*, *408*(1), 135–146.

Kapinos, L. E., Schumacher, J., Mucke, N., Machaidze, G., Burkhard, P., & Aebi, U. (2010). Characterization of the head-to-tail overlap complexes formed by human lamin A, B1 and B2 "half-minilamin" dimers. *Journal of Molecular Biology, 396*(3), 719–731.

Kokorin, A. I. (2012). *Forty years of the d1/d parameter. In A. I. Kokorin (Ed.), Nitroxides—Theory, experiment and applications* (pp. 113–164).Rijeka, Croatia: InTech. Available from, http://www.intechopen.com/books/nitroxides-theory-experiment-and-applications/forty-years-of-the-d1-d-parameter.

Krimm, I., Ostlund, C., Gilquin, B., Couprie, J., Hossenlopp, P., & Mornon, J. P. (2002). The Ig-like structure of the C-terminal domain of lamin a/c, mutated in muscular dystrophies, cardiomyopathy, and partial lipodystrophy. *Structure (Camb), 10*(6), 811–823.

Krissinel, E., & Henrick, K. (2007). Inference of macromolecular assemblies from crystalline state. *Journal of Molecular Biology, 372*(3), 774–797.

Krug, M., Weiss, M. S., Heinemann, U., & Mueller, U. (2012). XDSAPP: A graphical user interface for the convenient processing of diffraction data using XDS. *Journal of Applied Crystallography, 45*(3), 568–572.

Langen, R., Isas, J. M., Luecke, H., Haigler, H. T., & Hubbell, W. L. (1998). Membrane-mediated assembly of annexins studied by site-directed spin labeling. *Journal of Biological Chemistry, 273*(35), 22453–22457.

Lee, C. H., Kim, M. S., Chung, B. M., Leahy, D. J., & Coulombe, P. A. (2012). Structural basis for heteromeric assembly and perinuclear organization of keratin filaments. *Nature Structural & Molecular Biology, 19*(7), 707–715.

Likhtenshtein, G. I. (1993). *Biophysical labeling methods in molecular biology*. New York: Cambridge University Press.

Luft, J. R., Wolfley, J. R., Said, M. I., Nagel, R. M., Lauricella, A. M., & Smith, J. L. (2007). Efficient optimization of crystallization conditions by manipulation of drop volume ratio and temperature. *Protein Science, 16*(4), 715–722.

Lupas, A. N., & Gruber, M. (2005). The structure of alpha-helical coiled coils. *Advances in Protein Chemistry, 70*, 37–78.

Lupas, A., Van Dyke, M., & Stock, J. (1991). Predicting coiled coils from protein sequences. *Science, 252*(5009), 1162–1164.

McCoy, A. J., Grosse-Kunstleve, R. W., Adams, P. D., Winn, M. D., Storoni, L. C., & Read, R. J. (2007). Phaser crystallographic software. *Journal of Applied Crystallography, 40*(Pt. 4), 658–674.

McPherson, A., & Cudney, B. (2014). Optimization of crystallization conditions for biological macromolecules. *Acta Crystallographica Section F, Structural Biology Communications, 70*(Pt. 11), 1445–1467.

Meier, M., Padilla, G. P., Herrmann, H., Wedig, T., Hergt, M., & Patel, T. R. (2009). Vimentin coil 1A-A molecular switch involved in the initiation of filament elongation. *Journal of Molecular Biology, 390*(2), 245–261.

Mucke, N., Wedig, T., Burer, A., Marekov, L. N., Steinert, P. M., & Langowski, J. (2004). Molecular and biophysical characterization of assembly-starter units of human vimentin. *Journal of Molecular Biology, 340*(1), 97–114.

Nicolet, S., Herrmann, H., Aebi, U., & Strelkov, S. V. (2010). Atomic structure of vimentin coil 2. *Journal of Structural Biology, 170*(2), 369–376.

Otwinowski, Z., & Minor, W. (1997). Processing of X-ray diffraction data collected in oscillation mode. *Methods in Enzymology, 277*(Part A), 307–325.

Panjikar, S., Parthasarathy, V., Lamzin, V. S., Weiss, M. S., & Tucker, P. A. (2005). Auto-rickshaw: An automated crystal structure determination platform as an efficient tool for the validation of an X-ray diffraction experiment. *Acta Crystallographica. Section D: Biological Crystallography, 61*(Pt 4), 449–457.

Peters, J., Baumeister, W., & Lupas, A. (1996). Hyperthermostable surface layer protein tetrabrachion from the archaebacterium Staphylothermus marinus: Evidence for the presence of a right-handed coiled coil derived from the primary structure. *Journal of Molecular Biology, 257*(5), 1031–1041.

Petersen, B., Petersen, T. N., Andersen, P., Nielsen, M., & Lundegaard, C. (2009). A generic method for assignment of reliability scores applied to solvent accessibility predictions. *BMC Structural Biology, 9*, 51.

Pittenger, J. T., Hess, J. F., Budamagunta, M. S., Voss, J. C., & Fitzgerald, P. G. (2008). Identification of phosphorylation-induced changes in vimentin intermediate filaments by site-directed spin labeling and electron paramagnetic resonance. *Biochemistry, 47*(41), 10863–10870.

Powell, H. R., Johnson, O., & Leslie, A. G. (2013). Autoindexing diffraction images with iMosflm. *Acta Crystallographica. Section D: Biological Crystallography, 69*(Pt. 7), 1195–1203.

Rhodes, G. (2006). *Crystallography made crystal clear* (3rd ed.). Portland, Maine: Elsevier.

Rogers, K. R., Herrmann, H., & Franke, W. W. (1996). Characterization of disulfide crosslink formation of human vimentin at the dimer, tetramer, and intermediate filament levels. *Journal of Structural Biology, 117*(1), 55–69.

Ruan, J., Xu, C., Bian, C., Lam, R., Wang, J. P., & Kania, J. (2012). Crystal structures of the coil 2B fragment and the globular tail domain of human lamin B1. *FEBS Letters, 586*(4), 314–318.

Sale, K., Song, L., Liu, Y. S., Perozo, E., & Fajer, P. (2005). Explicit treatment of spin labels in modeling of distance constraints from dipolar EPR and DEER. *Journal of the American Chemical Society, 127*(26), 9334–9335.

Sheldrick, G. M. (2010). Experimental phasing with SHELXC/D/E: Combining chain tracing with density modification. *Acta Crystallographica. Section D: Biological Crystallography, 66*(Pt. 4), 479–485.

Strelkov, S. V., & Burkhard, P. (2002). Analysis of alpha-helical coiled coils with the program TWISTER reveals a structural mechanism for stutter compensation. *Journal of Structural Biology, 137*(1–2), 54–64.

Strelkov, S. V., Herrmann, H., Geisler, N., Lustig, A., Ivaninskii, S., & Zimbelmann, R. (2001). Divide-and-conquer crystallographic approach towards an atomic structure of intermediate filaments. *Journal of Molecular Biology, 306*(4), 773–781.

Strelkov, S. V., Herrmann, H., Geisler, N., Wedig, T., Zimbelmann, R., & Aebi, U. (2002). Conserved segments 1A and 2B of the intermediate filament dimer: Their atomic structures and role in filament assembly. *EMBO Journal, 21*(6), 1255–1266.

Strelkov, S. V., Kreplak, L., Herrmann, H., & Aebi, U. (2004). Intermediate filament protein structure determination. *Methods in Cell Biology, 78*, 25–43.

Strelkov, S. V., Schumacher, J., Burkhard, P., Aebi, U., & Herrmann, H. (2004). Crystal structure of the human lamin A coil 2B dimer: Implications for the head-to-tail association of nuclear lamins. *Journal of Molecular Biology, 343*(4), 1067–1080.

Studier, F. W. (2005). Protein production by auto-induction in high density shaking cultures. *Protein Expression and Purification, 41*(1), 207–234.

Szeverenyi, I., Cassidy, A. J., Chung, C. W., Lee, B. T., Common, J. E., & Ogg, S. C. (2008). The human intermediate filament database: Comprehensive information on a gene family involved in many human diseases. *Human Mutation, 29*(3), 351–360.

Taylor, G. L. (2010). Introduction to phasing. *Acta Crystallographica. Section D: Biological Crystallography, 66*(Pt. 4), 325–338.

Vagin, A. A., Steiner, R. A., Lebedev, A. A., Potterton, L., McNicholas, S., & Long, F. (2004). REFMAC5 dictionary: Organization of prior chemical knowledge and guidelines for its use. *Acta Crystallographica. Section D: Biological Crystallography, 60*(Pt. 12 Pt. 1), 2184–2195.

Vagin, A., & Teplyakov, A. (1997). MOLREP: An automated program for molecular replacement. *Journal of Applied Crystallography*, *30*, 1022–1025.

Weeks, S. D., Drinker, M., & Loll, P. J. (2007). Ligation independent cloning vectors for expression of SUMO fusions. *Protein Expression and Purification*, *53*(1), 40–50.

Winn, M. D., Ballard, C. C., Cowtan, K. D., Dodson, E. J., Emsley, P., & Evans, P. R. (2011). Overview of the CCP4 suite and current developments. *Acta Crystallographica. Section D: Biological Crystallography*, *67*(4), 235–242.

CHAPTER TWO

Mechanical Properties of Intermediate Filament Proteins

Elisabeth E. Charrier, Paul A. Janmey[1]
Institute for Medicine and Engineering, University of Pennsylvania, Philadelphia, Pennsylvania, USA
[1]Corresponding author: e-mail address: janmey@mail.med.upenn.edu

Contents

Abstract

Purified intermediate filament (IF) proteins can be reassembled *in vitro* to produce polymers closely resembling those found in cells, and these filaments form viscoelastic gels. The cross-links holding IFs together in the network include specific bonds between polypeptides extending from the filament surface and ionic interactions mediated by divalent cations. IF networks exhibit striking nonlinear elasticity with stiffness, as quantified by shear modulus, increasing an order of magnitude as the networks are deformed to large strains resembling those that soft tissues undergo *in vivo*. Individual IFs can be stretched to more than two or three times their resting length without breaking. At least 10 different rheometric methods have been used to quantify the viscoelasticity of IF networks over a wide range of timescales and strain magnitudes. The mechanical roles of different classes of cytoplasmic IFs on mesenchymal and epithelial cells in culture have also been studied by an even wider range of microrheological methods. These studies have documented the effects on cell mechanics when IFs are genetically or pharmacologically disrupted or when normal or mutant IF proteins are exogenously expressed in cells. Consistent with *in vitro* rheology, the mechanical role of IFs is more apparent as cells are subjected to larger and more frequent deformations.

Methods in Enzymology, Volume 568
ISSN 0076-6879
http://dx.doi.org/10.1016/bs.mie.2015.09.009

ABBREVIATIONS

AFM atomic force microscope
GFP green fluorescent protein
IF intermediate filament
KO knockout
MT microtubule
NF neurofilament
Pa Pascal (Newton/m^2)
SPC sphingosylphosphorylcholine
WT wild-type

1. INTRODUCTION

Intermediate filaments (IFs) provide the major structural support for many noncellular materials such as hair, nails, and the slime surrounding hagfish. The mechanical properties of intracellular IFs are hypothesized to be essential for the normal function of many soft tissues, and mutations in distinct IF proteins lead to human diseases such as cardiomyopathies and skin blistering disorders that are characterized by a failure of affected tissues to withstand mechanical stress. The structures of IF proteins and the manner by which they assemble into filaments are highly distinct from those of the other cytoskeletal filaments F-actin and microtubules (MTs), and the mechanical properties of IF also diverge strongly from the rest of the cytoskeleton. The viscoelasticity of IF networks *in vitro*, and their contribution to the viscoelasticity of cells are increasing well characterized by a wide range of different techniques. These studies are beginning to show how the unusual structures of IFs contribute to the normal function of a large number of different cell types.

2. VISCOELASTICITY OF PURIFIED IFs *IN VITRO*

The mechanical properties of individual IFs of different types have been measured directly by applying forces to them and imaging their deflection or have been inferred from images assuming that the polymer contours are deformed by thermal energy. The viscoelastic properties of IF networks constituted *in vitro* either as homogeneous networks or as composite network copolymerized with F-actin have been measured by a number of rheologic methods. The unique mechanical properties of IFs are related to two major structural differences between IFs and the other cytoskeletal polymers F-actin and MTs. As shown in Fig. 1, IFs are much more flexible than either

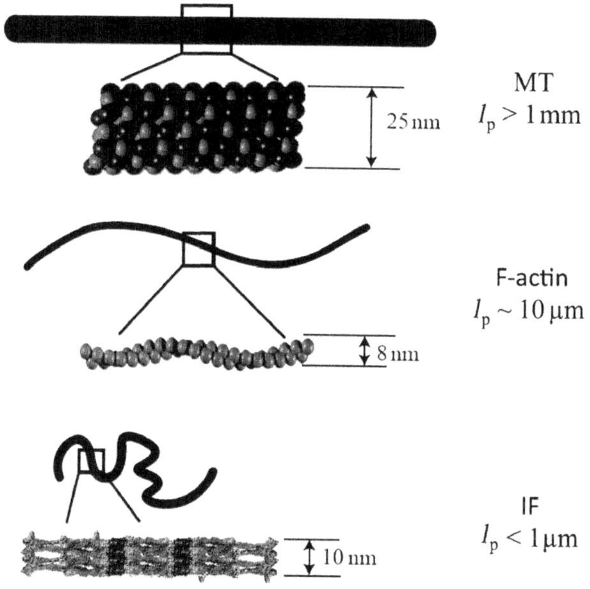

Figure 1 Schematic diagram of approximate diameter, subunit packing, and filament configuration of each of the three cytoskeletal polymer types: microtubules (MTs), F-actin, and intermediate filaments (IFs). The black filament outline represents the configuration of each filament in solution at 37 °C due to the thermal fluctuations acting on 10 μm long filaments with the persistence lengths l_p listed on the right.

MTs or actin filaments. This flexibility differs from the other cytoskeletal polymers by orders of magnitude and is quantified by the persistence length l_p, a measure of the distance over which a filament appears approximately straight. More precisely, l_p is defined by the expression $\langle \cos\theta(s) \rangle = e^{-s/l_p}$ where $\langle \cos\theta(s) \rangle$ is an ensemble average of the angle θ formed by two tangents drawn at distances s along the contour. The persistence length is related to the elastic bending constant of the filament K by the expression $K = l_p/k_B T$ where $k_B T$ is the thermal energy. This great flexibility is likely to be related to the greater degree of disorder and open hydrated space within IFs compared to actin or tubulin polymers. How precisely the subunit packing and higher-order structure of IFs allows them to be so flexible and resistant to breakage is not fully understood, but many different kinds of measurements reveal that IFs can potentially provide mechanical support to cells and tissues that cannot be achieved by the other polymer types. A representative, although not exhaustive, summary of the methods by which different types of IF have been characterized *in vitro* and the major findings of these studies are summarized in Table 1.

Table 1 Methods to Characterize IF Mechanical Properties *In Vitro*

IF Type	Method	Concentration	Time Scale	Main Result
NF/glial IFs (Leterrier & Eyer, 1987)	Falling ball viscometry	1–5 mg/ml	Seconds to minutes	NFs form gels by crossbridging. Divalent ions affect gelation
NF Vimentin (Leterrier, Kas, Hartwig, Vegners, & Janmey, 1996)	Oscillatory shear rheometry Parallel plate	3 mg/ml	10 ms to 1000 s	NF networks strain stiffen G' increases from <100 Pa to >kPa Modified by phosphorylation
Desmin Keratin NF (Kreplak, Bar, Leterrier, Herrmann, & Aebi, 2005)	AFM	Single filaments	Seconds to hours	IFs withstand stretching to >200% without rupture
NF NF–F-actin (Wagner et al., 2007)	Oscillatory shear rheometry	4 mg/ml	Seconds to minutes	NF gels rupture at high strain but rapidly reform. NF–F-actin composites lose recovery after large strain
Keratin (Leitner et al., 2012)	Single-bead thermal fluctuation microrheometry	0.5 mg/ml	0.5 ms to 1 s	$G' = 0.5$ Pa with 2 mM Mg^{2+}. Divalent ions stabilize networks
Vimentin (Janmey, Euteneuer, Traub, & Schliwa, 1991)	Torsion pendulum	0.3–10 mg/ml	10 ms to 100 s	Vimentin networks strain stiffen. Gels withstand >80% strain

Keratin Vimentin (Pawelzyk, Mucke, Herrmann, & Willenbacher, 2014)	Macroscopic shear rheometry and optical microrheometry	0.1–2 mg/ml	50 ms to 10 s	IF have attractive interactions due to hydrophobic and H bonds
Keratin Vimentin (Yamada, Wirtz, & Coulombe, 2003)	Shear rheometry Couette and cone-plate geometries	1 mg/ml	50 ms to 10 s	Apparent G' on order of 1–10 Pa affected by interfacial tensions. Weak frequency dependence
Vimentin (Lin, Broedersz, et al., 2010)	Parallel plate shear rheometry	0.2–1 mg/ml	300 ms to 50 s	Elastic response mainly entropic. Divalent ions act as cross-linkers
Desmin (Schopferer et al., 2009)	(1) Oscillatory squeeze flow (2) Cone-plate shear rheometry	1–2 mg/ml	(1) 50 μs to 1 s (2) 1 s	Strain stiffening but not always initial gelation is altered by disease-causing mutations
Vimentin and NF (Lin, Yao, et al., 2010)	Cone-plate shear rheometry	0.3–3 mg/ml	0.03–1000 s	Elasticity and strain-stiffening fit by theory for semiflexible polymer networks
Desmin Vimentin (Schopferer et al., 2009)	1. Oscillatory squeeze flow 2. Cone-plate shear rheometry	0.4–2.8 mg/ml	50 μs to 1 s	Desmin ($l_p \approx 900$ nm) is stiffer than vimentin ($l_p \approx 400$ nm) both electrostatics and binding affect network stiffness

Continued

Table 1 Methods to Characterize IF Mechanical Properties *In Vitro*—cont'd

IF Type	Method	Concentration	Time Scale	Main Result
Vimentin Vimentin + actin (Esue, Carson, Tseng, & Wirtz, 2006)	Cone-plate shear rheometry	0.04–0.4 mg/ml	1 ms to 5 s	Vimentin C-terminal tail binds F-actin to increase elastic modulus
Vimentin (Guzman et al., 2006)	AFM deflection	Single filaments	Seconds	Bending modulus of single IFs between 300 and 400 MPa
Vimentin (Mucke et al., 2004)	EM and AFM imaging	Single filaments	Static	Persistence length 1 μm
Keratin (Bousquet et al., 2001)	Cone-plate shear rheometry	0.5–1 mg/ml	Seconds	K14 C-terminal tail binds filament side to form cross-link
Keratin (Chou & Buehler, 2012)	Molecular dynamics	Single dimer	<20 ns	All atom simulation predicts force–extension of keratin dimer
NF (Janmey et al., 2007)	Parallel plate shear rheometry	2 mg/ml	Seconds	Shear deformations generate negative normal stress in NF networks

Several clear features unique to IF network mechanics emerge from these studies, and some issues related to the magnitude of IF network stiffness and the nature of interfilament links remains to be clarified. Unlike other elements of the cytoskeleton, individual IFs and the networks they form can withstand large deformations that would rupture F-actin or MTs (Guzman et al., 2006; Janmey et al., 1991; Kreplak et al., 2005). Not only do IF networks not rupture at large strain, but their elastic moduli increase, so that the incremental stiffness of vimentin, neurofilaments (NF), and other IF types can be 10 times larger at 100% strain that in the limit of low strain (Bertaud, Qin, & Buehler, 2010; Janmey et al., 1991; Leterrier et al., 1996; Lin, Yao, et al., 2010; Pawelzyk et al., 2014; Schopferer et al., 2009). The dependence of IF networks' elastic moduli on protein connection is also different from that of other biopolymer gels. Whereas the shear moduli of fibrin and actin networks scales with at least the square of the protein concentration, the shear moduli of vimentin and desmin networks increase much more gradually with power law exponents as low as 0.5 (Janmey et al., 1991; Lin, Broedersz, et al., 2010; Schopferer et al., 2009) relating elastic modulus to concentration. The reason for this discrepancy between IF and other biopolymer gels is not known.

The molecular mechanisms that link IFs together so that they form mechanically resistant networks are also not well understood. Specific cross-linking proteins do not appear to be required for network formation, and several bonds between IF subunit C-terminal extensions and the sides of other filaments have been reported (Bousquet et al., 2001; Esue et al., 2006; Pawelzyk et al., 2014). Complementary attractive interactions between NF sidearms are also implicated in linking these IFs to each other (Gou, Gotow, Janmey, & Leterrier, 1998). The most common method to create IF networks *in vitro* is to add divalent cations, usually Mg^{2+}, to several millimolar concentrations. The mechanisms by which divalent ions cross-link IFs is not fully characterized but has been hypothesized to involve either specific metal-binding bonds (Lin, Broedersz, et al., 2010) or polyelectrolyte effects that depend on the high surface charge of all IFs (Huisman et al., 2011; Janmey, Slochower, Wang, Wen, & Cebers, 2014). Identifying the molecular mechanisms for IF cross-linking and bundle formation remains a major challenge to defining this system with the same detail as currently available for network formation by other biopolymers.

3. IFs AND THE MECHANICAL PROPERTIES OF CELLS

The unique mechanical properties of IFs *in vitro*, characterized by strain stiffening of networks and the capacity of IFs to withstand very large extensions, have motivated recent studies to determine the roles of IFs in the mechanical properties of cells. Diverse studies have shown the effects of specific IF types in cell migration, adhesion, and mechanotransduction (Chung, Rotty, & Coulombe, 2013; Ivaska, Pallari, Nevo, & Eriksson, 2007; Pallari & Eriksson, 2006; Sakamoto, Boeda, & Etienne-Manneville, 2013; Wang & Stamenovic, 2000). The large diversity of cellular IF types, which are often integrated with actin and MTs networks, lead to a range of cellular effects when different IF types are genetically or pharmacologically disrupted or when they are overexpressed. Biochemical and genetic methods used to alter IF expression or assembly in cells are summarized in Table 2. Table 3 summarizes the methods used to characterize IF impact on cell mechanical properties and the main conclusions about their contribution to cell mechanics.

3.1 Type III IF: Vimentin and Desmin

Vimentin and desmin are the most represented filaments in this subgroup. Investigating the mechanical influence of type III IF on cell mechanical properties has led to divergent results largely because the results of mechanical measurements are strongly dependent on the way of probing the cell and the magnitude of the strain the cell is subjected to during the experiments (Table 3). The most commonly employed current method to characterize the cortical stiffness of cells is by indenting their surface with a conical tip or a colloidal bead attached to an atomic force microscope (AFM) cantilever, as depicted in Fig. 2. The principle of AFM and its application to reveal the mechanical effects of IFs in cells are summarized in Fig. 2 and Table 4. This method allows stiffness differences between cell types to be measured at various degrees of deformation, often with simultaneous imaging. Calculation of the absolute magnitude of elastic moduli from AFM as well as other microrheological techniques is difficult because numerous assumptions about contact geometry, material homogeneity, and volume conservation need to be made.

3.1.1 Knockout Models of Vimentin

Several biomechanical studies have been performed on mesenchymal cells with a disrupted or no vimentin network (Holwell, Schweitzer, & Evans,

Table 2 Biological Tools Allowing to Modify the Properties of IF Networks Used for Biomechanical Studies

Type of IF	Cell Type	Tools for Modifying Network	Effect on IF Network Morphology
Vimentin (Haudenschild et al., 2011; Wang & Stamenovic, 2000)	Endothelial cells and primary human articular chondrocytes	Acrylamide targets directly IF network, but has other effects that can obscure interpretation	Perinuclear condensation of the vimentin network
Vimentin (Gladilin, Gonzalez, & Eils, 2014)	Natural killer cells	Withaferin A targets directly vimentin network	Disruption of the vimentin network and aggregate formation
Vimentin (Brown, Hallam, Colucci-Guyon, & Shaw, 2001)	T lymphocytes	Calyculin A targets vimentin phosphatases, but also other enzymes that can indirectly affect IFs	Formation of a condensed juxtanuclear aggregate of vimentin
Vimentin (Rathje et al., 2014)	Immortalized human skin fibroblasts	Simian virus 40 large T antigen	Condensation of the vimentin network in the perinuclear area and retraction of thin peripheral filaments
Vimentin (Plodinec et al., 2011)	Rat-2 fibroblasts	Mutated desmin L345P targets directly vimentin or desmin network	Perinuclear aggregation of the vimentin inducing network total disruption
Desmin (Bonakdar et al., 2012)	Primary human myoblasts from patients carrying desmin mutations	Mutated desmin targets directly vimentin or desmin network	Not described
Keratin (Beil et al., 2003)	Human pancreatic epithelial tumor cells	Sphingosylphosphorylcholine to induce keratins phosphorylation	Perinuclear reorganization of the keratin network

Table 3 Summary of the Investigations of the Mechanical Role of IF Networks at the Cellular Level

Type of IF	Cell Type	Technique	Cellular Elements Probed	Effect of the Lack or the Disruption of the Vimentin Network at the Cellular Scale
Vimentin (Eckes et al., 1998)	Primary fibroblasts from vimentin KO rats	Rotational force magnetic twisting cytometer (Wang & Ingber, 1994)	Cell cortex submitted to large strains	Cortical rigidity lower by 40%
		Collagen lattice contraction (Mendez, Restle, & Janmey, 2014)	Cells contractile machinery	Contraction forces developed by vim-/- cells significantly reduced
Vimentin (Wang & Stamenovic, 2000)	Primary fibroblasts from vimentin KO rats Primary fibroblasts from WT rats and endothelial cells acrylamide treated	Rotational force magnetic twisting cytometer (Wang & Ingber, 1994)	Cell cortex submitted to different ranges of strain	Reduced ability to stiffen the cortex in response to applied forces and global cortex stiffness lower. These effects are amplified when the magnitude of the cell strain increases
Vimentin (Guo et al., 2013)	Primary fibroblasts from vimentin KO mice	Optical magnetic twisting cytometry (Fabry et al., 2001)	Cell cortex submitted to low strains	No effect on cells' cortical rigidity
		Optical tweezers	Cytoplasm	Intracytoplasmic rigidity of cell reduced by about a factor of 2

Vimentin (Gladilin et al., 2014)	Natural killer cells treated with withaferin A	Microfluidic optical stretcher (Guck et al., 2001)	Whole cell submitted to large strain	Global cell softening of about 20%
Vimentin (Brown et al., 2001)	T lymphocytes treated with Calyculin A	High *g*-force centrifugation (Mege, Capo, Benoliel, Foa, & Bongrand, 1985)	Whole cell submitted to large strain	Whole cell deformability increased by about 40%
Vimentin (Haudenschild et al., 2011)	Primary human articular chondrocytes	Straining of cells embedded in alginate gels	Whole cell submitted to large strain	Softening of the entire cell by a factor of 3
Vimentin (Rathje et al., 2014)	Immortalized human skin fibroblasts expressing simian virus 40 large T antigen	Colloidal probe force-mode AFM (Ducker, Senden, & Pashley, 1991)	Local cortex or cytoplasm as function of indentation depth	Cytoplasmic Young's modulus increased locally by two times
Vimentin (Plodinec et al., 2011)	Rat-2 fibroblasts expressing L345P-mutated desmin	AFM	Local cortex or cytoplasm as function of indentation depth	Perinuclear stiffening of the cytoplasm
Desmin (Bonakdar et al., 2012)	Primary human fibroblasts from patients carrying the R350P desmin mutation	Magnetic tweezers (Kollmannsberger & Fabry, 2007)	Cell cortex locally submitted to different ranges of strain	Cortical stiffness increased by factor of two Cortical stiffening three times lower after repeated straining of the cell

Continued

Table 3 Summary of the Investigations of the Mechanical Role of IF Networks at the Cellular Level—cont'd

Type of IF	Cell Type	Technique	Cellular Elements Probed	Effect of the Lack or the Disruption of the Vimentin Network at the Cellular Scale
Keratins (Beil et al., 2003)	Human pancreatic epithelial tumor cells treated with sphingosylphosphorylcholine	Parallel microplate cell stretcher	Whole cell	Cell elastic moduli decreased by 40%
		Migration through size-limited pores	Whole cell	Cell deformability significantly increased
Keratins (Seltmann, Fritsch, Kas, & Magin, 2013)	Primary keratinocytes from KO mice lacking all keratins	Optical stretcher (Lincoln et al., 2007)	Whole cell	Cell deformability increased by about 60%
Keratins (Sivaramakrishnan, DeGiulio, Lorand, Goldman, & Ridge, 2008)	Keratinocyte cell line KtyII −/−	AFM	Cytoplasm	Cytoplasmic Young modulus above cell nucleus is lowered by about 40%
		Magnetic tweezers	Cytoplasm	Cytoplasmic viscosity is 40% lower
Neurofilaments (Grevesse, Dabiri, Parker, & Gabriele, 2015)	Primary rat cortical neurons	Magnetic tweezers	Neurites versus soma	NF-rich neurites are both stiffer and more viscous than the soma

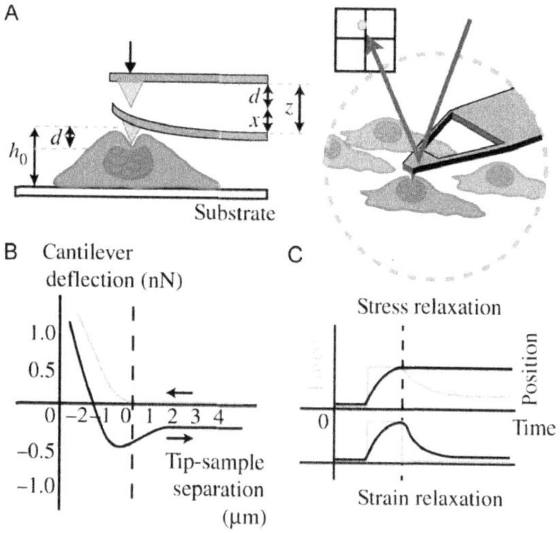

Figure 2 AFM modes of measurement. (A) AFM can be used to precisely apply compressive strains apically to cells within their aqueous environment. A laser deflected from the back of the AFM cantilever is measured by a photosensitive detector (PSD) to quantify cantilever deflection. (B) AFM force–indentation curves are often used to measure cellular elasticity, by fitting the approach curve (yellow (light gray in the print version)) to the Hertz model of contact mechanics. The retraction curve (blue (gray in the print version)) often shows a hysteresis and can be used to analyze adhesion and dissipation. (C) Stress and strain relaxation curves are often used to measure time-dependent cellular response. Following an applied strain on a cell, the cantilever can be kept at a constant height, and measurements of cellular force onto the cantilever can be made. Alternatively, following an initial strain, changes in the height of the cantilever as the cell relaxes can be measured. Modified cantilevers are also useful for measuring binding/unbinding forces between ligands and receptors (Haase & Pelling, 2015).

1997; Klymkowsky, 1981). The first demonstration of a role of vimentin in cells mechanical properties employed a rotational force magnetic twisting cytometer (Wang & Ingber, 1994) to study primary fibroblasts from vimentin knockout (KO) mice (Eckes et al., 1998). Vimentin null cells exhibit a lower cortical rigidity than wild-type (WT) cells and disturbed migratory abilities suggesting a role of vimentin in the stabilization of actin and MT networks in cells. A subsequent study of fibroblasts from vimentin KO mice quantified the mechanical impact of vimentin on cells (Wang & Stamenovic, 2000). This study, using the same magnetic twisting device, showed that the mechanical alteration of those fibroblasts due to the lack of vimentin, i.e., softer cortex and reduced ability to stiffen, is detectable only when cells are submitted to large strain. Studies using AFM stiffness mapping showed that the cortical stiffness of vimentin null cells measured

Table 4 Using Atomic Force Microscopy to Determine Cell Stiffness, with Emphasis on Methods to Detect Mechanical Effects of IFs

1. *AFM cantilever calibration.* Before each experiment, the cantilever is moved down onto a rigid surface such as the bottom of the glass or plastic plate while measuring the bending of the cantilever, as assessed by the laser beam deflected from its surface and the vertical displacement of the base of the cantilever as determined by the piezoelectric device. Since the rigid surface cannot be deformed by the relatively soft AFM cantilevers used for cells, any difference between the vertical displacement of the cantilever tip and base is due to deflection of the cantilever. The cantilever stiffness is calibrated by taking the slope of the cantilever deflection versus piezodisplacement curve. The spring constant of the cantilever can also be determined by measuring its resonance frequency in liquid, as discussed in detail elsewhere (Levy & Maaloum, 2002)
2. *Culture cells on standard glass or plastic dishes or on substrates of adjustable stiffness.* The substrates need to be rigidly held in a container that is large enough in diameter and deep enough to allow the AFM column (often call the head) which holds the cantilever, to be immersed into medium above the cell. Typically, the width of the dish is >20 mm and the depth of liquid above the cell is several mm. The piezoelectric devices that move the AFM probe vertically have limited range, so the depth of liquid cannot be much larger than mm
3. *Identifying the point of contact between AFM probe and cell surface.* Usually, the probe is moved near the cell surface using the microscope stage and imaging the focal plane of the AFM probe relative to that of the cell's apical surface. Once near enough to allow the piezoelectric drive to span the remaining distance (generally several microns), the final movements are made by the AFM software and hardware. The probe can be moved slowly until the deflection of the laser beam indicates that the cantilever is beginning to bend, presumably because it has touched the tip of the cell. Other methods based on changes in resonance can also determine the point at which contact is made
4. *Indentation of the cell surface.* As the AFM tip descends farther into the cell or gel, the cantilever will become increasingly bent (unless something breaks or slips) and the result is a force–extension curve where force is calculated from the measured bending of the cantilever, and extension from the vertical displacement of the AFM tip
5. *Force–displacement measurements.* The force–displacement data derived from the initial indentation into the sample are generally limited to a few hundred nm, over a time on the order of a second, depending on the shape of the probe, the material properties of interest, and the capabilities of the instrument hardware and software. Indentation is usually followed immediately by retraction. Perfect superposition of the indentation and retraction curves is expected for a purely elastic material to which the probe does not adhere. In reality, there is usually a difference between the indentation and retraction curves. The area between the

Table 4 Using Atomic Force Microscopy to Determine Cell Stiffness, with Emphasis on Methods to Detect Mechanical Effects of IFs—cont'd

curves is a measure of energy dissipated during the deformation, and often termed the plasticity index, but it is not simply related to a material constant such as a viscosity. This quantity is particularly dependent on the depth to which cells are indented and changes as a result of differences in IF expression
6. *Calculation of elastic modulus.* Quantifying the cell stiffness requires calculating an elastic modulus (usually the Young's modulus, a material property) from the force–extension curve (an experimental system-specific set of values). Conversion of force–indentation curves to absolute values of stiffness is perhaps the most challenging aspect of AFM measurements and the one most likely to lead to errors. Generally, a formula like that derived by Hertz is used to calculate elastic moduli from force measurements by accounting for the size and shape of the AFM probe and making assumptions about the nature of the sample's surface. For a spherical or hemi-spherical shape AFM tip, the Hertz relation is:
$f = kd = \frac{4}{3}\frac{ER^{1/2}\delta^{3/2}}{(1-\nu^2)}$
where f is the force applied to the cell, k is the spring constant of the cantilever, d is the deflection of the cantilever, E is the Young's modulus of the cell or other sample, R is the radius of the bead, δ is the indentation into the cell, and ν is the Poisson's ratio of the cell (a value related to the extent to which the sample maintains constant volume when deformed and often assumed to be near 0.5 for full volume conservation). For different geometries of the AFM tip the form of the Hertz relation varies, as detailed in several recent reports (Guz, Dokukin, Kalaparthi, & Sokolov, 2014; Melzak & Toca-Herrera, 2015; Thomas, Burnham, Camesano, & Wen, 2013)
7. *IF-specific AFM methods and results.* Detecting the mechanical effects of changes in IF expression by AFM depends on the way the cell is deformed. For example, loss of keratin leads to softening detected by small amplitude deformation of the cell surface, but often loss of vimentin does not. However, when cells are repeatedly deformed or deformed to greater depths, loss of vimentin becomes evident by changes in elastic modulus or plasticity index

at small strains, is not altered compared to WT cells, but that the absence of vimentin becomes apparent as cells are deformed more strongly or repeatedly. Loss of vimentin also increases the viscous loss during a cycle of cell deformation (Mendez et al., 2014). Another study on fibroblasts from KO vimentin mice, probed by optical magnetic twisting cytometry, demonstrated that the cortical rigidity of cells is not affected by the lack of vimentin, but the strain intensity applied to the cells during experiments was not reported (Guo et al., 2013). However, the same study showed that the interior cytoplasmic rigidity of fibroblasts null for vimentin, measured by active bead microrheometry, is reduced by a factor of 2, leading to an increased velocity of vesicular trafficking, suggesting that vimentin filaments are important to stabilize organelles position in the cytoplasm (Guo et al., 2013).

3.1.2 Drugs and Proteins Disrupting Vimentin

The contribution of vimentin networks to cell mechanical properties can also be assessed by using drugs specifically targeting vimentin or the enzymes that alter its phosphorylation, and inducing the disruption, aggregation, or depolymerization of its network. Withaferin A treatment induces disruption of the vimentin network and leads to its aggregation (Thaiparambil et al., 2011). Incubation of suspended natural killer (NK) cells with withaferin A induces a global cell softening of about 20%, when probed at large strains with a microfluidic optical stretcher (Gladilin et al., 2014). Calyculin A is a drug targeting vimentin phosphatases, and inducing the disruption of the vimentin network (Eriksson et al., 1992; Eriksson, Toivola, Sahlgren, Mikhailov, & Harmala-Brasken, 1998). In T lymphocytes, the vimentin network is organized as a cortical cage maintaining the mechanical integrity of the cell since the collapse of this network, with calyculin A, increases cell deformability by about 40% and presumably softens the cell, as quantified by high *g*-force centrifugation onto adhesive substrate followed by morphological analysis (Brown et al., 2001). Calyculin A is an inhibitor of vimentin protein phosphatases, but this molecule can also inhibit other cellular phosphatases like the one affecting myosin light chain and thus alter acto-myosin contractility of the cell. Acrylamide can also be used to induce the depolymerization and aggregation of vimentin IFs (Durham, Pena, & Carpenter, 1983; Eckert, 1985). Characterization of acrylamide-treated primary human articular chondrocytes shows that the loss of the vimentin network integrity induce a threefold softening of the entire cells as evaluated by applying large strains to cells embedded in alginate gels (Haudenschild et al., 2011). Acrylamide is used at low concentration in this study (4 m*M*) to limit its effect on other cytoskeletal components but 4 m*M* of acrylamide is enough to affect nuclear lamina architecture and mitochondrial homeostasis (Hay & De Boni, 1991). Moreover, at higher concentration, acrylamide disrupts actin and MT networks (Sager, 1989) so an effect of acrylamide on other cellular elements than vimentin cannot be excluded.

The overexpression of an oncogene protein (simian virus 40 large T antigen) is another way to induce a perinuclear reorganization of the vimentin network in human fibroblasts. This spatial reorganization of the network induces a twofold increase of the cytoplasmic young modulus, when quantified with colloidal probe force-mode AFM (Ducker et al., 1991), in the region where the vimentin density is increased (Rathje et al., 2014).

3.1.3 Mutated Desmin to Disrupt Vimentin

Another possible tool to disrupt the vimentin network is the expression of WT or mutated desmin in cells that originally express only vimentin. Desmin can copolymerize with vimentin and some dominant negative desmin mutants can induce the collapse of the vimentin network. The expression of the L345P desmin mutant in rat fibroblasts induces a perinuclear aggregation of the vimentin network correlated to a local stiffening of the cytoplasm in those areas as measured by AFM (Plodinec et al., 2011; Fig. 3). The effect of exogenous expression of three different desmin mutants on the elastic moduli of fibroblasts is shown in Fig. 3D. Expression of green fluorescent protein (GFP)-fused WT desmin has a small softening effect on the cell, possibly due to destabilization of the entire IF network by the GFP–desmin fusion protein. In contrast, the *des*A213V point mutant of desmin, which can form filaments, stiffens the cell not only possibly by increasing total IF content but also potentially by altering the mechanics of the IFs into which this mutant incorporates. The nonfilament forming desmin mutant *des*L345P has a complex effect on cell stiffness. It collapses the endogenous vimentin network around the nucleus, thereby strongly stiffening the perinuclear region, leaving the rest of the cell slightly less stiff than normal.

3.1.4 Mutated Desmin to Disrupt Desmin

Desmin is a type III IF specifically expressed in muscle cells. The study of primary myoblasts from a patient carrying the desmin mutation R350P by magnetic tweezers shows that the cortical rigidity of these cells is twice that of cells from a patient without this mutation, and their cortical stiffening under repeated stretching is threefold lower than for healthy human myoblasts (Bonakdar et al., 2012).

3.2 Keratins

Epithelial cells specifically express keratin belonging to type I (keratins 9–20) and II (keratins 1–8) IFs. The shear modulus of the keratin component of the isolated epithelial cell cytoskeleton ranges from approximately 34 Pa near

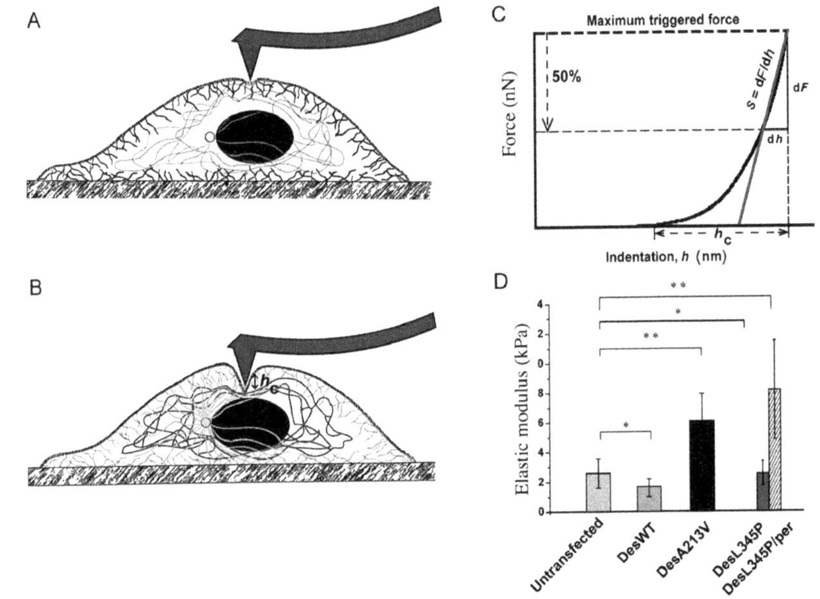

Figure 3 AFM indentation method for analyzing desmin and vimentin IF nanomechanics in cells. (A) When indenting a cell, the AFM tip first encounters the actin cytoskeleton (blue) below the plasma membrane and then (B) the intermediate filament network (red). (C) The retracting AFM force curve specifies the cell's response to the force *F* applied to indent the cell to a depth h_c. The force curve can be divided into two main segments. The lower segment corresponds to the response of the actin cytoskeleton beneath the plasma membrane, whereas the upper segment of the curve predominantly represents the response of the deeper intermediate filament network. A linear fit to the upper 50% of the force curve (red) is used to determine the elastic modulus. (D) Elastic modulus (E_s) of untransfected cells (rat-2 fibroblasts), cells transfected with WT desmin–GFP (DesWT) and cells expressing two types of desmin point mutants, DesA213V, which forms filaments and DesL345P, which does not. Solid bars denote the average stiffness of the whole cell or the region away from the nucleus, and the cross-hatched bar denotes the perinuclear area. The statistical analysis shows mean values of E_s and standard deviation ($*p < 0.05$, $**p < 0.0001$) (Plodinec et al., 2011). (See the color plate.)

the perinuclear area to 10 Pa near the cell edge (Sivaramakrishnan et al., 2008). This finding is consistent with studies of vimentin listed above that report a larger effect of disrupting IFs when cells are indented closer to the perinuclear region.

The involvement of keratins in cell mechanical properties has also been studied by drug-induced reorganization of the keratin network using sphingosylphosphorylcholine (SPC) to induce a perinuclear reorganization of keratin filaments in human pancreatic epithelial tumor cells (Beil et al., 2003). SPC treatment reduced the elastic moduli of treated cells, characterized with a parallel microplate cell stretcher, by 40%, and enhanced these cells ability to squeeze through small pores in a size-limited migration assay.

Double optical trapping experiments performed on suspended murine keratinocytes without keratin filaments show that deformability of these cells is increased about 60% (Seltmann et al., 2013). Another study on keratinocytes lacking keratin networks showed that the cell body Young's modulus of these cells is lowered by 40% as measured with AFM, and their intracytoplasmic viscosity is 40% lower as assessed with magnetic tweezers (Ramms et al., 2013).

3.3 Neurofilaments

Because NFs are localized to thin projections of neurons such as the axon of mature cells or neurite of developing cells or those reported *in vitro*, direct measurements of NF contributions to cell mechanics are less extensive than those of keratins or type III IFs. A recent study used magnetic tweezers to apply force to magnetic beads attached either to the cell body or the neurites of neurons cultured on adhesion lines patterned on a substrate. Creep curves similar to those in Fig. 2C showed that the high concentration of NFs in neurites caused them to be both stiffer than the cell body but also more viscous (Grevesse et al., 2015). The increased viscosity is attributed to the many transient interactions between NF sidearms that provide resistance to abrupt deformation but that can reorganize in response to prolonged forces, consistent with the rapidly reforming gels of NFs *in vitro* after they are disrupted by large strains (Wagner et al., 2007).

4. CONCLUSION

Macrorheological methods applied to purified networks formed by multiple types of IFs as well as AFM imaging and deformation of single IFs have revealed viscoelastic properties that differ from other biopolymers and that have potential to many biological functions. Measurements of the effects of IF disruption or deletion in cultured cells have revealed significant changes in cell mechanics, but thus far, usually more modest effects on cell stiffness than are produced, for example, by disrupting the actin network. The largest effects of IF disruption appear to be in keratin-containing cells and suspended cells such as lymphocytes. Future studies of systems that are closer to the *in vivo* context, such as confluent monolayers and cells in 3D culture might reveal additional mechanical effects of IF in many different cell types.

ACKNOWLEDGMENTS

We are grateful to Fitzroy Byfield for advice on AFM methods. This work was supported by grant GM096971 from the US National institutes of Health.

REFERENCES

Beil, M., Micoulet, A., von Wichert, G., Paschke, S., Walther, P., Omary, M. B., et al. (2003). Sphingosylphosphorylcholine regulates keratin network architecture and visco-elastic properties of human cancer cells. *Nature Cell Biology*, *5*, 803–811.

Bertaud, J., Qin, Z., & Buehler, M. J. (2010). Intermediate filament-deficient cells are mechanically softer at large deformation: A multi-scale simulation study. *Acta Biomaterialia*, *6*, 2457–2466.

Bonakdar, N., Luczak, J., Lautscham, L., Czonstke, M., Koch, T. M., Mainka, A., et al. (2012). Biomechanical characterization of a desminopathy in primary human myoblasts. *Biochemical and Biophysical Research Communications*, *419*, 703–707.

Bousquet, O., Ma, L. L., Yamada, S., Gu, C. H., Idei, T., Takahashi, K., et al. (2001). The nonhelical tail domain of keratin 14 promotes filament bundling and enhances the mechanical properties of keratin intermediate filaments in vitro. *Journal of Cell Biology*, *155*, 747–753.

Brown, M. J., Hallam, J. A., Colucci-Guyon, E., & Shaw, S. (2001). Rigidity of circulating lymphocytes is primarily conferred by vimentin intermediate filaments. *Journal of Immunology*, *166*, 6640–6646.

Chou, C. C., & Buehler, M. J. (2012). Structure and mechanical properties of human trichocyte keratin intermediate filament protein. *Biomacromolecules*, *13*, 3522–3532.

Chung, B. M., Rotty, J. D., & Coulombe, P. A. (2013). Networking galore: Intermediate filaments and cell migration. *Current Opinion in Cell Biology*, *25*, 600–612.

Ducker, W. A., Senden, T. J., & Pashley, R. M. (1991). Direct measurement of colloidal forces using an atomic force microscope. *Nature*, *353*, 239–241.

Durham, H. D., Pena, S. D., & Carpenter, S. (1983). The neurotoxins 2,5-hexanedione and acrylamide promote aggregation of intermediate filaments in cultured fibroblasts. *Muscle & Nerve*, *6*, 631–637.

Eckert, B. S. (1985). Alteration of intermediate filament distribution in PtK1 cells by acrylamide. *European Journal of Cell Biology*, *37*, 169–174.

Eckes, B., Dogic, D., Colucci-Guyon, E., Wang, N., Maniotis, A., Ingber, D., et al. (1998). Impaired mechanical stability, migration and contractile capacity in vimentin-deficient fibroblasts. *Journal of Cell Science*, *111*, 1897–1907.

Eriksson, J. E., Brautigan, D. L., Vallee, R., Olmsted, J., Fujiki, H., & Goldman, R. D. (1992). Cytoskeletal integrity in interphase cells requires protein phosphatase activity. *Proceedings of the National Academy of Sciences of the United States of America*, *89*, 11093–11097.

Eriksson, J. E., Toivola, D. M., Sahlgren, C., Mikhailov, A., & Harmala-Brasken, A. S. (1998). Strategies to assess phosphoprotein phosphatase and protein kinase-mediated regulation of the cytoskeleton. *Methods in Enzymology*, *298*, 542–569.

Esue, O., Carson, A. A., Tseng, Y., & Wirtz, D. (2006). A direct interaction between actin and vimentin filaments mediated by the tail domain of vimentin. *Journal of Biological Chemistry*, *281*, 30393–30399.

Fabry, B., Maksym, G. N., Butler, J. P., Glogauer, M., Navajas, D., & Fredberg, J. J. (2001). Scaling the microrheology of living cells. *Physical Review Letters*, *87*, 148102.

Gladilin, E., Gonzalez, P., & Eils, R. (2014). Dissecting the contribution of actin and vimentin intermediate filaments to mechanical phenotype of suspended cells using high-throughput deformability measurements and computational modeling. *Journal of Biomechanics*, *47*, 2598–2605.

Gou, J. P., Gotow, T., Janmey, P. A., & Leterrier, J. F. (1998). Regulation of neurofilament interactions in vitro by natural and synthetic polypeptides sharing Lys-Ser-Pro sequences with the heavy neurofilament subunit NF-H: Neurofilament crossbridging by antiparallel sidearm overlapping. *Medical & Biological Engineering & Computing*, *36*, 371–387.

Grevesse, T., Dabiri, B. E., Parker, K. K., & Gabriele, S. (2015). Opposite rheological properties of neuronal microcompartments predict axonal vulnerability in brain injury. *Scientific Reports*, *5*, 9475.

Guck, J., Ananthakrishnan, R., Mahmood, H., Moon, T. J., Cunningham, C. C., & Kas, J. (2001). The optical stretcher: A novel laser tool to micromanipulate cells. *Biophysical Journal*, *81*, 767–784.

Guo, M., Ehrlicher, A. J., Mahammad, S., Fabich, H., Jensen, M. H., Moore, J. R., et al. (2013). The role of vimentin intermediate filaments in cortical and cytoplasmic mechanics. *Biophysical Journal*, *105*, 1562–1568.

Guz, N., Dokukin, M., Kalaparthi, V., & Sokolov, I. (2014). If cell mechanics can be described by elastic modulus: Study of different models and probes used in indentation experiments. *Biophysical Journal*, *107*, 564–575.

Guzman, C., Jeney, S., Kreplak, L., Kasas, S., Kulik, A. J., Aebi, U., et al. (2006). Exploring the mechanical properties of single vimentin intermediate filaments by atomic force microscopy. *Journal of Molecular Biology*, *360*, 623–630.

Haase, K., & Pelling, A. E. (2015). Investigating cell mechanics with atomic force microscopy. *Journal of the Royal Society Interface*, *12*, 20140970.

Haudenschild, D. R., Chen, J. F., Pang, N. N., Steklov, N., Grogan, S. P., Lotz, M. K., et al. (2011). Vimentin contributes to changes in chondrocyte stiffness in osteoarthritis. *Journal of Orthopaedic Research*, *29*, 20–25.

Hay, M., & De Boni, U. (1991). Chromatin motion in neuronal interphase nuclei: Changes induced by disruption of intermediate filaments. *Cell Motility and the Cytoskeleton*, *18*, 63–75.

Holwell, T. A., Schweitzer, S. C., & Evans, R. M. (1997). Tetracycline regulated expression of vimentin in fibroblasts derived from vimentin null mice. *Journal of Cell Science*, *110*(Pt 16), 1947–1956.

Huisman, E. M., Wen, Q., Wang, Y. H., Cruz, K., Kitenbergs, G., Erglis, K., et al. (2011). Gelation of semiflexible polyelectrolytes by multivalent counterions. *Soft Matter*, *7*, 7257–7261.

Ivaska, J., Pallari, H. M., Nevo, J., & Eriksson, J. E. (2007). Novel functions of vimentin in cell adhesion, migration, and signaling. *Experimental Cell Research*, *313*, 2050–2062.

Janmey, P. A., Euteneuer, U., Traub, P., & Schliwa, M. (1991). Viscoelastic properties of vimentin compared with other filamentous biopolymer networks. *Journal of Cell Biology*, *113*, 155–160.

Janmey, P. A., McCormick, M. E., Rammensee, S., Leight, J. L., Georges, P. C., & MacKintosh, F. C. (2007). Negative normal stress in semiflexible biopolymer gels. *Nature Materials*, *6*, 48–51.

Janmey, P. A., Slochower, D. R., Wang, Y. H., Wen, Q., & Cebers, A. (2014). Polyelectrolyte properties of filamentous biopolymers and their consequences in biological fluids. *Soft Matter*, *10*, 1439–1449.

Klymkowsky, M. W. (1981). Intermediate filaments in 3T3 cells collapse after intracellular injection of a monoclonal anti-intermediate filament antibody. *Nature, 291*, 249–251.

Kollmannsberger, P., & Fabry, B. (2007). High-force magnetic tweezers with force feedback for biological applications. *The Review of Scientific Instruments, 78*, 114301.

Kreplak, L., Bar, H., Leterrier, J. F., Herrmann, H., & Aebi, U. (2005). Exploring the mechanical behavior of single intermediate filaments. *Journal of Molecular Biology, 354*, 569–577.

Leitner, A., Paust, T., Marti, O., Walther, P., Herrmann, H., & Beil, M. (2012). Properties of intermediate filament networks assembled from keratin 8 and 18 in the presence of Mg(2+). *Biophysical Journal, 103*, 195–201.

Leterrier, J. F., & Eyer, J. (1987). Properties of highly viscous gels formed by neurofilaments in vitro. A possible consequence of a specific inter-filament cross-bridging. *The Biochemical Journal, 245*, 93–101.

Leterrier, J. F., Kas, J., Hartwig, J., Vegners, R., & Janmey, P. A. (1996). Mechanical effects of neurofilament cross-bridges—Modulation by phosphorylation, lipids, and interactions with F-actin. *Journal of Biological Chemistry, 271*, 15687–15694.

Levy, R., & Maaloum, M. (2002). Measuring the spring constant of atomic force microscope cantilevers: Thermal fluctuations and other methods. *Nanotechnology, 13*, 33–37.

Lin, Y. C., Broedersz, C. P., Rowat, A. C., Wedig, T., Herrmann, H., MacKintosh, F. C., et al. (2010). Divalent cations crosslink vimentin intermediate filament tail domains to regulate network mechanics. *Journal of Molecular Biology, 399*, 637–644.

Lin, Y. C., Yao, N. Y., Broedersz, C. P., Herrmann, H., MacKintosh, F. C., & Weitz, D. A. (2010). Origins of elasticity in intermediate filament networks. *Physical Review Letters, 104*, 058101.

Lincoln, B., Schinkinger, S., Travis, K., Wottawah, F., Ebert, S., Sauer, F., et al. (2007). Reconfigurable microfluidic integration of a dual-beam laser trap with biomedical applications. *Biomedical Microdevices, 9*, 703–710.

Mege, J. L., Capo, C., Benoliel, A. M., Foa, C., & Bongrand, P. (1985). Study of cell deformability by a simple method. *Journal of Immunological Methods, 82*, 3–15.

Melzak, K. A., & Toca-Herrera, J. L. (2015). Atomic force microscopy and cells: Indentation profiles around the AFM tip, cell shape changes, and other examples of experimental factors affecting modeling. *Microscopy Research and Technique, 78*, 626–632.

Mendez, M. G., Restle, D., & Janmey, P. A. (2014). Vimentin enhances cell elastic behavior and protects against compressive stress. *Biophysical Journal, 107*, 314–323.

Mucke, N., Kreplak, L., Kirmse, R., Wedig, T., Herrmann, H., Aebi, U., et al. (2004). Assessing the flexibility of intermediate filaments by atomic force microscopy. *Journal of Molecular Biology, 335*, 1241–1250.

Pallari, H. M., & Eriksson, J. E. (2006). Intermediate filaments as signaling platforms. *Science's STKE, 2006*, pe53.

Pawelzyk, P., Mucke, N., Herrmann, H., & Willenbacher, N. (2014). Attractive interactions among intermediate filaments determine network mechanics in vitro. *PLoS One, 9*, e93194.

Plodinec, M., Loparic, M., Suetterlin, R., Herrmann, H., Aebi, U., & Schoenenberger, C. A. (2011). The nanomechanical properties of rat fibroblasts are modulated by interfering with the vimentin intermediate filament system. *Journal of Structural Biology, 174*, 476–484.

Ramms, L., Fabris, G., Windoffer, R., Schwarz, N., Springer, R., Zhou, C., et al. (2013). Keratins as the main component for the mechanical integrity of keratinocytes. *Proceedings of the National Academy of Sciences of the United States of America, 110*, 18513–18518.

Rathje, L. S. Z., Nordgren, N., Pettersson, T., Ronnlund, D., Widengren, J., Aspenstrom, P., et al. (2014). Oncogenes induce a vimentin filament collapse mediated

by HDAC6 that is linked to cell stiffness. *Proceedings of the National Academy of Sciences of the United States of America, 111*, 1515–1520.
Sager, P. R. (1989). Cytoskeletal effects of acrylamide and 2,5-hexanedione: Selective aggregation of vimentin filaments. *Toxicology and Applied Pharmacology, 97*, 141–155.
Sakamoto, Y., Boeda, B., & Etienne-Manneville, S. (2013). APC binds intermediate filaments and is required for their reorganization during cell migration. *Journal of Cell Biology, 200*, 249–258.
Schopferer, M., Bar, H., Hochstein, B., Sharma, S., Mucke, N., Herrmann, H., et al. (2009). Desmin and vimentin intermediate filament networks: Their viscoelastic properties investigated by mechanical rheometry. *Journal of Molecular Biology, 388*, 133–143.
Seltmann, K., Fritsch, A. W., Kas, J. A., & Magin, T. M. (2013). Keratins significantly contribute to cell stiffness and impact invasive behavior. *110*, 18507–18512.
Sivaramakrishnan, S., DeGiulio, J. V., Lorand, L., Goldman, R. D., & Ridge, K. M. (2008). Micromechanical properties of keratin intermediate filament networks. *Proceedings of the National Academy of Sciences of the United States of America, 105*, 889–894.
Thaiparambil, J. T., Bender, L., Ganesh, T., Kline, E., Patel, P., Liu, Y., et al. (2011). Withaferin A inhibits breast cancer invasion and metastasis at sub-cytotoxic doses by inducing vimentin disassembly and serine 56 phosphorylation. *International Journal of Cancer, 129*, 2744–2755.
Thomas, G., Burnham, N. A., Camesano, T. A., & Wen, Q. (2013). Measuring the mechanical properties of living cells using atomic force microscopy. *Journal of Visualized Experiments. 76*, http://dx.doi.org/10.3791/50497.
Wagner, O. I., Rammensee, S., Korde, N., Wen, Q., Leterrier, J. F., & Janmey, P. A. (2007). Softness, strength and self-repair in intermediate filament networks. *Experimental Cell Research, 313*, 2228–2235.
Wang, N., & Ingber, D. E. (1994). Control of cytoskeletal mechanics by extracellular matrix, cell shape, and mechanical tension. *Biophysical Journal, 66*, 2181–2189.
Wang, N., & Stamenovic, D. (2000). Contribution of intermediate filaments to cell stiffness, stiffening, and growth. *American Journal of Physiology. Cell Physiology, 279*, C188–C194.
Yamada, S., Wirtz, D., & Coulombe, P. A. (2003). The mechanical properties of simple epithelial keratins 8 and 18: Discriminating between interfacial and bulk elasticities. *Journal of Structural Biology, 143*, 45–55.

CHAPTER THREE

Multidimensional Monitoring of Keratin Intermediate Filaments in Cultured Cells and Tissues

Nicole Schwarz[1], Marcin Moch[1], Reinhard Windoffer, Rudolf E. Leube[2]
Institute of Molecular and Cellular Anatomy, RWTH Aachen University, Aachen, Germany
[2]Corresponding author: e-mail address: rleube@ukaachen.de

Contents

1. Introduction 60
2. 3D Imaging of Keratin Intermediate Filaments in Cultured Cells 62
   2.1 Microscopes 64
   2.2 Preparation of Imaging Medium 65
   2.3 Surface Coating 65
   2.4 Preparation of Cells for Imaging 66
   2.5 Image Acquisition 67
   2.6 Keratin Network Normalization 67
   2.7 Measuring Keratin Movement 68
   2.8 Keratin Bulk Flow Analysis 70
   2.9 Keratin Turnover Measured by FRAP 72
3. 3D Imaging of Keratin Intermediate Filaments in Murine Preimplantation Embryos 75
   3.1 Embryo Collection and Cultivation 75
   3.2 Imaging of Preimplantation Embryos 76
   3.3 Image Processing and Analysis 78
4. Outlook 80
Acknowledgments 81
References 81

Abstract

Keratin filaments are a hallmark of epithelial differentiation. Their cell type-specific spatial organization and dynamic properties reflect and support epithelial function. To study this interdependency, imaging of fluorescently tagged keratins is a widely used method by which the temporospatial organization and behavior of the keratin intermediate filament network can be analyzed in living cells. Here, we describe methods that have been adapted and optimized to dissect and quantify keratin intermediate filament network dynamics in vital cultured cells and functional tissues.

[1] These authors contributed equally to this work.

Methods in Enzymology, Volume 568
ISSN 0076-6879
http://dx.doi.org/10.1016/bs.mie.2015.07.034

1. INTRODUCTION

The intermediate filament network is a complex three-dimensional (3D) scaffold in metazoan cells contributing for the most part to the mechanical stability of cells and was therefore originally considered static in nature. This view was abandoned when different stimuli were found to induce profound organizational changes. During mitosis, for example, keratin intermediate filament networks are rapidly disassembled and appear as aggregates, followed by their reassembly in daughter cells (Franke, Schmid, Grund, & Geiger, 1982; Lane, Goodman, & Trejdosiewicz, 1982). The precise sequel of changes in the organization of the intermediate filament cytoskeleton, however, was difficult to deduce from still images. Subsequently, methods became available, which allowed monitoring of intermediate filaments in living cells.

Early approaches were based on the injection of fluorescently tagged proteins and antibodies (Mittal, Sanger, & Sanger, 1989; Okabe, Miyasaka, & Hirokawa, 1993; Vikstrom, Lim, Goldman, & Borisy, 1992), which revealed details of network reorganization during mitosis and aspects of filament turnover during interphase. Major problems in these studies were the precise regulation of expression and rapid photobleaching, which prevented long-term live cell imaging.

With the introduction of green fluorescent protein (GFP) as a marker for expression in 1994 (Chalfie, Tu, Euskirchen, Ward, & Prasher, 1994) and subsequent usage as a protein tag (Cubitt et al., 1995), imaging of protein distribution in living cells has been revolutionized. Initial studies on GFP-vimentin revealed an intrinsically dynamic and shape-changing network (Ho, Martys, Mikhailov, Gundersen, & Liem, 1998; Yoon, Moir, Prahlad, & Goldman, 1998). During spreading and adhesion of cells, three structural forms of intermediate filaments were distinguished: motile small particles, short nonconnected filamentous structures referred to as squiggles, and long intermediate filaments organized in a network (Prahlad, Yoon, Moir, Vale, & Goldman, 1998). It was proposed that the particles give rise to squiggles, which are incorporated into longer filaments (Prahlad et al., 1998). In neuronal cells, rapidly moving particles were identified by live cell imaging of GFP-tagged neurofilaments which intercalate into neurofilament bundles (Wang, Ho, Sun, Liem, & Brown, 2000; Yuan et al., 2009). Interestingly, short filaments termed keratin precursors were identified in the periphery of keratin-containing epithelial cells (Windoffer & Leube, 1999).

In case of the keratin intermediate filaments, a turnover cycle was proposed based on time-lapse fluorescence imaging, fluorescence recovery after photobleaching (FRAP) experiments, and photoactivation studies (Kölsch, Windoffer, Würflinger, Aach, Leube, 2010; Leube, Moch, Kölsch, & Windoffer, 2011; Windoffer, Beil, Magin, & Leube, 2011; Windoffer, Wöll, Strnad, & Leube, 2004; Yoon et al., 2001). Thus, keratin filament precursors nucleate in the cell periphery and constantly move inward, while they elongate and integrate into the peripheral keratin network. The network either matures into a stable nuclear cage or disassembles into soluble and highly diffusible subunits. These soluble subunits are reutilized in the cell periphery for another cycle of assembly and disassembly. Movie 1 (http://dx.doi.org/10.1016/bs.mie.2015.07.034) and corresponding Fig. 1 highlight features of the keratin cycle.

The studies made so far face two major challenges. On one hand, the subtle and physiologically most relevant alterations in keratin dynamics are often missed and/or difficult to measure (Baribault, Blouin, Bourgon, & Marceau, 1989; Chung, Murray, Eyk, & Coulombe, 2012; Keski-Oja, Lehto, & Virtanen, 1981; Ku & Omary, 1997). To overcome this, we developed a method to quantitatively map keratin intermediate filament motility and turnover at subcellular resolution (Moch, Herberich, Aach, Leube, & Windoffer, 2013). On the other hand, the additional expression of the fluorescently tagged keratin filaments may perturb network dynamics and organization (Windoffer & Leube, 1999). To mimic

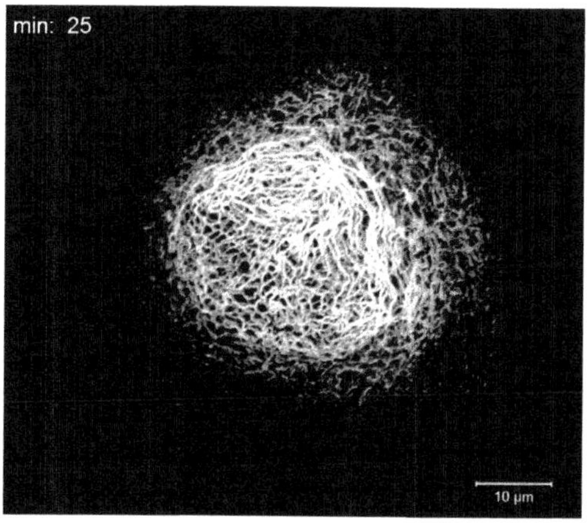

Movie 1 Time-lapse fluorescence recording of keratin 13-EGFP in AK13-pax1 cells (corresponding Fig. 1). Major steps of the keratin cycle, namely nucleation, elongation, integration, and bundling, are seen.

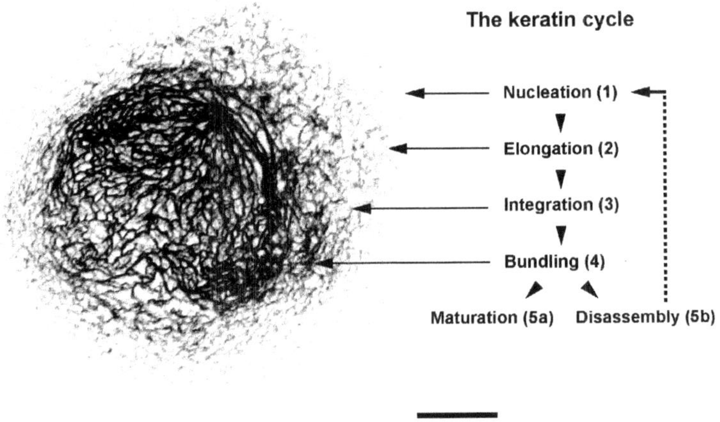

Figure 1 Keratin cycle of assembly and disassembly. The figure shows a representative still image of a time-lapse recording of keratin 13-EGFP fluorescence (inverse presentation) in vulvar carcinoma-derived AK13-pax1 cells (Leube, Moch, & Windoffer, 2015). Cells were plated on a laminin 332-rich matrix 20 h prior to confocal microscopy. The confocal plane shown represents the bottom part of the cell below the nucleus. On the right, the major steps of the keratin cycle, which can be best appreciated in Movie 1, are denoted: nucleation of keratin particles in the cell periphery (1), elongation of these particles (2), subsequent integration of the inward-moving and growing particles into the peripheral network (3), bundling of inward-moving filaments (4), and either maturation into a stable perinuclear cage-like structure (5a) or disassembly into rapidly diffusible subunits (5b) that can be reutilized for another round of assembly and disassembly. Scale bar: 10 μm.

the wild-type situation as closely as possible, we generated a knock-in mouse model that produces fluorescently tagged keratin 8 instead of the endogenous keratin 8 (Schwarz, Windoffer, Magin, & Leube, 2015). This model allows the analysis of keratin dynamics in the physiological 3D context of the preimplantation embryo and facilitates the examination of *de novo* network formation.

2. 3D IMAGING OF KERATIN INTERMEDIATE FILAMENTS IN CULTURED CELLS

Imaging keratins in cells producing fluorescently tagged keratins has become routine in several laboratories (Beriault et al., 2012; Fois et al., 2013; Liovic, Mogensen, Prescott, & Lane, 2003; Rolli, Seufferlein, Kemkemer, & Spatz, 2010; Windoffer & Leube, 1999; Yoon et al., 2001).

Conditions have been described to monitor keratin dynamics in 3D during the entire cell cycle (Windoffer & Leube, 2004). Examples are presented in Fig. 2 and corresponding Movie 2 (http://dx.doi.org/10.1016/bs.mie.2015.07.034). They show that the majority of the keratin network is located in the basal part of cultured cells adjacent to the extracellular matrix. It surrounds the nucleus in a cage-like structure and thins out toward the apical cytoplasm (Fig. 2A). Rapid breakdown of the network is detected at the onset of mitosis resulting in multiple granular aggregates, which are used for reassembly of the network in the daughter cells after completion of cell division (Windoffer & Leube, 2001, 2004). The sequence shown in Fig. 2B and Movie 2 demonstrates that the distribution of keratins can be monitored in 3D for long periods without disturbing cell viability and photobleaching. But keep in mind that scanning cells in three dimensions considerably increases cell stress by

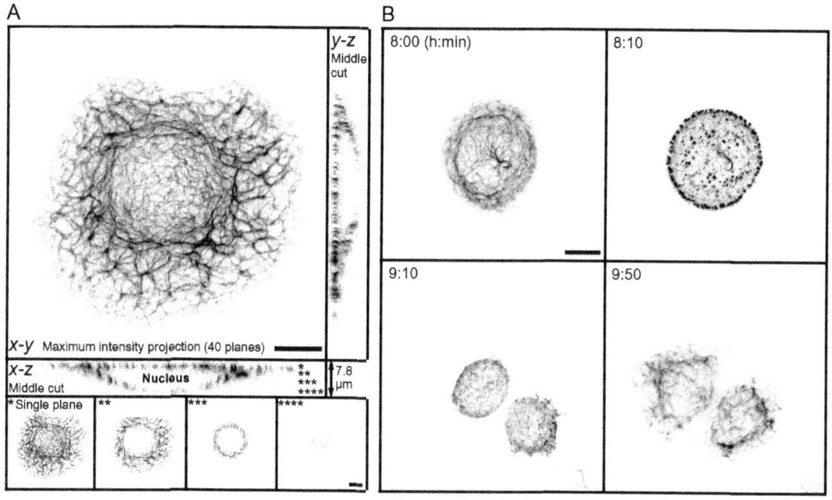

Figure 2 Keratin network morphology in interphase and during cell division. AK13-1 cells stably expressing keratin 13-EGFP were cultivated on a laminin 332-rich matrix without FCS for 48 h (A) and for 60 h (B). The fluorescence recorded by confocal laser scanning microscopy in the cell shown in (A) at high resolution depicts the arrangement of the keratin network during interphase as a maximum intensity projection of 40 planes and transverse sections on top and as single plane recordings at the bottom. The keratin network forms a cage around the nucleus and is mostly concentrated at the bottom plane of the cell (single star). The images in (B) are taken from Movie 2 that was recorded at lower spatial resolution (13 planes, 0.88 μm steps). In this instance, background was removed by Gaussian filtering using Fiji software. The inverse fluorescence micrographs show stages of keratin network disassembly before cell division and network reassembly thereafter. These processes occur within minutes and are independent of protein degradation and biosynthesis. Scale bars: 10 μm.

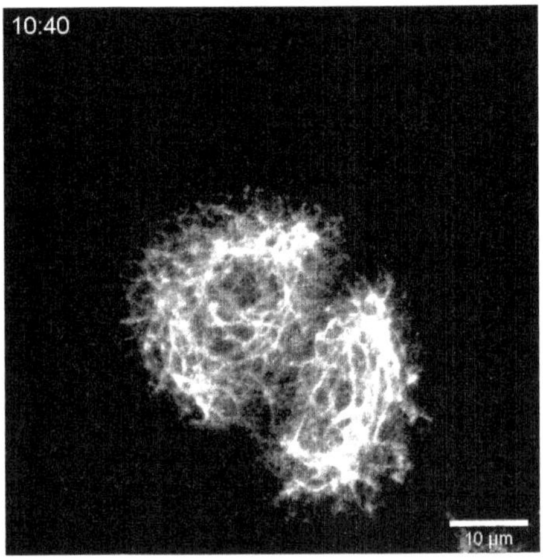

Movie 2 Maximum intensity projections of keratin 13-EGFP fluorescence detecting keratin network reorganization during cell division (corresponding Fig. 2B).

generating reactive oxygen species. The damage is not solely restricted to the imaged section but affects the entire cell. The stress is substantial when fluorescence is recorded repeatedly at high spatial resolution. Addition of antioxidants (e.g., 4 m*M* *N*-acetyl-L-cysteine; Strnad, Windoffer, & Leube, 2003) may alleviate the problem, although we never observed beneficial effects for the protocols presented here.

In the following sections, we will focus on new methodology to measure keratin dynamics in living cells under standardized conditions.

2.1 Microscopes

The methods described here were developed for a conventional confocal laser scanning microscope (LSM 710 Duo; Carl Zeiss). Excitation of EGFP or EYFP is performed at 488 nm using an Argon-ion laser (laser module LGK 7872 ML8). Images are acquired with a 63 × 1.4 N.A. Differential interference contrast (DIC) oil immersion objective. Focus drift in time series is corrected with the Definite Focus system (Carl Zeiss). The entire microscope is encased in a Plexiglas incubation chamber, and stable temperature is maintained with the help of a heating unit (Incubator XL LSM 710 S1). The complete microscope is mounted on a Vibration Isolated Workstation (Newport Corporation). Alternatively, other confocal laser scanning microscopes can be used. But it should be kept in mind that the keratin motion analysis software described below was not developed for charge-coupled device (CCD) cameras that are used, for example, in confocal spinning disc microscopy.

2.2 Preparation of Imaging Medium

For optimal fluorescence signal quality, experiments should be performed in cell culture medium without phenol red to avoid autofluorescence. If the microscope is not equipped with a CO_2 climate chamber, commercially available HEPES-buffered cell medium or self-made Hanks-HEPES imaging medium can be added prior to experiments. Be aware that several commercial media are often yellowish, which will increase in intensity upon addition of fetal calf serum (FCS). In our experience, however, this does not affect the signal quality of green and yellow fluorophores:

1. Prepare 500 ml Hanks-HEPES imaging medium by adding 50 ml Hanks' Balanced Salt Solution (HBSS, 10 ×; 14185052, Life Technologies), 5 ml nonessential amino acids (100 ×; 1114-035, Life Technologies), 10 ml essential amino acids (50 ×; 11130-036, Life Technologies), 5 ml GlutaMAX™ (35050-038, Life Technologies), 12.5 ml HEPES (1 *M*; 15630-056, Life Technologies), 5 ml *N*-acetyl-L-cysteine (0.48 *M*, freshly dissolved; A7250, Sigma-Aldrich), and 0.175 g sodium bicarbonate (HN01, Carl Roth).
2. Adjust volume to 500 ml using distilled H_2O (Aqua B. Braun).
3. Adjust pH to 7.4 using HCl.
4. Sterile filtrate through a 0.22-μm filter (Merck Millipore).
5. Store up to a month at 4 °C or for long-term at −20 °C.

2.3 Surface Coating

For optical imaging, cells are typically grown on glass. For cell adhesion and spreading, the glass surface should be coated with appropriate matrix proteins. The choice of matrix proteins affects cell shape and intracellular signaling. Ideally, single cells should be circular, flat, and stably attached. A number of commercial matrix proteins may fulfill these requirements. For the epithelial cell lines that we routinely use, we prefer a laminin 332-rich matrix preparation from 804G cells (Langhofer, Hopkinson, & Jones, 1993):

1. Grow rat bladder carcinoma-derived 804G cells in Dulbecco's Modified Eagle Medium (DMEM; D0819, Sigma-Aldrich) supplemented with 10% FCS (Life Technologies) to confluence in a glass bottom dish (35 mm Petri dish with 14 mm glass surface diameter and glass thickness no. 1.5; P35G-0.17-14-C, MatTek).
2. "Deroof" cells 1 day after reaching confluence by adding ice-cold 20 m*M* NH_4OH and incubating at room temperature for 10–15 min.

3. Discard NH_4OH and flush off remaining debris using distilled H_2O (Aqua B. Braun). This procedure will leave behind extracellular proteins on the glass surface.
4. Wash surface two times with distilled H_2O and two times with PBS (D8537, Sigma-Aldrich).
5. Remove PBS.
6. Coated vessels can be stored up to 2 months at −20 °C.

2.4 Preparation of Cells for Imaging

It is important to carefully select the conditions for cell culture to achieve reproducible kinetics of spreading and adhesion as well as growth and cell densities at defined times after passaging. Even then, heterogeneity is still a major challenge in quantitative imaging of intracellular structures in single cells of clonal cell lines. Another concern is biochemical stress. We therefore cultivate our cells in the absence of antibiotics and selective drugs. For cell lines that are difficult to grow at low cell density, we recommend preconditioned culture medium. This medium is obtained from dense cell cultures and is enriched in secreted factors. It is mixed with fresh cell culture medium at ratios that are empirically determined:

1. Seed ≈25,000 A431 cells in DMEM without FCS on glass bottom dishes coated with laminin 332-rich matrix (see Protocol 2.3).
2. Cells are typically kept in the incubator for 19–57 h prior to use.
3. Prewarm the microscope to 37 °C. Note that the objective will take longer until it is warm when it has no additional heating. It is advisable to test the temperature with a thermometer in a small water reservoir. Thorough adjustment and equilibration of the temperature is crucial for focal stability during image acquisition.
4. Prewarm Hanks-HEPES imaging medium (see Protocol 2.2) to 37 °C.
5. Remove cell culture medium from the glass bottom dish and add 1 ml Hanks-HEPES imaging medium.
6. Select suitable cells for time-lapse recording of fluorescence and phase contrast. Important criteria are that the selected cells are representative for all cells in a given culture dish and that most of the fluorescence signal is restricted to a single confocal plane. This is usually the case in extremely flat cells. For generation of topological maps of keratin dynamics, selected cells should have a similar size and a pancake-like circular shape with the nucleus in the cell center. Cells with more than one nucleus should be excluded from analysis.

2.5 Image Acquisition

The motion analysis method described here is optimized for confocal laser scanning microscopy where a photomultiplier tube (PMT) is used for signal detection. In comparison to images obtained with CCDs, such recordings provide a higher-than-average noise to signal ratio and eliminate efficiently out of focus signals. The slow laser scanning speed, which is in the range of several seconds per frame, is not problematic because keratins within the network move rather slowly. Of note, the motility of keratin precursors outside of the keratin network is not reliably measured by the method described here, because many of these rare particles (less than 1% of all keratins) are removed by the denoising during image analysis:

1. The scanning time per 67 μm × 67 μm should be less than 30 s to avoid distortion of moving structures (22 s in the examples shown here).
2. The images should be acquired at a resolution of 1024 × 1024 pixel. The signal detection range should be 16 bit.
3. The signal should be detected in a single scan for best accuracy without averaging of multiple scans into one image. It is also recommended to use unidirectional scanning instead of faster bidirectional scanning because of possible misalignment of scanned lines.
4. A laser intensity should be selected with minimal bleaching of fluorophores. In general, the signal does not have to be very strong as long as the background is low. This aspect is also relevant for choosing the optimal detector gain.
5. The pinhole can be slightly opened to increase the amount of detected light (2 airy units are optimal). In this way, the thickness of the focal plane is increased and thereby allows a more complete detection of the keratin network.
6. It is essential that the focus remains stable during the entire imaging period. Since manual adjustment requires extreme patience and diligence, an automated focus stabilization system is highly recommended. Some older microscopes can be upgraded by the manufacturer with such a device.
7. Time series are routinely done at intervals of 30 s for 10–15 min.

2.6 Keratin Network Normalization

Cells that are analyzed for longer time periods often show cell shape changes, especially when treated with different modulators. These changes affect the keratin network morphology and result in keratin displacement that is not related to active keratin transport inside the cell. These anomalies can be corrected in two different ways. When the changes occur only in a small part of the cell, it is possible to remove the regions from analysis by cutting them out or blackening. However, the faster and more elegant solution is to normalize the keratin network into a circular shape with a defined diameter. This can be done with the KeraMove software (http://www.moca.rwth-aachen.de/pubs/2013_01/softwarepackage_KeraMove_KeraDyn.zip). In this process, the border of the keratin network (which does not correspond to the cell border) is extended in every frame to the border of the predefined shape. This procedure sometimes induces errors but can help to analyze and compare data obtained from differently sized and shaped cells. The size of the newly produced networks is not related in a linear way to the data before transformation, but the keratin networks in these images can be tracked efficiently by the KeraMove software. This method is recommended for cells that show cell shape changes less than 10% of the total cell area as assessed by measuring the extension of the fluorescently tagged keratin network. It can yield almost perfectly symmetrical heat maps of keratin movement and bulk flow when results from many cells are overlaid:

1. Run the open source software KeraMove (http://www.moca.rwth-aachen.de/pubs/2013_01/softwarepackage_KeraMove_KeraDyn.zip).
2. Select "perform shape normalization."

2.7 Measuring Keratin Movement

The motion of fluorescently tagged keratin networks from confocal time-lapse recordings can be detected with the open source software KeraMove (Moch et al., 2013). The software and a user manual are provided at http://www.moca.rwth-aachen.de/pubs/2013_01/softwarepackage_KeraMove_KeraDyn.zip:

1. Start KeraMove as described in the manual. Set "maximum sought displacement" to 1200 nm. The software generates maps of mean keratin movement per pixel from analyzed recordings (output file: magOfMeanMotion.raw). The speed is calculated at the level of single pixels from the registered filaments. Therefore, the displacement of every of these pixels is tracked independently of its direction as long as it occurs in the 2D plane of the cell bottom. If keratin network normalization is needed, proceed as described in Section 2.6.
2. Import output file into Fiji software (freely available at http://fiji.sc/Fiji) as a heat map (submenu: File → import → raw; image type: 32 bit real,

little-endian byte order). The speed of movement is coded in 32-bit gray values in the same unit as in the analyzed recordings. The calculated speed corresponds to the sum of speed calculated for all frames divided by the number of frames. Do not forget that the speed has to be multiplied by the appropriate factor to obtain the speed per minute. For example, if images were acquired at 30 s intervals, the speed has to be multiplied by 2. Next, a threshold must be determined to remove noise caused by, e.g., Brownian motion and vibrations of the system. The threshold has to be removed from the calculated speed (gray values) for every pixel. We usually use a threshold of 100 nm min^{-1}. The following script performs all of these calculations in Fiji software and measures the total mean speed as gray value:

```
run("Multiply...", "value=2");
setThreshold(100, 500000);
wait(100);
run("Create Selection");
run("Measure");
```

If the mean gray value is not shown, it has to be activated in the following submenu: Analyze → Set Measurements → Mean gray value.

3. Prepare histograms by importing the measurements into GraphPad Prism (GraphPad Software).
4. The image displayed at the end of step 2 shows a heat map of mean keratin movement.
5. Note that the saved image still includes the background, which can be permanently removed (submenu: Process → Math → Min).
6. The differences in keratin movement can be visualized by addition of a color LUT (submenu: Image → Lookup Tables). For presentation of the heat maps, the removed gray values have to be considered in the lettering of the color scale.
7. Heat maps of keratin movement from multiple recordings of different cells can be overlaid and averaged (submenu: Image → Stacks → Z-Project → Average Intensity).

The motion analysis process is illustrated in Fig. 3. Figure 4A shows the averaged results for multiple recordings as both heat maps and histograms. A comparison of nonnormalized versus normalized image data highlights similarities in the overall pattern, i.e., no net movement in the cell center and peaks of keratin motility in a ring-shaped domain of the peripheral cytoplasm.

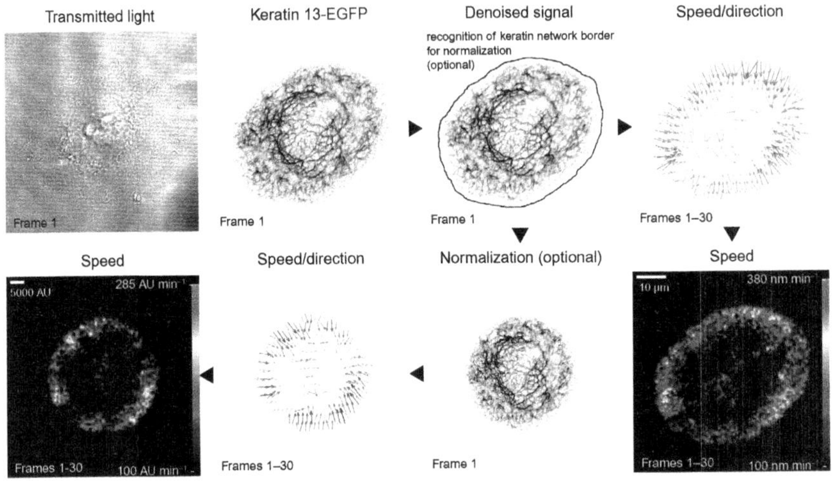

Figure 3 Calculation of keratin speed from time-lapse recordings. The pictures show steps of recording and analyzing keratin motion in a single AK13-1 cell. AK13-1 cells producing keratin 13-EGFP were plated on a laminin 332-rich matrix for 52 h prior to imaging. Phase contrast (transmitted light) and keratin 13-EGFP fluorescence (inverse presentation) were recorded for 15 min every 30 s in the bottom plane. The background was reduced by Anscombe curvelet transform-based denoising (cf. Moch et al., 2013). Optionally, the overall keratin network shape was automatically delineated (black outline) for normalization into a standard circular shape with a defined diameter. A comparison of the non-normalized and the normalized recordings is shown in Movie 3 (http://dx.doi.org/10.1016/bs.mie.2015.07.034). The results from motion analysis were then used to prepare vector maps to depict the direction of movement (corresponding to direction of arrows, which is mostly toward the cell interior). The vectors furthermore show the speed of keratin movement, which corresponds to length and thickness of the vectors. In addition, the speed is shown in detail in heat maps by a color scale as indicated in the images (the speed values correspond to mean speed per pixel). (See the color plate.)

2.8 Keratin Bulk Flow Analysis

During interphase, the keratin network is continuously moving from the cell periphery toward the cell interior. But this movement does not lead to noticeable keratin accumulation in the cell center because keratin filaments are disassembled into soluble subunits (sinks) that are reused for filament formation in the cell periphery (sources). The resulting keratin bulk flow can be measured from time-lapse recordings of fluorescently tagged keratins with the open source software KeraDyn (Moch et al., 2013). The software determines keratin filament mass from fluorescence intensity (see Protocol 2.7). The algorithm only measures the assembly and disassembly of moving keratin filaments, because nonmotile keratins are not registered by KeraMove:

1. Run the open source software KeraDyn (http://www.moca.rwth-aachen.de/pubs/2013_01/softwarepackage_KeraMove_KeraDyn.zip). The

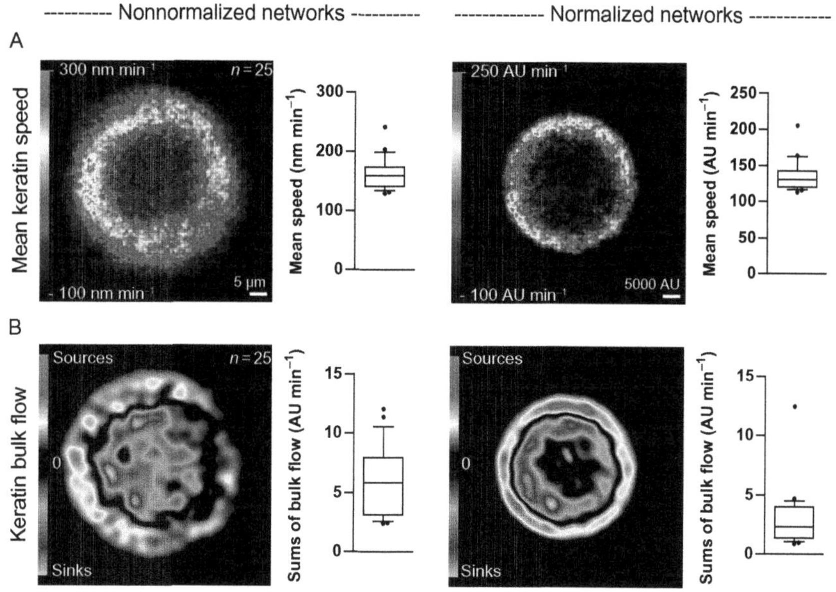

Figure 4 Quantitative measurement of keratin speed (A) and bulk flow (B) in nonnormalized and normalized fluorescence recordings of AK13-1 cells expressing keratin 13-EGFP. Fluorescence was recorded in the bottom plane of the cells by confocal microscopy at 30 s intervals for 15 min. The data show the compiled results of standardized measurements of 25 single cell recordings. They are presented as heat maps with subcellular resolution and as whisker box plots (10–90% percentiles). In (A), the heat maps reveal that keratins move faster in the cell periphery than in the cell center underneath the nucleus. The diagrams show that keratin filaments are moving with a median speed of 160 nm min^{-1} before normalization and with 131 AU min^{-1} after normalization of the network shape. In (B), the heat maps show that keratin filaments are primarily assembled in the cell periphery and disassembled in the perinuclear area. Between these two zones keratin is transported and no net keratin assembly or disassembly is detected. The keratin bulk flow can also be described in diagrams as shown in AU. Note that the AU results in diagrams are only of value when different populations/conditions are compared with each other or when data distribution is of interest (e.g., Gaussian distribution). (See the color plate.)

software generates a map with keratin bulk flow per pixel from analyzed time-lapse recordings of keratin motion (output file: mean_turnover.raw).

2. Import the output file into Fiji software as a heat map in the same way as described for keratin motion maps (see Protocol 2.7). Positive gray values correspond to keratin assembly and negative gray values to keratin disassembly in arbitrary units (AU).
3. For calculation of keratin bulk flow, it is sufficient to measure only the keratin assembly, because sources and sinks balance each other at equilibrium. Therefore, positive gray values are determined by setting the

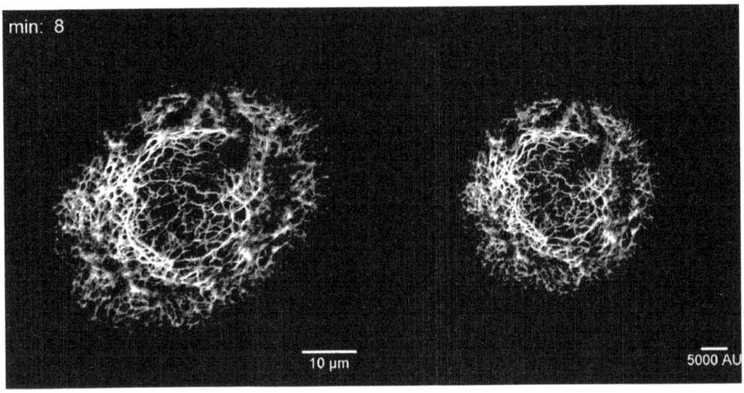

Movie 3 Denoised recordings of an AK13-1 cell expressing keratin 13-EGFP in a non-normalized (left) and normalized (right) format (corresponding Fig. 3). Note that normalization leads to occasional jumps because of extension/shrinkage of the overall network (time points 13 and 14 min).

threshold to zero. In this way, values equal to zero or smaller are not measured. The following script performs all of these calculations in Fiji software and measures the total bulk flow as integrated densitometry:

```
setThreshold(0.0000000, 1000000000000);
run("NaN Background");
wait(100);
run("Measure");
```

If the integrated densitometry is not shown, it has to be activated in the following submenu: Analyze → Set Measurements → Integrated density.

4. The heat maps of keratin bulk flow can be presented in the same way as described for keratin motion (see Protocol 2.7). Do not remove the background, because only sources would be displayed.
5. Heat maps of keratin bulk flow from multiple recordings of different cells can be overlaid and averaged (submenu: Image → Stacks → Z-Project → Average Intensity).

Figure 4B presents the bulk flow analysis of 25 recordings. Sources of keratin filament fluorescence are detected in the cell periphery (pink–yellow), sinks in the perinuclear cytoplasm (green–blue). Both regions are separated by an area without net turnover (black).

2.9 Keratin Turnover Measured by FRAP

Photobleaching is the photochemical destruction of fluorescence capabilities of a fluorophore. Fluorophores in specific cell regions can be bleached with lasers that are especially developed for this purpose or in confocal microscopy by the scanning laser at higher intensities. A major problem in FRAP experiments is phototoxic cell damage. In addition, care should be taken to apply bleaching intensities that do not induce cell retraction or other morphological changes.

The method described here is optimized for bleaching keratin filaments in single cells that are not connected to other cells. In these cells, half of the keratin network area is partially bleached which leads to a decrease in total fluorescence. It is important that the bleached network-half has similar organization and brightness (±10%) in comparison to the unbleached part. This will mostly be the case when the keratin cage surrounding the nucleus is localized in the cell center:

1. Choose a cell that fits in a 67 μm × 67 μm area and bleach half of the cell by repeated scanning with 100% laser power for 10 s. It is not necessary to bleach at maximum efficiency. 10–50% is sufficient. It is important, however, to be able to detect the bleached region reliably to account for cell shape changes during the fluorescence recovery time. It is also not necessary to bleach every sample with the same efficiency.
2. Fluorescence transfer from the unbleached to the bleached part is recorded at 0.2% laser intensity in three dimensions at 512 × 512 pixel (≈30 s per projection of complete cell). The chosen scanning speed (image quality) should be sufficient to distinguish the border between the unbleached and bleached regions. The recording of fluorescence recovery should be started directly after bleaching. The fluorescence transfer is recorded for 25 min at time intervals of 5 min.
3. The fluorescence recovery is calculated as the percentage of fluorescence flux from the unbleached to the bleached network part. The ideal choice for control is the unbleached half of the keratin network. The z-stacks from every time interval are opened in Fiji software, and 32-bit sum projections are generated. The background is measured by integrated densitometry at four arbitrary locations of 9.82 μm × 9.82 μm for every time point outside of the cell. It is averaged per pixel and subtracted from the corresponding sum projection.
4. The bleached and unbleached cell halves are outlined by hand with Polygon selections tool for every time point, and corresponding sums of gray values are measured as integrated densitometry. When necessary the regions are corrected at other time points.
5. The fluorescence in the bleached area is defined as $I^b_{t=n}$, and in the unbleached control area as $I^u_{t=n}$ (I=fluorescence intensity, t=time point). For calculation of keratin turnover, $I^b_{t=0}$ is subtracted from $I^b_{t=n}$

and $I^{u}_{t=n}$ for every time point, and the resulting difference corresponds to the keratin flux from the bleached to unbleached part. For example, at time point zero after bleaching the fluorescence in the unbleached part will always be 100% and in the bleached part 0%.

Figure 5 presents the results of FRAP experiments using the parameters described above.

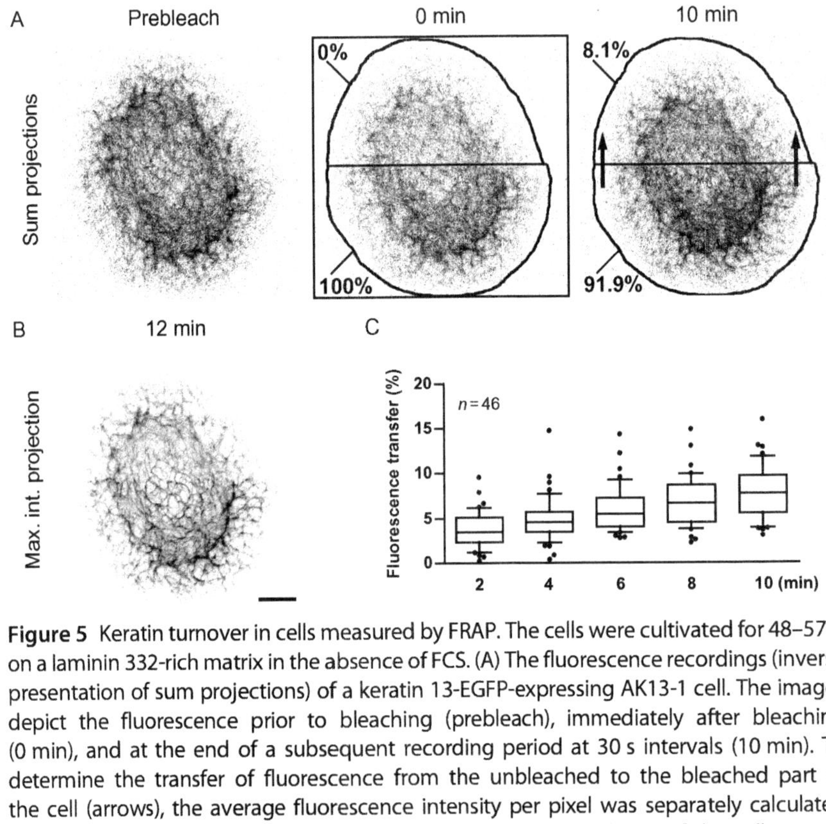

Figure 5 Keratin turnover in cells measured by FRAP. The cells were cultivated for 48–57 h on a laminin 332-rich matrix in the absence of FCS. (A) The fluorescence recordings (inverse presentation of sum projections) of a keratin 13-EGFP-expressing AK13-1 cell. The images depict the fluorescence prior to bleaching (prebleach), immediately after bleaching (0 min), and at the end of a subsequent recording period at 30 s intervals (10 min). To determine the transfer of fluorescence from the unbleached to the bleached part of the cell (arrows), the average fluorescence intensity per pixel was separately calculated for both parts. Next, the fluorescence intensity in the bleached part of the cell was set as 0% at time point 0 min. This value was subtracted from the fluorescence intensity in the unbleached part of the cell, and the resulting fluorescence intensity was defined as 100%. After 10 min, 8.1% of the initial fluorescence in the lower part had been translocated to the upper part. (B) A high-resolution scan as a maximum intensity projection (max. int. projection) of the cell in (A). Note that the border between the bleached and unbleached cell halves is still well demarcated. (C) A histogram of the results of fluorescence transfer measurements as exemplified in (A). The measurements were done in 46 single cells, and the results are shown for time points 2, 4, 6, 8, and 10 min postbleach. The whiskers are 10–90% percentiles. Scale bar: 10 μm.

3. 3D IMAGING OF KERATIN INTERMEDIATE FILAMENTS IN MURINE PREIMPLANTATION EMBRYOS

In the recently described mouse model, which carries a knock-in allele coding for a fluorescently tagged keratin 8 (Krt8-YFP), monitoring the *de novo* synthesis and dynamics of the keratin intermediate filament network has become possible in a physiological 3D context (Schwarz et al., 2015). The type II keratins 8 and 7 are the first intermediate filament proteins expressed in developing murine embryos as early as 2.5 days post conception (dpc) prior to the onset of keratin type I production (Jackson et al., 1980; Lu, Hesse, Peters, & Magin, 2005; Oshima, Howe, Klier, Adamson, & Shevinsky, 1983). Imaging of embryos thus allows following the events that eventually lead to a fully formed keratin intermediate filament network in the trophectodermal layer.

3.1 Embryo Collection and Cultivation

1. Breed homozygous Krt8-YFP mice naturally and sacrifice plugged female mice at 2.5 dpc by cervical dislocation according to national guidelines for use of laboratory animals.
2. Open the abdominal cavity and push the intestine to one side. Locate the uterus and grasp it with fine forceps. Gently pull the uterus, oviduct, and ovary away from the body cavity (Fig. 6A). Using forceps remove the mesometrium, which is now exposed (Fig. 6B). Locate the oviduct and ovary and grasp the uterus next to the oviduct (Fig. 6C). Gently slide scissors between the oviduct and ovary (do not cut) until they part. Doing so will make flushing easier as the infundibulum will be preserved. Cut next to the forceps, leaving at least 0.5 cm of the uterus and put the oviduct into a culture dish with PBS (D8537, Sigma-Aldrich). Repeat on the other side.
3. Prepare a needle (BD Microlance™ 3 Hypodermic Needle 30G × 1/2″, 304000, Becton-Dickinson) by blunting it with sandpaper (Grade: A400) in order to avoid puncturing the oviduct. Fill the syringe (BD Plastipak™ U-40, 300026, Becton-Dickinson) with M2 medium (M7167, Sigma-Aldrich) and make sure that it is free of air bubbles. Place the oviduct on the dry lid of the culture dish under a stereomicroscope and locate the infundibulum (Fig. 6D). Insert the needle and hold it in place with forceps (Fig. 6E). Gently flush the oviduct with ≈0.2 ml M2 medium.

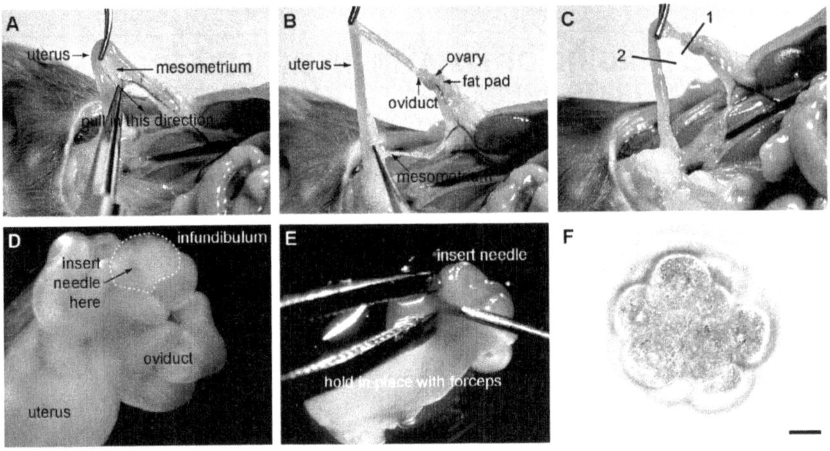

Figure 6 Critical steps of the isolation of preimplantation embryos. After opening the peritoneal cavity, the uterus is held with forceps to pull the mesometrium away (A) and to remove it as well as possible with forceps for exposure of the oviduct and ovary (B). Oviduct and ovary are separated (1 in C), and the uterus is cut as shown (2 in C). The infundibulum of the oviduct is located (D), and a blunted needle is inserted while it is held in place with forceps (E). (F) A healthy looking eight-cell embryo that is ready for further investigation. Scale bar: 20 μm.

4. At 2.5 dpc, eight-cell stage embryos are expected (Fig. 6F). Blastomeres of morphologically healthy looking embryos are round, similar in size, and fill the surrounding zona pellucida. No debris of dissolved cells should be visible. Pick up embryos using a mouth pipette system (embryo handling pipette, 441282, Reproline medical) and wash them through several drops of M2 medium to remove debris. Prepare a microdrop culture dish with 50 μl drops of M16 medium (M7292, Sigma-Aldrich) overlaid with mineral oil (M5310, Sigma-Aldrich). Place embryos individually into microdrops and keep at 37 °C and 5% CO_2.
5. Let embryos recover for 1 h for studies on the first time appearance of keratin intermediate filaments. From eight-cell stage to blastocyst stage, it will take ≈24 h in culture.

3.2 Imaging of Preimplantation Embryos

1. Prewarm the microscope chamber to 37 °C.
2. For imaging, prepare a microdrop culture in a glass bottom dish (P35G-0.17-14-C, MatTek) using 50 μl M2 medium overlaid with mineral oil. The mineral oil prevents evaporation of the medium.
3. Transfer a single embryo into the microdrop.

4. Record images with a Zeiss LSM 710 Duo microscope. For fluorescence detection, use the 488 nm line of the argon-ion laser and a 63 × 1.4 N.A. DIC M27 oil immersion objective. Monitor the emitted light between 500 and 590 nm with a pinhole set at 1–2 AU and a laser intensity of 5%. Additionally, record the transmitted light with the T-PMT detector.
5. For long-term imaging, the reduction of phototoxicity is of utmost importance. Therefore, intervals between recordings should be as long as possible. We routinely use 15–30 min between imaging of Z-stacks. The embryo can be monitored at a depth of 50 μm without loss in spatial resolution by confocal sectioning (a blastocyst has a diameter of about 100 μm). Therefore, imaging of the whole embryo is not advisable.

A representative recording is presented in Fig. 7 and corresponding Movie 4 (http://dx.doi.org/10.1016/bs.mie.2015.07.034). It shows the appearance of diffuse Krt8-YFP fluorescence at the late morula stage with small aggregates appearing upon compaction. In addition, a later stage of keratin network formation is shown in a blastocyst (Fig. 8).

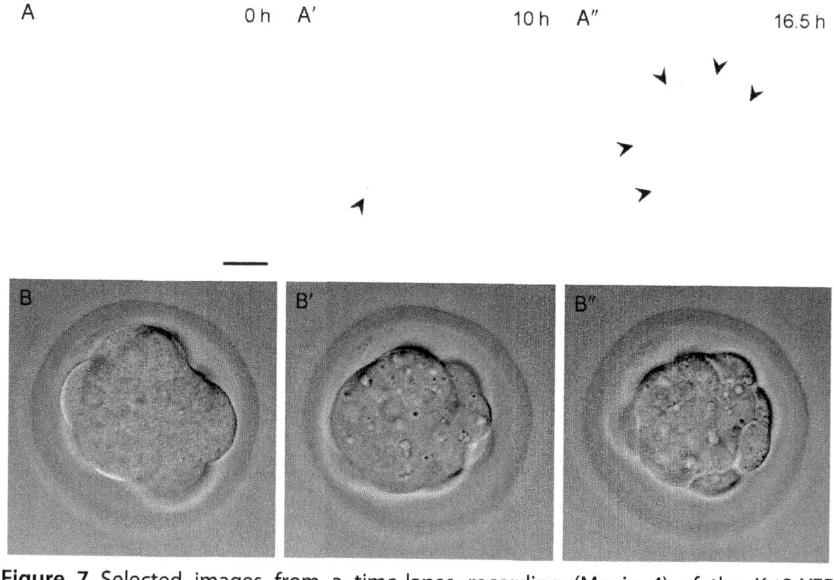

Figure 7 Selected images from a time-lapse recording (Movie 4) of the Krt8-YFP fluorescence (A) and phase contrast optics (B) of a compacting morula. At the beginning, no Krt8-YFP signal is observed. Within 8–10 h, diffuse fluorescence and small fluorescent dots appear (arrow in A′), which increase over time (arrows in A″). Scale bar: 20 μm.

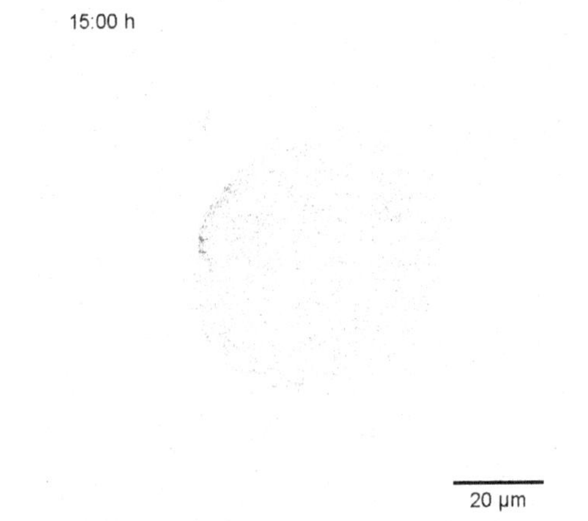

Movie 4 Time-lapse fluorescence recording of a compacted Krt8-YFP embryo shows the first appearance of keratins 10 h after compaction (corresponding Fig. 7).

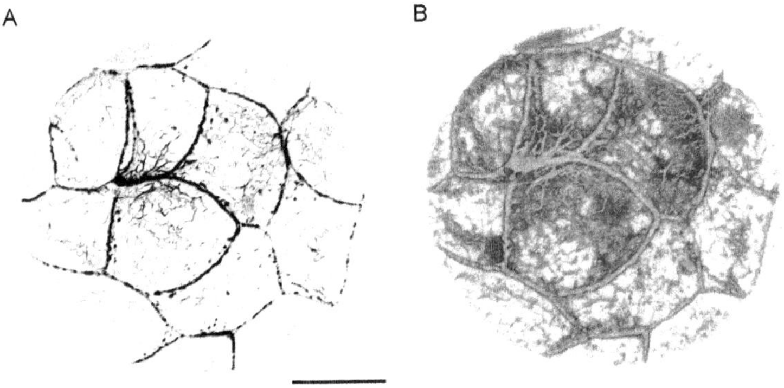

Figure 8 Detection of Krt8-YFP in a late blastocyst by confocal fluorescence microscopy. The maximum intensity projection (A) and 3D reconstruction (B) of the recorded fluorescence (25 focal planes, 1 µm steps) show an extensive network throughout the trophectoderm layer. Relative fluorescence intensity in (B) is color coded with light green being the strongest and deep blue being the weakest signal. An animation of the reconstruction is presented in Movie 5. Scale bar: 20 µm. (See the color plate.)

3.3 Image Processing and Analysis

We usually visualize keratin intermediate filament networks in murine preimplantation embryos either by maximum intensity projection or by 3D reconstruction.

Maximum intensity projection using Fiji software:

1. Run Fiji software (freely available at http://fiji.sc/Fiji).
2. Import data into Fiji (Submenu: File → Open).
3. Split channels, if bright-field image is included in data file (Submenu: Image → Color → Split channels).
4. Perform Z-projection for fluorescence channel data (Submenu: Image → Stacks → Z-Project (projection type: max intensity)).
5. If you want to show inverse fluorescence micrographs, choose a gray LUT (Submenu: Image → Lookup Table → Grays) and invert the gray values (Submenu: Image → Lookup Table → Invert LUT).

3D reconstruction using Amira software

1. Run Amira software (version 5.5.0, FEI).
2. Open data and load the fluorescence channel in the following submenu.
3. Highlight dataset in Pool window and choose "volren."
4. Set the following properties in Properties window: color: Volrengreen 1000–50,000; mode: VRT; Shading: none.
5. Create a TIFF file by clicking snapshot and choose tif format.
6. Save file.

Figure 8 shows a projection view and a colored 3D reconstruction (corresponding Movie 5 (http://dx.doi.org/10.1016/bs.mie.2015.07.034)) of the keratin network in a late blastocyst.

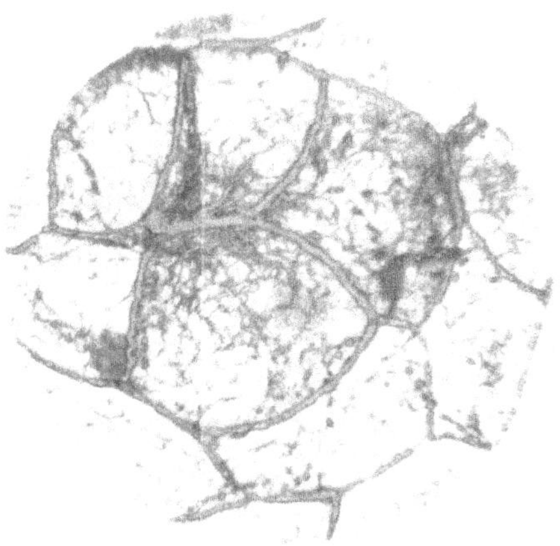

Movie 5 Animation of a 3D reconstruction of a Krt8-YFP blastocyst (corresponding Fig. 8).

4. OUTLOOK

With the given tools at hand, quantitative assessment of factors influencing keratin intermediate filament network dynamics is now achievable. This will help to understand the complexity of keratin network regulation, its impact on the keratin cycle, and the underlying mechanisms, i.e., signaling pathways, mechanics, and local activity of kinases and phosphatases. By averaging and compiling data of many cells, even subtle alterations become visible. The computational methods presented here can be used to study the dynamics of other intermediate filament networks with some restrictions. The filaments that are formed need to have a certain size and form in order to be ignored by the denoising algorithm (e.g., small round aggregates are filtered out). Furthermore, the algorithms are optimized for particles that are moving rather slowly (100–1000 nm min^{-1}). For filaments that are moving faster either the recording speed has to be increased in order to apply motion analysis or the code has to be adjusted to detect moving structures over longer distances. It remains to be shown whether the algorithm for bulk flow analysis is applicable for other systems that show a similar behavior.

The Krt8-YFP knock-in mouse line allows tracking of intermediate filament network formation and dynamics and its spatiotemporal determinants in a physiological 3D context. The murine preimplantation embryo is easy to handle and manipulate, and *in vivo* responses of the intermediate filaments to chemical and physical impacts can be faithfully monitored. It is necessary to experimentally test whether the generation of knock-in alleles coding for other intermediate filament - fluorescent protein chimeras is tolerable for the animal and does not cause any perturbations in network structure or function.

Further developments in standardization (e.g., printing procedures), image analysis (e.g., advanced modeling for simulation of network dynamics), monitoring and local regulation of enzymatic activities (e.g., Gardiner et al., 2002; Karginov, Ding, Kota, Dokholyan, & Hahn, 2010; Strickland et al., 2012; Regot, Hughey, Bajar, Carrasco, & Covert, 2014), detection of protein–protein interaction (e.g., by Förster resonance energy transfer, bimolecular fluorescence complementation), and 3D resolution (e.g., light sheet microscopy, super-resolution microscopy) will open exciting new possibilities in the analysis of the plasticity and function of the keratin cytoskeleton. This will help to understand the contribution of the keratin system to epithelial–mesenchymal transition during carcinogenesis and its dysfunction in skin disease.

ACKNOWLEDGMENTS

The work was supported by the German Research Council (LE 566/18, LE 566/20, WI 1731/6, and WI 1731/8).

REFERENCES

Baribault, H., Blouin, R., Bourgon, L., & Marceau, N. (1989). Epidermal growth factor-induced selective phosphorylation of cultured rat hepatocyte 55-kD cytokeratin before filament reorganization and DNA synthesis. *The Journal of Cell Biology, 109*, 1665–1676.

Beriault, D. R., Haddad, O., McCuaig, J. V., Robinson, Z. J., Russell, D., Lane, E. B., et al. (2012). The mechanical behavior of mutant K14-R125P keratin bundles and networks in NEB-1 keratinocytes. *PLoS One*, 7. e31320.

Chalfie, M., Tu, Y., Euskirchen, G., Ward, W. W., & Prasher, D. C. (1994). Green fluorescent protein as a marker for gene expression. *Science, 263*, 802–805.

Chung, B.-M., Murray, C. I., Eyk, J. E. V., & Coulombe, P. A. (2012). Identification of novel interaction between annexin A2 and keratin 17 evidence for reciprocal regulation. *The Journal of Biological Chemistry, 287*, 7573–7581.

Cubitt, A. B., Heim, R., Adams, S. R., Boyd, A. E., Gross, L. A., & Tsien, R. Y. (1995). Understanding, improving and using green fluorescent proteins. *Trends in Biochemical Sciences, 20*, 448–455.

Fois, G., Weimer, M., Busch, T., Felder, E. T., Oswald, F., von Wichert, G., et al. (2013). Effects of keratin phosphorylation on the mechanical properties of keratin filaments in living cells. *The FASEB Journal, 27*, 1322–1329.

Franke, W. W., Schmid, E., Grund, C., & Geiger, B. (1982). Intermediate filament proteins in nonfilamentous structures: Transient disintegration and inclusion of subunit proteins in granular aggregates. *Cell, 30*, 103–113.

Gardiner, E. M., Pestonjamasp, K. N., Bohl, B. P., Chamberlain, C., Hahn, K. M., & Bokoch, G. M. (2002). Spatial and temporal analysis of Rac activation during live neutrophil chemotaxis. *Current Biology, 12*, 2029–2034.

Ho, C. L., Martys, J. L., Mikhailov, A., Gundersen, G. G., & Liem, R. K. (1998). Novel features of intermediate filament dynamics revealed by green fluorescent protein chimeras. *Journal of Cell Science, 111*, 1767–1778.

Jackson, B. W., Grund, C., Schmid, E., Bürki, K., Franke, W. W., & Illmensee, K. (1980). Formation of cytoskeletal elements during mouse embryogenesis. *Differentiation, 17*, 161–179.

Karginov, A. V., Ding, F., Kota, P., Dokholyan, N. V., & Hahn, K. M. (2010). Engineered allosteric activation of kinases in living cells. *Nature Biotechnology, 28*, 743–747.

Keski-Oja, J., Lehto, V. P., & Virtanen, I. (1981). Keratin filaments of mouse epithelial cells are rapidly affected by epidermal growth factor. *The Journal of Cell Biology, 90*, 537–541.

Kölsch, A., Windoffer, R., Würflinger, T., Aach, T., & Leube, R. E. (2010). The keratin-filament cycle of assembly and disassembly. *Journal of Cell Science, 123*, 2266–2272.

Ku, N. O., & Omary, M. B. (1997). Phosphorylation of human keratin 8 in vivo at conserved head domain serine 23 and at epidermal growth factor-stimulated tail domain serine 431. *The Journal of Biological Chemistry, 272*, 7556–7564.

Lane, E. B., Goodman, S. L., & Trejdosiewicz, L. K. (1982). Disruption of the keratin filament network during epithelial cell division. *The EMBO Journal, 1*, 1365–1372.

Langhofer, M., Hopkinson, S. B., & Jones, J. C. (1993). The matrix secreted by 804G cells contains laminin-related components that participate in hemidesmosome assembly in vitro. *Journal of Cell Science, 105*, 753–764.

Leube, R. E., Moch, M., Kölsch, A., & Windoffer, R. (2011). Panta rhei. *BioArchitecture, 1*, 39–44.

Leube, R. E., Moch, M., & Windoffer, R. (2015). Intermediate filaments and the regulation of focal adhesion. *Current Opinion in Cell Biology, 32*, 13–20.

Liovic, M., Mogensen, M. M., Prescott, A. R., & Lane, E. B. (2003). Observation of keratin particles showing fast bidirectional movement colocalized with microtubules. *Journal of Cell Science, 116*, 1417–1427.

Lu, H., Hesse, M., Peters, B., & Magin, T. M. (2005). Type II keratins precede type I keratins during early embryonic development. *European Journal of Cell Biology, 84*, 709–718.

Mittal, B., Sanger, J. M., & Sanger, J. W. (1989). Visualization of intermediate filaments in living cells using fluorescently labeled desmin. *Cell Motility and the Cytoskeleton, 12*, 127–138.

Moch, M., Herberich, G., Aach, T., Leube, R. E., & Windoffer, R. (2013). Measuring the regulation of keratin filament network dynamics. *Proceedings of the National Academy of Sciences of the United States of America, 110*, 10664–10669.

Okabe, S., Miyasaka, H., & Hirokawa, N. (1993). Dynamics of the neuronal intermediate filaments. *The Journal of Cell Biology, 121*, 375–386.

Oshima, R. G., Howe, W. E., Klier, F. G., Adamson, E. D., & Shevinsky, L. H. (1983). Intermediate filament protein synthesis in preimplantation murine embryos. *Developmental Biology, 99*, 447–455.

Prahlad, V., Yoon, M., Moir, R. D., Vale, R. D., & Goldman, R. D. (1998). Rapid movements of vimentin on microtubule tracks: Kinesin-dependent assembly of intermediate filament networks. *The Journal of Cell Biology, 143*, 159–170.

Regot, S., Hughey, J. J., Bajar, B. T., Carrasco, S., & Covert, M. W. (2014). High-sensitivity measurements of multiple kinase activities in live single cells. *Cell, 157*, 1724–1734.

Rolli, C. G., Seufferlein, T., Kemkemer, R., & Spatz, J. P. (2010). Impact of tumor cell cytoskeleton organization on invasiveness and migration: A microchannel-based approach. *PLoS One, 5*. e8726.

Schwarz, N., Windoffer, R., Magin, T. M., & Leube, R. E. (2015). Dissection of keratin network formation, turnover and reorganization in living murine embryos. *Scientific Reports, 5*.

Strickland, D., Lin, Y., Wagner, E., Hope, C. M., Zayner, J., Antoniou, C., et al. (2012). TULIPs: Tunable, light-controlled interacting protein tags for cell biology. *Nature Methods, 9*, 379–384.

Strnad, P., Windoffer, R., & Leube, R. E. (2003). Light-induced resistance of the keratin network to the filament-disrupting tyrosine phosphatase inhibitor orthovanadate. *The Journal of Investigative Dermatology, 120*, 198–203.

Vikstrom, K. L., Lim, S. S., Goldman, R. D., & Borisy, G. G. (1992). Steady state dynamics of intermediate filament networks. *The Journal of Cell Biology, 118*, 121–129.

Wang, L., Ho, C., Sun, D., Liem, R. K. H., & Brown, A. (2000). Rapid movement of axonal neurofilaments interrupted by prolonged pauses. *Nature Cell Biology, 2*, 137–141.

Windoffer, R., Beil, M., Magin, T. M., & Leube, R. E. (2011). Cytoskeleton in motion: The dynamics of keratin intermediate filaments in epithelia. *The Journal of Cell Biology, 194*, 669–678.

Windoffer, R., & Leube, R. E. (1999). Detection of cytokeratin dynamics by time-lapse fluorescence microscopy in living cells. *Journal of Cell Science, 112*, 4521–4534.

Windoffer, R., & Leube, R. E. (2001). De novo formation of cytokeratin filament networks originates from the cell cortex in A-431 cells. *Cell Motility and the Cytoskeleton, 50*, 33–44.

Windoffer, R., & Leube, R. E. (2004). Imaging of keratin dynamics during the cell cycle and in response to phosphatase inhibition. *Methods in Cell Biology, 78*, 321–352.

Windoffer, R., Wöll, S., Strnad, P., & Leube, R. E. (2004). Identification of novel principles of keratin filament network turnover in living cells. *Molecular Biology of the Cell, 15*, 2436–2448.

Yoon, M., Moir, R. D., Prahlad, V., & Goldman, R. D. (1998). Motile properties of vimentin intermediate filament networks in living cells. *The Journal of Cell Biology, 143*, 147–157.

Yoon, K. H., Yoon, M., Moir, R. D., Khuon, S., Flitney, F. W., & Goldman, R. D. (2001). Insights into the dynamic properties of keratin intermediate filaments in living epithelial cells. *The Journal of Cell Biology*, *153*, 503–516.

Yuan, A., Sasaki, T., Rao, M. V., Kumar, A., Kanumuri, V., Dunlop, D. S., et al. (2009). Neurofilaments form a highly stable stationary cytoskeleton after reaching a critical level in axons. *The Journal of Neuroscience*, *29*, 11316–11329.

CHAPTER FOUR

Phospho-Specific Antibody Probes of Intermediate Filament Proteins

Hidemasa Goto*[,†], Hiroki Tanaka*, Kousuke Kasahara*[,‡], Masaki Inagaki*[,†,1]

*Division of Biochemistry, Aichi Cancer Center Research Institute, Nagoya, Aichi, Japan
[†]Department of Cellular Oncology, Graduate School of Medicine, Nagoya University, Nagoya, Aichi, Japan
[‡]Department of Oncology, Graduate School of Pharmaceutical Sciences, Nagoya City University, Nagoya, Aichi, Japan
[1]Corresponding author: e-mail address: minagaki@aichi-cc.jp

Contents

1. Introduction 86
2. Production of Site- and Phosphorylation State-Specific Antibodies 91
   2.1 Designing a Synthetic Phosphopeptide 91
   2.2 Conjugation of a Phosphopeptide to a Carrier Protein 97
3. Characterization of Site- and Phosphorylation State-Specific Antibodies 97
   3.1 ELISA 98
   3.2 Immunoblotting 99
4. Immunocytochemistry 100
   4.1 Overview for Immunostaining Methods 100
   4.2 Vimentin Phosphorylation in Mitosis 101
   4.3 Vimentin Phosphorylation in Signal Transduction 102
5. Other Applications 105
6. Conclusions 105
Acknowledgments 106
References 106

Abstract

Intermediate filaments (IFs) form one of the major cytoskeletal systems in the cytoplasm or beneath the nuclear membrane. Accumulating data have suggested that IF protein phosphorylation dramatically changes IF structure/dynamics in cells. For the production of an antibody recognizing site-specific protein phosphorylation (a site- and phosphorylation state-specific antibody), we first employed a strategy to immunize animals with an *in vitro*-phosphorylated polypeptide or a phosphopeptide (corresponding to a phosphorylated residue and its surrounding sequence of amino acids), instead of a phosphorylated protein. Our established methodology not only improves the chance of obtaining a phospho-specific antibody but also has the advantage that one can

Methods in Enzymology, Volume 568
ISSN 0076-6879
http://dx.doi.org/10.1016/bs.mie.2015.07.010

predesign a targeted phosphorylation site. It is now applied to the production of an antibody recognizing other types of site-specific posttranslational modification, such as acetylation or methylation. The use of such an antibody in immunocytochemistry enables us to analyze spatiotemporal distribution of site-specific IF protein phosphorylation. The antibody is of great use to identify a protein kinase responsible for *in vivo* IF protein phosphorylation and to monitor intracellular kinase activities through IF protein phosphorylation. Here, we present an overview of our methodology and describe stepwise approaches for the antibody characterization. We also provide some examples of analyses for IF protein phosphorylation involved in mitosis and signal transduction.

1. INTRODUCTION

Intermediate filaments (IFs), together with microtubules and actin filaments, form the cytoskeletal framework in the cytoplasm of virtually all vertebrates (Fig. 1). IFs are also found as major components of nuclear lamina, a filamentous layer associated with inner nuclear membrane. Compared with other two major cytoskeletal elements, there are much more

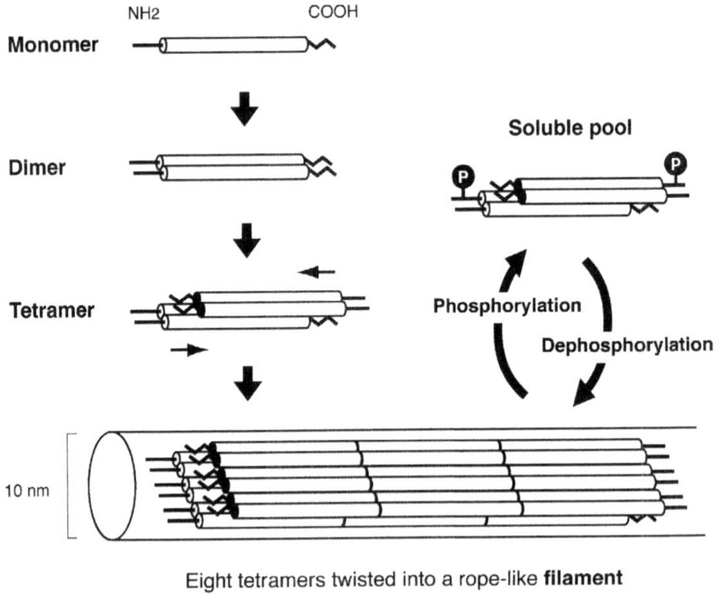

Figure 1 Schema showing phosphoregulation of IF structure. IF proteins form homo- or heterodimer. Two dimers associate with one tetramer, which assembles into 10 nm filaments. The phosphorylation of IF protein(s) induces the disassembly of IF structure at least *in vitro*. The balance of activities between IF kinases and phosphatases likely regulates IF structure *in vivo*.

(about 70) genes coding for IF component proteins (Coulombe, Ma, Yamada, & Wawersik, 2001; Eriksson et al., 2009; Eriksson, Opal, & Goldman, 1992; Fuchs & Weber, 1994; Herrmann & Aebi, 2004; Herrmann, Hesse, Reichenzeller, Aebi, & Magin, 2003; Parry & Steinert, 1992). Among them, vimentin is expressed in all mesenchymal cells with the eye lens being the tissue with by far the highest levels (Ivaska, Pallari, Nevo, & Eriksson, 2007; Song et al., 2009). Vimentin expression is also elevated during development or tumor progression (e.g., epithelial–mesenchymal transition; Eriksson et al., 2009; Ivaska et al., 2007; Kidd, Shumaker, & Ridge, 2014; Satelli & Li, 2011).

In 1987, an *in vitro* study first demonstrated direct evidence supporting IF reorganization by IF protein phosphorylation (Inagaki, Nishi, Nishizawa, Matsuyama, & Sato, 1987). The protein kinase A-mediated phosphorylation of vimentin resulted in complete disassembly of reconstituted vimentin filaments (Inagaki et al., 1987). Since then, a variety of protein kinases have been reported to phosphorylate almost all major types of IF proteins, which results in similar IF disassembly *in vitro* (Fig. 1; Inagaki et al., 1996). The above *in vitro* observations raise the question of whether IF remodeling is also regulated by IF protein phosphorylation *in vivo*.

Labeling of cells with radioactive phosphate was a widely used strategy to monitor *in vivo* phosphorylation of proteins in the past. Application of this technique to IFs led to the discovery that vimentin is phosphorylated during mitosis or cell differentiation (Evans & Fink, 1982). Recently, mass spectrometry (MS) analysis is a convenient alternative to the above method because MS analysis does not require radioisotopes to identify *in vivo* phosphorylation sites. However, the MS analysis also has several difficulties to analyze spatial and temporal change of site-specific phosphorylation (Goto & Inagaki, 2014a). For example, it is impossible to obtain clear images of subcellular distribution of protein phosphorylation in cells because the MS analysis generally requires cell lysis. SILAC (stable isotope labeling using amino acids) is often required to compare the level of site-specific phosphorylation between two groups.

In 1983, Sternberger's group reported that four of five antineurofibrillar monoclonal antibodies recognized specifically phosphorylated forms of neurofilaments (IFs expressed in neuronal cells) but not nonphosphorylated forms whereas the remaining one immunoreacted with both forms (Sternberger & Sternberger, 1983). The study also demonstrated that the use of each phospho-epitope-specific antibody in immunocytochemistry could lead to visualization of the intracellular distribution of IF protein phosphorylation (Sternberger & Sternberger, 1983). Later, these phospho-epitope-specific antibodies were reported to recognize a similar phosphorylated sequence of tau (Ksiezak-Reding, Dickson, Davies, & Yen, 1987; Lichtenberg-Kraag et al., 1992; Nukina, Kosik, & Selkoe, 1987). Therefore, for generating an antibody that can recognize a protein phosphorylated specifically at targeted residue(s), immunizing with a phosphorylated whole protein has little chance of being successful (Goto & Inagaki, 2014a).

For efficient production of an antibody recognizing site-specific phosphorylation on a target protein, we first established a novel methodology that utilizes an *in vitro*-phosphorylated polypeptide (corresponding to a phosphorylated residue and its surrounding sequence of amino acids) as an antigen in 1990 (Matsuoka et al., 1992; Nishizawa et al., 1991; Yano et al., 1991). Now, a chemically synthesized phosphopeptide is a convenient alternative to a peptide phosphorylated by a protein kinase *in vitro* (Nagata, Izawa, & Inagaki, 2001). Our strategy to design a phosphopeptide as an antigen has both a greater chance of obtaining a phospho-epitope-specific antibody and the advantage that one can predesign a targeted phosphorylation site (Goto & Inagaki, 2007, 2014a; Izawa & Inagaki, 2006; Nagata et al., 2001). A phosphoSer-, phosphoThr-, or phosphoTyr-specific antibody is now commercially available but quite different in the following aspect. This type of antibody recognizes one phosphorylated residue (Ser, Thr, or Tyr) on proteins, whereas the antibody based on our methodology simultaneously recognizes both phosphorylated residue(s) and the surround sequence of amino acids. The detection of a phosphorylated protein of interest by the former antibody absolutely requires the fractionation or purification from cells, even in the case of an abundant protein such as an IF protein (Barberis et al., 2009; Cohen et al., 2009). On the other hand, the latter can easily detect site-specific phosphorylation on a target protein in cells. Thus, we call it as a site- and phosphorylation state-specific antibody in order to distinguish between two types of phospho-epitope-specific antibodies. Since we established this methodology, many research groups have succeeded in the production of site- and phosphorylation state-specific antibodies. Table 1 summarizes the reported antibodies for IF proteins. Our methodology to immunize animals with a modified peptide is also applied to the production of antibodies that specifically recognize the other types of site-specific protein modification, such as acetylation and methylation (Goto & Inagaki, 2007, 2014a; Izawa & Inagaki, 2006; Nagata et al., 2001). Nowadays, such antibodies are commercially available: Table 2 shows the representative antibodies recognizing site-specific IF phosphorylation(s).

Table 1 Representative Site- and Phosphorylation State-Specific Antibodies for IF Proteins

IF Protein	P-site Human	P-site Mice	Animal	mAb/pAb	Ab Name	Refs.
Keratin 8	S8[a]	S8[a]	Rabbit	pAb		Akita et al. (2007)
	S23[a]	S23[a]	Rabbit	pAb		
	(−)	S46[a]	Rabbit	pAb		
	S73[a]	S79[a]	Mouse	mAb	LJ4	Liao, Ku and Omary (1997)
	Y267	Y273	Rabbit	pAb	5277	Snider, Park, and Omary (2013)
	S431[a]	S437[a]	Mouse	mAb	5B3	Ku and Omary (1997)
Keratin 18	S33[a]	S34[a]	Rabbit	pAb	8250	Ku, Liao, and Omary (1998)
			Mouse	mAb	IB4	
	S52[a]	(−)	Rabbit	pAb	3055	Liao, Lowthert, Ku, Fernandez, and Omary (1995)
Vimentin	S6[a]	S6[a]	Mouse	mAb	MO6	Yasui et al. (2001)
	S33[a]	S33[a]	Mouse	mAb	YT33	Ogawara et al. (1995)
	S38[a]	S38[a]	Rabbit	pAb	GK38	Inagaki et al. (1997)
			Rat	mAb	TM38	Kosako et al. (1999)
	S50[a]	S50[a]	Rat	mAb	TM50	Takai et al. (1996)
	S55[a]	S55[a]	Mouse	mAb	4A4	Tsujimura et al. (1994)
	S71[a]	S71[a]	Rabbit	pAb	GK71	Goto et al. (1998)
			Rat	mAb	TM71	Kosako et al. (1999)
	S72[a]	S72[a]	Rabbit	pAb	YG72	Yasui et al. (2001)
			Rat	mAb	TM72	Oguri et al. (2006)
	S82[a]	S82[a]	Mouse	mAb	MO82	Ogawara et al. (1995)
	S459	S459	Rabbit	pAb		Kotula et al. (2013)

Continued

Table 1 Representative Site- and Phosphorylation State-Specific Antibodies for IF Proteins—cont'd

IF Protein	P-site Human	P-site Mice	Animal	mAb/pAb	Ab Name	Refs.
Desmin	S11[a]	S11[a]	Rabbit	pAb	αPD11	Inada, Togashi, et al. (1999)
	T16[a]	T16[a]	Rabbit	pAb	αPD16	Kawajiri et al. (2003)
	S59[a]	S59[a]	Rabbit	pAb	αPD59	
	T75[a]	T75[a]	Rabbit	pAb	αPD75	Inada, Togashi, et al. (1999)
	T76[a]	T76[a]	Rabbit	pAb	αPD76	
GFAP	T7	T7	Rabbit	pAb	pG1-T	Matsuoka et al. (1992)
			Rat	mAb	TMG7	Goto et al. (1999)
	S8	S8	Mouse	mAb	YC10	Yano et al. (1991)
	S13	S12	Rabbit	pAb	pG1-II	Nishizawa et al. (1991)
			Mouse	mAb	KT13	Sekimata et al. (1996)
	S38[b]	S35[b]	Rabbit	pAb	pG2	Nishizawa et al. (1991)
			Mouse	mAb	KT34	Sekimata et al. (1996)
Lamin A/C	S22	S22	Rabbit	pAb		Kochin et al. (2014)
	S404	S404	Rabbit	pAb	αpS404	Cenni et al. (2008)
Lamin B2	T14	T12	Mouse	mAb		Kuga, Nozaki, Matsushita, Nomura, and Tomonaga (2010)
	S17	S15	Mouse	mAb		
	S385	S383	Mouse	mAb		
	S387	S385	Mouse	mAb		
	S401	S399	Mouse	mAb		
NFH	S503[c]	S500[c]	Rabbit	pAb	αpS493	Sasaki et al. (2002)
NFL	S26[a]	S26[a]	Rabbit	pAb	abNFL26	Hashimoto et al. (2000)
	S33[a]	S33[a]	Rabbit	pAb	abNFL33	
	S41[a]	S41[a]	Rabbit	pAb	abNFL41	
	S51[a]	S51[a]	Rabbit	pAb	abNFL51	
	S55[a]	S55[a]	Rabbit	pAb	abNFL55	Nakamura et al. (2000)

Table 1 Representative Site- and Phosphorylation State-Specific Antibodies for IF Proteins—cont'd

IF Protein	P-site Human	P-site Mice	Animal	mAb/pAb	Ab Name	Refs.
	S57[a]	S57[a]	Rabbit	pAb	abNFL57	Hashimoto et al. (2000)
	S66[a]	S66[a]	Rabbit	pAb	abNFL66	
	S472	S473	Rabbit	pAb	ab-pNFL473	Nakamura et al. (1999)
Peripherin	S59	S66	Rat	mAb	2C2	Konishi et al. (2007)

[a]In each amino acid position, the first Met is uncounted.
[b]This site is corresponding to S34 on porcine GFAP.
[c]This site is corresponding to S493 on rat NFH.
Each phosphorylation site (P-site) is indicated according to original publications.
Monoclonal or polyclonal antibody is indicated as mAb or pAb, respectively.

Here, we give an overview of our methodology to produce a site- and phosphorylation state-specific antibody with a particular focus on vimentin. We also provide some examples of ELISA (enzyme-linked immunosorbent assay), immunoblotting, and immunocytochemistry using antibodies recognizing site-specific vimentin phosphorylation. Readers are also referred to recent reviews on IF phosphorylation (Goto & Inagaki, 2014b; Hyder, Pallari, Kochin, & Eriksson, 2008; Sihag, Inagaki, Yamaguchi, Shea, & Pant, 2007; Snider & Omary, 2014).

2. PRODUCTION OF SITE- AND PHOSPHORYLATION STATE-SPECIFIC ANTIBODIES

We outline the production of a site- and phosphorylation state-specific antibody using a phosphopeptide as an antigen in Fig. 2. Among these procedures, we mainly focus on the design/production of an antigen phosphopeptide because we have already described methods to purify a polyclonal site- and phosphorylation state-specific antibody from antisera (Goto & Inagaki, 2007) and to select clone(s) secreting a desirable monoclonal one (Goto & Inagaki, 2014a). Readers are also referred to a general guidebook for antibody production (Greenfield, 2013).

2.1 Designing a Synthetic Phosphopeptide

We usually design a phosphopeptide to contain targeted phosphorylation site(s) and the flanking five amino acids at both sides, because five or six amino acid residues are considered to constitute an antigen epitope

Table 2 Commercially Available, Representative Anti-Phospho-IF Antibodies

IF Protein	P-site Human	P-site Mice	Vendor	Code	Animal	mAb/pAb	Remarks[a]
Keratin 8	S23[b]	S23[b]	Abcam	ab76584	Rabbit	mAb	EP1629Y
			Novus	NB100-79929	Rabbit	mAb	EP1629Y
	S73[b]	S79[b]	Abcam	ab32579	Rabbit	mAb	E431-2
				ab51150	Rabbit	pAb	
			S.C.	sc-293099	Rabbit	pAb	
	S431[b]	S437[b]	Abcam	ab109452	Rabbit	mAb	EP1630
				ab59434	Rabbit	pAb	
Keratin 18	S33[b]	S34[b]	Abcam	ab75747	Rabbit	pAb	
				ab51149	Rabbit	pAb	
			S.C.	sc-101727	Rabbit	pAb	
				sc-17031	Goat	pAb	
	S52[b]	(−)	Abcam	ab63393	Rabbit	pAb	
Vimentin	S6[b]	S6[b]	MBL	D096-3	Mouse	mAb	MO6
			S.C.	sc-57579	Mouse	mAb	MO6
	S33[b]	S33[b]	Abcam	ab115191	Mouse	mAb	YT33
			MBL	D099-3	Mouse	mAb	YT33
	S38[b]	S38[b]	CST	#13614	Rabbit	pAb	
			Abcam	ab115150	Rat	mAb	TM38
				ab52942	Rabbit	mAb	EP1069Y
			MBL	D094-3	Rat	mAb	TM38
	S50[b]	S50[b]	Abcam	ab115185	Rat	mAb	TM50
			MBL	D122-3	Rat	mAb	TM50
	S55[b]	S55[b]	CST	#7391	Rabbit	mAb	D5H2
				#3877	Rabbit	pAb	
			Abcam	ab61802	Rabbit	pAb	
				ab22651	Mouse	mAb	4A4
			MBL	D076-3	Mouse	mAb	4A4

Table 2 Commercially Available, Representative Anti-Phospho-IF Antibodies—cont'd

IF Protein	P-site Human	P-site Mice	Vendor	Code	Animal	mAb/pAb	Remarks
	S71[b]	S71[b]	Abcam	ab115189	Rat	mAb	TM71
			MBL	D093-3	Rat	mAb	TM71
	S72[b]	S72[b]	Abcam	ab52944	Rabbit	mAb	EP1070Y
			Novus	NB110-57647	Rabbit	mAb	EP1070Y
	S82[b]	S82[b]	CST	#12569	Rabbit	mAb	D5A2D
				#3878	Rabbit	pAb	
			Abcam	ab79189	Rabbit	pAb	
				ab52943	Rabbit	mAb	EP1071Y
			MBL	D095-3	Mouse	mAb	MO82
Desmin	T16[b]	T16[b]	S.C.	sc-130593	Rabbit	pAb	
	S59[b]	S59[b]	Abcam	ab111382	Rabbit	pAb	
	T75 and T76[b]	T75 and T76[b]	Abnova	PAB12619	Rabbit	pAb	
GFAP	S8	S8	Abcam	ab115898	Mouse	mAb	YC10
	S13	S12	S.C.	sc-32955	Rabbit	pAb	
				sc-32956	Rabbit	pAb	
	S38	S35	Abcam	ab62479	Rabbit	pAb	
Lamin A/C	S22	S22	CST	#13448	Rabbit	mAb	D2B2E
				#2026	Rabbit	pAb	
	S392	S392	Abcam	ab58528	Rabbit	pAb	Kong et al. (2012)
				ab192458	Rabbit	pAb	
NFM	S615 and S620	S605 and S610	Abcam	ab68142	Rabbit	mAb	EPR580(2)Y

[a]These remarks provide the information about the clone name of each monoclonal antibody or the publication in which the antibody was utilized for the detection of IF phosphorylation at the indicated site. Also refer to Table 1.
[b]In each amino acid position, the first Met is uncounted.
Abcam, Abcam Biotechnology (Cambridge, MA); Abnova, Abnova Corporation (Taipei, Taiwan); CST, Cell Signaling Technology (Beverly, MA); MBL, Medical & Biochemical Laboratories (Nagoya, Japan); Novus, Novus Biologicals (Littleton, CO); S.C., Santa Cruz (Santa Cruz, CA).

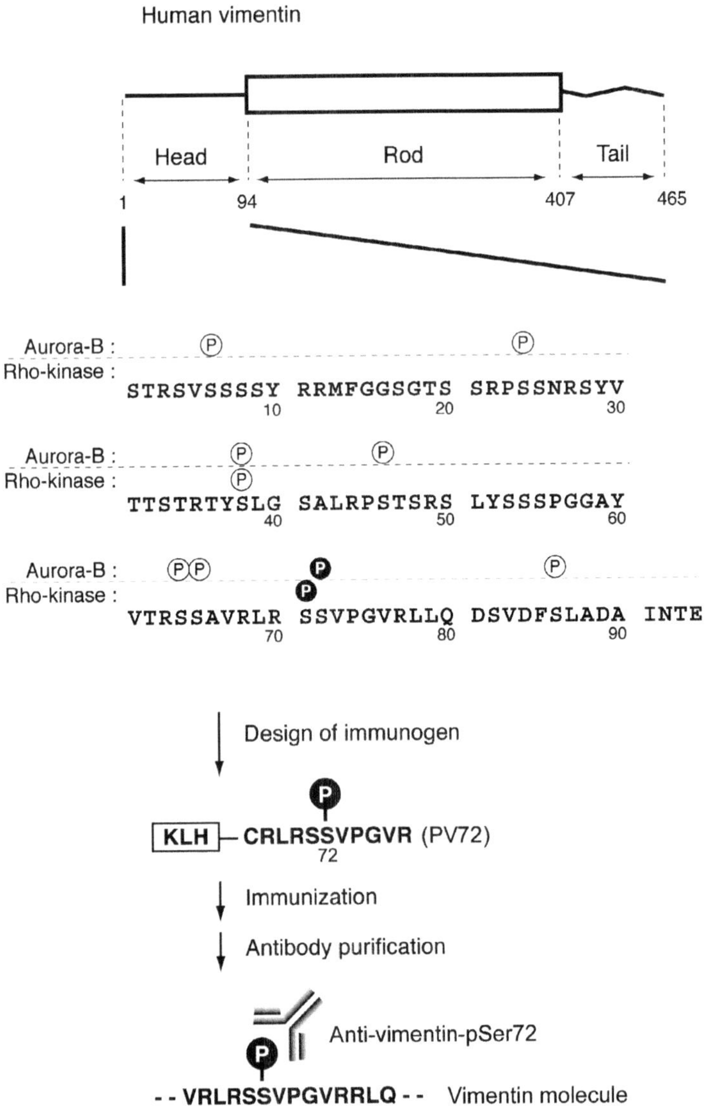

Figure 2 Overview of production and characterization of a site- and phosphorylation state-specific antibody. As an example, we indicate a rabbit polyclonal antibody specifically recognizing vimentin phosphorylated at Ser72 (anti-vimentin-pSer72; referred to as YG72). An upper map shows vimentin phosphorylation sites for Aurora-B and Rho-kinase (indicated as Ps in the circles). Each P in the black-colored circle indicates the site specific to Aurora-B or Rho-kinase among known vimentin kinases.

recognized by an antibody molecule. In order to conjugate it to a carrier protein (see below), we usually introduce a cysteine (C) residue at the amino- or carboxyl-terminal side of the synthetic peptide. As examples, we show synthetic peptides (total 12 amino acids in length) for the production of antibodies against phosphoSer55 (Fig. 3A) and phosphoSer72 (Figs. 2 and 3B) on vimentin. Up to 22 amino acids can be used, but smaller peptides are generally better because longer peptides may elevate the possibility of the production of antibodies against nonphosphorylated epitope. Simultaneously, we produce the same sequence of a nonphosphorylated peptide (Fig. 3A and B) because it is required for the removal of antibodies recognizing nonphospho-epitopes from antisera (Goto & Inagaki, 2007), ELISA assays (see below), etc. The above method in the peptide design can be also applicable to the production of an antibody that can specifically recognize the other types of site-specific protein modification, such as acetylation and methylation (Goto & Inagaki, 2007, 2014a; Izawa & Inagaki, 2006; Nagata et al., 2001).

Before the peptide production, it is ideal to check interspecies alignment of protein sequence around a phosphorylation site, using public databases such as PhosphoSitePlus® (http://www.phosphosite.org). In some cases, the sequence may not be completely conserved among species. For example, vimentin is phosphorylated at Ser55 by Cdk1 (Chou, Ngai, & Goldman, 1991; Kusubata et al., 1992), but the 53rd and the 59th amino acids on vimentin are not conserved among human and mouse (see Fig. 3A). In order to establish an antibody recognizing phosphoSer55 on human and mouse vimentin, we previously produced two types of phosphopeptides (HPV55 and MPV55 corresponding to human and mouse vimentin, respectively) and their corresponding nonphosphorylated peptides (HV55 and MV55; Fig. 3A). We then immunized mice with either phosphopeptide. From mice immunized with MPV55, we established a monoclonal antibody (4A4) recognizing HPV55 and MPV55 but not HV55 and MV55 (Fig. 3A; also see Table 1; Tsujimura et al., 1994). On the other hand, by the immunization with HPV55, we unexpectedly obtained a monoclonal antibody (1B8) reacting with HPV55 and HV55 to a similar extent (but not with MPV55 and MV55; data not shown); 1B8 is used as an antibody recognizing vimentin in a human-specific manner (Tsujimura et al., 1994). The above strategy is not always successful for the production of the antibody working in several species. If six or more (ideally 11) of consecutive amino acids including a phosphorylation site are conserved among species, we alternatively utilize them for the peptide production.

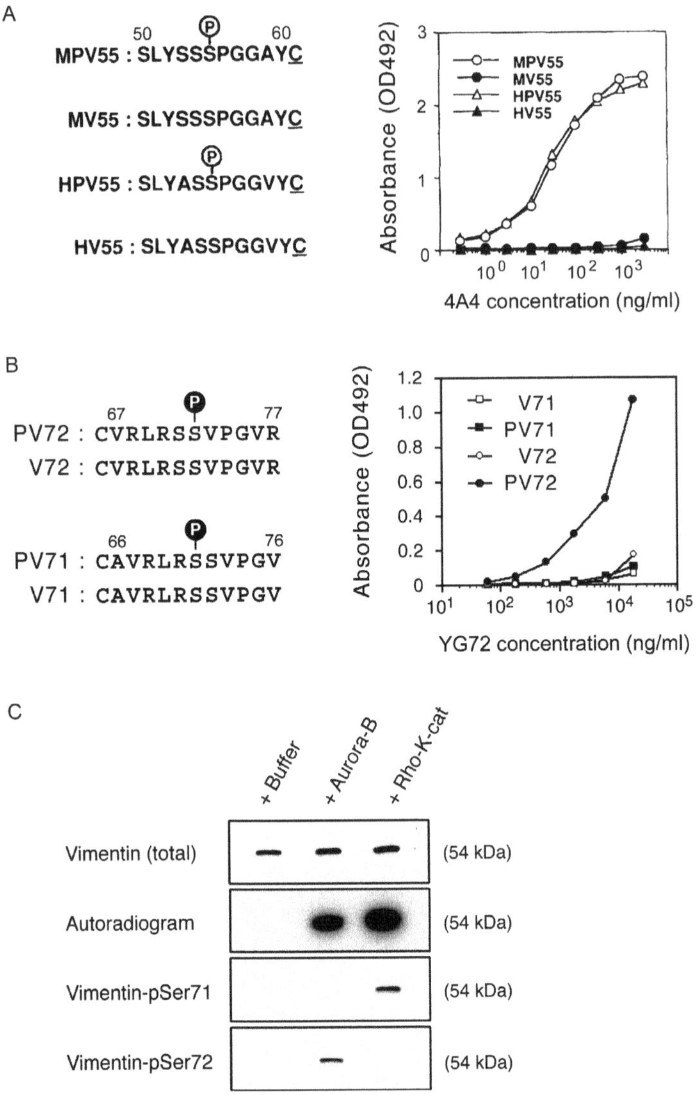

Figure 3 Characterization of site- and phosphorylation state-specific antibodies using ELISA (A and B) or immunoblotting (C). In (C), recombinant vimentin was incubated with or without [γ-^{32}P] ATP in the presence of recombinant GST-Aurora-B or GST-Rho-kinase catalytic domain (Rho-K-cat). As a negative control, vimentin was incubated without any kinases. *Reproduced from ©Tsujimura et al. (1994) (originally published in* The Journal of Biological Chemistry, *269: 31097–31106; A), ©Yasui et al. (2001) (originally published in* Oncogene, *http://dx.doi.org/10.1038/sj.onc.1204407; B), or ©Yokoyama, Goto, Izawa, Mizutani, and Inagaki (2005) (originally published in* Genes to Cells, *http://dx.doi.org/10.1111/j.1365-2443.2005.00824.x; C).*

Nowadays, many companies can chemically synthesize the modified peptides: the usage of highly pure peptide (over 90% purity; higher is better) is the key to the production of excellent antibodies. We usually order 15 mg nonmodified peptide and 25 mg phosphopeptide from Peptide Institute Inc. (Osaka, Japan).

The mutation at a phosphorylation site to Asp or Glu can generally mimic the change in electrical charge by IF protein phosphorylation (Kawajiri et al., 2003; Yamaguchi et al., 2005). However, Asp or Glu is structurally different from a phosphorylated residue. Therefore, the peptide in which a phosphorylation site is changed to Asp or Glu is generally inappropriate as an antigen for a site- and phosphorylation state-specific antibody.

2.2 Conjugation of a Phosphopeptide to a Carrier Protein

Maleimide linkage is the most reliable method for the conjugation of a phosphopeptide to a carrier protein. Therefore, we add a cysteine (C) in the amino- or carboxyl-terminus of a phosphopeptide (Figs. 2 and 3). Many companies also perform the conjugation to a carrier protein such as keyhole limpet hemocyanin (KLH). On ordering peptides, we usually request Peptide Institute Inc. to conjugate 5 mg of phosphopeptide to KLH. We also request to leave the peptide-conjugated KLH aqueous, because lyophilized KLH is difficult to be dissolved in the aqueous buffer such as phosphate-buffered saline (PBS). The conjugated protein solution is stored in aliquots at −80 °C before use.

3. CHARACTERIZATION OF SITE- AND PHOSPHORYLATION STATE-SPECIFIC ANTIBODIES

Using a phosphopeptide as an antigen, several types of antibodies are simultaneously produced in the animal. Some antibodies may recognize nonphospho-epitope(s) in the antigen. Some antibodies recognizing a phosphopeptide may not immunoreact with a protein phosphorylated at a target site. In order to exclude these undesirable antibodies, the purification steps from serum (Goto & Inagaki, 2007) or hybridoma selection steps (Goto & Inagaki, 2014a) are often required for the generation of a polyclonal or monoclonal antibody, respectively. After the antibody production, the first step is to examine whether the antibody recognizes site-specific protein phosphorylation. An indirect ELISA and an immunoblotting assay are often used for this characterization.

3.1 ELISA

In general, an indirect ELISA is one of the quickest and the most reliable assays to measure the antibody titer and assess lot-to-lot variability. In order to characterize an antibody raised against a phosphopeptide, it is critical to simultaneously perform the assays using at least two different types of ELISA plates; one is coated with an antigen phosphopeptide and the other with the corresponding, nonphosphorylated peptide. In order to rule out the antibody recognizing only a phosphorylated residue, it is desirable to use a different sequence of phosphopeptide as an additional negative control. We describe ELISA protocols, using a purified rabbit polyclonal anti-vimentin-phosphoSer72 (referred to as YG72; also see Table 1; Yasui et al., 2001) as an example (Fig. 3B). PV72 and PV71 are phosphopeptides corresponding to vimentin phosphorylated at Ser72 and Ser71, respectively (Fig. 3B). V72 and V71 are nonphosphorylated version of the above peptide (Fig. 3B).

1. Dilute each peptide to 1 μg/ml with 0.1 *M* sodium phosphate buffer ($Na_2HPO_4 \cdot NaH_2PO_4$; pH 7.4). Before use, check the pH of the peptide solution, which should be 7.0–7.5. If not, use 50 m*M* sodium carbonate buffer ($Na_2CO_3 \cdot NaHCO_3$; pH 9.0) for dilution.
2. Add 60 μl of the peptide solution into each well of ELISA 96-well plates (Nunc-Immuno Plates Maxisorb, Thermo Fisher Scientific Inc., Roskilde, Denmark) and then incubates for 2 h at room temperature (RT) or overnight at 4 °C.
3. Remove the solution from each well and wash with 100 μl of PBS per well 3 ×.
4. Add 300 μl of the ELISA blocking solution (10 m*M* $Na_2HPO_4 \cdot NaH_2PO_4$ (pH 8.0), 5% (w/v) BSA, 5% (w/v) sucrose, and 0.1% NaN_3) into each well. Incubate for 2–4 h at 37 °C or overnight at 4 °C.
5. Remove the solution. If the plates are not immediately used, leave the plates to air-dry for 5–10 min and then store them at 4 °C. In most cases, they can be stored at 4 °C for years.
6. Serially dilute YG72 (Yasui et al., 2001) at 1:10, 1:30, 1:100, 1:300, 1:1000, 1:3000, or 1:10,000 with PBS containing 1% (w/v) BSA, 1% (w/v) sucrose, and 0.1% NaN_3. Add 100 μl of each dilution per well: use two or three wells in order to determine the ELISA activity of each dilution against one type of peptide. Incubate for 1 h at 37 °C.
7. Remove the solution and then wash with 100 μl of PBS 5 ×.

8. Dilute an HRP (horseradish peroxidase)-conjugated anti-rabbit IgG (the second antibody) at 1:1000 in the ELISA secondary antibody buffer (10 m*M* $Na_2HPO_4{\cdot}NaH_2PO_4$ (pH 8.0), 100 m*M* NaCl, 1% (w/v) BSA, 0.1% (w/v) *p*-hydroxy phenylacetic acid (Sigma-Aldrich, St. Louis, MO) and 0.025% (w/v) thimerosal (Sigma-Aldrich); use the latter two chemicals as preservatives because sodium azide (NaN_3) inhibits HRP activity). Add 100 μl of the solution into each well and then incubate for 1 h at 37 °C.
9. Remove the solution and then wash with 100 μl of PBS 5 ×.
10. Freshly prepare the ELISA reaction solution (0.4 mg/ml *o*-phenylenediamine, 5% (v/v) methanol, and 0.01% (v/v) H_2O_2). Add 100 μl of the solution into each well and then incubate for 10–30 min at RT.
11. Stop the reaction by the adding 100 μl of 2 N H_2SO_4 and measure the absorbance at 492 nm with an ELISA plate reader. As you can see in Fig. 3B, ELISA activity toward PV72 is significantly higher than toward V72, V71, or PV71, which indicates high specificity of the antibody.

3.2 Immunoblotting

Immunoblotting is one of the standard techniques for the characterization of a polyclonal antibody purified from serum, a monoclonal antibody existing in hybridoma supernatant, etc. Regardless of whether a protein of interest is purified or mixed with other proteins (as in cell lysates and tissue samples), both positive and negative control samples are required to characterize a site- and phosphorylation state-specific antibody by immunoblotting. It is absolutely necessary for positive control samples to contain the protein highly phosphorylated at least at a target site. On the other hand, it is ideal that the target site is kept nonphosphorylated in the negative control sample. The target protein should be loaded equally between positive and negative control samples.

It is important to optimize experimental conditions, such as loading amounts of samples, transfer method, solutions for blocking or antibody dilution, chemiluminescent reagents, etc. In general, IF proteins are difficult to be electrically transferred onto an immunoblotting membrane. We prefer to use semi-dry-transfer because it is quicker and often more complete than wet transfer. In this case, we set voltage to 25 V (actual starting voltage is around 6–7 V), limit current to 1.5–2 mA/cm^2 gel area, and transfer proteins for 1.5–2 h at RT. Readers are also referred to our protocols for immunoblotting (Goto & Inagaki, 2014a).

We provide an example for immunoblotting using a site- and phosphorylation state-specific antibody for Ser71 (Goto et al., 1998) or Ser72 (Yasui et al., 2001) on vimentin (referred to as GK71 or YG72, respectively; also see Table 1) in Fig. 3C. Since vimentin is phosphorylated at Ser71 by Rho-associated kinase (Rho-kinase; also called ROK or ROCK; Goto et al., 1998) or at Ser72 by Aurora-B (Goto et al., 2003; also see Fig. 2), we use recombinant vimentin incubated with or without each purified active kinase *in vitro*: we load 50 ng (~1 pmol) of each recombinant vimentin per lane (Fig. 3C). If the same *in vitro* reactions are simultaneously performed with [γ-^{32}P] ATP in different experimental sets, it is easy to estimate the extent of vimentin phosphorylation in these immunoblotting samples (see "Autoradiogram" in Fig. 3C). As you can see in Fig. 3C, each antibody can specifically recognize vimentin phosphorylation specifically at the target site but not at its neighboring phosphorylation site. Readers are also referred to the protocols for IF protein purification from *Escherichia coli* and *in vitro* IF protein phosphorylation (Inada, Nagata, Goto, & Inagaki, 1999; Kawajiri & Inagaki, 2004).

4. IMMUNOCYTOCHEMISTRY

Here, we outline immunostaining methods using a site- and phosphorylation state-specific antibody and provide some examples.

4.1 Overview for Immunostaining Methods

It is critical to optimize experimental conditions, such as solutions for cell fixation or permeabilization. With regard to the fixative, we first test formaldehyde. However, it may not work in some cases because it masks some epitopes from recognition of some antibodies. In this case, we use ice-cold methanol or 10% trichloroacetic acid as an alternative. Except for methanol fixation, it is required to permeabilize cell membrane after the fixation. Since various permeabilization reagents (e.g., Triton X-100, Saponin, methanol, acetone, etc.) perforate the cell membrane by different mechanisms, it is also important to determine which is most appropriate. After the formalin fixation, we usually use ice-cold methanol for the permeabilization. Readers are also referred to the protocols as we described previously (Goto & Inagaki, 2014a; Inada, Nagata, et al., 1999; Kawajiri & Inagaki, 2004).

4.2 Vimentin Phosphorylation in Mitosis

Before the age of site- and phosphorylation state-specific antibodies, an *in vivo* kinase was generally identified by the comparison of phosphorylation sites on a protein between *in vivo* and *in vitro* using two-dimensional analysis of radioactive phosphopeptide digested by a protease such as trypsin. Application of this technique to vimentin led to the discovery that vimentin is phosphorylated by several protein kinases including cyclin-dependent protein kinase 1 (Cdk1) in mitotic cells (Chou, Bischoff, Beach, & Goldman, 1990). However, this method has several technical limitations. It is difficult to analyze the *in vivo* protein phosphorylation by each kinase separately when several kinases are simultaneously activated, such as in mitosis. It generally requires a lot of cells, which means that cell cycle synchronization is necessary to analyze the protein phosphorylation in a cell cycle-specific manner. However, it is often difficult to completely synchronize cells in a particular cell cycle phase. In addition, it is not easy to detect protein phosphorylation elevated only at the localized area.

The use of a site- and phosphorylation state-specific antibody in the immunocytochemistry enables us to analyze protein phosphorylation by each kinase in a single cell. Among the phosphorylation sites identified *in vitro*, there are sites specifically phosphorylated by a single kinase. In the case of vimentin, Ser55, Ser71, and Ser72 are specific phosphorylation sites for Cdk1 (Chou et al., 1991; Kusubata et al., 1992), Rho-kinase (Goto et al., 1998), and Aurora-B (Goto et al., 2003), respectively (also see Fig. 2). We target them for the production of antibodies. Using a monoclonal antibody (4A4) recognizing vimentin-Ser55 phosphorylation (also see Fig. 3A and Table 1; Tsujimura et al., 1994), we clearly demonstrated that Ser55 phosphorylation begins in prometaphase of mitosis, remains until metaphase, and declines gradually thereafter (Fig. 4). This Cdk1-induced phosphorylation is observed diffusely throughout the cytoplasm (Tsujimura et al., 1994; Fig. 4). On the other hand, vimentin phosphorylation at Ser71 (Goto et al., 1998) or Ser72 (Yasui et al., 2001) is spatially and temporally distinct (Fig. 4). It increases in anaphase of mitosis, maintains until telophase, and decreases at the exit of mitosis, and it is detected specifically at the cleavage furrow (between daughter nuclei; Goto et al., 1998; Yasui et al., 2001; Fig. 4). By the combination of RNA interference (RNAi) technology, we clearly demonstrated that Ser71 or Ser72 phosphorylation at the cleavage furrow is responsible for Rho-kinase or Aurora-B, respectively (Yokoyama et al., 2005; Fig. 5). Therefore, site- and phosphorylation state-specific

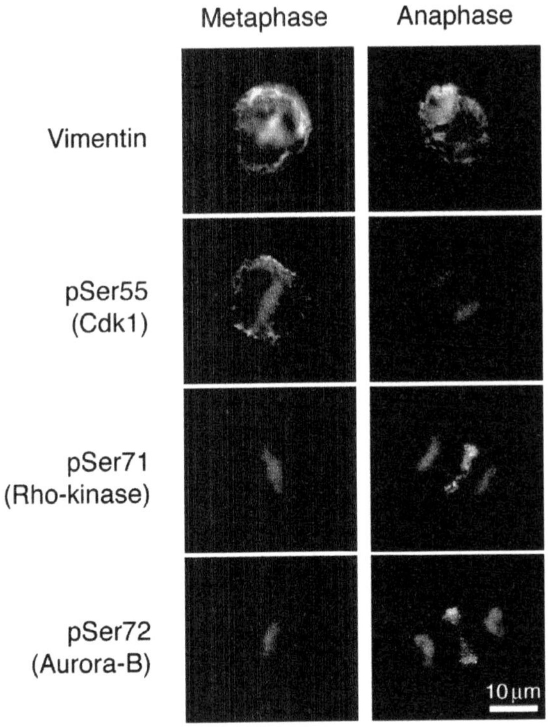

Figure 4 Site-specific vimentin phosphorylation during mitosis. U251 glioma cells were stained with anti-vimentin, anti-vimentin-pSer55, anti-vimentin-pSer71, or anti-vimentin-pSer72 (green). Chromosomes were also stained with propidium iodide (red). *Reproduced from ©Yamaguchi et al. (2005) (originally published in* The Journal of Cell Biology, *http://dx.doi.org/10.1083/jcb.200504091).* (See the color plate.)

antibodies are also of great use to identify each kinase responsible for *in vivo* protein phosphorylation.

4.3 Vimentin Phosphorylation in Signal Transduction

We also provide an example of vimentin phosphorylation downstream of receptor-activated phosphoinositide (PI) hydrolysis. For this purpose, we produced two monoclonal antibodies that recognize vimentin phosphorylations at Ser33 and Ser82 (Ogawara et al., 1995), which are relatively specific vimentin phosphorylation sites for protein kinase C (PKC) and Ca^{2+}/calmodulin-dependent protein kinase II (CaMKII), respectively (Inagaki et al., 1996). Simultaneous activation of CaMKII and PKC by PI

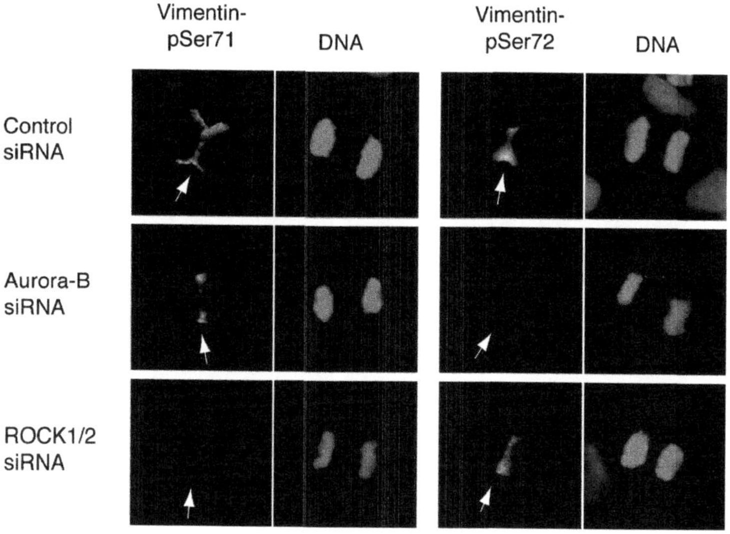

Figure 5 Cleavage furrow-specific vimentin phosphorylation at Ser71 or Ser72 is impaired in HeLa cells treated with small interfering RNA(s) (siRNA) specific to Rho-kinase or Aurora-B, respectively. Arrows indicate the position of cleavage furrow. *Reproduced from ©Yokoyama et al. (2005) (originally published in* Genes to Cells, *http://dx.doi.org/10.1111/j.1365-2443.2005.00824.x).* (See the color plate.)

hydrolysis with receptor stimulation leads to vimentin phosphorylation at Ser82 (a CaMKII site) but not at Ser33 (a PKC site) in astrocytes (Ogawara et al., 1995; also see Fig. 6). On the other hand, mitotic reorganization of intracellular membrane systems enables active PKC to interact with vimentin, which facilitates Ser33 phosphorylation predominantly during metaphase and anaphase (Takai et al., 1996). Therefore, CaMKII phosphorylates vimentin when it is activated by cell signaling whereas PKC-induced vimentin phosphorylation occurs only when intracellular membrane systems are reorganized, such as in mitosis.

In astrocytes, global elevation of Ca^{2+} concentration induces global vimentin phosphorylation at Ser82 by CaMKII, whereas local elevation does one localized in the same area (Fig. 6; Inagaki et al., 1997). Since vimentin is localized diffusely throughout the cytoplasm, these results indicate that cytoplasmic CaMKII activities can be monitored by immunocytochemical detection of vimentin-Ser82 phosphorylation. In other words, site- and phosphorylation state-specific antibodies can be used as an alternative method to monitor kinase activities in cells (Inagaki, Ito, Nakano, & Inagaki, 1994).

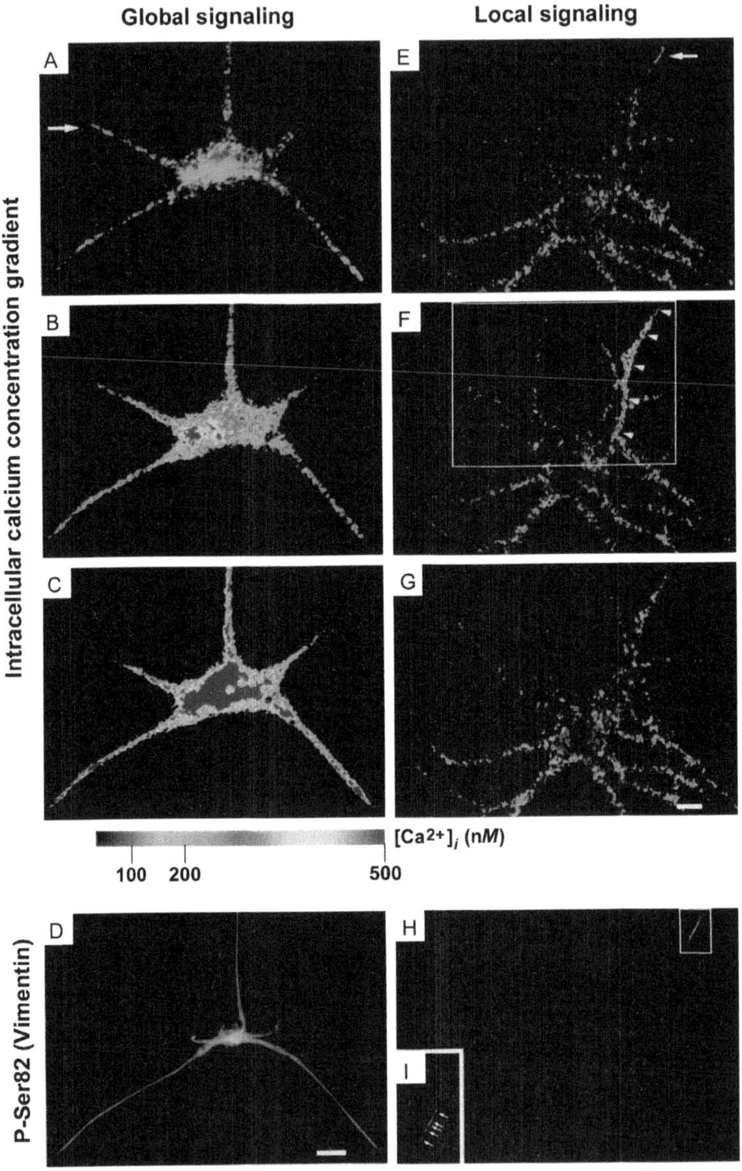

Figure 6 Local and global signaling of CaMKII defined by the area of Ca^{2+} signals. Intracellular distribution of Ca^{2+} signaling in an astrocyte was monitored by fura-2-based Ca^{2+} microscopy (A–C and E–F). $[Ca^{2+}]_i$ in an astrocyte before (A and E), or at 0.5 min (B and F), 1.5 min (C), or 4 min (G) after the local application of 10 μM $PGF_{2\alpha}$ for 0.25 min. At 5 min after the $[Ca^{2+}]_i$ measurement, each cell was fixed and then stained with anti-vimentin-pSer82 (D, H, and I; green). A magnified image in a rectangle (H) are shown in (I). Arrows (A and E) indicate sites of $PGF_{2\alpha}$ application. Arrowheads indicate the process showing Ca^{2+} signaling (F), or the area in which vimentin-Ser82 phosphorylation is elevated (I). Bars, 20 μm. *Reproduced from ©Inagaki et al. (1997) (originally published in* The Journal of Biological Chemistry, *272: 25195–25199).* (See the color plate.)

5. OTHER APPLICATIONS

Site- and phosphorylation state-specific antibodies can be used for an immunoprecipitation assay. The use in immunoprecipitation enables us to condense a site-specifically phosphorylated protein of interest from cell extracts (Ikegami et al., 2008; Kasahara et al., 2010). In addition to the antibody specificity, the use of correct affinity matrix (Protein A-, G-coated resin, etc.) is required for the successful immunoprecipitation. Since the affinity to Protein A or G varies among the subclasses of antibodies (Greenfield, 2013), it is desirable to determine the subtype of each monoclonal antibody before the assay.

Site- and phosphorylation state-specific antibodies were previously utilized for a microinjection assay to inhibit the function of a site-specifically phosphorylated protein in living cells (Dutertre et al., 2004). These antibodies also have the potential to monitor site-specific protein phosphorylation in living cells (Hayashi-Takanaka, Yamagata, Nozaki, & Kimura, 2009), but general use in this application requires technical advances (Jeong et al., 2013).

6. CONCLUSIONS

We provide a practical overview for the production of a site- and phosphorylation state-specific antibody as well as a theoretical framework for designing a phosphopeptide as an antigen. We outline how to characterize the antibody by ELISA and immunoblotting. We also highlight examples of how site- and phosphorylation state-specific antibodies are used in immunocytochemistry. However, not all the antibodies can be used for immunocytochemistry because some antibodies show nonspecific immunoreactivity (against nucleus, other intracellular organelles, microtubules, actin filaments, etc.) and/or cross-reactivity against other (phosphorylated) proteins. Therefore, the specificity of each antibody in immunocytochemistry should be also confirmed by additional assays, such as coimmunostaining with an antibody against an IF protein, the immunoblotting using cell lysates, the immunostaining after the knockdown of IF protein or a responsible kinase, etc.

As you can see in Figs. 4 and 5, site- and phosphorylation state-specific antibodies clearly show that vimentin phosphorylation by Rho-kinase (Goto et al., 1998) or Aurora-B (Goto et al., 2003) is restricted to the cleavage furrow from anaphase to telophase. Since vimentin loses its filament forming ability by Rho-kinase- or Aurora-B-induced phosphorylation *in vitro* (Goto et al., 1998, 2003), it is easily speculated that the cleavage furrow-specific vimentin phosphorylation is necessary for efficient separation of IFs to two daughter cells. Indeed, we observed unusual long bridge-like IF structures (referred to as IF-bridge) formed between daughter cells expressing phosphorylation-compromised vimentin mutants (Goto et al., 2003; Yamaguchi et al., 2005; Yasui et al., 2001). A mouse model we recently generated strongly confirms that cleavage furrow-specific phosphorylation is of critical importance both in the efficient IF separation and in the completion of cytokinesis (Matsuyama et al., 2013; Tanaka et al., 2015). Therefore, site- and phosphorylation state-specific antibodies are also of great use to speculate biological function(s) of IF phosphorylation by each kinase.

ACKNOWLEDGMENTS

This work was supported in part by Grants-in-Aid for Scientific Research from the Japan Society for the Promotion of Science and from the Ministry of Education, Science, Technology, Sports and Culture of Japan; by the Uehara Memorial Foundation; by Princess Takamatsu Cancer Research Fund; and by the Takeda Science Foundation. We sincerely apologize to all our colleagues whose contributions were unable to cite due to space limitations.

REFERENCES

Akita, Y., Kawasaki, H., Imajoh-Ohmi, S., Fukuda, H., Ohno, S., Hirano, H., et al. (2007). Protein kinase C epsilon phosphorylates keratin 8 at Ser8 and Ser23 in GH4C1 cells stimulated by thyrotropin-releasing hormone. *FEBS Journal*, *274*, 3270–3285.

Barberis, L., Pasquali, C., Bertschy-Meier, D., Cuccurullo, A., Costa, C., Ambrogio, C., et al. (2009). Leukocyte transmigration is modulated by chemokine-mediated PI3Kgamma-dependent phosphorylation of vimentin. *European Journal of Immunology*, *39*, 1136–1146.

Cenni, V., Bertacchini, J., Beretti, F., Lattanzi, G., Bavelloni, A., Riccio, M., et al. (2008). Lamin A Ser404 is a nuclear target of Akt phosphorylation in C2C12 cells. *Journal of Proteome Research*, *7*, 4727–4735.

Chou, Y. H., Bischoff, J. R., Beach, D., & Goldman, R. D. (1990). Intermediate filament reorganization during mitosis is mediated by p34cdc2 phosphorylation of vimentin. *Cell*, *62*, 1063–1071.

Chou, Y. H., Ngai, K. L., & Goldman, R. (1991). The regulation of intermediate filament reorganization in mitosis. p34cdc2 phosphorylates vimentin at a unique N-terminal site. *The Journal of Biological Chemistry*, *266*, 7325–7328.

Cohen, S., Brault, J. J., Gygi, S. P., Glass, D. J., Valenzuela, D. M., Gartner, C., et al. (2009). During muscle atrophy, thick, but not thin, filament components are degraded by MuRF1-dependent ubiquitylation. *The Journal of Cell Biology*, *185*, 1083–1095.

Coulombe, P. A., Ma, L., Yamada, S., & Wawersik, M. (2001). Intermediate filaments at a glance. *Journal of Cell Science*, *114*, 4345–4347.

Dutertre, S., Cazales, M., Quaranta, M., Froment, C., Trabut, V., Dozier, C., et al. (2004). Phosphorylation of CDC25B by Aurora-A at the centrosome contributes to the G2-M transition. *Journal of Cell Science*, *117*, 2523–2531.

Eriksson, J. E., Dechat, T., Grin, B., Helfand, B., Mendez, M., Pallari, H. M., et al. (2009). Introducing intermediate filaments: From discovery to disease. *The Journal of Clinical Investigation*, *119*, 1763–1771.

Eriksson, J. E., Opal, P., & Goldman, R. D. (1992). Intermediate filament dynamics. *Current Opinion in Cell Biology*, *4*, 99–104.

Evans, R. M., & Fink, L. M. (1982). An alteration in the phosphorylation of vimentin-type intermediate filaments is associated with mitosis in cultured mammalian cells. *Cell*, *29*, 43–52.

Fuchs, E., & Weber, K. (1994). Intermediate filaments: Structure, dynamics, function, and disease. *Annual Review of Biochemistry*, *63*, 345–382.

Goto, H., & Inagaki, M. (2007). Production of a site- and phosphorylation state-specific antibody. *Nature Protocols*, *2*, 2574–2581.

Goto, H., & Inagaki, M. (2014a). Method for the generation of antibodies specific for site and posttranslational modifications. In V. Ossipow & N. Fischer (Eds.), *Methods in molecular biology: Vol. 1131. Monoclonal antibodies: Methods and protocols* (2nd ed., pp. 21–31). New York, NY: Humana Press.

Goto, H., & Inagaki, M. (2014b). New insights into roles of intermediate filament phosphorylation and progeria pathogenesis. *IUBMB Life*, *66*, 195–200.

Goto, H., Kosako, H., Tanabe, K., Yanagida, M., Sakurai, M., Amano, M., et al. (1998). Phosphorylation of vimentin by Rho-associated kinase at a unique amino-terminal site that is specifically phosphorylated during cytokinesis. *The Journal of Biological Chemistry*, *273*, 11728–11736.

Goto, H., Tomono, Y., Ajiro, K., Kosako, H., Fujita, M., Sakurai, M., et al. (1999). Identification of a novel phosphorylation site on histone H3 coupled with mitotic chromosome condensation. *The Journal of Biological Chemistry*, *274*, 25543–25549.

Goto, H., Yasui, Y., Kawajiri, A., Nigg, E. A., Terada, Y., Tatsuka, M., et al. (2003). Aurora-B regulates the cleavage furrow-specific vimentin phosphorylation in the cytokinetic process. *The Journal of Biological Chemistry*, *278*, 8526–8530.

Greenfield, E. A. (2013). *Antibodies: A laboratory manual* (2nd ed.). Cold Spring Harbor, NY: Cold Spring Harbor Laboratory Press.

Hashimoto, R., Nakamura, Y., Komai, S., Kashiwagi, Y., Tamura, K., Goto, T., et al. (2000). Site-specific phosphorylation of neurofilament-L is mediated by calcium/calmodulin-dependent protein kinase II in the apical dendrites during long-term potentiation. *Journal of Neurochemistry*, *75*, 373–382.

Hayashi-Takanaka, Y., Yamagata, K., Nozaki, N., & Kimura, H. (2009). Visualizing histone modifications in living cells: Spatiotemporal dynamics of H3 phosphorylation during interphase. *The Journal of Cell Biology*, *187*, 781–790.

Herrmann, H., & Aebi, U. (2004). Intermediate filaments: Molecular structure, assembly mechanism, and integration into functionally distinct intracellular scaffolds. *Annual Review of Biochemistry*, *73*, 749–789.

Herrmann, H., Hesse, M., Reichenzeller, M., Aebi, U., & Magin, T. M. (2003). Functional complexity of intermediate filament cytoskeletons: From structure to assembly to gene ablation. *International Review of Cytology*, *223*, 83–175.

Hyder, C. L., Pallari, H. M., Kochin, V., & Eriksson, J. E. (2008). Providing cellular signposts—Post-translational modifications of intermediate filaments. *FEBS Letters*, *582*, 2140–2148.

Ikegami, Y., Goto, H., Kiyono, T., Enomoto, M., Kasahara, K., Tomono, Y., et al. (2008). Chk1 phosphorylation at Ser286 and Ser301 occurs with both stalled DNA replication

and damage checkpoint stimulation. *Biochemical and Biophysical Research Communications, 377*, 1227–1231.
Inada, H., Nagata, K., Goto, H., & Inagaki, M. (1999). Regulation of intermediate filament dynamics: A novel approach using site-and phosphorylation state-specific antibodies. In K. L. Carraway & C. A. C. Carraway (Eds.), *Practical approach series: Vol. 221. Cytoskeleton: Signalling and cell regulation. A practical approach* (1st ed., pp. 183–207). Oxford: Oxford University Press.
Inada, H., Togashi, H., Nakamura, Y., Kaibuchi, K., Nagata, K., & Inagaki, M. (1999). Balance between activities of Rho kinase and type 1 protein phosphatase modulates turnover of phosphorylation and dynamics of desmin/vimentin filaments. *The Journal of Biological Chemistry, 274*, 34932–34939.
Inagaki, N., Goto, H., Ogawara, M., Nishi, Y., Ando, S., & Inagaki, M. (1997). Spatial patterns of Ca^{2+} signals define intracellular distribution of a signaling by Ca^{2+}/calmodulin-dependent protein kinase II. *The Journal of Biological Chemistry, 272*, 25195–25199.
Inagaki, N., Ito, M., Nakano, T., & Inagaki, M. (1994). Spatiotemporal distribution of protein kinase and phosphatase activities. *Trends in Biochemical Sciences, 19*, 448–452.
Inagaki, M., Matsuoka, Y., Tsujimura, K., Ando, S., Tokui, T., Takahashi, T., et al. (1996). Dynamic property of intermediate filaments: Regulation by phosphorylation. *Bioessays, 18*, 481–487.
Inagaki, M., Nishi, Y., Nishizawa, K., Matsuyama, M., & Sato, C. (1987). Site-specific phosphorylation induces disassembly of vimentin filaments in vitro. *Nature, 328*, 649–652.
Ivaska, J., Pallari, H. M., Nevo, J., & Eriksson, J. E. (2007). Novel functions of vimentin in cell adhesion, migration, and signaling. *Experimental Cell Research, 313*, 2050–2062.
Izawa, I., & Inagaki, M. (2006). Regulatory mechanisms and functions of intermediate filaments: A study using site- and phosphorylation state-specific antibodies. *Cancer Science, 97*, 167–174.
Jeong, H. J., Ohmuro-Matsuyama, Y., Ohashi, H., Ohsawa, F., Tatsu, Y., Inagaki, M., et al. (2013). Detection of vimentin serine phosphorylation by multicolor Quenchbodies. *Biosensors and Bioelectronics, 40*, 17–23.
Kasahara, K., Goto, H., Enomoto, M., Tomono, Y., Kiyono, T., & Inagaki, M. (2010). 14-3-3gamma mediates Cdc25A proteolysis to block premature mitotic entry after DNA damage. *The EMBO Journal, 29*, 2802–2812.
Kawajiri, A., & Inagaki, M. (2004). Approaches to study phosphorylation of intermediate filament proteins using site-specific and phosphorylation state-specific antibodies. In M. B. Omary & P. A. Coulombe (Eds.), *Intermediate filament cytoskeleton: Vol. 78. Methods in cell biololgy* (pp. 353–371). Amsterdam: Elsevier.
Kawajiri, A., Yasui, Y., Goto, H., Tatsuka, M., Takahashi, M., Nagata, K., et al. (2003). Functional significance of the specific sites phosphorylated in desmin at cleavage furrow: Aurora-B may phosphorylate and regulate type III intermediate filaments during cytokinesis coordinatedly with Rho-kinase. *Molecular Biology of the Cell, 14*, 1489–1500.
Kidd, M. E., Shumaker, D. K., & Ridge, K. M. (2014). The role of vimentin intermediate filaments in the progression of lung cancer. *American Journal of Respiratory Cell and Molecular Biology, 50*, 1–6.
Kochin, V., Shimi, T., Torvaldson, E., Adam, S. A., Goldman, A., Pack, C. G., et al. (2014). Interphase phosphorylation of lamin A. *Journal of Cell Science, 127*, 2683–2696.
Kong, L., Schafer, G., Bu, H., Zhang, Y., Zhang, Y., & Klocker, H. (2012). Lamin A/C protein is overexpressed in tissue-invading prostate cancer and promotes prostate cancer cell growth, migration and invasion through the PI3K/AKT/PTEN pathway. *Carcinogenesis, 33*, 751–759.
Konishi, H., Namikawa, K., Shikata, K., Kobatake, Y., Tachibana, T., & Kiyama, H. (2007). Identification of peripherin as a Akt substrate in neurons. *The Journal of Biological Chemistry, 282*, 23491–23499.

Kosako, H., Goto, H., Yanagida, M., Matsuzawa, K., Fujita, M., Tomono, Y., et al. (1999). Specific accumulation of Rho-associated kinase at the cleavage furrow during cytokinesis: Cleavage furrow-specific phosphorylation of intermediate filaments. *Oncogene, 18*, 2783–2788.

Kotula, E., Faigle, W., Berthault, N., Dingli, F., Loew, D., Sun, J. S., et al. (2013). DNA-PK target identification reveals novel links between DNA repair signaling and cytoskeletal regulation. *PLoS One, 8*, e80313.

Ksiezak-Reding, H., Dickson, D. W., Davies, P., & Yen, S. H. (1987). Recognition of tau epitopes by anti-neurofilament antibodies that bind to Alzheimer neurofibrillary tangles. *Proceedings of the National Academy of Sciences of the United States of America, 84*, 3410–3414.

Ku, N. O., Liao, J., & Omary, M. B. (1998). Phosphorylation of human keratin 18 serine 33 regulates binding to 14-3-3 proteins. *The EMBO Journal, 17*, 1892–1906.

Ku, N. O., & Omary, M. B. (1997). Phosphorylation of human keratin 8 in vivo at conserved head domain serine 23 and at epidermal growth factor-stimulated tail domain serine 431. *The Journal of Biological Chemistry, 272*, 7556–7564.

Kuga, T., Nozaki, N., Matsushita, K., Nomura, F., & Tomonaga, T. (2010). Phosphorylation statuses at different residues of lamin B2, B1, and A/C dynamically and independently change throughout the cell cycle. *Experimental Cell Research, 316*, 2301–2312.

Kusubata, M., Tokui, T., Matsuoka, Y., Okumura, E., Tachibana, K., Hisanaga, S., et al. (1992). p13suc1 suppresses the catalytic function of p34cdc2 kinase for intermediate filament proteins, in vitro. *The Journal of Biological Chemistry, 267*, 20937–20942.

Liao, J., Ku, N. O., & Omary, M. B. (1997). Stress, apoptosis, and mitosis induce phosphorylation of human keratin 8 at Ser-73 in tissues and cultured cells. *The Journal of Biological Chemistry, 272*, 17565–17573.

Liao, J., Lowthert, L. A., Ku, N. O., Fernandez, R., & Omary, M. B. (1995). Dynamics of human keratin 18 phosphorylation: Polarized distribution of phosphorylated keratins in simple epithelial tissues. *The Journal of Cell Biology, 131*, 1291–1301.

Lichtenberg-Kraag, B., Mandelkow, E. M., Biernat, J., Steiner, B., Schroter, C., Gustke, N., et al. (1992). Phosphorylation-dependent epitopes of neurofilament antibodies on tau protein and relationship with Alzheimer tau. *Proceedings of the National Academy of Sciences of the United States of America, 89*, 5384–5388.

Matsuoka, Y., Nishizawa, K., Yano, T., Shibata, M., Ando, S., Takahashi, T., et al. (1992). Two different protein kinases act on a different time schedule as glial filament kinases during mitosis. *The EMBO Journal, 11*, 2895–2902.

Matsuyama, M., Tanaka, H., Inoko, A., Goto, H., Yonemura, S., Kobori, K., et al. (2013). Defect of mitotic vimentin phosphorylation causes microophthalmia and cataract via aneuploidy and senescence in lens epithelial cells. *The Journal of Biological Chemistry, 288*, 35626–35635.

Nagata, K., Izawa, I., & Inagaki, M. (2001). A decade of site- and phosphorylation state-specific antibodies: Recent advances in studies of spatiotemporal protein phosphorylation. *Genes to Cells, 6*, 653–664.

Nakamura, Y., Hashimoto, R., Kashiwagi, Y., Aimoto, S., Fukusho, E., Matsumoto, N., et al. (2000). Major phosphorylation site (Ser55) of neurofilament L by cyclic AMP-dependent protein kinase in rat primary neuronal culture. *Journal of Neurochemistry, 74*, 949–959.

Nakamura, Y., Hashimoto, R., Kashiwagi, Y., Wada, Y., Sakoda, S., Miyamae, Y., et al. (1999). Casein kinase II is responsible for phosphorylation of NF-L at Ser-473. *FEBS Letters, 455*, 83–86.

Nishizawa, K., Yano, T., Shibata, M., Ando, S., Saga, S., Takahashi, T., et al. (1991). Specific localization of phosphointermediate filament protein in the constricted area of dividing cells. *The Journal of Biological Chemistry, 266*, 3074–3079.

Nukina, N., Kosik, K. S., & Selkoe, D. J. (1987). Recognition of Alzheimer paired helical filaments by monoclonal neurofilament antibodies is due to crossreaction with tau protein. *Proceedings of the National Academy of Sciences of the United States of America*, *84*, 3415–3419.

Ogawara, M., Inagaki, N., Tsujimura, K., Takai, Y., Sekimata, M., Ha, M. H., et al. (1995). Differential targeting of protein kinase C and CaM kinase II signalings to vimentin. *The Journal of Cell Biology*, *131*, 1055–1066.

Oguri, T., Inoko, A., Shima, H., Izawa, I., Arimura, N., Yamaguchi, T., et al. (2006). Vimentin-Ser82 as a memory phosphorylation site in astrocytes. *Genes to Cells*, *11*, 531–540.

Parry, D. A., & Steinert, P. M. (1992). Intermediate filament structure. *Current Opinion in Cell Biology*, *4*, 94–98.

Sasaki, T., Taoka, M., Ishiguro, K., Uchida, A., Saito, T., Isobe, T., et al. (2002). In vivo and in vitro phosphorylation at Ser-493 in the glutamate (E)-segment of neurofilament-H subunit by glycogen synthase kinase 3beta. *The Journal of Biological Chemistry*, *277*, 36032–36039.

Satelli, A., & Li, S. (2011). Vimentin in cancer and its potential as a molecular target for cancer therapy. *Cellular and Molecular Life Sciences*, *68*, 3033–3046.

Sekimata, M., Tsujimura, K., Tanaka, J., Takeuchi, Y., Inagaki, N., & Inagaki, M. (1996). Detection of protein kinase activity specifically activated at metaphase-anaphase transition. *The Journal of Cell Biology*, *132*, 635–641.

Sihag, R. K., Inagaki, M., Yamaguchi, T., Shea, T. B., & Pant, H. C. (2007). Role of phosphorylation on the structural dynamics and function of types III and IV intermediate filaments. *Experimental Cell Research*, *313*, 2098–2109.

Snider, N. T., & Omary, M. B. (2014). Post-translational modifications of intermediate filament proteins: Mechanisms and functions. *Nature Reviews. Molecular Cell Biology*, *15*, 163–177.

Snider, N. T., Park, H., & Omary, M. B. (2013). A conserved rod domain phosphotyrosine that is targeted by the phosphatase PTP1B promotes keratin 8 protein insolubility and filament organization. *The Journal of Biological Chemistry*, *288*, 31329–31337.

Song, S., Landsbury, A., Dahm, R., Liu, Y., Zhang, Q., & Quinlan, R. A. (2009). Functions of the intermediate filament cytoskeleton in the eye lens. *The Journal of Clinical Investigation*, *119*, 1837–1848.

Sternberger, L. A., & Sternberger, N. H. (1983). Monoclonal antibodies distinguish phosphorylated and nonphosphorylated forms of neurofilaments in situ. *Proceedings of the National Academy of Sciences of the United States of America*, *80*, 6126–6130.

Takai, Y., Ogawara, M., Tomono, Y., Moritoh, C., Imajoh-Ohmi, S., Tsutsumi, O., et al. (1996). Mitosis-specific phosphorylation of vimentin by protein kinase C coupled with reorganization of intracellular membranes. *The Journal of Cell Biology*, *133*, 141–149.

Tanaka, H., Goto, H., Inoko, A., Makihara, H., Enomoto, A., Horimoto, K., et al. (2015). Cytokinetic failure-induced tetraploidy develops into aneuploidy, triggering skin aging in phosphovimentin-deficient mice. *The Journal of Biological Chemistry*, *290*, 12984–12998.

Tsujimura, K., Ogawara, M., Takeuchi, Y., Imajoh-Ohmi, S., Ha, M. H., & Inagaki, M. (1994). Visualization and function of vimentin phosphorylation by cdc2 kinase during mitosis. *The Journal of Biological Chemistry*, *269*, 31097–31106.

Yamaguchi, T., Goto, H., Yokoyama, T., Sillje, H., Hanisch, A., Uldschmid, A., et al. (2005). Phosphorylation by Cdk1 induces Plk1-mediated vimentin phosphorylation during mitosis. *The Journal of Cell Biology*, *171*, 431–436.

Yano, T., Taura, C., Shibata, M., Hirono, Y., Ando, S., Kusubata, M., et al. (1991). A monoclonal antibody to the phosphorylated form of glial fibrillary acidic protein:

Application to a non-radioactive method for measuring protein kinase activities. *Biochemical and Biophysical Research Communications, 175*, 1144–1151.

Yasui, Y., Goto, H., Matsui, S., Manser, E., Lim, L., Nagata, K., et al. (2001). Protein kinases required for segregation of vimentin filaments in mitotic process. *Oncogene, 20*, 2868–2876.

Yokoyama, T., Goto, H., Izawa, I., Mizutani, H., & Inagaki, M. (2005). Aurora-B and Rho-kinase/ROCK, the two cleavage furrow kinases, independently regulate the progression of cytokinesis: Possible existence of a novel cleavage furrow kinase phosphorylates ezrin/radixin/moesin (ERM). *Genes to Cells, 10*, 127–137.

CHAPTER FIVE

Assays for Posttranslational Modifications of Intermediate Filament Proteins

Natasha T. Snider*[,1], **M. Bishr Omary**[†,‡]
*Department of Cell Biology and Physiology, University of North Carolina, Chapel Hill, North Carolina, USA
[†]Department of Molecular & Integrative Physiology, Department of Medicine, University of Michigan, Ann Arbor, Michigan, USA
[‡]VA Ann Arbor Healthcare System, Ann Arbor, Michigan, USA
[1]Corresponding author: e-mail address: ntsnider@med.unc.edu

Contents

1. Introduction 114
 1.1 Posttranslational Modifications of Intermediate Filament Proteins 114
 1.2 Available Tools and Major Limitations for the Study of IF Protein PTMs 115
 1.3 Cross-Talk Between PTMs on IF Proteins 117
 1.4 Chemical and Pharmacological Approaches to Study Context-Specific IF Protein PTMs 117
2. Extraction of IF Proteins from Tissues and Cells for Biochemical Analysis of IF PTMs 120
 2.1 Materials and Reagents for Isolation of IF Proteins 120
 2.2 High Salt Extraction of IF Proteins 121
 2.3 Immunoprecipitation of Detergent-Soluble IF Proteins 124
3. Methods for Monitoring Specific PTMs on IF Proteins 126
 3.1 Analysis of IF Protein Phosphorylation 126
 3.2 Analysis of IF Protein Sumoylation 129
 3.3 Analysis of IF Protein Lysine Acetylation 130
 3.4 Analysis of IF Protein *O*-Linked Glycosylation 133
 3.5 Analysis of IF Protein Transamidation 134
4. Conclusions 135
Acknowledgments 135
References 135

Abstract

Intermediate filament (IF) proteins are known to be regulated by a number of posttranslational modifications (PTMs). Phosphorylation is the best-studied IF PTM, whereas ubiquitination, sumoylation, acetylation, glycosylation, ADP-ribosylation, farnesylation, and transamidation are less understood in functional terms but are known to regulate specific IFs under various contexts. The number and diversity of IF PTMs is certain to grow along with rapid advances in proteomic technologies. Therefore, the need for

Methods in Enzymology, Volume 568
ISSN 0076-6879
http://dx.doi.org/10.1016/bs.mie.2015.09.005

a greater understanding of the implications of PTMs to the structure, organization, and function of the IF cytoskeleton has become more apparent with the increased availability of data from global profiling studies of normal and diseased specimens. This chapter will provide information on established methods for the isolation and monitoring of IF PTMs along with the key reagents that are necessary to carry out these experiments.

ABBREVIATIONS

AcK acetyl-lysine
BHK-21 baby hamster kidney cells
dd double distilled
HSE high salt extract
IF intermediate filament
i.p. immunoprecipitation
K keratin
MS/MS tandem mass spectrometry
O-GlcNAc β-D-*N*-acetylglucosamine
PTM posttranslational modification
pY phosphotyrosine
SUMO small ubiquitin-like modifier
TG2 transglutaminase-2

1. INTRODUCTION

1.1 Posttranslational Modifications of Intermediate Filament Proteins

Intermediate filament (IF) proteins are important for the maintenance of cellular function in the basal state, and are particularly important under stress and in disease states (Davidson & Lammerding, 2014; Gruenbaum & Aebi, 2014; Homberg & Magin, 2014; Omary, 2009; Toivola, Boor, Alam, & Strnad, 2015). IFs are major structural components of the cell cytoskeleton, but through their dynamic behavior and under varying cellular conditions, they have also been demonstrated to impact virtually every aspect of cellular function, including gene transcription, signaling pathways, and cellular survival (Chung, Rotty, & Coulombe, 2013; Herrmann, Strelkov, Burkhard, & Aebi, 2009; Toivola, Strnad, Habtezion, & Omary, 2010). The assembly and disassembly dynamics of IF proteins, as well as their associations with other cellular components are regulated by various posttranslational modifications (PTMs), summarized in Table 1, and a myriad of enzymes that carry out specific PTM on/off reactions (Hyder, Pallari, Kochin, & Eriksson, 2008; Omary, Ku, Tao, Toivola, & Liao, 2006; Snider & Omary, 2014).

Table 1 Posttranslational Modifications of IF Proteins

IF Protein	Major PTMs
Lamins	Phosphorylation, farnesylation, ubiquitination, sumoylation
Keratins	Phosphorylation, *O*-linked glycosylation, ubiquitination, sumoylation, acetylation, transamidation
Vimentin	Phosphorylation, *O*-linked glycosylation, ubiquitination, sumoylation, ADP-ribosylation
Peripherin	Phosphorylation, ubiquitination[a]
α-internexin	Phosphorylation[a]
Neurofilament (-L, -M, -H)	Phosphorylation, *O*-linked glycosylation, ubiquitination
GFAP	Phosphorylation
Nestin	Phosphorylation
Desmin	Phosphorylation, ubiquitination, ADP-ribosylation
Synemin	Phosphorylation[a]
Syncoilin	Phosphorylation[a]
Filensin (BFSP1)	Phosphorylation[a]
Phakinin (BFSP2)	Phosphorylation[a]

[a]The modification has only been detected in high-throughput studies.
BFSP, beaded filament structural protein.

1.2 Available Tools and Major Limitations for the Study of IF Protein PTMs

The extent of functional understanding regarding the role of each PTM on IF protein function is highly dependent on the availability of tools to study the particular PTM of interest. For example, phosphorylation (Roux & Thibault, 2013) and ubiquitination (Sylvestersen, Young, & Nielsen, 2013) can be analyzed using mass spectrometry with relative ease, whereas sumoylation (Gareau & Lima, 2010), which has relatively low stoichiometry and is not easily analyzed by mass spectrometric means, is more difficult to probe. Therefore, the system-level PTM data currently available are skewed to highlight those PTMs that can be readily tracked using proteomic platforms (Choudhary & Mann, 2010; Hennrich & Gavin, 2015). The combination of global proteomic data with PTM databases that catalog experimentally determined and site-specific modifications, or that use computational approaches to predict and quantify PTMs (Table 2), has resulted

Table 2 Databases[a] That Curate Experimentally Determined or Predicted PTMs on Various Proteins

Database	Web Link	Modifications	References
PhosphoSitePlus	www.phosphosite.org	Phosphorylation, acetylation, ubiquitination, methylation, succinylation	Hornbeck et al. (2015)
PTM-SD	http://www.dsimb.inserm.fr/dsimb_tools/PTM-SD/	Many	Craveur, Rebehmed, and de Brevern (2014)
PTMCode	http://ptmcode.embl.de/	Many	Minguez, Letunic, Parca, and Bork (2012)
dbPTM	http://dbptm.mbc.nctu.edu.tw/	Phosphorylation, ubiquitination, acetylation, glycosylation, succinylation, nitrosylation, glutathionylation	Lu et al. (2013)
PTMCurator	http://selene.princeton.edu/PTMCuration/	Many (generates quantitative computational predictions)	Khoury, Baliban, and Floudas (2011)
PHOSIDA	http://www.phosida.com/	Phosphorylation, acetylation, *N*-glycosylation	Gnad, Gunawardena, and Mann (2011)

[a]Shown is a partial list of the available PTM databases. We commonly use PhosphoSitePlus.

in a wealth of information on modified residues on IF proteins. However, most of these modifications await functional assignment. For most IF protein PTMs, the use of molecular approaches (e.g., site-directed mutagenesis of modification sites), biochemical tools (pan- or site-specific PTM antibodies), chemical probes (inhibitors or activators of PTM enzymes), and transgenic mouse models, in combination with enrichment of the IF protein fraction from cells and tissues, has yielded useful insight into some of the functional roles of PTMs, although much more remains to be learned. The relative insolubility of IF proteins (particularly epidermal keratins) in nondenaturing detergent-containing buffers can be an impediment to the study of PTMs, although these limitations can be surmounted, as was shown for the type I keratin K17 (Pan, Kane, Van Eyk, & Coulombe, 2011).

1.3 Cross-Talk Between PTMs on IF Proteins

PTMs participate in complex cross-talk mechanisms to regulate IF function. The balance of various modified forms of IF proteins is dictated by cellular conditions, such as mitosis, cell migration, stress, and apoptosis. The key to resolving the information encoded by IF PTMs is to determine which PTM signatures are prevalent under a given condition and how altering the stoichiometry of IF PTMs alters IF function, distribution, interactions, and ultimately, cellular fate.

Using the database PhosphoSitePlus (Hornbeck et al., 2015), we conducted a search for PTMs on human keratin 8 (K8) that have been reported by at least one low-throughput study, or those that appear in at least five high-throughput studies/records (Table 3). In this case, low-throughput refers to data generated via amino acid sequencing, site-directed mutagenesis, or the use of specific antibodies, whereas high throughput refers to studies using unbiased discovery-mode mass spectrometry. This example analysis revealed that 16% (76/483) of residues on K8 are modified by either phosphorylation, acetylation, ubiquitination, sumoylation, or methylation. The majority of the PTM sites on K8 (47/76) are modified by phosphorylation (pSer > pTyr > pThr). It can also be appreciated that most of the acetylation sites are also targets for ubiquitination, and one residue (Lys-285) appears to be capable of undergoing sumoylation, acetylation, or ubiquitination. These observations suggest that, depending on the cellular conditions, the various PTMs may directly compete for a given site. In addition to direct competition, there is experimental evidence for functional cross-talk between the different PTMs, such as between phosphorylation and ubiquitination, sumoylation, and acetylation (Snider & Omary, 2014).

1.4 Chemical and Pharmacological Approaches to Study Context-Specific IF Protein PTMs

Regulation of IF proteins by PTMs occurs under specific cellular conditions, which can be modeled using chemical and pharmacological agents. An abbreviated list of representative reagents is shown in Table 4. Hyperphosphorylation of IF proteins can be induced by treatment with phosphatase inhibitors, such as okadaic acid or microcystin-LR. Phosphorylation, in

Table 3 Example of Phospho-Site Database Search for PTMs on Human K8

No. of Studies/ Records				No. of Studies/ Records			
LTP	HTP	Residue	Type of Modification	LTP	HTP	Site	Type of Modification
2	24	9	Phosphorylation (serine)	0	85	25	Phosphorylation (tyrosine)
1	5	13		0	8	143	
0	44	21		1	117	204	
0	25	22		1	440	267	
5	114	24		0	106	282	
0	33	27		0	49	286	
0	19	31		0	6	419	
1	48	34		0	19	427	
0	102	35		0	49	437	
0	11	36		0	6	11	Ubiquitination (lysine)
1	27	37		0	6	96	
0	23	39		0	9	101	
1	74	43		0	10	108	
0	21	44		0	9	117	
25	12	74		0	7	122	
0	23	104		0	10	130	
0	5	142		0	6	264	
0	19	253		0	5	285	
0	12	258		0	8	304	
0	9	274		0	22	325	
0	8	280		0	7	347	
1	6	291		0	5	352	
0	5	315		0	59	393	
0	10	330		0	11	472	

Table 3 Example of Phospho-Site Database Search for PTMs on Human K8—cont'd

No. of Studies/ Records				No. of Studies/ Records			
LTP	**HTP**	**Residue**	**Type of Modification**	**LTP**	**HTP**	**Site**	**Type of Modification**
0	8	410		2	1	11	Acetylation (lysine)
1	2	417		1	32	101	
20	27	432		0	7	108	
0	7	436		0	6	122	
0	8	442		1	4	207	
0	8	445		0	5	285	
0	9	456		1	2	393	
0	17	457		2	5	472	
0	20	475		3	38	483	
0	11	478		1	0	285	Sumoylation (lysine)
0	24	26	Phosphorylation (threonine)	1	0	364	
0	31	305		0	5	23	
1	1	431		0	7	32	Methylation (arginine)
0	7	455		0	9	47	

LTP, low throughput; HTP, high throughput.
PhosphoSitePlus (Hornbeck et al., 2015) was used to conduct a search for PTMs on human keratin 8 (K8). The search was limited to modified residues that have been reported by at least one low-throughput study (LTP), or those that appear in at least five high-throughput studies/records (HTP). LTP refers to data generated via amino acid sequencing, site-directed mutagenesis, or the use of specific antibodies; HTP refers to studies using unbiased discovery-mode mass spectrometry.

Table 4 Commonly Used Strategies to Induce IF Protein PTMs

Condition	Inducer(s)	PTM(s)
Oxidative stress	Hydrogen peroxide	Phosphorylation, sumoylation
Apoptosis	Anisomycin, Fas ligand	Phosphorylation, sumoylation
Phosphatase inhibition	Okadaic acid, microcystin-LR	Phosphorylation, sumoylation
Nutrient stimulation	Glucose	Lysine acetylation
Metabolic stress	PUGNAc	O-linked glycosylation
Proteasome inhibition	MG-132	Ubiquitination

Shown is an abbreviated list of reagents. There are many options for eliciting these types of responses for *in vitro* and *in vivo* monitoring of IF protein PTMs.

turn, can promote IF protein sumoylation, so these two modifications are generally observed together, particularly under conditions of oxidative stress and apoptosis (Snider, Weerasinghe, Iniguez-Lluhi, Herrmann, & Omary, 2011). The latter can be modeled in cells using hydrogen peroxide and anisomycin, respectively, among numerous other reagents that can induce similar responses. Increased acetylation of K8 in epithelial cells is dose–responsive to glucose levels *in vitro* and *in vivo*, and can be induced more robustly if the cells undergo transient glucose starvation prior to restimulation (Snider, Leonard, et al., 2013). Increased *O*-glycosylation of K18 *in vitro* and *in vivo* is observed in response to PUGNAc (Ku, Toivola, Strnad, & Omary, 2010), an inhibitor of the enzyme that removes *O*-linked *N*-acetylglucosamine (*O*-GlcNAc) groups (*O*-GlcNAcase). Since many IF proteins (Rogel, Jaitovich, & Ridge, 2010) are turned over by ubiquitination and proteasomal degradation, treatment with the proteasome inhibitor MG-132 promotes the accumulation of poly-ubiquitinated IF proteins. The sections that follow describe the general methods for isolation of IF proteins from cells and tissues, followed by a summary of specific tools and approaches to study more common PTMs on IF proteins.

2. EXTRACTION OF IF PROTEINS FROM TISSUES AND CELLS FOR BIOCHEMICAL ANALYSIS OF IF PTMs

2.1 Materials and Reagents for Isolation of IF Proteins

***Source of IF protein*:**
- Human or animal tissue
- Primary cells (e.g., freshly isolated hepatocytes)
- Cultured cells expressing endogenous IF proteins
- Cultured cells overexpressing IF protein(s) of interest

***Antibodies and immunoprecipitation beads*:**
- Anti-IF protein antibodies
- Anti-PTM antibodies (pan- or site-specific)
- IgG isotype control
- Dynabeads magnetic beads (or similar)

***Buffers*:**
- *TXB*: Triton-X Buffer
  - 1% Triton-X-100
  - 5 m*M* EDTA
  - Bring up volume in PBS, pH 7.4
- *TXB + PPI*: Triton-X Buffer with Protease and Phosphatase Inhibitors

Immediately before use, add protease and phosphatase inhibitors (follow manufacturer instructions) to the desired volume of TXB to be used in the experiment.

- *HSB*: High Salt Buffer:
  - 10 m*M* Tris–HCl, pH 7.6
  - 140 m*M* NaCl
  - 1.5 *M* KCl
  - 5 m*M* EDTA
  - 0.5% Triton-X-100
  - Bring up volume in double distilled (dd) H_2O
- *HSB + PPI*: High Salt Buffer with Protease and Phosphatase Inhibitors

  Immediately before use, add protease and phosphatase inhibitors (follow manufacturer instructions) to the desired volume of HSB to be used in the experiment.
- *PBS/EDTA*:

  5 m*M* EDTA in 1 × PBS, pH 7.4. The EDTA is included to protect from calcium-activated protease-related degradation (Chou, Riopel, Rott, & Omary, 1993).
- 2 × SDS Sample Buffer
- Coomassie-based stain compatible with mass spectrometry applications (e.g., Gel Code Blue)

2.2 High Salt Extraction of IF Proteins

1a (*If using tissue*): Cut 10–25 mg of freshly isolated or snap-frozen tissue from liquid nitrogen storage and place directly into 1 mL of ice-cold TXB + PPI on ice. Dounce (50 strokes) to a homogeneous suspension using a Potter-Elvehjem PTFE pestle and glass tube homogenizer (7 mL working volume size).

1b (*If using cells*): Collect cells by scraping (if adherent) and centrifugation at 500 *g* for 5 min. For cells grown to 80–95% confluency on a 100 mm dish (~60 cm^2 growth surface area vessel) use 2 mL of 1 × PBS to collect the cells. For a cell pellet with approximate volume of 50–100 μL add 1–2 mL of ice-cold TXB + PPI and solubilize the pellet by pipetting. Leave tubes sitting on ice for 5–10 min.

2. Centrifuge tissue or cell lysate (14,000 rpm) for 10 min at 4 °C.

3. Collect the supernatant fraction. This is the TXB-soluble fraction, which can be used to immunoprecipitate (i.p.) the detergent-soluble pool of IF proteins (Section 2.3).

Note: Under basal conditions ~5% of the cellular K8/K18 pool can be extracted from the human colonic cell line HT-29 in a detergent-free buffer, while a total of ~20% can be extracted using the nonionic detergent Nonidet P-40 which is similar to the TXB buffer (Chou et al., 1993; Omary, Ku, Liao, & Price, 1998). The K8/K18 soluble fraction increases during mitosis (Chou et al., 1993) but is much smaller for epidermal keratins but likely higher for vimentin and other type III IF proteins (Lowthert, Ku, Liao, Coulombe, & Omary, 1995; Omary et al., 1998). However, only 0.18–0.39% of vimentin in the rat cell line RVF-SMC was reported to be soluble (Soellner, Quinlan, & Franke, 1985), although this may represent an underestimate based on how vimentin was isolated using single-stranded DNA-cellulose columns from [^{35}S]-methionine pulse-labeled cells. In terms of keratin content in simple-type epithelial cells and tissues, K8/K18 make up 5% of total cellular protein in HT-29 cells (Chou et al., 1993) and 0.2–0.5% of total tissue protein in mouse pancreas, liver, and small intestine (Zhong et al., 2004); while human epidermal keratins make up 25–35% of total cellular proteins in human keratinocytes (Sun & Green, 1978) and 17–27% of newborn mouse keratinocytes (Feng, Zhang, Margolick, & Coulombe, 2013).

4. Resuspend the cell pellet in 1 mL of HSB + PPI by douncing (100 strokes) and leave on the shaker in the cold room for 1 h.

   Note: The 1 h shaking step is important for obtaining clean high salt extracts. In general, the High Salt Extraction (HSE) purity ranks: cell line overexpressed IFs > cell line endogenous IFs > tissue IFs. Therefore, this step is the most critical when working with tissues.
5. Centrifuge the samples at 14,000 rpm for 20 min at 4 °C. Discard the supernatant, which contains histones and other nuclear proteins that are not solubilized in TXB (Step 1).
6. Homogenize the pellet in 1 mL of ice-cold PBS/EDTA buffer and centrifuge at 14,000 rpm for 10 min at 4 °C to obtain the IF protein-rich high salt extract (HSE).
7. Dissolve the HSE pellets in 200–400 μL of 2 × reducing or nonreducing SDS sample buffer that has been preheated to 95 °C. Break up the pellet initially by pipetting/vortexing, then heat the samples for 2–5 min at 95 °C. Vortex and pipet as needed to ensure the pellet is dissolved.

   Note: The temperature and the volume of the buffer are critical for ensuring that the entire pellet goes into solution. Dissolving the entire pellet may require 5–15 min of vortexing and pipetting per sample, depending on the pellet size.

8. Proceed with downstream analysis:
   - Analyze the HSEs by SDS–PAGE followed by Coomassie stain to check the enrichment of IF proteins (an abbreviated protocol and example gel are shown in Fig. 1).

     Note: The pellets used in the gels that are shown in Fig. 1 were completely solubilized prior to loading on the gels. We have obtained HSEs of vimentin and GFAP with similar purity by following this extraction protocol (e.g., vimentin can be readily seen as noted in Fig. 3B).
   - Analyze the HSE samples by Western blot using PTM antibodies (pan- or site-specific)
     - Useful antibodies for specific PTMs are noted in each subsection.
   - If sufficient material is present on Coomassie-stained gel, excise the IF bands for proteomic analysis by mass spectrometry.
     - *Note*: Special precautions must be taken in order to avoid contamination: gels must be handled with clean gloves and incubated in clean containers, which should only be washed using ddH_2O

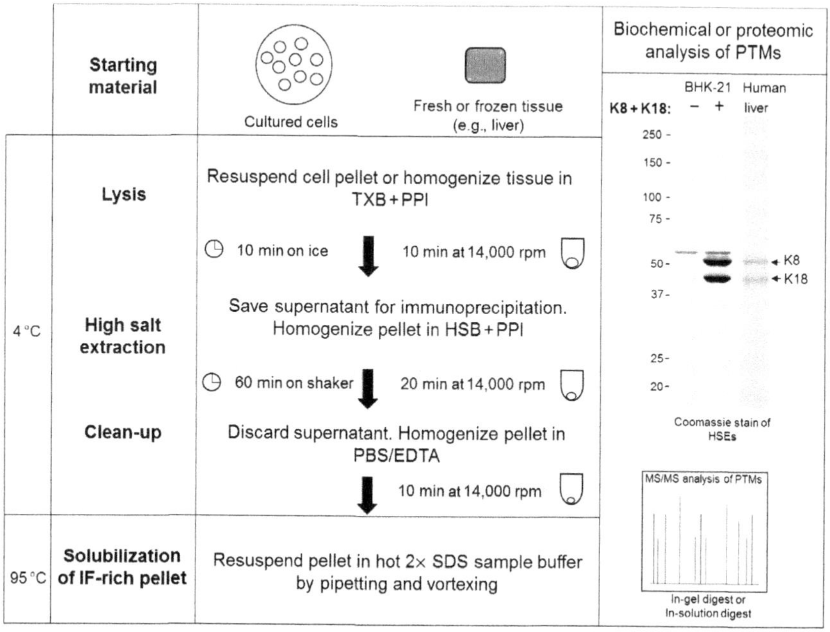

Figure 1 Isolation of intermediate filament proteins using high salt extraction. Shown is an abbreviated version of the protocol outlined in Section 2.2. In the panel on the right, BHK cells are transfected with human K8 and K18 (+) or vector alone (−) and the HSE is compared from these cells is shown in parallel with the HSE from normal human liver.

(no soap); note that keratins are typically considered common "contaminants" of mass spectrometry experiments.

- Alternatively, proteomic analysis may be performed on the total HSE (in-solution digest).
  - *Note*: A consultation with a proteomics expert is necessary to discuss the various needs and options for a given experiment prior to initiating a study.

2.3 Immunoprecipitation of Detergent-Soluble IF Proteins

1. Aliquot 50 μL of Dynabeads (e.g., Dynabeads protein-G) into an Eppendorf tube, place on the magnet, and aspirate storage solution.
2. Resuspend the beads in the antibody solution (1–10 μg of antibody in 200 μL of PBS + 0.02% Tween-20) and incubate on rotator (end-over-end) at room temperature for 20 min.
3. Aspirate antibody solution and wash beads once with 200 μL of PBS + 0.02% Tween-20. Add 600 μL of the cell or tissue lysate (save a small fraction of the original lysate to check the protein levels in the pre-i.p. input, as described in Step 8).
   - *Note*: Lysis buffers containing different detergents (e.g., NP-40 or Empigen-BB) may be used. Empigen-containing lysis buffers solubilize a larger pool of IF protein (Lowthert et al., 1995), although they may disrupt certain interactions with IF-binding partners.
4. Incubate for 3 h with rotation in a cold room.
5. Place samples on magnet and save the post-i.p. fraction for quality control in Step 8.
6. Wash the beads five times with 200 μL of PBS + 0.02% Tween-20. For the last wash step, collect the beads in 100 μL of PBS (without Tween-20) and transfer to a new tube.
7. Remove the PBS and add 60–100 μL of hot nonreducing sample buffer, heat the samples to 95 °C for 2–5 min and separate the i.p. fraction from the beads on the magnet and collect it into a new tube.
   - *Note*: It is necessary to use nonreducing buffer so that the antibody IgG heavy chain (which, when reduced, migrates at ~50 kDa) does not obscure the IF protein bands on an SDS–PAGE gel. Alternatively, the antibody can be cross-linked to the beads to avoid coelution (follow manufacturer guidelines for the specific cross-linking procedure).
8. As a quality check, analyze equivalent amounts of the pre-i.p. lysate (from Step 3), post-i.p. lysate (from Step 5), and the i.p. fraction (from

Step 7) on an SDS–PAGE gel and probe for the presence of the IF protein of interest. If significant amounts of protein remain in the post-i.p. fraction, vary the ratio of antibody to lysate in order to achieve immunodepletion.

- *Note*: We generally use 3 μg of anti-human keratin 18 (L2A1 or DC10 clones; see table 1 in Ku et al. (2004) for list of antibodies to simple epithelial keratins) antibody per 600 μL of HT-29 (human colon cancer cell line) lysate. We optimize this step based on the type and starting number of cells, rather than total protein concentration, because IF protein amounts will be highly variable depending on the starting material.

9. Proceed with downstream analysis:

- Perform Coomassie stain to check the enrichment of IF proteins in the HSE fraction; excise the IF band of interest for proteomic analysis by mass spectrometry.
  - An abbreviated protocol and example gel are shown in Fig. 2.
- Perform the immunoblot using antibodies to the IF protein/PTM of interest (pan- or site-specific).

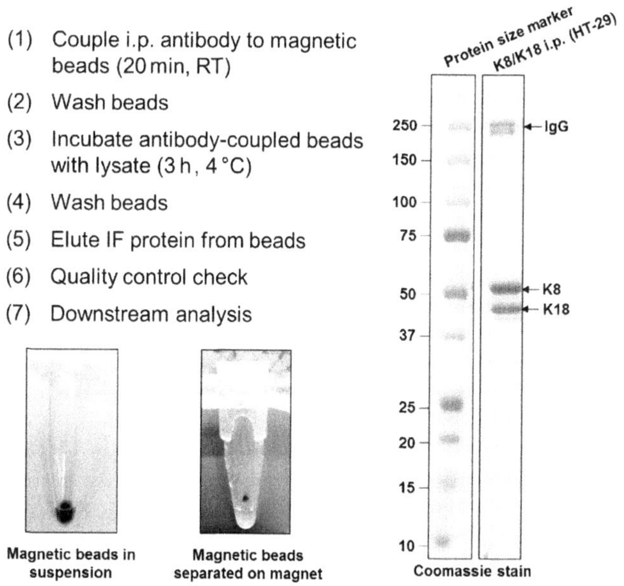

Figure 2 Immunoprecipitation of intermediate filament proteins from detergent-soluble fractions. Shown is an abbreviated version of the protocol outlined in Section 2.3. Representative Coomassie-stained gels of transfected cell- and liver tissue-extracted K8/K18 are shown on the right.

- Perform proteomic analysis of the total i.p. fraction (in-solution digest).
  - *Note*: Antibody bead cross-linking is ideal for an in-solution digest in order to avoid signal interference from the immunoglobulins.

3. METHODS FOR MONITORING SPECIFIC PTMs ON IF PROTEINS

The sections below describe the functional significance and the available tools and methods to study several IF protein PTMs: phosphorylation, sumoylation, acetylation, glycosylation, and transamidation. For specific methods for the analysis of other PTMs, such as farnesylation (Hannoush & Sun, 2010), ubiquitination (Lill & Wertz, 2014), and ADP-ribosylation (Moyle & Muir, 2010), we direct the readers to general protocols and previous applications of these modifications in the context of IF proteins.

3.1 Analysis of IF Protein Phosphorylation

Phosphorylation is the best-characterized PTM of IF proteins (Omary et al., 2006; Snider & Omary, 2014). Most of the known phosphorylation sites on IF proteins are Ser and Thr residues found in the globular head and tail domains, although proteomic studies have also revealed numerous phosphorylation sites in the rod domain that only recently began to undergo functional characterization (Snider, Park, & Omary, 2013). One major function of IF phosphorylation is to facilitate the reorganization, and the assembly–disassembly dynamics of IF proteins (Izawa & Inagaki, 2006; Sihag, Inagaki, Yamaguchi, Shea, & Pant, 2007; Snider & Omary, 2014). With the exception of neurofilaments, basal phosphorylation of IF proteins is minimal but becomes significantly upregulated during stress, mitosis, disease states, and in the presence of mutations that have structural effects on IF proteins. Multiple approaches can be used to study IF phosphorylation, including chemical inhibitors and activators, phospho-specific antibodies, mass spectrometry, and transgenic mice. A list of phospho-site-specific antibodies targeting different human IF proteins available through major suppliers is shown in Table 5. Additionally, pan-pSer/Thr and pan-pTyr antibody are widely available and may be used for biochemical detection of IF protein phosphorylation sites on IF protein-enriched samples in combination with proper experimental controls. As an example, the

Table 5 Phospho-Site-Specific Antibodies

IF Protein	Phosphorylation Sites	Commercial Source(s)
Keratin 8	pSer-24, pSer-74, pSer-432	Abcam
Keratin 18	pSer-34, pSer-53	Abcam
Keratin 17	pSer-44	Cell Signaling
Vimentin	pSer-34, pTyr-38, pSer-39, pSer-51, pSer-55, pSer-56, pTyr-61, pSer-72, pSer-83	Abcam, Cell Signaling, Sigma, Thermo Scientific
Desmin	pSer-60	Abcam, Sigma
GFAP	pSer-8, pSer-38	Abcam, Sigma
NF-M	pSer-615/620	Abcam
NF-H	KpSP repeat (NAP4 clone)	Thermo Scientific
Lamin-A/C	pSer-22, pSer-392	Abcam, Cell Signaling, Sigma, Thermo Scientific
Lamin-B1	pSer-575	Abcam

This is a list of IF protein phospho-antibodies available from four major antibody vendors. Note that it is not an exhaustive list, as it does not include investigator-generated antibodies or antibodies from smaller suppliers. Notably, the nomenclature may differ depending on the antibody source. For example, K8 pSer-24 may be catalogued as K8 pSer-23 in cases where the initiator methionine is not included in the count.

pan-phospho-Tyr antibody pY100 (Cell Signaling) recognizes human K8 phosphorylation at Tyr-267 (Snider, Park & Omary, 2013).

3.1.1 Specific Materials Needed to Study IF Protein Phosphorylation

- Kinase or phosphatase inhibitors
  - Commercially available with varying degree of target selectivity.
  - *Note*: To study epithelial keratin phosphorylation, we frequently use okadaic acid and microcystin-LR (Gehringer, 2004) for *in vitro* and *in vivo* studies, respectively (Toivola, Zhou, English, & Omary, 2002).
- Expression plasmids or siRNA against various kinases or phosphatases
  - Useful for examining PTM regulation mechanisms
- Phospho-site-specific or pan-phospho antibodies
  - Table 5: Most antibodies can be used for immunoblotting and some antibodies are also suitable for immunohistochemistry.
- HSE fraction or IF immunoprecipitates obtained by following Protocol 2.2 and 2.3

- Samples are prepared after chemical or genetic manipulation to alter kinase or phosphatase activity or induce stress conditions.
- *Note*: Phosphatase inhibitor treatment will increase the presence of IF proteins in the detergent-soluble pool, whereas some stress conditions (e.g., oxidative stress) will lead to IF protein aggregation. Therefore, both the insoluble (HSE) and soluble (i.p.) fractions should always be examined.

Figure 3 provides two different examples for monitoring K8 phosphorylation. Phosphorylation of K8 Ser-74 (Panel A) serves as a phosphate "sponge" during stress (Ku & Omary, 2006). Phosphorylation of K8 Tyr-267 (Panel B) is important for filament organization, since the

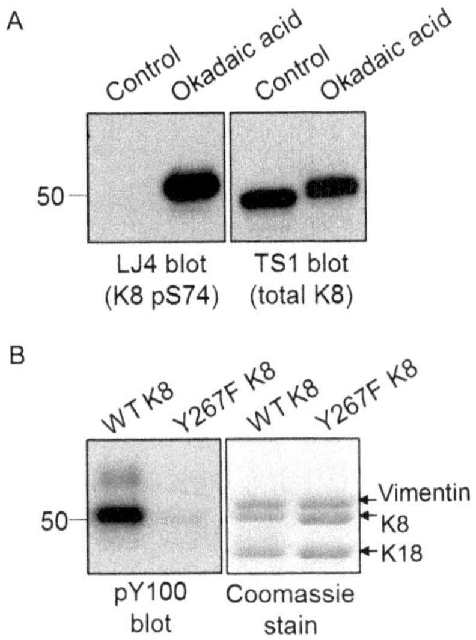

Figure 3 Biochemical analysis of K8 serine and tyrosine phosphorylation. (A) The phosphatase inhibitor okadaic acid was used to upregulate K8 Ser/Thr phosphorylation, as detected in total HepG2 cell lysates by a site-specific antibody (LJ4, which recognizes human K8 pSer-74). Total K8 was detected with the TS1 (mouse anti-human K8) antibody. Note the slight retardation in migration, which is typical of hyperphosphorylated proteins. (B) The pan-phosphotyrosine antibody pY100 antibody (Cell Signaling) was used to demonstrate that human K8 is phosphorylated at Tyr-267. In this case, the HSE fractions of BHK-21 cells expressing wild-type (WT) or the phosphorylation-deficient Y267F mutant of human K8 were analyzed, and shown are the pY100 immunoblot and Coomassie stain as a loading control. WT K18 was cotransfected in each case in order to allow stabilization of the partner K18; otherwise there would be degradation of the individual keratin by the proteasome (Ku & Omary, 2000). The Coomassie-stained band that migrates above K8 (arrow) corresponds to vimentin which is the major IF protein expressed in the transfected cells.

phospho-deficient Y267F mutation results in short and mostly perinuclear filaments and the phospho-mimetic Y267D mutation leads to K8/K18 aggregation (Snider, Park & Omary, 2013).

3.2 Analysis of IF Protein Sumoylation

Cytoplasmic IFs and nuclear lamins are targets for sumoylation, which entails the covalent addition of small ubiquitin-like modifier (SUMO) proteins to lysine residues on specific targets (Gareau & Lima, 2010). Monosumoylation of lamin-A under basal conditions appears to be important for the proper organization of the nuclear lamina (Zhang & Sarge, 2008). On the other hand, epithelial K8, K18, and K19 are modified by polymeric SUMO-2/3 chains primarily under conditions where the filament structures are altered, such as during stress or in the context of disease-associated mutations, and in these cases sumoylation likely functions (at least in part) to modulate the solubility of keratin IFs (Snider et al., 2011). Sumoylation of mammalian IF proteins can be studied in an *in vitro* reconstituted system or in cells and tissues. However, given that the fraction of sumoylated proteins is very small, experimental conditions that alter the stoichiometry of SUMO conjugation/deconjugation reactions are required in most cases to detect this modification on IF proteins. Sumoylation generally occurs within the classic consensus linear motif $\Psi KX[D/E]$ (Ψ and X denote a hydrophobic residue and any residue, respectively), although there are exceptions (Gareau & Lima, 2010). Sequence-based predictions should be performed as a first step using the freely available programs, such as SUMOplot (Abgent), GPS-SUMO, SUMOsp, and SUMOFI to aid with the experimental analysis.

3.2.1 Specific Materials Needed to Study IF Protein Sumoylation

- *In vitro sumoylation components*:
  - Purified IF protein
  - Recombinant E1 and E2 SUMO ligases
  - SUMO proteins (SUMO-1, -2, or -3)
  - ATP
    - *Note*: For beginning investigations, these components can be purchased as a SUMOylation kit (Enzo Life Sciences) and the reactions carried out following published procedures or manufacturer specifications. As controls, reactions lacking ATP or substrate should be performed routinely.

- SUMO antibodies
  - Available from multiple commercial sources. We have used a rabbit polyclonal antibody to SUMO-2/3 (Abcam; ab3742).
  - Can be used for immunoblotting, immunocytochemistry, or immunoprecipitation.
  - *Note*: Immunoprecipitation using anti-SUMO antibodies may prove challenging in the context of hypersumoylation due to masking of antibody epitopes.
- HSE fraction or IF immunoprecipitates obtained by following the protocols 2.2 and 2.3
  - *N*-ethylmaleimide (20 m*M*) should be added to the lysis buffer (e.g., TXB + PPI) to inhibit desumoylation during the sample processing.
  - Monosumoylation will result in a ~15 kDa increase in the molecular mass of the protein, whereas hypersumoylation appears as a smear of high molecular mass complexes, some of which might be trapped at the top of the gel. Therefore, the entire membrane should be probed during immunoblotting.

Figure 4 provides examples for monitoring IF protein sumoylation biochemically and by immunofluorescence analysis. Note that the sumoylation sites on K8, K18, and K19 are located in the rod domains (Snider et al., 2011). Therefore, coexpression of K8/K18 or K8/K19 pairs results in significantly diminished sumoylation relative to expression of each protein individually, likely due to limited access of the relevant enzyme in the context of heterodimer formation that involves the rod domains of the keratin pairs.

3.3 Analysis of IF Protein Lysine Acetylation

Acetylation of lysine residues is a common modification on numerous cellular proteins, and is best known for histones and enzymes involved in energy metabolism (Choudhary, Weinert, Nishida, Verdin, & Mann, 2014). IF proteins represent a separate category of targets for this modification, with a large number of acetylation sites determined by mass spectrometry proteomic studies. However, experimental validation has only been reported on human K8, where a single residue (Lys-207) was found to be acetylated in response to glucose stimulation, inhibition, or knockdown of the NAD-dependent deacetylase SIRT2, and in the presence of the NAD synthesis inhibitor FK-866 (Snider, Leonard, et al., 2013). The acetylation-deficient mutant of K8 (K207R) forms filaments but lacks the ability to assemble into a dense perinuclear network that is seen with WT

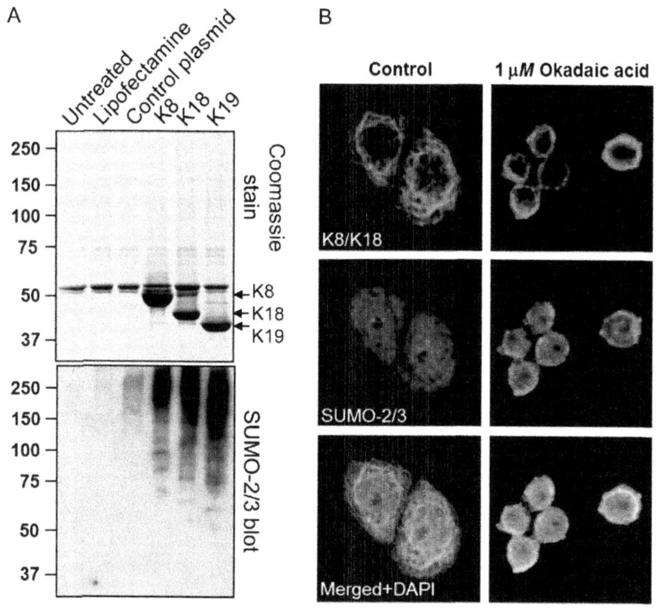

Figure 4 Biochemical and immunofluorescence analysis of keratin sumoylation. (A) Extensive poly-sumoylation (observed as ubiquitin-like smears on a SUMO-2/3 immunoblot) is observed when K8, K18, and K19 are transfected individually (i.e., without their obligate type I/II partner). This is due to the fact that sumoylation sites on K8/K18/K19 are located in the rod domains and are inaccessible to the SUMO proteins and their associated enzymes under basal conditions (in the context of intact heteropolymers). Note that large amounts of protein are needed to detect a signal, which is consistent with the low stoichiometry of sumoylation (as seen by the Coomassie-stained gel in the top part of the panel). (B) Colocalization between an IF protein (K8, green (dark gray in the print version)) and SUMO-2/3 (red (gray in the print version)) is shown in the presence of the phosphatase inhibitor okadaic acid (1 μm, 45 min).

K8, and shows diminished phosphorylation at Ser-74 (Snider, Leonard, et al., 2013). Therefore, lysine acetylation on K8 may couple the metabolic conditions to K8 phosphorylation and filament reorganization. Since there are no known consensus motifs for lysine acetylation, mass spectrometry can be used to identify acetylated sites on IFs and numerous acetylated residues on IF proteins have already been catalogued in databases like PhosphoSitePlus.

3.3.1 Specific Materials Needed to Study IF Protein Lysine Acetylation

- Acetyl-lysine antibodies
  - *Note*: There are a number of these available from commercial sources. However, the reactivity profiles of these antibodies vary significantly,

as shown by immunoblot comparison of three antibodies using normal mouse liver lysates (Fig. 5A). Therefore, multiple antibodies should be tested for their ability to recognize the IF protein of interest. We used a rabbit polyclonal anti-acetyl-lysine antibody (Abcam; ab80178) to monitor human K8 acetylation.

- Inhibitor of NAD synthesis (to test involvement of NAD-dependent sirtuins)
  - FK-866

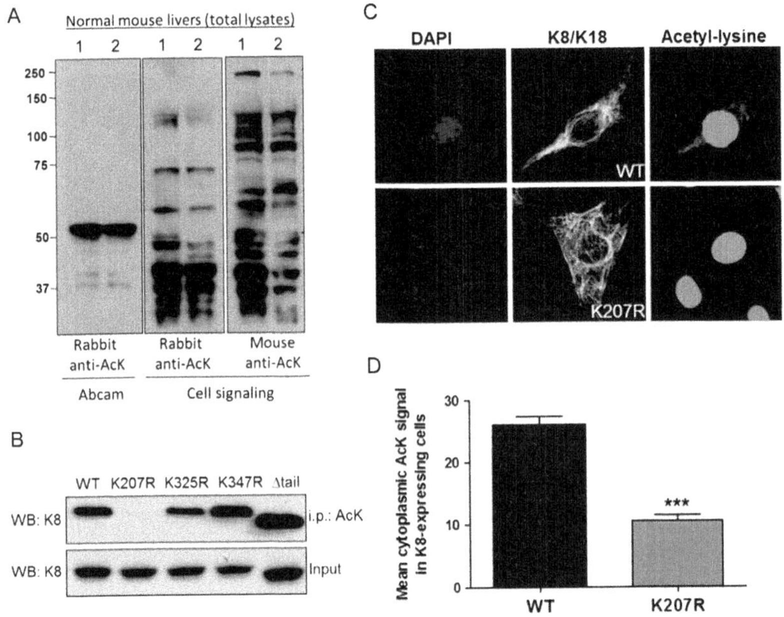

Figure 5 Monitoring lysine acetylation on IF proteins using pan-acetyl-lysine (AcK) antibodies. (A) Comparison of three different pan-AcK antibodies by immunoblot on total liver lysates from two control mouse livers. Note the different protein profiles detected by each antibody. (B) Identification of Lys-207 as an acetylation site on human K8 after immunoprecipitation using an anti-AcK antibody (Abcam antibody from Panel A) and immunoblotting with TS1 antibody against K8. This panel is from Snider, Leonard, et al. (2013) and is displayed under the terms of a Creative Commons License (Attribution-Noncommercial-Share Alike 3.0 Unported license, as described at http://creativecommons.org/licenses/by-nc-sa/3.0/). (C) Detection of K8 acetylation using immunofluorescence analysis of WT K8 and the K207R K8 mutant. Note the absence of cytoplasmic signal in the acetylation-deficient mutant. Blue (gray in the print version), DAPI; green (dark gray in the print version), K8; red (light gray in the print version), pan-AcK (Abcam antibody from Panel A). (D) Quantification of cytoplasmic AcK signal ($n=4$). $p<0.0001$, unpaired t-test.

- Inhibitors of sirtuin deacetylase enzymes
  - SIRT1/2 inhibitor: salermide
  - SIRT2 inhibitor: AGK-2
- pan-histone deacetylase (HDAC) inhibitor: trichostatin A

Currently, the best-available methods for detecting lysine acetylation on IF proteins are immunoprecipitation with a pan-acetyl-lysine antibody followed by mass spectrometry analysis to identify the acetylated residues. Site-directed mutagenesis to substitute candidate lysines to nonacetylatable arginines or acetylation-mimetic glutamines can narrow down the major sites, as shown for K8 (Fig. 5B). Immunofluorescence staining analysis may also be useful in some cases. Note that the nuclear acetyl-lysine (AcK) staining (acetylated histones) is usually the major signal detected, although the presence/absence of cytoplasmic staining can also be examined in the context of acetylation site mutants (Fig. 5C and D).

3.4 Analysis of IF Protein *O*-Linked Glycosylation

Several IF proteins (K8, K13, K18, vimentin, and NFs) are known to be regulated by *O*-linked glycosylation (Snider & Omary, 2014), a modification that involves the covalent attachment of β-D-*N*-acetylglucosamine (O-GlcNAc) to Ser/Thr residues (Hardiville & Hart, 2014). Functionally, *O*-linked glycosylation of K18 (Chou, Smith, & Omary, 1992; Ku & Omary, 1995) protects epithelial cells from injury by interfering with the phosphorylation and activation of stress kinases (Ku et al., 2010). In general, *O*-linked glycosylation is important in the maintenance of metabolic homeostasis and therefore, its broader functional significance to the regulation of IF protein function may fully come to light by examining various conditions of altered metabolism or metabolic stress.

3.4.1 Specific Materials Needed to Study IF Protein O-Linked Glycosylation

- Antibodies to *O*-linked *N*-acetylglucosamine
  - RL2, HGAC85 (Abcam, Pierce, Novus)
- Inhibitors of GlcNAc-β-*N*-acetylglucosaminidase
  - PUGNAc, Thiamet G (Cayman Chemical, Sigma, Tocris)
- Inducers of metabolic stress
  - Streptozotocin (Cayman Chemical, Sigma, Tocris)
- IF protein glycosylation-deficient mice (K18 Gly−/−)
  - mice that overexpress glycosylation-deficient human K18 (Ser30, Ser31, and Ser49 converted to Ala)

- Reagents for *in vitro* *O*-glycosylation
  - May be performed on IF immunoprecipitates or HSEs by incubation with radiolabeled UDP-(4, 5-^{3}H)-galactose and recombinant galactosyltransferase (Ku et al., 2004) to label terminal *O*-GlcNAc residues.
  - Nonradioactive methods for *in vitro* labeling of O-GlcNac modification include "click" chemistry approaches, such as the commercially available Click-iT O-GlcNAc Enzymatic Labeling System (Life Technologies).

3.5 Analysis of IF Protein Transamidation

Transamidation is a covalent PTM that involves formation of amide bonds between the ε-amino group of Lys and the γ-carboxyl group of Gln residues (Iismaa, Mearns, Lorand, & Graham, 2009). Epidermal and epithelial keratins are substrates of transglutaminase-2 (TG2), the calcium-dependent enzyme that catalyzes these reactions (Strnad et al., 2007). Transglutamination of type II epidermal keratins is involved in the maintenance of skin barrier function, whereas in the liver transglutamination of K8/K18 occurs in pathologic settings to promote the formation of the K8/K18-containing aggregates called Mallory-Denk bodies (Strnad et al., 2007). Although TG2-mediated cross-linking of keratins can be monitored *in vitro*, methods for *in vivo* detection of this modification and the specific residues involved are currently lacking. Strong consensus sequences for transglutamination have not been identified; however, phage display studies (Hitomi, Kitamura, & Sugimura, 2009; Sugimura et al., 2006) have revealed preferred sequences around the reactive glutamines to be: *Qx*(*P*,*Y*) ∅*D*(*P*) and *pQx*(*PTS*)*l*, where *x*=any, ∅=hydrophobic, *p*=polar, and *l*=aliphatic residue.

3.5.1 *Specific Materials Needed to Study IF Protein Transamidation*

- Plasmids encoding WT IF proteins or Gln→Asn and Lys→Arg substituted mutants.
- Purified peptides containing potential transglutamination sites.
- Recombinant transglutaminase enzyme.
- Antibody that recognizes the isopeptide bond (antibody 81D1C2, available from Abcam and Santa Cruz). However, there may be an issue with specificity of such antibodies (Johnson & LeShoure, 2004).

4. CONCLUSIONS

There are several hundred known PTMs across the proteome, and we are just beginning to understand the information that they encode in order to promote adaptable functions of proteins under specific physiological and pathophysiological conditions. IF proteins are known to be regulated by a number of PTMs, but with the improvement in detection methods we will be able to uncover more details about the known IF PTMs as well as identify novel IF-targeting PTMs. Generation of new site-specific antibodies will be necessary in order to study PTM dynamics in physiologically relevant systems. For some modifications, such as sumoylation, pharmacologic inhibitors of the conjugating and deconjugating enzymes are needed to further advance our understanding of this important modification. PTMs that target conserved IF protein domains are likely to have important roles on multiple IF protein types, and may provide insight into novel cellular functions of IFs and how they fit into the greater function of the cell cytoskeleton. From a disease perspective, various pathologic states (in humans and in animal models) display a mix of cellular stress responses where changes in IF protein PTMs may serve as disease markers, especially given the abundance of IFs. Finally, pharmacological strategies to alter PTM regulation of IF proteins through inhibiting the activity of various enzymes may also prove to be important from a human disease perspective. The field is wide open in terms the pursuit of functional aspects of IF PTMs, and it is hoped that the tools described herein will promote this line of investigation.

ACKNOWLEDGMENTS

The authors' work is supported by US National Institutes of Health (NIH) grants DK47918 and DK52951, and the Department of Veterans Affairs (M.B.O.); NIH grants DK093776 and DK102450 (N.T.S.); and NIH institutional grant DK34933 to the University of Michigan. M.B.O. expresses gratitude to his laboratory coworkers who have made significant contributions to our understanding of IF PTMs including Carrie Riopel, Chih-Fong Chou, Diana Toivola, Guo-Zhong Tao, Jian Liao, Li Feng, Lori Lowthert, Masaru Harada, Nam-On Ku, Natasha Snider, Pavel Strnad, Qin Zhou, Raymond Kwan, Shinichiro Hanada, and Xiangjun Zhou.

REFERENCES

Chou, C. F., Riopel, C. L., Rott, L. S., & Omary, M. B. (1993). A significant soluble keratin fraction in 'simple' epithelial cells. Lack of an apparent phosphorylation and glycosylation role in keratin solubility. *Journal of Cell Science*, *105*(Pt 2), 433–444.

Chou, C. F., Smith, A. J., & Omary, M. B. (1992). Characterization and dynamics of O-linked glycosylation of human cytokeratin 8 and 18. *The Journal of Biological Chemistry*, *267*(6), 3901–3906.
Choudhary, C., & Mann, M. (2010). Decoding signalling networks by mass spectrometry-based proteomics. *Nature Reviews. Molecular Cell Biology*, *11*(6), 427–439.
Choudhary, C., Weinert, B. T., Nishida, Y., Verdin, E., & Mann, M. (2014). The growing landscape of lysine acetylation links metabolism and cell signalling. *Nature Reviews. Molecular Cell Biology*, *15*(8), 536–550.
Chung, B. M., Rotty, J. D., & Coulombe, P. A. (2013). Networking galore: Intermediate filaments and cell migration. *Current Opinion in Cell Biology*, *25*(5), 600–612.
Craveur, P., Rebehmed, J., & de Brevern, A. G. (2014). PTM-SD: A database of structurally resolved and annotated posttranslational modifications in proteins. *Database (Oxford)*, http://dx.doi.org/10.1093/database/bau041. pii: bau041. Print 2014.
Davidson, P. M., & Lammerding, J. (2014). Broken nuclei—Lamins, nuclear mechanics, and disease. *Trends in Cell Biology*, *24*(4), 247–256.
Feng, X., Zhang, H., Margolick, J. B., & Coulombe, P. A. (2013). Keratin intracellular concentration revisited: Implications for keratin function in surface epithelia. *The Journal of Investigative Dermatology*, *133*(3), 850–853.
Gareau, J. R., & Lima, C. D. (2010). The SUMO pathway: Emerging mechanisms that shape specificity, conjugation and recognition. *Nature Reviews. Molecular Cell Biology*, *11*(12), 861–871.
Gehringer, M. M. (2004). Microcystin-LR and okadaic acid-induced cellular effects: A dualistic response. *FEBS Letters*, *557*(1–3), 1–8.
Gnad, F., Gunawardena, J., & Mann, M. (2011). PHOSIDA 2011: The posttranslational modification database. *Nucleic Acids Research*, *39*(Database issue), D253–D260. http://dx.doi.org/10.1093/nar/gkq1159. Epub 2010 Nov 16.
Gruenbaum, Y., & Aebi, U. (2014). Intermediate filaments: A dynamic network that controls cell mechanics. *F1000Prime Reports*, *6*, 54.
Hannoush, R. N., & Sun, J. (2010). The chemical toolbox for monitoring protein fatty acylation and prenylation. *Nature Chemical Biology*, *6*(7), 498–506.
Hardiville, S., & Hart, G. W. (2014). Nutrient regulation of signaling, transcription, and cell physiology by O-GlcNAcylation. *Cell Metabolism*, *20*(2), 208–213.
Hennrich, M. L., & Gavin, A. C. (2015). Quantitative mass spectrometry of posttranslational modifications: Keys to confidence. *Science Signaling*, *8*(371), re5.
Herrmann, H., Strelkov, S. V., Burkhard, P., & Aebi, U. (2009). Intermediate filaments: Primary determinants of cell architecture and plasticity. *The Journal of Clinical Investigation*, *119*(7), 1772–1783.
Hitomi, K., Kitamura, M., & Sugimura, Y. (2009). Preferred substrate sequences for transglutaminase 2: Screening using a phage-displayed peptide library. *Amino Acids*, *36*(4), 619–624.
Homberg, M., & Magin, T. M. (2014). Beyond expectations: Novel insights into epidermal keratin function and regulation. *International Review of Cell and Molecular Biology*, *311*, 265–306.
Hornbeck, P. V., Zhang, B., Murray, B., Kornhauser, J. M., Latham, V., & Skrzypek, E. (2015). PhosphoSitePlus, 2014: Mutations, PTMs and recalibrations. *Nucleic Acids Research*, *43*(Database issue), D512–D520.
Hyder, C. L., Pallari, H. M., Kochin, V., & Eriksson, J. E. (2008). Providing cellular signposts—Post-translational modifications of intermediate filaments. *FEBS Letters*, *582*(14), 2140–2148.
Iismaa, S. E., Mearns, B. M., Lorand, L., & Graham, R. M. (2009). Transglutaminases and disease: Lessons from genetically engineered mouse models and inherited disorders. *Physiological Reviews*, *89*(3), 991–1023.

Izawa, I., & Inagaki, M. (2006). Regulatory mechanisms and functions of intermediate filaments: A study using site- and phosphorylation state-specific antibodies. *Cancer Science, 97*(3), 167–174.

Johnson, G. V., & LeShoure, R., Jr. (2004). Immunoblot analysis reveals that isopeptide antibodies do not specifically recognize the epsilon-(gamma-glutamyl)lysine bonds formed by transglutaminase activity. *Journal of Neuroscience Methods, 134*(2), 151–158.

Khoury, G. A., Baliban, R. C., & Floudas, C. A. (2011). Proteome-wide post-translational modification statistics: Frequency analysis and curation of the Swiss-Prot database. *Scientific Reports, 1*. Article number: 90.

Ku, N. O., & Omary, M. B. (1995). Identification and mutational analysis of the glycosylation sites of human keratin 18. *The Journal of Biological Chemistry, 270*(20), 11820–11827.

Ku, N. O., & Omary, M. B. (2000). Keratins turn over by ubiquitination in a phosphorylation-modulated fashion. *The Journal of Cell Biology, 149*(3), 547–552.

Ku, N. O., & Omary, M. B. (2006). A disease- and phosphorylation-related nonmechanical function for keratin 8. *The Journal of Cell Biology, 174*(1), 115–125.

Ku, N. O., Toivola, D. M., Strnad, P., & Omary, M. B. (2010). Cytoskeletal keratin glycosylation protects epithelial tissue from injury. *Nature Cell Biology, 12*(9), 876–885.

Ku, N. O., Toivola, D. M., Zhou, Q., Tao, G. Z., Zhong, B., & Omary, M. B. (2004). Studying simple epithelial keratins in cells and tissues. *Methods in Cell Biology, 78*, 489–517.

Lill, J. R., & Wertz, I. E. (2014). Toward understanding ubiquitin-modifying enzymes: From pharmacological targeting to proteomics. *Trends in Pharmacological Sciences, 35*(4), 187–207.

Lowthert, L. A., Ku, N. O., Liao, J., Coulombe, P. A., & Omary, M. B. (1995). Empigen BB: A useful detergent for solubilization and biochemical analysis of keratins. *Biochemical and Biophysical Research Communications, 206*(1), 370–379.

Lu, C. T., Huang, K. Y., Su, M. G., Lee, T. Y., Bretaña, N. A., Chang, W. C., et al. (2013). DbPTM 3.0: An informative resource for investigating substrate site specificity and functional association of protein post-translational modifications. *Nucleic Acids Research, 41*(Database issue), D295–D305. http://dx.doi.org/10.1093/nar/gks1229. Epub 2012 Nov 27.

Minguez, P., Letunic, I., Parca, L., & Bork, P. (2013). PTMcode: A database of known and predicted functional associations between post-translational modifications in proteins. *Nucleic Acids Research, 41*(Database issue), D306–D311. http://dx.doi.org/10.1093/nar/gks1230. Epub 2012 Nov 28.

Moyle, P. M., & Muir, T. W. (2010). Method for the synthesis of mono-ADP-ribose conjugated peptides. *Journal of the American Chemical Society, 132*(45), 15878–15880.

Omary, M. B. (2009). "IF-pathies": A broad spectrum of intermediate filament-associated diseases. *The Journal of Clinical Investigation, 119*(7), 1756–1762.

Omary, M. B., Ku, N. O., Liao, J., & Price, D. (1998). Keratin modifications and solubility properties in epithelial cells and in vitro. *Sub-Cellular Biochemistry, 31*, 105–140.

Omary, M. B., Ku, N. O., Tao, G. Z., Toivola, D. M., & Liao, J. (2006). "Heads and tails" of intermediate filament phosphorylation: Multiple sites and functional insights. *Trends in Biochemical Sciences, 31*(7), 383–394.

Pan, X., Kane, L. A., Van Eyk, J. E., & Coulombe, P. A. (2011). Type I keratin 17 protein is phosphorylated on serine 44 by p90 ribosomal protein S6 kinase 1 (RSK1) in a growth- and stress-dependent fashion. *The Journal of Biological Chemistry, 286*(49), 42403–42413.

Rogel, M. R., Jaitovich, A., & Ridge, K. M. (2010). The role of the ubiquitin proteasome pathway in keratin intermediate filament protein degradation. *Proceedings of the American Thoracic Society, 7*(1), 71–76.

Roux, P. P., & Thibault, P. (2013). The coming of age of phosphoproteomics—From large data sets to inference of protein functions. *Molecular & Cellular Proteomics, 12*(12), 3453–3464.

Sihag, R. K., Inagaki, M., Yamaguchi, T., Shea, T. B., & Pant, H. C. (2007). Role of phosphorylation on the structural dynamics and function of types III and IV intermediate filaments. *Experimental Cell Research*, *313*(10), 2098–2109.

Snider, N. T., Leonard, J. M., Kwan, R., Griggs, N. W., Rui, L., & Omary, M. B. (2013). Glucose and SIRT2 reciprocally mediate the regulation of keratin 8 by lysine acetylation. *The Journal of Cell Biology*, *200*(3), 241–247.

Snider, N. T., & Omary, M. B. (2014). Post-translational modifications of intermediate filament proteins: Mechanisms and functions. *Nature Reviews. Molecular Cell Biology*, *15*(3), 163–177.

Snider, N. T., Park, H., & Omary, M. B. (2013). A conserved rod domain phosphotyrosine that is targeted by the phosphatase PTP1B promotes keratin 8 protein insolubility and filament organization. *The Journal of Biological Chemistry*, *288*(43), 31329–31337.

Snider, N. T., Weerasinghe, S. V., Iniguez-Lluhi, J. A., Herrmann, H., & Omary, M. B. (2011). Keratin hypersumoylation alters filament dynamics and is a marker for human liver disease and keratin mutation. *The Journal of Biological Chemistry*, *286*(3), 2273–2284.

Soellner, P., Quinlan, R. A., & Franke, W. W. (1985). Identification of a distinct soluble subunit of an intermediate filament protein: Tetrameric vimentin from living cells. *Proceedings of the National Academy of Sciences of the United States of America*, *82*(23), 7929–7933.

Strnad, P., Harada, M., Siegel, M., Terkeltaub, R. A., Graham, R. M., Khosla, C., et al. (2007). Transglutaminase 2 regulates mallory body inclusion formation and injury-associated liver enlargement. *Gastroenterology*, *132*(4), 1515–1526.

Sugimura, Y., Hosono, M., Wada, F., Yoshimura, T., Maki, M., & Hitomi, K. (2006). Screening for the preferred substrate sequence of transglutaminase using a phage-displayed peptide library: Identification of peptide substrates for TGASE 2 and factor XIIIA. *The Journal of Biological Chemistry*, *281*(26), 17699–17706.

Sun, T. T., & Green, H. (1978). Keratin filaments of cultured human epidermal cells. Formation of intermolecular disulfide bonds during terminal differentiation. *The Journal of Biological Chemistry*, *253*(6), 2053–2060.

Sylvestersen, K. B., Young, C., & Nielsen, M. L. (2013). Advances in characterizing ubiquitylation sites by mass spectrometry. *Current Opinion in Chemical Biology*, *17*(1), 49–58.

Toivola, D. M., Boor, P., Alam, C., & Strnad, P. (2015). Keratins in health and disease. *Current Opinion in Cell Biology*, *32*, 73–81.

Toivola, D. M., Strnad, P., Habtezion, A., & Omary, M. B. (2010). Intermediate filaments take the heat as stress proteins. *Trends in Cell Biology*, *20*(2), 79–91.

Toivola, D. M., Zhou, Q., English, L. S., & Omary, M. B. (2002). Type II keratins are phosphorylated on a unique motif during stress and mitosis in tissues and cultured cells. *Molecular Biology of the Cell*, *13*(6), 1857–1870.

Zhang, Y. Q., & Sarge, K. D. (2008). Sumoylation regulates lamin A function and is lost in lamin A mutants associated with familial cardiomyopathies. *The Journal of Cell Biology*, *182*(1), 35–39.

Zhong, B., Zhou, Q., Toivola, D. M., Tao, G. Z., Resurreccion, E. Z., & Omary, M. B. (2004). Organ-specific stress induces mouse pancreatic keratin overexpression in association with NF-kappaB activation. *Journal of Cell Science*, *117*(Pt 9), 1709–1719.

CHAPTER SIX

Immunofluorescence and Immunohistochemical Detection of Keratins

Cornelia Stumptner, Margit Gogg-Kamerer, Christian Viertler, Helmut Denk, Kurt Zatloukal[1]

Institute of Pathology, Medical University of Graz, Graz, Austria

[1]Corresponding author: e-mail address: kurt.zatloukal@medunigraz.at

Contents

Abstract

Reliable detection of keratins in tissues is important for investigating their physiological role and for using keratin expression as a biomarker in medical diagnostics. A particular challenge for the detection of keratins by immunofluorescence microscopy or immunohistochemistry relates to the fact that keratin intermediate filaments are obligatory heteropolymers, which may result in dissociation between RNA and protein expression levels in the event that the homeostasis of the expression of the proper keratin partners is disturbed. Furthermore, variable accessibility of epitopes on keratin polypeptides due to conformational changes may lead to false negative results. Preanalytical effects, such as warm/cold ischemia, fixation, tissue processing, and embedding may result in false negative or inappropriate reactions. An experimental design for how to systematically

Methods in Enzymology, Volume 568
ISSN 0076-6879
http://dx.doi.org/10.1016/bs.mie.2015.08.003

test preanalytical effects and to validate immunohistochemistry protocols is presented. This kind of evaluation should be performed for each antigen and antibody since the various epitopes recognized by antibodies may behave differently. In this context, one has to be aware that different cell structures may be affected or modified differently by various preanalytical procedures and may thus require different preanalytical and staining protocols.

ABBREVIATIONS

FFPE formalin-fixed paraffin-embedded
IHC immunohistochemistry
IF(s) intermediate filament(s)
IIF indirect immunofluorescence
Krt keratin gene
MDB Mallory-Denk body
PFPE PAXgene-fixed paraffin-embedded
wt wild type

1. INTRODUCTION

1.1 Types and Functions of Keratins

In this section, we will focus on IFs and in particular on the keratins, which are major objects of studies aimed at elucidation of IF properties, dynamics, regulation and diagnostic significance as markers in tumor pathology.

The family of IF proteins includes ~75 different genes and is further divided into six subtypes which are, at least in part, expressed in a cell type-, compartment- and differentiation-dependent manner. Individual IF proteins consist of a conserved central coiled-coil alpha-helical rod domain (interrupted by linkers), which is flanked by N-terminal (head) and C-terminal (tail) domains (Goldman, Cleland, Murthy, Mahammad, & Kuczmarski, 2012). The N- and C-terminal domains are major sites of post-translational modifications with phosphorylation being the best characterized one and contribute to structural heterogeneity and presumed tissue-specific functions (Omary, Ku, Tao, Toivola, & Liao, 2006). Site-specific phosphorylation of IFs can also cause segregation to specific compartments within a cell that might represent the apical or basolateral membrane area in polarized epithelia. Changes in phosphorylation status are responsible for IF dynamics, solubility and organization. In addition to posttranslational modifications, IF function is modified and complemented through interaction with a variety of IF-associated proteins (Strnad, Stumptner, Zatloukal, & Denk, 2008).

All these modifications and interactions may influence the binding of antibodies used for immune-detection techniques. Keratins can be divided into two classes, the acidic type I and the basic type II keratins. For IFs assembly, hetero-oligomerization of partners of both classes is obligatory (Herrmann & Aebi, 2000). Tissue-, compartment-, and cell type-specific distribution of keratins suggest unique functional roles. Keratins are highly dynamic and reorganize in response to stimuli, such as growth factor stimulation, mitosis, apoptosis, and a variety of cell stresses (drugs, virus infection, high temperature) with posttranslational modifications, such as phosphorylation, glycosylation, cross-linking, and proteolysis. Their functions depend on the ability to act as scaffolds for various molecules involved in cell signaling. Many of the emerging interactions relate to stress tolerance and cell survival. It was shown (Omary, Coulombe, & McLean, 2004) that (especially type II) keratins protect cells from injury by acting as a "phosphate sponge/sink" absorbing stress-induced kinase activity.

1.2 Keratins and Disease

Keratin mutations underlie many epithelium-specific diseases in man and animal models (Ku, Zhou, Toivola, & Omary, 1999; Omary et al., 2004; Toivola, Boor, Alam, & Strnad, 2015). Moreover, keratin fragments have been detected in sera of patients with epithelial tumors and antibodies to keratins may occur in association with several chronic diseases (e.g., autoimmune hepatitis), and thus have diagnostic potential.

Based on the fact that epithelial tumors largely maintain specific keratin expression associated with the respective cell type of origin (Moll, Divo, & Langbein, 2008) and the availability of specific antibodies, keratins are now extensively used as immunohistochemical markers in diagnostic tumor pathology and (together with other IF proteins) provide an important adjunct in tumor typing. Keratin profiling is especially valuable for identification of poorly differentiated carcinomas and metastases of unknown primary site (Karantza, 2011; Moll et al., 2008). For detailed information on the diagnostic utility of keratin and other cytoskeletal proteins in tumor pathology, special literature, e.g., AFIP Atlas of Tumor Pathology, series 4, should be consulted (Miettinen, Fetsch, Antonescu, Folpe & Wakely, 2014).

However, one has to be aware of inherent diagnostic pitfalls. IFs alone are not reliable markers of tumor histogenesis and should be used as part of a larger panel of antibodies. For example, IF proteins (e.g., keratin and vimentin) may be coexpressed in carcinomas (e.g., neuroendocrine carcinomas, papillary carcinomas of the thyroid, renal, endometrial, ovarian, and lung carcinomas), thymomas and mesotheliomas. Soft tissue tumors without true epithelioid differentiation may contain keratin-positive cells, usually as focal finding. Furthermore, keratin immunohistochemistry (IHC) has been positive in mesenchymal tissues and tumors, such as myometrium, leimyomas, leimyosarcomas (mostly in frozen sections, but only in a very low percentage of routinely processed samples; Brown, Theaker, Banks, Gatter, & Mason, 2002), fibrosarcomas, chondrosarcomas, and angiosarcomas. Moreover, Ewing sarcomas, primitive neuroectodermal tumors, and alveolar rhabdomyosarcomas might be positive for keratin (Azumi & Battifora, 1987; Banks, Jansen, Oberle, & Davey, 1995; Miettinen et al., 2014). Primary tumors of the central nervous system (astrocytomas, oligodendrogliomas, ependymomas, chorioid plexus papillomas, meningiomas) may also display keratin positivity and have thus to be distinguished on the basis of morphology and IHC from metastatic carcinomas. Keratins may further serve as prognostic markers. Expression of Keratin 17 and/or Keratins 5/6 in breast carcinoma cells was associated with a poor clinical outcome. The presence of Keratin 10 and Keratin 19 in hepatocellular, of Keratin 20 in pancreatic and the loss of Keratin 5/Keratin 6 in uterus cancers indicate poor prognosis (van de Rijn et al., 2002). All gastrointestinal stroma tumors with keratin expression carried a high risk for aggressive behavior (Sing, Ramdial, Ramburan, & Sewram, 2014). Most melanomas typically express vimentin, but Keratin 8/Keratin 18 coexpression in ocular melanomas is restricted to retino-invasive malignant melanomas. A correlation between coexpression of vimentin with Keratin 8 and Keratin 18 and the invasive and metastatic behavior of three human melanoma cell lines has also been observed (Hendrix et al., 1992).

The molecular mechanism of keratin expression in mesenchymal tumors is unknown but loose transcriptional control of *Krt8* and *Krt18* gene expression is implicated. In support, a transient expression of Keratin 8 and Keratin 18 has been observed in smooth and striated muscle cells during human embryogenesis and in transformed fibroblasts in cell culture. However, potential cross-reactivity of keratin antibodies with nonkeratin antigens or the nonspecific adsorption of "leaked" keratins to the cell surface is not excluded or assessed in some of the studies.

2. IMPACT OF MOLECULAR STRUCTURE ON THE DETECTION OF KERATINS IN FROZEN TISSUE SAMPLES

In addition to the biological variations in keratin expression, effects related to the tertiary and quaternary structure of keratins and IFs have to be considered in the detection of keratins by IHC or immunofluorescence. A pitfall in keratin IHC could be caused by masking or inaccessibility of the antigenic determinant by attached proteins or conformational changes induced by the pathologic process (Hazan, Denk, Franke, Lackinger, & Schiller, 1986).

2.1 Effects of Different Keratin Pair Formation and Lack of a Partner

As demonstrated *in vitro* as well as *in vivo* single keratin proteins are unable to form IFs, are unstable and rapidly degraded (Domenjoud, Jorcano, Breuer, & Alonso, 1988; Kulesh, Cecena, Darmon, Vassear, & Oshima, 1989; Magin, Bader, Freudenmann, & Franke, 1990).

In *krt8*$^{-/-}$ mice, mRNA of its type I partner Keratin 18 persists in hepatocytes whereas Keratin 8 protein is absent. Conversely, in *krt18*$^{-/-}$ mice, mRNA of its type II partner Keratin 8 persists in hepatocytes in the absence of Keratin 18 protein (Fig. 1).

Bile duct epithelia contain Keratins 8, 18, 7, and 19 (Fig. 1A (a) and (d)). However, they are expressed in a compartment-dependent manner as revealed by double-label indirect immunofluorescence (IIF). Whereas in wild-type (wt) mice, the whole cytoplasm of cholangiocytes contains intermediate filaments composed of Keratins 8, 18, and 19, Keratin 7 is restricted to a narrow subluminal cytoplasmic compartment in association with Keratin 18 and/or Keratin 19.

Note that in *krt8*$^{-/-}$ mice bile duct epithelia become Keratin 19 negative (Fig. 1A (e)) indicating that in bile ducts Keratin 19 is the preferred partner of Keratin 8, whereas Keratin 7 is still present in association with Keratin 18 (Fig. 1A (b)) (Zatloukal et al., 2000). In analogy, in *krt18*$^{-/-}$ mice bile duct epithelia become Keratin 7 negative, indicating Keratin 7 as partner of Keratin 18. Similar findings have been reported for *krt18*$^{-/-}$ mice where the knockout of *krt18* leads to a secondary loss of Keratin 7 in several tissue types and to formation of Keratin 8/Keratin 19 filaments (Haybaeck et al., 2012; Magin et al., 1998).

2.1.1 Protocol for Double-Label IIF Staining of Cryo-Sections

Antibodies for detection of keratins on cryo-sections of (mouse)tissue and their sequence of application in double-label IIF are summarized in Table 1.

Antibodies are diluted in phosphate-buffered saline (PBS) (50 mM potassium phosphate, 150 mM NaCl, pH 8.0–8.5). Diluted fluorochrome-conjugated secondary antibodies are centrifuged at 16,000 $\times g$ for 5 min

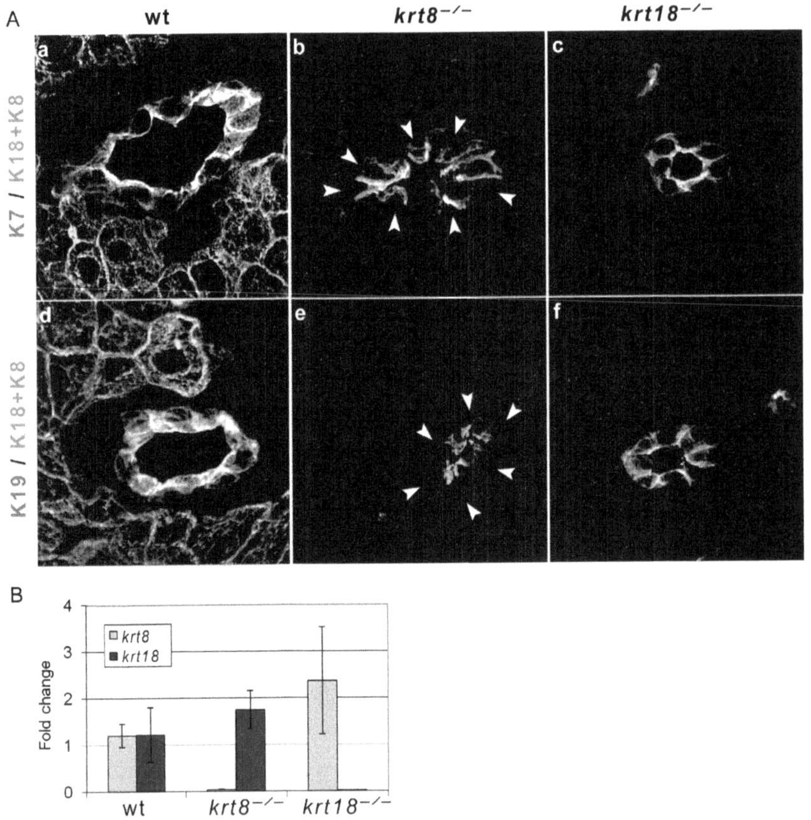

Figure 1 Expression (A and B) and localization (A) of keratins in *krt8*$^{-/-}$, *krt18*$^{-/-}$ and wild-type (wt) mice. (A) Double-label IIF microscopy of mouse hepatocytes and bile ducts with Keratin 7 (K7 red)/Keratin 8 or Keratin 18 (K8 or K18 green) (a–c) and with Keratin 19 (K19 red)/Keratin 8 or Keratin 18 (K8 or K18 green) antibodies (d–f). In wt mice (a and d) hepatocytes show a regular Keratin 8/18 network which is lost in hepatocytes of keratin knockout mice (b, c, e, f) in which one keratin partner is missing. In bile duct epithelia of wt mice (a and d) Keratin 7 and Keratin 19 (both red) and Keratin 8 and Keratin 18 (both green) form a cytoplasmic filament network. Note that in *krt8*$^{-/-}$ mice only Keratin 7 (b) but not Keratin 19 (e) is detectable and is restricted to the apical and lateral portion of the bile duct cells (arrowheads indicate the basal cell portion of bile ducts in b, e). (B) Quantitative real time RT-PCR analysis of *krt8* and *krt18* expression in livers of *krt8*$^{-/-}$, *krt18*$^{-/-}$ and wt mice performed using TaqMan® probe-based gene expression analysis. Keratin expression values are normalized to the house-keeping gene TATAbox binding protein and represent mean values of three to five mice each. (See the color plate.)

Table 1 Antibodies and Their Sequence of Application in Double-Label IIF

1st Primary Antibody	1st Secondary Antibody	2nd Primary Antibody	2nd Secondary Antibody
Keratin 7/Keratin 8–18:			
Monoclonal mouse anti-K7 (Clone RCK105 Monosan MON3007, 1:10)	Goat anti-mouse IgG, ALEXA-488-conjugated (Live Technologies, cat. no. A-11029) 1:100	50K160, polyclonal rabbit anti-human/mouse K8/18 (produced in our laboratory; Stumptner, Fuchsbichler, Heid, Zatloukal, & Denk, 2002) 1:50	Swine anti-rabbit Ig, TRITC-conjugated (Dako, cat. no. R0156) 1:50
Keratin 19/Keratin 8–18:			
Troma III, monoclonal rat anti-human/mouse K19 (Developmental Studies Hybridoma Bank, cat. no. TROMA-III) 1.100	Goat anti-rat IgG, Rhodamine (TRITC)-conjugated (Jackson ImmunoResearch Inc., cat. no. 112-025-167) 1:100	50K160, polyclonal rabbit anti-human/mouse K8/18 (Stumptner et al., 2002) 1:50	Swine anti-rabbit Ig, FITC-conjugated (Dako, cat. no. F0205) 1:50

prior to application onto the sections in order to remove aggregates, and protected from light during storage and use.

1. 4 μm cryo-sections are prepared from snap-frozen tissue using a cryocut (Leica CM3050, Leica) and air-dried before fixation in precooled acetone for 10 min at 20 °C. For staining of chromatin and chromatin-associated proteins, sections can be alternatively fixed in PBS-buffered 4% (v/v) formaldehyde solution for 10 min at room temperature, rinsed in PBS for 5 min, followed by fixation in precooled methanol for 5 min and in acetone for 3 min, both at −20 °C.
2. Sections are air-dried after fixation or rinsed in PBS.
3. For antibody incubation, slides are placed into a humid chamber under light protection and antibodies are applied to the slides separately and sequentially as indicated in Table 1 for 30 min at room temperature. Alternatively, primary antibodies can be applied for 1–2 h at room temperature or overnight at 4 °C. Furthermore, the two primary antibodies and two

secondary antibodies may be applied together to save time. However, this short protocol could favor interference (competition) of binding of antibodies to epitopes. An overview on keratin antibodies is provided in Zatloukal, Stumptner, Fuchsbichler, Janig, & Denk (2004).

After each antibody the sections are washed 2 × 5 min in PBS. Before applying the next antibody, excess liquid is removed by wiping around the section with a towel.

For negative controls, primary antibodies are replaced by PBS, PBS-diluted preimmune serum, or isotype-matched immunoglobulins, respectively.

4. After the last antibody incubation and washing step, slides are rinsed with distilled water, then with 100% ethanol for a few seconds before being left to air-dry.
5. Specimens are mounted with commercially available mounting medium (e.g., Fluorescence Mounting Medium, Dako, cat. no. S3023) or with Mowiol 4-88 (Sigma-Aldrich, cat. no. 81381) mounting medium (for preparation, see Zatloukal et al., 2004).
6. Double-label IIF staining is analyzed with a LSM510 laser-scanning microscope (Zeiss) and images acquired using a multitrack modus.
7. Slides are stored at 4 °C under light protection.

2.2 Conformation-Dependent Keratin Epitopes

The monoclonal antibody KM 54-5 produced against Mallory-Denk bodies (MDBs) (Zatloukal et al., 2007) reacted, *in vivo*, with MBDs but not with keratin IF in normal liver (Fig. 2). By immunoblotting KM 54-5 reacted with Keratin 8 and Keratin 18. In dot blot assay it reacted with individual Keratin 8 and Keratin 18 but not with heterotypic tetramer reconstituted from these polypeptides in 4 *M* urea. This indicates that antigenic determinants of keratin polypeptides can be masked in the heterotypic tetramer subunit and in the keratin IFs in normal cells, and that epitopes which are hidden in normal IFs may become accessible in the pathologic process of MDB formation (Hazan et al., 1986).

3. KERATIN IMMUNOHISTOCHEMISTRY ON PARAFFIN-EMBEDDED TISSUE

3.1 Optimizing Keratin Immunohistochemistry in Formalin-Fixed Paraffin-Embedded Tissues

Fixation of tissues in formalin (10% formaldehyde solution containing 3.7% by mass [corresponding to 4.0% by volume] formaldehyde buffered to

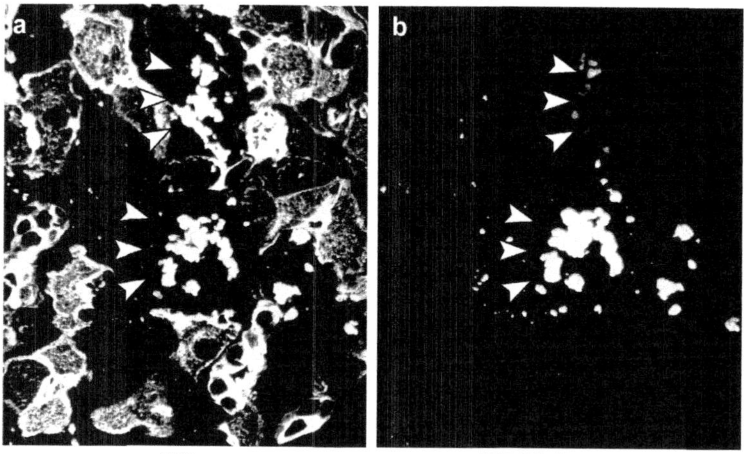

Figure 2 Epitope-specific staining of keratin in liver of the griseofulvin MDB mouse model. Double-label IIF microscopy of liver cryo-sections of 4 months griseofulvin-treated mice with two different keratin antibodies: (A) polyclonal rabbit antibody 577 (produced in our laboratory) and (B) monoclonal mouse antibody KM 54-5, which detect different epitopes on Keratin 8/18 generated and characterized by Hazan et al. (1986), as antibodies recognizing K8/18 (not commercially available). Arrowheads (A and B) indicate MDBs. Note that in contrast to the antibody 577 the antibody KM 54-5 does not stain the Keratin 8/18 network in hepatocytes and the keratin in some MDBs indicating that KM 54-5 recognizes a keratin epitope that is not accessible in the Keratin 8/18 filaments and present to a variable degree in MDBs. *Panel (B): the antibody may be obtained from W.W. Franke, Heidelberg.*

pH 6.8–7.2) and embedding in paraffin (formalin-fixed paraffin-embedded, FFPE) is the gold standard of tissue preparation for histopathological diagnosis (Fox, Johnson, Whiting, & Roller, 1985).

Formalin stabilizes and preserves tissues by modifying biomolecules, such as formation of methylol adducts or via Schiff bases formation of cross-links (e.g., protein–protein, protein–DNA, and DNA–DNA cross-links) (Do & Dobrovic, 2015; Fox et al., 1985). In aqueous solutions formaldehyde becomes rapidly hydrated to form methylene glycol which is present in equilibrium with formaldehyde. While methylene glycol is responsible for rapid penetration of the tissue, formaldehyde induces the actual covalent chemical reactions leading to tissue fixation. Formaldehyde might become oxidized to formic acid which could result in a drop of pH with deleterious effects on nucleic acids (e.g., by enhancing depurination or deamination) and formation of so-called formalin pigments, birefringent crystals resulting from the reaction of formic acid with blood

(Fox et al., 1985). Therefore, formalin should be prepared fresh, and pH and concentration monitored at a regular basis.

Formalin fixation causes denaturation of molecules and leads to conformational changes resulting in the masking of epitopes which can impair immunohistochemical detection of proteins (Otali et al., 2009). Consequently, to recover the formalin-modified antigenicity for IHC, antigen retrieval methods have been developed (Shi, Cote, & Taylor, 1997). Today, a variety of antigen retrieval techniques exist and are mainly based on the application of high temperature (heat-induced epitope retrieval, HIER, to revert modifications induced by formalin and to facilitate refolding), buffers of different pH (to modify charges of proteins to facilitate refolding) or enzymatic digestion (proteolytic-induced epitope retrieval (PIER) to introduce cleavages which compensate for cross-links that may inhibit refolding). All of these methods aim to reverse the conformational changes caused by tissue processing and paraffin embedding (Table 2).

In order to achieve reproducible IHC results, it is important to standardize fixation, tissue processing, in particular water content of alcohol and xylene, as well as the type and temperature of paraffin as embedding material. Tissue processing by replacing the water in tissues by paraffin in several steps of increasing concentrations of alcohol, xylene, and finally molten paraffin results in unfolding of hydrophobic amino acid residues of proteins with conformational changes of epitopes in FFPE tissues. The proper removal of water in the tissue seems to be critical for stability of the tissue and its biomolecules upon storage (unpublished observation).

In addition to the effects of fixation, processing and embedding, also warm and cold ischemia of tissues may impact on preservation of biomolecules and antigens. It is, therefore, important to document and, if possible, to standardize (i) the time period from interruption of blood supply of an organ (typically during surgery when blood vessels are clamped which is the start of warm ischemia time), (ii) the time between removal of tissue from the body and shift to ambient temperature (start of cold ischemia), and (iii) the start and the end of fixation of the tissue.

All these preanalytical parameters (e.g., warm/cold ischemia, fixation, tissue processing, and embedding) might have a major impact on the analysis. The European Committee for Standardization (www.cen.eu) is publishing Technical Specifications on molecular *in vitro* diagnostic examinations for preexamination processes. These Technical Specifications provide background information on preanalytical factors and guidance how to avoid them. In order to establish a reproducible protocol for IHC it is

Table 2 Examples of Antigen Retrieval Methods

Antigen Retrieval (AR)
Heat-induced epitope retrieval (HIER)
Device: Microwave, water bath, pressure cooker, vegetable steamer
Time/Watt/temperature: e.g.,
Microwave: e.g., 5–40 min; e.g., 160–270 W
Water bath: e.g., 60 °C ON, 98.5 °C for 20–40 min
Pressure cooker: e.g., 3 min
Buffer and pH: e.g.,
Tris–HCl, 10 m*M*, pH 9.5 + 5% urea
Sodium citrate, 10 m*M*, pH 6.0
Tris 10 m*M*/EDTA, 1 m*M* with 0.05% Tween (pH 9.0)
Dako REAL™ Target Retrieval Solution, pH 9.0 (Dako, cat. no. S2367)
Cell Conditioning 1 (Ventana, cat. no. 950-124)
Cell Conditioning Solution CC2 (Ventana, cat. no. 950-123)
Enzymatic or proteolytic-induced epitope retrieval (PIER)
Enzyme: e.g.,
Protease type XXIV (e.g., Sigma, cat. no. P 8038-1G)
Proteinase K (e.g., Dako S3004 or S3020)
Trypsin (e.g., Abcam, cat. no. ab970)
Pepsin (e.g., Abcam, cat. no. ab64201)
Concentration/time/temperature: e.g.,
0.1% Protease Type XXIV in PBS for 10 min at room temperature
50 m*M* Proteinase K in 15 m*M* Tris–HCl (pH 7.5) for 3–6 min at room temperature

necessary to evaluate the influence of these preanalytical parameters on the analytical technique (i.e., antigen retrieval and detection by IHC) (for overview, see Table 3; Engel & Moore, 2011; Grube, 2004; Shi & Taylor, 2010).

For investigation of preanalytical parameters impacting on the reliability of IHC protocols it is recommended to expose aliquots of a tissue sample to different preanalytical conditions (i.e., different ischemia and fixation times

Table 3 Key Variables that may Impact IHC Results on Paraffin-Embedded Tissue

Key preanalytical variables	
Tissue origin	• Organ of origin, disease, patient to patient variability
Ischemia time	• Warm ischemia (clamping of the vessel to removal of the tissue from the body) • Cold ischemia (removal of tissue from the body to start of fixation of tissue)
Fixation of tissue	• Size of tissue sample for fixation • Type of fixative: e.g., cross-linking such as formalin or non-cross-linking such as PAXgene • Condition of fixative: e.g., concentration, pH, or age of fixative • Tissue to fixative volume ratio • Fixation time • Fixation temperature: e.g., room temperature or 4 °C
Tissue processing	• Type and conditions of reagents for dehydration: e.g., percentage and freshness of ethanols • Type and conditions of reagents for clearing • Type and condition of paraffin for tissue impregnation: e.g., type/melting point paraffin and freshness
Tissue embedding	• Type of paraffin for embedding: e.g., type/melting point paraffin
Storage of paraffin tissue blocks	• Storage temperature: e.g., room temperature, 4 °C, temperature-controlled • Humidity during storage: e.g., amount and control of humidity • Storage time until use of tissue for IHC
Key IHC variables	
Tissue sections	• Thickness of sections • Drying of sections before IHC • Storage time and storage conditions of sections until IHC: e.g., immediate use versus storage at room temperature or 4 °C, covered with paraffin for long-term storage
Antigen retrieval method	• Heat- or proteolytic-induced epitope retrieval: e.g., device, enzyme, time, temperature, buffer type, pH
Antibody	• Type of antibody • Concentration and dilution buffer • Incubation time and temperature
Antigen detection system	• Biotin or polymer detection system • Incubation times • Chromogen: e.g., AEC or DAB

as shown in Fig. 3). It is further recommended to test different antigen retrieval protocols, different antibody concentrations and different detection systems in order to select the most reliable protocol.

Figure 3A shows the experimental design of a study to evaluate the impact of preanalytical factors (i.e., different tissue ischemia and fixation times) and presents an example of validating keratin IHC of human liver for detection of ballooned hepatocytes in the context of histopathologic diagnosis of steatohepatitis (Lackner et al., 2008). In Fig. 3, it is demonstrated that depending on the antibody or IHC protocol a positive reaction is only seen in some cases under certain conditions whereas other cases are falsely negative (compare Fig. 3B and C). Furthermore, depending on the antibody or IHC protocol there is a strong positive reaction with bile duct epithelia whereas the reaction with the keratin IF cytoskeleton of hepatocytes is variable (some hepatocytes reveal the typical cytoplasmic staining; in some hepatocytes only the submembranous cytoplasm is stained; some hepatocytes are falsely negative; Fig. 3C). There are also antibodies recognizing epitopes which are essentially not affected by the preanalytical variables tested (i.e., different ischemia and fixation times) using optimized IHC protocols (Fig. 3B). Such antibodies and IHC protocols result in robust, specific staining at various preanalytical conditions typically occurring in routine health care and are, therefore, recommended for diagnostic applications.

3.1.1 Optimized Protocol for Immunohistochemical Staining of Keratin 8/18 in FFPE Tissue Sections *(Results Are Shown in Fig. 3B)*

1. FFPE tissue microarray sections (3 μm) are deparaffinised in xylene (2 × 10 min) and rehydrated in graded ethanol (100%, 90%, 80%, 70%, 50% ethanol) and Aqua dest. for 3 min each.
2. Antigen retrieval is achieved by "slow cooking" microwave treatment of sections for 40 min at 160 W (one cuvette, 300 W for two, 450 W for three, or 600 W for four cuvettes) in Dako REAL™ Target Retrieval Solution pH 9.0 (cat. no. S2367). Since this microwave treatment results in evaporation of liquid the cuvette should be refilled with Aqua dest. after approximately 25 min.

   (Alternative antigen retrieval methods that can be used to optimize IHC protocols for other antibodies are shown in Table 2 and include HIER in (i) a water bath (e.g., at 98.5 °C for 20–40 min) or (ii) pressure cooker, or with (iii) PIER by incubating sections with

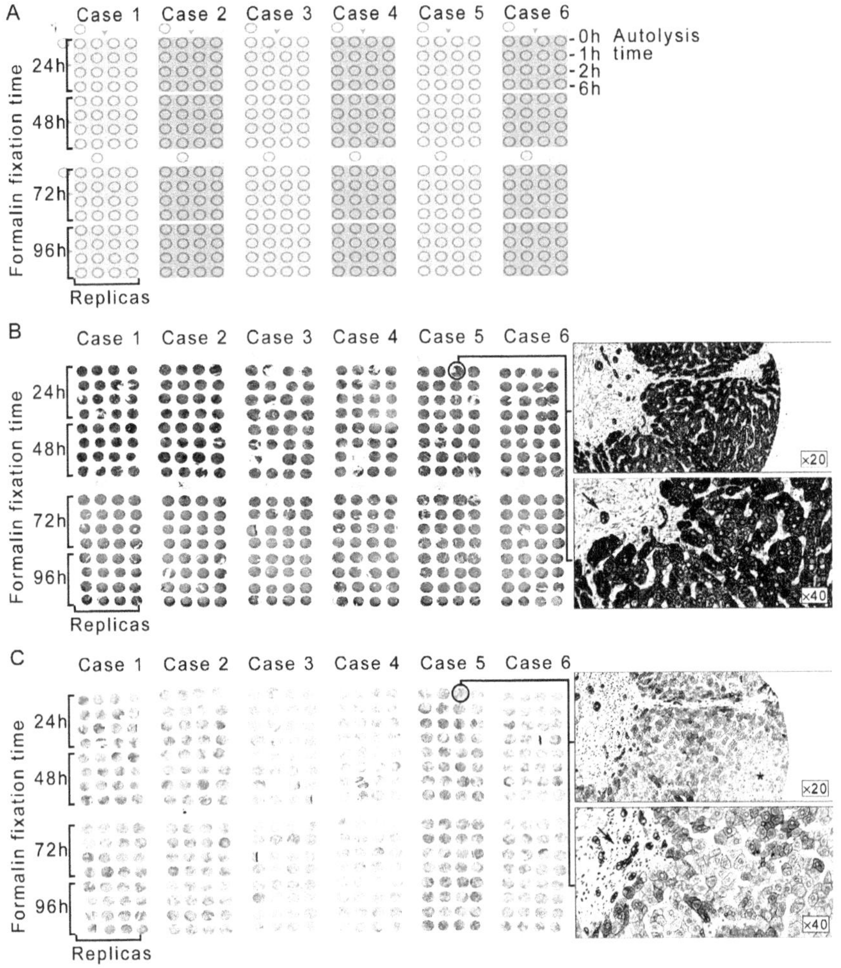

Figure 3 Example of a study design for evaluation of the impact of preanalytical factors and different staining protocols on IHC results. (A) Design of a tissue microarray representing liver samples from six different cases (four replicas each) exposed to different preanalytical conditions (0, 1, 2, and 6 h of cold ischemia and 24, 48, 72, and 96 h formalin fixation, respectively). (B) Example of a FFPE tissue microarray section showing an optimized IHC protocol for Keratin 8/18 that provides specific, positive results for all cases and all conditions tested. The protocol is described in Section 3.1. (C) Example of an IHC staining protocol for Keratin 8/18 which shows specific, positive results only for some cases and some conditions. Arrows indicate bile ducts (B and C). Asterisk indicates area in which hepatocytes are not stained by a Keratin 8/18 antibody (C). Rectangular areas in B (×20) and C (×20) indicate areas shown in ×40 magnification below.

enzymes such as 0.1% protease type XXIV solution (Sigma, cat. no. P 8038) in PBS for 5 min at room temperature in a humid chamber).

3. After cooling down for 20 min at room temperature, the sections are washed in PBS.
4. Endogenous peroxidase is blocked with Dako REAL™ Blocking Solution (cat. no. S2023) for 10 min at room temperature in a humid chamber.

   For labeling of Keratin 8/18 a mix of the primary monoclonal mouse antibodies to Keratin 8 (Clone TS1) (Leica, Novocastra™ cat. no. NCL-CK8-TS1) and to Keratin 18 (Ab-1 Clone DC 10) (NeoMarkers, cat. no. MS-142 P) is prepared by diluting both antibodies 1:50 in Dako REAL™ Antibody Diluent (cat. no. S2022). For negative controls, primary antibodies are omitted and replaced by diluted preimmune serum or isotype-matched immunoglobulins.
5. Sections are incubated with the Keratin 8/18 antibody mix in a humid chamber for 2 h at room temperature, followed by washing in PBS.
6. Antigen detection is performed with Dako REAL™ Detection System Peroxidase/DAB+, Rabbit/Mouse (cat. no. K5001). In brief,
   a. Sections are incubated with Dako REAL™ Link, Biotinylated Secondary Antibodies (AB2) ready-to-use for 30 min in a humid chamber.
   b. After 3 × washing in PBS, slides are incubated with Dako REAL™ Streptavidin Peroxidase (HRP) for 30 min at room temperature in a humid chamber.
   c. Detection is performed by incubation of sections for 10 min at room temperature in a humid chamber with a freshly prepared DAB-containing Substrate Working Solution (prepared by mixing thoroughly 20 μL of Dako REAL™ DAB+ Chromogen and 1 mL Dako REAL™ HRP Substrate Buffer).
   d. Slides are gently rinsed in warm tap water, counterstained with Mayer's haematoxylin, rinsed again in warm tap water and mounted with Aquatex (Merck, cat. no.1.08562.0050).
   e. Tissue microarray sections are scanned with an Aperio ScanScope AT and pictures taken with the Aperio Image Scope software (version V12.1.0.5029).

Examples of commercially available keratin antibodies and the conditions commonly used for immunohistochemical staining of FFPE tissue sections in diagnostic pathology are listed in Table 4.

Table 4 Examples of Antibodies Used for Immunohistochemical Staining of Human FFPE Tissue Sections

Antibodies	Clones	Possible Sources	Antibody Dilution	Detection System/ Chromogen	Examples of Positive Control Tissue
m-a-h keratin (K5/6/8/17/19)	MNF 116	Dako (cat. no. M821)	1:100	CM/AEC or DAB	Gut
m-a-h pan keratin	Lu 5	Biocarta (cat. no. CM 043)	1:50	CM/AEC or DAB	Skin, gut
m-a-h keratin (K1/5/6/8/10/14/18)	LP34/34ßE12/ßH11	Dako (cat. no. N1589)	Ready-to-use	CM/AEC or DAB	Skin, gut
m-a-h keratin large spectrum	KL 1	Immunotech (cat. no. 2128)	1:50	CM/AEC or DAB	Skin, gut
m-a-h epidermal keratin	AE1/AE3	Dako (cat. no. M 3515)	1:50	CM/AEC or DAB	Mamilla
m-a-h keratin, high-molecular-weight keratin (K1/5/10/14)	34ßE12	Dako (cat. no. M630)	1:50	CM/AEC or DAB	Prostate, mamma, adenocarcinoma
m-a-h keratin, low-molecular-weight keratin (K8/18)	5D3	Novocastra (cat. no. NCL-5D3)	1:50	CM/AEC or DAB	Mamilla
m-a-h keratin 5/6	D5/16B4	Dako (cat. no. M7237)	1:50	CM/AEC or DAB	Mamma carcinoma, mesothelioma
m-a-h keratin 5/8	RCK 102	ICN (cat. no. 10521)	1:10	CM/AEC or DAB	Skin, cervix
m-a-h keratin 8/18 (comparable with CAM 5.2)	TS1/DC10	Novocastra (cat. no. NCL-L-CK8-TST) or NeoMarker (cat. no. MS-142-PO)	1:50 each	CM/AEC or DAB	Renal cell carcinoma, hepatocellular carcinoma
m-a-h keratin 1	34ßB4	Novocastra (cat. no. NCL-CK1)	1:10	CM/AEC or DAB	Skin

m-a-h keratin 4	6B10	Monosan (cat. no. MON 3015)	1:5	CM/AEC or DAB	Parotid gland, hypophysis
m-a-h keratin 5	EP42	Epitomics (cat. no. AC-0181)	1:100	ENV/AEC	Skin, gut, lung carcinoma
m-a-h keratin 6	K_s6.KA12	Progen (cat. no. 65190)	Ready-to-use	CM/AEC or DAB	Skin
m-a-h keratin 7	OV-TL 12/30	Dako (cat. no. M7018)	1:100	CM/AEC or DAB	Mamilla
m-a-h keratin 8	TS1	Novocastra (cat. no. NCL-CK8-PS1)	1:50	CM/AEC or DAB	Skin, hepatocellular carcinoma
m-a-h keratin 9	Ks9.7/Ks9.12	Progen (cat. no. 651104)	1:20	CM/AEC or DAB	Skin
m-a-h keratin 10	DE-K10	Dako (cat. no. M7002) or NeoMarker (cat. no. MS-611-P)	1:100 1:200	CM/AEC or DAB	Skin, mamilla
m-a-h keratin 10/13 (detects K10 only when omitting antigen/epitope retrieval)	DE-K13	Dako (cat. no. M7003)	1:50	CM/AEC or DAB	Tonsil
m-a-h keratin 10/13 (detects K10 with heat-induced antigen/epitope retrieval, e.g., water bath)	DE-K13	Dako (cat. no. M7003)	1:50	CM/AEC or DAB	Skin, tonsil
m-a-h keratin 14	LL002	Novocastra (cat. no. NCL-LL002) or NeoMarker (cat. no. MS-115-PO)	1:50	CM/AEC or DAB	Skin

Continued

Table 4 Examples of Antibodies Used for Immunohistochemical Staining of Human FFPE Tissue Sections—cont'd

Antibodies	Clones	Possible Sources	Antibody Dilution	Detection System/ Chromogen	Examples of Positive Control Tissue
m-a-h keratin 15	LHK15	NeoMarker (cat. no. MS-1068-PO)	1:50	CM/AEC or DAB	Skin
m-a-h keratin 16	LL025	Novocastra (cat. no. NCL-CK16)	1:10	CM/AEC or DAB	Basal cell carcinoma
m-a-h keratin 17	E3	Dako (cat. no. M7046)	1:20	CM/AEC or DAB	Prostate, mamma
m-a-h keratin 18	DC 10	NeoMarker (cat. no. MS-142-PO) Dako (cat. no. M7010)	1:50 1:10	CM/AEC or DAB	Skin, mamma carcinoma
m-a-h keratin 19	RCK 108	Dako (cat. no. M888)	1:100	CM/AEC or DAB	Gut, mamilla
m-a-h keratin 20	K_s20.8	Dako (cat. no. M7019)	1:100	CM/AEC or DAB	Gut, mamilla
r-a-h keratin 7/m-a-h keratin 20 (mix of both for double immunolabeling)	BC1/K_s20.8	Biocare (cat. no. CRM 339 A)/Dako (M7019)	1:100 1:50	Mach2/Vulcan Fast Red and CM/DAB	Gut, mamma

+P, Protease (for antigen retrieval: incubation of sections in 0.1% Protease XXIV (Sigma, cat. no. P8038) diluted in PBS for 5 min at room temperature in a humid chamber); AEC, 3-amino-9-ethylcarbazole substrate chromogen for detection of IHC reaction (Dako, cat. no. K3464); CM, Dako REAL™ Detection System, Peroxidase/DAB+, Rabbit/Mouse (Dako, cat. no. K5001) (for detection of IHC reaction procedure, see optimized protocol for immunohistochemical staining in Section 3.1); DAB, 3,3′-diaminobenzidine tetrahydrochloride substrate chromogen for detection of IHC reaction (Dako REAL™ DAB+ Chromogen from Dako REAL™ kit cat. no. K5001); ENV, Dako EnVision™ Detection Systems, Peroxidase/DAB, Rabbit/Mouse (Dako, cat. no. K5007); m-a-h/r-a-h, mouse-anti-human; rabbit-anti-human; Vulcan Fast Red, Vulcan Fast Red chromogen from Vulcan Fast Red Chromogen Kit 2 (Biocare, cat. no. 022515) for detection of IHC reaction; no AR, no antigen retrieval; WB, water bath (for antigen retrieval: performed in cuvettes in Epitope Retrieval Solution from the HercepTest™ for the Dako Autostainer (Dako, cat. no. K5207) at 98.5 °C temperature for 20–40 min); MW 9.0, microwave (for antigen retrieval: performed in cuvettes in Target Antigen Retrieval Solution pH 9.0 at 160 W for 1 cuvette for 40 min); MACH2, Biocare Medical (Detection ohne Farbstoff für DF) Mach 2 Double Stain2 MRCT525L.

3.2 Keratin Detection in PAXgene-Fixed Paraffin-Embedded Tissue

Although there are many antibodies available that can be applied on FFPE tissues there are still antibodies for which no appropriate protocol for FFPE tissues exists. For some epitopes, the combined effect of formalin-induced cross-linking and unfolding during tissue processing and paraffin embedding are detrimental.

Therefore, several fixatives which are not based on formaldehyde and do not induce cross-links have been developed (Viertler et al., 2012). The benefit of non-cross-linking fixatives on antigen preservation in tissue is demonstrated using the example of PAXgene (PreAnalytiX, Hombrechtikon, Switzerland), which is based on a combination of alcohols, an acid, and a soluble organic compound. PAXgene has been demonstrated to preserve tissue morphology similarly to formalin (Gündisch et al., 2014) and, at the same time, preserves RNA almost as well as snap freezing (Viertler et al., 2012). Furthermore, proteins and in particular phosphoproteins can be readily investigated in PAXgene-fixed paraffin-embedded (PFPE) tissues (Ergin et al., 2010). The PAXgene tissue system preserves phosphoproteins in human tissue specimens and enables comprehensive protein biomarker research (Gündisch et al., 2013). The fixation of tissues without cross-linking may also reduce the requirement for (heat-induced) antigen retrieval for some antibodies as shown in Fig. 4 in which IHC of K8/18 on the same liver tissue is compared after FFPE and PFPE.

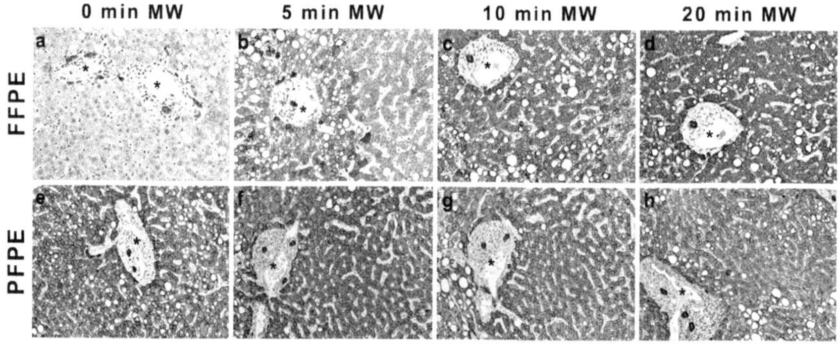

Figure 4 Effect of different tissue fixatives on keratin IHC. Tissue samples from a human liver were fixed in parallel with formalin (a–d) and PAXgene (e–h) for 24 h before further tissue processing and paraffin embedding. Five micrometer thick sections were used for IHC staining using the antibodies to Keratin 8/18 according to the optimized protocol as shown in Fig. 3B. Different times of microwave (MW) treatment of sections (0, 5, 10, and 20 min; pH 9.0) were tested for antigen retrieval. Asterisks indicate portal tracts (a–h). (See the color plate.)

3.2.1 Protocol for PAXgene Fixation, Embedding of Tissue, and IHC Staining of PFPE Sections *(Results Are Shown in Fig. 4)*

1. Fresh normal human liver is cut into small pieces (max. 4 × 15 × 15 mm in size), placed into standard histocassettes, and fixed with PAXgene Tissue FIX. This can either be done by using the PAXgene Tissue System (PreAnalytiX, cat. no. 765112, i.e., a dual-chamber container prefilled with two reagents, the fixative PAXgene Tissue FIX and the PAXgene Tissue Stabilizer; capacity: one cassette per container) or the PAXgene Tissue FIX Container (PreAnalytiX, cat. no. 765312; i.e., a single-chamber container prefilled with 50 mL PAXgene tissue fixative only; capacity for four cassettes).
2. Tissue samples are fixed for 24 h at room temperature; fixation time can range from 2 to 24 h without negative effect on IHC results. Even fixation (up to 72 h) seems to be possible without compromising the quality.
3. Fixation is terminated by transferring the cassette from the PAXgene fixative into the stabilizer (PAXgene Tissue System, cat. no. 765112: the cassette is transferred into the second chamber containing the stabilizer; PAXgene Tissue FIX Container, cat. no. 765312: fixative is replaced by PAXgene Tissue Stabilizer (PreAnalytiX, cat. no. 765512).)
4. Tissue samples are kept in PAXgene Tissue Stabilizer for 24 h at room temperature (alternatively, tissue can be stored in the stabilizer for a minimum of 7 days at room temperature or for a minimum of 4 weeks at 2–8 °C without negative effects on morphology.
5. Tissue processing of PAXgene-fixed and -stabilized samples is performed in an automated tissue processor (Sacura, S/AG VIP-2000 Floor 4622F) which is not used for tissue processing of formalin-fixed tissue, in order to avoid exposure of PAXgene-fixed and -stabilized tissue to even trace amounts of formalin present in the processing reagents. The tissue processing steps include 70% ethanol (2 × 15 min), 80% ethanol (1 × 30 min), 90% ethanol (1 × 1 h), 99% ethanol (2 × 1 h), followed by isopropanol (2 × 1 h), xylene (2 × 1 h), and low-melting paraffin (4 × 45 min at 55 °C; Paraffinum solidum F 43/46 Grad C; ACM, cat. no. S/N 2549662 30). Tissue processing reagents are replaced on a regular basis to avoid reagent dilution and incomplete dehydration of the tissue samples.
6. Embedding is performed using Histo-Comp® paraffin (Sanova, cat. no. 200856).

7. IHC on sections of PFPE tissue is performed as described in Section 3.1.1 with the exception that the microwave treatment is either omitted, or performed for 5, 10, or 20 min in the same buffer as described above (at 270 W).

4. CONCLUSIONS

Detection of keratins in tissues gained relevance not only in elucidating their biological role but also as diagnostic marker in medical diagnostics. However, immunological detection by immunofluorescence microscopy or IHC has to consider several factors impacting on the reliability and proper interpretation of staining results. Therefore, a solid understanding of the biology of keratins is mandatory. The impact of preanalytical factors is often underestimated and requires careful evaluation of all relevant issues ranging from ischemia times to stabilization techniques (e.g., snap freezing and formalin fixation) to storage and antigen retrieval. This is of particular importance when tissue samples are analyzed that have been processed according to different protocols. Only protocols that are tested for reliability for all the expected preanalytical variables should be used.

5. PEARLS AND PITFALLS

The following key issues should be considered for detecting keratins in tissues:

- Aberrant/loosely regulated expression may result in keratin positivity in nonepithelial cell types.
- Conformation-dependent epitopes may result in false negative reaction, particularly in case of using monoclonal antibodies.
- Dissociation of RNA and protein expression can be observed if the homeostasis of the expression of the proper keratin partner is disturbed.
- Preanalytical modifications caused by warm/cold ischemia, fixation, tissue processing and embedding may result in false negative or inappropriate reactions. In this context, it might be misleading that different cell structures could behave differently (e.g., cholangiocytic epithelia and hepatocytes). Inappropriate antigen retrieval protocols (e.g., too long microwave treatment) or applying solely routine protocols instead of testing different IHC protocol may result in false positive or negative reactions. Preanalytical effects have to be tested for each antibody and

antigen since the various epitopes recognized by antibodies may behave differently.
- Prolonged and inappropriate storage of FFPE tissues may affect antigenicity.

Standardization of preanalytical procedure, robustness of staining protocols and quality control together are the basis for reliable results both in the research and diagnostic settings. The less standardized preanalytical procedure are the more emphasis has to be placed on the robustness of the staining protocol (i.e., results are not affected by preanalytical variables) and the quality control for each tissue analyzed (e.g., staining of reference antigens [positive control] and negative controls).

ACKNOWLEDGMENTS

This work was supported by the following grants: European Union Seventh Framework Programme, project SPIDIA (grant no. 222916), the Austrian Genome Program GEN-AU (J20964), the Christian Doppler Laboratory BRBT, and the Austrian Biobanking and BioMolecular Research Infrastructure (BBMRI.at) funded by the Austrian Federal Ministry of Science, Research and Economy (BMWFW GZ 10.470/0016-II/3/2013). The funders had no role in study design, data collection and analysis, decision to publish, or preparation of the chapter. We thank Oridis Biomed for tissue microarray production and Penelope Kungl (Institute of Pathology, Medical University of Graz, Graz, Austria) who assisted in the proof-reading of the chapter.

Conflict of Interest: The authors have no competing interests to disclose. No writing assistance was utilized in the production of this chapter. K.Z. receives research funding from Qiagen.

REFERENCES

Azumi, N., & Battifora, H. (1987). The distribution of vimentin and keratin in epithelial and nonepithelial neoplasms. A comprehensive immunohistochemical study on formalin- and alcohol-fixed tumors. *American Journal of Clinical Pathology*, *88*(3), 286–296.

Banks, E. R., Jansen, J. F., Oberle, E., & Davey, D. D. (1995). Cytokeratin positivity in fine-needle aspirates of melanomas and sarcomas. *Diagnostic Cytopathology*, *12*(3), 230–233.

Brown, D. C., Theaker, J. M., Banks, P. M., Gatter, K. C., & Mason, D. Y. (2002). Cytokeratin expression in smooth muscle and smooth muscle tumours. *Histopathology*, *41*(3A), 85–94.

Do, H., & Dobrovic, A. (2015). Sequence artifacts in DNA from formalin-fixed tissues: Causes and strategies for minimization. *Clinical Chemistry*, *61*(1), 64–71.

Domenjoud, L., Jorcano, J. L., Breuer, B., & Alonso, A. (1988). Synthesis and fate of keratins 8 and 18 in nonepithelial cells transfected with cDNA. *Experimental Cell Research*, *179*(2), 352–361.

Engel, K. B., & Moore, H. M. (2011). Effects of preanalytical variables on the detection of proteins by immunohistochemistry in formalin-fixed, paraffin-embedded tissue. *Archives of Pathology & Laboratory Medicine*, *135*(5), 537–543.

Ergin, B., Meding, S., Langer, R., Kap, M., Viertler, C., Schott, C., et al. (2010). Proteomic analysis of PAXgene-fixed tissues. *Journal of Proteome Research*, *9*(10), 5188–5196.

Fox, C. H., Johnson, F. B., Whiting, J., & Roller, P. P. (1985). Formaldehyde fixation. *Journal of Histochemistry and Cytochemistry*, *33*, 845–853.

Goldman, R. D., Cleland, M. M., Murthy, S. N., Mahammad, S., & Kuczmarski, E. R. (2012). Inroads into the structure and function of intermediate filament networks. *Journal of Structural Biology*, *177*(1), 14–23.

Grube, D. (2004). Constants and variables in immunohistochemistry. *Archives of Histology and Cytology*, *67*(2), 115–134.

Gündisch, S., Schott, C., Wolff, C., Tran, K., Beese, C., Viertler, C., et al. (2013). The PAXgene® tissue system preserves phosphoproteins in human tissue specimens and enables comprehensive protein biomarker research. *PLoS One*, *8*(3). e60638.

Gündisch, S., Slotta-Huspenina, J., Verderio, P., Ciniselli, C. M., Pizzamiglio, S., Schott, C., et al. (2014). Evaluation of colon cancer histomorphology: A comparison between formalin and PAXgene tissue fixation by an international ring trial. *Virchows Archiv*, *465*(5), 509–519.

Haybaeck, J., Stumptner, C., Thueringer, A., Kolbe, T., Magin, T. M., Hesse, M., et al. (2012). Genetic background effects of keratin 8 and 18 in a DDC-induced hepatotoxicity and Mallory-Denk body formation mouse model. *Laboratory Investigation*, *92*(6), 857–867.

Hazan, R., Denk, H., Franke, W. W., Lackinger, E., & Schiller, D. L. (1986). Change of cytokeratin organization during development of Mallory bodies as revealed by a monoclonal antibody. *Laboratory Investigation*, *54*, 543–553.

Hendrix, M. J., Seftor, E. A., Chu, Y. W., Seftor, R. E., Nagle, R. B., McDaniel, K. M., et al. (1992). Coexpression of vimentin and keratins by human melanoma tumor cells: Correlation with invasive and metastatic potential. *Journal of the National Cancer Institute*, *84*(3), 165–174.

Herrmann, H., & Aebi, U. (2000). Intermediate filaments and their associates: Multi-talented structural elements specifying cytoarchitecture and cytodynamics. *Current Opinion in Cell Biology*, *12*(1), 79–90.

Karantza, V. (2011). Keratins in health and cancer: More than mere epithelial cell markers. *Oncogene*, *30*(2), 127–138.

Ku, N. O., Zhou, X., Toivola, D. M., & Omary, M. B. (1999). The cytoskeleton of digestive epithelia in health and disease. *American Journal of Physiology*, *277*(6 Pt. 1), G1108–G1137.

Kulesh, D. A., Cecena, G., Darmon, Y. M., Vassear, M., & Oshima, R. G. (1989). Posttranslational regulation of keratins: Degradation of mouse and human keratins 18 and 8. *Molecular and Cellular Biology*, *9*, 553–1565.

Lackner, C., Gogg-Kamerer, M., Zatloukal, K., Stumptner, C., Brunt, E. M., & Denk, H. (2008). Ballooned hepatocytes in steatohepatitis: The value of keratin immunohistochemistry for diagnosis. *Journal of Hepatology*, *48*(5), 821–828.

Magin, T. M., Bader, B. L., Freudenmann, M., & Franke, W. W. (1990). De novo formation of cytokeratin filaments in calf lens cells and cytoplasts after transfection with cDNAs or microinjection with mRNAs encoding human cytokeratins. *European Journal of Cell Biology*, *53*, 333–348.

Magin, T. M., Schröder, R., Leitgeb, S., Wanninger, F., Zatloukal, K., Grund, C., et al. (1998). Lessons from keratin 18 knockout mice: Formation of novel keratin filaments, secondary loss of keratin 7 and accumulation of liver-specific keratin 8-positive aggregates. *Journal of Cell Biology*, *140*(6), 1441–1451.

Miettinen, M., Fetsch, J. F., Antonescu, C. R., Folpe, A. I., & Wakely, P. E. (2014). Inflammatory myxohyaline tumor od distal extremities (acral myxoinflammatory fibroblastic sarcoma). In S. G. Silverberg (Ed.), *Atlas of tumor pathology, tumors of the soft tissues: Vol. 4* (pp. 190–192). Maryland: Silver Spring. American Registry of Pathology.

Moll, R., Divo, M., & Langbein, L. (2008). The human keratins: Biology and pathology. *Histochemistry and Cell Biology*, *129*(6), 705–733.

Omary, M. B., Coulombe, P. A., & McLean, W. H. (2004). Intermediate filament proteins and their associated diseases. *New England Journal of Medicine, 351*(20), 2087–2100.

Omary, M. B., Ku, N. O., Tao, G. Z., Toivola, D. M., & Liao, J. (2006). Heads and tails of intermediate filament phosphorylation: Multiple sites and functional insights. *Trends in Biochemical Sciences, 31*, 383–394.

Otali, D., Stockard, C. R., Oelschlager, D. K., Wan, W., Manne, U., Watts, S. A., et al. (2009). The combined effects of formalin fixation and individual steps in tissue processing on immunorecognition. *Biotechnic & Histochemistry, 84*(5), 223–247.

Shi, S.-R., Cote, R. J., & Taylor, C. R. (1997). Antigen retrieval immunohistochemistry: Past, present, and future. *Histochemistry and Cytochemistry, 45*, 327–343.

Shi, S.-R., & Taylor, C. R. (2010). Standardization of immunohistochemistry based on antigen retrieval technique. In S.-R. Shi & C. R. Taylor (Eds.), *Antigen retrieval immunohistochemistry based research and diagnostics* (pp. 75–86). Hoboken, NJ: John Wiley.

Sing, Y., Ramdial, P. K., Ramburan, A., & Sewram, V. (2014). Cytokeratin expression in gastrointestinal stromal tumors: Morphology, meaning, and mimicry. *Indian Journal of Pathology & Microbiology, 57*(2), 209–216.

Strnad, P., Stumptner, C., Zatloukal, K., & Denk, H. (2008). Intermediate filament cytoskeleton of the liver in health and disease. *Histochemistry and Cell Biology, 129*, 735–749.

Stumptner, C., Fuchsbichler, A., Heid, H., Zatloukal, K., & Denk, H. (2002). Mallory body—A disease-associated type of sequestosome. *Hepatology, 35*(5), 1053–1062.

Toivola, D. M., Boor, P., Alam, C., & Strnad, P. (2015). Keratins in health and disease. *Current Opinion in Cell Biology, 32*, 73–81.

van de Rijn, M., Perou, C. M., Tibshirani, R., Haas, P., Kallioniemi, O., Kononen, J., et al. (2002). Expression of cytokeratins 17 and 5 identifies a group of breast carcinomas with poor clinical outcome. *American Journal of Pathology, 161*(6), 1991–1996.

Viertler, C., Groelz, D., Gündisch, S., Kashofer, K., Reischauer, B., Riegman, P. H., et al. (2012). A new technology for stabilization of biomolecules in tissues for combined histological and molecular analyses. *Journal of Molecular Diagnostics, 14*(5), 458–466.

Zatloukal, K., French, S. W., Stumptner, C., Strnad, P., Harada, M., Toivola, D. M., et al. (2007). From Mallory to Mallory-Denk bodies: What, how and why? *Experimental Cell Research, 313*(10), 2033–2049.

Zatloukal, K., Stumptner, C., Fuchsbichler, A., Janig, E., & Denk, H. (2004). Intermediate filament protein inclusions. *Methods in Cell Biology, 78*, 205–228.

Zatloukal, K., Stumptner, C., Lehner, M., Denk, H., Baribault, H., Eshkind, L. G., et al. (2000). Cytokeratin 8 protects from hepatotoxicity, and its ratio to cytokeratin 18 determines the ability of hepatocytes to form Mallory bodies. *American Journal of Pathology, 156*(4), 1263–1274.

CHAPTER SEVEN

High-Throughput Screening for Drugs that Modulate Intermediate Filament Proteins

Jingyuan Sun[*,†,‡,§,¶], Vincent E. Groppi[||], Honglian Gui[*,†,‡,#], Lu Chen[#], Qing Xie[#], Li Liu[§,¶], M. Bishr Omary[*,†,‡,1]

[*]Department of Molecular & Integrative Physiology, University of Michigan, Ann Arbor, Michigan, USA
[†]Department of Medicine, University of Michigan, Ann Arbor, Michigan, USA
[‡]VA Ann Arbor Healthcare System, Ann Arbor, Michigan, USA
[§]Hepatology Unit, Department of Infectious Diseases, Nanfang Hospital, Southern Medical University, Guangzhou, PR China
[¶]Department of Radiation Oncology, Nanfang Hospital, Southern Medical University, Guangzhou, PR China
[||]Department of Pharmacology, The Center for Chemical Genomics, University of Michigan, Ann Arbor, Michigan, USA
[#]Department of Infectious Diseases, Ruijin Hospital, Jiaotong University School of Medicine, Shanghai, PR China
[1]Corresponding author: e-mail address: mbishr@umich.edu

Contents

1. Overview of Intermediate Filaments and Their Associated Diseases 164
2. Current Targeted Therapeutic Approaches for IF-Pathies 165
3. Unbiased Drug Screening to Target IF Mutations 170
4. High-Throughput Steps to Identify Drugs that Target Intermediate Filaments 171
   4.1 Methods for Drug Screening 171
   4.2 Validation of Initial Drug Screening Results 176
   4.3 Assessment of the Mechanism of Action for Compounds of Interest 178
5. Available Libraries and Vendors for Drug Screening 179
6. Pearls and Pitfalls 182

Acknowledgments 183

References 183

Abstract

Intermediate filament (IF) proteins have unique and complex cell and tissue distribution. Importantly, IF gene mutations cause or predispose to more than 80 human tissue-specific diseases (IF-pathies), with the most severe disease phenotypes being due to mutations at conserved residues that result in a disrupted IF network. A critical need for the entire IF-pathy field is the identification of drugs that can ameliorate or cure these diseases, particularly since all current therapies target the IF-pathy complication, such as diabetes or cardiovascular disease, rather than the mutant IF protein or gene. We describe a high-throughput approach to identify drugs that can normalize disrupted IF

Methods in Enzymology, Volume 568
ISSN 0076-6879
http://dx.doi.org/10.1016/bs.mie.2015.09.029

proteins. This approach utilizes transduction of lentivirus that expresses green fluorescent protein-tagged keratin 18 (K18) R90C in A549 cells. The readout is drug "hits" that convert the dot-like keratin filament distribution, due to the R90C mutation, to a wild-type-like filamentous array. A similar strategy can be used to screen thousands of compounds and can be utilized for practically any IF protein with a filament-disrupting mutation, and could therefore potentially target many IF-pathies. "Hits" of interest require validation in cell culture then using *in vivo* experimental models. Approaches to study the mechanism of mutant IF normalization by potential drugs of interest are also described. The ultimate goal of this drug screening approach is to identify effective and safe compounds that can potentially be tested for clinical efficacy in patients.

ABBREVIATIONS

DAPI 4′,6-diamidino-2-phenylindole
DMSO dimethyl sulfoxide
EBS epidermolysis bullosa simplex
FDA Food and Drug Administration
GFAP glial fibrillary acidic protein
GFP green fluorescent protein
IF intermediate filament(s)
IF-pathies intermediate filament-associated diseases
K keratin(s)
NMHC-IIA nonmuscle myosin heavy chain-IIA
PBS phosphate-buffered saline
PFA paraformaldehyde

1. OVERVIEW OF INTERMEDIATE FILAMENTS AND THEIR ASSOCIATED DISEASES

Intermediate filament (IF) proteins make up one of the three major components of the cytoskeleton, with the other two major groups being microfilaments (i.e., actins) and microtubules (i.e., tubulins) (Ku, Zhou, Toivola, & Omary, 1999). IF proteins, as contrasted with actins and tubulins, have several distinct properties that include being the largest in terms of its numbers of family members (e.g., the keratin (K) subgroup of IFs is encoded by 54 genes; Schweizer et al., 2006), relative insolubility, diverse structures, preferential expression in higher eukaryotes (e.g., they are not found in yeast), and extensive disease association (Fuchs & Weber, 1994; Omary, Coulombe, & McLean, 2004). Another distinctive feature of IF proteins is their tissue- and cell-type selective expression. For example, keratins are the IFs of epithelial cells, desmin is found in muscle, neurofilaments in neuronal cells, glial fibrillary acidic protein (GFAP) in glial cells, and vimentin in mesenchymal cells. All these examples are cytoplasmic IF, as contrasted with lamins which reside in the inner aspect of the nuclear membrane of nucleated cells (Fuchs & Weber, 1994; Osmanagic-Myers, Dechat, & Foisner, 2015; Schreiber & Kennedy, 2013).

In terms of human disease, IF mutations cause or predispose to >80 IF-associated human tissue-specific diseases (IF-pathies) (Omary, 2009; Worman & Schirmer, 2015) that can affect practically every organ in the body depending on the distribution of the IF (Fuchs & Weber, 1994; Omary et al., 2004; Szeverenyi et al., 2008). The first IF mutation found to be directly linked to any human disease involved keratin 14 (K14) (Bonifas, Rothman, & Epstein, 1991; Coulombe et al., 1991), which then led to multiple discoveries collectively showing that a broad range of human Mendelian-inherited diseases are caused by mutations in IF genes. Most of the known IF mutations are highly penetrant autosomal dominant, though some of the IF gene mutations predispose to, rather than cause, disease *per se* (Omary et al., 2004; Usachov et al., 2015). For example, K14 mutations cause the blistering skin disease epidermolysis bullosa simplex (EBS); GFAP mutations cause Alexander disease (Brenner et al., 2001); and K8 or K18 mutations predispose to the progression of several acute or chronic liver diseases (Ku, Gish, Wright, & Omary, 2001; Strnad et al., 2010; Usachov et al., 2015). Most disease-causing mutations found in IFs occur in the more conserved central portion of the protein, which is a coiled-coil α-helical stretch of 310–350 amino acids termed the "rod" domain (Fig. 1). Mutations in ultraconserved regions at the beginning or end of the rod domain result in disruption of the IF network from extended filaments into dots and short filaments (Fig. 1), and generally, lead to a more severe form of an IF-pathy (Coulombe, Kerns, & Fuchs, 2009; Lane & McLean, 2004).

2. CURRENT TARGETED THERAPEUTIC APPROACHES FOR IF-PATHIES

Mutations in most IF genes, with a few exceptions (e.g., α-internexin and a few of the keratins), have been linked to a human disease. The most pressing current obstacle in the IF field is that there is not a single direct cure or even partial therapy for any of the human IF-pathies. As such, the only current management of such diseases relates to lifestyle remedies such as prevention of skin trauma in the case of EBS (Gonzalez, 2013), or to treating

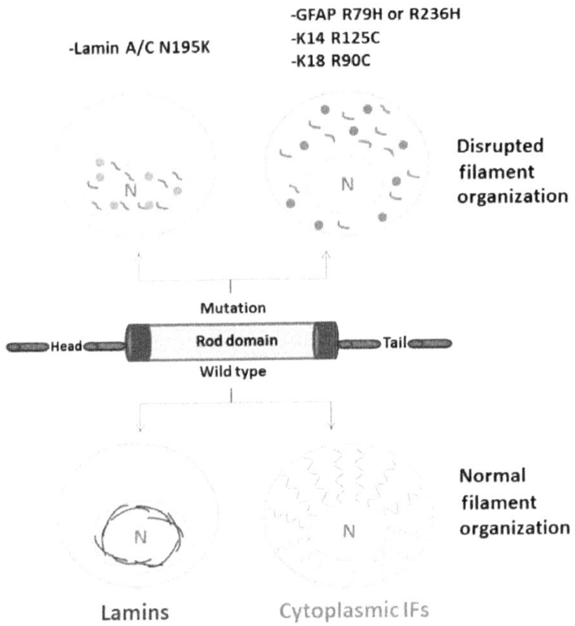

Figure 1 Prototype IF protein domains and consequences of IF mutation on filament organization. The schematic shows the three IF protein domains: a central α-helical coiled-coil relatively conserved "rod" domain (310–350 amino acids) that is flanked by N- and C-terminal non-α-helical "head" and "tail" domains (of variable length depending on the IF protein) which, in turn, provide the exceptional structural diversity among IFs. The mutations responsible for the most severe IF-pathy phenotypes are typically located at ultraconserved helix-initiation and helix-termination motifs at the beginning and end of the rod domain, and cause disruption of the filamentous organization into short filaments or dots. Examples of IF mutations that cause the type of filament disruption that is schematically shown include lamin A/C N195K mutation (causes Emery–Dreifuss muscular dystrophy) (Ostlund, Bonne, Schwartz, & Worman, 2001), GFAP R79H or R236H mutation (both cause Alexander disease) (Mignot et al., 2007; Wang, Colodner, & Feany, 2011), K14 R125C mutation (causes EBS) (Bonifas et al., 1991; Coulombe et al., 1991), and K18 R90C (predisposes to liver injury) (Ku, Michie, Oshima, & Omary, 1995). N, nucleus. (See the color plate.)

end-organ damage such as diabetes or cardiac complications as is the case for some of the lamin disorders (Lu, Muchir, Nagy, & Worman, 2011). However, several genetic (e.g., gene therapy or allele-specific silencing) and pharmacologic approaches have been attempted in experimental systems (Table 1). For example, the natural product sulforaphane, which activates Nrf2-dependent transcription and upregulates K16 and K17, ameliorated skin blistering in K14-null mice (Kerns et al., 2007, 2010). Similarly, forced overexpression of Nrf2 in astrocytes of R236H GFAP mutant mice

Table 1 Experimental Therapeutic Approaches for IF-Pathies

Interventions	Test Systems	Comments	References
Genetic approaches			
K16–14 hybrid transgene	• K14-deficient mice	Expressing a hybrid K16–14 transgene in the epidermis of mice null for K14 restored a wild-type phenotype to newborn epidermis. Of note, K18 expression was not protective	Hutton et al. (1998)
Ectopic expression of desmin	• $K5^{+/-}$ mice crossed with $Desmin^{tg/+}$ transgenic mice	Stable expression of desmin rescued K5 null mice, which in turn served as a model for severe EBS	Kirfel, Peters, Grund, Reifenberg, and Magin (2002)
siRNA-(K9 R163Q)	• AD293 cells • CD1 female mice	A allele-specific siRNA for mutant K9, which can lead to epidermolytic palmoplantar keratoderma, inhibits mutant allele expression *in vitro* and in mice	Leslie Pedrioli et al. (2012)
siRNA-(K12 R135T)	• AD293 cells	A potent allele-specific siRNA reduce the expression of mutant K12 which in turn can cause Meesmann epithelial corneal dystrophy	Allen et al. (2013)
siRNA-(K6 N171K)	• 293FT cells • Swiss wobster mice	Specific siRNA for mutant K6, which causes pachyonychia congenital, inhibits mutant but not WT K6 expression	Hickerson et al. (2015)

Continued

Table 1 Experimental Therapeutic Approaches for IF-Pathies—cont'd

Interventions	Test Systems	Comments	References
Pharmacologic approaches			
Sulforaphane	• K14-deficient mice	Sulforaphane ameliorates skin blistering by inducing K16/K17 in epidermis via Nrf2-dependent and independent pathways	Kerns, DePianto, Dinkova-Kostova, Talalay, and Coulombe (2007) and Kerns, DePianto, Yamamoto, and Coulombe (2010)
Doxycycline	• $K5^{-/-}$ mice	Doxycycline extended survival of neonatal $K5^{-/-}$ mice from <1 to up to 8 h via suppression of inflammatory cytokines	Lu et al. (2007)
PKC412	• A549 cells • K18 R90C mice	PKC412 normalizes K18 Arg90Cys mutation-induced filament disruption and disorganization by enhancing keratin association with NMHC-IIA in a myosin dephosphorylation-regulated manner	Kwan et al. (2015)
Celastrol	• $SW13^{-IF}$ cells transfected with mutant NF-L	Celastrol, an inducer of chaperone proteins, induced HSPA1 in motor neurons and prevented formation of NF inclusions and mitochondrial shortening induced by the expression of NF-L Q333P mutation	Gentil, Mushynski, and Durham (2013)

Farnesyltransferase inhibitors (FTIs)	• $Lmna^{HG/HG}$ mice	ATB-100 improves disease phenotypes in mice with Hutchinson–Gilford progeria syndrome, and R115777 prevents the onset and progression of existing cardiovascular disease	Yang et al. (2006) and Capell et al. (2008)
SP600125	• LmnaH222P/H222P mice	The JNK pathway inhibitor, SP600125 prevents cardiomyopathy caused by *LMNA* mutation	Wu, Shan, Bonne, Worman, and Muchir (2010)
Selumetinib	• LmnaH222P/H222P mice	The ERK1/2 pathway inhibitor, selumetinib preserves cardiac function and improves survival in mice with lamin A/C gene mutation	Muchir et al. (2012)

improved the pathologic features of the GFAP mutation and restored mouse body weights to wild-type levels (LaPash Daniels et al., 2012). The antibiotic doxycycline was also used in K5-null mice and extended the survival of neonatal mice from less than 1 to nearly 8 h, possibly via downregulation of inflammatory cytokines (Lu et al., 2007).

Other siRNA-related approaches, such as specific knockdown of mutant allele expression and clear mutant protein aggregation (Allen et al., 2013; Hickerson et al., 2015; Leslie Pedrioli et al., 2012). Furthermore, chaperone protein inducers and gene replacement therapy offer alternative approaches (Gentil et al., 2013). In addition, kinase signaling pathway inhibitors, such as JNK inhibitor SP600125 and ERK1/2 inhibitor selumetinib, benefited mice with cardiomyopathy due to *LMNA* mutation (Muchir et al., 2012; Wu et al., 2010). In these latter cases, specific kinases that are known to be activated in the context of lamin mutations were selectively and effectively targeted (Muchir & Worman, 2016). Lastly, Withaferin A, which inhibits NF-κB and may manifest some of its biologic effect by binding directly to the soluble fraction of vimentin, has antiangiogenic roles (Mohan & Bargagna-Mohan, 2016) that could impact some of the IF-pathies but this remains to be formally tested.

3. UNBIASED DRUG SCREENING TO TARGET IF MUTATIONS

Given that IF mutations, particularly those that result in more severe phenotypes, generally destabilize and disrupt the IF organization and network distribution, one potential general approach is to screen for compounds that stabilize IFs and potentially serve as direct or indirect chemical chaperones. Such compounds might function similar to taxol which stabilizes microtubules (Rohena & Mooberry, 2014). Other potential positive consequences of an unbiased screen is to identify compounds that target signaling pathways that are aberrantly activated or deactivated as a consequence of the mutation, or to even define novel pathways that are modulated by IF proteins that had not been previously appreciated. However, there are some potential drawbacks for an unbiased screen to keep in mind when planning experiments. For example, the screen is only as good as the assay that is being utilized for screening, so the design of the assay is critical, and some potentially powerful assays (e.g., *in vitro* filament assembly to identify potential compounds that bind directly to IF proteins) may not be readily suitable for high-throughput screening. In addition, compounds of potential interest might have multiple functions or might have undesirable side effects, and it might also be difficult to define their function.

Utilization of such an approach was successfully carried out by screening a kinase inhibitor library in a 384-well plate (Kwan et al., 2015). This approach identified the pan-kinase inhibitor PKC412 as a compound that reverts disrupted K18 R90C filaments into a wild-type-like extended filament network (Kwan et al., 2015). Assessment of the mechanism of action of PKC412 on filament organization showed that it promotes dephosphorylation of nonmuscle myosin heavy chain-IIA (NMHC-IIA), thereby enhancing NMHC-IIA association with K8/K18 which in turn was shown to stabilize the keratin filament network (Kwan et al., 2015). The following section highlights details of the approach used by Kwan et al. that can be implemented to screen for compounds that target any IF protein and can also be used for other non-IF proteins that have a fluorescently tractable cellular organization.

4. HIGH-THROUGHPUT STEPS TO IDENTIFY DRUGS THAT TARGET INTERMEDIATE FILAMENTS

4.1 Methods for Drug Screening

The initial critical elements for drug screening include: (i) developing an assay with a clear high-throughput friendly readout that is biologically meaningful and useful; (ii) deciding on the libraries to test. Some of the different options are covered in Section 5 but, in general, using a Food and Drug Administration (FDA)-approved library as an initial step is practical (because it allows the potential repurposing of already approved and likely relatively safe drugs). In addition, the size of such libraries is generally manageable (500–2000 compounds); and (iii) linking with an institutional core or a private company to carry out the screening. The cost is also a factor that needs to be considered.

For our purposes, we used A549 cells (human lung carcinoma cell line) because in preliminary experiments, we noted that they are readily transducible by GFP-K18 lentiviruses (green fluorescent protein (GFP) is fused with the N-terminus of K18), and they provide a clearly discernable dot pattern for the K18 R90C mutant or a well-defined filamentous pattern for K18 WT. We opted to use a simple epithelial cell line (i.e., A549 cells) that expresses keratins though we also tested mouse NIH-3T3 fibroblast cells that express vimentin but not keratins, and Chinese hamster ovary cells that are similar to NIH-3T3 cells in terms of their IF expression profile. The

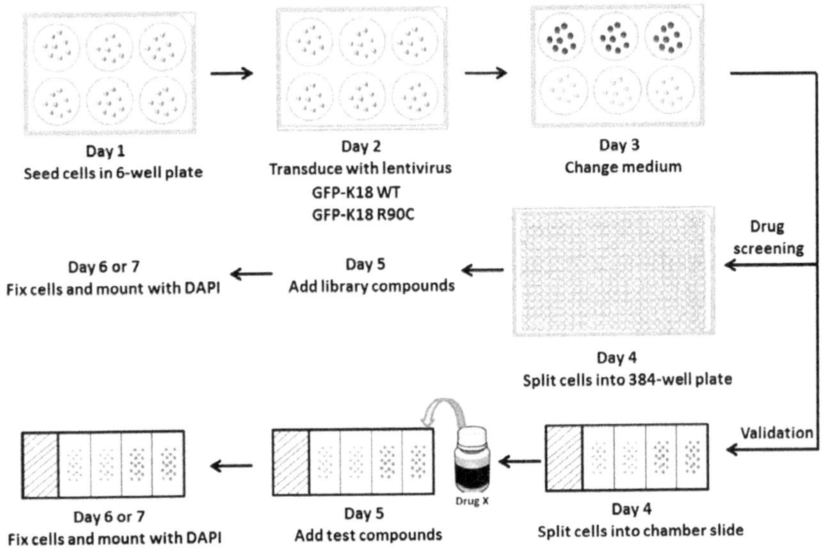

Figure 2 Flow chart of the cell transduction and validation drug screening setup. A549 cells (or any cell culture system that is used for the screening assay) are seeded into culture plates. After allowing cells to adhere for 1 day, they are transduced with GFP-WT K18 or GFP-R90C mutant K18, together with WT K8 (which result in normal control filament organization or in disrupted filaments, respectively, as exemplified in the schematic in Fig. 1 or in the immunofluorescence staining shown in Fig. 4). It is important to cotransfect with both type I and type II keratins to allow filament formation. Fresh media is then added on Day 3 followed on Day 4 by transfer (after trypsinization) to the 384 plates for drug screening (see also Fig. 3) or to chamber slides for specific drug validation. For validation, the test compounds are added into the chamber slide followed by 1 or 2 days of drug treatment then cell fixation and visualization by microscopy.

general approach we utilized is summarized in Fig. 2 and details of the apparatus setup are summarized in Fig. 3.

For selection of the cell system to be used for screening, it is also possible to use stable cell lines. One potential concern is the durability (and level) of expression of the IF protein of interest in the "stable" cell lines on hand. Also, in the case of keratin stable overexpression, another potential limitation to keep in mind is the induction of expression of compensatory keratins (or other IFs). For example, even in our system, we observed that the percentage of cells with dots decreased gradually from ~80% cells with dots on the second day after transduction to ~50% on the fourth day (due to WT keratin overexpression). Another alternative system is to derive mutant IF-expressing cells from cells of animal tissue or from human primary cells, as described in the latter case by immortalizing keratinocytes from patients

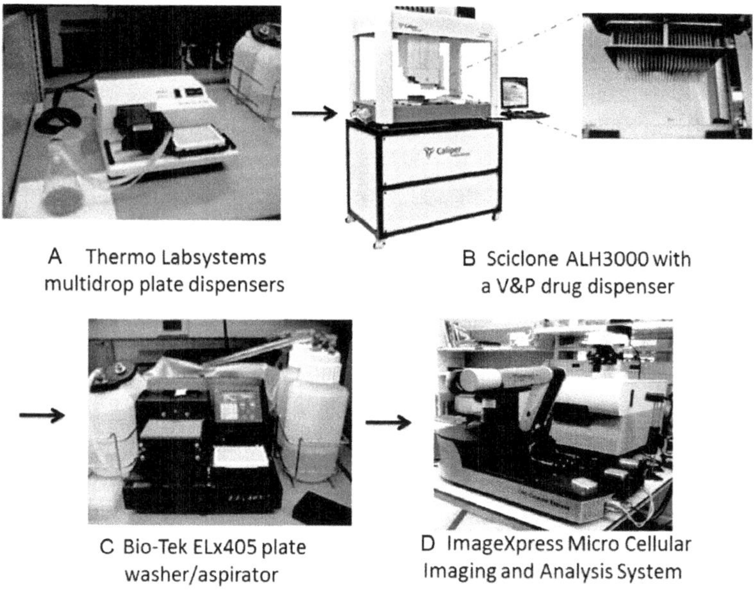

Figure 3 The drug screening high-throughput apparatus. The cells are seeded into 384-well plates using the Multidrop plate dispenser (A) followed by addition of the test compounds using the drug dispenser (B). After 1 or 2 days of incubation in the presence of the drugs, cells are fixed, permeabilized, and mounted with DAPI using an automated washer/aspirator system (C). Images are then acquired using an ImageXpress system (D) that is coupled to a computer to allow data analysis.

(with EBS due to K5 or K14 mutation) using simian virus 40T antigen or papillomavirus 16 (Morley et al., 2003).

4.1.1 Materials

- A549 cells
- GFP-tagged K18-WT lentivirus (positive control)
- GFP-tagged K18-R90C lentivirus
- Tissue culture 6-well plates (Falcon 353046)
- Perkin Elmer View Plate-384 Plates
- Phosphate-buffered saline (PBS)
- Complete medium
- 0.25% Trypsin with EDTA
- 4% Paraformaldehyde (PFA)
- 0.1% Triton X-100
- 1:5000 4′,6-Diamidino-2-phenylindole (DAPI) (Invitrogen H3570)

4.1.2 Methods

1. The day before transduction, trypsinize and count the A549 cells. Plate 0.6×10^6 cells per well (6-well plate) in 3–5 ml of complete growth medium. Cell density should be 50–80% confluent on the day of transduction.
2. Transduce cells with GFP-WT or GFP-R90C K18 lentivirus. Spin the plate at 2500 rpm for 30 min at 22 °C (6-well plates are used in allow to allow their centrifugation in order to enhance the transduction efficiency), then change the medium after 6 h or next day by simple aspiration. Incubate the cells at 37 °C in a 5% CO_2 incubator for 48 h.
3. Trypsinize and count the cells, adjust to 400,000 cells/ml.
4. Seed the cells 10μl/well into the 384 plates using the automated Multidrop and add compounds from the drug library by using the pintool (Fig. 3). Incubate the plates at 37 °C in a 5% CO_2 incubator for 24 or 48 h (depending on the duration that a biologic effect is anticipated).
5. Aspirate the medium from the plate and add 4% PFA with Multidrop, incubate for 15 min (22 °C). It is possible to also use other fixatives such as methanol. However, methanol can interfere with the GFP signal (Straight, 2007; Wang, Miller, Shaw, & Shaw, 1996), though we have not found this to be the case for GFP-K18. We opted to use PFA mainly because it works well for our purposes and for convenience (i.e., no need to maintain the methanol at −20 °C).
6. Wash the cells with PBS (three times) using the Bio-tek plate washer.
7. Add 20 μl of 0.1% Triton X-100 using the Multidrop, incubate for 5 min (22 °C), and then wash with PBS (three times).
8. Add 40 μl DAPI, incubate for 15 min (22 °C), and then wash with PBS (three times). We stain the nucleus with DAPI for two reasons. First, it helps determine the cell density in each well, so we can calculate the percentage of cells with dots by the software to generate the heat map (Fig. 4D). Second, it provides useful information regarding cell integrity, by defining the morphology of the nucleus.
9. Add 20 μl PBS, seal the plates, and store at 4 °C until imaging.
10. Image the plates using the ImageXpress Micro system.

For the analysis of the drug effect on IF organization, it is possible that a software program may need to be designed depending on the aim of the screening (e.g., in our case, our readout was disappearance of dots which is an easier "read" for a camera that appearance of filaments). The IXM uses MetaXpress Software Custom Modules software to allow image analysis and quantification. This software includes algorithms that the user can select to achieve the

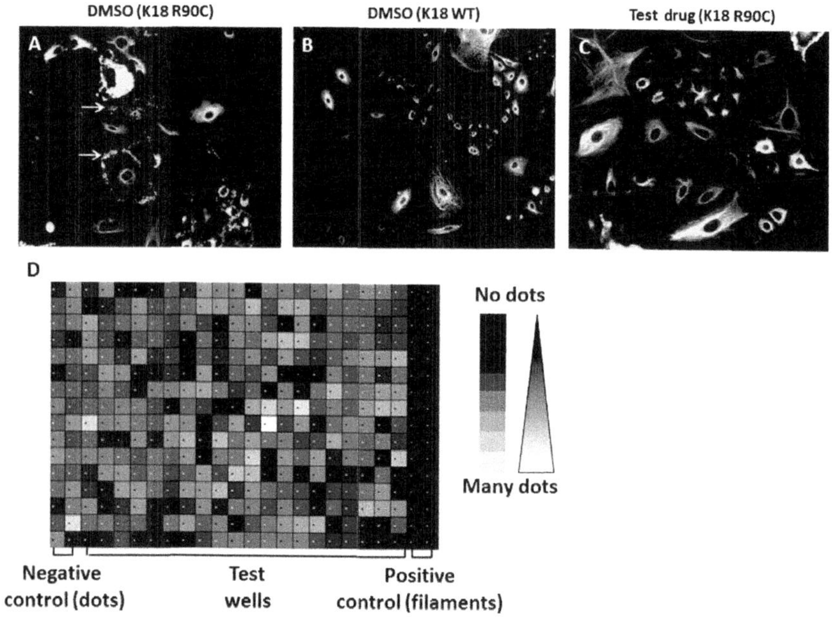

Figure 4 Examples of drug screening phenotypes and a typical heat map as visualized by ImageExpress. (A) Negative control cells which were transduced with GFP-K18 R90C lentivirus and treated with DMSO for 48 h (arrows highlight the dots). (B) Positive control cells which were transduced with GFP-K18 WT lentivirus and treated with DMSO for 48 h. (C) GFP-K18 R90C lentivirus-transduced cells treated with a compound that results in normalization of the disrupted keratins. (D) The first two columns (far left) of the 384-well plat represent GFP-mutant IF-transduced cells treated with DMSO for 48 h as a negative control (primarily red, or non-green color), while the two columns on the far right represent GFP-WT IF-transduced cells treated with DMSO as a positive control. The middle columns represent GFP-mutant IF-transduced cells with different drug library compounds after 48 h. Green indicates few dots/well, while red indicates mainly dots/well. Several wells in the middle columns are bright green, indicating the added compound may have corrected the IF mutant phenotype or is too toxic in those wells (i.e., dead cells will also have few dots and represent false positives). (See the color plate.)

greatest discrimination of phenotypes. In our case, the two phenotypes are "puncta" or "dots" versus "filaments." We used the algorithms to achieve the greatest discrimination between negative control (dots) and positive control (filaments) phenotypes. The algorithms offer discrimination for features that include size, shape, and intensity distribution.

The imager will typically acquire five pictures/well (see Fig. 4A–C, for examples of captured pictures taken from the screened wells), and the software will compare the profile of the pictures with the images from the wells with dots (which are treated with dimethyl sulfoxide (DMSO)) and positive control wells (which are K18-WT-transduced cells treated with DMSO, Fig. 4C). The "heat map" (Fig. 4D) can provide general information about which wells and their corresponding compounds might decrease the numbers of dots/cell/well, but visual inspection is also essential not only for quality control but also to be sure that what appears to be a positive well is not simply because of cell death that effectively provides loss of dots (i.e., a pseudo-positive hit). It is possible that drugs may increase the presence of dots (e.g., by increasing the phosphorylation of the IF). We have not encountered this, except in few rare cases and, if one wishes to pursue such compounds then one potential follow-up test would be to assess their effect on normal keratins (or other IFs).

4.2 Validation of Initial Drug Screening Results

The next important step is to validate the findings of the high-throughput screening by immunofluorescence staining (Fig. 2) using commercial sources of the positive hit compounds. Another aspect that may be tested at this stage is whether the test compounds can also correct the disorganized filaments from another mutant keratin (e.g., K14 R125C) or another IF mutation (e.g., in GFAP, lamin etc.). Different doses of the compound should be tried based on the dose used in drug screening and IC50 (half maximal inhibitory concentration), and different time points of exposure to the drug should be checked. After the compounds are added to the cells, the change in cell proliferation and morphology should be monitored. Furthermore, animal models can subsequently be used to validate the test compound, but the dose and duration and treatment plan would need to be optimized, depending on what is known about the compound(s) of interest.

4.2.1 *Cell System Materials*

The materials here are identical to those shown in Section 4.1.1 except that chamber slides are utilized to visualize the GFP fluorescence and to quantify the extent of filament normalization.

4.2.2 *Cell System Methods*

1. Follow steps 1 and 2 described in Section 4.1.2.
2. After the 48 h of transduction with the lentivirus constructs, trypsinize and count the cells and seed the cells into 4-well chamber slides (400,000 cells/ml; 1 ml per well).
3. Add the compound the next day. Incubate slides at 37 °C in a 5% CO_2 incubator for 24 or 48 h.

4. Aspirate the medium from the slides and add 4% PFA, incubate at room temperature for 15 min, and then proceed as described in Section 4.1.2 (steps 6–8).
5. Image the slides by fluorescent microscopy. It is possible to also use immune staining instead of GFP-tagged proteins, but there are some advantages and disadvantages for such an approach. GFP-tagged proteins offer convenience and save the steps of having to immunostain. The GFP tag also allows following the efficiency of transduction. However, some compounds may have autofluorescence that can mimic GFP or other fluorescent signals, and this should be tested for potential false positive hits. There is also the possibility of losing the GFP signal, depending on the fixation method as mentioned in Section 4.1.2.
6. Count the numbers of transfected cells with dots, dots and filaments, or mainly filaments; then calculate the percentage of cells with the three types of dots and/or filaments. This is done in triplicates to assess reproducibility and whether there is a significant increase in the number of cells with filaments rather than dots.

4.2.3 *In Vivo* System Materials

- K18 WT mice (that overexpress human K18-WT)
- K18 R90C mutant mice (that overexpress human K18-R90C)

The K18 WT mice (Abe & Oshima, 1990) and the K18 R90C mice (Ku et al., 1995) have been used extensively in numerous publications (Ku, Strnad, Zhong, Tao, & Omary, 2007; Strnad et al., 2008) and are available upon request. The K18 R90C mice offer several advantages including being an *in vivo* model for keratin mutation-induced liver injury and a model for mutation-induced keratin filament disruption (Ku et al., 2007). Similarly mice that express other IF mutants would need to be utilized for any preclinical *in vivo* testing of drugs that target other IF-pathies.

4.2.4 *In Vivo* System Methods

1. Choose age and gender matched mice that express the mutant IF. In initial experiments, it is ideal to test both male and female mice since the effect may differ depending on the metabolism of the test drug. Mice are administered the test drug or carrier (DMSO, saline, other). Caveats that go into consideration include the mode and frequency of delivery (oral, intraperitoneal, intravenous). This is why it is essential to carry out a detailed literature search to identify prior studies that may have tested

the compound of interest in mice which could save a tremendous amount of experimental ground work.

2. Another important caveat is the potential expense to obtain sufficient compound for *in vivo* experiments. In general, this is unlikely to be a significant issue unless the compound of interest is unique or is a natural product that is available in limited amounts.
3. The *in vivo* readout will be correction of the abnormal filament organization which would then require subsequent testing to determine if the underlying IF-pathy phenotype is corrected or whether a predisposition to a disease phenotype can be ameliorated. In the case of the K18 R90C mice, the compound PKC412 normalized the disrupted keratin filaments in hepatocytes (similar to what was seen in cell culture, Fig. 4A–C) after 4 days, after daily intraperitoneal administration, and significantly protected from Fas ligand-mediated liver injury (Kwan et al., 2015).

4.3 Assessment of the Mechanism of Action for Compounds of Interest

This is an important component that follows the identification of any potential compound of interest. The following questions need to be addressed, which represents that general approach used by Kwan and colleagues (Kwan et al., 2015) in defining the mechanism of action of PKC412 in normalizing mutant K18 filaments:

a. Does the drug upregulate a compensatory keratin or other IF, or does it decrease the levels of the mutant IF: this should be one of the first questions to answer and can be address easily by assessing the levels of the mutant IF. In addition, the potential overexpression of a compensatory IF can be examined by carrying out high-salt extraction (Snider & Omary, 2016) which is a simple and convenient method to assess whether new proteins now partition with the relatively insoluble IF fraction. Furthermore, a proteomic approach can be utilized to identify any new up- or down- regulated proteins if simple testing for potential IFs is unrevealing.
b. Is the compound of interest a known enzyme inhibitor or activator, or does it have a known function: compounds with known functions can help narrow down the hunt for the mechanism of action of the compound of interest. For example, PKC412, which normalizes mutant K18 R90C filaments, is a known pan-kinase inhibitor which helped focus its likely mechanism of action to changes in phosphorylation in K18 or its binding partner K8, or in the phosphorylation of an associated protein (Kwan et al., 2015).

c. Does the drug upregulate a keratin stabilizing protein or promote the binding of such protein to the mutant IF: this may relate to questions a and b (above). For example, in the case of PKC412, this compound leads to dephosphorylation of NMHC-II which in turn promotes its binding to K8 and helps stabilize the keratin filaments (Kwan et al., 2015) via an unknown mechanism. As mentioned above, a proteomic approach can help identify the presence of such a putative protein. In addition, expression profiling may be used to help identify up- or down-regulated proteins that may shed light on unique altered pathways.

d. Does the drug work on more than one mutant IF: this is also relevant to know since a potential compound that affects more than IF protein is likely to work via a conserved mechanism.

e. Does the drug bind directly to the IF protein: an analogy here would be a taxol-like effect, and this is presently a missing holy grail for IF-pathies and the IF field in general since no such compounds are presently known. Direct binding can be tested using several modalities that include: (i) examining the effect of the putative drug on *in vitro* filament assembly of purified IFs (protocols for IF *in vitro* assembly can be found in Herrmann, Kreplak, & Aebi, 2004), and (ii) testing the binding of a radiolabeled or derivatized form of the compound (if available) to the mutant and WT IF.

5. AVAILABLE LIBRARIES AND VENDORS FOR DRUG SCREENING

There is a wide range of commercial compound libraries that are available to purchase (Table 2). In addition, some universities may have drug screening cores that are available for use by their faculty investigators. In general terms, it is ideal to select libraries that include already approved drugs or drugs that are ongoing clinical trials because of the usually detailed available information on their pharmacodynamics and safety profile. Examples include the FDA-approved drugs that are part of the BioFocus NCC and Prestwick libraries (Table 2). In addition, some vendors including Enzo Life Sciences, ChemDiv, and Chembridge, can provide focused libraries that can target specific pathways, such as kinases, or biologic processes such as autophagy. Other vendors (e.g., Maybridge) provide libraries such as the HitFinder Library whereby compounds are selected to represent the overall diversity of the screening collection using a clustering algorithm. Such design libraries may shorten the drug discovery process and save resources

Table 2 Available Libraries and Vendors for Drug Screening

Vendors	Library Names	Library Size	Comments	Web Sites
Evotec (NIH Molecular Libraries Small Molecule Repository)	BioFocus NCC	446	NIH collection of FDA-approved drugs that have a history of use in human clinical trials	http://nihsmr.evotec.com/evotec/
Prestwick Chemical	Prestwick	1280	100% approved drugs (FDA, EMA, and other agencies), high chemical and pharmacological diversity, known bioavailability and safety in humans	http://www.prestwickchemical.com/index.html
Sigma-Aldrich	LOPAC (Library of Pharmacologically Active Compounds)	1280	A collection of molecules that span a broad range of cell signaling and neuroscience areas	https://www.sigmaaldrich.com/united-states.html
MicroSource	Spectrum	2400	Drugs that reached clinical trial stages; known biologically active compounds; natural products	http://www.msdiscovery.com/
Boston and Kansas University	Chemical Methodology Library Development (CMLD)	3000	Boston and Kansas University Collection	http://cmd.bu.edu/ https://hts.ku.edu/resources/compounds
Enzo Life Sciences	Multiple	4000	Toxicity (487), drug repurposing (786), pathway targeting (215), chemical genomics (801), receptor deorphaning (1134), natural products (502)	http://www.enzolifesciences.com/

ChemDiv	Multiple	1.5×10^6	Targeted diversity Libraries; Focus Libraries; Design Libraries	http://www.chemdiv.com/
Chembridge	Multiple	1×10^6	Diversity libraries(preselected diversity libraries) Targeted and focused libraries (kinase, GPCR, ion channel, CNS, and nuclear receptor libraries)	http://www.chembridge.com/index.php
Maybridge	Multiple	53,000	Hitfinder (representatives of the overall diversity of the screening collection); HitDiscover (the entire screening collection)	http://www.maybridge.com/

and time. Another cost-saving measure is multiplexing the screening such that up three or four compounds may be tested per well.

6. PEARLS AND PITFALLS

There are several caveats to keep in mind:

a. The design of the assay used for drug screening is key predictor of success. What is described in this chapter is a method to screen for compounds that normalize major organizational alterations in IFs. The hypothesis is that normalization of such severe IF disorganization will ameliorate or even cure the underlying IF-pathy that is caused by the mutant IF and its consequent filament alterations.

b. One limitation of the approach we describe is that it is not designed for more subtle IF mutations that can still cause significant patient morbidity and can lead to many IF-pathies. Assays with alternate selective readouts would need to be tailored for such situations.

c. One potential benefit of an unbiased IF-related drug screening approach is that it could illuminate novel signaling pathways or interacting proteins, as was the case for identifying NMHC-II binding to keratins (Kwan et al., 2015).

d. Once a hit compound is identified and validated in cell culture, it is important to use a systematic approach to define its mechanism of action, initially in cell culture, and then determine whether it is likely to work *in vivo*. This is the aspect that requires the most effort because of multiple potential mechanisms that may be involved. Also, some drugs may work in cell culture but may not be effective in animal studies for a number of reasons (e.g., pharmacodynamics, tissue/cell penetration), or may also impart protection independent of an IF mutation.

e. One potential concern is whether a given screen will result in too many potential compounds to pursue. However, this has not been an issue for us and we are presently working on only two additional compounds (aside from PKC412) after screening several libraries totaling more than 3000 compounds and ending up with nearly 15 potential compounds that were then narrowed down to two. For example, triaging generally includes (i) eliminating compounds that are known to be toxic or to have chemically reactive functional groups, (ii) selecting compounds that appear to be most effective in normalizing the mutant filament disruption phenotype, and (iii) selecting compounds that have a single or at most a limited number of known functions.

ACKNOWLEDGMENTS

Our work is supported by US National Institutes of Health (NIH) grants DK47918 and DK52951, and the Department of Veterans Affairs (M.B.O.); and NIH institutional grant DK34933 to the University of Michigan. We thank Dr. Martha Larsen, Nicholas Santoro, and Steve Swaney for their tremendous assistance with the high-throughput drug screening. We also thank Dr. Paula Gedraitis from Molecular Devices for her excellent technical support.

REFERENCES

Abe, M., & Oshima, R. G. (1990). A single human keratin 18 gene is expressed in diverse epithelial cells of transgenic mice. *The Journal of Cell Biology*, *111*, 1197–1206.

Allen, E. H., Atkinson, S. D., Liao, H., Moore, J. E., Leslie Pedrioli, D. M., Smith, F. J., et al. (2013). Allele-specific siRNA silencing for the common keratin 12 founder mutation in Meesmann epithelial corneal dystrophy. *Investigative Ophthalmology & Visual Science*, *54*, 494–502.

Bonifas, J. M., Rothman, A. L., & Epstein, E. H., Jr. (1991). Epidermolysis bullosa simplex: Evidence in two families for keratin gene abnormalities. *Science (New York)*, *254*, 1202–1205.

Brenner, M., Johnson, A. B., Boespflug-Tanguy, O., Rodriguez, D., Goldman, J. E., & Messing, A. (2001). Mutations in GFAP, encoding glial fibrillary acidic protein, are associated with Alexander disease. *Nature Genetics*, *27*, 117–120.

Capell, B. C., Olive, M., Erdos, M. R., Cao, K., Faddah, D. A., Tavarez, U. L., et al. (2008). A farnesyltransferase inhibitor prevents both the onset and late progression of cardiovascular disease in a progeria mouse model. *Proceedings of the National Academy of Sciences of the United States of America*, *105*, 15902–15907.

Coulombe, P. A., Hutton, M. E., Letai, A., Hebert, A., Paller, A. S., & Fuchs, E. (1991). Point mutations in human keratin 14 genes of epidermolysis bullosa simplex patients: Genetic and functional analyses. *Cell*, *66*, 1301–1311.

Coulombe, P. A., Kerns, M. L., & Fuchs, E. (2009). Epidermolysis bullosa simplex: A paradigm for disorders of tissue fragility. *The Journal of Clinical Investigation*, *119*, 1784–1793.

Fuchs, E., & Weber, K. (1994). Intermediate filaments: Structure, dynamics, function, and disease. *Annual Review of Biochemistry*, *63*, 345–382.

Gentil, B. J., Mushynski, W. E., & Durham, H. D. (2013). Heterogeneity in the properties of NEFL mutants causing Charcot-Marie-Tooth disease results in differential effects on neurofilament assembly and susceptibility to intervention by the chaperone-inducer, celastrol. *The International Journal of Biochemistry & Cell Biology*, *45*, 1499–1508.

Gonzalez, M. E. (2013). Evaluation and treatment of the newborn with epidermolysis bullosa. *Seminars in Perinatology*, *37*, 32–39.

Herrmann, H., Kreplak, L., & Aebi, U. (2004). Isolation, characterization, and in vitro assembly of intermediate filaments. *Methods in Cell Biology*, *78*, 3–24.

Hickerson, R. P., Speaker, T. J., Lara, M. F., Gonzalez-Gonzalez, E., Flores, M. A., Contag, C. H., et al. (2015). Non-invasive intravital imaging of siRNA-mediated mutant keratin gene repression in skin. *Molecular Imaging and Biology*, 1–9.

Hutton, E., Paladini, R. D., Yu, Q. C., Yen, M., Coulombe, P. A., & Fuchs, E. (1998). Functional differences between keratins of stratified and simple epithelia. *The Journal of Cell Biology*, *143*, 487–499.

Kerns, M. L., DePianto, D., Dinkova-Kostova, A. T., Talalay, P., & Coulombe, P. A. (2007). Reprogramming of keratin biosynthesis by sulforaphane restores skin integrity

in epidermolysis bullosa simplex. *Proceedings of the National Academy of Sciences of the United States of America, 104*, 14460–14465.
Kerns, M., DePianto, D., Yamamoto, M., & Coulombe, P. A. (2010). Differential modulation of keratin expression by sulforaphane occurs via Nrf2-dependent and -independent pathways in skin epithelia. *Molecular Biology of the Cell, 21*, 4068–4075.
Kirfel, J., Peters, B., Grund, C., Reifenberg, K., & Magin, T. M. (2002). Ectopic expression of desmin in the epidermis of transgenic mice permits development of a normal epidermis. *Differentiation; Research in Biological Diversity, 70*, 56–68.
Ku, N. O., Gish, R., Wright, T. L., & Omary, M. B. (2001). Keratin 8 mutations in patients with cryptogenic liver disease. *The New England Journal of Medicine, 344*, 1580–1587.
Ku, N. O., Michie, S., Oshima, R. G., & Omary, M. B. (1995). Chronic hepatitis, hepatocyte fragility, and increased soluble phosphoglycokeratins in transgenic mice expressing a keratin 18 conserved arginine mutant. *The Journal of Cell Biology, 131*, 1303–1314.
Ku, N. O., Strnad, P., Zhong, B. H., Tao, G. Z., & Omary, M. B. (2007). Keratins let liver live: Mutations predispose to liver disease and crosslinking generates Mallory–Denk bodies. *Hepatology, 46*, 1639–1649.
Ku, N. O., Zhou, X., Toivola, D. M., & Omary, M. B. (1999). The cytoskeleton of digestive epithelia in health and disease. *The American Journal of Physiology, 277*, G1108–G1137.
Kwan, R., Chen, L., Looi, K., Tao, G. Z., Weerasinghe, S. V., Snider, N. T., et al. (2015). PKC412 normalizes mutation-related keratin filament disruption and hepatic injury in mice by promoting keratin-myosin binding. *Hepatology, 62*, 1858–1869.
Lane, E. B., & McLean, W. H. (2004). Keratins and skin disorders. *The Journal of Pathology, 204*, 355–366.
LaPash Daniels, C. M., Austin, E. V., Rockney, D. E., Jacka, E. M., Hagemann, T. L., Johnson, D. A., et al. (2012). Beneficial effects of Nrf2 overexpression in a mouse model of Alexander disease. *The Journal of Neuroscience, 32*, 10507–10515.
Leslie Pedrioli, D. M., Fu, D. J., Gonzalez-Gonzalez, E., Contag, C. H., Kaspar, R. L., Smith, F. J., et al. (2012). Generic and personalized RNAi-based therapeutics for a dominant-negative epidermal fragility disorder. *The Journal of Investigative Dermatology, 132*, 1627–1635.
Lu, H., Chen, J., Planko, L., Zigrino, P., Klein-Hitpass, L., & Magin, T. M. (2007). Induction of inflammatory cytokines by a keratin mutation and their repression by a small molecule in a mouse model for EBS. *The Journal of Investigative Dermatology, 127*, 2781–2789.
Lu, J. T., Muchir, A., Nagy, P. L., & Worman, H. J. (2011). LMNA cardiomyopathy: Cell biology and genetics meet clinical medicine. *Disease Models & Mechanisms, 4*, 562–568.
Mignot, C., Delarasse, C., Escaich, S., Della Gaspera, B., Noe, E., Colucci-Guyon, E., et al. (2007). Dynamics of mutated GFAP aggregates revealed by real-time imaging of an astrocyte model of Alexander disease. *Experimental Cell Research, 313*, 2766–2779.
Mohan, R., & Bargagna-Mohan, P. (2016). The use of withaferin A to study intermediate filaments. *Methods in Enzymology, 568*, 187–218.
Morley, S. M., D'Alessandro, M., Sexton, C., Rugg, E. L., Navsaria, H., Shemanko, C. S., et al. (2003). Generation and characterization of epidermolysis bullosa simplex cell lines: Scratch assays show faster migration with disruptive keratin mutations. *The British Journal of Dermatology, 149*, 46–58.
Muchir, A., Reilly, S. A., Wu, W., Iwata, S., Homma, S., Bonne, G., et al. (2012). Treatment with selumetinib preserves cardiac function and improves survival in cardiomyopathy caused by mutation in the lamin A/C gene. *Cardiovascular Research, 93*, 311–319.
Muchir, A., & Worman, H. J. (2016). Targeting mitogen-activated protein kinase signaling in mouse models of cardiomyopathy caused by lamin A/C gene mutations. *Methods in Enzymology, 568*, 557–580.

Omary, M. B. (2009). "IF-pathies": A broad spectrum of intermediate filament-associated diseases. *The Journal of Clinical Investigation, 119*, 1756–1762.

Omary, M. B., Coulombe, P. A., & McLean, W. H. (2004). Intermediate filament proteins and their associated diseases. *The New England Journal of Medicine, 351*, 2087–2100.

Osmanagic-Myers, S., Dechat, T., & Foisner, R. (2015). Lamins at the crossroads of mechanosignaling. *Genes & Development, 29*, 225–237.

Ostlund, C., Bonne, G., Schwartz, K., & Worman, H. J. (2001). Properties of lamin A mutants found in Emery-Dreifuss muscular dystrophy, cardiomyopathy and Dunnigan-type partial lipodystrophy. *Journal of Cell Science, 114*, 4435–4445.

Rohena, C. C., & Mooberry, S. L. (2014). Recent progress with microtubule stabilizers: New compounds, binding modes and cellular activities. *Natural Product Reports, 31*, 335–355.

Schreiber, K. H., & Kennedy, B. K. (2013). When lamins go bad: Nuclear structure and disease. *Cell, 152*, 1365–1375.

Schweizer, J., Bowden, P. E., Coulombe, P. A., Langbein, L., Lane, E. B., Magin, T. M., et al. (2006). New consensus nomenclature for mammalian keratins. *The Journal of Cell Biology, 174*, 169–174.

Snider, N. T., & Omary, M. B. (2016). Assays for posttranslational modifications of intermediate filaments proteins. *Methods in Enzymology, 568*, 113–138.

Straight, A. F. (2007). Fluorescent protein applications in microscopy. *Methods in Cell Biology, 81*, 93–113.

Strnad, P., Tao, G. Z., Zhou, Q., Harada, M., Toivola, D. M., Brunt, E. M., et al. (2008). Keratin mutation predisposes to mouse liver fibrosis and unmasks differential effects of the carbon tetrachloride and thioacetamide models. *Gastroenterology, 134*, 1169–1179.

Strnad, P., Zhou, Q., Hanada, S., Lazzeroni, L. C., Zhong, B. H., So, P., et al. (2010). Keratin variants predispose to acute liver failure and adverse outcome: Race and ethnic associations. *Gastroenterology, 139*, 828–835. 835 e821–823.

Szeverenyi, I., Cassidy, A. J., Chung, C. W., Lee, B. T., Common, J. E., Ogg, S. C., et al. (2008). The human intermediate filament database: Comprehensive information on a gene family involved in many human diseases. *Human Mutation, 29*, 351–360.

Usachov, V., Urban, T. J., Fontana, R. J., Gross, A., Iyer, S., Omary, M. B., et al. (2015). Prevalence of genetic variants of keratins 8 and 18 in patients with drug-induced liver injury. *BMC Medicine, 13*, 196.

Wang, L., Colodner, K. J., & Feany, M. B. (2011). Protein misfolding and oxidative stress promote glial-mediated neurodegeneration in an Alexander disease model. *The Journal of Neuroscience, 31*, 2868–2877.

Wang, D. S., Miller, R., Shaw, R., & Shaw, G. (1996). The pleckstrin homology domain of human beta I sigma II spectrin is targeted to the plasma membrane in vivo. *Biochemical and Biophysical Research Communications, 225*, 420–426.

Worman, H. J., & Schirmer, E. C. (2015). Nuclear membrane diversity: Underlying tissue-specific pathologies in disease? *Current Opinion in Cell Biology, 34*, 101–112.

Wu, W., Shan, J., Bonne, G., Worman, H. J., & Muchir, A. (2010). Pharmacological inhibition of c-Jun N-terminal kinase signaling prevents cardiomyopathy caused by mutation in LMNA gene. *Biochimica et Biophysica Acta, 1802*, 632–638.

Yang, S. H., Meta, M., Qiao, X., Frost, D., Bauch, J., Coffinier, C., et al. (2006). A farnesyltransferase inhibitor improves disease phenotypes in mice with a Hutchinson-Gilford progeria syndrome mutation. *The Journal of Clinical Investigation, 116*, 2115–2121.

CHAPTER EIGHT

The Use of Withaferin A to Study Intermediate Filaments

Royce Mohan[1], Paola Bargagna-Mohan

Department of Neuroscience, University of Connecticut Health Center, Farmington, Connecticut, USA
[1]Corresponding author: e-mail address: mohan@uchc.edu

Contents

1. Introduction 188
2. WFA: A Novel Chemical Tool for IF Biology and Pharmacology 194
   2.1 WFA Properties and Uses in Cell Culture 195
   2.2 WFA Formulation for Animal Studies 195
   2.3 WFA Distribution in Rabbit Eyes 196
   2.4 WFA Measurement in Mice 197
3. Targeting IFs 197
   3.1 Targeting IFs *In Vitro* 197
   3.2 Targeting IFs *In Vivo* 205
4. Primary Cell Culture Systems 207
   4.1 Primary Cell Cultures from Ocular Tissues 207
5. Ocular Injury Model of Fibrosis 209
   5.1 Alkali Burn Corneal Injury in Mouse 209
   5.2 Tissue Harvesting 210
6. Techniques 211
   6.1 Immunohistochemistry Analysis 211
   6.2 Western Blot Analysis (Soluble and Insoluble Extracts) 212
7. Perspectives 213
Acknowledgments 214
References 214

Abstract

Withaferin A (WFA), initially identified as a compound that inhibits experimental angiogenesis, has been shown to bind to soluble vimentin (sVim) and other type III intermediate filament (IF) proteins. We review WFA's dose-related activities (Section 1), examining nanomolar concentrations effects on sVim in cell proliferation and submicromolar effects on lamellipodia and focal adhesion formation. WFA effects on polymeric IFs are especially interesting to the study of cell migration and invasion that depend on IF mechanical contractile properties. WFA interferes with NF-κB signaling, though this anti-inflammatory mechanism may occur via perturbation of sVim–protein complexes, and possibly also via targeting IκB kinase β directly. However, micromolar

Methods in Enzymology, Volume 568
ISSN 0076-6879
http://dx.doi.org/10.1016/bs.mie.2015.09.025

concentrations that induce vimentin cleavage to promote apoptosis may increasingly show off-target effects via targeting other IFs (neurofilaments and keratin) and non-IFs (tubulin, heat-shock proteins, proteasome). Thus, in Section 2, we describe our studies combining cell cultures with animal models of injury to validate relevant type III IF-targeting mechanisms of WFA. In Section 3, we illuminate from investigating myofibroblast differentiation how sVim phosphorylation may govern cell type-selective sensitivity to WFA, offering impetus for exploring vimentin phosphorylation isoforms as targets and biomarkers of fibrosis. These different WFA targets and activities are listed in a summary table.

ABBREVIATIONS

DAPI 4′,6-diamidino-2-phenylindole
DMSO dimethyl sulfoxide
epox epoxomicin
ERK extracellular signal-regulated kinase
IκB-α I kappa B alpha
IκB kinase β I kappa B kinase beta subunit
IFs intermediate filaments
i.p. intraperitoneal
NF-κB nuclear factor kappa B
PBS phosphate-buffered saline
PFA paraformaldehyde
pSer38Vim serine 38 phosphorylated vimentin
RbCF2 rabbit corneal fibroblasts passage 2 cells
RbCF8 rabbit corneal fibroblasts passage 8 cells
RbCFs rabbit corneal fibroblasts
RbTCFs rabbit Tenon's capsule fibroblasts
sVim soluble vimentin
TGF-β transforming growth factor beta
TNF-α tumor necrosis factor alpha
Vim KO vimentin knockout
WFA withaferin A
WT wild type
α-SMA alpha-smooth muscle actin

1. INTRODUCTION

Chemical genetic approaches have borne out many useful biologically active natural products (Newman & Cragg, 2012) that serve as useful probes of protein function (Crews & Mohan, 2000; Crews & Splittgerber, 1999). To identify novel inhibitors of angiogenesis, we investigated the medicinal plant *Withania somnifera* used in Ayurveda for treatment for rheumatoid arthritis and endometriosis (Mishra, Singh, & Dagenais, 2000), which led to our unraveling of the low-dose antiangiogenic mechanism of natural product withaferin A (WFA; Fig. 1A) (Mohan et al., 2004). The unique reactive pharmacophore warhead (α,β-unsaturated ketone and epoxide on the conjugated A and B rings) predicts that WFA covalently modifies its intracellular protein target(s). Thus, we developed a cell-permeable biotinylated WFA affinity probe (Yokota, Bargagna-Mohan, Ravindranath, Kim, & Mohan, 2006) that afforded us the identification of soluble vimentin (sVim) as WFA's binding target in endothelial cells (Bargagna-Mohan et al., 2007). WFA forms a covalent bond with the unique cysteine residue in the 2B rod domain of vimentin (Bargagna-Mohan et al., 2007). Using X-ray crystal structure information (Strelkov et al., 2001), a three-dimensional molecular model of the part of tetrameric vimentin's 2B domain bound to WFA along with molecular dynamics simulation was employed to show that WFA stably occupies a novel surface-binding pocket formed between two antiparallel dimers (Fig. 1B). WFA likely employs a two-step binding process to target sVim (Bargagna-Mohan et al., 2013). In the first step, both hydrogen bonding and hydrophobic interactions help to stabilize WFA within the binding pocket (Bargagna-Mohan et al., 2007). In the second step, held in this favorable pose, the reactive epoxide of WFA undergoes a stereospecific

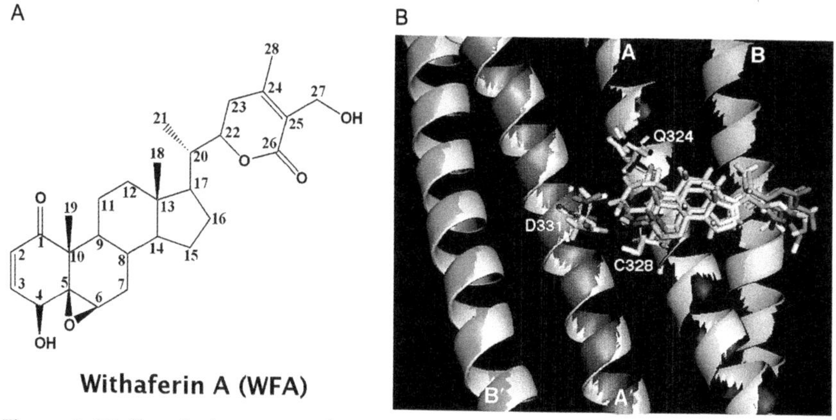

Figure 1 (A) Chemical structure of WFA. (B) Molecular model of the WFA-binding site in the 2B rod domain of tetrameric type III IFs. Shown is a multicolor composite overlap of WFA docked individually in the binding pockets of vimentin, GFAP, and desmin protein structures, respectively, that identifies a similar binding mode for WFA in these IFs. The key amino acid residues cysteine (C328), aspartic acid (D331), and glutamine (Q324) are numbered with respect to vimentin, for simplicity (Bargagna-Mohan et al., 2010, 2012). *Reproduced with permission as a modified version from Bargagna-Mohan et al. (2007).* (See the color plate.)

nucleophilic attack by the thiol group of cysteine to form the covalent bond. This covalent modification of sVim by WFA accounts for its potency, because a cysteine-mutant vimentin is perturbed by WFA only at 9 μ*M* (Grin et al., 2012), which is significantly higher than the dose (1–2 μ*M*) normally required for most cells (Bargagna-Mohan et al., 2013, 2007; Lahat et al., 2010; Thaiparambil et al., 2011). Moreover, the cysteine residue through zinc coordination promotes vimentin filament extension and subunit exchange (Perez-Sala et al., 2015), and hence, electrophilic prostaglandins that target sVim (Gharbi, Garzon, Gayarre, Timms, & Perez-Sala, 2007) also show reduced potency for perturbation of cysteine-mutated vimentin (Perez-Sala et al., 2015). Thus, cysteine modification appears to be a common mechanism of several different vimentin-binding ligands (Bargagna-Mohan et al., 2007; Burikhanov et al., 2014; Perez-Sala et al., 2015).

WFA also exhibits NF-κB inhibitory activity (IC_{50} 500 n*M*) by blocking NF-κB nuclear translocation through inhibiting IκB-α degradation, a mechanism accounting for its inhibition of TNF-α secretion *in vivo* (Bargagna-Mohan, Ravindranath, & Mohan, 2006; Mohan et al., 2004). NF-κB regulation is vimentin dependent as shown in an injury-induced corneal fibrosis model. Healing corneas of vimentin knockout (Vim KO) mice, similar to wild-type (WT) mice treated with WFA having downregulated vimentin, show recovered IκB-α expression levels and attenuated nuclear staining for p65 RelA (Bargagna-Mohan et al., 2012). Thus, WFA's inhibition of NF-κB occurs, in part, through perturbation of sVim (Che et al., 2011), which is localized in multiprotein complexes containing NF-κB, IκB-α, and tubulin (Chi, Bo, Wu, Jong, & Huang, 2012). NF-κB also activates vimentin transcription (Chi et al., 2012), which would explain WFA's downregulation of vimentin expression. This would be relevant for chronic inflammatory disease or injury paradigms (Bargagna-Mohan et al., 2012; Lahat et al., 2010; Lee, Hahm, Marcus, & Singh, 2013). WFA is also being tested in animal models of amyotrophic lateral sclerosis (Patel, Julien, & Kriz, 2015), but it is not known whether its NF-κB-antagonistic activity occurs independent of type III intermediate filaments (IFs), or alternatively, related to WFA's binding to IκB kinase β (Heyninck, Lahtela-Kakkonen, Van der Veken, Haegeman, & Vanden Berghe, 2014). Taken together, we posit that WFA's antagonism of a central chaperonic function of sVim that links to multiple stress-regulated kinase signaling pathways, as recently shown with pERK (Bargagna-Mohan et al., 2015; Kim et al., 2010; Kumar et al., 2007; Perez-Sala et al., 2015; Perlson et al., 2006), may be one of its principal inhibitory mechanisms by which low-dose WFA (see Table 1) blunts IF-associated signaling axes *in vivo*.

Table 1 Summary of WFA Targets and Activities

Target	Cellular Assay (WFA Effective Dose)	Phenotype	Comments
sVim	Proliferation of endothelial cells (IC_{50}: 12 n*M*; Mohan et al., 2004), astrocytes, and fibroblasts (IC_{50}: 25 n*M*; Bargagna-Mohan et al., 2010, 2012)	G_0/G_1 or G_2/M growth arrest	sVim downregulation; decreased cyclin D_1 in endothelial cells; decreased cyclin D_3, PCNA, and Skp2; and increased $p27^{Kip1}$ in astrocytes and fibroblasts
sVim	Proliferation of cancer cell lines (IC_{50}: 200–500 n*M*; Mohan et al., 2004)	G_2/M growth arrest	sVim downregulation
sVim	Cell spreading (250 n*M* to 1 μ*M*; Bargagna-Mohan et al., 2015)	Attenuation of cytoplasmic growth	Phosphorylation of sVim at serine 38 (pSer38Vim) with delayed formation of lamellipodia and focal adhesions; blockade of pERK nuclear localization
Vimentin IFs	Cell migration and invasion (250 n*M* to 2 μ*M*; Bargagna-Mohan et al., 2013, 2015, 2006; Mohan et al., 2004; Thaiparambil et al., 2011)	Loss of cell polarity and inhibition of cell migration and endothelial sprouting	Vimentin IF retraction, perinuclear condensation, and deploymerization into dots and squiggles
Vimentin IFs	Collagen contraction (500 n*M* to 1 μ*M*; Bargagna-Mohan et al., 2013, 2015)	Inhibition of cell contractile function	Differences between corneal fibroblasts and myofibroblasts due to pSer38Vim phosphorylation
sVim	Acute TNF-α induction in endothelial cells (IC_{50}: 500 n*M*; Mohan et al., 2004)	Anti-inflammatory	IκB-α degradation prevented with loss of NF-κB nuclear entry; possible mechanism is

Continued

Table 1 Summary of WFA Targets and Activities—cont'd

Target	Cellular Assay (WFA Effective Dose)	Phenotype	Comments
			perturbation of multiprotein complex containing sVim–IκB-α–NF-κB
sVim and sGFAP	Chronic TNF-α stimulation in astrocytes (IC_{50}: 200 n*M*; Bargagna-Mohan et al., 2010)	Anti-inflammatory	sVim and sGFAP downregulation; sVim is more sensitive than sGFAP
Vimentin IFs	Apoptosis (above 2 μ*M*; Bargagna-Mohan et al., 2007; Grin et al., 2012; Mohan et al., 2004)	Gradual loss of cell adhesion due to severity of effect on cytoskeleton and cell rounding	Vimentin is fragmented into sizes known to be cleavage products of caspases and blocked by inhibition of caspase activity. F-actin is found complexed with vimentin dots
Vimentin and GFAP	Ocular injury (2 mg/kg/day for 7 days; Bargagna-Mohan et al., 2010; Paranthan, Bargagna-Mohan, Lau, & Mohan, 2011)	Inhibition of retinal gliosis	sVim and sGFAP downregulation and fragmentation of GFAP filaments in Muller glia; cyclin D3 and PCNA downregulation; and $p27^{Kip1}$ increase
Vimentin and desmin	Ocular injury (2 mg/kg/day for 14 days; Bargagna-Mohan et al., 2013, 2012)	Inhibition of fibrosis in cornea and Tenon's capsule	Downregulation of sVim, sDesmin, vimentin IFs, and desmin IFs
Vimentin	Ocular injury (2 mg/kg/day for 7 and 14 days; Bargagna-Mohan et al., 2012)	Anti-inflammatory activity in the cornea	Restoration of injury-induced degradation of IκB-α with downregulation of vimentin
Peripherin	Filament targeting (above 1 μ*M*; Grin et al., 2012)	Filament retraction	Conserved WFA-binding site in type III IFs

Table 1 Summary of WFA Targets and Activities—cont'd

Target	Cellular Assay (WFA Effective Dose)	Phenotype	Comments
β-Tubulin	Proliferation in breast cancer cells (2 μ*M*; Antony et al., 2014)	Inhibition of cell growth	Binding to cysteine residue of β-tubulin
IκB kinase β	Binding activity in HEK293T cells (Heyninck et al., 2014)	Anti-inflammatory activity	Binding to cysteine residue of IκB kinase β
Neurofilaments	Filament targeting (above 4 μ*M*; Grin et al., 2012)	Filament retraction	Binding site not known
Keratin	Filament targeting (above 6 μ*M*; Grin et al., 2012)	Filament retraction	Binding site not known

Molecular models of glial fibrillary acidic protein (GFAP) (Bargagna-Mohan et al., 2010) and desmin tetramers (Bargagna-Mohan et al., 2012) reveal an evolutionary conservation of the WFA-binding site among all type III IFs. Given the overlapping roles for vimentin and GFAP in astrocytes and Muller glia, and similarly, vimentin and desmin in myofibroblasts, WFA effectively targets individual and copolymerized IF partners and down-regulates their expression during reactive gliosis and fibrosis (Bargagna-Mohan et al., 2010, 2012; Paranthan et al., 2011). However, WFA's selectivity among the type III IFs cannot be gleaned based on molecular modeling. Considering that IFs are dynamic and show heterogeneous iso-form expression (see other chapters), we believe that important aspects, such as biological context-dependent regulation of soluble IFs and their phos-phorylation status, are critical aspects of cell type-specific and differentiation-dependent sensitivities to WFA (Bargagna-Mohan et al., 2015; Perez-Sala et al., 2015; Robert, Rossow, Hookway, Adam, & Gelfand, 2015). Importantly, cytotoxic concentrations (above 2 μ*M*) per-turb also other classes of IFs (neurofilaments and keratin) (Grin et al., 2012) and non-IF proteins (Vanden Berghe, Sabbe, Kaileh, Haegeman, & Heyninck, 2012) (see Table 1). Thus, interrogating the methods employed to discover WFA's binding targets, such the proteasome and annexin II, reveals how differences in WFA–biotin probe construction and use in cell extracts versus *in vivo* and the WFA dose can lead to quite different results (Leslie & Hergenrother, 2008). We ruled out annexin II (Falsey et al., 2006) as a WFA-binding protein in the cornea (Bargagna-Mohan et al., 2012), and further showed that both actin and tubulin are reversibly coisolated with vimentin in WFA–biotin affinity purifications (Bargagna-Mohan et al., 2007). Several of WFA's claimed binding targets (reviewed in Vanden Berghe et al., 2012) represent components of cellular stress, where high doses of WFA that induce protein aggregates contain vimentin, heat-shock proteins, actin, plectin, and tubulin that can also become oxidatively cross-linked (Perez-Sala et al., 2015). Hence, while β-tubulin is targeted by WFA at 2 μM in cancer cells, whether this occurs independent of vimentin was not reported (Antony et al., 2014). Multiple mechanisms govern early steps of vimentin IF assembly (Grin et al., 2012; Robert et al., 2015) and, as such, could be differentially controlled in a cell type-specific manner affected by WFA. We believe such mechanisms may govern selective effects of WFA on malignant cells over fibroblasts (Nishikawa et al., 2015). Thus, vimentin represents a class of target that has remained undruggable for several decades, despite the numerous studies illuminating the disease-related roles of IFs (Eriksson et al., 2009; Kidd, Shumaker, & Ridge, 2014). Our discovery has provided an unprecedented opportunity to exploit WFA as a pharmacological probe of type III IFs in cell biology and disease (Bargagna-Mohan et al., 2007, 2015, 2010, 2012; Gladilin, Gonzalez, & Eils, 2014; Grin et al., 2012; Lahat et al., 2010; Lynch, Lazar, Iskratsch, Zhang, & Sheetz, 2013; Menko et al., 2014; Miyatake, Sheehy, Ikeshita, Hall, & Kasahara, 2015; Thaiparambil et al., 2011). In this chapter, we will describe some of the salient findings we have made with the use of WFA to investigate cell behavior that are tied to critical IF functions (see also Table 1 for summary of WFA activity on IFs and non-IFs).

2. WFA: A NOVEL CHEMICAL TOOL FOR IF BIOLOGY AND PHARMACOLOGY

Since the discovery of WFA as a potent angiogenesis inhibitor in our laboratory (Mohan et al., 2004), our major focus has centered on the relevance of WFA's low-dose pharmacological activities. This direction has guided our investigations to exploit WFA as an inhibitor of angiogenesis (Bargagna-Mohan et al., 2007), gliosis (Bargagna-Mohan et al., 2010), and fibrosis (Bargagna-Mohan et al., 2013, 2015, 2012). In this chapter, we will focus on WFA's antifibrotic activities.

2.1 WFA Properties and Uses in Cell Culture

WFA is available from several commercial vendors. We purchase ours from Chromadex (Irvine, CA) as dry powder and store it at −20 °C. Stock solutions prepared in dimethyl sulfoxide (DMSO) at 50 μ*M* are stored frozen in small amounts (5–10 μl) to be typically used within 1 year. Dilutions are prepared in cell culture medium on the day of use keeping the final concentration of DMSO under 0.2%. WFA's effects on vimentin show cell-type selectivity that is dose related. At low concentrations (under 250 n*M*), WFA affects sVim but spares the filamentous forms of vimentin (Bargagna-Mohan et al., 2013, 2007). WFA exerts antiproliferative activity with endothelial cells being highly sensitive (IC_{50} 12 n*M*) (Mohan et al., 2004). This cytostatic activity is due to downregulation of sVim with G_0/G_1 arrest being demonstrated in endothelial cells and astrocytes (Bargagna-Mohan et al., 2010; Mohan et al., 2004) or G_2/M in fibroblasts (Bargagna-Mohan et al., 2013, 2012). At concentrations between 250 n*M* and 1 μ*M*, increased serine 38 phosphorylation of vimentin (pSer38Vim) causes the retraction of vimentin IFs from cell membrane and its juxtanuclear localization, producing pSer38Vim particles and squiggles within 1–3 h (Grin et al., 2012). Interestingly, corneal myofibroblasts contain higher pSer38Vim levels in the soluble pool compared to fibroblasts. When treated with WFA, pSer38Vim becomes hyperphosphorylated, possibly by WFA causing a structural perturbation of sVim to activate a kinase to further phosphorylate pSer38Vim (see Section 3.1.3). As a consequence of hyperphosphorylation, these soluble precursors do not become incorporated into vimentin IFs in myofibroblasts, and focal adhesion formation is affected to a greater extent than in fibroblasts similarly treated with WFA (Bargagna-Mohan et al., 2015). While effects of WFA up to 500 n*M* on vimentin IFs are reversible, the continued incubation with this inhibitor can induce cell senescence (Grin et al., 2012). Concentrations at and above 1 μ*M* WFA show dose-dependent effects on cytoskeletal vimentin IFs causing perinuclear condensation (Bargagna-Mohan et al., 2007). Exceeding 2 μ*M*, WFA causes apoptosis and secondary vimentin cleavage within 2–4 h (Bargagna-Mohan et al., 2013, 2007; Grin et al., 2012; Lahat et al., 2010).

2.2 WFA Formulation for Animal Studies

For mouse studies, an *in vivo* delivery dose between 2 and 4 mg/kg WFA has been quite widely employed (Bargagna-Mohan et al., 2007; Lahat et al.,

2010; Nagalingam, Kuppusamy, Singh, Sharma, & Saxena, 2014; Thaiparambil et al., 2011). WFA is prepared in DMSO in advance for the entire course of injections (7-, 14-, or 21-day treatment period) and aliquots kept frozen at −20 °C. Each day a vial is thawed for drug injection. No toxicity has been reported with this treatment regimen (Bargagna-Mohan et al., 2007, 2010, 2012), but formal studies need to be performed. For rabbit ocular subconjunctival injections, we employed a suspension of WFA in saline. This method of delivery was chosen to study the efficacy of WFA to attenuate fibrosis in Tenon's capsule, which occurs as a side effect in humans who have undergone trabeculectomy to enhance outflow of aqueous humor in glaucoma (Cordeiro, Siriwardena, Chang, & Khaw, 2000). For this application, 50 m*M* WFA stock in DMSO was diluted into preservative-free sterile 0.9% sodium chloride solution (Hospira, Inc., Lake Forest, IL) to final concentrations of either 15 or 150 μ*M*. These diluted suspensions of WFA were prepared fresh on each day of injection and a volume of 100 μl provided by subconjunctival injection on day of surgery, 4 days postsurgery, and followed by weekly schedule (5 injections total). As controls, the contralateral eye of each rabbit was injected with vehicle solution (Bargagna-Mohan et al., 2013).

2.3 WFA Distribution in Rabbit Eyes

In experiments conducted in rabbits (see Section 2.2), we also measured WFA concentrations in different eye tissues. Using the lower-dose treatment group, enucleated eyes were dissected in the anterior aspects (cornea, aqueous humor, iris, conjunctiva) and tissues were thoroughly homogenized in buffer and extracted with ethyl acetate. Solvent extracts were dried down and suspended in methanol for quantification by liquid chromatography mass spectrometry as described (Bargagna-Mohan et al., 2013). These studies revealed that the conjunctiva retained the highest levels of WFA (28.8 n*M*), with decreasing levels in iris–ciliary body (18.6 n*M*), cornea (12.9 n*M*), and aqueous humor (7.6 n*M*), revealing antiproliferative concentrations being retained in the eye (Bargagna-Mohan et al., 2013). This study also identified that the lower WFA dose caused sVim levels to be downregulated but filamentous vimentin staining was not disrupted in Tenon's capsule, whereas the higher dose caused both sVim downregulation and filamentous vimentin to be disrupted, which was consistent with our cell culture dose–response studies in Tenon's capsule fibroblasts (TCFs) (Bargagna-Mohan et al., 2013). These data provide a useful guide for ocular delivery of WFA to the anterior segment of the eye and illuminate distinct dose effects of WFA. Remarkably, there was no toxicity in this WFA delivery paradigm for either dose. Further pharmacokinetic studies need to be performed to assess the ocular peak concentrations, half-life, and the dose-ranging toxicity.

2.4 WFA Measurement in Mice

Systemic delivery of WFA is typically employed for mice. Using a 10 mg/kg intraperitoneal (i.p.) injection, we showed that the circulating levels of WFA in serum after 6 h are 3 μg/ml. From these data, we estimated that the five-fold lower dose (2 mg/kg) we employ for mouse studies (Bargagna-Mohan et al., 2007, 2015, 2012) would deliver a daily peak concentration of ~1 μ*M* WFA (Bargagna-Mohan et al., 2010), considering that others showed a single 4 mg/kg i.p. injection of WFA produced peak concentrations of 2 μ*M* in plasma with a half-life of 1.36 h (Thaiparambil et al., 2011). Since WFA becomes undetectable in plasma after 24 h, the rapid clearance (0.15 l/h) has an advantage in that it affords reducing drug burden without sacrificing efficacy. Moreover, because prolonged action of covalent drugs reduces the need for frequent treatment in chronic regimens, the possibility to develop drug resistance may also be lower treatment (Bauer, 2015). We recommend maintaining a dose and treatment schedule not to exceed peak plasma concentrations of 2 μ*M* (2–4 mg/kg) for systemic delivery. However, in cancer settings this may be different. Since the lethal dose of a single WFA injection is around 80 mg/kg in mice (Sharada, Solomon, Devi, Udupa, & Srinivasan, 1996), the recommended doses are not likely to be toxic as treatment up to a month in mice showed no effects either on their health or behaviors (our personal observations).

3. TARGETING IFs

3.1 Targeting IFs *In Vitro*

It is widely known that the type III IF structure rapidly becomes altered by phosphorylation–dephosphorylation events (Izawa & Inagaki, 2006; Snider & Omary, 2014). Thus, the dynamic properties of this class of IF protein have been challenging for small-molecule targeting as multiple isoforms and structural variants can coexist in the same cell. However, WFA has offered a solution to this problem. WFA binds sVim at low concentrations delaying sVim incorporation into vimentin IFs. Second, by binding sVim,

WFA also regulates sVim-bound pERK's nuclear translocation, which could affect signaling functions (Bargagna-Mohan et al., 2015) (see Section 3.1.3). Currently, our data do not differentiate whether increasing WFA doses that perturb vimentin is due to direct IF targeting, or whether an excess of WFA-modified sVim subunits that exchange with unit-length filaments by becoming incorporated into vimentin causes IF destabilization. As such, the phosphorylation status of vimentin changes when cells engage in growth and migration, and thus, we hypothesize that vimentin is vulnerable to WFA when sVim becomes available during the reorganization of vimentin IFs (Grin et al., 2012; Robert et al., 2015), and with altered states of differentiation (Bargagna-Mohan et al., 2015) or cellular redox status (Perez-Sala et al., 2015).

3.1.1 Myofibroblast Transformation and Collagen Contractile Activity

Vimentin regulates wound repair by facilitating the transition of fibroblasts to myofibroblasts with force generation required for tissue contraction (Eckes et al., 1998; Herrmann, Strelkov, Burkhard, & Aebi, 2009). In this respect, it is not known, however, whether fibroblasts from different ocular tissues regulate vimentin differently when undergoing differentiation into myofibroblasts. We therefore investigated fibroblasts from two ocular anatomically distinct locations for their fibrotic response to growth factor TGF-β stimulation (Jester, Huang, Petroll, & Cavanagh, 2002). Corneal keratocytes derived from the avascular stroma differentiate into fibroblasts in presence of serum, and with serial cell passaging, differentiate into myofibroblasts (Jester, Petroll, & Cavanagh, 1999; Masur, Conors, Cheung, & Antohi, 1999). Stromal fibroblasts from Tenon's capsule, a vascularized tissue that is frequently involved in glaucoma filtration surgical scarring (Cordeiro et al., 2000), when similarly cultured, develop into myofibroblasts showing unique cell characteristics different from corneal cells (Bargagna-Mohan et al., 2013). For instance, under chronic low-dose TGF-β1 stimulation, rabbit corneal fibroblasts (RbCFs) are 6–8 times more sensitive to WFA treatment than rabbit Tenon's capsule fibroblasts (RbTCFs) in downregulating sVim expression (Fig. 2A). As such, this trend was also found in the greater sensitivity of RbCF cells compared to RbTCF cells to WFA in downregulating α-smooth muscle actin (α-SMA) expression. These findings suggest that Tenon's capsule myofibroblasts may possess vimentin isoforms that are not easily perturbed by WFA. Indeed, in TGF-β1-stimulated collagen contraction assays, a higher WFA dose (above 500 n*M*) is required to inhibit collagen contraction in RbTCF cells (Fig. 2D; Bargagna-Mohan et al., 2013), compared to RbCF cells (Bargagna-Mohan et al., 2015). These data also corroborate findings in a rabbit model of Tenon's capsule fibrosis that showed a high WFA dose was required to cause vimentin IF dismantling *in vivo* (Bargagna-Mohan et al., 2013). Although the molecular underpinnings for the differential sensitivities of ocular myofibroblasts are not understood at this time, we believe that underlying differences in vimentin phosphorylation between cell types may contribute to sensitizing sVim to WFA (see Section 3.1.3).

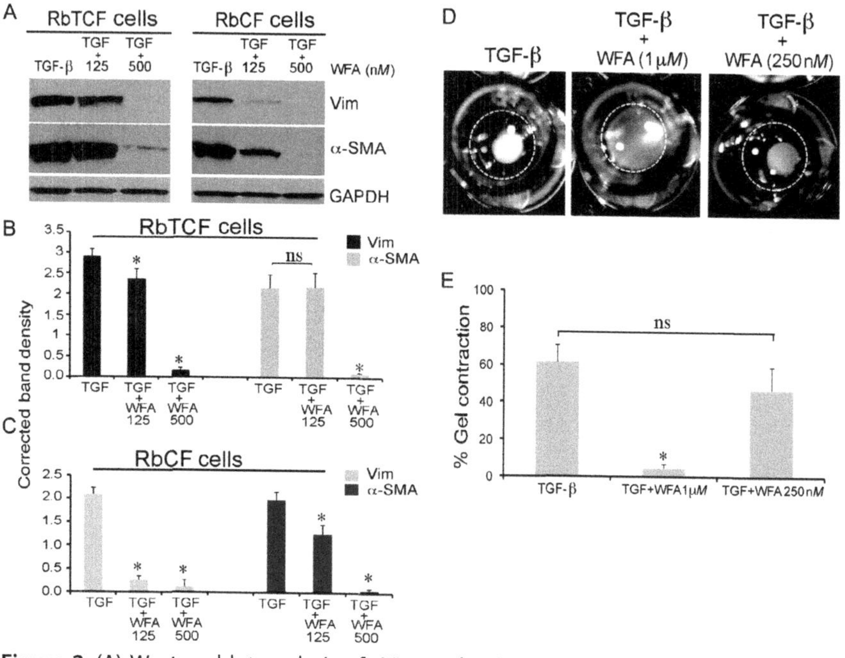

Figure 2 (A) Western blot analysis of sVim and α-SMA expression in RbTCF cells (left panel) and RbCF cells (right panel). (B,C) Densitometric quantification of vimentin (black and orange (light gray in the print version)) and α-SMA (gray and blue (black in the print version)) in RbTCF cells and RbCF cells, respectively, normalized to GAPDH using ImageJ software. Antibodies (mouse anti-vimentin monoclonal V9 from Santa Cruz Biotechnology; mouse anti-GAPDH monoclonal from Abcam; mouse anti-α-SMA antibody from Dako) were employed. (D) Representative images of polymerized collagen gels containing RbTCF cells ($n=6$ gels per treatment group) treated with TGF-β1 (2 ng/ml) in presence and absence of different doses of WFA for 3 days (10× magnification). (E) Quantification of percentage of gel contraction normalized to the initial size of the gel. Data are the mean of two independent experiments; dotted circles represent the well's area. *$p<0.05$ TGF-β versus WFA-treated cells. *This research was originally published in Bargagna-Mohan et al. (2013).*

Required materials, chemicals, and equipment

- 8-well glass cell culture chamber slides for IHC analysis and 60-mm and 96-well culture plates (Thermo Fisher Scientific, Pittsburg, PA)
- Phosphate-buffered saline (PBS) (Life Technology, Grand Island, NY)
- RbTCFs
- RbCFs
- TGF-β1 (2 ng/ml) (R&D Systems, Minneapolis, MN)
- WFA (Chromadex) stocks prepared at 50 m*M* in DMSO and diluted in medium for cell treatment
- Collagen type I from rat tail, formulated as previously described (Bargagna-Mohan et al., 2013)

3.1.1.1 Procedure for Collagen Contraction

1. Prepare 10 ml of ice-cold collagen solution and add 10^4 RbTCFs or RbCFs. Mix well.
2. Distribute 100 µl of cell–collagen mix into 96-well plates. Incubate gels for 2 h in an incubator at 37 °C with 5% CO_2 to allow polymerization. Then add TGF-β1 in cDMEM/F12 medium and culture for 2 days in presence or absence of WFA.
3. Analyze gels under an inverted microscope and take bright-field pictures to obtain images of the whole well.
4. Gel contractile activity is calculated as the percent area of collagen at 2 days from its initial area in the well (area measured using NIH ImageJ software).

3.1.1.2 Procedure for Myofibroblast Transformation

1. Plate RbTCF and RbCF cells in 60-mm cell culture plates and allow them to reach 95–100% confluency. Wash cells with 1 × PBS solution and add serum-free DMEM. Incubate cells for 48 h.

   Caveat: Plate cells one passage before the desired passage number and culture in serum-free medium for 48 h.
2. Remove serum-free medium, wash, and add 1 ml of warm trypsin/EDTA to each plate for 5 min.
3. Collect detached cells into a 15-ml tube and neutralize trypsin with 10% fetal bovine serum (FBS). Centrifuge at 800 rpm for 5 min and remove supernatant. At this point, cells can be used either for IHC or Western blot analysis.

Western blot analysis

1. Resuspend cell pellet in cDMEM/F12 medium and plate in 60-mm plates at a cell density of approximately 4.0×10^5 cells/plate. Incubate for 3 h.
2. Wash cells with 1 × PBS solution and culture with TGF-β1 in presence or absence of WFA for 3 days.
3. Proceed to Western blot analysis (see Section 6.2).

3.1.2 Cell Migration Inhibition

Cell migration is an obligatory step during wound healing, where vimentin IFs govern cell shape changes to drive directional motility (Helfand et al., 2011). Here, we describe the use of WFA to study the role of sVim in cell migration using RbTCFs. WFA exhibits an effective dose-related inhibition of cell migration between 250 n*M* and 2 μ*M* (Fig. 3) in the absence of toxicity

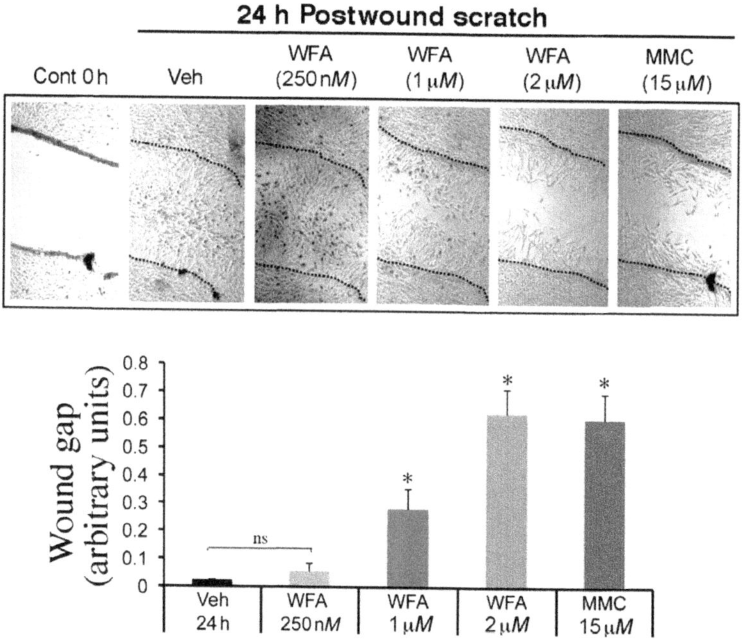

Figure 3 Cell migration in scratch wound assay. (A) Representative phase-contrast images of RbTCF cells immediately after a cross-stripe scrape injury on a confluent cell layer (0 h), 24 h postwounding (Veh), and after treatment with different concentrations of WFA or mitomycin C (MMC; positive control). (B) Arbitrary values of wound gap at 24 h normalized to size of initial wound gap and calculated from two independent experiments ($n = 3$ for each treatment group). *$p < 0.05$ vehicle versus drug treated; ns = nonsignificant. *This research was originally published in Bargagna-Mohan et al. (2013).*

(Bargagna-Mohan et al., 2013), which is a dose range that depolymerizes vimentin IFs in lamellipodia (Bargagna-Mohan et al., 2013). WFA can potently affect cell shape and migration over this dose range (Grin et al., 2012), and inhibits cancer cell migration through induction of vimentin phosphorylation that causes vimentin reorganization at lamellipodia (Thaiparambil et al., 2011) leading to loss of cell polarity.

Required materials, chemicals, and equipment

- PBS, 4% paraformaldehyde (PFA), and 8-well glass cell culture chamber slides
- RbTCFs cultured in cDMEM/F12 medium

Procedure

1. Plate RbTCF cells in an 8-well glass slide at 95–100% confluency in cDMEM/F12 medium and incubate for 24 h. Using a 100-μl micropipette tip, make a cross-stripe scrape in each well. Wash out debris with 1 × PBS solution.
2. Incubate adherent cells with different concentration of WFA for desired time. When control cells show 95% of wound closure, stop cell migration assay. Wash with 1 × PBS solution and fix cells with 4% PFA for 5 min. Measure wound gap in each sample and compare with controls.

3.1.3 Cell Spreading Inhibition

Cell spreading is also widely employed to study IF cytoskeleton dynamics (Prahlad, Yoon, Moir, Vale, & Goldman, 1998) during early stages of cell motility where vimentin is thought to regulate focal adhesion formation (Kim et al., 2010). We have adopted cell trypsinization as a method to induce dismantling of vimentin IFs into sVim so as to study their repolymerization when replated cells undergo cell spreading (Prahlad et al., 1998). To model the behavior of a corneal wound where contributions from both repair fibroblasts and myofibroblasts regulate the healing process (Fini & Stramer, 2005; Jester et al., 1999), we caused the differentiation of fibroblasts into myofibroblasts by serial cell passaging in serum. This produces an incremental, passage-dependent, accumulation of myofibroblasts initially representing the repair phenotype in passage 2 cells (RbCF2; 80% fibroblasts and 20% myofibroblasts) that is converted into a fibrotic phenotype by passage 8 (RbCF8; 80% myofibroblasts and 20% fibroblasts). This cell-culturing paradigm would allow one to interrogate the behavior of such mixed cell populations, and identify the dominant phenotype at any desired stage of cell differentiation. We found that sVim becomes increasingly phosphorylated during fibroblast (RbCF2) to myofibroblast (RbCF8) differentiation, accumulating pSer38Vim in the soluble pool as cytoplasmic dot-like structures (Fig. 4A). During cell spreading, pSer38Vim-containing species in myofibroblasts become hyperphosphorylated in response to drug treatment (Fig. 4B). As opposed to RbCF2 cells, RbCF8 cells do not efficiently incorporate hyperphosphorylated pSer38Vim into growing vimentin IFs and show delayed cell spreading. WFA-treated RbCF2 repair fibroblasts do not trigger this hyperphosphorylation step and efficiently incorporate pSer38Vim into filaments (Fig. 4B), which may involve an inherent mechanism by which fibroblasts are relatively desensitized to WFA. These differences

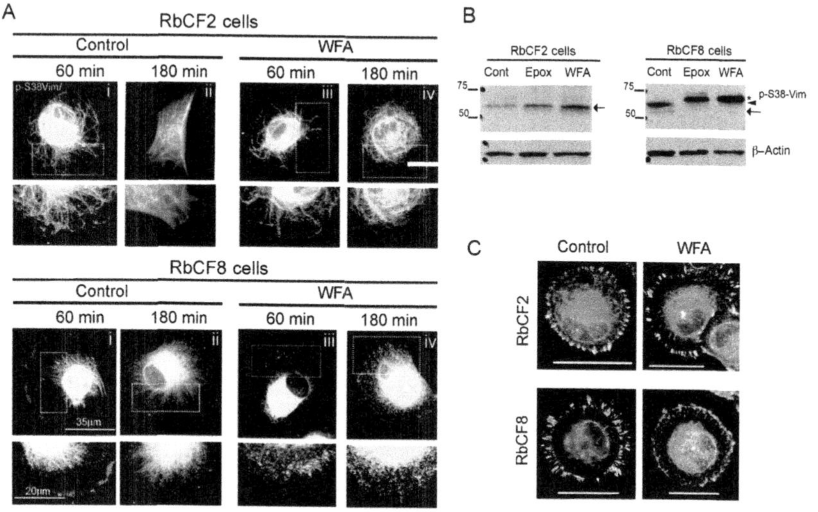

Figure 4 (A) Immunohistochemistry of pSer38Vim staining (green) in RbCF2 and RbCF8 cells. Scale bar = 35 μm. Inset panels represent 60× magnification of selected areas. Scale bar = 20 μm. (B) Western blot analysis of pSer38Vim in soluble extracts from RbCF2 and RbCF8 cells treated in presence or absence of WFA for 30 min. Asterisk marks the 67-kDa high molecular weight pSer38Vim hyperphosphorylated species, and the arrow and arrowheads represent the 57- and 61-kDa pSer38Vim bands, respectively. (C) Immunohistochemistry of RbCF2 and RbCF8 cells plated for 30 min followed by treatment in presence or absence of 1 μ*M* WFA for 1 h. Cells were fixed and stained with paxillin (green), vimentin (red), and DAPI (blue). Scale bar = 35 μm. Antibodies (rabbit anti-pSer38Vim antibody (A and B) and mouse anti-vimentin monoclonal V9 (C) from Santa Cruz Biotechnology; mouse anti-β-actin monoclonal from Sigma (B); rabbit anti-paxillin monoclonal (C) from Abcam) were employed. *Modified version from original published images in Bargagna-Mohan et al. (2015).* (See the color plate.)

are not due to differential uptake of WFA as both cell types show rapid response to the chemical compound in live cell imaging experiments (personal observations). As the proteasome inhibitor epoxomicin (Epox) also induces sVim hyperphosphorylation in RbCF8 cells, we showed the mechanism of WFA is not proteasome mediated, because only Epox induces accumulation of ubiquitinated species in this assay (Bargagna-Mohan et al., 2015). Next, we also examined the coalition between vimentin IFs and focal adhesions (Kim et al., 2010; Tomasek et al., 2002), and found that focal adhesions do occur in both RbCF2 and RbCF8 cells (Fig. 4C). However, RbCF8 cells display an increase in mature FAs (0.32 ± 0.054 vs. 0.22 ± 0.05; $p < 0.0001$) compared to RbCF2 cells. WFA selectively reduces the extension of vimentin IFs in the lamella zone in RbCF8 cells (0.32 ± 0.054 vs. 0.24 ± 0.05; $p < 0.0001$). WFA was also shown to induce pSer38Vim phosphorylation in the human BJ-5ta fibroblast cell line, and similar to corneal fibroblasts, they do not undergo pSer38Vim hyperphosphorylation (Grin et al., 2012). The dynamic recruitment of sVim into growing vimentin IFs was also investigated in endoplasmic spreading, which is blocked by WFA through targeting sVim (Lynch et al., 2013). Considering that phosphorylation has such profound effects on the dynamics of sVim, it is interesting that in breast cancer cells WFA induces serine 56 phosphorylation to cause the dismantling of vimentin IFs at lamellipodia in cell migration assays (Thaiparambil et al., 2011). Together, these studies identify that WFA affects different vimentin-signaling kinases to phosphorylate unique vimentin sites, which could result from a conformational change induced in sVim by WFA binding that triggers the activation of preassociated kinase(s) or via recruitment of other kinase(s). Since tetrameric sVim subunits exchange from unit-length filaments in a very rapid manner and this is ATP dependent, a posttranslational mechanism(s), which is believed to be other than pSer38Vim phosphorylation, may govern this early dynamic step (Robert et al., 2015). It is likely that WFA affects several mechanisms of vimentin dynamics according to when and where sVim becomes available (Chernoivanenko, Matveeva, Gelfand, Goldman, & Minin, 2015; Helfand et al., 2011), the contextual activation of vimentin-binding kinases and their cross-talk with other signaling components in subcellular organelles (Toda et al., 2012; Wortzel & Seger, 2011), and the magnitude to which vimentin is perturbed (Bargagna-Mohan et al., 2013, 2015; Grin et al., 2012). Thus, factors such as dose, treatment period, and posttranslational state of sVim will all define the nature of cellular response to WFA.

Required materials, chemicals, and equipment

- PBS, 0.05% trypsin–EDTA solution, and trypsin-inhibitor medium (Life Technology, Grand Island, NY)
- RbCFs cultured in cDMEM/F12 medium plated separately at passage 1 or passage 7

Procedure

Follow plating instruction as described earlier in Section 3.1.1.

Measurement of focal adhesion is performed according to the published study of Lynch et al. (2013).

Immunohistochemistry analysis

1. Resuspend cell pellet in 10 ml of DMEM. Distribute 250 μl of cell suspension in each glass well and gently tap on each side of the glass slide to allow equal distribution of cells into the well.

   Caveat: Now cells are passage 2 (RbCF2) and passage 8 (RbCF8).
2. Incubate cells at 37 °C for 30 min to allow cell attachment. Next, carefully add to each well WFA and incubate cells at 37 °C in a humidified chamber maintained under 5% CO_2 for 1 h. Carefully remove medium by aspiration and wash once wells with 1 × PBS solution.
3. Proceed to IHC analysis (see Section 6.1).

Western blot analysis

1. Resuspend cell pellet in complete DMEM. Plate cells in 60-mm plates at a cell density of approximately 4.0×10^5 cells/plate and incubate cells for 30 min.
2. Add WFA treatment and incubate cells at 37 °C for 30 min. Carefully remove medium by aspiration and wash once wells with 1 × PBS solution.
3. Proceed to Western blot analysis (see Section 6.2).

3.2 Targeting IFs *In Vivo*

3.2.1 Antifibrotic Activity In Vivo

Vimentin and desmin become overexpressed after an injury in corneal fibroblasts and continue to remain elevated in myofibroblasts to drive corneal fibrosis (Bargagna-Mohan et al., 2012; Chaurasia, Kaur, de Medeiros, Smith, & Wilson, 2009). In fact, Vim KO mice upon alkali injury have reduced corneal fibrosis compared to WT mice, but retain some haze, which we showed is from persistent expression of desmin in myofibroblasts. WFA treatment of WT mice significantly reduces corneal fibrosis through reduction of both vimentin and desmin expression, but some haze remains that is similar to that found in Vim KO mouse corneas. When Vim KO mice are treated with WFA for 14 days, the reduction in desmin expression in corneal myofibroblasts and reduction of α-SMA (Bargagna-Mohan et al., 2012) correlate with maximal improvement in corneal clarity (Fig. 5). These findings identify that WFA exerts its antifibrotic activities and protects against haze

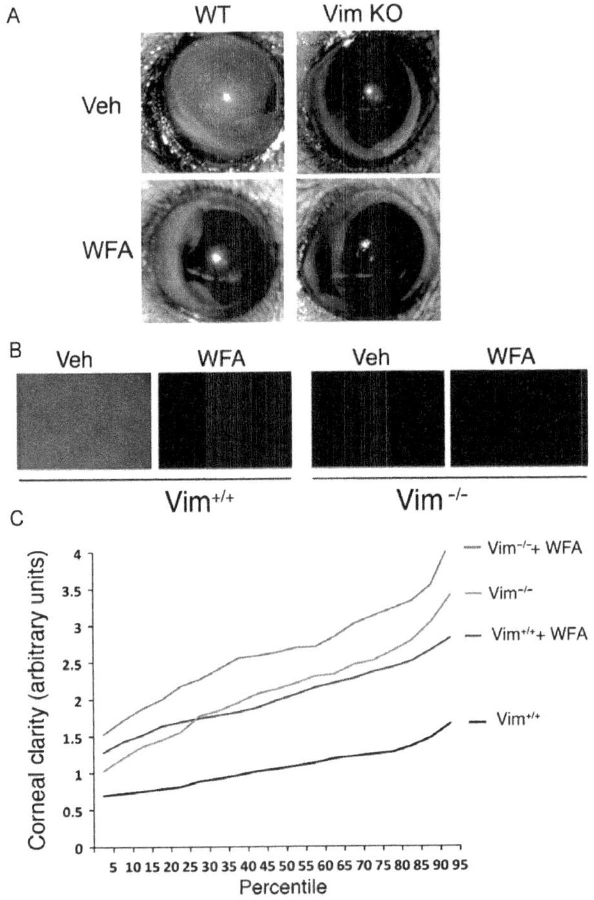

Figure 5 (A) Representative images of WT and Vim KO mice injured for 14 days showing levels of corneal opacity in mice that have vimentin expression downregulated either pharmacologically (WFA) or genetically (Vim KO). (B) Computer-aided imaging analysis of corneal transparency. Representative images of injured (vehicle; *Veh*) and WFA-treated corneas of WT and Vim KO mice showing opacity that ranked at the 50th percentile for corneal clarity. (C) Corneal clarity values for WT and Vim KO mice treated with and without WFA ($n > 150$ images/group) were plotted as a function of their percentile rank distributions to reveal the trends in healing for each group. Maximal clarity scores are at 4.0. *This research was originally published in Bargagna-Mohan et al. (2012). © The American Society for Biochemistry and Molecular Biology.*

through both vimentin- and desmin-targeted mechanisms in corneal myofibroblasts (Bargagna-Mohan et al., 2012).

Imaging Procedure

1. Anesthetize mice with 80% dose of ketamine/xylazine by i.p. injection and obtain multiple (3–4) color digital images of corneas of both eyes. As controls, image also noninjured mice from both genetic backgrounds used in the study.
2. We developed a highly sensitive, computer-based, colorimetric method to score corneal opacity that allows for an objective analysis with very high level of sensitivity. This novel colorimetric imaging method has been described (Bargagna-Mohan et al., 2012) and requires the user to run the software on Matlab.

4. PRIMARY CELL CULTURE SYSTEMS

4.1 Primary Cell Cultures from Ocular Tissues

4.1.1 Rabbit Corneal Fibroblasts

RbCFs have been extensively used as a cell culture model to study the mechanisms controlling myofibroblast transformation and fibrotic events in the cornea.

Required materials, chemicals, and equipment

- Freshly isolated eyes ($n=50$) from New Zealand white rabbits of both sexes approximately 4.75–5.75 lb (Pel-Freez Biologicals, Rogers, AR) can be obtained on wet-ice in glass containers filled with saline and antibiotic–antimytotic drugs and shipped from vendor by overnight parcel service
- DMEM/F12 basal medium (Life Technology, Grand Island, NY)
- FBS (Life Technology, Grand Island, NY)
- Cocktail of antibiotic/antimycotic solution (5000 U/ml penicillin, 50 μg/ml streptomycin, and 100 μg/ml neomycin) (Life Technology, Grand Island, NY)
- Complete DMEM/F12 medium (cDMEM/F12): DMEM/F12 basal medium, 10% FBS, cocktail of antibiotic/antimycotic solution
- Collagenase powder (1000–3000 CDU/mg solid) (Sigma-Aldrich, St. Louis, MO)
- Sterile surgical scissors and forceps

Procedure

1. Wash rabbit eyes three times with 1 × PBS containing antibiotic–antimycotic cocktail mix, and under sterile conditions make an incision

at the limbal/conjunctiva border with scalpel blade. With curved dissecting scissors, cut 9–10-mm central cornea buttons and place corneas in ice-cold sterile 1 × PBS solution.

2. Under a dissecting microscope, place cornea buttons endothelial-side-up, and with forceps remove the endothelium from each tissue using a tearing motion.
3. Place endothelium-free corneas in 60-mm cell culture plates (5–6 corneas/each plate) epithelium-side-up, and add 0.05% trypsin–EDTA. Incubate at 4 °C overnight.
4. Under sterile conditions, scrape off epithelium with a scalpel and digest remaining stromal tissue with collagenase (3 mg/ml) for 1 h in incubator.
5. Neutralize collagenase with cDMEM/F12 medium (2:1 ratio) and dissociate cells by gentle pipetting.
6. Centrifuge at 8000 rpm for 15 min, remove supernatants, and quickly mix cell pellets with fresh cDMEM/F12 medium.
7. Add equal amount of cell suspension in 60-mm plates and incubate cells for 2 days in incubator.
8. Replenish fresh cDMEM/F12 medium every 2 days. RbCF cells will reach confluency after 7–10 days. These cells will be labeled as RbCF0.

4.1.2 Rabbit Tenon's Capsule Fibroblasts

Glaucoma filtration surgery often fails because of the fibrotic reaction from TCFs (Esson et al., 2004). This paragraph illustrates a protocol to isolate and culture TCFs from rabbit eyes.

Required materials, chemicals, and equipment

- Fresh isolated eyes from New Zealand white rabbits ($n=10$ eyes) (Pel-Freez Biologicals, Rogers, AR)
- Sterile surgical scissors and forceps
- Dissecting microscope
- Sterile round glass coverslips (Thermo Fisher Scientific, Pittsburg, PA)
- cDMEM/F12 medium (as described in Section 4.1.1)

Procedure

1. Wash eyes three times with 1 × PBS solution, containing the antibiotic–antimycotic cocktail mix.
2. Under sterile conditions, make a conjunctival flap incision at approximately 1 mm from the limbus, and by blunt dissection, open the Tenon's capsule space.

3. Excise 10–15 mm^2 Tenon's capsule tissue and transfer into 60-mm cell culture plates.
4. Place 5–8 explants in each plate and gently press each tissue against the bottom of the plate. Allow the tissues to air-dry, under cell culture hood, for 15–20 min.

   Caveat: Do not skip this step. Air-dry time allows a stronger adherence of the explant to the plate.
5. Place a sterile round glass coverslip on top of each tissue. Gently press the glass against the tissue.
6. Add 6 ml of DMEM complete medium to the plate and place in incubator for 48 h.
7. After 48 h, carefully remove medium and replace it with fresh cDMEM/F12 medium.
8. After 12–15 days in culture, a significant number of RbTCF cells will migrate out from the explants. Carefully remove explants with sterile forceps.
9. Change medium every 2 days and allow RbTCF cells to reach confluency (7–10 days from removal of explants).

5. OCULAR INJURY MODEL OF FIBROSIS

All animal experiments were conducted in accordance with the Institutional Animal Care and Use Committee of our research institutions. Animals were housed in pathogen-free cages, under a 12-h light–dark cycle with food and water available *ad libitum*.

5.1 Alkali Burn Corneal Injury in Mouse

Required materials, chemicals, and equipment

- Dissecting microscope and warming pads
- 4–6-week-old mice in 129/Svev or C57Bl6 background from commercial vendors
- Blunt Tooke corneal knife
- 0.15 N NaOH diluted from 5 N stock (Sigma-Aldrich, St. Louis, MO)
- Sterile PBS
- Ketamine (Ketajet 100 mg/ml stock, 10 ml; Mylan Institutional LLC, Galway, Ireland)
- Xylazine (100 mg/ml stock, 50 ml; Lloyd Laboratories, Shenandoah, Iowa)

- Proparacaine hydrochloride ophthalmologic solution (0.5%; Akron, Lake Forest, IL)
- Atropine sulfate ophthalmic solution (1%; Bausch & Lomb)
- Tobramycin ophthalmic solution (0.3%; Bausch & Lomb)
- Erythromycin ophthalmic ointment (0.5%; Bausch & Lomb)
- DMSO. Use this as vehicle control for i.p. injections
- WFA (stock solution made in DMSO). Inject the requisite volume of 1 mg/ml WFA stock solution by i.p. to deliver 2 mg/kg body weight

Procedure

1. Weigh mice and anesthetize by i.p. injection of a diluted solution of ketamine and xylazine according to body weight. A toe-pinch test is employed to ensure total anesthesia.
2. Topically anesthetize mouse cornea with a drop of proparacaine ophthalmic solution. Wait for 2–3 min and check the cornea for ocular reflex by gently touching the cornea with a sterile blunt instrument.
3. Place the mouse on a warming pad placed under the microscope to visualize the eye. Next apply 1 μl drop of dilute 0.15 *M* sodium hydroxide to the cornea for 1 min. Perform one eye at a time. Wash each eye immediately with 10 ml sterile PBS solution.
4. Using a blunt Tooke corneal knife, gently remove corneal and limbal epithelium until corneal epithelium is completely denuded. Apply atropine eye drop followed by a drop of tobramycin antibiotic.
5. Apply copious amounts of erythromycin antibiotic eye ointment over the eye. This ointment also ensures that the cornea remains protected from dehydration.
6. Allow mouse to recover from injury on warming pads until fully recovered from anesthesia and begin grooming.
7. WFA solution (2 mg/kg) or vehicle (DMSO) is provided by i.p. injection and mice returned to animal housing facilities.
8. Mice are treated with tobramycin eye drops the day following injury by which time the conjunctival epithelium should have resurfaced the cornea.
9. Subsequent injections of WFA and vehicle are provided each day as per experimental paradigm.

5.2 Tissue Harvesting

5.2.1 Isolation of Mouse Cornea

- Enucleate mouse eye by carefully cutting through the extraocular muscles using a curved corneal scissors.

- Be careful when removing the eye so as not to pull on the optic nerve.
- Wash the eye with ice-cold 1 × PBS solution and proceed to isolate the cornea or freeze at − 80 °C.

For immunohistochemistry analysis

- Embed fresh eye in Tissue-Tek OCT medium (VWR, Radnor, PA) and freeze immediately at −80 °C.
- Proceed to cryosection of samples (10–12 μm thickness).

For Western blot analysis

- Under a dissecting microscope, make an incision with a sharp blade at the limbus/conjunctiva border. Insert small scissors into the space and carefully cut along the limbus/conjunctiva border to separate the cornea from the posterior chamber.
- Collect corneas in new tubes.

  Caveat: In order to avoid protein degradation, this process must be completed in less than 1–2 min/eye.
- Tissues can be frozen at −80 °C or processed immediately for protein analysis (see Section 6.2).

6. TECHNIQUES

6.1 Immunohistochemistry Analysis

Required materials, chemicals, and equipment

- Permeabilization/blocking buffer: 1 × PBS buffer containing 0.2% Triton X-100 (Sigma-Aldrich, St. Louis, MO), 5% goat serum (Life Technology, Grand Island, NY), and 1% bovine serum albumin (Sigma-Aldrich, St. Louis, MO)
- Primary antibodies: goat anti-rabbit vimentin (1:100; Abcam, Boston, MA), goat anti-rabbit p-Ser38-vimentin (1:100; Santa Cruz Biotechnology, Dallas, TX), goat anti-rabbit GFAP (1:100; Abcam, Boston, MA)
- Secondary antibodies: goat anti-rabbit Alexa 488 (1:1000; Life Technology, Grand Island, NY), goat anti-rabbit Alexa 455 (1:1000; Life Technology, Grand Island, NY), goat anti-mouse Alexa 555 (1:500; Life Technology, Grand Island, NY)
- 300 n*M* DAPI (4′,6-diamidino-2-phenylindole) (Life Technology, Grand Island, NY)
- Background reduction solution (Cat # S3022, Dako, Carpinteria, CA)
- 4% PFA (Sigma-Aldrich, St. Louis, MO)

6.1.1 In Vitro *Experiments*

For cell plating density, cell culture plates, and treatment, refer to Section 2. In brief, cells are plated and treated in 8-well glass chambers for IHC analysis.

Procedure

1. Remove medium from wells, rinse with 1 × PBS solution, and fix cells with 4% PFA at room temperature for 5 min and wash twice with 1 × PBS solution.
2. Incubate slides with permeabilization/blocking buffer at 4 °C overnight or at 37 °C for 1 h.
3. Wash with 1 × PBS and apply primary antibody of choice diluted in the Dako background solution at 4 °C overnight. The day after, rinse wells twice with 1 × PBS solution and incubate with secondary antibody of choice diluted in the Dako background solution, at room temperature for 1 h.
4. Wash wells three times with 1 × PBS solution, 10 min each wash. Apply DAPI for 5 min, and wash once with 1 × PBS solution. Keep slides in 1 × PBS solution until ready to be analyzed.

6.1.2 In Vivo *Experiments*

Tissue of interest is isolated from the host and then imbedded in OCT medium and immediately frozen at −80 °C.

Procedure

1. Air-dry frozen tissue sections on slides under a ventilated hood for 45 min.
2. Follow steps used for staining cells.

6.2 Western Blot Analysis (Soluble and Insoluble Extracts)

Required materials, chemicals, and equipment

- Primary antibodies: anti-rabbit vimentin (1:100; Abcam), anti-rabbit p-S38-vimentin (1:100; Santa Cruz Biotechnology, Dallas, TX), mouse α-SMA antibody (1:100; DAKO)
- Secondary antibodies: goat anti-rabbit (1:1000; Santa Cruz Biotechnology, Dallas, TX), goat anti-mouse (1:1000; Santa Cruz Biotechnology, Dallas, TX)
- Soluble buffer: Tris-buffered saline, 1% Nonidet P-40, 200 m*M* NaCl, and a protease inhibitor mixture (Roche, Indianapolis, IN). Phosphatase inhibitors are not included
- Insoluble buffer: Laemmli sample buffer

6.2.1 In Vitro *Experiments*

Procedure

1. Place 60-mm plates on top of an ice bed and carefully remove medium.
2. Add 500 μl of ice-cold PBS solution into each plate and, using a cell scraper, collect cells into 1.5-ml tubes.
3. Centrifuge for 3 min at 800 rpm at 4 °C.
4. Remove supernatants and add 100 μl of ice-cold soluble buffer. Gently pipette cells into solution. Incubate cells on ice for 15 min.
5. Add 6 μl of 10% Triton X-100 in each tubes and vortex for 30 s, then immediately centrifuge for additional 30 s.

   Caveat: This step must be done within 1 min time frame and tubes must be kept at all times on ice.
6. Transfer supernatant in a new tube. This represents the soluble fraction.
7. Resuspend pellets in 50 μl of insoluble buffer. This represents the insoluble fraction.
8. Freeze soluble and insoluble extracts at −20 °C until ready for Western blot analysis.

6.2.2 In Vivo *Experiments*

Obtain either freshly isolated or previously frozen eyes.

Procedure

Cornea extraction

1. Mince tissues thoroughly with scissors until tissue is uniformly chopped and add 100 μl of ice-cold soluble buffer. Use a pipette to aspirate up and down to produce a homogenous mixture. Incubate for 45 min on ice with occasional mixing.
2. Centrifuge for 5 min at 800 rpm at 4 °C.
3. Collect supernatant in new tubes. This represents the soluble fraction.
4. Resuspend pellets in 100 μl of insoluble buffer. This represents the insoluble fraction.
5. Freeze soluble and insoluble extracts at −20 °C until ready for Western blot analysis.

7. PERSPECTIVES

The discovery of WFA as a binding ligand for type III IFs has opened up a plethora of opportunities to probe this important class of IFs. Because the WFA ligand-binding site is evolutionary conserved, WFA can be exploited as a probe for all type III IFs. However, the theoretical model developed for tetrameric vimentin bound to WFA does not consider post-translational modifications, which could affect the protein conformation and thus binding characteristics of this ligand *in vivo*. As such, this may be a limitation in our understanding of how many IFs structures are expressed *in vivo* at any particular time and their liability to WFA targeting. Since WFA has potent pharmacological activity, the quest to identify and target the disease-related IF isoforms is then highly pertinent, as some IF isoforms may also have protective functions (Verardo et al., 2008). Some insight into this challenging task comes from our recent study that shows vimentin phosphorylation can play an important role in sensitizing sVim to WFA. That phosphorylation differences (e.g., in fibroblasts and myofibroblasts) have illuminated pSer38Vim; it is possible that pSer38Vim is a biomarker of corneal fibrosis due to its persistence in fibrotic corneas of mice and highly reduced expression in WFA-treated mice that recover corneal clarity (Bargagna-Mohan et al., 2015). These scientific insights into the use of WFA will hopefully continue to show promise to target and study type III IFs as well as define the doses that avoid off-target effects.

ACKNOWLEDGMENTS

This work was supported in part by grants from National Eye Institute (R01 EY016782), Fight for Sight Foundation, and John A. and Florence Mattern Solomon Endowed Chair in Vision Biology and Eye Diseases to R.M.

REFERENCES

Antony, M. L., Lee, J., Hahm, E. R., Kim, S. H., Marcus, A. I., Kumari, V., et al. (2014). Growth arrest by the antitumor steroidal lactone withaferin A in human breast cancer cells is associated with down-regulation and covalent binding at cysteine 303 of beta-tubulin. *The Journal of Biological Chemistry*, *289*(3), 1852–1865.

Bargagna-Mohan, P., Deokule, S. P., Thompson, K., Wizeman, J., Srinivasan, C., Vooturi, S., et al. (2013). Withaferin A effectively targets soluble vimentin in the glaucoma filtration surgical model of fibrosis. *PLoS One*, *8*(5), e63881.

Bargagna-Mohan, P., Hamza, A., Kim, Y. E., Khuan Abby Ho, Y., Mor-Vaknin, N., Wendschlag, N., et al. (2007). The tumor inhibitor and antiangiogenic agent withaferin A targets the intermediate filament protein vimentin. *Chemistry & Biology*, *14*(6), 623–634.

Bargagna-Mohan, P., Lei, L., Thompson, A., Shaw, C., Kasahara, K., Inagaki, M., et al. (2015). Vimentin phosphorylation underlies myofibroblast sensitivity to withaferin A *in vitro* and during corneal fibrosis. *PLoS One*, *10*(7), e0133399.

Bargagna-Mohan, P., Paranthan, R. R., Hamza, A., Dimova, N., Trucchi, B., Srinivasan, C., et al. (2010). Withaferin A targets intermediate filaments glial fibrillary acidic protein and vimentin in a model of retinal gliosis. *The Journal of Biological Chemistry*, *285*(10), 7657–7669.

Bargagna-Mohan, P., Paranthan, R. R., Hamza, A., Zhan, C. G., Lee, D. M., Kim, K. B., et al. (2012). Corneal anti-fibrotic switch identified in genetic and pharmacological deficiency of vimentin. *The Journal of Biological Chemistry*, *287*(2), 989–1006.

Bargagna-Mohan, P., Ravindranath, P. P., & Mohan, R. (2006). Small molecule anti-angiogenic probes of the ubiquitin proteasome pathway: Potential application to choroidal neovascularization. *Investigative Ophthalmology & Visual Science*, *47*(9), 4138–4145.

Bauer, R. A. (2015). Covalent inhibitors in drug discovery: From accidental discoveries to avoided liabilities and designed therapies. *Drug Discovery Today*, *20*(9), 1061–1073.

Burikhanov, R., Sviripa, V. M., Hebbar, N., Zhang, W., Layton, W. J., Hamza, A., et al. (2014). Arylquins target vimentin to trigger Par-4 secretion for tumor cell apoptosis. *Nature Chemical Biology*, *10*(11), 924–926.

Chaurasia, S. S., Kaur, H., de Medeiros, F. W., Smith, S. D., & Wilson, S. E. (2009). Dynamics of the expression of intermediate filaments vimentin and desmin during myofibroblast differentiation after corneal injury. *Experimental Eye Research*, *89*(2), 133–139.

Che, X., Chi, F., Wang, L., Jong, T. D., Wu, C. H., Wang, X., et al. (2011). Involvement of IbeA in meningitic Escherichia coli K1-induced polymorphonuclear leukocyte transmigration across brain endothelial cells. *Brain Pathology*, *21*(4), 389–404.

Chernoivanenko, I. S., Matveeva, E. A., Gelfand, V. I., Goldman, R. D., & Minin, A. A. (2015). Mitochondrial membrane potential is regulated by vimentin intermediate filaments. *FASEB Journal*, *29*(3), 820–827.

Chi, F., Bo, T., Wu, C. H., Jong, A., & Huang, S. H. (2012). Vimentin and PSF act in concert to regulate IbeA+ E. coli K1 induced activation and nuclear translocation of NF-kappaB in human brain endothelial cells. *PLoS One*, 7(4), e35862.

Cordeiro, M. F., Siriwardena, D., Chang, L., & Khaw, P. T. (2000). Wound healing modulation after glaucoma surgery. *Current Opinion in Ophthalmology*, *11*(2), 121–126.

Crews, C. M., & Mohan, R. (2000). Small-molecule inhibitors of the cell cycle. *Current Opinion in Chemical Biology*, *4*(1), 47–53.

Crews, C. M., & Splittgerber, U. (1999). Chemical genetics: Exploring and controlling cellular processes with chemical probes. *Trends in Biochemical Sciences*, *24*(8), 317–320.

Eckes, B., Dogic, D., Colucci-Guyon, E., Wang, N., Maniotis, A., Ingber, D., et al. (1998). Impaired mechanical stability, migration and contractile capacity in vimentin-deficient fibroblasts. *Journal of Cell Science*, *111*(Pt. 13), 1897–1907.

Eriksson, J. E., Dechat, T., Grin, B., Helfand, B., Mendez, M., Pallari, H. M., et al. (2009). Introducing intermediate filaments: From discovery to disease. *The Journal of Clinical Investigation*, *119*(7), 1763–1771.

Esson, D. W., Neelakantan, A., Iyer, S. A., Blalock, T. D., Balasubramanian, L., Grotendorst, G. R., et al. (2004). Expression of connective tissue growth factor after glaucoma filtration surgery in a rabbit model. *Investigative Ophthalmology & Visual Science*, *45*(2), 485–491.

Falsey, R. R., Marron, M. T., Gunaherath, G. M., Shirahatti, N., Mahadevan, D., Gunatilaka, A. A., et al. (2006). Actin microfilament aggregation induced by withaferin A is mediated by annexin II. *Nature Chemical Biology*, *2*(1), 33–38.

Fini, M. E., & Stramer, B. M. (2005). How the cornea heals: Cornea-specific repair mechanisms affecting surgical outcomes. *Cornea*, *24*(8 Suppl.), S2–S11.

Gharbi, S., Garzon, B., Gayarre, J., Timms, J., & Perez-Sala, D. (2007). Study of protein targets for covalent modification by the antitumoral and anti-inflammatory prostaglandin PGA1: Focus on vimentin. *Journal of Mass Spectrometry*, *42*(11), 1474–1484.

Gladilin, E., Gonzalez, P., & Eils, R. (2014). Dissecting the contribution of actin and vimentin intermediate filaments to mechanical phenotype of suspended cells using high-throughput deformability measurements and computational modeling. *Journal of Biomechanics*, *47*(11), 2598–2605.

Grin, B., Mahammad, S., Wedig, T., Cleland, M. M., Tsai, L., Herrmann, H., et al. (2012). Withaferin A alters intermediate filament organization, cell shape and behavior. *PLoS One*, *7*(6), e39065.

Helfand, B. T., Mendez, M. G., Murthy, S. N., Shumaker, D. K., Grin, B., Mahammad, S., et al. (2011). Vimentin organization modulates the formation of lamellipodia. *Molecular Biology of the Cell*, *22*(8), 1274–1289.

Herrmann, H., Strelkov, S. V., Burkhard, P., & Aebi, U. (2009). Intermediate filaments: Primary determinants of cell architecture and plasticity. *The Journal of Clinical Investigation*, *119*(7), 1772–1783.

Heyninck, K., Lahtela-Kakkonen, M., Van der Veken, P., Haegeman, G., & Vanden Berghe, W. (2014). Withaferin A inhibits NF-kappaB activation by targeting cysteine 179 in IKKbeta. *Biochemical Pharmacology*, *91*(4), 501–509.

Izawa, I., & Inagaki, M. (2006). Regulatory mechanisms and functions of intermediate filaments: A study using site- and phosphorylation state-specific antibodies. *Cancer Science*, *97*(3), 167–174.

Jester, J. V., Huang, J., Petroll, W. M., & Cavanagh, H. D. (2002). TGFbeta induced myofibroblast differentiation of rabbit keratocytes requires synergistic TGFbeta, PDGF and integrin signaling. *Experimental Eye Research*, *75*(6), 645–657.

Jester, J. V., Petroll, W. M., & Cavanagh, H. D. (1999). Corneal stromal wound healing in refractive surgery: The role of myofibroblasts. *Progress in Retinal and Eye Research*, *18*(3), 311–356.

Kidd, M. E., Shumaker, D. K., & Ridge, K. M. (2014). The role of vimentin intermediate filaments in the progression of lung cancer. *American Journal of Respiratory Cell and Molecular Biology*, *50*(1), 1–6.

Kim, H., Nakamura, F., Lee, W., Shifrin, Y., Arora, P., & McCulloch, C. A. (2010). Filamin A is required for vimentin-mediated cell adhesion and spreading. *American Journal of Physiology. Cell Physiology*, *298*(2), C221–C236.

Kumar, N., Robidoux, J., Daniel, K. W., Guzman, G., Floering, L. M., & Collins, S. (2007). Requirement of vimentin filament assembly for beta3-adrenergic receptor activation of ERK MAP kinase and lipolysis. *The Journal of Biological Chemistry*, *282*(12), 9244–9250.

Lahat, G., Zhu, Q. S., Huang, K. L., Wang, S., Bolshakov, S., Liu, J., et al. (2010). Vimentin is a novel anti-cancer therapeutic target; insights from *in vitro* and *in vivo* mice xenograft studies. *PLoS One*, *5*(4), e10105.

Lee, J., Hahm, E. R., Marcus, A. I., & Singh, S. V. (2013). Withaferin A inhibits experimental epithelial-mesenchymal transition in MCF-10A cells and suppresses vimentin protein level *in vivo* in breast tumors. *Molecular Carcinogenesis*, *54*, 417–429.

Leslie, B. J., & Hergenrother, P. J. (2008). Identification of the cellular targets of bioactive small organic molecules using affinity reagents. *Chemical Society Reviews*, *37*(7), 1347–1360.

Lynch, C. D., Lazar, A. M., Iskratsch, T., Zhang, X., & Sheetz, M. P. (2013). Endoplasmic spreading requires coalescence of vimentin intermediate filaments at force-bearing adhesions. *Molecular Biology of the Cell*, *24*(1), 21–30.

Masur, S. K., Conors, R. J., Jr., Cheung, J. K., & Antohi, S. (1999). Matrix adhesion characteristics of corneal myofibroblasts. *Investigative Ophthalmology & Visual Science*, *40*(5), 904–910.

Menko, A. S., Bleaken, B. M., Libowitz, A. A., Zhang, L., Stepp, M. A., & Walker, J. L. (2014). A central role for vimentin in regulating repair function during healing of the lens epithelium. *Molecular Biology of the Cell*, *25*(6), 776–790.

Mishra, L. C., Singh, B. B., & Dagenais, S. (2000). Scientific basis for the therapeutic use of Withania somnifera (ashwagandha): A review. *Alternative Medicine Review*, *5*(4), 334–346.

Miyatake, Y., Sheehy, N., Ikeshita, S., Hall, W. W., & Kasahara, M. (2015). Anchorage-dependent multicellular aggregate formation induces CD44 high cancer stem cell-like

ATL cells in an NF-kappaB- and vimentin-dependent manner. *Cancer Letters*, *357*(1), 355–363.

Mohan, R., Hammers, H. J., Bargagna-Mohan, P., Zhan, X. H., Herbstritt, C. J., Ruiz, A., et al. (2004). Withaferin A is a potent inhibitor of angiogenesis. *Angiogenesis*, *7*(2), 115–122.

Nagalingam, A., Kuppusamy, P., Singh, S. V., Sharma, D., & Saxena, N. K. (2014). Mechanistic elucidation of the antitumor properties of withaferin A in breast cancer. *Cancer Research*, *74*(9), 2617–2629.

Newman, D. J., & Cragg, G. M. (2012). Natural products as sources of new drugs over the 30 years from 1981 to 2010. *Journal of Natural Products*, *75*(3), 311–335.

Nishikawa, Y., Okuzaki, D., Fukushima, K., Mukai, S., Ohno, S., Ozaki, Y., et al. (2015). Withaferin A induces cell death selectively in androgen-independent prostate cancer cells but not in normal fibroblast cells. *PLoS One*, *10*(7), e0134137.

Paranthan, R. R., Bargagna-Mohan, P., Lau, D. L., & Mohan, R. (2011). A robust model for simultaneously inducing corneal neovascularization and retinal gliosis in the mouse eye. *Molecular Vision*, *17*, 1901–1908.

Patel, P., Julien, J. P., & Kriz, J. (2015). Early-stage treatment with withaferin A reduces levels of misfolded superoxide dismutase 1 and extends lifespan in a mouse model of amyotrophic lateral sclerosis. *Neurotherapeutics*, *12*(1), 217–233.

Perez-Sala, D., Oeste, C. L., Martinez, A. E., Carrasco, M. J., Garzon, B., & Canada, F. J. (2015). Vimentin filament organization and stress sensing depend on its single cysteine residue and zinc binding. *Nature Communications*, *6*, 7287.

Perlson, E., Michaelevski, I., Kowalsman, N., Ben-Yaakov, K., Shaked, M., Seger, R., et al. (2006). Vimentin binding to phosphorylated ERK sterically hinders enzymatic dephosphorylation of the kinase. *Journal of Molecular Biology*, *364*(5), 938–944.

Prahlad, V., Yoon, M., Moir, R. D., Vale, R. D., & Goldman, R. D. (1998). Rapid movements of vimentin on microtubule tracks: Kinesin-dependent assembly of intermediate filament networks. *The Journal of Cell Biology*, *143*(1), 159–170.

Robert, A., Rossow, M. J., Hookway, C., Adam, S. A., & Gelfand, V. I. (2015). Vimentin filament precursors exchange subunits in an ATP-dependent manner. *Proceedings of the National Academy of Sciences of the United States of America*, *112*(27), E3505–E3514.

Sharada, A. C., Solomon, F. E., Devi, P. U., Udupa, N., & Srinivasan, K. K. (1996). Antitumor and radiosensitizing effects of withaferin A on mouse Ehrlich ascites carcinoma *in vivo*. *Acta Oncologica*, *35*(1), 95–100.

Snider, N. T., & Omary, M. B. (2014). Post-translational modifications of intermediate filament proteins: Mechanisms and functions. *Nature Reviews. Molecular Cell Biology*, *15*(3), 163–177.

Strelkov, S. V., Herrmann, H., Geisler, N., Lustig, A., Ivaninskii, S., Zimbelmann, R., et al. (2001). Divide-and-conquer crystallographic approach towards an atomic structure of intermediate filaments. *Journal of Molecular Biology*, *306*(4), 773–781.

Thaiparambil, J. T., Bender, L., Ganesh, T., Kline, E., Patel, P., Liu, Y., et al. (2011). Withaferin A inhibits breast cancer invasion and metastasis at sub-cytotoxic doses by inducing vimentin disassembly and serine 56 phosphorylation. *International Journal of Cancer*, *129*, 2744–2755.

Toda, M., Kuo, C. H., Borman, S. K., Richardson, R. M., Inoko, A., Inagaki, M., et al. (2012). Evidence that formation of vimentin mitogen-activated protein kinase (MAPK) complex mediates mast cell activation following FcepsilonRI/CC chemokine receptor 1 cross-talk. *The Journal of Biological Chemistry*, *287*(29), 24516–24524.

Tomasek, J. J., Gabbiani, G., Hinz, B., Chaponnier, C., & Brown, R. A. (2002). Myofibroblasts and mechano-regulation of connective tissue remodelling. *Nature Reviews Molecular Cell Biology*, *3*(5), 349–363.

Vanden Berghe, W., Sabbe, L., Kaileh, M., Haegeman, G., & Heyninck, K. (2012). Molecular insight in the multifunctional activities of withaferin A. *Biochemical Pharmacology, 84*(10), 1282–1291.

Verardo, M. R., Lewis, G. P., Takeda, M., Linberg, K. A., Byun, J., Luna, G., et al. (2008). Abnormal reactivity of muller cells after retinal detachment in mice deficient in GFAP and vimentin. *Investigative Ophthalmology & Visual Science, 49*(8), 3659–3665.

Wortzel, I., & Seger, R. (2011). The ERK cascade: Distinct functions within various subcellular organelles. *Genes & Cancer, 2*(3), 195–209.

Yokota, Y., Bargagna-Mohan, P., Ravindranath, P. P., Kim, K. B., & Mohan, R. (2006). Development of withaferin A analogs as probes of angiogenesis. *Bioorganic & Medicinal Chemistry Letters, 16*(10), 2603–2607.

CHAPTER NINE

Assays to Study Consequences of Cytoplasmic Intermediate Filament Mutations: The Case of Epidermal Keratins

Tong San Tan, Yi Zhen Ng, Cedric Badowski, Tram Dang, John E.A. Common, Lukas Lacina[1], Ildikó Szeverényi, E. Birgitte Lane[2]
Institute of Medical Biology, Singapore
[2]Corresponding author: e-mail address: birgit.lane@imb.a-star.edu.sg

Contents

1. Introduction 221
   1.1 The Link Between Human Diseases and Keratin Mutations 221
   1.2 *In Vivo* Consequences of Keratin Mutations 222
2. Identification of Keratin Mutations 223
   2.1 Skin Biopsies to Identify Keratinopathies 223
   2.2 Diagnosis and Candidate Gene Identification by Immunohistochemistry 224
   2.3 Deoxyribonucleic Acid Collection and Molecular Identification of Keratin Mutations 226
3. Developing Experimental Model Systems 227
   3.1 Disease Modeling in Simple Culture Systems 229
   3.2 Fluorescence Time-Lapse Imaging of Cells Expressing Mutant Keratins 234
   3.3 Disease Modeling in 3D Cultures 235
   3.4 The Use of Mouse Models to Study Effects of Keratin Mutations 237
4. Keratin Mutations in Stress Assays 238
   4.1 Heat Stress 238
   4.2 Osmotic Stress 240
   4.3 Cell Shape Changes: Cell Spreading 240
   4.4 Cell Shape Changes: Cell Motility and Migration 241
   4.5 Mechanical Stress 244
5. Conclusion, Pearls, and Pitfalls 246

Acknowledgments 248
References 248
Further reading 253

[1] Current address: Charles University in Prague, First Faculty of Medicine, Institute of Anatomy & Department of Dermatology, U Nemocnice 3, 12800 Prague 2.

Methods in Enzymology, Volume 568
ISSN 0076-6879
http://dx.doi.org/10.1016/bs.mie.2015.09.030

Abstract

The discovery of the causative link between keratin mutations and a growing number of human diseases opened the way for a better understanding of the function of the whole intermediate filament families of cytoskeleton proteins. This chapter describes analytical approaches to identification and interpretation of the consequences of keratin mutations, from the clinical and diagnostic level to cells in tissue culture. Intermediate filament pathologies can be accurately diagnosed from skin biopsies and DNA samples. The Human Intermediate Filament Database collates reported mutations in intermediate filament genes and their diseases, and can help clinicians to establish accurate diagnoses, leading to disease stratification for genetic counseling, optimal care delivery, and future mutation-aligned new therapies.

Looking at the best-studied keratinopathy, epidermolysis bullosa simplex, the generation of cell lines mimicking keratinopathies is described, in which tagged mutant keratins facilitate live-cell imaging to make use of today's powerful enhanced light microscopy modalities. Cell stress assays such as cell spreading and cell migration in scratch wound assays can interrogate the consequences of the compromised cytoskeletal network. Application of extrinsic stresses, such as heat, osmotic, or mechanical stress, can enhance the differentiation of mutant keratin cells from wild-type cells. To bring the experiments to the next level, 3D organotypic human cultures can be generated, and even grafted onto the backs of immunodeficient mice for greater *in vivo* relevance. While development of these assays has focused on mutant K5/K14 cells, the approaches are often applicable to mutations in other intermediate filaments, reinforcing fundamental commonalities in spite of diverse clinical pathologies.

ABBREVIATIONS

cDNA complementary deoxyribonucleic acid
DAPI 4′,6-diamidine-2′-phenylindole dihydrochloride
DebRA dystrophic epidermolysis bullosa research association
DMEM Dulbecco's modified eagle's medium
DNA deoxyribonucleic acid
EB epidermolysis bullosa
EBS epidermolysis bullosa simplex
EBS, gen-non-DM epidermolysis bullosa simplex, other generalized
EBS-DM epidermolysis bullosa simplex, Dowling-Meara
EBS-loc epidermolysis bullosa simplex, localized
EGF epidermal growth factor
EGFP enhanced green fluorescent protein
EI epidermolytic ichthyosis
FACS fluorescent-activated cell sorting
FBS fetal bovine serum
FFPE formalin-fixed, paraffin-embedded
GFP green fluorescent protein
HEK293 human embryonic kidney 293 cells
HIFD Human Intermediate Filament Database

JEB junctional epidermolysis bullosa
K14 keratin 14 protein
KRT14 keratin 14 gene
M.O.I. multiplicity of infection
MAPK mitogen-activated protein kinase
PBS phosphate-buffered saline
PBS-T phosphate-buffered saline, with tween-20
PC pachyonychia congenital
SEI superficial epidermolytic ichthyosis

1. INTRODUCTION

1.1 The Link Between Human Diseases and Keratin Mutations

The causative link between mutations in keratin intermediate filaments and a wide range of human epithelial pathologies was first recognized in the early 1990s, with the association between the rare skin fragility disorder *epidermolysis bullosa simplex* (EBS) and mutations in basal keratin genes keratin 5 (*KRT5*) and keratin 14 (*KRT14*) (Bonifas, Rothman, & Epstein, 1991; Coulombe, Hutton, Letai, et al., 1991; Lane et al., 1992). Mutations in other keratins associated with epithelial fragility disorders were then quickly uncovered. Keratins (with other intermediate filaments) are a large family of proteins expressed in clear tissue-specific patterns (Moll, Dhouailly, & Sun, 1989), and antibodies to intermediate filaments are widely used in diagnostic pathology, especially for cancer (Fischer, Altmannsberger, Weber, & Osborn, 1987). Yet, the functions of these proteins were hard to access until the link with skin diseases revealed their essential role in tissue function.

There are now 99 different distinct clinical conditions associated with intermediate filaments. Keratins, the type I and type II intermediate filament proteins, account for 54 of the 70 intermediate filament genes in the human genome, and keratin-associated disorders are numerous. Known collectively as *keratinopathies*, most of them affect skin and associated structures. There are 25 reported disorders associated with mutations in type I keratins and 36 disorders with mutations in type II keratins (http://www.interfil.org), although many of these overlap as type I and type II keratins only assemble as obligate heteropolymers and the two proteins of each coexpressed pair of keratins are therefore functionally interdependent.

The Human Intermediate Filament Database (HIFD) (http://www.interfil.org) retrieves data on published sequence variants (mutations) and allelic variants (polymorphisms) identified in human intermediate filament genes (Szeverenyi et al., 2008). Half of all reported disease-associated mutations in intermediate filaments are in keratins. To date, over 1200 unrelated keratin mutations and 871 polymorphisms have been published and entered in the HIFD. The accurate recognition of keratin mutations, and their consequences, both *in vivo* and *in vitro*, is an essential step toward understanding the mechanisms of the pathology. With this understanding, effective and practical routes to therapy can be devised for those affected with these to date incurable conditions.

1.2 *In Vivo* Consequences of Keratin Mutations

The clinical consequences of keratin mutations, and the phenotypes resulting from mutations, usually suggest a structural fragility in the epithelial cells expressing the mutated protein. The integument or skin (epidermis and appendages) is the site of the most diverse expression of keratins, and the organ in which pathological changes can most easily be seen, on the outside of the body. Skin is under constant and formidable stress and strain—mechanical, chemical, and physiological, and the epidermis undergoes steady high cell turnover throughout life. The consequences of keratin mutations present clinically as tissue fragility or instability, specifically in the cell subpopulation expressing the mutant keratin as a major component of its cytoskeleton (Fig. 1). Assessing the clinical presentation of a patient is therefore a key first step to identifying the nature of the keratin mutation.

The earliest recognized and most intensively studied keratinopathy was the rare skin fragility disorder EBS. This disorder is characterized by skin blistering, induced by quite mild physical trauma to the skin, and is caused by mutant keratins 5 or 14 (K5 or K14). EBS dramatically demonstrates the link between the expression pattern of the keratins and the cell fragility resulting from mutations. K5 and K14 proteins are the primary, and predominant, pair of keratins expressed in basal layer keratinocytes, becoming less prominent in suprabasal epidermal cell layers where K1 and K10 are synthesized. Thus, the regional restriction of the intraepidermal fragility caused by K5/K14 mutations can be seen histologically as a specific splitting of cells in the basal layer upon physical trauma in EBS, while mutations in K1/K10 or K2 lead to suprabasal (epidermolytic ichthyosis, EI) or upper spinous layer (superficial epidermolytic ichthyosis, SEI) cell rupture, respectively (see Fig. 1).

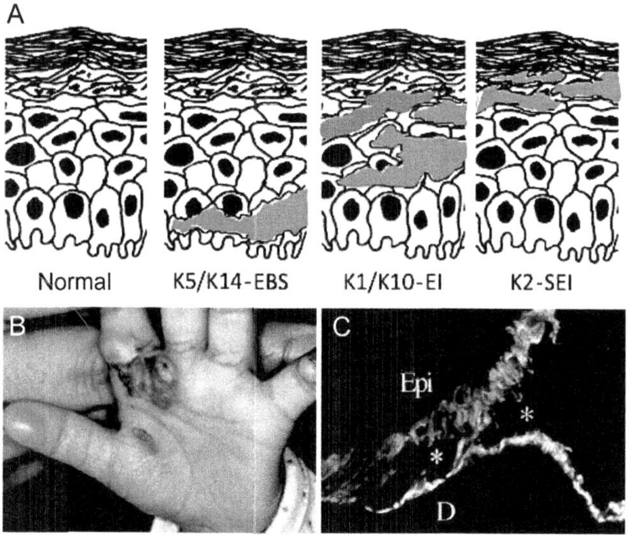

Figure 1 The *in vivo* consequences of keratin mutations. (A) Diagrammatic examples of planes of epidermal fracture caused by different keratin mutations (EBS: epidermolysis bullosa simplex, EI: epidermolytic ichthyosis, SEI: superficial epidermolytic ichthyosis). Cell fragility (red) is closely correlated with expression range of the mutated keratin and is diagnostically indicative. (B) Clinical presentation of EBS, Dowling-Meara type (EBS-DM). (C) Histological frozen section of EBS-DM biopsy showing immunofluorescence staining of blister area using monoclonal antibody RCK107 to K14 (white). Cleavage plane leaves K14 material on both upper and lower aspects of blister. Epi: Epidermis; D: Dermis; * Intracellular cleavage of basal keratinocytes. *Panel (B) and (C): from Mark Koh, Kandang Kerbau Women's and Children's Hospital, Singapore.* (See the color plate.)

Mutations in other keratin genes give rise to very different phenotypes. Sequence variants in *KRT6A/B/C*, *KRT16*, and *KRT17* genes cause pachyonychia congenita (thick nails, plantar keratoderma), mutations in corneal keratins *KRT3* and *KRT12* link to Meesmann corneal dystrophy, and similar mutations in mucosal keratins *KRT4* or *KRT13* lead to white sponge nevus, and so on (Toivola, Boor, Alam, & Strnad, 2015).

2. IDENTIFICATION OF KERATIN MUTATIONS

2.1 Skin Biopsies to Identify Keratinopathies

Collecting clinical material to study skin fragility disorders associated with intermediate filaments, such as EBS, is a vital part of understanding the mechanisms of pathology. Many different skin disorders present with blistering, and histological examination of a biopsy of a small fresh blister can help to distinguish keratinopathies from viral, autoimmune, vascular, or metabolic causes of blistering. The "cleavage plane" at which the epidermis splits is informative as splitting can occur within the keratinocytes of the basal layer (in EBS), in the suprabasal layers (in EI), or below the stratum corneum (in SEI), or can be subepidermal in nonkeratin diseases, with cleavage within or below the basal lamina (in junctional or dystrophic forms of epidermolysis bullosa, EB). All produce fluid-filled blisters that are hard to distinguish without a biopsy.

For optimal observation of epithelial damage, a new blister should be generated for biopsy. The skin is stressed by holding a clean disinfected pencil eraser vertical on the skin, pressing down, and twisting back and forth, initially 20 times. Mild erythema should appear within a few minutes, and the biopsy is taken after a further 5 min to allow blister formation. In milder cases, the blister can take longer to appear and may need additional mechanical stress. If any skin damage or tearing is seen immediately, mechanical stress is stopped and the biopsy taken (Intong & Murrell, 2010).

2.2 Diagnosis and Candidate Gene Identification by Immunohistochemistry

2.2.1 Immunofluorescence Histochemistry for EB Diagnosis

For optimum discrimination of immunohistochemistry, samples should be snap-frozen as soon as possible after excision. Alternatively, tissue can be transported in Michel's fixative at ambient temperature and processed by snap freezing after delivery to laboratory.

The availability of specific monoclonal antibodies to nearly all the keratins implicated in genetic diseases (see www.interfil.org), as well as most of the proteins involved in genetic blistering disorders, has made immunohistochemistry the preferred first strategy for keratinopathy diagnostic pathology. Although frozen sections are preferred for immunohistochemistry, there are many good strong antibodies that work well on formalin-fixed paraffin-embedded (FFPE) material after enzyme or heat treatment antigen retrieval procedures. Table 1 lists some antibodies used routinely in histopathology for EB subtype identification. Note that for clinically validated diagnoses, research immunohistochemistry findings will need to be confirmed by a formally accredited histopathology laboratory.

1. Frozen sections are cut at 5–7 μm thickness using a cryostat at predetermined optimal cutting temperature. (Freshly cut sections are always

Table 1 Antibodies Used for Immunomapping of Major EB Subtypes

Target Protein Antigen	Major EB Group Classification	Commercial Monoclonal Antibody and Distributor	Works in FFPE Sections Also?	Antibody References
Keratin 5	EBS	XM26 (DAKO)	Yes	Moll, Franke, Schiller, Geiger, and Krepler (1982)
Keratin 14	EBS	LL001 (Abcam)	Yes	Purkis et al. (1990), Wetzels et al. (1989), and Wetzels et al. (1991)
		LL002 (Abcam)	Yes	
		RCK107 (Abcam)	No	
Type IV Collagen	JEB	CIV 22 (DAKO)	Yes	Odermatt, Lang, Ruttner, Winterhalter, and Trueb (1984)
Type VII Collagen	Recessive dystrophic epidermolysis bullosa (RDEB)	LH7.2 (Leica)	No	Heagerty, Kennedy, Leigh, Purkis, and Eady (1986)
Type XVII Collagen	JEB	NC16A-3 (Abcam)	Not evaluated	Kawachi et al. (1996)
Integrin α6	JEB	NKI-GoH3 (AbD Serotec)	Not evaluated	Sonnenberg, Modderman, and Hogervorst (1988)
Integrin β4	JEB	BD 611233 (Becton Dickinson)	No	Giancotti, Stepp, Suzuki, Engvall, and Ruoslahti (1992) and Kennel et al. (1990)
Plectin	EBS	HD121 (currently no commercial supplier)	Not evaluated	Owaribe, Nishizawa, and Franke (1991)
Laminin 332	JEB	D4B5 (Millipore)	No	Mizushima et al. (1998)

optimal, but antigenic reactivity can be well preserved for many weeks at −80 °C.)

2. Sections are brought to room temperature for staining and rehydrated in phosphate-buffered saline (PBS) with 0.05% Tween-20 (PBS-T) for 10 min. Brief fixation with cold methanol/acetone (1:1) at −20 °C, or 4% paraformaldehyde at ambient temperature, can be carried out at this stage.
3. Nonspecific staining is blocked by immersion in 10% serum for 30 min at ambient temperature, using serum from same species as that of secondary antibody. Excess serum is removed by light blotting with tissue paper or cotton bud. Sections should not be washed at this point.
4. Primary antibody (antigen specific) is applied at predetermined optimal dilution and sections incubated for 60 min at room temperature or preferably overnight at 4 °C. After this, they are washed for 3 × 5 min in PBS-T or in running tap water.
5. Secondary antibody (primary antibody specific with fluorescent label) is applied at predetermined optimal dilution and incubated for 30–60 min at room temperature, and the slides are washed for 3 × 10 min in PBS-T or running water.
6. Slides are counterstained with 4′,6-diamidine-2′-phenylindole dihydrochloride (DAPI) and then washed again for 3 × 5 min in PBS-T or running water. Excess liquid is removed by gently blotting and the coverslip mounted in suitable mounting medium. Sections are then observed by fluorescence microscopy.

2.3 Deoxyribonucleic Acid Collection and Molecular Identification of Keratin Mutations

Having confirmed a candidate gene by histopathology, deoxyribonucleic acid (DNA) sequencing is carried out to identify a specific mutation. DNA from both affected and unaffected family members is analyzed to allow downstream validation of the pathological significance of any allelic variants identified. Again, collection of any patient material is always dependent on prior ethical approval from the hospital and laboratory ethics committees and upon fully informed patient consent.

DNA is usually analyzed from a blood sample. Use of any of the good commercial extraction kits is recommended for isolation of DNA from peripheral blood lymphocytes. As a guide, it is recommended to collect 3 mL or more of peripheral whole blood, so that multiple downstream genetic analyses can be performed without going back to the family for further material. Alternatively, DNA analysis from saliva samples is well validated, and a useful route when collecting samples from difficult to access patients, and where long distance sample transport is only possible at room temperature. However, some techniques requiring more DNA, such as whole-genome sequencing, are not yet validated for saliva samples.

Approximately 80% of EBS patients have been found to carry a *KRT5* or *KRT14* mutation. Complete sequencing of the coding region for these two genes with specifically designed oligonucleotide primer pairs is required, to avoid sequencing highly related pseudogenes and other keratin family members (Jerabkova et al., 2010; Rugg et al., 2007). We would recommend the primers used by Rugg et al. for amplifying keratin 5 and 14 gene exons reliably (Rugg et al., 2007).

Numerous research groups have published validated methods for keratin gene sequencing. Several laboratories offer clinically accredited molecular diagnosis as a fee-for-service, and world-wide patient advocacy organizations such as dystrophic epidermolysis bullosa research association (DebRA) International (http://www.debra-international.org/homepage.html) or Pachyonychia Congenita Project (http://www.pachyonychia.org) can provide information for researchers and patients looking for diagnostics services.

The HIFD (http://www.interfil.org) collates the wide spectrum of disease-causing keratin mutations reported to date. Figure 2 shows the most severe mutation hotspots at the ends of the rod domains (Fig. 2A) and specific mutations and their diseases (Fig. 2B). There is clear evidence for a phenotype–genotype correlation in keratinopathies, with the most detrimental mutations occurring within these helix boundary peptides at either end of the α-helical rod domain of the proteins. These ends are thought to be key interaction sites in assembly and disassembly of the keratin polymers, such that alterations in their properties due to mutation (as in Dowling-Meara-type EBS) will interfere with the keratin assembly kinetics. For the mutation cluster sites seen in less severe forms of EBS, the reader is referred to other published reviews.

3. DEVELOPING EXPERIMENTAL MODEL SYSTEMS

With the causative mutations identified, one can analyze the consequences of mutations by mimicking aspects of the disease phenotype in an experimentally accessible format. The fact that most keratinopathies affect the skin and other barrier epithelia now becomes a huge advantage: human skin cells can be accessed through biopsies, cultured, engineered, and even

A

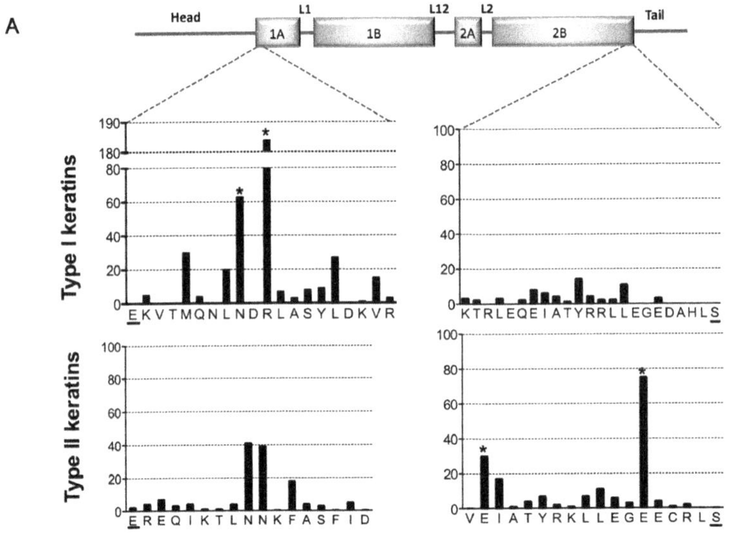

B

Type I keratins	Asn (N) mutations		Arg (R) mutations		Major disease groups
	Position	Reported cases	Position	Reported cases	
K9	N161	14	R163	39	Epidermolytic palmoplantar keratoderma, EPPK
K10	N154	1	R156	50	Epidermolytic ichthyosis, EI
K12	N133	1	R135	5	Meesmann corneal dystrophy, MCD
K13	N112	1	R114	3	White sponge nevus, WSN
K14	N123	8	R125	67	Generalized severe epidermolysis bullosa simplex, EBS-DM
K16	N125	14	R127	11	Pachyonychia congenita, PC-K16
K17	N92	24	R94	9	Pachyonychia congenita, PC-K17

Type II keratins	Glu (E) mutations				Major disease groups
	Position	Reported cases	Position	Reported cases	
K1	E478	4	E489	1	Epidermolytic ichthyosis, EI
K2	E476	2	E487	17	Superficial epidermolytic ichthyosis, SEI
K3	E498	1	E509	1	Meesmann corneal dystrophy, MCD
K4	E509	1	E520	1	White sponge nevus, WSN
K5	E466	2	E477	13	Generalized severe epidermolysis bullosa simplex, EBS-DM
K6a	E461	2	E472	6	Pachyonychia congenita, PC-K6a
K6b	E461	1	E472	5	Pachyonychia congenita, PC-K6b
K6c	E461	0	E472	3	Focal non-epidermolytic palmoplantar keratoderma, FNEPPK
K81	E402	6	E413	2	Monilethrix
K83	E401	1	E418	0	Monilethrix
K86	E402	10	E413	25	Monilethrix

Figure 2 Patterns of keratin mutations causing human disease. (A) Most mutations occur within the conserved amino-terminal rod end of type I keratins and the carboxy-terminal rod end of type II keratins. The amino acids marking the end of the head domains (Glu, E) and the start of the tail domains (Ser, S) are underlined. The two most frequently mutated positions are labeled with asterisks (*). (B) The specific sequence variants reported at these highly mutated positions (* in A) are listed for all the keratins, together with their principal-associated diseases. These residues are highly conserved within type I (equivalent positions to N123 and R125 of K14) and type II (equivalent positions to E466 and E477 of K5) keratins. *Data extracted from HIFD (http://www.interfil.org).*

transplanted back under full experimental observation. Several laboratories are now working toward an end goal of developing cell and gene therapy for skin fragility disorders (Droz-Georget Lathion et al., 2015; Mavilio et al., 2006).

3.1 Disease Modeling in Simple Culture Systems

Experimental models of skin disease can be generated with different degrees of complexity, all with advantages and disadvantages, and the system of choice must be selected strategically (Table 2). *In vitro* cell models with reproducible phenotypes are essential tools for analyzing disease mechanisms and will become essential for evaluation of new therapeutic strategies for keratinopathies as they become available. Primary keratinocytes can be passaged through several generations but are easily lost to terminal differentiation. Immortalized cell lines have therefore been generated to overcome the limited supply of primary cells; while these should retain characteristic features of keratinocytes *in situ*, they can lose ability to undergo terminal differentiation (Morley et al., 2003). Potential problems associated with immortalized cell lines also include genetic drift in tissue culture and intrinsic variability due to a random immortalization process.

3.1.1 Generation of Pathomimetic Cell Lines

Two approaches are used to generate stable model cell lines with keratin mutations (Table 3). The most direct one is to immortalize keratinocytes derived from keratinopathy patients. Different methods of immortalization have been adopted, from SV40 large T antigen, to HPV16 E6/E7 (Morley et al., 2003) to the now-preferred telomerase-based systems (Dickson et al., 2000). Pairs of cell lines should be generated whenever possible, one from a patient paired with one from an unaffected close relative (e.g., NEB-1 is a control cell line (unaffected parent) for KEB-1 derived from a patient with epidermolysis bullosa simplex, Dowling-Meara, EBS-DM). In spite of interindividual genetic variability in the background, experiments using cell lines with keratin mutations demonstrate significant consistency, regardless of the methods adopted for cell line generation. Cell lines mimicking severe EBS (e.g., KEB-1, KEB-2, and KEB-7) all show keratin aggregates, even in resting states and especially in subconfluent or stress-activated states. Cell lines with milder mutations (e.g., KEB-3 and KEB-4) only show significant aggregates in stress assays.

A second approach, which avoids reliance on patient-derived cells and so broadens the range of model lines available, is transfection of mutant or

Table 2 Different Experimental Systems to Study Effects of Keratin Mutations

Approach	Advantages	Disadvantages
Simple monolayer tissue culture, using transfected epithelial cell lines	Simple, fast, and reproducible	Immortalized cells usually have limited differentiation capacity
Simple monolayer culture, using primary cells from patients	No genome scrambling from immortalization	Primary keratinocytes are fiddly to handle, harder to transfect. Limited tissue availability, complex to obtain material. Donor variability can hinder controls
Mixed monolayer culture using primary patient cells and fibroblast feeders	Cultures have longer life span, more clonogenic cells preserved	Very limited primary tissue availability. Mixture of cell types. Best practice uses mouse feeders
"Organotypic" or "skin-in-a-dish" cultures	Best culture-generated differentiation	Limited lifespan with standard protocols
Ex vivo organ culture	Closest culture option to *in vivo* skin. Multiple cell types	No patient tissue availability. No blood, nerve supply, and limited immune cells. Short lifespan
Genetically engineered nonhuman animal models, e.g., mouse	Can observe specific tissue effects over time through development and differentiation	Major species differences in skin can give misleading results. Care needed with interpretation
Human-to-mouse grafting of culture-generated human skin	Possibly fully differentiated. A long-term experimental model of human skin	Slow, technically complex preclinical strategy
Clinical trials	Most relevant. Orphan disease status may accelerate progress to clinical trials	Clinical trials are complex and expensive. Extensive and expensive preparation needed for legal and ethical approval

wild-type keratin constructs into a "parental" isogenic background of normal (wild-type keratin) cells, generating lines which express additional mutant or wild-type keratins in addition to their normal proteins (D'Alessandro et al., 2011; Liovic et al., 2009). These cells can also be tagged

Table 3 Published Human Cell Line Models of Keratin Diseases

Cell Line	Immortalization Route	Keratin Mutation	Associated Disease	References	Remarks
Patient-derived immortalized cell lines					
EB21 EB22	HPV16 E6/E7	K5 p.V186L	EBS, gen-nonDM	Chamcheu et al. (2009)	EB21 established in serum-free media, EB22 in serum-containing media
EBDM-1	HPV16 E6/E7	K14 p.R125H	EBS-DM	Lettner et al. (2013)	
EH11	HPV16 E6/E7	K1 p.V176_K197del	EI (least severe)	Chamcheu et al. (2011)	EH11, EH21, and EH31 were derived from patients with increasing clinical severity (EH11 < EH21 < EH31)
EH21	HPV16 E6/E7	K10 p.R156G	EI (more severe)		
EH31	HPV16 E6/E7	K10 p.L161_N162del	EI (most severe)		
KEB-1	SV40 T antigen	K5 p.E475G	EBS-DM	Morley et al. (2003)	Clinical severity less than KEB-7
KEB-2	SV40 T antigen	K5 p.E475G	EBS-DM	Morley et al. (2003)	
KEB-3	SV40 T antigen	K14 p.V270M	EBS-loc	Morley et al. (2003)	
KEB-4	HPV16 E6/E7	K14 p.V270M	EBS-loc	Morley et al. (2003)	
KEB-7	HPV16 E6/E7	K14 p.R125P	EBS-DM	Morley et al. (2003)	Aggregates in resting condition
KEB-11	HPV16 E6/E7	K14 c.314delGC	EBS, gen-nonDM	D'Alessandro, Coats, Jonkmann, Leigh, and Lane (2011)	K14-null

Continued

Table 3 Published Human Cell Line Models of Keratin Diseases—cont'd

Cell Line	Immortalization Route	Keratin Mutation	Associated Disease	References	Remarks
KEB-13	HPV16 E6/E7	K14. c.526-2A > C	EBS-DM	D'Alessandro et al. (2011)	K14-null
PC-10_K6a_N171K		K6a p.N171K	PC	Leachman et al. (2008)	
Pathomimetic cell lines					
NEB-1 EGFP-K14 R125P	HPV16 E6/E7	K14 p.R125P	EBS-DM	Liovic et al. (2009)	Immortalized normal wild-type keratinocytes transfected with EBS-mutant K14 or K5 sequences, tagged with EGFP
NEB-1 EGFP-K14 V270M	HPV16 E6/E7	K14 p.V270M	EBS-loc	Liovic et al. (2009)	
NEB-1 EGFP-K5 E475G	HPV16 E6/E7	K5 p.E475G	EBS-DM	Liovic et al. (2009)	
NEB-1 EGFP-K14 wt	HPV16 E6/E7	None	None	Liovic et al. (2009)	Transfected with EGFP-tagged wild-type K14

with green fluorescent protein (GFP, enhanced green fluorescent protein (EGFP)) for live-cell imaging of keratins. To the best of our knowledge, these lines are not commercially available, but the technology for their generation is fairly standard.

1. Keratinocyte cell lines (e.g., NEB-1 (wild type) and KEB-7 (EBS-DM) line, Table 3) are cultured in 75% Dulbecco's modified eagle's medium (DMEM) with 25% Ham's F12 medium, containing 1% L-glutamine, 1% penicillin/streptomycin, and 10% fetal bovine serum (FBS), with additional growth factors such as epidermal growth factor (EGF) (10 ng/mL), insulin (5 μg/mL), hydrocortisone (0.4 μg/mL), adenine (1.9×10^{-4} *M*), transferrin (5 μg/mL), and lyothyronine (2×10^{-11} *M*). Human embryonic kidney 293 (HEK293)T cells are cultured in DMEM supplemented with 10% FBS and 1% penicillin/streptomycin. Primary keratinocytes, if used, are cultured according to standard protocols (Rheinwald & Green, 1975). All cells are cultured at 37 °C in 5% CO_2 atmosphere.
2. The coding region of human K14 complementary deoxyribonucleic acid (cDNA) (NM_000526) is cloned into the EGFP-C1 vector (Clontech). The keratin mutation (e.g., K14 R125P) is introduced into the wild-type sequence to create an EGFP-tagged construct (EGFP-K14 R125P), using QuikChange site-directed mutagenesis (Stratagene) (Liovic et al., 2009). The EGFP tag is cloned into the N-terminal end of the keratin as it has been reported that C-terminal GFP tags are more likely to interfere with intermediate filament assembly efficiency (Herrmann, Hesse, Reichenzeller, Aebi, & Magin, 2003). The tagged K14 wild-type and mutant cDNAs are excised from these constructs and recloned into the pLVX-EF1α-AcGFP1-C1 lentiviral expression vector using the *BstBI* and *BamHI* restriction sites (www.clontech.com).
3. Nonreplicative lentiviruses are produced by triple transfection of HEK293T cells with 1.5 μg pLVX-EF1α-AcGFP1-C1 lentiviral expression vector, 1 μg pHR-CMV 8.2 deltaR packaging vector, and 1 μg pCMV VSV-G envelope vector (Davidson et al., 2010) using Effectene transfection reagent (Qiagen, Germany). Virus particle-containing supernatants are collected after 48 and 72 h, filtered through a 0.45-μm filter, ultracentrifuged at 19,600 × *g*, and the pellet resuspended in DMEM and frozen in aliquots.
4. Keratinocytes are seeded into a 24-well plate for 16 h before cells are infected with virus-packaged constructs at titrations of 10^{-2}, 10^{-3}, $10^{-3}/2$ dilutions (of each construct) and incubated in the presence of

8 μg/mL of polybrene. (8 μg/mL is normally recommended but a higher polybrene concentration can be used to improve transduction efficiency of keratinocytes.) After 72 h, cells are trypsinized, centrifuged, and resuspended in PBS for fluorescent-activated cell sorting (FACS) analysis.

5. Viral titer for each construct is derived from the percentage of GFP-positive cells in the population of infected cells at a specific dilution, and determined as the number of viral particles per mL. Multiplicity of infection (M.O.I.) = (viral particles/cell number). Viral particles needed = (No. of cells to be infected × desired M.O.I.). Volume of viral supernatant to be added = (viral particles needed/viral titer) = [(No. of cells to be infected × desired M.O.I.)/viral particles per mL].
6. To generate stable cell lines, various virus-packaged constructs (M.O.I. = 1.0) are used to infect immortalized keratinocyte cells in suspension in 8 μg/mL of polybrene. Cells are preincubated at 37 °C for 1 h before plating onto 10 cm dishes.
7. After 72 h, virus-containing supernatant is replaced with fresh culture medium. GFP-positive cells are selected by FACS (at 1 week after transduction) or by antibiotic resistance (in medium containing 2 μg/mL of puromycin for over 2 weeks). Cells are then ready for use. N.B. Prior to antibiotic selection, any antibiotic must be first titrated to determine the lowest effective concentration with tolerable toxicity.

3.2 Fluorescence Time-Lapse Imaging of Cells Expressing Mutant Keratins

Provided that sufficient controls are examined in parallel to exclude the possibility of toxic effects of the EGFP moiety, the use of fluorescence tags cloned into mutant keratins in cell model lines allows direct real-time microscopic monitoring of target protein in the cell. This has been widely used to examine keratin dynamics in keratinocytes (Moch, Herberich, Aach, Leube, & Windoffer, 2013).

1. 10,000 cells are grown on a 35-mm WillCo-dish® glass-bottomed dish (WillCo wells, Amsterdam).
2. Images of cells expressing the fluorescence constructs are collected every 15 s for a period of 30 min. For this, we use an Olympus UPlanApo/IX70 100× (N.A. 1.35) oil immersion objective lens in a Deltavision epifluorescence inverted microscope (Applied Precision, USA), equipped with a fully motorized Z stage (Applied Precision, USA) and linked to Photometrics CCD camera (CoolSNAP HQ2) using the SEDAT filter set.

3. The sequences of images are then converted into movies and subsequently analyzed using SoftWoRx software (Applied Precision, USA). The image series in Fig. 3, presented as inverse fluorescence micrographs, are taken from time-lapse recordings generated in this way.

3.3 Disease Modeling in 3D Cultures

To mimic *in vivo* physiological conditions more closely, 3D "organotypic" cultures are used, in which keratinocytes are grown on a gel matrix (typically rat-tail type I collagen) embedded with fibroblasts. Organotypics generated using immortalized EBS patient cells can recapitulate some of the basal cell cytolysis reminiscent of actual EBS blistered skin (Chamcheu et al., 2009). Similar organotypics grown using immortalized EI-mimetic keratinocyte cell lines (with mutant K10) demonstrated cytolysis in the suprabasal layers (Chamcheu et al., 2011).

1. 8 mL type I rat-tail collagen (Corning, #354236) and 1 mL 10× DMEM (minimal essential medium as a pH indicator, Sigma) is added to a chilled 50 mL tube. Air bubbles are avoided during mixing and the tube kept immersed in ice. Collagen/DMEM mixture turns yellow due to acetic acid used to dissolve collagen.
2. Solution is brought to neutral pH by adding 1 *M* sodium hydroxide dropwise until solution turns orange. The sodium hydroxide must be well dissolved by stirring. Once neutralized, collagen sets quickly.
3. Add 1 mL fibroblast suspension (1×10^6 cells in FBS) to the neutralized collagen. 2 mL of collagen mixture containing fibroblasts is added onto insert platforms (Greiner) in a 6-well plate and plates incubated at 37 °C for 1–3 h to allow collagen to gel completely.
4. The next day, keratinocytes (500 μL of keratinocyte suspension at 1×10^6 cells/mL) are seeded onto the collagen gels after aspirating away the medium from inside and outside the inserts. Cultures are incubated at 37 °C for 3 h (minimum) to allow keratinocyte attachment onto gels. 2 mL of keratinocyte medium is then added into each insert and 1 mL outside the insert.
5. Upon keratinocyte confluence (typically around 4 days), inserts are raised to the air–liquid interface. All medium is carefully aspirated, and each culture insert is transferred into deep well plates (Greiner). Sufficient medium is added to the outside of each insert such that medium just touches the underside of culture insert. Medium is changed every 3–4 days. Organotypic cultures can be harvested and processed after 2–4 weeks.

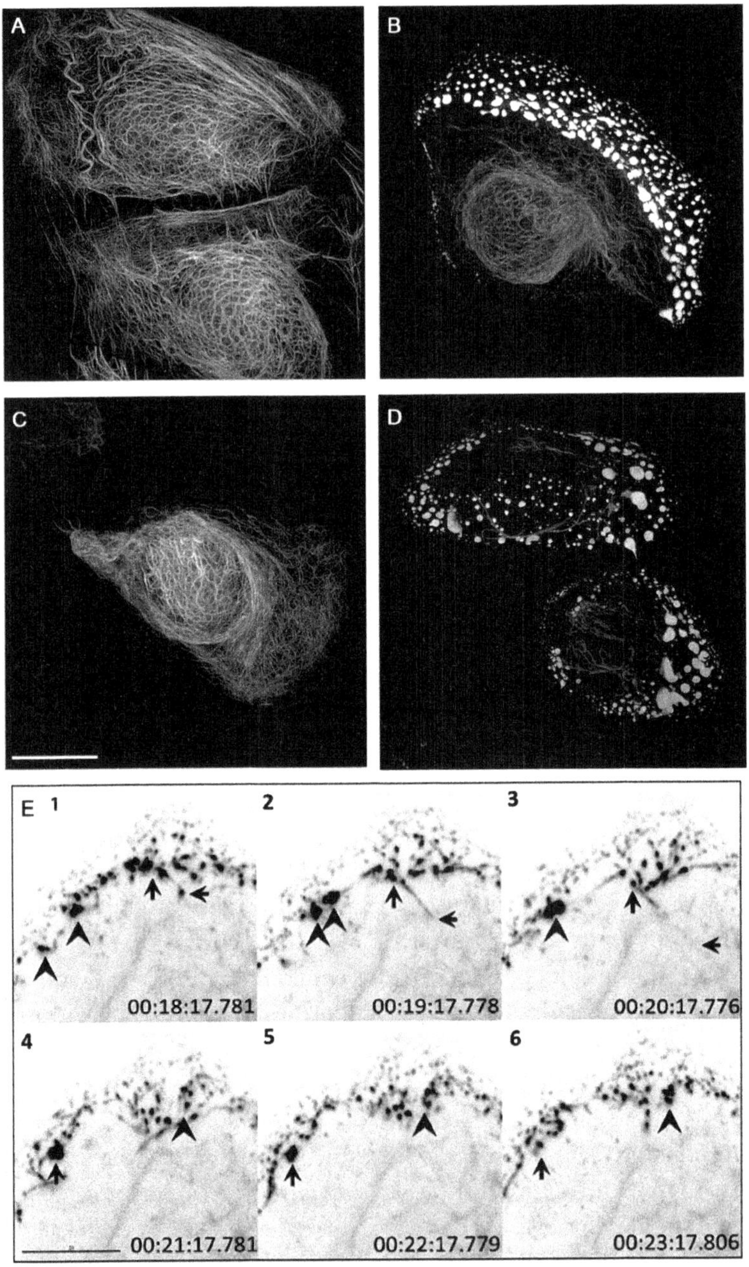

Figure 3 Human disease-mimetic cell lines expressing GFP-tagged mutant keratins. (A and C) Cell lines with wild-type keratin show filaments only, whereas keratin aggregates (misfolded protein) are seen in (B) EBS-DM (K14 R125P) and (D) PC-mimetic cell lines (K16 L132P) (Quan, T. and Common, J., unpublished) generated as described in Section 3.1.1. Scale bar, 10 μm. (E) Time-lapse imaging of EGFP-K14 R125P cell at 15 s intervals for 30 min. The montage showed keratin aggregate movements at 1 min intervals during acquisition from 18th to 23rd min. Small peripheral keratin aggregates coalesce (arrowheads), disassemble, and intercalate into preexisting keratin filaments (arrows). Scale bar, 10 μm. *Panel (E): Reproduced from Tan (2012).*

N.B. Cell numbers can be adjusted for experimental needs. Time between fibroblast and keratinocyte seedings can also vary. Fibroblasts may remodel the gel and cause it to contract. Keratinocyte media formulations used will depend on cell type (primary or immortalized).

3.4 The Use of Mouse Models to Study Effects of Keratin Mutations

In routine culture conditions, organotypic cultures can be sustained for a few weeks. For longer term studies, they can also be grafted onto immunodeficient mice and maintained for several months, as was done with K6 mutant cells derived from pachyonychia congenital (PC) patients (Garcia et al., 2011). With this approach, "*in vivo*" and "human" aspects of the disease can be combined in an experimentally tractable format. However, the defective immune system of these mice may in itself distort the pathology of the disease as there is evidence that inflammation may play a role in the disease (Lu et al., 2007; Wally et al., 2013).

Transgenic mouse models of keratinopathies have been made by many labs, and the reader is referred to other reviews for details (Arin & Roop, 2004; Chen & Roop, 2005; Vijayaraj, Sohl, & Magin, 2007). Disease mimetic models have been generated for EHK (EI) (Arin, Longley, Wang, & Roop, 2001; Fuchs, Esteves, & Coulombe, 1992; Reichelt & Magin, 2002), EBS (Coulombe, Hutton, Vassar, & Fuchs, 1991; Lloyd et al., 1995; Peters, Kirfel, Bussow, Vidal, & Magin, 2001; Vassar, Coulombe, Degenstein, Albers, & Fuchs, 1991), and PC (Lessard & Coulombe, 2012) and these mice duly develop site-specific keratinopathy phenotypes such as paw blisters, or hyperkeratosis on mechanical stress. They have also been powerfully used to identify unexpected consequences of modulating skin keratins, such as protective upregulation of K17 by sulforaphane (Kerns, DePianto, Dinkova-Kostova, Talalay, & Coulombe, 2007). However, they have drawbacks, such as lethality in mice of mutations that humans can sustain for life, which necessitate construction of more complex mouse models with postnatally induced mutations (Arin & Roop, 2004; Cao, Longley, Wang, & Roop, 2001). Sometimes, more radical engineering was needed to reproduce the human phenotype (Coulombe, Hutton, Vassar, et al., 1991; Lloyd et al., 1995; Peters et al., 2001; Vassar et al., 1991). Furthermore, mouse skin is histologically and anatomically different from human skin (for example, mouse epidermis is thinner than human, and the dermis has a layer of smooth muscle (the *panniculus carnosus*) that is not seen in human skin) such that caution is needed when extrapolating from mouse experiments. Increasing societal pressure to move away from animal work towards nonanimal alternatives is now directing the focus on human systems and *in vitro* models of disease where possible.

4. KERATIN MUTATIONS IN STRESS ASSAYS

There is currently no practical cure for any of the keratinopathies. To design strategies for therapy, an understanding of the mechanisms underlying tissue failure is needed. Yet, when cultured EBS-DM keratinocytes (Section 3) are grown to confluence, only a fraction of the cells show spontaneous keratin aggregates, although all cells carry the mutation. This parallels the clinical observations that most of a patient's skin is intact most of the time and indicates that mutation-associated skin fragility is not a total failure of tissue resilience. Thus, effort has been directed into designing assays that increase the stress on the cells to variable degrees and can mimic, *in vitro*, aspects of the stresses that lead to blistered skin *in vivo*.

Keratin aggregates and misfolded protein are common features of keratinopathies, especially prominent in EBS-DM as first recognized by Anton-Lamprecht and Schnyder (1982). Keratin aggregates can also appear in nonpathological situations, such as during mitosis (Franke, Schmid, Grund, & Geiger, 1982; Horwitz, Kupfer, Eshhar, & Geiger, 1981; Lane, Goodman, & Trejdosiewicz, 1982; Plancha, Carmo-Fonseca, & David-Ferreira, 1991). This was subsequently shown to be due to hyperphosphorylation of keratins (such as K8 pSer73, K5 pThr150, and K6 pThr145) associated with stress, mitosis, and apoptosis (Liao, Ku, & Omary, 1997; Omary, Ku, Tao, Toivola, & Liao, 2006; Toivola, Zhou, English, & Omary, 2002). *In vitro*, the effects of keratin mutations can be assayed in various ways, but invariably some form of stress is needed to trigger a differential response in the mutant cells.

4.1 Heat Stress

Elevated ambient temperature is often reported by EBS patients to worsen their susceptibility to blister formation (Horn & Tidman, 2000). Elevated temperature has been used on keratinocyte-derived lines expressing EBS keratin mutations to demonstrate the greater instability of the mutant keratin networks (Morley et al., 1995; Sorensen et al., 2003). EBS cells were subjected to transient elevated temperature after which keratin aggregates formed spontaneously, reaching a maximum at 15 min into the recovery period and disappearing about 60 min of recovery. Aggregates were inducible this way in several EBS-mimetic cell lines (KEB-1, KEB-3, KEB-4, and KEB-7) but not in control keratinocytes (NEB-1, KT, SVK14, and TR146) (Morley et al., 1995, 2003). This was the first assay to demonstrate the reversible instability of mutant keratins in EBS cells. In K14 null cells (KEB-11), aggregates were not induced by heat shock, evidence that it is the mutant protein, rather than a general defect of the keratin network, that leads to formation of keratin aggregates (D'Alessandro et al., 2011).

1. Cells are grown to near full confluence on coverslips in culture dishes for 2–3 days, and then the 37 °C culture medium is replaced with prewarmed 43 °C medium.
2. Culture dishes are placed in a 47 °C water bath for 15 min, taking readings of the medium temperature every 5 min (expected mean temperature 43 °C). At 15 min, the medium is replaced with fresh 37 °C medium and the dish is returned back to the 37 °C incubator.
3. Coverslips are removed from the dish every 15 min and fixed in methanol/acetone (1:1) over a time period of 1 h.
4. The fixed cells are processed for immunocytochemistry to visualize the keratin network (and other cytoskeleton elements as desired) using appropriate antibodies, and imaged by fluorescence microscopy.
5. Cell viability is determined from the cells remaining on the culture dish at 24 h after treatment, after removal of all the coverslips. Remaining cells are stained with Nigrosin or Trypan blue, which will stain dead cells only, as living cells exclude the dye (Fig. 4).

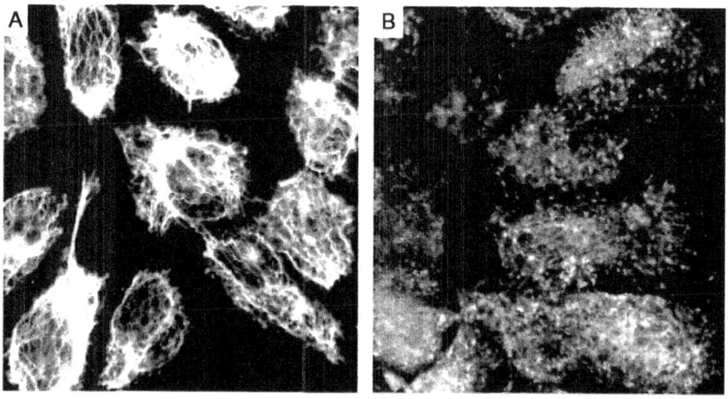

Figure 4 Keratin filament disruption in mutant EBS cells after heat shock. The heat shock treated cells were then fixed and stained after 15 min of recovery at 37 °C. (A) KT cells (immortalized wild-type keratinocytes). (B) KEB-1 cells (immortalized EBS-DM keratinocytes, Table 3). Keratin network was stained using pan-keratin antibody LP34. *Reproduced from Morley et al. (1995).*

4.2 Osmotic Stress

Osmotic swelling can frequently occur *in vivo* following a transient energy deficit (as caused by transient ischemia) which inactivates the cell membrane ion pumps so that osmotic balance in the cell is not maintained. Keratinocytes subjected to hypoosmotic shock undergo rapid cell swelling that transiently disrupts cytoskeletal components such as actin and microtubules. It is thought that keratins may provide stability to epithelial cells during the turbulent shape changes of osmotic stress, as keratin assembly is not so energy dependent as actin and tubulin. *In vitro*, cells respond to hypoosmotic stress by undergoing sudden rapid cell swelling, but this is quickly followed by a return to their original volume within minutes (Hoffmann, 1987; Kimelberg, 1991), by a regulatory volume decrease involving efflux of osmotically active molecules such as taurine from the cells (Moran, Maar, & Pasantes-Morales, 1994). In tissue culture, urea can be used to induce hypoosmotic stress (Kucerova & Strbak, 2001). Urea is a small permeable molecule that can rapidly diffuse into the cells, increasing the cell's osmolality which draws water in across the membrane and so causes cell swelling (DeFelice, 1991). By allowing cells to recover from this hypoosmotic shock, the stress responses during the recovery phase can be compared between cell lines (D'Alessandro et al., 2011; D'Alessandro, Russell, Morley, Davies, & Lane, 2002; Liovic et al., 2008).

1. 1×10^6 cells are seeded onto a 10 cm petri dish and cultured to 80% confluence for maximum effect of the osmotic stress.
2. Keratinocytes are either left unstressed or subjected to hypoosmotic shock by immersion in 150 m*M* urea for 5 min at 37 °C. Cell swelling occurs as water passes into the cell, followed by the regulatory volume decrease within 5 min of exposure to urea treatment.
3. Keratinocytes are then allowed to recover in fresh tissue culture medium for varying periods of time before being harvested for immunoblotting analysis.

Stress kinases such as JNK or p38 mitogen-activated protein kinase (MAPK) are often activated during osmotic stress response, and their activated phosphorylated forms can be detected biochemically using immunoblotting from cells harvested during recovery from osmotic stress.

4.3 Cell Shape Changes: Cell Spreading

Other types of stress on the epithelial cytoskeleton can be induced by events that caused dramatic cell shape change, as these will all require cytoskeleton remodeling. The most obvious of these is mitosis. Although no significant evidence of altered cell proliferation rates related to expression of mutant keratins has been found (D'Alessandro et al., 2011; Morley et al., 1995), Morley and colleagues observed slower postmitotic cell spreading in mutant cells (Fig. 5A). This finding predicts that epidermis with a higher mitotic index contains more cells in an unstable state, possibly explaining the greater fragility in fast-growing infants. The efficiency of cell spreading was measured in newly replated cells to confirm that EBS-derived keratinocytes (KEB-1 and KEB-3) remained rounded for longer than wild-type keratin cell lines (Morley et al., 1995). By 24 h, however, all cell lines were equally spread (Morley et al., 1995), indicating that mutant cells retain the ability to spread but that the process is less efficient.

1. Keratinocytes are grown to confluence under culture conditions described in Section 3.1.1. Cells are dissociated with trypsin as for passaging, resuspended in culture medium to obtain single-cell suspensions, and filtered through 70 μm cell strainers.
2. Cells are plated at high density (1×10^6 cells/mL) on coverslips in culture dish and left to settle for 1 h, rinsed with PBS, and fresh medium is added. Coverslips are removed at hourly intervals up to 24 h and fixed in methanol/acetone (1:1).
3. Cells are stained with Toluidine Blue for 15 min and rinsed in alcohol and xylene. Coverslips are inverted onto drops of DPX mounting medium on microscope slides and imaged at low magnification (10×, bright field). Images are analyzed using ImageJ to calculate cell size. Cell area of 100 cells (ranked according to size) at every timepoint is displayed logarithmically (Fig. 5A).

4.4 Cell Shape Changes: Cell Motility and Migration

Cell migration also involves profound cell shape changes in epidermal keratinocytes, as seen in the initiation of wound healing *in vivo*, as well as during embryonic development. Most studies of cell migration have focused on actin and tubulin systems, but some studies have implicated a role for intermediate filaments. The type III intermediate filament protein vimentin is associated with the invasive properties during epithelial–mesenchymal transition (EMT) (Mendez, Kojima, & Goldman, 2010) and required for epithelial wound repair (Rogel et al., 2011). Another study reported that cells expressing K14 lead the collective invasion of breast cancer and that subsequent knockdown of K14 can prevent this invasive behavior

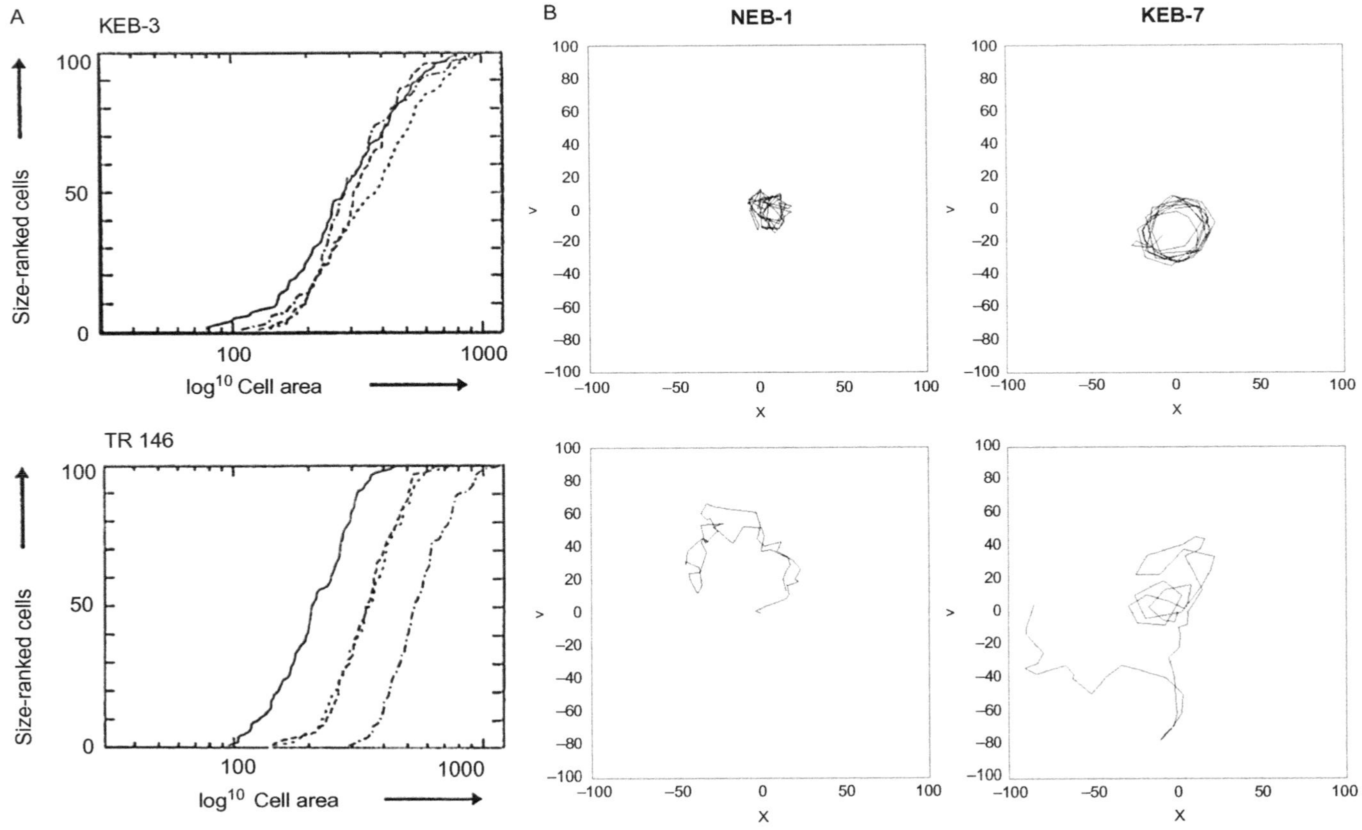

Figure 5 Effects of keratin mutation on cell spreading and motility. (A) Graphs showing cell spreading with increasing time following suspension and replating of KEB-3 (EBS) cells (top) and TR146 (control) cells (bottom). *X*-axis = log 10 of cell area (μm), *Y*-axis = cumulative number of cells. Each curve shows the distribution of sizes of 100 cells counted at one time point. Measurements were taken at 1 h (continuous line), 4 h (dashed line), 8 h (dotted line), and 12 h (dot-dashed line) after replating. (B) EBS-DM keratinocytes travel further than wild-type keratinocytes in single-cell migration. Live-cell DIC images of single cell from each cell line are obtained, images are segmented, and centroids are plotted with ImageJ macro. Representative migration tracks of a cell over a period of 5 h for NEB-1 and KEB-7 are shown. Figure axes are

(Cheung, Gabrielson, Werb, & Ewald, 2013). In melanoma cell lines, invasive and metastatic behavior was associated with expression of keratins K8 and K18 (Hendrix et al., 1992) and that this observation could be reproduced by coexpressing vimentin and K8/K18 experimentally (Chu, Seftor, Romer, & Hendrix, 1996). It is quite possible that keratin mutations could affect cell migration, since many intermediate filaments now seem to play a role in cell migration (Leduc & Etienne-Manneville, 2015).

4.4.1 Single-Cell Movement

Recent advances in microscopy tools such as digital imaging and sophisticated software have allowed studies of cell motility and migration to be done by time-lapse filming of cell behavior and shape changes over time. Single-cell movement can be monitored periodically and cell tracks can be used to analyze motility and translocation (Hamill, Hopkinson, Jonkman, & Jones, 2011).

1. 100–300 cells are seeded onto 4-well Lab-Tek II #1.5 chambered coverslips.
2. A single cell is selected with the "mark and visit" tool of SoftWoRx (Applied Precision, USA) for each cell line.
3. Repetitive images are collected every 5 min for a total of 5 h using an Olympus UApo/340 20× (N.A. 0.75) objective lens, to generate movies of randomly migrating live cells. A minimum of ten individual fields is recorded for each cell line.
4. Differential interference contrast (DIC) image stacks of single-cell migration are analyzed and processed using several macros written with ImageJ and their corresponding migration paths are plotted. Representative results are seen in Fig. 5B.

4.4.2 Collective Cell Movement

Most keratinocyte cell movement is collective rather than singular, as epithelial cells are well attached to each other by cell–cell adhesion structures. Collective migration can be stimulated by disrupting a confluent cultured monolayer epithelial sheet with a scratch to remove a strip of cells. This stimulates cells at the scratch edge to become migratory and to close up the gap over a period of time, mimicking cell migration during wound healing *in vivo*. "Scratch wound" assays are useful to examine epidermal cell activation and migration without the complication of dermal responses. The assay can be scaled up by using the commercially available 96-well WoundMaker Tool, with its 96-well format of PTFE pin tips. The scratched 96-well plate can then be tracked by IncuCyte phase contrast live-cell imaging to measure progressive wound closure in each well over time. Scratch wound experiments have been used to compare cell migration rates between several mutant cell lines and their corresponding wild-type keratinocytes (Morley et al., 2003; Osmanagic-Myers et al., 2006; Seltmann et al., 2013). Scratch wound closure is found to be consistently faster in EBS cells than in wild-type keratinocytes (D'Alessandro et al., 2011; Morley et al., 2003).

1. 20,000 cells are seeded onto each well of an Essen ImageLock 96-well plate and grown until 2 days postconfluence. N.B. The state of confluence should be standardized, as it affects the wound response through the maturation of desmosomes after confluence (Garrod, Berika, Bardsley, Holmes, & Tabernero, 2005).
2. Confluent wells are scratched (through the culture medium) with a 96-well WoundMaker Tool, which gently remove cells from the confluent monolayer using a 96 arrays of PTFE pin tips.
3. Cells are then washed twice with sterile culture medium and the plate is placed inside an IncuCyte imaging system for phase contrast live-cell imaging of scratch wounds. Wound images are collected hourly for 26 h. The rate of wound closure will depend on the cell line used.
4. Data are processed using ImageJ. Representative results are seen in Fig. 6.

4.5 Mechanical Stress

The physical resilience of epidermal keratinocytes is largely due to the keratin intermediate filaments they express, as shown by the pathological skin fragility resulting from mutations in skin keratins (Fig. 1), which rupture upon mild mechanical stress. The keratin-desmosome cytoskeleton system is ideally placed to act as a mechanosensory system in the epidermis. Understanding the physical forces acting on keratinocytes, and the consequences of disturbing keratin integrity, is crucial for understanding the pathogenesis of keratin diseases as well as for understanding wound healing regulation. Stretching keratinocytes are known to stimulate DNA synthesis (Brunette, 1984), and to increase cell proliferation through MAPK activation pathways (Kippenberger et al., 2000), but the stretch type has different consequences. Russell and colleagues showed that oscillating mechanical stress differentiates between wild-type keratinocytes and mutant EBS keratinocytes in culture (Russell, Andrews, James, & Lane, 2004), whereas mutant cells behaved the same as wild-type cells in a sustained stretch, to which they were remarkably resistant (Beriault et al., 2012). Mechanical stretch also induces ERK activation, which contributes to the apoptosis

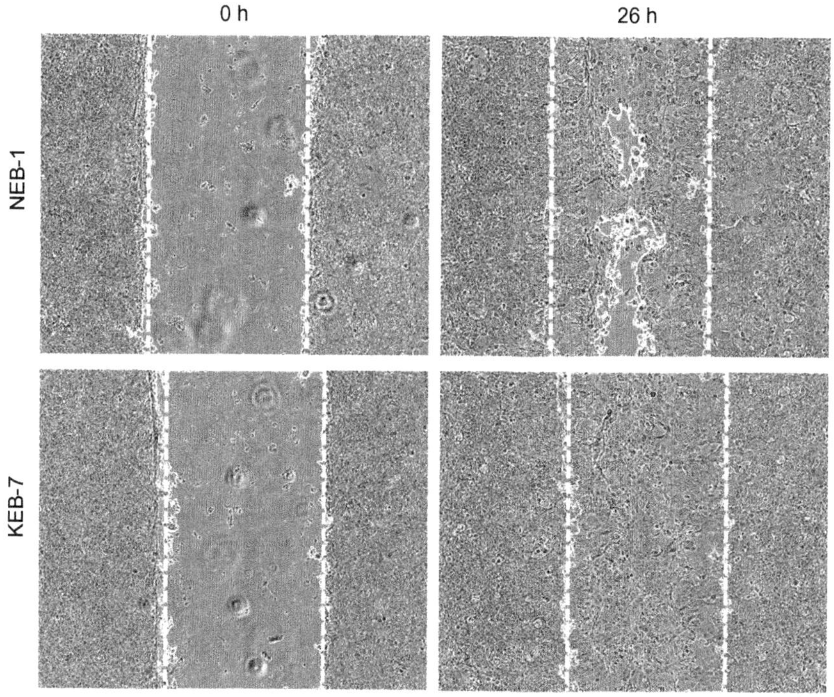

Figure 6 Effects of keratin mutation on collective cell migration. EBS-DM keratinocyte sheets close scratch wounds faster than wild-type keratinocytes. Scratch wounds are generated in confluent keratinocyte monolayers and left to close over time. White dashed line denotes the denuded area at the start (0 h); white line denotes the remaining uncovered area after 26 h. *Reproduced from Tan (2012).*

resistance seen in mutant keratin cells (Russell, Ross, & Lane, 2010). Thus, mechanical stretch studies could reveal a role for keratin intermediate filaments in cell survival.

Mechanical stress can be applied at the basal surfaces of cultured cells by stretching and deforming the substrate to which they are attached. The Flexcell Cell Stretcher (Flexcell International, Hillsborough, NC) is a commercial apparatus for mechanical stretching cultured cells that uses vacuum to apply constant or cyclical stretch/relaxation at variable force to cells grown on a deformable membrane. Oscillating stretch at 12% amplitude and 4 Hz frequency demonstrated clear differences between EBS-mimetic and wild-type keratinocytes (Russell et al., 2004, 2010).

1. 200,000 cells are seeded onto collagen IV-coated silicon membranes in a 6-well BioFlex flexible culture plate and grown to 80% confluence.

2. The FX-4000T Cell Stretcher (Flexcell International) is used to stretch cells with oscillation frequency of 4 Hz and effective amplitude of 12% for times up to 180 min at 37 °C in 5% CO_2 atmosphere. Control wells (unstretched) are isolated from the vacuum.
3. After stretch, silicone membranes are excised from the culture plate with a sharp scalpel and fixed in cold methanol/acetone (1:1) for 5 min, then washed twice with cold PBS.
4. Cells are processed for immunofluorescence as desired and the silicone membranes are mounted onto glass microscope slides using CitiFluor before assessment by fluorescence microscopy (Fig. 7).

5. CONCLUSION, PEARLS, AND PITFALLS

The consequences of mutations in keratins are turning out to be more complex than predicted from the early model "box in a balloon" model of tissue failure (Lane, 1994) and early views on the altered dynamics of mutant keratins (Ma, Yamada, Wirtz, & Coulombe, 2001). Many of the higher order consequences are likely to be secondary due to the importance of maintaining the integrity of any epithelial sheet tissue that by definition forms the biological barrier between two different environments in the body. Nevertheless, several key indicator processes, such as resistance to distortion of the cell and its keratin cytoskeleton, can be measured in the quite simple tissue culture assays described here. Taking advantage of the phenotypic hallmarks of EBS-DM (keratin aggregates), these assays can now be adapted for drug screening in the search for therapy.

Diseases caused by keratin mutations are however rare, and this will always pose challenges to doctors and intermediate filament researchers. The challenges range from accurate diagnosis (which may be overcome by networked telediagnosis, as used by the International PC Consortium) to patient numbers for future trials (although patient advocacy networks foster great enthusiasm to volunteer for participation, and governments fast-track rare disease therapies based on their "orphan disease" status). The biggest obstacle is always the cost of new therapy for a very rare disorder. Thinking about keratinopathies as one group rather than many different diseases will probably be helpful, especially if generic treatment strategies can be devised that are not dependant on private mutations, but rather on processes shared with many or all of these disorders, such as protein misfolding, assembly dynamics, or cell stress. The judicious selection of assays, models, and controls to give maximum informative preclinical data will all help make therapeutics for keratinopathies a reality.

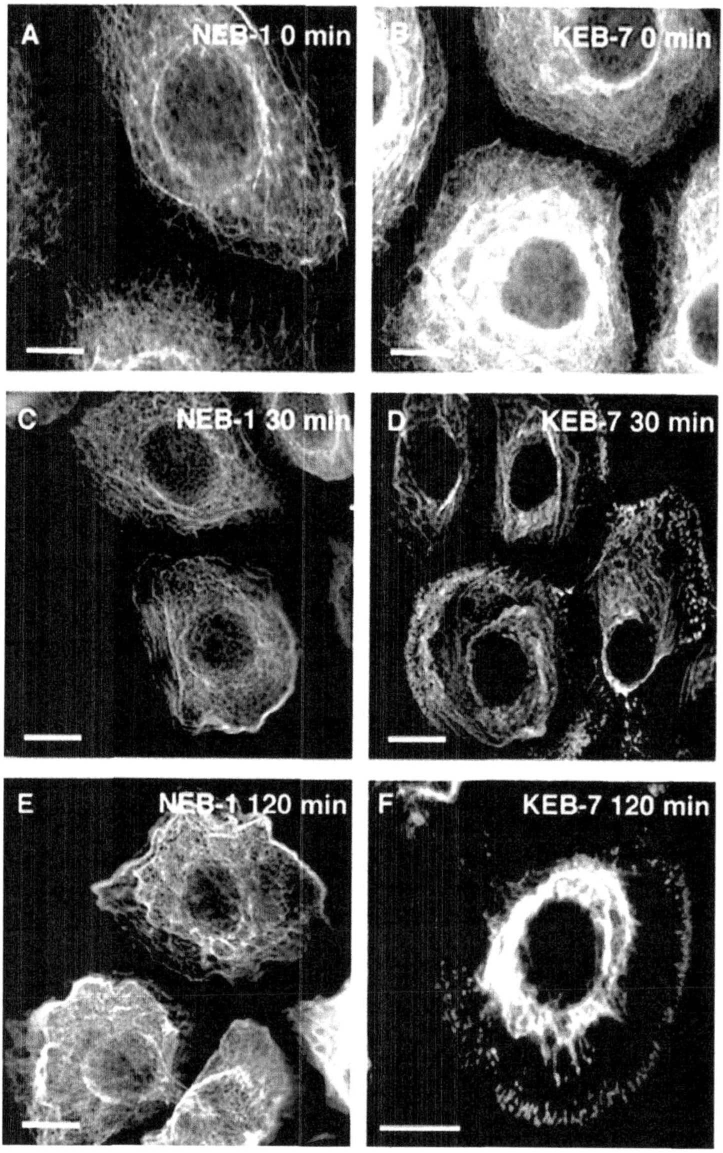

Figure 7 Mechanical stretch induces keratin fragmentation in EBS-DM keratinocytes. (A, C, E) Control and (B, D, F) EBS-DM keratin mutant cells were subjected to oscillating stretch at 4 Hz and 12% amplitude. Concentric compaction and wrinkling of keratin filaments are seen in all cells, but only EBS cells show keratin aggregates after stretching. After 2 h of stretching, EBS cells show severely disrupted keratin network with filaments concentrated around the nucleus. Scale bars = 10 μm. *Reproduced from Russell et al. (2004).*

ACKNOWLEDGMENTS

Our thanks to the patients whose courage has inspired research on keratin diseases, and to their doctors and carers who make the research possible. The authors are supported by the Biomedical Sciences Research Council of Singapore and DebRA International.

REFERENCES

Anton-Lamprecht, I., & Schnyder, U. W. (1982). Epidermolysis bullosa herpetiformis Dowling-Meara. Report of a case and pathomorphogenesis. *Dermatologica, 164*, 221–235.

Arin, M. J., Longley, M. A., Wang, X. J., & Roop, D. R. (2001). Focal activation of a mutant allele defines the role of stem cells in mosaic skin disorders. *The Journal of Cell Biology, 152*, 645–649.

Arin, M. J., & Roop, D. R. (2004). Inducible mouse models for inherited skin diseases: Implications for skin gene therapy. *Cells, Tissues, Organs, 177*, 160–168.

Beriault, D. R., Haddad, O., McCuaig, J. V., Robinson, Z. J., Russell, D., Lane, E. B., et al. (2012). The mechanical behavior of mutant K14-R125P keratin bundles and networks in NEB-1 keratinocytes. *PloS One, 7*, e31320.

Bonifas, J. M., Rothman, A. L., & Epstein, E. H., Jr. (1991). Epidermolysis bullosa simplex: Evidence in two families for keratin gene abnormalities. *Science, 254*, 1202–1205.

Brunette, D. M. (1984). Mechanical stretching increases the number of epithelial cells synthesizing DNA in culture. *Journal of Cell Science, 69*, 35–45.

Cao, T., Longley, M. A., Wang, X. J., & Roop, D. R. (2001). An inducible mouse model for epidermolysis bullosa simplex: Implications for gene therapy. *The Journal of Cell Biology, 152*, 651–656.

Chamcheu, J. C., Lorie, E. P., Akgul, B., Bannbers, E., Virtanen, M., Gammon, L., et al. (2009). Characterization of immortalized human epidermolysis bullosa simplex (KRT5) cell lines: Trimethylamine N-oxide protects the keratin cytoskeleton against disruptive stress condition. *Journal of Dermatological Science, 53*, 198–206.

Chamcheu, J. C., Pihl-Lundin, I., Mouyobo, C. E., Gester, T., Virtanen, M., Moustakas, A., et al. (2011). Immortalized keratinocytes derived from patients with epidermolytic ichthyosis reproduce the disease phenotype: A useful in vitro model for testing new treatments. *The British Journal of Dermatology, 164*, 263–272.

Chen, J., & Roop, D. R. (2005). Mouse models in preclinical studies for pachyonychia congenita. *The Journal of Investigative Dermatology. Symposium Proceedings, 10*, 37–46.

Cheung, K. J., Gabrielson, E., Werb, Z., & Ewald, A. J. (2013). Collective invasion in breast cancer requires a conserved basal epithelial program. *Cell, 155*, 1639–1651.

Chu, Y. W., Seftor, E. A., Romer, L. H., & Hendrix, M. J. (1996). Experimental coexpression of vimentin and keratin intermediate filaments in human melanoma cells augments motility. *The American Journal of Pathology, 148*, 63–69.

Coulombe, P. A., Hutton, M. E., Letai, A., Hebert, A., Paller, A. S., & Fuchs, E. (1991a). Point mutations in human keratin 14 genes of epidermolysis bullosa simplex patients: Genetic and functional analyses. *Cell, 66*, 1301–1311.

Coulombe, P. A., Hutton, M. E., Vassar, R., & Fuchs, E. (1991b). A function for keratins and a common thread among different types of epidermolysis bullosa simplex diseases. *The Journal of Cell Biology, 115*, 1661–1674.

D'Alessandro, M., Coats, S. E., Jonkmann, M. F., Leigh, I. M., & Lane, E. B. (2011). Keratin 14-null cells as a model to test the efficacy of gene therapy approaches in epithelial cells. *The Journal of Investigative Dermatology, 131*, 1412–1419.

D'Alessandro, M., Russell, D., Morley, S. M., Davies, A. M., & Lane, E. B. (2002). Keratin mutations of epidermolysis bullosa simplex alter the kinetics of stress response to osmotic shock. *Journal of Cell Science, 115*, 4341–4351.
Davidson, L., Maccario, H., Perera, N. M., Yang, X., Spinelli, L., Tibarewal, P., et al. (2010). Suppression of cellular proliferation and invasion by the concerted lipid and protein phosphatase activities of PTEN. *Oncogene, 29*, 687–697.
DeFelice, L. J. (1991). Channels, carriers, and pumps: An introduction to membrane transport: By W. D. Stein. San Diego: Academic Press (1990). *Cell, 66*, 13–14.
Dickson, M. A., Hahn, W. C., Ino, Y., Ronfard, V., Wu, J. Y., Weinberg, R. A., et al. (2000). Human keratinocytes that express hTERT and also bypass a p16 (INK4a)-enforced mechanism that limits life span become immortal yet retain normal growth and differentiation characteristics. *Molecular and Cellular Biology, 20*, 1436–1447.
Droz-Georget Lathion, S., Rochat, A., Knott, G., Recchia, A., Martinet, D., Benmohammed, S., et al. (2015). A single epidermal stem cell strategy for safe ex vivo gene therapy. *EMBO Molecular Medicine*, 7, 380–393.
Fischer, H. P., Altmannsberger, M., Weber, K., & Osborn, M. (1987). Keratin polypeptides in malignant epithelial liver tumors. Differential diagnostic and histogenetic aspects. *The American Journal of Pathology, 127*, 530–537.
Franke, W. W., Schmid, E., Grund, C., & Geiger, B. (1982). Intermediate filament proteins in nonfilamentous structures: Transient disintegration and inclusion of subunit proteins in granular aggregates. *Cell, 30*, 103–113.
Fuchs, E., Esteves, R. A., & Coulombe, P. A. (1992). Transgenic mice expressing a mutant keratin 10 gene reveal the likely genetic basis for epidermolytic hyperkeratosis. *Proceedings of the National Academy of Sciences of the United States of America, 89*, 6906–6910.
Garcia, M., Larcher, F., Hickerson, R. P., Baselga, E., Leachman, S. A., Kaspar, R. L., et al. (2011). Development of skin-humanized mouse models of pachyonychia congenita. *The Journal of Investigative Dermatology, 131*, 1053–1060.
Garrod, D. R., Berika, M. Y., Bardsley, W. F., Holmes, D., & Tabernero, L. (2005). Hyperadhesion in desmosomes: Its regulation in wound healing and possible relationship to cadherin crystal structure. *Journal of Cell Science, 118*, 5743–5754.
Giancotti, F. G., Stepp, M. A., Suzuki, S., Engvall, E., & Ruoslahti, E. (1992). Proteolytic processing of endogenous and recombinant beta 4 integrin subunit. *The Journal of Cell Biology, 118*, 951–959.
Hamill, K. J., Hopkinson, S. B., Jonkman, M. F., & Jones, J. C. (2011). Type XVII collagen regulates lamellipod stability, cell motility, and signaling to Rac1 by targeting bullous pemphigoid antigen 1e to alpha6beta4 integrin. *The Journal of Biological Chemistry, 286*, 26768–26780.
Heagerty, A. H., Kennedy, A. R., Leigh, I. M., Purkis, P., & Eady, R. A. (1986). Identification of an epidermal basement membrane defect in recessive forms of dystrophic epidermolysis bullosa by LH 7:2 monoclonal antibody: Use in diagnosis. *The British Journal of Dermatology, 115*, 125–131.
Hendrix, M. J., Seftor, E. A., Chu, Y. W., Seftor, R. E., Nagle, R. B., McDaniel, K. M., et al. (1992). Coexpression of vimentin and keratins by human melanoma tumor cells: Correlation with invasive and metastatic potential. *Journal of the National Cancer Institute, 84*, 165–174.
Herrmann, H., Hesse, M., Reichenzeller, M., Aebi, U., & Magin, T. M. (2003). Functional complexity of intermediate filament cytoskeletons: From structure to assembly to gene ablation. *International Review of Cytology, 223*, 83–175.
Hoffmann, E. K. (1987). Volume regulation in cultured cells. In K. Arnost (Ed.), *Current topics in membranes and transport: Vol. 30* (pp. 125–180). San Diego: Academic Press.
Horn, H. M., & Tidman, M. J. (2000). The clinical spectrum of epidermolysis bullosa simplex. *The British Journal of Dermatology, 142*, 468–472.

Horwitz, B., Kupfer, H., Eshhar, Z., & Geiger, B. (1981). Reorganization of arrays of prekeratin filaments during mitosis. Immunofluorescence microscopy with multiclonal and monoclonal prekeratin antibodies. *Experimental Cell Research, 134*, 281–290.

Intong, L. R., & Murrell, D. F. (2010). How to take skin biopsies for epidermolysis bullosa. *Dermatologic Clinics, 28*(197–200), vii.

Jerabkova, B., Marek, J., Buckova, H., Kopeckova, L., Vesely, K., Valickova, J., et al. (2010). Keratin mutations in patients with epidermolysis bullosa simplex: Correlations between phenotype severity and disturbance of intermediate filament molecular structure. *The British Journal of Dermatology, 162*, 1004–1013.

Kawachi, Y., Ikegami, M., Hashimoto, T., Matsumura, K., Tanaka, T., & Otsuka, F. (1996). Autoantibodies to bullous pemphigoid and epidermolysis bullosa acquisita antigens in an infant. *The British Journal of Dermatology, 135*, 443–447.

Kennel, S. J., Epler, R. G., Lankford, T. K., Foote, L. J., Dickas, V., Canamucio, M., et al. (1990). Second generation monoclonal antibodies to the human integrin alpha 6 beta 4. *Hybridoma, 9*, 243–255.

Kerns, M. L., DePianto, D., Dinkova-Kostova, A. T., Talalay, P., & Coulombe, P. A. (2007). Reprogramming of keratin biosynthesis by sulforaphane restores skin integrity in epidermolysis bullosa simplex. *Proceedings of the National Academy of Sciences of the United States of America, 104*, 14460–14465.

Kimelberg, H. K. (1991). Swelling and volume control in brain astroglial cells. In R. Gilles, E. K. Hoffmann, & L. Bolis (Eds.), *Advances in comparative and environmental physiology: Vol. 9.* (pp. 81–117). Berlin, Heidelberg: Springer.

Kippenberger, S., Bernd, A., Loitsch, S., Guschel, M., Muller, J., Bereiter-Hahn, J., et al. (2000). Signaling of mechanical stretch in human keratinocytes via MAP kinases. *The Journal of Investigative Dermatology, 114*, 408–412.

Kucerova, J., & Strbak, V. (2001). The osmotic component of ethanol and urea action is critical for their immediate stimulation of thyrotropin-releasing hormone (TRH) release from rat brain septum. *Physiological Research/Academia Scientiarum Bohemoslovaca, 50*, 309–314.

Lane, E. B. (1994). Keratin diseases. *Current Opinion in Genetics & Development, 4*, 412–418.

Lane, E. B., Goodman, S. L., & Trejdosiewicz, L. K. (1982). Disruption of the keratin filament network during epithelial cell division. *The EMBO Journal, 1*, 1365–1372.

Lane, E. B., Rugg, E. L., Navsaria, H., Leigh, I. M., Heagerty, A. H., Ishida-Yamamoto, A., et al. (1992). A mutation in the conserved helix termination peptide of keratin 5 in hereditary skin blistering. *Nature, 356*, 244–246.

Leachman, S. A., Hickerson, R. P., Hull, P. R., Smith, F. J., Milstone, L. M., Lane, E. B., et al. (2008). Therapeutic siRNAs for dominant genetic skin disorders including pachyonychia congenita. *Journal of Dermatological Science, 51*, 151–157.

Leduc, C., & Etienne-Manneville, S. (2015). Intermediate filaments in cell migration and invasion: The unusual suspects. *Current Opinion in Cell Biology, 32C*, 102–112.

Lessard, J. C., & Coulombe, P. A. (2012). Keratin 16-null mice develop palmoplantar keratoderma, a hallmark feature of pachyonychia congenita and related disorders. *The Journal of Investigative Dermatology, 132*, 1384–1391.

Lettner, T., Lang, R., Klausegger, A., Hainzl, S., Bauer, J. W., & Wally, V. (2013). MMP-9 and CXCL8/IL-8 are potential therapeutic targets in epidermolysis bullosa simplex. *PloS One, 8*, e70123.

Liao, J., Ku, N. O., & Omary, M. B. (1997). Stress, apoptosis, and mitosis induce phosphorylation of human keratin 8 at Ser-73 in tissues and cultured cells. *The Journal of Biological Chemistry, 272*, 17565–17573.

Liovic, M., D'Alessandro, M., Tomic-Canic, M., Bolshakov, V. N., Coats, S. E., & Lane, E. B. (2009). Severe keratin 5 and 14 mutations induce down-regulation of junction proteins in keratinocytes. *Experimental Cell Research, 315*, 2995–3003.

Liovic, M., Lee, B., Tomic-Canic, M., D'Alessandro, M., Bolshakov, V. N., & Lane, E. B. (2008). Dual-specificity phosphatases in the hypo-osmotic stress response of keratin-defective epithelial cell lines. *Experimental Cell Research*, *314*, 2066–2075.

Lloyd, C., Yu, Q. C., Cheng, J., Turksen, K., Degenstein, L., Hutton, E., et al. (1995). The basal keratin network of stratified squamous epithelia: Defining K15 function in the absence of K14. *The Journal of Cell Biology*, *129*, 1329–1344.

Lu, H., Chen, J., Planko, L., Zigrino, P., Klein-Hitpass, L., & Magin, T. M. (2007). Induction of inflammatory cytokines by a keratin mutation and their repression by a small molecule in a mouse model for EBS. *The Journal of Investigative Dermatology*, *127*, 2781–2789.

Ma, L., Yamada, S., Wirtz, D., & Coulombe, P. A. (2001). A "hot-spot" mutation alters the mechanical properties of keratin filament networks. *Nature Cell Biology*, *3*, 503–506.

Mavilio, F., Pellegrini, G., Ferrari, S., Di Nunzio, F., Di Iorio, E., Recchia, A., et al. (2006). Correction of junctional epidermolysis bullosa by transplantation of genetically modified epidermal stem cells. *Nature Medicine*, *12*, 1397–1402.

Mendez, M. G., Kojima, S., & Goldman, R. D. (2010). Vimentin induces changes in cell shape, motility, and adhesion during the epithelial to mesenchymal transition. *FASEB Journal: Official Publication of the Federation of American Societies for Experimental Biology*, *24*, 1838–1851.

Mizushima, H., Koshikawa, N., Moriyama, K., Takamura, H., Nagashima, Y., Hirahara, F., et al. (1998). Wide distribution of laminin-5 gamma 2 chain in basement membranes of various human tissues. *Hormone Research*, *50*(Suppl. 2), 7–14.

Moch, M., Herberich, G., Aach, T., Leube, R. E., & Windoffer, R. (2013). Measuring the regulation of keratin filament network dynamics. *Proceedings of the National Academy of Sciences of the United States of America*, *110*, 10664–10669.

Moll, R., Dhouailly, D., & Sun, T. T. (1989). Expression of keratin 5 as a distinctive feature of epithelial and biphasic mesotheliomas. An immunohistochemical study using monoclonal antibody AE14. *Virchows Archiv B Cell Pathology Including Molecular Pathology*, *58*, 129–145.

Moll, R., Franke, W. W., Schiller, D. L., Geiger, B., & Krepler, R. (1982). The catalog of human cytokeratins: Patterns of expression in normal epithelia, tumors and cultured cells. *Cell*, *31*, 11–24.

Moran, J., Maar, T., & Pasantes-Morales, H. (1994). Cell volume regulation in taurine deficient cultured astrocytes. *Advances in Experimental Medicine and Biology*, *359*, 361–367.

Morley, S. M., D'Alessandro, M., Sexton, C., Rugg, E. L., Navsaria, H., Shemanko, C. S., et al. (2003). Generation and characterization of epidermolysis bullosa simplex cell lines: Scratch assays show faster migration with disruptive keratin mutations. *The British Journal of Dermatology*, *149*, 46–58.

Morley, S. M., Dundas, S. R., James, J. L., Gupta, T., Brown, R. A., Sexton, C. J., et al. (1995). Temperature sensitivity of the keratin cytoskeleton and delayed spreading of keratinocyte lines derived from EBS patients. *Journal of Cell Science*, *108*, 3463–3471.

Odermatt, B. F., Lang, A. B., Ruttner, J. R., Winterhalter, K. H., & Trueb, B. (1984). Monoclonal antibodies to human type IV collagen: Useful reagents to demonstrate the heterotrimeric nature of the molecule. *Proceedings of the National Academy of Sciences of the United States of America*, *81*, 7343–7347.

Omary, M. B., Ku, N. O., Tao, G. Z., Toivola, D. M., & Liao, J. (2006). "Heads and tails" of intermediate filament phosphorylation: multiple sites and functional insights. *Trends in Biochemical Sciences*, *31*, 383–394.

Osmanagic-Myers, S., Gregor, M., Walko, G., Burgstaller, G., Reipert, S., & Wiche, G. (2006). Plectin-controlled keratin cytoarchitecture affects MAP kinases involved in cellular stress response and migration. *The Journal of Cell Biology*, *174*, 557–568.

Owaribe, K., Nishizawa, Y., & Franke, W. W. (1991). Isolation and characterization of hemidesmosomes from bovine corneal epithelial cells. *Experimental Cell Research, 192*, 622–630.

Peters, B., Kirfel, J., Bussow, H., Vidal, M., & Magin, T. M. (2001). Complete cytolysis and neonatal lethality in keratin 5 knockout mice reveal its fundamental role in skin integrity and in epidermolysis bullosa simplex. *Molecular Biology of the Cell, 12*, 1775–1789.

Plancha, C. E., Carmo-Fonseca, M., & David-Ferreira, J. F. (1991). Cytokeratin in early hamster embryogenesis and parthenogenesis: Reorganization during mitosis and association with clusters of interchromatinlike granules. *Differentiation, 48*, 67–74.

Purkis, P. E., Steel, J. B., Mackenzie, I. C., Nathrath, W. B., Leigh, I. M., & Lane, E. B. (1990). Antibody markers of basal cells in complex epithelia. *Journal of Cell Science, 97*(Pt. 1), 39–50.

Reichelt, J., & Magin, T. M. (2002). Hyperproliferation, induction of c-Myc and 14-3-3 sigma, but no cell fragility in keratin-10-null mice. *Journal of Cell Science, 115*, 2639–2650.

Rheinwald, J. G., & Green, H. (1975). Serial cultivation of strains of human epidermal keratinocytes: The formation of keratinizing colonies from single cells. *Cell, 6*, 331–343.

Rogel, M. R., Soni, P. N., Troken, J. R., Sitikov, A., Trejo, H. E., & Ridge, K. M. (2011). Vimentin is sufficient and required for wound repair and remodeling in alveolar epithelial cells. *FASEB Journal: Official Publication of the Federation of American Societies for Experimental Biology, 25*, 3873–3883.

Rugg, E. L., Horn, H. M., Smith, F. J., Wilson, N. J., Hill, A. J., Magee, G. J., et al. (2007). Epidermolysis bullosa simplex in Scotland caused by a spectrum of keratin mutations. *The Journal of Investigative Dermatology, 127*, 574–580.

Russell, D., Andrews, P. D., James, J., & Lane, E. B. (2004). Mechanical stress induces profound remodelling of keratin filaments and cell junctions in epidermolysis bullosa simplex keratinocytes. *Journal of Cell Science, 117*, 5233–5243.

Russell, D., Ross, H., & Lane, E. B. (2010). ERK involvement in resistance to apoptosis in keratinocytes with mutant keratin. *The Journal of Investigative Dermatology, 130*, 671–681.

Seltmann, K., Roth, W., Kroger, C., Loschke, F., Lederer, M., Huttelmaier, S., et al. (2013). Keratins mediate localization of hemidesmosomes and repress cell motility. *The Journal of Investigative Dermatology, 133*, 181–190.

Sonnenberg, A., Modderman, P. W., & Hogervorst, F. (1988). Laminin receptor on platelets is the integrin VLA-6. *Nature, 336*, 487–489.

Sorensen, C. B., Andresen, B. S., Jensen, U. B., Jensen, T. G., Jensen, P. K., Gregersen, N., et al. (2003). Functional testing of keratin 14 mutant proteins associated with the three major subtypes of epidermolysis bullosa simplex. *Experimental Dermatology, 12*, 472–479.

Szeverenyi, I., Cassidy, A. J., Chung, C. W., Lee, B. T., Common, J. E., Ogg, S. C., et al. (2008). The Human Intermediate Filament Database: Comprehensive information on a gene family involved in many human diseases. *Human Mutation, 29*, 351–360.

Tan, T. S. (2012). *Keratin remodelling in stress*. Singapore: National University of Singapore. Ph.D. thesis.

Toivola, D. M., Boor, P., Alam, C., & Strnad, P. (2015). Keratins in health and disease. *Current Opinion in Cell Biology, 32*, 73–81.

Toivola, D. M., Zhou, Q., English, L. S., & Omary, M. B. (2002). Type II keratins are phosphorylated on a unique motif during stress and mitosis in tissues and cultured cells. *Molecular Biology of the Cell, 13*, 1857–1870.

Vassar, R., Coulombe, P. A., Degenstein, L., Albers, K., & Fuchs, E. (1991). Mutant keratin expression in transgenic mice causes marked abnormalities resembling a human genetic skin disease. *Cell, 64*, 365–380.

Vijayaraj, P., Sohl, G., & Magin, T. M. (2007). Keratin transgenic and knockout mice: Functional analysis and validation of disease-causing mutations. *Methods in Molecular Biology, 360*, 203–251.
Wally, V., Lettner, T., Peking, P., Peckl-Schmid, D., Murauer, E. M., Hainzl, S., et al. (2013). The pathogenetic role of IL-1beta in severe epidermolysis bullosa simplex. *The Journal of Investigative Dermatology, 133*, 1901–1903.
Wetzels, R. H., Holland, R., van Haelst, U. J., Lane, E. B., Leigh, I. M., & Ramaekers, F. C. (1989). Detection of basement membrane components and basal cell keratin 14 in noninvasive and invasive carcinomas of the breast. *The American Journal of Pathology, 134*, 571–579.
Wetzels, R. H., Kuijpers, H. J., Lane, E. B., Leigh, I. M., Troyanovsky, S. M., Holland, R., et al. (1991). Basal cell-specific and hyperproliferation-related keratins in human breast cancer. *The American Journal of Pathology, 138*, 751–763.

FURTHER READING

Lane, E. B., & McLean, W. H. I. (2008). Broken bricks and cracked mortar - epidermal diseases arising from genetic abnormalities. *Drug Discovery Today: Disease Mechanisms, 5*, e93–e101.

CHAPTER TEN

Using Data Mining and Computational Approaches to Study Intermediate Filament Structure and Function

David A.D. Parry[1]

Institute of Fundamental Sciences and Riddet Institute, Massey University, Palmerston North, New Zealand

[1]Corresponding author: e-mail address: d.parry@massey.ac.nz

Contents

1. Introduction	256
2. Methodology	259
2.1 Sequences and Preliminary Characterization	259
2.2 Sequence Comparisons	260
2.3 Secondary Structure	261
2.4 Structural and Functional Motifs	262
2.5 Assembly of Chains into Molecules	262
2.6 Assembly of Molecules	263
2.7 Tertiary Structure	266
2.8 The Effects of Mutations on Structure	266
2.9 Model Structures	267
2.10 Imaging Techniques	268
3. Summary	270
References	273

Abstract

Experimental and theoretical research aimed at determining the structure and function of the family of intermediate filament proteins has made significant advances over the past 20 years. Much of this has either contributed to or relied on the amino acid sequence databases that are now available online, and the data mining approaches that have been developed to analyze these sequences. As the quality of sequence data is generally high, it follows that it is the design of the computational and graphical methodologies that are of especial importance to researchers who aspire to gain a greater understanding of those sequence features that specify both function and structural hierarchy. However, these techniques are necessarily subject to limitations and it is important that these be recognized. In addition, no single method is likely to be successful in solving a particular problem, and a coordinated approach using a suite of

Methods in Enzymology, Volume 568
ISSN 0076-6879
http://dx.doi.org/10.1016/bs.mie.2015.07.011

methods is generally required. A final step in the process involves the interpretation of the results obtained and the construction of a working model or hypothesis that suggests further experimentation. While such methods allow meaningful progress to be made it is still important that the data are interpreted correctly and conservatively. New data mining methods are continually being developed, and it can be expected that even greater understanding of the relationship between structure and function will be gleaned from sequence data in the coming years.

1. INTRODUCTION

Data mining relates not to the extraction of data *per se* but, instead, to the computational process of discovering regularities/common features in large datasets, in this case the amino acid sequences of the intermediate filament (IF) chains. A subsequent step involves interpreting these observations in terms of structure and/or function, and providing a model or hypothesis that stimulates further experimentation. Success can be achieved only if the data are of sufficiently high quality and quantity, and the limitations of the methodology for analyzing the data are both understood and appreciated.

Consider the nature of data in general. It is conventional wisdom that data represent knowledge and that knowledge endows the recipient with power, which in this context implies a greater understanding of the structure and function of IFs. It is less commonly appreciated that it is more often than not the quality of the data, as distinct from their number, that are likely to allow a meaningful conclusion to be drawn. There are, of course, many examples in everyday life where controversial decisions are made on the basis of selected or incomplete datasets. It is fundamental to the scientific approach, however, that the researcher first considers all of the data available and then recognizes which of these are likely to be of high quality (and thus informative) and which of them may be less reliable (and hence potentially misleading). It is bad practice to use all available data, thereby weighting them equally, simply because they are there, without firstly having a realistic appreciation of which are relevant to the problem in hand. As Francis Crick once said, in one of his many insightful comments, "A theory that fits all the facts is bound to be wrong, as some of the facts will be wrong" (Wolpert, 1992). It is worth emphasizing that an uncritical approach to the nature of the data themselves and to the methods of their analysis has the potential to invalidate the conclusions made.

It is also pertinent to point out that no amount of computational analysis will yield a meaningful result from poor data. The potential dangers of misinterpreting data, using a partial dataset or oversimplifying the interpretation of the data are always with the research worker. Knowing which data to ignore, i.e., determining those data that are outliers, for example, or those that are incompatible with other experimentally derived information, is fraught with danger. A classic example refers to Maurice Wilkins' work on DNA structure. He states that "Our main mistake was to pay too much attention to experimental evidence. Nelson won the Battle of Copenhagen by putting his blind eye to the telescope so that he did not see the signal to stop fighting. In the same way, scientists sometimes should use the Nelson Principle and ignore experimental evidence" (Wilkins, 2003). It remains a hazardous occupation to do so, of course, but those with the intuition, as well as experience and knowledge, to do so correctly are often those that make the most significant breakthroughs.

In addition to the issue of data quality, it is also important to recognize that large sequence datasets do allow various statistical analyses to be performed. These too can result in common features being identified, often with a very high degree of reliability. With hundreds of IF sequences now accessible from a wide range of species, the statistical approach has become invaluable.

Since datasets can only reflect the current status of the field they must, by definition, be incomplete, thereby forcing the researcher to work with imperfect information. This can lead to problems. For example, the insistence of Rosalind Franklin that the density measurements for DNA were consistent only with a three-chain structure severely restricted the thinking of the King's College group at a critical stage of their research work on the structure of DNA. This is an example not of incorrect data *per se* but of an incorrect interpretation of those data and an over-reliance of her colleagues on a single piece of information. Data obtained from different researchers using different techniques but which yield the same conclusion are, of course, inherently more reliable than any single piece of information will ever be.

None of these points should dissuade researchers from using the data available in order to advance their field. Each model or hypothesis produced has value and represents another step forward in our understanding of a biological system. More importantly, it forms a basis for future experimentation and research that will cause the ideas to be refined, amended, or perhaps even radically changed. However, it is important that those involved in such work should appreciate the limitations of their work and have the courage to respond positively to developments that challenge the status quo when new data do emerge. Again, as quoted in the New Yorker (25 April 2011) Francis Crick was reported to have said "The dangerous man is the one who has only one idea, because then he'll fight and die for it." The realization that the models for structure and function that we present to our colleagues (and in which we have often invested so much personal effort) merely represent but one stage in an on-going process (and are thus imperfect) is not always an easy one to accept. A great scientist will readily appreciate that point but the average scientist frequently will not.

It is not only the quality and number of data that present a challenge to the researcher. The use of computational approaches too, without due regard for their limitations, can easily lead to a blind belief in their outputs. Data mining relies, for example, on the hydropathy scale chosen, the secondary structure prediction method selected, the heptad recognition software used, the time used in molecular dynamics simulations, the water content incorporated, and the effect of the sequence on the tertiary structure adopted. There are, of course, numerous other factors involved besides those listed here, and each will give a different result though, hopefully, these will not differ too greatly from one another. Thus, it follows that a degree of caution must be associated with all data mining methodology and the results it produces. Frequently, a battery of different but related methods will need to be employed if a result is to be substantiated. It cannot be denied, of course, that computational methods have revolutionized the field of structural biophysics and have allowed the researcher to analyze enormous datasets, especially in the area of X-ray crystallography and molecular dynamics, that would have been impossible a generation or so ago. As stressed earlier, the output of any computational technique remains critically dependent upon the quality of the input data and, in some cases, their number too if a statistically meaningful conclusion is to be drawn. Yet another Crickism noted that determining the structure of a protein was akin to badger-hunting in that it required "a knowledge of the animal's habits and a certain amount of low cunning" (Crick & Kendrew, 1957). The truth of this, especially in the area of fibrous proteins, is beyond question but, of course, such a sentiment cannot readily be incorporated in data mining software.

Presenting an idea via one of the many display programs now available can lead to stunning images or video clips. In recent times, scientists have also increasingly sought to produce their own animated videos to illustrate aspects of their research. This has involved not only writing an appropriate script but also gaining the skills in the production process. It is certainly not difficult to be seduced by the beauty of such presentations. For example, the three-dimensional conformations of proteins can be viewed from various angles, and possible mechanisms of docking with other molecules investigated directly. Such presentations have yielded great insight into aspects of both structure and function and form an integral part of the armory of techniques used conventionally by bioinformaticists.

2. METHODOLOGY

2.1 Sequences and Preliminary Characterization

There are now many hundreds of IF chain sequences available encompassing the Type I and Type II trichocyte and epithelial keratins; the Type III vimentin, desmin, peripherin, glial fibrillary acidic protein, and syncoilin chains; the Type IV neurofilament NF-L, NF-M, and NF-H chains, α-internexin, and nestin; and the Type V nuclear lamins. There are also other IF chains, such as synemin, paranemin, phakinin, and filensin (for example), that are less easily characterized in terms of chain type but they are, of course, of no less importance. The sequences of the human keratin IF chains are all located within a single database (Szeverenyi et al., 2008, http://www.interfil.org). Genome sequences have also been published for a number of other vertebrate species, including mouse (http://www.informatics.jax.org), rat (Shimoyama et al., 2015, http://www.rgd.mcw.edu), and dog (http://www.broadinstitute.org). All of these are readily accessed via a Web-based search engine or through an online library search. Often it is also valuable to calculate the volumes and molecular weights of the IF protein chains or fragments thereof, and the Web facility (http://www.basic.northwestern.edu/biotools/proteincalc.html) has proved useful in this regard.

Since there are now large numbers of sequences representing many different species in the various sequence databases, it is a relatively simple computational matter to consider the characteristics of the IF chain, or specific regions of it, to ascertain whether differences occur at statistically meaningful levels. Such features relate to structural and/or functional roles including the mechanical attributes bestowed upon the tissue in question. An example of this approach was undertaken by Strnad et al. (2011), who investigated the compositional differences among hair, epidermal, and simple-type keratin chains. Some of these manifested themselves in the following manner: (1) cysteines are very common in hair keratins, but only a few exist in epidermal keratins, and almost none are found in simple-type keratins, (2) the heads and tails of hair keratins are proline-rich, those of epidermal keratins are glycine and phenylalanine rich but alanine poor, and the simple-type keratins are rich in charged residues. Sequence and amino acid composition differences in these keratin types are believed to relate to differing functions, specifically features such as structural flexibility, rigidity, and solubility.

2.2 Sequence Comparisons

Having accessed the sequence data, there are many routines that allow the researcher to investigate the similarities and differences that exist between family members. Methods most commonly used include ClustalW, BLAST, or PSI-BLAST in, for example, ExPASy (Gasteiger et al., 2003,http://www.expasy.org/proteomics) or Predict Protein (https://www.predictprotein.org). There are many other Web sites, however, that also allow similar approaches to be used. The results may be displayed using JalView (The Barton Group, University of Dundee, Scotland, UK) or equivalent program. Sequence similarities will indicate conserved features, whereas differences will reveal those aspects of the structure and/or function that are unique to that particular chain.

The methods noted above are also easily able to identify those residues or motifs that are conserved across IF family members. Two regions of high homology that have been shown to be largely conserved across all IF chains are the helix initiation motif and the helix termination motif that lie near to the N- and at the C-terminal ends of the rod domain, respectively (Parry & Steinert, 1999), though the initial observation of these motifs arose from direct observation. Another example was the conserved nine-residue sequence found in the head domains of desmin, vimentin, peripherin, and the light chain of neurofilaments. This has been shown experimentally to be important in allowing regular assembly to occur (Herrmann, Hofmann, & Franke, 1992). In addition to this type of repeat, there are some IF chains that exhibit internal sequence repeats, thereby indicating a common structure and function for each element. Such repeats are confined to the head or tail domains, and examples include (a) 22-residue (human) and 44-residue repeats (hamster and rat) in the tail domain of nestin (Parry & Steinert, 1999; Steinert, Chou, et al., 1999), (b) 11-residue motif repeated 51 times contiguously in the tail domain of mouse keratin K78 (Langbein et al., 2015), and (c) degenerate KSP-containing repeats in the NF-M and NF-H chains for human (and other species) that serve as phosphorylation sites (Omary, Ku, Tao, Toivola, & Liao, 2006). Numerous methods are now available to identify such structures. These include the commonly used DOTPLOT (Gibbs & McIntyre, 1970) but now some 20 other methods are also accessible online (see Luo & Nijveen, 2013, for a comprehensive list).

2.3 Secondary Structure

Once any protein sequence has been completed, it is often a quick first step to get a feel for its likely structure and function. The initial approach, especially in the past, was to assess its likely α- and β-contents using several of the numerous techniques available online (see, for example, PSIPRED, McGuffin, Bryson, & Jones, 2000, http://www.globin.bio.warwick.ac.uk/psipred/; JPred3, Cole, Barber, & Barton, 2008, http://www.compbio.dundee.ac.uk/jpred; Phyre2, Kelley, Mezulis, Yates, Wass, & Sternberg, 2015, http://www.sbg.bio.ic.ac.uk/phyre2; RaptorX, Källberg et al., 2012, http://www.raptorX.uchicago.edu). This method tends to be much less important these days in the field of IF proteins, where the tri-domain character of the molecule (head, rod, and tail domains) is now well established and easily identified by comparison with IF homologues. Where this methodology still has some limited value, however, is in a study of the head and tail domains, where no three-dimensional crystal data, other than that for the C-terminal domain of lamin B1, are currently available.

Prediction techniques are based on conformational features observed in globular crystalline proteins, and not the fibrous ones of particular interest here. Indeed, there are indications that the methods that have been developed result in significant differences between the success rates for predicting secondary structure in globular proteins and fibrous ones. An example of the problem is that of PSIPRED, which is a complete failure when used on repetitive sequences as in *Bombyx mori* silk fibroin. This is an archetypical β-protein but PSIPRED predicts no residues whatsoever in a β-conformation (Fraser & Parry, 2011). Also, the probability of successfully predicting an α-helical structure in a fibrous protein remains markedly higher than that for a β-structure. This is largely a consequence of the percentage reduction in specifying the ends of the long α-helices compared to that for the far shorter β-strands generally found. Indeed, in fibrous proteins, the lengths of α-helix are frequently several hundred residues long whereas the β-strands may be as little as 5–10 residues in extent.

2.4 Structural and Functional Motifs

It is good practice to ascertain whether the IF chain under study contains any structural or functional motifs that have previously been characterized in other proteins. There are now more than 200 such features, and these include sites of posttranslational modifications (such as phosphorylation and glycosylation sites), ligand-binding sites, proteolytic cleavage sites, and nuclear localization signals. The standard method by which these and other features are identified is ELM (Dinkel et al., 2014, http://elm.eu.org). As the motifs thus found are frequently quite short (in some cases just 3–10 residues), there is a high probability that false positives will be identified. For this reason, it is necessary to use such information in a conservative manner and then actively seek out other data that support or refute the role of the short sequence motif that has been identified. Undertaking additional experimental work based on the information gleaned from the analysis of the data represents an appropriate way forward.

2.5 Assembly of Chains into Molecules

A number of programs exist to analyze sequences and predict those regions with a heptad substructure. These include COILS (Lupas, Van Dyke, & Stock, 1991, http://www.ch.embnet.org/software/COILS_form.html) and MARCOIL (Delorenzi & Speed, 2002, http://www.isrec.isb-sib.ch.BCF/Delorenzi/Marcoil/index). Note also that Lupas and Gruber (2005) have listed a number of other methods that are also used to delineate regions of potential coiled-coil structure. Without exception, however, these (and other) methods fail to consistently predict either short regions with these features or those regions that suffer breaks in heptad/hendecad continuity. They also fail to recognize sequence features that have the potential to disrupt the coiled-coil structure locally. The "problem" thus remains that someone with an experienced eye can still spot an anomalous feature in a sequence that the programs are currently unable to detect.

Another technique, NetSurfP predictor (Petersen, Petersen, Andersen, Nielsen, & Lundegaard, 2009) allows the solvent exposure of residues to be determined as a function of position along a sequence. This can be related to the packing of the chains within the IF molecule. The hydropathy index (Kyte & Doolittle, 1982) of residues as a function of position can also yield information on the packing of chains within the molecule.

Once a heptad/hendecad region has been identified, it is possible to calculate computationally the potential number of interchain ionic interactions as a function of both the relative polarity of the chains involved in the molecular structure and the relative stagger of one chain with respect to its neighbor(s) (McLachlan & Stewart, 1976). This approach indicated a heterodimer structure for the keratin IF molecule (Parry, Crewther, Fraser, & MacRae, 1977) well before any experimental data were available (Coulombe & Fuchs, 1990; Hatzfeld & Weber, 1990; Parry, Steven, & Steinert, 1985; Steinert, 1990). The theoretical method involving the calculation of the potential number of interchain ionic interactions also indicated that the chains were both parallel to one another and in axial register. The latter predictions were subsequently verified using cross-linking techniques (Steinert, Marekov, Fraser, & Parry, 1993; Steinert, Marekov & Parry, 1993a, 1993b, 1999) and X-ray crystallography and/or SDSL-EPR (Aziz et al., 2012, Chernyatina, Guzenko, & Strelkov, 2015; Chernyatina, Nicolet, Aebi, Herrmann, & Strelkov, 2012; Chernyatina & Strelkov, 2012; Lee, Kim, Chung, Leahy, & Coulombe, 2012; Meier et al., 2009; Nicolet, Herrmann, Aebi, & Strelkov, 2010; Ruan et al., 2012; Strelkov et al., 2002; Strelkov, Schumacher, Burkhard, Aebi, & Herrmann, 2004). Interestingly, experimentally-derived crystal structures of coiled-coil proteins have shown that not all of the predicted interchain ionic interactions actually occur *in vivo*. The theoretical approach thus derives a maximum number for this type of interaction. *In vivo*, the actual number observed will generally be less.

2.6 Assembly of Molecules

All fibrous proteins that assemble into filamentous structures contain periodic distributions of their charged and/or apolar residues. These periodicities are normally detected by a fast Fourier transform technique (Parry, 1975; Stewart & McLachlan, 1975). The sequence is first digitized into a linear array with unit values (1) assigned for each one of the amino acid(s) of interest, and zero values (0) for all of the others. This array, which is of length equal to that of the amino acid sequence, is then embedded in a longer array of zeroes to give an overall length for the array of 2^n (typically 1024, 2048, or 4096). The array is then Fourier transformed and scaled so that the statistical significance of a peak with an intensity I is given by the expression exp $(-I)$ (McLachlan & Stewart, 1976). The rationale for the embedding is that it naturally leads to a more accurate determination of the periodicity present. This method (Parry et al., 1977; Parry & Fraser, 1985) revealed a near out-of-phase periodicity in the linear distributions of the acidic residues and the basic residues in segment 1B (period about 9.55 residues) and in segment 2 (period about 9.85 residues). The small difference in these two periods, allied to the fact that the periods are not exactly out-of-phase, predisposes a pair of 1B segments to aggregate in an antiparallel (rather than parallel) manner via the maximization of intermolecular ionic interactions (A_{11} mode), as it does with a pair of 2 segments (A_{22} mode). It does not, however, preclude a less significant interaction between (antiparallel) 1B and 2 segments (A_{12} mode). The different modes of assembly of IF molecules within the filament (A_{11}, A_{22}, A_{12}, and A_{CN}) were defined by Steinert, Marekov, Fraser, et al. (1993) on the basis of cross-linking results but were based on the original suggestions of Crewther, Dowling, Steinert, and Parry (1983).

It has been established by cross-linking studies on the "reduced" and "oxidized" forms of the hair keratin IF that the structures differ significantly from one another (Fig. 1) and that this arises as a consequence of an axial shift between antiparallel 1B segments in adjacent molecules (Wang et al., 2000). An explanation for this observation, plus the facts that a strong interaction between antiparallel 1B segments has been implicated in unit-length filament formation (a precursor to IF formation, Herrmann & Aebi, 1998, 2004; Mücke et al., 2004) and in stabilizing an enzyme-resistant four-chain fragment from trichocyte keratin (Ahmadi & Speakman, 1978; Crewther & Harrap, 1967; Dobb, Millward, & Crewther, 1972; Suzuki, Crewther, Fraser, MacRae, & McKern, 1973), was sought by Fraser and Parry (2014). In particular, they studied the spatial relationship of residues that favor dimer formation for the 1B segment (Bahadur, Chakabarti, Rodier, & Janin, 2003). In this work, they showed that the sequence differences that exist between the Type I and Type II chains that constitute the keratin heterodimer resulted in a single line of contact that involved each of the keratin chains in turn. A similar analysis was undertaken for vimentin and neurofilament homodimers. The conclusion was that the transition from the reduced to the oxidized states in hair keratin did not involve any rotation of one segment with respect to the other but occurred purely as a simple linear shift between the two antiparallel 1B segments. The reduced form corresponds to molecules that are loosely packed (node-to-node) whereas the oxidized structure contains molecules that are more compact (node-to-antinode). This is consistent with the radial compaction of the IF that is known to occur at this stage (Fraser & Parry, 2007). In addition, the cysteine residues in segment 1B (previously not aligned axially in the reduced form) become closely in-register axially after molecular slippage has occurred. This facilitates disulfide bond formation, as observed (Fraser & Parry, 2007, 2012; Parry, 1996).

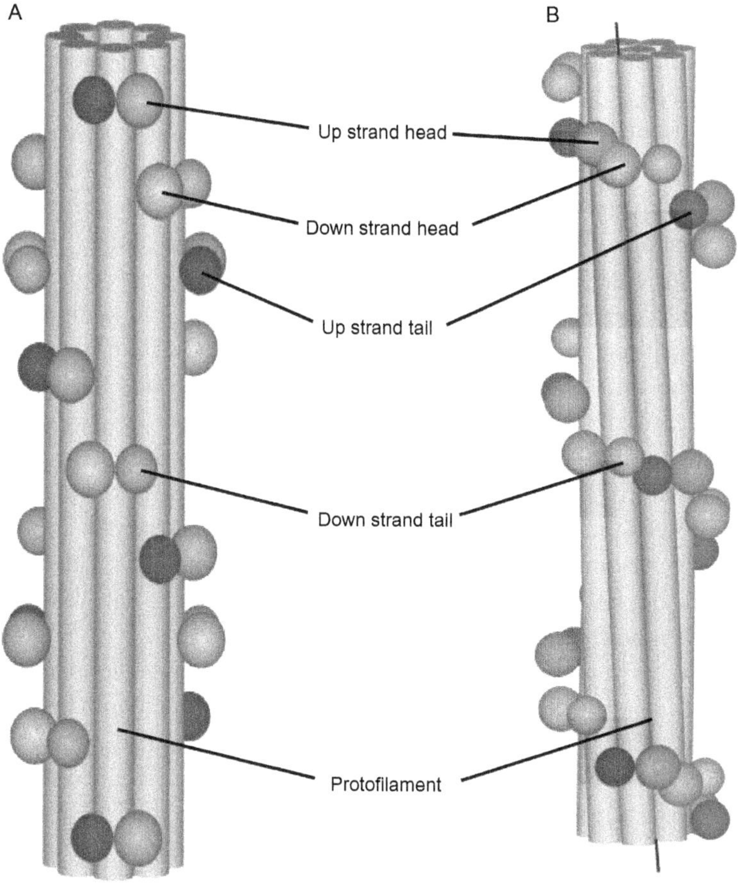

Figure 1 Diagrammatic representation of the structures of (A) "reduced" hair keratin IF and (B) "oxidized" hair keratin IF. The α-helical coiled-coil rod domains of the protofilaments, which are composed of antiparallel molecular strands, are colored yellow. The head domains are represented by green and blue spheres for the up and down strands, respectively, and the tail domains are represented by red and orange spheres in the up and down strands, respectively. The dispositions of the head and tail domains differ significantly between the two structures. In the "reduced" structure, the head and tail domains lie on a two-start left-hand helix that would appear at low resolution to give a diagonal banding pattern of spacing 22 nm. In contrast, in the "oxidized" structure, the head and tail domains are in much closer spatial proximity and appear to be distributed on a helix of pitch length 23.5 nm. Also, there is some radial compaction of the "oxidized" structure relative to the "reduced" one. Details of the radial, axial, and azimuthal coordinates of the protofibrils are given in the text. *Reprinted from Fraser and Parry (2005) with permission from Elsevier.* (See the color plate.)

2.7 Tertiary Structure

A more common approach these days, thanks to the vast number of protein sequences now determined, is to ascertain whether any sequence with similarity to the IF one under investigation, has been reported previously. If it has, then its tertiary structure might also have been determined by either protein X-ray crystallography or NMR spectroscopy techniques. The protein crystallography database (PDB) (Berman et al., 2000, http://www.rcsb.org) then becomes of considerable value. There are now, of course, a number of rod domain fragments from vimentin, keratin, and lamin that have been crystallized, their three-dimensional conformations determined (see list of references noted earlier) and their coordinates deposited in the PDB.

As regards the crystal data (or high-field NMR data) there seems, at first sight, no potential for concern as to their reliability as the data reveal but a single structure (or multiple related structures in the case of the NMR studies). The possibility exists, nonetheless, that small assemblies in a highly aqueous environment as studied by X-ray crystallography will not adopt an identical conformation to that in the IF, where the density of molecular packing is much higher and the water content is much lower. So far, we have data on the former scenario but no data on the latter so the extent of the problem, if indeed one exists, is unknown. It is noted, nonetheless, that the cross-linking studies on vimentin IF structures and the crystal/EPR results on small aggregates suggest a small but significant difference in the relative axial stagger of antiparallel 1B segments. Which of these structures best represents the *in vivo* situation remains to be seen.

2.8 The Effects of Mutations on Structure

The effect of mutations on the likely structure of IF segments has also been investigated using a molecular dynamics approach. For example, in the case of the keratin K5/K14 heterodimer (Smith, Steinert, & Parry, 2004), a model for segment 1A was first derived from the structure of Liovic et al. (2001). The known mutations in keratin K5/K14 for Epidermolysis Bullosa Simplex were inserted one at a time in the appropriate places in the K5 or K14 sequences. Molecular dynamics was then performed on segment 1A in an aqueous environment using AMBER (Case et al., 2005, http://www.ambermd.org). This resulted in "refined" structures where the conformational changes resulting from the mutation could be identified. These fell into five groups: (1) the change in structure is localized to one or two turns of α-helix in the chain that suffered the mutation, (2) the changes in structure is localized to one or two turns of α-helix in the partner chain immediately opposite to the site of the mutation, (3) the structural changes occur locally in both chains at the site of the mutation, (4) the change in conformation extends over a considerable distance in the chain with the mutation, and (5) no visible change in conformation occurred as a result of the mutation. The use of AGADIR (Muñoz & Serrano, 1997) was also used to assess the effect of the mutations on the stability of the α-helical structure.

2.9 Model Structures

To produce models of the coiled-coil rod domain of IF molecules in order to use these to study its function or three-dimensional structure, it is first necessary to use either the mathematically derived coordinates using the Crick formulae (1953)—thereby producing a highly regular coiled-coil—or an *in silico* construction of a model based on crystallographically or SDSL-EPR-derived structures (see list of references noted earlier). Crystal data of coiled-coil proteins indicate that there are significant variations locally in both the radius of the coiled-coil and its pitch length, and it is presumed that these arise from the local amino acid sequence (see, for example, Chernyatina et al., 2012; Seo & Cohen, 1993; Strelkov & Burkhard, 2002). This indicates that mathematically derived models may suffer from the problem of being unrealistically regular. However, the crystal-derived models may suffer from a different problem in that they represent the conformation for that particular sequence (and *only* that particular sequence) in a highly aqueous and low-density environment that differs significantly from that experienced *in vivo* in the IF. Currently, both approaches have their merits depending on the situation being investigated, but the potential problem of using an incorrect conformation for a model structure should not be disregarded.

Model structures of IF molecules other than vimentin, keratin, or lamin (for which some crystal data are available) can also be constructed and studied by using COOT (Emsley, Lohkamp, Scott, & Cowtan, 2010, http://www2.mrc-lmb.cam.ac.uk/personal/pemsley/coot) or TURBO-FRODO (http://www.csb.yale.edu). This allows residues in the crystal structure to be changed into those present in the IF molecule that one wishes to study. The assumption is that this molecule and the crystalline one on which it is based will have identical three-dimensional conformations. This may or may not be true but it does, nonetheless, represent a reasonable starting structure.

Electrostatic surfaces of the real or model molecules can be displayed using PYMOL (http://www.sourceforge.net) with, for example, the acidic, basic, and apolar residues being colored red, blue, and green, respectively. Using this approach, Smith and Parry (2008) showed that one face of segment 1A (face A) in K35/K85 trichocyte keratin was considerably more apolar than was the opposite one (face B) (Fig. 2A and B), and also that the differences in the charged residues that occurred between the Type I and Type II chains were largely confined to face A (Fig. 2C and D). In addition, it was shown that one face of segment 1B in K10/K1 epidermal keratin (face C) displayed the bulk of the apolar residues as compared to its opposite face(D) (Fig. 2E and F). Face C also contained the bulk of the charged residue differences that occurred between the two chains (Fig. 2G and H). These results indicated the probable internal face of these segments and those sequence features that play an especially important part in assembly.

2.10 Imaging Techniques

Presenting data in a graphical form can often reveal information that would otherwise be difficult to visualize, especially so if it relates to a three-dimensional structure. A good example is that of the structural change of the hair keratin IF that occurs between its initial formation in the cell and the final form that occurs after cell death. Models based on the dimensions of the IF molecule and the surface lattice structures of the IF in the two cases graphically illustrate the effect of the slippage of molecules within the protofilament that arises from the altered packing of the 1B segments (Fig. 1A and B; Fraser & Parry, 2005). The radial compaction of the IF is also evident. The resulting alignment of the cysteine residues that permit disulfide bond formation to occur gives hair keratin its required mechanical properties. In the initial (reduced) state, the IF has eight protofilaments arranged on a ring of diameter 3.5 nm. Each protofilament is related to its predecessor by a rotation of 45° and an axial displacement of 16.85 nm. The head and tail domains lie on a two-start left-hand helix that would appear at low resolution to give a diagonal banding pattern of spacing 22 nm (Fig. 1B). In contrast, in the final (oxidized) form of the hair keratin IF, there are seven protofilaments arranged on a ring of radius 3.0 nm with one central straight one located at the ring's center. Each protofilament is related to its predecessor by a rotation of 24.3° and an axial displacement of 19.82 nm. The head and tail domains are now in much closer spatial proximity and appear to be distributed

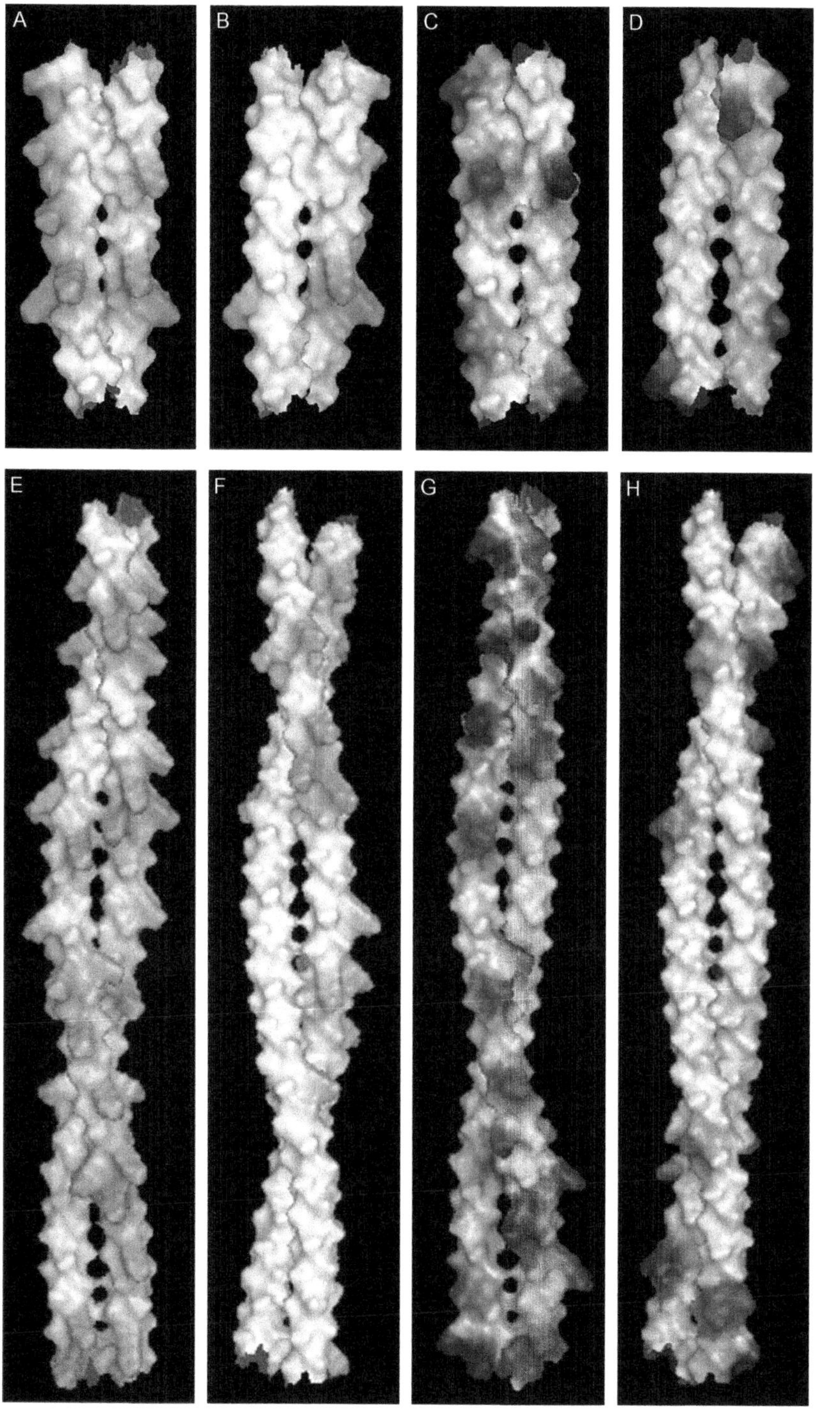

Figure 2 See legend on next page.

on a helix of pitch length 23.5 nm. It would have been very difficult to imagine these important structural changes merely from the tabular presentation of the data.

3. SUMMARY

Rather than attempting an all-encompassing review of every data mining application that is now available (>100), the decision was made to list only the most commonly used techniques in the field of IF and to give examples of the contributions that these have made. The Web site addresses relating to this selection are summarized in Table 1. Consequently, the data mining methodologies available to those seeking to understand IF structure and function described here reflect a degree of personal preference, and hence it is not unlikely that some other workers will prefer different methods. An attempt has also been made to describe the limitations of the techniques used. What has become very clear is that data mining techniques have provided a wealth of information about both the structure and function of IF. Where such methods have conspicuously failed thus far is in describing more of the probable conformation present in the head and tail domains of IF in general. Of course, the ability of the researcher to feed in a sequence to a program and generate a full three-dimensional structure remains a holy grail for structural biophysicists, and while progress has been made for some simple globular proteins there appears to be a long way to go before such successes are repeated for the heads and tails of IF proteins. Nonetheless, crystal structure data for the rod domains allied to cross-linking

Figure 2—Con't One face of segment 1A (face A) in K35/K85 trichocyte keratin (A) is more apolar than the opposite one (face B) (B), and also the differences in the charged residues that occurred between the Type I and Type II chains were largely confined to face A (C) rather than face B (D). In addition, in (E), it was shown that one face of segment 1B in K10/K1 epidermal keratin (face C) displayed the bulk of the apolar residues when compared to that present in its opposite face D (F). Face C also contained the bulk of the charged residue differences that occurred between the two chains (G) compared to face D (H). These results indicate the probable internal face of these segments and those sequence features that play an especially important part in assembly. Details of the methods used (PYMOL) and the color scheme employed are given in Smith and Parry (2008). Apolar residues are colored green, and difference profiles for acidic and for basic residues are colored red and blue, respectively. All other residues are represented by alanines (white). *Reprinted from Smith and Parry (2008) with permission from Elsevier.* (See the color plate.)

Table 1 Summary of a Small Selection of Data Mining Methodologies: IF Proteins

Technique/ Database	Web Site Address	References	Comments
Protein sequences			
Human amino acid sequences	http://www.interfil.org	Szeverenyi et al. (2008)	Easy access to all human keratin IF protein sequences
Genome sequences			
Mouse	http://www.informatics.jax.org		Many genome sequences are now available for comparative purposes and those listed here are examples only
Rat	http://www.reg.mcw.edu	Shimoyama et al. (2015)	
Dog	http://www.broadinstitute.org		
Molecular parameters			
Volume and molecular weight determination	http://www.basic.northwestern.edu/biotools/proteincalc.html		Simple, requires protein sequence data as input
Sequence analysis algorithms			
ExPASy	http://www.expasy.org/proteomics	Gasteiger et al. (2003)	Wide range of algorithms available to aid sequence comparison and determination of structural features
Predict protein	https://www.predictprotein.org		
Secondary structure prediction			
PSIPRED	http://www.compbio.dundee.ac.uk/jpred	McGuffin et al. (2000)	OK for globular domains, poor for β in fibrous ones
Jpred3	http://www.compbio.dundee.ac.uk/jpred	Cole et al. (2008)	Higher success rates for α-helix than for β-strands as α-helical regions are often much longer i.e. the end effects are much smaller than for the β-strands
Phyre2	http://www.sbg.bio.ic.ac.uk/phyre2	Kelley et al. (2015)	
RaptorX	http://www.raptorX.uchicago.edu	Källberg et al. (2012)	
Structural motifs			
Recognition of structural motifs	http://www.elm.eu.org	Dinkel et al. (2014)	Indicative only, may produce false positives

Continued

Table 1 Summary of a Small Selection of Data Mining Methodologies: IF Proteins—cont'd

Technique/ Database	Web Site Address	References	Comments
Recognition of heptad substructure			
COILS	http://www.ch.embnet.org/software/COILS_form.html	Lupas et al. (1991)	Good for long heptad regions but often misses discontinuities and short heptad-containing regions
MARCOIL	http://www.isrec.isb-sib.ch.BCF/Delorenzi?Marcoil/index	Delorenzi and Speed (2002)	
Solvent exposure			
NetSurfP	http://www.cbs.dtu.dk	Petersen et al. (2009)	Assesses solvent exposure of residues, chain packing
Structural database			
Protein crystal structure database	http://www.rcsb.org	Berman et al. (2000)	3D structure database, numerous high-quality details
Molecular dynamics			
AMBER	http://www.ambermd.org	Case et al. (2005)	Water essential, sufficient refinement time vital
Mutations			
AGADIR		Muñoz and Serrano (1997)	Estimates effect of mutations on α-helical structure
Model building			
COOT	http://www2.mrc-lmb.cam.ac.uk/personal/pemsley/coot	Emsley et al. (2010)	Very commonly used techniques, user-friendly and have wide applications in all model-building studies
TURBO-FRODO	http://www.csb.yale.edu		
PYMOL	http://www.sourceforge.net		

studies and other data mining methods have been very successful in allowing the researcher real insight into the assembly of IF molecules into fully functional IF.

REFERENCES

Ahmadi, B., & Speakman, P. T. (1978). Suberimidate crosslinking shows that a rod-shaped, low cystine, high helix protein prepared by limited proteolysis of reduced wool has four protein chains. *FEBS Letters*, *94*, 365–367.

Aziz, A., Hess, J. F., Budamagunta, M. S., Voss, J. C., Kuzin, A. P., Huang, Y. J., et al. (2012). The structure of vimentin linker 1 and rod 1b domains characterized by site-directed spin-labeling electron paramagnetic resonance (SDSL-EPR) and X-ray crystallography. *Journal of Biological Chemistry*, *287*, 28349–28361.

Bahadur, R. P., Chakabarti, P., Rodier, F., & Janin, J. (2003). Dissecting subunit interfaces in homodimeric proteins. *Proteins: Structure, Function, and Bioinformatics*, *53*, 708–719.

Berman, H. M., Westbrook, J., Feng, Z., Gilliland, G., Bhat, T. N., Weissig, H., et al. (2000). The Protein Data Bank. *Nucleic Acids Research*, *28*, 235–242.

Case, D. A., Cheatham, T. E., III, Darden, T., Gohlke, H., Luo, R., Merz, K. M., Jr., et al. (2005). The Amber biomolecular simulation programs. *Journal of Computational Chemistry*, *26*, 1668–1688.

Chernyatina, A. A., Guzenko, D., & Strelkov, S. V. (2015). Intermediate filament structure: The bottom-up approach. *Current Opinion in Cell Biology*, *32*, 65–72.

Chernyatina, A. A., Nicolet, S., Aebi, U., Herrmann, H., & Strelkov, S. V. (2012). Atomic structure of the vimentin central α-helical domain and its implications for intermediate filament assembly. *Proceedings of the National Academy of Sciences of the United States of America*, *109*, 13620–13625.

Chernyatina, A. A., & Strelkov, S. V. (2012). Stabilization of vimentin coil 2 fragment via an engineered disulfide. *Journal of Structural Biology*, *177*, 46–53.

Cole, C., Barber, J. D., & Barton, G. J. (2008). The Jpred3 secondary structure prediction server. *Nucleic Acids Research*, *35*, W197–W201.

Coulombe, P. A., & Fuchs, E. (1990). Elucidating the early stages of keratin filament assembly. *Journal of Cell Biology*, *111*, 153–169.

Crewther, W. G., Dowling, L. M., Steinert, P. M., & Parry, D. A. D. (1983). Structure of intermediate filaments. *International Journal of Biological Macromolecules*, *5*, 267–274.

Crewther, W. G., & Harrap, B. S. (1967). Preparation and properties of a helix-rich fraction obtained by partial proteolysis of low sulfur S-carboxymethylkerateine from wool. *Journal of Biological Chemistry*, *242*, 4310–4319.

Crick, F. H. C. (1953). The Fourier transform of a coiled-coil. *Acta Crystallographica*, *6*, 685–689.

Crick, F. H. C., & Kendrew, J. C. (1957). X-ray analysis and protein structure. *Advances in Protein Chemistry*, *12*, 133–214.

Delorenzi, M., & Speed, T. (2002). An HMM model for coiled-coil domains and a comparison with PSSM-based predictions. *Bioinformatics*, *18*, 617–625.

Dinkel, H., Van Roey, K., Michael, S., Davey, N. E., Weatheritt, R. J., Born, D., et al. (2014). The eukaryotic linear motif resource ELM: 10 years and counting. *Nucleic Acids Research*, *42*, D259–D266.

Dobb, M. G., Millward, G. R., & Crewther, W. G. (1972). Examination in electron microscope of S-carboxymethylkerateine-A and of helix-rich fraction obtained from it by partial proteolysis. *Journal of the Textile Institute*, *64*, 374–385.

Emsley, P., Lohkamp, B., Scott, W. G., & Cowtan, K. (2010). Features and development of COOT. *Acta Crystallographica*, *D66*, 486–501.

Fraser, R. D. B., & Parry, D. A. D. (2005). The three-dimensional structure of trichocyte (hard α-) keratin intermediate filaments: Features of the molecular packing deduced from the sites of induced crosslinks. *Journal of Structural Biology, 151*, 171–181.
Fraser, R. D. B., & Parry, D. A. D. (2007). Structural changes in the trichocyte intermediate filaments accompanying the transition from the reduced to the oxidized form. *Journal of Structural Biology, 159*, 36–45.
Fraser, R. D. B., & Parry, D. A. D. (2011). Structural basis of the filament-matrix texture in the avian/reptilian group of hard β-keratins. *Journal of Structural Biology, 173*, 391–405.
Fraser, R. D. B., & Parry, D. A. D. (2012). The role of disulfide bond formation in the structural transition observed in the intermediate filaments of the developing hair. *Journal of Structural Biology, 180*, 117–124.
Fraser, R. D. B., & Parry, D. A. D. (2014). Keratin intermediate filaments: Differences in the sequences of the type I and type II chains explain the origin of the stability of an enzyme-resistant four-chain fragment. *Journal of Structural Biology, 185*, 317–326.
Gasteiger, E., Gattiker, A., Hoogland, C., Ivanyi, I., Appel, R. D., & Bairoch, A. (2003). ExPasy: The proteomics server for in-depth protein knowledge and analysis. *Nucleic Acids Research, 31*, 3784–3788.
Gibbs, A. J., & McIntyre, G. A. (1970). The diagram, a method for comparing sequences. Its use with amino acid and nucleotide sequences. *European Journal of Biochemistry, 16*, 1–11.
Hatzfeld, M., & Weber, K. (1990). The coiled-coil of in vitro assembled keratin filaments is a heterodimer of type I and II keratin: Use of site-specific mutagenesis and recombinant protein expression. *Journal of Cell Biology, 110*, 1199–1210.
Herrmann, H., & Aebi, U. (1998). Intermediate filament assembly: Fibrillogenesis is driven by decisive dimer-dimer interactions. *Current Opinion in Structural Biology, 8*, 177–185.
Herrmann, H., & Aebi, U. (2004). Intermediate filament assembly: Molecular structure, assembly mechanism, and integration into functionally distinct intracellular scaffolds. *Annual Review of Biochemistry, 73*, 749–789.
Herrmann, H., Hofmann, I., & Franke, W. W. (1992). Identification of a nonapeptide motif in the vimentin head domain involved in intermediate filament assembly. *Journal of Molecular Biology, 223*, 637–650.
Källberg, M., Wang, H., Wang, S., Peng, J., Wang, Z., Lu, H., et al. (2012). Template-based protein structure modeling using the RaptorX web server. *Nature Protocols, 7*, 1511–1522.
Kelley, L. A., Mezulis, S., Yates, C. M., Wass, M. N., & Sternberg, M. J. E. (2015). The Phyre2 webportal for protein modeling, prediction and analysis. *Nature Protocols, 10*, 845–858.
Kyte, J., & Doolittle, R. F. (1982). A simple method for displaying the hydropathic character of a protein. *Journal of Molecular Biology, 157*, 105–132.
Langbein, L., Eckhart, L., Fischer, H., Rogers, M. A., Praetzel-Wunder, S., Parry, D. A. D., et al. (2015). Localisation of keratin K78 in the basal layer and the first suprabasal layers of stratified epithelia completes the expression catalog of type II keratins and provides new insights into sequential keratin expression. *Cell and Tissue Research*, (submitted).
Lee, C. H., Kim, M. S., Chung, B. M., Leahy, D. J., & Coulombe, P. A. (2012). Structural basis for heteromeric assembly and perinuclear organization of keratin filaments. *Nature Structural and Molecular Biology, 19*, 707–715.
Liovic, M., Stojan, J., Bowden, P. E., Gibbs, D., Vahlquist, A., & Lane, E. B. (2001). Novel keratin 5 mutation (K5V186L) in a family with EBS-K: A conservative substitution can lead to development of different disease phenotypes. *Journal of Investigative Dermatology, 116*, 964–969.
Luo, H., & Nijveen, H. (2013). Understanding and identifying amino acid repeats. *Briefings in Bioinformatics, 15*, 582–591.

Lupas, A. N., & Gruber, M. (2005). The structure of α-helical coiled coils. *Advances in Protein Chemistry, 70*, 37–78.

Lupas, A., Van Dyke, M., & Stock, J. (1991). Predicting coiled coils from protein sequences. *Science, 252*, 1162–1164.

McGuffin, L., Bryson, K., & Jones, D. T. (2000). The PSIPRED protein structure prediction server. *Bioinformatics, 16*, 404–405.

McLachlan, A. D., & Stewart, M. (1976). The 14-fold periodicity in α-tropomyosin and the interaction with actin. *Journal of Molecular Biology, 103*, 271–298.

Meier, M., Padilla, G. P., Herrmann, H., Wedig, T., Hergt, M., Patel, T. R., et al. (2009). Vimentin coil 1A—A molecular switch involved in the initiation of filament elongation. *Journal of Molecular Biology, 390*, 245–261.

Mücke, N., Wedig, T., Bürer, A., Marekov, L. N., Steinert, P. M., Langowski, J., et al. (2004). Molecular and biophysical characterization of assembly-starter units of human vimentin. *Journal of Molecular Biology, 340*, 97–114.

Muñoz, V., & Serrano, L. (1997). Development of the multiple sequence approximation within the AGADIR model of α-helix formation. Comparison with Zimm-Bragg and Lifson-Roig formalisms. *Biopolymers, 42*, 495–509.

Nicolet, S., Herrmann, H., Aebi, U., & Strelkov, S. V. (2010). Atomic structure of vimentin coil 2. *Journal of Structural Biology, 170*, 369–376.

Omary, M. B., Ku, N.-O., Tao, G.-F., Toivola, D. M., & Liao, J. (2006). "Heads and tails" of intermediate filament phosphorylation: Multiple sites and functional insights. *Trends in Biochemical Sciences, 31*, 383–394.

Parry, D. A. D. (1975). Analysis of the primary sequence of α-tropomyosin from rabbit skeletal muscle. *Journal of Molecular Biology, 98*, 519–535.

Parry, D. A. D., & Fraser, R. D. B. (1985). Intermediate filament structure. 1. Analysis of IF protein sequence data. *International Journal of Biological Macromolecules, 7*, 203–213.

Parry, D. A. D. (1996). Hard α-keratin intermediate filaments: An alternative explanation of the low-angle equatorial X-ray diffraction pattern, and the axial disposition of putative disulphide bonds in the intra- and inter-protofilamentous networks. *International Journal of Biological Macromolecules, 19*, 45–50.

Parry, D. A. D., Crewther, W. G., Fraser, R. D. B., & MacRae, T. P. (1977). Structure of α-keratin: Structural implication of the amino acid sequences of the type I and type II chain segments. *Journal of Molecular Biology, 113*, 449–454.

Parry, D. A. D., & Steinert, P. M. (1999). Intermediate filaments: Molecular architecture, assembly, dynamics and polymorphism. *Quarterly Reviews of Biophysics, 32*, 99–187.

Parry, D. A. D., Steven, A. C., & Steinert, P. M. (1985). The coiled-coil molecules of intermediate filaments consist of two parallel chains in exact axial register. *Biochemical and Biophysical Research Communications, 127*, 1012–1018.

Petersen, B., Petersen, T. N., Andersen, P., Nielsen, M., & Lundegaard, C. (2009). A generic method for assignment of reliability scores applied to solvent accessibility predictions. *BMC Structural Biology, 9*, 51. http://dx.doi.org/10.1186/1472-6807-9-51.

Ruan, J., Xu, C., Bian, C., Lam, R., Wang, J. P., Kania, J., et al. (2012). Crystal structures of the coil 2B fragment and the globular tail domain of human lamin B1. *FEBS Letters, 586*, 314–318.

Seo, J., & Cohen, C. (1993). Pitch diversity in alpha-helical coiled coils. *Proteins: Structure, Function, and Genetics, 15*, 223–234.

Shimoyama, M., De Pons, J., Hayman, G. T., Laulederkind, S. J., Liu, W., Nigam, R., et al. (2015). The Rat Genome Database 2015: Genomic, phenotypic and environmental variations and disease. *Nucleic Acids Research, 43*, D743–D750.

Smith, T. A., & Parry, D. A. D. (2008). Three-dimensional modeling of interchain sequence similarities and differences in the coiled-coil segments of keratin intermediate filament heterodimers highlight features important in assembly. *Journal of Structural Biology, 162*, 139–151.

Smith, T. A., Steinert, P. M., & Parry, D. A. D. (2004). Modeling effects of mutations in coiled-coil structures: Case study using epidermolysis Bullosa Simplex mutations in segment 1A of K5/K14 intermediate filaments. *Proteins: Structure, Function, and Bioinformatics, 55*, 1043–1052.

Steinert, P. M. (1990). The two-chain coiled-coil molecule of native epidermal keratin intermediate filaments is a type I–type II heterodimer. *Journal of Biological Chemistry, 265*, 8766–8774.

Steinert, P. M., Chou, Y.-H., Prahlad, V., Parry, D. A. D., Marekov, L. N., Wu, K. C., et al. (1999). A high molecular weight intermediate filament-associated protein in BHK-21 cells is nestin, a type VI intermediate filament protein: Limited co-assembly in vitro to form heteropolymers with type III vimentin and type IV α-internexin. *Journal of Biological Chemistry, 274*, 1657–1666.

Steinert, P. M., Marekov, L. N., Fraser, R. D. B., & Parry, D. A. D. (1993). Keratin intermediate filament structure: Crosslinking studies yield quantitative information on molecular dimensions and mechanism of assembly. *Journal of Molecular Biology, 230*, 436–452.

Steinert, P. M., Marekov, L. N., & Parry, D. A. D. (1993a). Conservation of the structure of keratin intermediate filaments: Molecular mechanism by which different keratin molecules integrate into pre-existing keratin intermediate filaments during differentiation. *Biochemistry, 32*, 10046–10056.

Steinert, P. M., Marekov, L. N., & Parry, D. A. D. (1993b). Diversity of intermediate filament structure: Evidence that the alignment of coiled-coil molecules in vimentin is different from that in keratin intermediate filaments. *Journal of Biological Chemistry, 268*, 24916–24925.

Steinert, P. M., Marekov, L. N., & Parry, D. A. D. (1999). Molecular parameters of type IV α-internexin and type IV-type III α-internexin-vimentin copolymer intermediate filaments. *Journal of Biological Chemistry, 274*, 1657–1666.

Stewart, M., & McLachlan, A. D. (1975). Fourteen actin-binding sites on tropomyosin? *Nature (London), 257*, 331–333.

Strelkov, S. V., & Burkhard, P. (2002). Analysis of α-helical coiled coils with the program TWISTER reveals a structural mechanism for stutter compensation. *Journal of Structural Biology, 137*, 54–64.

Strelkov, S. V., Herrmann, H., Geisler, N., Wedig, T., Zimbelmann, R., & Aebi, U. (2002). Conserved segments 1A and 2B of the intermediate filament dimer: Their atomic structures and role in filament assembly. *EMBO Journal, 21*, 1255–1266.

Strelkov, S. V., Schumacher, J., Burkhard, P., Aebi, U., & Herrmann, H. (2004). Crystal structure of the human lamin A coil 2B dimer: Implications for the head-to-tail association of nuclear lamins. *Journal of Molecular Biology, 343*, 1067–1080.

Strnad, P., Usachov, V., Debes, C., Gräter, F., Parry, D. A. D., & Omary, M. B. (2011). Unique amino acid signatures that are evolutionarily conserved distinguish simple-type, epidermal and hair keratins. *Journal of Cell Science, 124*, 4221–4232.

Suzuki, E., Crewther, W. G., Fraser, R. D. B., MacRae, T. P., & McKern, N. M. (1973). X-ray diffraction and infrared studies of an α-helical fragment from α-keratin. *Journal of Molecular Biology, 73*, 275–278.

Szeverenyi, I., Cassidy, A. J., Chung, C. W., Lee, B. T., Common, J. E., Ogg, S. C., et al. (2008). The human intermediate filament database: Comprehensive information on a gene family involved in many human diseases. *Human Mutation, 29*, 351–360.

Wang, H., Parry, D. A. D., Jones, L. N., Idler, W. W., Marekov, L. N., & Steinert, P. M. (2000). In vitro assembly and structure of trichocyte keratin intermediate filaments: A novel role for stabilization by disulfide bonding. *Journal of Cell Biology, 151*, 1459–1468.

Wilkins, M. H. F. (2003). *The third man of the double helix*. Oxford: Oxford University Press.

Wolpert, L. (1992). Estimating Popper's impact. *Nature, 360*, 202–204.

PART II

Mammalian IF Proteins

CHAPTER ELEVEN

Isolation and Analysis of Keratins and Keratin-Associated Proteins from Hair and Wool

Santanu Deb-Choudhury[1], Jeffrey E. Plowman, Duane P. Harland
Food and Bio-Based Products Group, AgResearch Ltd., Christchurch, New Zealand
[1]Corresponding author: e-mail address: santanu.deb-choudhury@agresearch.co.nz

Contents

1. Introduction 280
2. Chemical Extraction of Whole Fibers 281
   2.1 Alkaline Extraction 281
   2.2 Differential Extraction of Keratins 282
3. Isolation and Digestion of Other Components of the Fiber 286
   3.1 Acidic Extraction 286
   3.2 Proteolytic Digestion of Whole Fiber 288
   3.3 Isolation of the a-Layer 288
   3.4 Chemical Digestion of Resistant Membranes 289
4. Combination of Sodium Deoxycholate and C18 Empore™ for an Efficient Extraction of Keratin Peptides from Gel-Resolved Keratin Proteins 290
5. Ionic Liquid-Assisted Extraction of Fiber Keratins 294
   5.1 Isolation of KAPs 294
   5.2 Isolation of Keratins 296
   5.3 Digestion with $BMIM^+Cl^-$ 296
   5.4 IEF Prefractionation 296
6. Conclusions 299
References 300

Abstract

The presence of highly cross-linked protein networks in hair and wool makes them very difficult substrates for protein extraction, a prerequisite for further protein analysis and characterization. It is therefore imperative that these cross-links formed by disulfide bridges are first disrupted for the efficient extraction of proteins. Chaotropes such as urea are commonly used as efficient extractants. However, a combination of urea and thiourea not only improves recovery of proteins but also results in improved resolution of the keratins in 2DE gels. Reductants also play an important role in protein dissolution. Dithiothreitol effectively removes keratinous material from the cortex, whereas phosphines, like Tris(2-carboxyethyl)phosphine, remove material from the exocuticle. The

Methods in Enzymology, Volume 568
ISSN 0076-6879
http://dx.doi.org/10.1016/bs.mie.2015.07.018

relative extractability of the keratins and keratin-associated proteins is also dependent on the concentration of chaotropes, reductants, and pH, thus providing a means to preferentially extract these proteins. Ionic liquids such as 1-butyl-3-methylimidazolium chloride ($BMIM^{+}[Cl]^{-}$) are known to solubilize wool by disrupting noncovalent interactions, specifically intermolecular hydrogen bonds. $BMIM^{+}[Cl]^{-}$ proved to be an effective extractant of wool proteins and complementary in nature to chaotropes such as urea and thiourea for identifying unique peptides of wool proteins using mass spectrometry (MS). Successful identification of proteins resolved by one- or two-dimensional electrophoresis and MS is highly dependent on the optimal recovery of its protease-digested peptides with an efficient removal of interfering substances. The detergent sodium deoxycholate used in conjunction with Empore™ disks improved identification of proteins by mass spectrometry leading to higher percentage sequence coverage, identification of unique peptides and higher score.

ABBREVIATIONS

2DE two-dimensional electrophoresis
AcN acetonitrile
$BMIM^{+}[Cl]^{-}$ 1-butyl-3-methylimidazolium chloride
CMC cell membrane complex
DTT dithiothreitol
HGTP high-glycine–tyrosine protein
HSP high-sulfur protein
IPG immobilized pH gradient
KAP keratin-associated protein
MS mass spectrometry
NTCB 2-nitro-5-thiocyano-benzoic acid
SDC sodium deoxycholate
TCEP tris(2-carboxyethyl)phosphine
TFA trifluoroacetic acid
Tris tris(hydroxymethyl)aminomethane

1. INTRODUCTION

Trichocyte keratins from hair, wool, and fibers from other species are among some of the most difficult proteins to extract from any animal tissue. This is largely because of a highly cross-linked network that is formed by disulfide bridges between the major protein components of the fiber, namely, keratins and keratin-associated proteins (KAPs). Keratins can have anything from 17 to 32 cysteines, whereas KAPs can have an even higher number of cysteines.

Extractions of trichocyte keratins involve disruption of these cross-linked networks using a chaotropic agent and a reductant. Typically, these solutions are buffered at pHs greater than the p*K*a of the thiol group (pH 8.0) to ensure that it is deprotonated; however, the cysteine residues in wool are also prone to degradation if the pH is too high, being partially converted to lanthionine (Maclaren & Milligan, 1981). A variety of reductants have been trialed, including thioglycollic acid, tributylphosphine, and β-mercaptoethanol (Maclaren & Milligan, 1981). More recently, dithiothreitol (DTT) has found favor as a reductant because it has a less objectionable odor than mercaptoethanol and is a more powerful reductant (redox potentials, DTT: 0.33 V; β-mercaptoethanol: 0.26 V). A solution of 8 *M* urea, 50 m*M* tris(hydroxymethyl)aminomethane (Tris), and 50 m*M* DTT has been previously used to extract proteins from wool (Woods & Orwin, 1987). In this study, they found that more protein was extracted from the follicle end of the fiber with solubility ranging between 66.6% and 96.7%, whereas the solubility of the tip end ranged between 23.3% and 40.0%. This illustrates a key point, namely, that environmental effects, be it natural weathering or dye or hair treatments, often markedly reduce extractability; thus it pays to clearly understand the history of fiber samples.

Since the introduction of these protocols, we have established a range of techniques to improve the extraction efficiency of wool proteins as described in the following sections. Furthermore, despite the differences in protein profiles of the fibers of other species (Thomas et al., 2012), these protocols are readily applicable to the study of keratins from them as well.

Prior to extraction approximately one-third of the tip end of the wool or other fiber is removed. It is cleaned or scoured first with 0.15% Teric GN9 at 60 °C, then 40 °C for 2 min each, after which it is washed with water at 40 and 60 °C. After allowing it to dry overnight, it is washed twice with dichloromethane for 30 s each, followed by two washes with ethanol and two washes with water. After this, the cleaned fibers can either be crushed to a powder in liquid nitrogen or cut into snippets of less than 1 mm in length with scissors.

2. CHEMICAL EXTRACTION OF WHOLE FIBERS

2.1 Alkaline Extraction

Thiourea, when used in conjunction with high concentrations of urea, has led to significant improvements in the recovery of proteins from the nucleus, integral membranes, and tubulin compared to classical solubilization solutions as demonstrated by two-dimensional electrophoresis (2DE) (Rabilloud, Adessi, Giraudel, & Lunardi, 1997). The standard extraction conditions for wool (Woods & Orwin, 1987) were thus modified:

1. A solution of 7 *M* urea, 2 *M* thiourea, 50 m*M* Tris, and 50 m*M* DTT at pH 9.3 is added to the fiber at an extraction fiber:solution ratio of 10 mg:1 mL.
2. The wool suspension is shaken on a reciprocal action shaker for 18 h.
3. The extract is spun down at 14,000 rpm for 15 min and the supernatant removed with a micropipette.

Alkaline extraction with DTT primarily removes proteins from the macrofibrils of the cortex, leaving behind most or all of the nonkeratinous components of the cortex. Details of how extraction affects specific fiber structures can be visualized, and measured, using a combination of transmission electron microscopy (Woods & Orwin, 1987) and staining chemistry that allows clear differentiation of keratinous and nonkeratinous features (Harland, Vernon, Walls, & Woods, 2011). In transmission electron micrographs of cross-sections of intact wool (Fig. 1A), most nonkeratinous features stain darkly, including the thin network of intermacrofibrillar material surrounding small keratin macrofibrils, the cell membrane complex (CMC), the lumen-like cytoplasmic remnants (composed of the debris of cytoplasmic nuclear and organelle material), and also the cuticle, which contains keratinous and nonkeratinous components. Following extraction (Fig. 1B), micrographs clearly indicate that the previously light gray-stained keratin structures are gone and only the darkly stained nonkeratinous material remains. The use of 7 *M* urea in combination with 2 *M* thiourea results in better resolution of spots on 2DE gels (Deb-Choudhury et al., 2010), in particular, the lower pH portion of the Type II keratin region (i.e., K81, K83, and K86). The Type I keratins were also better resolved with three spots observed for K33a and K33b and four spots for K31.

2.2 Differential Extraction of Keratins

Extractability of the keratins and KAPs is highly dependent on both the concentration of chaotrope and reductant, as well as pH (Plowman et al., 2010).

2.2.1 Effect of Chaotrope Concentration

Reducing the concentration of urea from 8 to 1 *M* results in the decrease in the amount of the keratins extracted, a significant reduction being noted around 5 *M* urea, while no keratins are extracted at urea concentrations

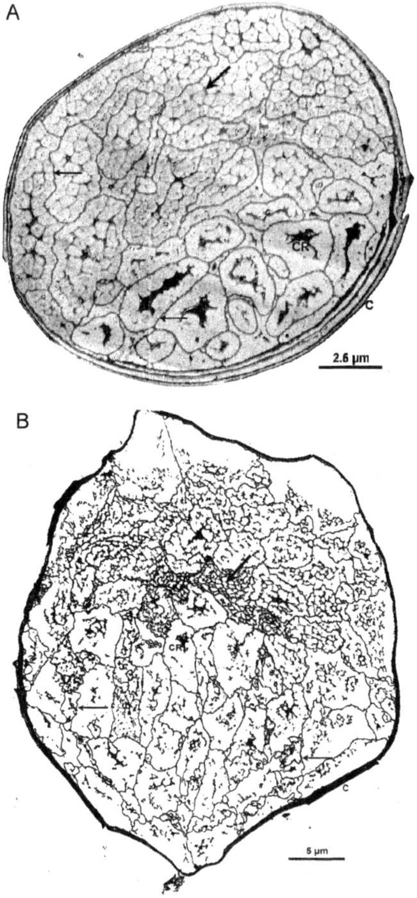

Figure 1 A transmission electron microscopy (TEM) micrograph of (A) an intact wool fiber cross-section and (B) wool extracted with DTT as the reductant showing cell membrane complex (thin arrow), cytoplasmic remnants (CR), intermacrofibrillar material (thick arrow), and cuticle (C) remaining after extraction of the keratins of the macrofibrils. *Reproduced with permission from Woods and Orwin (1987).*

below this (Fig. 2A). The extractability of some KAPs around 25 kDa is partly affected by the reduction in urea concentration and are extractable below 4 *M* urea, while the extractability of some other KAPs around 15 kDa decreases steadily as the urea concentration decreases. The extractability of one KAP remains constant until 5 *M* urea whereupon it is not extractable. The high-glycine-tyrosine proteins (HGTPs) below 10 kDa are not extractable below urea concentrations of 6 *M*. The remaining KAPs appeared to be unaffected by urea concentration above 2 *M*.

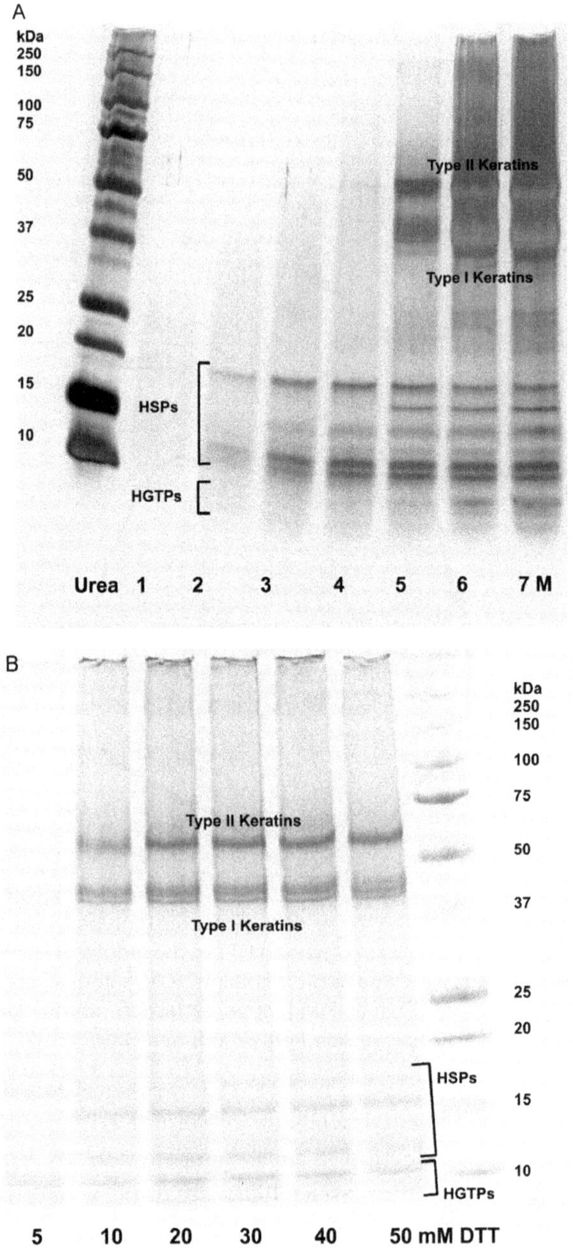

Figure 2 The effect of (A) urea concentration in molarity (*M*); (B and C) DTT concentration in milli-molarity (m*M*) on the extractability of proteins from Merino wool. (D) The effect of pH on the extractability of proteins from Merino wool fiber using an extraction buffer composed of 8 *M* urea, 50 m*M* Tris, and 50 m*M* DTT. *Reproduced with permission from Plowman et al. (2010).*

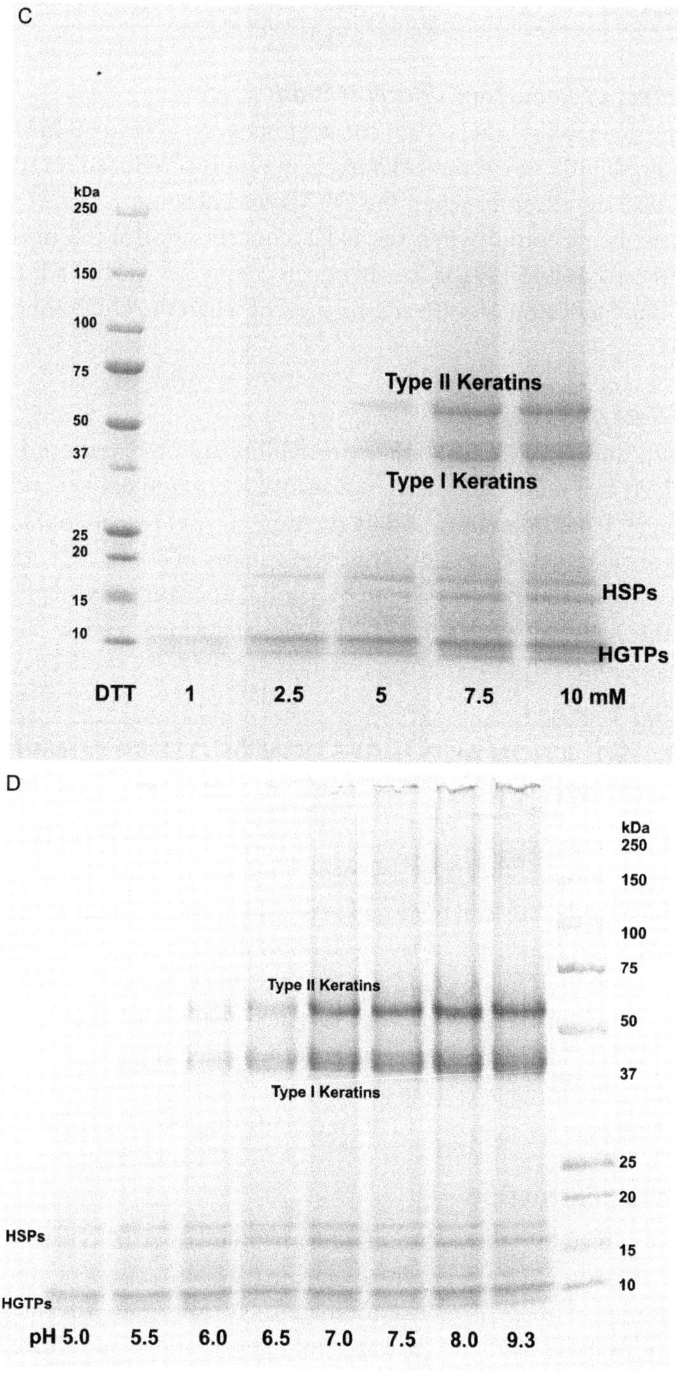

Figure 2—Cont'd

2.2.2 Effect of Reductant Concentration

Keratin extractability is relatively constant between 20 and 50 m*M* of DTT, decreasing slightly when the DTT concentration is reduced to 10 m*M* and much more significantly when the DTT concentration is 5 m*M* (Fig. 2B), with none being found when the DTT concentration is 2.5 m*M*. At least one KAP around 15 kDa is lost between 5 and 2.5 m*M* DTT (Fig. 2C), while some HGTPs at 10 kDa are still present when the DTT concentration is 1 m*M*.

2.2.3 Effect of Solution pH

No significant changes in keratin extractability are observed until pH 6.5 is reached. At that point, there is a considerable reduction in the concentration of the Type I and II keratins, and by the time the pH reaches 5.0, no further keratins are extracted (Fig. 2D). The extractability of the KAPs is less affected by pH, with the relative proportion remaining unchanged until pH 5.0, after which the high-sulfur proteins (HSPs) above 15 kDa tend to be less well extracted.

3. ISOLATION AND DIGESTION OF OTHER COMPONENTS OF THE FIBER

Wool and other hair fibers are surrounded by a sheath of flattened and overlapping cuticle cells. Each cuticle cell has a laminar structure composed of, on the outside, the thin but highly cross-linked keratinous a-layer (sometimes called the exocuticle a-layer). Below, this is the rest of the also keratinous exocuticle. The innermost layer is the endocuticle which is supposed to be nonkeratinous. The cells are attached to one another by a cuticle-specific CMC. These subcomponents of the cuticle appear to be unaffected by standard chemical extraction procedures, the a-layer being the most resistant to digestion by enzymes or strong acids (Fig. 3A).

3.1 Acidic Extraction

Tris(2-carboxyethyl)phosphine (TCEP) is useful as an alternative to DTT because it is stoichiometric with regard to the reduction of thiol groups and less toxic than β-mercaptoethanol. It also has advantages over tributylphosphine in that it is water soluble, odorless, and more stable in solution, the latter only being stable in urea solutions for 2 h (Chan, Lo, & Hodgkins, 2002). It is sold in the form of a hydrochloride, and as a result, the extraction solution is acidic. The extraction conditions used are

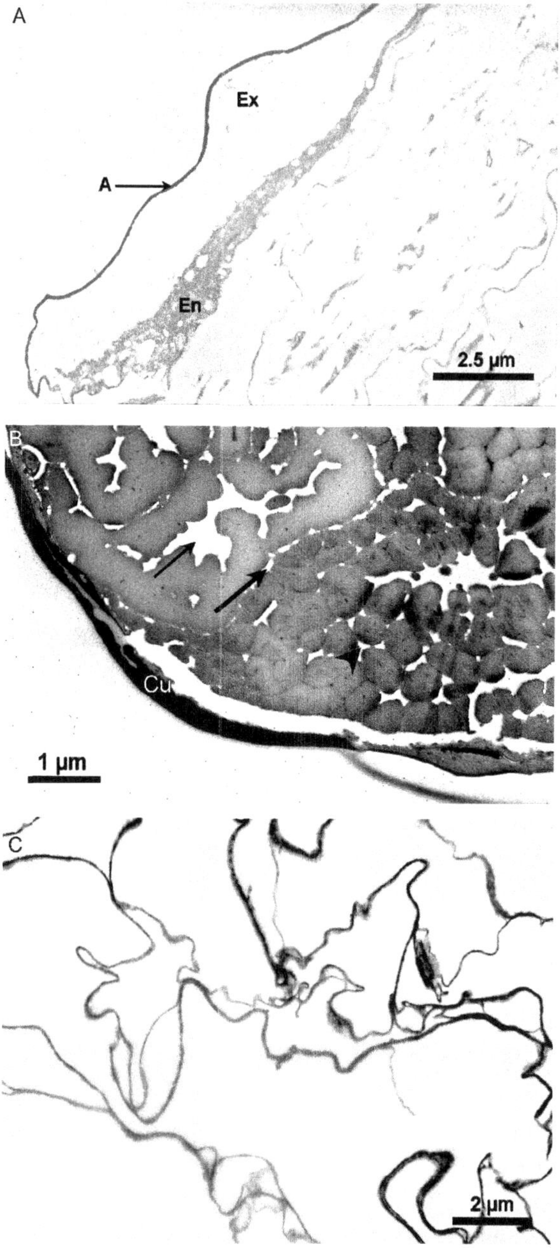

Figure 3 TEM micrographs illustrating layers of the wool cuticle and partial cross-sections of Merino wool (A) after reductive extraction with TCEP, showing the a-layer (A), exocuticle (Ex), and endocuticle (En); (B) after proteolytic digestion, showing the cuticle (Cu), cell membrane complex (thick arrow), cytoplasmic remnant (thin arrow), and intermacrofibrillar material (arrow head); and (C) enriched a-layer preparations sectioned end-on after reductive extraction and proteolytic digestion. *Reproduced with permission from Bringans et al. (2007).*

1. 7 *M* urea, 2 *M* thiourea, 50 m*M* Tris, and 50 m*M* TCEP at pH 4.3 at a wool to extraction solution ratio of 10 mg:1 mL.
2. The wool suspension is shaken on a reciprocal action shaker for 18 h, spun down at 14,000 rpm for 15 min, and the supernatant removed.

The use of acid conditions, when TCEP is the reductant, in addition to removal of macrofibrillar material, removes proteins from the exocuticle layer in the cuticle but not the a-layer, endocuticle, CMCs, or cytoplasmic remnants (Bringans et al., 2007).

The above extraction conditions result in a number of new spots in a 2DE map. Among them three strings of protein spots around 50 kDa below the Type II keratins, eight strings between 37 and 50 kDa below the Type I keratins and two strings between 25 and 30 kDa above the KAP1 HSP family are detected (Fig. 4B).

3.2 Proteolytic Digestion of Whole Fiber

Specific elements of the fiber can be removed by enzymic digestion (Bringans et al., 2007). The procedure includes digestion with Pronase E in 1% ammonium acetate buffer at a ratio of buffer:enzyme:substrate of 2000:1:20 at 37 °C for 7 days (Bringans et al., 2007).

Proteolytic digestion results in the incomplete removal of inter-macrofibrillar material, cytoplasmic remnants, and endocuticle. Extraction is most readily observed in the orthocortex with often less apparent removal of these components from the paracortex cells (Fig. 3B). The keratinous components of the cuticle and cortex are not extracted, and the CMC is damaged (where adjacent to nonkeratinous material) but not extracted. A similar result was also noted when trypsin was used to digest wool.

3.3 Isolation of the a-Layer

To obtain a purified isolate of the a-layer, the following protocol was used (Bringans et al., 2007):

1. Acidic extraction with TCEP as the reductant (as per Section 3.1).
2. Enzymatic digestion with Pronase E (as per Section 3.2).

Acidic extraction followed by proteolytic digestion of Merino wool fibers results in the complete removal of the cortex, endocuticle, and exocuticle leaving a "string-like" residue (Fig. 3C). The width of this residue is on average 50 nm and at the widest 130 nm. As the former is close to the average width of the a-layer (54 nm reported for Merino wool), this approach is the most effective way of removing protein material from the fiber to isolate a-layer.

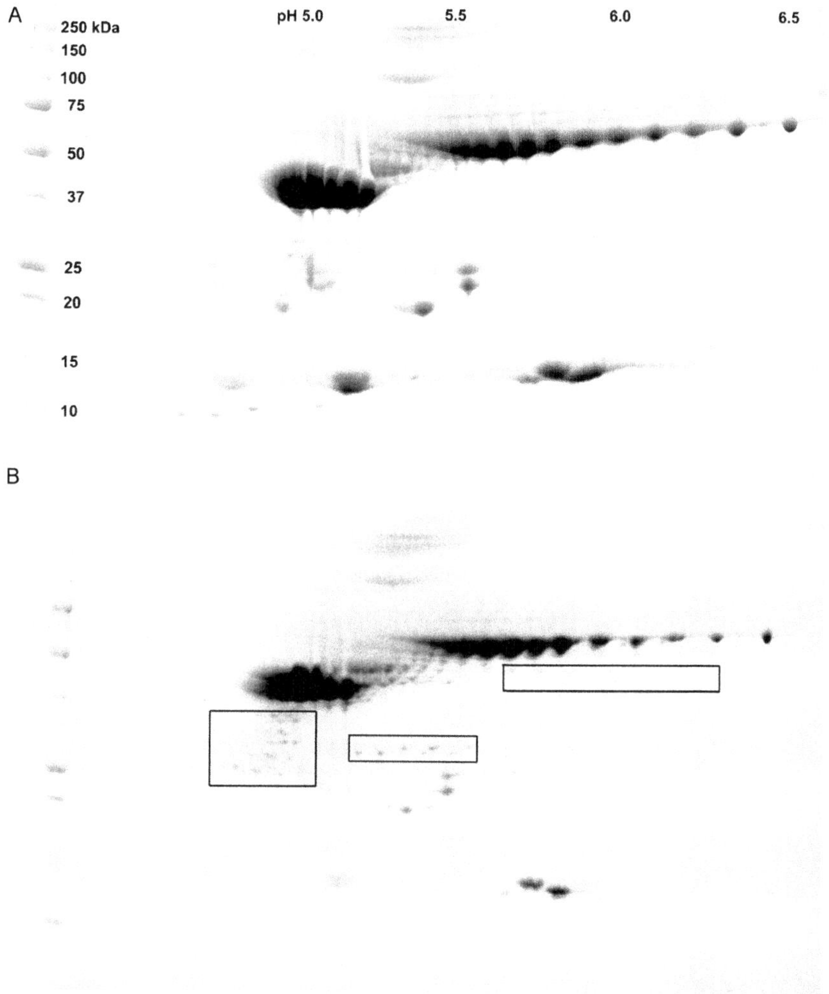

Figure 4 2DE maps (pH 4–7) of the wool extracted when (A) DTT and (B) TCEP were the reductants. The additional proteins extracted with TCEP are highlighted in the boxed regions.

3.4 Chemical Digestion of Resistant Membranes

2-Nitro-5-thiocyano-benzoic acid (NTCB) cleaves proteins on the N-terminal side of the cysteine residue resulting in its conversion to an iminothiazolidinyl carboxyl residue (Degani & Patchornik, 1974; Stark, 1977; Tang & Speicher, 2004) with an absorbance at 220 nm, which enables easy identification of cysteines by HPLC and mass spectrometry (MS). Because of the high number of cysteine residues in keratins and KAPs in hair fibers compared to other eukaryote organs and tissues, this method was evaluated as follows:

1. Cyanylation is performed by suspending the sample in a solution of 10 m*M* DTT, 4 *M* guanidine HCl, 1 *M* NH_4OH buffer solution, pH 8.0, then adding NTCB in a 20-fold molar excess and incubating at 37 °C for 1 h.
2. Protein cleavage is carried out by raising the pH to 9.0 with 1 *M* NaOH and incubating for 16 h
3. The reaction is terminated by incubation with 3 m*M* β-mercaptoethanol at 24 °C for 15 min.

This method digests up to 60% of the intractable a-layer as revealed by the identification of cuticular ultra-high-sulfur proteins by MS, in particular sheep KAP5.5 and proteins similar to human HSPs, KAP10 and KAP12 from the cuticle and the HGTP KAP19.2 found in both the cuticle and the cortex (Bringans et al., 2007).

4. COMBINATION OF SODIUM DEOXYCHOLATE AND C18 EMPORE™ FOR AN EFFICIENT EXTRACTION OF KERATIN PEPTIDES FROM GEL-RESOLVED KERATIN PROTEINS

Successful identification of proteins resolved by one-dimensional electrophoresis or 2DE and MS, as in Fig. 5, is highly dependent on the optimal recovery of its protease-digested peptides with an efficient removal of interfering substances. Iterative organic solvent extractions, followed by micropipette tip sample clean-up, are labor intensive and results in significant sample loss. Procedures that include the use of acid-labile surfactants, known to improve digestion efficiency, as well as the use of C18 Empore™ disks, which have been demonstrated to yield efficient peptide extraction are already present. An alternative method using surfactants in combination with C18 Empore™ disks was developed (Koehn et al., 2011) and compared against a standard protocol (Speicher, Kolbas, Harper, & Speicher, 2000), an SDC-assisted digestion protocol (Lin et al., 2008; Zhou et al., 2006), and an Empore™-assisted peptide extraction protocol (Meng et al., 2008).

The combined Empore™/SDC protocol consists of the following steps:

1. 4 μL of 1% sodium deoxycholate (SDC) in 50 m*M* NH_4HCO_3 is first added to the gel pieces that had undergone reduction with 50 m*M* TCEP and alkylation with 360 m*M* acrylamide.

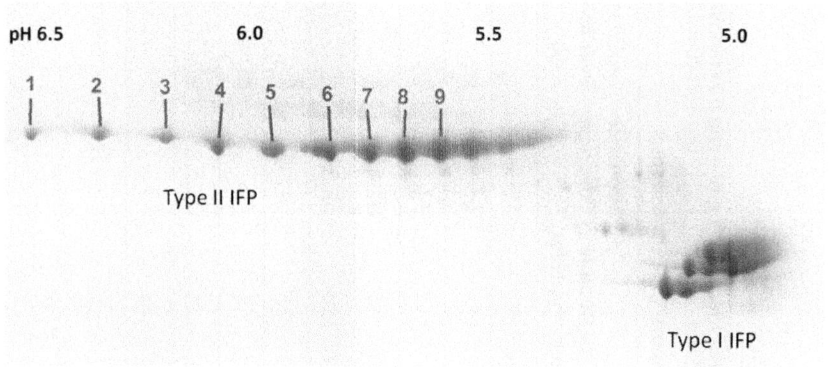

Figure 5 A 2DE gel of Merino wool proteins showing the distribution of Type II and Type I keratins. The labeled spots were excised for this study. The spot numbers refer to the numbers in Tables 1 and 2. *Reproduced with permission from Koehn et al. (2011).*

2. The incubation in the presence of SDC is carried out for 10 min at 37 °C after which the excess SDC solution is removed using a pipette.
3. The gel pieces are then trypsin digested as per standard protocol.
4. Small Empore™ disks are wetted with acetonitrile (AcN) for 1 min, conditioned with methanol for 1 min, and are then added to the gel digest and incubated for 3 h at room temperature with light vortexing.
5. The disks are removed and washed with 0.1% trifluoroacetic acid (TFA) for 5 min, and the bound peptides are eluted from the disks by lightly vortexing in 10 μL 75% AcN, 0.1% TFA over 30 min (Meng et al., 2008).
6. The eluent is then concentrated down to ~5 μL before adding 10 μL of 0.1% TFA for MS analysis.

Based on peptide mass fingerprinting, the Empore™-assisted and combined SDC + Empore™-assisted peptide protocols prove clearly superior to the standard protocol with respect to the most critical factors in protein identification: sequence coverage (%), unique peptides, and score. While the results of Empore™-assisted and combined SDC + Empore™-assisted extraction both produces good overall results, the combined approach is especially efficient in obtaining higher scores and better sequence coverage in some spots (Table 1).

Based on MS/MS data, both Empore™-assisted methods proved better suited for the identification of peptides of masses over 2000 Da compared to the standard method (Table 2; Fig. 6). Factors contributing to the identification of additional peptides possibly include surfactant-enhanced protein

Table 1 Comparison of Standard, Empore™, and Empore™ + SDC Peptide Extraction Procedures for the Identification of Type II Keratins from Merino Wool Protein Extracts

		Standard				Empore™ Assisted				Empore™ + SDC Assisted			
Spot#	Protein	Score	Unique Peptides	Sequence Coverage %	Peptides Over *m/z* 2000	Score	Unique Peptides	Sequence Coverage %	Peptides Over *m/z* 2000	Score	Unique Peptides	Sequence Coverage %	Peptides Over *m/z* 2000
1	K85	236	23	48.1	3	197	28	51.7	5	258	31	55.3	6
2	K85	240	26	50.1	4	325	33	59.4	8	171	27	52.3	6
3	K85	222	26	50.1	4	206	29	52.9	7	248	33	54.1	9
4	K85	211	26	50.3	4	162	25	47.7	6	205	31	58.6	6
5	K81	155	21	40.4	3	245	31	53.3	6	227	29	55.2	6
6	K81	193	28	50.1	5	236	30	61.9	9	231	31	54.4	7
7	K86	181	32	55.4	5	263	34	61	6	222	31	59	6
8	K81	173	30	55.4	6	181	23	46	7	200	28	52.5	5
9	K83	120	22	40.3	4	200	28	56.8	8	209	29	55.2	5

The spot numbers refer to the numbers in Fig. 5.

Table 2 Peptide Identification Performance of Standard, Empore™, and Combined Empore™ + SDC Peptide Extraction Procedures for the MS/MS Identification of Higher Mass Peptides from Merino Wool Keratin 2DE Spots

	Standard	Empore™	Empore™ + SDC	Peptide ID
m/z 2285 Spot 1	No peak detected	Score: 89	Score: 102	K85 199–220
m/z 2344 Spot 6	No match	Score: 97	Score: 107	K81 144–162 K86 161–179
m/z 3685 Spot 7	No peak detected	Score: 45	Score: 43	K81 66–102 K86 83–119
m/z 1844 Spot 9	No peak detected	Score: 68	Score: 100	K83 179–194

The spot numbers refer to the numbers in Fig. 5.

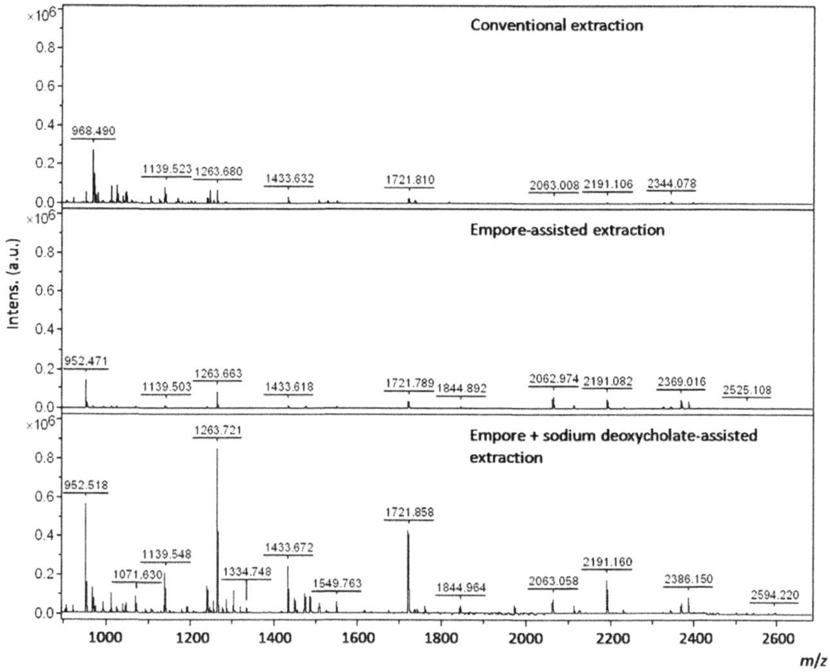

Figure 6 MALDI-TOF MS spectra of Type II keratin tryptic digests (spot 9) using the conventional, Empore™, and Empore™ + SDC procedures. *Reproduced with permission from Koehn et al. (2011).*

dissolution and avoidance of the use of organic solvents for better diffusion of the peptides from the gel matrix.

5. IONIC LIQUID-ASSISTED EXTRACTION OF FIBER KERATINS

The ionic liquid 1-butyl-3-methylimidazolium chloride ($BMIM^+[Cl]^-$) has been reported to solubilize wool by disrupting noncovalent interactions, specifically intermolecular hydrogen bonds (Xie, Li, & Zhang, 2005). To explore this potentially useful characteristic, a microscopic evaluation of wool dissolved in ($BMIM^+[Cl]^-$) was contrasted this with cryo-crushed wool powder. Differential interference contrast (Normarski) imaging microscopy of the cryo-crushed wool powder revealed discrete, dense wool fragments (Fig. 7A). After the powder was exposed to 99 °C for 18 h in the presence of $BMIM^+[Cl]^-$, a viscous solution formed in which wool powder was no longer visible by eye, but microscopic examination revealed fiber swelling and apparent disintegration of fibers with loss of cohesion between cells, leading, in some cases, to an intriguing "unwinding" effect of cuticle (Fig. 7B). Subcellular fragments that included individual macrofibrils were also observed at high magnification. These observations suggest that $BMIM^+Cl^-$ opens up the fiber structure by disrupting structural connections between cells and possibly between macrofibrils. Similar effects were observed with human scalp hair (Fig. 7C).

A MS compatible method using ionic liquid-assisted dissolution of wool proteins was developed and contrasted with the existing urea/thiourea extraction method (see Section 2.1). The following fractionation steps to isolate wool protein families are first carried out (Plowman et al., 2014).

5.1 Isolation of KAPs

1. Wool is extracted in 2 *M* urea, 50 m*M* Tris, and 50 m*M* DTT at pH 9.3, with a wool:extraction buffer ratio of 1:100 (w/v) in a reciprocal action shaker for 18 h at ambient temperature.
2. The extract is centrifuged at 20,000 × *g* for 15 min and the supernatant removed to give Fraction 1 (Plowman et al., 2010).
3. The supernatant is dialyzed against eight changes of water with a 3500 Da MWCO membrane and freeze-dried.

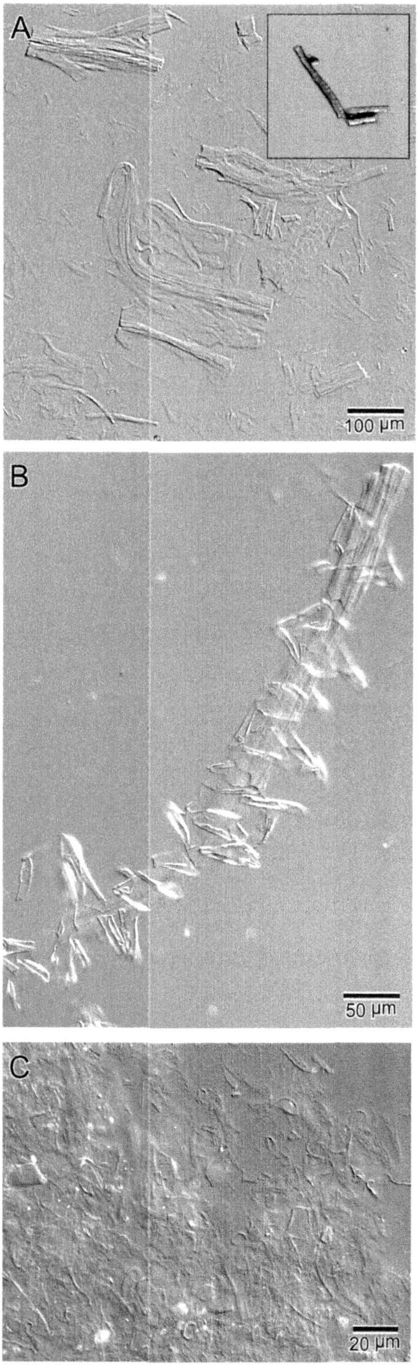

Figure 7 Effects of ionic liquid ($BMIM^{+}[Cl]^{-}$) on keratin fiber structure. (A) Liquid nitrogen ground wool, (B) following processing in ionic liquid, and (C) effect on human hair. *Reproduced with permission from Plowman et al. (2014).*

5.2 Isolation of Keratins

1. The fiber residue from the first extraction is further extracted for 18 h in 8 *M* urea, 50 m*M* tris, and 50 m*M* DTT at pH 9.3 at a residue to extraction buffer ratio of 1:100 (w/v).
2. The extract is centrifuged and the supernatant dialyzed for 18 h against eight changes of water and freeze-dry to give Fraction 2. The residue from this step is termed as Fraction 3.

5.3 Digestion with $BMIM^+Cl^-$

1. Fractions 1, 2, and 3 are individually mixed with $BMIM^+Cl^-$ (w/w ratio: 1:20) and heated with agitation at 99 °C for 18 h in 1.5 mL sample tubes.
2. After cooling, a reextraction of all the various fractions is performed by vortexing for 18 h in 7 *M* urea, 2 *M* thiourea, and 50 m*M* DTT (v/v, ratio: 1:50).
3. The proteinaceous material is precipitated from the extracts using chloroform/methanol (Wessel & Flügge, 1984).
4. The resulting pellet is resuspended by sonication for 10 min in 50 m*M* ammonium bicarbonate.
5. The resuspended pellet is reduced with 50 m*M* TCEP at 56 °C for 45 min and alkylated by adding 150 m*M* iodoacetamide and vortexed for 30 min.
6. The alkylated protein extract is then digested with trypsin for 18 h at 37 °C at an enzyme:substrate ratio of 1:50.

Fractions 1, 2, and 3 are also extracted by 7 *M* urea, 2 *M* thiourea, and 50 m*M* DTT without the initial $BMIM^+Cl^-$ step, alkylated and subjected to tryptic digestion as described above.

5.4 IEF Prefractionation

An IEF prefractionation is performed to increase the probability of identifying a higher number of proteins. Tryptic peptides generated from the various extraction procedures are prefractionated by isoelectric focusing using immobilized pH gradient (IPG) gel strips as follows:

1. Tryptic peptides from each fraction are redissolved in a solution of 8 *M* urea, 50 m*M* DTT, and 0.5% Pharmalyte buffer and used to rehydrate 11 cm, pH 3–11, NL IPG strips.

2. The IPG strips are electrofocused for 50 kVh. Approximately 1 cm lengths of gel were scraped off the surface of the backing plastic into 1.5-mL microfuge tubes.
3. The peptides from the gel pieces are progressively extracted with solutions containing 10%, 50%, and 80% AcN and 0.5% TFA and then concentrated and cleaned using Empore disks (Koehn et al., 2011).
4. Peptides are eluted from the Empore disks with 50% AcN, dried down, and reconstituted in 35 μL of 0.1% FA.

MS analysis of Fraction 1 revealed more identified peptides in the $BMIM^+Cl^-$/(urea/thiourea) extract compared to the urea/thiourea only extraction (control). The number of proteins identified in Fractions 1 and 2 was, however, similar in both the urea/thiourea control and the $BMIM^+Cl^-$ extractions. Both extraction procedures generated peptides mostly in the mass range m/z 1000–1500 (Fig. 8). Interestingly, the urea/thiourea only extraction generated more peptides with $m/z > 1500$, while $BMIM^+Cl^-$ extraction generated more peptides of $m/z < 1000$ (Fig. 8).

A similar number of trichocyte keratin proteins were identified by both extraction procedures across all fractions (Table 3). However, the ionic liquid procedure favored identification of both the KAPs and the cytokeratins in the first fraction. A total of 78 proteins were identified using both procedures (Fig. 9A). Of these, 9 were found to be uniquely identified by

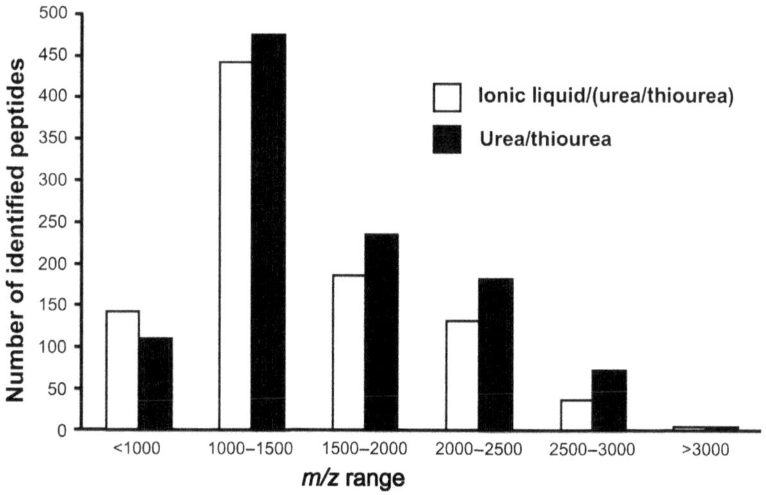

Figure 8 Distribution of identified peptides (from both keratinous and nonkeratinous proteins) by mass spectrometry in the combined fractions of the urea/thiourea and ionic liquid/(urea/thiourea) extractions.

Table 3 Comparison of the Total Number of Keratins Identified in Each Fraction Using Ionic Liquid/(Urea/Thiourea) and Urea Extraction

	Fraction 1		Fraction 2		Fraction 3	
	Urea/Thiourea	Ionic Liquid/(Urea/Thiourea)	Urea/Thiourea	Ionic Liquid/(Urea/Thiourea)	Urea/Thiourea	Ionic Liquid/(Urea/Thiourea)
Trichocyte keratins	14	14	14	12	13	13
KAPs	18	21	21	13	18	6
Epithelial keratins	1	7	2	1	8	5
Total	33	42	37	26	39	24

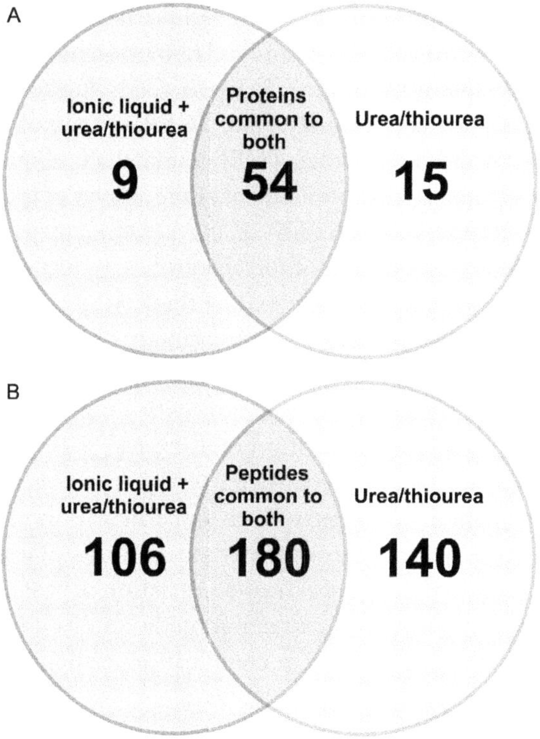

Figure 9 Unique (A) keratins and (B) keratin peptides identified in the combined ionic liquid/(urea/thiourea) and urea/thiourea procedures. *Reproduced with permission from Plowman et al. (2014).*

the ionic liquid procedure and 15 by the urea/thiourea only extraction method, leaving 54 common to both. Furthermore, 106 peptides were found to be unique to the ionic liquid procedure and 140 to the urea/thiourea only extraction procedure, while 180 were common to both (Fig. 9B). These results clearly revealed the complementary nature of the procedures, because the unique peptides generated by these extraction procedures resulted in a larger set of identified proteins than was possible with either of them on their own (Plowman et al., 2014).

6. CONCLUSIONS

The keratinization process in fibers results in a highly cross-linked network of proteins, which are themselves relatively insoluble in aqueous solution. Breaking this network involves the use of reductants with chaotropes being required to solubilize the proteins. While the use of urea is standard, the addition of thiourea results in improved resolution of the keratins in 2DE gels. Reductants, such as DTT, are effective in removing keratinous material from the cortex, whereas phosphines, like TCEP, can remove material from the exocuticle. Furthermore, when TCEP is used in combination with proteolytic digestion, a relatively pure isolate of the a-layer of the cuticle can be obtained. The relative extractability of the keratins, in particular, is also dependent on the concentration of the chaotropes, reductant, and pH, thus providing a means to preferentially extract the KAPs.

When it comes to improving the extractability of keratins and KAPs for mass spectroscopic analysis, there are a number of approaches available. The detergent SDC when used in conjunction with Empore™ disks has been found to improve identification of proteins by MS with enhancements in the critical factors leading to higher percent sequence coverage, identification of unique peptides and higher score. In particular, more peptides over 2000 Da were matched than by other methods. The use of ionic liquids adds another dimension to the mass spectral identification of proteins because it is complementary to the standard extraction method in its use of a chaotrope and reductant, the use of the two in combination leading to the identification of more proteins than either one alone.

REFERENCES

Bringans, S. D., Plowman, J. E., Dyer, J. M., Clerens, S., Vernon, J. A., & Bryson, W. G. (2007). Characterization of the exocuticle a-layer proteins of wool. *Experimental Dermatology, 16*(11), 951–960.

Chan, L. L., Lo, S. C., & Hodgkins, I. J. (2002). Proteomic study of a model causative agent of harmful red tide, *Prorocentrum triestinum* I: Optimization of sample preparation methodologies for analyzing with two-dimensional electrophoresis. *Proteomics, 2*(9), 1169–1186.

Deb-Choudhury, S., Plowman, J. E., Thomas, A., Krsinic, G. L., Dyer, J. M., & Clerens, S. (2010). Electrophoretic mapping of highly homologous keratins: A novel marker peptide approach. *Electrophoresis, 31*(17), 2894–2902.

Degani, Y., & Patchornik, A. (1974). Cyanylation of sulfhydryl groups by 2-nitro-5-thiocyanobenzoic acid. High-yield modification and cleavage of peptides at cysteine residues. *Biochemistry, 13*(1), 1–11.

Harland, D. P., Vernon, J. A., Walls, R. J., & Woods, J. L. (2011). Transmission electron microscopy staining methods for the cortex of human hair: A modified osmium method and comparison with other stains. *Journal of Microscopy, 243*(2), 184–196.

Koehn, H., Lau, B., Clerens, S., Plowman, J. E., Dyer, J. M., Ramli, U. S., et al. (2011). Combination of acid labile detergent and C18 Empore™ disks for improved identification and sequence coverage of in-gel digested proteins. *Analytical and Bioanalytical Chemistry, 400*(2), 415–421. (Technical note).

Lin, Y., Zhou, J., Bi, D., Chen, P., Wang, X., & Liang, S. (2008). Sodium-deoxycholate-assisted tryptic digestion and identification of proteolytically resistant proteins. *Analytical Biochemistry*, *377*(2), 259–266.

Maclaren, J. A., & Milligan, B. (1981). *Wool science—The chemical reactivity of the wool fibre*. Marrickville, NSW 2204, Australia: Science Press, pp. 1–18.

Meng, W., Zhang, H., Guo, T., Pandey, C., Zhu, Y., Kon, O. L., et al. (2008). One-step procedure for peptide extraction from in-gel digestion sample for mass spectrometric analysis. *Analytical Chemistry*, *80*(24), 9797–9805. (Technical notes).

Plowman, J. E., Clerens, S., Lee, E., Harland, D. P., Dyer, J. M., & Deb-Choudhury, S. (2014). Ionic liquid-assisted extraction of wool keratin proteins as an aid to MS identification. *Analytical Methods*, *6*, 7305–7311.

Plowman, J. E., Deb-Choudhury, S., Thomas, A., Clerens, S., Cornellison, C. D., Grosvenor, A. J., et al. (2010). Characterisation of low abundance wool proteins through novel differential extraction techniques. *Electrophoresis*, *31*(12), 1937–1946. http://dx.doi.org/10.1002/elps.201000053. (Research article).

Rabilloud, T., Adessi, C., Giraudel, A., & Lunardi, J. (1997). Improvement of the solubilization of proteins in two-dimensional electrophoresis with immobilized pH gradients. *Electrophoresis*, *18*, 307–316.

Speicher, K., Kolbas, O., Harper, S., & Speicher, D. (2000). Systematic analysis of peptide recoveries from in-gel digestions for protein identifications in proteome studies. *Journal of Biomolecular Techniques*, *11*(2), 74–86.

Stark, G. R. (1977). Cleavage at cysteine after cyanylation. *Methods in Enzymology*, *47*, 129–132.

Tang, H. Y., & Speicher, D. W. (2004). Identification of alternative products and optimization of 2-nitro-5-thiocyanatobenzoic acid cyanylation and cleavage at cysteine residues. *Analytical Biochemistry*, *334*(1), 48–61.

Thomas, A., Harland, D. P., Clerens, S., Deb-Choudhury, S., Vernon, J. A., Krsinic, G. K., et al. (2012). Interspecies comparison of morphology, ultrastructure and proteome of mammalian keratin fibres of similar diameter. *Journal of Agricultural and Food Chemistry*, *60*(10), 2434–2446.

Wessel, D., & Flügge, U. I. (1984). A method for the quantitative recovery of protein in dilute solution in the presence of detergents and lipids. *Analytical Biochemistry*, *138*(1), 141–143.

Woods, J. L., & Orwin, D. F. G. (1987). Wool proteins of New Zealand Romney sheep. *Australian Journal of Biological Sciences*, *40*, 1–14.

Xie, H., Li, S., & Zhang, S. (2005). Ionic liquids as novel solvents for the dissolution and blending of wool keratin fibers. *Green Chemistry*, *7*, 606–608.

Zhou, J., Zhou, T., Cao, R., Liu, Z., Shen, J., Chen, P., et al. (2006). Evaluation of the application of sodium deoxycholate to proteomic analysis of rat hippocampal plasma membrane. *Journal of Proteome Research*, *5*(10), 2547–2553.

CHAPTER TWELVE

Skin Keratins

Fengrong Wang*, Abigail Zieman*, Pierre A. Coulombe*,†,‡,§,1
*Department of Biochemistry and Molecular Biology, Bloomberg School of Public Health, Johns Hopkins University, Baltimore, Maryland, USA
†Department of Biological Chemistry, School of Medicine, Johns Hopkins University, Baltimore, Maryland, USA
‡Department of Dermatology, School of Medicine, Johns Hopkins University, Baltimore, Maryland, USA
§Department of Oncology, School of Medicine, Johns Hopkins University, Baltimore, Maryland, USA
[1]Corresponding author: e-mail address: coulombe@jhsph.edu

Contents

1. Introduction 305
   1.1 General Features of Keratin Genes, Proteins, and Filaments 305
   1.2 Keratin Gene Expression in Skin 309
   1.3 Functions of Keratin Filaments and Their Associated Skin Disorders 310
2. Collection of Mouse Skin Tissue for Analysis 311
   2.1 Isolation of Mouse Skin Samples for Morphological Studies 311
   2.2 Harvest Mouse Skin Samples for RNA and Proteins 318
3. Cell Culture Studies 319
   3.1 Materials 319
   3.2 Isolation of Keratinocytes from Newborn Mouse Skin 320
   3.3 Conditions for the Primary Culture of Skin Keratinocytes 322
   3.4 Culture of Skin Explants *Ex Vivo* 325
   3.5 Immunofluorescence Staining of Cells in Culture 325
   3.6 Isolation and Analysis of RNA and Proteins from Keratinocytes in Culture 326
   3.7 Immunoprecipitation of Keratin Proteins from Cell Culture to Study Their Interacting Partners 329
4. *In Vitro* Methods to Study Keratin Proteins 330
   4.1 Materials 330
   4.2 Expression and Purification of Keratin Proteins 331
   4.3 *In Vitro* Assembly of Keratin Filaments 333
   4.4 High-Speed Sedimentation Assay and TEM 334
   4.5 Low-Speed Sedimentation Assay 335
   4.6 Far-Western Assay and Cosedimentation Assay to Study Direct Interactions 335
5. Pearls and Pitfalls 337
   5.1 Collection of Mouse Skin Tissue 337
   5.2 Morphological Studies 337
   5.3 Cell Culture Studies 338
   5.4 Isolation and Analysis of Proteins and RNA from Skin Cells and Tissues 338
   5.5 *In Vitro* Studies with Recombinant Keratin Proteins 339

Methods in Enzymology, Volume 568
ISSN 0076-6879
http://dx.doi.org/10.1016/bs.mie.2015.09.032

Abstract

Keratins comprise the type I and type II intermediate filament-forming proteins and occur primarily in epithelial cells. They are encoded by 54 evolutionarily conserved genes (28 type I, 26 type II) and regulated in a pairwise and tissue type-, differentiation-, and context-dependent manner. Keratins serve multiple homeostatic and stress-enhanced mechanical and nonmechanical functions in epithelia, including the maintenance of cellular integrity, regulation of cell growth and migration, and protection from apoptosis. These functions are tightly regulated by posttranslational modifications as well as keratin-associated proteins. Genetically determined alterations in keratin-coding sequences underlie highly penetrant and rare disorders whose pathophysiology reflects cell fragility and/or altered tissue homeostasis. Moreover, keratin mutation or misregulation represents risk factors or genetic modifiers for several acute and chronic diseases. This chapter focuses on keratins that are expressed in skin epithelia, and details a number of basic protocols and assays that have proven useful for analyses being carried out in skin.

ABBREVIATIONS

BSA bovine serum albumin
Co-IP coimmunoprecipitation
DAPI 4′,6-diamidino-2-phenylindole
DMEM Dulbecco's modified Eagle's medium
DNA deoxyribonucleic acid
DNase deoxyribonuclease
DTT DL-dithiothreitol
ECL enhanced chemiluminescence
E. coli *Escherichia coli*
EDTA ethylenediaminetetraacetic acid
EGF epidermal growth factor
EGTA ethylene glycol-bis(2-aminoethylether)-N,N,N′,N′-tetraacetic acid
EM electron microscopy
FPLC fast protein liquid chromatography
H&E hematoxylin and eosin
HRP horseradish peroxidase
IF intermediate filament
IP immunoprecipitation
IPTG isopropyl β-D-1-thiogalactopyranoside
K keratin (protein)
Krt keratin (gene)
LB Luria broth

NGS normal goat serum
OCT optimal cutting temperature
PBS phosphate-buffered saline
PFA paraformaldehyde
PIC protease inhibitor cocktail
PIPES 1,4-piperazinediethanesulfonic acid
PMSF phenylmethylsulfonyl fluoride
RNA ribonucleic acid
SDS sodium dodecyl sulfate
SDS-PAGE sodium dodecyl sulfate polyacrylamide gel electrophoresis
TBST Tris-buffered saline with Tween 20
TEM transmission electron microscopy

1. INTRODUCTION

1.1 General Features of Keratin Genes, Proteins, and Filaments

Keratins represent a major subclass within the large family of intermediate filament (IF) proteins, which self-assemble into 10-nm-wide filaments (Fuchs & Weber, 1994). The genes encoding the 28 type I and 26 type II keratins are, respectively, clustered on chromosomes 17q21.2 and 12q13.13, with the exception of the type I keratin 18 (*Krt18*), which is located next to *Krt8* on the type II gene cluster (Hesse, Zimek, Weber, & Magin, 2004; Fig. 1). The molecular features of keratin genes, such as size, number and positions of introns/exon junctions, transcriptional orientation, and positions relative to other family members are largely conserved in mammals, suggesting that keratin genes arose from duplication of an ancestral gene during evolution (Coulombe & Bernot, 2004; Hesse et al., 2004). Keratins share the tripartite structure of all IF proteins: a highly conserved central α-helical rod domain flanked by highly variable nonhelical N- and C-terminal head and tail domains (Fuchs & Weber, 1994; Steinert, Steven, & Roop, 1985; Fig. 2A). The N-terminal head and C-terminal tail domains are dynamically subjected to numerous posttranslational modifications (Omary, Ku, Liao, & Price, 1998; Pan, Hobbs, & Coulombe, 2013; Snider & Omary, 2014), which affect keratin filament assembly, organization, and interactions with other proteins (Haines & Lane, 2012; Snider & Omary, 2014; Toivola, Boor, Alam, & Strnad, 2015).

In vitro studies have shown that keratin filament assembly begins with the formation of heterodimers from type I and type II keratin monomers, with

A

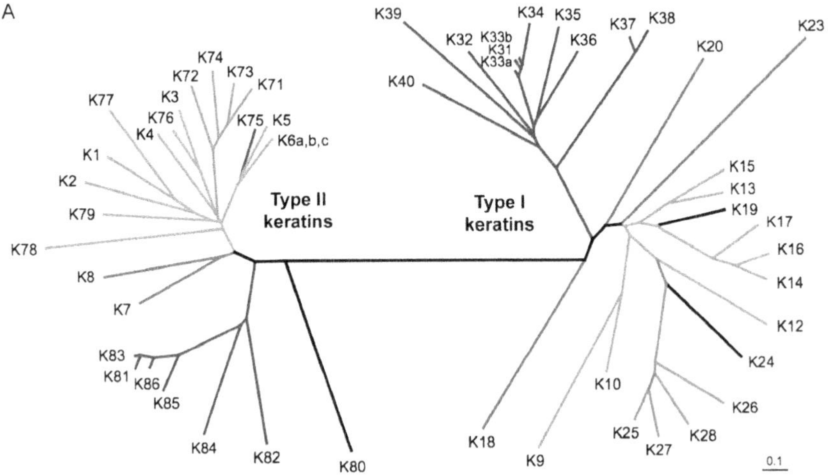

B

Type I keratin gene cluster—human chromosome 17q21.2

centromere ... *KRT222P KRT24 KRT223P KRT25 KRT26 KRT27 KRT28 KRT10 KRT12 KRT20 KRT23 KRT39 KRT40 KRT33A KRT33B KRT34 KRT31 KRT41P KRT37 KRT38 KRT221P KRT32 KRT35 KRT36 KRT13 KRT15 KRT19 KRT9 KRT14 KRT16 KRT17 KRT42P* ... telomere

Type II keratin gene cluster—human chromosome 12q13.13

centromere ... *KRT80 KRT7 KRT121P KRT122P KRT81 KRT86 KRT23 KRT123P KRT85 KRT84 KRT82 KRT124P KRT75 KRT6B KRT6C KRT6A KRT5 KRT71 KRT74 KRT72 KRT73 KRT125P KRT2 KRT1 KRT77 KRT126P KRT127P KRT128P KRT76 KRT3 KRT4 KRT79 KRT78 KRT8 KRT18* ... telomere

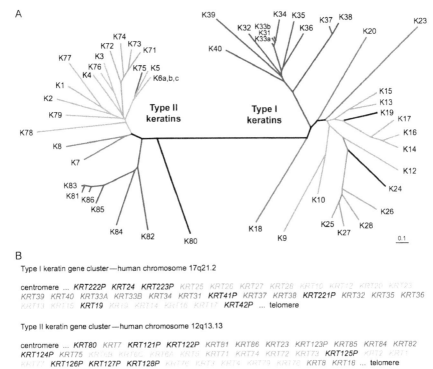

Figure 1 The keratin gene family. (A) Comparison of the primary structure of human keratins using the ClustalW and TreeView softwares. Sequence relatedness is inversely correlated with the length of the lines connecting sequences, and number and position of branch points. This comparison makes use of the sequences from the head and central rod domain for each keratin. Two major branches are seen, corresponding exactly to the known partitioning of keratin genes into type I and type II sequences. Beyond this dichotomy, each subtype is further segregated into major subgroupings. (B) Location and organization of type I and type II keratin genes in the human genome. All functional type I keratin genes, except *Krt18*, are clustered on the long arm of human chromosome 17, while all functional type II keratin genes are located on the long arm of chromosome 12. *Krt18*, a type I gene, is located at the telomeric (Tel) boundary of the type II cluster. The suffix P identifies keratin pseudogenes. As highlighted by the color code used in frames A and B, individual type I and type II keratin genes belonging to the same subgroup, based on the primary structure of their protein products, tend to be clustered in the genome. Moreover, highly homologous keratin proteins (e.g., K5 and K6 paralogs; also, K14, K16, and K17) are often encoded by neighboring genes, pointing to the key role of gene duplication in generating keratin diversity. These features of the keratin family are virtually identical in mouse (not shown). *Adapted from Coulombe, Bernot, and Lee (2013), figure 1.*

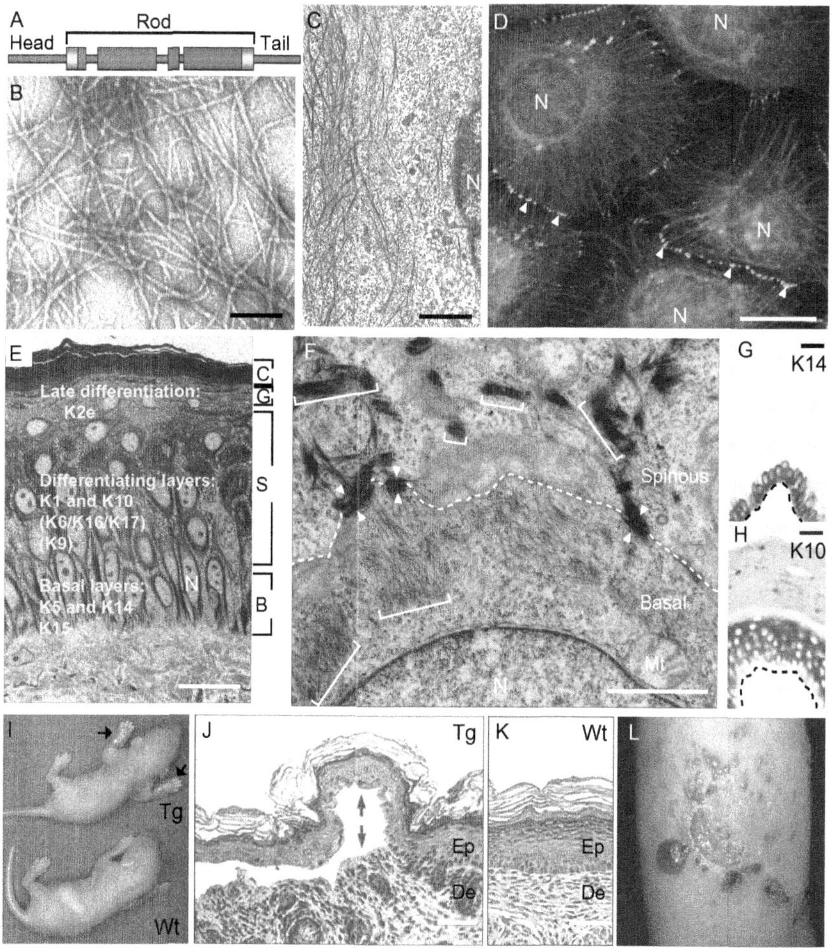

Figure 2 Attributes, differential regulation, and disease association of keratins. (A) Tripartite domain structure shared by all keratin and other intermediate filament (IF) proteins. A central α-helical "rod" domain acts as a key determinant of self-assembly and is flanked by nonhelical "head" and "tail" domains at the N-terminus and C-terminus, respectively. The ends of the rod domain contain 15–20 amino acid regions, here shown is yellow that are highly conserved among all IFs. (B) Visualization of filaments, reconstituted *in vitro* from purified K5 and K14, by negative staining and electron microscopy. Bar, 125 nm. (C) Ultrastructure of the cytoplasm of epidermal cells in primary culture as shown by transmission electron microscopy. Keratin filaments are abundant and tend to be organized in large bundles of loosely packed filaments in the cytoplasm. Bar, 5 μm. (D) Triple-labeling for keratin (red) and desmoplakin (green), a desmosome component, and DNA (blue) by indirect immunofluorescence of epidermal cells in culture. Keratin filaments are organized in a network that spans the entire cytoplasm

(Continued)

the central rod domain S aligned in parallel and in register (Coulombe & Fuchs, 1990; Hatzfeld & Weber, 1990a). These heterodimers then interact along their lateral surfaces with an axial stagger and an antiparallel orientation, giving rise to structurally apolar tetramers. The latter further interact in an end-to-end and lateral fashion to give rise to 10-nm-wide filaments that are apolar, display a smooth surface, and contain on average 16 coiled-coil dimers in cross section (Herrmann, Wedig, Porter, Lane, & Aebi, 2002; Fig. 2B). How IF assembly proceeds in living cells is poorly understood at present (Fig. 2C). A model based upon observations from time-lapse imaging of live epithelial cells in culture proposes that filament nucleation occurs at the cell periphery, with filament assembly proceeding as part of a centripetal flow culminating in the formation of a dense filament network that surrounds the nucleus (Windoffer, Beil, Magin, & Leube, 2011). In its present form, this model does not account for how keratin filaments form stable arrays anchored at desmosome (cell–cell) and hemidesmosome

Figure 2—Cont'd and are attached to desmosomes at points of cell–cell contacts (arrowheads). Bar, 30 μm. N, nucleus. (E) Histological cross section of resin-embedded human trunk epidermis, revealing the basal (B), spinous (S), granular (G), and cornified (C) compartments. The differentiation-dependent distribution of keratin proteins in the epidermis is indicated. Bar, 50 μm. N, nucleus. (F) Ultrastructure of the boundary between the basal and suprabasal cells in mouse trunk epidermis as seen by routine transmission electron microscopy. The sample, from which this micrograph was taken, is oriented in the same manner as (E). Organization of keratin filaments as loose bundles (brackets in basal cell) correlates with the expression of K5–K14 in basal cells, whereas the formation of much thicker and electron-dense filament bundles (brackets in spinous cell) reflects the onset of K1–K10 expression in early differentiating keratinocytes. Arrowheads point to desmosomes. Bar, 1 μm. N, nucleus. (G and H) Differential distribution of keratin epitopes on human skin tissue cross sections (similar to E) as visualized by an antibody-based detection method. K14 occurs in the basal layer, where the epidermal progenitor cells reside (G). K10 primarily occurs in the differentiating suprabasal layers of epidermis (H). Dashed line, basal lamina. Bar, 100 μm. (I) Newborn mouse littermates. The top mouse is transgenic (Tg) and expresses a mutated form of K14 in its epidermis. Unlike the control pup below (Wt), this transgenic newborn shows extensive blistering of its front paws (arrows). (J and K) Hematoxylin and eosin (H&E)-stained histological cross section through paraffin-embedded newborn mouse skin similar to those shown in (I). Compared with the intact skin of a control littermate (K, Wt), the epidermis of the K14 mutant expressing transgenic pup (J, Tg) shows intraepidermal cleavage within the basal layer, where the mutant keratin is expressed (opposing arrows). Bar, 100 μm. (L) Leg skin in a patient with the Dowling-Meara form of epidermolysis bullosa simplex. Several skin blisters are grouped in a herpetiform pattern. *Reproduced from Coulombe & Bernot, 2004.* (See the color plate.)

(cell–matrix) junctions (Fig. 2D)—yet, these elements are crucial to the structural support and cytoarchitectural functions of keratin IFs.

1.2 Keratin Gene Expression in Skin

More than half of keratin genes are expressed in mature mammalian skin tissue. Keratin gene expression exquisitely reflects the type of epithelium (e.g., hair vs. epidermis), the differentiation state of epithelial cells and is subjected to striking modulation upon wounding, infection, or disease (Fuchs & Green, 1980; Takahashi, Yan, Yamanishi, Imamura, & Coulombe, 1998; Toivola et al., 2015; Woodcock-Mitchell, Eichner, Nelson, & Sun, 1982). The interfollicular epidermis is a stratified epithelium consisting of a proliferative basal layer (progenitor status) that gives rise to multiple suprabasal layers (spinous, granular, and cornified) through a programmed differentiation process (Fuchs, 1995; Fig. 2E). Basal keratinocytes are mitotically active and express keratin 5 (K5; type II), K14, and minor amounts of K15 (both type I) (Fuchs & Green, 1980; Lloyd et al., 1995; Nelson & Sun, 1983; Fig. 2E–G). Upon their commitment to terminal differentiation, basal keratinocytes stop dividing and start expressing K1 (type II) and K10 (type I) concurrent with their migration upward into the suprabasal compartment (Byrne, 1997; Fuchs & Green, 1980; Woodcock-Mitchell et al., 1982; Fig. 2E, F, and H). As they differentiate, epidermal keratinocytes undergo a dramatic flattening and begin accumulating keratohyalin granules, forming the granular layer (Holbrook & Wolff, 1993). Additionally, expression of K2e, yet another type II keratin, is induced in the upper spinous and granular layers (Collin, Moll, Kubicka, Ookayoun, & Franke, 1992; Fig. 2E).

Keratin gene expression in palmar-plantar epidermis is far more complicated than that of interfollicular epidermis. In this case, K9 (type I) is prominently expressed in the suprabasal, differentiating layers of stress-bearing regions, presumably to increase mechanical resilience (Langbein, Heid, Moll, & Franke, 1993; Swensson et al., 1998; Fig. 2E). Further, K6 (type II), K16, K17, and K19 (all type I) are also expressed in specific spatial patterns (McGowan & Coulombe, 1998b; Michel et al., 1996; Swensson et al., 1998). Interestingly, K6, K16, and K17 are rapidly and robustly induced upon various challenges to the interfollicular epidermis (e.g., wounding, infection), as well as in the setting of chronic hyperproliferative diseases (e.g., psoriasis and cancer) (Freedberg, Tomic-Canic, Komine, & Blumenberg, 2001; McGowan & Coulombe, 1998b; Paladini, Takahashi, Bravo, & Coulombe, 1996; Stoler, Kopan, Duvic, & Fuchs, 1988; Weiss, Eichner, & Sun, 1984; Fig. 2E).

1.3 Functions of Keratin Filaments and Their Associated Skin Disorders

Keratin filaments impact cytoarchitecture and endow epithelial cells with their remarkable mechanical resilience and ability to withstand various forms of stresses (Coulombe, Hutton, Vassar, & Fuchs, 1991; Fuchs, Esteves, & Coulombe, 1992; Ma, Yamada, Wirtz, & Coulombe, 2001; Seltmann, Fritsch, Kas, & Magin, 2013). *In vitro* biophysical studies have shown that keratin filaments possess unique and remarkable viscoelastic properties that support their involvement in a structural support capacity (Ma et al., 2001). Mutations in keratin genes can reduce the strength of keratin filaments by disrupting filament formation, altering the properties and dynamics of keratin assemblies, and/or destabilizing junction proteins (Haines & Lane, 2012; Omary, Coulombe, & McLean, 2004) and ultimately result in disease. Examples of these skin fragility disorders include epidermolysis bullosa simplex (caused by mutations in K5 or K14), the first IF disorder to be discovered (Bonifas, Rothman, & Epstein, 1991; Coulombe, Hutton, Letai, et al., 1991; Lane et al., 1992; Fig. 2I–L), epidermolytic hyperkeratosis (mutations in K1, K10, or K9) (Lane & McLean, 2004), and monilethrix (mutations in K81, K83, or K86) (McLean & Moore, 2011).

Additionally, keratin filaments act as platforms that modulate cellular and molecular events including migration (Rotty & Coulombe, 2012), innate immunity (Lessard et al., 2013; Roth et al., 2012; Tam, Mun, Evans, & Fleiszig, 2012), hair cycling (Tong & Coulombe, 2006), and tumor progression (Chung et al., 2015; Depianto, Kerns, Dlugosz, & Coulombe, 2010; Hobbs et al., 2015) as well as other fundamental cell processes including growth and programed cell death (e.g., apoptosis) (Gilbert, Loranger, Daigle, & Marceau, 2001; Inada et al., 2001; Kim & Coulombe, 2007; Kim, Wong, & Coulombe, 2006; Ku & Omary, 2006). Mutations that possibly disrupt both structural and nonstructural roles of select keratins (K6, K16, or K17) underlie the cutaneous disorders pachyonychia congenita and steatocystoma multiplex (K17) (McLean & Moore, 2011, McLean et al., 1995).

A keratin mutation database summarizing keratin variants and associated disorders is maintained at the Center for Molecular Medicine and the Bioinformatics Institute in Singapore and can be assessed online at http://www.interfil.org (Szeverenyi et al., 2008).

2. COLLECTION OF MOUSE SKIN TISSUE FOR ANALYSIS

2.1 Isolation of Mouse Skin Samples for Morphological Studies

Skin tissue is easily accessible for the study of keratins in skin development and homeostasis. The epidermis of different body regions is not identical in architecture and morphological details; therefore, skin samples must be harvested from the same location (Montanez et al., 2007; Muller-Rover et al., 2001; Paus et al., 1999). Dorsal back skin is traditionally chosen for routine histology in mouse—another popular site for study is the ear. Animal euthanasia must be performed using an institutionally approved protocol.

2.1.1 Materials

Carbon dioxide euthanasia chamber (Nalgene); razor blades (VWR, cat. no. 55411-050); dissecting scissors; two forceps; sterile bacterial Petri dishes; isoflurane; propylene glycol; AcuPunch (Acuderm, Inc.); trimmer (Wahl® MiniARCO™); liquid nitrogen; Bouin's solution (Sigma, cat. no. HT10132); small glass vials; ethanol; xylene; paraplast paraffin (Tyco Healthcare/Kendall, Hampshire, UK); microtome; Mayer's hematoxylin solution (Thermo Scientific, cat. no. TA-125-MH); eosin solution (Thermo Scientific, cat. no. 71504); Permount (Fisher Scientific, cat. no. SP15-100); sodium citrate; heat-resistant rack; normal goat serum (NGS); 4′,6-diamidino-2-phenylindole (DAPI); 1 × phosphate-buffered saline (PBS); ClearMount mounting media (Thermo Scientific, cat. no. 00-8110); optimal cutting temperature (OCT) compound (Sakura, Finetek); plastic embedding molds (Ted Pella, Inc., cat. no. 27112); cryostat (Thermo Scientific, Microm HM550); formaldehyde; paraformaldehyde (PFA); sucrose; EM-grade glutaraldehyde; sodium cacodylate; osmium tetroxide.

2.1.2 Harvesting Adult and Newborn Mice Skin Samples

For adult mice, if only a small skin sample (up to 5 mm punch biopsy) is required, anesthetize the animal with 20% (v/v) isoflurane in propylene glycol, and verify unconsciousness by gently pinching the footpad. If the animal responds, extend the anesthesia time or augment the anesthetic dosage. Use an AcuPunch to acquire a circular full-thickness skin biopsy (epidermis, dermis, and subcutaneous fat). If experiments call for a larger area of skin, or skin tissue from areas such as paws or whisker pads, sacrifice the animals with CO_2 in a suitable chamber, and confirm euthanasia with cervical dislocation. Prior to tissue harvesting, remove the fur with a trimmer at the sites of interest. Lift the skin and make a full-thickness incision with scissors. Cut along both sides of the midline to obtain a rectangular sample of the back skin, while noting tissue orientation relative to the main body axis.

To obtain skin tissue from neonatal mice, sacrifice them by decapitation with surgical scissors. Remove the tail, genital area, and limbs with scissors. Cut along the dorsal surface from tail to neck and then peel off the skin from the body as one piece.

If skin samples are to be used for protein and ribonucleic acid (RNA) extraction, immediately snap-freeze them with liquid nitrogen if protein or RNA extraction cannot be performed at the time of harvest. If the tissue processing requires fixation, temporarily spread the skin tissue flat in a sterile Petri dish. This enables the tissue to adopt and maintain a flat shape during fixation and subsequent embedding. Trim the tissue to the proper size with a razor blade.

2.1.3 Preparation of Skin Tissue for Routine Morphological Study

For histological analyses, tissues are sliced into thin (5–10 μm) sections with either a microtome or a cryostat and stained to distinguish between different tissue components either with hematoxylin and eosin (H&E) staining or with immunohistochemistry. Hematoxylin, a basic dye, stains acidic structures purple or blue (e.g., nuclear deoxyribonucleic acid (DNA), ribosomal RNA, and rough endoplasmic reticulum). Eosin, an acidic dye, stains basic structures pink (e.g., cytoplasmic components) (Fischer, Jacobson, Rose, & Zeller, 2008). Both paraffin-embedded tissues and fresh-frozen tissues can be used for H&E staining (Fig. 3A) or immunohistochemistry (Fig. 3B). Keratin antibodies used for immunofluorescence staining in our laboratory are summarized in Table 1.

Preparation of paraffin-embedded tissues:

Fixation:

(1) To prevent skin curling during fixation, spread it flat in a weigh boat with epidermis facing up and cover with Bouin's solution (Sigma) for 10–15 min.

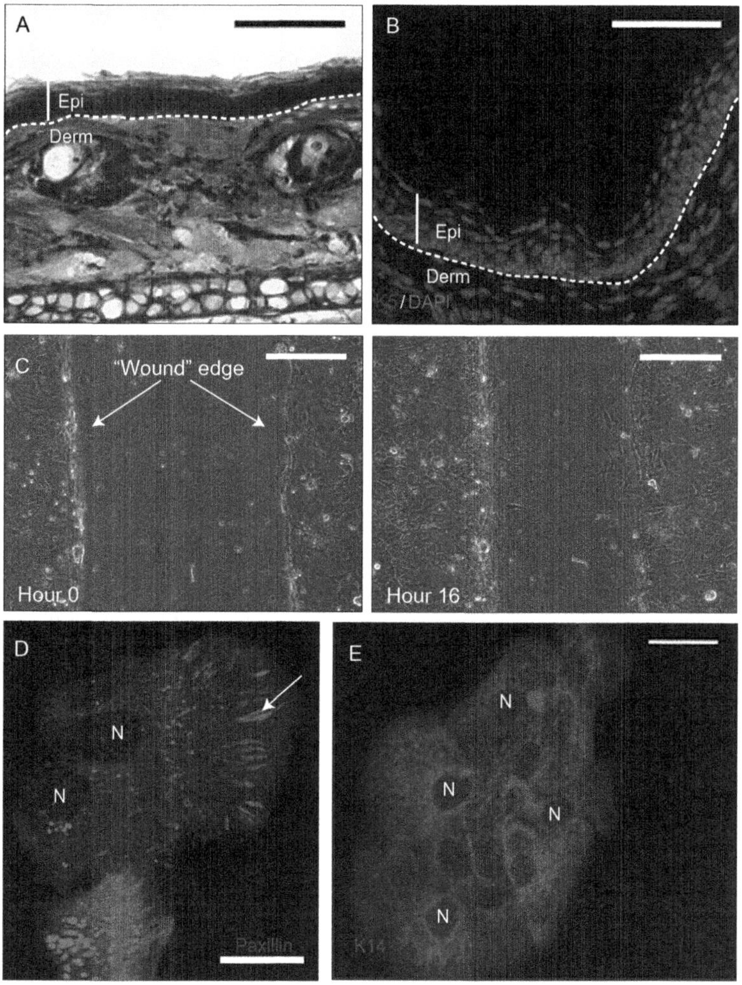

Figure 3 Various analyses of skin keratins utilizing mouse tissue and cultured primary keratinocytes. (A) Hematoxylin and eosin stain of fresh-frozen adult mouse ear tissue (4 months old). Dotted line marks the boundary between the epidermis (Epi) and dermis (Derm). Bar, 50 μm. (B) Fresh-frozen front paw tissue of a 2-month-old mouse processed for immunofluorescence of basal keratin K5. Note the restriction of K5 to the basal (progenitor) layer of keratinocytes. Dotted line marks the boundary between the epidermis (Epi) and dermis (Derm). Bar, 50 μm. (C) Live cell images of "wounding assay" using WT mouse keratinocytes in primary culture. Freshly isolated keratinocytes were plated in chamber slides with culture inserts. The "wound" was introduced by removing culture inserts when cells were 100% confluent. Phase contrast imaging was performed with a Zeiss Axio Observer Z1 microscope equipped with Zeiss EC Plan-Neofluar 10 ×/0.3 Ph1 objective for 16 h. Bar, 200 μm. (D) Transient expression of mCherry fluorescence protein (mCherry)-tagged paxillin in mouse primary keratinocytes. Freshly isolated

(Continued)

(2) Transfer the skin into a small glass vial. Fix skin samples (0.5–1 cm^3) in Bouin's solution overnight at 4 °C.
(3) After overnight fixation, remove excess fixative with multiple 70% ethanol washes.
(4) Trim tissue to the proper size, paying attention to the orientation of the specimen relative to the main body axis (rostral–caudal axis vs. dorsal–ventral axis).
(5) Store samples temporarily at 4 °C in 70% ethanol before further processing.

Dehydration and embedding:

(1) Place tissue samples in 50% ethanol for 30 min.
(2) Transfer tissue to 75% ethanol for 30 min.
(3) Transfer tissue to 95% ethanol for 30 min.
(4) Wash twice with 100% ethanol for 15 min.
(5) Wash twice with xylene for 10 min.
(6) Place tissue samples in molten paraffin at 60 °C while stirring in a beaker for 20–30 min. Repeat three times.
(7) Add a small amount of fresh molten paraffin to the embedding mold. Cool the mold on ice to solidify.
(8) Place the sample in the mold in the correct orientation. Fill the rest of the mold with molten paraffin to cover the sample and allow it to harden.
(9) Section 5–10 μm thick samples with a microtome.

H&E staining of paraffin sections:

(1) Remove paraffin by incubating the slides in 65 °C for 30 min and wash three times for 5 min with xylene.
(2) Rehydrate slides with two washes of 100% ethanol, one wash each of 95%, 70%, and 50% ethanol and a final wash with ddH_2O. All washes are 2.5 min each.

Figure 3—Cont'd keratinocytes were transfected with a plasmid encoding mCherry-tagged paxillin using the nucleofection method before plating in chamber slides with culture inserts. After removing the culture inserts, keratinocytes were allowed to migrate for at least 8 h before imaging using a Zeiss Axio Observer Z1 fluorescence microscope equipped with Zeiss EC Plan-Neofluar 40× objective. Keratinocytes shown in this panel were located at the leading edge in "wounding assay." Arrow points to a paxillin-positive focal adhesion. Bar, 20 μm. N, nucleus. (E) Immunofluorescence staining of keratin K14 in mouse skin keratinocytes in primary culture. Keratinocytes were isolated from newborn mouse pups and cultured in mKer media for 2 days. They were then fixed with 4% PFA and permeabilized with 0.5% Triton/PBS. Bar, 50 μm. N, nucleus. (See the color plate.)

Table 1 Summary of antibodies found to be useful for the detection of select keratins by western blotting and indirect immunofluorescence on fixed cells or tissue sections

Keratins	Name	Antibody	Source	Reactivity	Antigen	Dilution (Western blot)	Dilution (Immunofluorescence Staining)	References
K1	AF109	Rabbit	Covance	Mouse, rat	VKFVSTSYSRGTK-COOH	1:1000	1:100	Roop et al. (1984)
K5	AF138	Rabbit	Covance	Human, mouse	C-terminus peptide	1:10,000	1:1000	
K6a/b	4th bleed	Rabbit	Coulombe Lab	Human, mouse	CSSTIKYTT-COOH	1:2000	1:250	McGowan and Coulombe (1998a)
K10		Rabbit	Covance	Human, mouse	GGGDQSSKGPRY-COOH	1:1000	1:100	Roop et al. (1984)
K14	AF64	Rabbit	Covance	Human, mouse	KVVSTHEQVLRTKN-COOH	1:10,000	1:1000	Roop et al. (1984)
K16	5th bleed	Rabbit	Coulombe Lab	Human, mouse	CSTSFSQSQSQSSRD-COOH	1:2000	1:250	Bernot, Coulombe, and McGowan (2002)
K17	3rd bleed 2250	Rabbit	Coulombe Lab	Human, mouse	CSSREQVHQTTR-COOH	1:5000	1:500	McGowan and Coulombe (1998a)

(3) Stain in Mayer's hematoxylin solution 5–10 min.
(4) Wash with running tap water 20 min.
(5) Counterstain with eosin solution 1 min.
(6) Perform two 2 min 95% ethanol washes.
(7) Perform two 2 min 100% ethanol washes.
(8) Perform two 2 min xylene washes.
(9) Mount slides with Permount (Fisher Scientific) and glass coverslips.

Immunohistochemistry of paraffin sections:

(1) Deparaffinize and rehydrate slides following steps 1–2 in the H&E staining protocol.
(2) Perform antigen retrieval as needed. Boil ~500 mL of 10 m*M* sodium citrate buffer (pH 6.0) (see Appendix). Remove the boiling solution from heat source and immediately add the slides to this solution in a heat-resistant rack. Incubate 5–10 min.
(3) Let the slides cool 30 min.
(4) Perform three 5 min 1 × PBS washes.
(5) Block unspecific binding with 5% NGS/1 × PBS for 30 min at room temperature.
(6) Incubate with primary antibody (diluted in 2.5% NGS/1 × PBS) for 1 h at room temperature or overnight at 4 °C.
(7) Perform three 5 min 1 × PBS washes.
(8) Incubate with secondary antibody (diluted in 2.5% NGS/1 × PBS) for 1 h at room temperature.
(9) Wash the slide(s) once with 1 × PBS containing DAPI; (1:10,000).
(10) Perform three 5 min 1 × PBS washes.
(11) Mount slides with ClearMount mounting media (Life Technologies) and glass coverslips.

Preparation of Fresh Frozen Tissue Samples:

(1) Fill plastic mold with enough OCT compound (Sakura, Finetek) to cover the tissue.
(2) Place harvested tissue into OCT in the desired orientation. Avoid generating bubbles in the OCT. Bubbles will interfere with subsequent sectioning.
(3) Transfer the mold onto dry ice to solidify the OCT. Tissues frozen in OCT can be stored at −80 °C for several months.
(4) Cut tissues to 5–10 μm sections using a cryostat machine. Equilibrate tissue blocks in cryostat at least 30 min before sectioning. Tissue sections can be stored at −20 °C.

To process for various staining protocols, thaw tissue sections and air dry 10 min. For H&E staining (Fig. 3A), fix the tissue with 10% formalin for 5 min on ice, wash three times with ddH_2O and then follow steps 3–9 for H&E staining of paraffin-embedded sections. For immunostaining and/or fluorescence staining (Fig. 3B), fix the tissue with 4% PFA for 10 min at room temperature, and then follow the steps 4–11 for immunohistochemistry staining of paraffin-embedded sections.

2.1.4 Preparation of Skin Tissue for *In Situ* Hybridization

(1) To prevent RNA degradation, fix tissue samples immediately in freshly made, RNAse-free 4% PFA in RNAse-free PBS overnight at 4 °C.
(2) Drain PFA and wash tissue with cold RNAse-free PBS (4 °C) three times for 10–15 min.
(3) Incubate tissue in RNAse-free 30% sucrose in RNAse-free PBS (w/v) overnight at 4 °C.
(4) Embed samples in OCT (Sakura, Finetek) and freeze them at least overnight at −80 °C.
(5) Make sections of 10–15 μm thickness.
(6) Let slides air dry and store at −80 °C before hybridization process.

The procedures for the preparation of probes and their application for detecting specific keratin mRNAs in skin tissue sections have been described elsewhere (Tong & Coulombe, 2004; Wang, Wong, Langbein, Schweizer, & Coulombe, 2003).

2.1.5 Preparation of Skin Tissue for Transmission Electron Microscopy

Compared with light microscopy, transmission electron microscopy (TEM) provides a significantly higher resolution view of biological elements. TEM can be used as a diagnostic and research tool for the characterization of the fine structure and organization of skin tissue and detect morphological defects at the subcellular level. A buffered solution containing PFA and glutaraldehyde is used as a primary fixative (Hayat, 2000).

(1) Freshly prepare fixative solutions and prechill on ice.
(2) Harvest tissue and fix overnight with 2% TEM-grade glutaraldehyde, 1% PFA prepared in 0.1 *M* sodium cacodylate (pH 7.2) at 4 °C.
(3) On the following day, wash three times with 0.1 *M* sodium cacodylate (pH 7.2) at room temperature.
(4) In a well-ventilated hood, postfix the tissue samples with 1% osmium tetroxide for 1 h. Osmium tetroxide is hazardous to humans so wear gloves, facemask, and dispose of waste according to institutional

guidelines for hazardous disposal. Note: tissue samples will become a dark brown/black color upon osmication, leading to a loss of histological cues.

(5) Wash samples three times with 0.1 *M* sodium cacohylate (pH 7.2).

(6) Rinse samples three times with ddH_2O.

(7) Transfer tissues to 50% ethanol and store at 4 °C before embedding (usually in an epoxy resin medium) using a standard TEM embedding protocol (Alvarado & Coulombe, 2014; Lessard et al., 2013).

2.2 Harvest Mouse Skin Samples for RNA and Proteins

2.2.1 Materials

Razor blades (VWR, cat. no. 55411-050); TRIzol® Reagent (Life Technologies); 15 mL falcon tubes; Polytron hand-held homogenizer (Kinematica AG, Switzerland); chloroform; RNAse-free water; RNeasy kit (Qiagen); Bio-Rad protein assay dye reagent concentrate (Bio-Rad; cat. no. 500-0006); Albumin standard (Thermo Scientific, cat. no. 23209).

2.2.2 Preparation and Analysis of RNA and Protein Samples from Tissues

RNA and proteins from skin tissues are extracted using TRIzol® (TRIzol® Reagent, Life Technologies), which can be used to sequentially precipitate RNA, DNA, and proteins from the same sample.

RNA extraction:

(1) Cut the tissue into small pieces with a razor blade.

(2) Transfer the tissue pieces into a 15-mL falcon tube containing 1 mL of TRIzol® Reagent (Life Technologies).

(3) Homogenize these tissue pieces for 1 min at room temperature.

(4) Incubate the homogenate for 15 min at room temperature.

(5) Add 0.2 mL of chloroform.

(6) Shake the tube vigorously, by hand, for 15 s.

(7) Incubate the homogenate for 2–3 min at room temperature.

(8) Centrifuge homogenate at 12,000 × *g* for 15 min at 4 °C.

(9) Collect the upper aqueous phase with RNA. Save the organic phase for DNA and protein extraction.

RNA precipitation:

(10) Isolate RNA according to RNA isolation protocol (see TRIzol® protocol RNA Isolation procedure).

(11) Resuspend RNA pellet in RNAse-free ddH_2O.

(12) RNA should be purified by using an RNA clean-up kit (RNeasy kit; Lessard et al., 2013).
(13) The final RNA samples can be used for real-time quantitative reverse transcription polymerase chain reaction, Northern Blot analysis, etc.

DNA precipitation: (must be completed before protein isolation)

(1) Refer to DNA precipitation procedure in TRIzol® protocol.

Protein Isolation:

(1) Follow protein isolation procedure in TRIzol® protocol until final resuspension step.
(2) After 100% ethanol wash, resuspend protein pellet in 6.5 *M* Urea buffer (see Appendix) by adding 500 μL of Urea buffer.
(3) To optimize protein resuspension, rotate samples at 4 °C for 20 min to overnight.
(4) Centrifuge samples at 10,000 × *g* for 10 min at 4 °C to sediment any insoluble material.
(5) Transfer supernatant containing protein to a new tube and measure protein concentration using Bradford assay. Isolated proteins can be subjected to western blot analysis. Keratin antibodies used for western blotting in our laboratory are summarized in Table 1.

3. CELL CULTURE STUDIES

Compared with *in vivo* studies, *ex vivo* cell culture provides a less complex setting that also enables the study specific cell populations, such as keratinocytes or fibroblasts, with potential for mechanistic insight. Thanks to continuous refinements in cell culture assays over the past 35 years, our understanding of the cellular and molecular basis underlying the proliferation, differentiation, migration, and death of epidermal keratinocytes has been greatly advanced (Lichti, Anders, & Yuspa, 2008). The procedures described here can be routinely applied to isolate and culture keratinocytes harvested from newborn mouse skin.

3.1 Materials

Newborn (P0–P3) mice; two pair of forceps; dissecting scissors; povidone-iodine solution (Ricca Chemical Company, cat. no. 3955-16); ethanol; sterile bacteria Petri dish; 0.25% trypsin (Corning, cat. no. 25-053-Cl); Dulbecco's modified Eagle's medium (DMEM; Gibco, cat. no. 11885-084); Lymphoprep™ (cat. no. 1114545); Cell scrapers; hemocytometer; Collagen type I (Corning, cat. no. 354236); 15 mL conical tubes; CnT-57 (CELLnTEC);

KBM-2 (Lonza, cat. no. CC-3103); Chelex-100 resin (Bio-Rad, cat. no. 142-2842); 1 × PBS; Whatman filter paper; calcium-free low glucose DMEM (US Biologicals, cat. no. D9800-10); sodium bicarbonate; penicillin-streptomycin (GIBCO/Invitrogen, cat. no. 15140-122); calcium chloride; chamber slides (Lab-Tek, cat. no. 155380); culture insert (Ibidi, cat. no. 80209); Amaxa™ 4D-Nucleofector™ (Lonza); P1 Primary Cell 4D-Nucleofector™ X Kit (Lonza); AcuPunch (Acuderm, Inc.); RNeasy kit (Qiagen); Nanophotometer (Implen); Bioruptor (Diagenode, UCD200); Bio-Rad protein assay dye reagent concentrate (Bio-Rad, cat. no. 500-0006); Albumin standard (Thermo Scientific, cat. no. 23209); Empigen BB (Calbiochem, cat. no. 324690); TrueBlot® Anti-Rabbit Ig IP Beads (Rockland, cat. no. 00-8800-25); TrueBlot® Anti-Rabbit-HRP (Rockland, 18-8816-31); β-mercaptoethanol.

3.2 Isolation of Keratinocytes from Newborn Mouse Skin

Generally, newborn pups between ages P0 (birth) and P3 are used as a source of skin keratinocytes for primary culture. If mice older than 3 days after birth are used, the keratinocyte yield will be low due to difficulties separating the epidermis from dermis due to hair follicle development (note that the epidermis gets thinner as the hair coat emerges at the surface of the skin).

(1) On the first day, sacrifice pups via decapitation with scissors. Bathe them in iodine solution 2 min, wash twice with ddH_2O, and then rinse twice with 70% ethanol. All of the following steps should be performed inside a tissue culture hood using sterilized surgical tools.

(2) Place pups into 10 cm bacterial Petri dishes. Hold mouse pup with a pair of forceps, cut off the front limbs at the ankle joint, and the hind limbs just below the joint, leaving visible stumps. This ensures that the skin harvested remains whole and intact, optimizing cell yield. Remove the tail close to the skin, leaving a hole. Keep a segment of the tail for genotyping.

(3) Carefully insert scissors into the hole and cut skin along the dorsal surface of the mouse pup towards where the head was.

(4) Use two pairs of forceps to carefully peel the skin off the mouse pup.

(5) Place the skin onto a 6-cm sterile Petri dish, dermis side down, and carefully spread the skin flat so that the edges are not curled. If a large litter is being harvested and the skins need to rest for more than 45 min, place dishes on ice until the next step.

(6) Float skins (dermis facing down) on 4 mL of 0.25% trypsin in a 6-cm dish overnight (approximately 18 ~ 24 h) at 4 °C.

(7) Extract DNA from the tail snips and genotype by polymerase chain reaction (if relevant).

(8) On day 2, recover the floating skin samples from the trypsin solution with forceps, and place the skin, dermis side up, on the inner surface of the Petri dish lid.

(9) Spread the skin flat, and remove the dermis by gently lifting it up straight from the epidermis. The dermis can be retained for fibroblast isolation.

(10) Gently scrape keratinocytes away from the cornified layer of the epidermis with a (new) sterile cell scraper.

(11) Add 3 mL of DMEM supplemented with 10% serum or mKER media (see Appendix) to collect keratinocytes, and transfer the resuspended cells into a 15-mL conical tube. Combine cells from (up to 3) mice of the same genotype.

(12) Pellet keratinocytes at 1200 rpm for 5 min at 4 °C, remove the supernatant and resuspend cells in 4 mL of fresh media. Thorough resuspension of keratinocytes is critical for a good yield.

(13) Gently pipette cell suspension on top of 10 mL of Lymphoprep without disturbing the Lymphoprep/media interface.

(14) Spin at 1800 rpm for 30 min at 4 °C. A white hazy layer enriched with keratinocytes can be seen after centrifugation.

(15) Starting at the interface, collect all the media, along with the top 4 mL of Lymphoprep. Avoid disturbing the pellet of unwanted cells at the bottom.

(16) Transfer the solution containing isolated keratinocytes into a fresh 15-mL conical tube, and spin at 1200 rpm, 5 min at 4 °C.

(17) Aspirate the supernatant and resuspend cells in the desired culture medium (0.5 mL of media per pup).

(18) Count cells using a hemocytometer and plate cells in sterile culture dishes. Precoating culture dishes with extracellular matrix proteins (e.g., Collagen type I) can facilitate cell attachment to the plate. Plating cells at 1.0×10^6 per 60 mm dish will yield around 80% confluence after 72 h.

(19) Change media the following day to remove dead cells and their secreted toxins.

3.3 Conditions for the Primary Culture of Skin Keratinocytes

Consistent with the existence of a Ca^{2+} gradient across the epidermis in skin tissue *in situ*, primary keratinocytes can be stimulated to differentiate by increasing Ca^{2+} concentration to >0.07 m*M* *in vitro* (Lichti et al., 2008). Accordingly, much consideration needs to be taken when choosing culture media for experiments conducted on keratinocytes in primary culture.

3.3.1 Growth Media for Keratinocytes in Primary Culture

For long-term maintenance, keratinocytes need to be grown on collagen-coated plates in fibroblast-conditioned, low Ca^{2+} media. Reducing Ca^{2+} concentration facilitates the maintenance of keratinocytes in a basal layer like, relatively undifferentiated state. Media commonly used in our laboratory to culture keratinocytes for a limited period without passaging are mKer, CnT-57(CELLnTEC), FAD (Reichelt & Haase, 2010; see Appendix), and KBM-2 (Lonza). mKer medium promotes proliferation of keratinocytes while inhibiting fibroblast growth, but because it has a high Ca^{2+} concentration, the majority of keratinocytes will differentiate within a few days postplating. Consequently, alternative media with low to negligible levels of Ca^{2+} are increasingly applied in routine studies in the laboratory. The CnT-57 medium (0.07 m*M* Ca^{2+}) is commercially available and formulated to retain the basal state of progenitor cells and delay their differentiation. FAD medium has similar supplements as mKer medium, but uses calcium-free DMEM and calcium-depleted fetal bovine serum (FBS). KBM-2 medium (0.15 m*M* Ca^{2+}), without additional growth factors, can also be used to culture keratinocytes for the short term, but keratinocytes proliferate very slowly in this medium. Low calcium media such as CnT-57 and FAD are recommended if the intent is to promote the proliferation and expansion of keratinocytes while minimizing differentiation.

3.3.2 Calcium Switch Protocol to Induce Keratinocyte Differentiation

Keratinocytes usually proliferate readily under low calcium conditions (0.05 m*M*) but will commit to terminal differentiation when the calcium concentration is >0.07 m*M*. The differentiation status of keratinocytes can be assessed by morphological changes (flattening, increased surface area), formation of tight cell–cell adhesions, and expression of differentiation markers (e.g., K1, K10, involucrin, filaggrin, and loricrin) (Candi, Schmidt, & Melino, 2005; Yuspa, Kilkenny, Steinert, & Roop, 1989). This section provides procedures for preparing media with varying calcium concentrations, as well as performing the calcium switch (Lichti et al., 2008).

Preparation of calcium-depleted serum:

(1) Suspend 25 g Chelex-100 resin in 500 mL of ddH_2O.
(2) Stir 5 min and gradually adjust pH to ~7.4.
(3) Let the resin settle for 30 min without stirring.
(4) Pour out ddH_2O.
(5) Wash resin once with 1 L of fresh ddH_2O.
(6) Wash resin twice with 1 L of 1 × PBS.
(7) In the last wash of PBS, adjust pH to ~7.4. Pour off PBS.
(8) Add the resin to 500 mL of FBS, stir mixture overnight at 4 °C.
(9) Filter the chelated FBS with Whatman filter paper.
(10) Discard resin. Sterilize FBS in tissue culture hood with a 0.22-μ m filter.

Preparation of calcium switch media:

(1) Dissolve 4.885 g calcium-free low glucose DMEM powder (US Biologicals) in 500 mL ddH_2O.
(2) Add 1.85 g sodium bicarbonate.
(3) Sterilize media by filtration in tissue culture hood.
(4) Supplement with chelated serum (8% final) and penicillin-streptomycin (0.5% final).
(5) Use a sterile concentrated calcium chloride stock solution to obtain either low (0.05 m*M*), moderate (0.12–0.20 m*M*), or high (1.2–2.0 m*M*) calcium media.
(6) Calcium concentration can be verified using atomic absorption spectroscopy (Hennings, Holbrook, & Yuspa, 1983).

Calcium switch:

(1) Plate keratinocytes in moderate calcium (0.12–0.20 m*M*) overnight.
(2) On the next day, wash cells three times with calcium-free PBS, and then add low calcium medium (0.05 m*M*).
(3) Culture cells for 48 h or until confluent, and then induce them to differentiate using either moderate (0.12–0.20 m*M*) or high (1.2–2.0 m*M*) calcium media.
(4) Continue culturing cells for another 24–48 h before harvesting cells for analysis.

3.3.3 Scratch Wounding Assay for Primary Keratinocytes

Scratch wounding, performed in confluent or near-confluent cultures of cells, represents a simple means to evaluate the migration properties of keratinocytes upon the appearance of the equivalent of a wound. Such wounds are created either by scratching a confluent monolayer of cells with a pipette tip or by plating cells into a chamber slide with a culture insert (Ibidi) and removing the insert when cells reach complete confluence, thus generating a gap. Cells at the newly generated "wound edge" will polarize and initiate migration to close the wound (Fig. 3C). Time-lapse imaging can capture cell migration at specific time points and ImageJ software can be used to quantitate migration velocity, directionality, and distance.

3.3.4 Gene Transfer Protocols for Keratinocytes in Primary Culture

Transfection of primary keratinocytes is generally challenging because conventional methods such as liposome-mediated gene delivery result in low transfection efficiency while the calcium phosphate-based strategy induces keratinocytes to differentiate. Adenovirus can achieve nearly 100% transfection efficiency for transient expression in keratinocytes in primary culture while lentivirus- and retrovirus-based strategies have lower transfection efficiencies. A detailed viral transfection protocol has been described by Li (2013). This segment details a procedure involving the Lonza Primary Cell P1 kit, which gives rise to ~50–60% transfection efficiency in mouse primary keratinocytes (Fig. 3D; Distler et al., 2005, Feng & Coulombe, 2015).

(1) Resuspend primary keratinocytes in culture media after isolation. Count cells.
(2) Turn on the Amaxa™ 4D-Nucleofector™, and select the Keratinocyte Mouse Primary program (CM-102).
(3) Mix 16.4 μL of nucleofection solution and 3.6 μL of supplement solution from P1 Primary Cell 4D-Nucleofector™ X Kit with 0.5 μg of plasmid DNA.
(4) Pellet $\sim 5 \times 10^5$ keratinocytes by centrifugation at $0.5 \times 1000g$ for 5 min at 4 °C.
(5) Gently resuspend cell pellets in nucleofection mixture (see step 3).
(6) Transfer the entire mixture into one well in the 16-well strip (included in nucleofection kit).
(7) Place the strip into the nucleofector. Begin program.
(8) After electrical pulse, wait 5–10 min. Add 80 μL of culture media to the well. Gently transfer the cell suspension into a culture plate.
(9) Wait 24–48 h after transfection before using cells for experimentation.

3.4 Culture of Skin Explants *Ex Vivo*

The *in vitro* scratch-wounding assay can yield reproducible, sophisticated insight about cell migration but cannot mimic the cross talk that occurs between various cell types during epithelialization as it occurs *in vivo*. Our laboratory adapted a simple skin explant culture assay *ex vivo* that allows migration to proceed from a stratified epithelium, and otherwise, allows for interactions between keratinocytes and fibroblasts. A comprehensive description of this assay can be accessed in Mazzalupo, Wawersik, and Coulombe (2002).

(1) Obtain the skin of newborn pups as described (see isolation of skin keratinocytes for primary culture).
(2) Place the skin, dermis side down, onto a sterile Petri dish and spread flat.
(3) Use a biopsy punch device to generate circular, full-thickness skin pieces and transfer biopsies to a 24-well tissue culture dish. Lay the biopsies flat, dermal side down.
(4) Allow ~10–15 min for explants to adhere to the plate. Add 250 μL of mKer medium to moisten the tissue edges, but not cover the center of the explants.
(5) On the next day, submerge explants by adding an additional 250 μL of mKer media to each well. Keratinocytes will grow from the explant edges, outward as a stratified sheet, over the next several days.
(6) Change media every 2 days for duration of the experiment, usually 6–8 days (see Mazzalupo et al., 2002).
(7) Cells making up the outgrowth can either be used for immunofluorescence staining or be harvested for RNA and protein analysis, after the original biopsy piece has been removed. These protocols are described in the following two sections.

3.5 Immunofluorescence Staining of Cells in Culture

The fixation and permeabilization method used for keratinocytes in primary culture is dependent upon the antigen of interest and the antibody being used. Our laboratory commonly uses two methods: methanol-fixation, and 4% PFA fixation coupled with 0.5% Triton/PBS permeabilization (Fig. 3E). Methanol, an organic solvent, precipitates proteins to preserve cellular structure. It works best for detecting intermediate filaments and their associated proteins (Ma & Lorincz, 1988). Immunostainings of cytosolic proteins are usually fixed with

PFA. It is a cross-linking reagent that yields a better preservation of cell morphology, but may hinder the access of antibody to its epitope(s) on target protein(s).

Methanol fixation:

(1) Fix cells in methanol for 5 min at −20°C

(2) Wash cells three times with PBS, 5 min each.

(3) Perform steps 5–11 from Immunofluorescence of Adult Tissues protocol.

PFA fixation:

(1) Fix cells in 4% PFA for 10 min at room temperature.

(2) Wash cells three times with PBS, 5 min each.

(3) Permeabilize cells with 0.5% Triton/PBS for 5 min.

(4) Wash cells three times with PBS, 5 min each.

(5) Perform steps 5–11 from Immunofluorescence of Adult Tissues protocol

3.6 Isolation and Analysis of RNA and Proteins from Keratinocytes in Culture

For both RNA and protein assays, it is important to keep the density of cells between experiments relatively consistent as confluence is a critical factor that regulates cell differentiation and alters gene expression. RNA extraction can be performed using an RNA Isolation kit following the manufacturer's protocol (such as RNeasy kit; Lessard et al., 2013). After RNA isolation and clean-up, samples should be kept on ice to prevent RNA degradation. RNA concentration should be determined using Nanophotometer (Implen). If the RNA is not to be used immediately, it should be kept frozen at −20 °C (short-term) or −70 °C (long-term storage).

Cellular proteins can be extracted as a whole cell lysate, or as subcellular fractions (e.g., soluble vs. insoluble pools (Fig. 4A); cytosolic vs. membrane proteins (Fig. 4B)). Protease and phosphatase inhibitors are added to the lysis buffer to preserve the size and posttranslational modification of proteins.

Preparing whole cell lysate:

(1) Culture cells to proper confluence.

(2) Aspirate media and wash once with PBS.

(3) Add whole-cell extraction buffer (6.5 *M* Urea) to lyse cells (200 μL per well of a 6-well dish).

(4) Collect cell lysates by scraping cells with cell scraper and transfer to a new tube.

(5) Incubate 20 min on a shaker at 4 °C.

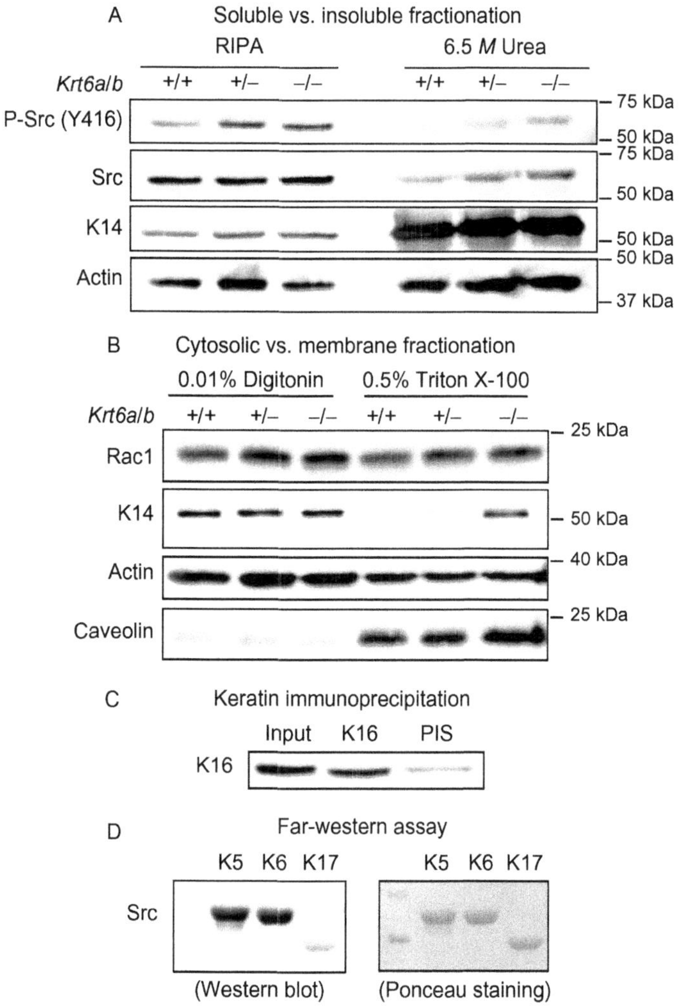

Figure 4 Subcellular fractionation methods and assays for studying keratin-interacting proteins. (A) Western blot analysis of Src activity (Y416 phosphorylated epitope) in WT, *Krt6a/b*$^{+/-}$, and *Krt6a/b*$^{-/-}$ skin explant keratinocyte protein lysates. Cellular outgrowths from skin explants cultured for 6 days were pooled and solubilized with RIPA buffer, and then with 6.5 *M* Urea buffer. Use of an antibody to K14 for western blotting reveals the relative amount of keratin protein occurring at the RIPA and urea extraction steps. β-Actin was used as a loading control for both RIPA-soluble fraction and RIPA-insoluble fraction (urea-soluble fraction). (B) Western blot analysis of subcellular localization of Rac1 in WT, *Krt6a/b*$^{+/-}$, and *Krt6a/b*$^{-/-}$ skin explant keratinocyte protein lysates. Cellular outgrowths from skin explants cultured for 6 days were pooled and solubilized with 0.01% Digitonin buffer and then with 0.5% Triton X-100 buffer. Loading was assessed using β-actin (digitonin-soluble fraction/cytosolic fraction) and caveolin (triton-soluble fraction/membrane fraction). Use of an antibody to K14 for western blotting reveals the

(Continued)

(6) Sonicate lysate 5 min (alternating 30 s ON/30 s OFF) on "high" setting with Bioruptor (Diagenode).
(7) Pellet sonicated samples at maximum speed (16.1 × 1000*g*) for 20 min at 4 °C.
(8) Collect supernatant and determine concentration of proteins using Bradford assay.

Soluble versus insoluble protein fractionation (Fig. 4A):
(1) Lyse cells in RIPA buffer (see Appendix; Wong & Coulombe, 2003).
(2) Incubate lysates 30 min on a shaker at 4 °C.
(3) Centrifuge at maximum speed (16.1 × 1000*g*) for 10 min at 4 °C.
(4) Collect supernatant as the "RIPA-soluble fraction."
(5) Wash the remaining pellet with RIPA buffer three times.
(6) Add 6.5 *M* Urea buffer and incubate 20 min on a shaker at 4 °C.
(7) Sonicate the lysate 5 min (alternating 30 s ON/30 s OFF) on "high" setting Bioruptor (Diagenode).
(8) Pellet at maximum speed (16.1 × 1000*g*) for 20 min at 4 °C. Collect supernatant as "insoluble fraction," which contains many cytoskeletal proteins.
(9) Measure protein concentration with Bradford assay.

Cytosolic versus membrane protein fractionation (Fig. 4B):
(1) Lyse cells in 0.01% Digitonin buffer (see Appendix; Rotty & Coulombe, 2012).
(2) Incubate samples 10 min on a shaker at 4 °C.
(3) Centrifuge at maximum speed (16.1 × 1000*g*) for 1 min.
(4) Collect supernatant as "cytosolic fraction."
(5) Wash remaining pellet three times with Digitonin buffer.

Figure 4—Cont'd relative amount of keratin protein occurring at the RIPA and urea extraction steps. (C) Immunoprecipitation (IP) of keratin 16 (K16) from mouse keratinocytes in primary culture. Keratinocytes were isolated from newborn mouse pups and cultured to confluence before IP. "Input" represents the whole cell lysate, which serves as a positive control, "K16" refers to the K16 antibody (see Table 1) used for IP, and "PIS" refers to preimmune serum, used as a negative control. (D) Far-western assay to study the (direct) interaction between keratin proteins and, in this case, Src protein. Recombinant K5, K6, and K17 proteins were purified using HiTrap Q column. 5 μg of each of these keratin proteins were run on 10% SDS-PAGE and transferred to a nitrocellulose membrane. Ponceau staining was done to assess loading of these proteins. The membrane was then incubated with recombinant Src protein (150 ng/mL; Abcam) for 4 h at room temperature. The association of Src protein with keratin proteins on the nitrocellulose membrane was next detected via conventional western blotting for Src protein. *See Rotty and Coulombe (2012), for details.*

(6) Resuspend pellet in 0.5% Triton X-100 buffer (see Appendix).
(7) Incubate 20 min on a shaker at 4 °C.
(8) Centrifuge at maximum speed (16.1 × 1000*g*) for 1 min.
(9) Collect supernatant as "membrane fraction."
(10) The remaining pellet can be resuspended in 6.5 *M* Urea buffer to collect the "insoluble fraction."

3.7 Immunoprecipitation of Keratin Proteins from Cell Culture to Study Their Interacting Partners

As keratins typically are the least soluble proteins in keratinocytes, immunoprecipitation (IP) of keratin proteins has been a challenge. Lowthert, Ku, Liao, Coulombe, and Omary (1995) have successfully developed a method for immunoprecipitating keratin proteins using 2% Empigen BB or 1% NP-40 buffer. Both detergents can maintain some protein–protein interactions and solubilize a fraction of keratins. Our laboratory has adapted the Lowthert protocol over time and currently uses 1% Triton X-100 containing 2% Empigen BB for antigen extraction towards IP assays (Chung et al., 2015) (see Fig. 4C for an example of K16 coimmunoprecipitation (Co-IP) results). The main steps involved (specific to K17 Co-IPs) are as follows:

(1) Grow two 10-cm plates of keratinocytes in primary culture to about 90% confluence.
(2) Aspirate the media. Perform the following steps on ice.
(3) Wash cells with cold PBS.
(4) Use 500 μL of 1% Triton X-100 lysis buffer containing 2% Empigen BB (see Appendix) to lyse cells in each plate. Scrape cells into a new tube.
(5) Incubate cell lysates for 5 min on a shaker at 4 °C.
(6) Pellet at maximum speed (16.1 × 1000*g*) for 5 min at 4 °C.
(7) Combine the supernatants from both plates. Measure protein concentration using Bradford assay using bovine serum albumin (BSA) as a standard.
(8) Save 40 μL of lysate and add 40 μL 2 × sodium dodecyl sulfate (SDS) sample buffer (with 5% β-mercaptoethanol) as input. Boil on heat block (95 °C) for 5 min.
(9) Transfer the combined supernatants to a new tube and add 20 μL of TrueBlot® Anti-Rabbit Ig IP Beads to the lysates. Incubate for 45 min on a shaker at 4 °C to "preclear" beads and prevent nonspecific binding.

(10) Centrifuge until maximal speed (16.1 × 1000*g*) at 4 °C in a benchtop microfuge.

(11) Aliquot 500 μL of the supernatant into a 1.5-mL Eppendorf tube labeled "IP control" and another 500 μL into another tube labeled "K17 IP."

(12) Add 0.5 μL of preimmune serum into the "IP control" sample and add 0.5 μL of K17 antibody (McGowan & Coulombe, 1998a) into the "K17 IP." Generally use 1 μL of K17 antibody for 1 mg of lysate.

(13) Incubate the mixture for 3 h on a shaker at 4 °C. Add 25 μL of TrueBlot® Anti-Rabbit Ig IP Beads into each tube. Incubate 45 min on a shaker at 4 °C.

(14) Centrifuge until maximum speed (16.1 × 1000*g*) at 4 °C.

(15) Wash the beads in each tube with 500 μL of 1% Triton X-100 lysis buffer three times.

(16) Add 20 μL of 2 × sample buffer (with 5% β-mercaptoethanol) to the pelleted beads in each tube. Boil on a heat block (95 °C) for 5 min.

(17) Analyze samples by western blotting. Use TrueBlot® Anti-Rabbit-HRP secondary antibody when probing with rabbit primary antibodies.

4. *IN VITRO* METHODS TO STUDY KERATIN PROTEINS

In addition to skin tissue and cell culture settings, recombinant keratin proteins can be used to investigate the properties of keratins *in vitro*. Keratin proteins readily concentrate in inclusion bodies when expressed in recombinant forms in *Escherichia coli*, facilitating their subsequent purification (see below; Coulombe & Fuchs, 1990). 10-nm keratin filaments can be reconstituted *in vitro* from purified type I and II keratins through a simple series of dialysis steps. By introducing mutations into such recombinant keratin proteins, the amino acid residues and protein domains that are crucial for proper filament formation can be assessed (Coulombe, Chan, Albers, & Fuchs, 1990; Hatzfeld & Weber, 1990a, 1990b). Furthermore, one can study direct interactions between keratin proteins, or keratin filaments, with target proteins using far-western or the cosedimentation assays.

4.1 Materials

BL21 (DE3) competent cells; Luria Broth (LB) agar plates; isopropyl β-D-1-thiogalactopyranoside (IPTG; Sigma, cat. no. 16758); lysozyme (Sigma, cat. no. L6876); magnesium chloride; manganese chloride; deoxyribonuclease

(Roche, cat. no. 10104159001); 0.22 μm filter (Millex-GV, cat. no. SLGV033RS); 60 mL Syringe (BD, cat. no. 309653); fast protein liquid chromatography (FPLC; AKTA purifier); HiTrap Q column (GE Healthcare, cat. no. 17-1154-01); Coomassie blue dye; Mono Q column (GE Healthcare, cat. no. 17-5166-01); centrifugal filter units (Millipore, cat. no. UFC801096); cellulose dialysis tubing (Fisher Scientific); air-driven ultracentrifuge (Beckman Coulter); Parafilm; paper 1% uranyl acetate; electron microscopy (EM) grid (Electron Microscopy Sciences, cat. no. CF400CU50); TEM; sodium chloride; 1.5 mL Eppendorf tubes; nitrocellulose membrane (Bio-Rad, cat. no. 162-0115); Ponceau S solution (Sigma, cat. no. P7170); BSA powder; Tween 20 (Amresco, cat. no. M147); enhanced chemiluminescence (ECL) solution (GE Healthcare, cat. no. RPN2235); Fluorchem Q MultiImage® III.

4.2 Expression and Purification of Keratin Proteins

Expression of recombinant keratin proteins:

(1) Day 1: Transform the plasmid encoding the keratin protein of interest into a suitable strain of *E. coli* bacteria. Generally, we use BL21 (DE3) plysS competent cells for type II keratins (K1, K5, K6) and BL21 (DE3) competent cells for type I keratins (K10, K14, K16, K17). Plate bacteria freshly transformed with the relevant plasmid on LB agar plate containing desired antibiotic and incubate overnight at 37 °C.

(2) Day 2: Toward the end of the day, inoculate 10 mL of LB medium supplemented with antibiotic (for positive selection) with a colony from step 1, to initiate a liquid culture. Let the starter culture grow overnight in a 37 °C shaker.

(3) Day 3: Inoculate 1 L of LB media (with selective antibiotic) with a 1:100 dilution of the starter culture (step 2). Grow bacteria until OD_{600} reaches ~0.6.

(4) Induce expression of keratin proteins by adding 1 mL of 200 mg/mL IPTG and allow for protein expression for 4–5 h in a 37 °C shaker.

(5) Pellet cells by centrifuging at 4000 rpm for 20 min at 4 °C. The cell pellets can be stored at −20 °C until the next step.

Preparation of inclusion body fractions:

(1) Resuspend the cells in 60 mL of Bacteria Lysis Buffer (see Appendix).

(2) Add 15 mL of 10 mg/mL lysozyme, mix well, and incubate 30 min on ice.

(3) Freeze the cells on dry ice for 30 min and then thaw cells (preferably at room temperature, or in a 37 °C incubator).
(4) Repeat the freeze–thaw cycle twice (step 3).
(5) Prepare DNase solution by mixing 450 μL of 1 *M* $MgCl_2$ and 45 μL of 1 *M* $MnCl_2$ with 1 mL of 10 mg/mL DNase.
(6) Add the DNase solution (step 5) to the lysates from step 4. Incubate the mixture on a shaker at room temperature for 1 h. Then, transfer the mixture to a shaker in a cold room (4 °C) and incubate overnight.
(7) On the next day, add 150 mL of Detergent I solution (see Appendix), mix well and incubate the mixture on a shaker in a cold room (4 °C) for 10 min.
(8) Centrifuge at 7000 rpm for 30 min. Discard the supernatant.
(9) Resuspend the pellet in 150 mL of Detergent II solution (see Appendix), incubate the mixture in a shaker at 4 °C for 10 min and centrifuge at 8000 rpm for 30 min.
(10) Wash the pellet again with Detergent II solution (repeat step 9).
(11) Store the pellets (inclusion body) at −20 °C (up to 2 months) until purification.

Purification of keratin proteins:

(1) Add 8 mL of Urea Buffer A solution (see Appendix) to the inclusion body prep. Incubate on a shaker in a cold room (4 °C) for 1–2 h or overnight.
(2) Centrifuge at 14,000 rpm for 10 min. Filter the supernatant through a 0.22-μm filter syringe(s).
(3) Purify keratin proteins via "FPLC" (AKTA purifier) with a HiTrap Q column (GE Healthcare) using Urea Buffer A and Urea Buffer B (see Appendix). Keratin proteins are separated from other proteins by gradually increasing the percentage of Urea Buffer B at the expense of Urea Buffer A (Coulombe & Fuchs, 1990; Wawersik, Paladini, Noensie, & Coulombe, 1997).
(4) Run collected fractions on 10% sodium dodecyl sulfate polyacrylamide gel electrophoresis (SDS-PAGE) gel and stain with Coomassie blue dye to evaluate the purity and integrity of proteins.
(5) To increase purity, combine desired fractions containing keratin proteins from HiTrap Q purification (step 3) and purify again via FPLC with a Mono Q column (GE Healthcare). Before Mono Q purification, make sure to dilute the sample(s) 1/3 with Urea Buffer A (to decrease salt concentration) and adjust pH to 8.5 with NaOH (Coulombe & Fuchs, 1990; Wawersik et al., 1997).

(6) Collect fractions, run them on 10% SDS-PAGE gel, and stain with Coomassie blue to evaluate the purity and integrity of proteins. Protein concentration can be measured using the Bradford assay.

4.3 *In Vitro* Assembly of Keratin Filaments

Formation of type I–type II keratin heterotypic complexes:

(1) Mix an approximately equal molar ratio of purified type I and type II keratins. The preferred concentration of proteins is 0.2 μg/μL (or above).
(2) Incubate the mixture at room temperature for 1 h.
(3) Dilute the sample 1/3 with 6 *M* Urea Buffer A and add one drop of 10 N NaOH.
(4) Purify heterotypic complexes via FPLC using a Mono Q column (GE healthcare) as described above (Coulombe & Fuchs, 1990; Wawersik et al., 1997). Such complexes will elute at a higher salt concentration (higher percentage of 6 *M* Urea Buffer B) relative to type II keratin monomers, and type I keratin monomers.
(5) Collect fractions, run them on 10% SDS-PAGE gel, and stain with Coomassie blue to evaluate the purity and integrity of proteins. For the purpose of *in vitro* filament assembly assays, set aside (and pool) the fractions with highest purity and protein concentration (>0.5 μg/μL). As needed, proteins can be concentrated by using centrifugal filter units (Minipore) and centrifuging samples at 2500 rpm at 4 °C until the sample reaches a suitable concentration.

This step enables the recovery of keratin heterotypic complexes with type I and II proteins in a perfect 1:1 molar ratio and, as a bonus, allows for the elimination of remaining protein impurities if present.

Reconstitution of keratin intermediate filaments from purified proteins *in vitro*:

(1) Rinse dialysis tubing (Fisher Scientific) in dH_2O for at least 10 min. Place 0.5–1 mL of sample (>0.25 μg/μL or above) into the dialysis tubing. For the following dialysis procedures, the buffer needs to be >1000 × volume of the sample.
(2) Dialyze the sample in 9 *M* Urea-containing assembly buffer (see Appendix; Feng & Coulombe, 2015; Lee & Coulombe, 2009) for 4 h at room temperature.
(3) Dialyze the sample in 2 *M* Urea-containing assembly buffer (see Appendix) for 1 h at room temperature.

(4) Dialyze the sample in 0 *M* Urea-containing final assembly buffer (see Appendix) in a cold room (4 °C) overnight.

4.4 High-Speed Sedimentation Assay and TEM

High-speed sedimentation assay is used to assess the polymerization efficiency of keratin proteins. The ultrastructure of keratin filaments can be examined by negative staining (1% uranyl acetate) and with TEM (Coulombe & Fuchs, 1990; Ma et al., 2001).

High-speed sedimentation assay:

(1) Take 100 μL of sample after the last dialysis step (step 5 from filament assembly) and centrifuge at 150,000 × *g* for 30 min using an Air-Driven tabletop Ultracentrifuge (Beckman Coulter).

(2) Transfer supernatant (soluble pool) to a new tube and resuspend pellet (10-nm filaments) in 100 μL of 2 × SDS sample buffer.

(3) Mix 5 μL of input (before dialysis) with 5 μL of 2 × SDS. Mix 5 μL of supernatant with 5 μL of 2 × SDS. Boil these samples at 95 °C for 5 min. Additionally, boil 5 μL of resuspended pellet solution at 95 °C for 5 min.

(4) Run samples from step 3 on 10% SDS-PAGE gel and stain with Coomassie blue dye.

(5) Intensity of the "supernatant" and "pellet" fractions can be quantified to evaluate filament formation efficiency. For K5-K14 samples, for example, >95% of the input is typically retrieved in the pellet fraction, reflecting high assembly efficiency.

TEM (Fig. 2B)**:**

(6) Apply 10 μL of the dialyzed sample (filament assembly procedure) on a piece of Parafilm paper.

(7) Place a drop of dH_2O and two drops of 1% uranyl acetate separately on the Parafilm paper.

(8) Place a carbon-coated, 400-mesh EM grid (Electron Microscopy Sciences) on the upper surface of the sample drop for 2 min.

(9) Place the grid on the surface of the dH_2O drop for 5–10 s.

(10) Place the grid on the surface of 1% uranyl acetate drop for 1 min.

(11) Place the grid on the surface of another 1% uranyl acetate drop for 1 min.

(12) Remove residual solution from the grid using filter paper, and let the grid dry.

(13) Place the grid into a grid box.

(14) Examine the grid by TEM using Philips BioTwin CM120 (FEI Company) or Hitachi 7600 (Hitachi) instruments (Lee & Coulombe, 2009) to assess filament ultrastructure.

4.5 Low-Speed Sedimentation Assay

The formation of cross-linked keratin filaments can be assessed by low-speed sedimentation assay.

Low-speed sedimentation assay:

(1) To promote bundling of keratin filament assemblies, add 10 m*M* NaCl to the 0 *M* Urea-containing, final buffer (filament assembly procedure) or lower the pH to 7.0. Dialyze the sample in a 4 °C cold room overnight.

(2) Place 100 μL of dialyzed sample in to a 1.5-mL Eppendorf tube. Centrifuge at 8000 × *g* for 30 min.

(3) Transfer the supernatant into a new 1.5-mL Eppendorf tube and mix it with 100 μL of 2 × SDS sample buffer. Use 100 μL of 2 × SDS buffer to resuspend the pellet. Boil both the "supernatant" and "pellet" samples at 95 °C for 5 min.

(4) Run 10 μL of the "supernatant" sample and 5 μL of the "pellet" sample on 10% SDS-PAGE gel and stain with Coomassie blue dye.

(5) Quantify the intensity of the "supernatant" and the "pellet" fractions. For K5-K14 samples, for example, >50% of the input is typically retrieved in the pellet fraction, indicating a high degree of stable-filament–filament interactions (e.g., bundling).

As a complement to this say, keratin assemblies can be visualized using digital interference contrast light microscopy as described (Lee & Coulombe, 2009).

4.6 Far-Western Assay and Cosedimentation Assay to Study Direct Interactions

While Co-IP assay provides a way to identify keratin protein(s) interacting partners, one of its limitations is that it does not distinguish between direct or indirect interactions. Moreover, the bulk of keratin that is amenable to immunoprecipitation from cell lysates occur in the soluble pool, so the interaction of target proteins with keratin filaments cannot be studied using Co-IP. This section introduces the far-western assay and cosedimentation assay, which enable the investigation of direct interactions between target proteins with keratin proteins or keratin filaments, respectively (Lee & Coulombe, 2009; Rotty & Coulombe, 2012):

Far-western assay (Fig. 4D):

(1) Run 5 μg of each bait protein (e.g., purified recombinant keratin proteins: K5, K6, and K17) on a 10% SDS-PAGE gel.
(2) Transfer the gel to a nitrocellulose membrane (0.45 μm pore size).
(3) Stain the membrane with Ponceau solution (Sigma) to assess protein loading.
(4) Block the membrane in 5% milk in 1 × Tris-buffered saline with Tween 20 (TBST) for 30 min.
(5) Wash the membrane briefly with 1 × TBST.
(6) Incubate the membrane with the target protein, e.g., Src protein (150 ng/mL), in 5% BSA in 1 × TBST on a shaker at room temperature for 4 h.
(7) Wash with 1 × TBST several times.
(8) Block the membrane in 5% milk in 1 × TBST for 30 min.
(9) Incubate the membrane with primary antibody for the target protein (Src, 1:1000 dilution) in 5% BSA in 1 × TBST on a shaker in a cold room (4 °C) overnight.
(10) Perform three 5-min washes with 1 × TBST.
(11) Incubate the membrane with HRP-conjugated secondary antibody (1:2000 dilution) in 5% milk in 1 × TBST on a shaker at room temperature for 1 h.
(12) Perform three 5 min washes with 1 × TBST.
(13) Apply ECL solution (GE Healthcare) and image the membrane with FluorchemQ system.

***In vitro* cosedimentation assay (Rotty & Coulombe, 2012):**

(1) Dilute the protein of interest, e.g., Src protein (Abcam) in a suitable buffer (for Src: 60 m*M* HEPES, pH 7.5, 5 m*M* $MgCl_2$, and 5 m*M* $MnCl_2$) to a 10 ng/μL final concentration.
(2) After last step of dialysis with 0 *M* Urea-containing assembly buffer (see filament assembly protocol), incubate 25 μg of the assembled keratin filaments with 100 ng of target protein in a volume of final 100 μL.
(3) As a negative control, incubate 25 μg BSA with 100 ng Src protein (Abcam) in a volume of 100 μL of (10 μL of Src protein in Src kinase buffer, into 90 μL of BSA in 0 *M* Urea buffer).
(4) After mixing, immediately centrifuge the samples from steps 1 and 2 at 160,000 × *g* for 30 min at room temperature.
(5) Transfer the supernatant from each sample to a new separate tube and mix it with 100 μL of 2 × SDS sample buffer. Boil at 95 °C for 5 min.

(6) Resuspend the pellet from each sample with 100 μL of 2 × SDS sample buffer and boil the mixture at 95 °C for 5 min.
(7) Run 20 μL of the "supernatant" fraction and 10 μL of the "pellet" fraction from each sample on 10% SDS-PAGE gel. Detect target protein (e.g., Src) via western blotting. If Src protein interacts with keratin filaments, a significant fraction of the Src protein should cosediment with filaments into the "pellet" fraction (Rotty & Coulombe, 2012).

5. PEARLS AND PITFALLS

5.1 Collection of Mouse Skin Tissue

When harvesting skin from adult mice, the fur needs to be trimmed to avoid various complications related to its density and properties; care must be given to avoid wounding the skin. The time interval between skin isolation and fixation or snap freezing should be minimized to best preserve the original attributes of the tissue and its constituents.

5.2 Morphological Studies

The orientation of skin samples relative to the main body axis needs to be tracked during collection, embedding, and sectioning. Sectioning in a plane parallel to the rostral-caudal axis allows the viewing of single hair follicles along their long axis. Further, hair follicles cycle through growth (anagen), regression (catagen), and rest (telogen) phases, and therefore comparative experiments require skin tissue from the same phase (Montanez et al., 2007; Muller-Rover et al., 2001; Paus et al., 1999).

Chemical fixation and paraffin embedding enable long-term storage of tissue samples with good preservation of tissue morphology, but may mask epitopes of target antigens of interest. Even with antigen retrieval, the use of paraffin sections for keratin antigen immunostaining is less sensitive than fresh-frozen sections. Fresh-frozen sections preserve most epitopes and are widely used for immunohistochemistry, but may not be ideal for morphological preservation because water frozen in the tissue could result in morphological changes (Stoppacciaro & Ruco, 1999).

When preparing samples for TEM, it is important to fix tissue samples quickly because delayed and/or incomplete fixation may result in cellular damage (e.g., blebbing of the plasma membrane, swelling of extranuclear space, and abnormal mitochondrial cristae) (Graham & Orenstein, 2007), thereby interfering with the assessment of the specific impact of genetic or other types of manipulation.

5.3 Cell Culture Studies

The viable cells recovered from dissected epidermal sheets represent a heterogeneous population consisting of basal keratinocytes, keratinocytes that originate from the upper segment of hair follicles (infundibulum and outer root sheath in particular; see Kamimura, Lee, Baden, Brissette, & Dotto, 1997), melanocytes, and dermal fibroblasts. Though mKer medium can inhibit the growth of fibroblasts, fibroblasts generally have a higher proliferation rate than keratinocytes and can dominate the plate if keratinocytes are plated at low density. Keratinocytes do not grow well if they are seeded at low density.

As calcium is an important regulator of keratinocyte differentiation and other aspects of their biology, the concentration of this divalent cation in media needs to be carefully monitored. If interested in monitoring progression through terminal differentiation, a "moderate" amount of calcium in the medium is preferred to "high" calcium (Yuspa et al., 1989).

Keratinocytes form tight cell–cell adhesions and can detach as a sheet during scratch-wounding assays, thus generating "wounds" of uneven characteristics so care should be taken when performing these assays. When preparing *ex vivo* skin explant culture, skin punches should be placed directly onto the dish. Moving the explant after placement on the dish will increase fibroblast outgrowth from the skin punch, and ultimately, their dominance in the dish.

5.4 Isolation and Analysis of Proteins and RNA from Skin Cells and Tissues

Tissues and cells should be processed quickly without going through multiple freeze–thaw cycles to prevent RNA and protein degradation. For RNA isolation, it is crucial to maintain an RNAse-free environment as RNAses are very stable, ubiquitous enzymes that do not require cofactors to function. Gloves must be worn at all times and changed frequently. Aerosol-barrier tips and disposable tubes are highly recommended throughout the procedure. Nondisposable containers and glassware need to be pretreated with diethyl pyrocarbonate, a strong RNAse inhibitor, before usage. For protein isolation, protease and phosphatase inhibitors should be added freshly every time to prevent degradation and preserve posttranslational modifications.

Keratins exist as very stable type I–type II heterotypic complexes *in vivo* and *in vitro* even in the presence of Empigen BB, which partially solubilizes keratins. So when immunoprecipitating K17 from keratinocytes in culture other keratins such as K5, K6, and K14 will coimmunoprecipitate. The immunoglobulin G heavy chain (~55 kDa) falls in within the size range of most keratins (40–60 kDa), so to avoid interference from the heavy chain we use TrueBlot® Anti-Rabbit Ig IP Beads to immunoprecipitate keratins using rabbit antibodies and TrueBlot® Anti-Rabbit-HRP secondary antibody to detect proteins by western blotting (Chung et al., 2015). Alternatively, cross-linking antibody to the beads can circumvent this problem, because the antibody will remain attached to the beads during protein elution (Bernot, Coulombe, & Wong, 2004).

5.5 *In Vitro* Studies with Recombinant Keratin Proteins

The DNase digestion step is critical during the inclusion body preparation. If the digestion is not complete, the inclusion body will not pellet well during the Detergent I and II wash steps. A white viscous and fluffy pellet, instead of a solid white/amber pellet, will form as a result. Consequently, a large amount of the inclusion body will flow away when decanting the detergent after centrifugation, thereby significantly decreasing the keratin protein yield.

Serially decreasing the urea concentration from 9 to 0 *M* during keratin filament assembly is needed to allow for the successful formation of long and disperse 10 nm keratin filaments with high efficiency complexes (Herrmann et al., 2002). Concentration of keratin protein monomers should be higher than 0.2 μg/μL for best outcomes.

For the far-western assay, incubating the membrane with primary antibody longer than overnight, or washing with TBST solution for an excessive amount of time, may result in decreased signal of target proteins. One caveat for this assay is that the association tested may not reflect an *in vivo* interaction because the purified recombinant proteins immobilized on the membrane were first denatured during the preparation step by boiling in SDS sample buffer at 95 °C. Although these proteins partially renature when transferred onto the nitrocellulose membrane and incubated in gentler buffers, the resulting protein conformation may not approximate the native confirmation of the protein in the cell. Therefore, additional methods to test direct interactions are needed to support findings obtained from far-western assays.

6. CONCLUSIONS

Mouse skin represents a powerful model system to study the properties, regulation, and function of keratins in complex epithelia, whether normal or diseased. Further, skin tissue is readily amenable to clinical assessment, topical treatment, and survival surgery. Expression of keratin genes and proteins in skin follows established and evolutionary conserved patterns under healthy normal conditions, while disease processes are invariably accompanied by abnormal keratin regulation. Skin tissue thus makes it possible to investigate the roles of keratins in the complex setting of an intact, heterogeneous tissue. Cell culture, on the other hand, enables the study of the response of a single population of cells to a specific treatment and is amenable to gaining mechanistic insight. Keratinocytes represent the predominant cell type in primary culture, while fibroblasts also figure prominently in skin explant cultures. Insight gained from studies of keratinocytes in culture *ex vivo* can be readily related to their properties and functions *in vivo*. Additionally, recombinant keratin proteins can be used to study filament assembly as well as test for direct interactions between keratin proteins, or filaments, with other cellular proteins. *In vitro* studies using recombinant keratin proteins provide insight into how keratin proteins and filaments are functioning *in vivo*. We hope the methods and assays described here will prove useful for researchers interested in keratins and skin tissue.

ACKNOWLEDGMENTS

We would like to thank Dr. Ryan P. Hobbs for proofreading this chapter and providing the *in situ* hybridization protocol. We are grateful to Dr. Jeremy Rotty for sharing the cosedimentation protocol and to Dr. Xia Feng for the negative staining protocol. Efforts in the laboratory are supported by grants AR042047, AR044232, and CA160255, from the National Institutes of Health, to P.A.C.

APPENDIX

Citrate buffer for antigen retrieval

0.1 *M* Citric acid stock ($C_6H_8O_7 \cdot H_2O$)	21.0 g in 1 L
0.1 *M* Sodium citrate stock ($C_6H_5Na_3O_7 \cdot 2H_2O$)	29.4 g in 1 L
For 500 mL of 0.01 *M* citrate buffer, pH 6.0, add 9 mL of 0.1 *M* citric acid, 41 mL of 0.1 *M* sodium citrate, and bring volume up to 500 mL.	

6.5 M Urea buffer (for whole cell lysates and "insoluble fraction")
50 m*M* Tris (pH 7.5)
1 m*M* EGTA
6.5 *M* Urea
2 m*M* DTT
1 m*M* PMSF
1 × PIC1
1 × PIC2
50 m*M* NaF
1 m*M* Na_3VO_4
RIPA buffer (for "RIPA-soluble fractionation")
50 m*M* Tris (pH 7.5)
150 m*M* NaCl
0.5 m*M* EDTA
1 m*M* EGTA
1% Triton X-100
1% sodium deoxycholate
0.1% SDS
1 m*M* PMSF
1 × PIC1
1 × PIC2
50 m*M* NaF
1 m*M* Na_3VO_4
0.01% Digitonin buffer (for "cytosolic fraction")
0.01% Digitonin
10 m*M* PIPES (pH 6.8)
300 m*M* Sucrose
100 m*M* NaCl
3 m*M* $MgCl_2$
5 m*M* EDTA
0.2 m*M* DTT
1 m*M* PMSF
1 × PIC1
1 × PIC2
50 m*M* NaF
1 m*M* Na_3VO_4
0.5% Triton buffer (for "membrane fraction")
0.5 % Triton X-100
10 m*M* PIPES (pH 7.4)

300 m*M* Sucrose
100 m*M* NaCl
3 m*M* $MgCl_2$
3 m*M* EDTA
0.2 m*M* DTT
1 m*M* PMSF
1 × PIC1
1 × PIC2
50 m*M* NaF
1 m*M* Na_3VO_4

1% Triton X-100 buffer containing 2% Empigen BB (for IP keratins)

40 m*M* HEPES (pH 7.5)
120 m*M* NaCl
1 m*M* EDTA
1% Triton X-100
2% Empigen BB
10 m*M* sodium pyrophosphate
1 m*M* PMSF
1 × PIC1
1 × PIC2
50 m*M* NaF
1 m*M* Na_3VO_4

Phenylmethylsulfonyl fluoride (PMSF)

Make 200 m*M* stock. Use at 1–2 m*M* concentration. (174.2 g/mol) Dissolve in isopropanol and store at −20 °C. Warm stock to room temperature to resuspend precipitate immediately before use, PMSF has a half-life of <1 h in aqueous solutions. Use extreme caution, as PMSF is highly toxic.

Protease inhibitor cocktail 1 1000× (PIC1)

2 mg/mL	Antipain
10 mg/mL	Aprotinin
10 mg/mL	Benzamidine
1 mg/mL	Leupeptin
Dissolve in sterile ddH_2O, aliquot, and store at −20 °C until use.	

Protease inhibitor cocktail 2 1000× (PIC2)

1 mg/mL	Cymostatin
1 mg/mL	Pepstatin-A
Dissolve in dimethyl sulfoxide, aliquot, and store at −20 °C until use.	

mKer media

DMEM Low Glucose	3 parts
Ham's F-12	1 part
Fetal Bovine Serum	10%
Cholera toxin	1 n*M*
Insulin	5 μg/mL
EGF	10 ng/mL
Gentamicin	25 μg/mL
Hydrocortisone	400 ng/mL
Transferrin	5 μg/mL
3,3′,5-Triiodo-L-thyronine	2 n*M*
Penicillin	60 μg/mL

FAD media

Calcium-free DMEM	3 parts
Ham's F-12	1 part
Fetal bovine serum (chelex-treated)	10%
Cholera toxin	1 n*M*
Insulin	5 μg/mL
EGF	10 ng/mL
Gentamicin	25 μg/mL
Hydrocortisone	400 ng/mL
Penicillin	60 μg/mL

Adenine	0.18 m*M*
Glutamine	2 m*M*
Pyruvate	1 m*M*

Calcium switch media
DMEM without Ca^{2+}
8% Chelated FBS
0.5% Penicillin–streptomycin
Add desired amount of Ca^{2+} in the form of $CaCl_2$:

Low	0.05 m*M*
Moderate	0.12–0.20 m*M*
High	1.2–2.0 m*M*

Bacterial lysis buffer

Sucrose	70 m*M*
Tris–HCl (pH 8.0)	50 m*M*
EDTA (pH 8.0)	1 m*M*

Detergent I

Tris–HCl (pH 7.5)	20 m*M*
EDTA (pH 8.0)	2 m*M*
NaCl	0.2 *M*
Sodium deoxycholate	1%
Igepal CA-630 (or NP-40)	1%

Detergent II

EDTA (pH 8.0)	1 m*M*
Triton X-100	0.5%

Urea Buffer A (for keratin protein purification)

Tris–HCl (pH 8.0)	50 m*M*
EGTA	1 m*M*
Urea	6.5 *M*
DTT (add freshly)	2 m*M*
PMSF (30 mg/mL in isopropanol; add freshly)	Use at 1/100 dilution
Adjust pH to 8.5 before use.	

Urea Buffer B (for keratin protein purification)

Urea buffer A	300 mL
Guanidine HCl	0.5 *M*

9 M Urea buffer (for filament assembly)

Urea	9 *M*
Tris–HCl (pH 7.4)	25 m*M*
β-Mercaptoethanol	25 m*M*

2 M Urea buffer (for filament assembly)

Urea	2 *M*
Tris–HCl (pH 7.4)	5 m*M*
β-Mercaptoethanol	5 m*M*

0 M Urea buffer (for filament assembly)

Tris–HCl (pH 7.5)	5 m*M*
β-Mercaptoethanol	5 m*M*

REFERENCES

Alvarado, D. M., & Coulombe, P. A. (2014). Directed expression of a chimeric type II keratin partially rescues keratin 5-null mice. *The Journal of Biological Chemistry*, *289*, 19435–19447.

Bernot, K. M., Coulombe, P. A., & McGowan, K. M. (2002). Keratin 16 expression defines a subset of epithelial cells during skin morphogenesis and the hair cycle. *The Journal of Investigative Dermatology*, *119*, 1137–1149.

Bernot, K. M., Coulombe, P. A., & Wong, P. (2004). Skin: An ideal model system to study keratin genes and proteins. *Methods in Cell Biology*, *78*, 453–487.

Bonifas, J. M., Rothman, A. L., & Epstein, E. H., Jr. (1991). Epidermolysis bullosa simplex: Evidence in two families for keratin gene abnormalities. *Science*, *254*, 1202–1205.

Byrne, C. (1997). Regulation of gene expression in developing epidermal epithelia. *Bioessays*, *19*, 691–698.

Candi, E., Schmidt, R., & Melino, G. (2005). The cornified envelope: A model of cell death in the skin. *Nature Reviews. Molecular Cell Biology*, *6*, 328–340.

Chung, B. M., Arutyunov, A., Ilagan, E., Yao, N., Wills-Karp, M., & Coulombe, P. A. (2015). Regulation of C-X-C chemokine gene expression by keratin 17 and hnRNP K in skin tumor keratinocytes. *The Journal of Cell Biology*, *208*, 613–627.

Collin, C., Moll, R., Kubicka, S., Ooukayoun, J. P., & Franke, W. W. (1992). Characterization of human cytokeratin 2, an epidermal cytoskeletal protein synthesized late during differentiation. *Experimental Cell Research*, *202*, 132–141.

Coulombe, P. A., & Bernot, K. M. (2004). Keratins and the skin. In W. J. Lennarz & M. D. Lane (Eds.), *Encyclopedia of biological chemistry: Vol. 2* (pp. 497–504). Oxford: Elsevier.

Coulombe, P. A., Bernot, K. M., & Lee, C. H. (2013). Keratins and the skin. In W. J. Lennarz & M. D. Lane (Eds.), *Encyclopedia of biological chemistry* (pp. 665–671). Oxford: Elsevier.

Coulombe, P. A., Chan, Y.-M., Albers, K., & Fuchs, E. (1990). Deletions in epidermal keratins that lead to alterations in filament organization and assembly: *In vivo* and *in vitro* studies. *The Journal of Cell Biology*, *111*, 3049–3084.

Coulombe, P. A., & Fuchs, E. (1990). Elucidating the early stages of keratin filament assembly. *The Journal of Cell Biology*, *111*, 153–169.

Coulombe, P. A., Hutton, M. E., Letai, A., Hebert, A., Paller, A. S., & Fuchs, E. (1991a). Point mutations in human keratin 14 genes of epidermolysis bullosa simplex patients: Genetic and functional analyses. *Cell*, *66*, 1301–1311.

Coulombe, P. A., Hutton, M. E., Vassar, R., & Fuchs, E. (1991b). A function for keratins and a common thread among different types of epidermolysis bullosa simplex diseases. *The Journal of Cell Biology*, *115*, 1661–1674.

Depianto, D., Kerns, M. L., Dlugosz, A. A., & Coulombe, P. A. (2010). Keratin 17 promotes epithelial proliferation and tumor growth by polarizing the immune response in skin. *Nature Genetics*, *42*, 910–914.

Distler, J. H., Jungel, A., Kurowska-Stolarska, M., Michel, B. A., Gay, R. E., Gay, S., et al. (2005). Nucleofection: A new, highly efficient transfection method for primary human keratinocytes. *Experimental Dermatology*, *14*, 315–320.

Feng, X., & Coulombe, P. A. (2015). A role for disulfide bonding in keratin intermediate filament organization and dynamics in skin keratinocytes. *The Journal of Cell Biology*, *209*, 59–72.

Fischer, A. H., Jacobson, K. A., Rose, J., & Zeller, R. (2008). Hematoxylin and eosin staining of tissue and cell sections. *CSH Protocols*, *2008*. prot4986.

Freedberg, I. M., Tomic-Canic, M., Komine, M., & Blumenberg, M. (2001). Keratins and the keratinocyte activation cycle. *The Journal of Investigative Dermatology*, *116*, 633–640.

Fuchs, E. (1995). Keratins and the skin. *Annual Review of Cell and Developmental Biology, 11*, 123–153.
Fuchs, E., Esteves, R. A., & Coulombe, P. A. (1992). Transgenic mice expressing a mutant keratin 10 gene reveal the likely genetic basis for epidermolytic hyperkeratosis. *Proceedings of the National Academy of Sciences of the United States of America, 89*, 6906–6910.
Fuchs, E., & Green, H. (1980). Changes in keratin gene expression during terminal differentiation of the keratinocyte. *Cell, 19*, 1033–1042.
Fuchs, E., & Weber, K. (1994). Intermediate filaments: Structure, dynamics, function, and disease. *Annual Review of Biochemistry, 63*, 345–382.
Gilbert, S., Loranger, A., Daigle, N., & Marceau, N. (2001). Simple epithelium keratins 8 and 18 provide resistance to Fas-mediated apoptosis. The protection occurs through a receptor-targeting modulation. *The Journal of Cell Biology, 154*, 763–773.
Graham, L., & Orenstein, J. M. (2007). Processing tissue and cells for transmission electron microscopy in diagnostic pathology and research. *Nature Protocols, 2*, 2439–2450.
Haines, R. L., & Lane, E. B. (2012). Keratins and disease at a glance. *Journal of Cell Science, 125*, 3923–3928.
Hatzfeld, M., & Weber, K. (1990a). The coiled coil of *in vitro* assembled keratin filaments is a heterodimer of type I and II keratins: Use of site-specific mutagenesis and recombinant protein expression. *The Journal of Cell Biology, 110*, 1199–1210.
Hatzfeld, M., & Weber, K. (1990b). Tailless keratins assemble into regular intermediate filaments in vitro. *Journal of Cell Science, 97*, 317–324.
Hayat, M. A. (2000). *Principles and techniques of electron microscopy: Biological applications*. Cambridge, UK: Cambridge University Press.
Hennings, H., Holbrook, K. A., & Yuspa, S. H. (1983). Factors influencing calcium-induced terminal differentiation in cultured mouse epidermal cells. *Journal of Cellular Physiology, 116*, 265–281.
Herrmann, H., Wedig, T., Porter, R. M., Lane, E. B., & Aebi, U. (2002). Characterization of early assembly intermediates of recombinant human keratins. *Journal of Structural Biology, 137*, 82–96.
Hesse, M., Zimek, A., Weber, K., & Magin, T. M. (2004). Comprehensive analysis ofkeratin gene clusters in humans and rodents. *European Journal of Cell Biology, 83*, 19–26.
Hobbs, R. P., DePianto, D. J., Jacob, J. T., Han, M. C., Chung, B. M., Batazzi, A. S., et al. (2015). Keratin-dependent regulation of Aire and gene expression in skin tumor keratinocytes. *Nature Genetics, 47*, 933–938.
Holbrook, K. A., & Wolff, K. (1993). The structure and development of skin. In T. B. Fitzpatrick, A. Z. Eisen, K. WolV, I. M. Freedberg, & M. D. Austen (Eds.), *Dermatology in general medicine* (pp. 97–144). New York: McGraw-Hill.
Inada, H., Izhawa, I., Nishizawa, M., Fujita, E., Kiyono, T., Takahashi, T., et al. (2001). Keratin attenuates tumor necrosis factor-induced cytotoxicity through association with TRADD. *The Journal of Cell Biology, 155*, 415–426.
Kamimura, J., Lee, D., Baden, H. P., Brissette, J., & Dotto, G. P. (1997). Primary mouse keratinocyte cultures contain hair follicle progenitor cells with multiple differentiation potential. *The Journal of Investigative Dermatology, 109*, 534–540.
Kim, S., & Coulombe, P. A. (2007). Intermediate filament scaffolds fulfill mechanical, organizational, and signaling functions in the cytoplasm. *Genes & Development, 21*, 1581–1597.
Kim, S., Wong, P., & Coulombe, P. A. (2006). A keratin cytoskeletal protein regulates protein synthesis and epithelial cell growth. *Nature, 441*, 362–365.
Ku, N. O., & Omary, M. B. (2006). A disease- and phosphorylation-related nonmechanical function for keratin 8. *The Journal of Cell Biology, 174*, 115–125.
Lane, E. B., & McLean, W. H. (2004). Keratins and skin disorders. *The Journal of Pathology, 204*, 355–366.

Lane, E. B., Rugg, E. L., Navsaria, H., Leigh, I. M., Heagerty, A. H., Ishida-Yamamoto, A., et al. (1992). A mutation in the conserved helix termination peptide of keratin 5 in hereditary skin blistering. *Nature, 356*, 244–246.

Langbein, L., Heid, H. W., Moll, I., & Franke, W. W. (1993). Molecular characterization of the body site-specific human epidermal cytokeratin 9: cDNA cloning, amino acid sequence, and tissue specificity of gene expression. *Differentiation, 55*, 57–71.

Lee, C. H., & Coulombe, P. A. (2009). Self-organization of keratin intermediate filaments into cross-linked networks. *The Journal of Cell Biology, 186*, 409–421.

Lessard, J. C., Pina-Paz, S., Rotty, J. D., Hickerson, R. P., Kaspar, R. L., Balmain, A., et al. (2013). Keratin 16 regulates innate immunity in response to epidermal barrier breach. *Proceedings of the National Academy of Sciences of the United States of America, 110*, 19537–19542.

Li, L. (2013). Mouse epidermal keratinocyte culture. *Methods in Molecular Biology, 945*, 177–191.

Lichti, U., Anders, J., & Yuspa, S. H. (2008). Isolation and short-term culture of primary keratinocytes, hair follicle populations and dermal cells from newborn mice and keratinocytes from adult mice for *in vitro* analysis and for grafting to immunodeficient mice. *Nature Protocols, 3*, 799–810.

Lloyd, C., Yu, Q. C., Cheng, J., Turksen, K., Degenstein, L., Hutton, E., et al. (1995). The basal keratin network of stratified squamous epithelia: Defining K15 function in the absence of K14. *The Journal of Cell Biology, 129*, 1329–1344.

Lowthert, L. A., Ku, N. O., Liao, J., Coulombe, P. A., & Omary, M. B. (1995). Empigen BB: A useful detergent for solubilization and biochemical analysis of keratins. *Biochemical and Biophysical Research Communications, 206*, 370–379.

Ma, A. S. P., & Lorincz, A. L. (1988). Immunofluorescence localization of peripheral proteins in cultured human keratinocytes. *The Journal of Investigative Dermatology, 90*, 331–335.

Ma, L., Yamada, S., Wirtz, D., & Coulombe, P. A. (2001). A 'hot-spot' mutation alters the mechanical properties of keratin filament networks. *Nature Cell Biology, 3*, 503–506.

Mazzalupo, S., Wawersik, M. J., & Coulombe, P. A. (2002). An *ex vivo* assay to assess the potential of skin keratinocytes for wound epithelialization. *The Journal of Investigative Dermatology, 118*, 866–870.

McGowan, K. M., & Coulombe, P. A. (1998a). Onset of keratin 17 expression coincides with the definition of major epithelial lineages during skin development. *The Journal of Cell Biology, 143*, 469–486.

McGowan, K. M., & Coulombe, P. A. (1998b). The wound repair associated keratins 6, 16, and 17: Insights into the role of intermediate filaments in specifying cytoarchitecture. In J. R. Harris & H. Herrmann (Eds.), *Subcellular biochemistry: Intermediate filaments* (pp. 141–165). London: Plenum Publishing Corp.

McLean, W. H., & Moore, C. B. (2011). Keratin disorders: From gene to therapy. *Human Molecular Genetics, 20*, R189–R197.

McLean, W. H., Rugg, E. L., Lunny, D. P., Morley, S. M., Lane, E. B., Swensson, O., et al. (1995). Keratin 16 and keratin 17 mutations cause pachyonychia congenita. *Nature Genetics, 9*, 273–278.

Michel, M., Torok, N., Godbout, M. J., Lussier, M., Gaudreau, P., Royal, A., et al. (1996). Keratin 19 as a biochemical marker of skin stem cells *in vivo* and *in vitro*: Keratin 19 expressing cells are differentially localized in function of anatomic sites, and their number varies with donor age and culture stage. *Journal of Cell Science, 109*, 1017–1028.

Montanez, E., Piwko-Czuchra, A., Bauer, M., Li, S., Yurchenco, P., & Fassler, R. (2007). Analysis of integrin functions in peri-implantation embryos, hematopoietic system, and skin. *Methods in Enzymology, 426*, 239–289.

Muller-Rover, S., Handjiski, B., van der Veen, C., Eichmuller, S., Foitzik, K., McKay, I. A., et al. (2001). A comprehensive guide for the accurate classification of murine hair follicles in distinct hair cycle stages. *The Journal of Investigative Dermatology*, *117*, 3–15.
Nelson, W. G., & Sun, T. T. (1983). The 50- and 58-kdalton keratin classes as molecular markers for stratified squamous epithelia: Cell culture studies. *The Journal of Cell Biology*, *97*, 244–251.
Omary, M. B., Coulombe, P. A., & McLean, W. H. (2004). Intermediate filament proteins and their associated diseases. *The New England Journal of Medicine*, *351*, 2087–2100.
Omary, M. B., Ku, N. O., Liao, J., & Price, D. (1998). Keratin modifications and solubility properties in epithelial cells and *in vitro*. *Sub-Cellular Biochemistry*, *31*, 105–140.
Paladini, R. D., Takahashi, K., Bravo, N. S., & Coulombe, P. A. (1996). Onset of re-epithelialization after skin injury correlates with a reorganization of keratin filaments in wound edge keratinocytes: Defining a potential role for keratin 16. *The Journal of Cell Biology*, *132*, 381–397.
Pan, X., Hobbs, R. P., & Coulombe, P. A. (2013). The expanding significance of keratin intermediate filaments in normal and diseased epithelia. *Current Opinion in Cell Biology*, *25*, 47–56.
Paus, R., Muller-Rover, S., Van Der Veen, C., Maurer, M., Eichmuller, S., Ling, G., et al. (1999). A comprehensive guide for the recognition and classification of distinct stages of hair follicle morphogenesis. *The Journal of Investigative Dermatology*, *113*, 523–532.
Reichelt, J., & Haase, I. (2010). Establishment of spontaneously immortalized keratinocyte lines from wild type and mutant mice. *Methods in Molecular Biology*, *585*, 59–69.
Roop, D. R., Cheng, C. K., Titterington, L., Meyers, C. A., Stanley, J. R., Steinert, P. M., et al. (1984). Synthetic peptides corresponding to keratin subunits elicit highly specific antibodies. *The Journal of Biological Chemistry*, *259*, 8037–8040.
Roth, W., Kumar, V., Beer, H. D., Richter, M., Wohlenberg, C., Reuter, U., et al. (2012). Keratin 1 maintains skin integrity and participates in an inflammatory network in skin through interleukin-18. *Journal of Cell Science*, *125*, 5269–5279.
Rotty, J. D., & Coulombe, P. A. (2012). A wound-induced keratin inhibits Src activity during keratinocyte migration and tissue repair. *The Journal of Cell Biology*, *197*, 381–389.
Seltmann, K., Fritsch, A. W., Kas, J. A., & Magin, T. M. (2013). Keratins significantly contribute to cell stiffness and impact invasive behavior. *Proceedings of the National Academy of Sciences of the United States of America*, *110*, 18507–18512.
Snider, N. T., & Omary, M. B. (2014). Post-translational modifications of intermediate filament proteins: Mechanisms and functions. *Nature Reviews. Molecular Cell Biology*, *15*, 163–177.
Steinert, P. M., Steven, A. C., & Roop, D. R. (1985). The molecular biology of intermediate filaments. *Cell*, *42*, 411–420.
Stoler, A., Kopan, R., Duvic, M., & Fuchs, E. (1988). Use of monospecific antisera and cRNA probes to localize the major changes in keratin expression during normal and abnormal epidermal differentiation. *The Journal of Cell Biology*, *107*, 427–446.
Stoppacciaro, A., & Ruco, L. P. (1999). Detection of adhesion molecules by immunohistochemistry on human and murine tissue sections. *Methods in Molecular Biology*, *96*, 93–106.
Swensson, O., Langbein, L., McMillan, J. R., Stevens, H. P., Leigh, I. M., McLean, W. H., et al. (1998). Specialized keratin expression pattern in human ridged skin as an adaptation to high physical stress. *The British Journal of Dermatology*, *139*, 767–775.
Szeverenyi, I., Cassidy, A. J., Chung, C. W., Lee, B. T., Common, J. E., Ogg, S. C., et al. (2008). The human intermediate filament database: Comprehensive information on a gene family involved in many human diseases. *Human Mutation*, *29*, 351–360.

Takahashi, K., Yan, B., Yamanishi, K., Imamura, S., & Coulombe, P. A. (1998). The two functional type II keratin 6 genes of mouse show a differential regulation and evolved independently from their human orthologs. *Genomics*, *53*, 170–183.

Tam, C., Mun, J. J., Evans, D. J., & Fleiszig, S. M. (2012). Cytokeratins mediate epithelial innate defense through their antimicrobial properties. *The Journal of Clinical Investigation*, *122*, 3665–3677.

Toivola, D. M., Boor, P., Alam, C., & Strnad, P. (2015). Keratins in health and disease. *Current Opinion in Cell Biology*, *32*, 73–81.

Tong, X., & Coulombe, P. A. (2004). A novel mouse type I intermediate filament gene, keratin 17n (K17n), exhibits preferred expression in nail tissue. *The Journal of Investigative Dermatology*, *122*, 965–970.

Tong, X., & Coulombe, P. A. (2006). Keratin 17 modulates hair follicle cycling in a TNFα-dependent fashion. *Genes & Development*, *20*, 1353–1364.

Wang, Z., Wong, P., Langbein, L., Schweizer, J., & Coulombe, P. A. (2003). Type II epithelial keratin 6hf (K6hf) is expressed in the companion layer, matrix, and medulla in anagen-stage hair follicles. *The Journal of Investigative Dermatology*, *121*, 1276–1282.

Wawersik, M., Paladini, R. D., Noensie, E., & Coulombe, P. A. (1997). A proline residue in the a-helical rod domain of type I keratin 16 destabilizes keratin heterotetramers and influences incorporation into filaments. *The Journal of Biological Chemistry*, *272*, 32557–32565.

Weiss, R. A., Eichner, R., & Sun, T. T. (1984). Monoclonal antibody analysis of keratin expression in epidermal diseases: A 48- and 56-kdalton keratin as molecular markers for hyperproliferative keratinocytes. *The Journal of Cell Biology*, *98*, 1397–1406.

Windoffer, R., Beil, M., Magin, T. M., & Leube, R. E. (2011). Cytoskeleton in motion: The dynamics of keratin intermediate filaments in epithelia. *The Journal of Cell Biology*, *194*, 669–678.

Wong, P., & Coulombe, P. A. (2003). Loss of keratin 6 (K6) proteins reveals a function for intermediate filaments during wound repair. *The Journal of Cell Biology*, *163*, 327–337.

Woodcock-Mitchell, J., Eichner, R., Nelson, W. G., & Sun, T. T. (1982). Immunolocalization of keratin polypeptides in human epidermis using monoclonal antibodies. *The Journal of Cell Biology*, *95*, 580–588.

Yuspa, S. H., Kilkenny, A. E., Steinert, P. M., & Roop, D. R. (1989). Expression of murine epidermal differentiation markers is tightly regulated by restricted extracellular calcium concentrations *in vitro*. *The Journal of Cell Biology*, *109*, 1207–1217.

CHAPTER THIRTEEN

Simple Epithelial Keratins

Pavel Strnad*[,1], Nurdan Guldiken*, Terhi O. Helenius[†,‡], Julia O. Misiorek[†,‡], Joel H. Nyström[†,‡], Iris A.K. Lähdeniemi[†,‡], Jonas S.G. Silvander[†,‡], Deniz Kuscuoglu*, Diana M. Toivola[†,‡,1]

*Department of Internal Medicine III and IZKF, University Hospital Aachen, Aachen, Germany
[†]Faculty of Science and Engineering, Department of Biosciences, Cell Biology, Åbo Akademi University, Turku, Finland
[‡]Turku Center for Disease Modeling, University of Turku, Turku, Finland
[1]Corresponding authors: e-mail address: pstrnad@ukaachen.de; dtoivola@abo.fi

Contents

1. Introduction 352
2. Isolation of SEKs 354
3. Studying SEKs in Cell Culture 357
   3.1 Overexpression of SEKs in Cells 358
   3.2 Downregulation of SEKs in Cells 362
4. Genetic Mouse Models of SEKs 365
   4.1 SEK KO Mouse Models 365
   4.2 SEK WT Overexpressor Mice 370
   4.3 Transgenic Mice Overexpressing SEK Variants 371
5. MDBs—Keratin Aggregates in Liver Disease 372
   5.1 Methods to Induce MDBs 373
   5.2 Imaging and Quantification of MDBs 374
   5.3 Biochemical Analysis of MDBs 377
6. SEK Variant Detection in Humans 378
7. Conclusions 379
Acknowledgments 380
References 380

Abstract

Simple epithelial keratins (SEKs) are the cytoplasmic intermediate filament proteins of single-layered and glandular epithelial cells as found in the liver, pancreas, intestine, and lung. SEKs have broad cytoprotective functions, which are facilitated by dynamic posttranslational modifications and interaction with associated proteins. SEK filaments are composed of obligate heteropolymers of type II (K7, K8) and type I (K18–K20, K23) keratins. The multifaceted roles of SEKs are increasingly appreciated due to findings obtained from transgenic mouse models and human studies that identified SEK variants in several digestive diseases. Reorganization of the SEK network into aggregates called Mallory–Denk bodies (MDBs) is characteristic for specific liver disorders such as alcoholic and nonalcoholic steatohepatitis. To spur further research on SEKs, we here review the

Methods in Enzymology, Volume 568
ISSN 0076-6879
http://dx.doi.org/10.1016/bs.mie.2015.08.004

methods and potential caveats of their isolation as well as possibilities to study them in cell culture. The existing transgenic SEK mouse models, their advantages and potential drawbacks are discussed. The tools to induce MDBs, ways of their visualization and quantification, as well as the possibilities to detect SEK variants in humans are summarized.

ABBREVIATIONS

ALD alcoholic liver disease
DDC 3,5-diethoxycarbonyl-1,4-dihydrocollidine
DHPLC denaturing high-performance liquid chromatography
FRET fluorescence resonance energy transfer
HSE high salt extraction
IF intermediate filament
IP immunoprecipitation
K keratin
KO knockout
MDBs Mallory–Denk bodies
MEF mouse embryonic fibroblast
NASH nonalcoholic steatohepatitis
PBS phosphate buffer saline
PTM posttranslational modification
SDS sodium dodecyl sulfate
SEK simple epithelial keratin
shRNA short hairpin RNA
siRNA short interfering RNA
TX-100 Triton X-100
WT wild type

1. INTRODUCTION

Intermediate filaments (IFs) are, unlike the ubiquitous microfilaments and microtubules, expressed in a tissue- and cell-dependent manner (Ku, Zhou, Toivola, & Omary, 1999). Within the IF family of over 70 proteins, the 54 keratin (K) family members compose the most abundant subgroup (Omary, Ku, Strnad, & Hanada, 2009). Keratins are divided in acidic type I and basic to neutral type II proteins that form obligate heteropolymers (Schweizer et al., 2006). Keratins are further divided into hair keratins, epidermal and simple epithelial keratins (SEKs), the latter being expressed in simple epithelia, such as liver, lung, intestine, pancreas, and most glandular epithelia and comprise K7, K8, K18–K20, and K23 (Moll, Divo, & Langbein, 2008; Omary et al., 2009; Strnad et al., 2011). The epithelial cells of some tissues only express one type I/type II pair, e.g., hepatocytes which express K8 and K18, whereas intestinal epithelial cells express all SEKs. Within the epithelial tissues, SEKs can be expressed in a layer, cell, or cellular compartment-specific manner, generating a large variation and functional diversity that is yet not well understood. For example, in colonic crypts K20 is only expressed in the uppermost differentiated cells, whereas K7 is found only in the lower parts of the crypt (Zhou et al., 2003). In murine acinar cells of the pancreas, K19 is located only at the apico-lateral cell membrane under basal conditions, while K8 and K18 form filaments all over the cytoplasm (Toivola, Baribault, Magin, Michie, & Omary, 2000). The differential expression patterns are useful in pathological analysis for subtyping tumors, studying circulating cancer cells, and determining the origin of metastases of unknown primary tumor origin (Omary, 2009; Toivola, Boor, Alam, & Strnad, 2015). K8 and K18 constitute the most abundant SEKs that are expressed early on in embryonic development and are also found at low levels in stratified epithelia and even nonepithelial tissues such as heart muscle (Omary et al., 2009).

SEKs display both important differences and similarities with the other keratins. For the latter, SEKs function as important stress protectors and are upregulated at the mRNA and protein levels during various stress situations (Toivola, Strnad, Habtezion, & Omary, 2010). In transgenic animals, a defect in SEKs leads to a predominantly hepatic phenotype, whereas the involvement of other tissues seems to be more variable, a notion that needs to be further characterized in future studies (Ku, Strnad, Zhong, Tao, & Omary, 2007; Omary et al., 2009). Similarly, K8/K18 variants predispose to development and progression of various human liver diseases, while their importance in other disorders is less clear (Toivola et al., 2015). With regard to their unique properties, SEKs are less exposed to mechanical stress than keratins from hair and stratified epithelia, and they display a distinct amino acid composition with an almost complete lack of cysteine (Strnad et al., 2011). SEKs are also more soluble than the other keratin subtypes which is in a good agreement with their more pliable networks (Omary et al., 2009). In specific liver diseases, such as alcoholic liver disease (ALD) or nonalcoholic steatohepatitis (NASH), the SEK network becomes reorganized into inclusions termed as Mallory–Denk bodies (MDBs; Strnad et al., 2013; Zatloukal, Stumptner, Fuchsbichler, Janig, & Denk, 2004).

Due to the lack of endogenous SEKs in lower organisms, such as bacteria and the fruit fly, as well as the importance of keratin variants in human disease (Toivola et al., 2015), most research on SEK has been done in mammalian systems, i.e., cell culture, mouse models, and human samples (Ku et al., 2004; Omary et al., 2009). To reflect these model systems, this chapter aims to provide a guide how to (i) isolate SEKs (Section 2), (ii) manipulate them in cell culture (Section 3), (iii) analyze them in transgenic animals (Section 4), (iv) induce and evaluate the formation of MDBs (Section 5), and (v) detect keratin variants in humans (Section 6). For a more detailed manual for how to study the biochemical properties of keratins including their posttranslational modifications (PTMs) (this issue of Methods in Enzymology (MIE) chapter by Snider and Omary, "Assays for Post-translational Modifications of Intermediate Filament Proteins"), their dynamic behavior in living cells (MIE) chapter by Schwarz et al., "Multidimensional monitoring of keratin intermediate filaments in cultured cells and tissues" and how to visualize keratins (Ku et al., 2004) (this issue of MIE Chapter by Stumptner et al., "Immunofluorescence and immunohistochemical detection of keratins"), the readers should refer to the appropriate chapters of this book.

2. ISOLATION OF SEKs

Keratin filaments are dynamic proteins that exist both in insoluble and soluble forms (Windoffer, Beil, Magin, & Leube, 2011). SEKs are significantly more soluble than both epidermal and hair keratins, as approximately 5% of K8 and K18 can be dissolved in nonionic detergents such as Triton X-100 (TX-100) (Omary, Ku, Liao, & Price, 1998). Detailed protocols for keratin isolation/enrichment via high salt extraction (HSE) and immunoprecipitation (IP) from the soluble keratin pool have been previously published (Ku et al., 2004; Snider & Omary, 2014). HSE takes advantage of the relative insolubility of keratins in 0.5–1% TX-100 as well as in high salt buffers containing 1.5 *M* KCl; the latter being used to remove nucleic acids (Ku et al., 2004). At the end of the procedure, keratins are solubilized in 2–4% (w/v) sodium dodecyl sulfate (SDS)-containing Laemmli buffer (Ku et al., 2004). Alternatively, buffers with high concentrations of urea (8–9.5 *M*, 25 m*M* Tris–HCl, pH 8.0) (Eichner & Kahn, 1990) or guanidium hydrochloride (2 *M* guanidium–HCl, 10 m*M* Tris–HCl, pH 7.5) may be used; however, the guanidium chloride-containing buffers cannot be easily loaded on a gel (Hatzfeld & Weber, 1990). Of note, recombinant keratins have been shown to polymerize even in 6 *M* urea (Coulombe & Fuchs,

1990), which is why SDS-containing buffers are typically used to extract keratins, whereas urea-containing buffers are employed to study keratin polymerization as well as for proteomic analysis (Herrmann, Kreplak, & Aebi, 2004; Liao, Ku, & Omary, 1996; Plowman, 2007). It is important to note that in the HSE extract protocol, the TX-100 soluble keratin fraction will not be present in the final pellet, whereas IP collects only soluble keratins. The zwitterionic detergent Empigen B represents a useful alternative that can solubilize up to 50% of SEKs, however, may disrupt some protein–protein interactions (Lowthert, Ku, Liao, Coulombe, & Omary, 1995). Of note, TX-100-insoluble, Empigen B soluble fraction yields highly enriched keratins; however, this fraction has not been systematically analyzed.

Using the above-mentioned methods combined with size separation (Table 1) by 1D or 2D gel electrophoresis and Western blotting, the soluble properties of SEKs and SEK-binding proteins can be analyzed using SEK antibodies (Table 2). A challenge is to study the direct interaction of other proteins with the nonsoluble keratin pool since SEK IP from this fraction is not easily performed. The uses of advanced microscopy techniques such as fluorescence resonance energy transfer (FRET) or proximity *in situ* ligation assays are promising tools to address these questions (Zatloukal et al., 2014).

In agreement with their role as cytoprotective proteins, keratins are often increased during stress and the subsequent regeneration process (Alam et al., 2013; Toivola et al., 2010). On the other hand, keratins particularly of the epidermal and hair subtype constitute a common contamination that represents a challenge in large-scale proteomics studies (Petrak et al., 2008); however, this does not justify the elimination of SEKs from the

Table 1 Molecular Weights of Human and Mouse Keratins

	Type II Keratin		**Type I Keratin**			
Species	**K7**	**K8**	**K18**	**K19**	**K20**	**K23**
Human	51.3	53.7	48.1	44.1	48.5	48.1
Mouse	50.6	54.1	47.4	44.4	48.9	47.9

The estimated molecular weights (presented as kDa values in the table) are based on amino acid sequences and the relative weight of individual amino acids (Szeverenyi et al., 2008). Note that based on Coomassie and Western blot analysis performed on keratin high salt extracts (HSE) (Ku et al., 2004), mouse K18 runs at a higher molecular weight than the human K18 isoform, and human K20 is larger than the mouse K20 isoform, in contrast to the estimated molecular weights. The molecular weight of keratins in a PAGE gel may also change dependent on the level of PTM, such that hyperphosphorylated keratins could shift to a higher molecular weight (depending on the site and extent of phosphorylation), and caspase-mediated K18 cleavage results in low-molecular-weight fragments.

Table 2 SEK Antibodies

Keratin	Antibody Name	Host	Species Reactivity	Applications	Supplier
K7	RCK105*	Mouse	m, hu, r, c, h, p	WB, ICC, IHC, FCM	Progen, Abcam, Santa Cruz biotech.**, Thermo Scientific, Millipore
K8	Troma I*	Rat	m, hu	WB, ICC, IHC, IP, FCM	DSHB
	M20*	Mouse	hu, r, rb	WB, ICC, IHC, IP, FCM	Abcam, Thermo Scientific, Santa Cruz biotech.**
	E3264	Rabbit	Hu	WB, IHC	Spring bioscience
	Ks8.7*	Mouse	hu, m, r	WB, ICC, IHC	Progen
K18	DC10*	Mouse	hu	WB, ICC, IHC, IP	Abcam, Santa Cruz biotech., Thermo Scientific**
	L2A1*	Mouse	hu	WB, ICC, IHC, IP	Thermo Scientific
	Ks18.04*	Mouse	hu, m, r, h, c, p, s, z	WB, ICC, IHC, FCM	Progen
K19	Troma III*	Rat	m, hu (ICC, IHC)	WB, ICC, IHC, IP	DSHB
	RCK108*	Mouse	hu, z	WB, ICC, IHC, FCM	Thermo Scientific Abcam, Biorbyt, Novus Biol.**
K20	Ks20.8*	Mouse	hu, m	WB, IHC	Millipore, Thermo Scientific**
	ITKs20.10*	Mouse	hu, r, p	WB, ICC, IHC	Progen
	Q6*	Mouse	hu	WB, IHC, IP	Thermo Scientific

Table 2 SEK Antibodies—cont'd

Keratin	Antibody Name	Host	Species Reactivity	Applications	Supplier
Pan Keratin	C-11*	Mouse	hu, r, mo, r, mm	WB, ICC, IHC, FCM, IP	Cell Signaling, BioLegend**, United states biological
	Z0622	Rabbit	hu	IHC	Dako

The table lists commonly used, commercially available SEK antibodies, their applications and suppliers. m, mouse; r, rat; rb, rabbit; g, goat; h, hamster; c, canine; f, feline; hu, human; p, porcine; z, zebrafish; mo, monkey; mm, mammalian; WB, Western blot; ICC, immunocytochemistry; IHC, immunohistochemistry; FCM, flow cytometry; IP, immunoprecipitation; DSHB, Developmental Studies Hybridoma Bank; *, monoclonal antibody; **, Companies offering competitive pricing and/or smaller aliquots that might be useful for pilot experiments. Note that most SEK antibodies do not perform well on formaldehyde-fixed samples for immunohistochemical or immunofluorescence staining.

analysis. For rising to this challenge, methods that minimize potential contaminations need to be implemented (Xu et al., 2011), and keratin-related "hits" should be validated by an alternative approach whenever possible.

3. STUDYING SEKs IN CELL CULTURE

Many SEK properties and functions have been discovered through biochemical analysis of HT-29 or Caco-2 colon cancer cell lines (Weerasinghe, Ku, Altshuler, Kwan, & Omary, 2014), HepG2 liver cells (Fortier, Asselin, & Cadrin, 2013), breast epithelial cell lines (Iyer et al., 2013), human oral squamous cell carcinoma AW13516 cells (Alam et al., 2011), and human lung adenocarcinoma A549 cells (Sivaramakrishnan, Schneider, Sitikov, Goldman, & Ridge, 2009) to mention a few. However, these cell lines are either transformed or derived from cancers and therefore behave significantly different than differentiated cells. For example, colon epithelial health and energy metabolism are fueled by short-chain fatty acids (produced by the luminal microbiota), a process where keratins have a modulatory role (Helenius et al., 2015), while the same short-chain fatty acids cause cell death in cultured cancer cells (Fung, Cosgrove, Lockett, Head, & Topping, 2012). Primary cultures can help solve this issue, and lung alveolar epithelial cells (Sivaramakrishnan, DeGiulio, Lorand, Goldman, & Ridge, 2008) and hepatocytes (Toivola, Goldman, Garrod, & Eriksson, 1997) have been very useful, e.g., in studying the role of K8 and K18 in protection from apoptosis and in their regulation by PTMs (Snider & Omary, 2014). Under careful isolation conditions, even the very fragile K8 knockout (K8 KO) hepatocytes can be isolated and cultured (Marceau, Gilbert, & Loranger, 2004). For other primary cells, such as pancreatic acini (Toivola et al., 2010), islets of Langerhans (Alam et al., 2013), and colonocytes (Toivola, Krishnan, Binder, Singh, & Omary, 2004), culture/incubation times limited to a few hours are feasible only in some cases. Epithelial cells are very dependent on their 3D microenvironment architecture, and consequently, the recently described 3D organoid culture systems grown from organ-specific stem cells will likely revolutionize the study of SEKs in many epithelial organs, e.g., liver (Huch et al., 2015), small intestine, and colon (Sato & Clevers, 2013). Lastly, mice lacking all keratins have recently been generated, and cell cultures derived from these animals constitute an impressive tool for future research (Loschke, Seltmann, Bouameur, & Magin, 2015; Seltmann, Fritsch, Kas, & Magin, 2013).

To study the effects of specific human SEK variants or PTM alterations on *de novo* assembly of keratin networks, overexpression of keratins in cell lines that lack keratins has been helpful (see Sections 3.1). Generation of cells stably expressing keratin constructs has also been described, e.g., introduction of human K8 variants in HT-29 colon cancer cells (Zupancic et al., 2014). In addition, stable overexpression of fluorescently tagged SEKs allows for *in vivo* imaging of SEK dynamics (Windoffer & Leube, 2004) (see chapter MIE by Schwarz et al.). However, since a long-term overexpression may introduce artifacts, the use of inducible constructs should be considered for future studies (Wald, Oriolo, Casanova, & Salas, 2005). It is also important to acknowledge that some epithelial cells lines may express only low levels of keratins, particularly if cells have undergone epithelial–mesenchymal transition (EMT), where the mesenchymal IF protein vimentin replaces keratins, e.g., in invasive breast cancer cell lines (Iyer et al., 2013). Finally, some epithelial cell lines may even contain SEK nucleotide deletions or insertions (Kwan, Looi, & Omary, 2015).

3.1 Overexpression of SEKs in Cells

Since keratins are obligate heteropolymers, overexpression of SEKs in cells using SEK-containing plasmids (Table 3) requires transfection of equal amounts of type I and type II keratin DNA. For example, to study the effect of human liver disease keratin variant K8G62C, cells must be transfected with the type II K8G62C together with a type I keratin, such as K18 (Ku et al., 2005). Many overexpression studies utilized transient transfection of BHK-21 (baby hamster kidney cells), NIH-3T3 fibroblasts, or CHO (Chinese hamster ovary) cells (Ku & Omary, 2000; Ku, Azhar, et al., 2002;

Table 3 Overexpression Plasmids for Nontagged SEK Expression
A. Wild-Type SEK Constructs

Gene Insert	Construct Name	References
Type I		
KRT18	human K18	Ku and Omary (1994)
KRT18	Keratin 18 (pcDNA3)	Ku and Omary (1994)
KRT18	pGC1853	Kulesh and Oshima (1988)
KRT18 (m)	PUC9B71A*	Singer, Trevor, and Oshima (1986)
KRT19	human K19	Zhou, Liao, Hu, Feng, and Omary (1999)
KRT20	pcDNA3.1-K20	Zhou et al. (2003)
Type II		
KRT8	Keratin 8 (pcDNA3)	Ku, Azhar, Omary (2002)
KRT8	human K8	Ku, Azhar, et al. (2002), and Kulesh, Cecena, Darmon, Vasseur, and Oshima (1989)
KRT8	pK812	Kulesh et al. (1989)

B. Human Constructs for SEK Variants

Variant	Disease Context	Phenotype of the Variant	References
Type I			
KRT18 R90C	Epidermal hot spot mutation	Filament collapse	Ku and Omary (1994)
KRT18 H128L	Cryptogenic liver cirrhosis	Altered *in vitro* filament assembly	Ku, Wright, Terrault, Gish, and Omary (1997)
KRT18 S230T	IBD, liver disease	Increased permeability in colon cell culture	Zupancic et al. (2014)
KRT20 R80H	Epidermal hot spot mutation	Filament collapse	Zhou et al. (2003)

Continued

Table 3 Overexpression Plasmids for Nontagged SEK Expression—cont'd
B. Human Constructs for SEK Variants

Variant	Disease Context	Phenotype of the Variant	References
Type II			
KRT8 Y54H	Various diseases	Disruption of keratin network under stress conditions	Ku, Gish, Wright, and Omary (2001)
KRT8 G62C	Various diseases	Predisposing to liver injury and apoptosis Keratin crosslinking	Ku and Omary (2006)
KRT8 R341H	Liver disease	Predisposing to acetaminophen-induced liver injury and keratin network disruption	Guldiken, Zhou et al. (2015)
KRT8 R454C	Liver disease	Keratin crosslinking	Ku et al. (2005)
KRT8 G434S	Liver disease	Decreased phosphorylation at S432	Ku et al. (2005)
KRT8 K464N	IBD variant	Increased permeability in colon cell culture	Zupancic et al. (2014)

(A) The table presents mammalian untagged expression vectors of wild-type SEKs (except for the P-UC9B71A* vector, which is used for bacterial expression) that are commercially available through Addgene (www.addgene.org). For fluorescently tagged plasmids, please see Addgene or MIE chapter by Schwarz et al. (B) The table presents human constructs for SEK disease variants. Note that the amino acid number listed reflects the current nomenclature including the first amino acid methionine, while the construct/variant name on vendor webpages and in original literature may reflect the old nomenclature which is the amino acid number −1. The phenotype of the corresponding human variant is indicated. KRT, keratin gene; m, mouse; A, Alanine; C, Cysteine; G, Glycine; H, Histidine; K, Lysine; L, Leucine; N, Aspargine; R, Arginine; S, Serine; T, Threonine; Y, Tyrosine. IBD, inflammatory bowel disease. For constructs related to PTMs, please see MIE chapter by Snider and Omary and www.addgene.org.

Weerasinghe et al., 2014). BHK cells display high transfection efficiency and are therefore useful for biochemical assays, while NIH-3T3 fibroblasts have been preferred for morphological analyses of SEK networks (Ku, Azhar, et al., 2002). Of note, fibroblasts do not basally express epithelial-specific proteins and are not polarized like glandular simple-type epithelial cells, thus limiting their usefulness for studies on epithelial biology. To exclude potential confounding effects of other IF proteins, IF KO cell lines can be used, such as mouse embryonic fibroblasts isolated from vimentin-null mice

(MEFvimKO) (Hyder et al., 2015) or keratin-free keratinocytes from animals lacking the type II keratin cluster (Vijayaraj et al., 2009).

3.1.1 Protocol for K8 and K18 Overexpression in Cultured Cells

Here, we describe a protocol for K8 + K18 overexpression in MEFvimKO cells. In this particular cell line, electroporation-based transfection provides good results with 50–60% transfection efficiency, while Lipofectamin 3000 resulted in 20–30% efficiency. The best mode of transfection varies based on the employed cell line and should be tested.

1. Cells are grown in a 100 × 100 mm cell culture dish in DMEM culture media (Sigma-Aldrich, cat no. D6171) supplemented with 10% fetal bovine serum (FBS), 2 m*M* L-glutamine, and 100 U/ml penicillin/streptomycin (Supplemented DMEM) at 37 °C in 95% air and 5% CO_2 atmosphere.
2. Cells are grown until they reach 80% confluence, then the medium is removed and cells are washed three times with sterile 1 × phosphate buffer saline (PBS; Biowest, cat no. L0615). Cells are detached with 2 ml trypsin containing 0.25% ethylenediaminetetraacetic acid (EDTA) (Biowest, cat no. L0931) at 37 °C for 5 min in the incubator and are collected into 10 ml of supplemented DMEM.
3. Cells are centrifuged at 1000 rpm for 5 min. After centrifugation, cells are resuspended in 400 μl Opti-MEM I Reduced-Serum Medium (Life Technologies, cat no. 11058-021) and placed in an electroporation cuvette (BTX, cat no. 45-0126) where 10 μg of the plasmids, for example, [wild-type (WT) K8 and WTK18] for normal filaments, [K8S74A+WTK18] or [WTK8 and K18R90C] for keratin variants has been placed. The plasmids are pipetted into different corners of the electroporation cuvette before addition of the cells. An empty vector should be used as a negative control.
4. The cells and plasmids are gently mixed by tapping the cuvette, while avoiding bubbles.
5. The cells with plasmids are electroporated with a Gene Pulser (Bio-Rad) at 975 μF and 260 V for 30–40 ms.
6. The electroporated cells are mixed in 13 ml Supplemented DMEM, and 2 ml/well is applied in a 6-well plate with cover slips for immunostaining purposes and incubated as described above for 24–48 h.
7. Keratin immunofluorescence staining (Fig. 1) or RNA studies can be performed. When larger quantities of material are needed (such as for biochemical analyses), cells should be suspended in 10 ml Supplemented DMEM in a 100 × 100 mm cell culture dish after electroporation.

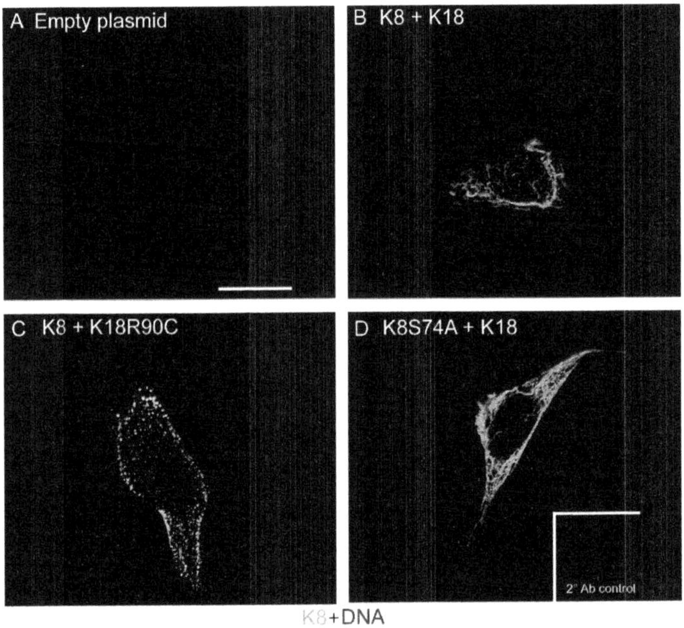

Figure 1 Overexpression of WT SEKs and SEK variants in cells. Immunofluorescence staining for K8 (Troma I, 1:1500, Developmental Studies Hybridoma Bank, University of Iowa, IA) was performed in MEFvimKO cells transfected with empty plasmid (A), WT K8/K18 (B) and K8/18 variants together with their WT partner keratin (C) and (D) as indicated in the figure. The transfection of cells was carried out via electroporation, and the cells were fixed in acetone at −20 °C for 10 min. DNA was stained with Draq5 (1:5000), and the cells were visualized with a Leica TCS SP5 confocal microscope. The insert in panel (D) shows signals from cells stained with secondary antibody and Draq5. Scale bar = 20 μm. (See the color plate.)

3.2 Downregulation of SEKs in Cells

Downregulation of SEKs in cells can be achieved through the use of keratin-specific short interfering RNAs (siRNAs) or short hairpin RNAs (shRNAs). siRNAs convey a fast, effective, and short-term gene silencing, while shRNA-mediated silencing is long-term and stable. In contrast to keratin overexpression, where a type I and type II keratin is needed for filament stability, it can be sufficient to specifically downregulate only one major SEK, such as type II K8 (Helenius et al., 2015), to also decrease partner keratin protein levels (Fig. 2). However, future studies are needed to fully uncover the consequences of the resulting "dysregulated" SEK production. Keratins have successfully been downregulated in a number of studies

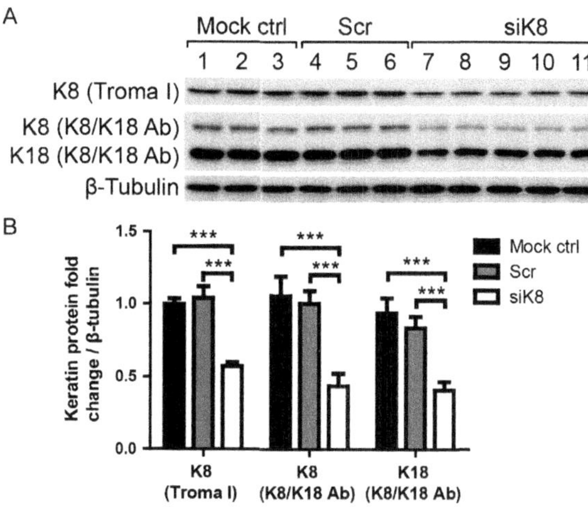

Figure 2 Downregulation of K8 in Caco-2 cells via K8 siRNA. Caco-2 cells were treated with K8 siRNA (siK8), scrambled negative control siRNA (Scr), or with no siRNA but with the transfection agents (Mock Ctrl) using Lipofectamine 2000 as described in the Section 3.1.1. Western blotting (A) was performed on total cell lysates collected 96 h after transfection. The following antibodies were used: K8 (antibody Troma I, Developmental Studies Hybridoma Bank, University of Iowa), K8/K18 (antibody K8/K18 (Toivola et al., 1997); upper band K8, lower band K18), and β-tubulin (antibody T8328, clone AA2, Sigma-Aldrich). The results were quantified by densitometry, and the level of downregulation was determined (B) as fold change keratin decrease normalized to β-tubulin levels, which were used as loading controls. ***$p < 0.001$.

through both siRNA- and shRNA-mediated gene silencing: K7 (Hong, Jiang, Kim, Li, & Lee, 2014; Sano et al., 2010), K8 (Alam et al., 2011; Fortier et al., 2013; Helenius et al., 2015; Mashukova et al., 2009; Uchiumi, Yamashita, & Katagata, 2012), K18 (Duan et al., 2009; Helenius et al., 2015; Uchiumi et al., 2012), K19 (Ju et al., 2013; Uchiumi et al., 2012), K20 (Min et al., 2012), and K23 (Birkenkamp-Demtroder et al., 2013). SEK-specific gene silencing has successfully been used to confirm and to further extend the *in vivo* findings in an *in vitro* setting. For instance, Mashukova and coworkers used both different keratin-null mouse models and shRNA-mediated K8 knockdown in Caco-2 cells to demonstrate the importance of K8 in the regulation of atypical PKC expression and localization (Mashukova et al., 2009). Helenius et al. have reported that the loss of K8 leads to downregulation of key proteins in colon energy metabolism (e.g., MCT1 and HMGCS2), in colonocytes of K8KO mice as well as in K8-siRNA-treated Caco-2 cells (Helenius et al., 2015).

It is important to bear in mind that siRNA transfection is not trivial and requires optimization to obtain high transfection efficiencies. In that respect, viral constructs (e.g., of lenti- or adenoviral origin) represent one of the most efficient and versatile tools that can transduce nearly any mammalian cell type (Shearer & Saunders, 2015). The Clustered Regularly Interspaced Short Palindromic Repeats (CRISPR) Type II system (Larson et al., 2013) offers another recently established method to permanently disrupt genes by introduction of inserts or deletions; however, its usefulness for SEKs remains to be proven. When working with knockdown systems, it is often useful to reintroduce the missing SEK to determine whether the biological phenotype can be rescued. For example, reintroduction of K8 by retroviral vectors to K8KO hepatocytes abrogated their susceptibility to Fas-induced apoptosis (Gilbert, Loranger, Daigle, & Marceau, 2001).

3.2.1 Protocol for SEK siRNA Transfection with Lipofectamine 2000

A robust shRNA-mediated gene silencing of K8 was shown in Caco-2 cells with the K8 shRNA clone ID TRCN0000062384 from the TRC library database (Mashukova et al., 2009). Using this shRNA as a template, a K8 siRNA with the sequence 5′-GCCUCCUUCAUAGACAAGGUA (dTdT)-3′ was synthesized (siMAX siRNA, Eurofins Genomics). The siRNA was dissolved in 1× siMAX universal buffer prepared from sterile water and 5× siMAX universal buffer (30 m*M* HEPES, 100 m*M* KCl, 1 m*M* MgCl2, pH 7.3) stock solution (Eurofins Genomics) in order to obtain a 100 pmol/μl siRNA solution. The reconstituted K8 siRNA was introduced into Caco-2 cells with Lipofectamine 2000, resulting in a 40–60% decrease in protein levels of both K8 and the type I partners K18 (Fig. 2) and K19 (not shown). Controls include mock (all reagents except siRNA added) and negative (scrambled sequence of target siRNA sequence or a universal nontargeting sequence).

1. For a 24-well plate assay, Caco-2 colon cancer cells are plated in 500 μl of DMEM growth medium (Sigma-Aldrich, cat no. D6171; supplemented with 10% fetal calf serum (FCS), 2 m*M* L-glutamine, 100 U/ml penicillin/streptomycin) and grown at 37 °C in 95% air and 5% CO_2 atmosphere 1–2 days to obtain 30–40% confluency. On the day of transfection, the cells are briefly washed with PBS (Biowest, cat no. L0615) and fresh DMEM (with or without serum/antibiotics) is added.

2. K8 siRNA-Lipofectamine 2000 (Life Technologies, cat no. 11668) complexes for transfection are prepared. For one well in a 24-well plate, 50 pmol (0.5 μl of 100 pmol/μl siRNA stock solution) of K8 siRNA (siMAX siRNA, Eurofins Genomics) is added to 50 μl Opti-MEM I Reduced-Serum Medium (Life Technologies, cat no. 11058-021) in a sterile 1.5-ml Eppendorf tube. The solution is mixed gently by pipetting up and down or by tapping the tube and kept at room temperature until needed.
3. In another 1.5-ml Eppendorf tube, 1.5 μl Lipofectamine 2000 is combined with 50 μl Opti-MEM I Reduced-Serum Medium and mixed gently by pipetting the solution up and down or by tapping the tube. The mixture is incubated for 5 min at room temperature. The next step should be started within 25 min.
4. The siRNA solution prepared in step 2 is added to the Lipofectamine 2000 solution prepared in step 3, mixed gently, and incubated for 20 min at room temperature to allow siRNA-Lipofectamine 2000 complexes to form.
5. The siRNA-Lipofectamine 2000 complexes are added to each sample well and mixed gently by rocking the plate back and forth.
6. Cells are incubated for at least 72–96 h (as keratins have a long half-life). Medium may be replaced 4–6 h after transfection with normal DMEM growth medium with no loss of transfection activity.
7. The effect of gene silencing can be assayed by measuring SEK target gene protein (and mRNA) levels.

In-depth guidelines and tools for the design and selection of siRNAs and shRNAs have been described elsewhere. It is, however, worth mentioning here that it is vital to make sure that the siRNAs/shRNAs selected for SEK gene silencing are highly specific and effective and to identify potential off-target effects (Boudreau et al., 2013).

4. GENETIC MOUSE MODELS OF SEKs

4.1 SEK KO Mouse Models

For studying SEKs, mouse models (Table 4) have proven to be exceptionally important, as many SEK functions are not evident on cellular level. The K8KO mouse was the first SEK and IF KO and demonstrated that the function of SEKs may depend on genetic background. In that respect, K8KOs in the C57Bl/6 background display 94% embryonic lethality (Baribault et al., 1993), whereas only ~50% embryolethality is seen in the

Table 4 List of SEK Transgenic Mouse Strains

Keratin	Genetic Modification	Phenotype	References
K7	KO; C57Bl/6 background	Hyperproliferation of the bladder urothelium	Sandilands et al. (2013)
K8	KO C57Bl/6 background	Embryonic lethality (95%) due to internal bleeding and growth retardation	Baribault, Price, Miyai, and Oshima (1993)
K8*	KO	Embryonic lethality (50%). In survivors, mild hepatitis, increased susceptibility to liver toxins and injury; colorectal hyperproliferation, colitis and abnormal colonic energy metabolism, and intestinal protein targeting; abnormal insulin production, glucose signaling, thymus architecture, and susceptibility to apoptosis; compensatory K7 in colon brushborder cells	Alam et al. (2013), Ameen, Figueroa, and Salas (2001), Baribault, Penner, Iozzo, and Wilson-Heiner (1994), Habtezion, Toivola, Butcher, and Omary (2005), Helenius et al. (2015), Mathew, Loranger, Gilbert, Faure, and Marceau (2013), Odaka et al. (2013), Omary et al. (2009), Toivola et al. (1998), and Toivola et al. (2004)
K8	Mouse WT overexpression	Increased MDB formation upon MDB induction or aging; increased neoexpression of K19/K20 and K18 hyperphosphorylation in pancreatic acini	Nakamichi et al. (2005) and Toivola et al. (2008)
K8	Human WT overexpression	Pancreatitis; epidermal and hair follicle dysplasia; disrupted intestinal brushborder	Casanova et al. (1999), Casanova et al. (2004), and Wald et al. (2005)
K8	Human WT overexpression	Increased liver injury and MDB formation after feeding with high fat diet	Ku and Omary (2006), and Kucukoglu et al. (2014)

Table 4 List of SEK Transgenic Mouse Strains—cont'd

Keratin	Genetic Modification	Phenotype	References
K8	K8-YFP; C57Bl/6 background	Fluorescent YFP-K8 expression	Schwarz, Windoffer, Magin, and Leube (2015)
K8	Human G62C	Predisposition to liver toxins (Fas, microcystin-LR, acetaminophen) and inhibition of phosphorylation at K8 S74, reduced ability to form MDBs	Guldiken, Zhou, et al. (2015), Ku and Omary (2006), Kwan et al. (2012), and Omary et al. (2009)
K8	Human S74A	Predisposition liver injury and apoptosis; reduced ability to form MDBs	Ku and Omary (2006) and Kwan et al. (2012)
K8	Human R341C	Predisposition to hepatotoxicity by acetaminophen	Guldiken, Zhou, et al. (2015)
K8	Human R341H	Predisposition to hepatotoxicity by acetaminophen	Guldiken, Zhou, et al. (2015)
K18*	KO; mixed background	Mild liver injury, increased sensitivity to apoptosis and liver toxins; spontaneous MDB formation in old mice	Caulin, Ware, Magin, and Oshima (2000), Leifeld et al. (2009), and Magin et al. (1998)
K18	Human WT overexpression	Protected from MDB formation	Harada, Strnad, Resurreccion, Ku, and Omary (2007)
K8/K18	Mouse K8/ human K18 overexpression	Spontaneous exocrine pancreas injury	Nakamichi et al. (2005) and Toivola et al. (2008)
K18	Human R90C overexpression	Hepatitis; liver, pancreas, and intestinal keratin filament disruption; susceptibility to hepatotoxicity	Ku, Michie, Oshima, and Omary (1995), Ku et al. (1996), Omary et al. (2009), Strnad et al. (2008), Toivola et al. (1998), and Zhou et al. (2005)

Continued

Table 4 List of SEK Transgenic Mouse Strains—cont'd

Keratin	Genetic Modification	Phenotype	References
K18	Human S34A overexpression	Accumulation of abnormal hepatocyte mitotic bodies after partial hepatectomy	Ku, Michie, et al. (2002)
K18	Human S53A overexpression	Predisposition to griseofulvin- or microcystin-LR-induced hepatotoxic injury	Ku et al. (1998)
K18	Human S30/31/49A overexpression (K18-Gly(-))	Susceptible to induced apoptosis, liver, and pancreas injury	Ku, Toivola, Strnad, and Omary (2010)
K18	Human D238E and D397E overexpression	Increased FAS-mediated liver damage	Weerasinghe et al. (2014)
K19	KO	Mild myopathy, compensatory overexpression of K18 and K20 in gallbladder, attenuated ductal reaction but stronger cholestatic liver injury	Chen et al. (2015), Stone et al. (2007), Tamai et al. (2000), and Tao et al. (2003)
K19	Human WT overexpression	None	Bader and Franke (1990)
K20	Human WT overexpression	No phenotype but missing expression in pancreas and stomach	Zhou et al. (2003)
K20	Human R80H overexpression	Filament organization defect in enterocytes of small intestine partially rescued by WT human K18 overexpression	Zhou et al. (2003)
hK20/K18	Human K20 R80H/K18 R90C overexpression	Filament organization defect in enterocytes of small intestine partially rescued by WT human K18 or K20 overexpression	Zhou et al. (2003)

Table 4 List of SEK Transgenic Mouse Strains—cont'd

Keratin	Genetic Modification	Phenotype	References
K8/K19	Double KO; mixed background	Embryonic lethality due to defects in placenta	Tamai et al. (2000)
K18/K19	Double KO; mixed background	Embryonic lethality due to defects in placental trophoblast giant cells	Hesse, Franz, Tamai, Taketo, and Magin (2000)
K18/K19	K18 R90C overexpression in K19KO/K18KO mice	Embryonic lethality, which can be rescued when the mutation is replaced by one endogenous K18WT copy but still causing keratin aggregates	Hesse et al. (2007)
K type II	KO of the entire type II gene cluster; C57Bl/6 background	Embryonic lethality due to severe growth retardation	Vijayaraj et al. (2009)

Unless otherwise stated (C57Bl/6 or mixed background), mouse models were generated in the FVB/n background. * Indicates strains that are commercially available at Jackson laboratory (www.jax.org). Note that all original references and complete phenotypes have not been included due to space limitations. Note also that amino acids numbering includes the first methionine and may thus vary with +1 compared to the mouse name described in the original references. WT, wild type; KO, knockout.

FVB/n background (Baribault et al., 1994). However, these animals develop colitis, hyperplasia in cecum, colon and rectum, and mild hepatitis (Baribault et al., 1994; Habtezion et al., 2005; Loranger et al., 1997; Toivola et al., 2004). Another important feature that was uncovered in K8KOs, and subsequently confirmed in various SEK-transgenic strains, is their susceptibility to different hepatic insults (Ku et al., 2007; Toivola et al., 1998). K8KOs also harbor abnormalities in the endocrine pancreas (Alam et al., 2013), but no obvious phenotype in gallbladder and exocrine pancreas (Tao et al., 2003; Toivola et al., 2000), while other tissues were not studied in detail. Analysis of heterozygous K8 mice demonstrated that K8 protein levels decreased to 50%, which was not sufficient to fully maintain colon homeostasis (Asghar et al., 2015). Of note, the multitissue phenotype of K8KOs makes the interpretation of some of the data difficult, as e.g., chronic diarrhea seen in the animals may affect both their endocrine pancreas and liver given the importance of the gut–liver axis (Schnabl & Brenner, 2014). Accordingly, development of tissue-specific KOs is highly recommended to fully dissect the tissue-specific importance of SEKs. As the first such model, conditional type II keratin KO mice (that lack all type II keratins) have been generated (Vijayaraj et al., 2009) and used to study the importance of keratins in lung cancer (Konig et al., 2013).

Similar to K8KOs, K18KO mice develop a mild hepatitis, but no embryonic lethality and no obvious phenotype in the intestine (Magin et al., 1998; Toivola et al., 2001). As a potential explanation for this finding, loss of K18 is likely compensated by K19 in most tissues (except in hepatocytes), while loss of K8 cannot be offset given the fairly restricted expression of its "substitute" K7. An exception to this is the K8KO colonic crypts, where the K7 expression pattern is expanded to the topmost cells of the crypt, which normally do not express K7 (Asghar et al., 2015). In support of this hypothesis, a combined loss of both K18 and K19 results in early embryonic lethality (Hesse et al., 2000). In that respect, generation of tissue-specific double KO animals would provide a useful tool to dissect the redundancy of both keratins in different tissues. Further support for the redundancy of SEKs is that an ablation of less abundant keratins K7 and K19 results in no or only a saddled phenotype, respectively (Sandilands et al., 2013; Stone et al., 2007; Tamai et al., 2000). For the latter, K19KOs display a mild skeletal myopathy (Stone et al., 2007) and an attenuated ductal/ductular reaction that represents a specific form of hepatobiliary stress response (Chen et al., 2015). Last but not least, the double K18 R90C/K20 R80H mice show disrupted intestinal keratin filaments that are not seen in the single mutant mice when the partner WT protein is overexpressed (Zhou et al., 2003).

4.2 SEK WT Overexpressor Mice

In addition to KO animals, multiple strains overexpressing various SEKs have been generated (Table 4). These include two different lines overexpressing human K8 and one line overexpressing mouse K8 (Casanova et al., 1999; Ku & Omary, 2006; Nakamichi et al., 2005). While lower K8 overexpression levels were well tolerated under basal conditions, mice with higher levels of transgenic human K8 demonstrated progressive alterations in the exocrine pancreas (Casanova et al., 1999; Kucukoglu et al., 2014; Nakamichi et al., 2005). These data suggest that overexpression of keratins need to be used with caution since it may result in a spontaneous gain-of-function phenotype (Casanova et al., 1999; Kucukoglu et al., 2014; Nakamichi et al., 2005).

Even at lower levels, K8 overexpression predisposed animals to development of MDBs and high-fat-diet-induced liver injury (Casanova et al., 1999; Kucukoglu et al., 2014; Nakamichi et al., 2005). K18 overexpression seemed to be better tolerated; however, when combined with increased K8 production, it further increases keratin levels that likely precipitate the development of a spontaneous pancreatic phenotype (Toivola et al., 2008).

4.3 Transgenic Mice Overexpressing SEK Variants

Despite their limitations, mice overexpressing SEK variants have revealed important insights into SEK biology (Table 4). As the first one, animals overexpressing an epidermolysis bullosa simplex-like variant of K18 (K18 R90C) displayed a disruption of the hepatocellular keratin network, mild hepatitis, and marked susceptibility to various hepatic stresses, i.e., a phenotype similar to K18KOs (Ku et al., 2007; Toivola et al., 1998). An analogous mutation in K20 (K20 R80H) similarly led to keratin network disruption but no tissue damage (Zhou et al., 2003). To study the importance of SEK phosphorylation and glycosylation, several mouse lines were generated. K8 S74A mice were susceptible to Fas-induced hepatocellular apoptosis, K18 S34A displayed abnormal regeneration and mitotic bodies after partial hepatectomy, whereas K18 S53A mice were predisposed to microcystin-LR (i.e., hyperphosphorylation)-driven hepatocellular damage (Ku et al., 2007). Mice with largely abolished K18 glycosylation, due to mutation of three major glycosylation sites, were also described but did not have an obvious phenotype under basal conditions (Ku et al., 2010). On the other hand, they were more susceptible to hyperglycosylation-associated liver and pancreatic injury and apoptosis (Ku et al., 2010).

In the latest generation of SEK overexpressing animals, mice overexpressing commonly found variants of human K8, such as G62C or R341H, have been studied. These animals have no obvious phenotype under basal conditions but are predisposed to certain stresses, such as acetaminophen or Fas-induced liver injury (Guldiken, Zhou, et al., 2015; Ku & Omary, 2006). Since the liver represents the organ most obviously affected by SEK variants, several transgenic mice were preferentially analyzed for their hepatic phenotype and an analysis of additional organs may therefore reveal so far unknown phenotypes.

As mentioned above, overexpression of SEKs may lead to a gain-of-function phenotype, and therefore, a parallel analysis of nontransgenic mice as well as mice overexpressing similar levels of wild-type and modified SEKs is crucial. A more convenient and a less error-prone alternative may be the generation of knock-in mouse lines. Accordingly, K8-YFP knock-in mice have recently been described and offer a great opportunity to improve our understanding of SEK organization *in vivo* and to track K8-expressing cells (Schwarz et al., 2015).

5. MDBs—KERATIN AGGREGATES IN LIVER DISEASE

MDBs are protein aggregates that are found in specific liver disorders, such as ALD and NASH (Strnad et al., 2013). K8 and K18 are the major, specific constituents of MDBs and because of that MDBs represent the most prominent, disease-associated SEK aggregates (Strnad et al., 2013; Zatloukal et al., 2007). MDBs can be induced in mice, and the rodent model was instrumental in revealing MDB pathogenesis (Strnad et al., 2013; Zatloukal et al., 2007). In particular, it revealed a crucial role of K8/K18, their PTMs and stoichiometric ratio for the aggregate formation (Strnad et al., 2013). While both griseofulvin and 3,5-diethoxycarbonyl-1,4-dihydrocollidine (also termed as 1,4-dihydro-1,4,6-trimethyl-3,5-pyridinedicarboxylate; DDC) are known to induce MDBs, DDC is preferred in current studies because of its reproducible MDB formation and lower toxicity (Stumptner, Fuchsbichler, Lehner, Zatloukal, & Denk, 2001). One important caveat for both DDC and griseofulvin is that they do not mimic the histological features of NASH or ALD. Instead, they induce hepatic porphyria, sclerosing cholangitis, and biliary fibrosis with massive ductal cell proliferation (Fickert et al., 2007; Ku et al., 2007). MDBs are also formed in ferrochelatase KO mice that serve as a genetic model of human porphyria, in K18KOs and in K8-overexpressing mice, in particular after administration of high fat diet (Kucukoglu et al., 2014; Magin et al., 1998; Singla et al., 2012). The latter model is likely of human relevance given that excess K8 production has been found in ALD (Guldiken, Usachov, et al., 2015). In conclusion, DDC administration represents the gold standard of MDB induction in rodents, while the ferrochelatase KO mice, K8 overexpressors and K18KOs may yield important additional information. As a potential caveat, MDBs develop spontaneously in aging mouse K8-overexpressing mice (Nakamichi et al., 2005), whereas an additional physiologic stress seems to be needed to induce MDB formation in human K8-overexpressing mice (Kucukoglu et al., 2014).

5.1 Methods to Induce MDBs

To induce MDBs, DDC (0.1% w/w) can be either added to a powdered Formulab 5008 diet (Dean's Animal Feeds, Redwood City, CA) or it can be supplemented in form of commercially prepared pellets (Kucukoglu et al., 2014; Strnad et al., 2007). For the latter, the D12450B diet from Research Diets (New Brunswick, NJ) can be used. While both methods are suitable to induce MDBs, the latter offers an easier and more comfortable handling. We use DDC from Sigma-Aldrich (cat # 137030), and our preliminary data indicate that the source of DDC may affect the efficiency of MDB formation. The time needed for development of MDBs varies between 2 and 3.5 months (Strnad et al., 2007; Stumptner et al., 2001). This variability may be due to varying susceptibility to MDB induction in different genetic backgrounds (Hanada, Strnad, Brunt, & Omary, 2008) but also due to differences in experimental conditions including the fat content of the employed diet (Kucukoglu et al., 2014). Of note, MDBs form during a later stage of DDC intoxication, while the cholestatic injury peaks around week 4 and decreases thereafter (Fickert et al., 2007). Another important feature of MDBs is their reversibility; MDBs largely disappear when mice recover on a DDC-free diet for 4 weeks. The resulting animals are termed "primed mice" and are highly susceptible to MDB reinduction. Formation of MDBs in primed mice only takes 5–7 days and can be induced by a variety of unspecific insults including heat shock, common bile duct ligation, or treatment with, among others, thioacetamide, ethanol, and okadaic acid (Zatloukal et al., 2007).

In addition to mouse models, cellular approaches have been used to study MDB pathogenesis. Among them, isolation of primary hepatocytes from DDC-primed mice (Wu et al., 2005) represents the most established technique. In this model, MDBs reform spontaneously within a few days in culture, and the efficiency of their formation is highest when laminin-coated plates are used (Wu et al., 2005). When coupled with knockdown of a gene of interest, this method may be useful for screening of factors involved in MDB pathogenesis. However, future studies that should include a subsequent validation in an *in vivo* mouse model are needed to confirm the usefulness of this approach. As another potentially interesting research tool, overexpression of K8 or K18 together with p62 and ubiquitin was shown to induce MDB formation in cell culture (Stumptner, Fuchsbichler, Zatloukal, & Denk, 2007). Despite these attractive preliminary data, cell culture-based MDB models are still not well established and their similarity to *in vivo* MDBs is not clear.

5.2 Imaging and Quantification of MDBs

Keratin network alterations observed in human liver disease and in corresponding experimental models can be divided into two major subtypes: (i) characteristic, large MDBs that are irregularly shaped, often exceed the size of the nucleus, and are found in the hepatocyte perinuclear area and (ii) "empty cells" that largely lack any keratin staining but may contain small aggregates that are often located in the cell periphery (Molnar, Haybaeck, Lackner, & Strnad, 2011; Strnad et al., 2013; Fig. 3). Additionally, cells with partially disrupted SEK networks and small aggregates are occasionally seen and likely represent an early step of the MDB formation process (Molnar et al., 2011; Zatloukal et al., 2007).

Recent studies suggest that a predominant formation of "large MDBs" versus "empty cells" might be influenced by genetic background and/or type of employed stress (Hanada et al., 2008; Kucukoglu et al., 2014). While there are no established standards for categorizing different forms of MDB-associated keratin network remodeling, previously published studies may provide some guidance (Hanada et al., 2008; Kucukoglu et al., 2014). These studies defined large MDBs either as aggregates of >5 μm in diameter or inclusions exceeding half the size of a nucleus. Both the amount of cells with large MDBs and either the number of empty cells or cells with visible disruptions in SEK network have been analyzed, the latter constituted the "total MDB count." Quantitative (i.e., number of cells per microscope field) or semiquantitative analyses performed with the help of 10 or 20 magnification lens were described (Hanada et al., 2008; Kucukoglu et al., 2014). Given the focal distribution of MDBs, at least 5 fields at 10 magnification or 20 fields at 20 magnification should be analyzed per specimen.

Histological staining with hematoxylin and eosin (H&E) represents the most widely used technique to detect MDBs, particularly in the clinic, but is of limited sensitivity and specificity (Strnad et al., 2013; Zatloukal et al., 2007). Alternative histological staining methods may provide a better MDB resolution; however, they have not been systematically analyzed (Strnad et al., 2013; Zatloukal et al., 2007). In the research setting, immunohistochemistry and/or immunofluorescence labeling constitutes the best validated method for MDB detection and allows discrimination between MDB subtypes. In our hands, a double labeling of the abundant MDB components, K8/K18 and p62 (Fig. 3), yields the most consistent results (see Table 5 for details). The use of double labeling is crucial and allows to distinguish MDBs from various p62 aggregates such as intracellular hyaline

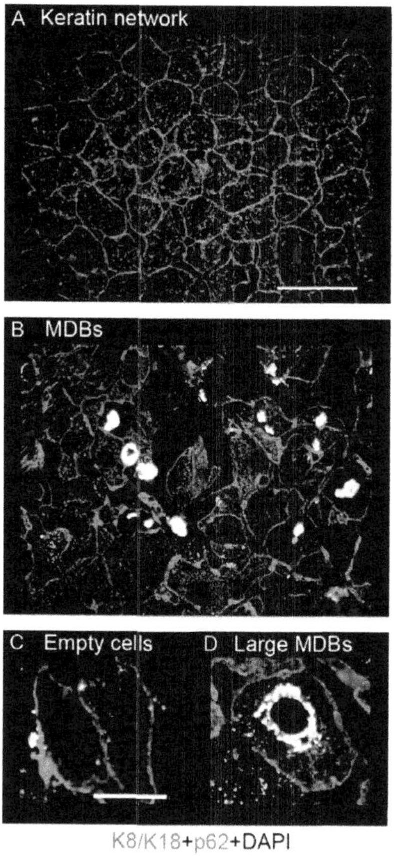

Figure 3 Keratin network reorganization in a mouse model of Mallory–Denk body (MDB) formation. Double immunofluorescence staining for K8/K18 (antibody 8592-red) and p62 (antibody GP62-C, Progen; green; see Table 5 for details) depicts the normal appearance of the SEK network in untreated C57BL/6 mice (A), as well as the SEK reorganization in animals fed with 0.1% 3,5-diethoxycarbonyl-1,4-dihydrocollidine (DDC) for 10 weeks (B)–(D). Note that some cells display only residual nonfilamentous K8/K18 signal in the cell periphery (empty cells; C); while others contain large MDBs that appear as yellow inclusions due to the double-labeling with K8/K18 and p62 (D). ProLong Gold mounting medium containing DAPI (Invitrogen) was used to visualize the nuclei. Scale bars: A,B = 50 μm; C,D = 20 μm. (See the color plate.)

bodies or DDC-induced p62 accumulations (Zatloukal et al., 2007). On the other hand, small keratin aggregates occurring during apoptosis are p62 negative (Zatloukal et al., 2007). Other potentially useful tools include luminescent conjugated oligothiophenes, which recognize cross ß-sheet protein

Table 5 Overview of Antibodies and Markers Used for Visualization of Mallory–Denk Bodies (MDBs)

Detected Proteins	Antibody/ Marker Name	References	Applications
K8/K18	**8592**[a]	Strnad et al. (2007)	IF-A
	50K160[b]	Fickert et al. (2002) and Stumptner, Fuchsbichler, Heid, Zatloukal, and Denk (2002)	IF-A, IHC
K8	M20[a]	Li et al. (2008) Fitzgerald Industries, Cat No 10R-2286	IF-A, IF-P
	Ks 8.7[b]	Stumptner et al. (2002) and Zatloukal et al. (2002) Progen Cat No 61038	IF-A
	Troma I[a]	Yuan, Marceau, French, Fu, and French (1996) Developmental Studies Hybridoma Bank Cat No AB_531826	IF-A
K18	Ks18.04[b]	Stumptner et al. (2002) and Zatloukal et al. (2002) Progen Cat No 61028	IF-A
Ubiquitin	662099[a]	Bardag-Gorce et al. (2008) Calbiochem Inc Cat No 662099	IF-A
	Z0458[b]	Stumptner et al. (2002) Dako Cat No Z0458	IF-A,IHC
	P4D1[a]	Strnad et al. (2007) Santa Cruz Cat No sc-8017	IF-A

Table 5 Overview of Antibodies and Markers Used for Visualization of Mallory-Denk Bodies (MDBs)—cont'd

Detected Proteins	Antibody/ Marker Name	References	Applications
p62 (sequestosome 1)	**P62CT**[b]	Haybaeck et al. (2012) and Zatloukal et al. (2002) Progen Cat no GP62-C	IF-A, IHC
High-molecular weight nonkeratin component	**M_M120–1**[b]	Zatloukal et al. (1990)	IF-A, IHC
ß-Sheet containing misfolded proteins	p-HTAA[c] p-FTAA[c] (luminescent-conjugated oligothiophenes (LCOs))	Mahajan et al. (2011)	Direct labeling of sections without fixation or of acetone/ paraformaldehyde fixed sections

[a]Antibodies that are mostly used for MDB detection in mice.
[b]Antibodies that are used for detection of both mouse and human MDBs.
[c]Reagents that label both mouse and human MDBs.
Antibodies that are most frequently used for MDB detection are highlighted in bold. IF-A, immunofluorescence on frozen, acetone-fixed sections; IF-P, immunofluorescence on frozen, paraformaldehyde-fixed sections; IHC, immunohistochemistry.

structures, and antibodies against ubiquitin (Mahajan et al., 2011; Zatloukal et al., 2004). Of note, MDB scores obtained by histological and immunological staining may vary substantially, and histological scores may reflect the amount of large MDBs (Hanada et al., 2008). Electron microscopy represents a well-established method for analysis of MDB ultrastructure, allowing their further subclassification (Zatloukal et al., 2007). Consequently, for reliable evaluation of whether an unknown structure represents an MDB or not, we recommend that both double immunofluorescence staining and electron microscopy are performed.

5.3 Biochemical Analysis of MDBs

Biochemical analysis constitutes another important aspect of MDB research. MDBs can be isolated by a combination of sucrose gradient centrifugation and subsequent isolation of insoluble proteins, as described in detail previously (Zatloukal et al., 2004). Given the rapidly expanding knowledge about keratin PTMs (Snider & Omary, 2014), careful consideration should be given regarding addition of appropriate inhibitors to the isolation buffers in order to maintain *in vivo* conditions. FACS-based MDB isolation represents another attractive tool to obtain pure aggregates; however, it has not been routinely used (Zatloukal, Bock, Rainer, Denk, & Weber, 1991). Lastly, the amount of p62, high-molecular weight K8 or ubiquitinated proteins (for list of useful antibodies, see Table 5) within the insoluble protein fraction, determined via immunoblotting, are increasingly being used as a tool to quantify the extent of MDB formation (Strnad et al., 2007).

6. SEK VARIANT DETECTION IN HUMANS

Inherited variations in keratins have been linked to at least 60 different human diseases (Toivola et al., 2015). Among them, K8/K18/K19 variants predispose to development and adverse outcomes of various liver disorders, while the involvement of SEK variants in other diseases is not as well documented (Omary, 2009; Toivola et al., 2015). While multiple well-established methods may be used for detection of genetic variants, several potential pitfalls need to be taken into consideration. Up to date, SEK variants have been mostly analyzed by a combination of denaturing high-performance liquid chromatography (DHPLC) and subsequent sequencing of preselected samples (Strnad et al., 2010). However, DHPLC needs to be used with caution, and inclusion of positive controls (i.e., samples with a known variant) is highly recommended to determine the appropriate analysis conditions, particularly the correct denaturing temperature (Strnad et al., 2006). Of note, DHPLC is well suited for detection of heterozygous, but not homozygous variants (Strnad et al., 2006). For the latter, samples can be mixed with a known WT sequence at an equimolar ratio and denatured to allow formation of heteroduplexes (Strnad et al., 2006).

It is also important to keep in mind that a direct comparison of results obtained by different analytical methods may be misleading, as these methods may differ in their sensitivity/specificity. Another major caveat is the selection of an appropriate control group. In that respect, SEK variants have been shown to display racial and ethnic variations, and their frequencies may differ even among different nationalities of the same ethnicity (Strnad et al., 2010; Treiber et al., 2006). This problem is particularly relevant since the amount of control subjects that have been analyzed for presence of SEK variants is rather limited. Publicly available databases such as Exome Variant Server (http://evs.gs.washington.edu/EVS/) or the 1000 Genomes Project data (http://browser.1000genomes.org/index.html) provide help in this regard (Usachov et al., 2015).

Genome-wide association studies and next-generation sequencing represent other useful tools to evaluate the importance of SEK variants in human disease, for instance, as shown for K5 G138E variant in basal cell carcinoma (Stacey et al., 2009). Unfortunately, the currently used SNP chips, such as Affymetrix 6.0 SNP array, Illumina Human 1M Duo-BeadChip, Human CNV370-QuadV3, and Illumina Human610-Quad v1.0 BeadChip, do not contain the common SEK variants. To check whether a particular SEK variant is available on a chip of interest, one can take advantage of public databases such as http://gvs.gs.washington.edu/GVS138 and http://www.gwascentral.org.

7. CONCLUSIONS

In the present review, we summarize the current approaches to isolate SEKs and tools to study them in cell culture and transgenic rodent models. In addition, approaches to study the reorganization of SEK networks into hepatic MDBs and to detect SEK variants in human disease patients are also addressed. This potpourri of methods reflects the versatile functions of SEKs and their broad importance in human disease.

Despite the great progress during the last decade, there is still a long way to go. While we have learned a great deal about K8/K18 as the primary SEKs, much less in known about the less abundant SEK family members, in particular K23, for which the tissue expression pattern and biological function remains almost completely unknown. In addition, the importance of K8/K18 in tissues other than the liver needs to be further explored, and the findings from cell culture and animal models should spur further studies addressing the role of SEKs in human disease.

While the relative insolubility of SEKs represents a challenge for the biochemical studies of interaction with other proteins, several new techniques such as *in situ* proximity ligation assay or FRET may help to circumvent this problem. An ongoing improvement in proteomic approaches should lead to a better appreciation of SEKs as stress-inducible proteins rather than simple contaminants. Genetic tools that allow manipulation of SEKs in a time- and cell-type-specific manner should help us to dissect the importance of SEKs in complex processes, including tumor development and metastasis.

ACKNOWLEDGMENTS

Many of the techniques described here were adopted by P.S. and D.M.T. during the postdoctoral training in the laboratory of Bishr Omary at Stanford University and University of Michigan. Our work was supported by the COST Action BM1002 Nanonet; German Research Foundation Grants STR 1095/4-1, 1095/5-1 and SFB TRR57, the Interdisciplinary Center for Clinical Research (IZKF) in Aachen (to P.S.), the Academy of Finland #140759/12616, Sigrid Juselius Foundation, EU FP7 IRG KIFREO, ÅAU Center of Excellence of Cell Stress and Molecular Aging, Finnish Diabetes Research Foundation, Diabetes Wellness, and Liv och Hälsa foundation (to D.M.T.), Victoriastiftelsen-foundation (to T.O.H.), Swedish Cultural Foundation (to I.A.K.L. and J.H.N.), Turku Doctoral Programme in Molecular Biosciences at ÅAU (to I.A.K.L. and J.S.G.S.), Turku Doctoral Program of Biomedical Sciences, Åbo Akademi University foundation and Oskar Öflund's Foundation (to J.O.M.), and Makarna Agneta and Carl-Erik Olins foundation (J.H.N.) are acknowledged for financial support. The funding sources were not involved in the study design; the collection, analysis, and interpretation of data, writing of the report, nor in the decision to submit the article for publication. We thank all the scientists who have contributed to the work summarized here and apologize for the studies we have not been able to include due to space limitations.

Conflict of interest: The authors have no competing interests to disclose. No writing assistance was utilized in the production of this manuscript.

REFERENCES

Alam, H., Gangadaran, P., Bhate, A. V., Chaukar, D. A., Sawant, S. S., Tiwari, R., et al. (2011). Loss of keratin 8 phosphorylation leads to increased tumor progression and correlates with clinico-pathological parameters of OSCC patients. *PLoS One*, *6*(11), e27767.

Alam, C. M., Silvander, J. S., Daniel, E. N., Tao, G. Z., Kvarnstrom, S. M., Alam, P., et al. (2013). Keratin 8 modulates beta-cell stress responses and normoglycaemia. *Journal of Cell Science*, *126*(Pt. 24), 5635–5644.

Ameen, N. A., Figueroa, Y., & Salas, P. J. (2001). Anomalous apical plasma membrane phenotype in CK8-deficient mice indicates a novel role for intermediate filaments in the polarization of simple epithelia. *Journal of Cell Science*, *114*(Pt. 3), 563–575.

Asghar, M. N., Silvander, J. S., Helenius, T. O., Lahdeniemi, I. A., Alam, C., Fortelius, L. E., et al. (2015). The amount of keratins matters for stress protection of the colonic epithelium. *PLoS One*, *10*(5), e0127436.

Bader, B. L., & Franke, W. W. (1990). Cell type-specific and efficient synthesis of human cytokeratin 19 in transgenic mice. *Differentiation; Research in Biological Diversity*, *45*(2), 109–118.

Bardag-Gorce, F., Oliva, J., Villegas, J., Fraley, S., Amidi, F., Li, J., et al. (2008). Epigenetic mechanisms regulate mallory denk body formation in the livers of drug-primed mice. *Experimental and Molecular Pathology*, *84*(2), 113–121.

Baribault, H., Penner, J., Iozzo, R. V., & Wilson-Heiner, M. (1994). Colorectal hyperplasia and inflammation in keratin 8-deficient FVB/N mice. *Genes & Development*, *8*(24), 2964–2973.

Baribault, H., Price, J., Miyai, K., & Oshima, R. G. (1993). Mid-gestational lethality in mice lacking keratin 8. *Genes & Development*, 7(7A), 1191–1202.

Birkenkamp-Demtroder, K., Hahn, S. A., Mansilla, F., Thorsen, K., Maghnouj, A., Christensen, R., et al. (2013). Keratin23 (KRT23) knockdown decreases proliferation and affects the DNA damage response of colon cancer cells. *PLoS One*, *8*(9), e73593.

Boudreau, R. L., Spengler, R. M., Hylock, R. H., Kusenda, B. J., Davis, H. A., Eichmann, D. A., et al. (2013). siSPOTR: A tool for designing highly specific and potent siRNAs for human and mouse. *Nucleic Acids Research*, *41*(1), e9.

Casanova, M. L., Bravo, A., Martinez-Palacio, J., Fernandez-Acenero, M. J., Villanueva, C., Larcher, F., et al. (2004). Epidermal abnormalities and increased malignancy of skin tumors in human epidermal keratin 8-expressing transgenic mice. *FASEB Journal: Official Publication of the Federation of American Societies for Experimental Biology*, *18*(13), 1556–1558.

Casanova, M. L., Bravo, A., Ramirez, A., Morreale de Escobar, G., Were, F., Merlino, G., et al. (1999). Exocrine pancreatic disorders in transsgenic mice expressing human keratin 8. *The Journal of Clinical Investigation*, *103*(11), 1587–1595.

Caulin, C., Ware, C. F., Magin, T. M., & Oshima, R. G. (2000). Keratin-dependent, epithelial resistance to tumor necrosis factor-induced apoptosis. *The Journal of Cell Biology*, *149*(1), 17–22.

Chen, Y., Guldiken, N., Spurny, M., Mohammed, H. H., Haybaeck, J., Pollheimer, M. J., et al. (2015). Loss of keratin 19 favours the development of cholestatic liver disease through decreased ductular reaction. *The Journal of Pathology*, in press.

Coulombe, P. A., & Fuchs, E. (1990). Elucidating the early stages of keratin filament assembly. *The Journal of Cell Biology*, *111*(1), 153–169.

Duan, S., Yao, Z., Zhu, Y., Wang, G., Hou, D., Wen, L., et al. (2009). The Pirh2-keratin 8/18 interaction modulates the cellular distribution of mitochondria and UV-induced apoptosis. *Cell Death and Differentiation*, *16*(6), 826–837.

Eichner, R., & Kahn, M. (1990). Differential extraction of keratin subunits and filaments from normal human epidermis. *The Journal of Cell Biology*, *110*(4), 1149–1168.

Fickert, P., Stoger, U., Fuchsbichler, A., Moustafa, T., Marschall, H. U., Weiglein, A. H., et al. (2007). A new xenobiotic-induced mouse model of sclerosing cholangitis and biliary fibrosis. *The American Journal of Pathology*, *171*(2), 525–536.

Fickert, P., Trauner, M., Fuchsbichler, A., Stumptner, C., Zatloukal, K., & Denk, H. (2002). Bile acid-induced mallory body formation in drug-primed mouse liver. *The American Journal of Pathology*, *161*(6), 2019–2026.

Fortier, A. M., Asselin, E., & Cadrin, M. (2013). Keratin 8 and 18 loss in epithelial cancer cells increases collective cell migration and cisplatin sensitivity through claudin1 up-regulation. *The Journal of Biological Chemistry*, *288*(16), 11555–11571.

Fung, K. Y., Cosgrove, L., Lockett, T., Head, R., & Topping, D. L. (2012). A review of the potential mechanisms for the lowering of colorectal oncogenesis by butyrate. *The British Journal of Nutrition*, *108*(5), 820–831.

Gilbert, S., Loranger, A., Daigle, N., & Marceau, N. (2001). Simple epithelium keratins 8 and 18 provide resistance to fas-mediated apoptosis. Tprotection occurs through a receptor-targeting modulation. *The Journal of Cell Biology*, *154*(4), 763–773.

Guldiken, N., Usachov, V., Levada, K., Trautwein, C., Ziol, M., Nahon, P., et al. (2015). Keratins 8 and 18 are type II acute-phase responsive genes overexpressed in human liver disease. *Liver International: Official Journal of the International Association for the Study of the Liver*, *35*(4), 1203–1212.

Guldiken, N., Zhou, Q., Kucukoglu, O., Rehm, M., Levada, K., Gross, A., et al. (2015). Human keratin 8 variants promote mouse acetaminophen hepatotoxicity coupled with c-jun amino-terminal kinase activation and protein adduct formation. *Hepatology*, *62*, 876–886.

Habtezion, A., Toivola, D. M., Butcher, E. C., & Omary, M. B. (2005). Keratin-8-deficient mice develop chronic spontaneous Th2 colitis amenable to antibiotic treatment. *Journal of Cell Science, 118*(Pt. 9), 1971–1980.

Hanada, S., Strnad, P., Brunt, E. M., & Omary, M. B. (2008). The genetic background modulates susceptibility to mouse liver mallory-denk body formation and liver injury. *Hepatology (Baltimore, Md.), 48*(3), 943–952.

Harada, M., Strnad, P., Resurreccion, E. Z., Ku, N. O., & Omary, M. B. (2007). Keratin 18 overexpression but not phosphorylation or filament organization blocks mouse mallory body formation. *Hepatology (Baltimore, Md.), 45*(1), 88–96.

Hatzfeld, M., & Weber, K. (1990). The coiled coil of in vitro assembled keratin filaments is a heterodimer of type I and II keratins: Use of site-specific mutagenesis and recombinant protein expression. *The Journal of Cell Biology, 110*(4), 1199–1210.

Haybaeck, J., Stumptner, C., Thueringer, A., Kolbe, T., Magin, T. M., Hesse, M., et al. (2012). Genetic background effects of keratin 8 and 18 in a DDC-induced hepatotoxicity and mallory-denk body formation mouse model. *Laboratory Investigation; a Journal of Technical Methods and Pathology, 92*(6), 857–867.

Helenius, T. O., Misiorek, J. O., Nystrom, J. H., Fortelius, L. E., Habtezion, A., Liao, J., et al. (2015). Keratin 8 absence down-regulates colonocyte HMGCS2 and modulates colonic ketogenesis and energy metabolism. *Molecular Biology of the Cell, 26*(12), 2298–2310.

Herrmann, H., Kreplak, L., & Aebi, U. (2004). Isolation, characterization, and in vitro assembly of intermediate filaments. *Methods in Cell Biology, 78*, 3–24.

Hesse, M., Franz, T., Tamai, Y., Taketo, M. M., & Magin, T. M. (2000). Targeted deletion of keratins 18 and 19 leads to trophoblast fragility and early embryonic lethality. *The EMBO Journal, 19*(19), 5060–5070.

Hesse, M., Grund, C., Herrmann, H., Brohl, D., Franz, T., Omary, M. B., et al. (2007). A mutation of keratin 18 within the coil 1A consensus motif causes widespread keratin aggregation but cell type-restricted lethality in mice. *Experimental Cell Research, 313*(14), 3127–3140.

Hong, S. W., Jiang, Y., Kim, S., Li, C. J., & Lee, D. K. (2014). Target gene abundance contributes to the efficiency of siRNA-mediated gene silencing. *Nucleic Acid Therapeutics, 24*(3), 192–198.

Huch, M., Gehart, H., van Boxtel, R., Hamer, K., Blokzijl, F., Verstegen, M. M., et al. (2015). Long-term culture of genome-stable bipotent stem cells from adult human liver. *Cell, 160*(1–2), 299–312.

Hyder, C. L., Kemppainen, K., Isoniemi, K. O., Imanishi, S. Y., Goto, H., Inagaki, M., et al. (2015). Sphingolipids inhibit vimentin-dependent cell migration. *Journal of Cell Science, 128*(11), 2057–2069.

Iyer, S. V., Dange, P. P., Alam, H., Sawant, S. S., Ingle, A. D., Borges, A. M., et al. (2013). Understanding the role of keratins 8 and 18 in neoplastic potential of breast cancer derived cell lines. *PLoS One, 8*(1), e53532.

Ju, J. H., Yang, W., Lee, K. M., Oh, S., Nam, K., Shim, S., et al. (2013). Regulation of cell proliferation and migration by keratin19-induced nuclear import of early growth response-1 in breast cancer cells. *Clinical Cancer Research: An Official Journal of the American Association for Cancer Research, 19*(16), 4335–4346.

Konig, K., Meder, L., Kroger, C., Diehl, L., Florin, A., Rommerscheidt-Fuss, U., et al. (2013). Loss of the keratin cytoskeleton is not sufficient to induce epithelial mesenchymal transition in a novel KRAS driven sporadic lung cancer mouse model. *PLoS One, 8*(3), e57996.

Ku, N. O., Azhar, S., & Omary, M. B. (2002). Keratin 8 phosphorylation by p38 kinase regulates cellular keratin filament reorganization: Modulation by a keratin 1-like disease causing mutation. *The Journal of Biological Chemistry, 277*(13), 10775–10782.

Ku, N. O., Gish, R., Wright, T. L., & Omary, M. B. (2001). Keratin 8 mutations in patients with cryptogenic liver disease. *The New England Journal of Medicine, 344*(21), 1580–1587.

Ku, N. O., Lim, J. K., Krams, S. M., Esquivel, C. O., Keeffe, E. B., Wright, T. L., et al. (2005). Keratins as susceptibility genes for end-stage liver disease. *Gastroenterology, 129*(3), 885–893.

Ku, N. O., Michie, S., Oshima, R. G., & Omary, M. B. (1995). Chronic hepatitis, hepatocyte fragility, and increased soluble phosphoglycokeratins in transgenic mice expressing a keratin 18 conserved arginine mutant. *The Journal of Cell Biology, 131*(5), 1303–1314.

Ku, N. O., Michie, S., Resurreccion, E. Z., Broome, R. L., & Omary, M. B. (2002). Keratin binding to 14-3-3 proteins modulates keratin filaments and hepatocyte mitotic progression. *Proceedings of the National Academy of Sciences of the United States of America, 99*(7), 4373–4378.

Ku, N. O., Michie, S. A., Soetikno, R. M., Resurreccion, E. Z., Broome, R. L., & Omary, M. B. (1998). Mutation of a major keratin phosphorylation site predisposes to hepatotoxic injury in transgenic mice. *The Journal of Cell Biology, 143*(7), 2023–2032.

Ku, N. O., Michie, S. A., Soetikno, R. M., Resurreccion, E. Z., Broome, R. L., Oshima, R. G., et al. (1996). Susceptibility to hepatotoxicity in transgenic mice that express a dominant-negative human keratin 18 mutant. *The Journal of Clinical Investigation, 98*(4), 1034–1046.

Ku, N. O., & Omary, M. B. (1994). Identification of the major physiologic phosphorylation site of human keratin 18: Potential kinases and a role in filament reorganization. *The Journal of Cell Biology, 127*(1), 161–171.

Ku, N. O., & Omary, M. B. (2000). Keratins turn over by ubiquitination in a phosphorylation-modulated fashion. *The Journal of Cell Biology, 149*(3), 547–552.

Ku, N. O., & Omary, M. B. (2006). A disease- and phosphorylation-related nonmechanical function for keratin 8. *The Journal of Cell Biology, 174*(1), 115–125.

Ku, N. O., Strnad, P., Zhong, B. H., Tao, G. Z., & Omary, M. B. (2007). Keratins let liver live: Mutations predispose to liver disease and crosslinking generates mallory-denk bodies. *Hepatology (Baltimore, Md.), 46*(5), 1639–1649.

Ku, N. O., Toivola, D. M., Strnad, P., & Omary, M. B. (2010). Cytoskeletal keratin glycosylation protects epithelial tissue from injury. *Nature Cell Biology, 12*(9), 876–885.

Ku, N. O., Toivola, D. M., Zhou, Q., Tao, G. Z., Zhong, B., & Omary, M. B. (2004). Studying simple epithelial keratins in cells and tissues. *Methods in Cell Biology, 78*, 489–517.

Ku, N. O., Wright, T. L., Terrault, N. A., Gish, R., & Omary, M. B. (1997). Mutation of human keratin 18 in association with cryptogenic cirrhosis. *The Journal of Clinical Investigation, 99*(1), 19–23.

Ku, N. O., Zhou, X., Toivola, D. M., & Omary, M. B. (1999). The cytoskeleton of digestive epithelia in health and disease. *The American Journal of Physiology, 277*(6 Pt. 1), G1108–G1137.

Kucukoglu, O., Guldiken, N., Chen, Y., Usachov, V., El-Heliebi, A., Haybaeck, J., et al. (2014). High-fat diet triggers mallory-denk body formation through misfolding and crosslinking of excess keratin 8. *Hepatology (Baltimore, Md.), 60*(1), 169–178.

Kulesh, D. A., Cecena, G., Darmon, Y. M., Vasseur, M., & Oshima, R. G. (1989). Posttranslational regulation of keratins: Degradation of mouse and human keratins 18 and 8. *Molecular and Cellular Biology, 9*(4), 1553–1565.

Kulesh, D. A., & Oshima, R. G. (1988). Cloning of the human keratin 18 gene and its expression in nonepithelial mouse cells. *Molecular and Cellular Biology, 8*(4), 1540–1550.

Kwan, R., Hanada, S., Harada, M., Strnad, P., Li, D. H., & Omary, M. B. (2012). Keratin 8 phosphorylation regulates its transamidation and hepatocyte mallory-denk body formation. *FASEB Journal: Official Publication of the Federation of American Societies for Experimental Biology*, *26*(6), 2318–2326.

Kwan, R., Looi, K., & Omary, M. B. (2015). Absence of keratins 8 and 18 in rodent epithelial cell lines associates with keratin gene mutation and DNA methylation: Cell line selective effects on cell invasion. *Experimental Cell Research*, *335*(1), 12–22.

Larson, M. H., Gilbert, L. A., Wang, X., Lim, W. A., Weissman, J. S., & Qi, L. S. (2013). CRISPR interference (CRISPRi) for sequence-specific control of gene expression. *Nature Protocols*, *8*(11), 2180–2196.

Leifeld, L., Kothe, S., Sohl, G., Hesse, M., Sauerbruch, T., Magin, T. M., et al. (2009). Keratin 18 provides resistance to fas-mediated liver failure in mice. *European Journal of Clinical Investigation*, *39*(6), 481–488.

Li, J., Bardag-Gorce, F., Dedes, J., French, B. A., Amidi, F., Oliva, J., et al. (2008). S-adenosylmethionine prevents mallory denk body formation in drug-primed mice by inhibiting the epigenetic memory. *Hepatology (Baltimore, Md)*, *47*(2), 613–624.

Liao, J., Ku, N. O., & Omary, M. B. (1996). Two-dimensional gel analysis of glandular keratin intermediate filament phosphorylation. *Electrophoresis*, *17*(11), 1671–1676.

Loranger, A., Duclos, S., Grenier, A., Price, J., Wilson-Heiner, M., Baribault, H., et al. (1997). Simple epithelium keratins are required for maintenance of hepatocyte integrity. *The American Journal of Pathology*, *151*(6), 1673–1683.

Loschke, F., Seltmann, K., Bouameur, J. E., & Magin, T. M. (2015). Regulation of keratin network organization. *Current Opinion in Cell Biology*, *32*, 56–64.

Lowthert, L. A., Ku, N. O., Liao, J., Coulombe, P. A., & Omary, M. B. (1995). Empigen BB: A useful detergent for solubilization and biochemical analysis of keratins. *Biochemical and Biophysical Research Communications*, *206*(1), 370–379.

Magin, T. M., Schroder, R., Leitgeb, S., Wanninger, F., Zatloukal, K., Grund, C., et al. (1998). Lessons from keratin 18 knockout mice: Formation of novel keratin filaments, secondary loss of keratin 7 and accumulation of liver-specific keratin 8-positive aggregates. *The Journal of Cell Biology*, *140*(6), 1441–1451.

Mahajan, V., Klingstedt, T., Simon, R., Nilsson, K. P., Thueringer, A., Kashofer, K., et al. (2011). Cross beta-sheet conformation of keratin 8 is a specific feature of mallory-denk bodies compared with other hepatocyte inclusions. *Gastroenterology*, *141*(3). 1080-1090.e1-7.

Marceau, N., Gilbert, S., & Loranger, A. (2004). Uncovering the roles of intermediate filaments in apoptosis. *Methods in Cell Biology*, *78*, 95–129.

Mashukova, A., Oriolo, A. S., Wald, F. A., Casanova, M. L., Kroger, C., Magin, T. M., et al. (2009). Rescue of atypical protein kinase C in epithelia by the cytoskeleton and Hsp70 family chaperones. *Journal of Cell Science*, *122*(Pt. 14), 2491–2503.

Mathew, J., Loranger, A., Gilbert, S., Faure, R., & Marceau, N. (2013). Keratin 8/18 regulation of glucose metabolism in normal versus cancerous hepatic cells through differential modulation of hexokinase status and insulin signaling. *Experimental Cell Research*, *319*(4), 474–486.

Min, Y. S., Yi, E. H., Lee, J. K., Choi, J. W., Sim, J. H., Kang, J. S., et al. (2012). CK20 expression enhances the invasiveness of tamoxifen-resistant MCF-7 cells. *Anticancer Research*, *32*(4), 1221–1228.

Moll, R., Divo, M., & Langbein, L. (2008). The human keratins: Biology and pathology. *Histochemistry and Cell Biology*, *129*(6), 705–733.

Molnar, A., Haybaeck, J., Lackner, C., & Strnad, P. (2011). The cytoskeleton in nonalcoholic steatohepatitis: 100 years old but still youthful. *Expert Review of Gastroenterology & Hepatology*, *5*(2), 167–177.

Nakamichi, I., Toivola, D. M., Strnad, P., Michie, S. A., Oshima, R. G., Baribault, H., et al. (2005). Keratin 8 overexpression promotes mouse mallory body formation. *The Journal of Cell Biology*, *171*(6), 931–937.

Odaka, C., Loranger, A., Takizawa, K., Ouellet, M., Tremblay, M. J., Murata, S., et al. (2013). Keratin 8 is required for the maintenance of architectural structure in thymus epithelium. *PLoS One*, *8*(9), e75101.
Omary, M. B. (2009). "IF-pathies": A broad spectrum of intermediate filament-associated diseases. *The Journal of Clinical Investigation*, *119*(7), 1756–1762.
Omary, M. B., Ku, N. O., Liao, J., & Price, D. (1998). Keratin modifications and solubility properties in epithelial cells and in vitro. *Sub-Cellular Biochemistry*, *31*, 105–140.
Omary, M. B., Ku, N. O., Strnad, P., & Hanada, S. (2009). Toward unraveling the complexity of simple epithelial keratins in human disease. *The Journal of Clinical Investigation*, *119*(7), 1794–1805.
Petrak, J., Ivanek, R., Toman, O., Cmejla, R., Cmejlova, J., Vyoral, D., et al. (2008). Deja vu in proteomics. A hit parade of repeatedly identified differentially expressed proteins. *Proteomics*, *8*(9), 1744–1749.
Plowman, J. E. (2007). The proteomics of keratin proteins. *Journal of Chromatography. B. Analytical Technologies in the Biomedical and Life Sciences*, *849*(1–2), 181–189.
Sandilands, A., Smith, F. J., Lunny, D. P., Campbell, L. E., Davidson, K. M., MacCallum, S. F., et al. (2013). Generation and characterisation of keratin 7 (K7) knockout mice. *PLoS One*, *8*(5), e64404.
Sano, M., Aoyagi, K., Takahashi, H., Kawamura, T., Mabuchi, T., Igaki, H., et al. (2010). Forkhead box A1 transcriptional pathway in KRT7-expressing esophageal squamous cell carcinomas with extensive lymph node metastasis. *International Journal of Oncology*, *36*(2), 321–330.
Sato, T., & Clevers, H. (2013). Growing self-organizing mini-guts from a single intestinal stem cell: Mechanism and applications. *Science (New York, N.Y.)*, *340*(6137), 1190–1194.
Schnabl, B., & Brenner, D. A. (2014). Interactions between the intestinal microbiome and liver diseases. *Gastroenterology*, *146*(6), 1513–1524.
Schwarz, N., Windoffer, R., Magin, T. M., & Leube, R. E. (2015). Dissection of keratin network formation, turnover and reorganization in living murine embryos. *Scientific Reports*, *5*, 9007.
Schweizer, J., Bowden, P. E., Coulombe, P. A., Langbein, L., Lane, E. B., Magin, T. M., et al. (2006). New consensus nomenclature for mammalian keratins. *The Journal of Cell Biology*, *174*(2), 169–174.
Seltmann, K., Fritsch, A. W., Kas, J. A., & Magin, T. M. (2013). Keratins significantly contribute to cell stiffness and impact invasive behavior. *Proceedings of the National Academy of Sciences of the United States of America*, *110*(46), 18507–18512.
Shearer, R. F., & Saunders, D. N. (2015). Experimental design for stable genetic manipulation in mammalian cell lines: Lentivirus and alternatives. *Genes to Cells: Devoted to Molecular & Cellular Mechanisms*, *20*(1), 1–10.
Singer, P. A., Trevor, K., & Oshima, R. G. (1986). Molecular cloning and characterization of the endo B cytokeratin expressed in preimplantation mouse embryos. *The Journal of Biological Chemistry*, *261*(2), 538–547.
Singla, A., Moons, D. S., Snider, N. T., Wagenmaker, E. R., Jayasundera, V. B., & Omary, M. B. (2012). Oxidative stress, Nrf2 and keratin up-regulation associate with mallory-denk body formation in mouse erythropoietic protoporphyria. *Hepatology (Baltimore, Md.)*, *56*(1), 322–331.
Sivaramakrishnan, S., DeGiulio, J. V., Lorand, L., Goldman, R. D., & Ridge, K. M. (2008). Micromechanical properties of keratin intermediate filament networks. *Proceedings of the National Academy of Sciences of the United States of America*, *105*(3), 889–894.
Sivaramakrishnan, S., Schneider, J. L., Sitikov, A., Goldman, R. D., & Ridge, K. M. (2009). Shear stress induced reorganization of the keratin intermediate filament network requires phosphorylation by protein kinase C zeta. *Molecular Biology of the Cell*, *20*(11), 2755–2765.

Snider, N. T., & Omary, M. B. (2014). Post-translational modifications of intermediate filament proteins: Mechanisms and functions. *Nature Reviews. Molecular Cell Biology*, *15*(3), 163–177.

Stacey, S. N., Sulem, P., Masson, G., Gudjonsson, S. A., Thorleifsson, G., Jakobsdottir, M., et al. (2009). New common variants affecting susceptibility to basal cell carcinoma. *Nature Genetics*, *41*(8), 909–914.

Stone, M. R., O'Neill, A., Lovering, R. M., Strong, J., Resneck, W. G., Reed, P. W., et al. (2007). Absence of keratin 19 in mice causes skeletal myopathy with mitochondrial and sarcolemmal reorganization. *Journal of Cell Science*, *120*(Pt. 22), 3999–4008.

Strnad, P., Harada, M., Siegel, M., Terkeltaub, R. A., Graham, R. M., Khosla, C., et al. (2007). Transglutaminase 2 regulates mallory body inclusion formation and injury-associated liver enlargement. *Gastroenterology*, *132*(4), 1515–1526.

Strnad, P., Lienau, T. C., Tao, G. Z., Ku, N. O., Magin, T. M., & Omary, M. B. (2006). Denaturing temperature selection may underestimate keratin mutation detection by DHPLC. *Human Mutation*, *27*(5), 444–452.

Strnad, P., Nuraldeen, R., Guldiken, N., Hartmann, D., Mahajan, V., Denk, H., et al. (2013). Broad spectrum of hepatocyte inclusions in humans, animals, and experimental models. *Comprehensive Physiology*, *3*(4), 1393–1436.

Strnad, P., Tao, G. Z., Zhou, Q., Harada, M., Toivola, D. M., Brunt, E. M., et al. (2008). Keratin mutation predisposes to mouse liver fibrosis and unmasks differential effects of the carbon tetrachloride and thioacetamide models. *Gastroenterology*, *134*(4), 1169–1179.

Strnad, P., Usachov, V., Debes, C., Grater, F., Parry, D. A., & Omary, M. B. (2011). Unique amino acid signatures that are evolutionarily conserved distinguish simple-type, epidermal and hair keratins. *Journal of Cell Science*, *124*(Pt. 24), 4221–4232.

Strnad, P., Zhou, Q., Hanada, S., Lazzeroni, L. C., Zhong, B. H., So, P., et al. (2010). Keratin variants predispose to acute liver failure and adverse outcome: Race and ethnic associations. *Gastroenterology*, *139*(3), 828–835. 835.e1-3.

Stumptner, C., Fuchsbichler, A., Heid, H., Zatloukal, K., & Denk, H. (2002). Mallory body—A disease-associated type of sequestosome. *Hepatology (Baltimore, Md.)*, *35*(5), 1053–1062.

Stumptner, C., Fuchsbichler, A., Lehner, M., Zatloukal, K., & Denk, H. (2001). Sequence of events in the assembly of mallory body components in mouse liver: Clues to the pathogenesis and significance of mallory body formation. *Journal of Hepatology*, *34*(5), 665–675.

Stumptner, C., Fuchsbichler, A., Zatloukal, K., & Denk, H. (2007). In vitro production of mallory bodies and intracellular hyaline bodies: The central role of sequestosome 1/p62. *Hepatology (Baltimore, Md.)*, *46*(3), 851–860.

Szeverenyi, I., Cassidy, A. J., Chung, C. W., Lee, B. T., Common, J. E., Ogg, S. C., et al. (2008). The human intermediate filament database: Comprehensive information on a gene family involved in many human diseases. *Human Mutation*, *29*(3), 351–360.

Tamai, Y., Ishikawa, T., Bosl, M. R., Mori, M., Nozaki, M., Baribault, H., et al. (2000). Cytokeratins 8 and 19 in the mouse placental development. *The Journal of Cell Biology*, *151*(3), 563–572.

Tao, G. Z., Toivola, D. M., Zhong, B., Michie, S. A., Resurreccion, E. Z., Tamai, Y., et al. (2003). Keratin-8 null mice have different gallbladder and liver susceptibility to lithogenic diet-induced injury. *Journal of Cell Science*, *116*(Pt. 22), 4629–4638.

Toivola, D. M., Baribault, H., Magin, T., Michie, S. A., & Omary, M. B. (2000). Simple epithelial keratins are dispensable for cytoprotection in two pancreatitis models. *American Journal of Physiology. Gastrointestinal and Liver Physiology*, *279*(6), G1343–G1354.

Toivola, D. M., Boor, P., Alam, C., & Strnad, P. (2015). Keratins in health and disease. *Current Opinion in Cell Biology*, *32*, 73–81.

Toivola, D. M., Goldman, R. D., Garrod, D. R., & Eriksson, J. E. (1997). Protein phosphatases maintain the organization and structural interactions of hepatic keratin intermediate filaments. *Journal of Cell Science*, *110*(Pt. 1), 23–33.
Toivola, D. M., Krishnan, S., Binder, H. J., Singh, S. K., & Omary, M. B. (2004). Keratins modulate colonocyte electrolyte transport via protein mistargeting. *The Journal of Cell Biology*, *164*(6), 911–921.
Toivola, D. M., Nakamichi, I., Strnad, P., Michie, S. A., Ghori, N., Harada, M., et al. (2008). Keratin overexpression levels correlate with the extent of spontaneous pancreatic injury. *The American Journal of Pathology*, *172*(4), 882–892.
Toivola, D. M., Nieminen, M. I., Hesse, M., He, T., Baribault, H., Magin, T. M., et al. (2001). Disturbances in hepatic cell-cycle regulation in mice with assembly-deficient keratins 8/18. *Hepatology (Baltimore, Md.)*, *34*(6), 1174–1183.
Toivola, D. M., Omary, M. B., Ku, N. O., Peltola, O., Baribault, H., & Eriksson, J. E. (1998). Protein phosphatase inhibition in normal and keratin 8/18 assembly-incompetent mouse strains supports a functional role of keratin intermediate filaments in preserving hepatocyte integrity. *Hepatology (Baltimore, Md.)*, *28*(1), 116–128.
Toivola, D. M., Strnad, P., Habtezion, A., & Omary, M. B. (2010). Intermediate filaments take the heat as stress proteins. *Trends in Cell Biology*, *20*(2), 79–91.
Treiber, M., Schulz, H. U., Landt, O., Drenth, J. P., Castellani, C., Real, F. X., et al. (2006). Keratin 8 sequence variants in patients with pancreatitis and pancreatic cancer. *Journal of Molecular Medicine (Berlin, Germany)*, *84*(12), 1015–1022.
Uchiumi, A., Yamashita, M., & Katagata, Y. (2012). Downregulation of keratins 8, 18 and 19 influences invasiveness of human cultured squamous cell carcinoma and adenocarcinoma cells. *Experimental and Therapeutic Medicine*, *3*(3), 443–448.
Usachov, V., Urban, T. J., Fontana, R. J., Gross, A., Iyer, S., Omary, M. B., et al. (2015). Prevalence of genetic variants of keratins 8 and 18 in patients with drug-induced liver injury. *BMC Medicine*, 13, 196.
Vijayaraj, P., Kroger, C., Reuter, U., Windoffer, R., Leube, R. E., & Magin, T. M. (2009). Keratins regulate protein biosynthesis through localization of GLUT1 and -3 upstream of AMP kinase and raptor. *The Journal of Cell Biology*, *187*(2), 175–184.
Wald, F. A., Oriolo, A. S., Casanova, M. L., & Salas, P. J. (2005). Intermediate filaments interact with dormant ezrin in intestinal epithelial cells. *Molecular Biology of the Cell*, *16*(9), 4096–4107.
Weerasinghe, S. V., Ku, N. O., Altshuler, P. J., Kwan, R., & Omary, M. B. (2014). Mutation of caspase-digestion sites in keratin 18 interferes with filament reorganization, and predisposes to hepatocyte necrosis and loss of membrane integrity. *Journal of Cell Science*, *127*(Pt. 7), 1464–1475.
Windoffer, R., Beil, M., Magin, T. M., & Leube, R. E. (2011). Cytoskeleton in motion: The dynamics of keratin intermediate filaments in epithelia. *The Journal of Cell Biology*, *194*(5), 669–678.
Windoffer, R., & Leube, R. E. (2004). Imaging of keratin dynamics during the cell cycle and in response to phosphatase inhibition. *Methods in Cell Biology*, *78*, 321–352.
Wu, Y., Nan, L., Bardag-Gorce, F., Li, J., French, B. A., Wilson, L. T., et al. (2005). The role of laminin-integrin signaling in triggering MB formation. an in vivo and in vitro study. *Experimental and Molecular Pathology*, *79*(1), 1–8.
Xu, B., Zhang, Y., Zhao, Z., Yoshida, Y., Magdeldin, S., Fujinaka, H., et al. (2011). Usage of electrostatic eliminator reduces human keratin contamination significantly in gel-based proteomics analysis. *Journal of Proteomics*, *74*(7), 1022–1029.
Yuan, Q. X., Marceau, N., French, B. A., Fu, P., & French, S. W. (1996). Mallory body induction in drug-primed mouse liver. *Hepatology (Baltimore, Md.)*, *24*(3), 603–612.
Zatloukal, K., Bock, G., Rainer, I., Denk, H., & Weber, K. (1991). High molecular weight components are main constituents of mallory bodies isolated with a fluorescence

activated cell sorter. *Laboratory Investigation; a Journal of Technical Methods and Pathology, 64*(2), 200–206.
Zatloukal, K., Denk, H., Spurej, G., Lackinger, E., Preisegger, K. H., & Franke, W. W. (1990). High molecular weight component of mallory bodies detected by a monoclonal antibody. *Laboratory Investigation; a Journal of Technical Methods and Pathology, 62*(4), 427–434.
Zatloukal, K., French, S. W., Stumptner, C., Strnad, P., Harada, M., Toivola, D. M., et al. (2007). From mallory to mallory-denk bodies: What, how and why? *Experimental Cell Research, 313*(10), 2033–2049.
Zatloukal, B., Kufferath, I., Thueringer, A., Landegren, U., Zatloukal, K., & Haybaeck, J. (2014). Sensitivity and specificity of in situ proximity ligation for protein interaction analysis in a model of steatohepatitis with mallory-denk bodies. *PLoS One, 9*(5), e96690.
Zatloukal, K., Stumptner, C., Fuchsbichler, A., Heid, H., Schnoelzer, M., Kenner, L., et al. (2002). p62 is a common component of cytoplasmic inclusions in protein aggregation diseases. *The American Journal of Pathology, 160*(1), 255–263.
Zatloukal, K., Stumptner, C., Fuchsbichler, A., Janig, E., & Denk, H. (2004). Intermediate filament protein inclusions. *Methods in Cell Biology, 78*, 205–228.
Zhou, Q., Ji, X., Chen, L., Greenberg, H. B., Lu, S. C., & Omary, M. B. (2005). Keratin mutation primes mouse liver to oxidative injury. *Hepatology (Baltimore, Md.), 41*(3), 517–525.
Zhou, X., Liao, J., Hu, L., Feng, L., & Omary, M. B. (1999). Characterization of the major physiologic phosphorylation site of human keratin 19 and its role in filament organization. *The Journal of Biological Chemistry, 274*(18), 12861–12866.
Zhou, Q., Toivola, D. M., Feng, N., Greenberg, H. B., Franke, W. W., & Omary, M. B. (2003). Keratin 20 helps maintain intermediate filament organization in intestinal epithelia. *Molecular Biology of the Cell, 14*(7), 2959–2971.
Zupancic, T., Stojan, J., Lane, E. B., Komel, R., Bedina-Zavec, A., & Liovic, M. (2014). Intestinal cell barrier function in vitro is severely compromised by keratin 8 and 18 mutations identified in patients with inflammatory bowel disease. *PLoS One, 9*(6), e99398.

CHAPTER FOURTEEN

Methods for Determining the Cellular Functions of Vimentin Intermediate Filaments

Karen M. Ridge[*,†,‡,1,2], Dale Shumaker[*,†,1], Amélie Robert[†,1], Caroline Hookway[†,1], Vladimir I. Gelfand[†,1], Paul A. Janmey[§,¶,1], Jason Lowery[*,1], Ming Guo[||,1], David A. Weitz[||,#,1], Edward Kuczmarski[†,1], Robert D. Goldman[*,†,1]

[*]Division of Pulmonary and Critical Care Medicine, Chicago, Illinois, USA
[†]Department of Cell and Molecular Biology, Northwestern University, Feinberg School of Medicine, Chicago, Illinois, USA
[‡]Veterans Administration, Chicago, Illinois, USA
[§]Institute for Medicine and Engineering, University of Pennsylvania, Philadelphia, Pennsylvania, USA
[¶]Departments of Physiology and Physics & Astronomy, University of Pennsylvania, Philadelphia, Pennsylvania, USA
[||]School of Engineering and Applied Sciences, Harvard University, Cambridge, Massachusetts, USA
[#]Department of Physics, Harvard University, Cambridge, Massachusetts, USA
[2]Corresponding author: e-mail address: kridge@northwestern.edu

Contents

1. Introduction 391
2. Disruption of Vimentin IFs 391
   - 2.1 Microinjection of Full-Length Vimentin and Mimetic Peptides 393
   - 2.2 Transfection with Green Fluorescent Protein-Labeled Full-Length and Dominant-Negative Forms of Vimentin 395
   - 2.3 Silencing the Expression of Vimentin in Cells 396
   - 2.4 Withaferin A 397
   - 2.5 Gigaxonin as an Efficient Tool for Reducing and Eliminating Vimentin IFs from Cells 398
3. Analysis of Vimentin Dynamics Using Photoactivatable and Photoconvertible Protein Tags 398
   - 3.1 Creating Cells That Express Photoactivatable and Photoconvertible Protein Fusions with Vimentin 399
   - 3.2 Imaging Vimentin Dynamics Using Photoactivatable or Photoconvertible Proteins 400
   - 3.3 Quantification of Vimentin Dynamics 403
4. Investigating Vimentin–Protein Interactions 405
   - 4.1 Biolayer Interferometry 406
   - 4.2 Soluble Bead Binding Assay Between Vimentin and Protein Extracts 409

[1] Authors contributed equally.

Methods in Enzymology, Volume 568
ISSN 0076-6879
http://dx.doi.org/10.1016/bs.mie.2015.09.036

5. Investigating the Mechanical Properties of Vimentin IF Networks 411
5.1 Three-Dimensional Substrate Studies 411
5.2 Collagen Gel Contraction Studies 413
6. Investigating the Role of Vimentin IFs in Cell Mechanics 415
6.1 Investigating the Role of Vimentin in Cytoplasmic Mechanics Using Optical-Tweezer Active Microrheology 415
6.2 Investigating the Role of Vimentin in Intracellular Dynamics 417
6.3 Investigating the Contribution of Vimentin to the Aggregate Intracellular Forces 418
7. Conclusion 420
Acknowledgments 421
References 421

Abstract

The type III intermediate filament protein vimentin was once thought to function mainly as a static structural protein in the cytoskeleton of cells of mesenchymal origin. Now, however, vimentin is known to form a dynamic, flexible network that plays an important role in a number of signaling pathways. Here, we describe various methods that have been developed to investigate the cellular functions of the vimentin protein and intermediate filament network, including chemical disruption, photoactivation and photoconversion, biolayer interferometry, soluble bead binding assay, three-dimensional substrate experiments, collagen gel contraction, optical-tweezer active microrheology, and force spectrum microscopy. Using these techniques, the contributions of vimentin to essential cellular processes can be probed in ever further detail.

ABBREVIATIONS

AFI average fluorescence intensity
BLI biolayer interferometry
BMDM bone marrow-derived macrophage
DMEM Dulbecco-modified Eagle medium
EMT epithelial–mesenchymal transition
FSM force spectrum microscopy
GAN giant axonal neuropathy
GFP green fluorescent protein
IF intermediate filament
mEF mouse embryonic fibroblast
MSU monosodium urate
NLRP3 NACHT, LRR, and PYD domains-containing protein 3
PAA polyacrylamide
shRNA short hairpin RNA
siRNA small-interfering RNA
TIRF total internal reflection fluorescence
ULF unit-length filament
Vim$^{-/-}$ Vimentin$^{-/-}$
WFA Withaferin A
WT wild-type

1. INTRODUCTION

Vimentin is a type III intermediate filament (IF) cytoskeletal protein expressed in cells of mesenchymal origin. It serves as a canonical marker of epithelial–mesenchymal transition (EMT) and is involved in a number of diseases and conditions, including cancer, inflammation, and congenital cataracts (Dos Santos et al., 2015; Kidd, Shumaker, & Ridge, 2014; Muller et al., 2009; Stevens et al., 2013). In the past, IF proteins, including vimentin, were assumed to form static structures, until evidence of a dynamic exchange of IF subunits came to light (Eriksson et al., 2009). Changes in the shapes and assembly states of IFs were also observed, revealing dynamic and flexible cytoskeletal networks (Eriksson et al., 2009).

The basic structure of vimentin consists of a central α-helical rod domain flanked by unstructured head and tail domains (Eriksson et al., 2009). Vimentin monomers pair up into coiled-coil dimers, which then align in a staggered, antiparallel fashion to form tetramers; groups of eight tetramers make up the unit-length filaments (ULFs) that join end-to-end and subsequently undergo a radial compaction to form the mature vimentin IFs (Herrmann et al., 1996; Hess, Budamagunta, Voss, & FitzGerald, 2004; Mucke et al., 2004; Steinert, Marekov, & Parry, 1993). The dynamics of the IF network dictate the structural and mechanical properties of the cell and its organelles. For example, vimentin IFs modulate lamellipodia formation during cell migration and mitochondrial movement within the cytoplasm (Helfand et al., 2011; Nekrasova et al., 2011). Vimentin also acts as a scaffold for important signaling molecules and even mediates the activation of a variety of signaling pathways (Barberis et al., 2009; Dos Santos et al., 2015; Stevens et al., 2013; Tzivion, Luo, & Avruch, 2000).

The diverse cellular functions of vimentin IFs lend themselves to analysis by a wide assortment of experimental techniques using various reagents (see Table 1). In this chapter, we describe a variety of methods that have been developed to analyze the cellular functions of vimentin IFs.

2. DISRUPTION OF VIMENTIN IFs

No reliable drugs or natural products have been sufficiently characterized with respect to their disruption of the assembly states of vimentin IFs in cells, in contrast to the readily available inhibitors of microtubules

Table 1 Vimentin-Related Reagents

Reagent	Type	Reference or Manufacturer	Application Notes
Biotinylated vimentin	Labeled protein	Vikstrom, Miller, and Goldman (1991)	Incorporates into existing IF network when microinjected into cells, detected with fluorescent antibiotin antibody
Chicken antivimentin polyclonal antibody	Antibody	#919101, BioLegend, San Diego, CA	Recognizes human, mouse, and rat vimentin
Dominant-negative mutant vimentin$_{(1-138)}$	Mutant protein	Kural et al. (2007)	Disrupts vimentin IF network, causing retraction to perinuclear region, when expressed in living cells
GFP-tagged vimentin	Fluorescent protein	Yoon, Moir, Prahlad, and Goldman (1998)	Fluorescent form of vimentin expressed in living cells
Gigaxonin	Protein inhibitor	Mahammad et al. (2013)	Virtually all vimentin IFs are cleared when this E3 ligase adaptor protein is overexpressed in living cells
Mimetic peptide 104–138 (1A peptide)	Mimetic peptide	Goldman, Khuon, Chou, Opal, and Steinert (1996)	Causes complete disassembly of vimentin IFs
Mimetic peptide 355–412 (2B2 peptide)	Mimetic peptide	Helfand et al. (2011)	Causes disassembly of vimentin IFs to the stage of ULFs
PQCXIP-mEos3.2-Vimentin	Retroviral plasmid	Hookway et al. (2015)	Photoconvertible vimentin
PQCXIP-mEos3.2-VimentinY117L	Retroviral plasmid	Robert, Rossow, Hookway, Adam, and Gelfand (2015)	Photoconvertible ULF vimentin mutant
PQCXIP-mMaple3-Vimentin	Retroviral plasmid	Robert et al. (2015)	Photoconvertible vimentin

Table 1 Vimentin-Related Reagents—cont'd

Reagent	Type	Reference or Manufacturer	Application Notes
PQCXIP-PAGFP-Vimentin	Retroviral plasmid	Hookway et al. (2015)	Photoactivatable vimentin
Rhodamine-vimentin	Labeled protein	Vikstrom et al. (1991)	Fluorescent form of vimentin, incorporates into existing IF network when microinjected into cells
Silencer *VIM* siRNA	siRNA	#AM16708, Life Technologies, Grand Island, NY	Up to 90% knockdown of vimentin protein expression in mammalian cells
Withaferin A	Small molecule inhibitor	#ASB-00023250, ChromaDex, Irvine, CA	Extracted from *Withania somnifera*, reorganizes vimentin IFs into a perinuclear aggregate

GFP, green fluorescent protein; IF, intermediate filament; siRNA, small-interfering RNA; ULF, unit-length filament.

(e.g., nocodazole and vinblastine) and microfilaments (F-actin; e.g., cytochalasin and latrunculin). For this reason, different approaches and methodologies have been developed for disrupting vimentin IFs in order to determine their cellular functions.

2.1 Microinjection of Full-Length Vimentin and Mimetic Peptides

Microinjection of biotinylated vimentin or vimentin directly conjugated to rhodamine permits the tracking of unpolymerized subunits as they assemble into endogenous vimentin IF networks (Vikstrom, Borisy, & Goldman, 1989). Importantly, rhodamine-conjugated vimentin can also be used for photobleaching experiments (Vikstrom, Lim, Goldman, & Borisy, 1992). More recently, the microinjection of vimentin has been used to study the impact of its assembly in EMT (Mendez, Kojima, & Goldman, 2010). These techniques permit the analysis of the immediate steps of vimentin polymerization within cells.

The development and use of vimentin mimetic peptides designed to perturb the function of vimentin IFs in cells has provided insights into their structure and function. When these peptides are microinjected into cells, they induce IF disassembly or disrupt IF organization. The advantage of the microinjection technique is that cells can be studied immediately following the introduction of the peptides, which begin to disrupt IF assembly within minutes after injection. Prior to microinjection, it is essential to demonstrate the efficacy of these peptides *in vitro*. For example, when a mimetic peptide with a sequence derived from the helix initiation 1A domain of vimentin (amino acid residues 104–138; see Fig. 1A) is mixed with fully polymerized vimentin IFs at 1:1 molar ratios, it causes disassembly into vimentin monomers and dimers within 30 min at room temperature (Goldman et al., 1996). Thus, the peptide disrupts the interactions among vimentin subunits, causing disassembly that is presumably due to competitive inhibition with the helix 1A domains of full-length IFs. When this peptide is microinjected into live fibroblasts, polymerized IF networks are disassembled and the cells round up and lose their adhesions, demonstrating a role for IFs in maintaining cell shape and mechanical integrity. As the cells round up, they show extensive loss of microtubules and microfilaments, demonstrating the important role of vimentin IFs in stabilizing these other cytoskeletal systems. The effects of this peptide are reversible and the cells recover their normal shapes within a few hours (Straube-West, Loomis, Opal, & Goldman, 1996).

The vimentin 2B2 mimetic peptide (residues 355–412; see Fig. 1A) is derived from the C-terminal end of the α-helical rod domain, and also functions in a dominant-negative fashion, causing the disassembly of polymerized vimentin IFs. *In vitro*, the 2B2 peptide causes vimentin IFs to disassemble at molar ratios of 1:10 or less as determined by specific viscosity and negative stain electron microscopy. However, the disassembly stops at the ULF stage (Strelkov et al., 2002), not the monomers and dimers seen in the more catastrophic disassembly induced by the 1A peptide described above (Herrmann et al., 1996). Within living fibroblasts, the microinjection of the 2B2 peptide at concentrations of less than 5 μg/mL causes the vimentin IF network to begin to disassemble into short IFs within seconds; after longer periods, the majority of these short IFs disassemble into ULF-like structures. Under these conditions, the cells do not completely round up as in cells injected with the 1A peptide (see above), but rather can be used to induce the local disassembly of vimentin IFs. This peptide has been used to demonstrate the importance of vimentin IF disassembly in the formation of lamellipodia and cell polarity during fibroblast motility (Helfand et al., 2011).

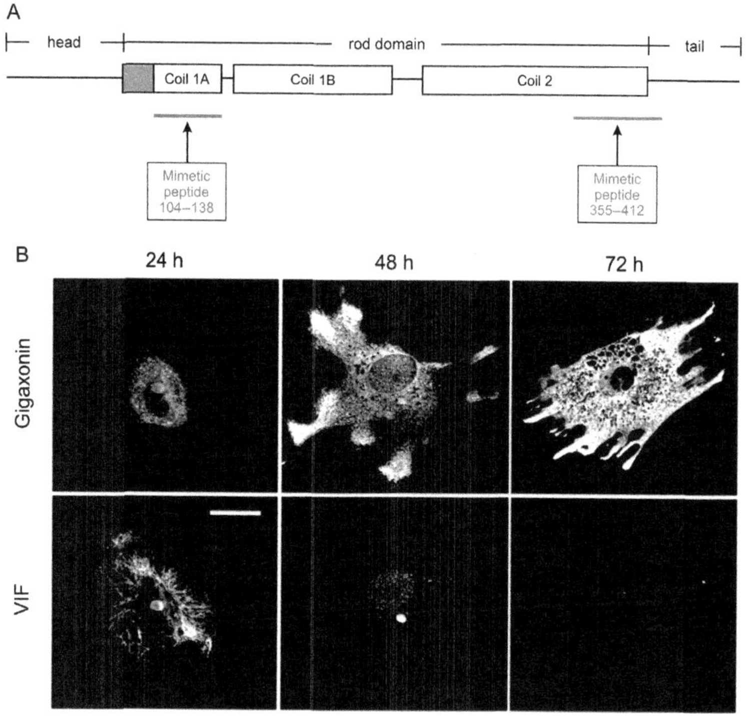

Figure 1 Use of mimetic peptides and gigaxonin to study the cell biology of vimentin intermediate filaments (IFs). (A) Location of the mimetic peptides relative to the organization of a vimentin monomer. The monomer is divided into three major domains: head, α-helical rod, and tail. The rod domain is further segmented into coil 1 and coil 2. The diagram shows the secondary segmentation of coil 1 into 1A and 1B. Mimetic peptide 104–138 spans almost the entire coil 1A region. Mimetic peptide 355–412 includes part of the carboxy-terminus of coil 2 and seven amino acids of the tail. Green (gray in the print version) indicates the precoil domain. (B) Gigaxonin as an efficient tool for reducing and eliminating vimentin IFs from cells. Fibroblasts derived from a patient with giant axonal neuropathy were transfected with a mammalian expression vector-containing FLAG-tagged wild-type gigaxonin, fixed at the indicated times, and double-labeled with anti-FLAG (*top row*) and antivimentin (*bottom row*) antibodies. Twenty-four hours after induction of gigaxonin expression, vimentin IFs (VIF) were still observed. By 48 h, only short filaments and nonfilamentous vimentin particles remained; and by 72 h, there was no detectable vimentin. Scale bar, 10 μm. *Images in (B) adapted from Mahammad et al. (2013).*

2.2 Transfection with Green Fluorescent Protein-Labeled Full-Length and Dominant-Negative Forms of Vimentin

To study the dynamic properties of vimentin IF networks in live cells over prolonged periods and to study the impact of dominant-negative vimentin mutants, constructs labeled with green fluorescent protein (GFP) or Emerald have been employed (Helfand et al., 2011; Hookway et al., 2015; Mendez et al., 2010; Yoon et al., 1998). Emerald is a GFP derivative which has improved photostability and brightness (Shaner, Patterson, & Davidson, 2007). For live-cell imaging techniques such as fluorescence recovery after photobleaching, the use of Emerald is not optimal and enhanced GFP is more advantageous. These fluorescently labeled constructs are easily transfected into cells for either transient or stable expression of wild-type (WT) or dominant-negative vimentin mutants. Using full-length GFP-vimentin, one can demonstrate that vimentin IFs are highly dynamic and undergo constant changes in their shapes, assembly and disassembly, translocation, and length over relatively brief periods (Ho, Martys, Mikhailov, Gundersen, & Liem, 1998; Kajita et al., 2014; Martys, Ho, Liem, & Gundersen, 1999; Prahlad, Yoon, Moir, Vale, & Goldman, 1998; Yoon et al., 1998).

Dominant-negative mutants of vimentin have been used to reorganize IF networks in living cells. For example, the dominant-negative vimentin$_{(1-138)}$ mutant is an effective tool for disrupting vimentin IF networks, causing them to accumulate mainly in the juxtanuclear regions instead of dispersing throughout cells. The expression of this mutant results in the inhibition of cell locomotion, suggesting that vimentin IFs play a critical role in motility (Helfand et al., 2011). In melanophores, this mutant was used to more accurately determine the step size of the molecular motors kinesin-2, dynein, and myosin V in living cells (Kural et al., 2007) and to demonstrate the role of vimentin IFs in positioning and anchoring pigment granules (Chang et al., 2009).

2.3 Silencing the Expression of Vimentin in Cells

Small-interfering RNA (siRNA) and short hairpin RNA (shRNA) have been very useful tools to knockdown vimentin protein expression in mammalian cells, reducing vimentin protein levels by up to 90% (Chernoivanenko, Matveeva, Gelfand, Goldman, & Minin, 2015; Tezcan & Gunduz, 2014). This approach has been employed in experiments aimed at determining the function of vimentin IFs. For example, knockdown of vimentin in fibroblasts causes changes in cell shape, cell motility, and mitochondrial membrane potential (Chernoivanenko et al., 2015; Helfand et al., 2011; Mendez et al., 2010). In addition, the utility of vimentin-null fibroblasts from the vimentin knockout mouse has been key in uncovering the functional significance of vimentin IF networks in various cellular processes including cell mechanics, cell motility, and organelle positioning (Guo et al., 2013; Helfand et al., 2011; Mendez et al., 2010).

Vimentin siRNA is now commercially available (e.g., Life Technologies, Grand Island, NY; see Table 1) and siRNA to specific targets can also be ordered (e.g., Integrated DNA Technologies, Coralville, IA). When designing primers for siRNA, the oligonucleotide sequence length should be 19–25 nucleotides and contain a two-nucleotide overhang at both 3′ ends. If the experiment involves both knocking down endogenous vimentin and expression of an exogenous vimentin, then the siRNA should be designed to target 5′ and 3′ untranslated regions.

1. Plate cells onto a 6-well, 12-well, or 24-well plate at least 24 h before transfection so that they are 50% confluent for transfection. For a 6-well dish, start with ~150,000 cells per well. Alter the number of cells based on well size to achieve 50% confluence. When using a smaller well size, scale the number of cells plated relative to the area of the well compared to a 6-well dish. The protocol below is based on the Mission siRNA transfection reagent from Sigma-Aldrich (St. Louis, MO).
2. For a 6-well dish, combine 12 pmol of siRNA duplex with 200 μL of serum-free medium. The final siRNA concentration will be ~5.45 n*M*.
3. Add 8 μL of transfection reagent to the siRNA, serum-free medium solution.
4. Mix the sample gently by pipetting ~10 times, then incubate for 10 min at room temperature.
5. Replace the medium in the wells with 2 mL of fresh medium containing serum.
6. Add the siRNA mixture with transfection agent to the cells in a dropwise manner.
7. Incubate the cells for ~72 h for the maximum vimentin knockdown.
8. Use the vimentin knocked down cells for downstream applications.

2.4 Withaferin A

Withaferin A (WFA), a steroidal lactone extracted from *Withania somnifera*, induces the reorganization of vimentin IFs. In fibroblasts treated with WFA at concentrations of 0.5–2.0 μ*M*, vimentin IF networks retract from the cell periphery toward the nucleus, leaving behind a small number of nonfilamentous vimentin particles and short IFs. This reorganization of IFs into large juxtanuclear aggregates is dose and time dependent and is followed by changes in cell shape and motility. After 3 h of treatment with 2 μ*M* WFA, fibroblasts change from an asymmetric elongated shape to a more rounded shape typical of epithelial cells. Time-lapse imaging reveals cell migration is significantly slower in cells treated with WFA compared with controls. The effects of WFA are reversible; following its removal, a majority of cells reestablish a normal vimentin IF network. The use of this small molecule as a specific inhibitor of vimentin IF structure and function remains to be determined. Despite these reservations regarding the specificity of WFA, there are numerous studies employing it as a specific probe for vimentin IF functions, especially in metastatic cells (Bargagna-Mohan et al., 2007; Satelli & Li, 2011; Shirahata & Hibi, 2014; Thaiparambil et al., 2011). WFA is discussed in further detail by Mohan and Bargagna-Mohan in Chapter 8 of this volume.

2.5 Gigaxonin as an Efficient Tool for Reducing and Eliminating Vimentin IFs from Cells

Giant axonal neuropathy (GAN) is a rare disease of children causing mainly neurological disorders and leading to death in the second or third decade of life (Mahammad et al., 2013). The pathological hallmark of GAN is the formation of large aggregates and bundles of polymerized vimentin IFs in dermal fibroblasts and in different types of neurons. This disease is caused by mutations in the *GAN* gene which encodes gigaxonin, a predicted E3-ligase adaptor protein that targets vimentin IFs for degradation through the ubiquitin-proteasome pathway (Mahammad et al., 2013). Overexpression of gigaxonin in fibroblasts causes vimentin IFs to disassemble and subsequently become completely degraded and cleared, with no obvious effects on the microtubule and microfilament cytoskeletal systems (Fig. 1B). Thus, gigaxonin is a useful tool for helping to dissect the structure and function of vimentin IFs.

3. ANALYSIS OF VIMENTIN DYNAMICS USING PHOTOACTIVATABLE AND PHOTOCONVERTIBLE PROTEIN TAGS

Once considered to form merely static structures, vimentin is now known to be dynamic and undergo active movement and rearrangement in cells. The use of live-cell imaging combined with photoconvertible protein tagging facilitates the study of the dynamic properties of vimentin.

3.1 Creating Cells That Express Photoactivatable and Photoconvertible Protein Fusions with Vimentin

It is important that the native vimentin IF network is not perturbed by the fluorescently tagged vimentin. In this regard, the choices of the photoconvertible probe, the linker between the probe and vimentin, and the position of the fusion protein on the N- or C-terminus are important. We have found that fusion of mMaple3, mEos3.2, or PA-GFP to the N-terminus of vimentin works well to create vimentin networks with a normal distribution of IFs (Hookway et al., 2015). Others have shown that vimentin organization also appears normal when mEos3.2 is fused to the C-terminus of vimentin (Wang, Moffitt, Dempsey, Xie, & Zhuang, 2014). Usually, it is preferable to use a photoconvertible rather than a photoactivatable probe for two reasons: first, the entire network is visible in the nonconverted channel before conversion. This facilitates cell positioning, focusing before conversion, and selection of an area for conversion. Second, photoactivatable probes have a low basal fluorescence before activation. For this reason, the signal-to-noise ratio after conversion will be higher for photoconvertible proteins than for photoactivatable ones. However, sometimes the use of photoactivatable tags may be advantageous because these probes require only one channel. Therefore, the other channels remain free to simultaneously image other fluorescently tagged proteins.

It is preferable to express vimentin-fusion proteins in cells that normally express vimentin, such as fibroblasts (e.g., NIH-3T3 and BJ-5ta), adenocarcinoma cells (vimentin-positive clones of SW13), and retinal pigment epithelial cells. This is because the tagged protein can incorporate nicely into the endogenous network, whereas it can be difficult for normal-looking networks to form when vimentin-fusion proteins are expressed in vimentin-free cell lines. In addition, many tags, especially those that oligomerize (but also monomeric dendra2), cause abnormal accumulation and/or aggregation of vimentin. Therefore, it is imperative to compare the pattern of localization of the vimentin-fusion protein to that of endogenous vimentin, which can be revealed by immunofluorescent labeling with a vimentin antibody (e.g., chicken polyclonal antivimentin antibody from BioLegend, San Diego, CA; see Table 1). Since overexpression of vimentin often causes filament aggregation, subcloning of cells expressing vimentin-fusion proteins to select cells with low vimentin expression may be necessary. Finally, we typically use retroviral techniques to express vimentin-fusion proteins because the high efficiency of transduction facilitates subcloning and speeds the creation of stable cell lines. However, transient transfection may also be used to express vimentin-fusion proteins. The transfection reagent used and the efficiency of expression will depend on the cell line and should be experimentally determined.

To study dynamics in vimentin filament precursors, the Y117L point mutant of vimentin should be used (Meier et al., 2009) instead of WT vimentin. Assembly of this mutant is blocked at the ULF stage. This construct should be expressed in cells that do not contain endogenous vimentin (such as vimentin-negative clones of SW13 or fibroblasts from a vimentin knockout mouse) because the Y117L mutant can copolymerize with endogenous vimentin and become incorporated into the endogenous IF network (Robert, Herrmann, Davidson, & Gelfand, 2014). Usually, tags that work well with WT vimentin behave well after fusion with the Y117L mutant.

3.2 Imaging Vimentin Dynamics Using Photoactivatable or Photoconvertible Proteins

Vimentin IFs are difficult to image since they are fine structures (of only ~10 nm in diameter) that form a very dense network. We recommend using total internal reflection fluorescence (TIRF) microscopy to image filaments close to the ventral side of the cell to increase the signal-to-noise ratio. However, we have found spinning disc confocal microscopy to be sufficient to image ULFs. We do not recommend wide-field epifluorescent imaging because out-of-focus light due to the density of vimentin creates high background.

A high-powered source of 405 nm light is necessary for photoactivation or photoconversion. To this end, we have successfully used LED (405 Heliophor, 89 North, Burlington, VT) and mercury (HBO 100 W/2) light illuminators with a filter cube (LF405B000; Semrock, Rochester, NY) in the epifluorescent light path as well as 405 nm laser light. In order to study dynamics, photoconversion should be limited to a small area of the cytoplasm. We have achieved this with LED or mercury light by replacing the field diaphragm in the microscope's epifluorescent light path with a removable pinhole. The projected region should be 10–20 μm in diameter (the pinhole size will vary depending on the microscope and magnification). This setup restricts the light used for conversion, but not laser illumination used for image collection in the TIRF or confocal modes. If 405 nm laser light is used to photoactivate/convert, the laser must be confined to a region of interest. Before conducting experiments, imaging parameters such as laser power and exposure time should be determined as described in the following section.

3.2.1 Optimizing Imaging and Photoconversion Parameters

1. Determine the exposure time required for optimal photoconversion: Photoconvert the entire field by removing any restriction to 405 nm illumination (e.g., by removing the pinhole). Take an image in the converted channel (red for photoconversion, green for photoactivation). Compare the intensity in the converted channel against different 405 nm light exposure times (and laser intensity if using a 405 nm laser) such that cells are maximally converted but not bleached by the conversion light. Repeat with many cells, since cells will vary in expression level. We find that 3–5 s of LED or mercury light exposure for photoconversion (mEos3.2 and mMaple3) and 10–20 s of exposure for photoactivation (PA-GFP) works well on our setup, but the time will vary for different light sources.
2. Once the parameters for conversion have been set, determine the bleaching rate in the photoactivated/converted channel. Use an exposure time and laser power that will yield an appropriate number of frames for the desired experiment. (For example, if only two frames are needed, a higher laser power and longer exposure time may be used, but the same setting may bleach the sample too rapidly for a different image series that requires many frames.) Be aware that some cell treatments can affect the bleaching. Thus, the bleaching rate should be determined for each experimental condition. Under control conditions, we find we can collect about 10 frames after mEos3.2 conversion and at least double that for mMaple3 before significant bleaching occurs. Once the number of frames and the exposure times have been determined, collect an image sequence and measure the average intensity for each frame to determine the photobleaching rate specific to the given experiment.
3. If two-color image series are to be collected after photoconversion, repeat step 2 for the nonconverted channel (in the absence of photoconversion).

3.2.2 Using Photoactivation/Conversion to Image Filament Transport

1. One day before the experiment, plate cells in their regular medium on glass bottom Petri dishes at about 50–80% confluence.
2. Thirty to sixty minutes before performing the imaging experiments, prepare the microscope for live-cell imaging by warming the stage to 37 °C

and creating a humid, 5% CO_2 environment for the cells (usually achieved using a stage-top incubator). Optional: Switch medium on cells to pre-warmed and CO_2-equilibrated medium optimized for imaging such as FluoroBrite Dulbecco-modified Eagle medium (DMEM; Life Technologies, Carlsbad, CA) or phenol-red free versions of the same medium used to culture the cells (with usual supplements, e.g., 10% fetal bovine serum).

3. Place cells on the stage for 10 min prior to imaging to allow the microscope and chamber stabilize. For long-term imaging, add mineral oil over the cell medium to prevent evaporation.
4. Use transmitted light microscopy to locate cells for imaging. If a photoconvertible probe is being used, the green channel (488 nm laser) can be used to check network distribution and focus on filaments before conversion. However, since bleaching in the green channel will decrease the protein available for photoconversion (or photoactivation in the case of PA-GFP), exposure to 488 nm light should be kept to a minimum.
5. Before photoactivation/conversion, take a background image using the same laser intensity and exposure time that will be used after photoactivation/conversion.
6. Photoactivate/convert using the parameters determined from Section 3.2.1 using a pinhole to restrict the light to a small region of the cell.
7. Immediately after conversion, collect an image series with frames taken every 15–20 s for at least 3 min in the appropriate channel (red for photoconvertible proteins and green for PA-GFP). Use exposure settings as determined in Section 3.2.1. An example of a typical photoconversion experiment to observe filament transport is shown in Fig. 2A.

3.2.3 Using Photoactivation/Conversion to Image Filament Severing and Reannealing

Use the same procedure described in Section 3.2.2 except with a photoconvertible (not photoactivatable) vimentin probe. In addition, step 7 of Section 3.2.2 should be modified such that frames (in both red and green channels) are collected over at least a 3 h period to allow time for filaments to sever and re-anneal. The number of frames collected during that time should be adjusted to account for the bleaching rate of the photoconvertible protein (see step 2 of Section 3.2.1).

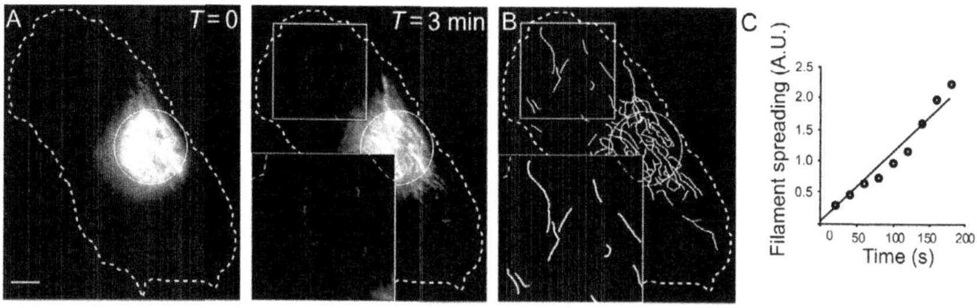

Figure 2 Study of vimentin filament transport using photoconversion of mEos3.2-vimentin. (A) Images taken immediately following photoconversion of the area marked with a white circle ($T = 0$) and 3 min later ($T = 3$). Many converted filaments can be seen outside the region of conversion after 3 min (*inset*). A log filter was applied and gamma was adjusted to 1.3. (B) Identified filaments for quantification. Inset shows detail for comparison to inset in (A). (C) Filament transport quantified as described in Section 3.3.1, step 4. Scale bar, 5 μm. *Figure adapted from Hookway et al. (2015).*

3.2.4 Using Photoconversion to Image Subunit Exchange in ULFs

Use the same procedure described in Section 3.2.2 except with a photoconvertible (not photoactivatable) probe fused to vimentin with the Y117L point mutation. In addition, step 7 of Section 3.2.2 should be modified such that images in both the red and green channels are collected following photoconversion. See Fig. 3 for an example of a photoconversion experiment analyzing ULF subunit exchange.

3.3 Quantification of Vimentin Dynamics

3.3.1 Measuring Vimentin Filament Transport

1. Correct for photobleaching in each image sequence by histogram matching (this can be performed in Fiji, an ImageJ processing package).
2. Identify filaments in each frame that have emerged from the region of photoactivation/conversion. This can be done by filament segmentation as described elsewhere (Hookway et al., 2015). The result is a binary representation of filaments so that they can be quantified by the number of pixels used to represent them (Fig. 2B).
3. Normalize the filament counts by the value of the sum intensity measured in the region of photoactivation/conversion in the first frame of the time series. This accounts for variation in the initial amount of converted vimentin.
4. Plot normalized filament counts versus time and calculate the slope, which we define as the rate of filament transport in the given image series (for series of less than 5 min, this range is linear) (Fig. 2C).

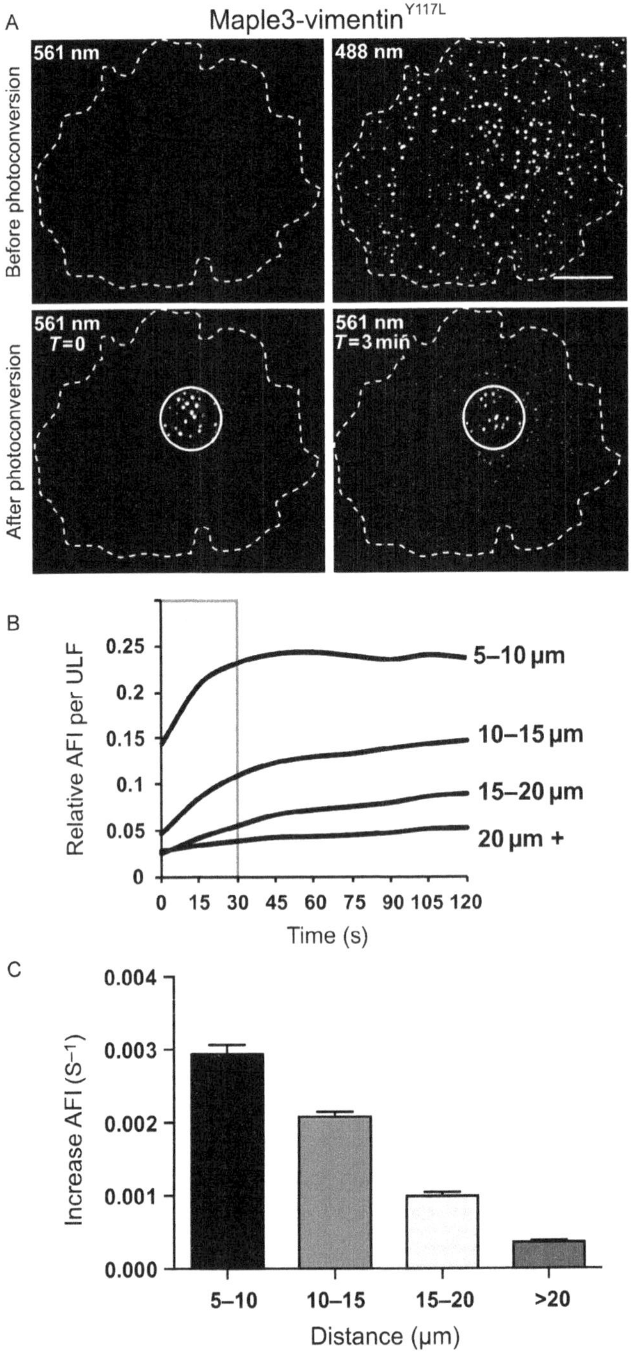

Figure 3 See legend on opposite page.

3.3.2 Measuring Subunit Exchange in ULFs

1. Track particles in the green channel using DiaTrack software (http://diatrack.org/index.html).
2. Using the coordinates of the particles generated in DiaTrack, measure the intensity of the particles in the red channel as a function of time.
3. Subtract background from each measurement using the image in the red channel taken before photoconversion.
4. Correct the data for photobleaching (described in step 2 of Section 3.2.1). Note that mMaple3 is more photostable than mEos3.2 and these data do not always require bleach correction.
5. Divide the measured red intensity of each ULF by the average intensity measured in the photoconverted region of the first frame to account for variation in photoconversion between cells.
6. To observe how the distance between a ULF and the converted region impacts the rate of fluorescence increase, group ULFs as a function of their distance from the center of the converted region.
7. Plot the average normalized intensities versus time for every group (Fig. 3B). Take the slope over the linear range to get the initial rate of exchange (Fig. 3C).

4. INVESTIGATING VIMENTIN–PROTEIN INTERACTIONS

A growing body of evidence has shown that vimentin interacts directly with a number of important signaling proteins (Barberis et al., 2009; Dos Santos et al., 2015; Stevens et al., 2013; Tzivion et al., 2000). Various methods can be used to investigate these vimentin–protein interactions.

Figure 3 Study of unit-length filament (ULF) subunit exchange using photoconversion of Maple3-vimentinY117L. (A) Images taken before photoconversion show ULF particles formed by Maple3-vimentinY117L are only visible in the green channel (488 nm) but not in the red channel (561 nm). Images taken after photoconversion in the area marked with a circle show that red fluorescence is initially confined inside the photoconverted zone ($T=0$) but accumulates in ULFs located outside the photoconverted area within 3 min ($T=3$). Subunit exchange was quantified as described in Section 3.3.2. The graph in (B) shows the normalized average fluorescence intensities (AFIs) versus time for individual ULFs grouped according to their distance from the photoconverted area. The slope of the curves was taken in their linear range (gray box on the graph) to determine the initial rate of exchange (increase in AFI per second) shown in (C).

4.1 Biolayer Interferometry

A new method to examine the binding affinity between vimentin and expected interacting proteins is biolayer interferometry (BLI) (Dos Santos et al., 2015). BLI is a dip-and-read, real-time assay system that evaluates changes in interference patterns of reflected white light between an internal reference and biomolecules attached to the tip of the sensor. Changes in the interference pattern are measured in nanometers and indicate that protein has bound to the sensor. The system can determine association rate (k_a), dissociation rate (k_d), and affinity constants (K_D) as well as protein concentration (Sultana & Lee, 2015).

BLI instruments are commercially available from ForteBio in two systems: the BLItz, which can analyze one sample at a time, and the Octet models, which are automated and can perform 8–96 assays in parallel using 96- or 384-well plates. The BLItz has an affinity range of 1–0.1 n*M* and analyzes proteins larger than 10 kDa. The Octet systems have a larger affinity range, 0.1 m*M*–10 p*M*, and can analyze samples that are larger than 150 Da.

Direct analysis of protein binding works best with purified proteins. BLI can be used to determine whether a protein is present in a sample by means of specific antibodies. Purification of vimentin and vimentin fragments has been described elsewhere (Strelkov et al., 2001); expression and purification strategies can also be found elsewhere (Palmer & Wingfield, 2012; Wingfield, Palmer, & Liang, 2014). Purification of proteins tagged with His_6 or GST is much simpler than purification of untagged proteins (Palmer & Wingfield, 2012; Wingfield et al., 2014).

1. Decide which protein should be attached to the biosensor as the bait. This choice should be made based on size, cost, and/or availability of the protein. The interference measured by BLI is related to the amount of protein on the sensor. A more substantial interference change during the association step can be observed when the protein with the larger molecular weight is used as the analyte.
2. Hydrate the biosensor for at least 10 min in BLI kinetics buffer (ForteBio). Alternatively, a buffer consisting of 1 × phosphate-buffered saline, 0.5% (w/v) bovine serum albumin, and 0.02% (v/v) Tween-20 can also be used. The biosensors can be hydrated for 24 h, so all biosensors expected be used for the experiment can be hydrated at the start of the experiment. If the biosensor is not completely hydrated, a noisy sensorgram (trace) will be recorded.
3. Set up the experiment in the BLItz Pro software (Fig. 4A).

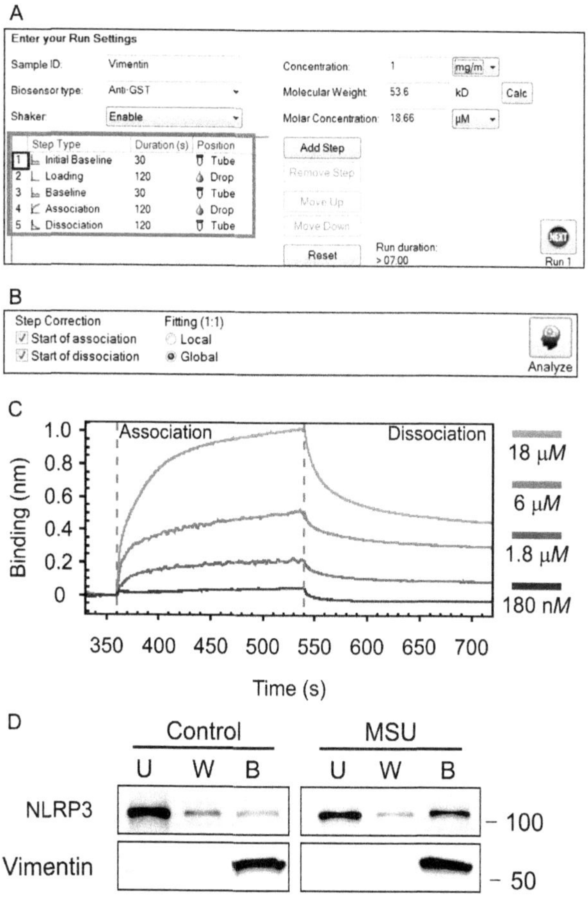

Figure 4 Biolayer interferometric analysis of interaction between vimentin and NLRP3 (NACHT, LRR, and PYD domains-containing protein 3). (A) Screenshot of the BLItz Pro software in which the settings for an experiment are shown. The only value that needs to be provided is the Molar Concentration, which can be altered for each run. If the concentration and molecular weight of the protein being used as the analyte is provided, the Molar Concentration value will be calculated. The red square shows where the Step Type, Duration, and Position can be specified for the experiment. (B) Settings for binding analysis: Start of Association and Start of Dissociation checked, and Global Fitting selected. (C) Determination of association and dissociation rates between a bait protein (NLRP3) attached to a biosensor and various concentrations of vimentin diluted in BLI kinetics buffer. (D) Examination of binding between NLRP3 and vimentin. Cell extracts were prepared from vimentin-null bone marrow-derived macrophages (BMDMs). Some cells were treated with monosodium urate (MSU) to activate the BMDMs, while others were treated with saline (control). S-tag, His-tag vimentin was added to these extracts and binding of NLRP3 was examined (U, unbound; W, washed; B, bound). Binding was evaluated by chemiluminescence. (See the color plate.)

4. Place the biosensor onto the biosensor mount and ensure a tight fit by twisting the biosensor a quarter turn while pushing the biosensor onto the mount.
5. Pipette 250 μL of BLI kinetics buffer into a 0.5 mL black opaque microcentrifuge tube and place the tube into the tube holder. Move the slider to the correct position so the biosensor lines up with the microcentrifuge tube.
6. Run the experiment. For the BLItz system, the slider will have to be moved manually and BLI kinetics buffer, ligand (bait), and analyte added when necessary.
   a. Lower the biosensor into the BLI kinetics buffer and determine the initial baseline.
   b. Add 4 μL of bait, diluted with BLI kinetics buffer, to the drop holder and move the slider to the correct position. The concentration of the bait protein should be 10–50 μg/mL. A starting concentration of 30 μg/mL is recommended.
   c. Lower the biosensor into the sample and determine the loading of bait onto the biosensor.
   d. Place the biosensor back into the microcentrifuge tube and determine the baseline.
   e. Add 4 μL of the analyte, diluted in BLI kinetics buffer, to the drop holder and slide the sample into position to determine the association.
   f. Move the slider back to the tube holder and determine the dissociation.
   g. Discard the biosensor.
   h. Clean the drop holder with either a swab or a Kimwipe and rinse three times with sample buffer.
7. Repeat the experiment for all dilutions of the analyte. ForteBio recommends at least four dilutions per analyte. For best results, the concentration of one of the analyte dilutions should be ~10-fold higher than the expected K_D.
8. Analyze the data using the BLItz Pro software.
   a. Examine the sensorgram and uncheck any runs that did not perform correctly. These would include very noisy sensorgrams, runs in which the baseline was not stable, or runs for which the loading curve indicates that protein was not loaded onto the biosensor.
   b. Select which run will be used as a reference trace (no protein in the analyte).

c. Check the boxes for step correction for both Start of Association and Start of Dissociation, and click the radio button for Global Fitting (Fig. 4B).
b. Save the data set, copy the analysis to a spreadsheet, and save the figure (Fig. 4C).

4.2 Soluble Bead Binding Assay Between Vimentin and Protein Extracts

Interactions between vimentin and other proteins in a cell can be examined in solution through the addition of exogenous protein or protein fragments to a cell lysate (Dos Santos et al., 2015; Shumaker et al., 2008). This method can indicate which domain is required for interaction with a protein or a protein complex. The assay in Fig. 4D shows the interaction between vimentin and NACHT, LRR, and PYD domains-containing protein 3 (NLRP3), a NOD-like receptor protein which forms a large multiprotein complex that activates interleukin 1β (Dos Santos et al., 2015). Here, NLRP3 is used as the binding partner with vimentin as a demonstration of the efficacy of the assay for examining protein–protein interactions with IF proteins (Dos Santos et al., 2015; Shumaker et al., 2008).

1. Clone the complementary DNA sequence for vimentin or vimentin fragments into a bacterial expression vector that contains a tag, for example, S-tag, GST, or His_6.
2. Bacterially express and purify vimentin or vimentin fragments (Strelkov et al., 2001).
3. Prepare cell extract by lysing cells. For adherent cells, wash cells twice with ice-cold phosphate-buffered saline. Place the dishes on ice for 5 min, then add 1 mL of prechilled lysis buffer (4 °C) for a 100 mm dish; alter the volume relative to the surface area of the dish.
   a. Start with a basic lysis buffer: 50 m*M* Tris (pH 7.4), 150 m*M* NaCl, and 1% NP-40. If the basic lysis buffer does not release the protein of interest into a soluble form, the salt concentration and detergents may be altered. Salt concentrations of 50–250 m*M* can be used, or NP-40 can be replaced with Triton X-100. Other nondenaturing detergents such as CHAPS or deoxycholate can also be used.
   b. Protease inhibitors should be included in the lysis buffer. Boehringer sells protease inhibitor cocktail tablets. Protease inhibitors that are commonly used include aprotinin (1 μg/mL), pepstatin (1 μg/mL), leupeptin (1 μg/mL), PMSF (50 μg/mL), and tosyl phenylalanyl chloromethyl ketone (10 μg/mL).

   c. If the phosphorylation state of the cellular proteins is important, use Na_3VO_4 (100 μ*M*), NaF (25 m*M*), or 40 m*M* β-glycerol phosphate. EDTA (1 m*M*) can also be used but it may interfere with downstream assays, for example, if Ni-NTA (nickel agarose) is being used instead of S-protein agarose.
   d. To stabilize the proteins, up to 5% glycerol or 300 m*M* sucrose can be used in the lysis buffer. Lysosomal lysis can be diminished with ~250 m*M* sucrose.
4. Lyse cells on ice for 5–10 min depending on cell type. During lysis, rock the plate to distribute the lysis solution.
5. Scrape the cells to one side of the dish. The DNA can be sheared by pulling the lysate through a 30-gauge needle about 10 times. Alternatively the lysates can be sonicated using a small tip on the sonicator.
6. Cell supernatant can be clarified by pelleting cell membranes and large complexes at 14,000 × *g* for 10 min at 4 °C.
7. Measure the concentration of the lysate. The lysate can be flash frozen in liquid nitrogen and stored at −80 °C until needed. It is preferable to use the lysate immediately.
8. Mix 2 mg of lysate with 20 μg of S-tag vimentin in a microcentrifuge tube and then rotate the mixture at 4 °C for at least 4 h.
9. Add 30 μL of washed S-protein agarose beads (50% slurry) to the protein mix. The binding capacity for the S-protein agarose beads is approximately 2000 μg/mL of beads.
10. Incubate at 25 °C for 30 min while rotating.
11. Pellet the beads at 500 × *g* for 10 min at 4 °C.
12. Remove the supernatant and retain to evaluate binding of vimentin to the S-protein agarose.
13. Wash the beads with either the 1 × binding/wash solution that comes with the beads or the lysis buffer.
14. Pellet the beads at 500 × *g* for 10 min at 4 °C.
15. Remove the supernatant. Retain a portion to evaluate binding of vimentin and proteins of interest. The rest can be discarded.
16. Repeat steps 12–14 two times.
17. The bound protein can be eluted from the beads either with a low pH buffer (0.2 *M* citrate, pH 2.0), 0.3 *M* $MgCl_2$, or with 1 × binding/wash buffer containing 3 *M* guanidine thiocyanate. Incubate the beads with the buffer of choice for 10 min at 25 °C. Pellet the beads at 500 × *g* for 10 min and retain the supernatant, which contains the bound proteins. If the downstream application is only a Western blot, add sodium

dodecyl sulfate lysis buffer and prepare the sample for Western blotting. The unbound, washed, and bound samples can be examined thusly for specific bound proteins (Fig. 4D).

5. INVESTIGATING THE MECHANICAL PROPERTIES OF VIMENTIN IF NETWORKS

The unusual rheological properties of vimentin networks *in vitro* have stimulated studies to determine how vimentin networks alter the mechanical properties of cells and tissues. Studies of purified vimentin networks have led to a consensus that they are very soft at low shear deformations, compared with the shear modulus of cross-linked actin, for example, but also show dramatic stiffening with increasing shear strains at which actin and microtubule networks fail (Guzman et al., 2006; Janmey, Euteneuer, Traub, & Schliwa, 1991; Lin et al., 2010; Schopferer et al., 2009). Studies of cells and tissues in which vimentin levels or distribution have been altered have, in contrast, led to divergent conclusions about the contribution of vimentin to the overall mechanical properties of the cell (these results are discussed by Charrier and Janmey in Chapter 2 of this volume).

5.1 Three-Dimensional Substrate Studies

An important limitation of most cell studies *in vitro* is their reliance on rigid two-dimensional surfaces designed to optimize imaging, but not to reproduce the mechanical properties of the compliant tissues in which vimentin-expressing cells function *in vivo*. Additional recent emphasis on using three-dimensional substrates with tunable stiffness to study mesenchymal cells reveals effects of vimentin not evident in single-cell, two-dimensional cultures. For example, mesenchymal stem cells, endothelial cells, and fibroblasts exhibit biphasic changes in vimentin detergent solubility when cultured on three-dimensional substrates of different stiffness, whereas the vimentin remains largely insoluble when these cells are cultured on glass or plastic substrates (Murray, Mendez, & Janmey, 2014).

Soft substrates with elastic modulus in the range of 100–50,000 Pa are generally made from polyacrylamide (PAA) or other covalently cross-linked hydrogel networks to which specific ligands for integrins or other receptors are covalently attached. A typical formulation suitable for studies of vimentin-expressing fibroblasts (Mendez, Restle, & Janmey, 2014; Murray et al., 2014), endothelial cells (Galie, van Oosten, Chen, & Janmey, 2015), or glioblastoma cells (Pogoda et al., 2014) is listed below and discussed in more detail in other reports (Engler et al., 2004; Wang & Pelham, 1998; Yeung et al., 2005).

Substrates of 100 μm thickness on 22 mm^2 coverslips are prepared using the following mixtures:

- For 6 kPa gels: 150 μL of 7.5% (w/v) PAA plus 106 μL of 2% bis-acrylamide (BioRad, Hercules, CA).
- For 36 kPa gels: 150 μL of 12% (w/v) PAA plus 196 μL of 2% bis-PAA (BioRad, Hercules, CA).

The elastic modulus of these substrates is varied by altering either the total amount of acrylamide or the ratio of bis-acrylamide cross-linker to acrylamide monomer. Once the monomer solutions are mixed, polymerization is initiated by ammonium persulfate and Tetramethylethylenediamine (TEMED) by the standard method used to make PAA gels for electrophoresis and described in more detail elsewhere (Kandow, Georges, Janmey, & Beningo, 2007; Wang & Pelham, 1998). Polymerized gels are most commonly activated for attachment to amine- or sulfhydryl-containing proteins or other ligands by Sulfo-SANPAH (Thermo Scientific, Waltham, MA) followed by incubation typically in 100 μg/mL fibronectin or other adhesion proteins.

PAA gels have several advantages for such studies because the gels they form are optically transparent. Their network structure is also isotropic, with a relatively uniform mesh size throughout the sample, and their viscoelastic properties remain constant over a large range of strains and timescales. The local deformation of the network also closely matches the global sample strain (Basu et al., 2011).

Changing the stiffness, as quantified by the elastic modulus, of such substrates has a large effect on the state of vimentin assembly as well as global aspects of cell morphology, motility, and other features. For example, when attached to stiff 30 kPa substrates, nearly all of the vimentin in fibroblasts, endothelial cells, and mesenchymal stem cells is insoluble after detergent extraction, consistent with previous reports of the relatively stable vimentin network (Murray et al., 2014). However, as seen in Fig. 5, solubility increases to over 50% when cells adhere to substrates with elastic moduli of a few kilopascals, similar to the stiffness of many mesenchymal tissues. The soluble pool of vimentin does not appear to be enriched in vimentin tetramer subunits, but rather in ULFs that are thought to be fundamental assembly intermediates in IF formation (Portet et al., 2009). The amount of soluble vimentin coincides with the fraction of the cells undergoing active ruffling (Murray et al., 2014) and is consistent with previous reports that

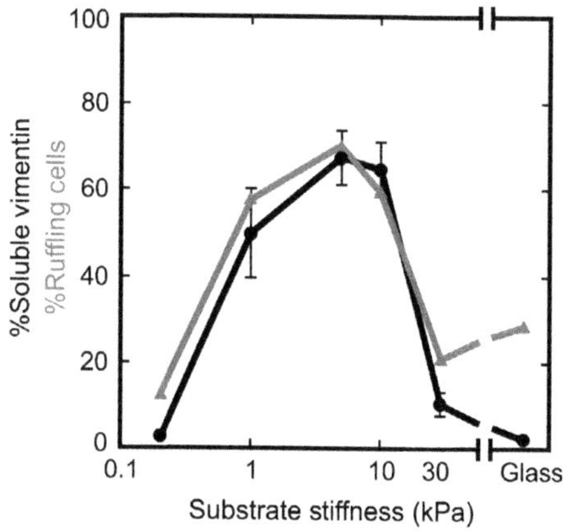

Figure 5 Effect of substrate stiffness on vimentin assembly state. Shown is the Triton X-100 soluble fraction of vimentin (*black line*) and fraction of cell undergoing active ruffling (*red line* (gray in the print version)) in human mesenchymal stem cells obtained from Lonza (Walkersville, MD) and cultured on fibronectin-coated substrates with various degrees of stiffness.

dynamic disassembly of vimentin networks is required for protrusion of the cell edge (Helfand et al., 2011).

5.2 Collagen Gel Contraction Studies

Three-dimensional constructs of cells embedded within polymer networks can reveal effects for cellular vimentin that are not evident from studies of cells on surfaces. For example, the role of vimentin on the forces generated by cells adhered to an extracellular matrix was first documented by measuring the contraction of collagen gels containing fibroblasts prepared from vimentin-null mice or their WT littermates. The original studies of these cells reported that vimentin-null cells were less capable of contracting collagen gels than were normal fibroblasts, suggesting that the lack of vimentin lessens either the contractile force of the cells or their ability to remodel the matrix (Eckes et al., 1998). Later studies not only confirmed this result as long as the cell density was relatively low but also showed that at high cell density at which cell–cell contact was significant compared to cell–matrix contact, cells lacking vimentin were ultimately able to contract the gels more rapidly and to a greater extent (Mendez et al., 2014).

Three-dimensional collagen gels are prepared by suspending pelleted cells to produce a controlled density of cells in medium containing 1 ×

DMEM (diluted from 5 ×; Life Technologies, Grand Island, NY), 10% fetal bovine serum, and 2 mg/mL collagen I in a 3 mL volume, cultured within a 35 mm dish. Other adhesion proteins such as 0.1 μg/mL fibronectin can be added, and the collagen concentration can be varied from 1 μg/mL to as high as solubility allows to alter the adhesive density and elastic moduli of the gels. The ability of cells to contract the matrix is most conveniently measured by freeing the collagen matrix from the sides of the dish by running a pipette tip around the circumference at a given time, then imaging over a time course of minutes to days as the gel diameter decreases as the result of active contraction by cells.

Figure 6 shows a typical time course of collagen gel contraction by embedded fibroblasts isolated from WT or vimentin-null mice. At low cell densities the vimentin-null cells are less able to contact the gels, in agreement with earlier studies (Eckes et al., 1998), but at high cell densities at which cell–cell contacts become significant, the vimentin-null cells are more active in gel contraction (Mendez et al., 2014).

An important limitation of gel contraction studies such as those in Fig. 6 is that they are not a direct measurement of the contractile forces generated by the cells. The change in gel diameter depends in part of the contractile work done by the cells, but also on the effects of the cells on the gel stiffness, as well as their ability to remodel the collagen networks by secretion and activation of proteases. As a result, the net effect of force generation by the cells is quantified by these studies, but the direct role of vimentin on the forces generated by individual cells within three-dimensional matrices remains to be quantified.

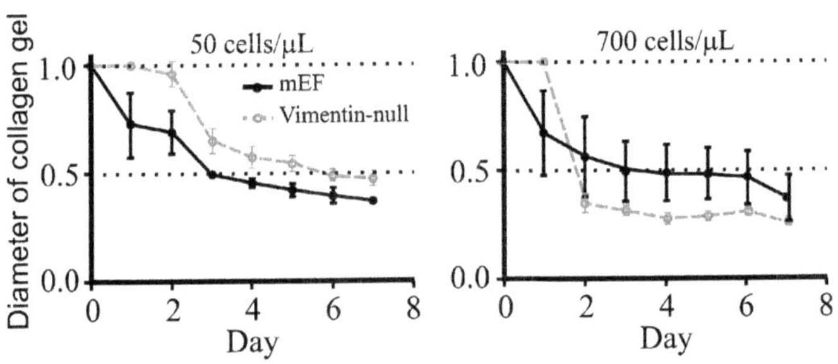

Figure 6 Variation in contractility of vimentin-null cells with cell density and gel composition. Normal or vimentin-null mouse embryonic fibroblasts (mEFs) were cultured in 2 mg/mL collagen gels for 7 days at densities of 50 (A) or 700 (D) cells/μL. Gels were freed from the dish edge 24 h after plating.

6. INVESTIGATING THE ROLE OF VIMENTIN IFs IN CELL MECHANICS

Cytoplasmic IFs, such as vimentin, typically form beautiful structures in the cell; however, their network configurations in cellular architecture are often altered in IF-related diseases due to incorrect polymerization or organization (Omary, Coulombe, & McLean, 2004). This suggests dramatic alterations in that the mechanical properties of the IF networks, in addition to possible changes in biochemical functionality, and may play a role in the development of IF-related diseases. Therefore, characterizing the role of cytoplasmic IF networks in determining the mechanical property of cells is essential to understand the fundamental function of IFs and their related diseases. Recent advances in direct characterization of intracellular mechanics enable the mechanical role of IFs and its consequence in regulating intracellular dynamics to be revealed (Guo et al., 2013, 2014).

6.1 Investigating the Role of Vimentin in Cytoplasmic Mechanics Using Optical-Tweezer Active Microrheology

To measure cytoplasmic mechanics, active microrheology is performed using optical tweezers on single 500 nm diameter polystyrene particles that have been endocytosed by WT or *Vimentin*$^{-/-}$ (*Vim*$^{-/-}$) mouse embryonic fibroblasts (mEFs) (Mendez et al., 2010). These particles are covered with lipid layers during endocytosis and can thus be transported along microtubules; however, most of the time these particles display random movement and are randomly distributed within the cytoplasm. These WT and *Vim*$^{-/-}$ mEFs are generated according to the protocol in Section 2.3. To focus on the contribution of vimentin IFs to cytoplasmic mechanics, measurements are performed with particles located away from both the thin lamellar region and the nucleus, which avoids these mechanically distinct regions of the cell. About 8 h after adding particles, cytoplasmic mechanical properties are measured by active microrheology using optical tweezers to apply an oscillating force F on the trapped particle to deform the cytoplasm (Fig. 7A), with the following setup (Guo et al., 2013).

1. To optically trap and manipulate beads in the cytoplasm, the beam from a variable power Nd:YAG solid-state laser (4 W, 1064 nm; Spectra Physics, Mountain View, CA) is steered through a series of Keplerian beam expanders to overfill the back aperture of a 100× and 1.3 numerical aperture microscope objective (Nikon S-fluor; Nikon, Tokyo, Japan).

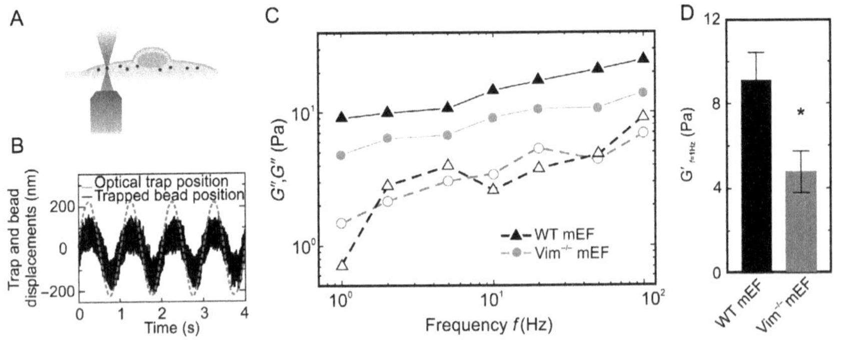

Figure 7 Optical tweezers measurement of intracellular mechanics. (A) Schematic of the optical tweezer experiment. (B) Typical displacements of the trapped bead and the optical trap oscillating at 1 Hz. (C) Frequency-dependent cytoplasmic elastic moduli G' (*filled symbols*) and loss moduli G'' (*open symbols*) of the wild type (WT) and *Vimentin*$^{-/-}$ (*Vim*$^{-/-}$) mouse embryonic fibroblasts (mEFs). The cytoplasm of the WT mEFs (*triangles*) is stiffer than that of the *Vim*$^{-/-}$ mEFs (*circles*). (D) Cytoplasmic elastic moduli in the WT and *Vim*$^{-/-}$ mEFs at 1 Hz. Error bars, SEM ($*p < 0.05$).

2. To steer the beam and manipulate the trapped bead, two acousto-optic deflectors are used (NEOS Technologies, Melbourne, FL). Using a custom-written Labview program (National Instruments, Austin, TX), the acousto-optic deflectors are manipulated to control the beam in the plane of the microscope glass slide.
3. For detection, the bead is centered on a high-resolution position detection quadrant detector (MBPS; Spectral Applied Research, Richmond Hill, ON, Canada) and illuminated using bright-field illumination from a 75 W xenon lamp. The linear region of the detector is calibrated by trapping a bead identical to those used in the cells in water and moving it across the detector using the acousto-optic deflectors in known step sizes.
4. The trap stiffness is calibrated from the mean squared Brownian motion of a trapped bead in water at various laser power settings using the principle of energy equipartition as described elsewhere (Veigel, Bartoo, White, Sparrow, & Molloy, 1998). Once calibrated, the laser trap is used to optically trap and manipulate beads intracellularly. For measurements, in the cytoplasm of mEFs, a trap stiffness of 0.05 pN/nm is used.
5. Trapped beads are oscillated across a frequency range of 1–100 Hz, and the laser position and bead displacement are recorded simultaneously, from which the elastic and viscous shear moduli are determined. By measuring the resulting displacement of the bead, $x(\omega)$, subjected to

an applied sinusoidal trap oscillation with a force F at frequency ω, the effective spring constant, $K(\omega) = F(\omega)/x(\omega)$, can be extracted for a given intracellular environment.

6. For purely elastic materials, displacement and force are in phase; for materials with dissipation, the displacement and force are not in phase, which results in a complex spring constant. For a homogeneous, incompressible viscoelastic material, this spring constant is related to a complex modulus, $G = G' + iG''$, through a generalization of the Stokes relation $K = 3\pi Gd$ (Mizuno, Tardin, Schmidt, & MacKintosh, 2007), where d is the bead diameter.

Active microrheology measurements show that the cytoplasm of mEFs is an elastic gel instead of a viscous fluid, when measured on submicron length scales. Furthermore, both the elastic modulus G' and the loss modulus G'' increase with frequency, following a power-law form, $|G(\omega)| \sim \omega\beta$, with $\beta \approx 0.25$ (Fig. 7B). Although both WT and $Vim^{-/-}$ mEFs show similar frequency-dependent behavior, the cytoplasmic elastic modulus, G', of WT mEFs is larger than that of $Vim^{-/-}$ mEFs, as shown in Fig. 7C. Specifically, at 1 Hz the cytoplasm of WT mEFs is twice as stiff as that of $Vim^{-/-}$ mEFs; thus, the presence of vimentin increases the cytoplasmic elastic modulus from approximately 5–9 Pa (Fig. 7D). However, the loss modulus G'' is not significantly different between the WT and $Vim^{-/-}$ cells over the investigated frequency range; the loss tangent, as defined by G''/G', which represents the relative dissipation of materials, is roughly twice as large for the $Vim^{-/-}$ cells, indicating that the presence of vimentin also reduces energy dissipation in the cytoplasm. The significant difference in cytoplasmic moduli between WT and $Vim^{-/-}$ mEFs reflects the contribution of vimentin IFs to the intracellular stiffness, suggesting that vimentin is a crucial structural cellular component within the cytoplasm.

6.2 Investigating the Role of Vimentin in Intracellular Dynamics

The vimentin IFs also affect intracellular activity. To investigate how intracellular dynamics are influenced by cytoplasmic mechanics due to the vimentin IF network, the movements of endogenous vesicles and protein complexes are tracked in WT and $Vim^{-/-}$ mEFs (Guo et al., 2013). These refractive objects are visualized by bright-field microscopy using a 633-nm laser and a 63 × and 1.2 numerical aperture water immersion objective on a Leica TSC SP5 microscope. To avoid cell-boundary effects, trajectories from the thin actin-rich lamellar region and the mechanically distinct nucleus are excluded, and instead trajectories greater than ~ 1 μm deep within the cell are analyzed, where vimentin IFs are typically distributed. The trajectories of the vesicles and protein complexes are recorded every 18 ms for 30 s. Vesicle and protein complex centers are determined by calculating the centroid of the object's brightness distributions in each image with an accuracy of 20 nm using custom-written particle tracking software in IDL. Object trajectories are tracked in order to calculate the time- and ensemble-averaged mean squared displacement (MSD), $\langle \Delta r^2(\tau) \rangle$, where $\Delta r(\tau) = r(t+\tau) - r(t)$. The MSD of the probe particles is nearly constant in time at short timescales ($t \leq 0.1$ s), and is about an order of magnitude greater than the noise floor. Occasionally, the motion is clearly directed, with objects moving along a straight path at a constant velocity, reflecting vectorial transport along microtubules by motors. However, the majority of the motion appears to be random, and the MSD increases linearly in time, reflecting the diffusive-like nature of the motion (Lau, Hoffman, Davies, Crocker, & Lubensky, 2003). While the trajectories of vesicles and protein complexes in both WT and $Vim^{-/-}$ mEFs indicate random movements, these organelles in $Vim^{-/-}$ mEFs move farther over the same timescale, as shown in Fig. 8A and B. Quantifying the trajectories by plotting the MSD of these organelles reveals that while both of them increase linearly with time, the vesicles and protein complexes move an order of magnitude faster in the $Vim^{-/-}$ mEFs compared with the control WT mEFs (Fig. 8C). This increased movement in $Vim^{-/-}$ cells is consistent with previous observations of the movements of mitochondria (Nekrasova et al., 2011), melanosomes (Chang et al., 2009), the Golgi apparatus (Gao & Sztul, 2001; Gao, Vrielink, MacKenzie, & Sztul, 2002), and other organelles (Styers, Kowalczyk, & Faundez, 2005; Styers et al., 2004), indicating that vimentin IFs contribute to the localization of a variety of different organelles.

6.3 Investigating the Contribution of Vimentin to the Aggregate Intracellular Forces

The fluctuating motion of intracellular organelles reflects the average random fluctuations due to the aggregate motor activity in the cell. If both the fluctuating motion and the cytoplasmic viscoelasticity are measured, the spectrum of the average fluctuating force due to these motors, which drives this motion, can be directly determined, through $\langle f^2(v) \rangle = |K(v)|^2 \langle x^2(v) \rangle$, using a method called force spectrum microscopy (FSM) (Guo et al., 2014). This average force is due to the aggregate,

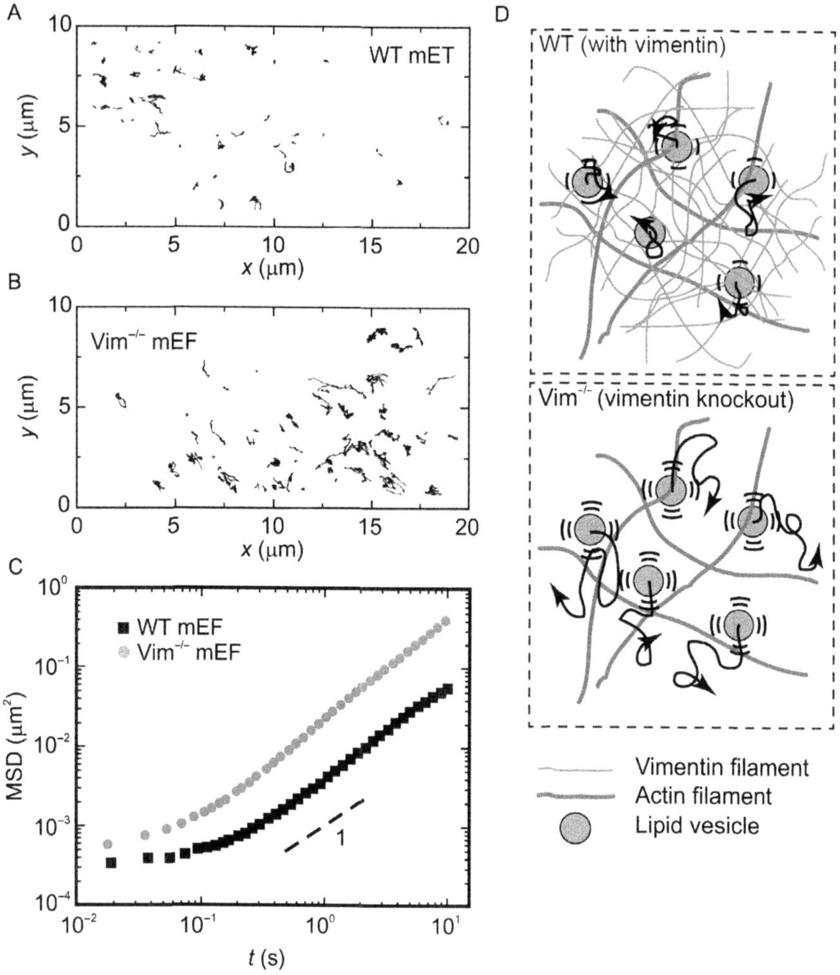

Figure 8 Intracellular movement of endogenous vesicles and protein complexes inside wild-type (WT) and *Vimentin*$^{-/-}$ (*Vim*$^{-/-}$) mouse embryonic fibroblasts (mEFs). (A, B) Ten-second trajectories of endogenous vesicles and protein complexes in the cytoplasm of (A) WT mEFs and (B) *Vim*$^{-/-}$ mEFs. These refractive objects are visualized by bright-field microscopy. (C) Calculation of the mean squared displacement of vesicles and protein complexes shows that these organelles move faster in the *Vim*$^{-/-}$ mEFs than in the WT mEFs. (D) Illustration of random organelle movement in networks with and without vimentin. In the WT cells, the vimentin network constrains the diffusive-like movement of organelles; in the *Vim*$^{-/-}$ cells, organelles move more freely. (See the color plate.)

yet random, effects of all active processes in the cell. Although these forces are inherently time-dependent, it is more convenient to describe their frequency-dependent spectrum. FSM provides a way to investigate the contribution of specific cellular components to the level of aggregate intracellular forces. For example, FSM can be applied to characterize the role of vimentin in the aggregate of intracellular forces in WT and *Vim*$^{-/-}$ mEFs, simply combining the cytoplasmic viscoelasticity measurement in Section 6.1 and intracellular movement measurement in Section 6.2. Interestingly, no significant difference in the intracellular force spectrum between WT and *Vim*$^{-/-}$ mEFs is found. In contrast, the force spectrum is markedly reduced when actin filaments are depolymerized with 5 μg/mL cytochalasin D in WT mEFs. These results suggest that vimentin IFs are mainly structural polymers that are an important contributor to the internal stiffness of cells, but do not affect the aggregate intracellular forces.

7. CONCLUSION

Vimentin IFs were once thought to be static, merely structural proteins, but they are now appreciated as dynamic structures that participate in essential cellular processes. These processes link vimentin to a number of diseases and conditions, including cancer, inflammation, and even cataract formation (Dos Santos et al., 2015; Kidd et al., 2014; Muller et al., 2009; Stevens et al., 2013). To study the cellular functions of vimentin IFs, various methods have been developed which should be applicable to other types of IF. In general, one does not have to be aware of any special caveats when applying the techniques described for disrupting vimentin IFs to other IF systems. With respect to mimetic peptides, for example, it has been possible to design peptides that disrupt keratin IF filaments (see references in Goldman et al., 1996). Overexpression of gigaxonin not only clears vimentin but also eliminates the neurofilament proteins peripherin and NF-L. The effect of gigaxonin overexpression on other types of IF protein remains to be determined. The dynamics of vimentin IFs can be investigated by means of photoactivatable and photoconvertible protein fusions combined with live-cell imaging. These techniques should not be restricted to the study of vimentin, but in fact have been used successfully to characterize dynamic properties of keratin (Kolsch, Windoffer, Wurflinger, Aach, & Leube, 2010) and neurofilaments (Colakoglu & Brown, 2009; Uchida, Colakoglu, Wang, Monsma, & Brown, 2013). The direct interactions between vimentin and other proteins can be examined by specialized binding assays such as BLI, a recently developed technique for examining protein–protein interactions. To our knowledge, the only IF protein that has been examined by BLI is vimentin (Dos Santos et al., 2015), but both BLI and soluble bead binding assays are applicable to all IF proteins.

Altogether, these methods hold promise for continued discovery and investigation of vimentin's cellular functions in the future.

ACKNOWLEDGMENTS

The authors wish to thank Ms. Jennifer Davis for the editing of the Chapter; Melissa Mendez, Maria Murray for their respective contributions in the development of the techniques outlined in this chapter.

Funding: R.D.G. is supported by grants from the National Institute of General Medical Sciences (P01GM09697), National Institutes of Health and the Hannah's Hope Fund. V.I.G. is supported by grants from the National Institute of General Medical Sciences (P01GM09697 and GM052111), National Institutes of Health. K.M.R. is supported by the National Heart, Lung, and Blood Institute (HL71643; HL124664), Department of Veterans Affairs (MERIT Award). P.A.J. is supported by grants from the National Institutes of health (GM096971 and EB017753). D.A.W. is supported by the NIH (PO1GM096971), the Harvard Materials Research Science and Engineering Center (DMR-0820484).

REFERENCES

Barberis, L., Pasquali, C., Bertschy-Meier, D., Cuccurullo, A., Costa, C., Ambrogio, C., et al. (2009). Leukocyte transmigration is modulated by chemokine-mediated PI3Kgamma-dependent phosphorylation of vimentin. *European Journal of Immunology, 39*(4), 1136–1146. http://dx.doi.org/10.1002/eji.200838884.

Bargagna-Mohan, P., Hamza, A., Kim, Y. E., Khuan Abby Ho, Y., Mor-Vaknin, N., Wendschlag, N., et al. (2007). The tumor inhibitor and antiangiogenic agent withaferin A targets the intermediate filament protein vimentin. *Chemistry & Biology, 14*(6), 623–634. http://dx.doi.org/10.1016/j.chembiol.2007.04.010.

Basu, A., Wen, Q., Mao, X., Lubensky, T. C., Janmey, P. A., & Yodh, A. G. (2011). Non-affine displacements in flexible polymer networks. *Macromolecules, 44*(6), 1671–1679. http://dx.doi.org/10.1021/ma1026803.

Chang, L., Barlan, K., Chou, Y. H., Grin, B., Lakonishok, M., Serpinskaya, A. S., et al. (2009). The dynamic properties of intermediate filaments during organelle transport. *Journal of Cell Science, 122*(Pt 16), 2914–2923. http://dx.doi.org/10.1242/jcs.046789. jcs.046789 [pii].

Chernoivanenko, I. S., Matveeva, E. A., Gelfand, V. I., Goldman, R. D., & Minin, A. A. (2015). Mitochondrial membrane potential is regulated by vimentin intermediate filaments. *FASEB Journal, 29*(3), 820–827. http://dx.doi.org/10.1096/fj.14-259903.

Colakoglu, G., & Brown, A. (2009). Intermediate filaments exchange subunits along their length and elongate by end-to-end annealing. *Journal of Cell Biology, 185*(5), 769–777. http://dx.doi.org/10.1083/jcb.200809166.

Dos Santos, G., Rogel, M. R., Baker, M. A., Troken, J. R., Urich, D., Morales-Nebreda, L., et al. (2015). Vimentin regulates activation of the NLRP3 inflammasome. *Nature Communications, 6*, 6574. http://dx.doi.org/10.1038/ncomms7574.

Eckes, B., Dogic, D., Colucci-Guyon, E., Wang, N., Maniotis, A., Ingber, D., et al. (1998). Impaired mechanical stability, migration and contractile capacity in vimentin-deficient fibroblasts. *Journal of Cell Science, 111*(Pt 13), 1897–1907.

Engler, A., Bacakova, L., Newman, C., Hategan, A., Griffin, M., & Discher, D. (2004). Substrate compliance versus ligand density in cell on gel responses. *Biophysical Journal, 86*(1 Pt 1), 617–628. http://dx.doi.org/10.1016/S0006-3495(04)74140-5. S0006-3495(04)74140-5 [pii].

Eriksson, J. E., Dechat, T., Grin, B., Helfand, B., Mendez, M., Pallari, H. M., et al. (2009). Introducing intermediate filaments: From discovery to disease. *Journal of Clinical Investigation, 119*(7), 1763–1771. http://dx.doi.org/10.1172/JCI38339.

Galie, P. A., van Oosten, A., Chen, C. S., & Janmey, P. A. (2015). Application of multiple levels of fluid shear stress to endothelial cells plated on polyacrylamide gels. *Lab on a Chip, 15*(4), 1205–1212. http://dx.doi.org/10.1039/c4lc01236d.

Gao, Y., & Sztul, E. (2001). A novel interaction of the Golgi complex with the vimentin intermediate filament cytoskeleton. *Journal of Cell Biology, 152*(5), 877–894.

Gao, Y. S., Vrielink, A., MacKenzie, R., & Sztul, E. (2002). A novel type of regulation of the vimentin intermediate filament cytoskeleton by a Golgi protein. *European Journal of Cell Biology, 81*(7), 391–401. http://dx.doi.org/10.1078/0171-9335-00260.

Goldman, R. D., Khuon, S., Chou, Y. H., Opal, P., & Steinert, P. M. (1996). The function of intermediate filaments in cell shape and cytoskeletal integrity. *Journal of Cell Biology, 134*(4), 971–983.

Guo, M., Ehrlicher, A. J., Jensen, M. H., Renz, M., Moore, J. R., Goldman, R. D., et al. (2014). Probing the stochastic, motor-driven properties of the cytoplasm using force spectrum microscopy. *Cell, 158*(4), 822–832. http://dx.doi.org/10.1016/j.cell.2014.06.051.

Guo, M., Ehrlicher, A. J., Mahammad, S., Fabich, H., Jensen, M. H., Moore, J. R., et al. (2013). The role of vimentin intermediate filaments in cortical and cytoplasmic mechanics. *Biophysical Journal, 105*(7), 1562–1568. http://dx.doi.org/10.1016/j.bpj.2013.08.037.

Guzman, C., Jeney, S., Kreplak, L., Kasas, S., Kulik, A. J., Aebi, U., et al. (2006). Exploring the mechanical properties of single vimentin intermediate filaments by atomic force microscopy. *Journal of Molecular Biology, 360*(3), 623–630. http://dx.doi.org/10.1016/j.jmb.2006.05.030.

Helfand, B. T., Mendez, M. G., Murthy, S. N., Shumaker, D. K., Grin, B., Mahammad, S., et al. (2011). Vimentin organization modulates the formation of lamellipodia. *Molecular Biology of the Cell, 22*(8), 1274–1289. http://dx.doi.org/10.1091/mbc.E10-08-0699.

Herrmann, H., Haner, M., Brettel, M., Muller, S. A., Goldie, K. N., DFedtke, B., et al. (1996). Structure and assembly properties of the intermediate filament protein vimentin: The role of its head, rod and tail domains. *Journal of Molecular Biology, 264*, 933–953.

Hess, J. F., Budamagunta, M. S., Voss, J. C., & FitzGerald, P. G. (2004). Structural characterization of human vimentin rod 1 and the sequencing of assembly steps in intermediate filament formation in vitro using site-directed spin labeling and electron paramagnetic resonance. *Journal of Biological Chemistry, 279*(43), 44841–44846. http://dx.doi.org/10.1074/jbc.M406257200.

Ho, C. L., Martys, J. L., Mikhailov, A., Gundersen, G. G., & Liem, R. K. (1998). Novel features of intermediate filament dynamics revealed by green fluorescent protein chimeras. *Journal of Cell Science, 111*(Pt 13), 1767–1778.

Hookway, C., Ding, L., Davidson, M. W., Rappoport, J. Z., Danuser, G., & Gelfand, V. I. (2015). Microtubule-dependent transport and dynamics of vimentin intermediate filaments. *Molecular Biology of the Cell, 26*(9), 1675–1686. http://dx.doi.org/10.1091/mbc.E14-09-1398.

Janmey, P. A., Euteneuer, U., Traub, P., & Schliwa, M. (1991). Viscoelastic properties of vimentin compared with other filamentous biopolymer networks. *Journal of Cell Biology, 113*(1), 155–160.

Kajita, M., Sugimura, K., Ohoka, A., Burden, J., Suganuma, H., Ikegawa, M., et al. (2014). Filamin acts as a key regulator in epithelial defence against transformed cells. *Nature Communications, 5*, 4428. http://dx.doi.org/10.1038/ncomms5428.

Kandow, C. E., Georges, P. C., Janmey, P. A., & Beningo, K. A. (2007). Polyacrylamide hydrogels for cell mechanics: Steps toward optimization and alternative uses. In Y. L. Wang & D. E. Discher (Eds.), *Cell mechanics: Vol. 83* (p. 29). Waltham, MA, USA: Academic Press.

Kidd, M. E., Shumaker, D. K., & Ridge, K. M. (2014). The role of vimentin intermediate filaments in the progression of lung cancer. *American Journal of Respiratory Cell and Molecular Biology, 50*(1), 1–6. http://dx.doi.org/10.1165/rcmb.2013-0314TR.

Kolsch, A., Windoffer, R., Wurflinger, T., Aach, T., & Leube, R. E. (2010). The keratin-filament cycle of assembly and disassembly. *Journal of Cell Science, 123*(Pt 13), 2266–2272. http://dx.doi.org/10.1242/jcs.068080.

Kural, C., Serpinskaya, A. S., Chou, Y. H., Goldman, R. D., Gelfand, V. I., & Selvin, P. R. (2007). Tracking melanosomes inside a cell to study molecular motors and their interaction. *Proceedings of the National Academy of Sciences of the United States of America, 104*(13), 5378–5382. http://dx.doi.org/10.1073/pnas.0700145104.

Lau, A. W. C., Hoffman, B. D., Davies, A., Crocker, J. C., & Lubensky, T. C. (2003). Microrheology, stress fluctuations, and active behavior of living cells. *Physical Review Letters, 91*(19). http://dx.doi.org/10.1103/PhysRevLett.91.198101.

Lin, Y. C., Broedersz, C. P., Rowat, A. C., Wedig, T., Herrmann, H., Mackintosh, F. C., et al. (2010). Divalent cations crosslink vimentin intermediate filament tail domains to regulate network mechanics. *Journal of Molecular Biology, 399*(4), 637–644. http://dx.doi.org/10.1016/j.jmb.2010.04.054.

Lin, Y. C., Yao, N. Y., Broedersz, C. P., Herrmann, H., Mackintosh, F. C., & Weitz, D. A. (2010). Origins of elasticity in intermediate filament networks. *Physical Review Letters, 104*(5), 058101.

Mahammad, S., Murthy, S. N., Didonna, A., Grin, B., Israeli, E., Perrot, R., et al. (2013). Giant axonal neuropathy-associated gigaxonin mutations impair intermediate filament protein degradation. *Journal of Clinical Investigation, 123*(5), 1964–1975. http://dx.doi.org/10.1172/JCI66387.

Martys, J. L., Ho, C. L., Liem, R. K., & Gundersen, G. G. (1999). Intermediate filaments in motion: Observations of intermediate filaments in cells using green fluorescent protein-vimentin. *Molecular Biology of the Cell, 10*(5), 1289–1295.

Meier, M., Padilla, G. P., Herrmann, H., Wedig, T., Hergt, M., Patel, T. R., et al. (2009). Vimentin coil 1A-A molecular switch involved in the initiation of filament elongation. *Journal of Molecular Biology, 390*(2), 245–261. http://dx.doi.org/10.1016/j.jmb.2009.04.067.

Mendez, M. G., Kojima, S., & Goldman, R. D. (2010). Vimentin induces changes in cell shape, motility, and adhesion during the epithelial to mesenchymal transition. *FASEB Journal, 24*(6), 1838–1851. http://dx.doi.org/10.1096/fj.09-151639.

Mendez, M. G., Restle, D., & Janmey, P. A. (2014). Vimentin enhances cell elastic behavior and protects against compressive stress. *Biophysical Journal, 107*(2), 314–323. http://dx.doi.org/10.1016/j.bpj.2014.04.050.

Mizuno, D., Tardin, C., Schmidt, C. F., & MacKintosh, F. C. (2007). Nonequilibrium mechanics of active cytoskeletal networks. *Science, 315*(5810), 370–373. http://dx.doi.org/10.1126/science.1134404.

Mucke, N., Wedig, T., Burer, A., Marekov, L. N., Steinert, P. M., Langowski, J., et al. (2004). Molecular and biophysical characterization of assembly-starter units of human vimentin. *Journal of Molecular Biology, 340*(1), 97–114. http://dx.doi.org/10.1016/j.jmb.2004.04.039.

Muller, M., Bhattacharya, S. S., Moore, T., Prescott, Q., Wedig, T., Herrmann, H., et al. (2009). Dominant cataract formation in association with a vimentin assembly disrupting mutation. *Human Molecular Genetics, 18*(6), 1052–1057. http://dx.doi.org/10.1093/hmg/ddn440.

Murray, M. E., Mendez, M. G., & Janmey, P. A. (2014). Substrate stiffness regulates solubility of cellular vimentin. *Molecular Biology of the Cell*, *25*(1), 87–94. http://dx.doi.org/10.1091/mbc.E13-06-0326.

Nekrasova, O. E., Mendez, M. G., Chernoivanenko, I. S., Tyurin-Kuzmin, P. A., Kuczmarski, E. R., Gelfand, V. I., et al. (2011). Vimentin intermediate filaments modulate the motility of mitochondria. *Molecular Biology of the Cell*, *22*(13), 2282–2289. http://dx.doi.org/10.1091/mbc.E10-09-0766.

Omary, M. B., Coulombe, P. A., & McLean, W. H. I. (2004). Mechanisms of disease: Intermediate filament proteins and their associated diseases. *New England Journal of Medicine*, *351*(20), 2087–2100. http://dx.doi.org/10.1056/NEJMra040319.

Palmer, I., & Wingfield, P. T. (2012). Preparation and extraction of insoluble (inclusion-body) proteins from Escherichia coli. *Current Protocols in Protein Science*. http://dx.doi.org/10.1002/0471140864.ps0603s70. Chapter 6, Unit 6.3.

Pogoda, K., Chin, L. K., Georges, P. C., Byfield, F. J., Bucki, R., Kim, R., et al. (2014). Compression stiffening of brain and its effect on mechanosensing by glioma cells. *New Journal of Physics*, *16*, 075002. http://dx.doi.org/10.1088/1367-2630/16/7/075002.

Portet, S., Mucke, N., Kirmse, R., Langowski, J., Beil, M., & Herrmann, H. (2009). Vimentin intermediate filament formation: In vitro measurement and mathematical modeling of the filament length distribution during assembly. *Langmuir*, *25*, 8817–8823.

Prahlad, V., Yoon, M., Moir, R. D., Vale, R. D., & Goldman, R. D. (1998). Rapid movements of vimentin on microtubule tracks: Kinesin-dependent assembly of intermediate filament networks. *Journal of Cell Biology*, *143*(1), 159–170.

Robert, A., Herrmann, H., Davidson, M. W., & Gelfand, V. I. (2014). Microtubule-dependent transport of vimentin filament precursors is regulated by actin and by the concerted action of Rho- and p21-activated kinases. *FASEB Journal*, *28*(7), 2879–2890. http://dx.doi.org/10.1096/fj.14-250019.

Robert, A., Rossow, M. J., Hookway, C., Adam, S. A., & Gelfand, V. I. (2015). Vimentin filament precursors exchange subunits in an ATP-dependent manner. *Proceedings of the National Academy of Sciences of the United States of America*, *112*(27), E3505–E3514. http://dx.doi.org/10.1073/pnas.1505303112.

Satelli, A., & Li, S. (2011). Vimentin in cancer and its potential as a molecular target for cancer therapy. *Cellular and Molecular Life Sciences*, *68*(18), 3033–3046. http://dx.doi.org/10.1007/s00018-011-0735-1.

Schopferer, M., Bar, H., Hochstein, B., Sharma, S., Mucke, N., Herrmann, H., et al. (2009). Desmin and vimentin intermediate filament networks: Their viscoelastic properties investigated by mechanical rheometry. *Journal of Molecular Biology*, *388*(1), 133–143. http://dx.doi.org/10.1016/j.jmb.2009.03.005.

Shaner, N. C., Patterson, G. H., & Davidson, M. W. (2007). Advances in fluorescent protein technology. *Journal of Cell Science*, *120*(Pt 24), 4247–4260. http://dx.doi.org/10.1242/jcs.005801.

Shirahata, A., & Hibi, K. (2014). Serum vimentin methylation as a potential marker for colorectal cancer. *Anticancer Research*, *34*(8), 4121–4125.

Shumaker, D. K., Solimando, L., Sengupta, K., Shimi, T., Adam, S. A., Grunwald, A., et al. (2008). The highly conserved nuclear lamin Ig-fold binds to PCNA: Its role in DNA replication. *Journal of Cell Biology*, *181*(2), 269–280. http://dx.doi.org/10.1083/jcb.200708155.

Steinert, P. M., Marekov, L. N., & Parry, D. A. (1993). Diversity of intermediate filament structure. Evidence that the alignment of coiled-coil molecules in vimentin is different from that in keratin intermediate filaments. *Journal of Biological Chemistry*, *268*(33), 24916–24925.

Stevens, C., Henderson, P., Nimmo, E. R., Soares, D. C., Dogan, B., Simpson, K. W., et al. (2013). The intermediate filament protein, vimentin, is a regulator of NOD2 activity. *Gut*, *62*(5), 695–707. http://dx.doi.org/10.1136/gutjnl-2011-301775.

Straube-West, K., Loomis, P. A., Opal, P., & Goldman, R. D. (1996). Alterations in neural intermediate filament organization: Functional implications and the induction of pathological changes related to motor neuron disease. *Journal of Cell Science*, *109*(Pt 9), 2319–2329.

Strelkov, S. V., Herrmann, H., Geisler, N., Lustig, A., Ivaninskii, S., Zimbelmann, R., et al. (2001). Divide-and-conquer crystallographic approach towards an atomic structure of intermediate filaments. *Journal of Molecular Biology*, *306*(4), 773–781. http://dx.doi.org/10.1006/jmbi.2001.4442.

Strelkov, S. V., Herrmann, H., Geisler, N., Wedig, T., Zimbelmann, R., Aebi, U., et al. (2002). Conserved segments 1A and 2B of the intermediate filament dimer: Their atomic structures and role in filament assembly. *EMBO Journal*, *21*(6), 1255–1266. http://dx.doi.org/10.1093/emboj/21.6.1255.

Styers, M. L., Kowalczyk, A. P., & Faundez, V. (2005). Intermediate filaments and vesicular membrane traffic: The odd couple's first dance? *Traffic*, *6*(5), 359–365. http://dx.doi.org/10.1111/j.1600-0854.2005.00286.x.

Styers, M. L., Salazar, G., Love, R., Peden, A. A., Kowalczyk, A. P., & Faundez, V. (2004). The endo-lysosomal sorting machinery interacts with the intermediate filament cytoskeleton. *Molecular Biology of the Cell*, *15*(12), 5369–5382. http://dx.doi.org/10.1091/mbc.E04-03-0272.

Sultana, A., & Lee, J. E. (2015). Measuring protein-protein and protein-nucleic acid interactions by biolayer interferometry. *Current Protocols in Protein Science*, *79*, 19.25.11–19.25.26. http://dx.doi.org/10.1002/0471140864.ps1925s79.

Tezcan, O., & Gunduz, U. (2014). Vimentin silencing effect on invasive and migration characteristics of doxorubicin resistant MCF-7 cells. *Biomedicine & Pharmacotherapy*, *68*(3), 357–364. http://dx.doi.org/10.1016/j.biopha.2014.01.006.

Thaiparambil, J. T., Bender, L., Ganesh, T., Kline, E., Patel, P., Liu, Y., et al. (2011). Withaferin A inhibits breast cancer invasion and metastasis at sub-cytotoxic doses by inducing vimentin disassembly and serine 56 phosphorylation. *International Journal of Cancer*, *129*(11), 2744–2755. http://dx.doi.org/10.1002/ijc.25938.

Tzivion, G., Luo, Z. J., & Avruch, J. (2000). Calyculin A-induced vimentin phosphorylation sequesters 14-3-3 and displaces other 14-3-3 partners in vivo. *Journal of Biological Chemistry*, *275*(38), 29772–29778. http://dx.doi.org/10.1074/jbc.M001207200.

Uchida, A., Colakoglu, G., Wang, L., Monsma, P. C., & Brown, A. (2013). Severing and end-to-end annealing of neurofilaments in neurons. *Proceedings of the National Academy of Sciences of the United States of America*, *110*(29), E2696–E2705. http://dx.doi.org/10.1073/pnas.1221835110.

Veigel, C., Bartoo, M. L., White, D. C. S., Sparrow, J. C., & Molloy, J. E. (1998). The stiffness of rabbit skeletal actomyosin cross-bridges determined with an optical tweezers transducer. *Biophysical Journal*, *75*(3), 1424–1438.

Vikstrom, K. L., Borisy, G. G., & Goldman, R. D. (1989). Dynamic aspects of intermediate filament networks in BHK-21 cells. *Proceedings of the National Academy of Sciences of the United States of America*, *86*(2), 549–553.

Vikstrom, K. L., Lim, S. S., Goldman, R. D., & Borisy, G. G. (1992). Steady state dynamics of intermediate filament networks. *Journal of Cell Biology*, *118*(1), 121–129.

Vikstrom, K. L., Miller, R. K., & Goldman, R. D. (1991). Analyzing dynamic properties of intermediate filaments. *Methods in Enzymology*, *196*, 506–525.

Wang, S., Moffitt, J. R., Dempsey, G. T., Xie, X. S., & Zhuang, X. (2014). Characterization and development of photoactivatable fluorescent proteins for single-molecule-based

superresolution imaging. *Proceedings of the National Academy of Sciences of the United States of America*, *111*(23), 8452–8457. http://dx.doi.org/10.1073/pnas.1406593111.
Wang, Y. L., & Pelham, R. J. (1998). Preparation of a flexible, porous polyacrylamide substrate for mechanical studies of cultured cells. *Molecular Motors and the Cytoskeleton, Part B*, *298*, 489–496. http://dx.doi.org/10.1016/s0076-6879(98)98041-7.
Wingfield, P. T., Palmer, I., & Liang, S. M. (2014). Folding and purification of insoluble (inclusion body) proteins from Escherichia coli. *Current Protocols in Protein Science*, *78*, 6.51–6.530. http://dx.doi.org/10.1002/0471140864.ps0605s78.
Yeung, T., Georges, P. C., Flanagan, L. A., Marg, B., Ortiz, M., Funaki, M., et al. (2005). Effects of substrate stiffness on cell morphology, cytoskeletal structure, and adhesion. *Cell Motility and the Cytoskeleton*, *60*(1), 24–34. http://dx.doi.org/10.1002/cm.20041.
Yoon, M., Moir, R. D., Prahlad, V., & Goldman, R. D. (1998). Motile properties of vimentin intermediate filament networks in living cells. *Journal of Cell Biology*, *143*(1), 147–157.

CHAPTER FIFTEEN

Strategies to Study Desmin in Cardiac Muscle and Culture Systems

Antigoni Diokmetzidou[1], Mary Tsikitis[1], Sofia Nikouli, Ismini Kloukina, Elsa Tsoupri, Stamatis Papathanasiou, Stelios Psarras, Manolis Mavroidis, Yassemi Capetanaki[2]

Center of Basic Research, Biomedical Research Foundation, Academy of Athens, Athens, Greece

[2]Corresponding author: e-mail address: ycapetanaki@bioacademy.gr

Contents

1. Introduction: Desmin's Scaffold—The Fine-Tuning Machinery of Striated Muscle 428
2. Cell Systems Used for Desmin Studies 430
   2.1 Primary Culture Systems 430
   2.2 Generation of Cardiomyocytes from Neonatal Fibroblast Culture 434
   2.3 Embryoid Bodies as a Model for Differentiation 436
   2.4 Cell Line Systems 437
3. Methods for Desmin Detection (Expression and Localization) 440
   3.1 Immunofluorescence 440
   3.2 Immunoelectron Microscopy 442
   3.3 Subcellular Fractionation and Immunoblotting 444
   3.4 Detection of Desmin-Expressing Cells with Flow Cytometry Analysis 445
   3.5 Desmin Expression in Whole Mouse Embryos Using Reporter Genes 445
4. Methods for Desmin Isolation and Assembly 447
   4.1 Desmin Isolation from Cells and Tissues 448
   4.2 Production and Purification of GST-Fused Desmin Protein and Desmin Deletion Mutants from Bacteria 449
   4.3 *In Vitro* Assembly 450
   4.4 *In Vivo* Assembly 450
5. Model Animals for Desmin Mutation Studies 451
   5.1 Knockout, Knockin, and Transgenic Mice 451
   5.2 Zebrafish Morphants and Knock-In Lines 452
6. Identification of Desmin-Associated Proteins 452
   6.1 Yeast 2-Hybrid System 452
   6.2 GST Pull-Down Assay 453
   6.3 Immunoprecipiation 453
7. Pearls and Pitfalls 454
References 454

[1] These authors contributed equally to this work.

Methods in Enzymology, Volume 568
ISSN 0076-6879
http://dx.doi.org/10.1016/bs.mie.2015.09.026

Abstract

Intermediate filament (IF) cytoskeleton comprises the fine-tuning cellular machinery regulating critical homeostatic mechanisms. In skeletal and cardiac muscle, deficiency or disturbance of the IF network leads to severe pathology, particularly in the latter. The three-dimensional scaffold of the muscle-specific IF protein desmin interconnects key features of the cardiac muscle cells, including the Z-disks, intercalated disks, plasma membrane, nucleus, mitochondria, lysosomes, and potentially sarcoplasmic reticulum. This is crucial for the highly organized striated muscle, in which effective energy production and transmission as well as mechanochemical signaling are tightly coordinated among the organelles and the contractile apparatus. The role of desmin and desmin-associated proteins in the biogenesis, trafficking, and organelle function, as well as the development, differentiation, and survival of the cardiac muscle begins to be enlightened, but the precise mechanisms remain elusive. We propose a set of experimental tools that can be used, *in vivo* and *in vitro*, to unravel crucial new pathways by which the IF cytoskeleton facilitates proper organelle function, homeostasis, and cytoprotection and further understand how its disturbance and deficiency lead to disease.

1. INTRODUCTION: DESMIN'S SCAFFOLD—THE FINE-TUNING MACHINERY OF STRIATED MUSCLE

Intermediate filaments (IFs) along with microtubules and actin microfilaments comprise the filamentous cytoskeleton of higher eukaryotes. Desmin is the major IF protein specifically expressed in all muscle types, whereas the nonmuscle-specific synemin, paranemin, and syncoilin are much less abundant. It is expressed in embryonic cardiac and skeletal muscle progenitor cells, as well as in the corresponding adult stem cells, the satellite cells, and adult cardiac side population (CSP) stem cells, respectively (for review, see Capetanaki, Papathanasiou, Diokmetzidou, Vatsellas, & Tsikitis, 2015). Desmin is important for skeletal muscle differentiation in both C2C12 cells and embryonic stem cells (ESCs) (Li et al., 1994; Weitzer, Milner, Kim, Bradley, & Capetanaki, 1995) and promotes cardiogenesis in ESC-derived embryonic bodies (Hollrigl, Hofner, Stary, & Weitzer, 2007).

In mature cardiac muscle, desmin IFs form a three-dimensional scaffold which extends from the Z-disks to most membranous compartments and organelles, including the costameres and intercalated disks (IDs) of the sarcolemma, mitochondria, lysosomes, and potentially sarcoplasmic reticulum (SR) and T-tubules (reviewed in Capetanaki, Bloch, Kouloumenta, Mavroidis, & Psarras, 2007; Tsoupri & Capetanaki, 2013). This continuous network of desmin IFs could coordinate vital processes of the muscle, including force transmission and mechanochemical signaling, as well as trafficking and organelle biogenesis and function.

The generation of desmin-deficient mice (des$^{-/-}$; Li et al., 1996; Milner, Mavroidis, Weisleder, & Capetanaki, 2000; Milner, Weitzer, Tran, Bradley, & Capetanaki, 1996) highlighted the importance of desmin in muscle maintenance and function. Des$^{-/-}$ mice show defects in all muscle types with the heart being more severely affected by extensive cell death leading to inflammation, fibrosis, and myocardial degeneration and eventually dilated cardiomyopathy and heart failure (Li et al., 1996; Mavroidis & Capetanaki, 2002; Mavroidis et al., 2015; Milner et al., 1996; Psarras et al., 2012; Thornell, Carlsson, Li, Mericskay, & Paulin, 1997).

Mitochondrial structural and functional abnormalities are the major and earliest observed defects in des$^{-/-}$ myocardium (Milner et al., 2000, 1996; Papathanasiou et al., 2015). Furthermore, in des$^{-/-}$ myocardium, lysosomes lose their proper shape and perinuclear distribution, potentially through the disturbance of the desmin-associated protein myospryn, a TRIM-like protein that interacts with dysbindin, a component of the BLOC-1 complex involved in protein trafficking and organelle biogenesis (Kouloumenta, Mavroidis, & Capetanaki, 2007; Tsoupri & Capetanaki, 2013).

In addition to mitochondria and lysosomes, desmin filaments extending from Z-disks toward the nuclear pores seem to affect, potentially through LINC complex and lamins, nuclear shape, positioning, and mechanotransduction (Nikolova et al., 2004; Ralston et al., 2006; Shah et al., 2004). Indeed, the muscle-specific genes myoD and myogenin are abnormally expressed in C2C12 myogenic cells expressing reduced levels of desmin and in des$^{-/-}$ ES cells (Li et al., 1994; Weitzer et al., 1995). Furthermore, mutations of desmin interfere with cardiomyogenesis and lead to downregulation of brachyury, goosecoid, nkx2.5, and mef2c, while ectopic expression of desmin in ES cells has the opposite effects (Hofner, Hollrigl, Puz, Stary, & Weitzer, 2007).

The important role of desmin in disease is manifested by the severe pathologies arising from desmin and desmin-binding protein deficiencies, mutations, and posttranslational modifications (reviewed in Capetanaki et al., 2015). In this context, the transgenic mouse technologies endorsing such alterations are invaluable tools toward our efforts to unravel the multifunctional role of desmin scaffold in health and pathology (Capetanaki et al., 2015). However, the description of such well-established approaches is beyond the scope of this work. Herein, we focus on presenting a toolbox of cell types, animal models, and methodologies to analyze the expression, assembly pattern, and subcellular distribution of wild type and mutant desmin *in vitro* and *in vivo*, at the cellular, subcellular, and organ levels. We also include methods to identify and characterize desmin-binding proteins.

2. CELL SYSTEMS USED FOR DESMIN STUDIES

2.1 Primary Culture Systems

To study the expression and function of desmin network *in vitro* and to understand its contribution in heart pathophysiology, primary cultures of neonatal and adult cardiomyocytes and fibroblasts can be used for relevant differentiation, pharmacology, toxicology, and functional studies.

2.1.1 Neonatal Cardiomyocyte Culture

Cultured neonatal cardiomyocytes are used to study contraction, ischemia, hypoxia, hypertrophy, and the toxicity of different compounds. We provide an example (Supplemental Movies S1 and S2) demonstrating the effect of the nonselective β-adrenergic agonist isoproterenol on cardiomyocyte beating.

The protocol provided is intended for the isolation from one–two litter(s) but can be scaled up.

2.1.1.1 Procedure

Hearts (×10) are harvested from 0- to 3-day-old neonatal mice and transferred immediately into ice-cold HBSS [Ca^{2+}/Mg^{2+}-free HBSS (Hanks' Balanced Salt Solution/Gibco)]. Atria are removed and ventricles are subjected to trypsin digestion [100 μg/ml in 10 ml HBSS (Gibco)] for 16–18 h at 4 °C. Digestion is stopped by the addition of *trypsin inhibitor* [200 μg/ml in HBSS (Gibco)] and incubation for 10 min/37 °C. Further digestion with *collagenase digestion medium* [500 μg/ml collagenase type II (CLS-2, Worthington) in 10 ml L-15 (Leibovitz's/Gibco)] is followed, and tissue is incubated for 20–25 min/37 °C with periodic gentle agitation. Tissue fragments are gently triturated ×10, allowed to settle for 10 min, and the supernatant is filtered through a 70-μm cell strainer (BD). The filtered cells are left to settle for 20 min and centrifuged for 5 min/50–100 ×*g*. Cell pellet is suspended in *NCM culture medium* [Ham's-F10 (Biochrom AG) supplemented with 5% FBS (Fetal Bovine Serum/Gibco), 10% Horse Serum (Gibco), 1% penicillin/streptomycin (Gibco)], plated in 10 cm cell culture dish, and incubated at 5% CO_2/37 °C for 1–3 h allowing the fibroblasts to seed. Afterwards, nonadherent cardiomyocytes are collected and plated at 2×10^5 cells/cm^2 on collagen-A-coated dishes (Biochrom AG). To minimize fibroblasts' growth, 10 μ*M* cytosine β-D-arabinofuranoside (AraC, Sigma) is added to *NCM culture medium*. After incubation for 24 h, cardiomyocytes should adhere to the dish and optimally will spontaneously contract. NCM culture medium is renewed daily (for 4–5 days).

2.1.2 Adult Cardiomyocyte Culture

Isolation of adult cardiomyocytes is a tricky process as it requires dissociation of their strong and complex interconnections, affecting their yield and performance. Conventionally, the Langendorff retrograde perfusion system is used (Zhou et al., 2000), an expensive procedure requiring specific training and equipment. We describe a low cost, homemade system routinely yielding up to 1 million high-quality rod-shaped myocytes per heart of a 1- to 2-month-old mouse, which can be further used for numerous downstream applications, e.g., immunostaining, calcium measurements and signaling, fluorescence-activated cell sorting (FACS) analysis, live-cell imaging, etc.

2.1.2.1 Procedure

Caution: All buffers should be prepared with ddH_2O and filtered (0.22 μm). All glassware should be thoroughly washed with ddH_2O and autoclaved.

Setup

Dishes are coated with laminin (Invitrogen; 5 μg/ml in PBS) on the day of use and maintained at 37 °C/2% CO_2/2 h. *Stock perfusion buffer* (113 m*M* NaCl, 4.7 m*M* KCl, 0.6 m*M* KH_2PO_4, 0.6 m*M* Na_2HPO_4, 1.2 m*M* $MgSO_4 \cdot 7H_2O$, 0.032 m*M* phenol red, 12 m*M* $NaHCO_3$, 10 m*M* $KHCO_3$, 10 m*M* HEPES, 30 m*M* taurine) is stored at 4 °C for up to 1 month. The *perfusion buffer (PB)* [98% (v/v) stock perfusion buffer, 10 m*M* BDM (2,3-butanedionemonoxime), 5.5 m*M* glucose] and *digestion buffer (DB)* (87.5% (v/v) perfusion buffer, 12.5 μ*M* $CaCl_2$) are prepared fresh and prewarmed at 37 °C before use. *Cardiomyocyte plating medium (CPM)* [91% (v/v) MEM, 5% (v/v) BCS, 10 m*M* BDM, 100 U/ml Penicillin, 2 m*M* L-glutamine; prepared fresh] and *cardiomyocyte culture medium (CCM)* [98% (v/v) MEM, 100 U/ml penicillin, 2 m*M* L-glutamine, 0.1 mg/ml bovine serum albumin; store at 4 °C for limited time] are prepared and equilibrated at 37 °C/2% CO_2/2 h.

Setup of the Perfusion Apparatus (Fig. 1)
A Petri dish is placed on heating block at 38–39 °C to collect perfusion wastes. A handmade chamber (made of a 50-ml conical plastic tube cut in half with a round cut of 1 cm diameter in the conical end) is placed on the collection dish to preserve the temperature and air around the perfused

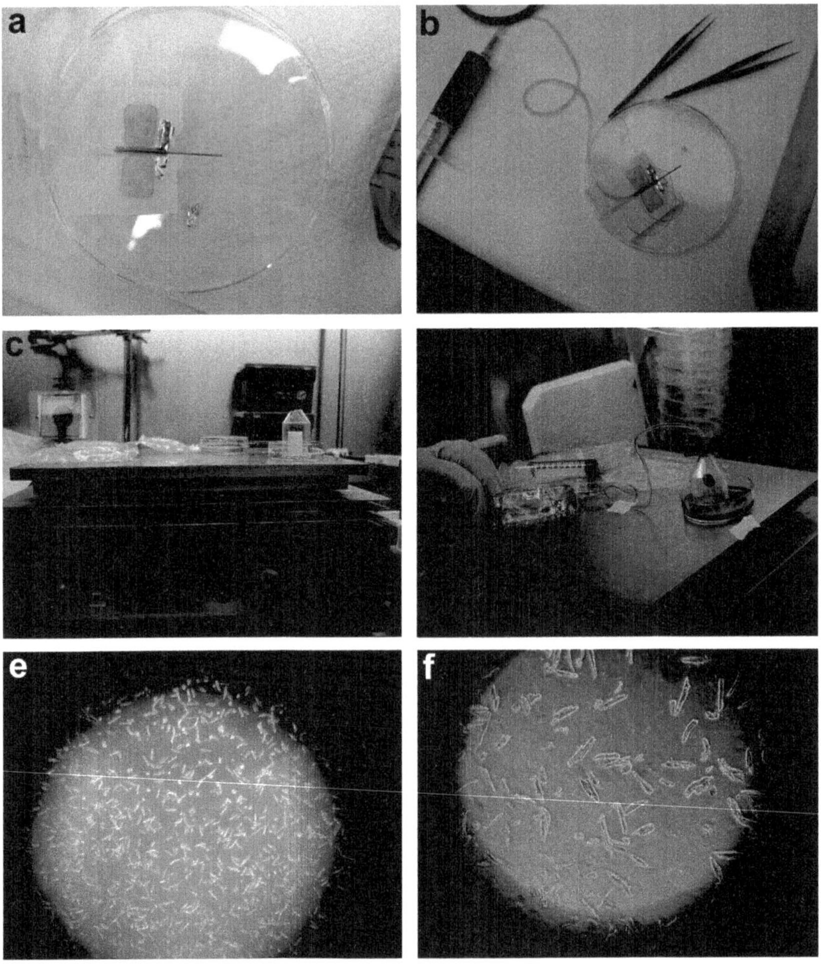

Figure 1 Experimental setup for the isolation of adult cardiomyocytes. (A) The butterfly needle of the perfusion system is immobilized on the back of a Petri dish. (B) The silk suture is tightened around the needle to secure the heart as soon as it is cannulated. (C) The heating block is set at 38–40 °C, and the water jackets are prewarmed. (D) A Petri dish is used to collect and reuse the enzyme solution. The heart is placed inside the experimental devise to be protected from cooling and is perfused by the use of syringes. (E and F) Adult cardiomyocytes isolated by this system.

heart. The cannulated heart will be placed through the cut inside the chamber. The perfusion system consists of a 21G butterfly needle and two 10 ml syringes (one to perfuse and another to collect and reuse the digestion buffer). The point of the butterfly needle is cut to avoid tissue damage. Use prewarmed water jackets to keep the perfusion system warm.

Heart Perfusion and Cardiomyocyte Dissociation

The heart of a 4- to 8-week-old mouse is quickly removed, placed in a collection dish with PB at RT, and cannulated using fine-tip forceps to slide the aorta onto the tip of the butterfly needle and 6–0 silk suture [nonsterile, Deknatel® (F.S.T.)]. Perfusion is started immediately with 3 ml/min constant flow rate until it is cleared from blood. The PB is changed to prewarmed DB (total volume 7.5 ml) supplemented with 1 mg/ml Collagenase D (Roche) and 0.25% Trypsin (Gibco).

Caution: Digestion lasts for 8–10 min with the exact termination point determined by the stiffness of the heart.

As soon as the heart tissue becomes soft and watery in appearance, it is transferred to a $35 \times 10\ mm^2$ culture dish containing 2.5 ml of the used *DB*, minced into 1–2 mm pieces, and mechanically dissociated by pipetting. 2.5 ml of *cardiomyocyte stopping buffer 1* [90% (v/v) PB, 10% (v/v) BCS, 12.5 μM $CaCl_2$; prepared fresh] is added, and the small undigested pieces are let to settle in a 15-ml tube. The supernatant is transferred to a new 15-ml tube. The rod-shaped and round cardiomyocytes are counted. Routinely, $1–1.5 \times 10^6$ cells/heart can be isolated (~70% are rod shaped).

Calcium Reintroduction

The supernatant is centrifuged at $180 \times g$/1 min, and the pellet is resuspended in 10 ml of *cardiomyocyte digestion stopping buffer 2* [95% (v/v) perfusion buffer, 5% (v/v) BCS, 12.5 μM $CaCl_2$; prepared fresh] and transferred to a culture dish. Calcium is gradually reintroduced every 4 min at RT as follows: 50 µl 10 mM $CaCl_2$, 50 µl 10 mM $CaCl_2$, 100 µl 10 mM $CaCl_2$, 30 µl 100 mM $CaCl_2$, 50 µl 100 mM $CaCl_2$. The rod-shaped cardiomyocytes are counted.

Cardiomyocyte Plating and Culture

The pelleted cardiomyocytes ($180 \times g$/1 min) are resuspended in prewarmed *CPM*, plated onto 5 µg/ml laminin-coated dishes (5×10^3 rod-shaped cells/cm^2), and incubated at 37 °C/2% CO_2 for 2 h to attach. The *CPM* is changed to *CCM* containing 10% BCS and 10 mM BDM.

Adult cardiomyocytes can remain in culture for up to a week, changing the culture medium daily.

2.2 Generation of Cardiomyocytes from Neonatal Fibroblast Culture

Recently, "next-generation" cellular reprogramming approaches have been developed achieving direct conversion, known as transdifferentiation, of one cell type into another by using combinations of lineage-specific factors. Three transcription factors, Gata4, Tbx5, and Mef2c (GMT), are sufficient for direct reprogramming cardiac fibroblasts into induced cardiomyocytes (iCMs) (Ieda et al., 2010; Qian et al., 2012; Song et al., 2012), which can serve as tools for cardiac regenerative approaches and studies of reprogramming mechanism(s).

Neonatal cardiac fibroblasts (NCFs) comprise more than half of the cardiac cell population. They contribute to structural, biochemical, mechanical, and electrical properties of the myocardium. They do not express desmin; nonetheless when they are transdifferentiated into iCM, desmin functional studies can be performed.

We provide a short protocol for NCF isolation, and some suggestions for retroviral GMT delivery to convert cardiac fibroblasts to cardiomyocytes (see Fig. 2 for overview). The researcher is referred to Qian, Berry, Fu, Ieda, and Srivastava (2013) for detailed protocol.

2.2.1 Isolation of NCFs

2.2.1.1 Procedure

Hearts harvested from 0- to 3-day-old neonatal mice are placed to an empty dish on ice, minced, and then transferred to a 15-ml conical tube. Tissue fragments are broken by gently triturating (×12–15) with 5 ml prewarmed

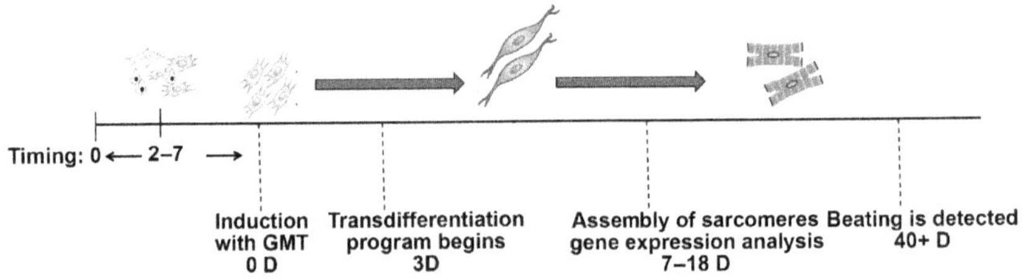

Figure 2 Timetable of NCF transdifferentiation into iCM. The red (gray in the print version) line indicates days after transduction of fibroblasts with reprogramming factors. Potential analysis per time point is listed.

NCF digestion buffer [1 mg/ml collagenase (Type II, Worthington), 1 mg/ml trypsin (Gibco), in L-15 medium (Gibco)] and incubated for 8 min/37 °C. Another cycle of trituration and incubation is followed, and the supernatant containing cells is collected on ice. The remaining undigested tissue undergoes the same process, and the additional supernatant is pooled with the previous one, filtered through 70-μm cell strainer and centrifuged at 600 × *g*/5 min. Cell pellet is suspended in *NCF medium* [Iscove's modified Dulbecco's medium (IMDM, Gibco) supplemented with Gibco 20% FBS, 1% nonessential amino acids (NEAA), and 1% penicillin/streptomycin], plated into 10-cm dish, and incubated 5% CO_2/37 °C/3–6 h. The attached NCFs are gently washed with PBS to eliminate cardiomyocyte clumps, and NCF medium is added. Fibroblasts can be used the following day or cultured for long periods.

2.2.2 Direct Transition of Cardiac Fibroblasts to Cardiomyocyte-Like Cells

A detailed transdifferentiation protocol is described in Qian et al. (2013). However, some additional suggestions may help the procedure.

Suggestions for virus generation

1. High-titer virus can be obtained by the use of Plat-E cells [Platinum-E Retroviral Packaging Cell Line, Ecotropic (Plat-E), CELL BIOLABS, RV-101] cultured as per manufacturer instructions. Freshly thawed Plat-E cells may take time to attach and thus do not change media till most cells are attached. Cells should be cultured in media with selection antibiotics to eliminate cells not expressing the packaging proteins. They should be split when they reach 60–70% confluency. Keep low passage of the cells for higher viral titer.
2. Transfection can be performed with different reagents; however, an inexpensive one is PEI [Polyethylenimine, linear (Polysciences); stock 2.5 mg/ml]. Dilute DNA (refer to Ieda et al., Cell 2010) in Serum Free Medium (SFM, DMEM, or OptiMEM). Add PEI at ratio DNA: PEI 1:3. Apply solution dropwise into cells and move gently the plate to mix well.

2.2.2.1 NCF Transduction into iCM

Suggestions for viral transduction:

1. The day before infection, fibroblasts are split at $1 \times 10^4/cm^2$. Freshly isolated fibroblasts result in better infection efficiencies. NCF isolation protocol is shown in Section 2.2.1.

2. Infection occurs over 24 h. Infect once by harvesting virus after 48 h of transfection. However, multiple boosts of infection can be performed, e.g., harvest and infect every 6 h (4 × /24 h), 8 h (3 × /24 h), or 12 h (2 × /24 h).

2.2.2.2 Evaluate Transdifferentiation Efficiency

3. Transdifferentiation begins by day 3 and peak on day 15 (10–20%).
4. Transdifferentiated cells expressing desmin and other cardiac markers can be detected using flow cytometry (FC500, BD), according to Section 3.4. When using a transgenic mouse expressing an endogenous fluorescence protein upon cardiac induction, the iCMs can be isolated with FACS (FACSAria IIU, BD) and collected in a very rich medium (e.g., pure FBS).
5. After collection, the cells can be either cultured longer or directly analyzed for cardiac sarcomeric marker expression by qPCR, immunohistochemistry, immunoblotting, etc.

2.3 Embryoid Bodies as a Model for Differentiation

EBs are three-dimensional aggregates of ESCs which form tissue-like spheroids in suspension culture. EBs can undergo differentiation and cell specification of all the three germ lineages (Doetschman, Eistetter, Katz, Schmidt, & Kemler, 1985; Itskovitz-Eldor et al., 2000). EB differentiation recapitulates aspects of early embryogenesis presenting an excellent system to study the role of tissue-specific IFs, such as desmin, during muscle development. Indeed, studies using EBs have shown that desmin is necessary for skeletal muscle development along with enhanced expression of skeletal and cardiac regulators (Hofner et al., 2007; Weitzer et al., 1995).

A protocol for mouse EB generation and differentiation into mouse cardiomyocytes is described, based on the hanging drop method (Weitzer et al., 1995) which generates a homogeneous fraction of EBs, ensuring reproducibility.

2.3.1 Procedure

ESCs are cultured [e.g., AB2.2 (Soriano, Montgomery, Geske, & Bradley, 1991)] on 10 μg/ml mitomycin C (Acros Organics)-inactivated feeder fibroblasts [SNL 76/7 (McMahon & Bradley, 1990)] and plated in *ESC medium* [high-glucose (4.5 g/l) DMEM (Gibco), 15% FBS (HyClone), 2 m*M* L-glutamine (Gibco), NEAA (Gibco), 100 μ*M* β-mercaptoethanol

(Sigma), 10% penicillin/streptomycin (Gibco), and 1000 units/l LIF (Sigma)] till forming colonies.

Caution: For reproducible EB differentiation, ESCs should be passaged every 24–48 h on freshly inactivated feeder cells. Cell density, quality of feeders, source, and batch of FBS can severely affect differentiation potential.

Colonies are trypsinized [(Trypsin–EDTA (Gibco)] for 30 s–2 min and detached from the feeders. Trypsinization is stopped with *EB differentiation medium (EBDM)* [high-glucose DMEM, 15% FBS (Sigma F7524), 2 m*M* L-glutamine, 1% NEAA, 100 μ*M* β-mercaptoethanol, and 10% penicillin/streptomycin]. ES colonies are collected, triturated to achieve single-cell suspension, and plated on 0.1% (w/v) gelatin-coated tissue culture dish for 60 min/37 °C for absorption of remaining feeder cells. The remaining ESCs in the supernatant are collected and spun at 1000 rpm for 5 min at RT. After suspending the pellet in *EBDM*, cells are counted and diluted to 4×10^4 cells/ml. Single drops of 20 μl (containing approximately 600 ESCs) are plated onto the inner face of the lids of 10 cm dishes containing 10 ml PBS and gently turned upside down on the plates. After incubation for 4.5 days in 37 °C/5% CO_2, the cells should aggregate and form one EB per drop.

Note: Check that EBs assume a regular round-shaped morphology with a smooth peripheral cell layer and a dense (neuroectodermal) layer inside.

The aggregates are washed from the lid with *EBDM* and distributed evenly by pipetting gently onto gelatin (Sigma)-coated tissue culture plates ($n = 90 \pm 10$ EBs/10 cm plate). Medium is changed gently every 3 days after aggregation until day 33.

Caution: The aggregates should not be disturbed! The cells are monitored for beating areas. EBs can be dissociated to individual cardiomyocytes for further studies (see Fig. 3 for overview).

2.4 Cell Line Systems

In addition to isolated cardiomyocytes and fibroblasts, there are a few cell lines that can be used to study desmin.

2.4.1 Skeletal Muscle: C2C12

For studying desmin in skeletal muscle, a commonly used cell line is C2C12. C2C12 cells were generated by Blau, Chiu, and Webster (1983) as a subclone of the C2 cell line isolated by Yaffe and Saxel (1977). C2C12 cells were originally derived from satellite cells from the thigh muscle of a

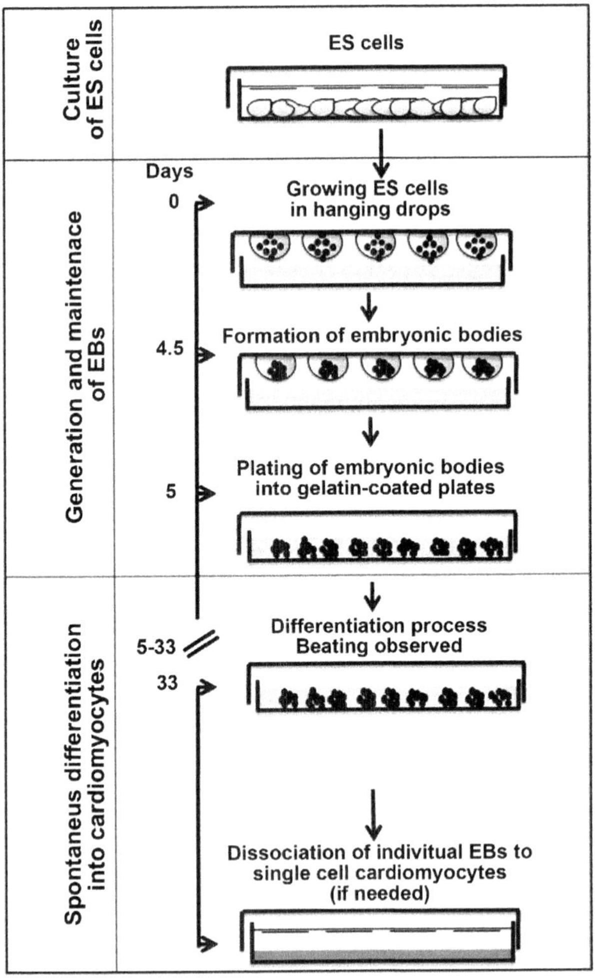

Figure 3 Generation of cardiomyocytes through EBs. Undifferentiated feeder-free ES cells are transformed into EBs by cultivation in *hanging drops* for 4.5 days. Under EB differentiation medium EBs spontaneously differentiate into cardiomyocytes. Grow for 28.5 more days after aggregation.

2-month-old female C3H mouse donor 70 h after a crush injury. They readily proliferate in high-serum conditions, and differentiate and fuse in low-serum conditions. Proliferating C2C12 expresses desmin which is upregulated upon differentiation. Lack of desmin inhibits C2C12 differentiation and interferes with the expression of MyoD and myogenin (Li et al., 1994).

C2C12 grow as undifferentiated myoblasts in growth medium, *GM* [DMEM—high glucose no sodium pyruvate (Gibco), 10–20% FBS (HyClone or Gibco), 1% penicillin/streptomycin, 1% L-glutamine (final 2 m*M*)] at 37 °C 5% CO_2. *GM* should be replaced every 2 days, and cells should not be allowed to reach >70% confluency (ideally split at 50–60%), as they quickly fuse when they are in contact with one another, resulting in a selective depletion. Differentiation of C2C12 cells is achieved by replacing *GM* to differentiation media, *DM* [DMEM—high glucose no sodium pyruvate (Gibco), 2% horse serum (Gibco), 1% glutamine (Gibco), 1% pen/strep (Gibco)]. After 24 h in *DM*, fused cells should be visible. *DM* should be changed every 48 h. Mature myotubes could be collected from day 4 to 7 with mild trypsinization.

Note

(1) The rate of differentiation can be increased, if *DM* is supplemented with insulin/transferrin/selenium.

(2) C2C12 myoblast cells will begin to express myogenin upon reaching 100% confluency, and it will be upregulated after switch to *DM*.

(3) Keep C2C12 cells at low passage number for avoiding reduced differentiation rate.

2.4.2 Cardiomyocyte Cell Lines: HL-1 and H9c2

It has been very difficult to generate cell lines from cardiac muscle; therefore, more studies are done in primary cardiomyocytes. However, there are two cell lines, HL-1 and H9c2, that are used occasionally.

HL-1 is a cardiac muscle cell line, derived from AT-1 mouse artial cardiomyocyte tumor lineage (Claycomb et al., 1998). These cells have the morphology of differentiated cardiomyocytes, maintaining the biochemical and electrophysiological properties and the ability to contract, while they can be passaged several times. A healthy culture of these cells demands precoating of culture flasks with 0.5% fibronectin in 0.02% gelatine solution overnight at 37°C/5% CO_2, and maintenance of the cells in *Claycomb Medium* (JRH Biosciences) supplemented with 10% FBS (Life Technologies), 4 m*M* L-glutamine (Life Technologies), 10 μ*M* norepinephrine (Sigma-Aldrich), and 1% penicillin–streptomycin (Life Technologies). The cultures are kept at 37 °C/5% CO_2.

H9c2 is another cardiac cell line, a subclone of the original clonal cell line derived from embryonic BD1X rat heart tissue (Kimes & Brandt, 1976). However, this rat cell line exhibits several skeletal muscle cell characteristic and thus may not be appropriate for cardiac studies.

3. METHODS FOR DESMIN DETECTION (EXPRESSION AND LOCALIZATION)

Different methodologies have been used to study the cellular localization of desmin protein, including immunofluorescence protocols, subcellular fractionation combined with immunoblotting, and immunogold labeling techniques. In this section, we briefly describe the most extensively used protocols in our laboratory.

3.1 Immunofluorescence

For immunofluorescence staining of desmin in both skeletal and cardiac tissue as well as cells, we recommend the following antibodies: goat anti-desmin Y20, Santa Cruz, sc-7559, 1:50 dilution; rabbit anti-desmin, H-76, Santa Cruz, sc-14026, 1:50 dilution; mouse anti-desmin DE-U-10, Sigma-Aldrich, D1033, 1:20 dilution; and rabbit anti-desmin, Abcam, ab-8592, 1:100 dilution.

3.1.1 OCT Embedded Tissue

This protocol is used for any mammalian cardiac or skeletal muscle. Tissue is washed in ice-cold 1 × PBS, dried, and placed in 15 × 15 × 5 base mold filled with O.C.T. compound (VWR). The mold is placed on the top of a metal base in liquid nitrogen containing polysterene box in order for the OCT-embedded tissue to freeze slowly and stored at −80 °C until use. Sections are taken at 7–10 μm using a cryostat and placed on Superfrost(R) Plus 25 × 75 × 1 mm^3 slides (Thermo Scientific). The cryosections are fixed with ice-cold 70% methanol–30% ethanol for 30 min at −20 °C or 100% acetone for 30 min at −20 °C. They are washed twice with 1 × PBS for 5 min and permeabilized with 1% Triton X-100 in 1 × PBS at RT. Then, the tissue is washed × 3 with 1 × PBS for 5 min and incubated with blocking buffer (5% BSA in 1 × PBS) for 1 h at RT. The anti-desmin antibody is applied in 2.5% BSA–PBST (0.01% Tween-20 in PBS) overnight at 4 °C. The tissue is washed ×3 with PBST and incubated for 1 h at RT, in the dark, with Alexa Fluor secondary antibody conjugated with fluorescent dyes (Molecular Probes) diluted 1:1500 in 2.5% BSA–PBST. Then, it is washed 3 × with 1 × PBST and incubated with 0,1–1 μg DAPI (Sigma) in 1 × PBS for 2min at RT. The section are washed twice with 1 × PBS and mounted with mounting medium (DAKO) (Fig. 4).

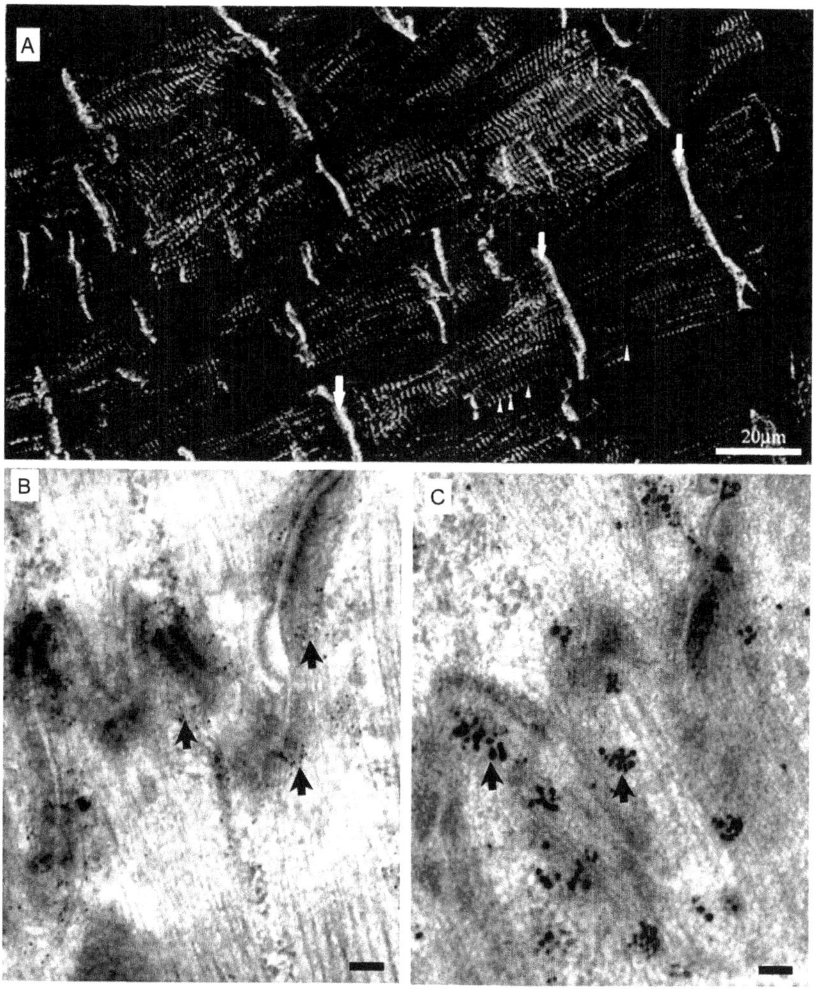

Figure 4 Localization of desmin in the myocardium by immunofluorescence (A) and immunoelectron microscopy (B and C). (A) Frozen sections of human myocardium are stained with anti-desmin. White arrows point to Intercalated Disks and arrowheads to Z-disks. (B and C) Ultrathin sections of mouse myocardium stained with anti-desmin and decorated by 15 nm colloidal gold particles (A) or labeling with silver amplification (B), both pointed by white arrows. Only a part of an Intercalated Disk is shown in both cases. Scale bar: 2 μm. (See the color plate.)

3.1.2 Adherent Cells

For immunofluorescence of adherent cells, the above protocol can be followed with slight modifications: Cells are fixed for 10 min in 100% acetone or 70% methanol–30% ethanol at −20 °C, or 3.7% formaldehyde in

1 × PBS for 10 min. Fixed cells are permeabilized with 1 × PBS–Tween 0.2% for 5 min and blocked for 30 min.

3.2 Immunoelectron Microscopy

Electron microscopic immunocytochemistry is a very powerful technique since it combines the specificity and flexibility of immunocytochemistry with the resolution at high magnification of electron microscopy. Here, we describe a postembedding immunogold labeling of desmin using low-temperature embedding in Lowicryl resin and also a preembedding immunogold labeling protocol on cryostat sections.

3.2.1 Tissue Preparation

Mouse heart is perfused with cold fixative solution (4% paraformaldehyde and 1% glutaraldehyde in 0.1 *M* phosphate buffer (PB), pH 7.4), excised, and dissected into 1-mm cubes, which are postfixed in the same fixative for 2 h and rinsed in 0.1 *M* PB. Dehydration and embedding are carried out according to the progressive lowering of temperature (PLT) method, which improves the efficiency of immunolocalization, using the Leica EM AFS (Leica Microsystems, Vienna, Austria), according to Robertson, Monaghan, Clarke, and Atherton (1992). Samples are dehydrated in successive alcohols with PLT to −50 °C and embedded in Lowicryl HM20 resin. The blocks are polymerized using UV light for 48 h at −50 °C followed by 24 h at 0 °C. Ultrathin sections are cut on ultramicrotome and collected onto 300-mesh Formvar-coated nickel grids for immunolabeling.

3.2.2 Postembedding Immunogold Labeling

All incubations are carried out using the immunogold labeling system Leica EM IGL, in which reagents are automatically applied on ultrathin sections at RT, except for the primary antibody step, which is carried out at 4 °C, according to Havaki et al. (2003). Before use, all the solutions are filtered with Millipore filters (0.2-μm pore size).

3.2.3 Procedure

Grids are floated section side on 30 μl aliquots of Tris–HCl buffer 0.05 *M*, pH 7.4 and then transferred to blocking buffer (0.05 *M* Tris–HCl, pH 7.4 supplemented with 0.2% BSA + 0.1% coldwater fish gelatin, 5% normal goat serum, 0.1% Tween-20) for 30 min. Samples are incubated in primary antibody (desmin rabbit polyclonal, 1:20, Abcam, ab-8592) diluted in Tris–HCl buffer 0.05 *M*, pH 7.4, containing 1% BSA, in Terasaki plate overnight at

4 °C. After several washings with Tris–HCl/BSA, the grids are incubated for 1 h in goat anti-mouse IgG conjugated to 15 nm gold particles (Aurion, The Netherlands) [1:40] diluted in Tris–HCl/BSA and centrifuged (450 g/20 min). Alternatively, IgG conjugated to nanogold particles can be used [1:200] with silver amplification (Nanoprobes) for 5 min. Grids are then rinsed ×3 with Tris–HCl/BSA, ×5 with Tris–HCl, and ×5 with dH_2O. Finally, grids are dried and counterstained with 7.5% ethanolic uranyl acetate for 25 min and 0.4% lead citrate for 3 min. Sections are examined with a Philips 420 transmission electron microscope. As controls for immunogold specificity, primary antibody is omitted, while secondary gold conjugates are applied.

3.2.4 Double Immunogold Labeling

Postembedding techniques allow the simultaneous localization of two antigens in the same grid. The current protocol can be adapted for double immunogold labeling with incubation of the grids in primary antibodies mixture followed by incubation in a mixture of secondary colloidal gold antibodies with two different gold particle sizes.

3.2.5 Preembedding Immunogold Labeling

A preembedding immunogold labeling on cryostat sections, embedded in Epon, of mouse hearts has been recently described (Papathanasiou et al., 2015).

Cryosections are fixed in PBS containing 2% formaldehyde for 5–7 min and permeabilized with PBS containing 0.1% saponin (3–5 min). Incubation with primary antibody follows for 2 h. Specifically for desmin, a-Desmin D9 (Progen Biotechnik, 10519) can be used. After three washing steps, the samples are incubated with secondary antibody conjugated with 1.4 nm gold particles (Nanogold, Biotrend) for 4 h. Samples are postfixed for 15 min at RT with 2.5% glutaraldehyde in sodium cacodylate buffer, briefly rinsed in the same buffer, and twice incubated in buffer containing 200 m*M* sucrose and 50 m*M* HEPES (pH 5.8) for 10 min. This step is followed by silver enhancement (Nanoprobes) for 7–10 min, two washes in 250 m*M* sodium thiosulfate, buffered with 50 m*M* HEPES (pH 5.8; 5 min each), and two short washes in dH_2O.

After postfixation with 0.2% OsO_4 in cacodylate buffer for 30 min on ice, samples are dehydrated, embedded in Epon, sectioned, and stained with 2% uranyl acetate in methanol for 15 min.

3.3 Subcellular Fractionation and Immunoblotting

Caution should be taken when analyzing the localization of desmin by subcellular fractionation. The buffers used to obtain the different cellular compartments should have low ionic strength and pH 8 and above to avoid sedimentation of desmin during the first centrifugation step. Fractionation of the heart is performed by differential centrifugation (for detailed protocols refer to Frezza, Cipolat, & Scorrano, 2007; Wieckowski, Giorgi, Lebiedzinska, Duszynski, & Pinton, 2009). All procedures are carried out at 4 °C or on ice. Protease inhibitors should be used in all steps to avoid desmin proteolysis.

3.3.1 Procedure

Hearts from two to three mice are perfused with ice-cold PBS, isolated, and minced in ice-cold MSE (225 m*M* mannitol, 75 m*M* sucrose, 1 m*M* EGTA, 1 m*M* Tris–HCl pH 8, 0.2 m*M* PMSF, 37 μg/ml TLCK, and *protease inhibitors*). The minced tissue is homogenized by up to 10 strokes of Teflon head at 1500 rpm on a Glas-Col homogenizer. The homogenate is centrifuged at 500 × *g* for 10 min to remove nuclei. The resulting supernatant is centrifuged at 8000 × *g*/30 min to give crude mitochondria at the pellet and cytoplasm in the supernatant. Cytoplasm is centrifuged at 20,000 × *g*/30 min to pellet membrane contaminations and then at 100,000 × *g*/1 h to give SR and cytosol. Crude mitochondria are suspended in *MRB* (250 m*M* mannitol, 5 m*M* HEPES pH 7.4, and 0.5 m*M* EGTA), placed on the top of *Percoll medium* (225 m*M* mannitol, 25 m*M* HEPES pH 7.4, 1 m*M* EGTA, and 30% Percoll), and centrifuged at 95,000 × *g*/30 min/4 °C using aTH640 (Sorvall) swinging rotor. The mitochondrial and mitochondria-associated membrane (MAM) fractions are collected, resuspended in *MRB*, and washed twice at 6300 × *g* for 10 min at 4 °C. The MAM supernatant is centrifuged at 100,000 × *g* for 1 h at 4 °C, and then the pellet is resuspended in *MRB*. The mitochondrial pellet is resuspended in *MRB*.

Immunoblotting is performed according to standard procedures. The same desmin antibodies used for immunofluorescence can also be used for Western blotting in a dilution range 1:500–1:1000. Desmin is detected in the cytoplasm, crude mitochondria, MAMs, and SR fractions. Due to the ionic strength of the buffer used in this protocol desmin sediments in the nuclear fraction.

3.4 Detection of Desmin-Expressing Cells with Flow Cytometry Analysis

An alternative approach for studying desmin expression in a cell population is Flow Cytometry (FC) analysis. FC analysis is based on the detection of a population that expresses a specific cell membrane marker using fluorochrome-conjugated antibodies. However, it can also be used for the detection of cells expressing intracellular markers following immunofluorescence staining of the cells. Thus, FC analysis is a useful method to determine the percentage of desmin-positive cells in a population of noncardiomyocytes isolated from neonatal or adult heart, such as the cardiac side population stem cells (Goodell, Brose, Paradis, Conner, & Mulligan, 1996; Oyama et al., 2007; Pfister et al., 2005), or the transdifferentiated cardiac neonatal fibroblasts (see Section 2). Isolation procedures for CSP cells can be found elsewhere (Pfister, Oikonomopoulos, Sereti, & Liao, 2010).

3.4.1 Procedure

Cultured cells are detached from the culture plate with standard trypsinization method, counted, and stained according to immunofluorescence protocol described in Section 3.1.1, with 100 μl anti-desmin antibody (Y20, Santa Cruz, sc-75591, dilution 1:50) in blocking buffer per 10^6 cells. A secondary antibody conjugated with Alexa-488 fluorochrome (dilution 1:1500) is used. Following the staining procedure, cells are resuspended at 10^6 cells/500 ml FC buffer (PBS, 2 m*M* EDTA) and analyzed with flow cytometer equipped with lasers able to excite at 488 nm (e.g., FC500, BD) and to collect emissions at 530 nm (detector FL1) (Fig. 5).

Note: (a) Caution should be taken after cell fixation between immunofluorescence staining steps as fixation makes cells lighter, and there is a possibility of cell lost between washes and centrifugations. (b) During FC analysis, cells are kept in dark. (c) Nuclei staining for cell counting is not required during FC analysis.

3.5 Desmin Expression in Whole Mouse Embryos Using Reporter Genes

Desmin is one of the first muscle-specific proteins to appear during mammalian embryonic development. In addition to immunofluorescence, to unravel the most precise pattern of the desmin, gene expression throughout development is the use of transgenic mouse embryos expressing a reporter

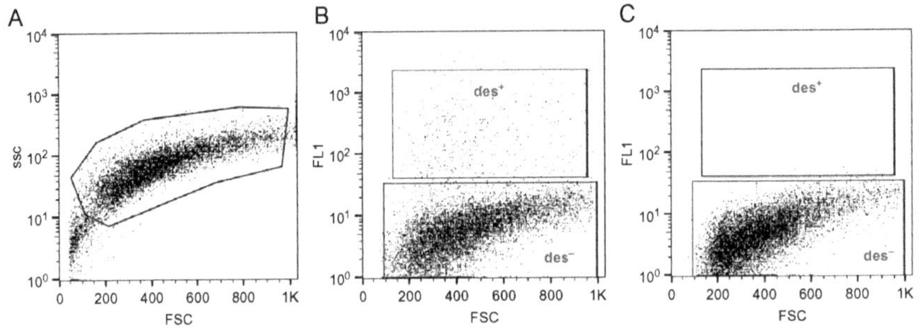

Figure 5 FC analysis of differentiated cardiac cells expressing desmin. (A) Differentiated cardiac cells are identified and gated according to the cell size (forward scatter, FSC) and granularity (side scatter, SSC). (B) Within the FSC/SSC gate, the FL1 versus FSC profile is displayed in a new graph and desmin-expressing cells (desmin+) are gated separately from the nonexpressing cells (desmin−). (C) Unstained differentiated cardiac cells are used as negative control sample.

gene such as lacZ under the control of the desmin gene regulatory elements (Kuisk, Li, Tran, & Capetanaki, 1996). It has been shown that the *cis*-acting DNA sequences within 1-kb 5′ flanking region of the mouse desmin gene are sufficient to direct appropriate temporal transcription in both heart and skeletal muscle during mouse embryogenesis (Kuisk et al., 1996). On the other hand, a 5.1-kb region is required for expression in all three muscle types (N. Weisleder and Y. Capetanaki, unpublished).

3.5.1 Procedure

For the generation of desmin-nlacZ transgenic mice, fragments containing the corresponding 5′ flanking region (1 kb, 5.1 kb, or even longer) plus 200 bp of desmin exon 1 sequences cloned to the MCS of the nlacZ reporter vector 46.21 (Fire, Harrison, & Dixon, 1990), carrying the nuclear localization signal, can be used following the standard protocols of transgenic mice generation (Capetanaki, Smith, & Heath, 1989; Kuisk et al., 1996). For β-galactosidase staining, whole embryos are fixed for 1 h in 2% paraformaldehyde, 0.2% glutaraldehyde, washed with PBS, and then stained in a PBS solution containing 5 m*M* X-gal, 5 m*M* potassium ferrocyanide, and 2 m*M* $MgCl_2$ at RT overnight. The embryos are then rinsed ×2 with PBS containing 3% DMSO, ×3 with 70% ethanol and stored until assayed (Fig. 6) or embedded in paraffin, and sectioned for histological counterstaining (Macgregor, Nolan, Fiering, Roederer, & Herzenberg, 1991).

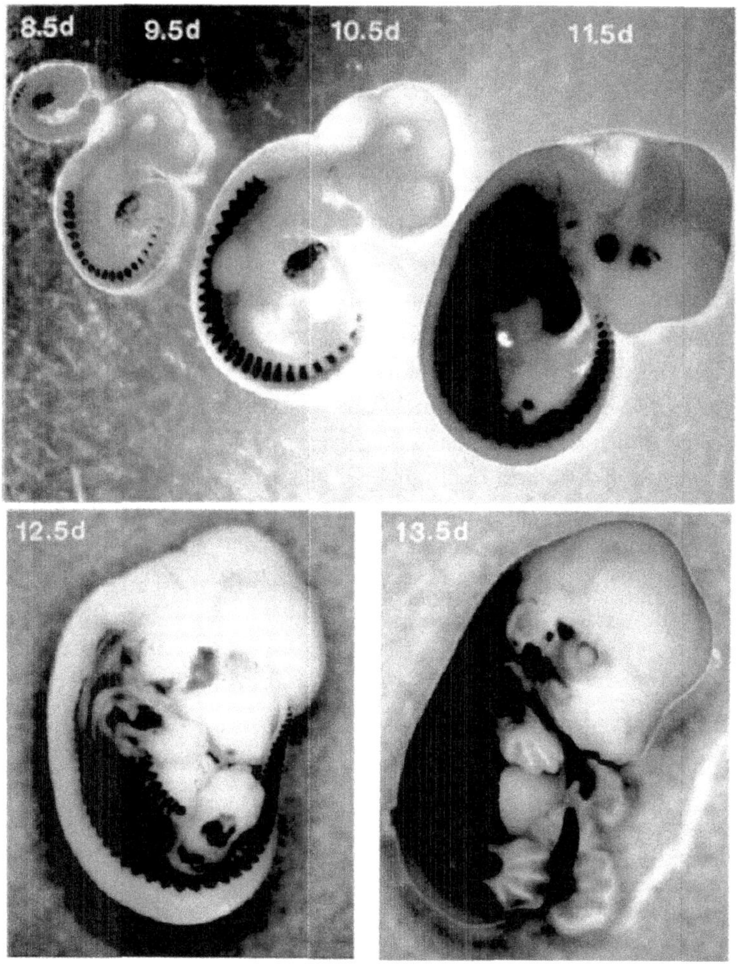

Figure 6 Visualization of desmin gene expression during development using desmin–LacZ reporter. Whole mount staining for LacZ activity of desmin–LacZ transgenic embryos from 8.5 to 13.5 dpc.

4. METHODS FOR DESMIN ISOLATION AND ASSEMBLY

The most recent methodologies for the isolation, purification, and assembly of desmin filaments are extensively described in Herrmann, Kreplak, and Aebi (2004) and Robson, Huiatt, and Bellin (2004), and the references within. Briefly, purified desmin remains soluble in buffers of very low ionic strength (lower of 10 m*M* Tris) and pH above 8. Assembly of filaments is achieved by increasing the ionic strength and/or by lowering the pH. Most studies on filament assembly can be performed in assembly buffer consisting of 10–20 m*M* Tris pH 7–7.5, 100–150 m*M* KCl or NaCl, and 2–5 m*M* $MgCl_2$ (Yuan, Huiatt, Liao, Robson, & Graves, 1999).

4.1 Desmin Isolation from Cells and Tissues

For the isolation of high-purity desmin for *in vitro* assembly experiments or other use (as atomic structure analysis), we take advantage of the fact that IFs and a few other proteins are insoluble in buffers containing moderate levels of nonionic detergents and high concentrations of monovalent ions, and in such buffers almost complete solubilization of the rest cellular material is achieved. As a source of desmin IFs, cardiac or skeletal muscle tissue and also cultures of transformed bacteria overexpressing desmin could be used. The procedure for isolation of high-purity desmin from bacteria is essentially the same as described in Section 4.2. The procedure for desmin isolation from skeletal muscle is extensively described in Robson et al. (2004) and the references within and from porcine smooth muscle in Vorgias and Traub (1983). In Fig. 7, the major fractions obtained during desmin

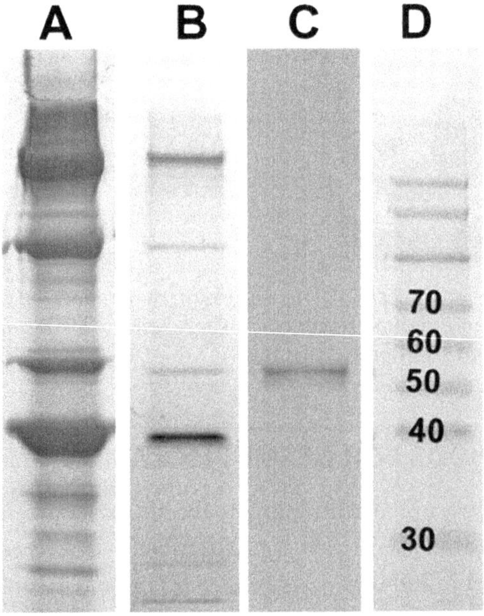

Figure 7 SDS-PAGE analysis of the different fractions obtained during high-purity desmin isolation from porcine cardiac muscle. See text for details. Lane D:molecular weight protein markers (kDa).

purification from porcine cardiac muscle are depicted by SDS-PAGE analysis. Briefly, porcine cardiac muscle is pulverized well with a pestle–mortar liquid nitrogen, and subsequently 5 g of tissue powder is homogenized with 30 ml of *extraction buffer* (5 m*M* Tris pH 7.5, 0.6 *M* KCl, 1% Triton X-100, 1 m*M* EDTA, 1 m*M* DTT, protease inhibitor cocktail/Sigma P8340, 2 m*M* PMSF, 0.2 m*M* Na_2VO_3, and 1 m*M* NaF) using an electric homogenizer (polytron). The homogenate is then sonicated for 3 × 3 min on ice and is centrifuged for 30 min at 4000 × *g*. The pellet is dissolved in 30 ml *extraction buffer*, sonicated for 3 × 3 min on ice, and then centrifuged for 30 min at 4000 × *g*. The pellet is washed with H_2O and then resuspended in 10 ml of 5 m*M* Tris pH 7.5, 1 m*M* EDTA, 1 m*M* DTT, 8 *M* urea. The dissolved material is dialyzed overnight against another batch of the same solution containing 6 *M* urea and then centrifuged at 15,000 × *g* for 30 min. The supernatant (crude desmin; see Fig. 7, lane B) is subjected to anion exchange chromatography in the same solution and eluted with a linear 0–500 m*M* potassium chloride gradient. A homemade 15 ml column, AccellPlus QMA resin (Waters) connected to AKTA purifier 900 (Amersham Biosciences) FPLC system is used. Desmin is eluted in 160 m*M* KCl fraction. The fractions enriched in desmin are collected (partially purified desmin; see Fig. 7, lane B), dialyzed against 5 m*M* Tris pH 8.4, 1 m*M* EDTA, 1 m*M* DTT supplied with 8*M* urea, concentrated, and subjected to exclusion chromatography using the Shodex KW804 column connected to the AKTA purifier. The fractions are subjected to SDS-PAGE, and those fractions containing desmin, without contaminants, are pooled as together (see Fig. 7, lane C). Purity is verified by mass spectroscopy.

4.2 Production and Purification of GST-Fused Desmin Protein and Desmin Deletion Mutants from Bacteria

4.2.1 Procedure

Desmin cDNA, synthesized from RNA isolated from mouse hearts, and desmin mutants: DesHead (H) aa 1–108 (1–324 bp), HeadRod (HR) aa 1–412 (1–1236 bp), Rod (R) aa 109–412 (324–1236 bp), RodTail (RT) aa 109–470 (324–1410 bp), Tail (T) aa 413–470 (1237–1410 bp) are PCR amplified and subcloned into pGEX-4T-3. The plasmids are transformed in BL21 bacteria (New England Biolabs), and the recombinant polypeptides are expressed by induction with 1 m*M* isopropyl-1-thio-β-D-galactopyranoside for 5 h. The bacteria are harvested and resuspended in *lysis buffer* (20 m*M* Tris pH 8, 5 m*M* EDTA, 5 m*M* DTT, 0.5% NP-40, 5% glycerol, protease inhibitor cocktail, 2 m*M* phenylmethylsulfonylfluoride).

Lysozyme is added (Sigma) to 5 mg/ml and incubated for 30 min at 4 °C. The lysate is sonicated and centrifuged at 12,000 rpm for 30 min. GST desmin and GST-desmin mutants, except for GST-tail, are accumulated in occlusion bodies which are dissolved in *urea buffer* (6 *M* urea, 5 m*M* Tris–HCl pH 6.8, 1 m*M* EDTA, 10 m*M* sodium acetate pH 6.3, 0.5% Triton, 1 × protease inhibitor cocktail, 2 m*M* phenylmethylsulfonylfluoride). The lysates are sonicated, centrifuged as above, and dialyzed against *dialysis buffer 1* (1 *M* urea, 10 m*M* Tris–HCl pH 8, 5 m*M* EDTA, 20 m*M* NaCl, 0.1% NP-40, 0.1% DOC, 5% glycerol, *protease inhibitor cocktail*, 2 m*M* phenylmethylsulfonylfluoride) overnight at 4 °C and *dialysis buffer 2* (dialysis buffer 1 without urea) for 8–10 h at 4 °C. The GST fused proteins are affinity purified using glutathione sepharose 4B (GE Healthcare). The sepharose-bound proteins can be used directly for GST pull-down assay. Otherwise, the GST-tag can be removed by site-specific proteolysis using thrombin (GE Healthcare, 27-0846-01).

4.3 *In Vitro* Assembly

Detailed methodologies for assembly of desmin filaments are described in Herrmann et al. (2004), and an extensive analysis of the impact of disease mutations on desmin filament assembly process can be found in Bar et al. (2006, 2005) and Bar, Strelkov, Sjoberg, Aebi, and Herrmann (2004).

4.3.1 *Procedure*

Briefly, for *in vitro* reconstitution of purified recombinant desmin, 0.5–1.0 mg/ml of protein is dialyzed overnight into a buffer containing 5 m*M* Tris–HCl (pH 8.4), 1 m*M* EDTA, 0.1 m*M* EGTA, and 1 m*M* DTT. Assembly is started by addition of equal amount of *assembly buffer* (45 m*M* Tris–HCl, pH 7.0, 100 m*M* NaCl). Assembly is stopped by the addition of 0.2% glutaraldehyde (final concentration) in a time-lapse manner (usually 10 s, 1 min, 5 min, and 1 h), and electron microscopy of negatively stained samples is performed.

4.4 *In Vivo* Assembly

Transfection of cell lines with plasmids expressing wild type or mutated desmin has been a very helpful method to study IF dynamics, assembly, interactions, and function (Schweitzer et al., 2001).

Desmin is a type III IF protein, like vimentin, glial fibrillary acidic protein (GFAP), and peripherin. Type III IF proteins can form coiled-coil dimers with themselves, with other type III IF members and with type IV IF proteins, such as syncoilin, nestin, neurofilaments, and synemins, but not with the keratins (Herrmann & Aebi, 2000). Thus, the assembly properties of desmin transfected in a cell line also depend on the IF proteins that are already expressed by the given cell type.

Different cell lines that already express other IF proteins, such as vimentin, keratins, or even desmin (as 3T3 fibroblasts, MCF-7, and C2C12), have been used for desmin transfection experiments. Nevertheless, the SW-13 cell line (Sarria, Lieber, Nordeen, & Evans, 1994; Sarria, Nordeen, & Evans, 1990), which is IF free, has been widely used for the analysis of the impact of a disease causing mutation in the formation of filaments (Bar, Goudeau, et al., 2007; Bar, Mucke, Katus, Aebi, & Herrmann, 2007).

However, caution should be taken when studying desmin in systems devoid of all IFs, as the formation of extended desmin filament networks requires the presence of either paranemin or vimentin (Schweitzer et al., 2001). Eventually, what became apparent from these experiments is that the severity of disease and the degree of the assembly defect of a mutant desmin variant cannot be correlated directly (Bar et al., 2005).

Common procedures for cell line transfection with the given plasmid are used, as the calcium phosphate method (Sarria et al., 1994) or commercially available transfection reagents (as Fugene 6, Roche or Lipofectamine-2000, ThermoFisher Scientific) according to the manufacturer's protocol. After transfection, cells are incubated for 24 or 48 h before being processed for immunofluorescence microscopy, eventually as described in Section 3.1.1.

5. MODEL ANIMALS FOR DESMIN MUTATION STUDIES

5.1 Knockout, Knockin, and Transgenic Mice

All studies for desmin function *in vivo* have been done with desmin-null mice generated in two independent laboratories (Li et al., 1996; Milner et al., 1996). For *in vivo* studies aiming to elucidate the consequences of desmin point or deletion mutations in the desmin network and muscle structure, differentiation, and function, two strategies have been used: (a) the generation of transgenic mice carrying the corresponding desmin mutation, such as Arg173-Glu179del (Wang et al., 2001), Ile451Met (Mavroidis, Panagopoulou, Kostavasili, Weisleder, & Capetanaki, 2008), and Leu345Pro (Kostareva et al., 2008) under the control of a cardiac-specific promoter (MHC) and (b) a knock-in strategy where the WT desmin allele is replaced by the mutant allele, as in the case of desmin R350P (Clemen et al., 2015). The TNF-α-overexpressing heart failure model is also a good model to be used for studies on desmin aggregate formation mediated by caspase-induced desmin cleavage (Panagopoulou et al., 2008). All the above methodologies are standard and widely used, thus not described here. The knock-in strategy generates a better model with equal amounts of wt and mutant protein, whereas in the transgenic models the expression level of the transgene depends on copy number and site of integration.

5.2 Zebrafish Morphants and Knock-In Lines

Another vertebrate model used for *in vivo* studies of desmin deficiency or network disruption (Ramspacher et al., 2015) is zebrafish, which possesses two desmin genes, desma and desmb, expressed in both cardiac and skeletal muscle and sharing high homology with mammalian desmin (Li, Andersson-Lendahl, Sejersen, & Arner, 2013; Loh, Chan, Gong, Lim, & Chua, 2000). Two strategies have been used, morpholino (MO)-mediated knockdown (Li et al., 2013; Vogel et al., 2009) and genetically engineered zebrafish knock-in lines (Ramspacher et al., 2015). In desma^{ct122aGt} (or ct122aGt) line, obtained by a gene trap screen, citrine fluorescent protein has been inserted between amino acids 460 and 461 in the C-terminus of desma (Trinh le et al., 2011). Following Flip Trap, this line was converted by Cre–lox recombination to a mCherry fusion that generated a stop codon leading to the second line, desma^{ct122aRGt} (or ct122aRGt), with truncation of the last 13 aa of desma tail. Desma aggregation and defects were observed in both cardiac and skeletal muscle in both lines. Given the fact that the insertion of the fluorescence tag, and not a desminopathy related specific mutation, leads to desma aggregation, one must be cautious with these lines. Nevertheless, they might be very useful for screening strategies for the identification of chemical compounds inhibiting desmin aggregation (Ramspacher et al., 2015). Finally, a desmin knockout line, desma$^{sa5-/-}$ (Kettleborough et al., 2013), obtained from Sanger Institute ENU-mutagenesis screen, which carries a point mutation that leads to premature stop codon, has been studied and showed similar defects to those obtained with the MO knockdowns and the aggregations lines. The detailed widely used methodologies for all these strategies are not included here.

6. IDENTIFICATION OF DESMIN-ASSOCIATED PROTEINS

6.1 Yeast 2-Hybrid System

Desmin exerts its multiple functions through direct and indirect binding to various other cellular molecules. A technology that can be used to identify novel protein interactions, confirm suspected interactions, and define interacting domains is the yeast two-hybrid assay. We have successfully used this technology to identify novel proteins interacting with desmin (Kouloumenta et al., 2007).

6.1.1 Procedure

Briefly, the entire desmin or specific domains, such as head, road, or tail, can be used as bait. The bait cDNA is inserted downstream of the GAL4 DNA-binding domain (DNA-BD) in the pGBKT7 plasmid vector (Clontech), while a cDNA library derived from any muscle of interest, in our case from cardiac muscle (Clontech), is expressed as a fusion to the GAL4 activation domain (AD) in the pGADT7-Rec plasmid vector. When bait and library fusion proteins interact in a yeast reporter strain, the DNA-BD and AD are brought into proximity and activate transcription of four reporter genes which are the selection markers. The interaction positive clones are selected in selection medium: (SD/Trp/Leu/Ade/His X—galactosidase plates). The pGADT7-Rec and the pGBKT7 plasmid vector contain c-Myc and HA epitope tags, bacterial selection markers, and T7 promoters to help expedite the discovery and verification of the new interactions.

6.2 GST Pull-Down Assay

The production and purification of GST protein, GST-desmin, and GST-desmin mutants are described in Section 4.2. 5 mg heart or muscle lysates are precleared for 3 h at 4 °C with glutathione sepharose 4B (GE Healthcare) and then added to the immobilised GST fused proteins and incubated for 3 h at 4 °C. The flow through is collected, and the columns are washed 5 × with wash buffer (20 m*M* Tris–HCl pH 8, 200 m*M* NaCl, 1 m*M* EDTA, 0.5% Nonident-P40, 1 m*M* phenylmethylsulfonyl fluoride, 1 × protease inhibitor cocktail). The bound proteins are eluted by incubation for 30 min at RT with 0.5 ml of 20 m*M* reduced glutathione in 50 m*M* Tris–HCl, pH 8 per ml bed volume of Glutathione Sepharose 4B and then by heating at 97 °C for 10 min in 2 × electrophoresis loading buffer. The eluates are analyzed by SDS-PAGE, transferred to PVDF membrane, and probed with the desired antibodies and anti-GST antibody (Santa Cruz) as control.

6.3 Immunoprecipiation

Mouse heart tissue is isolated and homogenized by 10 strokes of Teflon head at 500 rpm on a Glas-Col homogenizer in solubilization buffer (20 m*M* Bis–Tris, 500 m*M* aminocaproic acid, 20 m*M* NaCl, 2 m*M* EDTA pH 8, 10%

glycerol, pH 7, 2 m*M* phenylmethylsulfonyl fluoride, 1 × protease inhibitor cocktail) and incubated with 4 g digitonin/g protein for 30 min on ice. Five to seven milligrams of the solubilized heart tissue is precleared with protein A/G Ultralink resin (Thermo Scientific) for 3 h at 4 °C and then incubated with desmin antibody (20 μl Santa Cruz Y-20 or 10 μl Abcam, ab-8592) or BSA overnight at 4 °C rotating. One hundred microliter protein A/G Ultralink resin (slurry washed twice with 25 m*M* Tris, 150 m*M* NaCl, pH 7.2) is added to the antibody-lysate and incubated for 2 h at RT. The unbound material is collected and the beads are washed ×10 with wash buffer (20 m*M* Tris–HCl pH 8, 200 m*M* NaCl, 1 m*M* EDTA, 0.5% Nonident-P40, 1 m*M* phenylmethylsulfonyl fluoride, 1 × protease inhibitor cocktail). The immune complex is eluted by adding 50 μl 1 × electrophoresis loading buffer and incubating for 10 min at 97 °C.

7. PEARLS AND PITFALLS

We have tried to address critical issues in our individual sections instead of this one, due to the high number of methodologies described. We have prompted the researcher to consider our suggestions, notes, and cautions.

REFERENCES

Bar, H., Goudeau, B., Walde, S., Casteras-Simon, M., Mucke, N., Shatunov, A., et al. (2007). Conspicuous involvement of desmin tail mutations in diverse cardiac and skeletal myopathies. *Human Mutation, 28*(4), 374–386. http://dx.doi.org/10.1002/humu.20459.

Bar, H., Kostareva, A., Sjoberg, G., Sejersen, T., Katus, H. A., & Herrmann, H. (2006). Forced expression of desmin and desmin mutants in cultured cells: Impact of myopathic missense mutations in the central coiled-coil domain on network formation. *Experimental Cell Research, 312*(9), 1554–1565. http://dx.doi.org/10.1016/j.yexcr.2006.01.021. S0014-4827(06)00022-X [pii].

Bar, H., Mucke, N., Katus, H. A., Aebi, U., & Herrmann, H. (2007). Assembly defects of desmin disease mutants carrying deletions in the alpha-helical rod domain are rescued by wild type protein. *Journal of Structural Biology, 158*(1), 107–115. http://dx.doi.org/10.1016/j.jsb.2006.10.029. S1047-8477(06)00351-0 [pii].

Bar, H., Mucke, N., Kostareva, A., Sjoberg, G., Aebi, U., & Herrmann, H. (2005). Severe muscle disease-causing desmin mutations interfere with in vitro filament assembly at distinct stages. *Proceedings of the National Academy of Sciences of the United States of America, 102*(42), 15099–15104. http://dx.doi.org/10.1073/pnas.0504568102. 0504568102 [pii].

Bar, H., Strelkov, S. V., Sjoberg, G., Aebi, U., & Herrmann, H. (2004). The biology of desmin filaments: How do mutations affect their structure, assembly, and organisation? *Journal of Structural Biology, 148*(2), 137–152. http://dx.doi.org/10.1016/j.jsb.2004.04.003. S1047847704000978 [pii].

Blau, H. M., Chiu, C. P., & Webster, C. (1983). Cytoplasmic activation of human nuclear genes in stable heterocaryons. *Cell*, *32*(4), 1171–1180. 0092-8674(83)90300-8 [pii].
Capetanaki, Y., Bloch, R. J., Kouloumenta, A., Mavroidis, M., & Psarras, S. (2007). Muscle intermediate filaments and their links to membranes and membranous organelles. *Experimental Cell Research*, *313*(10), 2063–2076. http://dx.doi.org/10.1016/j.yexcr.2007.03.033. S0014-4827(07)00151-6 [pii].
Capetanaki, Y., Papathanasiou, S., Diokmetzidou, A., Vatsellas, G., & Tsikitis, M. (2015). Desmin related disease: A matter of cell survival failure. *Current Opinion in Cell Biology*, *32C*, 113–120. http://dx.doi.org/10.1016/j.ceb.2015.01.004. S0955-0674(15)00006-X [pii].
Capetanaki, Y., Smith, S., & Heath, J. P. (1989). Overexpression of the vimentin gene in transgenic mice inhibits normal lens cell differentiation. *The Journal of Cell Biology*, *109*(4 Pt. 1), 1653–1664.
Claycomb, W. C., Lanson, N. A., Jr., Stallworth, B. S., Egeland, D. B., Delcarpio, J. B., Bahinski, A., et al. (1998). HL-1 cells: A cardiac muscle cell line that contracts and retains phenotypic characteristics of the adult cardiomyocyte. *Proceedings of the National Academy of Sciences of the United States of America*, *95*(6), 2979–2984.
Clemen, C. S., Stockigt, F., Strucksberg, K. H., Chevessier, F., Winter, L., Schutz, J., et al. (2015). The toxic effect of R350P mutant desmin in striated muscle of man and mouse. *Acta Neuropathologica*, *129*(2), 297–315. http://dx.doi.org/10.1007/s00401-014-1363-2.
Doetschman, T. C., Eistetter, H., Katz, M., Schmidt, W., & Kemler, R. (1985). The in vitro development of blastocyst-derived embryonic stem cell lines: Formation of visceral yolk sac, blood islands and myocardium. *Journal of Embryology and Experimental Morphology*, *87*, 27–45.
Fire, A., Harrison, S. W., & Dixon, D. (1990). A modular set of lacZ fusion vectors for studying gene expression in *Caenorhabditis elegans*. *Gene*, *93*(2), 189–198. 0378-1119(90)90224-F [pii].
Frezza, C., Cipolat, S., & Scorrano, L. (2007). Organelle isolation: Functional mitochondria from mouse liver, muscle and cultured fibroblasts. *Nature Protocols*, *2*(2), 287–295. http://dx.doi.org/10.1038/nprot.2006.478. nprot.2006.478 [pii].
Goodell, M. A., Brose, K., Paradis, G., Conner, A. S., & Mulligan, R. C. (1996). Isolation and functional properties of murine hematopoietic stem cells that are replicating in vivo. *The Journal of Experimental Medicine*, *183*(4), 1797–1806.
Havaki, S., Kittas, C., Marinos, E., Dafni, U., Sotiropoulou, C., Voloudakis-Baltatzis, I., et al. (2003). Ultrastructural immunostaining of infiltrating ductal breast carcinomas with the monoclonal antibody H: A comparative study with cytokeratin 8. *Ultrastructural Pathology*, *27*(6), 393–407. KVJ6M6HVAC772QE4 [pii].
Herrmann, H., & Aebi, U. (2000). Intermediate filaments and their associates: Multi-talented structural elements specifying cytoarchitecture and cytodynamics. *Current Opinion in Cell Biology*, *12*(1), 79–90. S0955-0674(99)00060-5 [pii].
Herrmann, H., Kreplak, L., & Aebi, U. (2004). Isolation, characterization, and in vitro assembly of intermediate filaments. *Methods in Cell Biology*, *78*, 3–24.
Hofner, M., Hollrigl, A., Puz, S., Stary, M., & Weitzer, G. (2007). Desmin stimulates differentiation of cardiomyocytes and up-regulation of brachyury and nkx2.5. *Differentiation*, *75*(7), 605–615. http://dx.doi.org/10.1111/j.1432-0436.2007.00162.x. S0301-4681(09)60154-7 [pii].
Hollrigl, A., Hofner, M., Stary, M., & Weitzer, G. (2007). Differentiation of cardiomyocytes requires functional serine residues within the amino-terminal domain of desmin. *Differentiation*, *75*(7), 616–626. http://dx.doi.org/10.1111/j.1432-0436.2007.00163.x. S0301-4681(09)60155-9 [pii].

Ieda, M., Fu, J. D., Delgado-Olguin, P., Vedantham, V., Hayashi, Y., Bruneau, B. G., et al. (2010). Direct reprogramming of fibroblasts into functional cardiomyocytes by defined factors. *Cell, 142*(3), 375–386. http://dx.doi.org/10.1016/j.cell.2010.07.002. S0092-8674(10)00771-3 [pii].

Itskovitz-Eldor, J., Schuldiner, M., Karsenti, D., Eden, A., Yanuka, O., Amit, M., et al. (2000). Differentiation of human embryonic stem cells into embryoid bodies compromising the three embryonic germ layers. *Molecular Medicine, 6*(2), 88–95.

Kettleborough, R. N., Busch-Nentwich, E. M., Harvey, S. A., Dooley, C. M., de Bruijn, E., van Eeden, F., et al. (2013). A systematic genome-wide analysis of zebrafish protein-coding gene function. *Nature, 496*(7446), 494–497. http://dx.doi.org/10.1038/nature11992. nature11992 [pii].

Kimes, B. W., & Brandt, B. L. (1976). Properties of a clonal muscle cell line from rat heart. *Experimental Cell Research, 98*(2), 367–381.

Kostareva, A., Sjoberg, G., Bruton, J., Zhang, S. J., Balogh, J., Gudkova, A., Hedberg, B., Edstrom, L., Westerblad, H., & Sejersen, T. (2008). Mice expressing L345P mutant desmin exhibit morphological and functional changes of skeletal and cardiac mitochondria. *Journal of Muscle Research and Cell Motility, 29*(1), 25–36. http://dx.doi.org/10.1007/s10974-008-9139-8.

Kouloumenta, A., Mavroidis, M., & Capetanaki, Y. (2007). Proper perinuclear localization of the TRIM-like protein myospryn requires its binding partner desmin. *The Journal of Biological Chemistry, 282*(48), 35211–35221. http://dx.doi.org/10.1074/jbc.M704733200. M704733200 [pii].

Kuisk, I. R., Li, H., Tran, D., & Capetanaki, Y. (1996). A single MEF2 site governs desmin transcription in both heart and skeletal muscle during mouse embryogenesis. *Developmental Biology, 174*(1), 1–13. http://dx.doi.org/10.1006/dbio.1996.0046. S0012-1606(96)90046-9 [pii].

Li, M., Andersson-Lendahl, M., Sejersen, T., & Arner, A. (2013). Knockdown of desmin in zebrafish larvae affects interfilament spacing and mechanical properties of skeletal muscle. *The Journal of General Physiology, 141*(3), 335–345. http://dx.doi.org/10.1085/jgp.201210915. jgp.201210915 [pii].

Li, H., Choudhary, S. K., Milner, D. J., Munir, M. I., Kuisk, I. R., & Capetanaki, Y. (1994). Inhibition of desmin expression blocks myoblast fusion and interferes with the myogenic regulators MyoD and myogenin. *The Journal of Cell Biology, 124*(5), 827–841.

Li, Z., Colucci-Guyon, E., Pincon-Raymond, M., Mericskay, M., Pournin, S., Paulin, D., et al. (1996). Cardiovascular lesions and skeletal myopathy in mice lacking desmin. *Developmental Biology, 175*(2), 362–366. http://dx.doi.org/10.1006/dbio.1996.0122. S0012-1606(96)90122-0 [pii].

Loh, S. H., Chan, W. T., Gong, Z., Lim, T. M., & Chua, K. L. (2000). Characterization of a zebrafish (*Danio rerio*) desmin cDNA: An early molecular marker of myogenesis. *Differentiation, 65*(5), 247–254. http://dx.doi.org/10.1046/j.1432-0436.2000.6550247.x. S0301-4681(09)60428-X [pii].

Macgregor, G. R., Nolan, G. P., Fiering, S., Roederer, M., & Herzenberg, L. A. (1991). Use of *Escherichia coli* (*E. coli*) lacZ (beta-galactosidase) as a Reporter Gene. *Methods in Molecular Biology*, 7, 217–235. http://dx.doi.org/10.1385/0-89603-178-0:217.

Mavroidis, M., & Capetanaki, Y. (2002). Extensive induction of important mediators of fibrosis and dystrophic calcification in desmin-deficient cardiomyopathy. *The American Journal of Pathology, 160*(3), 943–952. http://dx.doi.org/10.1016/S0002-9440(10)64916-4. S0002-9440(10)64916-4 [pii].

Mavroidis, M., Davos, C. H., Psarras, S., Varela, A., Athanasiadis, N. C., Katsimpoulas, M., et al. (2015). Complement system modulation as a target for treatment of arrhythmogenic cardiomyopathy. *Basic Research in Cardiology, 110*(3), 485. http://dx.doi.org/10.1007/s00395-015-0485-6.

Mavroidis, M., Panagopoulou, P., Kostavasili, I., Weisleder, N., & Capetanaki, Y. (2008). A missense mutation in desmin tail domain linked to human dilated cardiomyopathy promotes cleavage of the head domain and abolishes its Z-disc localization. *The FASEB Journal*, *22*(9), 3318–3327. http://dx.doi.org/10.1096/fj.07-088724. fj.07-088724 [pii].

McMahon, A. P., & Bradley, A. (1990). The Wnt-1 (int-1) proto-oncogene is required for development of a large region of the mouse brain. *Cell*, *62*(6), 1073–1085. http://dx.doi.org/0092-8674(90)90385-R [pii].

Milner, D. J., Mavroidis, M., Weisleder, N., & Capetanaki, Y. (2000). Desmin cytoskeleton linked to muscle mitochondrial distribution and respiratory function. *The Journal of Cell Biology*, *150*(6), 1283–1298.

Milner, D. J., Weitzer, G., Tran, D., Bradley, A., & Capetanaki, Y. (1996). Disruption of muscle architecture and myocardial degeneration in mice lacking desmin. *The Journal of Cell Biology*, *134*(5), 1255–1270.

Nikolova, V., Leimena, C., McMahon, A. C., Tan, J. C., Chandar, S., Jogia, D., et al. (2004). Defects in nuclear structure and function promote dilated cardiomyopathy in lamin A/C-deficient mice. *The Journal of Clinical Investigation*, *113*(3), 357–369. http://dx.doi.org/10.1172/JCI19448.

Oyama, T., Nagai, T., Wada, H., Naito, A. T., Matsuura, K., Iwanaga, K., et al. (2007). Cardiac side population cells have a potential to migrate and differentiate into cardiomyocytes in vitro and in vivo. *The Journal of Cell Biology*, *176*(3), 329–341. http://dx.doi.org/10.1083/jcb.200603014. jcb.200603014 [pii].

Panagopoulou, P., Davos, C. H., Milner, D. J., Varela, E., Cameron, J., Mann, D. L., & Capetanaki, Y. (2008). Desmin mediates TNF-alpha-induced aggregate formation and intercalated disk reorganization in heart failure. *Journal of Cell Biology*, *181*(5), 761–775. http://dx.doi.org/10.1083/jcb.200710049jcb.200710049 [pii].

Papathanasiou, S., Rickelt, S., Soriano, M. E., Schips, T. G., Maier, H. J., Davos, C. H., et al. (2015). Tumor necrosis factor-alpha confers cardioprotection through ectopic expression of keratins K8 and K18. *Nature Medicine*, *21*(9), 1076–1084. http://dx.doi.org/10.1038/nm.3925. nm.3925 [pii].

Pfister, O., Mouquet, F., Jain, M., Summer, R., Helmes, M., Fine, A., et al. (2005). CD31 – but not CD31 + cardiac side population cells exhibit functional cardiomyogenic differentiation. *Circulation Research*, *97*(1), 52–61. http://dx.doi.org/10.1161/01.RES.0000173297.53793.fa. 01.RES.0000173297.53793.fa [pii].

Pfister, O., Oikonomopoulos, A., Sereti, K. I., & Liao, R. (2010). Isolation of resident cardiac progenitor cells by Hoechst 33342 staining. *Methods in Molecular Biology*, *660*, 53–63. http://dx.doi.org/10.1007/978-1-60761-705-1_4.

Psarras, S., Mavroidis, M., Sanoudou, D., Davos, C. H., Xanthou, G., Varela, A. E., et al. (2012). Regulation of adverse remodelling by osteopontin in a genetic heart failure model. *European Heart Journal*, *33*(15), 1954–1963. http://dx.doi.org/10.1093/eurheartj/ehr119. ehr119 [pii].

Qian, L., Berry, E. C., Fu, J. D., Ieda, M., & Srivastava, D. (2013). Reprogramming of mouse fibroblasts into cardiomyocyte-like cells in vitro. *Nature Protocols*, *8*(6), 1204–1215. http://dx.doi.org/10.1038/nprot.2013.067. nprot.2013.067 [pii].

Qian, L., Huang, Y., Spencer, C. I., Foley, A., Vedantham, V., Liu, L., et al. (2012). In vivo reprogramming of murine cardiac fibroblasts into induced cardiomyocytes. *Nature*, *485*(7400), 593–598. http://dx.doi.org/10.1038/nature11044. nature11044 [pii].

Ralston, E., Lu, Z., Biscocho, N., Soumaka, E., Mavroidis, M., Prats, C., et al. (2006). Blood vessels and desmin control the positioning of nuclei in skeletal muscle fibers. *Journal of Cellular Physiology*, *209*(3), 874–882. http://dx.doi.org/10.1002/jcp.20780.

Ramspacher, C., Steed, E., Boselli, F., Ferreira, R., Faggianelli, N., Roth, S., et al. (2015). Developmental alterations in heart biomechanics and skeletal muscle function in desmin mutants suggest an early pathological root for desminopathies. *Cell Reports, 11*(10), 1564–1576. http://dx.doi.org/10.1016/j.celrep.2015.05.010. S2211-1247(15)00522-7 [pii].

Robertson, D., Monaghan, P., Clarke, C., & Atherton, A. J. (1992). An appraisal of low-temperature embedding by progressive lowering of temperature into Lowicryl HM20 for immunocytochemical studies. *Journal of Microscopy, 168*(Pt. 1), 85–100.

Robson, R. M., Huiatt, T. W., & Bellin, R. M. (2004). Muscle intermediate filament proteins. *Methods in Cell Biology, 78*, 519–553.

Sarria, A. J., Lieber, J. G., Nordeen, S. K., & Evans, R. M. (1994). The presence or absence of a vimentin-type intermediate filament network affects the shape of the nucleus in human SW-13 cells. *Journal of Cell Science, 107*(Pt. 6), 1593–1607.

Sarria, A. J., Nordeen, S. K., & Evans, R. M. (1990). Regulated expression of vimentin cDNA in cells in the presence and absence of a preexisting vimentin filament network. *The Journal of Cell Biology, 111*(2), 553–565.

Schweitzer, S. C., Klymkowsky, M. W., Bellin, R. M., Robson, R. M., Capetanaki, Y., & Evans, R. M. (2001). Paranemin and the organization of desmin filament networks. *Journal of Cell Science, 114*(Pt. 6), 1079–1089.

Shah, S. B., Davis, J., Weisleder, N., Kostavassili, I., McCulloch, A. D., Ralston, E., et al. (2004). Structural and functional roles of desmin in mouse skeletal muscle during passive deformation. *Biophysical Journal, 86*(5), 2993–3008. http://dx.doi.org/10.1016/S0006-3495(04)74349-0. S0006-3495(04)74349-0 [pii].

Song, K., Nam, Y. J., Luo, X., Qi, X., Tan, W., Huang, G. N., et al. (2012). Heart repair by reprogramming non-myocytes with cardiac transcription factors. *Nature, 485*(7400), 599–604. http://dx.doi.org/10.1038/nature11139. nature11139 [pii].

Soriano, P., Montgomery, C., Geske, R., & Bradley, A. (1991). Targeted disruption of the c-src proto-oncogene leads to osteopetrosis in mice. *Cell, 64*(4), 693–702. http://dx.doi.org/0092-8674(91)90499-O [pii].

Thornell, L., Carlsson, L., Li, Z., Mericskay, M., & Paulin, D. (1997). Null mutation in the desmin gene gives rise to a cardiomyopathy. *Journal of Molecular and Cellular Cardiology, 29*(8), 2107–2124. http://dx.doi.org/S0022282897904466 [pii].

Trinh le, A., Hochgreb, T., Graham, M., Wu, D., Ruf-Zamojski, F., Jayasena, C. S., et al. (2011). A versatile gene trap to visualize and interrogate the function of the vertebrate proteome. *Genes & Development, 25*(21), 2306–2320. http://dx.doi.org/10.1101/gad.174037.111. 25/21/2306 [pii].

Tsoupri, E., & Capetanaki, Y. (2013). Muyospryn: A multifunctional desmin-associated protein. *Histochemistry and Cell Biology, 140*(1), 55–63. http://dx.doi.org/10.1007/s00418-013-1103-z.

Vogel, B., Meder, B., Just, S., Laufer, C., Berger, I., Weber, S., et al. (2009). In-vivo characterization of human dilated cardiomyopathy genes in zebrafish. *Biochemical and Biophysical Research Communications, 390*(3), 516–522. http://dx.doi.org/10.1016/j.bbrc.2009.09.129. S0006-291X(09)01946-9 [pii].

Vorgias, C. E., & Traub, P. (1983). Isolation, purification and characterization of the intermediate filament protein desmin from porcine smooth muscle. *Preparative Biochemistry, 13*(3), 227–243. http://dx.doi.org/10.1080/00327488308064250.

Wang, X., Osinska, H., Dorn, G. W., 2nd., Nieman, M., Lorenz, J. N., Gerdes, A. M., et al. (2001). Mouse model of desmin-related cardiomyopathy. *Circulation, 103*(19), 2402–2407.

Weitzer, G., Milner, D. J., Kim, J. U., Bradley, A., & Capetanaki, Y. (1995). Cytoskeletal control of myogenesis: A desmin null mutation blocks the myogenic pathway during embryonic stem cell differentiation. *Developmental Biology, 172*(2), 422–439. http://dx.doi.org/10.1006/dbio.1995.8070. S0012-1606(85)78070-0 [pii].

Wieckowski, M. R., Giorgi, C., Lebiedzinska, M., Duszynski, J., & Pinton, P. (2009). Isolation of mitochondria-associated membranes and mitochondria from animal tissues and cells. *Nature Protocols*, *4*(11), 1582–1590. http://dx.doi.org/10.1038/nprot.2009.151. nprot.2009.151 [pii].

Yaffe, D., & Saxel, O. (1977). Serial passaging and differentiation of myogenic cells isolated from dystrophic mouse muscle. *Nature*, *270*(5639), 725–727.

Yuan, J., Huiatt, T. W., Liao, C. X., Robson, R. M., & Graves, D. J. (1999). The effects of mono-ADP-ribosylation on desmin assembly-disassembly. *Archives of Biochemistry and Biophysics*, *363*(2), 314–322. http://dx.doi.org/S0003986198910967 [pii].

Zhou, Y. Y., Wang, S. Q., Zhu, W. Z., Chruscinski, A., Kobilka, B. K., Ziman, B., et al. (2000). Culture and adenoviral infection of adult mouse cardiac myocytes: Methods for cellular genetic physiology. *American Journal of Physiology Heart and Circulatory Physiology*, *279*(1), H429–H436.

CHAPTER SIXTEEN

Genetic Manipulation of Neurofilament Protein Phosphorylation

Maria R. Jones[*,†], **Eric Villalón**[*,†], **Michael L. Garcia**[*,†,1]
[*]Department of Biological Sciences, University of Missouri, Columbia, Missouri, USA
[†]C.S. Bond Life Sciences Center, University of Missouri, Columbia, Missouri, USA
[1]Corresponding author: e-mail address: garciaml@missouri.edu

Contents

1. Introduction 462
2. Genetic Manipulation of Mice 464
3. Steps for Creating a Gene-Targeted Mutant of a Mouse 465
   3.1 Site-Directed Mutagenesis of Gene of Interest 465
   3.2 Generation of Targeting Construct 466
   3.3 Targeting of ES Cells Via Homologous Recombination 466
   3.4 Injecting Blastocyst into Psuedopregnant Females, and Breeding Selection 468
   3.5 Generating a Line of Mice/Genotyping 468
   3.6 Analysis of Neurofilament Phosphorylation Utilizing Immunohistochemistry 469
4. Conclusions and Applications 471
References 473

Abstract

Neurofilament biology is important to understanding structural properties of axons, such as establishment of axonal diameter by radial growth. In order to study the function of neurofilaments, a series of genetically modified mice have been generated. Here, we describe a brief history of genetic modifications used to study neurofilaments, as well as an overview of the steps required to generate a gene-targeted mouse. In addition, we describe steps utilized to analyze neurofilament phosphorylation status using immunoblotting. Taken together, these provide comprehensive analysis of neurofilament function *in vivo*, which can be applied to many systems.

ABBREVIATIONS

CNS central nervous system
C-terminal carboxy terminal
DNA deoxyribose nucleic acid
dNTP deoxynucleotide triphosphate

Methods in Enzymology, Volume 568
ISSN 0076-6879
http://dx.doi.org/10.1016/bs.mie.2015.07.027

ES cell embryonic stem cell
KSP lysine, serine, proline
NFs neurofilaments
NF-H neurofilament heavy
NF-L neurofilament light
NF-M neurofilament medium
PCR polymerase chain reaction
Pfu *Pyrococcus furiosus*
pGK-Neo phosphoglycerate kinase promoter-neomycin phosphotransferase gene
PMSF phenylmethanesulfonylfluoride
PNS peripheral nervous system
PVDF polyvinylidine difluoride
RT room temperature
UTR untranslated region

1. INTRODUCTION

Rapid impulse propagation allows complex nervous systems to operate quickly and efficiently. A key structural property that contributes to the rate of impulse propagation is axonal diameter (Rushton, 1951; Waxman, 1980), which is established by myelin-dependent radial growth (de Waegh, Lee, & Brady, 1992). Axonal radial growth is also dependent upon neurofilaments (de Waegh et al., 1992; Eyer & Peterson, 1994; Sakaguchi, Okada, Kitamura, & Kawasaki, 1993; Zhu, Couillard-Despres, & Julien, 1997).

Neurofilaments (NFs) are the main cytoskeletal proteins of large myelinated axons. The composition of NFs varies with development and location (central nervous system, CNS vs. peripheral nervous system, PNS). In mouse utero, NFs are composed primarily of neurofilament light (NF-L) and medium (NF-M) (Carden, Trojanowski, Schlaepfer, & Lee, 1987; Cochard & Paulin, 1984; Pachter & Liem, 1984; Shaw & Weber, 1982). This composition is maintained for about 3 months during postnatal development when neurofilament heavy (NF-H) is incorporated into the filament (Pachter & Liem, 1984; Shaw & Weber, 1982). Recent evidence suggests that a fourth subunit is incorporated into NFs (Yuan et al., 2006, 2012), which varies depending upon location. Within the CNS, α-internexin is the fourth subunit (Yuan et al., 2006). However, within the PNS, peripherin is incorporated into NF filaments (Yuan et al., 2012).

The nonvariant NF subunit proteins in adults (NF-L, NF-M, and NF-H) are subdivided into three main domains. All three have a highly conserved 310 amino acid central rod domain, which is necessary for copolymerization (Fuchs & Weber, 1994; Geisler, Kaufmann, Fischer, Plessmann, & Weber, 1983). The amino termini, referred to as the head domain, are less conserved, and are necessary for NF formation (Heins et al., 1993; Herrmann & Aebi, 2004; Omary, Ku, Tao, Toivola, & Liao, 2006). The carboxy termini (C-termini), referred to as the tail domain, are the most variable domains. The tail domains of NF-M and NF-H have a subregion of lysine–serine–proline (KSP) repeats, which are phosphorylated at the serine *in vivo* (Julien & Mushynski, 1982, 1983) upon myelination (de Waegh et al., 1992).

Genetic manipulation of the mouse embryo has allowed us to investigate the role NF subunit proteins in mediating axonal radial growth. Subunit composition is important for radial growth. Altering NF subunit composition by altering the expression of any single subunit leads to reduction of axonal diameters (Collard, Cote, & Julien, 1995; Cote, Collard, & Julien, 1993; Lee, Xu, Wong, & Cleveland, 1993; Monteiro, Hoffman, Gearhart, & Cleveland, 1990; Xu et al., 1996). Gene targeting has revealed the role of each individual subunit as well as the function of the carboxy terminal (C-terminal) tail domains. Deletion of the NF-L subunit resulted in loss NF formation and prevented axonal radial growth (Zhu et al., 1997). Moreover, deleting NF-M (Elder, Friedrich, Bosco, et al., 1998) and NF-H (Elder, Friedrich, Kang, et al., 1998; Rao et al., 1998; Zhu et al., 1997; Zhu, Lindenbaum, Levavasseur, Jacomy, & Julien, 1998) identified NF-M as the critical subunit for radial growth. Delayed expression of NF-H might allow for nerves to develop in a distal to proximal gradient (Shen, Barry, & Garcia, 2010). Utilizing gene replacement, NF-M C-terminus was found to be the critical domain for mediating radial growth (Garcia et al., 2003; Rao et al., 2003), whereas the C-terminal tail domain of NF-H regulates the rate of radial growth (Garcia et al., 2003; Rao et al., 2002).

Gene targeting has also been utilized to provide mechanistic insights into how NF-M C-terminus mediates radial axonal growth. Prior to these studies, the prevailing hypothesis was myelin-dependent phosphorylation of KSP repeats within NF-M C-terminus (de Waegh et al., 1992; Elder, Friedrich, Bosco, et al., 1998; Elder, Friedrich, Kang, et al., 1998; Garcia et al., 2003; Rao et al., 2003, 2002, 1998; Yin et al., 1998). However, substitution of KSP serine residues with alanine prevented myelin-dependent phosphorylation without altering radial growth (Garcia et al., 2009). Generation of a chimeric NF-M protein, which was composed of mouse head and rod domains and cow C-terminal tail domain, effectively lengthened the C-terminal tail domain (Barry et al., 2012). Lengthening NF-M C-terminus increased axonal diameter without increasing impulse propagation (Barry et al., 2012).

The ability to manipulate the mouse genome has resulted in significant progress in determining the mechanisms mediating axonal radial growth. Both neurofilament accumulation (Zhu et al., 1997) and myelination (de Waegh et al., 1992) have been established to be critical for radial growth. NF-M and NF-H both have multiple KSP repeats on the C-terminal tail domain, and these become heavily phosphorylated within internodes (Hsieh et al., 1994). Deletion of NF-M (Elder, Friedrich, Bosco, et al., 1998), or truncation of the NF-M C-terminus (Garcia et al., 2003), resulted in reduced axonal caliber. Similar experiments deleting NF-H (Elder, Friedrich, Kang, et al., 1998; Rao et al., 1998), or truncation of the C-terminal tail domain (Rao et al., 2002), did not result in similar reductions in axonal caliber. Thus, it was hypothesized that myelin-dependent phosphorylation of NF-M KSP sites may contribute to establishing axonal diameter (Garcia et al., 2003). To directly test this hypothesis, site-directed mutagenesis was utilized to mutate all seven KSP serine residues to alanine to prevent phosphorylation. This mutant allele was targeted generating NF-M$^{S\rightarrow A}$ mutant mice (Garcia et al., 2009). In this chapter, we provide a detailed procedure on generating mutant alleles with site-directed mutagenesis; targeting mutant alleles in embryonic stem cells (ES cells); and how to analyze neurofilament expression and phosphorylation utilizing commercially available resources.

2. GENETIC MANIPULATION OF MICE

Multiple scientific advances came together in the 1980s to allow for creation of the first gene-targeted mouse line. In 1981, Sir Martin Evans isolated mouse ES cells (Evans & Kaufman, 1981). Subsequently, Oliver Smithies and Mario Capecchi discovered that homologous recombination in ES cells could be utilized to target genes for generation of genetically modified mice (Capecchi, 1989a, 1989b; Koller et al., 1989; Mansour, Thomas, Deng, & Capecchi, 1990). Since the first knockout mouse in 1989, modifying the mouse genome has become a standard approach to studying gene function. Currently, multiple techniques exist for genetic manipulation of mice including transgenic mice, gene deletion, gene targeted, and conditional deletion.

Transgenics and gene targeting have been employed for studying the role of neurofilaments. This chapter will provide an outline for creating a gene-targeted mouse using the NF-M$^{S \rightarrow A}$ mouse mutant mouse as an example.

3. STEPS FOR CREATING A GENE-TARGETED MUTANT OF A MOUSE

1. Site-directed mutagenesis of gene of interest
2. Generation of targeting construct
3. Targeting of ES cells via homologous recombination
4. Injecting clone into blastocysts
5. Generating a line of mice
6. Analysis of neurofilament phosphorylation utilizing immunohistochemistry

3.1 Site-Directed Mutagenesis of Gene of Interest

1. Obtain a genomic clone of the gene of interest. Analyze genomic sequence to identify locations of endonuclease recognition sequences. If possible, identify unique sites flanking gene region that contains codons that will be mutated.
2. Digest genomic clone with unique endonucleases (example *Acc*I-*Bcl*I for NF-M$^{S \rightarrow A}$ mouse) to subclone smaller fragment for mutagenesis.
3. Primers for site-directed mutagenesis should be designed such that the mutated codon is flanked by at least 25 bases on each side of the mutations. Example of primer used to mutate lysine-serine-aspartate to lysine-alanine-aspartate: agaggaggaagatgaaggtgtcaagGCCgaccaggcaga agaggggggatctg.
4. Site-directed mutagenesis using polymerase chain reaction (PCR) approach:

   100 ng deoxyribose nucleic acid (DNA) template
   2.5 μL of 10 × *Pyrococcus furiosus* (Pfu) buffer
   2.5 μL phosphorylated mutagenic primer
   0.2 μL 25 m*M* Adenosine triphosphate
   0.25 μL 10 m*M* deoxynucleotide triphosphate (dNTP)
   0.25 μL Pfu polymerase
   0.5 μL Pfu ligase
   X μL sterile water
   25 μL total volume reaction
   PCR parameters:

Denature: 95 °C 1 min.
30 cycles:
95 °C 1 min.
55 °C 1 min
65–72 °C (1–2 min/kb)
Then: 4 °C hold

For standard Pfu polymerase, extension time is 2 min/kb. For Pfu turbo polymerase, extension time is 1 min/Kb for 0–4 kb templates, and 2 min/kb for templates greater than 4 kb.

5. Template was digested with *Dpn*I, which digests methylated DNA, at 37 °C for 1 h, and single stranded plasmid was transfected into competent cells.
6. Sequence PCR products to determine that desired sites were mutated and that PCR error did not introduce undesired mutations.
7. Clone fragment back into full gene.

3.2 Generation of Targeting Construct

To increase the probability of successful homologous recombination, the mutated gene must contain at least 8 kb of sequence identity with the gene of interest. The general strategy for generating a targeting construct of the NF-M$^{S\rightarrow A}$ allele was to create a long arm (6 kb) and short arm (2 kb) that flanked exon three (exon three contains all mutated codons). Additionally, a cassette containing the NF-L 3′ untranslated region (UTR) and phosphoglycerate kinase promoter-neomycin phosphotransferase gene (pGK-Neo) was cloned into the 3′UTR of NF-M. NF-L 3′UTR provided an upstream polyadenylation signal, which was necessary as pGK-Neo was cloned in the same reading direction as *nefm*. pGK-Neo gene was utilized for positive selection. Furthermore, a negative selection factor was cloned 3′ of the short arm and outside the region of identity. For the NF-M$^{S\rightarrow A}$ allele, the herpes simplex virus thymidine kinase gene was utilized for negative selection (Garcia et al., 2009). Diphtheria toxin alpha gene with polyadenylation (Yanagawa et al., 1999) was utilized in subsequent NF-M alleles such as NF-M^{BovineTail} (Barry et al., 2012). The full targeting construct for the NF-M$^{S\rightarrow A}$ mouse is shown in Fig. 1A.

3.3 Targeting of ES Cells Via Homologous Recombination

Once the desired targeting construct has been made, it is necessary to incorporate the desired DNA into mouse ES cells.

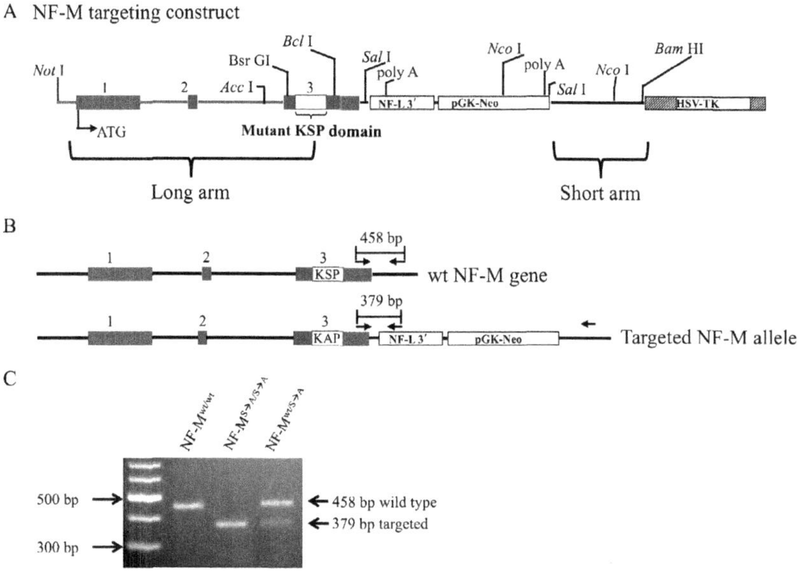

Figure 1 Overview of targeting constructs and genotyping strategy for NF-M$^{S\rightarrow A}$ mouse. (A) Schematic of the long and short arm of the NF-M targeting construct, used to promote homologous recombination. KSP serine residues were replaced with alanine to prevent phosphorylation. (B) Schematic of primer annealing sites for wild-type and targeted *nefm* alleles with approximate size of PCR products (458 and 379 bp product for the endogenous and targeted genes, respectively). (C) Genotyping results using PCR and electrophoretic analysis of genomic DNA isolated from tail biopsies.

1. The targeting construct was linearized using a unique restriction site that was localized outside the region of identity. The linearized construct was then electroporated into ES cells.
2. Electroporated cells were treated with G418 (250 μg/mL) for positive selection and ganciclovir (2 μ*M*) for negative selection. ES cells were treated for 1 week.
3. Colonies that survive selection were harvested, and were placed in a single well in a 96-well plate. Colonies were maintained under selective pressure, and were further expanded by placement in successively larger wells.
4. After expansion, colonies were split into two groups. One group was stored at −80 °C while genomic DNA was isolated from the remaining cells. ES cells were screened on the 3′ end utilizing PCR. PCR was performed with a 5′ primer anchored in within the pGK-Neo gene and a 3′ primer that was localized outside the short arm within the genomic sequence.

5. PCR positive clones were further expanded, and genomic DNA was isolated for screening of the 5′ end by southern blot.
6. As an additional confirmation, PCR and southern positive clones were sequenced to ensure that all mutant codons were contained within the targeted allele.

3.4 Injecting Blastocyst into Psuedopregnant Females, and Breeding Selection

1. Positive RI ES cells were injected into blastocysts isolated from C57Bl/6 females. Injected blastocysts were subsequently implanted into C57Bl/6 pseudopregnant females. Culturing ES cells and implanting into blastocysts requires good timing and coordination, so we have provided a reference to a thorough protocol (Longenecker & Kulkarni, 2009).
2. Coat color will reflect random incorporation of the targeted ES cells into the offspring. Select for male mice with high levels of brown fur, and mate with C57Bl/6 animals. Offspring of this initial mating will have brown fur when utilizing RI ES cells.

3.5 Generating a Line of Mice/Genotyping

1. Coat color was utilized for the first round of screening from chimeric-C57Bl/6 breeding. Positive pups were all brown. PCR screening was utilized for confirmation and for screening of all subsequent mice. Genomic DNA was isolated from tail biopsies. PCR screens were performed with a primer anchored in the 3′ end of exon three. Two three prime primers were included. One primer was common to both endogenous and targeted allele. Amplification of the endogenous allele resulted in a ~458 bp product. However, incorporation of the NF-L 3′UTR/pGK-Neo cassette into the mutant allele prevented amplification with this primer pair due to size. A primer located within the 5′ end of NF-L 3′UTR generated a ~379 bp product, which was confirmation of the targeted allele.
2. Genotyping NF-$M^{S\rightarrow A}$ mice

   Primers

   NF-M Exon3: AGTGGTGGTCACCAAGAAGG

   NF-L 3′UTR: GTTGACCTGATTGGGGAGAA

   NF-M 3′UTR: TGGCTCAGTTGGTACTCTGC

   Successful PCR resulted in a ~379 bp fragment for the targeted band and ~458 bp fragment for the endogenous band. A schematic of primer

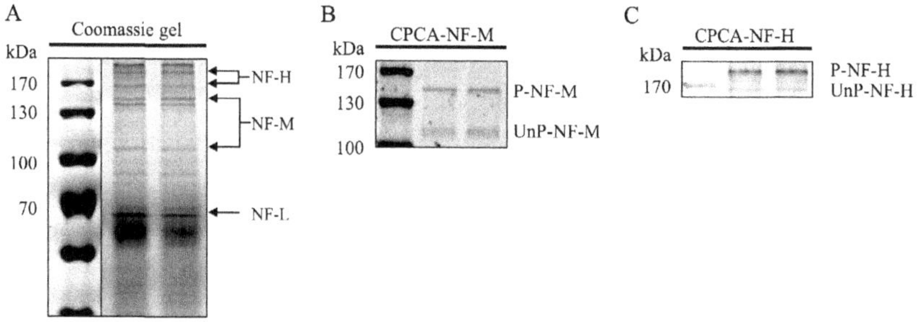

Figure 2 Overview of neurofilament analysis utilizing immunoblotting. (A) Coomassie stain of an SDS-PAGE gel of sciatic nerve extracts indicating approximate locations of NF-H, NF-M, and NF-L proteins. Notice two locations for NF-H and NF-M depending on phosphorylation states. Immunoblotting identifies both phosphorylated and unphosphorylated forms of (B) NF-M and (C) NF-H from sciatic nerve extracts isolated from wild-type mice.

annealing sites and PCR products are shown in Fig. 2B and C, respectively. The following PCR conditions were used:

100 ng DNA template
1 μL NF-M Exon 3 primer (10 μ*M* stock)
1 μL NF-L 3′UTR primer (10 μ*M* stock)
1 μL NF-M 3′UTR primer (10 μ*M* stock)
1 μL dNTP (*Takara*)
5 μL 10 × buffer (*Invitrogen*)
1.5 μL 50 m*M* $MgCl_2$ (*Invitrogen*)
0.5 μL Taq polymerase (*Invitrogen*)
X μL sterile water
50 μL total volume reaction
PCR parameters:
Denature: 94 °C 1 min
35 cycles
94 °C 30 s
60 °C 30 s
72 °C 30 s
Then 4 °C hold

3.6 Analysis of Neurofilament Phosphorylation Utilizing Immunohistochemistry

After a mouse line was created, assessment of neurofilament expression and phosphorylation was accomplished through immunoblotting.

A plethora of antibodies are available to detect both phosphorylated and unphosphorylated forms of Neurofilament Medium, and Heavy. Furthermore, phosphorylation of NF-M and NF-H alter the migration of these proteins in SDS-PAGE (Fig. 2A–C) allowing for two potential mechanisms to determine the phosphorylation status of NF-M.

Solutions

Homogenization buffer:

50 m*M* Tris (pH 7.5)
150 m*M* NaCl
5 m*M* Ethylenediaminetetraacetic acid
1 m*M* Leupeptin
1 m*M* Aprotinin
1 m*M* Chymostatin
Phenylmethanesulfonylfluoride (PMSF) (20%, w/v)

Lysis buffer:

50 m*M* Tris (pH 7.5)
150 m*M* NaCl
1% NP-40
1% deoxycholate
2% Sodium dodecyl sulfate
1 m*M* Leupeptin
1 m*M* Aprotinin
1 m*M* Chymostatin
PMSF (20%, w/v)

1. Homogenize spinal cord or sciatic nerve in appropriate amount of homogenization buffer. Suggested volume for spinal cord is 500 μL, and sciatic nerves 50 μL per nerve.
2. Add equal amount of lysis buffer, and continue to homogenize tissue.
3. Sonicate sample at 30% duty cycle, and output control setting of 1 for 10–15 strokes. Wash sonicator with ethanol and ddH_2O to avoid contamination between samples.
4. Boil extract for 10 min in 100 °C sand bath. Centrifuge at 4 °C at full speed for 10 min.
5. Transfer supernatant to new tube.
6. Determine protein concentration by standard Pierce™ BCA Protein Assay Kit or Pierce Coomassie Plus (Bradford) Assay Kit (both kits available at http://www.lifetechnologies.com, catalogue 23227 and 23236, respectively).

7. Add Laemmli buffer to protein samples, mix by pipetting up and down. Boil samples for 3–5 min.
8. Separate on 7.5% polyacrylamide gel.
9. Transfer to nitrocellulose membrane. Polyvinylidine difluoride (PVDF) can be substituted if preferred. However, given the abundance of NF proteins in spinal tissue and sciatic nerves, we have not found the increased binding capacity associated with PVDF membranes beneficial. We suggest transferring overnight at 25 V at 4 °C, or for 3 h at 50 V at 4 °C. Cut membrane at 100 kDa marker if you wish to separately block for NF-H or NF-M and NF-L.
10. Rinse membrane with ddH_2O.
11. Block in 5% nonfat dry milk in TBS-T for 2 h at room temperature (RT).
12. Rinse in TBS-T three times for 5 min at RT
13. Incubate in primary antibody for 2 h at RT or overnight at 4 °C. EnCor Biotechnology anti-NF-M and anti-NF-H antibodies detect both phosphorylated and nonphosphorylated forms of NF-M and NF-H, respectively, eliminating the need for multiple blots.

    Suggested dilutions for chicken polyclonal antibodies (EnCor Biotechnology):

    NF-L: 1:10,000
    NF-M: 1:10,000
    NF-H: 1:1,000,000

    However, oftentimes these antibodies are detectable at even more dilute levels depending on concentration within samples.
14. Rinse in TBS-T three times for 5 min at RT.
15. Incubate in secondary antibody for 1 h at RT or 2 h 4 °C.

    We suggest using anti-chicken secondary antibody (Rockland, Gilbertsonville, PA, USA) at 1:5000.
16. Image using infrared scanner or film.

4. CONCLUSIONS AND APPLICATIONS

Interestingly, preventing NF-M KSP phosphorylation, through the development of phospho-incompetent NF-M, did not affect radial growth. This was a surprising finding considering the prevailing hypothesis predicted that neurofilament phosphorylation was required for radial growth. These results generated two immediate questions for the field: (1) How is axonal diameter developmentally determined? (2) What is the role for NF KSP phosphorylation? Question 1 has been partially addressed through DNA sequence analysis of NF-M (Barry et al., 2010) and through the generation of mice expressing NF-M with an elongated C-terminus (Barry et al., 2012). Together, these studies suggested that the length of NF-M C-terminal domain plays an important role in axonal diameter, yet the mechanism by which it does so remains unclear. However, question 2 remains unanswered *in vivo*.

Phosphorylation is a key posttranslational modification for NF subunit proteins as well as other intermediate filament proteins. Phosphorylation has been shown to play a key role in intermediate filament biology, and multiple sites on the head and tail domains of intermediate filament proteins can be phosphorylated. Site-directed mutagenesis has been a key tool in providing insights into the role of phosphorylation *in vivo*. Transgenic expression of the phospho-mimetic NF-L^{S55D} resulted in early and robust neuropathology that included accumulation of NF aggregates within the cell body (Gibb, Brion, Brownless, Anderton & Miller, 1998). Moreover, site-directed mutagenesis has been utilized to analyze the role of multiple phosphorylation sites in unique combinations. Utilizing site-directed mutagenesis to prevent phosphorylation of all five head domain phosphorylation sites of glial fibrillary acid protein (GFAP) reduced filament formation *in vivo* (Takemura, Gomi, Colucci-Guyon, & Itohara, 2002). Interestingly, preventing phosphorylation of subsets of sites on the head domain of GFAP resulted in different phenotypes suggesting that each phosphorylation site plays a unique role in regulating filament assembly (Takemura et al., 2002). Taken together, these results suggest that phosphorylation of sites located within the head domain of intermediate filaments altered filament formation.

While it has been shown that phosphorylation of the head domain plays a key role in filament assembly and transport (Omary et al., 2006), the role of phosphorylation within the tail domain of NF-M or NF-H remains unclear. Deletion of the tail domain of NF-M (NF-$M^{\Delta Tail}$) (Garcia et al., 2003) did not alter NF transport or filament formation, but there were alterations in NF spacing, clustering, and reduced radial growth, which resulted in reduced conduction velocity. However, taken together with the information gained from the NF-$M^{S\rightarrow A}$, reduced axonal diameter observed in NF-$M^{\Delta Tail}$ mice was not due to deletion of NF-M tail domain phosphorylation sites. Rather, the reduction in radial growth was likely due to the fact that the length of NF-M tail domain contributed to establishing axon diameter (Barry et al., 2010, 2012). Deletion of the NF-H tail domain (NF-$H^{\Delta Tail}$) resulted in reduced rates of radial growth, but diameters of all but the very largest motor axons were similar to control by 6 months (Rao et al., 2003).

Given the relatively mild phenotypes observed in NF-M$^{S \rightarrow A}$ and NF-H$^{\Delta Tail}$ mice, it is interesting that preventing the phosphorylation of a single NF subunit resulted in compensatory phosphorylation of the remaining phospho-competent subunit. Phosphorylation of NF-H was increased in NF-M$^{\Delta Tail}$ (Garcia et al., 2003) and NF-M$^{S \rightarrow A}$ (Garcia et al., 2009) mice. Deletion of the tail domain of NF-H resulted in an increase in NF-M phosphorylation (Rao et al., 2002), and simultaneous deletion of both NF-H and NF-M C-termini resulted in an increase in Tau phosphorylation (Lobsiger, Garcia, Ward & Cleveland, 2005). With the current set of genetically modified mice, it is not possible to determine the role for NF C-terminal phosphorylation. However, compensatory phosphorylation has been observed in all four lines of mice (NF-H$^{\Delta Tail}$, NF-M$^{\Delta Tail}$, NF-(M/H)$^{\Delta Tail}$, and NF-M$^{S \rightarrow A}$) expressing phospho-incompetent NF subunits suggesting an important physiological role.

Two strategies will need to be employed to address the role of NF C-terminal phosphorylation during nerve development. NF phosphorylation is dependent upon the formation of compact myelin (de Waegh et al., 1992), and NF phosphorylation is increased in myelinated internodes relative to unmyelinated regions (de Waegh et al., 1992; Hsieh et al., 1994; Yin et al., 1998). Therefore, it would be interesting to develop gene-targeted mice expressing a constitutively expressed phospho-mimetic NF subunit. This would alter the spatial-temporal dynamics of NF phosphorylation by uncoupling NF phosphorylation and myelination. Moreover, generating inducible forms of NF phospho-mutants could provide key insights into any potential short-term consequences of altering NF C-terminal phosphorylation. Both methods will address fundamental questions important to understanding nerve development, and both methods will require the use of site-directed mutagenesis.

REFERENCES

Barry, D. M., Carpenter, C., Yager, C., Golik, B., Barry, K. J., Shen, H., et al. (2010). Variation of the neurofilament medium KSP repeat sub-domain across mammalian species: Implications for altering axonal structure. *Journal of Experimental Biology, 213*(1), 128–136.

Barry, D. M., Stevenson, W., Bober, B. G., Wiese, P. J., Dale, J. M., Barry, G. S., et al. (2012). Expansion of neurofilament medium C terminus increases axonal diameter independent of increases in conduction velocity or myelin thickness. *Journal of Neuroscience, 32*(18), 6209–6219.

Capecchi, M. R. (1989a). Altering the genome by homologous recombination. *Science*, *244*(4910), 1288–1292.
Capecchi, M. R. (1989b). The new mouse genetics: Altering the genome by gene targeting. *Trends in Genetics*, *5*(3), 70–76.
Carden, M. J., Trojanowski, J. Q., Schlaepfer, W. W., & Lee, V. M.-Y. (1987). Two-Stage expression of neurofilament polypeptides during rat neurogenesis with early establishment of adult phosphorylation patterns. *Journal of Neuroscience*, 7(11), 3489–3504.
Cochard, P., & Paulin, D. (1984). Initial expression of neurofilaments and vimentin in the central and peripheral nervous system of the mouse embryo *in vivo*. *Journal of Neuroscience*, *4*(8), 2080–2094.
Collard, J. F., Cote, F., & Julien, J. P. (1995). Defective axonal transport in a transgenic mouse model of amyotrophic lateral sclerosis. *Nature*, *375*(6526), 61–64.
Cote, F., Collard, J. F., & Julien, J. P. (1993). Progressive neuropathy in transgenic mice expressing the human neurofilament heavy gene: A mouse model of amyotrophic lateral sclerosis. *Cell*, *73*, 35–46.
de Waegh, S. M., Lee, V. M.-Y., & Brady, S. T. (1992). Local modulation of neurofilament phosphorylation, axonal caliber, and slow axonal transport by myelinating Schwann cells. *Cell*, *68*, 451–463.
Elder, G. A., Friedrich, V. L., Jr., Bosco, P., Kang, C., Gourov, A., Tu, P. H., et al. (1998). Absence of the mid-sized neurofilament subunit decreases axonal calibers, levels of light neurofilament (NF-L), and neurofilament content. *Journal of Cell Biology*, *141*(3), 727–739.
Elder, G. A., Friedrich, V. L., Jr., Kang, C., Bosco, P., Gourov, A., Tu, P.-H., et al. (1998). Requirement of heavy neurofilament subunit in the development of axons with large calibers. *Journal of Cell Biology*, *143*(1), 195–205.
Evans, M. J., & Kaufman, M. H. (1981). Establishment in culture of pluripotent cells from mouse embryos. *Nature*, *272*, 154–156.
Eyer, J., & Peterson, A. (1994). Neurofilament-deficient axons and perikaryal aggregates in viable transgenic mice expressing a neurofilament-beta-galactosidase fusion protein. *Neuron*, *12*(2), 389–405.
Fuchs, E., & Weber, C. (1994). Intermediate filaments: Structure, dynamics, function, and disease. *Annual Review of Biochemistry*, *63*, 345–382.
Garcia, M. L., Lobsiger, C. S., Shah, S. B., Deerinck, T. J., Crum, J., Young, D., et al. (2003). NF-M is an essential target for the myelin-directed "outside-in" signaling cascade that mediates radial axonal growth. *The Journal of Cell Biology*, *163*(5), 1011–1020.
Garcia, M. L., Rao, M. V., Fujimoto, J., Garcia, V. B., Shah, S. B., Crum, J., et al. (2009). Phosphorylation of highly conserved neurofilament medium KSP repeats is not required for myelin-dependent radial axonal growth. *Journal of Neuroscience*, *29*(5), 1277–1284.
Geisler, N., Kaufmann, E., Fischer, S., Plessmann, U., & Weber, K. (1983). Neurofilament architecture combines structural principles of intermediate filaments with carboxy-terminal extensions increasing in size between triplet proteins. *The EMBO Journal*, *2*(8), 1295–1302.
Gibb, J. M., Brion, J. P., Brownless, J., Anderton, B. H., & Miller, C. C. J. (1998). Neuropathological abnormalities in transgenic mice harbouring a phosphorylation mutant neurofilament transgene. *Journal of Neurochemistry*, *70*(2), 492–500.
Heins, S., Wong, P. C., Muller, S., Goldie, K., Cleveland, D., & Aebi, U. (1993). The rod domain of NF-L determines neurofilament architecture, whereas the end domains specify filament assembly and network formation. *Journal of Cell Biology*, *123*(6), 1517–1533.
Herrmann, H., & Aebi, U. (2004). Intermediate filaments: Molecular structure, assembly mechanism, and integration into functionally distinct intracellular Scaffolds. *Annual Review of Biochemistry*, *73*, 749–789.

Hsieh, S., Kidd, G. J., Crawford, T. O., Xu, Z., Lin, W., Trapp, B. D., et al. (1994). Regional modulation of neurofilament organization by myelination in normal axons. *Journal of Neuroscience*, *14*(11), 6392–6401.

Julien, J. P., & Mushynski, W. E. (1982). Multiple phosphorylation sites in mammalian neurofilament polypeptides. *The Journal of Biological Chemistry*, *257*(17), 10467–10470.

Julien, J. P., & Mushynski, W. E. (1983). The distribution of phosphorylation sites among identified proteolytic fragments of mammalian neurofilaments. *The Journal of Biological Chemistry*, *258*(6), 4019–4025.

Koller, B. H., Hagemann, L. J., Doetschman, T., Hagaman, J. R., Huang, S., Williams, P. J., et al. (1989). Germ-line transmission of a planned alteration made in a hypoxanthine phosphoribosyltransferase gene by homologous recombination in embryonic stem cells. *Proceedings of the National Academy of Sciences of the United States of America*, *86*(22), 8927–8931.

Lee, M. K., Xu, Z., Wong, P. C., & Cleveland, D. W. (1993). Neurofilaments are obligate heteropolymers in vivo. *The Journal of Cell Biology*, *122*(6), 1337–1350.

Lobsiger, C. S., Garcia, M. L., Ward, C. M., & Cleveland, D. W. (2005). Altered axonal architecture by removal of the heavily phosphorylated neurofilament tail domains strongly slows superoxide dismutase 1 mutant-mediated ALS. *Proceedings of the National Academy of Sciences of the United States of America*, *102*(29), 10351–10356.

Longenecker, G., & Kulkarni, A. B. (2009). Generation of gene knockout mice by ES cell microinjection. *Current Protocols in Cell Biology*, 1–36. Chapter 19, Unit 19.14.

Mansour, S. L., Thomas, K. R., Deng, C. X., & Capecchi, M. R. (1990). Introduction of a lacZ reporter gene into the mouse int-2 locus by homologous recombination. *Proceedings of the National Academy of Sciences of the United States of America*, *87*(19), 7688–7692.

Monteiro, M. J., Hoffman, P. N., Gearhart, J. D., & Cleveland, D. W. (1990). Expression of NF-L in both neuronal and nonneuronal cells of transgenic mice: Increased neurofilament density in axons without affecting caliber. *The Journal of Cell Biology*, *111*(4), 1543–1557.

Omary, M. B., Ku, N. O., Tao, G. Z., Toivola, D. M., & Liao, J. (2006). "Heads and tails" of intermediate filament phosphorylation: Multiple sites and functional insights. *Trends in Biochemical Sciences*, *31*(7), 383–394.

Pachter, J. S., & Liem, R. K. (1984). The differential appearance of neurofilament triplet polypeptides in the developing rat optic nerve. *Developmental Biology*, *103*(1), 200–210.

Rao, M. V., Campbell, J., Yuan, A., Kumar, A., Gotow, T., Uchiyama, Y., et al. (2003). The neurofilament middle molecular mass subunit carboxyl-terminal tail domains is essential for the radial growth and cytoskeletal architecture of axons but not for regulating neurofilament transport rate. *The Journal of Cell Biology*, *163*(5), 1021–1031.

Rao, M. V., Garcia, M. L., Miyazaki, Y., Gotow, T., Yuan, A., Mattina, S., et al. (2002). Gene replacement in mice reveals that the heavily phosphorylated tail of neurofilament heavy subunit does not affect axonal caliber or the transit of cargoes in slow axonal transport. *The Journal of Cell Biology*, *158*(4), 681–693.

Rao, M. V., Houseweart, M. K., Williamson, T. L., Crawford, T. O., Folmer, J., & Cleveland, D. W. (1998). Neurofilament-dependent radial growth of motor axons and axonal organization of neurofilaments does not require the neurofilament heavy subunit (NF-H) or its phosphorylation. *The Journal of Cell Biology*, *143*(1), 171–181.

Rushton, W. A. H. (1951). A theory of the effects of fibre size in medullated nerve. *The Journal of Physiology*, *115*, 101–122.

Sakaguchi, T., Okada, M., Kitamura, T., & Kawasaki, K. (1993). Reduced diameter and conduction velocity of myelinated fibers in the sciatic nerve of a neurofilament-deficient mutant quail. *Neuroscience Letters*, *153*(1), 65–68.

Shaw, G., & Weber, K. (1982). Differential expression of neurofilament triplet proteins in brain development. *Nature*, *298*(5871), 277–279.

Shen, H., Barry, D. M., & Garcia, M. L. (2010). Distal to proximal development of peripheral nerves requires the expression of neurofilament heavy. *Neuroscience*, *170*(1), 16–21.

Takemura, M., Gomi, H., Colucci-Guyon, E., & Itohara, S. (2002). Protective role of phosphorylation in turnover of glial fibrillary acidic protein in mice. *Journal of Neuroscience*, *22*(16), 6972–6979.

Waxman, S. G. (1980). Determinants of conduction velocity in myelinated nerve fibers. *Muscle & Nerve*, *3*(2), 141–150.

Xu, Z., Marszalek, J. R., Lee, M. K., Wong, P. C., Folmer, J., Crawford, T. O., et al. (1996). Subunit composition of neurofilaments specifies axonal diameter. *The Journal of Cell Biology*, *133*(5), 1061–1069.

Yanagawa, Y., Kobayashi, T., Ohnishi, M., Kobayashi, T., Tamura, S., Tsuzuki, T., et al. (1999). Enrichment and efficient screening of ES cells containing a targeted mutation: The use of DT-A gene with the polyadenylation signal as a negative selection maker. *Transgenic Research*, *8*(3), 215–221.

Yin, X., Crawford, T. O., Griffin, J. W., Tu, P., Lee, V. M., Li, C., et al. (1998). Myelin-associated glycoprotein is a myelin signal that modulates the caliber of myelinated axons. *Journal of Neuroscience*, *18*(6), 1953–1962.

Yuan, A., Rao, M. V., Sasaki, T., Chen, Y., Kumar, A., Veeranna, et al. (2006). Alpha-internexin is structurally and functionally associated with the neurofilament triplet proteins in the mature CNS. *Journal of Neuroscience*, *26*(39), 10006–10019.

Yuan, A., Sasaki, T., Kumar, A., Peterhoff, C. M., Rao, M. V., Liem, R. K., et al. (2012). Peripherin is a subunit of peripheral nerve neurofilaments: Implications for differential vulnerability of CNS and peripheral nervous system axons. *Journal of Neuroscience*, *32*(25), 8501–8508.

Zhu, Q., Couillard-Despres, S., & Julien, J. P. (1997). Delayed maturation of regenerating myelinated axons in mouse lacking neurofilaments. *Experimental Neurology*, *148*(1), 299–316.

Zhu, Q., Lindenbaum, M., Levavasseur, F., Jacomy, H., & Julien, J. P. (1998). Disruption of the NF-H gene increases axonal microtubule content and velocity of neurofilament transport: Relief of axonopathy resulting from the toxin beta, beta′-iminodipropionitrile. *The Journal of Cell Biology*, *143*(1), 183–193.

CHAPTER SEVENTEEN

α-Internexin and Peripherin: Expression, Assembly, Functions, and Roles in Disease

Jian Zhao, Ronald K.H. Liem[1]
Department of Pathology and Cell Biology, Taub Institute for Research on Alzheimer's Disease and the Aging Brain, Columbia University College of Physicians and Surgeons, New York, USA
[1]Corresponding author: e-mail address: rkl2@cumc.columbia.edu

Contents

1. α-Internexin 478
   1.1 Introduction: α-Internexin is a Type IV Neuronal Intermediate Filament Protein 478
   1.2 α-Internexin: Structure and Assembly 479
   1.3 α-Internexin: Tissue and Developmental Expression 481
   1.4 Functions of α-Internexin 487
   1.5 α-Internexin in Disease 488
2. Peripherin 489
   2.1 Introduction: Peripherin is a Type III Neuronal IF Protein 489
   2.2 Peripherin: Structure and Assembly 490
   2.3 Peripherin: Tissue and Developmental Expression 490
   2.4 Functions of Peripherin 493
   2.5 Role of Peripherin in Disease 494
3. Conclusion 497
References 498

Abstract

α-Internexin and peripherin are neuronal-specific intermediate filament (IF) proteins. α-Internexin is a type IV IF protein like the neurofilament triplet proteins (NFTPs, which include neurofilament light chain, neurofilament medium chain, and neurofilament high chain) that are generally considered to be the primary components of the neuronal IFs. However, α-internexin is often expressed together with the NFTPs and has been proposed as the fourth subunit of the neurofilaments in the central nervous system. α-Internexin is also expressed earlier in the development than the NFTPs and is a maker for neuronal IF inclusion disease. α-Internexin can self-polymerize *in vitro* and in transfected cells and it is present in the absence of the NFTP in development and in granule cells in the cerebellum. In contrast, peripherin is a type III IF protein. Like α-internexin, peripherin is specific to the nervous system, but it is expressed predominantly in the peripheral nervous system (PNS). Peripherin can also self-assemble both *in vitro* and

Methods in Enzymology, Volume 568
ISSN 0076-6879
http://dx.doi.org/10.1016/bs.mie.2015.09.012

in transfected cells. It is as abundant as the NFTPs in the sciatic nerve and can be considered a fourth subunit of the neurofilaments in the PNS. Peripherin has multiple isoforms that arise from intron retention, cryptic intron receptor site or alternative translation initiation. The functional significance of these isoforms is not clear. Peripherin is a major component found in inclusions of patients with amyotrophic lateral sclerosis (ALS) and peripherin expression is upregulated in ALS patients.

ABBREVIATIONS

ALS amyotrophic lateral sclerosis
FGF fibroblast growth factor
HTLV-1 human T-cell leukemia virus type 1
IL-6 interleukin-6
IF intermediate filament
IFA intermediate filament antigen
LIF leukemia inhibitory factor
NGF nerve growth factor
NFTPs neurofilament triplet proteins
NFL neurofilament light chain
NFM neurofilament medium chain
NFH neurofilament high chain
NIFID neuronal intermediate filament inclusion disease
NPSLE neuropsychiatric system lupus erythematosus
NOD nonobese diabetic
PFA paraformaldehyde
PBS phosphate buffered saline
PBST phosphate buffered saline with 0.1% Triton X-100
PNS peripheral nervous system
SAXS small angle X-ray scattering
ULF unit-length-filament

1. α-INTERNEXIN

1.1 Introduction: α-Internexin is a Type IV Neuronal Intermediate Filament Protein

α-Internexin protein was first identified during a procedure to purify neurofilaments from rat optic nerve and spinal cord (Pachter & Liem, 1985). This 66 kDa protein was recognized by the anti-IFA antibody against a conserved epitope present on most intermediate filament (IF) proteins (Pruss, 1985). α-Internexin was at first thought to be an IF-associated protein (Pachter & Liem, 1985; Steinert & Roop, 1988), but when its cDNA was cloned and sequenced, its predicted amino acid sequence showed all the characteristics of an IF protein (Fliegner, Ching, & Liem, 1990). Further studies, including RNA and protein distribution, *in vitro* polymerization and axonal transport, showed that α-internexin is a neuronal IF protein found in most neurons of the CNS (Fliegner et al., 1990; Kaplan, Chin, Fliegner, & Liem, 1990). It was also identical to a 66 kDa neuronal IF protein previously described (Chiu et al., 1989). Subsequently, the gene encoding α-internexin was cloned using low stringency hybridization and a neurofilament medium chain (NFM) cDNA probe (Ching & Liem, 1991). The sequence data and exon–intron organization of the gene established α-internexin as a member of the type IV IF gene family along with the neurofilament triplet proteins (NFTPs), neurofilament light chain (NFL), NFM, and neurofilament heavy chain (NFH) (Ching & Liem, 1991; Julien et al., 1987, 1988; Levy, D'Eustachio, Cowan, & Liem, 1987; Lewis & Cowan, 1986).

Protocol for the purification of α-internexin (Pachter & Liem, 1985):

1. Isolate IFs from spinal cord by the axonal flotation method, which will contain a mixture of neurofilaments, GFAP and α-internexin (Liem, Yen, Salomon, & Shelanski, 1978). Alternatively, use a bacterial expression system for α-internexin to produce the protein (Kaplan et al., 1990).
2. Dissolve the insoluble protein in 10 m*M* phosphate buffer, pH 7.4 made up in freshly deionized 8 *M* urea. Clear the insoluble debris by centrifugation at 10,000 × *g* for 30 min.
3. Apply the solubilized material to a hydroxylapatite column (BioRad HTP, Bio-Rad Laboratories, Richmond, CA), equilibrated with the same buffer.
4. Elute the column with 0.1 *M* phosphate buffer, pH 7.0 in 8 *M* urea. The α-internexin protein will be eluted in these fractions.
5. α-Internexin can be reassembled from this material by dialyzing against PBS, centrifuged at 100,000 × *g* for 1 h and the pellet can be dissolved in 10 m*M* phosphate buffer, pH 7.0 in 8 *M* urea. Analyze reassembled protein by SDS-PAGE for purity.
6. If necessary, α-internexin can be further purified by applying the solubilized protein to a DEAE-cellulose column (DE52 from Whatman) and eluted with a linear gradient of 10 m*M*–0.1 *M* phosphate buffer, pH 7.0 in 8 *M* urea. Fractions should be analyzed by SDS-PAGE for purity.

1.2 α-Internexin: Structure and Assembly

As an IF protein, α-internexin shares the common tripartite molecular structure, which consists of a highly conserved central coiled-coil rod domain flanked by amino-terminal head and carboxyl-terminal tail domains. Similar to NFL, α-internexin has a short tail region compared to NFM and NFH. The tail region of α-internexin contains a high content of glutamate, lysine, and serine residues, which are commonly found in the tails of NFTPs (Fliegner et al., 1990).

A common assembly mechanism exists for vertebrate neuronal IF proteins. The first step is the alignment of α-internexin or NFL monomer with any other NF monomer, through association of the conserved rod domains, to form parallel and coiled-coil dimers. Two dimers line up side-by-side in a half-staggered manner to form an antiparallel tetramer. Rapid aggregation of tetramers leads to the formation of 55 nm long unit-length-filament (ULF) (Herrmann, Häner, Brettel, Ku, & Aebi, 1999). In the second step, ULFs anneal axially to elongate and form immature filaments of about 16 nm wide and several microns long. The third step is a radial compaction that involves close packing of molecular strands to eventually form mature IFs (Herrmann & Aebi, 1998, 2004). However, in one study, ULFs are reported to be the first assembly products for α-internexin under conditions favored by other IF proteins to form tetramers and other small oligomers (Abumuhor, Spencer, & Cohlberg, 1998).

Most of the data on IF structure and assembly has been obtained by a variety of methods including X-ray crystallography, small angle X-ray scattering (SAXS), electron cryotomography, computational analysis, and modeling (Parry, Strelkov, Burkhard, Aebi, & Herrmann, 2007). Studying atomic structures will lead to a better understanding of how α-internexin and other IF proteins organize into filaments and interact with other proteins. Currently, the information of IF crystal structures is mostly from vimentin, lamin, and keratin (Aziz et al., 2012; Chernyatina, Nicolet, Aebi, Herrmann, & Strelkov, 2012; Parry et al., 2007). Thus, data of the structure and assembly of α-internexin have only been obtained from indirect studies. A number of studies have demonstrated the existence of α-internexin homopolymers. Purified α-internexin proteins can self-assemble into 10-nm filament by itself *in vitro* after removing urea by dialysis (Chiu et al., 1989; Kaplan et al., 1990). α-Internexin also self-assembled into filamentous networks in transfected SW13vim^{-} cells lacking a preexisting filament network (Ching & Liem, 1993). In addition, by using immunocytochemistry and *in situ* hybridization, expression levels of neuronal IFs in developing mouse cerebellum and cultured granule neurons have been determined, and the results showed that α-internexin is the only neuronal IF in migrating granule cells and in mature granule neuron neurites, presumably forming a homopolymeric filamentous network (Chien, Mason, & Liem, 1996).

α-Internexin forms heteropolymers with NFTPs and can be considered as a fourth neurofilament subunit in the mature CNS. Yuan et al. showed that α-internexin copurified with the NFTPs in a fixed stoichiometry of 4:2:2:1 (NFL:α-internexin:NFM:NFH) in adult mouse CNS (Yuan et al., 2006). α-Internexin and NFTPs can coassemble into a single filamentous network in transfected SW13vim^{-} cells (Yuan et al., 2006, 2009). Ultrastructural colocalization studies using EM revealed that α-internexin colocalizes with NFM on the same NF in mouse optic axons (Yuan et al., 2006). Data from the measurements of transport and content levels of NF in axons of knockout mice missing one or two members of the four neuronal IF proteins reinforced the assumption that α-internexin is functionally associated with NFTPs in the adult CNS (Yuan, Rao, Kumar, Julien, & Nixon, 2003; Yuan et al., 2006). Moreover, α-internexin, together with NFL, NFM, and peripherin, has been shown to be an integral component of neurofilament polymers in axons of cultured mouse sympathetic neurons (Yan, Jensen, & Brown, 2007).

1.3 α-Internexin: Tissue and Developmental Expression

IF expression is not only tissue specific but also developmentally regulated (Ho & Liem, 1996). The expression level of α-internexin mRNA was first studied in the whole embryonic rat brain using northern blot analysis. This study showed that α-internexin mRNA levels are the highest on E16 and decline postnatally as a fraction of the total brain mRNA (Fliegner et al., 1990). Its expression in the developing rat nervous system was further characterized by *in situ* hybridization. These studies demonstrated that α-internexin is expressed in most neurons as differentiation begins, and it is the only IF gene expressed in the adult cerebellar granule cells. Moreover, expression of α-internexin precedes that of NFL and NFM in many neurons during development (Fliegner, Kaplan, Wood, Pintar, & Liem, 1994). Examples of immunostaining on embryonic mouse brain is shown in Fig. 1. The results in this figure show the difference in staining of α-internexin and NFL in a mouse with a knock-in mutation (N98S) in the NFL gene that corresponds to a mutation found in Charcot-Marie-Tooth type 2E disease (Adebola et al., 2015). As a result of this mutation, the NFL-positive neurons that show aggregates are apparent in Fig. 1. α-Internexin is also expressed in these aggregates, however, α-internexin is present in other processes in this embryonic cerebellum that are still NFL negative and the processes formed by α-internexin alone do not form aggregates.

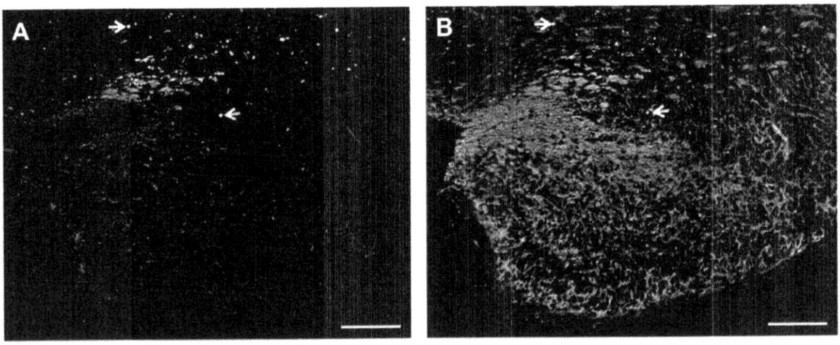

Figure 1 Double immunofluorescence of sections of embryonic brain (E16) of an animal with a N98SNFL knock-in mutation with antibodies specific for NFL (red) and α-internexin (green). NFL positive neurons show aggregates that also contain α-internexin (arrows). However, a number of α-internexin positive processes can be observed that are NFL-negative and are therefore not aggregated. Bars = 100 μm. (See the color plate.)

For these experiments, dissected embryos (E16–E18) were rinsed well in PBS and decapitated. Embryos were fixed in 4% paraformaldehyde (PFA) for 2 h at 4 °C. After fixation, embryos were washed in PBS twice for 10 min each and in 0.85% saline once for 10 min before they were incubated with 50% and 70% ethanol for 15 min each. Tissues can now be stored in 70% ethanol at 4 °C. To obtain paraffin-embedded sections, tissues were first immersed in 70% ethanol twice, 90% ethanol twice, 100% ethanol three times, and xylene three times for 30 min each. Tissues were then transferred to paraffin at 58 °C and aligned appropriately to achieve horizontal sections of the brain. Five micrometer thick sections were cut using a rotary microtome and floated in a 56 °C water bath and mounted onto gelatin-coated histological slides. The slides were dried overnight at room temperature.

For IHC staining, sections were incubated at 58 °C for 1 h and rehydrated (xylene 10 min ×2, 100% ethanol 10 min ×2, 95% ethanol 5 min, 75% ethanol 5 min, 50% ethanol 5 min, distilled H_2O 5 min, phosphate buffered saline with 0.1% Triton X-100 (PBST) 5 min ×2). For antigen retrieval, sections were placed in preheated Trilogy (Cell Marque, Rocklin, CA) in a steamer for 25 min. Sections were rinsed twice with PBST and were blocked with 10% goat serum in PBST for 30 min. Primary antibodies (mouse anti-NFL monoclonal from Abcam; anti-α-internexin rabbit serum; described below) were incubated overnight at 4 °C. The next day, sections were brought back to room temperature, rinsed with PBST twice for 8 min each, incubated with a mixture of Alexa Fluor 488 goat anti-rabbit and Alexa Fluor 568 goat anti-mouse secondary antibodies at

1:1000 for 1 h, and rinsed with PBST twice for 8 min each again. Slides were mounted using Aquamount (Lerner Laboratories, Pittsburgh, PA) and digital images were taken with a Zeiss Axioplan 2 microscope equipped with camera AxioCam with Axiovision software (Zeiss).

Detailed protocol for immunostaining of mouse embryos:

1. Dissect embryos. Wash extensively in PBS to remove as much blood as possible.
   Cut off the tail for genotyping. Decapitate the embryos.
2. Transfer embryos to 4% PFA. Incubate for 2 h (up to 24 h for E16–E18) at 4 °C.
3. Wash embryos extensively in cold PBS and then wash with PBS twice for 10 min with rotation.
4. Wash with 0.85% saline for 10 min with rotation (remove phosphate from PBS, which would precipitate during ethanol dehydration).
5. Incubate in 50% ethanol with rotation for 15 min.
6. Incubate in 70% ethanol with rotation for 15 min.
   Embryos can now be stored in 70% ethanol at 4 °C for several months.

Tissue preparation and paraffin infiltration

1. Dehydrate tissues:
   A. Immerse the tissue in 70% ethanol twice for 30 min each at room temperature.
   B. Immerse the tissue in 90% ethanol twice for 30 min each at room temperature.
   C. Immerse the tissue in 100% ethanol three times for 30 min each at room temperature.
   D. Immerse the tissue in xylene three times for 30 min each at room temperature.
2. Embed the tissue in paraffin at 58 °C.
3. Cut 5 μm thick tissue sections using a rotary microtome.
4. Float the sections in a 56 °C water bath.
5. Mount the sections onto gelatin-coated histological slides.
6. Dry the slides overnight at room temperature.

Immunostaining

1. Incubate paraffin sections at 58 °C for 1 h.
2. Rehydration:
   Xylene: 10 min ×2
   100% ethanol: 10 min ×2

95% ethanol: 5 min
75% ethanol: 5 min
50% ethanol: 5 min
H_2O: 5 min
(Put Trilogy (Cell Marque, Rocklin, CA) in a steamer to preheat)

3. Wash with PBST (PBS w/0.1% Triton) for 5 min twice.
4. Antigen retrieval: Transfer slides in preheated Trilogy in steamer for 25 min.
5. Remove the Trilogy container from the steamer and let it cool down in Trilogy to room temperature.
6. Wash with PBST for 5 min twice.
7. Blocking: 10% goat serum in PBST for 30 min.
8. Primary antibody incubation: Dilute α-internexin polyclonal antibody and monoclonal mouse anti-NFL (Abcam) at 1:1000 in 10% goat serum in PBST overnight at 4 °C.
9. On the next day, take out slides and let them sit at room temperature for 1 h.
   Wash with PBST for 8 min twice.
10. Secondary antibody incubation: Incubate slides with Alexa Fluor 488 goat anti-rabbit and Alexa Fluor 568 goat anti-mouse secondary antibodies for 1 h at 1:1000 (Life Technologies).
11. Wash with PBST for 8 min twice.
12. Affix cover slips over the slides using Aquamount (Lerner Laboratories, Pittsburgh, PA) and store in the dark.

Commercial antibodies to α-internexin and peripherin are shown in Tables 1 and 2. In addition, our laboratory has generated a series of polyclonal rabbit antibodies to the neuronal IF proteins. NFL-N antibody was made against a 20-mer synthetic peptide corresponding to the N-terminus of rat NFL with one amino acid difference. Human NFL and mouse NFL differs from rat NFL by one and two amino acids, respectively, on this region (Kaplan et al., 1991). The antibody recognizes all three species. Peripherin antibody (used in the next section) was generated against a C-terminal peptide of rat peripherin, and the sequence of this peptide is different from mouse and human by one and two amino acids, respectively (Angelastro, Ho, Frappier, Liem, & Greene, 1998). This antibody also recognizes all three species. An α-internexin polyclonal antibody was raised in the same manner as described for the production of a monoclonal antibody to α-internexin (Ching, Chien, Flores, & Liem, 1999): An 0.8 kb BglII–BamHI fragment encoding the C-terminal amino acid residues 340–505 of rat α-internexin containing the SV40 poly A signal was isolated from a rat α-internexin expression cDNA

Table 1 Commercial Anti-α-Internexin Antibodies

Company	Catalog Number (Clone)	Host	Clonality	Reacts with (Predicted to Work with)	Immunogen
Abcam	ab7259	Rabbit	Polyclonal	Rat (human, all mammals)	Recombinant full length rat protein
Abcam	ab108302 (EPR1529)	Mouse	Monoclonal	Mouse, rat, human	Synthetic peptide, C-terminus of human protein
Thermo Scientific Pierce Antibodies	PA1-10009	Chicken	Polyclonal	Mouse, rat, human	Recombinant rat protein
Bioss	Bs-2408R	Rabbit	Polyclonal	Mouse, rat, human	Synthetic peptide between 265 and 310 amino acids of human protein
Novus Biologicals	NBP2-29978 (257CT7.1.2)	Mouse	Monoclonal	Mouse (bovine, rat, human)	Synthetic peptide between 290 and 319 amino acids of human protein

vector (pRSV-α) (Ching & Liem, 1993). This fragment, designated α-BB was highly conserved and shared over 90% homology with human α-internexin and mouse α-internexin, and was cloned into a pET16b vector that also encoded an N-terminal histidine tag. After transformation into the bacterial strain BL21 (DE3), the production of the α-internexin C-terminal peptide was induced by 0.5 m*M* isopropyl-ß-D thiogalactopyranoside for 4 h and the histidine tag containing α-internexin fragment was purified as described (Ching et al., 1999). The purified α-internexin peptide was used as antigen for injection into rabbits. Booster injections were given every 3 weeks. The antibodies were tested against the α-BB fusion protein, as well as whole brain extracts to show specificity. High titer, specific α-internexin antibodies were obtained using this method. The antibody recognizes rat, human, and mouse α-internexin.

Table 2 Commercial Anti-peripherin Antibodies

Company	Catalog Number (Clone)	Host	Clonality	Reacts with (Predicted to Work with)	Immunogen
Abcam	ab4666	Rabbit	Polyclonal	Mouse, rat, rabbit, cat, human	Recombinant full length rat protein
Abcam	ab4573 (7C5)	Mouse	Monoclonal	Chicken (all mammals)	Recombinant full length rat protein
Thermo Scientific Pierce Antibodies	PA3-16723	Rabbit	Polyclonal	Mammals	Recombinant full length rat protein
EMD Millipore	AB1530	Rabbit	Polyclonal	Human, pig, bovine, rat, mouse	Fusion protein containing all but the 4 N-terminal amino acids of rat protein
Santa Cruz Biotechnology	sc-7604	Goat	Polyclonal	Human, rat, mouse, bovine	Synthetic peptide, C-terminus of human protein
Santa Cruz Biotechnology	sc-28539	Rabbit	Polyclonal	Human, rat, mouse, bovine, equine, canine, procine	Synthetic peptide between 21 and 90 amino acids near N-terminus of human protein
Sigma-Aldrich	P5117 (8G2)	Mouse	Monoclonal	Human, rat, mouse, rabbit, pig, feline, bovine	Recombinant full length rat protein

The rabbit polyclonal antibodies to all these neuronal IF proteins are available from our laboratory upon request.

The *in situ* hybridization results confirmed a previous immunohistochemical study that showed that α-internexin protein is distributed extensively in the rat embryo, remained mostly in the CNS in the adult, and is expressed earlier than NFL in the developing brain (Kaplan et al., 1990). In the adult CNS, the amount of α-internexin isolated is comparable to the NFTPs, which account for more than 10% of total protein (soluble and insoluble) in axoplasm (Morris & Lasek, 1982; Yuan et al., 2006).

1.4 Functions of α-Internexin

Most of our knowledge of the function of α-internexin comes from studies of the expression of α-internexin and other neuronal IFs in animal models and cultured cells. Although the precise physiological functions of α-internexin and the mechanisms employed remain to be discovered, several roles have emerged.

Several lines of evidence have suggested that α-internexin may be essential for axonal outgrowth. Gefiltin in zebrafish and xefiltin in *Xenopus laevis* are homologs of α-internexin and they share about 60% homology to human α-internexin. Expression of gefiltin and xefiltin are highly upregulated during retinal growth and optic axon regeneration (Glasgow, Druger, Fuchs, Lane, & Schechter, 1994; Leake, Asch, Canger, & Schechter, 1999; Zhao & Szaro, 1997). In lizard, when regenerating optic axons form synapses with inappropriate targets, expression of the α-internexin homolog is increased but never returns to normal (Rodger et al., 2001). Similarly, in injured rat motor neurons, α-internexin protein expression is dramatically increased and remains elevated when axons are hindered from reaching their targets (McGraw, Mickle, Shaw, & Streit, 2002). Manipulation of α-internexin expression in cultured cell lines provides further support for this role. Overexpression of α-internexin enhances neurite outgrowth of PC12 cells, whereas downregulation of α-internexin in NB2a/d1 neuroblastoma cells with antisense oligonucleotides and anti-α-internexin antibody inhibits axonal neurite elongation (Chien, Liu, Ho, & Lu, 2005; Shea & Beermann, 1999).

α-Internexin may also play a role in the formation and maintenance of dendritic structure as well as in postsynaptic signaling. α-Internexin is found in dendritic shafts, dendritic spines and postsynaptic densities in neurons by immunocytochemistry, immunohistochemistry, and immunoelectron microscopy (Benson, Mandell, Shaw, & Banker, 1996; Suzuki et al., 1997). It is possible that by interacting with other cytoskeletal components such as microtubules and actin, α-internexin may have a structural role in the formation and maintenance of dendrites (Benson et al., 1996). Recently, a novel role of α-internexin was identified from the study of NMDA receptor signaling, in which α-internexin forms a complex with phosphorylated Jacob and ERK after activation of synaptic NMDA receptors, and protects

Jacob from dephosphorylation during trafficking to the nucleus (Karpova et al., 2013). By using the yeast two-hybrid system and a pull-down assay, the binding region of α-internexin for Jacob is mapped to the C-terminal region (Karpova et al., 2013).

Unexpectedly, α-internexin knockout mice showed normal nervous system development and no overt phenotype (Levavasseur, Zhu, & Julien, 1999). However, it cannot be ruled out that subtle changes have not been detected and/or other proteins are compensating for the function(s) of α-internexin. Studies on a transgenic mouse model over-expressing α-internexin showed high levels of neuronal IF accumulations and progressive neurodegeneration (Ching et al., 1999). This result is consistent with the finding of overexpressing α-internexin in PC12 cells (Chien et al., 2005). Neuronal cell death of α-internexin-overexpressing PC12 cells can be rescued by kinase inhibitors that suppress NF hyperphosphorylation (Lee et al., 2012).

1.5 α-Internexin in Disease

1.5.1 Neuronal Intermediate Filament Inclusion Disease

Abnormal accumulation of α-internexin is a pathological hallmark of many human degenerative disorders. Neuronal intermediate filament inclusion disease (NIFID) is a neurodegenerative disease with frontotemporal dementia, pyramidal, and extrapyramidal signs (Bigio, Lipton, White, Dickson, & Hirano, 2003; Cairns et al., 2003; Josephs et al., 2003; Lee et al., 2012). Inclusions positive for α-internexin and NF triplet proteins have been found in many but not all NFID cases (Uchikado, Li, Lin, & Dickson, 2006; Yokota et al., 2008), and they may be an early event in the pathogenesis of NIFID (Cairns et al., 2004). However, the mechanism of NF aggregation in NIFID has yet to be identified. To study the specificity of α-internexin as a component of inclusions of other neurodegenerative diseases, immunohistochemiscal studies have been undertaken, and the results suggest that α-internexin is also involved in Alzheimer's disease, dementia with Lewy bodies and motor neuron disease (Cairns et al., 2004). It should be noted that no human mutations of α-internexin have been identified to be associated with any diseases.

1.5.2 Other Neurodegenerative Diseases

In another neurodegenerative disorder named tropical spastic paraparesis/ human T-cell leukemia virus type 1 (HTLV-1) associated myelopathy, HTLV-1 transcriptional transactivator Tax is found to interact with α-internexin using the yeast two-hybrid system (Reddy et al., 1998). By using truncated constructs and ß-galactosidase assays, the region of α-internexin interacting with Tax was found within the rod domain, and their interaction results in the disruption of the cytoskeletal network (Reddy et al., 1998). α-Internexin has been demonstrated to be helpful in the diagnosis and prognosis determination in several types of tumors (Ducray et al., 2009, 2011; Durand et al., 2011; Eigenbrod et al., 2011; Mokhtari et al., 2011; Nagaishi, Suzuki, Nobusawa, Yokoo, & Nakazato, 2014; Willoughby, Sonawala, Werlang-Perurena, & Donner, 2008). α-Internexin is also implicated in neuropsychiatric system lupus erythematosus (NPSLE) with signs including cognitive impairment and memory loss (Lu et al., 2010). In NPSLE, α-Internexin has been found to be a new pathogenetically relevant autoantigen by using a proteomics approach; higher titers of anti-α-internexin Abs in both the serum and the cerebrospinal fluid of NPSLE have been detected by ELISA, and the titer was inversely correlated with the cognitive status. In addition, data from mice immunized with α-internexin and cultured embryonic neurons treated with anti-α-internexin antibodies demonstrate that anti-α-internexin antibody inhibits neurite elongation and promotes neuron apoptosis (Lu et al., 2010).

2. PERIPHERIN

2.1 Introduction: Peripherin is a Type III Neuronal IF Protein

Peripherin was first identified as a 58 kDa Triton-X100-insoluble protein that is highly expressed in the peripheral nervous system (PNS). Like α-internexin, it was recognized by the anti-IFA antibody (Pruss, 1985) and it therefore contains the common antigenic determinant found in all IF proteins (Portier, de Néchaud, & Gros, 1983). Peripherin cDNA was first identified by differential screening of cDNA libraries from rat PC12 cells to identify mRNAs that were upregulated by long-term treatment with nerve growth factor (NGF) (Leonard, Gorham, Cole, Greene, & Ziff, 1988). Analyses of the predicted amino acid sequence and its exon–intron genomic organization of the isolated rat peripherin gene established it as a member of type III IF family (Thompson & Ziff, 1989). Peripherin and other type III IF family members, including vimentin, desmin, and GFAP, have >70% sequence homology, but only peripherin is neuronal specific (Fuchs & Weber, 1994; Thompson & Ziff, 1989). Peripherin has the same tripartite structure shared by all IF proteins. Unlike the NF triplet proteins, peripherin has a very short C-terminus without a glutamate rich region (Thompson & Ziff, 1989). Peripherin has been purified from differentiated PC12 cells (Aletta et al., 1988), as well as differentiated neuroblastoma cells (NE115 cells) (Portier, Croizat, & Gros, 1982). In both cell lines, peripherin is the primary IF present.

2.2 Peripherin: Structure and Assembly

The assembly mechanism of type III IF proteins is similar to other IFs and involves steps of lateral association of monomers into ULFs, longitudinal elongation of ULFs and radial compaction (Herrmann & Aebi, 2004). Transfection studies showed that expression of peripherin by itself could generate a filamentous network in cultured SW13vim$^-$ cells (Cui, Stambrook, & Parysek, 1995; Ho, Chin, Carnevale, & Liem, 1995). A disulfide dimer has been isolated from rat dorsal root ganglion and sciatic nerve (Chadan, Moya, Portier, & Filliatreau, 1994). This dimer was found to be a native form of peripherin instead of an artifact during sample preparation. The peripherin dimer itself may play some role in development and regeneration as the level of dimeric form of peripherin undergoes more dramatic changes than the monomer. Several lines of evidences also suggest that peripherin can form heteromers with neurofilament proteins, and that peripherin is actually a fourth subunit of NF in the adult PNS (Athlan & Mushynski, 1997; Beaulieu, Robertson, & Julien, 1999; Parysek, McReynolds, Goldman, & Ley, 1991; Yuan et al., 2012). Nixon and colleagues have shown by quantitative immunoblotting of adult mice sciatic nerve homogenates that peripherin is found to be as abundant as the NFTPs and exists with the NFTPs in a fixed ratio of 4:2:1:1 (NFL:NFM:Peripherin:NFH). Peripherin colocalizes with NFL on single neurofilaments of sciatic nerve as revealed by immunogold electron microscopy. NFL knockout mice have significantly less content of peripherin in sciatic nerves. In addition, transfection studies using SW13vim$^-$ cells showed that peripherin coassembles with NF triplet proteins into a single network (Yuan et al., 2012). As described below peripherin and NFL do not always colocalize in dorsal root ganglion cells.

2.3 Peripherin: Tissue and Developmental Expression

2.3.1 Expression of the Major Peripherin Isoform

The expression patterns of peripherin protein and mRNA during development and in the adult have been studied using immunohistochemical methods and *in situ* hybridization (Brody, Ley, & Parysek, 1989; Escurat, Djabali, Gumpel, Gros, & Portier, 1990; Gorham, Baker, Kegler, & Ziff, 1990; Leonard et al., 1988; Parysek & Goldman, 1988; Portier et al., 1983). In adults, expression of peripherin is confined to the PNS and a limited number of neuronal populations in the CNS (Brody et al., 1989; Leonard et al., 1988; Parysek & Goldman, 1988). Peripherin is expressed more widely in embryos, where it has been found in neurons with different functions (motor neurons, sensory neurons, and autonomic neurons), and its expression is associated with the onset of terminal differentiation (Escurat et al., 1990; Gorham et al., 1990). During development, environmental factors play a fundamental role in regulating gene expression and determining the morphological and biochemical differentiation of neurons. *In vitro* studies have shown that the expression of peripherin is induced by NGF (Leonard, Ziff, & Greene, 1987; Parysek & Goldman, 1987), leukemia inhibitory factor (LIF) (Lecomte, Basseville, Landon, Karpov, & Fauquet, 1998), interleukin-6 (IL-6) (Sterneck, Kaplan, & Johnson, 1996), and fibroblast growth factor (FGF) (Choi et al., 2001). Peripherin gene-regulating *cis-* and *trans-* elements have also been identified using *in vitro* studies and transgenic models (Adams, Choate, & Thompson, 1995; Chang & Thompson, 1996; Desmarais, Filion, Lapointe, & Royal, 1992; Lecomte et al., 1998; Leconte et al., 1996; Thompson, Lee, Lawe, Gizang-Ginsberg, & Ziff, 1992). Using immunocytochemistry and *in situ* hybridization, expression of peripherin has been studied *in vivo* in rat motor neurons after injury. It was shown that a retrograde transported inhibitory signal modulates the normal expression of peripherin *in vivo* (Terao, Janssens, van den Bosch de Aguilar, Portier, & Klosen, 2000).

Although peripherin is often expressed together with the NFTPs, the dorsal root ganglia show a particularly interesting pattern of expression (Fig. 2). By double staining of adult mouse dorsal root ganglia with antibodies to NFL (monoclonal) and peripherin (rabbit polyclonal), we can see that there are three different populations of cells: NFL (+)/peripherin (−); NFL (−)/peripherin (+) ;and NFL (+)/peripherin (+). The NFL (+)/peripherin (−) cells tend to be larger and most of the axonal staining is only NFL positive. Some smaller axons can be observed that are NFL (−)/peripherin (+), as well as cells with smaller cell bodies. There are a few cells (arrow) that are positive for both peripherin and NFL. To perform these experiments adult mouse DRG were immersion fixed in 4% PFA for 20 min and washed in PBS twice before they were dehydrated and embedded in paraffin. The remaining protocol is the same as described earlier in Section 1.3.

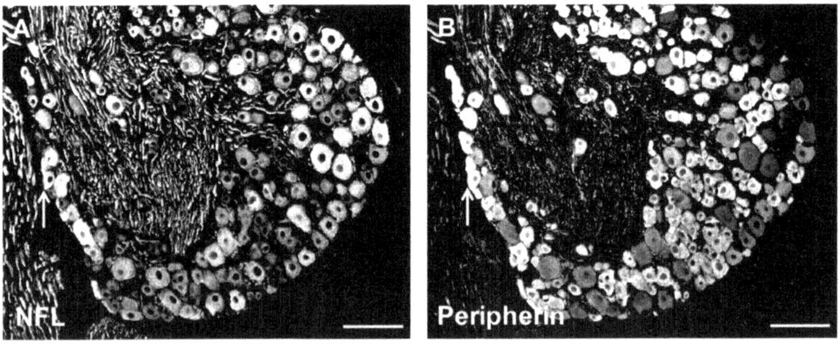

Figure 2 Double immunofluorescence of sections of dorsal root ganglia from adult mice using antibodies specific for NFL (red) and peripherin (green). Most neurons show strong staining of only one of NFL and peripherin and little staining of the other. Arrow depicts a neuron showing coexpression of both NFL and peripherin. Axons show preferential staining of NFL. Bars = 100 μm. (See the color plate.)

2.3.2 Expression of Other Peripherin Isoforms

In addition to the constitutively expressed peripherin protein with a molecular weight of 58 kDa (Per-58), several other isoforms have been identified from mouse (Per-45, Per-56, and Per-61) and human (Per-28 and Per-45) (Landon, Wolff, & de Néchaud, 2000; McLean, Xiao, Miyazaki, & Robertson, 2008; Xiao et al., 2008) (Fig. 3). These protein isoforms are the results of alternative splicing or translation initiation (Thompson & Ziff, 1989). Although the functional significance of these isoforms is not clear, several studies have revealed their distinct properties in IF network organization. Per-58 associates with Per-45, a truncated protein generated from alternative translation of an in-frame downstream initiation codon, to form an intra-isoform complex during IF network formation. *In vivo*, this intra-isoform ratio is often region specific and changes to this ratio affects the integrity of the peripherin filament network as reflected in neuronal injury and motor neuron diseases (McLean et al., 2008, 2010; Xiao et al., 2008). Per-61 is the result of an in-frame retention of intron 4 (Landon et al., 1989, 2000), and Per-28 is a C-terminal truncated protein derived from translation to a premature stop codon in the retained intron 3 (Xiao et al., 2008). Expression of Per-61 or Per-28 in $SW13vim^{-}$ cells induces protein aggregates, and they may contribute to the pathogenesis of a transgenic mouse model expressing mutant superoxide dismutase-1 ($SOD1^{G37R}$) and in amyotrophic lateral sclerosis (ALS) (Robertson et al., 2003; Xiao et al., 2008).

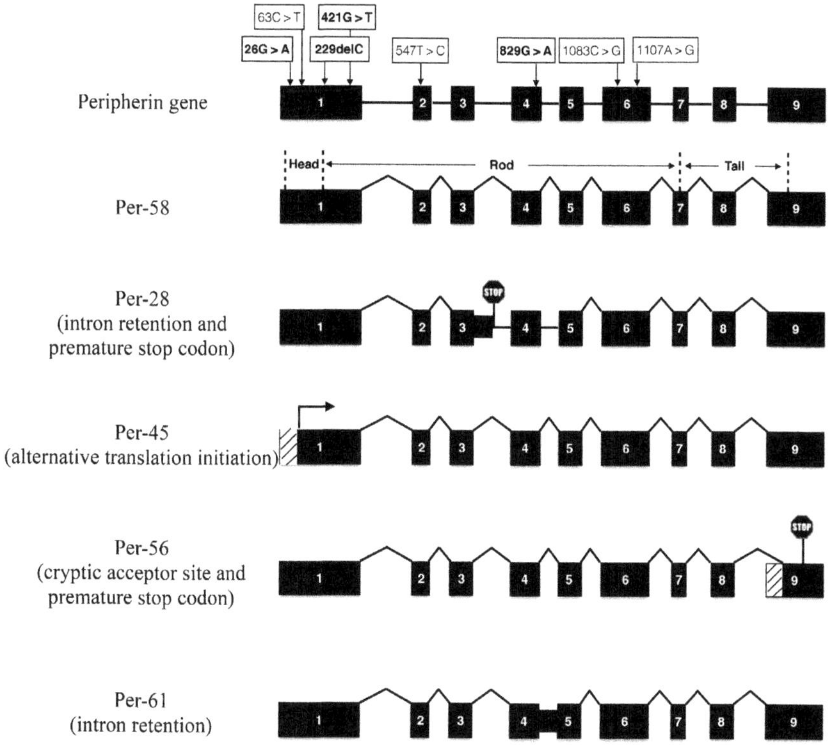

Figure 3 The peripherin gene contains nine exons. Variants in the *peripherin* gene exons are indicated. The variants causing an amino acid change are in bold. Alternate splicing and alternate initiation results in multiple isoforms. Per-58: complete intron splicing gives rise to Per-58 mRNA, which encodes the constitutively expressed 58 kDa peripherin. Per-28 mRNA is generated from alternative splicing with the retention of introns 3 and 4; a premature stop codon in intron 3 terminates the translation and leads to a truncated 28 kDa peripherin isoform. Per-45: Per-45 mRNA is generated from complete intron splicing, but translation initiates from an in-frame downstream start codon to generate a truncated protein of 45 kDa. Per-56 mRNA arises from alternative splicing using a cryptic acceptor site in exon 9. The truncated exon 9 is missing 62 nucleotides, resulting in a frame shift and early termination of translation. Per-61 mRNA results from the in-frame retention of intron 4. Translation of intron 4 leads to a 32 amino acid insertion within the rod domain.

2.4 Functions of Peripherin

Peripherin has been suggested to play an important role in some aspects of neurite growth and stability. This role is supported by the findings that expression of peripherin emerges early in neurite outgrowth during development (Escurat et al., 1990; Gorham et al., 1990; Troy, Brown, Greene, & Shelanski, 1990; Undamatla & Szaro, 2001), and that elevated peripherin expression has been found in PNS and CNS neurons after injury (Beaulieu, Kriz, & Julien, 2002; Oblinger, Wong, & Parysek, 1989; Troy, Muma, Greene, Price, & Shelanski, 1990; Wong & Oblinger, 1990). A study using antisense oligonucleotides in cultured PC12 cells suggested that peripherin is necessary for the stability of processes, but not neurite formation (Troy, Greene, & Shelanski, 1992). In another study using siRNA to deplete peripherin in PC12 cells, the initiation, extension and maintenance of neurites were inhibited, indicating a critical role of peripherin in determining the architecture of neurons (Helfand, Mendez, Pugh, Delsert, & Goldman, 2003).

Studies of the peripherin knockout mice demonstrated that peripherin is dispensable for growth of long myelinated neurons as the knockout mice developed normally with no overt phenotypes or anatomical abnormalities (Larivière, Nguyen, Ribeiro-da-Silva, & Julien, 2002). However, an increase in expression of α-internexin was observed in motor axons, possibly compensating for the function of peripherin. Moreover, the peripherin null mice also exhibited a reduced number of unmyelinated sensory axons in L5 dorsal roots, suggesting its role in the proper development of a subset of sensory neurons (Larivière et al., 2002).

2.5 Role of Peripherin in Disease

2.5.1 Amyotrophic Lateral Sclerosis

Peripherin is a major component found in inclusions of ALS patients (Corbo & Hays, 1992; Migheli, Pezzulo, Attanasio, & Schiffer, 1993; Tu et al., 1996). Currently, the mechanism underlying the formation of inclusions is still not clear. Neither it is known to what extent peripherin contributes to the onset of ALS. Interestingly, several peripherin mutations have been identified in sporadic ALS patients (Corrado et al., 2011; Gros-Louis et al., 2004; Leung et al., 2004). By associating with normal NF proteins, these peripherin mutants could cause deleterious effects on the neuronal IF network, which eventually could lead to the onset of ALS.

It has also been shown that peripherin expression is upregulated in ALS patients (Robertson et al., 2003). Thus, to study the contribution of peripherin overexpression to motor neuron degeneration in ALS, several transgenic mouse and *in vitro* models have been developed in which one or more NF genes have been overexpressed or knocked out. A transgenic mouse model overexpressing wild-type peripherin has been generated, and it developed a late-onset (about 2 years) motor neuron disease featuring perikaryal and axonal IF inclusions and selective loss of motor neurons (Beaulieu, Nguyen, & Julien, 1999). *In vitro*, peripherin overexpression not only induced the apoptotic death of cultured motor neurons but also of cultured DRG, providing the proinflammatory CNS environment (Robertson et al., 2001). In an axonal transport study in 6-month-old mice, impaired axonal transport of NF proteins in neurons overexpressing peripherin has been noted (Millecamps, Robertson, Lariviere, Mallet, & Julien, 2006).

Since there is a 60% reduction in NFL mRNA levels in motor neurons of ALS patients (Bergeron et al., 1994; Wong, He, & Strong, 2000), a double-transgenic mouse model has been developed that would better mimic the conditions in ALS by crossing mice overexpressing peripherin with NFL knockout mice (Per; $L^{-/-}$) (Beaulieu, Nguyen, et al., 1999; Beaulieu, Robertson, et al., 1999). The results of that cross revealed that the onset of motor neuron disease was accelerated from 2 years to about 6–8 months with a dramatic motor neuron loss seen at 5 month (Beaulieu, Nguyen, et al., 1999; Beaulieu, Robertson, et al., 1999). An *in vitro* study of cultured DRG neurons using time-lapse microscopy found a net retrograde transport of mitochondria in neurons from Per; $L^{-/-}$ mice, a phenomenon possibly contributing to the pathogenesis of some neurodegenerative disorders (Perrot & Julien, 2009).

To further explore the mechanisms causing motor neuron disease and interactions among NF members, another transgenic mouse model overexpressing both peripherin and NFH but deficient in NFL was generated (Beaulieu & Julien, 2003). Results showed that the motor neuron death mediated by peripherin overexpression in the absence of NFL is rescued by NFH overexpression, and this rescue effect may have originated from the sequestration of peripherin in the perikaryon of motor neurons to prevent the formation of axonal inclusions (Beaulieu & Julien, 2003). Indeed, mice overexpressing NFH and deficient for NFL show perikaryal inclusions but do not suffer motor neuron death (Beaulieu, Jacomy, & Julien, 2000).

The above results from transgenic and *in vitro* models highlight the importance of neuronal IF protein stoichiometry in the formation of inclusions, and maybe in the pathogenesis of some ALS patients.

2.5.2 Role of Peripherin in Other Neurodegenerative Diseases

Peripherin has been found to associate with protein partners involved in vesicle trafficking in recent studies using the yeast two-hybrid screen (Cogli et al., 2013; Gentil et al., 2014). A yeast two-hybrid screen using a dorsal root ganglia cDNA library has identified peripherin as an interacting partner of RAB7A encoded by *rab7a* gene, which is the causative gene of a neurodegenerative disease called Charcot-Marie-Tooth type 2B (Cogli et al., 2013). The biological importance of this interaction remains to be studied, but it has been suggested RAB7A may regulate peripherin assembly by altering the peripherin soluble/insoluble ratio. It is possible that disease-causing RAB7A mutant proteins affect neuronal IF assembly and axonal transport resulting in neurodegeneration (Cogli et al., 2013). Similarly, a yeast two-hybrid screen on a mouse brain cDNA library also helped identify a wide range of peripherin binding partners, among which SNAP25-interacting protein 30 (SIP30) was studied further in cell cultures (Gentil et al., 2014). By interacting through their coiled–coiled domains, SIP30 affects peripherin assembly and peripherin isoforms, Per-58 and Per-61, alter SIP30 and SNAP25 subcellular localization. Thus, peripherin may play a role in synaptic vesicle exocytosis and trafficking (Gentil et al., 2014).

2.5.3 Peripherin and Diabetes

Peripherin has also been reported to be involved in type 1 diabetes. Type 1 diabetes is an autoimmune disease that involves the selective destruction of insulin-producing pancreatic ß-cells (Lehuen, Diana, Zaccone, & Cooke, 2010). Autoantibodies to peripherin have been detected in patients with autoimmune neuropathies and endocrinopathies, as well as in the nonobese diabetic (NOD) mouse model (Boitard et al., 1992; Chamberlain et al., 2010). By using isolated islet-infiltrating B cells and recombinant protein techniques, the epitope was mapped to a fragment in the C-terminal tail of Per-58 and Per-61, but not Per-56, which does not share the identical C-terminal sequence seen in Per-58 and Per-61 (see Fig. 3) (Garabatos et al., 2014; Puertas et al., 2007). Peripherin is expressed in ß-cells during development, but it is not clear when and where it is presented to B cells, or whether or not these peripherin autoantibodies are attacking peripherin-expressing cells (Garabatos et al., 2014; Puertas et al., 2007). The association of the existence of peripherin autoantibodies and Type 1 diabetic patients with neuropathies suggests peripherin and autoreactive B cells may play a role in diabetic neuropathy.

3. CONCLUSION

α-Internexin and peripherin were both believed to form independent IF networks with unknown functions. Recent studies have provided strong evidence suggesting that α-internexin and peripherin are a fourth subunit of neurofilaments in the adult CNS and PNS, respectively. A better understanding of how these neurofilament subunits assemble and compose the neurofilament network will provide fresh insights into the functions of NF proteins in neuronal development and neurodegenerative diseases, and shed light upon findings hard to explain before. To aid in further studies, commercial antibodies to the two proteins are shown in Tables 1 and 2 and available cDNA clones (all from Origine), including ones with GFP tags are shown in Table 3. All the GFP tags are in the tails of the molecule. The involvement of α-internexin and peripherin in inclusions of several neurodegenerative diseases leads us to ask whether α-internexin and peripherin aggregations are key pathogenic factors or just early events in the pathogenesis. The factors and signaling pathways inducing the inclusions remain to be determined. Novel roles of α-internexin and peripherin isoforms are starting to emerge. Further studies of their roles in different signaling pathways may unfold physiological functions of α-internexin and peripherin.

Table 3 cDNA Clones of α-Internexin and Peripherin from Origene

	Category	Vector	Tag	SKU
α-Internexin	Human clone	pCMV6-XL5	No	SC104003
	Human clone	pCMV6-AC	No	SC324402
	Human lenti-ORF	pLenti-C-mGFP	mGFP	RC202877L2
	Human clone	pCMV6-AC-GFP	TurboGFP	RG202877
	Mouse clone	pCMV6-Kan/Neo	No	MC203999
	Mouse clone	pCMV6-AC-GFP	TurboGFP	MG208054
	mouse lenti-ORF	pLenti-C-mGFP	mGFP	MR208054L2
	Rat clone	pCMV6-Entry	No	RN215311
	Rat lenti-ORF	pLenti-C-mGFP	mGFP	RR215311L2

Continued

Table 3 cDNA Clones of α-Internexin and Peripherin from Origene—cont'd

	Category	Vector	Tag	SKU
Peripherin	Human clone	pCMV6-XL5	No	SC123831
	Human lenti-ORF	pLenti-C-mGFP	mGFP	RC207561L2
	Human clone	pCMV6-AC-GFP	TurboGFP	RG207561
	Mouse clone, variant 1	pCMV6-Entry	No	MC217143
	Mouse clone, variant 2	pCMV6-Entry	No	MC217144
	Mouse clone, variant 3	pCMV6-Entry	No	MC218098
	Mouse clone, variant 1	pCMV6-AC-GFP	TurboGFP	MG226366
	Mouse clone, variant 2	pCMV6-AC-GFP	TurboGFP	MG226367
	Mouse lenti-ORF, variant 1	pLenti-C-mGFP	mGFP	MR226366L2
	Mouse lenti-ORF, variant 2	pLenti-C-mGFP	mGFP	MR226367L2
	Mouse lenti-ORF, variant 3	pLenti-C-mGFP	mGFP	MR207614L2
	Rat clone	pCMV6-Entry	No	RN208519
	Rat lenti-ORF	pLenti-C-mGFP	mGFP	RR208519L2

REFERENCES

Abumuhor, I. A., Spencer, P. H., & Cohlberg, J. A. (1998). The pathway of assembly of intermediate filaments from recombinant α-internexin. *Journal of Structural Biology*, *123*(3), 187–198. http://dx.doi.org/10.1006/jsbi.1998.4040.

Adams, A. D., Choate, D. M., & Thompson, M. A. (1995). NF1-L is the DNA-binding component of the protein complex at the peripherin negative regulatory element. *Journal of Biological Chemistry*, *270*(12), 6975–6983. Retrieved from: http://www.jbc.org/content/270/12/6975.abstract.

Adebola, A. A., Di Castri, T., He, C. Z., Salvatierra, L. A., Zhao, J., Brown, K., et al. (2015). Neurofilament light polypeptide gene N98S mutation in mice leads to neurofilament network abnormalities and a Charcot-Marie-Tooth Type 2E phenotype. *Human Molecular Genetics*, *24*(8), 2163–2174. http://dx.doi.org/10.1093/hmg/ddu736.

Aletta, J. M., Angeletti, R., Liem, R. K., Purcell, C., Shelanski, M. L., & Greene, L. A. (1988). Relationship between the nerve growth factor-regulated clone 73 gene product and the 58-kilodalton neuronal intermediate filament protein (peripherin). *Journal of Neurochemistry*, *51*(4), 1317–1320.

Angelastro, J. M., Ho, C. L., Frappier, T., Liem, R. K., & Greene, L. A. (1998). Peripherin is tyrosine-phosphorylated at its carboxyl-terminal tyrosine. *Journal of Neurochemistry*, *70*(2), 540–549. Retrieved from: http://www.ncbi.nlm.nih.gov/pubmed/9453548.

Athlan, E. S., & Mushynski, W. E. (1997). Heterodimeric associations between neuronal intermediate filament proteins. *Journal of Biological Chemistry*, *272*(49), 31073–31078. http://dx.doi.org/10.1074/jbc.272.49.31073.

Aziz, A., Hess, J. F., Budamagunta, M. S., Voss, J. C., Kuzin, A. P., Huang, Y. J., et al. (2012). The structure of vimentin linker 1 and Rod 1B domains characterized by site-directed spin-labeling electron paramagnetic resonance (SDSL-EPR) and X-ray crystallography. *Journal of Biological Chemistry*, *287*(34), 28349–28361. http://dx.doi.org/10.1074/jbc.M111.334011.

Beaulieu, J.-M., Jacomy, H., & Julien, J.-P. (2000). Formation of intermediate filament protein aggregates with disparate effects in two transgenic mouse models lacking the neurofilament light subunit. *Journal of Neuroscience*, *20*(14), 5321–5328. Retrieved from: http://www.jneurosci.org/content/20/14/5321.abstract.

Beaulieu, J.-M., & Julien, J.-P. (2003). Peripherin-mediated death of motor neurons rescued by overexpression of neurofilament NF-H proteins. *Journal of Neurochemistry*, *85*(1), 248–256. http://dx.doi.org/10.1046/j.1471-4159.2003.01653.x.

Beaulieu, J.-M., Kriz, J., & Julien, J.-P. (2002). Induction of peripherin expression in subsets of brain neurons after lesion injury or cerebral ischemia. *Brain Research*, *946*(2), 153–161. http://dx.doi.org/10.1016/S0006-8993(02)02830-5.

Beaulieu, J.-M., Nguyen, M. D., & Julien, J.-P. (1999). Late onset death of motor neurons in mice overexpressing wild-type peripherin. *Journal of Cell Biology*, *147*(3), 531–544. Retrieved from: http://www.ncbi.nlm.nih.gov/pmc/articles/PMC2151189/.

Beaulieu, J. M., Robertson, J., & Julien, J. P. (1999). Interactions between peripherin and neurofilaments in cultured cells: Disruption of peripherin assembly by the NF-M and NF-H subunits. *Biochemistry and Cell Biology*, 77(1), 41–45. http://dx.doi.org/10.1139/bcb-77-1-41.

Benson, D., Mandell, J., Shaw, G., & Banker, G. (1996). Compartmentation of alpha-internexin and neurofilament triplet proteins in cultured hippocampal neurons. *Journal of Neurocytology*, *25*(1), 181–196. http://dx.doi.org/10.1007/BF02284795.

Bergeron, C., Beric-Maskarel, K., Muntasser, S., Weyer, L., Somerville, M. J., & Percy, M. E. (1994). Neurofilament light and polyadenylated mRNA levels are decreased in amyotrophic lateral sclerosis motor neurons. *Journal of Neuropathology & Experimental Neurology*, *53*(3), 221–230. Retrieved from: http://www.ncbi.nlm.nih.gov/pubmed/7909836.

Bigio, E. H., Lipton, A. M., White, C. L., Dickson, D. W., & Hirano, A. (2003). Frontotemporal and motor neurone degeneration with neurofilament inclusion bodies: Additional evidence for overlap between FTD and ALS. *Neuropathology and Applied Neurobiology*, *29*(3), 239–253. http://dx.doi.org/10.1046/j.1365-2990.2003.00466.x.

Boitard, C., Villa, M. C., Becourt, C., Gia, H. P., Huc, C., Sempe, P., et al. (1992). Peripherin: An islet antigen that is cross-reactive with nonobese diabetic mouse class II gene products. *Proceedings of the National Academy of Sciences of the United States of America*, *89*(1), 172–176. Retrieved from: http://www.ncbi.nlm.nih.gov/pmc/articles/PMC48198/.

Brody, B. A., Ley, C. A., & Parysek, L. M. (1989). Selective distribution of the 57 kDa neural intermediate filament protein in the rat CNS. *Journal of Neuroscience: The Official Journal of the Society for Neuroscience*, *9*(7), 2391–2401. Retrieved from: http://europepmc.org/

abstract/MED/2746334. http://www.jneurosci.org/cgi/content/abstract/9/7/2391; http://www.jneurosci.org/cgi/reprint/9/7/2391.pdf.

Cairns, N. J., Perry, R. H., Jaros, E., Burn, D., McKeith, I. G., Lowe, J. S., et al. (2003). Patients with a novel neurofilamentopathy: Dementia with neurofilament inclusions. *Neuroscience Letters, 341*(3), 177–180. http://dx.doi.org/10.1016/S0304-3940(03)00100-9.

Cairns, N. J., Uryu, K., Bigio, E. H., Mackenzie, I. R. A., Gearing, M., Duyckaerts, C., et al. (2004). α-Internexin aggregates are abundant in neuronal intermediate filament inclusion disease (NIFID) but rare in other neurodegenerative diseases. *Acta Neuropathologica, 108*(3), 213–223. http://dx.doi.org/10.1007/s00401-004-0882-7.

Chadan, S., Moya, K. L., Portier, M. M., & Filliatreau, G. (1994). Identification of a peripherin dimer: Changes during axonal development and regeneration of the rat sciatic nerve. *Journal of Neurochemistry, 62*(5), 1894–1905. http://dx.doi.org/10.1046/j.1471-4159.1994.62051894.x.

Chamberlain, J. L., Pittock, S. J., Oprescu, A.-M., Dege, C., Apiwattanakul, M., Kryzer, T. J., et al. (2010). Peripherin-IgG association with neurologic and endocrine autoimmunity. *Journal of Autoimmunity, 34*(4), 469–477. http://dx.doi.org/10.1016/j.jaut.2009.12.004.

Chang, L., & Thompson, M. A. (1996). Activity of the distal positive element of the peripherin gene is dependent on proteins binding to an Ets-like recognition site and a novel inverted repeat site. *Journal of Biological Chemistry, 271*(11), 6467–6475. http://dx.doi.org/10.1074/jbc.271.11.6467.

Chernyatina, A. A., Nicolet, S., Aebi, U., Herrmann, H., & Strelkov, S. V. (2012). Atomic structure of the vimentin central α-helical domain and its implications for intermediate filament assembly. *Proceedings of the National Academy of Sciences of the United States of America, 109*(34), 13620–13625. http://dx.doi.org/10.1073/pnas.1206836109.

Chien, C.-L., Liu, T.-C., Ho, C.-L., & Lu, K.-S. (2005). Overexpression of neuronal intermediate filament protein α-internexin in PC12 cells. *Journal of Neuroscience Research, 80*(5), 693–706. http://dx.doi.org/10.1002/jnr.20506.

Chien, C. L., Mason, C. A., & Liem, R. K. H. (1996). α-internexin is the only neuronal intermediate filament expressed in developing cerebellar granule neurons. *Journal of Neurobiology, 29*(3), 304–318. http://dx.doi.org/10.1002/(SICI)1097-4695(199603)29:3<304::AID-NEU3>3.0.CO;2-D.

Ching, G. Y., Chien, C.-L., Flores, R., & Liem, R. K. H. (1999). Overexpression of α-internexin causes abnormal neurofilamentous accumulations and motor coordination deficits in transgenic mice. *Journal of Neuroscience, 19*(8), 2974–2986. Retrieved from: http://www.jneurosci.org/content/19/8/2974.abstract.

Ching, G. Y., & Liem, R. K. (1991). Structure of the gene for the neuronal intermediate filament protein alpha-internexin and functional analysis of its promoter. *Journal of Biological Chemistry, 266*(29), 19459–19468. Retrieved from: http://www.jbc.org/content/266/29/19459.abstract.

Ching, G., & Liem, R. (1993). Assembly of type IV neuronal intermediate filaments in nonneuronal cells in the absence of preexisting cytoplasmic intermediate filaments. *Journal of Cell Biology, 122*(6), 1323–1335. http://dx.doi.org/10.1083/jcb.122.6.1323.

Chiu, F. C., Barnes, E. A., Das, K., Haley, J., Socolow, P., Macaluso, F. P., et al. (1989). Characterization of a novel 66 kd subunit of mammalian neurofilaments. *Neuron, 2*(5), 1435–1445.

Choi, D.-Y., Toledo-Aral, J.-J., Lin, H. Y., Ischenko, I., Medina, L., Safo, P., et al. (2001). Fibroblast growth factor receptor 3 induces gene expression primarily through Ras-independent signal transduction pathways. *Journal of Biological Chemistry, 276*(7), 5116–5122. http://dx.doi.org/10.1074/jbc.M002959200.

Cogli, L., Progida, C., Thomas, C., Spencer-Dene, B., Donno, C., Schiavo, G., et al. (2013). Charcot–Marie–Tooth type 2B disease-causing RAB7A mutant proteins show altered interaction with the neuronal intermediate filament peripherin. *Acta Neuropathologica, 125*(2), 257–272. http://dx.doi.org/10.1007/s00401-012-1063-8.

Corbo, M., & Hays, A. P. (1992). Peripherin and neurofilament protein coexist in spinal spheroids of motor neuron disease. *Journal of Neuropathology & Experimental Neurology, 51*(5), 531–537. Retrieved from: http://journals.lww.com/jneuropath/Fulltext/1992/09000/Peripherin_and_Neurofilament_Protein_Coexist_in.8.aspx.

Corrado, L., Carlomagno, Y., Falasco, L., Mellone, S., Godi, M., Cova, E., et al. (2011). A novel peripherin gene (PRPH) mutation identified in one sporadic amyotrophic lateral sclerosis patient. *Neurobiology of Aging, 32*(3), 552.e551–552.e556. http://dx.doi.org/10.1016/j.neurobiolaging.2010.02.011.

Cui, C., Stambrook, P. J., & Parysek, L. M. (1995). Peripherin assembles into homopolymers in SW13 cells. *Journal of Cell Science, 108*(Pt 10), 3279–3284.

Desmarais, D., Filion, M., Lapointe, L., & Royal, A. (1992). Cell-specific transcription of the peripherin gene in neuronal cell lines involves a cis-acting element surrounding the TATA box. *EMBO Journal, 11*(8), 2971–2980. Retrieved from: http://www.ncbi.nlm.nih.gov/pmc/articles/PMC556779/.

Ducray, F., Crinière, E., Idbaih, A., Mokhtari, K., Marie, Y., Paris, S., et al. (2009). α-Internexin expression identifies 1p19q codeleted gliomas. *Neurology, 72*(2), 156–161. http://dx.doi.org/10.1212/01.wnl.0000339055.64476.cb.

Ducray, F., Mokhtari, K., Crinière, E., Idbaih, A., Marie, Y., Dehais, C., et al. (2011). Diagnostic and prognostic value of alpha internexin expression in a series of 409 gliomas. *European Journal of Cancer, 47*(5), 802–808. http://dx.doi.org/10.1016/j.ejca.2010.11.031.

Durand, K., Guillaudeau, A., Pommepuy, I., Mesturoux, L., Chaunavel, A., Gadeaud, E., et al. (2011). Alpha-internexin expression in gliomas: Relationship with histological type and 1p, 19q, 10p and 10q status. *Journal of Clinical Pathology, 64*(9), 793–801. http://dx.doi.org/10.1136/jcp.2010.087668.

Eigenbrod, S., Roeber, S., Thon, N., Giese, A., Krieger, A., Grasbon-Frodl, E., et al. (2011). α-Internexin in the diagnosis of oligodendroglial tumors and association with 1p/19q status. *Journal of Neuropathology & Experimental Neurology, 70*(11), 970–978. http://dx.doi.org/10.1097/NEN.0b013e3182333ef5.

Escurat, M., Djabali, K., Gumpel, M., Gros, F., & Portier, M. (1990). Differential expression of two neuronal intermediate-filament proteins, peripherin and the low-molecular-mass neurofilament protein (NF-L), during the development of the rat. *Journal of Neuroscience, 10*(3), 764–784. Retrieved from: http://www.jneurosci.org/content/10/3/764.abstract.

Fliegner, K. H., Ching, G. Y., & Liem, R. K. (1990). The predicted amino acid sequence of alpha-internexin is that of a novel neuronal intermediate filament protein. *EMBO Journal, 9*(3), 749–755. Retrieved from: http://www.ncbi.nlm.nih.gov/pmc/articles/PMC551731/.

Fliegner, K. H., Kaplan, M. P., Wood, T. L., Pintar, J. E., & Liem, R. K. H. (1994). Expression of the gene for the neuronal intermediate filament protein α-internexin coincides with the onset of neuronal differentiation in the developing rat nervous system. *Journal of Comparative Neurology, 342*(2), 161–173. http://dx.doi.org/10.1002/cne.903420202.

Fuchs, E., & Weber, K. (1994). Intermediate filaments: Structure, dynamics, function and disease. *Annual Review of Biochemistry, 63*(1), 345–382. http://dx.doi.org/10.1146/annurev.bi.63.070194.002021.

Garabatos, N., Alvarez, R., Carrillo, J., Carrascal, J., Izquierdo, C., Chapman, H. D., et al. (2014). In vivo detection of peripherin-specific autoreactive B cells during type

1 diabetes pathogenesis. *Journal of Immunology, 192*(7), 3080–3090. http://dx.doi.org/10.4049/jimmunol.1301053.

Gentil, B. J., McLean, J. R., Xiao, S., Zhao, B., Durham, H. D., & Robertson, J. (2014). A two-hybrid screen identifies an unconventional role for the intermediate filament peripherin in regulating the subcellular distribution of the SNAP25-interacting protein, SIP30. *Journal of Neurochemistry, 131*(5), 588–601. http://dx.doi.org/10.1111/jnc.12928.

Glasgow, E., Druger, R. K., Fuchs, C., Lane, W. S., & Schechter, N. (1994). Molecular cloning of gefiltin (ON1): Serial expression of two new neurofilament mRNAs during optic nerve regeneration. *EMBO Journal, 13*(2), 297–305. Retrieved from: http://www.ncbi.nlm.nih.gov/pmc/articles/PMC394808/.

Gorham, J. D., Baker, H., Kegler, D., & Ziff, E. B. (1990). The expression of the neuronal intermediate filament protein peripherin in the rat embryo. *Developmental Brain Research, 57*(2), 235–248. http://dx.doi.org/10.1016/0165-3806(90)90049-5.

Gros-Louis, F., Larivière, R., Gowing, G., Laurent, S., Camu, W., Bouchard, J.-P., et al. (2004). A frameshift deletion in peripherin gene associated with amyotrophic lateral sclerosis. *Journal of Biological Chemistry, 279*(44), 45951–45956. http://dx.doi.org/10.1074/jbc.M408139200.

Helfand, B. T., Mendez, M. G., Pugh, J., Delsert, C., & Goldman, R. D. (2003). A role for intermediate filaments in determining and maintaining the shape of nerve cells. *Molecular Biology of the Cell, 14*(12), 5069–5081. http://dx.doi.org/10.1091/mbc.E03-06-0376.

Herrmann, H., & Aebi, U. (1998). Intermediate filament assembly: Fibrillogenesis is driven by decisive dimer-dimer interactions. *Current Opinion in Structural Biology, 8*(2), 177–185. http://dx.doi.org/10.1016/S0959-440X(98)80035-3.

Herrmann, H., & Aebi, U. (2004). Intermediate filaments: Molecular structure, assembly mechanism, and integration into functionally distinct intracellular scaffolds. *Annual Review of Biochemistry, 73*(1), 749–789. http://dx.doi.org/10.1146/annurev.biochem.73.011303.073823.

Herrmann, H., Häner, M., Brettel, M., Ku, N.-O., & Aebi, U. (1999). Characterization of distinct early assembly units of different intermediate filament proteins1. *Journal of Molecular Biology, 286*(5), 1403–1420. http://dx.doi.org/10.1006/jmbi.1999.2528.

Ho, C. L., Chin, S. S., Carnevale, K., & Liem, R. K. (1995). Translation initiation and assembly of peripherin in cultured cells. *European Journal of Cell Biology, 68*(2), 103–112. Retrieved from: http://www.ncbi.nlm.nih.gov/pubmed/8575457.

Ho, C.-L., & Liem, R. H. (1996). Intermediate filaments in the nervous system: Implications in cancer. *Cancer and Metastasis Reviews, 15*(4), 483–497. http://dx.doi.org/10.1007/BF00054014.

Josephs, K. A., Holton, J. L., Rossor, M. N., Braendgaard, H., Ozawa, T., Fox, N. C., et al. (2003). Neurofilament inclusion body disease: A new proteinopathy? *Brain, 126*(Pt 10), 2291–2303.

Julien, J.-P., Côté, F., Beaudet, L., Sidky, M., Flavell, D., Grosveld, F., et al. (1988). Sequence and structure of the mouse gene coding for the largest neurofilament subunit. *Gene, 68*(2), 307–314. http://dx.doi.org/10.1016/0378-1119(88)90033-9.

Julien, J.-P., Grosveld, F., Yazdanbaksh, K., Flavell, D., Meijer, D., & Mushynski, W. (1987). The structure of a human neurofilament gene (NF-L): A unique exon-intron organization in the intermediate filament gene family. *Biochimica et Biophysica Acta (BBA) - Gene Structure and Expression, 909*(1), 10–20. http://dx.doi.org/10.1016/0167-4781(87)90041-8.

Kaplan, M., Chin, S., Fliegner, K., & Liem, R. (1990). Alpha-internexin, a novel neuronal intermediate filament protein, precedes the low molecular weight neurofilament protein (NF-L) in the developing rat brain. *Journal of Neuroscience, 10*(8), 2735–2748. Retrieved from: http://www.jneurosci.org/content/10/8/2735.abstract.

Kaplan, M. P., Chin, S. S., Macioce, P., Srinawasan, J., Hashim, G., & Liem, R. K. (1991). Characterization of a panel of neurofilament antibodies recognizing N- terminal epitopes. *Journal of Neuroscience Research, 30*(3), 545–554.

Karpova, A., Mikhaylova, M., Bera, S., Bär, J., Reddy, P. P., Behnisch, T., et al. (2013). Encoding and transducing the synaptic or extrasynaptic origin of NMDA receptor signals to the nucleus. *Cell, 152*(5), 1119–1133. http://dx.doi.org/10.1016/j.cell.2013.02.002.

Landon, F., Lemonnier, M., Benarous, R., Huc, C., Fiszman, M., Gros, F., et al. (1989). Multiple mRNAs encode peripherin, a neuronal intermediate filament protein. *EMBO Journal, 8*(6), 1719–1726. Retrieved from: http://www.ncbi.nlm.nih.gov/pmc/articles/PMC401014/.

Landon, F., Wolff, A., & de Néchaud, B. (2000). Mouse peripherin isoforms. *Biology of the Cell, 92*(6), 397–407. http://dx.doi.org/10.1016/S0248-4900(00)01099-6.

Larivière, R. C., Nguyen, M. D., Ribeiro-da-Silva, A., & Julien, J. P. (2002). Reduced number of unmyelinated sensory axons in peripherin null mice. *Journal of Neurochemistry, 81*(3), 525–532. http://dx.doi.org/10.1046/j.1471-4159.2002.00853.x.

Leake, D., Asch, W. S., Canger, A. K., & Schechter, N. (1999). Gefiltin in zebrafish embryos: Sequential gene expression of two neurofilament proteins in retinal ganglion cells. *Differentiation, 65*(4), 181–189. http://dx.doi.org/10.1046/j.1432-0436.1999.6540181.x.

Lecomte, M. J., Basseville, M., Landon, F., Karpov, V., & Fauquet, M. (1998). Transcriptional activation of the mouse peripherin gene by leukemia inhibitory factor: Involvement of STAT proteins. *Journal of Neurochemistry, 70*(3), 971–982. Retrieved from: http://europepmc.org/abstract/MED/9489716.

Leconte, L., Santha, M., Fort, C., Poujeol, C., Portier, M.-M., & Simonneau, M. (1996). Cell type-specific expression of the mouse peripherin gene requires both upstream and intragenic sequences in transgenic mouse embryos. *Developmental Brain Research, 92*(1), 1–9. http://dx.doi.org/10.1016/0165-3806(95)00182-4.

Lee, W.-C., Kan, D., Chen, Y.-Y., Han, S.-K., Lu, K.-S., & Chien, C.-L. (2012). Suppression of extensive neurofilament phosphorylation rescues α-internexin/peripherin-overexpressing PC12 cells from neuronal cell death. *PLoS One*, 7(8), e43883. http://dx.doi.org/10.1371/journal.pone.0043883.

Lehuen, A., Diana, J., Zaccone, P., & Cooke, A. (2010). Immune cell crosstalk in type 1 diabetes. *Nature Reviews. Immunology, 10*(7), 501–513. Retrieved from: http://dx.doi.org/10.1038/nri2787.

Leonard, D. G., Gorham, J. D., Cole, P., Greene, L. A., & Ziff, E. B. (1988). A nerve growth factor-regulated messenger RNA encodes a new intermediate filament protein. *Journal of Cell Biology, 106*(1), 181–193. http://dx.doi.org/10.1083/jcb.106.1.181.

Leonard, D. G., Ziff, E. B., & Greene, L. A. (1987). Identification and characterization of mRNAs regulated by nerve growth factor in PC12 cells. *Molecular and Cellular Biology*, 7(9), 3156–3167. Retrieved from: http://www.ncbi.nlm.nih.gov/pmc/articles/PMC367950/.

Leung, C. L., He, C. Z., Kaufmann, P., Chin, S. S., Naini, A., Liem, R. K. H., et al. (2004). A pathogenic peripherin gene mutation in a patient with amyotrophic lateral sclerosis. *Brain Pathology, 14*(3), 290–296. http://dx.doi.org/10.1111/j.1750-3639.2004.tb00066.x.

Levavasseur, F., Zhu, Q., & Julien, J.-P. (1999). No requirement of α-internexin for nervous system development and for radial growth of axons. *Molecular Brain Research, 69*(1), 104–112. http://dx.doi.org/10.1016/S0169-328X(99)00104-7.

Levy, E., D'Eustachio, P., Cowan, N. J., & Liem, R. K. H. (1987). Structure and evolutionary origin of the gene encoding mouse NF-M, the middle-molecular-mass neurofilament protein. *European Journal of Biochemistry, 166*(1), 71–77. http://dx.doi.org/10.1111/j.1432-1033.1987.tb13485.x.

Lewis, S. A., & Cowan, N. J. (1986). Anomalous placement of introns in a member of the intermediate filament multigene family: An evolutionary conundrum. *Molecular and Cellular Biology, 6*(5), 1529–1534. Retrieved from: http://www.ncbi.nlm.nih.gov/pmc/articles/PMC367678/.

Liem, R. K., Yen, S. H., Salomon, G. D., & Shelanski, M. L. (1978). Intermediate filaments in nervous tissues. *Journal of Cell Biology, 79*(3), 637–645.

Lu, X.-y., Chen, X.-x, Huang, L.-d., Zhu, C.-q., Gu, Y.-y., & Ye, S. (2010). Anti-α-internexin autoantibody from neuropsychiatric lupus induce cognitive damage via inhibiting axonal elongation and promote neuron apoptosis. *PLoS One*, *5*(6), e11124. http://dx.doi.org/10.1371/journal.pone.0011124.

McGraw, T. S., Mickle, J. P., Shaw, G., & Streit, W. J. (2002). Axonally transported peripheral signals regulate α-internexin expression in regenerating motoneurons. *Journal of Neuroscience*, *22*(12), 4955–4963. Retrieved from: http://www.jneurosci.org/content/22/12/4955.abstract.

McLean, J., Liu, H.-N., Miletic, D., Weng, Y. C., Rogaeva, E., Zinman, L., et al. (2010). Distinct biochemical signatures characterize peripherin isoform expression in both traumatic neuronal injury and motor neuron disease. *Journal of Neurochemistry*, *114*(4), 1177–1192. http://dx.doi.org/10.1111/j.1471-4159.2010.06846.x.

McLean, J., Xiao, S., Miyazaki, K., & Robertson, J. (2008). A novel peripherin isoform generated by alternative translation is required for normal filament network formation. *Journal of Neurochemistry*, *104*(6), 1663–1673. http://dx.doi.org/10.1111/j.1471-4159.2007.05198.x.

Migheli, A., Pezzulo, T., Attanasio, A., & Schiffer, D. (1993). Peripherin immunoreactive structures in amyotrophic lateral sclerosis. *Laboratory Investigation: A Journal of Technical Methods and Pathology*, *68*(2), 185–191. Retrieved from: http://europepmc.org/abstract/MED/8441252.

Millecamps, S., Robertson, J., Lariviere, R., Mallet, J., & Julien, J.-P. (2006). Defective axonal transport of neurofilament proteins in neurons overexpressing peripherin. *Journal of Neurochemistry*, *98*(3), 926–938. http://dx.doi.org/10.1111/j.1471-4159.2006.03932.x.

Mokhtari, K., Ducray, F., Kros, J. M., Gorlia, T., Idbaih, A., Taphoorn, M., et al. (2011). Alpha-internexin expression predicts outcome in anaplastic oligodendroglial tumors and may positively impact the efficacy of chemotherapy. *Cancer*, *117*(13), 3014–3026. http://dx.doi.org/10.1002/cncr.25827.

Morris, J. R., & Lasek, R. J. (1982). Stable polymers of the axonal cytoskeleton: The axoplasmic ghost. *Journal of Cell Biology*, *92*(1), 192–198. Retrieved from: http://www.ncbi.nlm.nih.gov/pmc/articles/PMC2112002/.

Nagaishi, M., Suzuki, A., Nobusawa, S., Yokoo, H., & Nakazato, Y. (2014). Alpha-internexin and altered CIC expression as a supportive diagnostic marker for oligodendroglial tumors with the 1p/19q co-deletion. *Brain Tumor Pathology*, *31*(4), 257–264. http://dx.doi.org/10.1007/s10014-013-0168-7.

Oblinger, M. M., Wong, J., & Parysek, L. M. (1989). Axotomy-induced changes in the expression of a type III neuronal intermediate filament gene. *Journal of Neuroscience: The Official Journal of the Society for Neuroscience*, *9*(11), 3766–3775. Retrieved from: http://europepmc.org/abstract/MED/2585054. http://www.jneurosci.org/cgi/content/abstract/9/11/3766; http://www.jneurosci.org/cgi/reprint/9/11/3766.pdf.

Pachter, J. S., & Liem, R. K. (1985). Alpha-internexin, a 66-kD intermediate filament-binding protein from mammalian central nervous tissues. *Journal of Cell Biology*, *101*(4), 1316–1322. http://dx.doi.org/10.1083/jcb.101.4.1316.

Parry, D. A. D., Strelkov, S. V., Burkhard, P., Aebi, U., & Herrmann, H. (2007). Towards a molecular description of intermediate filament structure and assembly. *Experimental Cell Research*, *313*(10), 2204–2216. http://dx.doi.org/10.1016/j.yexcr.2007.04.009.

Parysek, L., & Goldman, R. (1987). Characterization of intermediate filaments in PC12 cells. *Journal of Neuroscience*, 7(3), 781–791. Retrieved from: http://www.jneurosci.org/content/7/3/781.abstract.

Parysek, L., & Goldman, R. (1988). Distribution of a novel 57 kDa intermediate filament (IF) protein in the nervous system. *Journal of Neuroscience*, *8*(2), 555–563. Retrieved from: http://www.jneurosci.org/content/8/2/555.abstract.

Parysek, L. M., McReynolds, M. A., Goldman, R. D., & Ley, C. A. (1991). Some neural intermediate filaments contain both peripherin and the neurofilament proteins.

Journal of Neuroscience Research, *30*(1), 80–91. http://dx.doi.org/10.1002/jnr.490300110.

Perrot, R., & Julien, J.-P. (2009). Real-time imaging reveals defects of fast axonal transport induced by disorganization of intermediate filaments. *FASEB Journal*, *23*(9), 3213–3225. http://dx.doi.org/10.1096/fj.09-129585.

Portier, M. M., Croizat, B., & Gros, F. (1982). A sequence of changes in cytoskeletal components during neuroblastoma differentiation. *FEBS Letters*, *146*(2), 283–288.

Portier, M. M., de Néchaud, B., & Gros, F. (1983). Peripherin, a new member of the intermediate filament protein family. *Developmental Neuroscience*, *6*(6), 335–344. Retrieved from: http://www.karger.com/DOI/10.1159/000112360.

Pruss, R. M. (1985). Efficient detection of intermediate filament proteins using a panspecific monoclonal antibody: Anti-IFA. *Journal of Neuroimmunology*, *8*(4–6), 293–299. Retrieved from: http://www.ncbi.nlm.nih.gov/pubmed/4040138.

Puertas, M. C., Carrillo, J., Pastor, X., Ampudia, R. M., Planas, R., Alba, A., et al. (2007). Peripherin is a relevant neuroendocrine autoantigen recognized by islet-infiltrating B lymphocytes. *Journal of Immunology*, *178*(10), 6533–6539. http://dx.doi.org/10.4049/jimmunol.178.10.6533.

Reddy, T. R., Li, X., Jones, Y., Ellisman, M. H., Ching, G. Y., Liem, R. K. H., et al. (1998). Specific interaction of HTLV tax protein and a human type IV neuronal intermediate filament protein. *Proceedings of the National Academy of Sciences of the United States of America*, *95*(2), 702–707. Retrieved from: http://www.ncbi.nlm.nih.gov/pmc/articles/PMC18484/.

Robertson, J., Beaulieu, J.-M., Doroudchi, M. M., Durham, H. D., Julien, J.-P., & Mushynski, W. E. (2001). Apoptotic death of neurons exhibiting peripherin aggregates is mediated by the proinflammatory cytokine tumor necrosis factor-α. *Journal of Cell Biology*, *155*(2), 217–226. http://dx.doi.org/10.1083/jcb.200107058.

Robertson, J., Doroudchi, M. M., Nguyen, M. D., Durham, H. D., Strong, M. J., Shaw, G., et al. (2003). A neurotoxic peripherin splice variant in a mouse model of ALS. *Journal of Cell Biology*, *160*(6), 939–949. http://dx.doi.org/10.1083/jcb.200205027.

Rodger, J., Bartlett, C. A., Harman, A. M., Thomas, C., Beazley, L. D., & Dunlop, S. A. (2001). Evidence that regenerating optic axons maintain long-term growth in the lizard Ctenophorus ornatus: Growth-associated protein-43 and gefiltin expression. *Neuroscience*, *102*(3), 647–654. http://dx.doi.org/10.1016/S0306-4522(00)00506-6.

Shea, T. B., & Beermann, M. L. (1999). Neuronal intermediate filament protein α-internexin facilitates axonal neurite elongation in neuroblastoma cells. *Cell Motility and the Cytoskeleton*, *43*(4), 322–333. http://dx.doi.org/10.1002/(SICI)1097-0169(1999)43:4<322::AID-CM5>3.0.CO;2-B.

Steinert, P. M., & Roop, D. R. (1988). Molecular and cellular biology of intermediate filaments. *Annual Review of Biochemistry*, *57*(1), 593–625. http://dx.doi.org/10.1146/annurev.bi.57.070188.003113.

Sterneck, E., Kaplan, D. R., & Johnson, P. F. (1996). Interleukin-6 induces expression of peripherin and cooperates with Trk receptor signaling to promote neuronal differentiation in PC12 cells. *Journal of Neurochemistry*, *67*(4), 1365–1374. http://dx.doi.org/10.1046/j.1471-4159.1996.67041365.x.

Suzuki, T., Mitake, S., Okumura-Noji, K., Shimizu, H., Tada, T., & Fujii, T. (1997). Excitable membranes and synaptic transmission: Postsynaptic mechanisms: Localization of α-internexin in the postsynaptic density of the rat brain. *Brain Research*, *765*(1), 74–80. http://dx.doi.org/10.1016/S0006-8993(97)00492-7.

Terao, E., Janssens, S., van den Bosch de Aguilar, P., Portier, M. M., & Klosen, P. (2000). In vivo expression of the intermediate filament peripherin in rat motoneurons: Modulation by inhibitory and stimulatory signals. *Neuroscience*, *101*(3), 679–688. http://dx.doi.org/10.1016/S0306-4522(00)00423-1.

Thompson, M. A., Lee, E., Lawe, D., Gizang-Ginsberg, E., & Ziff, E. B. (1992). Nerve growth factor-induced derepression of peripherin gene expression is associated with alterations in proteins binding to a negative regulatory element. *Molecular and Cellular Biology, 12*(6), 2501–2513. Retrieved from: http://www.ncbi.nlm.nih.gov/pmc/articles/PMC364443/.

Thompson, M. A., & Ziff, E. B. (1989). Structure of the gene encoding peripherin, an NGF-regulated neuronal-specific type III intermediate filament protein. *Neuron, 2*(1), 1043–1053. http://dx.doi.org/10.1016/0896-6273(89)90228-6.

Troy, C. M., Brown, K., Greene, L. A., & Shelanski, M. L. (1990). Ontogeny of the neuronal intermediate filament protein, peripherin, in the mouse embryo. *Neuroscience, 36*(1), 217–237.

Troy, C., Greene, L., & Shelanski, M. (1992). Neurite outgrowth in peripherin-depleted PC12 cells. *Journal of Cell Biology, 117*(5), 1085–1092. http://dx.doi.org/10.1083/jcb.117.5.1085.

Troy, C. M., Muma, N. A., Greene, L. A., Price, D. L., & Shelanski, M. L. (1990). Regulation of peripherin and neurofilament expression in regenerating rat motor neurons. *Brain Research, 529*(1–2), 232–238. http://dx.doi.org/10.1016/0006-8993(90)90832-V.

Tu, P. H., Raju, P., Robinson, K. A., Gurney, M. E., Trojanowski, J. Q., & Lee, V. M. (1996). Transgenic mice carrying a human mutant superoxide dismutase transgene develop neuronal cytoskeletal pathology resembling human amyotrophic lateral sclerosis lesions. *Proceedings of the National Academy of Sciences of the United States of America, 93*(7), 3155–3160. Retrieved from: http://www.ncbi.nlm.nih.gov/pmc/articles/PMC39778/.

Uchikado, H., Li, A., Lin, W.-L., & Dickson, D. W. (2006). Heterogeneous inclusions in neurofilament inclusion disease. *Neuropathology, 26*(5), 417–421. http://dx.doi.org/10.1111/j.1440-1789.2006.00709.x.

Undamatla, J., & Szaro, B. G. (2001). Differential expression and localization of neuronal intermediate filament proteins within newly developing neurites in dissociated cultures of Xenopus laevis embryonic spinal cord. *Cell Motility and the Cytoskeleton, 49*(1), 16–32. http://dx.doi.org/10.1002/cm.1017.

Willoughby, V., Sonawala, A., Werlang-Perurena, A., & Donner, L. R. (2008). A comparative immunohistochemical analysis of small round cell tumors of childhood: Utility of peripherin and α-internexin as markers for neuroblastomas. *Applied Immunohistochemistry & Molecular Morphology, 16*(4), 344–348. http://dx.doi.org/10.1097/PAI.0b013e318165fe78.

Wong, N. K., He, B. P., & Strong, M. J. (2000). Characterization of neuronal intermediate filament protein expression in cervical spinal motor neurons in sporadic amyotrophic lateral sclerosis (ALS). *Journal of Neuropathology & Experimental Neurology, 59*(11), 972–982. Retrieved from: http://www.ncbi.nlm.nih.gov/pubmed/11089575.

Wong, J., & Oblinger, M. M. (1990). Differential regulation of peripherin and neurofilament gene expression in regenerating rat DRG neurons. *Journal of Neuroscience Research, 27*(3), 332–341. http://dx.doi.org/10.1002/jnr.490270312.

Xiao, S., Tjostheim, S., Sanelli, T., McLean, J. R., Horne, P., Fan, Y., et al. (2008). An aggregate-inducing peripherin isoform generated through intron retention is upregulated in amyotrophic lateral sclerosis and associated with disease pathology. *Journal of Neuroscience, 28*(8), 1833–1840. http://dx.doi.org/10.1523/jneurosci.3222-07.2008.

Yan, Y., Jensen, K., & Brown, A. (2007). The polypeptide composition of moving and stationary neurofilaments in cultured sympathetic neurons. *Cell Motility and the Cytoskeleton, 64*(4), 299–309. http://dx.doi.org/10.1002/cm.20184.

Yokota, O., Tsuchiya, K., Terada, S., Ishizu, H., Uchikado, H., Ikeda, M., et al. (2008). Basophilic inclusion body disease and neuronal intermediate filament inclusion disease: A comparative clinicopathological study. *Acta Neuropathologica, 115*(5), 561–575. http://dx.doi.org/10.1007/s00401-007-0329-z.

Yuan, A., Rao, M. V., Kumar, A., Julien, J.-P., & Nixon, R. A. (2003). Neurofilament transport in vivo minimally requires hetero-oligomer formation. *Journal of Neuroscience*, *23*(28), 9452–9458. Retrieved from: http://www.jneurosci.org/content/23/28/9452.abstract.

Yuan, A., Rao, M. V., Sasaki, T., Chen, Y., Kumar, A., Veeranna, et al. (2006). α-Internexin is structurally and functionally associated with the neurofilament triplet proteins in the mature CNS. *Journal of Neuroscience*, *26*(39), 10006–10019. http://dx.doi.org/10.1523/jneurosci.2580-06.2006.

Yuan, A., Sasaki, T., Kumar, A., Peterhoff, C. M., Rao, M. V., Liem, R. K., et al. (2012). Peripherin is a subunit of peripheral nerve neurofilaments: Implications for differential vulnerability of CNS and PNS axons. *Journal of Neuroscience*, *32*(25), 8501–8508. http://dx.doi.org/10.1523/JNEUROSCI.1081-12.2012.

Yuan, A., Sasaki, T., Rao, M. V., Kumar, A., Kanumuri, V., Dunlop, D. S., et al. (2009). Neurofilaments form a highly stable stationary cytoskeleton after reaching a critical level in axons. *Journal of Neuroscience*, *29*(36), 11316–11329. http://dx.doi.org/10.1523/jneurosci.1942-09.2009.

Zhao, Y., & Szaro, B. G. (1997). Xefiltin, a Xenopus laevis neuronal intermediate filament protein, is expressed in actively growing optic axons during development and regeneration. *Journal of Neurobiology*, *33*(6), 811–824. http://dx.doi.org/10.1002/(SICI)1097-4695(19971120)33:6<811::AID-NEU8>3.0.CO;2-C.

CHAPTER EIGHTEEN

Studying Nestin and its Interrelationship with Cdk5

Julia Lindqvist*,†, Num Wistbacka*,†, John E. Eriksson*,†,1
*Cell Biology, Biosciences, Faculty of Science and Engineering, Åbo Akademi University, Turku, Finland
†Turku Centre for Biotechnology, University of Turku and Åbo Akademi University, Turku, Finland
[1]Corresponding author: e-mail address: john.eriksson@abo.fi

Contents

1. Introduction 510
   1.1 Dynamic Regulation of Nestin Expression During Development and Tissue Pathology 512
   1.2 Nestin Dynamics Is Under Strict Control of Posttranslational Modifications 514
   1.3 Truncation Mutants Give Insight into the Regulation of Nestin Protein Levels During Neuronal Cell Death 515
   1.4 The Solubility State of Nestin Determines Its Scaffolding Properties Towards Cdk5 517
2. Cellular Techniques to Study Nestin Functions in Cells 520
   2.1 Modulation of Nestin Expression in Cell Lines 521
   2.2 Tools for Analyzing the Signaling Properties of Nestin 525
3. *In Vivo* Mouse Models to Study Nestin Function 528
4. Conclusions 530
Acknowledgments 530
References 530

Abstract

Current research utilizes the specific expression pattern of intermediate filaments (IF) for identifying cellular state and origin, as well as for the purpose of disease diagnosis. Nestin is commonly utilized as a specific marker and driver for CNS progenitor cell types, but in addition, nestin can be found in several mesenchymal progenitor cells, and it is constitutively expressed in a few restricted locations, such as muscle neuromuscular junctions and kidney podocytes. Alike most other members of the IF protein family, nestin filaments are dynamic, constantly being remodeled through posttranslational modifications, which alter the solubility, protein levels, and signaling capacity of the nestin filaments. Through its interactions with kinases and other signaling executors, resulting in a complex and bidirectional regulation of cell signaling events, nestin has the potential to determine whether cells divide, differentiate, migrate, or stay in place. In this review, the broad and similar roles of IFs as dynamic signaling scaffolds, is exemplified by observations of nestin functions and its interaction with the cyclin-dependent kinase 5, the atypical kinase in the family of cyclin-dependent kinases.

Methods in Enzymology, Volume 568
ISSN 0076-6879
http://dx.doi.org/10.1016/bs.mie.2015.09.019

ABBREVIATIONS

ACh acetylcholine
AChR acetycholine receptor
Cdk5 cyclin-dependent kinase 5
CNS central nervous system
DN-Cdk5 dominant negative-Cdk5
GFP green fluorescent protein
IF intermediate filament
IP immunoprecipitation
MTJ myotendinous junction
Nes-640 GFP-tagged nestin deletion construct, amino acids 1–640
Nes-S nestin small-isoform
NMJ neuromuscular junction
PFA paraformaldehyde
PKCζ protein kinase C zeta
PTM posttranslational modification
shRNA small hairpin RNA
siRNA small interfering RNA

1. INTRODUCTION

Ever since the intermediate filament (IF) protein nestin was detected by the Rat-401 antibody in developing central nervous system (CNS) (Hockfield & McKay, 1985), it has been widely used as a marker for proliferating neuroepithelial progenitor cells. Due to similarities with the genes encoding neurofilaments, nestin is considered to be a type IV IF protein (Dahlstrand, Zimmerman, McKay, & Lendahl, 1992). Although the nestin protein is large (>1800 amino acids in rodents) mainly due to its unconventionally long tail domain, the N-terminal head domain is only eight amino acids long. These intrinsic features of the protein determine two unique features of nestin: firstly, the short head domain restricts the self-polymerization capacity of nestin, leaving its filament formation completely dependent upon interactions with other IF proteins. The long nestin tail is in turn the prime domain for interspecies variability (Yang et al., 2001). The tail contains multiple 11 amino acid repeats nearly identical in rats and humans, the number of which varies between species (41 copies in rodents vs. 18 in humans) (Dahlstrand et al., 1992). Although the repeat copy number varies, the repeats themselves are highly conserved. Alike most other IFs, the rod domain of nestin is the most conserved region (82% between humans and mice), while the C-terminus has evolved faster during evolution (55%), most likely due to the repetitive nature of the tail structure (Dahlstrand et al., 1992). The exceptional nestin tail, protruding from the filament structure is accessible to posttranslational modifications (PTMs) and protein interactions, and serves as a platform for cell signal integration. Nestin is expressed in mammals, but is found also in zebrafish (Chen, Yuh, & Wu, 2010), as well as birds (where the ortholog is termed transitin) (Jalouli et al., 2010; Wakamatsu, Nakamura, Lee, Cole, & Osumi, 2007).

Two nestin isoforms have been described so far: nestin short (Nes-S) and Nes-SΔ107-254 (Fig. 1), which are exclusively expressed in the otherwise nestin-negative dorsal root ganglia from postnatal day 5 (Su, Weng, et al., 2013; Wong et al., 2013). Nes-S incorporates into filamentous structures in rat postnatal dorsal root ganglia sensory and motor neurons and is not expressed elsewhere. Nes-S is alternatively spliced from the nestin gene so that exon 4, which encodes part of the C-terminus, is spliced out, giving rise to a 49.5 kDa protein with a 94 amino acid long tail, including a Nes-S-specific 32 amino acid tail domain (Su, Weng, et al., 2013). Nes-S-vimentin filament structures form in SW13 cells, while in dorsal root ganglia, Nes-S seems to incorporate into a filamentous network with the IFs peripherin and

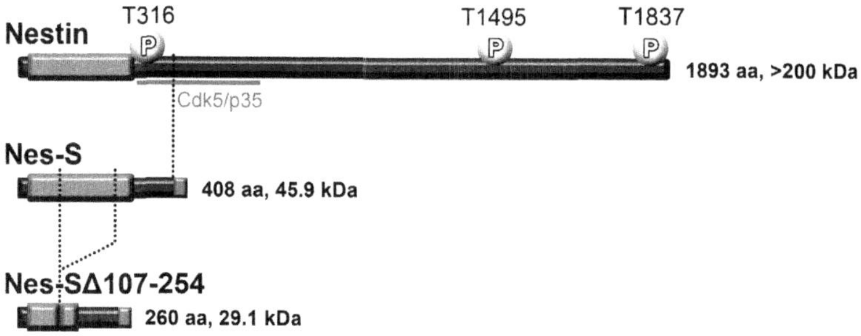

Figure 1 The protein structure of nestin and its splice variants. The full length rat nestin protein is 1893 amino acids in length and contains a long C-terminal tail, which constitutes the majority of the protein. The proposed Cdk5 phosphorylation sites T316, T1495 (Sahlgren et al., 2003), and T1837 (Contreras-Vallejos et al., 2014) all reside in the C-terminus, making this domain the primary site of PTM-mediated regulation. The Cdk5/p35 interacting domain on nestin (highlighted in red (dark gray in the print version)) has been mapped to reside between amino acids 316 and 640 (Sahlgren et al., 2006), but even Nes-S has been suggested to direct Cdk5 activity (Su, Chen, et al., 2013). Nes-S is a truncated nestin isoform, where the rod domain (orange (dark gray in the print version)) remains intact, but where most of the C-terminus (blue (black in the print version)) is deleted, and where the proximal C-terminus contains a 32 amino acids long frameshift region with a Nes-S-specific sequence (green (dark gray in the print version)). Nes-SΔ107-254 is a nestin splice variant lacking a large portion of the rod domain, and thereby being unable to form polymers.

neurofilament-H in perinuclear regions as well as in axonal processes (Su, Weng, et al., 2013). Nes-SΔ107-254, on the other hand, is missing most of its rod domain and cannot form filaments (Wong et al., 2013). Although present in dorsal root ganglia at low levels, its possible physiological function remains ambiguous.

Vimentin is the major polymerization partner of nestin in most cell types (Eliasson et al., 1999). In muscle cells, nestin copolymerizes mainly with desmin (Sjöberg, Jiang, Ringertz, Lendahl, & Sejersen, 1994) but, interestingly, nestin distribution seems to be unaltered at neuromuscular junctions (NMJs) and myotendinous junctions (MTJs) in desmin$^{-/-}$ muscle, suggesting that there are possibly other IFs compensating for desmin filaments *in vivo*, allowing nestin filament formation to occur in the absence of desmin (Carlsson, Li, Paulin, & Thornell, 1999). Nestin can also form filaments *in vitro* with the neuronal IF α-internexin (Steinert et al., 1999). Without the association of appropriate polymerization partners, nestin protein is not only incapable of forming filaments, but also prone to degradation at the protein level, as is the case in many vimentin null cell types (Eliasson et al., 1999). Intriguingly, also nestin seems to have a regulatory role in the vimentin–nestin heteropolymers: nestin–vimentin filament formation *in vitro* occurs optimally at a ratio of 1:4, while increasing nestin will ultimately have negative effects on vimentin assembly (Steinert et al., 1999). The same is true when nestin is overexpressed in mitotic cells: while overexpression of nestin can keep vimentin in a more soluble state, nestin downregulation seems to promote vimentin filament formation around the mitotic spindle. Therefore, depending on the cellular context, nestin incorporation into major IF networks may drastically change the nature of the IF cytoskeleton in terms of functionality, as detailed below.

1.1 Dynamic Regulation of Nestin Expression During Development and Tissue Pathology

The nestin promoter is located in the 5′-untranslated region and its expression is mainly driven by the zinc-finger transcription factors Sp1 and Sp3 (Cheng et al., 2004). Nestin expression is regulated in a tissue-specific fashion by enhancers that reside within two of its three introns: while nestin expression in the somites and myotome is regulated by the first intron, the second intron regulates CNS-specific nestin expression (Kawaguchi et al., 2001; Zimmerman et al., 1994). Nestin is best known for its expression in neuroepithelial progenitor cells in the CNS and peripheral nervous system starting from early development (Kawaguchi et al., 2001), but is in addition expressed in many other organs and cell types. In the embryonic muscle-forming somites, nestin is expressed from day E9.5 (Kachinsky, Dominov, & Miller, 1994; Sejersen & Lendahl, 1993). Besides these major organ systems, nestin expression has been repeatedly reported in adult tissue resident progenitor cells (mesenchymal stem cells) that assist in adult tissue repair and renewal. Nestin is also expressed in certain myofibroblast populations, such as breast myoepithelial cells, propria mucosa of colon and intestine as well as endometrium (Kishaba, Matsubara, & Niki, 2010). In addition to its progenitor cell-related expression, nestin is constitutively expressed in renal podocytes (Perry et al., 2007) as well as at NMJs and MTJs (Vaittinen et al., 1999).

With the induction of tissue stress and injury, nestin has been found to be reexpressed at least in regenerating muscle after injury and in myodegeneration (Sjöberg, Edström, Lendahl, & Sejersen, 1994; Vaittinen et al., 1999), in reactive astrocytes after CNS injury (Frisén, Johansson, Török, Risling, & Lendahl, 1995), in regenerating testes when vascular smooth muscle cells and pericytes differentiate to Leydig cells (Davidoff et al., 2004), in both mesangial cells and podocytes in healing kidney after nephritis (Daniel, Albrecht, Lüdke, & Hugo, 2008; Perry et al., 2007) as well as in the myocardium after infarction (Scobioala et al., 2008). In addition to this highly cell type-specific expression of nestin, it is also upregulated in neoangiogenetic blood vessels in remolding tissue during development, injury, and cancer (Mokrý et al., 2004). Nestin-driven green fluorescent protein (GFP) reporter mouse models have contributed to the knowledge of nestin regulation (although promoter activity does not necessarily correlate with protein levels) and are valuable tools for stem cell isolation as well as angiogenesis studies particularly in the cancer field (Amoh et al., 2005; Uehara et al., 2014).

The expression of nestin in tumors has recently evoked plenty of interest due to its potential use as a clinical marker and therapeutic target. Nestin expression has been found to correlate with poor disease prognosis in at least melanoma and breast cancer (Ishiwata, Matsuda, & Naito, 2011). The downregulation of nestin seems to negatively affect the cell migration/invasion in at least glioma, melanoma as well as pancreatic and lung cancer models (Akiyama, Matsuda, Ishiwata, Naito, & Kawana, 2013; Ishiwata, Teduka, et al., 2011; Matsuda et al., 2011; Takakuwa et al., 2013). While Kleeberger et al. (2007) found that nestin downregulation inhibited migration and invasion of prostate cancer cells, a later study argued that nestin downregulation promotes prostate cancer cell invasion, but not migration per se (Hyder et al., 2014). The lack of nestin was found to affect the turnover of focal adhesions through increasing membrane expression of specific integrins and modulation of phosphorylated focal adhesion kinase localization, providing an explanation at the molecular level for the specific effects of nestin on impaired invasion (Hyder et al., 2014). The downregulation of nestin also promotes melanoma cell line invasion and matrix-metalloprotease upregulation (Lee et al., 2014). Some reports suggest that nestin expression might be linked to epithelial-to-mesenchymal transition, at least in pancreatic (Su, Chen, et al., 2013) and breast (Zhao et al., 2014) cancer cell models. In many types of cancer cells, the knockdown of nestin interferes with cell growth and viability, including lung cancer cells (Takakuwa et al., 2013), breast cancer cells (Zhao et al., 2014), and melanoma (Akiyama et al., 2013). In liver cancer tumorigenesis, nestin expression is regulated by the p53–Sp1 signaling axis as an essential part of tumor initiation (Tschaharganeh et al., 2014), suggesting that nestin expression may in fact drive an oncogenic program in this particular setup and further proposing that nestin plays an active part in a cancerous dedifferentiation program. These and many other publications do support that nestin plays a role in cancer cell line-specific effects in tumorigenesis and cell motility, but the mechanisms are in many cases not understood well, if at all. Also the aspects of nestin as a potential regulator and target for tumor angiogenesis (Ishiwata, Matsuda, et al., 2011; Matsuda, Hagio, & Ishiwata, 2013) have showed growing interest in the past years, as nestin is highly expressed in pericytes and endothelial cells in neovasculature.

The specificity of nestin expression in mesenchymal stem cells, neuronal progenitors, and cancer have created an interest in the research community to understand its expression pattern in more detail, possibly even including its use as a clinical marker. There are, however, only a limited number of research articles that address in detail the functionality of nestin and the significance of its presence. Given its surprisingly abundant, but highly specific, expression, the knowledge of which has expanded in recent years, there is a more solid basis but also an increasing need to better understand the molecular functions of nestin.

1.2 Nestin Dynamics Is Under Strict Control of Posttranslational Modifications

Although the genetic regulation of nestin is naturally crucial for its timely expression in specific tissues, nestin is, alike most IF proteins, strictly regulated at the posttranslational level through at least phosphorylation and ubiquitination (Sahlgren et al., 2001, 2003, 2006). To date, no other PTMs of nestin have been described in literature, but it would be surprising if nestin was not modified in a similar complex manner, as other IFs (Snider & Omary, 2014). In addition, nestin antibodies often detect multiple differentially migrating bands by Western blotting, suggesting the prominent presence of nestin splicing-, degradation-, and/or PTM-isoforms. A series of reports by our laboratory have studied nestin phosphorylation. Through *in vitro* phosphorylation and mass spectrometry identification, cyclin-dependent kinase 5 (Cdk5) was shown to phosphorylate nestin on Thr316 and Thr1495 (both Ser/Thr-Pro sites and, thereby suitable as Cdk5 sites) with obvious effects on nestin cytoskeleton organization, sequestration, and solubility (Sahlgren et al., 2003). Nestin, however, contains multiple other potential phosphorylation sites that may be modified by Cdk5, as well as other kinases. Thr316 is also phosphorylated by Cdk1 during mitosis, causing mitotic reorganization of nestin filaments (Sahlgren et al., 2001). Additionally, the phosphorylation of nestin Ser1837 (a Cdk5 consensus site) was reported to be downregulated in E18.5 $Cdk5^{-/-}$ brain in a phosphoproteomic-study, possibly implying it may be a Cdk5 site *in vivo* (Contreras-Vallejos et al., 2014). The phosphoregulation of nestin has vast consequences on Cdk5 itself, and the signaling mechanism is discussed further in forthcoming sections. It is likely that in most model systems, nestin protein levels are regulated by simultaneous input from transcriptional activity as well as PTMs (Mellodew et al., 2004; Sahlgren et al., 2003; Vaittinen et al., 2001).

1.3 Truncation Mutants Give Insight into the Regulation of Nestin Protein Levels During Neuronal Cell Death

Nestin and Cdk5 were for the first time shown to interact in the rat-derived neuronal progenitor ST15A cells (Sahlgren et al., 2003) and the concept has been later been shown to be applicable in other model systems. Cdk5 is a kinase involved in both neuronal (Ohshima et al., 1996) and muscle development (Lazaro et al., 1997), and deregulation of its activity has been later linked to many diseases, including Alzheimer's disease, diabetes, and cancer. The activity of Cdk5 is strictly determined through association of activators, such as the noncyclin protein p35, which is important for neuronal differentiation and axonal maintenance, or the calpain cleavage product of p35, p25, which is linked to hyperactivation of the kinase and neuronal cell death. Regulation of the balance of p35/p25 is, therefore, crucial in correct targeting of Cdk5 activity in relation to its dual functions in neuronal maintenance and death.

In several models, nestin has been shown to regulate Cdk5 kinase activity through modulation of the p35/p25 balance, which in the context of dying neuronal progenitors translates into a cytoprotective function of the nestin scaffold (Sahlgren et al., 2006). When the ST15A cells were treated with hydrogen peroxide to induce apoptosis, nestin was found to undergo proteasomal degradation at the time of cell death. The direct involvement of nestin in balancing apoptosis was demonstrated by its downregulation, which markedly increased apoptosis, while overexpression of exogenous nestin had a strong cytoprotective effect (Sahlgren et al., 2006). Cdk5 was shown to be the executor of apoptosis in this particular model, confirmed by experiments showing that expression of dominant negative-Cdk5 (DN-Cdk5) rescued the increased apoptosis in nestin-downregulated cells (Sahlgren et al., 2006). In fact, expression of nestin reduced the calpain cleavage of p35 to p25 during apoptosis while protecting p35 from ubiquitination and proteasomal degradation. To determine the specific functions of different parts of nestin, expression of nestin truncations N-terminally tagged with GFP proved to be a powerful measure. By employing these constructs, it was possible to determine that the nestin tail contributes strongly to its own degradation, as the nestin truncation containing amino acids 1–640 of rat nestin (Nes-640) is more stable compared to full length forms (Sahlgren et al., 2006). Immunoprecipitation (IP) studies also revealed that Cdk5/p35 binds to the nestin C-terminus (between amino acids 314 and 640) next to the rod domain (Sahlgren et al., 2006). Finally, it was established that while expression of Nes-640 did not increase steady-state p35 protein levels as such, its intracellular localization (and thus Cdk5 activity) shifted towards a soluble pool. Thus, it is believed that a soluble pool of Nes-640, which is indeed prominent, sequesters p35 and Cdk5 activity to the cytoplasm and protects p35 from ubiquitination, thereby preventing cytotoxic Cdk5/p25 activity in the nucleus. Importantly, the nestin-Cdk5 signaling occurred in both directions: while nestin was shown to regulate p35 turnover, disruption of Cdk5 activity with roscovitine or DN-Cdk5 significantly retarded the H_2O_2-induced nestin degradation (Sahlgren et al., 2006). As Cdk5-mediated nestin phosphorylation promotes filament solubilization (Sahlgren et al., 2003), the evidence from both publications imply that Cdk5 itself acts to regulate the turnover of p35 through nestin to prevent unwanted kinase activity.

The novel Nes-S isoform also appeared to interact with Cdk5 when overexpressed in N2a neuroblastoma cells that are intrinsically Nes-S deficient (Su, Weng, et al., 2013). In support of nestin-Cdk5 interplay in ST15A cells, the expression of Nes-S promoted survival after H_2O_2 treatment (Su, Weng, et al., 2013). In addition, Nes-S downregulation seemed to impair primary dorsal root ganglia neuron viability. Although further evidence is required to understand the interplay between Nes-S and Cdk5, these results attractively support the nestin–Cdk5 interaction which has been under intense investigation previously. Also in zebrafish, the downregulation of nestin causes an increase in apoptosis of neuronal precursors during development of brain and eye, leading to severely malformed and small-sized brains as well as hydrocephalus (Chen et al., 2010).

1.4 The Solubility State of Nestin Determines Its Scaffolding Properties Towards Cdk5

The nestin–Cdk5 interaction was confirmed to occur also in differentiating myoblasts, where both nestin protein expression and Cdk5 activity are induced strongly upon induction of differentiation (Sahlgren et al., 2003). Myoblast differentiation is a multistep process involving activation and proliferation of myoblasts, followed by cell fusion, to allow for the formation of multinucleated myotubes. The differentiation process involves multiple morphological changes that require extensive remodeling of the cytoskeletal networks, including that of nestin. The initial perinuclear nestin filaments, which is typical of proliferating cells, reorganize during differentiation into long filaments that follow the shape of the elongated and fused myotubes (Sahlgren et al., 2003). Cdk5 acts upstream of nestin to regulate both nestin solubility through phosphorylation, as well as its protein levels, which are upregulated during the differentiation process. Although phospho-Thr316 seems to be one of the key Cdk5 phosphorylation sites on nestin that determine its solubility status in myoblasts, several other phosphosites, the functions of which remained uncharacterized, were induced by Cdk5, suggesting that nestin is heavily phosphorylated by the kinase (Sahlgren et al., 2003). Generation of phospho-specific antibodies against phospho-Thr316 revealed that phospho-nestin occurred exclusively in soluble fractions in differentiating myoblasts, suggesting that Cdk5-mediated nestin phosphorylation releases a pool of soluble nestin. Thus, the nestin filament reorganization during differentiation is indeed dependent on Cdk5 activity (Sahlgren et al., 2003). Later, the PKCζ-kinase which is an essential upstream regulator of Cdk5 activity during myoblast differentiation, was also demonstrated to be essential for nestin filament organization and induction of nestin protein levels in the system through Cdk5 (De Thonel et al., 2010).

Further experiments demonstrated that a soluble pool of nestin promotes Cdk5/p35 association and p35 stability (Pallari et al., 2011). This observation has direct consequences for myogenic differentiation: when nestin is depleted transiently with small interfering RNA (siRNA) in myoblasts and the cells are induced to differentiate, differentiation occurs at a faster rate due to the increased processing of p35 to p25 (Pallari et al., 2011). The cleavage of p25 during myoblast differentiation is crucial to allow the generation of correct differentiation-associated Cdk5 activity (De Thonel et al., 2010). In contrast, overexpression of the stable Nes-640 markedly delayed the process of differentiation and severely disturbed p25 generation (Pallari et al., 2011). Thus, the nestin-mediated scaffolding of Cdk5 during muscle differentiation closely resembles its cytoprotective function during neuronal apoptosis (Sahlgren et al., 2006); when nestin is depleted, p35 is uncontrollably processed to p25, leading to hyperactivation of Cdk5, with the outcome of increased cell death, or increased myogenic differentiation (Fig. 2A). Intriguingly, Cdk5 itself regulates the solubility of nestin through direct phosphorylation, allowing bidirectional crosstalk between the scaffold and the kinase. Therefore, Cdk5 regulates its own activity through phosphorylation of nestin.

In addition to being expressed in differentiating myoblasts, nestin is also localized directly underneath the postsynaptic (muscle) plasma membrane of the NMJ (Vaittinen et al., 1999, 2001) where the filaments extend into sarcomeres in the vicinity of the NMJs and surround the junctional myonuclei (Kang et al., 2007). Nestin expression at the postsynaptic NMJ is regulated by innervation of the muscle, which causes nestin mRNA to be transcribed in the postsynaptic myonuclei (Kang et al., 2007).

Moreover, nestin-mediated scaffolding of Cdk5 was found to be essential in determining Cdk5 activity at the NMJs. NMJ development and maintenance is actively balanced by signals on both the presynaptic (neuronal) and postsynaptic (muscle) side of the junction that act to stabilize or disperse the postsynaptic signaling apparatus in response to innervation. Cdk5 acts at several levels of NMJ development to regulate both positive (Cdk5 promotes the neuregulin-induced acetylcholine receptor [AChR] transcription) and negative signals (acetylcholine [ACh]-mediated Cdk5 activity disperses AChR clusters that are not stabilized by agrin) that act on AChR clustering (Fu et al., 2001, 2005; Lin et al., 2005). As a response to ACh agonists, which initiate a signaling cascade that destabilizes the NMJs, C2C12 myotubes upregulate both membrane-associated and total Cdk5 activity, induce p35 membrane recruitment and ultimately trigger nestin Thr316 phosphorylation

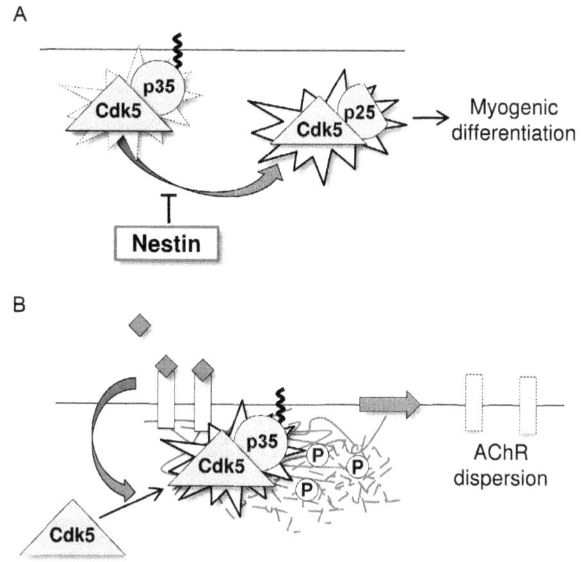

Figure 2 Nestin regulates Cdk5 in differentiating muscle cells and at NMJs. (A) The levels and solubility of nestin determines Cdk5 activity in differentiating myoblasts through adjusting the balance between p35 (membrane-bound trough a myristoyl group) and p25 (cytoplasmic/nuclear). If nestin is overexpressed, excess nestin sequesters and protects p35 from calpain cleavage, inhibiting the generation of p25, which in turn is required for differentiation. Therefore, nestin has an inhibitory effect on differentiation. Conversely, if nestin is downregulated by siRNA, p35 is uncontrollably processed and p25-generation is favored, resulting in increased Cdk5 activity and increased differentiation (Pallari et al., 2011). (B) Nestin is required for Cdk5/p35 activation at the NMJ membrane, where Cdk5 kinase activity is important for the dispersion of nonstabilized AChR clusters. First, ACh itself induces Cdk5 activity. Then, Cdk5/p35 is recruited to the membrane by nestin, where nestin is phosphorylated, further contributing to p35 stabilization, and AChR cluster dispersion. If nestin is removed from the system, or if phospho-deficient T316A nestin is expressed, Cdk5 is not activated as efficiently, resulting in increased AChR clustering, as observed in the nestin deficient mice (Mohseni et al., 2011; Yang et al., 2011).

(Yang et al., 2011). Nestin plays a crucial role in the ACh agonist-mediated Cdk5 activation, as the downregulation of nestin or expression of phospho-deficient T316A mutant nestin suppresses Cdk5 activation and reduces ACh-agonist dependent NMJ dispersion (Yang et al., 2011). These results show that (phosphorylated) nestin is required for AChR dispersion (Yang et al., 2011) (Fig. 2B). Indeed, T316A mutant nestin sequesters Cdk5 strongly but keeps it in a less active state. Nestin siRNA mice that were generated by Yang et al. (2011) did show an increase in AChR cluster number, indicating that nestin is required for optimal NMJ dispersion. The results were verified by a genetic nestin deletion mouse model, which similarly showed an increase in AChR cluster number and size, causing motor coordination problems in the nestin-deficient mice when measured by the Roto-Rod test (Mohseni et al., 2011).

The nestin-mediated scaffolding of Cdk5 activity seems to occur slightly differently at NMJs compared to differentiating myoblasts. Whereas nestin siRNA increases Cdk5 activity through increased p25 cleavage in differentiating myoblasts (Pallari et al., 2011), nestin depletion (or mutation of the Cdk5-site Thr316 to alanine) decreases both membrane bound and total Cdk5 activity at NMJs (Yang et al., 2011), which prevents NMJ dispersion. Thus, in both the differentiation and NMJ models, soluble nestin seems to be required for Cdk5/p35 stabilization. At NMJs, soluble nestin/Cdk5/p35 is required for NMJ stabilization, and depletion of nestin would therefore disturb Cdk5 activity. On the other hand, the calpain-mediated p25 generation is obligatory for muscle differentiation (De Thonel et al., 2010). When nestin is depleted from differentiating myoblasts, p35 is uncontrollably cleaved by calpains, causing an increase in Cdk5/p25-associated activity, which promotes differentiation (Pallari et al., 2011). When nestin is overexpressed using the Nes-640 construct, the balance of nestin filaments is severely hampered. A noticeable fraction of the overexpressed Nes-640 remains soluble (Pallari et al., 2011), and this particular fraction is likely to severely interfere with Cdk5 activator turnover. Thus, the activity of Cdk5 is dynamically regulated through nestin by the kinase itself, in a complex autoregulatory feedback loop.

2. CELLULAR TECHNIQUES TO STUDY NESTIN FUNCTIONS IN CELLS

In the following section, a selection of useful approaches and methods has been compiled, to facilitate researchers that attempt to study the roles of nestin in different cellular, developmental, as well as homeostasis and disease-related processes. Although the expression of nestin is restricted in most tissues, there are a number of cell lines that can be utilized to study the endogenous protein (Table 1). Many of the nestin-expressing cell lines are, in fact, cancer cells, of which numerous remain unlisted here. For example, many lung cancer cells express nestin (Chen et al., 2014; Takakuwa et al., 2013).

Table 1 Examples of Commercial Cell Lines with High Nestin Expression

Cell Line	Cell Type	Species	Reference
C2C12	Myoblast	Mouse	Pallari et al. (2011)
L6	Myoblast	Rat	Kachinsky et al. (1994)
ST15A	Neuronal progenitor	Rat	Sahlgren et al. (2006)
PC-12	Adrenal gland/ pheochromocytoma	Rat	Jia, Ji, Maillet, and Zhang (2010)
BHK-21	Kidney fibroblast	Hamster	Chou, Khuon, Herrmann, and Goldman (2003)
PC-3	Prostate adenocarcinoma	Human	Hyder et al. (2014)
U251 MG	Glioma	Human	Macarthur et al. (2014)
A2058 and A375	Melanoma	Human	Lee et al. (2014)

2.1 Modulation of Nestin Expression in Cell Lines

2.1.1 Transient siRNA Transfections

Transient downregulation of nestin by siRNA in mouse (C2C12) and human (PC3) cells is performed using Lipofectamine RNAiMAX transfection reagent (Invitrogen) according to instructions of the manufacturer in antibiotic-free cell culture media. A mixture of four siRNA oligonucleotides targeting nestin is purchased from Qiagen (FlexiTube GeneSolution GS18008 against mouse nestin and FlexiTube GeneSolution GS10763 for human cells). We have found that a mixture of four different siRNAs gives maximal downregulation of nestin protein, compared to the use of single siRNAs only. The AllStars Negative Control siRNA scrambled oligonucleotide is also purchased from Qiagen. All siRNA oligos are used at a final concentration of 50 n*M*. For transfection of one well in a 12-well plate, the following protocol is utilized in strictly RNAse free condition:

1. Cells are seeded the day before in antibiotic free media so that they reach 60–80% confluency the day of transfection. If needed, the cells can be divided after performing transfection according to experimental needs.
2. Five microliter Lipofectamine RNAiMAX (Invitrogen) is suspended in 100 μl Opti-MEM media (Life Technologies). A mixture of four nestin-targeting siRNA oligos is suspended in 100 μl Opti-MEM media, so that the final transfection concentration reaches 50 n*M*.

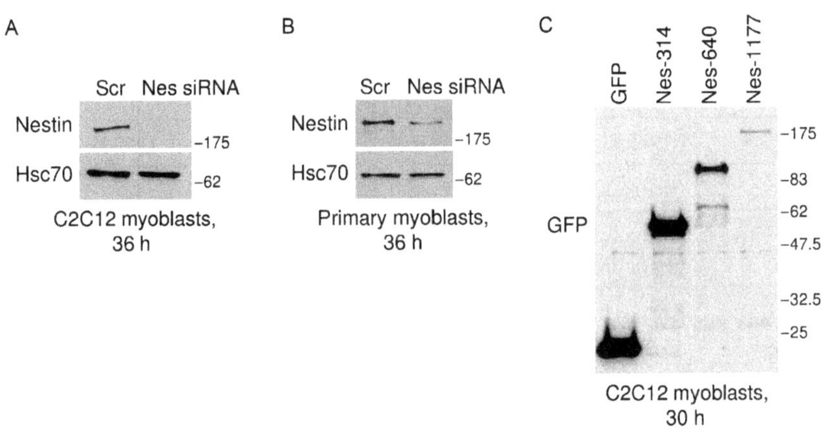

Figure 3 Transient downregulation and overexpression of nestin in myoblasts. (A) siRNA-mediated downregulation of nestin (Nes siRNA) in C2C12 myoblasts. Cells were seeded at 80% confluency, incubated in RNAi oligos for 12 h, after which proliferation medium (10% serum) was replaced with differentiation medium (1% serum). Samples were collected for Western blotting 36 h after induction of differentiation. Scr = scrambled oligos. (B) Downregulation of nestin in primary myoblasts is less efficient compared to C2C12 cells, but yet notable. (C) Overexpression of nestin truncation constructs becomes increasingly difficult with the increasing size of the constructs and compared to GFP plasmid. Here, cells were induced to differentiate 12 h after initiation of differentiation, and samples were collected after 30 h.

3. The siRNA mixture is added to the transfection reagent mixture and incubated for 20 min at room temperature.
4. The siRNA–liposome complexes (200 μl) are added to the cells and incubated 16–24 h.

The transfection efficiency reaches its maximum (50–90%) typically after 48 h, depending on the cell line used (Fig. 3A and B).

2.1.2 Transient Overexpression of Nestin

Previously, we have utilized several nestin truncation constructs to study the signaling properties of nestin (Sahlgren et al., 2006). Due to the large size of the nestin protein, it is difficult to reach good transfection efficiency using the full length nestin construct, as it is fairly unstable in the cells. Therefore, the N-terminally GFP tagged truncation containing the 640 first amino acids of rat nestin (Nes-640) has been utilized instead for the purpose of Cdk5 interaction studies. This particular deletion construct retains both the eight amino acid N-terminus as well as the rod domain (amino acids 9–314), and contains the interaction site for Cdk5. A schematic representation of the available GFP-tagged nestin constructs is highlighted in Fig. 4. Whereas rod-domain deletions do not form filaments, all other truncations do.

Another important consideration when studying overexpression of nestin is the presence of an endogenous filament network of a nestin polymerization partner in the target cell line, typically vimentin or desmin, which allow nestin filament organization to occur. On the contrary, the signaling properties of soluble and filamentous nestin (using nestin overexpression constructs) are easily investigated with the help of endogenously vimentin negative and positive cell lines, such as the respective clones of the adenocarcinoma cell line SW13 (Pallari et al., 2011).

2.1.3 Transient Overexpression of Nes-640 in C2C12 Cells Using JetPEI

Transient overexpression of nestin in sensitive cell lines which are generally difficult to transfect (such as the mouse C2C12 myoblasts) is performed by liposome transfection. We utilize the JetPEI transfection reagent (PolyPlus Transfection), which has the advantage of minimal antibiotic-induced toxicity, allowing the use of antibiotic-containing media during the course of transfection. Also Lipofectamine LTX with PLUS-reagent (Invitrogen) has proven to give good transfection efficiency when following the instructions of the manufacturer. For transfection of cells in one 12-well plate, the following protocol is used:

1. Cells are seeded the day before in regular media so that they reach 60–80% confluency the day of transfection.
2. Four microliter JetPEI transfection reagent (PolyPlus Transfection) is suspended in 50 μl sterile NaCl (Life Technologies). Two microgram DNA construct (Nes-640 or empty pEGFP vector [Clontech]) is suspended in sterile NaCl.
3. The liposome and DNA mixtures are combined and incubated in room temperature for 25 min.
4. The DNA–liposome complexes (in a volume of 100 μl) are added to the cells and incubated 48 h before further experimental use. Representative results are presented in Fig. 3C.

Alternatively, other overexpression techniques, such as electroporation (which yields higher cytotoxicity compared to liposome-mediated transfection) or lentiviral expression, can be considered as potential alternatives for liposome-mediated transfection of C2C12 myoblasts.

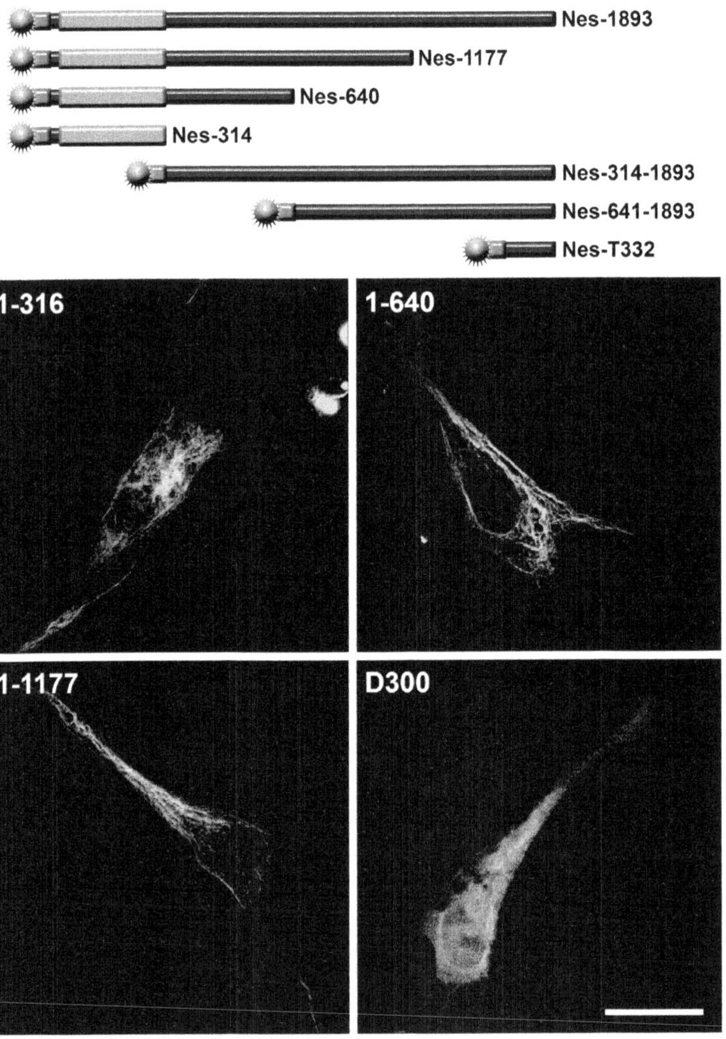

Figure 4 Nestin deletion mutant constructs. Differentially truncated, N-terminally GFP-tagged rat nestin constructs can be employed to effectively investigate the physiological roles of nestin. The position of the GFP-tag is highlighted in green and the rod domain in orange. The short nestin head (N-terminus) and long tail (C-terminus) are blue. The mutants include the following constructs: full-length nestin (Nes-1893), nestin tail truncations (Nes-1177, Nes-640), and nestin lacking the C-terminal tail completely (Nes-314). N-terminally truncated nestin constructs (lacking the rod domain) include Nes-314-1893, Nes-641-1893, Nes-T332, the latter construct containing only the last 332 amino acids of the C-terminus. The morphology obtained with these constructs is exemplified by images from BHK-21 cells that were transfected with four different representative nestin constructs. In cells transfected with the construct containing

2.1.4 Transient Overexpression of Nes-640 in SW13 Cells by Electroporation

Vimentin positive and negative SW13 cell clones were transfected using electroporation according to (Pallari et al., 2011).

1. For transfection of one 100 mm cell culture dish, the SW13 cells were grown to 80% confluency. The cells were trypsinized, collected, and resuspended in 200 μl Opti-MEM media (Life Technologies).
2. Ten microgram of Nes-640 plasmid was added to 0.4 cm electroporation cuvettes (BTX) together with the cells.
3. Cells were subjected to a single electric pulse (220 V, 975 μF) using a Bio-Rad Gene Pulser electroporator, followed by dilution in DMEM supplemented with 10% FBS, 2 m*M* L-glutamine, and antibiotics.
4. Cells were incubated for 24–48 h before experimental procedures.

2.2 Tools for Analyzing the Signaling Properties of Nestin

2.2.1 Considerations for Microscopy-Based Techniques

For immunolabeling of nestin, and colabeling with other proteins, a 3% paraformaldehyde (PFA) solution is typically used for fixation. In some cases, PFA fixation may compromise accurate filament structures during microscopy and may require the use and optimization of additional antigen retrieval protocols due to heavy crosslinking and epitope masking. With the use of the below-mentioned antibody, we have not recorded a need for antigen retrieval in the immunolabeling protocol (Fig. 5). To better enhance the filament structure, a 10 min fixation using acetone, methanol or 50:50 acetone–methanol solutions at −20 °C yields in general more accurate filaments for high-resolution microscopy. However, acetone or methanol-based fixation will disturb cellular membranes and may, thus, hamper the study of such nestin-interacting proteins that are membrane-associated or soluble. When using acetone-fixation, the permeabilization step can be disregarded. For immunocytochemical detection of nestin in C2C12 cells (or alternatively, any other cell line or primary cell type of interest) the following immunolabeling protocol is used:

Figure 4—Cont'd the head and rod domain of nestin, Nes-314, filaments can be seen to form throughout the cells. Likewise, the constructs Nes-640 and Nes-1177 also give rise to filamentous structures. Mutants lacking the rod-domain, like Nes-T332(D300), fail to form filamentous structures (similar morphology was obtained with all truncations lacking the rod domain). The images were acquired with Leica TCS SP5 confocal microscope and are presented as maximum projections of the GFP signal. Scale bar is 20 μm. (See the color plate.)

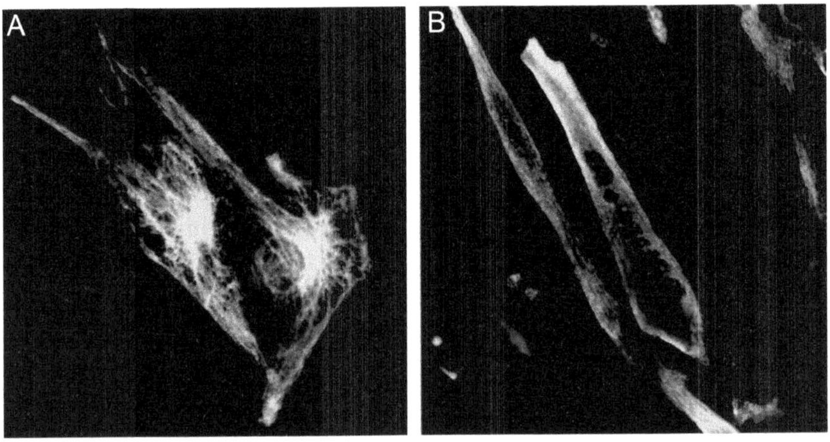

Figure 5 Nestin immunolabeling in myoblasts. Endogenous nestin (green (gray in the print version)) was immunolabeled with the protocol described in Section 2.2.1 in proliferating myoblasts (A) and 48 h after induction of differentiation, when myotubes are forming trough fusion (B). Nuclei were counterstained with Dapi (blue (black gray in the print version)). The images were acquired with Zeiss LSM510 confocal microscope.

1. Cells are plated on coverslips to reach 70% confluency and treated according to experimental requirements.
2. Cells are fixed 10 min in 3% PFA at room temperature. After fixation, the cells are thoroughly rinsed in phosphate buffered saline (PBS) 3 ×.
3. Cells are permeabilized with 0.1–0.2% Triton X-100 for 5–10 min. Samples are then washed once with PBS. The permeabilization conditions used for nestin immunolabeling in differentiating myotubes is 0.5% Triton X-100 for 10 min, as the myotubes tend to form thick and overlapping mats on the coverslips and requires stronger permeabilization.
4. Samples are blocked in 1% bovine serum albumin (BSA) in PBS for 60 min.
5. Samples are incubated in primary antibody against nestin for 1–2 h, but overnight incubation may enhance the signal. For immunostaining of mouse cells, the anti-nestin Rat-401 antibody (BD Pharmingen, Cat. nr. 556309) is typically used at a dilution of 1:200 in 1% BSA/PBS. For immunolabeling of human nestin, anti-nestin clone 10C2 (Millipore) has been used at a concentration of 1:200.
6. The coverslips are thoroughly washed 3 × by dipping the coverslip, and samples are incubated in secondary species-specific Alexa fluorescent probe-conjugated antibodies (Invitrogen), diluted 1:500–1:1000 for 1 h

7. After four consecutive washes, cells are mounted in ProLong Gold with DAPI (Life technologies) or Mowiol mounting solution (made in-house) and analyzed by confocal microscopy.

2.2.2 Biochemical Methods

The detection of IF-protein interactions by IP can be problematic due to the insoluble nature of the proteins. Therefore, the use of strong detergents in IP buffer is required to solubilize the IFs. Nestin, for example, occurs naturally in a soluble state (in addition to filaments), and this particular fraction is easy to extract. The soluble pool of nestin in a given cell type varies depending on the cell lines and types used, the differentiation state of the cells, whether endogenous or overexpressed proteins are investigated, and naturally, the phosphorylation status of the cells. Keeping in mind that all of these factors affect the solubility of nestin, the soluble nestin fraction typically reaches 10–20%, but can even constitute as much as 50% of total nestin. Regardless of the use of strong detergents in IP buffers, a large fraction of nestin (or any other IF in general) will often remain insoluble and will be discarded with the insoluble pellet-fraction. Uncleared cellular extracts or additional crosslinking of proteins prior to IP may be used to enhance the probability of detection of protein interactions, but may adversely contribute to detection of unspecific false-positives and higher background. Therefore, microscopy-based techniques, such as colocalization, proximity-ligation assay (Paul & Skalli, in press), Förster resonance energy transfer and fluorescence correlation spectroscopy, or *in vitro*-interaction analysis of purified proteins may enlighten the actual status of the interaction, even though many methods may be obscured by the requirement of protein overexpression. Therefore, IP does have the advantage of detection of naturally occurring endogenous interactions. For IP of nestin, the following protocol for C2C12 cells (and can be used for any other cell line of interest), which has been optimized for one 100 mm cell culture dish at 80% confluency, can be utilized for the analysis of interaction partners (such as Cdk5/p35) with the soluble nestin fraction:

1. Cells pellets are collected for IP, and lysed 30 min on ice. The following lysis buffer (1 ml per 100 mm plate) is used for IP of nestin (Pallari et al., 2011; Sahlgren et al., 2006): 50 m*M* Tris (pH 8.0), 150 m*M* NaCl, 1% Nonidet P-40, 0.5% deoxycholate, 0.05% SDS, 5 m*M* EDTA, 5 m*M* EGTA, Complete Protease Inhibitor Cocktail (Roche Diagnostics). The addition of phosphatase inhibitors should be considered.

2. Samples are centrifuged at 15,000 × *g* for 10 min at 4 °C for clearing, but may be adjusted as high-speed centrifugation will inevitably pellet filaments. The soluble supernatant fraction is recovered and used for the following steps.
3. Samples are taken for input controls, and each lysate is precleared using a 50% slurry of protein G-Sepharose beads (GE Healthcare) (in a volume of 25 μl) for 1 h at 4 °C under rotation.
4. Two microgram of antibody is added to the cell extracts, and incubated for 1 h. For immunoprecipitation of mouse nestin, the anti-nestin Rat-401 antibody (BD Pharmingen, Cat.nr. 556309) is used. For IP of human nestin, anti-nestin clone 10C2 (Millipore) is used instead.
5. The capture of immunocomplexes by protein G-Sepharose beads is allowed 4 h at 4 °C. Alternatively, magnetic A+G protein beads (Millipore) have been successfully utilized, and allows shorter incubation times.
6. Last, samples are washed 3× in 1 ml lysis buffer, and beads are resuspended in Laemmli lysis buffer and boiled.
7. Successful IP can be analyzed with Western blotting. For detection of nestin using Western blotting we use anti-nestin (BD Pharmingen, Cat. Nr. 611659) for mouse cells, and anti-nestin clone 10C2 (Millipore) or anti-nestin clone MCA-4D11 (EnCor Biotechnology) for human cells.

3. *IN VIVO* MOUSE MODELS TO STUDY NESTIN FUNCTION

The function of nestin has remained somewhat unclear with the only recent generation of three mice lacking nestin, showing phenotypic inconsistency. The first one to be published targeted the coding region of nestin exon 1 by homologous recombination and the mice showed perinatal lethality (Park et al., 2010). These transgenic mice died due to high apoptosis-rate of neuronal precursors during development (Park et al., 2010). Later, Mohseni and colleagues (2011) published a viable and grossly normal nestin knockout mouse, where most of exon 1 and part of the 5′-untranslated region were deleted by gene targeting. The knockout mouse by Mohseni et al. (2011) did, however, show aberrant NMJ structure and motor coordination problems. Mohseni and coauthors (2011) generated two separate and viable knockout mouse strains, which were further analyzed in two different genetic backgrounds, supporting that nestin deficiency would not be lethal. In line with Mohseni et al. (2011), a third nestin-deficient mouse was generated by ubiquitous lentiviral expression of nestin small hairpin RNA (shRNA) plasmid, causing a disruption of NMJ organization in nestin-downregulated mice (Yang et al., 2011). Several independent lines were created also in this case. Also the shRNA mice were viable and grossly normal. Although the knockdown was efficient, it was not complete in all tissues. There are several factors that may underlie the differing phenotypes in the transgenic animals, such as off-target effects due to recombination consequently leading to lethality in the first mouse (Park et al., 2010), or incomplete knockout in the second mouse model (Mohseni et al., 2011). Up to the present time, the mouse models are not available commercially.

The cellular models for isolation of nestin-expressing primary cells from mice are relatively limited. For cells of myogenic origin, primary myoblasts can be isolated from limb skeletal muscles of postnatal and adult mice according to Danoviz and Yablonka-Reuveni (2012). Briefly, the muscles are isolated, minced and digested in a solution of dispase II and collagenase A for 60 min. Myoblasts are released by serial trituration and filtration of the slurry. Cells are then centrifuged and reconstituted in growth media (20% fetal bovine serum, 10% horse serum, 1% chicken embryo extract in high glucose DMEM with standard supplements and 1 m*M* sodium pyruvate). Myoblasts can be plated on gelatin-coated cell culture dishes or growth factor-reduced Matrigel-coated coverslips. Differentiation of primary myoblast cultures is induced at 80% confluency by serum depletion (1% horse serum in DMEM including supplements). Given that the muscle tissue is directly minced, the result is a myoblast culture containing of a variety of cell types, such as fibroblasts and adipocytes. Alternatively, purer myoblast cultures can be achieved through the isolation of live myofibers (Keire, Shearer, Shefer, & Yablonka-Reuveni, 2013), to which satellite cells are still attached. In culture, the satellite cells can be allowed to migrate out from the fiber, or alternatively, the myofibers can be grown in floating conditions, allowing satellite cells to activate and proliferate on the surface of the myofibers in a more natural environment. While the myoblast cultures from minced muscles are characterized by the presence of other cell types, the method allows the isolation of great amounts of material suitable for biochemical analysis. On the contrary, the myofiber explant cultures are, due to their high purity, well suited for imaging-based methods. In our laboratory, both isolation techniques have proven to be highly functional for studying nestin. In addition, mice that express GFP under the control of regulatory elements of the gene encoding nestin have been in other laboratories successfully utilized for the isolation and identification progenitor cells of different origin, including muscle (Day, Shefer, Richardson, Enikolopov, & Yablonka-Reuveni, 2007). Besides myogenic cell cultures, nestin-expressing cells are particularly prominent in postnatally isolated neurospheres, that have in other laboratories been found to be well suited for analysis of proliferation, self-renewal, and differentiation of neuronal stem cells into various lineages (Gil-Perotín et al., 2013; Wilhelmsson et al., 2012).

4. CONCLUSIONS

The prominent PTMs of IFs have in recent years been of utmost interest in the field, as virtually every member of the family is subjected to PTMs (Snider & Omary, 2014). Phosphorylation along with other PTMs that affect the solubility, charges, and hydrophobicity play a major role in regulating polymerization and dynamics of nestin. Phosphorylation also regulates nestin interaction with partner proteins, such as the signaling complex of Cdk5, which was highlighted in this treatise. In this respect, it is evident that PTMs are not only required for regulation of IF assembly but also changes in solubility orchestrate cellular signaling pathways. Nestin levels are undergoing continuous regulation on the protein level, and ubiquitination, often in concert with other PTMs, has a prime role in regulation of nestin protein level.

ACKNOWLEDGMENTS

J.L. was supported by Turku Graduate School for Biomedical Sciences and Turku Doctoral Program in Molecular Biosciences. The described research has been supported by Sigrid Jusélius Foundation, the Academy of Finland, the Åbo Akademi Foundation, the Finnish Cancer Foundations, and the Magnus Ehrnrooth Foundation.

REFERENCES

Akiyama, M., Matsuda, Y., Ishiwata, T., Naito, Z., & Kawana, S. (2013). Inhibition of the stem cell marker nestin reduces tumor growth and invasion of malignant melanoma. *Journal of Investigative Dermatology*, *133*(5), 1384–1387. http://dx.doi.org/10.1038/jid.2012.508.

Amoh, Y., Yang, M., Li, L., Reynoso, J., Bouvet, M., Moossa, A. R., et al. (2005). Nestin-linked green fluorescent protein transgenic nude mouse for imaging human tumor angiogenesis. *Cancer Research*, *65*(12), 5352–5357. http://dx.doi.org/10.1158/0008-5472.CAN-05-0821.

Carlsson, L., Li, Z., Paulin, D., & Thornell, L. E. (1999). Nestin is expressed during development and in myotendinous and neuromuscular junctions in wild type and desmin knock-out mice. *Experimental Cell Research, 251*(1), 213–223. http://dx.doi.org/10.1006/excr.1999.4569.

Chen, Z., Wang, J., Cai, L., Zhong, B., Luo, H., Hao, Y., et al. (2014). Role of the stem cell-associated intermediate filament nestin in malignant proliferation of non-small cell lung cancer. *PLoS One, 9*(2), e85584. http://dx.doi.org/10.1371/journal.pone.0085584.

Chen, H.-L., Yuh, C.-H., & Wu, K. K. (2010). Nestin is essential for zebrafish brain and eye development through control of progenitor cell apoptosis. *PLoS One, 5*(2), e9318. http://dx.doi.org/10.1371/journal.pone.0009318.

Cheng, L., Jin, Z., Liu, L., Yan, Y., Li, T., Zhu, X., et al. (2004). Characterization and promoter analysis of the mouse nestin gene. *FEBS Letters, 565*(1–3), 195–202. http://dx.doi.org/10.1016/j.febslet.2004.03.097.

Chou, Y.-H., Khuon, S., Herrmann, H., & Goldman, R. D. (2003). Nestin promotes the phosphorylation-dependent disassembly of vimentin intermediate filaments during mitosis. *Molecular Biology of the Cell, 14*(4), 1468–1478. http://dx.doi.org/10.1091/mbc.E02-08-0545.

Contreras-Vallejos, E., Utreras, E., Bórquez, D. A., Prochazkova, M., Terse, A., Jaffe, H., et al. (2014). Searching for novel Cdk5 substrates in brain by comparative phosphoproteomics of wild type and Cdk5 −/− mice. *PLoS One, 9*(3), e90363. http://dx.doi.org/10.1371/journal.pone.0090363.

Dahlstrand, J., Zimmerman, L. B., McKay, R. D., & Lendahl, U. (1992). Characterization of the human nestin gene reveals a close evolutionary relationship to neurofilaments. *Journal of Cell Science, 103*(Pt 2), 589–597.

Daniel, C., Albrecht, H., Lüdke, A., & Hugo, C. (2008). Nestin expression in repopulating mesangial cells promotes their proliferation. *Laboratory Investigation; A Journal of Technical Methods and Pathology, 88*(4), 387–397. http://dx.doi.org/10.1038/labinvest.2008.5.

Danoviz, M. E., & Yablonka-Reuveni, Z. (2012). Skeletal muscle satellite cells: Background and methods for isolation and analysis in a primary culture system. *Methods in Molecular Biology (Clifton, NJ), 798*, 21–52. http://dx.doi.org/10.1007/978-1-61779-343-1_2.

Davidoff, M. S., Middendorff, R., Enikolopov, G., Riethmacher, D., Holstein, A. F., & Müller, D. (2004). Progenitor cells of the testosterone-producing Leydig cells revealed. *Journal of Cell Biology, 167*(5), 935–944. http://dx.doi.org/10.1083/jcb.200409107.

Day, K., Shefer, G., Richardson, J. B., Enikolopov, G., & Yablonka-Reuveni, Z. (2007). Nestin-GFP reporter expression defines the quiescent state of skeletal muscle satellite cells. *Developmental Biology, 304*(1), 246–259. http://dx.doi.org/10.1016/j.ydbio.2006.12.026.

De Thonel, A., Ferraris, S. E., Pallari, H.-M., Imanishi, S. Y., Kochin, V., Hosokawa, T., et al. (2010). Protein kinase C zeta regulates Cdk5/p25 signaling during myogenesis. *Molecular Biology of the Cell, 21*(8), 1423–1434. http://dx.doi.org/10.1091/mbc.E09-10-0847.

Eliasson, C., Sahlgren, C., Berthold, C.-H., Stakeberg, J., Celis, J. E., Betsholtz, C., et al. (1999). Intermediate filament protein partnership in astrocytes. *Journal of Biological Chemistry, 274*(34), 23996–24006. http://dx.doi.org/10.1074/jbc.274.34.23996.

Frisén, J., Johansson, C. B., Török, C., Risling, M., & Lendahl, U. (1995). Rapid, widespread, and longlasting induction of nestin contributes to the generation of glial scar tissue after CNS injury. *Journal of Cell Biology, 131*(2), 453–464.

Fu, A. K., Fu, W. Y., Cheung, J., Tsim, K. W., Ip, F. C., Wang, J. H., et al. (2001). Cdk5 is involved in neuregulin-induced AChR expression at the neuromuscular junction. *Nature Neuroscience, 4*(4), 374–381. http://dx.doi.org/10.1038/86019.

Fu, A. K. Y., Ip, F. C. F., Fu, W.-Y., Cheung, J., Wang, J. H., Yung, W.-H., et al. (2005). Aberrant motor axon projection, acetylcholine receptor clustering, and neurotransmission in cyclin-dependent kinase 5 null mice. *Proceedings of the National Academy of Sciences of the United States of America, 102*(42), 15224–15229. http://dx.doi.org/10.1073/pnas.0507678102.

Gil-Perotín, S., Duran-Moreno, M., Cebrián-Silla, A., Ramírez, M., García-Belda, P., & García-Verdugo, J. M. (2013). Adult neural stem cells from the subventricular zone: A review of the neurosphere assay. *Anatomical Record (Hoboken, N.J.: 2007)*, *296*(9), 1435–1452. http://dx.doi.org/10.1002/ar.22746.

Hockfield, S., & McKay, R. D. (1985). Identification of major cell classes in the developing mammalian nervous system. *Journal of Neuroscience: The Official Journal of the Society for Neuroscience*, *5*(12), 3310–3328.

Hyder, C. L., Lazaro, G., Pylvänäinen, J. W., Roberts, M. W. G., Qvarnström, S. M., & Eriksson, J. E. (2014). Nestin regulates prostate cancer cell invasion by influencing the localisation and functions of FAK and integrins. *Journal of Cell Science*, *127*(Pt 10), 2161–2173. http://dx.doi.org/10.1242/jcs.125062.

Ishiwata, T., Matsuda, Y., & Naito, Z. (2011). Nestin in gastrointestinal and other cancers: Effects on cells and tumor angiogenesis. *World Journal of Gastroenterology*, *17*(4), 409–418. http://dx.doi.org/10.3748/wjg.v17.i4.409.

Ishiwata, T., Teduka, K., Yamamoto, T., Kawahara, K., Matsuda, Y., & Naito, Z. (2011). Neuroepithelial stem cell marker nestin regulates the migration, invasion and growth of human gliomas. *Oncology Reports*, *26*(1), 91–99. http://dx.doi.org/10.3892/or.2011.1267.

Jalouli, M., Lapierre, L. R., Guérette, D., Blais, K., Lee, J.-A., Cole, G. J., et al. (2010). Transitin is required for the differentiation of avian QM7 myoblasts into myotubes. *Developmental Dynamics: An Official Publication of the American Association of Anatomists*, *239*(11), 3038–3047. http://dx.doi.org/10.1002/dvdy.22448.

Jia, L., Ji, S., Maillet, J.-C., & Zhang, X. (2010). PTEN suppression promotes neurite development exclusively in differentiating PC12 cells via PI3-kinase and MAP kinase signaling. *Journal of Cellular Biochemistry*, *111*(6), 1390–1400. http://dx.doi.org/10.1002/jcb.22867.

Kachinsky, A. M., Dominov, J. A., & Miller, J. B. (1994). Myogenesis and the intermediate filament protein, nestin. *Developmental Biology*, *165*(1), 216–228. http://dx.doi.org/10.1006/dbio.1994.1248.

Kang, H., Tian, L., Son, Y.-J., Zuo, Y., Procaccino, D., Love, F., et al. (2007). Regulation of the intermediate filament protein nestin at rodent neuromuscular junctions by innervation and activity. *Journal of Neuroscience: The Official Journal of the Society for Neuroscience*, *27*(22), 5948–5957. http://dx.doi.org/10.1523/JNEUROSCI.0621-07.2007.

Kawaguchi, A., Miyata, T., Sawamoto, K., Takashita, N., Murayama, A., Akamatsu, W., et al. (2001). Nestin-EGFP transgenic mice: Visualization of the self-renewal and multipotency of CNS stem cells. *Molecular and Cellular Neurosciences*, *17*(2), 259–273. http://dx.doi.org/10.1006/mcne.2000.0925.

Keire, P., Shearer, A., Shefer, G., & Yablonka-Reuveni, Z. (2013). Isolation and culture of skeletal muscle myofibers as a means to analyze satellite cells. *Methods in Molecular Biology (Clifton, N.J.)*, *946*, 431–468. http://dx.doi.org/10.1007/978-1-62703-128-8_28.

Kishaba, Y., Matsubara, D., & Niki, T. (2010). Heterogeneous expression of nestin in myofibroblasts of various human tissues. *Pathology International*, *60*(5), 378–385. http://dx.doi.org/10.1111/j.1440-1827.2010.02532.x.

Kleeberger, W., Bova, G. S., Nielsen, M. E., Herawi, M., Chuang, A.-Y., Epstein, J. I., et al. (2007). Roles for the stem cell associated intermediate filament Nestin in prostate cancer migration and metastasis. *Cancer Research*, *67*, 9199–9206.

Lazaro, J. B., Kitzmann, M., Poul, M. A., Vandromme, M., Lamb, N. J., & Fernandez, A. (1997). Cyclin dependent kinase 5, cdk5, is a positive regulator of myogenesis in mouse C2 cells. *Journal of Cell Science*, *110*(Pt 1), 1251–1260.

Lee, C.-W., Zhan, Q., Lezcano, C., Frank, M. H., Huang, J., Larson, A. R., et al. (2014). Nestin depletion induces melanoma matrix metalloproteinases and invasion. *Laboratory Investigation; A Journal of Technical Methods and Pathology*, *94*, 1382–1395. http://dx.doi.org/10.1038/labinvest.2014.130.

Lin, W., Dominguez, B., Yang, J., Aryal, P., Brandon, E. P., Gage, F. H., et al. (2005). Neurotransmitter acetylcholine negatively regulates neuromuscular synapse formation by a Cdk5-dependent mechanism. *Neuron*, *46*(4), 569–579. http://dx.doi.org/10.1016/j.neuron.2005.04.002.

Macarthur, K. M., Kao, G. D., Chandrasekaran, S., Alonso-Basanta, M., Chapman, C., Lustig, R. A., et al. (2014). Detection of brain tumor cells in the peripheral blood by a telomerase promoter-based assay. *Cancer Research*, *74*(8), 2152–2159. http://dx.doi.org/10.1158/0008-5472.CAN-13-0813.

Matsuda, Y., Hagio, M., & Ishiwata, T. (2013). Nestin: A novel angiogenesis marker and possible target for tumor angiogenesis. *World Journal of Gastroenterology*, *19*(1), 42–48. http://dx.doi.org/10.3748/wjg.v19.i1.42.

Matsuda, Y., Naito, Z., Kawahara, K., Nakazawa, N., Korc, M., & Ishiwata, T. (2011). Nestin is a novel target for suppressing pancreatic cancer cell migration, invasion and metastasis. *Cancer Biology & Therapy*, *11*(5), 512–523.

Mellodew, K., Suhr, R., Uwanogho, D. A., Reuter, I., Lendahl, U., Hodges, H., et al. (2004). Nestin expression is lost in a neural stem cell line through a mechanism involving the proteasome and Notch signalling. *Developmental Brain Research*, *151*(1–2), 13–23. http://dx.doi.org/10.1016/j.devbrainres.2004.03.018.

Mohseni, P., Sung, H.-K., Murphy, A. J., Laliberte, C. L., Pallari, H.-M., Henkelman, M., et al. (2011). Nestin is not essential for development of the CNS but required for dispersion of acetylcholine receptor clusters at the area of neuromuscular junctions. *Journal of Neuroscience: The Official Journal of the Society for Neuroscience*, *31*(32), 11547–11552. http://dx.doi.org/10.1523/JNEUROSCI.4396-10.2011.

Mokrý, J., Cízková, D., Filip, S., Ehrmann, J., Osterreicher, J., Kolár, Z., et al. (2004). Nestin expression by newly formed human blood vessels. *Stem Cells and Development*, *13*(6), 658–664. http://dx.doi.org/10.1089/scd.2004.13.658.

Ohshima, T., Ward, J. M., Huh, C. G., Longenecker, G., Veeranna, Pant, H. C., et al. (1996). Targeted disruption of the cyclin-dependent kinase 5 gene results in abnormal corticogenesis, neuronal pathology and perinatal death. *Proceedings of the National Academy of Sciences of the United States of America*, *93*(20), 11173–11178.

Pallari, H.-M., Lindqvist, J., Torvaldson, E., Ferraris, S. E., He, T., Sahlgren, C., et al. (2011). Nestin as a regulator of Cdk5 in differentiating myoblasts. *Molecular Biology of the Cell*, *22*(9), 1539–1549. http://dx.doi.org/10.1091/mbc.E10-07-0568.

Park, D., Xiang, A. P., Mao, F. F., Zhang, L., Di, C.-G., Liu, X.-M., et al. (2010). Nestin is required for the proper self-renewal of neural stem cells. *Stem Cells (Dayton, Ohio)*, *28*(12), 2162–2171. http://dx.doi.org/10.1002/stem.541.

Paul, M., & Skalli, O. (2016). Synemin: Tools to study and characterize binding partners by proximal ligation assay. *Methods in Enzymology*, (in press).

Perry, J., Ho, M., Viero, S., Zheng, K., Jacobs, R., & Thorner, P. S. (2007). The intermediate filament nestin is highly expressed in normal human podocytes and podocytes in glomerular disease. *Pediatric and Developmental Pathology: The Official Journal of the Society for Pediatric Pathology and the Paediatric Pathology Society*, *10*(5), 369–382. http://dx.doi.org/10.2350/06-11-0193.1.

Sahlgren, C. M., Mikhailov, A., Hellman, J., Chou, Y. H., Lendahl, U., Goldman, R. D., et al. (2001). Mitotic reorganization of the intermediate filament protein nestin involves phosphorylation by cdc2 kinase. *Journal of Biological Chemistry*, *276*(19), 16456–16463. http://dx.doi.org/10.1074/jbc.M009669200.

Sahlgren, C. M., Mikhailov, A., Vaittinen, S., Pallari, H.-M., Kalimo, H., Pant, H. C., et al. (2003). Cdk5 regulates the organization of Nestin and its association with p35. *Molecular and Cellular Biology*, *23*(14), 5090–5106.

Sahlgren, C. M., Pallari, H.-M., He, T., Chou, Y.-H., Goldman, R. D., & Eriksson, J. E. (2006). A nestin scaffold links Cdk5/p35 signaling to oxidant-induced cell death. *EMBO Journal, 25*(20), 4808–4819. http://dx.doi.org/10.1038/sj.emboj.7601366.

Scobioala, S., Klocke, R., Kuhlmann, M., Tian, W., Hasib, L., Milting, H., et al. (2008). Up-regulation of nestin in the infarcted myocardium potentially indicates differentiation of resident cardiac stem cells into various lineages including cardiomyocytes. *FASEB Journal: Official Publication of the Federation of American Societies for Experimental Biology, 22*(4), 1021–1031. http://dx.doi.org/10.1096/fj.07-8252com.

Sejersen, T., & Lendahl, U. (1993). Transient expression of the intermediate filament nestin during skeletal muscle development. *Journal of Cell Science, 106*(Pt 4), 1291–1300.

Sjöberg, G., Edström, L., Lendahl, U., & Sejersen, T. (1994). Myofibers from Duchenne/Becker muscular dystrophy and myositis express the intermediate filament nestin. *Journal of Neuropathology and Experimental Neurology, 53*(4), 416–423.

Sjöberg, G., Jiang, W. Q., Ringertz, N. R., Lendahl, U., & Sejersen, T. (1994). Colocalization of nestin and vimentin/desmin in skeletal muscle cells demonstrated by three-dimensional fluorescence digital imaging microscopy. *Experimental Cell Research, 214*(2), 447–458. http://dx.doi.org/10.1006/excr.1994.1281.

Snider, N. T., & Omary, M. B. (2014). Post-translational modifications of intermediate filament proteins: Mechanisms and functions. *Nature Reviews Molecular Cell Biology, 15*(3), 163–177. http://dx.doi.org/10.1038/nrm3753.

Snider, N. T., & Omary, M. B. (2016). Assays for posttranslational modifications of intermediate filament proteins. *Methods in Enzymology, 568*, 113–138.

Steinert, P. M., Chou, Y.-H., Prahlad, V., Parry, D. A. D., Marekov, L. N., Wu, K. C., et al. (1999). A high molecular weight intermediate filament-associated protein in BHK-21 cells is nestin, a type VI intermediate filament protein: Limited co-assembly in vitro to form heteropolymers with type III vimentin and type IV a-internexin. *Journal of Biological Chemistry, 274*(14), 9881–9890. http://dx.doi.org/10.1074/jbc.274.14.9881.

Su, P.-H., Chen, C.-C., Chang, Y.-F., Wong, Z.-R., Chang, K.-W., Huang, B.-M., et al. (2013). Identification and cytoprotective function of a novel nestin isoform, Nes-S, in dorsal root ganglia neurons. *Journal of Biological Chemistry, 288*(12), 8391–8404. http://dx.doi.org/10.1074/jbc.M112.408179.

Su, H.-T., Weng, C.-C., Hsiao, P.-J., Chen, L.-H., Kuo, T.-L., Chen, Y.-W., et al. (2013). Stem cell marker nestin is critical for TGF-β1-mediated tumor progression in pancreatic cancer. *Molecular Cancer Research, 11*(7), 768–779. http://dx.doi.org/10.1158/1541-7786.MCR-12-0511.

Takakuwa, O., Maeno, K., Kunii, E., Ozasa, H., Hijikata, H., Uemura, T., et al. (2013). Involvement of intermediate filament nestin in cell growth of small-cell lung cancer. *Lung Cancer (Amsterdam, the Netherlands), 81*(2), 174–179. http://dx.doi.org/10.1016/j.lungcan.2013.04.022.

Tschaharganeh, D. F., Xue, W., Calvisi, D. F., Evert, M., Michurina, T. V., Dow, L. E., et al. (2014). p53-dependent Nestin regulation links tumor suppression to cellular plasticity in liver cancer. *Cell, 158*(3), 579–592. http://dx.doi.org/10.1016/j.cell.2014.05.051.

Uehara, F., Tome, Y., Miwa, S., Hiroshima, Y., Yano, S., Yamamoto, M., et al. (2014). Osteosarcoma cells enhance angiogenesis visualized by color-coded imaging in the in vivo Gelfoam® assay. *Journal of Cellular Biochemistry, 115*(9), 1490–1494. http://dx.doi.org/10.1002/jcb.24799.

Vaittinen, S., Lukka, R., Sahlgren, C., Hurme, T., Rantanen, J., Lendahl, U., et al. (2001). The expression of intermediate filament protein nestin as related to vimentin and desmin in regenerating skeletal muscle. *Journal of Neuropathology and Experimental Neurology, 60*(6), 588–597.

Vaittinen, S., Lukka, R., Sahlgren, C., Rantanen, J., Hurme, T., Lendahl, U., et al. (1999). Specific and innervation-regulated expression of the intermediate filament protein nestin at neuromuscular and myotendinous junctions in skeletal muscle. *The American Journal of Pathology*, *154*(2), 591–600. http://dx.doi.org/10.1016/S0002-9440(10)65304-7.

Wakamatsu, Y., Nakamura, N., Lee, J.-A., Cole, G. J., & Osumi, N. (2007). Transitin, a nestin-like intermediate filament protein, mediates cortical localization and the lateral transport of Numb in mitotic avian neuroepithelial cells. *Development (Cambridge, England)*, *134*(13), 2425–2433. http://dx.doi.org/10.1242/dev.02862.

Wilhelmsson, U., Faiz, M., de Pablo, Y., Sjöqvist, M., Andersson, D., Widestrand, A., et al. (2012). Astrocytes negatively regulate neurogenesis through the Jagged1-mediated Notch pathway. *Stem Cells (Dayton, Ohio)*, *30*(10), 2320–2329. http://dx.doi.org/10.1002/stem.1196.

Wong, Z.-R., Su, P.-H., Chang, K.-W., Huang, B.-M., Lee, H., & Yang, H.-Y. (2013). Identification of a rod domain-truncated isoform of nestin, Nes-S$\Delta_{107\text{-}254}$, in rat dorsal root ganglia. *Neuroscience Letters*, *553*, 181–185. http://dx.doi.org/10.1016/j.neulet.2013.08.035.

Yang, J., Cheng, L., Yan, Y., Bian, W., Tomooka, Y., Shiurba, R., et al. (2001). Mouse nestin cDNA cloning and protein expression in the cytoskeleton of transfected cells. *BBA. Gene Structure and Expression*, *1520*(3), 251–254. http://dx.doi.org/10.1016/S0167-4781(01)00275-5.

Yang, J., Dominguez, B., de Winter, F., Gould, T. W., Eriksson, J. E., & Lee, K.-F. (2011). Nestin negatively regulates postsynaptic differentiation of the neuromuscular synapse. *Nature Neuroscience*, *14*(3), 324–330. http://dx.doi.org/10.1038/nn.2747.

Zhao, Z., Lu, P., Zhang, H., Xu, H., Gao, N., Li, M., et al. (2014). Nestin positively regulates the Wnt/ß-catenin pathway and the proliferation, survival, and invasiveness of breast cancer stem cells. *Breast Cancer Research: BCR*, *16*(4), 408. http://dx.doi.org/10.1186/s13058-014-0408-8.

Zimmerman, L., Parr, B., Lendahl, U., Cunningham, M., McKay, R., Gavin, B., et al. (1994). Independent regulatory elements in the nestin gene direct transgene expression to neural stem cells or muscle precursors. *Neuron*, *12*(1), 11–24.

CHAPTER NINETEEN

Synemin: Molecular Features and the Use of Proximity Ligation Assay to Study Its Interactions

Madhumita Paul, Omar Skalli[1]
Department of Biological Sciences, The University of Memphis, Memphis, Tennessee, USA
[1]Corresponding author: e-mail address: oskalli@memphis.edu

Contents

1. Synemin 538
   1.1 Structure and Isoforms 538
   1.2 Assembly Properties 539
   1.3 Tissue and Cell Distribution 540
   1.4 Roles and Interacting Partners 543
2. Reagents and Tools to Study Synemin 545
   2.1 Antibodies 545
   2.2 Synemin cDNAs, siRNA, shRNAs, and CRISPR Reagents 546
   2.3 Cell Lines and Tissues 546
   2.4 Purified Synemin and Immunoprecipitation Protocols 546
3. PLA for *In Situ* Detection of Synemin Interaction with Binding Partners 546
   3.1 Choosing Primary Antibodies and Optimizing Their Dilutions 547
   3.2 Performing the PLA Reaction 549
   3.3 Controls 551
4. Conclusions 552
References 552

Abstract

Synemin has three splice variants (α, β, and L) with identical head and rod domains but with tail domains of varying size. α- and β-Synemin are larger than most intermediate filament proteins (1565 and 1253 amino acids, respectively) but L-synemin is shorter (339 amino acids). Synemin isoforms do not self-assemble into filaments but can copolymerize with vimentin and desmin. Synemin is present in all muscle cell types, in a few neural cell types, and in various other nonepithelial cell types. Synemin expression is regulated, sometimes in an isoform-specific manner, during development of the nervous system, in brain and breast cancer cells and during injuries to the brain and liver. Mice-lacking synemin develop a myopathic phenotype, possibly due to synemin role in linking desmin filaments to costameres and sarcomeres. Synemin may play this role through its demonstrated binding to costameric and sarcolemmal proteins, such as

Methods in Enzymology, Volume 568
ISSN 0076-6879
http://dx.doi.org/10.1016/bs.mie.2015.08.005

α-actinin, vinculin, and members of the dystroglycan complex. In astrocytoma cells, synemin regulates proliferation by interacting with PP2A to modulate Akt phosphorylation status. Methods to identify synemin binding partners are central to understand the roles of this protein in diverse cell types. Here, we describe how to use proximal ligation assays (PLA) for this purpose. PLA complement biochemical methods such as immunoprecipitation by relying on the use of antibodies conjugated to oligonucleotide probes to visualize by fluorescence microscopy protein–protein interactions in cells and tissues.

1. SYNEMIN

1.1 Structure and Isoforms

Synemin was initially characterized as a high-molecular-weight protein copurifying with desmin intermediate filaments (IFs) from chicken smooth muscle and is less abundant than desmin (98:2 molar ratio of desmin to synemin) in this tissue (Granger & Lazarides, 1980). These early results suggested that synemin was an IF-associated protein rather than an IF protein and accordingly coined the name from the Greek *syn* (with) and *nema* (filament) (Granger & Lazarides, 1980). Sequencing of avian (Becker, Bellin, Sernett, Huiatt, & Robson, 1995; Bellin et al., 1999) and mammalian (Mizuno, Thompson, et al., 2001; Titeux et al., 2001; Xue et al., 2004) synemin cDNAs revealed, however, that synemin is a bona fide IF protein since it possesses structural features common to all IF proteins, i.e., a central α-helical domain of 310 amino acids interrupted at specific positions by short non-α-helical regions (linkers). Synemin sequence shares the most similarity with nestin providing the rationale to group these two proteins into a class VI IF proteins (Guérette, Khan, Savard, & Vincent, 2007). More often, however, synemin is classified as a type IV IF protein, together with nestin and neurofilament proteins (Peter & Stick, 2015). The human synemin gene localizes on chromosome 15q26.3 and displays some sequence polymorphism (Mizuno, Puca, O'Brien, Beggs, & Kunkel, 2001; Titeux et al., 2001). Thus far, no synemin mutation has been associated with diseases.

In mammals, three synemin isoforms (α, β, and L) arise from alternative splicing (Titeux et al., 2001; Xue et al., 2004; Fig. 1). Compared to most other IF proteins, the head domain of synemin isoforms is about 10 times shorter, whereas the tail domain of α- and β-synemin is about 10 times larger and that of L-synemin is shorter (Fig. 1). Human β-synemin was initially named desmulin (Mizuno, Thompson, et al., 2001) because at the time only avian synemin sequence was known, and it differed substantially from that of

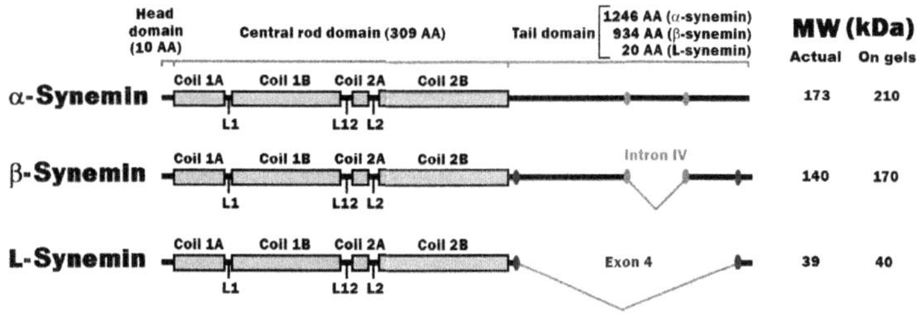

Figure 1 The three synemin isoforms have identical head and central rod domain, but differ in the tail domain. β-Synemin contains the sequence of all the five exons of the synemin gene, whereas α-synemin contains in addition the sequence encoded by intron IV. In L-synemin, exon 4 is spliced out and intron IV sequence is absent. The number of amino acids (AA) and molecular weight (MW) are those of human synemin isoforms. MW is indicated as calculated from the protein sequence (actual) and as it appears after SDS-polyacrylamide gel electrophoresis (on gels).

its human homolog. The name "desmulin," however, still often appears in "omics" databases and vendor's catalogs, and this should be taken into account when mining databases for information about synemin or purchasing reagents for synemin such as antibodies and siRNAs.

1.2 Assembly Properties

Unlike most IF proteins, purified α-synemin does not assemble into 10 nm filaments under physiological conditions but instead forms globular structures of ~30 nm in diameter (Bilak et al., 1998; Granger & Lazarides, 1982; Hirako et al., 2003). Similarly, in cells, α-synemin does not assemble into IFs but instead forms punctate structures as determined by transfecting α-synemin cDNA into SW13-C2 cells (a cell line lacking cytoplasmic IF proteins) (Bellin et al., 1999; Titeux et al., 2001). Similar experiments also established that β- and L-synemin do not polymerize into IFs (Khanamiryan, Li, Paulin, & Xue, 2008). Synemin inability to self-assemble into filaments is due to the combination of three factors: the short size of the head domain, the lack of a TAAL motif in helix 2A, and a modified TYRKLLEGEE polymerization motif in helix 2B (Khanamiryan et al., 2008). This was demonstrated by the finding that β- and L-synemin become self-assembly competent when their short 10 amino acid head domain is substituted by the ~100 amino acid long vimentin head domain, and a TAAL motif is inserted into helix 2A while the TYRALEGES sequence in helix 2B is replaced by the canonical IF protein polymerization motif TYRKLLEGEE (Khanamiryan et al., 2008).

In spite of its inability to self-polymerize, synemin can copolymerize with type III vimentin and desmin both *in vitro* and *in vivo* (Bellin et al., 1999; Jing et al., 2007; Titeux et al., 2001). In avian erythrocytes, synemin incorporation into vimentin IFs occurs at periodically spaced foci, and this spacing is developmentally regulated (Granger & Lazarides, 1982). At high synemin to desmin molar ratio (6:25), synemin acts as a capping protein preventing desmin filament elongation (Hirako et al., 2003). Synemin does not copolymerize with type III GFAP but does bind to preformed GFAP IF (Jing et al., 2007). Overlay assays have also shown that synemin binds to type II K5 and K6 but not to type I K14 (Hirako et al., 2003). However, whether synemin actually copolymerizes with these keratins has yet to be examined.

1.3 Tissue and Cell Distribution

In the normal adult, synemin is primarily found in contractile cells, namely striated and smooth muscle (Bilak et al., 1998; Granger & Lazarides, 1980) and myoepithelial cells (Hirako et al., 2003; Table 1). β-Synemin is the predominant isoform in contractile cells (Mizuno, Thompson, et al., 2001). In striated muscle cells, synemin localizes at the circumference of Z-discs (Bilak et al., 1998; Granger & Lazarides, 1980) and M lines (Prudner, Roy, Damron, & Russell, 2014). Synemin also associates with muscle junctional complexes including intercalated discs in cardiocytes (Bellin, Huiatt, Critchley, & Robson, 2001) as well as costameres and neuromuscular and myotendinous junctions in myocytes (Bellin et al., 2001; Carlsson et al., 2000; Mizuno et al., 2004). In desmin-null mice, synemin no longer associates with Z-discs but remains present in myotendinous and neuromuscular junctions, suggesting that synemin association with junctional complexes, but not with Z-discs, is desmin independent (Carlsson et al., 2000). This could be due to differential synemin isoforms distribution since α-synemin preferentially localizes in cell–cell junctions while β-synemin predominates in Z-discs (Lund et al., 2012).

Besides muscles, synemin is present in the adult nervous systemin subsets of neurons in the spinal cord and peripheral nervous system and in some glial cells including Schwann cells and astrocytes in the retina and optic nerve (Hirako et al., 2003; Izmiryan et al., 2006; Luna, Lewis, Banna, Skalli, & Fisher, 2010; Mizuno, Guyon, Okamoto, & Kunkel, 2007, 2009; Tawk et al., 2003; Table 1). Multipotent neural stem cells in the adult brain subventricular zone are also synemin positive (de Souza Martins et al., 2011).

Table 1 Synemin-Positive Cell Types Grouped According to the Cytoplasmic IF Proteins That They Contain in Addition To Synemin (Type of IF Protein is Indicated in Brackets After Protein Name)

Synemin-positive cell types	Myocytes Cadiomyocytes Smooth muscle cells	Myoblasts Cardiomyoblasts	Hepatic stellate cells	Lens cells Endothelial cells Avian erythrocytes	Myoepithelial cells	Retina and optic nerve astrocytes Schwann cells Ependymal cells	Astrocytes during development SVZ neural stem cells	Subpopulations of neurons in the pons and medulla Subpopulations of neurons in the anterior horn of the spinal cord Purkinje cells	Neurons in the peripheral nervous system
IF protein(s) coexpressed with synemin	Desmin (III)	Desmin (III) Vimentin (III) Nestin (IV)	Desmin (III) GFAP (III)	Vimentin (III)	CK14 (I), CK5/6 (II/II)	GFAP (III) Vimentin (III)	GFAP (III) Vimentin (III) Nestin (IV)	NF triplet proteins (IV)	Peripherin (III)

This table shows that synemin is always coexpressed with other cytoplasmic IF proteins and is present in cell types belonging to the five major lineages (muscle, mesenchymal, epithelial, glial, and neural).

α-Synemin is generally the predominant isoform in the nervous system (Sultana, Sernett, Bellin, Robson, & Skalli, 2000).

Some cell types other than muscle and neural cells also contain synemin. These include endothelial cells (Izmiryan, Franco, Paulin, Li, & Xue, 2009; Schmitt-Graeff, Jing, Nitschke, Desmoulière, & Skalli, 2006), lens cells (Granger & Lazarides, 1984; Tawk et al., 2003), hepatic stellate cells (Schmitt-Graeff et al., 2006; Uyama et al., 2006), and avian erythrocytes (Granger & Lazarides, 1982).

Synemin is developmentally regulated in the nervous system and, to a lesser extent, in muscle. During rat brain development, synemin protein level steadily increases from embryonic stages to peak 2 days after birth and to rapidly decrease thereafter (Sultana et al., 2000). This peak of synemin expression corresponds to the maturation of radial glial cells into astrocytes, a transition during which these cells coexpress synemin, GFAP, vimentin, and nestin (Sultana et al., 2000). While synemin expression is widespread in differentiating astrocytes, most mature astrocytes do not contain synemin, with few exceptions (Hirako et al., 2003). Synemin is developmentally regulated in an isoform-specific manner in most types of peripheral neurons and during the maturation of embryonic stem cells into various neural precursors (de Souza Martins et al., 2011; Izmiryan et al., 2006; Table 1).

In striated muscles, synemin is coexpressed with vimentin and nestin during early development and persists as myoblasts differentiate and replace nestin and vimentin with desmin (Lund et al., 2012; Price & Lazarides, 1983). Quantitative changes, however, take place during this process since both α- and β-synemin increase after birth in the heart (Lund et al., 2012).

Diseases also alter synemin expression. In contrast to most mature astrocytes, which are synemin negative, astrocytoma cells express synemin (Jing, Pizzolato, Robson, Gabbiani, & Skalli, 2005) and about 1/3 of glioblastomas (the most malignant type of astrocytomas) coexpress high levels of synemin, vimentin, GFAP, and nestin (Skalli et al., 2013). In astrocytoma cells, synemin distributes over the vimentin IF network and is also enriched in membrane domains involved in cell motility (Jing et al., 2005). In normal breast myoepithelial cells, synemin levels are elevated but drop dramatically in breast cancer due to the tumor-specific methylation of the synemin gene promoter (Noetzel et al., 2010). Lastly, synemin is present in reactive cells that develop in diseases involving physical trauma or metabolic dysfunction to the brain, retina, and liver (Jing et al., 2005, 2007; Luna et al., 2010; Pekny et al., 2014; Schmitt-Graeff et al., 2006; Uyama et al., 2006).

Myopathies modify synemin distribution in myocytes (Mizuno, Puca, et al., 2001; Olivé et al., 2003). In myopathies caused by desmin mutations, synemin associates with cytoplasmic desmin aggregates (Carlsson et al., 2002; Olivé et al., 2003). In nondesmin myopathies, synemin accumulates under the sarcolemma and/or forms diffuse cytoplasmic aggregates (Olivé et al., 2003).

1.4 Roles and Interacting Partners

Synemin role has been explored in striated muscle and in astrocytoma cells. Genetic ablation of synemin in mice induces a myopathic phenotype affecting mostly fast twitch muscles, with impairment in growth and regenerative capacity and increased susceptibility to injury (García-Pelagio et al., 2015; Li et al., 2014). Synemin-null mice also present defects in sarcomere organization (Li et al., 2014), although this may depend on genetic background (García-Pelagio et al., 2015). The structural integrity and mechanical properties of the sarcolemma are also impacted by the absence of synemin (García-Pelagio et al., 2015; Li et al., 2014). Similarly, siRNA-mediated synemin downregulation in neonatal cardiomyocytes disrupts cell–cell junctions and desmin IFs alignment with Z-lines (Lund et al., 2012).

The phenotype of the muscles of synemin-null mice may be attributed to the disruption of the interactions between synemin and its binding partners at the level of sarcomeres and sarcolemma. These partners include α-actinin, plectin, and zyxin (Bellin et al., 1999; Hijikata et al., 2008; Jing et al., 2005; Sun, Huiatt, Paulin, Li, & Robson, 2010), which bind both α- and β-synemin, and vinculin, talin, and titin, which bind β-synemin only (Bellin et al., 2001; Lund et al., 2012; Prudner et al., 2014; Sun, Critchley, Paulin, Li, & Robson, 2008). In addition, synemin interacts with α-dystrobrevin, dystrophin, and utrophin, which are components of the membrane dystroglycan complex in muscle cells (Bhosle, Michele, Campbell, Li, & Robson, 2006; Mizuno, Thompson, et al., 2001). Synemin is also an A-kinase-anchoring protein (AKAP) (Russell et al., 2006) and as such may influence muscle growth and repair by regulating the phosphorylation of proteins at the Z-disks and sarcolemma via protein kinase A (García-Pelagio et al., 2015). In addition, synemin also interacts with Akt (Pitre, Davis, Paul, Orr, & Skalli, 2012), a kinase important for the normal growth of fast twitch muscles (García-Pelagio et al., 2015).

In astrocytoma cells, synemin downregulation by shRNAs dramatically alters cell shape and decreases motility and proliferation (Pan, Jing, Pitre,

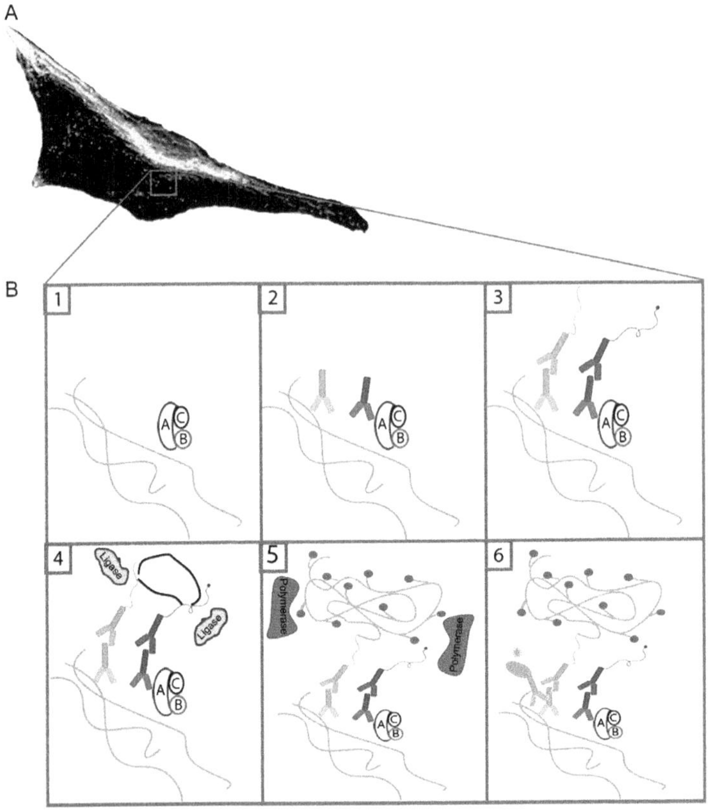

Figure 2 Diagram showing the different steps of a PLA assay. (A) Micrograph of U87 cells transfected with β-synemin, processed for PLA with anti-synemin and anti-PP2A (gray dots are the PLA reaction product) and counterstained for synemin (light gray filaments). (B) In panels 1–6, synemin is represented by light gray filaments and PP2A by its A–C subunits. Panel 1 shows the proximity of synemin with PP2A. Panel 2: the first step of the PLA reaction is the incubation of fixed and permeabilized cells with primary antibodies against synemin and PP2A A subunit. Panel 3: next, cells are incubated with secondary antibodies (PLA probes PLUS and MINUS) conjugated to oligonucleotides. Note that the PLA probe that will not be used for amplification in step 4 has a blocked 2′-*O*-methyl RNA (dark gray dot in secondary PLA probe binding to the primary for PP2A) at the 3′ end of its conjugated oligonucleotide. Panel 4: ligation step in which connector oligonucleotides (black lines) and ligase are added to the cells to ligate the oligonucleotides attached to the PLA probes into a circular nucleic acid structure. Panel 5: following ligation, amplification solution is added to the cells to initiate rolling circle amplification using as primer the oligonucleotide which does not have the 3′ blocked. Fluorescently labeled oligonucleotides (gray ovals) hybridize to complementary tag sequences in the amplified product. Panel 6: cells are counterstained with fluorescently labeled secondary antibody (light gray oval indicated by *) directed against the anti-synemin primary antibody.

Williams, & Skalli, 2008). This is accompanied by decreased filamentous actin and associated α-actinin, suggesting a role for synemin in actin dynamics (Pan et al., 2008). In addition, Akt activity is much lower in synemin-silenced astrocytoma cells, accounting for the decreased proliferation of these cells (Pitre et al., 2012). The decreased Akt activity is linked to increased activity of PP2A, the major phosphatase regulating Akt dephosphorylation. Synemin, in fact, binds PP2A and the effect of synemin downregulation on astrocytoma cell proliferation is rescued by PP2A inhibitors (Pitre et al., 2012).

Synemin interacting partners have been identified through various methods, including yeast two-hybrid screens, immunoprecipitation, *in vitro* binding assays with purified proteins, and more recently proximal ligation assays (PLA) (Pitre et al., 2012; Fig. 2). This latter method is the most recent, and it will be described because it enables to assess protein–protein interactions in fixed and permeabilized cells (Fredriksson et al., 2002) in contrast to immunoprecipitation which takes into account only proteins soluble in physiological buffer.

2. REAGENTS AND TOOLS TO STUDY SYNEMIN

Some reagents and tools to study synemin may be listed under the name desmuslin by vendors (see Section 1.1).

2.1 Antibodies

Our laboratory has produced and affinity purified rabbit and goat antibodies against synemin that are available upon request (Jing et al., 2005; Skalli et al., 2013). In addition, we have tested a series of commercial antibodies, including a rabbit polyclonal against α-desmulin (Novus Biological, Cat # NBP2-37892), mouse monoclonal A8 against desmulin (Santa Cruz Biotech, Cat # sc-374484), and goat polyclonal E15 against desmulin (Santa Cruz Biotech, Cat # sc-49649). This latter antibody should not be used for synemin studies as it yields an actin-like pattern by immunofluorescence and does not stain by Western blot. In contrast, monoclonal A8 yields superb immunofluorescence staining and only labels α- and β-synemin by Western blot. The rabbit polyclonal from Novus yields a diffuse synemin staining, and stains α- and β-synemin as well as other bands by Western blotting.

2.2 Synemin cDNAs, siRNA, shRNAs, and CRISPR Reagents

Molecular biology constructs to express, downregulate, or knock down synemin can be obtained from a variety of vendors such as Sigma, Origen, and Santa Cruz biotech.

2.3 Cell Lines and Tissues

Primary cell cultures expressing synemin include cardiomyocytes (Lund et al., 2012; Prudner et al., 2014), myotubes (Granger & Lazarides, 1980), embryonic stem cells (de Souza Martins et al., 2011), embryonic neurons (Izmiryan et al., 2009), and neonatal astrocytes (Sultana et al., 2000).

Cell lines expressing synemin include HeLa cells (Sun et al., 2010), several astrocytoma lines such as U373 and A172 cells (Jing et al., 2005), A10 vascular smooth muscle cells (Sun et al., 2008), C2C12 myoblast (Hijikata et al., 2008), and SW13 clone 1 cells (Bellin et al., 1999).

Tissues rich in synemin include heart, muscle, and embryonic brain.

2.4 Purified Synemin and Immunoprecipitation Protocols

Synemin can be purified from muscles (Bilak et al., 1998; Sandoval, Colaco, & Lazarides, 1983) or from bacterial expression systems (Jing et al., 2007).

Immunoprecipitation and pull-down protocols have been developed by different laboratories (Hijikata et al., 2008; Jing et al., 2005; Pitre et al., 2012; Sun et al., 2008).

3. PLA FOR *IN SITU* DETECTION OF SYNEMIN INTERACTION WITH BINDING PARTNERS

The first step of PLA is the incubation of the cells with a pair of primary antibodies, one against synemin and the other against a potential synemin-interacting partner (Fig. 2). After washes, cells are incubated with appropriate secondary antibodies each conjugated to oligonucleotides with a unique sequence (Fig. 2). Next, cells are incubated with two-circle-forming DNA oligonucleotides in the presence of a ligase. If the nucleotides attached to each of the secondary antibodies are in close proximity (less than 40 nm apart) (Trifilieff et al., 2011; Zatloukal et al., 2014), they will hybridize with the two-circle-forming DNA oligonucleotides, and the ligase will catalyze formation of a closed circle (Fig. 2). An amplification step is then performed in which the oligonucleotide arm of one of the secondary antibodies serves as primer for a rolling circle amplification reaction catalyzed by a polymerase and using the closed circle as template. This generates a long concatemeric copy of the closed DNA circle formed during the ligation reaction (Fig. 2). During amplification, fluorescently labeled oligonucleotides complementary to a tag sequence on the concatemer hybridize with the concatemer to generate fluorescent dots that can be imaged with a fluorescence microscope. The presence of such fluorescent dots indicates that synemin and the protein tested are close by, either because they bind directly to each other or because they are separated by a short distance in a multiprotein complex.

Probes and reagents to perform PLA can be obtained from Sigma Inc. (St. Louis, MO) either individually or as the Duolink *In Situ* Detection Reagent kit. The manufacturer provides information about some of the buffers comprised in the kit. This includes buffer A (1 ×): 0.01 *M* Tris, 0.15 *M* NaCl, and 0.05% Tween 20, pH 7.4, and buffer B (1 ×): 0.2 *M* Tris and 0.1 *M* NaCl, pH 7.5. Both solutions should be filtered through a 0.22 μm filter and stored at 4 °C. In addition, the ligation and amplification reaction reagents can be prepared as described by Jarvius et al. (2007) and Soderberg et al. (2006). While the manufacturer does not disclose the composition of the blocking solution and antibody diluent included in the kit, some laboratories prepare their own (Jarvius et al., 2007; Soderberg et al., 2006). Finally, users can custom design their PLA probes by conjugating amine modified oligonucleotides (Jarvius et al., 2007; Soderberg et al., 2006) to secondary antibodies of their choice with the Duolink Probemaker kit (Sigma Co., St. Louis, MO).

3.1 Choosing Primary Antibodies and Optimizing Their Dilutions

Similar to double immunofluorescence, the two primary antibodies used for PLA must be raised in two different species. In addition, it is critical to first control the specificity of these antibodies and their suitability for immunostaining and then to optimize their dilution before proceeding to PLA.

Step 1. Verifying antibody specificity

To obtain reliable PLA information, it is essential to first evaluate the specificity of the primary antibodies by Western blotting even if the antibodies' manufacturer claims that they are highly specific. These experiments should demonstrate that, in cell type(s) in which PLA is to be performed, the antibodies strongly label their target antigens and do not recognize any other proteins. Antibodies failing to demonstrate specificity should not be used for

PLA as they may generate false-positive signals by recognizing proteins other than those of interest.

Step 2. Verifying antibody suitability for immunostaining

After validation of the specificity of the primaries, double immunofluorescence staining of the cell type(s) to be used for PLA is performed to evaluate whether both primaries bind to their respective antigen under the same set of fixation and permeabilization conditions. After permeabilization, perform a 30 min blocking step at 37 °C in a humid chamber using the PLA blocking buffer provided by the manufacturer. All antibodies should be diluted in PLA diluent provided by the manufacturer. This will ensure that the blocking agents included in blocking and dilution buffers do not interfere with antibody binding and/or cause background artifacts. Antibody pairs yielding the expected staining patterns under these conditions are acceptable for PLA.

Step 3. Optimizing antibody dilutions

PLA is much more sensitive than immunofluorescence due to the rolling circle amplification step. Thus, dilutions of primary antibodies for PLA should generally be higher than those for immunofluorescence to avoid background and/or false-positive signals.

Each primary antibody should be optimized separately for PLA. This is done by using the fixation, permeabilization, and blocking conditions working for immunofluorescence staining, but with a dilution range of the primary antibody spanning from 10 to 200 times of the dilution for immunofluorescence staining. For instance, with a primary antibody diluted at 1:50 for immunofluorescence, PLA dilutions of 1:500, 1:1000, 1:2000, 1:4000, 1:8000, and 1:16,000 should be tested. The goal of this titration is to determine the dilution at which the PLA signal is extinguished and then to use half that dilution for the PLA experiment. For instance, if no or very few PLA fluorescent products are observed at 1:8000, the primary antibody should be diluted at 1:4000 in the actual experiment.

After incubation with the primary, the PLA procedure described below is followed. In this instance, the two secondary PLA probes used are conjugated to the minus and plus oligonucleotides and are against IgGs of the species in which the primary antibody was produced. Using the dilution determined for the primary, it is also useful to test three dilutions of the secondary, including 1:5 (recommended by the manufacturer), 1:10, and 1:20 and to choose the dilution that results in less than 200 dots per cell.

3.2 Performing the PLA Reaction

Unless otherwise stated, all the steps are carried out at room temperature (20–25 °C)

1. Grow cells of interest in appropriate culture medium on No. 1.5 sterile cover glasses.
2. On the day of the experiment, use fine point jeweler forceps to lift the cover glasses off the wells. Briefly wash cover glasses by gently dipping in and out a 100 ml beaker containing 70 ml PBS. From this point on, keep track of the side of the cover glass onto which the cells are attached.
3. Transfer cover glasses to a cover glass staining jar (Electron Microscopy Sciences, Cat # 72242-21) containing 4% formaldehyde in PBS (diluted from 20% formaldehyde solution, Tousimis, Cat # 1008A). Fix for 5 min.
4. Transfer cover glasses to a staining jar containing PBS and let sit 5 min.
5. Transfer cover glasses to a staining jar containing 0.1% NP-40 or Triton-X-100 (v/v, PBS) and incubate for 5 min. This solubilizes the plasma membrane and enables antibodies and other reagents to enter the cell interior in subsequent steps.
6. Transfer cover glasses to a staining jar containing PBS and let sit 5 min.
7. In a humid chamber, lay cover glasses flat, cells facing up and place 40 μl/cm^2 blocking buffer on the cells. Incubate for 30 min at 37 °C.
8. Rinse off blocking buffer by dipping cover glass two to three times in a beaker containing PBS.
9. Return cover glasses to the humid chamber and place 40 μl/cm^2 of the two primary antibodies diluted in dilution buffer at the dilution determined in preliminary experiments. Incubate for 1 h at room temperature.
10. Tap off the primary antibodies solution from the cover glasses and rinse them 2 × 5 min with wash buffer A included in the PLA kit.
11. Return cover glasses to the humid chamber, place 40 μl/cm^2 of complementary PLA probes diluted in dilution buffer at the dilution determined in preliminary experiments. Incubate for 1 h at 37 °C.

    Choice of the PLA probes: for instance, if one of the primaries was raised in rabbit and the other in mice, the PLA probes should be donkey anti-mouse IgG Plus probe and donkey anti-rabbit IgG Minus probe. The opposite combination of probes (donkey anti-mouse IgG Minus probe and donkey anti-rabbit IgG Plus probe) would work as well.

12. Tap off the PLA probe solution and rinse cover glasses 2 × 5 min with wash buffer A.
13. Return cover glasses to the humid chamber and incubate for 30 min at 37 °C with 40 μl/cm^2 of ligase diluted 1:40 in ligation buffer provided by the manufacturer.
14. Tap off ligation solution off the cover glasses and wash them in 1 × wash buffer A for 2 × 2 min.
15. Return cover glasses to the humid chamber and incubate for 100 min at 37 °C with 40 μl/cm^2 of polymerase diluted 1:80 in amplification solution provided by the manufacturer.
16. Tap off amplification solution from the cover glasses and wash them in 1 × wash buffer B (provided by the manufacturer) for 2 × 10 min.
17. Perform a 1 min high-stringency wash in 0.01 × wash buffer B.
18. At this stage, a counterstaining step can be performed. This can be done, for instance, by incubating the cells for 10 min at room temperature with fluorescently labeled phalloidin diluted 1:100 in PBS, or with a 1:50 dilution (in PBS) of a fluorescently labeled secondary antibody against one of the primaries (Fig. 2). The fluorescent conjugate attached to the antibody or phalloidin must emit light of a different wavelength than the fluorescent PLA dots. After this counterstaining step, wash cells 2 × 2 min in PBS.
19. Without allowing the cells to dry, mount cover glasses with Prolong Diamond Antifade with or without DAPI (Molecular Probes) and let the mounting medium harden overnight. Alternatively, cover glasses can be mounted using Duolink II mounting medium with DAPI. Note that DAPI is a nuclear DNA fluorescent dye.
20. Observe the cells with a wide-field fluorescent microscope rather than with a confocal scanning laser microscope (CSLM). Alternatively, use a CSLM to obtain optical sections z-stacks and make sure to use settings such that these sections will cover most of the cell volume. This will enable observation of PLA dots present throughout the entire cytoplasmic volume, which is especially important when carrying out quantification.
21. Any image analysis tool can be used for quantifying PLA signals as the number of fluorescent dots per cell. Typically, use the region of interest (ROI) tool of the analysis software and define the interior of the cells with the PLA signals to quantify. After defining the ROIs, proper thresholding should be performed to remove background noise and ensure that only PLA signals from within the defined ROIs are counted by the analysis software. Alternatively, quantify the number of PLA dots per cell with the Duolink Image Tool software (Olink Bioscience Co.).

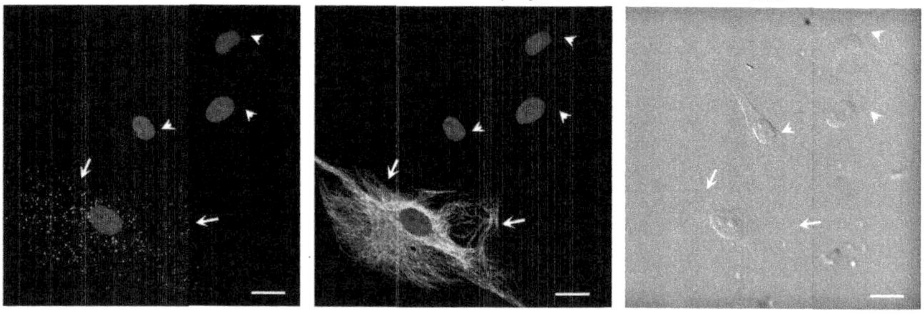

Figure 3 PLA demonstrating the interaction between β-synemin and PP2A A subunit in U87 human glioblastoma cells (obtained from the ATCC), which was previously reported by our laboratory (Pitre et al., 2012). U87 cells were transfected with β-synemin tagged with the V5 epitope and PLA was performed with mouse anti-V5 and rabbit anti-PP2A A subunit (A). Following the PLA reaction, cells were stained with anti-mouse IgG conjugated to Alexa Fluor 488 to identify the cells transfected with the synemin construct as they contain a filamentous synemin network (light gray filaments) (B). Nuclear DNA was stained with DAPI and appears dark gray on the micrographs (A and B). PLA dots in the cell transfected with synemin (arrow) demonstrate interaction between synemin and PP2A A subunit. The absence of PLA dots in the cells not expressing the synemin construct (arrowhead) indicates the specificity of the reaction. (C) DIC, differential interference contrast. Bars = 10 μm. (See the color plate.)

3.3 Controls

Control reactions should be performed to ascertain the specificity of PLA signals. First, one of the primary antibodies is omitted and the specimen is incubated with two appropriate PLA probes. In another control reaction, the specimen is incubated with only the PLA probes and no primary antibodies. Under both conditions, PLA dots should be absent.

If possible, PLA reactions should also be performed in cells in which one of the proteins under investigation is not expressed or is downregulated by siRNA, resulting in an absence of PLA reaction products. Perhaps the best control is obtained by performing PLA with cells transfected with the cDNA coding for one of the proteins of interest: evidence for specific PLA reaction is obtained by the presence of dots in transfected but not in nontransfected cells (Fig. 3). Validation of PLA results obtained with one set of primary antibodies can also be provided by showing positive PLA reaction with a different set of primary antibodies. Finally, biochemical methods such as immunoprecipitation and/or binding assays with purified proteins should be used to confirm PLA results. It should be kept in mind, however, that some positive PLA results may reflect interactions between proteins insoluble in immunoprecipitation buffer, in which case immunoprecipitation does not represent a practical validation method.

4. CONCLUSIONS

Current knowledge on synemin indicates that it regulates cellular behaviors pertaining to growth and survival. Synemin appears to function in these capacities by interacting with a surprising number of partners, a diversity enabled by the large size of the tail domain of α- and β-synemin. To further our understanding of synemin roles, it is important to characterize *in situ* the interaction of synemin with its partners in different cell types and to map the domains involved. Similar to other IF proteins, most of the cellular synemin is insoluble in immunoprecipitation buffers. Therefore, immunoprecipitation can only detect proteins interacting with the small pool of soluble synemin. PLA overcomes this limitation because it is performed on fixed cells thereby enabling to also analyze the pool of synemin insoluble in immunoprecipitation buffers. Compared to FRET, another method for tracking protein–protein interactions in cells, PLA is simpler to implement and can also be adapted to examine interactions between different proteins in tissue samples. FRET, however, affords studies of protein–protein interactions in live cells, whereas PLA can be performed only in fixed cells.

One of the most important technical considerations for obtaining valid PLA results is to demonstrate the specificity of the antibodies to be used in PLA protocols. It is also advised not to omit the optimization steps. When a reliable PLA positive signal is obtained, one should keep in mind that this indicates that the two proteins examined are in close proximity but not necessarily that they directly bind to each other. Direct binding can be demonstrated through *in vitro* binding assays with purified proteins.

REFERENCES

Becker, B., Bellin, R. M., Sernett, S. W., Huiatt, T. W., & Robson, R. M. (1995). Synemin contains the rod domain of intermediate filaments. *Biochemical and Biophysical Research Communications*, *213*, 796–802.

Bellin, R. M., Huiatt, T. W., Critchley, D. R., & Robson, R. M. (2001). Synemin may function to directly link muscle cell intermediate filaments to both myofibrillar Z-lines and costameres. *Journal of Biological Chemistry*, *276*, 32330–32337.

Bellin, R. M., Sernett, S. W., Becker, B., Ip, W., Huiatt, T. W., & Robson, R. M. (1999). Molecular characteristics and interactions of the intermediate filament protein synemin. Interactions with alpha-actinin may anchor synemin-containing heterofilaments. *Journal of Biological Chemistry*, *274*, 29493–29499.

Bhosle, R. C., Michele, D. E., Campbell, K. P., Li, Z., & Robson, R. M. (2006). Interactions of intermediate filament protein synemin with dystrophin and utrophin. *Biochemical and Biophysical Research Communications*, *346*, 768–777.

Bilak, S. R., Sernett, S. W., Bilak, M. M., Bellin, R. M., Stromer, M. H., Huiatt, T. W., et al. (1998). Properties of the novel intermediate filament protein synemin and its identification in mammalian muscle. *Archives of Biochemistry and Biophysics, 355*, 63–76.

Carlsson, L., Fischer, C., Sjöberg, G., Robson, R. M., Sejersen, T., & Thornell, L. E. (2002). Cytoskeletal derangements in hereditary myopathy with a desmin L345P mutation. *Acta Neuropathologica, 104*, 493–504.

Carlsson, L., Li, Z. L., Paulin, D., Price, M. G., Breckler, J., Robson, R. M., et al. (2000). Differences in the distribution of synemin, paranemin, and plectin in skeletal muscles of wild-type and desmin knock-out mice. *Histochemistry and Cell Biology, 114*, 39–47.

de Souza Martins, S. C., Agbulut, O., Diguet, N., Larcher, J. C., Paulsen, B. S., Rehen, S. K., et al. (2011). Dynamic expression of synemin isoforms in mouse embryonic stem cells and neural derivatives. *BMC Cell Biology, 12*, 51.

Fredriksson, S., Gullberg, M., Jarvius, J., Olsson, C., Pietras, K., Gústafsdóttir, S. M., et al. (2002). Protein detection using proximity-dependent DNA ligation assays. *Nature Biotechnology, 20*, 473–477.

García-Pelagio, K. P., Muriel, J., O'Neill, A., Desmond, P. F., Lovering, R. M., Lund, L., et al. (2015). Myopathic changes in murine skeletal muscle lacking synemin. *American Journal of Physiology. Cell Physiology, 308*, C448–C462.

Granger, B. L., & Lazarides, E. (1980). Synemin: A new high molecular weight protein associated with desmin and vimentin filaments in muscle. *Cell, 22*, 727–738.

Granger, B. L., & Lazarides, E. (1982). Structural associations of synemin and vimentin filaments in avian erythrocytes revealed by immunoelectron microscopy. *Cell, 30*, 263–275.

Granger, B. L., & Lazarides, E. (1984). Expression of the intermediate-filament-associated protein synemin in chicken lens cells. *Molecular Cell. Biology, 4*, 1943–1950.

Guérette, D., Khan, P. A., Savard, P. E., & Vincent, M. (2007). Molecular evolution of type VI intermediate filament proteins. *BMC Evolutionary Biology, 7*, 164.

Hijikata, T., Nakamura, A., Isokawa, K., Imamura, M., Yuasa, K., Ishikawa, R., et al. (2008). Plectin 1 links intermediate filaments to costameric sarcolemma through beta-synemin, alpha-dystrobrevin and actin. *Journal of Cell Science, 121*, 2062–2074.

Hirako, Y., Yamakawa, H., Tsujimura, Y., Nishizawa, Y., Okumura, M., Usukura, J., et al. (2003). Characterization of mammalian synemin, an intermediate filament protein present in all four classes of muscle cells and some neuroglial cells: Co-localization and interaction with type III intermediate filament proteins and keratins. *Cell and Tissue Research, 313*, 195–207.

Izmiryan, A., Cheraud, Y., Khanamiryan, L., Leterrier, J. F., Federici, T., Peltekian, E., et al. (2006). Different expression of synemin isoforms in glia and neurons during nervous system development. *Glia, 54*, 204–213.

Izmiryan, A., Franco, C. A., Paulin, D., Li, Z., & Xue, Z. (2009). Synemin isoforms during mouse development: Multiplicity of partners in vascular and neuronal systems. *Experimental Cell Research, 315*, 769–783.

Jarvius, M., Paulsson, J., Weibrecht, I., Leuchowius, K.-J., Andersson, A.-C., Wahlby, C., et al. (2007). In situ detection of phosphorylated platelet derived growth factor receptor beta using a generalized proximity ligation method. *Molecular & Cell Proteomics, 6*, 1500–1509.

Jing, R., Pizzolato, G., Robson, R. M., Gabbiani, G., & Skalli, O. (2005). Intermediate filament protein synemin is present in human reactive and malignant astrocytes and associates with ruffled membranes in astrocytoma cells. *Glia, 50*, 107–120.

Jing, R., Wilhelmsson, U., Goodwill, W., Li, L., Pan, Y., Pekny, M., et al. (2007). Synemin is expressed in reactive astrocytes in neurotrauma and interacts differentially with vimentin and GFAP intermediate filament networks. *Journal of Cell Science, 120*, 1267–1277.

Khanamiryan, L., Li, Z., Paulin, D., & Xue, Z. (2008). Self-assembly incompetence of synemin is related to the property of its head and rod domains. *Biochemistry*, *47*, 9531–9539.

Li, Z., Parlakian, A., Coletti, D., Alonso-Martin, S., Hourdé, C., Joanne, P., et al. (2014). Synemin acts as a regulator of signalling molecules during skeletal muscle hypertrophy. *Journal of Cell Science*, *127*, 4589–4601.

Luna, G., Lewis, G. P., Banna, C. D., Skalli, O., & Fisher, S. K. (2010). Expression profiles of nestin and synemin in reactive astrocytes and Müller cells following retinal injury: A comparison with glial fibrillar acidic protein and vimentin. *Molecular Vision*, *16*, 2511–2523.

Lund, L. M., Kerr, J. P., Lupinetti, J., Zhang, Y., Russell, M. A., Bloch, R. J., et al. (2012). Synemin isoforms differentially organize cell junctions and desmin filaments in neonatal cardiomyocytes. *The FASEB Journal*, *26*, 137–148.

Mizuno, Y., Guyon, J. R., Okamoto, K., & Kunkel, L. M. (2007). Synemin expression in brain. *Muscle and Nerve*, *36*, 497–504.

Mizuno, Y., Guyon, J. R., Okamoto, K., & Kunkel, L. M. (2009). Expression of synemin in the mouse spinal cord. *Muscle and Nerve*, *39*, 634–641.

Mizuno, Y., Guyon, J. R., Watkins, S. C., Mizushima, K., Sasaoka, T., Imamura, M., et al. (2004). Beta-synemin localizes to regions of high stress in human skeletal myofibers. *Muscle and Nerve*, *30*, 337–346.

Mizuno, Y., Puca, A. A., O'Brien, K. F., Beggs, A. H., & Kunkel, L. M. (2001). Genomic organization and single-nucleotide polymorphism map of desmuslin, a novel intermediate filament protein on chromosome 15q26.3. *BMC Genetics*, *2*, 8.

Mizuno, Y., Thompson, T. G., Guyon, J. R., Lidov, H. G., Brosius, M., Imamura, M., et al. (2001). Desmuslin, an intermediate filament protein that interacts with alpha-dystrobrevin and desmin. *Proceedings of the National Academy of Sciences of the United States of America*, *98*, 6156–6161.

Noetzel, E., Rose, M., Sevinc, E., Hilgers, R. D., Hartmann, A., Naami, A., et al. (2010). Intermediate filament dynamics and breast cancer: Aberrant promoter methylation of the synemin gene is associated with early tumor relapse. *Oncogene*, *29*, 4814–4825.

Olivé, M., Goldfarb, L., Dagvadorj, A., Sambuughin, N., Paulin, D., Li, Z., et al. (2003). Expression of the intermediate filament protein synemin in myofibrillar myopathies and other muscle diseases. *Acta Neuropathologica*, *106*, 1–7.

Pan, Y., Jing, R., Pitre, A., Williams, B. J., & Skalli, O. (2008). Intermediate filament protein synemin contributes to the migratory properties of astrocytoma cells by influencing the dynamics of the actin cytoskeleton. *The FASEB Journal*, *22*, 3196–3206.

Pekny, T., Faiz, M., Wilhelmsson, U., Curtis, M. A., Matej, R., Skalli, O., et al. (2014). Synemin is expressed in reactive astrocytes and Rosenthal fibers in Alexander disease. *Acta Pathologica, Microbiologica et Immunologica Scandinavica*, *122*, 76–80.

Peter, A., & Stick, R. (2015). Evolutionary aspects in intermediate filament proteins. *Current Opinion in Cell Biology*, *32C*, 48–55.

Pitre, A., Davis, N., Paul, M., Orr, A. W., & Skalli, O. (2012). Synemin promotes AKT-dependent glioblastoma cell proliferation by antagonizing PP2A. *Molecular Biology of the Cell*, *23*, 1243–1253.

Price, M. G., & Lazarides, E. (1983). Expression of intermediate filament-associated proteins paranemin and synemin in chicken development. *Journal of Cell Biology*, *97*, 1860–1874.

Prudner, B. C., Roy, P. S., Damron, D. S., & Russell, M. A. (2014). α-Synemin localizes to the M-band of the sarcomere through interaction with the M10 region of titin. *FEBS Letters*, *588*, 4625–4630.

Russell, M. A., Lund, L. M., Haber, R., McKeegan, K., Cianciola, N., & Bond, M. (2006). The intermediate filament protein, synemin, is an AKAP in the heart. *Archives of Biochemistry and Biophysics*, *456*, 204–215.

Sandoval, I. V., Colaco, C. A., & Lazarides, E. (1983). Purification of the intermediate filament-associated protein, synemin, from chicken smooth muscle. Studies on its physicochemical properties, interaction with desmin, and phosphorylation. *Journal of Biological Chemistry*, *258*, 2568–2576.

Schmitt-Graeff, A., Jing, R., Nitschke, R., Desmoulière, A., & Skalli, O. (2006). Synemin expression is widespread in liver fibrosis and is induced in proliferating and malignant biliary epithelial cells. *Human Pathology*, *37*, 1200–1210.

Skalli, O., Wilhelmsson, U., Orndahl, C., Fekete, B., Malmgren, K., Rydenhag, B., et al. (2013). Astrocytoma grade IV (glioblastoma multiforme) displays 3 subtypes with unique expression profiles of intermediate filament proteins. *Human Pathology*, *44*, 2081–2088.

Soderberg, O., Gullberg, M., Jarvius, M., Ridderstrale, K., Leuchowius, K.-J., Jarvius, J., et al. (2006). Direct observation of individual endogenous protein complexes in situ by proximity ligation. *Nature Methods*, *3*, 995–1000.

Sultana, S., Sernett, S. W., Bellin, R. M., Robson, R. M., & Skalli, O. (2000). Intermediate filament protein synemin is transiently expressed in a subset of astrocytes during development. *Glia*, *30*, 143–153.

Sun, N., Critchley, D. R., Paulin, D., Li, Z., & Robson, R. M. (2008). Identification of a repeated domain within mammalian alpha-synemin that interacts directly with talin. *Experimental Cell Research*, *314*, 1839–1849.

Sun, N., Huiatt, T. W., Paulin, D., Li, Z., & Robson, R. M. (2010). Synemin interacts with the LIM domain protein zyxin and is essential for cell adhesion and migration. *Experimental Cell Research*, *316*, 491–505.

Tawk, M., Titeux, M., Fallet, C., Li, Z., Daumas-Duport, C., Cavalcante, L. A., et al. (2003). Synemin expression in developing normal and pathological human retina and lens. *Experimental Neurology*, *183*, 499–507.

Titeux, M., Brocheriou, V., Xue, Z., Gao, J., Pellissier, J. F., Guicheney, P., et al. (2001). Human synemin gene generates splice variants encoding two distinct intermediate filament proteins. *European Journal of Biochemistry*, *268*, 6435–6449.

Trifilieff, P., Rives, M. L., Urizar, E., Piskorowski, R. A., Vishwasrao, H. D., Castrillon, J., et al. (2011). Detection of antigen interactions ex vivo by proximity ligation assay: Endogenous dopamine D2-adenosine A2A receptor complexes in the striatum. *Biotechniques*, *51*, 111–118.

Uyama, N., Zhao, L., Van Rossen, E., Hirako, Y., Reynaert, H., Adams, D. H., et al. (2006). Hepatic stellate cells express synemin, a protein bridging intermediate filaments to focal adhesions. *Gut*, *55*, 1276–1289.

Xue, Z. G., Cheraud, Y., Brocheriou, V., Izmiryan, A., Titeux, M., Paulin, D., et al. (2004). The mouse synemin gene encodes three intermediate filament proteins generated by alternative exon usage and different open reading frames. *Experimental Cell Research*, *298*, 431–444.

Zatloukal, B., Kufferath, I., Thueringer, A., Landegren, U., Zatloukal, K., & Haybaeck, J. (2014). Sensitivity and specificity of in situ proximity ligation for protein interaction analysis in a model of steatohepatitis with Mallory-Denk bodies. *PloS One*, *9*, e96690.

CHAPTER TWENTY

Targeting Mitogen-Activated Protein Kinase Signaling in Mouse Models of Cardiomyopathy Caused by Lamin A/C Gene Mutations

Antoine Muchir*, Howard J. Worman[†,‡,1]

*Center of Research in Myology, UPMC-Inserm UMR974, CNRS FRE3617, Institut de Myologie, G.H. Pitie Salpetriere, Paris Cedex, France

[†]Department of Medicine, College of Physicians and Surgeons, Columbia University, New York, USA

[‡]Department of Pathology and Cell Biology, College of Physicians and Surgeons, Columbia University, New York, USA

[1]Corresponding author: e-mail address: hjw14@columbia.edu

Contents

1. Introduction 558
2. Mouse Models of Cardiomyopathy Caused by *LMNA* Mutations 560
3. Altered MAP Kinase Signaling in Hearts of *Lmna*$^{H222P/H222P}$ Mice 563
   3.1 Cardiac Transcriptomic Analysis 563
   3.2 Immuoblotting Assessment of MAP Kinase Activities in Hearts 565
   3.3 Activated/Phosphorylated MAP Kinases Translocate into the Nucleus of Cardiomyocytes 567
4. MAP Kinase Inhibitor Treatment Studies in *Lmna*$^{H222P/H222P}$ Mice 568
   4.1 MEK1/2 Inhibitors 568
   4.2 JNK Inhibitors 569
   4.3 p38α Inhibitors 569
   4.4 Treatment Protocols 570
5. Assessment of *Lmna*$^{H222P/H222P}$ Mice After Treatment with MAP Kinase Inhibitors 570
   5.1 Echocardiography 573
   5.2 Inhibition of MAP Kinase Activities in Hearts 574
   5.3 Natriuretic Peptide A Secretion 574
   5.4 Cardiac Fibrosis 575
   5.5 Survival 576
6. Conclusions 576
Acknowledgments 577
References 577

Methods in Enzymology, Volume 568
ISSN 0076-6879
http://dx.doi.org/10.1016/bs.mie.2015.07.028

Abstract

The most frequently occurring mutations in the gene encoding nuclear lamin A and nuclear lamin C cause striated muscle diseases virtually always involving the heart. In this review, we describe the approaches and methods used to discover that cardiomyopathy-causing lamin A/C gene mutations increase MAP kinase signaling in the heart and that this plays a role in disease pathogenesis. We review different mouse models of cardiomyopathy caused by lamin A/C gene mutations and how transcriptomic analysis of one model identified increased cardiac activity of the ERK1/2, JNK, and p38α MAP kinases. We describe methods used to measure the activity of these MAP kinases in mouse hearts and then discuss preclinical treatment protocols using pharmacological inhibitors to demonstrate their role in pathogenesis. Several of these kinase inhibitors are in clinical development and could potentially be used to treat human subjects with cardiomyopathy caused by lamin A/C gene mutations.

ABBREVIATIONS

ERK1/2 extracellular-signal-regulated kinase 1/2
JNK c-Jun N-terminal kinase
MAP mitogen-activated protein
MEK1/2 mitogen-activated protein kinase kinase 1/2

1. INTRODUCTION

Lamin A and lamin C are two of the protein building blocks of the nuclear lamina, a meshwork of intermediate filaments on the inner aspect of the nuclear envelope inner membrane (Aebi, Cohn, Buhle, & Gerace, 1986; Fisher, Chaudhary, & Blobel, 1986; Goldman, Maul, Steinert, Yang, & Goldman, 1986; McKeon, Kirschner, & Caput, 1986). They are encoded by the lamin A/C gene (*LMNA*) on chromosome 1q21 and expressed in most differentiated somatic cells (Lin & Worman, 1993; Wydner, McNeil, Lin, Worman, & Lawrence, 1996). Human lamin A and lamin C are identical for the first 566 amino acids with lamin C having six unique carboxyl-terminal amino acids and prelamin A, the precursor of lamin A, having 98 unique carboxyl-terminal amino acids. Prelamin A contains a cysteine–aliphatic–aliphatic–any amino acid (CAAX) motif at its carboxyl-terminus, which signals the following series of posttranslational enzymatic modifications: (1) farnesylation of the cysteine, (2) cleavage of the –AAX, and (3) methylation of the carboxyl group of the newly exposed farnesylcysteine (Bergo, Wahlstrom, Fong, & Young, 2008). This modified protein is then recognized by the zinc metalloproteinase ZMPSTE24 and cleaved 15 amino acids from the farnesylated cysteine to yield lamin A (Rusiñol & Sinensky, 2006; Young, Meta, Yang, & Fong, 2006). Lamins and the nuclear lamina have been implicated in a wide range of functions from providing structural support to the nucleus to regulating transcription and DNA replication. More recent data also demonstrate that lamins are connected to the cytoskeleton via integral proteins of the inner and outer nuclear membranes, functioning in mechanotransduction and nuclear positioning (Chang, Worman, & Gundersen, 2015).

Scientific interest in lamin A and lamin C has significantly increased in the past decade and a half since mutations in *LMNA* have been connected to a broad range of inherited diseases often called laminopathies (Worman, Fong, Muchir, & Young, 2009). Depending upon the mutation, these diseases predominantly affect either (1) striated muscle, (2) adipose tissue, (3) peripheral nerve, or (4) multiple systems generating progeroid phenotypes. The most frequently occurring *LMNA* mutations lead to striated muscle diseases virtually always involving the heart. In 1999, Bonne et al. (1999) identified *LMNA* mutations causing autosomal dominant Emery–Dreifuss muscular dystrophy. Progressive muscle weakness and wasting, contractures of the elbows, ankles, and neck; and dilated cardiomyopathy with an early onset atrioventricular conduction block are the classical clinical features. Soon after, mutations in *LMNA* were shown to cause dilated cardiomyopathy without significant skeletal muscle involvement, limb-girdle muscular dystrophy type 1B, and cardiomyopathy with variable skeletal muscle involvement (Brodsky et al., 2000; Fatkin et al., 1999; Muchir et al., 2000). Based on the case series and reports published since these initial discoveries, we now know that the same mutations in *LMNA* can cause any one of these phenotypes, overlaps of these phenotypes and congenital muscular dystrophy, with dilated cardiomyopathy as a common feature (Lu, Muchir, Nagy, & Worman, 2011).

Various cellular signaling pathways are perturbed in diseases arising from mutations in genes encoding nuclear envelope proteins including *LMNA* (Dauer & Worman, 2009). We have used mouse models of cardiomyopathy caused by *LMNA* mutations to analyze alterations in cell signaling in affected heart. In particular, our research has focused on abnormal mitogen-activated protein (MAP) kinase signaling and AKT-mTOR signaling in the

Lmna$^{H222P/H222P}$ mouse model (Choi et al., 2012; Muchir et al., 2007; Muchir, Wu, et al., 2012). Here, we review methods underlying our discoveries of altered MAP kinase signaling in hearts of mice with *Lmna* mutations and its role in the pathogenesis of cardiomyopathy.

2. MOUSE MODELS OF CARDIOMYOPATHY CAUSED BY *LMNA* MUTATIONS

Several mouse models of human laminopathies, as well as mice with selective deletions of lamin A or lamin C and altered prelamin A processing, have been generated (Stewart, Kozlov, Fong, & Young, 2007; Zhang, Kieckhaefer, & Cao, 2013). While the heart is secondarily affected in some models of progeria, several knockout and knockin mice develop a primary dilated cardiomyopathy, sometimes with accompanying skeletal muscle disease resembling muscular dystrophy (Table 1).

The original *Lmna*$^{-/-}$ mice generated by Sullivan et al. develop dilated cardiomyopathy and muscular dystrophy at an early age, dying by 6–8 weeks (Nikolova et al., 2004; Sullivan et al., 1999). One report has suggested that these mice may not be complete knockouts but rather express a truncated prelamin A (Jahn et al., 2012). Nonetheless, these mice and embryonic fibroblasts derived from them have been widely used to obtain important insights regarding the functions of lamin A and lamin C. Another gene trap *Lmna* knockout line has a shorter lifespan and does not develop left ventricular dilatation prior to death (Kubben et al., 2011).

Lmna$^{N195K/N195K}$ mice, *Lmna*$^{H222P/H222P}$ mice, and homozygous mice expressing nonfarnesylated prelamin A without lamin C all develop cardiomyopathy; however, their phenotypes vary. *Lmna*$^{N195K/N195K}$ have a significantly shorter lifespan than *Lmna*$^{H222P/H222P}$ mice or homozygous mice expressing nonfarnesylated prelamin A without lamin C (Arimura et al., 2005; Davies et al., 2010; Mounkes et al., 2005). *Lmna*$^{H222P/H222P}$ mice also develop skeletal muscle disease. *Lmna*$^{\Delta 32/\Delta 32}$ mice have a lifespan of less than 3 weeks and skeletal muscle defects but do not develop left ventricular dilatation prior to death (Bertrand et al., 2012). *Lmna*$^{H222P/H222P}$ mice and homozygous mice expressing nonfarnesylated prelamin A without lamin C have sex differences in disease severity, with male mice more severely affected than female mice.

In humans, cardiomyopathy caused by *LMNA* mutations is virtually always an autosomal dominant disease. In contrast, heterozygous *Lmna* knockout and knockin mice generally have normal lifespans. An exception

Table 1 Knockout and Knockin Mouse Models of Cardiomyopathy Caused by *LMNA* Mutations

Strain	Homozygous	Heterozygous	References
Lmna$^{-/-}$ (deletion of exons 8–11)	Lifespan 6–8 weeks, growth retardation at 2 weeks, cardiac conduction defects and left ventricular dilatation at 4–6 weeks, regional muscular dystrophy	Apparently normal lifespan, cardiac conduction defects at 10 weeks, and left ventricular dilatation at 20 weeks (Wolf et al., 2008)	Sullivan et al. (1999), Nikolova et al. (2004), and Chandar et al. (2010)
Lmna$^{GT-/-}$ (promoter trap construct inserted into intron 2)	Lifespan 2–3 weeks, postnatal growth retardation, left ventricular hypertrophy but no dilatation or conduction defects, muscular dystrophy, decreased subcutaneous adipose tissue	Apparently normal lifespan, no significant cardiac abnormalities	Kubben et al. (2011)
Lmna$^{H222P/H222P}$	Male mice median survival 28 weeks, growth retardation at 12 weeks, progressive left ventricular dilatation starting at 8–10 weeks, conduction defects at 12 weeks, regional muscular dystrophy 16–20 weeks; female mice live longer and have later onset of abnormalities	Apparently normal lifespan, no reported cardiac abnormalities	Arimura et al. (2005)
Lmna$^{N195K/N195K}$	Lifespan 12–16 weeks, growth retardation at 4 weeks, left	Apparently normal lifespan, no reported cardiac abnormalities	Mounkes, Kozlov, Rottman, and Stewart (2005)

Continued

Table 1 Knockout and Knockin Mouse Models of Cardiomyopathy Caused by *LMNA* Mutations—cont'd

Strain	Homozygous	Heterozygous	References
	ventricular dilatation at 8 weeks (when diameter normalized to body mass), conduction defects by 8 weeks, mild skeletal muscle disease		
$Lmna^{nPLAO/nPLAO}$ (expresses nonfarnesylated lamin A only and no lamin C)	Male mice median survival 38.5 weeks, left ventricular dilatation (age of onset not specified), conduction not analyzed, no apparent skeletal muscle disease; female mice live longer and have less severe disease	Not characterized	Davies et al. (2010)
$Lmna^{\Delta 32/\Delta 32}$	Median survival 15 days (all die by 19 days), growth retardation at 5 days, decreased heart weight but no left ventricular dilatation at 14 days, skeletal myopathy disease at birth	Lifespan 35–70 weeks, left ventricular dilatation by 30–40 weeks, no conduction defects, no skeletal muscle disease	Bertrand et al. (2012) and Cattin et al. (2013)

is $Lmna^{\Delta 32/+}$ mice, which die between 30 and 70 weeks of age and develop dilated cardiomyopathy (Cattin et al., 2013). Although they apparently have a normal lifespan, the $Lmna^{+/-}$ mice generated by Sullivan et al. have been reported to develop cardiac conduction defects and late-onset left ventricular dilation at around 20 weeks (Chandar et al., 2010; Sullivan et al., 1999). It is unclear if other heterozygous lines develop late-onset heart abnormalities, as careful phenotyping of older mice has not been reported in the literature.

For most of our research, we have used *Lmna*$^{H222P/H222P}$ mice generated by Arimura et al. (2005). We have found this to be a useful model, as the mice do not develop heart abnormalities until 8–10 weeks of age and grow to a size, when they are much easier to analyze than those with earlier-onset disease and shorter lifespans. By 16 weeks of age, male *Lmna*$^{H222P/H222P}$ mice also have significant cardiac fibrosis, which occurs in humans with cardiomyopathy caused by *LMNA* mutations (Holmström et al., 2011; Raman, Sparks, Baker, McCarthy, & Wooley, 2007). In contrast to the mice expressing nonfarnesylated prelamin A without lamin C, *Lmna* H222P corresponds to a naturally occurring human disease-causing mutation. Because of the sex differences in disease severity, we have mostly utilized male *Lmna*$^{H222P/H222P}$ mice. The male mice develop clinically detectable cardiomyopathy at an earlier age than female mice, making it less time consuming and more cost-effective to carry out experiments. One drawback of male *Lmna*$^{H222P/H222P}$ mice is that at 16–20 weeks of age they develop myopathy that involves the diaphragm as well as other muscle groups, which makes it difficult to determine if cardiomyopathy is the only process responsible for their death.

3. ALTERED MAP KINASE SIGNALING IN HEARTS OF *Lmna*$^{H222P/H222P}$ MICE

3.1 Cardiac Transcriptomic Analysis

To determine if altered signal transduction occurs in cardiomyopathy caused by *LMNA* mutation, we carried out a transcriptomic analysis of hearts of *Lmna*$^{H222P/H222P}$ male mice at 10 weeks of age. This age is concurrent with the onset of left ventricular dysfunction (Muchir et al., 2007; Muchir, Wu, et al., 2012). We first examined similarities in transcription profiles between hearts from wild-type ($n=8$), *Lmna*$^{H222P/+}$ ($n=7$), and *Lmna*$^{H222P/H222P}$ ($n=6$) mice using hierarchical cluster analysis. This analysis revealed a strong consistency between replicates and yielded 104 probe sets in hearts from *Lmna*$^{H222P/+}$ mice and 114 in hearts from *Lmna*$^{H222P/H222P}$ mice with statistically significant differences in expressed compared to wild-type mouse hearts. There were 57 similar probe sets between hearts from *Lmna*$^{H222P/H222P}$ and *Lmna*$^{H222P/+}$ mice.

We then analyzed Gene Ontology terms applied to genes to identify functional related groups differentially expressed in hearts of mutant mice compared to control mice. This analysis revealed significant differences in expression of genes in several groups, including those encoding proteins in the extracellular-signal-regulated kinase 1/2 (ERK1/2), c-Jun N-terminal kinase (JNK), and p38α kinase branches of the MAP kinase pathway. This can be visualized using the KEGG-based pathway visualization tool (Arakawa, Kono, Yamada, Mori, & Tomita, 2005; Fig. 1). The stepwise protocol used for the transcriptomic analysis as follows:

I. Left ventricles from wild-type, $Lmna^{H222P/+}$, and $Lmna^{HH222P/H222P}$ male mice were isolated after cervical dislocation at 10 weeks of age.

II. A rotor-stator homogenizer (Omni International) was used to disrupt 20 mg of tissue in 350 μl of buffer RNeasy Lysis Buffer (Qiagen) containing 1% β-mercaptoethanol until the tissue was uniformly homogeneous. We centrifuged the lysate for 3 min at 13,000 × *g* and transferred the supernatant to a microcentrifuge tube. We then added 1 volume of 70% ethanol and mixed immediately by pipetting. We transferred up to 700 μl of the sample to an RNeasy spin column placed in a 2-ml

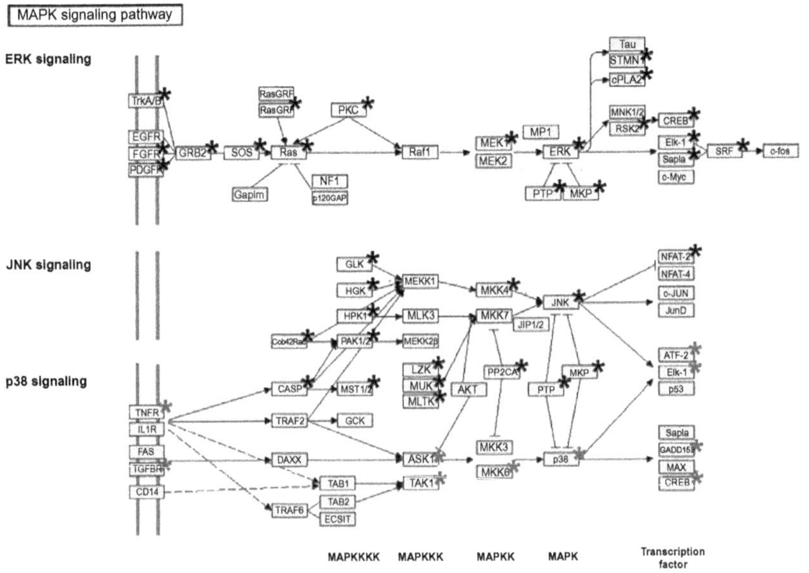

Figure 1 Genes in the MAP kinase signaling pathway (MAPK signaling pathway) identified using the KEGG-based pathway visualization tool. These include the ERK1/2 (ERK signaling), JNK (JNK signaling), and p38α (p38 signaling) branches within this GO term. Black asterisks indicate genes in the ERK1/2 and JNK branches with statistically significant differences ($q < 0.05$) in expression detected on Affymetrix Mouse Genome 430 2.0 Arrays in hearts of $Lmna^{H222P/H222P}$ mice compared to wild-type controls. Red asterisks (dark gray in the print version) indicate genes with statistically differences in expression genes in the p38α branch. *From Muchir, Wu, et al. (2012), by permission of Oxford University Press.*

collection tube, which was centrifuged for 15 s at 8000 × *g*. We discarded the flow-through. We add 700 μl of Buffer RNeasy Wash 1 to the RNeasy spin column and centrifuged for 15 s at 8000 × *g*. We discarded the flow-through. We then added 500 μl of Buffer RPE (Qiagen) and centrifuged for 15 s at 8000 × *g*. We discarded the flow-through. We again added 500 μl of Buffer RPE and centrifuged for 2 min at 8000 × *g*. We discarded the flow-through. We placed the RNeasy spin column in a 1.5-ml tube and added 30 μl of RNAse-free water and centrifuged for 1 min at 8000 × *g*. Adequacy and integrity of extracted RNA were determined by gel electrophoresis and concentrations were measured by ultraviolet absorption spectroscopy (OD 260 and 280 nm).

III. To analyze RNA transcripts, we used GeneChip Mouse Genome 430 2.0 Arrays (Affymetrix). Methods for target preparation, hybridization, fluidics station setup, array washing staining, and probe array scanning were those in the GeneChip Expression Analysis Technical Manual (http://media.affymetrix.com/support/downloads/manuals/expression_analysis_technical_manual.pdf).

IV. Image files were obtained through Affymetrix GeneChip software and analyzed by robust multichip analysis using Affymetrix microarray "*.cel*" image files and GeneTraffic 3.0 software (Stratagene). Genes were identified as being differentially expressed if they met a false discovery rate threshold of $q < 0.05$ in a two-tailed Student's *t*-test and showed at least a twofold difference in expression independent of absolute signal intensity.

V. Gene expression changes related to functional groups were analyzed using the class score method in ErmineJ (version 2.1.12; http://www.bioinformatics.ubc.ca/ermineJ/) (Lee, Braynen, Keshav, & Pavlidis, 2005) and the database for annotation, visualization, and integrated discovery program (http://david.abcc.ncifcrf.gov/) (Dennis et al., 2003). Alterations in expression of genes encoding component of the MAP kinase pathway, specifically the ERK1/2, JNK, and p38α branches, were visualized using the KEGG-based pathway visualization tool (Arakawa et al., 2005).

3.2 Immuoblotting Assessment of MAP Kinase Activities in Hearts

Transcriptomic data suggesting that expression of genes in the MAP kinase signaling pathway are altered in hearts of *Lmna*$^{H222P/H222P}$ mice at 10 weeks of age, led us to test whether the activities of the proteins were altered in heart tissue. We used immunoblotting analysis to directly demonstrate increased phosphorylation (activation) of ERK1/2, JNK, and p38α using the following protocol.

I. Hearts were excised from *Lmna*$^{H222P/H222P}$ and *Lmna*$^{+/+}$ mice. Left ventricles were dissected and snap-frozen in liquid nitrogen-cooled isopentane and then homogenized using a roto-stator homogenizer (Omni International) in extraction buffer (25 m*M* Tris–HCl [pH 7.4], 150 m*M* NaCl, 5 m*M* ethylenediaminetetraacetic acid, 10 m*M* sodium pyrophosphate, 1 m*M* Na_3VO_4, 1% SDS, 1 m*M* dithiothreitol) containing 25 mg/ml aprotinin and 10 mg/ml leupeptin.

II. Proteins in homogenates (20 μg) were separated by SDS–polyacrylamide gel electrophoresis (Laemmli, 1970), transferred to nitrocellulose membranes (0.45 μ*M*, Invitrogen), and blotted with optimal dilutions of primary antibodies against ERK1/2 (No sc94, Santa Cruz Biotechnology), phosphorylated ERK1/2 (No 9101, Cell Signaling Technology), JNK (No sc474, Santa Cruz Biotechnology), phosphorylated JNK (No 9251, Cell Signaling), p38α (No 9212, Cell Signaling), or phosphorylated p38α (No 9216, Cell Signaling) in Tris-buffered saline containing 0.1% polysorbate 20 and incubated at room temperature overnight. Antibodies against β-tubulin, β-actin, or GADPH were used as internal controls to normalize the amounts of protein between immunoblots. Blots were then washed in Tris-buffered saline containing 0.1% polysorbate 20, labeled with horseradish peroxidase-conjugated secondary antibodies, and washed again.

III. Recognized proteins were visualized by enhanced chemiluminescence (GE Healthcare Life Sciences). Band densities were calculated using Scion Imaging software (version alpha 4.0.3.2), and kinase activity was estimated by calculating the ratio of phosphorylated MAP kinase/total MAP kinase.

This analysis showed increased ratios of phosphorylated to total JNK, ERK1/2, and p38α in hearts of male *Lmna*$^{H222P/H222P}$ mice compared to those from *Lmna*$^{+/+}$ mice (Fig. 2). For ERK1/2, this increased activation occurs in mice as young as 4 weeks of age and for p38α in mice as young as 8 weeks (Choi et al., 2012; Muchir, Wu, et al., 2012).

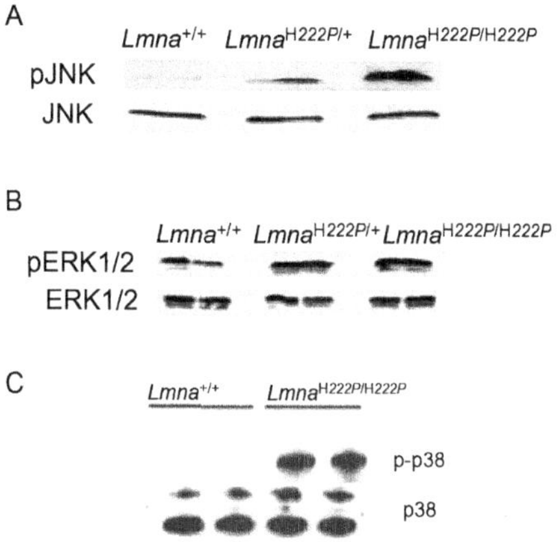

Figure 2 Increased phosphorylated to total ERK1/2, JNK, and p38α in hearts of male *Lmna*$^{H222P/H222P}$ mice. (A) Immunoblot showing phosphorylated JNK (pJNK) and total JNK in protein extracts of hearts from 10-week-old *Lmna*$^{+/+}$, *Lmna*$^{H222P/+}$, and *Lmna*$^{H222P/H222P}$ male mice. (B) Immunoblot showing phosphorylated ERK1/2 (pERK1/2) and total ERK1/2 in protein extracts of hearts from 10-week-old *Lmna*$^{+/+}$, *Lmna*$^{H222P/+}$, and *Lmna*$^{H222P/H222P}$ male mice. (C) Immunoblot showing phosphorylated p38α (p-p38) (pJNK) and total p38α (p38) in protein extracts of hearts from 8-week-old *Lmna*$^{+/+}$ and *Lmna*$^{H222P/H222P}$ male mice; samples from two hearts of each genotype are shown. *(A) and (B) are from Muchir et al. (2007), by permission of the American Society for Clinical Investigation and (C) is from Muchir, Wu, et al. (2012), by permission of Oxford University Press.*

3.3 Activated/Phosphorylated MAP Kinases Translocate into the Nucleus of Cardiomyocytes

Activated/phosphorylated MAP kinases translocate to the nucleus where they modulate the activities of target genes (Plotnikov, Zehorai, Procaccia, & Seger, 2011). We therefore examined nuclear translocation of ERK1/2 in mouse hearts by using an antibody that recognized phosphorylated ERK1/2 in tissue sections. We showed that the phosphorylated form of ERK1/2 is predominantly located in the nucleus of cardiomyocytes. Immunofluorescence microscopic analysis of heart sections from *Lmna*$^{+/+}$ mice using these antibodies revealed a faint, rather diffuse pattern, whereas in heart sections from *Lmna*$^{H222P/H222P}$ mice there was intense and predominantly nuclear fluorescence (Muchir et al., 2007). The protocol we used to analyze nuclear localization of ERK1/2 is as follows:

1. Hearts were excised from *Lmna*$^{+/+}$ mice and *Lmna*$^{H222P/H222P}$ mice and snap-frozen in liquid nitrogen-cooled isopentane; portions of these hearts can be stored in liquid nitrogen for other subsequent experiments.

II. Immunofluorescence labeling of phosphorylated ERK1/2 was performed on transverse frozen sections (8 μm) of left ventricular muscle. Sections were fixed in 3.7% formaldehyde in phosphate-buffered saline for 15 min and then blocked in 5% fetal goat serum in phosphate-buffered saline with 0.1% Triton X-100 for 1 h. This fixation was suitable for the anti-ERK1/2 antibodies used.

III. Sections were incubated in blocking solution with antiphosphorylated ERK1/2 monoclonal antibody (Cell Signaling Technology) overnight at 4 °C followed by washing in phosphate-buffered saline and incubation with Texas red-conjugated goat anti-mouse immunoglobulin G secondary antibody (Invitrogen). Sections were counterstained with 0.1 μg/ml 4′6-diamidino-2-phenylindole (Sigma-Aldrich) to label nuclei.

IV. Images were analyzed using a planar immunofluorescence microscope and intensity of phosphorylated ERK1/2 labeling of cardiomyocytes was measured using Scion Imaging software (version alpha 4.0.3.2; Scion Corp.).

4. MAP KINASE INHIBITOR TREATMENT STUDIES IN *Lmna*$^{H222P/H222P}$ MICE

The discovery that MAP kinase activities were abnormally elevated in hearts of *Lmna*$^{H222P/H222P}$ mice led us to hypothesize that decreasing their activities would lead to clinical improvement in the living animals. We therefore examined the effects of treatment with mitogen-activated protein kinase kinase 1/2 (MEK1/2), JNK, and p38α inhibitors on *Lmna*$^{H222P/H222P}$ mice.

4.1 MEK1/2 Inhibitors

Allosteric inhibitors of MEK1/2, the MAP kinase kinase that phosphorylates ERK1/2, have been synthesized and tested in humans, primary for oncology indications (Zhao & Adjei, 2014). Glaxo's trametinib was the first drug in this class to be approved by the U.S. Food and Drug Administration (Wright & McCormack, 2013). We have used several other MEK1/2 inhibitors to examine the effects of blocking ERK1/2 activity in hearts of *Lmna*$^{H222P/H222P}$ mice. Our initial published studies utilized PD098059, which was one of the first small molecule inhibitors of MEK1/2 synthesized by scientists at Parke-Davis Pharmaceutical Research Division of Warner-Lambert Co. (now Pfizer) (Dudley, Pang, Decker, Bridges, & Saltiel, 1995). Although PD098059 inhibits MEK1/2 *in vitro* and *in vivo* (Alessi, Cuenda, Cohen, Dudley, & Saltiel, 1995), it never advanced in clinical development. In subsequent studies, we therefore switched to selumetinib, a potent and selective MEK1/2 inhibitor discovered by scientists at Array BioPharma and licensed to AstraZeneca for clinical development; it is currently being studied in phase III clinical trials (Ciombor & Bekaii-Saab, 2015; Yeh et al., 2007).

4.2 JNK Inhibitors

Pharmacological inhibitors of JNK have been synthesized but their clinical development has been rather limited (Bogoyevitch & Arthur, 2008). They have been hypothesized to be potentially therapeutically useful in a wide range of conditions including neurodegeneration, inflammation, diabetes, viral infections, and fibrosis; however, there are no human clinical data to support these indications. One, PGL5001, has been studied in a phase II clinical trial for inflammatory endometriosis (Barnes, 2013). In our studies of $Lmna^{H222P/H222P}$ mice, we used the reversible ATP-competitive JNK inhibitor SP600125 (Bennett et al., 2001).

4.3 p38α Inhibitors

Several small molecule inhibitors of p38α have been synthesized (Coulthard, White, Jones, McDermott, & Burchill, 2009). Dilmapimod has been used in phase II clinical trials for neuropathic pain and chronic obstructive pulmonary disease (Anand et al., 2011; Betts et al., 2015; Singh, Smyth, Borrill, Sweeney, & Tal-Singer, 2010). The p38α inhibitor ARRY-371797, discovered at Array BioPharma, has been studied in a phase II trial for acute inflammatory pain (Coulthard et al., 2009). According to the company, its development for this indication was discontinued because the results were similar to other p38 inhibitors evaluated in rheumatoid arthritis (http://investor.arraybiopharma.com/phoenix.zhtml?c=123810&p=irol-newsArticle&ID=1305781). We used ARRY-371797 in our studies of $Lmna^{H222P/H222P}$ mice. Based on its beneficial effects on left ventricular diameters and fractional shortening in these mice (Muchir, Wu, et al., 2012), Array BioPharma has started a phase II trial of ARRY-371797 for patients with cardiomyopathy caused by *LMNA* mutations (https://clinicaltrials.gov/ct2/show/NCT02057341).

4.4 Treatment Protocols

I. A summary of the MAP kinase inhibitors used is provided in Table 2.

II. For MEK1/2 inhibition, PD098059 (Selleck Chemicals) or selumetinib (Selleck Chemicals) were delivered to male *Lmna*$^{H222P/H222P}$ mice by intraperitoneal injection using a 27-gauge 5/8-in. needle and syringe. For JNK inhibition, SP600125 (Calbiochem) was similarly delivered. The drugs were dissolved in dimethyl sulfoxide (Sigma-Aldrich) at a concentration of 1.0 or 0.5 mg/ml. Doses were 3 mg/kg/day for PD098059, 1 mg/kg/day for selumetinib, and 1 mg/kg for SP600125, given 5 days a week. To study the effects of an angiotensin-converting-enzyme inhibitor, benazepril (Sigma-Aldrich) was dissolved in dimethyl sulfoxide and similarly administered by intraperitoneal injection at a dose of 10 mg/kg/day. The placebo control consisted of dimethyl sulfoxide alone and was delivered similarly. During treatment, mice were fed a chow diet and housed in a barrier facility.

III. For p38α inhibition, ARRY-371797 (Array BioPharma) was delivered to male *Lmna*$^{H222P/H222P}$ mice by oral gavage. It was dissolved in Water for Injection (Gibco) at a concentration of 0.5 mg/ml and dosed at 30 mg/kg twice a day. The placebo consisted of the same volume of Water for Injection.

IV. To assess effects on heart, treatment was started either when mice were 8 weeks of age (asymptomatic) and continued until 16 weeks or, alternatively, started at 16 weeks of age (symptomatic) and continued until 20 weeks of age. At 16 or 20 weeks of age, mice were analyzed by echocardiography, then euthanized and hearts were excised for biochemical and histological analyses (see below).

V. To assess the effects of treatment on survival, selumetinib dissolved in dimethyl sulfoxide was diluted in drinking water to a concentration of 0.2 mg/ml; the same volume of dimethyl sulfoxide diluted in drinking water was used as placebo. Treatment was begun at 16 weeks of age and continued until the endpoints were reached (see below).

5. ASSESSMENT OF *Lmna*$^{H222P/H222P}$ MICE AFTER TREATMENT WITH MAP KINASE INHIBITORS

As summarized in Table 3, treatment of *Lmna*$^{H222P/H222P}$ mice with MAP kinase inhibitors has various beneficial effects. Most of our work

Table 2 MAP Kinase Inhibitors Used in Treatment Studies of *Lmna*$^{H222P/H222P}$ Mice

Drug	Target	Source	Stock (mg/ml)	Stability	Dose	References for Dose
PD98059	MEK1/2	Calbiochem	1.0	Up to 4 months −20 °C (protect from light)	3 mg/kg/day ip	Dudley et al. (1995)
Selumetinib	MEK1/2	Selleck Chemicals	0.5	Up to 8 months −80 °C	1 mg/kg/day ip ~1 mg/kg/day po[a]	Denton and Gustafson (2011)
SP600125	JNK1/2/3	Calbiochem	1.0	Up to 2 months −20 °C (protect from light)	3 mg/kg/day ip	Bennett et al. (2001)
ARRY-371797	p38α	Array BioPharma	0.5	Not disclosed by source	30 mg/kg/bid po[b]	Array BioPharma personal communication

[a]Drug added to drinking water to provide estimated dose of 1 mg/kg/day.
[b]By gastric gavage.
Ip, intraperitoneal; po, oral; bid, twice a day.

Table 3 Summary of Beneficial Effects of MAP Kinase Inhibitor Treatment on *Lmna*$^{H222P/H222P}$ Mice

Target	Drug	Treatment	LV Function[a]	Serum ANF[b]	Fibrosis	Survival	References
MEK1/2	PD98059	8–16 weeks	Significantly improved	NA	NA	NA	Muchir, Shan, Bonne, Lehnart, and Worman (2009)
MEK1/2	PD98059	16–20 weeks	Significantly improved	NA	Significantly decreased	NA	Wu, Muchir, Shan, Bonne, and Worman (2011)
MEK1/2	Selumetinib	16–20 weeks	Significantly improved	Decreased	Significantly decreased	NA	Muchir, Reilly, et al. (2012)
MEK1/2	Selumetinib	16 weeks to death/euthanasia		NA	Significantly decreased	Prolonged	Muchir, Reilly, et al. (2012)
JNK	SP600125	8–16 weeks	Significantly improved	NA	Significantly decreased	NA	Wu, Shan, Bonne, Worman, and Muchir (2010)
JNK	SP600125	16–20 weeks	Significantly improved	NA	Significantly decreased	NA	Wu et al. (2011)
P38α	ARRY-371797	16–20 weeks	Significantly improved	NA	No change	NA	Muchir, Wu, et al. (2012)

[a]Left ventricular function compared to placebo based on fractional shortening/ejection fraction determined by echocardiography.
[b]Natriuretic peptide A serum concentration.
NA, not assessed.

focused on MEK1/2 inhibitors because, as a class, they are further along in clinical development than JNK and p38α inhibitors. When treatment with a MEK1/2 inhibitor is begun at 8 weeks of age, before the onset of symptoms, male *Lmna*$^{H222P/H222P}$ mice have significantly smaller left ventricular end systolic and end diastolic diameters and a significantly greater left ventricular ejection fraction, with values similar to wild-type mice, at 16 weeks of age (Muchir et al., 2009). Treatment with the JNK inhibitor SP600125 and analysis at the same ages gives similar results (Wu et al., 2010). Male *Lmna*$^{H222P/H222P}$ mice treated with MEK1/2 inhibitors or a JNK inhibitor starting at 16 weeks of age, when heart function is already abnormal, have significantly greater left ventricular fractional shortening and ejection fraction, significantly smaller left ventricular diameters and significantly less left ventricular fibrosis than placebo-treated mice at 20 weeks of age (Muchir, Reilly, et al., 2012; Wu et al., 2011). MEK1/2 inhibitor treatment from 16 to 20 weeks also decreases serum concentrations of natriuretic peptide A, which is secreted in response to increased cardiac filling pressure (Muchir, Reilly, et al., 2012). Treatment with a MEK1/2 inhibitor also has synergistic benefits when combined with an angiotensin-converting enzyme inhibitor (Muchir, Wu, Sera, Homma, & Worman, 2014). Treatment with a p38α inhibitor from 16 to 20 weeks has similar beneficial effects on left ventricular diameters and fractional shortening but does not decrease the expression of genes encoding collagens responsible for fibrosis (Muchir, Wu, et al., 2012). Finally, treatment with a MEK1/2 inhibitor starting at 16 weeks of age significantly prolonged survival of male *Lmna*$^{H222P/H222P}$ mice (Muchir, Reilly, et al., 2012).

The protocols and methods we used to assess the effects of the MAP kinase inhibitors were as follows. Ideally, preclinical drug studies using such protocols such as these should, at a minimum, report on sample-size estimation, whether and how animals were randomized, whether investigators were blind to the treatment and the handling of data, as recommended by the U.S. National Institute of Neurological Disorders and Stroke (Landis et al., 2012).

5.1 Echocardiography

Echocardiography provides a convenient, noninvasive method to assess left ventricular diameters and functions in mice (Rottman, Ni, & Brown, 2007). Potential limitations have been related to the small size and rapid heart rate; however, the development of small probes operating at higher frequencies and faster frame rates has overcome these limitations. Care must be taken not to give too much anesthesia to maintain a physiologically relevant heart rate.

I. Mice were anesthetized with 1.5% isoflurane by inhalation and placed on a heating pad at 37 °C. We aimed at maintaining a heart rate of greater than 500/s.

II. Echocardiography was performed using a Visualsonics Vevo 770 ultrasound with a 30 MHz transducer applied to the chest wall. Cardiac ventricular dimensions were measured in two-dimensional mode and M-mode three times for each animal. A typical echocardiographic examination takes approximately 3–5 min.

III. Fractional shortening was calculated using the following formula: fractional shortening (%) = [(left ventricular end diastolic diameter − left ventricular end systolic diameter)/left ventricular end diastolic diameter] × 100. Left ventricular ejection fraction was calculated using the modified Simpson rule (Folland et al., 1979).

IV. The echocardiographer was always blind to mouse genotype and treatment.

5.2 Inhibition of MAP Kinase Activities in Hearts

After treatment with MAP kinase inhibitors, hearts were excised from drug-treated and placebo-treated mice to assess inhibition of ERK1/2 (which is activated by MEK1/2), JNK, and p38α. As described in Section 3.2 above, we used immunoblotting to measure their activities. Treatment with inhibitors gave approximately 30–70% decreases in the ratio of the phosphorylated MAP kinase to the total MAP kinase compared to treatment with placebo (Muchir, Reilly, et al., 2012; Muchir et al., 2009; Muchir, Wu, et al., 2012; Wu et al., 2011, 2010).

5.3 Natriuretic Peptide A Secretion

Natriuretic peptide A, also known as atrial natriuretic factor and atriopeptin, is secreted from the heart in response to increased intravascular volume and right atrial pressure. Its concentration is elevated in the blood in congestive heart failure (Tikkanen, Fyhrquist, Metsärinne, & Leidenius, 1985). We measured natriuretic peptide A in serum from $Lmna^{H222P/H222P}$ mice treated with selumetinib or placebo using the following methods.

I. Natriuretic peptide A is detected in mouse serum samples using a commercial competitive enzyme immunoassay (RayBiotech).

II. We added 100 μl antinatriuretic peptide A antibody solution to each well of the 96-well plate coated with rabbit secondary antibody and incubate for 1.5 h at room temperature.

III. Serum samples containing biotinylated natriuretic peptide A (10 pg/ml) was added to wells of the microplate and incubated for 2.5 h at room temperature.

IV. After washing four times in phosphate-buffered saline, we added 100 ml of horseradish peroxidase–streptavidin solution to each well and incubated for 45 min at room temperature. After washing and incubation for 30 min with 100 μl of RayBiotech 3,3′,5,5′-tetramethylbenzidine One-Step substrate reagent in the dark, absorbance was measured at 450 nm.

5.4 Cardiac Fibrosis

We assessed left ventricular fibrosis in excised hearts of *Lmna*$^{H222P/H222P}$ treated with either a MAP kinase inhibitor or a placebo using histological and biochemical methods. Histological assessment was a semiquantitative analysis of collagen staining. Biochemical analysis measured the expression of two genes encoding collagens.

I. Sections of left ventricles from *Lmna*$^{H222P/H222P}$ mice treated with a MAP kinase inhibitor or placebo were fixed in 4% formaldehyde for 48 h, embedded in paraffin, sectioned at 5 μm, and stained with Gomori trichrome, which stains collagen blue.

II. Representative stained sections were photographed using a Microphot SA (Nikon) light microscope attached to a Spot RT Slide camera (Diagnostic Instruments).

III. To quantify fibrosis, micrographs were processed (JMicroVision software) and blue-stained fibrotic tissue measured (ImageJ64 software).

IV. For biochemical analysis of collagen gene expression, primers corresponding to mouse *Col1a1* and *Co1a2* gene cDNAs (*Col1a1* forward 5′-agacggacagtactggatcg-3′ and reverse 5′-gcttcttttccttggggttc-3′; *Col1a2* forward 5′-ccgtgcttctcagaacatca-3′ and reverse 5′-gagcagccatcgactaggac-3′) are used from real-time RT-PCR.

V. Reactions contained HotStart-IT SYBR Green qPCR Master Mix (Affymetrix), 200 n*M* of each primer, and 0.2 ml of template in a 25-ml reaction volume.

VI. Amplification was carried out using the ABI 7300 Real-Time PCR System (Applied Biosystems) with an initial denaturation at 95 °C for 2 min followed by 50 cycles at 95 °C for 30 s and 62 °C for 30 s. Relative levels of mRNA expression were calculated using the $\Delta\Delta C_T$ method (Ponchel et al., 2003) and normalized to calculated Gapdh mRNA expression level.

5.5 Survival

We demonstrated a significantly prolonged survival of male *Lmna*$^{H222/H222P}$ mice treated with the MEK1/2 inhibitor selumetinib. We actually used a combined endpoint of survival and required euthanasia for mice that are in distress. This is a more compassionate protocol to limit pain and distress and was encouraged by our institution's Institutional Animal Care and Use Committee.

I. Selumetinib dissolved in dimethyl sulfoxide was diluted in drinking water to a concentration of 0.2 mg/ml; the same volume of dimethyl sulfoxide diluted in drinking water was used as placebo.

II. Treatment was started when mice were 16 weeks of age and continued until death or until a veterinarian blind to treatment group recommended euthanasia because of undue distress.

III. Criteria for euthanasia were (1) difficulty with normal ambulation, (2) failure to eat or drink, (3) loss of body mass of >20%, (4) rough or unkempt coat, and (5) respiratory distress.

IV. Data were analyzed using the Kaplan–Meier estimator (Kaplan & Meier, 1958) followed by a log-rank test with $p < 0.05$ considered statistically significant; GraphPad Prism software was used for this analysis.

6. CONCLUSIONS

The experiments and methods described in this review have demonstrated the utility of a mouse model of cardiomyopathy caused by *LMNA* mutation in identifying dysregulated cardiac MAP kinase signaling in the hearts of these animals. Pharmacological inhibitors of kinases in the ERK1/2, JNK, and p38α pathways were used to demonstrate a role for their abnormally increased signaling activity in pathogenesis. Several of these kinase inhibitors are in clinical development for other indications and could potentially be developed to treat human subjects with cardiomyopathy caused by *LMNA* mutations. Abnormally elevated ERK1/2 and p38α activity has indeed been demonstrated in hearts from affected human subjects (Muchir, Reilly, et al., 2012; Muchir, Wu, et al., 2012). A small clinical trial of a p38α inhibitor is already underway based on this research and others could be anticipated.

ACKNOWLEDGMENTS

The work described in this paper was supported by grants from the National Institutes of Health (AR048997), Muscular Dystrophy Association (MDA172222), Los Angeles Thoracic and Cardiovascular Foundation (CRV 2011-873R1), and the New York City Investment Fund.

REFERENCES

Aebi, U., Cohn, J., Buhle, L., & Gerace, L. (1986). The nuclear lamina is a meshwork of intermediate-type filaments. *Nature, 323*(6088), 560–564.

Alessi, D. R., Cuenda, A., Cohen, P., Dudley, D. T., & Saltiel, A. R. (1995). PD 098059 is a specific inhibitor of the activation of mitogen-activated protein kinase kinase *in vitro* and *in vivo*. *Journal of Biological Chemistry, 270*(46), 27489–27494.

Anand, P., Shenoy, R., Palmer, J. E., Baines, A. J., Lai, R. Y., Robertson, J., et al. (2011). Clinical trial of the p38 MAP kinase inhibitor dilmapimod in neuropathic pain following nerve injury. *European Journal of Pain, 15*(10), 1040–1048.

Arakawa, K., Kono, N., Yamada, Y., Mori, H., & Tomita, M. (2005). KEGG-based pathway visualization tool for complex omics data. *In Silico Biology, 5*(4), 419–423.

Arimura, T., Helbling-Leclerc, A., Massart, C., Varnous, S., Niel, F., Lacène, E., et al. (2005). Mouse model carrying H222P-Lmna mutation develops muscular dystrophy and dilated cardiomyopathy similar to human striated muscle laminopathies. *Human Molecular Genetics, 14*(1), 155–169.

Barnes, P. J. (2013). New anti-inflammatory targets for chronic obstructive pulmonary disease. *Nature Reviews. Drug Discovery, 12*(7), 543–559.

Bennett, B. L., Sasaki, D. T., Murray, B. W., O'Leary, E. C., Sakata, S. T., Xu, W., et al. (2001). SP600125, an anthrapyrazolone inhibitor of Jun N-terminal kinase. *Proceedings of the National Academy of Sciences of the United States of America, 98*(24), 13681–13686.

Bergo, M. O., Wahlstrom, A. M., Fong, L. G., & Young, S. G. (2008). Genetic analyses of the role of RCE1 in RAS membrane association and transformation. *Methods in Enzymology, 438*, 367–389.

Bertrand, A. T., Renou, L., Papadopoulos, A., Beuvin, M., Lacène, E., Massart, C., et al. (2012). DelK32-lamin A/C has abnormal location and induces incomplete tissue maturation and severe metabolic defects leading to premature death. *Human Molecular Genetics, 21*(5), 1037–1048.

Betts, J. C., Mayer, R. J., Tal-Singer, R., Warnock, L., Clayton, C., Bates, S., et al. (2015). Gene expression changes caused by the p38 MAPK inhibitor dilmapimod in COPD patients: Analysis of blood and sputum samples from a randomized, placebo-controlled clinical trial. *Pharmacology Research & Perspectives, 3*(1), e00094.

Bogoyevitch, M. A., & Arthur, P. G. (2008). Inhibitors of c-Jun N-terminal kinases: JuNK no more? *Biochimica et Biophysica Acta, 1784*(1), 76–93.

Bonne, G., Di Barletta, M. R., Varnous, S., Bécane, H. M., Hammouda, E. H., Merlini, L., et al. (1999). Mutations in the gene encoding lamin A/C cause autosomal dominant Emery-Dreifuss muscular dystrophy. *Nature Genetics, 21*(3), 285–288.

Brodsky, G. L., Muntoni, F., Miocic, S., Sinagra, G., Sewry, C., & Mestroni, L. (2000). Lamin A/C gene mutation associated with dilated cardiomyopathy with variable skeletal muscle involvement. *Circulation, 101*(5), 473–476.

Cattin, M. E., Bertrand, A. T., Schlossarek, S., Le Bihan, M. C., Skov Jensen, S., Neuber, C., et al. (2013). Heterozygous LmnadelK32 mice develop dilated cardiomyopathy through a combined pathomechanism of haploinsufficiency and peptide toxicity. *Human Molecular Genetics*, *22*(15), 3152–3164.

Chandar, S., Yeo, L. S., Leimena, C., Tan, J. C., Xiao, X. H., Nikolova-Krstevski, V., et al. (2010). Effects of mechanical stress and carvedilol in lamin A/C-deficient dilated cardiomyopathy. *Circulation Research*, *106*(3), 573–582.

Chang, W., Worman, H. J., & Gundersen, G. G. (2015). Accessorizing and anchoring the LINC complex for multifunctionality. *Journal of Cell Biology*, *208*(1), 11–22.

Choi, J. C., Muchir, A., Wu, W., Iwata, S., Homma, S., Morrow, J. P., et al. (2012). Temsirolimus activates autophagy and ameliorates cardiomyopathy caused by lamin A/C gene mutation. *Science Translational Medicine*, *4*(144), 144ra102.

Ciombor, K. K., & Bekaii-Saab, T. (2015). Selumetinib for the treatment of cancer. *Expert Opinion on Investigational Drugs*, *24*(1), 111–123.

Coulthard, L. R., White, D. E., Jones, D. L., McDermott, M. F., & Burchill, S. A. (2009). p38(MAPK): Stress responses from molecular mechanisms to therapeutics. *Trends in Molecular Medicine*, *15*(8), 369–379.

Dauer, W. T., & Worman, H. J. (2009). The nuclear envelope as a signaling node in development and disease. *Developmental Cell*, *17*(5), 626–638.

Davies, B. S., Barnes, R. H., 2nd., Tu, Y., Ren, S., Andres, D. A., Spielmann, H. P., et al. (2010). An accumulation of non-farnesylated prelamin A causes cardiomyopathy but not progeria. *Human Molecular Genetics*, *19*(13), 2682–2694.

Dennis, G., Sherman, B. T., Hosack, D. A., Yang, J., Gao, W., Lane, H. C., et al. (2003). DAVID: Database for annotation, visualization, and integrated discovery. *Genome Biology*, *4*, P3.

Denton, C. L., & Gustafson, D. L. (2011). Pharmacokinetics and pharmacodynamics of AZD6244 (ARRY-142886) in tumor-bearing nude mice. *Cancer Chemotherapy and Pharmacology*, *67*(2), 349–360.

Dudley, D. T., Pang, L., Decker, S. J., Bridges, A. J., & Saltiel, A. R. (1995). A synthetic inhibitor of the mitogen-activated protein kinase cascade. *Proceedings of the National Academy of Sciences of the United States of America*, *92*(17), 7686–7689.

Fatkin, D., MacRae, C., Sasaki, T., Wolff, M. R., Porcu, M., Frenneaux, M., et al. (1999). Missense mutations in the rod domain of the lamin A/C gene as causes of dilated cardiomyopathy and conduction-system disease. *The New England Journal of Medicine*, *341*(23), 1715–1724.

Fisher, D. Z., Chaudhary, N., & Blobel, G. (1986). cDNA sequencing of nuclear lamins A and C reveals primary and secondary structural homology to intermediate filament proteins. *Proceedings of the National Academy of Sciences of the United States of America*, *83*(17), 6450–6454.

Folland, E. D., Parisi, A. F., Moynihan, P. F., Jones, D. R., Feldman, C. L., & Tow, D. E. (1979). Assessment of left ventricular ejection fraction and volumes by real-time, two-dimensional echocardiography. A comparison of cineangiographic and radionuclide techniques. *Circulation*, *60*(4), 760–766.

Goldman, A. E., Maul, G., Steinert, P. M., Yang, H. Y., & Goldman, R. D. (1986). Keratin-like proteins that coisolate with intermediate filaments of BHK-21 cells are nuclear lamins. *Proceedings of the National Academy of Sciences of the United States of America*, *83*(11), 3839–3843.

Holmström, M., Kivistö, S., Heliö, T., Jurkko, R., Kaartinen, M., Antila, M., et al. (2011). Late gadolinium enhanced cardiovascular magnetic resonance of lamin A/C gene mutation related dilated cardiomyopathy. *Journal of Cardiovascular Magnetic Resonance*, *13*, 30.

Jahn, D., Schramm, S., Schnölzer, M., Heilmann, C. J., de Koster, C. G., Schütz, W., et al. (2012). A truncated lamin A in the Lmna-/- mouse line: Implications for the understanding of laminopathies. *Nucleus, 3*(5), 463–474.

Kaplan, E. L., & Meier, P. (1958). Nonparametric estimation from incomplete observations. *Journal of the American Statistical Association, 53*(282), 457–481.

Kubben, N., Voncken, J. W., Konings, G., van Weeghel, M., van den Hoogenhof, M. M., Gijbels, M., et al. (2011). Post-natal myogenic and adipogenic developmental: Defects and metabolic impairment upon loss of A-type lamins. *Nucleus, 2*(3), 195–207.

Laemmli, U. K. (1970). Cleavage of structural proteins during the assembly of the head of bacteriophage T4. *Nature, 227*(5259), 680–685.

Landis, S. C., Amara, S. G., Asadullah, K., Austin, C. P., Blumenstein, R., Bradley, E. W., et al. (2012). A call for transparent reporting to optimize the predictive value of preclinical research. *Nature, 490*(7419), 187–191.

Lee, H. K., Braynen, W., Keshav, K., & Pavlidis, P. (2005). ErmineJ: Tool for functional analysis of gene expression data sets. *BMC Bioinformatics, 6*, 269.

Lin, F., & Worman, H. J. (1993). Structural organization of the human gene encoding nuclear lamin A and nuclear lamin C. *Journal of Biological Chemistry, 268*(22), 16321–16326.

Lu, J. T., Muchir, A., Nagy, P. L., & Worman, H. J. (2011). LMNA cardiomyopathy: Cell biology and genetics meet clinical medicine. *Disease Models & Mechanisms, 4*(5), 562–568.

McKeon, F. D., Kirschner, M. W., & Caput, D. (1986). Homologies in both primary and secondary structure between nuclear envelope and intermediate filament proteins. *Nature, 319*(6053), 463–468.

Mounkes, L. C., Kozlov, S. V., Rottman, J. N., & Stewart, C. L. (2005). Expression of an LMNA-N195K variant of A-type lamins results in cardiac conduction defects and death in mice. *Human Molecular Genetics, 14*(15), 2167–2180.

Muchir, A., Bonne, G., van der Kooi, A. J., van Meegen, M., Baas, F., Bolhuis, P. A., et al. (2000). Identification of mutations in the gene encoding lamins A/C in autosomal dominant limb girdle muscular dystrophy with atrioventricular conduction disturbances (LGMD1B). *Human Molecular Genetics, 9*(9), 1453–1459.

Muchir, A., Pavlidis, P., Decostre, V., Herron, A. J., Arimura, T., Bonne, G., et al. (2007). Activation of MAPK pathways links LMNA mutations to cardiomyopathy in Emery-Dreifuss muscular dystrophy. *Journal of Clinical Investigation, 117*(5), 1282–1293.

Muchir, A., Reilly, S. A., Wu, W., Iwata, S., Homma, S., Bonne, G., et al. (2012). Treatment with selumetinib preserves cardiac function and improves survival in cardiomyopathy caused by mutation in the lamin A/C gene. *Cardiovascular Research, 93*(2), 311–319.

Muchir, A., Shan, J., Bonne, G., Lehnart, S. E., & Worman, H. J. (2009). Inhibition of extracellular signal-regulated kinase signaling to prevent cardiomyopathy caused by mutation in the gene encoding A-type lamins. *Human Molecular Genetics, 18*(2), 241–247.

Muchir, A., Wu, W., Choi, J. C., Iwata, S., Morrow, J., Homma, S., et al. (2012). Abnormal p38α mitogen-activated protein kinase signaling in dilated cardiomyopathy caused by lamin A/C gene mutation. *Human Molecular Genetics, 21*(19), 4325–4333.

Muchir, A., Wu, W., Sera, F., Homma, S., & Worman, H. J. (2014). Mitogen-activated protein kinase kinase 1/2 inhibition and angiotensin II converting inhibition in mice with cardiomyopathy caused by lamin A/C gene mutation. *Biochemical and Biophysical Research Communications, 452*(4), 958–961.

Nikolova, V., Leimena, C., McMahon, A. C., Tan, J. C., Chandar, S., Jogia, D., et al. (2004). Defects in nuclear structure and function promote dilated cardiomyopathy in lamin A/C-deficient mice. *Journal of Clinical Investigation, 113*(3), 357–369.

Plotnikov, A., Zehorai, E., Procaccia, S., & Seger, R. (2011). The MAPK cascades: Signaling components, nuclear roles and mechanisms of nuclear translocation. *Biochimica et Biophysica Acta, 1813*(9), 1619–1633.

Ponchel, F., Toomes, C., Bransfield, K., Leong, F. T., Douglas, S. H., Field, S. L., et al. (2003). Real-time PCR based on SYBR-Green I fluorescence: An alternative to the TaqMan assay for a relative quantification of gene rearrangements, gene amplifications and micro gene deletions. *BMC Biotechnology*, *3*, 18.

Raman, S. V., Sparks, E. A., Baker, P. M., McCarthy, B., & Wooley, C. F. (2007). Mid-myocardial fibrosis by cardiac magnetic resonance in patients with lamin A/C cardiomyopathy: Possible substrate for diastolic dysfunction. *Journal of Cardiovascular Magnetic Resonance*, *9*(6), 907–913.

Rottman, J. N., Ni, G., & Brown, M. (2007). Echocardiographic evaluation of ventricular function in mice. *Echocardiography*, *24*(1), 83–89.

Rusiñol, A. E., & Sinensky, M. S. (2006). Farnesylated lamins, progeroid syndromes and farnesyl transferase inhibitors. *Journal of Cell Science*, *119*(16), 3265–3272.

Singh, D., Smyth, L., Borrill, Z., Sweeney, L., & Tal-Singer, R. (2010). A randomized, placebo-controlled study of the effects of the p38 MAPK inhibitor SB-681323 on blood biomarkers of inflammation in COPD patients. *Journal of Clinical Pharmacology*, *50*(1), 94–100.

Stewart, C. L., Kozlov, S., Fong, L. G., & Young, S. G. (2007). Mouse models of the laminopathies. *Experimental Cell Research*, *313*(10), 2144–2156.

Sullivan, T., Escalante-Alcalde, D., Bhatt, H., Anver, M., Bhat, N., Nagashima, K., et al. (1999). Loss of A-type lamin expression compromises nuclear envelope integrity leading to muscular dystrophy. *Journal of Cell Biology*, *147*(5), 913–920.

Tikkanen, I., Fyhrquist, F., Metsärinne, K., & Leidenius, R. (1985). Plasma atrial natriuretic peptide in cardiac disease and during infusion in healthy volunteers. *Lancet*, *2*(8446), 66–69.

Wolf, C. M., Wang, L., Alcalai, R., Pizard, A., Burgon, P. G., Ahmad, F., et al. (2008). Lamin A/C haploinsufficiency causes dilated cardiomyopathy and apoptosis-triggered cardiac conduction system disease. *Journal of Molecular and Cellular Cardiology*, *44*(2), 293–303.

Worman, H. J., Fong, L. G., Muchir, A., & Young, S. G. (2009). Laminopathies and the long strange trip from basic cell biology to therapy. *Journal of Clinical Investigation*, *119*(7), 1825–1836.

Wright, C. J., & McCormack, P. L. (2013). Trametinib: First global approval. *Drugs*, *73*(11), 1245–1254.

Wu, W., Muchir, A., Shan, J., Bonne, G., & Worman, H. J. (2011). Mitogen-activated protein kinase inhibitors improve heart function and prevent fibrosis in cardiomyopathy caused by mutation in lamin A/C gene. *Circulation*, *123*(1), 53–61.

Wu, W., Shan, J., Bonne, G., Worman, H. J., & Muchir, A. (2010). Pharmacological inhibition of c-Jun N-terminal kinase signaling prevents cardiomyopathy caused by mutation in *LMNA* gene. *Biochimica et Biophysica Acta*, *1802*(7-8), 632–638.

Wydner, K. L., McNeil, J. A., Lin, F., Worman, H. J., & Lawrence, J. B. (1996). Chromosomal assignment of human nuclear envelope protein genes LMNA, LMNB1, and LBR by fluorescence in situ hybridization. *Genomics*, *32*(3), 474–478.

Yeh, T. C., Marsh, V., Bernat, B. A., Ballard, J., Colwell, H., Evans, R. J., et al. (2007). Biological characterization of ARRY-142886 (AZD6244), a potent, highly selective mitogen-activated protein kinase kinase 1/2 inhibitor. *Clinical Cancer Research*, *13*(5), 1576–1583.

Young, S. G., Meta, M., Yang, S. H., & Fong, L. G. (2006). Prelamin A farnesylation and progeroid syndromes. *Journal of Biological Chemistry*, *281*(52), 39741–39745.

Zhang, H., Kieckhaefer, J. E., & Cao, K. (2013). Mouse models of laminopathies. *Aging Cell*, *12*(1), 2–10.

Zhao, Y., & Adjei, A. A. (2014). The clinical development of MEK inhibitors. *Nature Reviews. Clinical Oncology*, *11*(7), 385–400.

CHAPTER TWENTY-ONE

In vivo, *Ex Vivo*, and *In Vitro* Approaches to Study Intermediate Filaments in the Eye Lens

Miguel Jarrin*, Laura Young†, Weiju Wu*, John M. Girkin†, Roy A. Quinlan*,[1]

*Integrative Cell Biology Laboratory, School of Biological and Biomedical Sciences, The University of Durham, Durham, United Kingdom

†Centre for Advanced Instrumentation and Biophysical Sciences Institute, Department of Physics, The University of Durham, Durham, United Kingdom

[1]Corresponding author: e-mail address: r.a.quinlan@dur.ac.uk

Contents

Abstract

The role of the eye lens is to focus light into the retina. To perform this unique function, the ocular lens must be transparent. Previous studies have demonstrated the expression of vimentin, BFSP1, and BFSP2 in the eye lens. These intermediate filament (IF) proteins are essential to the optical properties of the lens. They are also important to its biomechanical properties, to the shape of the lens fiber cells, and to the organization and function of the plasma membrane. The eye lens is an iconic model in developmental studies, as a result different vertebrate models, including zebrafish, have been developed to study lens formation. In the present chapter, we have summarized the new approaches and the more breakthrough models (e.g., iPSc) that can be used to study the function of IFs in the ocular lens. We have presented three different groups of models. The first group includes *in vitro* models, where IFs can be studied and manipulated in lens cell cultures. The second includes *ex vivo* models. These replicate better the complex lens cell differentiation processes and the role(s) played by IFs. The third class is the *in vivo*

Methods in Enzymology, Volume 568
ISSN 0076-6879
http://dx.doi.org/10.1016/bs.mie.2015.09.010

models, and here, we have focused on Zebrafish and new imaging approaches using selective plane illumination microscopy. Finally, we present protocols on how to use these lens models to study IFs.

ABBREVIATION

BFSP1 beaded filament structural protein 1
BFSP2 beaded filament structural protein 2
BMP4 bone morphogenetic protein 4
BMP7 bone morphogenetic protein 7
Cas CRISPR-associated protein
CFP cyan fluorescent protein
CRISPR clustered regularly interspaced short palindromic repeats
ED embryonic day
ES embryonic stem cell
FEP fluorinated ethylene propylene
FGF2 fibroblast growth factor 2
GFP green fluorescent protein
GOI gene of interest
IFs intermediate filaments
iPSc inducible pluripotent stem cells
mRNA messenger ribonucleic acid
NA numerical aperture
Pax6 paired box protein 6
RNA ribonucleic acid
RNAi interference ribonucleic acid
siRNA small interfering ribonucleic acid
SPIM selective plane illumination microscopy
WNT3a Wingless-type member 3a

1. INTRODUCTION

Here, we review the recent techniques and approaches used to study the role of the cytoskeleton, and specifically intermediate filaments (IFs), in eye lens homeostasis and optical function (Gokhin et al., 2012; Sandilands et al., 2003, 2004). Changes in lens IFs dramatically alter lens function (Gokhin et al., 2012; Sandilands et al., 2003, 2004; Song et al., 2009). One of the challenges for cell biology is to understand the molecular mechanisms that determine the dynamic cell shape changes and mechanical stability during tissue formation (Wei, Reidler, Shen, & Huang, 2013). Given the enormous complexity of the cytoskeletal systems involved and particularly the pivotal role played by IFs, a spectrum of different approaches is needed to understand the morphological changes as well as their functional consequence(s). The importance of these mechanisms and the regulation at both the cellular and tissue level of IFs is evidenced by their extensive links with human disease (Butin-Israeli, Adam, Goldman, & Goldman, 2012; Homberg & Magin, 2014; Landsbury, Perng, Pohl, & Quinlan, 2010; Lepinoux-Chambaud & Eyer, 2013; Pan, Hobbs, & Coulombe, 2013; Snider & Omary, 2014; Toivola, Boor, Alam, & Strnad, 2015; Wang, Zhang, Wu, & Zhang, 2013). It is therefore no surprise that to maintain this cellular homeostasis, a dynamic cytoskeletal network of IFs is required in eukaryotic cells to support the transitions needed during cell growth, proliferation, migration, and differentiation (Chung, Rotty, & Coulombe, 2013; Herrmann, Strelkov, Burkhard, & Aebi, 2009; Ingber, 2006; Koster, Weitz, Goldman, Aebi, & Herrmann, 2015; Quinlan, Bromley, & Pohl, 2015; Shabbir, Cleland, Goldman, & Mrksich, 2014; Toivola, Strnad, Habtezion, & Omary, 2010). These cytoskeletal networks provide the cells with specific structural and physical properties that maintain cell boundaries and their tissue integration upon exposure to environmental, mechanical, and metabolic stresses (Herrmann et al., 2009; Jensen, Morris, Goldman, & Weitz, 2014).

IFs were first identified in Howard Holtzer's laboratory in 1968 (Ishikawa, Bischoff, & Holtzer, 1968) as polymers of 10 nm diameter morphologically distinct from microfilaments (6 nm diameter) and microtubules (25 nm in diameter). They are the most resilient components of the cytoskeleton (Janmey, Euteneuer, Traub, & Schliwa, 1991) which initially focused attention onto the structural roles played by these filaments in cells and tissues (Herrmann et al., 2009). The different members of the IFs family play important roles in maintaining cell and tissue integrity, such as cytokeratins in the epidermis (Coulombe, Hutton, Vassar, & Fuchs, 1991; Ishida-Yamamoto et al., 1991), desmin in muscle, (Capetanaki & Milner, 1998; Capetanaki, Milner, & Weitzer, 1997), or beaded filament structural proteins (BFSP1, also known as filensin, and BFSP2, also known as either phakinin or CP49) in the eye lens (Fudge et al., 2011; Quinlan, Carte, Sandilands, & Prescott, 1996; Sandilands et al., 2003, 2004).

The role of the eye lens is to focus light onto retina (Fig. 1A). To achieve this, the lens requires a specialized cellular organization (Fig. 1B) and lifelong programme of differentiation (Wride, 2011). This involves the coordination of a complex program of cell proliferation, cell migration, and cell differentiation to generate the required cell shape, cytoarchitecture, and mechanical properties of lens tissue (Wride, 2011). This produces a mature

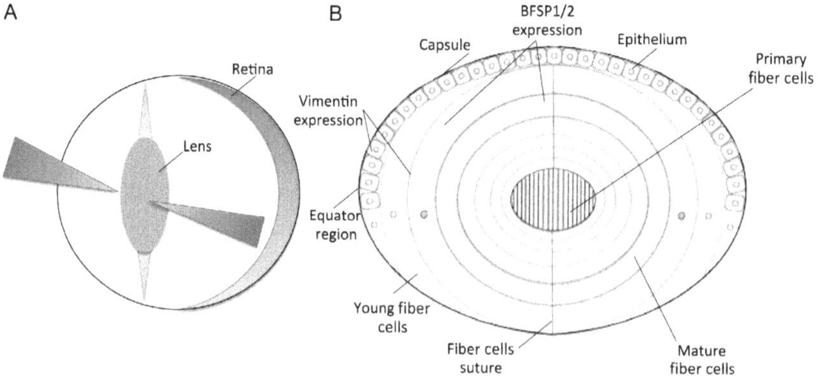

Figure 1 Diagram of the eye lens. The eye lens is a transparent and avascular tissue localized in the anterior segment of the eye. Its role is to focus the light onto the retina (A). The lens is contained within a capsule, one of the thickest basement membranes in the body (B). A single epithelial cell layer covers the anterior hemisphere of the lens. Those epithelial cells found in the equatorial region start to differentiate into lens fiber cells, exiting the lens epithelium via the meridional rows, which indicate the perimeter of the epithelium. Lens cell differentiation involves cell elongation and the expression of fiber cell markers. Differentiating lens cells elongate to form a characteristic suture pattern that varies with different species and the optical properties of the lens concerned. Lens cell differentiation includes the removal of organelles (e.g., mitochondria and ER) and nuclei. The oldest fiber cells form the core (nucleus) of the lens and are those (primary) fiber cells that first filled the lens vesicle. Vimentin is expressed in both lens epithelial and young fiber cells. The expression of BFSP1 and BFSP2 and their proteolytic fragments are found in all fiber cells.

lens that is transparent and with the required optical (Perng, Zhang, & Quinlan, 2007; Sandilands et al., 2003) and biomechanical properties (Won, Fudge, & Choh, 2015). Aging affects these lens properties and also the expression pattern and subcellular distribution of IFs (Alizadeh et al., 2002, 2004; Fudge et al., 2011; Gokhin et al., 2012; Sandilands et al., 2003, 2004).

The eye lens is avascular and is contained within the lens capsule (Fig. 1B), a transparent basement membrane comprising mainly type IV collagen and sulfated glycosaminoglycans. It is one of the thickest basement membranes in the body. A single epithelial cell layer covers the anterior hemisphere of the adult lens (Fig. 1B). The body of the lens comprises fiber cells, which have originated from the epithelial cells at the lens equator, a region where cell proliferation is concentrated (Fig. 1B; Wu et al., 2015). Many studies have analyzed the lens expression of BFSP1 and BFSP2 (Ireland et al., 2000), vimentin (Ellis, Alousi, Lawniczak, Maisel, &

Welsh, 1984; Sandilands, Prescott, Carter, et al., 1995), glial fibrilliary acidic protein (GFAP) (Bozanic, Bocina, & Saraga-Babic, 2006), simple epithelial keratins (Bozanic et al., 2006; Kasper & Viebahn, 1992), and nestin (Mokry & Nemecek, 1998; Yang, Bian, Gao, Chen, & Jing, 2000). Of these, vimentin, BFSP1, and BFSP2 are the most highly expressed and all cause cataract when mutated in humans (Muller et al., 2009; Perng et al., 2007). Vimentin is expressed in both lens epithelial and young fiber cells (Fig. 1B), although it is lost from the central (mature) fiber cells (Sandilands, Prescott, Carter, et al., 1995). The expression of BFSP1 and BFSP2 in the adult lens is restricted to fiber cells (Fig. 1B; Ireland et al., 2000; Perng et al., 2007; Sandilands, Prescott, Hutcheson, et al., 1995). The fiber cell plasma membranes are very precisely aligned to minimize intercellular spaces (Michael, van Marle, Vrensen, & van den Berg, 2003), and the cytoplasmic protein concentrations are extremely high (Bloemendal et al., 2004). It is the beaded filaments that help with the precise cellular organization of lens fiber cells (Landsbury et al., 2010; Sandilands et al., 2003) and also the stabilization of the high protein concentrations needed to achieve the high refractive index of the lens (Bloemendal et al., 2004; Slingsby, Wistow, & Clark, 2013).

Historically, the lens was used as a source of vimentin and later BFSP1 and 2 (Perng et al., 2004). The relative expression of vimentin in the lens is critical to its function as its overexpression disturbs fiber cell differentiation and causes lens cataract (Capetanaki, Smith, & Heath, 1989). The atopic expression of desmin in the lens also caused cataract (Dunia et al., 1990) reinforcing the point that IFs are critical to lens structure and function (Song et al., 2009). The identification of a vimentin mutation that caused inherited cataract (Muller et al., 2009) in humans was followed by engineered phosphorylation site mutations in mice that caused microophthalmia and cataract (Matsuyama et al., 2013). Vimentin is critical to the epithelial mesenchyme transition (Menko et al., 2014). This makes the lens a good model to study the dynamic role and behavior of IFs and their roles in cytokinesis, aneuploidy and transdifferentiation and cell motility. Microinjection, multifluorescence labeling, and *in vivo* imaging are all suitable for IF structure–function studies in the lens especially when combined with genetic manipulation approaches (Bibliowicz, Tittle, & Gross, 2011; Graw, 2009; Gross & Perkins, 2008; Watanabe et al., 2010).

IFs have been highly conserved during evolution, originating from a lamin-like progenitor (Karabinos, 2013; Kollmar, 2015; Peter & Stick, 2015). Beaded filaments are an iconic IF-based network in the lens, discovered by Maisel & Perry, (1972) and structurally and immunologically conserved from squid to man (FitzGerald & Casselman, 1991). The major structural proteins of these filaments, BFSP1 and 2, are found in mature fiber cells in different species as chicken, human, mouse, and zebrafish (Perng et al., 2007; Song et al., 2009). During their evolution, the C-terminal tail domain of BFSP2 has been lost in mammals, but is retained in fishes (Qu et al., 2012) and in birds, an extended helix IB was observed in a splice variant of BFSP2 (Wallace, Signer, Paton, Burt, & Quinlan, 1998). These lens IF proteins are considered as outliers within the IF family of vertebrates (Karabinos, 2013), and yet during their evolution, they manifest a key lamin-like feature, namely, an extended helix I (Tomarev, Zinovieva, & Piatigorsky, 1993; Wallace et al., 1998), suggesting that they too originated from a lamin precursor (Kollmar, 2015; Peter & Stick, 2015).

2. MODELS TO STUDY IFs IN THE EYE LENS

Xenopus (Hirsch et al., 2002), zebrafish (Greiling & Clark, 2012), chicken (Musil, 2012), mouse (Ogino, Ochi, Reza, & Yasuda, 2012) and rat (Tripathi, Tripathi, Borisuth, Dhaliwal, & Dhaliwal, 1991) as well as mole rats (Nikitina & Kidson, 2014), cave fish (Ma, Parkhurst, & Jeffery, 2014; Yamamoto, Stock, & Jeffery, 2004), and newt (Eguchi et al., 2011; Inoue et al., 2012) have all been used to study vertebrate eye lens development. Here, we focus on a few models to facilitate the study IFs and their structure–function relationship in the eye lens. We describe three different models. The first are *in vitro* models. IFs can be studied and manipulated in lens cell cultures (Ireland, Tran, & Mrock, 1993; Leonard, Chan, & Menko, 2008). Fiber cell differentiation is usually compromised or even absent in such systems (e.g., Krausz et al., 1996). The second class of models includes *ex vivo* models. These replicate better the complex cell differentiation processes that accompany lens development and allow to understand the role of IFs in a more diverse and complex environment more similar to *in vivo* models but keeping the main feature of cell culture. Finally, the third class of models collect the *in vivo* models.

2.1 *In Vitro* Lens Models

In vitro models based on established lens epithelial cell lines have provided a homogeneous population of cells that are transfectable but that are compromised due a very limited differentiation potential. The main advantage of such cells lines is that they can be synchronized, exposed to drug/small molecule treatments, or genetically manipulated in order to alter expression patterns and to probe disease mechanisms. We have listed the most commonly used lens epithelial cell lines (Table 1). The recent advance in stem cell technology and inducible pluripotent cells from somatic cells realizes the potential to use *in vitro* models to address the role of IFs in lens fiber cell differentiation and aging (Yang et al., 2010).

2.1.1 Efficient Lens Formation from Human Stem Cells

Stem cell technologies now support an efficient protocol to obtain differentiated lens cells (Yang et al., 2010). This model allows some aspects of lens fiber cell differentiation to be studied *in vitro*. The expression of the lens IFs proteins, BFSP1 and BFSP2, evidences successful cell differentiation in this system (Yang et al., 2010) and provides a valuable platform to address the role of IFs in lens development and epithelial cell differentiation.

The study used the WA01 (H1) human ES cells line (WiCell Research Institute, Madison, WI, USA). Cells were cultured on Matrigel-coated plate in DMEM/F-12 supplemented with 0.05% BSA (Sigma, USA), 1% (w/v) nonessential amino acids (Invitrogen, USA), 2 m*M* L-glutamine (Gibco, USA), Pen Strep (Gibco, USA), 1 × N2 supplements (Gibco, USA), 1 × B27 supplements (Gibco, USA). The culture medium was changed every other day, and the cells were passaged once a week mechanically.

The protocol for the generation of mature lens cells requires the sequential manipulation of the bone morphogenetic protein (BMP), fibroblast growth factor (FGF), and Wnt pathways. The decision to focus on these specific pathways was because of their essential roles in lens development. The strategy is a three-step process that can be adapted to any other stem cell line (Fig. 2 summarizes the main steps in the protocol).

1. Generation of lens progenitor cells.

   H1 human ES cell line was cultured supplemented with Noggin (100 ng/mL, R&D systems, USA). Noggin prevents BMP to bind to the receptors. After 6 days, Noggin is removed from media, and BMP4 (20 ng/mL, R&D systems, USA), BMP7 (20 ng/mL, R&D systems, USA), and FGF2 (100 ng/mL, R&D systems, USA) are added to the media to induce lens progenitors. At day 10, markers for lens progenitors: Pax6 and αB-Crystallin are expressed followed by αA-Crystallin (Fig. 2).

2. Increase of presence of lens progenitor cells.

   H1 human ES cell line was incubated with BMP4 (20 ng/mL), BMP7 (20 ng/mL), and FGF2 (100 ng/mL) from day 10 to day 18 (Fig. 2).

3. Formation of lentoid bodies.

Table 1 Mammalian Lens Epithelial Cell Lines Cited in the Literature

Cell Line	Species	Source	References
CCat1-2	Human	Control patients	Rhodes, Monckton, McAbney, Prescott, and Duncan (2006)
CD5A	Human	Lens epithelial cell line	Lengler et al. (2001)
DM1	Human	Myotonic dystrophy type 1 Adult patients	Rhodes et al. (2006)
FHL124	Human	Capsule-lens epithelial explants	Wormstone, Tamiya, Marcantonio, and Reddan (2000)
HLB-E3	Human	Retinopathy of prematurity Infant patients	Andley, Rhim, Chylack, and Fleming (1994) (ATCC® CRL-11421)
HLE-1	Human	Normal human lens epithelial	Chang et al. (2000)
H36CE2	Human	Central lens epithelium Postmortem infant	Tholozan et al. (2007)
SRA 01/04	Human	Lens epithelial cells	Ibaraki et al. (1998)
alpha-TN4	Mouse	Lens epithelial cells	Yamada, Nakamura, Westphal, and Russell (1990)
MLEC/ Hsf4−/−	Mouse	Lens epithelial cell lines with the genotype of Hsf4−/−	Zhang, Ma, Wang, Wang, and Hu (2013)
ML6	Mouse	Lens epithelial	Haque, Arora, Dikdan, Lysz, and Zelenka (1999)
NKR-11	Mouse	Nakano adult lens mice	Russell, Fukui, Tsunematsu, Huang, and Konoshita (1977)
21EM15	Mouse	Emory and age-matched cataract-resistant mice	Reddan et al. (1989)
17EM15	Mouse	Emory and age-matched cataract-resistant mice	Reddan et al. (1989)
B3	Rabbit	Rabbit lens epithelial cells	Haque et al. (1999)
LEP2	Rabbit	Rabbit lens epithelial cells	Haque et al. (1999)
N/N1003A	Rabbit	Rabbit lens epithelial cells	Andley, Hebert, Morrison, Reddan, and Pentland (1994)

Human, mouse, and rabbit cell lines have been developed and characterized by a variety of different labs, from whom these lines can be sourced as there is no commercial repository, with the exception of the HLE-B3 cell line, where these can be obtained. Such cell lines are useful models to explore the role of the IFs in the lens.

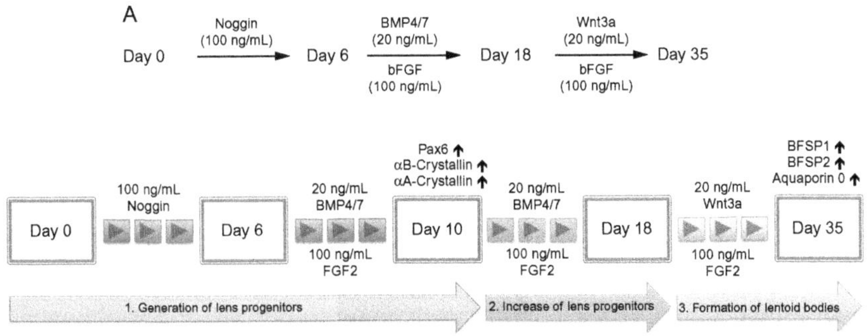

Figure 2 Efficient lentoid formation from embryonic human stem cells. The figure summarized the three-step differentiation procedure to the generation of lentoid bodies. The protocol includes a combination of cytokines followed by monitoring lens differentiation markers to confirm the progression of lens cell differentiation. Lens progenitors are first generated by incubating H1 human ES cells with Noggin (100 ng/mL) for 6 days. Noggin is then replaced with a cocktail of BMP4 (20 ng/mL), BMP7 (20 ng/mL), and FGF2 (100 ng/mL). Lens progenitors are confirmed by the expression of Pax6, αB- and αA-crystallins. The second step (expansion of the lens progenitor population) starts on day 10. The last step produces lentoid bodies by incubating with a second growth factor cocktail containing Wnt3a (20 ng/mL) and FGF2 (100 ng/mL). The expression of BFSP1, BFSP2, and Aquaporin 0 is evident from day 35. *Adapted from Yang et al. (2010).*

The last step in the protocol encourages the progenitor cells into the final differentiation phase as evidenced by the formation of lentoid bodies (Fig. 2). Lentoid bodies are 3D transparent and light refractive structures (O'Connor & McAvoy, 2007; Wagner & Takemoto, 2001). To form lentoids, BMP4 and BMP7 are removed at day 18, and cell cultures are then incubated with Wingless-type member 3a (Wnt3a) (20 ng/mL, PePro tech, USA) and FGF2 (100 ng/mL). At day 35, markers of fiber cell differentiation are expressed, for example, the IF proteins, BFSP1 and BFSP2, and the integral membrane protein Aquaporin 0.

The current procedure requires a deeper appreciation of the molecular mechanisms involved in lentoid body formation. Culture conditions have to be optimized in order to improve the success of the differentiation process in culture, but it is a unique model that allows human lens development and differentiation to be studied *in vitro*. It is therefore also an important emerging tool to study the role of IFs in lens development and fiber cell differentiation.

2.2 *Ex Vivo* Lens Models to Study IFs

The dissection of tissues and then their culture in a controlled tissue culture environment is a well-established approach to study tissue interactions and organogenesis (Resau, Sakamoto, Cottrell, Hudson, & Meltzer, 1991). It has been applied to the lens (Barnett, Lin, Akoyev, Willard, & Takemoto, 2008; Heys, Friedrich, & Truscott, 2007; Jarrin, Pandit, & Gunhaga, 2012; Mehta, Hull, & Lawrenson, 2015; Varma & Hegde, 2010), including a human lens capsular bag system (human lens epithelial cell proliferation in a protein-free medium; Wormstone et al., 1997). This model system maintains the integrity of the tissue, while allowing the effects of molecular activators and inhibitors to be studied. Once organ/tissue culture conditions have been established, this type of model makes it easier to observe and study specific morphological changes and correlate these with gene and protein expression changes. The issue with such models is that any intraocular signaling (Beebe, Holekamp, Siegfried, & Shui, 2011) or morphogen gradients (Wu et al., 2014) will be compromised.

2.2.1 Rat Lens Epithelium Explants

Lens epithelial explants (McAvoy, 1980) and their derivatives are a good model to study epithelial cell proliferation and migration. The critical aspect of this model is the isolation of the lens epithelium layer. Different variations of the explanting procedure have been described (West-Mays, Pino, & Lovicu, 2010; Zelenka, Gao, & Saravanamuthu, 2009).

Newborn rats are euthanized, eyes enucleated, and lenses isolated at the correct age. Lenses were placed in a 35-mm dish (Greiner, UK) with Medium 199 (Gibco, UK) and quickly cleaned of surrounding adherent eye tissue, e.g., ciliary body and iris. Lenses were placed with the posterior side facing upward under a dissecting microscope. Lens epithelium is tightly bound to the lens capsule (Wu et al., 2015). This feature allows the lens epithelial explant to be isolated by removing the lens capsule.

Several cuts are made in the posterior lens capsule, where there are no lens epithelial cells. The posterior lens capsule is peeled back and pinned down to reveal the body of the lens comprising the fiber cells. This fiber cell mass is removed in its entirety by a pair of curve tweezers. Explants are briefly washed once with Medium 199. Lens capsule with the epithelial cells left facing upward are placed onto cell culture plastic glass bottom Petri dishes, poly-D-lysine coated (Greiner, UK; 35 mm in diameter). Such samples can be cultivated *ex vivo* under adherent conditions with fresh Medium 199 supplemented with 10% (v/v) FBS and 1% (v/v) antibiotics (penicillin–streptomycin; Sigma) at 37 °C and 5% CO_2. The preparations were cultured until epithelial cells had recolonized the lens capsule before any treatments were considered. The anterior rat lens epithelium explant provides a homogeneous population of undifferentiated cells. Culture explants for different lengths of time should provide information of lens differentiation. The combination of these explants with different activators of inhibitors will allow to manipulate lens differentiation.

2.2.2 *Rat* Ex Vivo *Lens Fiber Cell Explants*

We describe a protocol to isolate lens fiber cells (Chandra et al., 2002). Eyes are dissected from young adult lenses from 200 to 250 g rats, intact lenses isolated and then transferred into a trypsin solution, and incubated for 15 min at 33 °C. The solution contains 0.6 mg of trypsin (Sigma, UK), 280 n*M* sucrose, 10 m*M* Na–EDTA, and 10 m*M* Hepes, pH 7.4. After removal of the solution, fresh trypsin solution is added and incubated at 32 °C for 6 min. After this, the temperature is increased gradually in 1 °C per minute of incubation until arrive to 37 °C, then incubated for additional 3 min. The capsule is then carefully removed with the attached lens epithelium. The fiber cell mass is then rotated for 20 min at a 1 revolution per second for 15–20 min at room temperature. Finally, the suspension is removed and single fiber cells plated onto a coverslip.

2.2.3 *Chick Lens-Dissociated Epithelial Model*

Culturing primary lens epithelial cells from the dissociated chick lens epithelium of embryonic chicks has been used to study fiber cell differentiation (Jarrin, Mansergh, Boulton, Gunhaga, & Wride, 2012; Menko, Klukas, & Johnson, 1984; Wride & Sanders, 1998). These cultures are capable of reproducing several of the main events of lens morphogenesis. Epithelial cells that differentiate into lentoid bodies have many characteristics in common with the lens fiber cells including the robust expression of delta-crystallin, the formation of adherens junctions and the expression of beaded filament proteins (Granger & Lazarides, 1984; Menko et al., 2014).

Thirty-six lenses were collected from embryonic day (ED) 10 chicken embryos. Lenses were dissected. The lenses were placed in a plastic Petri dish (Falcon, UK) in Tyrode's saline solution containing gentamycin (50 μg/mL). Vitreous and other tissues were removed from the lenses. The lenses were washed three times, and then the lenses were incubated at 37 °C for 15 min, with 1 mL prewarmed 2.5% (w/v) trypsin solution. The solution was passed through a 22 G needle and then centrifuged at 10,000 × *g* for 2 min. The pellet was resuspended by pipetting 300 μL of Medium 199 (Gibco, UK) containing 10% (v/v) fetal calf serum plus antibiotics and then filtered using a 40 μm Falcon cell strainer (BD Biosciences, UK).

Twenty-four-well plates were coated with 1.2 mg/mL of Matrigel (Invitrogen, UK). The wells were washed with Medium 199 before 5×10^5 cells/well were seeded. Incubation of cells was performed in a humid atmosphere at 37 °C in 5% (v/v) CO_2. Cells were allowed to attach and begin to spread for 24 h, and this was designated day 0 of culture. The medium was subsequently changed every day. The use of *in vitro* models or explants provides important information on epithelial cell proliferation and cell migration, but they are not ideal to study lens fiber cell differentiation. Although *in vivo* models overcome this limitation, the complexity of the *in vivo* models and difficulties associated with *in vivo* bioimaging techniques mean that there are still methodological obstacles to be overcome. The dissociated chick lens epithelial model is an alternative to study fiber cell differentiation events.

2.3 *In Vivo* Lens Models to Study IFs

Basic biomedical research relies on selecting the model most optimal for specific scientific questions and so it is rare that a single approach suffice. Rodent models have been the dominant *in vivo* model, but for those studies requiring real-time imaging of lens development, other models are better suited, such as the zebrafish.

2.3.1 Zebrafish

In the last few decades, the zebrafish (*Danio rerio*) model has grown in importance to the research community. The fact that zebrafish embryos develop externally within their own chorion provides the opportunity for numerous manipulations as soon as fertilization has taken place. These include the injection of tracer dyes, morpholinos, or of messenger ribonucleic acid (mRNA) and transplantation of donor cells (Bibliowicz et al., 2011). The zebrafish is a vertebrate with several unique features. Zebrafish development is fast, acquiring sexual maturation after 3 months. As many as 200 embryos per week per mating pair can be generated assisting the statistical analysis of experimental data (Kimmel, 1989). Their rapid development and the high number of embryos coupled with their transparency make the zebrafish a perfect model for *in vivo* imaging to study cell proliferation, migration, and organogenesis without having to sacrifice the animal. Zebrafish also absorb drugs and vital dyes dissolved in the tank water, a unique advantage for chemical/drug screening as well as reverse and forward genetic manipulation. Furthermore, the research community is supported by an excellent database of genetic, genomic, and developmental information: the Zebrafish Model Organism Database (Bradford et al., 2011).

The zebrafish genome is published and as ~70% of all human disease genes have functional homologs in zebrafish, they are proving to be a valuable model of human diseases (Langheinrich, 2003). The zebrafish has been widely used for chemical and insertional mutagenesis (Amsterdam & Hopkins, 2006). This approach has led to the identification of a myriad of mutants with disruption of conserved genes that correlate to human (Amsterdam & Hopkins, 2006; Haffter et al., 1996; Patton & Zon, 2001). Different large-scale forward genetic screens have been performed using zebrafish and the methylating agent ethylnitrosourea as a mutagen (Amsterdam & Hopkins, 2006; Muto et al., 2005), but probably the more common use of zebrafish is the reverse genetic manipulation to assess the consequence of the removal of specific genes in either transient or stable approaches.

These transient reverse genetic techniques permit a quick and easy probing of specific gene function *in vivo*, since in zebrafish, most of the organs have formed and are fully functional 5 days after hatching (Kimmel, 1989; Santoriello & Zon, 2012). Transient gene manipulation involves either the over or reduced expression of the gene of interests (GOIs). Transient overexpression is performed by microinjection of specific mRNA into embryos (Santoriello & Zon, 2012). Knockdown or reduced expression of GOI products requires the microinjection of modified antisense oligos (known as "morpholinos"). Morpholinos are injected into one- to two-cell stage embryos to knockdown expression of a target gene. These oligonucleotides can be designed to 5′ UTR or initiation codon to prevent translation (Nasevicius & Ekker, 2000). Targeting an exon–intron boundary will interfere with GOI splicing (Draper, Morcos, & Kimmel, 2001). The efficiency of morpholino-mediated mRNA knockdown is evaluated at the protein expression level using specific antibodies. Alternatively, injection of small interfering RNA (siRNA) is also possible (Dodd, Chambers, & Love, 2004; Liu et al., 2005). This technology cannot be used to study gene function in mice because antisense oligonucleotides are rapidly diluted during mouse development (Santoriello & Zon, 2012).

Transient genetic manipulations are commonly used, but when long-term effects on or the complete removal of a GOI is required, then stable mutants have to be generated by transgenesis or gene editing approaches. Several methods have been developed to perform permanent gene editing: custom zinc finger nucleases (Woods & Schier, 2008) and transcription activator-like effector nucleases (Huang et al., 2011). These techniques are relatively efficient to effect the knockout of a GOI, but the insertion of exogenous DNA sequences into a targeted genomic locus can be a challenge (Hisano et al., 2015).

Recently, the CRISPR/Cas system has been used to generate gene knockouts both cheaply and with very high frequency (75–99%; Hwang et al., 2013). CRISPR/Cas9 is also a useful tool to generate mutations using a donor vector harboring short homologous sequences (10–40 bp) flanking the genomic target locus (Hisano et al., 2015).

This makes the zebrafish an excellent model system, and as vision is highly conserved between vertebrates, it is ideal to study eye and lens development. It also has significant potential to study IF structure and function. Several cDNAs from members of the IFs family have been cloned in various fish species and have been demonstrated to be very similar to their mammalian homologs (Cerda, Conrad, Markl, Brand, & Herrmann, 1998; Herrmann, Munick, Brettel, Fouquet, & Markl, 1996). Vimentin and desmin have been already studied in the zebrafish (Cerda et al., 1998; Jones & Schechter, 1987; Jones et al., 1986; Li, Andersson-Lendahl, Sejersen, & Arner, 2013). Zebrafish are a viable alternative to mouse particularly with the availability of genetic tools, mutants, and transgenics to facilitate the analysis of mechanism in a living animal.

3. METHODS TO STUDY IFs

In this section, we focus first on gene manipulations, followed by bioimaging *in vivo*. The protocols have been written to cover some of the models introduced above, but they can easily be adapted to any other animal model.

3.1 Vimentin Knockdown RNAi *In Vitro* or *Ex Vivo* Cells Culture

RNA interference (RNAi) analysis is a routine approach to study gene function by reducing GOI expression in cell cultures. Small RNA duplexes of 21-nt with 2-nt 3′ overhangs are introduced by transfection. These duplexes (known as siRNA) trigger the degradation of mRNAs containing the same sequence.

siRNA can be designed by the researcher (see Pei & Tuschl, 2006 for guidance) or can be purchased as a validated siRNA from company sources. We have produced a summary of published human siRNA-targeting vimentin (Table 2).

Table 2 RNA Interference (RNAi) Has Been Used to Target Different Regions of the Vimentin Transcript

Name	Vector	Sequence	Transfection	References
Vim-1	p-SHING-G	5′-CTACATCGACAAGGTGCGC-3′	Oligofectamine	Harborth, Elbashir, Bechert, Tuschl, and Weber (2001) Kojima and Borisy (2014)
Vim-2	p-SHING-G	5′-TACCAAGACCTGCTCAATG-3′	Oligofectamine	Harborth et al. (2001) Kojima and Borisy (2014)
Vim-3	p-SHING-G	5′-GAATGGTACAAATCCAAGT-3′	Oligofectamine	Harborth et al. (2001) Kojima and Borisy (2014)
Vim-4	p-SHING-G/p-Silencer 3.0.H1	5′-GTACGTCAGCAATATGAAA-3′	Lipofectamine 2000	Ui-Tei et al. (2004) Kojima and Borisy (2014)
Vim-5	p-SHING-G/p-Silencer 3.0.H1	5′-GATGAGATTCAGAATATGA-3′	Lipofectamine 2000	Ui-Tei et al. (2004) Kojima and Borisy (2014)

RNAi tools can be purchased from a variety of companies as validated reagents or developed empirically. In this table, we have summarized different targets that have been used successfully. We also have included vector details suitable for the cloning of the RNAi if long-term expression is required.

3.1.1 Preparation and Storage of siRNAs

1. Resuspend the siRNAs with nuclease-free water to 100 μ*M*. Fluorescently labeled siRNAs can be used to assess siRNA stability and efficiency of the transfection.
2. Positive and negative (scrambled siRNA) siRNAs should be included as controls.
3. siRNA concentration is measuring absorbance at 260 nm using a nanodrop or equivalent instrument. Aliquot into 0.5-mL tubes and store at −20 °C.

3.1.2 Transfection of Cells with siRNA

1. Prepare cells under optimal culture conditions. Plate the cells in a 96-well plate. Ideally, the cell culture should be plated at 25–50% confluency. When the cultures reach 80% confluency, then they are ready for transfection.
2. Dilute stock siRNA solution 100 μ*M* to working solution of 10 μ*M* using Opti-MEM media (Gibco, UK).
3. Determine the concentration of siRNAs to use. The final concentrations to test in the culture are from 20 to 100 n*M*. Test at least three concentrations (e.g., 25, 50, and 80 n*M*). Total volume per well is 100 μL.
4. Dilute transfection reagent Lipofectamine 2000 (Invitrogen, UK) in Opti-MEM media. Recommended concentration is 2 μL in 50 μL.
5. Prepare the final transfection solution containing the final concentration of siRNA. For a 25 n*M* solution of siRNA per well, prepare as follow: 2 μL of Lipofectamine 2000 + 97.75 μL of Opti-MEM. Gently shake and incubate at room temperature for 5 min. Add 0.25 μL of 10 μ*M* siRNA solution diluted in Opti-MEM. Prepare also your positive and negative control.
6. Incubate 20 min at room temperature before adding to the cell culture medium.
7. Remove medium from the cells. Wash once with 150 μL of Opti-MEM.
8. Add 100 μL per well of the solution containing Lipofectamine 2000 and siRNA.
9. Incubate cells for 6 h at 37 °C and 5% of CO_2.
10. Remove transfection media containing siRNAs.
11. Gently wash cells with PBS three times and add fresh culture media to allow the cells to continue growing.
12. Analysis of the effects of siRNAs with antibodies or PCR before 72 h after transfection.

3.2 Transient Manipulation of Vimentin in Zebrafish by Morpholino Microinjection

Zebrafish are an attractive alternative model to the mouse for genetic manipulation. Transient genetic manipulation using microinjection is a quick and nonexpensive approach. Several hundred eggs can be microinjected for overexpression or knockdown of vimentin within an hour (Fig. 3).

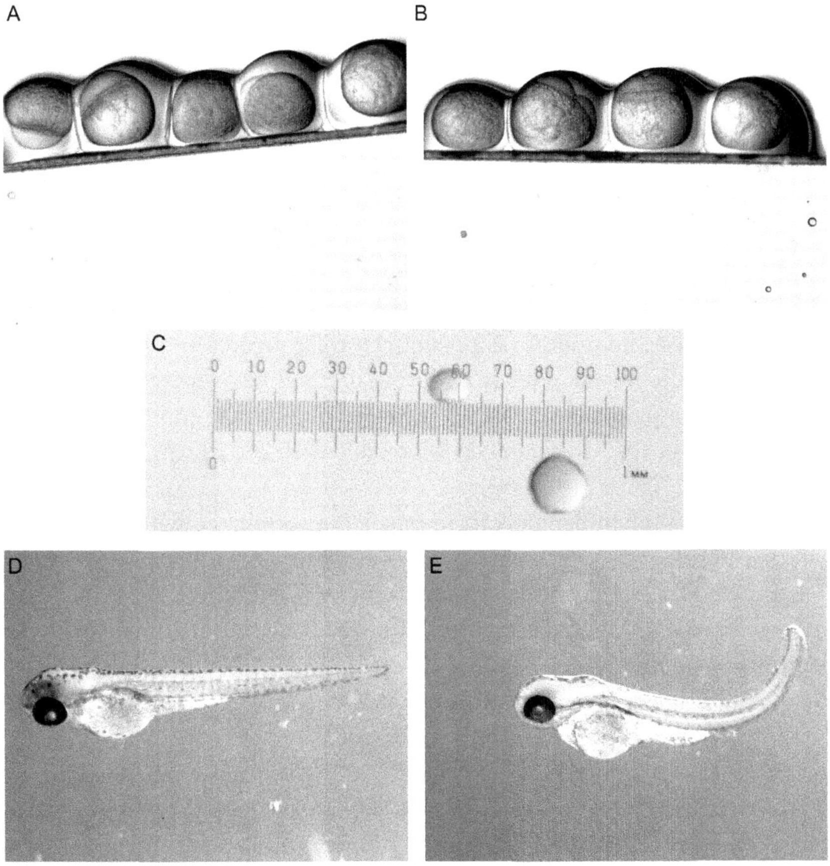

Figure 3 Preparation of zebrafish embryos for microinjection. (A) One-cell stage embryos were selected and transferred into a Petri dish and then aligned against a microscope slide prior to microinjection. (B) Two-cell stage embryos prepared for microinjection. (C) Calibration of the microinjection volume. The volume to be microinjected is measured using mineral oil and a micrometer. (D) Day 3 postfertilization control embryo with no morphological defects. (E) Day 3 microinjected embryos showing morphological defects. Effectiveness of the microinjection is confirmed by phenotypic analyses followed by protein expression analysis by immunoblotting. (See the color plate.)

3.2.1 Preparation of the Vimentin Morpholinos

Morpholino antisense oligonucleotides are common used to transient knockdown of gene expression by blocking translation of a targeted protein. Morpholinos can be designed by Gene tools (Philomath, OR).

1. 300 nmol of a morpholino oligo against vimentin, 100 nmol of positive and negative controls were ordered from Gene Tools, LLC (Philomath, OR).
2. Morpholinos were resuspended at 1 m*M* in nuclease-free water. Morpholinos and controls were heated at 65 °C for 10 min and briefly vortex. Keep at room temperature.
3. The concentration of the morpholinos was checked.
4. Morpholino stocks were diluted to a working concentration of 2 ng/nL in injection buffer (Danieau's buffer) with 1% phenol red. Different concentrations of morpholino can be achieved by varying the volume microinjected, e.g., 1 pulse: 2 ng/nL and 2 pulses: 4 ng/nL.

3.2.2 Preparation of Vimentin mRNA

Overexpression of vimentin is possible by microinjection of capped RNA into zebrafish embryos. This is a routine method to overexpress any GOI.

1. Linearize 5 μg of plasmid DNA by digestion with the appropriate restriction enzyme for 2 h. Then DNA is treated with 200 μg/mL in 0.5% (w/v) SDS at 50 °C for 30–60 min.
2. Samples are phenol–chloroform purified, precipitated with ethanol, and then resuspended in nuclease-free water.
3. Sample concentration determined using the nanodrop.
4. An *in vitro* CAP RNA transcription synthesis is performed using the mMessage mMachine SP6 kit (Ambion, UK).
5. RNA is phenol–chloroform extracted, ethanol precipitated, and then resuspended in nucleases-free water.
6. Samples are aliquoted and stored at −80 °C until is needed.
7. On the day of microinjection, mRNA stocks are diluted to the desired working concentration in Danieau's buffer with 0.5% (w/v) phenol red. All new mRNA preparations need to be titrated for efficacy. Several pulses can be injected to test different doses.
8. mRNA samples should be kept on ice at all times.

3.2.3 Preparation of the Embryos

1. Set up fish pairs the night before microinjection. Prepare breeding tanks with dividers to keep two females and one male separated in the same tank.

2. On the morning of the microinjection, prepare the sample, set up the instrument, load the needle, and connect to the microinjector rig after confirming the pulse/volume parameters.
3. After step 2 is completed, remove the dividers from 2–3 breeding tanks. Wait 10–15 min to allow eggs to be laid and fertilized. Do not remove the dividers of all available tanks because microinjection can only be done at the one- to two-cell embryo stage. Once the first set is microinjected, then remove the dividers of the next 2–3 tanks to ensure a steady supply of fertilized eggs.
4. Collect the eggs using a cell strainer. Rinse the eggs and place the eggs in a Petri dish with egg water media. Remove unfertilized eggs.

3.2.4 Microinjection

1. Place a glass slide in a 100-mm Petri dish. Transfer 20 eggs to the dish and line up the eggs against the slide. Remove excess of egg water.
2. One- to two-cell stages embryos are microinjected. Orientate the embryos with the cytoplasm visible under the dissecting microscope.
3. Introduce the tip of the micropipette via the chorion into the egg yolk and then microinject. Successful microinjection in the yolk is confirmed by the presence of phenol red. Microinjected material is absorbed into the embryo from the yolk. Withdraw carefully the micropipette without disturbing the embryo.
4. After microinjection, place the embryos into a Petri dish with egg water medium and incubate them at 28.5 °C for normal development.
5. After 24 h, remove any dead embryos.
6. Always microinject positive and negative controls to evaluate properly the phenotypes obtained.

3.3 Bioimaging in Living Zebrafish

Imaging *in vivo* means that cells can be tracked throughout development. This allows not only the recording of cell shape and size changes but also cell proliferation and migration. This technique is an invaluable tool to address the cellular changes after gene manipulation of vimentin, BFSP1, or BFSP2. In order to perform *in vivo* imaging, different approaches can be followed. Here, we have chosen to use selective plane illumination microscopy (SPIM). SPIM is a technique for three-dimensional imaging of tissue specimens *in vivo* by sectioning them optically using a light sheet. This technique reduces unnecessary exposure of the sample since it only illuminates the target plane, reducing phototoxic and photobleaching effects.

This technique is also fast since it records an entire plane in the sample in a single frame, which allows cells to be tracked with a high time resolution.

3.3.1 Material Required

a. Fluorescent stable lines. There are many zebrafish models available expressing stably integrated GFP-tagged proteins and their derivatives. We use the Q01 transgenic zebrafish, which express cyan fluorescent protein (CFP) fused to a membrane-targeting sequence of the Gap43 gene. This transgenic line was provided by Rachel O.L. Wong (University of Washington, Seattle). CFP expression is directed to those tissues regulated by Pax-6. The second transgenic zebrafish line that we use is H2A. This fish line expresses a fusion of a histone variant H2A to the green fluorescent protein (GFP), provided by David Raible (Department of Biological Structure, University of Washington, Seattle, USA).

b. Nonfluorescent fish can be labeled *in vivo* by adding fluorescent vital dyes to the water. Staining with BODIPY TR methyl ester dye (Invitrogen, UK) is common performed to label cellular membranes *in vivo*. We use also Coumarin 6, a membrane staining vital dye that is preferentially uptaken in the eye (Watanabe et al., 2010; Fig. 4).

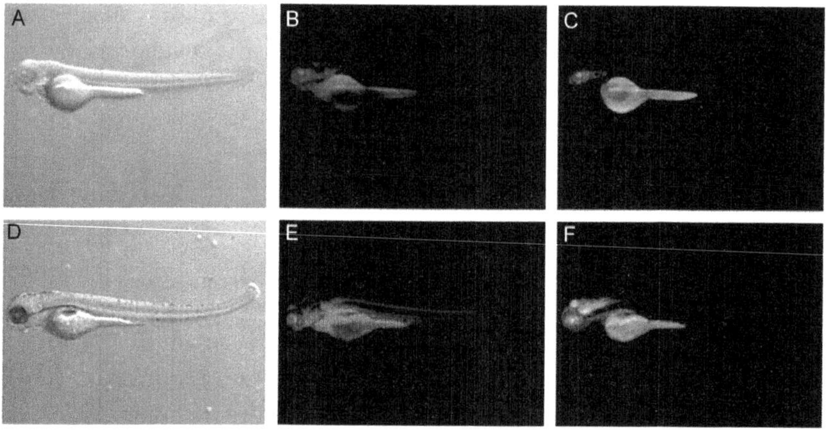

Figure 4 Staining of the embryos *in vivo* with vital dyes. The use of different vital dyes is an invaluable technique for bioimaging applicable to any zebrafish embryo. In this figure, we have summarized the representative staining of the day 3 embryos. (A)–(C) Control day 3 embryo. (D)–(F) Day 3 microinjected embryo. Embryos were staining for 1 h with coumarin 6 (C) and (F), and BODIPY TR (B) and (E) to detect cell membranes. (See the color plate.)

c. We use a custom-built SPIM, which uses a cylindrical lens to create a light sheet that is relayed through a 0.3 NA (numerical aperture) 10 × Nikon objective lens (the illumination objective, Nikon, Japan). The light sheet is imaged along an axis perpendicular to the direction of its propagation, using a 0.8 NA 16 × Nikon objective lens (Nikon, Japan). The sample is scanned through the light sheet to image planes at different depths. This technique is summarized in Fig. 5.

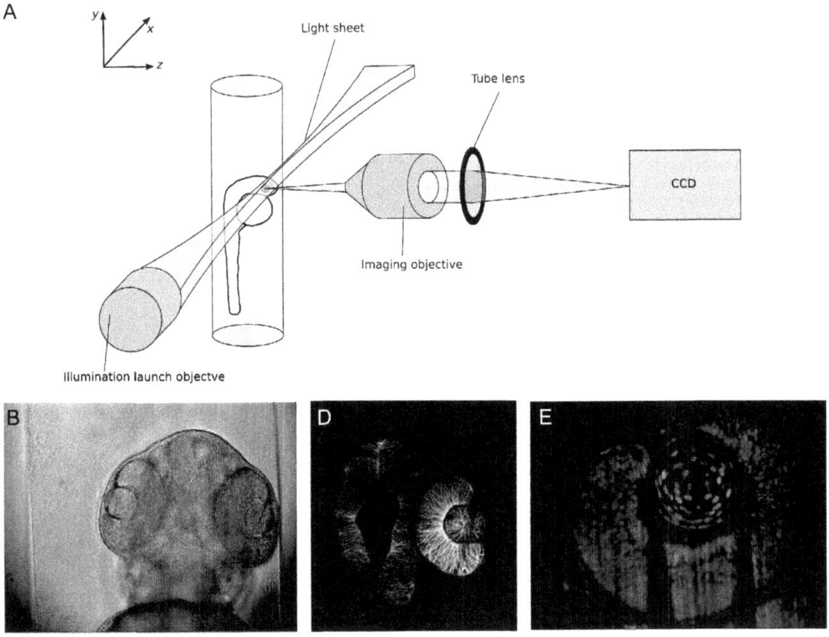

Figure 5 Selective plane illumination microscopy (SPIM). The SPIM technique allows optical sectioning of the sample using a light sheet. Q01 and H2A-eGFP embryos were used for imaging positioned within the light sheet by a fluorinated ethylene propylene (FEP) tube. A schematic of the SPIM technique for eye lens imaging is summarized in panel (A). The SPIM uses two Nikon water-dipping objective lenses. A light sheet is generated using a cylindrical lens and projected on to the back aperture of a 10 × Plan Fluor objective. This produces a light sheet that illuminates a thin (~3 μm) slice of the sample, and the resulting fluorescence is imaged perpendicular to the sheet via a 16 × LWD objective. A three-dimensional image of the sample can be built up, layer-by-layer, by scanning the sample through the light sheet. (B) Bright field image of day 2 Q01 embryo held in the FEP tube, captured during orientation of the sample. (C) Fluorescence image of a plane within the head of a day 1 Q01 embryo, recorded using SPIM. The image shows expression of CFP in the brain, retina and lens in a plane through the center of the eye lens. (D) Fluorescence image of a section through the center of the eye lens of a day 2 H2A-GFP embryo. GFP expression is targeted to all cell nuclei in the fish. *Panel (A) adapted from Young et al. (2014).*

3.3.2 Sample Preparation

1. Follow steps 1–4 in Section 3.2.3.
2. Split the fertilized eggs and early embryos into groups of 50 eggs per 100-mm dish with egg water. Remove the nonviable eggs. Put them in an incubator at 28.5 °C.
3. After 4 h, add 0.03 mg/mL 1-phenyl 2-thiourea (PTU; Sigma, UK) to each dish to prevent embryo pigmentation.
4. After 20 h, check again for nonviable embryos.
5. At the developmental stage required, select the fluorescent embryos or else select unlabelled fish for incubation with fluorescent (vital) dyes.
6. Transfer the selected fish to a 35-mm dish with fresh egg water. Add 0.05 mg/mL of tricaine (MS-222, Sigma, UK) and carefully dechorionate the embryo using two forceps under a dissecting microscope.
7. Transfer the embryos to fresh egg water and incubate 1 h before imaging.
8. If labeling is required, put up to 10 embryos in 1.5-mL tube with 500 μL of egg water for 1 h at 28.5 °C. Add the required dye. BODIPY TR methyl ester dye (50 m*M*) and Coumarin 6 (2 μg/mL) in 1% DMSO.
9. After labeling the fish, wash three times with egg water and then image.

3.3.3 Imaging Preparation

1. Cut the cleaned fluorinated ethylene propylene (FEP) tube into a piece of about 4 cm. Attach the FEP tube to a 1-mL syringe (Terumo, UK).
2. Introduce the fish into the syringe-attached FEP tube with the egg water and tricaine. This method allows imaging the fish for a short period of time (30–40 min).
3. Orientate the fish with the head upward and close to the wall of the tube. Check for air bubbles and make sure that the tail does not obscure the head.
4. A stack of images are taken across the entire width of the lens in each eye using an excitation wavelength of 405 nm for Q01, 488 nm for coumarin, 6 or 561 nm for BODIPY TR methyl ester dye.

3.3.4 Live Cell Imaging over the Course of Zebrafish Development

For time series, fish are embedded in 0.7% (w/v) low melting agarose in egg water containing 0.2 mg/mL of tricaine (MS-222, Sigma, UK), using the procedure outlined below.

1. Heat up the 0.7% (w/v) agarose solution to 65 °C in a heat block. Let it cool down to 40 °C and maintain this temperature until use.

2. Put the agarose in a dish at room temperature to cool down. Add tricaine (MS-222, Sigma, UK) until final concentration of 0.2 mg/mL. Select one dechorionated zebrafish embryo and transfer it to the dish.
3. Embed the fish using FEP tube and a plunger. Start by taking only the agarose solution with tricaine, then collect the fish with the head oriented upward, and finally add more agarose solution after the fish.
4. Allow 5 min for the agarose to set.
5. For imaging, push the embedded fish very gently out of the tubing using the plunger.

4. CONCLUDING REMARKS

The lens is a good model to study IFs structure and function (Fudge et al., 2011; Gokhin et al., 2012; Sandilands et al., 2003; Song et al., 2009). Bovine lenses are a good source of IF proteins when biochemical quantities are required. The mouse and zebrafish are complementary animal models, but the zebrafish is the system of choice for live cell imaging and dynamic studies in a living animal. Here, we have described different models and protocols to study IFs in the eye lens. Protocols to differentiate either stem cells or inducible pluripotential stem cells (iPSc) into lens cells are probably the most exciting cell-based modes given their potential to investigate gene networks and their disease-causing mutants through the generation of iPSc cell banks from patients with congenital cataract. Such a resource will help show how IFs mutations perturb and alter normal filament dynamics and their role in cataractogenesis. Zebrafish allow the study of IF structure and function *in vivo*, using GFP-fusions, reverse/forward genetics as well as drugs/small molecules to manipulate IFs. Here, new (SPIM) and old (2-photon) *in vivo* bioimaging technologies are continually being refined to improve resolution and are matched by segmentation tools to optimize data analysis.

Lens research provides a valuable system to reveal the role of IFs in organogenesis, but the ultimate aim is to understand how IFs are integrated into and facilitate the cell biology of the lens so that this can be translated into the lens regeneration strategies to combat presbyopia and cataract.

REFERENCES

Alizadeh, A., Clark, J., Seeberger, T., Hess, J., Blankenship, T., & FitzGerald, P. G. (2004). Characterization of a mutation in the lens-specific CP49 in the 129 strain of mouse. *Investigative Ophthalmology & Visual Science*, *45*(3), 884–891.

Alizadeh, A., Clark, J. I., Seeberger, T., Hess, J., Blankenship, T., Spicer, A., et al. (2002). Targeted genomic deletion of the lens-specific intermediate filament protein CP49. *Investigative Ophthalmology & Visual Science, 43*(12), 3722–3727.

Amsterdam, A., & Hopkins, N. (2006). Mutagenesis strategies in zebrafish for identifying genes involved in development and disease. *Trends in Genetics, 22*(9), 473–478. http://dx.doi.org/10.1016/j.tig.2006.06.011.

Andley, U. P., Hebert, J. S., Morrison, A. R., Reddan, J. R., & Pentland, A. P. (1994). Modulation of lens epithelial cell proliferation by enhanced prostaglandin synthesis after UVB exposure. *Investigative Ophthalmology & Visual Science, 35*(2), 374–381.

Andley, U. P., Rhim, J. S., Chylack, L. T., Jr., & Fleming, T. P. (1994). Propagation and immortalization of human lens epithelial cells in culture. *Investigative Ophthalmology & Visual Science, 35*(7), 3094–3102.

Barnett, M., Lin, D., Akoyev, V., Willard, L., & Takemoto, D. (2008). Protein kinase C epsilon activates lens mitochondrial cytochrome c oxidase subunit IV during hypoxia. *Experimental Eye Research, 86*(2), 226–234. http://dx.doi.org/10.1016/j.exer.2007.10.012.

Beebe, D. C., Holekamp, N. M., Siegfried, C., & Shui, Y. B. (2011). Vitreoretinal influences on lens function and cataract. *Philosophical Transactions of the Royal Society of London. Series B, Biological Sciences, 366*(1568), 1293–1300. http://dx.doi.org/10.1098/rstb.2010.0228.

Bibliowicz, J., Tittle, R. K., & Gross, J. M. (2011). Toward a better understanding of human eye disease insights from the zebrafish, Danio rerio. *Progress in Molecular Biology and Translational Science, 100*, 287–330. http://dx.doi.org/10.1016/b978-0-12-384878-9.00007-8.

Bloemendal, H., de Jong, W., Jaenicke, R., Lubsen, N. H., Slingsby, C., & Tardieu, A. (2004). Ageing and vision: Structure, stability and function of lens crystallins. *Progress in Biophysics and Molecular Biology, 86*(3), 407–485. http://dx.doi.org/10.1016/j.pbiomolbio.2003.11.012.

Bozanic, D., Bocina, I., & Saraga-Babic, M. (2006). Involvement of cytoskeletal proteins and growth factor receptors during development of the human eye. *Anatomy and Embryology, 211*(5), 367–377. http://dx.doi.org/10.1007/s00429-006-0087-z.

Bradford, Y., Conlin, T., Dunn, N., Fashena, D., Frazer, K., Howe, D. G., et al. (2011). ZFIN: Enhancements and updates to the zebrafish model organism database. *Nucleic Acids Research, 39*(Database issue), D822–829. http://dx.doi.org/10.1093/nar/gkq1077.

Butin-Israeli, V., Adam, S. A., Goldman, A. E., & Goldman, R. D. (2012). Nuclear lamin functions and disease. *Trends in Genetics, 28*(9), 464–471. http://dx.doi.org/10.1016/j.tig.2012.06.001.

Capetanaki, Y., & Milner, D. J. (1998). Desmin cytoskeleton in muscle integrity and function. *Subcellular Biochemistry, 31*, 463–495.

Capetanaki, Y., Milner, D. J., & Weitzer, G. (1997). Desmin in muscle formation and maintenance: Knockouts and consequences. *Cell Structure and Function, 22*(1), 103–116.

Capetanaki, Y., Smith, S., & Heath, J. P. (1989). Overexpression of the vimentin gene in transgenic mice inhibits normal lens cell differentiation. *The Journal of Cell Biology, 109*(4 Pt. 1), 1653–1664.

Cerda, J., Conrad, M., Markl, J., Brand, M., & Herrmann, H. (1998). Zebrafish vimentin: Molecular characterization, assembly properties and developmental expression. *European Journal of Cell Biology, 77*(3), 175–187. http://dx.doi.org/10.1016/s0171-9335(98)80105-2.

Chandra, D., Ramana, K. V., Wang, L., Christensen, B. N., Bhatnagar, A., & Srivastava, S. K. (2002). Inhibition of fiber cell globulization and hyperglycemia-induced lens opacification by aminopeptidase inhibitor bestatin. *Investigative Ophthalmology & Visual Science, 43*(7), 2285–2292.

Chang, P. Y., Bjornstad, K. A., Chang, E., McNamara, M., Barcellos-Hoff, M. H., Lin, S. P., et al. (2000). Particle irradiation induces FGF2 expression in normal human lens cells. *Radiation Research, 154*(5), 477–484.

Chung, B. M., Rotty, J. D., & Coulombe, P. A. (2013). Networking galore: Intermediate filaments and cell migration. *Current Opinion in Cell Biology*, *25*(5), 600–612. http://dx.doi.org/10.1016/j.ceb.2013.06.008.

Coulombe, P. A., Hutton, M. E., Vassar, R., & Fuchs, E. (1991). A function for keratins and a common thread among different types of epidermolysis bullosa simplex diseases. *The Journal of Cell Biology*, *115*(6), 1661–1674.

Dodd, A., Chambers, S. P., & Love, D. R. (2004). Short interfering RNA-mediated gene targeting in the zebrafish. *FEBS Letters*, *561*(1–3), 89–93. http://dx.doi.org/10.1016/s0014-5793(04)00129-2.

Draper, B. W., Morcos, P. A., & Kimmel, C. B. (2001). Inhibition of zebrafish fgf8 pre-mRNA splicing with morpholino oligos: A quantifiable method for gene knockdown. *Genesis*, *30*(3), 154–156.

Dunia, I., Pieper, F., Manenti, S., van de Kemp, A., Devilliers, G., Benedetti, E. L., et al. (1990). Plasma membrane-cytoskeleton damage in eye lenses of transgenic mice expressing desmin. *European Journal of Cell Biology*, *53*(1), 59–74.

Eguchi, G., Eguchi, Y., Nakamura, K., Yadav, M. C., Millan, J. L., & Tsonis, P. A. (2011). Regenerative capacity in newts is not altered by repeated regeneration and ageing. *Nature Communications*, *2*, 384. http://dx.doi.org/10.1038/ncomms1389.

Ellis, M., Alousi, S., Lawniczak, J., Maisel, H., & Welsh, M. (1984). Studies on lens vimentin. *Experimental Eye Research*, *38*(2), 195–202.

FitzGerald, P. G., & Casselman, J. (1991). Immunologic conservation of the fiber cell beaded filament. *Current Eye Research*, *10*(5), 471–478.

Fudge, D. S., McCuaig, J. V., Van Stralen, S., Hess, J. F., Wang, H., Mathias, R. T., et al. (2011). Intermediate filaments regulate tissue size and stiffness in the murine lens. *Investigative Ophthalmology & Visual Science*, *52*(6), 3860–3867. http://dx.doi.org/10.1167/iovs.10-6231.

Gokhin, D. S., Nowak, R. B., Kim, N. E., Arnett, E. E., Chen, A. C., Sah, R. L., et al. (2012). Tmod1 and CP49 synergize to control the fiber cell geometry, transparency, and mechanical stiffness of the mouse lens. *PLoS One*, 7(11), e48734. http://dx.doi.org/10.1371/journal.pone.0048734.

Granger, B. L., & Lazarides, E. (1984). Expression of the intermediate-filament-associated protein synemin in chicken lens cells. *Molecular and Cellular Biology*, *4*(10), 1943–1950.

Graw, J. (2009). Mouse models of cataract. *Journal of Genetics*, *88*(4), 469–486.

Greiling, T. M., & Clark, J. I. (2012). New insights into the mechanism of lens development using zebra fish. *International Review of Cell and Molecular Biology*, *296*, 1–61. http://dx.doi.org/10.1016/b978-0-12-394307-1.00001-1.

Gross, J. M., & Perkins, B. D. (2008). Zebrafish mutants as models for congenital ocular disorders in humans. *Molecular Reproduction and Development*, *75*(3), 547–555. http://dx.doi.org/10.1002/mrd.20831.

Haffter, P., Granato, M., Brand, M., Mullins, M. C., Hammerschmidt, M., Kane, D. A., et al. (1996). The identification of genes with unique and essential functions in the development of the zebrafish, Danio rerio. *Development*, *123*, 1–36.

Harborth, J., Elbashir, S. M., Bechert, K., Tuschl, T., & Weber, K. (2001). Identification of essential genes in cultured mammalian cells using small interfering RNAs. *Journal of Cell Science*, *114*(Pt 24), 4557–4565.

Haque, M. S., Arora, J. K., Dikdan, G., Lysz, T. W., & Zelenka, P. S. (1999). The rabbit lens epithelial cell line N/N1003A requires 12-lipoxygenase activity for DNA synthesis in response to EGF. *Molecular Vision*, *5*, 8.

Herrmann, H., Munick, M. D., Brettel, M., Fouquet, B., & Markl, J. (1996). Vimentin in a cold-water fish, the rainbow trout: Highly conserved primary structure but unique assembly properties. *Journal of Cell Science*, *109*(Pt. 3), 569–578.

Herrmann, H., Strelkov, S. V., Burkhard, P., & Aebi, U. (2009). Intermediate filaments: Primary determinants of cell architecture and plasticity. *The Journal of Clinical Investigation*, *119*(7), 1772–1783. http://dx.doi.org/10.1172/jci38214.

Heys, K. R., Friedrich, M. G., & Truscott, R. J. (2007). Presbyopia and heat: Changes associated with aging of the human lens suggest a functional role for the small heat shock protein, alpha-crystallin, in maintaining lens flexibility. *Aging Cell*, *6*(6), 807–815. http://dx.doi.org/10.1111/j.1474-9726.2007.00342.x.

Hirsch, N., Zimmerman, L. B., Gray, J., Chae, J., Curran, K. L., Fisher, M., et al. (2002). Xenopus tropicalis transgenic lines and their use in the study of embryonic induction. *Developmental Dynamics*, *225*(4), 522–535. http://dx.doi.org/10.1002/dvdy.10188.

Hisano, Y., Sakuma, T., Nakade, S., Ohga, R., Ota, S., Okamoto, H., et al. (2015). Precise in-frame integration of exogenous DNA mediated by CRISPR/Cas9 system in zebrafish. *Scientific Reports*, *5*, 8841. http://dx.doi.org/10.1038/srep08841.

Homberg, M., & Magin, T. M. (2014). Beyond expectations: Novel insights into epidermal keratin function and regulation. *International Review of Cell and Molecular Biology*, *311*, 265–306. http://dx.doi.org/10.1016/b978-0-12-800179-0.00007-6.

Huang, P., Xiao, A., Zhou, M., Zhu, Z., Lin, S., & Zhang, B. (2011). Heritable gene targeting in zebrafish using customized TALENs. *Nature Biotechnology*, *29*(8), 699–700. http://dx.doi.org/10.1038/nbt.1939.

Hwang, W. Y., Fu, Y., Reyon, D., Maeder, M. L., Tsai, S. Q., Sander, J. D., et al. (2013). Efficient genome editing in zebrafish using a CRISPR-Cas system. *Nature Biotechnology*, *31*(3), 227–229. http://dx.doi.org/10.1038/nbt.2501.

Ibaraki, N., Chen, S. C., Lin, L. R., Okamoto, H., Pipas, J. M., & Reddy, V. N. (1998). Human lens epithelial cell line. *Experimental Eye Research*, *67*(5), 577–585. http://dx.doi.org/10.1006/exer.1998.0551.

Ingber, D. E. (2006). Mechanical control of tissue morphogenesis during embryological development. *The International Journal of Developmental Biology*, *50*(2–3), 255–266. http://dx.doi.org/10.1387/ijdb.052044di.

Inoue, T., Inoue, R., Tsutsumi, R., Tada, K., Urata, Y., Michibayashi, C., et al. (2012). Lens regenerates by means of similar processes and timeline in adults and larvae of the newt Cynops pyrrhogaster. *Developmental Dynamics*, *241*(10), 1575–1583. http://dx.doi.org/10.1002/dvdy.23854.

Ireland, M. E., Tran, K., & Mrock, L. (1993). Beta-adrenergic mechanisms affect cell division and differentiation in cultured chick lens epithelial cells. *Experimental Eye Research*, *57*(3), 325–333. http://dx.doi.org/10.1006/exer.1993.1131.

Ireland, M. E., Wallace, P., Sandilands, A., Poosch, M., Kasper, M., Graw, J., et al. (2000). Up-regulation of novel intermediate filament proteins in primary fiber cells: An indicator of all vertebrate lens fiber differentiation? *The Anatomical Record*, *258*(1), 25–33.

Ishida-Yamamoto, A., McGrath, J. A., Chapman, S. J., Leigh, I. M., Lane, E. B., & Eady, R. A. (1991). Epidermolysis bullosa simplex (Dowling-Meara type) is a genetic disease characterized by an abnormal keratin-filament network involving keratins K5 and K14. *The Journal of Investigative Dermatology*, *97*(6), 959–968.

Ishikawa, H., Bischoff, R., & Holtzer, H. (1968). Mitosis and intermediate-sized filaments in developing skeletal muscle. *The Journal of Cell Biology*, *38*(3), 538–555.

Janmey, P. A., Euteneuer, U., Traub, P., & Schliwa, M. (1991). Viscoelastic properties of vimentin compared with other filamentous biopolymer networks. *The Journal of Cell Biology*, *113*(1), 155–160.

Jarrin, M., Mansergh, F. C., Boulton, M. E., Gunhaga, L., & Wride, M. A. (2012). Survivin expression is associated with lens epithelial cell proliferation and fiber cell differentiation. *Molecular Vision*, *18*, 2758–2769.

Jarrin, M., Pandit, T., & Gunhaga, L. (2012). A balance of FGF and BMP signals regulates cell cycle exit and Equarin expression in lens cells. *Molecular Biology of the Cell*, *23*(16), 3266–3274. http://dx.doi.org/10.1091/mbc.E12-01-0075.

Jensen, M. H., Morris, E. J., Goldman, R. D., & Weitz, D. A. (2014). Emergent properties of composite semiflexible biopolymer networks. *Bioarchitecture*, *4*(4–5), 138–143. http://dx.doi.org/10.4161/19490992.2014.989035.

Jones, P. S., & Schechter, N. (1987). Distribution of specific intermediate-filament proteins in the goldfish retina. *The Journal of Comparative Neurology*, *266*(1), 112–121. http://dx.doi.org/10.1002/cne.902660109.

Jones, P. S., Tesser, P., Keyser, K. T., Quitschke, W., Samadi, R., Karten, H. J., et al. (1986). Immunohistochemical localization of intermediate filament proteins of neuronal and nonneuronal origin in the goldfish optic nerve: Specific molecular markers for optic nerve structures. *Journal of Neurochemistry*, *47*(4), 1226–1234.

Karabinos, A. (2013). The cephalochordate Branchiostoma genome contains 26 intermediate filament (IF) genes: Implications for evolution of chordate IF proteins. *European Journal of Cell Biology*, *92*(8–9), 295–302. http://dx.doi.org/10.1016/j.ejcb.2013.10.004.

Kasper, M., & Viebahn, C. (1992). Cytokeratin expression and early lens development. *Anatomy and Embryology*, *186*(3), 285–290.

Kimmel, C. B. (1989). Genetics and early development of zebrafish. *Trends in Genetics*, *5*(8), 283–288.

Kojima, S., & Borisy, G. (2014). An image-based, dual fluorescence reporter assay to evaluate the efficacy of shRNA for gene silencing at the single-cell level. *F1000Research*, *3*, 60. http://dx.doi.org/10.12688/f1000research.3-60.v1.

Kollmar, M. (2015). Polyphyly of nuclear lamin genes indicates an early eukaryotic origin of the metazoan-type intermediate filament proteins. *Scientific Reports*, *5*, 10652. http://dx.doi.org/10.1038/srep10652.

Koster, S., Weitz, D. A., Goldman, R. D., Aebi, U., & Herrmann, H. (2015). Intermediate filament mechanics in vitro and in the cell: From coiled coils to filaments, fibers and networks. *Current Opinion in Cell Biology*, *32c*, 82–91. http://dx.doi.org/10.1016/j.ceb.2015.01.001.

Krausz, E., Augusteyn, R. C., Quinlan, R. A., Reddan, J. R., Russell, P., Sax, C. M., et al. (1996). Expression of Crystallins, Pax6, Filensin, CP49, MIP, and MP20 in lens-derived cell lines. *Investigative Ophthalmology & Visual Science*, *37*(10), 2120–2128.

Landsbury, A., Der Perng, M., Pohl, E., Roy, A., & Quinlan, R. (2010). Functional symbiosis between the intermediate filament cytoskeleton and small heat shock proteins. In S. Simon & A. P. Arigo (Eds.), *Small stress proteins and human disease*. New York, USA: Nova Science Publishers.

Langheinrich, U. (2003). Zebrafish: A new model on the pharmaceutical catwalk. *Bioessays*, *25*(9), 904–912. http://dx.doi.org/10.1002/bies.10326.

Lengler, J., Krausz, E., Tomarev, S., Prescott, A., Quinlan, R. A., & Graw, J. (2001). Antagonistic action of Six3 and Prox1 at the gamma-crystallin promoter. *Nucleic Acids Research*, *29*(2), 515–526.

Leonard, M., Chan, Y., & Menko, A. S. (2008). Identification of a novel intermediate filament-linked N-cadherin/gamma-catenin complex involved in the establishment of the cytoarchitecture of differentiated lens fiber cells. *Developmental Biology*, *319*(2), 298–308. http://dx.doi.org/10.1016/j.ydbio.2008.04.036.

Lepinoux-Chambaud, C., & Eyer, J. (2013). Review on intermediate filaments of the nervous system and their pathological alterations. *Histochemistry and Cell Biology*, *140*(1), 13–22. http://dx.doi.org/10.1007/s00418-013-1101-1.

Li, M., Andersson-Lendahl, M., Sejersen, T., & Arner, A. (2013). Knockdown of desmin in zebrafish larvae affects interfilament spacing and mechanical properties of skeletal muscle. *The Journal of General Physiology*, *141*(3), 335–345. http://dx.doi.org/10.1085/jgp.201210915.

Liu, W. Y., Wang, Y., Sun, Y. H., Wang, Y., Wang, Y. P., Chen, S. P., et al. (2005). Efficient RNA interference in zebrafish embryos using siRNA synthesized with SP6 RNA

polymerase. *Development, Growth & Differentiation, 47*(5), 323–331. http://dx.doi.org/10.1111/j.1440-169X.2005.00807.x.
Ma, L., Parkhurst, A., & Jeffery, W. R. (2014). The role of a lens survival pathway including sox2 and alphaA-crystallin in the evolution of cavefish eye degeneration. *EvoDevo, 5*, 28. http://dx.doi.org/10.1186/2041-9139-5-28.
Maisel, H., & Perry, M. M. (1972). Electron microscope observations on some structural proteins of the chick lens. *Experimental Eye Research, 14*(1), 7–12.
Matsuyama, M., Tanaka, H., Inoko, A., Goto, H., Yonemura, S., Kobori, K., et al. (2013). Defect of mitotic vimentin phosphorylation causes microophthalmia and cataract via aneuploidy and senescence in lens epithelial cells. *The Journal of Biological Chemistry, 288*(50), 35626–35635. http://dx.doi.org/10.1074/jbc.M113.514737.
McAvoy, J. W. (1980). Beta- and gamma-crystallin synthesis in rat lens epithelium explanted with neural retinal. *Differentiation, 17*(2), 85–91.
Mehta, V. V., Hull, C. C., & Lawrenson, J. G. (2015). The effect of varying glucose levels on the ex vivo crystalline lens: Implications for hyperglycaemia-induced refractive changes. *Ophthalmic & Physiological Optics, 35*(1), 52–59. http://dx.doi.org/10.1111/opo.12176.
Menko, A. S., Bleaken, B. M., Libowitz, A. A., Zhang, L., Stepp, M. A., & Walker, J. L. (2014). A central role for vimentin in regulating repair function during healing of the lens epithelium. *Molecular Biology of the Cell, 25*(6), 776–790. http://dx.doi.org/10.1091/mbc.E12-12-0900.
Menko, A. S., Klukas, K. A., & Johnson, R. G. (1984). Chicken embryo lens cultures mimic differentiation in the lens. *Developmental Biology, 103*(1), 129–141.
Michael, R., van Marle, J., Vrensen, G. F., & van den Berg, T. J. (2003). Changes in the refractive index of lens fibre membranes during maturation—Impact on lens transparency. *Experimental Eye Research*, 77(1), 93–99.
Mokry, J., & Nemecek, S. (1998). Immunohistochemical detection of intermediate filament nestin. *Acta Medica (Hradec Králové), 41*(2), 73–80.
Muller, M., Bhattacharya, S. S., Moore, T., Prescott, Q., Wedig, T., Herrmann, H., et al. (2009). Dominant cataract formation in association with a vimentin assembly disrupting mutation. *Human Molecular Genetics, 18*(6), 1052–1057. http://dx.doi.org/10.1093/hmg/ddn440.
Musil, L. S. (2012). Primary cultures of embryonic chick lens cells as a model system to study lens gap junctions and fiber cell differentiation. *The Journal of Membrane Biology, 245*(7), 357–368. http://dx.doi.org/10.1007/s00232-012-9458-y.
Muto, A., Orger, M. B., Wehman, A. M., Smear, M. C., Kay, J. N., Page-McCaw, P. S., et al. (2005). Forward genetic analysis of visual behavior in zebrafish. *PLoS Genetics, 1*(5), e66. http://dx.doi.org/10.1371/journal.pgen.0010066.
Nasevicius, A., & Ekker, S. C. (2000). Effective targeted gene 'knockdown' in zebrafish. *Nature Genetics, 26*(2), 216–220. http://dx.doi.org/10.1038/79951.
Nikitina, N. V., & Kidson, S. H. (2014). Eye development in the Cape dune mole rat. *Development Genes and Evolution, 224*(2), 107–117. http://dx.doi.org/10.1007/s00427-014-0468-x.
O'Connor, M. D., & McAvoy, J. W. (2007). In vitro generation of functional lens-like structures with relevance to age-related nuclear cataract. *Investigative Ophthalmology & Visual Science, 48*(3), 1245–1252. http://dx.doi.org/10.1167/iovs.06-0949.
Ogino, H., Ochi, H., Reza, H. M., & Yasuda, K. (2012). Transcription factors involved in lens development from the preplacodal ectoderm. *Developmental Biology, 363*(2), 333–347. http://dx.doi.org/10.1016/j.ydbio.2012.01.006.
Pan, X., Hobbs, R. P., & Coulombe, P. A. (2013). The expanding significance of keratin intermediate filaments in normal and diseased epithelia. *Current Opinion in Cell Biology, 25*(1), 47–56. http://dx.doi.org/10.1016/j.ceb.2012.10.018.

Patton, E. E., & Zon, L. I. (2001). The art and design of genetic screens: Zebrafish. *Nature Reviews. Genetics*, *2*(12), 956–966. http://dx.doi.org/10.1038/35103567.

Pei, Y., & Tuschl, T. (2006). On the art of identifying effective and specific siRNAs. *Nature Methods*, *3*(9), 670–676. http://dx.doi.org/10.1038/nmeth911.

Perng, M. D., Sandilands, A., Kuszak, J., Dahm, R., Wegener, A., Prescott, A. R., et al. (2004). The intermediate filament systems in the eye lens. *Methods in Cell Biology*, *78*, 597–624.

Perng, M. D., Zhang, Q., & Quinlan, R. A. (2007). Insights into the beaded filament of the eye lens. *Experimental Cell Research*, *313*(10), 2180–2188. http://dx.doi.org/10.1016/j.yexcr.2007.04.005.

Peter, A., & Stick, R. (2015). Evolutionary aspects in intermediate filament proteins. *Current Opinion in Cell Biology*, *32*, 48–55. http://dx.doi.org/10.1016/j.ceb.2014.12.009.

Qu, B., Landsbury, A., Schonthaler, H. B., Dahm, R., Liu, Y., Clark, J. I., et al. (2012). Evolution of the vertebrate beaded filament protein, Bfsp2; comparing the in vitro assembly properties of a "tailed" zebrafish Bfsp2 to its "tailless" human orthologue. *Experimental Eye Research*, *94*(1), 192–202. http://dx.doi.org/10.1016/j.exer.2011.12.001.

Quinlan, R. A., Bromley, E. H., & Pohl, E. (2015). A silk purse from a sow's ear-bioinspired materials based on alpha-helical coiled coils. *Current Opinion in Cell Biology*, *32c*, 131–137. http://dx.doi.org/10.1016/j.ceb.2014.12.010.

Quinlan, R. A., Carte, J. M., Sandilands, A., & Prescott, A. R. (1996). The beaded filament of the eye lens: An unexpected key to intermediate filament structure and function. *Trends in Cell Biology*, *6*(4), 123–126.

Reddan, J. R., Kuck, J. F., Dziedzic, D. C., Kuck, K. D., Reddan, P. R., & Wasielewski, P. (1989). Establishment of lens epithelial cell lines from Emory and cataract resistant mice and their response to hydrogen peroxide. *Lens and Eye Toxicity Research*, *6*(4), 687–701.

Resau, J. H., Sakamoto, K., Cottrell, J. R., Hudson, E. A., & Meltzer, S. J. (1991). Explant organ culture: A review. *Cytotechnology*, *7*(3), 137–149.

Rhodes, J. D., Monckton, D. G., McAbney, J. P., Prescott, A. R., & Duncan, G. (2006). Increased SK3 expression in DM1 lens cells leads to impaired growth through a greater calcium-induced fragility. *Human Molecular Genetics*, *15*(24), 3559–3568. http://dx.doi.org/10.1093/hmg/ddl432.

Russell, P., Fukui, H. N., Tsunematsu, Y., Huang, F. L., & Konoshita, J. H. (1977). Tissue culture of lens epithelial cells from normal and Nakano mice. *Investigative Ophthalmology & Visual Science*, *16*(3), 243–246.

Sandilands, A., Prescott, A. R., Carter, J. M., Hutcheson, A. M., Quinlan, R. A., Richards, J., et al. (1995). Vimentin and CP49/filensin form distinct networks in the lens which are independently modulated during lens fibre cell differentiation. *Journal of Cell Science*, *108*(Pt. 4), 1397–1406.

Sandilands, A., Prescott, A. R., Hutcheson, A. M., Quinlan, R. A., Casselman, J. T., & FitzGerald, P. G. (1995). Filensin is proteolytically processed during lens fiber cell differentiation by multiple independent pathways. *European Journal of Cell Biology*, *67*(3), 238–253.

Sandilands, A., Prescott, A. R., Wegener, A., Zoltoski, R. K., Hutcheson, A. M., Masaki, S., et al. (2003). Knockout of the intermediate filament protein CP49 destabilises the lens fibre cell cytoskeleton and decreases lens optical quality, but does not induce cataract. *Experimental Eye Research*, *76*(3), 385–391.

Sandilands, A., Wang, X., Hutcheson, A. M., James, J., Prescott, A. R., Wegener, A., et al. (2004). Bfsp2 mutation found in mouse 129 strains causes the loss of CP49' and induces vimentin-dependent changes in the lens fibre cell cytoskeleton. *Experimental Eye Research*, *78*(4), 875–889. http://dx.doi.org/10.1016/j.exer.2003.09.028.

Santoriello, C., & Zon, L. I. (2012). Hooked! Modeling human disease in zebrafish. *The Journal of Clinical Investigation, 122*(7), 2337–2343. http://dx.doi.org/10.1172/jci60434.

Shabbir, S. H., Cleland, M. M., Goldman, R. D., & Mrksich, M. (2014). Geometric control of vimentin intermediate filaments. *Biomaterials, 35*(5), 1359–1366. http://dx.doi.org/10.1016/j.biomaterials.2013.10.008.

Slingsby, C., Wistow, G. J., & Clark, A. R. (2013). Evolution of crystallins for a role in the vertebrate eye lens. *Protein Science, 22*(4), 367–380. http://dx.doi.org/10.1002/pro.2229.

Snider, N. T., & Omary, M. B. (2014). Post-translational modifications of intermediate filament proteins: Mechanisms and functions. *Nature Reviews. Molecular Cell Biology, 15*(3), 163–177. http://dx.doi.org/10.1038/nrm3753.

Song, S., Landsbury, A., Dahm, R., Liu, Y., Zhang, Q., & Quinlan, R. A. (2009). Functions of the intermediate filament cytoskeleton in the eye lens. *The Journal of Clinical Investigation, 119*(7), 1837–1848. http://dx.doi.org/10.1172/jci38277.

Tholozan, F. M., Gribbon, C., Li, Z., Goldberg, M. W., Prescott, A. R., McKie, N., et al. (2007). FGF-2 release from the lens capsule by MMP-2 maintains lens epithelial cell viability. *Molecular Biology of the Cell, 18*(11), 4222–4231. http://dx.doi.org/10.1091/mbc.E06-05-0416.

Toivola, D. M., Boor, P., Alam, C., & Strnad, P. (2015). Keratins in health and disease. *Current Opinion in Cell Biology, 32c*, 73–81. http://dx.doi.org/10.1016/j.ceb.2014.12.008.

Toivola, D. M., Strnad, P., Habtezion, A., & Omary, M. B. (2010). Intermediate filaments take the heat as stress proteins. *Trends in Cell Biology, 20*(2), 79–91. http://dx.doi.org/10.1016/j.tcb.2009.11.004.

Tomarev, S. I., Zinovieva, R. D., & Piatigorsky, J. (1993). Primary structure and lens-specific expression of genes for an intermediate filament protein and a beta-tubulin in cephalopods. *Biochimica et Biophysica Acta, 1216*(2), 245–254.

Tripathi, B. J., Tripathi, R. C., Borisuth, N. S., Dhaliwal, R., & Dhaliwal, D. (1991). Rodent models of congenital and hereditary cataract in man. *Lens and Eye Toxicity Research, 8*(4), 373–413.

Ui-Tei, K., Naito, Y., Takahashi, F., Haraguchi, T., Ohki-Hamazaki, H., Juni, A., et al. (2004). Guidelines for the selection of highly effective siRNA sequences for mammalian and chick RNA interference. *Nucleic Acids Research, 32*(3), 936–948. http://dx.doi.org/10.1093/nar/gkh247.

Varma, S. D., & Hegde, K. R. (2010). Kynurenine-induced photo oxidative damage to lens in vitro: Protective effect of caffeine. *Molecular and Cellular Biochemistry, 340*(1–2), 49–54. http://dx.doi.org/10.1007/s11010-010-0399-4.

Wagner, L. M., & Takemoto, D. J. (2001). PKCalpha and PKCgamma overexpression causes lentoid body formation in the N/N 1003A rabbit lens epithelial cell line. *Molecular Vision*, 7, 138–144.

Wallace, P., Signer, E., Paton, I. R., Burt, D., & Quinlan, R. (1998). The chicken CP49 gene contains an extra exon compared to the human CP49 gene which identifies an important step in the evolution of the eye lens intermediate filament proteins. *Gene, 211*(1), 19–27.

Wang, H., Zhang, T., Wu, D., & Zhang, J. (2013). A novel beaded filament structural protein 1 (BFSP1) gene mutation associated with autosomal dominant congenital cataract in a Chinese family. *Molecular Vision, 19*, 2590–2595.

Watanabe, K., Nishimura, Y., Oka, T., Nomoto, T., Kon, T., Shintou, T., et al. (2010). In vivo imaging of zebrafish retinal cells using fluorescent coumarin derivatives. *BMC Neuroscience, 11*, 116. http://dx.doi.org/10.1186/1471-2202-11-116.

Wei, Q., Reidler, D., Shen, M. Y., & Huang, H. (2013). Keratinocyte cytoskeletal roles in cell sheet engineering. *BMC Biotechnology, 13*, 17. http://dx.doi.org/10.1186/1472-6750-13-17.

West-Mays, J. A., Pino, G., & Lovicu, F. J. (2010). Development and use of the lens epithelial explant system to study lens differentiation and cataractogenesis. *Progress in Retinal and Eye Research*, *29*(2), 135–143. http://dx.doi.org/10.1016/j.preteyeres.2009.12.001.
Won, G. J., Fudge, D. S., & Choh, V. (2015). The effects of actomyosin disruptors on the mechanical integrity of the avian crystalline lens. *Molecular Vision*, *21*, 98–109.
Woods, I. G., & Schier, A. F. (2008). Targeted mutagenesis in zebrafish. *Nature Biotechnology*, *26*(6), 650–651. http://dx.doi.org/10.1038/nbt0608-650.
Wormstone, I. M., Liu, C. S., Rakic, J. M., Marcantonio, J. M., Vrensen, G. F., & Duncan, G. (1997). Human lens epithelial cell proliferation in a protein-free medium. *Investigative Ophthalmology & Visual Science*, *38*(2), 396–404.
Wormstone, I. M., Tamiya, S., Marcantonio, J. M., & Reddan, J. R. (2000). Hepatocyte growth factor function and c-Met expression in human lens epithelial cells. *Investigative Ophthalmology & Visual Science*, *41*(13), 4216–4222.
Wride, M. A. (2011). Lens fibre cell differentiation and organelle loss: Many paths lead to clarity. *Philosophical Transactions of the Royal Society of London. Series B, Biological Sciences*, *366*(1568), 1219–1233. http://dx.doi.org/10.1098/rstb.2010.0324.
Wride, M. A., & Sanders, E. J. (1998). Nuclear degeneration in the developing lens and its regulation by TNFalpha. *Experimental Eye Research*, *66*(3), 371–383. http://dx.doi.org/10.1006/exer.1997.0440.
Wu, W., Tholozan, F. M., Goldberg, M. W., Bowen, L., Wu, J., & Quinlan, R. A. (2014). A gradient of matrix-bound FGF-2 and perlecan is available to lens epithelial cells. *Experimental Eye Research*, *120*, 10–14. http://dx.doi.org/10.1016/j.exer.2013.12.004.
Wu, J. J., Wu, W., Tholozan, F. M., Saunter, C. D., Girkin, J. M., & Quinlan, R. A. (2015). A dimensionless ordered pull-through model of the mammalian lens epithelium evidences scaling across species and explains the age-dependent changes in cell density in the human lens. *Journal of the Royal Society Interface*, *12*(108), 20150391. http://dx.doi.org/10.1098/rsif.2015.0391.
Yamada, T., Nakamura, T., Westphal, H., & Russell, P. (1990). Synthesis of alpha-crystallin by a cell line derived from the lens of a transgenic animal. *Current Eye Research*, *9*(1), 31–37.
Yamamoto, Y., Stock, D. W., & Jeffery, W. R. (2004). Hedgehog signalling controls eye degeneration in blind cavefish. *Nature*, *431*(7010), 844–847. http://dx.doi.org/10.1038/nature02864.
Yang, J., Bian, W., Gao, X., Chen, L., & Jing, N. (2000). Nestin expression during mouse eye and lens development. *Mechanisms of Development*, *94*(1–2), 287–291.
Yang, C., Yang, Y., Brennan, L., Bouhassira, E. E., Kantorow, M., & Cvekl, A. (2010). Efficient generation of lens progenitor cells and lentoid bodies from human embryonic stem cells in chemically defined conditions. *The FASEB Journal*, *24*(9), 3274–3283. http://dx.doi.org/10.1096/fj.10-157255.
Young, L. K., Jarrin, M., Saunter, C., Nelson, C., Taylor, J., Quinlan, R., et al. (2014). Using light to study the development of the eye's lens. *Kaleidoscope*, *6*(1), 114–118.
Zhang, J., Ma, Z. Y., Wang, Y. L., Wang, J. Y., & Hu, Y. Z. (2013). Establishment of mouse lens epithelial cell lines with the genotype of Hsf4-/-. *[Zhonghua yan ke za zhi] Chinese Journal of Ophthalmology*, *49*(11), 1029–1031.
Zelenka, P. S., Gao, C. Y., & Saravanamuthu, S. S. (2009). Preparation and culture of rat lens epithelial explants for studying terminal differentiation. *Journal of Visualized Experiments*, (31). pii: 1519. http://dx.doi.org/10.3791/1519.

PART III

Non-Mammalian IF Protein Systems

CHAPTER TWENTY-TWO

Compartment-Specific Phosphorylation of Squid Neurofilaments

Philip Grant, Harish C. Pant[1]
CPR, NINDS, NIH, Bethesda, MD, USA
[1]Corresponding author: e-mail address: panth@ninds.nih.gov

Contents

1. Introduction 616
2. Squid Giant Fiber System 617
3. Isolation of Axoplasm 619
4. Squid NF Genes 621
5. Squid NF Antibodies 622
6. Developmental Regulation of NF Expression 622
   6.1 Fractionation of Axoplasm and Preparation of NFs 624
   6.2 SDS-PAGE and Western Blots 625
   6.3 Phosphorylation Assay 625
   6.4 Soma (GFL) Versus Axoplasm 627
7. NF-Associated Protein Kinases 627
8. Squid NFs Are Included in Compartment-Specific Multimeric Protein Complexes: P13suc1 Affinity Chromatography 628
References 630

Abstract

Studies of the giant axon and synapse of third-order neurons in the squid stellate ganglion have provided a vast literature on neuronal physiology and axon transport. Large neuronal size also lends itself to comparative biochemical studies of cell body versus axon. These have focused on the regulation of synthesis, assembly, posttranslational modification and function of neuronal cytoskeletal proteins (microtubules (MTs) and neurofilaments (NFs)), the predominant proteins in axoplasm. These contribute to axonal organization, stability, transport, and impulse transmission responsible for rapid contractions of mantle muscles underlying jet propulsion. Studies of vertebrate NFs have established an extensive literature on NF structure, organization, and function; studies of squid NFs, however, have made it possible to compare compartment-specific regulation of NF synthesis, assembly, and function in soma versus axoplasm. Since NFs contain over 100 eligible sites for phosphorylation by protein kinases, the compartment-specific patterns of phosphorylation have been a primary focus of

Methods in Enzymology, Volume 568
ISSN 0076-6879
http://dx.doi.org/10.1016/bs.mie.2015.09.033

biochemical studies. We have learned that NF phosphorylation is tightly compartmentalized; extensive phosphorylation occurs only in the axonal compartment in squid and in vertebrate neurons. This extensive phosphorylation plays a key role in organizing NFs, in association with microtubules (MTs), into a stable, dynamic functional lattice that supports axon growth, diameter, impulse transmission, and synaptic activity. To understand how cytoskeletal phosphorylation is topographically regulated, the kinases and phosphatases, bound to NFs isolated from cell bodies and axoplasm, have also been studied.

1. INTRODUCTION

For decades, since its discovery (Young, 1938), and its use in unraveling ionic fluxes underlying the action potential (Hodgkin & Huxley, 1952a,1952b,1952c), studies of the squid giant axon have contributed significantly to our understanding of the nature and function of neuronal ion pumps, Na^+, K^+, and Ca^{2+} exchangers and channels (Altamirano, Breitwieser, & Russell, 1999; Llinas, 1982; Westerfield & Joyner, 1982), synaptic transmission (Augustine, Charlton, & Smith, 1985; Clay, 1998; Clay & Kuzirian, 2000; Gilly & Armstrong, 1982; Rosenthal & Gilly, 2003), the mechanisms underlying molecular motors in axon transport (Allen, Metuzals, Tasaki, Brady, & Gilbert, 1982; Bearer, DeGiorgis, Bodner, Kao, & Reese, 1996; Brady, Lasek, & Allen, 1985), and the role of kinases and phosphatases in neuronal cytoskeletal organization and function (Cohen, Pant, House, & Gainer, 1987; Floyd, Grant, Gallant, & Pant, 1991; Gallant, Pant, Pruss, & Gainer, 1986; Grant, Diggins, & Pant, 1999; Grant & Pant, 2004; Pant, Gallant, & Gainer, 1986; Takahashi, Amin, Grant, & Pant, 1995). We have learned that what is true for the giant axon is true for all neurons.

Over the years, most biochemical studies of squid neurofilaments (NFs) have focused on the stellate ganglion and its giant axons because of the ease and simplicity of separating cell bodies in the giant fiber lobe (GFL) from pure axoplasm, uncontaminated by sheaths or associated glia. This is not possible in the more complicated vertebrate nervous system in which billions of small neurons and axons predominate. Hence the squid giant fiber system has made it possible to compare the biochemistry and molecular biology of two compartments, the soma, the site of gene expression, regulation, and protein synthesis, and the respective axoplasm to which products from the soma are transported, assembled, and organized into a functioning "organelle" specialized for neurotransmission and behavior. Microtubules

(MTs), microfilaments (MFs), and NFs are the predominant cytoskeletal proteins in axoplasm, the latter making up more than >60% of total protein (Brown & Lasek, 1993).

The organization of the dynamic axonal lattice is not well understood, although evidence of crossbridges among NF polymers and interactions with MTs has been reported (Abercrombie, Masukawa, Sjodin, & Livengood, 1981; Gotow & Tanaka, 1994; Gotow, Tanaka, Nakamura, & Takeda, 1994; Martin, 1996; Metuzals, Fishman, & Robb, 1995; Metuzals, Hodge, Lasek, & Kaiserman-Abramof, 1983). Posttranslational modifications of high molecular weight NF subunits, such as phosphorylation, induce crossbridges which affect the dynamic exchanges between a NF–MT core and oligomers that increase and stabilize axon diameter (Cohen et al., 1987; Floyd et al., 1991; Jaffe, Sharma, Grant, & Pant, 2001; Martin, 1996; Martin et al., 1999; Shea & Chan, 2008; Shea & Lee, 2011). Squid NF proteins (NF60, NF70, and NF220) are more similar to lamin-like intermediate proteins than to mammalian NFs, originating by alternative splicing from a single gene rather than from three independent genes (Szaro, Pant, Way, & Battey, 1991; Way et al., 1992). Only squid NF220 is phosphorylated endogenously at more than 80 sites (mostly proline-directed serines KSP) in the C-terminal tail domain (Jaffe et al., 2001) and like phosphorylation of mammalian NFM/H, it occurs primarily in the axon (Grant & Pant, 2000; Pant, Veeranna, & Grant, 2000). In fact, phosphorylation of most proteins in the giant axon is conspicuously compartmentalized; it is topographically regulated with most activity restricted to the axoplasm of the giant axon (Grant et al., 1999; Grant & Pant, 2004). Although many studies have focused on this problem, the factors responsible for compartmentalization are still not well understood. Localization of kinases, regulators, and phosphatases has been explored and, in general, found to be organized into compartment-specific large multimeric complexes; the mechanisms underlying their function, however, still elude us (Cohen et al., 1987; Grant & Pant, 2004; Takahashi et al., 1995; Tytell, Pant, Gainer, & Hill, 1990; Tytell, Zackroff, & Hill, 1988).

2. SQUID GIANT FIBER SYSTEM

The giant axon is the largest of 8–10 third-order giant axons arising in the stellate ganglion that innervate the circular muscles of the mantle, contractions of which underlie the rapid jet propulsion mode of squid swimming behavior (Fig. 1A). Each giant axon originates from the

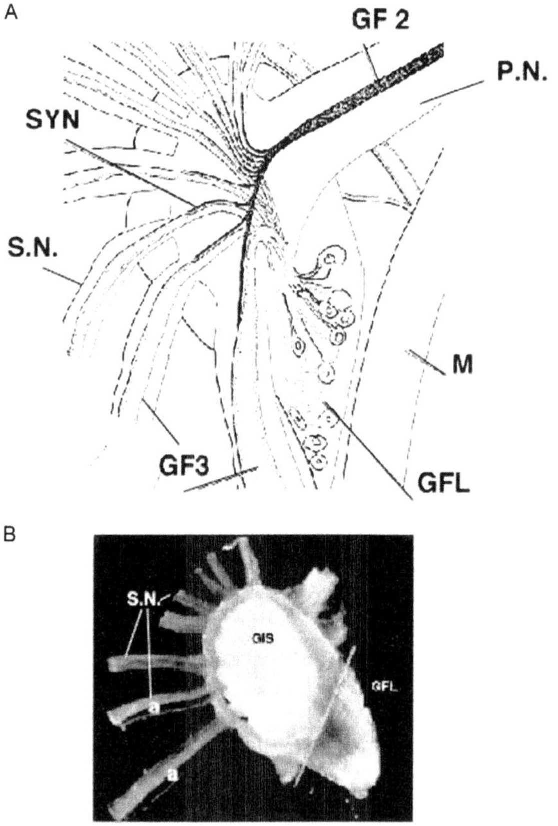

Figure 1 Squid giant fiber system: (A) Diagram of *Loligo* stellate ganglion showing giant axons, GF3, exiting the ganglion in stellate nerves (S.N.) to innervate mantle muscles (M). Axons of small cells in giant fiber lobe (GFL) fuse to form giant axons. Second-order giant fibers (GF2), from the brain, in the pallial nerve (P.N.) enter the ganglion where they synapse with giant axons in the giant synapse region (SYN). (B) Stellate ganglion with stubs of attached giant axons (a.) appearing as clear areas in stellate nerves (S.N.). GIS is neuropil region of the ganglion showing how cell bodies of the giant fiber lobe (GFL) are dissected from the body of the ganglion. *Panel (A) modified after Young (1939), fig. 11. Panel (B) modified after Tytell et al. (1990), fig. 1, Wiley-Liss.*

fusion of many axons of small neurons in the GFL (Martin, 1965; Young, 1939) thereby increasing axon diameter (up to 1 mm) to facilitate rapid impulse transmission essential to the escape response (Otis & Gilly, 1990; Young, 1938). Giant axons receive input via axon-axonal synapses within a giant synapse in the ganglion from second-order giant fibers in the squid brain (Martin & Miledi, 1975; Williamson & Chrachri, 2004; Young, 1973).

Three major cytoskeletal structures contribute to the cytoskeletal scaffolding underlying the axonal architecture, MTs, MFs, and NFs. In addition,

MAPs, myosins, and the various molecular motors kinesin and dynein contribute to the dynamics of axonal transport and function. Fast and slow axon transport include vesicular cargos of transmitters, modulators, receptor proteins, and ion channels, while larger oligomers of tubulin and NFs exchange slowly with a stable MT–NF lattice core (Adams & Gillespie, 1988; Allen et al., 1982; Brady, Lasek, & Allen, 1982; Brady et al., 1985; Galbraith, Reese, Schlief, & Gallant, 1999). Mitochondria are the largest organelles transported from cell body to the neuromuscular synapses in mantle muscle (Brinley & Tiffert, 1978; Pratt, 1986).

3. ISOLATION OF AXOPLASM

In most studies, the initial source of squid NFs comes from extractions of pure axoplasm collected from the largest of the giant axons exiting the stellate ganglion. The squid *Loligo pealei*, obtained live from Marine Biological Laboratory, Woods Hole, MA, are kept in a tank of running sea water (temp. 20–22 °C), usually for 1 to 2 days before use. The procedure for dissecting the stellate ganglion, isolating the giant axon, and extruding the axoplasm is based on a description in chapter 14 of the book, *Squid as Experimental Animals* (Brown & Lasek, 1993). The animal is decapitated, and dissection is carried out in running sea water on a seawater table illuminated from below through a glass window. The animal is eviscerated, leaving the large mantle muscle with stellate ganglion and axons attached dorsally (Figs. 1B and 2). The two hindmost and medial most giant axons are the axons of choice. Depending on the size of the squid, their length varies from 50 to 100 mm with diameters of 300–600 μm (Fig. 2A). The ends of the axons are ligated with thread prior to isolation, placed in a dish of ice-cold Ca-free artificial seawater. The axon is blotted to remove sea water and cleansed of surrounding small fin nerve and stellar nerve fibers before the axoplasm is extruded (Fig. 2A–C). For extrusion, the axon is extended on a square of parafilm, cut at one end, and a small capillary glass tube placed gently is rolled on the axon to squeeze the axoplasm to the cut end into a small blob which is then (Fig. 2B and C) collected in an Eppendorf tube embedded in ice until used in one of the buffers (see below). Alternatively, axoplasm may be frozen for future use; collected in an Eppendorf with or without 100 m*M* NaCl, 20 m*M* Hepes, pH 7.0, 10 m*M* $MgCl_2$, 1 m*M* EGTA, and 50% glycerol; and quickly transferred and stored at −80 °C. Buffers vary according to use: For an enriched NF preparation, buffer A: 10 m*M* HEPES, pH 7.4, 455 m*M* KF, 1 m*M* EDTA, 1 m*M* EGTA, 1%

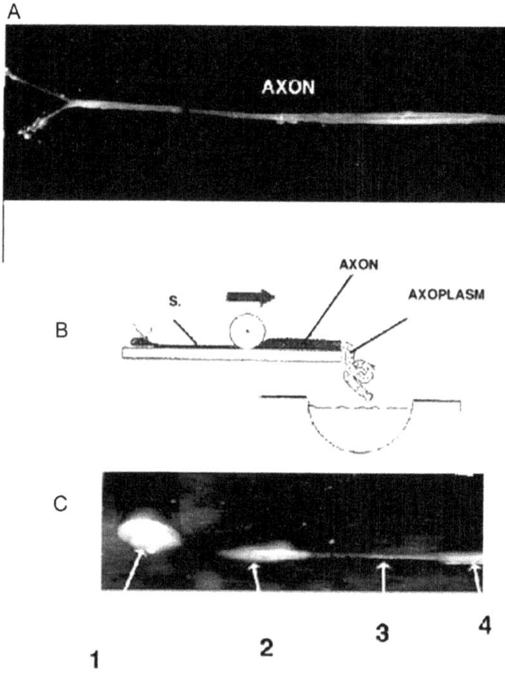

Figure 2 Isolation of axoplasm: (A) Photo of isolated giant axon attached to stellate ganglion. (B) Procedure for squeezing axoplasm from surrounding sheath as described in text. (C) Collecting axoplasm. 1. Blob of axoplasm from several axons, 2. axoplasm emerging from the sheath, 3. empty sheath, and 4. intact axon. *Panel (B) modified after Brown and Lasek (1993), chapter 14, fig. 3.*

Triton X-100, and protease inhibitor 10 μg/ml; extraction buffer B; phosphate-buffered saline (PBS) containing 0.5% Nonidet P-40, 5 m*M* $MgCl_2$, 10 m*M* *b*-phosphoglycerol, 10 m*M* KF plus protease inhibitor, 10 μg/ml, alternatively, extraction buffer C; 20 m*M* Tris–HCl, pH 7.4, 100 m*M* NaCl, 1% Nonidet NP40, 1 m*M* DTT, protease inhibitor pellets one-fourth pellet/ml (Roche). For kinase assays, phosphorylating buffer, 20 m*M* Tris, pH 7.4, 10 m*M* $MgCl_2$, 1 m*M* EGTA, 1 m*M* EDTA, 1 m*M* DTT, and 0.1 μ*M* okadaic acid. GFLs were dissected out and placed on ice (Fig. 1B) in one of the above buffers depending upon the experiment or frozen on dry ice and kept −80° C for future experiments. For comparative studies of soma (GFL) versus axoplasm, the small GFL can be cut from the ganglion and collected in a small tube (frozen at −80° C), or placed in ice cold buffer A, B, or C for comparative studies.

4. SQUID NF GENES

The squid NF subunits have been cloned and sequenced (Szaro et al., 1991; Way et al., 1992). They differ from the three vertebrate NFs (α-internexin, NFL, NFM, and NFH), resembling nuclear lamins instead. Moreover, in contrast to vertebrate NFs that are coded by three independent genes, the three squid subunits (NF60, NF70, and NF220) are derived by alternative splicing of a single gene (Fig. 3). The N-terminal and rod domain sequences of the three subunits are identical; only the C-terminal tail domains differ as to length (NF60 and NF70 being short), while the NF220 is considerably longer and highly phosphorylated on the many KSP (Lys-Ser-Pro motifs) repeats in the tail domain. Phosphorylation of these sites induces side arm extensions that interact with one another and with adjacent MTs to stabilize the cytoskeletal lattice. Here, the NF220 tail domain with its many KSP repeats resembles the tail domain of the vertebrate high NFH and mid NFM proteins that are also phosphorylated. All 88 prospective sites in the NF220 tail domain are endogenously phosphorylated (Jaffe et al., 2001), probably phosphorylated by proline-directed kinases such as Cdk5, Erk1, or GSK3. It is also notable that 78% (31 0f 40) of the serines in the region with repeats KAESEK and EKSARSP are also phosphorylated, suggesting nonproline-directed kinases like casein kinase 1 (CK1) are likely kinases (Floyd et al., 1991). This is consistent with the view that large macromolecules with many potential sites require sequential phosphorylation by multiple kinases.

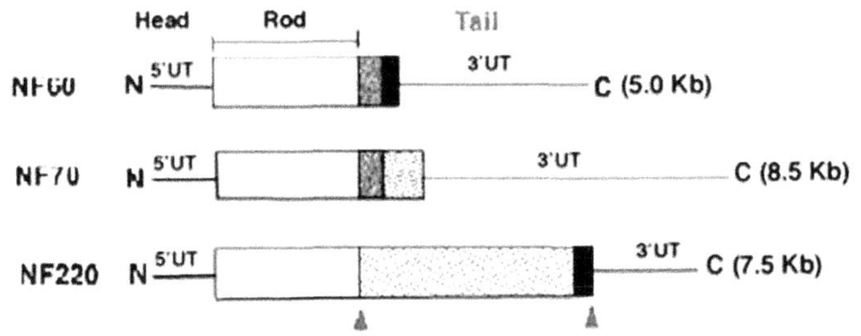

Figure 3 Alternative splicing of squid NF gene. Schematic drawing of cDNAs encoding NF60, NF70, and NF220 proteins. Coding regions are represented by rectangles: untranslated (UT) regions are straight lines. Identical amino acid residues in the rod region marked by open rectangles. An arrow marks the beginning of divergent tail domains (shaded and hatched boxes). Lengths of mRNA species are shown at right. *Modified from fig. 2 of Way et al. (1992).*

5. SQUID NF ANTIBODIES

A diagram showing the structure of the three NFs is seen in Fig. 3 (see Szaro et al., 1991; Way et al., 1992). Note that all share identical 5′ N-terminal and rod sequences but differ significantly in the 3′ C-terminal regions, except for a short region shared by the NF60 and NF70 proteins. An antibody (NF–NT) was prepared in rabbits to a common N-terminal sequence EISTTTTYEGESRPSS (aa's 9–24, Immunodynamics) (Fig. 4A). Another antibody (NF60/70) was prepared against a peptide EAEVLSTILTRSEGG (NF60/70, Peptide Technologies) shared by both the NF60 and NF70 proteins. A third antibody (NF70) was prepared against a peptide KGEDKANYTQNTWQ (AAs 601–615, Peptide Technologies) specific for the NF70 protein. An attempt to make an antibody specific for the NF220 was unsuccessful. Prebleed sera were also taken from donor rabbits for controls. All antibodies were prepared as 1:500 dilutions in 10% normal goat serum (NGS) in PBS, pH 7.4, 0.1% azide for immunohistochemical assays, and 1:1000 in 0.2% Tween-20–Tris-buffered saline, pH 7.5, 0.1% azide (TTBS) for immunoblots. Two mammalian-derived antibodies were also used, a commercially obtained monoclonal to 1-tubulin at 1:1000 dilution in 10% NGSiPBS or in TTBS for immunohistochemical or immunoblot assay, respectively, and Sternbergers' SMI 31 (the monoclonal antibody to phosphorylated mammalian NFH; Sternberger & Sternberger, 1983) at a 1:1000 dilution in the same diluants. This antibody (here called P-NFH) was used to detect the phosphorylated form of squid NF220. The Western blots and immunohistochemical data obtained using this antibody were similar to a monoclonal antibody prepared against the phosphorylated epitope of the squid NF220 (Cohen et al., 1987). The identification of these antibodies is shown in Fig. 4B.

6. DEVELOPMENTAL REGULATION OF NF EXPRESSION

NF phosphorylation is also developmentally regulated in the squid (Grant, Tseng, Gould, Gainer, & Pant, 1995). The smaller NFs (NF60 and NF70) appear early after stage 22 (Arnold, 1990) as axons develop in the nervous system and only later (stage 25) in the stellate ganglion does the NF220 appear already phosphorylated (see Fig. 5 of stellate ganglion

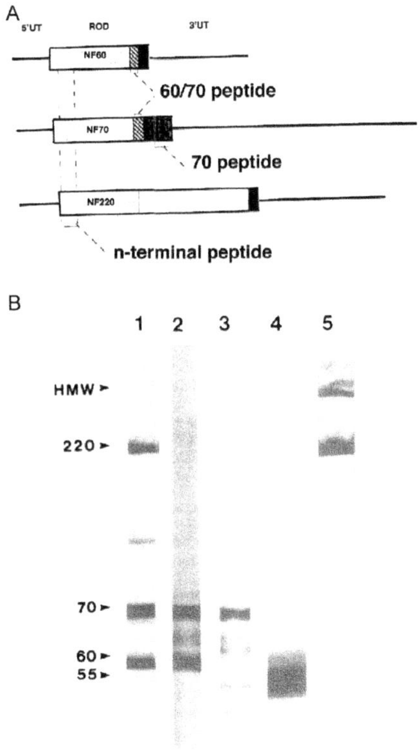

Figure 4 Neurofilament peptides for antibody production. (A) Structure of squid neurofilament proteins showing the regions from which individual peptide sequences were obtained for the production of polyclonal antibodies. The symbols (dots, hatching, etc.) represent regions of identical amino acid sequences. Dashed lines mark sites from which peptide sequences were obtained. For details, see Fig. 3 above. (B) Immunoblots of squid axoplasm from the giant axons showing patterns of NF protein expression with antibodies prepared as described in Fig. 3A above and the text. Lane 1, NF–N-terminal; Lane 2, NF60/70; Lane 3, NF70; Lane 4, α-tubulin; Lane 5, mP-NFH. Molecular weight markers are also shown. *Panel (A) modified from fig. 1 of Grant et al. (1995). Wiley-Liss, Inc. Panel (B) from fig. 2 of Grant et al. (1995), Wiley-Liss, Inc.*

and axon). This pattern also resembles that of developing vertebrates, smaller NFs appearing early before the NFM/NFH and phosphorylation. Early maturation of the NF cytoskeleton in the giant axon system is essential to initiate and sustain the jet propulsion swimming and escape responses. Respiration also regulated by mantle contractions. It is also essential prior to hatching. In subsequent studies after the squid antibodies were exhausted, we continued to use the commercial antibodies, particularly Sternberger's

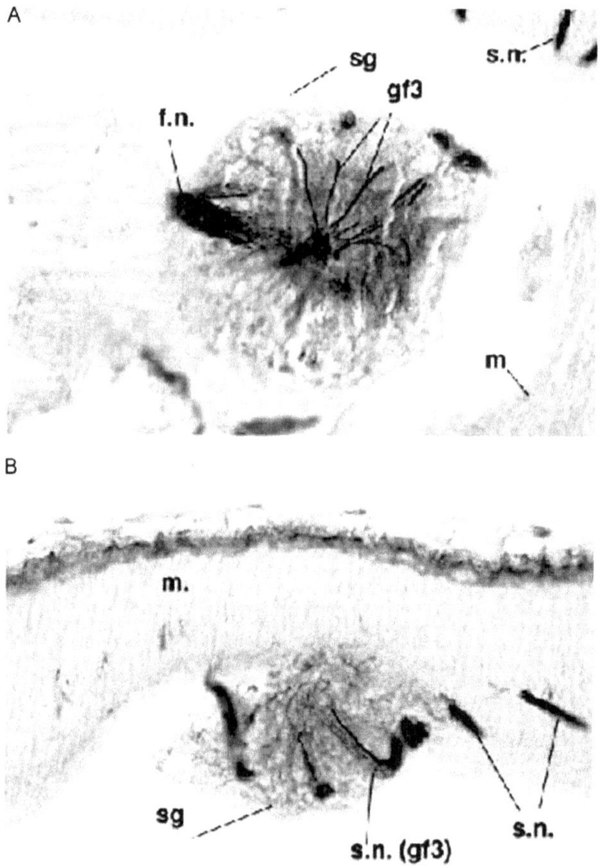

Figure 5 Immunohistochemical expression of NF proteins in stellate ganglion of squid hatchling. (A) Horizontal section of stage 30 squid hatchling stained with SMI 31, antibody which detects the NF220 phosphorylated NF. f.n., fin nerve; sg, stellate ganglion cell bodies; gf3 are giant axons in stellar nerves; s.n., stellar nerve in mantle muscle, m. (B) Horizontal section of stage 30 hatchling showing expression of NF–NT antibody which stains rod region shared by all NF proteins (NF60, NF70, and NF220), m, mantle muscle; sg, cell bodies in stellate ganglion; s.n., stellar nerve containing giant axon (gf3). (See the color plate.)

SMI 31 which continues to be appropriate for assay of NF220 and the high molecular weight phosphorylated HMW complex.

6.1 Fractionation of Axoplasm and Preparation of NFs

The extruded axoplasm (collected from eight axons (~40 μl) is suspended in 500 μl of buffer A and homogenized in a small conical glass homogenizer and vortexed for 5 min followed by centrifugation at 20,000 rpm for 10 min in the cold. Discard the supernatant and wash the NF pellet two times with 300 μl 100 m*M* HEPES, 1 m*M* DTT, plus protease inhibitor, with (1%) or without Triton X100. The washed pellet is resuspended in low salt phosphorylating buffer (D: 20 m*M* Tris, pH 7.4, 1 m*M* EDTA, 1 m*M* EGTA for SDS-PAGE, protein staining. Western blots (Fig. 6A) and kinase assays. A complete biochemical analysis of NF proteins isolated from GFL and axoplasm is described below.

6.2 SDS-PAGE and Western Blots (Fig. 6B–D)

SDS-PAGE is carried out according to procedure by Laemmli, Beguin, Gujer-Kellenberger (1970). Several gels may be used: 10% SDS-polyacrylamide, 10% precast Novex Tris–glycine, and 4–16%, 8–16%, 4–20% precast Novex Tris–glycine gradient gels. Then, 10–30 μl aliquots of each sample are loaded in each lane (~10–20 μg protein/lane) and electrophoresed according to standard procedures. Gels are stained with 0.2% Coomassie blue in 20% acetic acid, 50% methanol for 1 h then destained in changes of 10% acetic acid/50% methanol.

Western blots are performed after electrophoretic transfer of proteins from gels unto PVDF membranes (Immobilon). The transfer is carried out for 4 h at room temperature (RT) at 150 mA. The membranes are blocked for 2–3 h at RT in TTBS (20 m*M* Tris–HCl, pH 7.4, 150 m*M* NaCl, 0.2% Tween) and hung up to dry and stored for later use. Alternatively, for a quick immunoblot, membranes are blocked in TTBS for 0.5–1 h at RT and subsequently incubated in the primary antibody at 4 ° C overnight. Primary antibodies (see note about primary antibodies) are diluted in TTBS. After incubation, the membranes are washed four times in TTBS for 15 min each, alternating with 5-min rinses in dH_2O. The membranes are then incubated in a 1:5000 (or greater) dilution of horseradish-conjugated secondary antibody for 1 h at RT, washed four times as above, and developed with a commercial chemiluminescent reagent.

6.3 Phosphorylation Assay (Fig. 6B–D)

Comparison of endogenous and exogenous phosphorylation in GFL and axoplasm. The assay for axoplasm and GFL extracts is usually carried out with 5 μL of sample (~3 μg protein) placed in 50 μl phosphorylating buffer (20 m*M* Tris, pH 7.4, 10 m*M* $MgCl_2$, 1 m*M* EGTA, 1 m*M* EDTA, 1 m*M* DTT, and 0.1 μ*M* okadaic acid to inhibit serine/threonine phosphatases

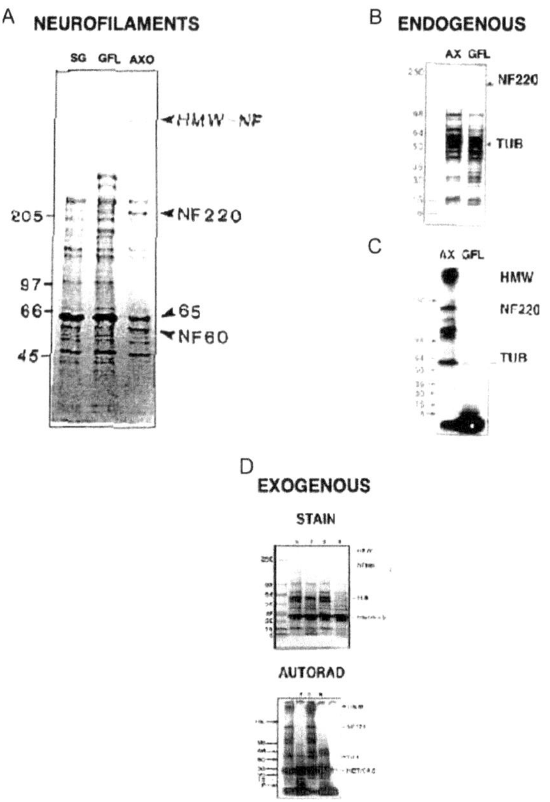

Figure 6 Preparation of neurofilaments and comparative axon–GFL phosphorylation patterns. (A) Coomassie stained gradient gel (3–15%) showing pattern of protein bands obtained from Triton-extracted pellets of ganglion isthmus region (IS), cell bodies of GFL, and the axoplasm (AX). Molecular weight markers at left, markers at right show three prominent subunits of axonal neurofilaments, HMW, NF220, NF60 and 65 kDa. Note that little if any of HMW and NF220 are found in GFL or in IS lanes. (B) and (C) Endogenous phosphorylation patterns. Coomassie stained gel (B), showing shared protein patterns and phosphorylation assays (C), autoradiogram of GFL and axoplasm (AX) extracts showing profound difference in the level of endogenous protein phosphorylation. Markers on right (B) show NF220 and TUB bands, while markers in (C) identify phosphorylated bands in axoplasm. Only low molecular weight fractions in GFL exhibit any phosphorylation. (D) Protein staining patterns and autoradiographs of exogenous proteins (histone and casein) phosphorylated by kinases in axoplasm and GFL lysates. Staining patterns in upper panel show similar protein bands. Lower panels: autoradiogram showing substrate phosphorylations are similar in both compartments indicating that they share similar kinases. Phosphorylation differences are most apparent in endogenous proteins. *Panel (A) modified from fig. 2 of Tytell et al. (1988). Panels (B) and (C) modified from fig. 2 of Grant et al. (1999). Panel (D) modified from fig. 3 of Grant et al. (1999).*

(0.1 μ*M* vanadate may also be included to inhibit tyrosine phosphatases). An exogenous substrate may be added such as histone H1 or casein at 1 mg/ml to compare endogenous with exogenous activity. The reaction is initiated by the addition of 10 μl of 0.01 uCi (γ-^{32}P) ATP (100 μ*M*) and incubated for 1 h at RT. An aliquot (20 μl) is taken and placed on a small square of P-81 phosphocellulose paper. All phosphorylated compounds bind to the paper. Papers are washed three times in 75 m*M* phosphoric acid followed by 95% alcohol washes before drying in air. The papers are then placed in scintillation fluid-filled vials for counting to determine P^{32} incorporation. The endogenous counts (without exogenous substrate) are substracted from those with exogenous substrates to determine activity. This assay is used to identify specific kinases that phosphorylate NFs, or other specific proteins that may be involved in the regulation of cytoskeletal protein phosphorylation. Specific kinase substrate proteins are included in the assay with the addition of appropriate regulators or specific inhibitors.

6.4 Soma (GFL) Versus Axoplasm

Squid NFs are also highly phosphorylated: in the presence of radioactive or nonradioactive ATP in a phosphorylating buffer, NFs, particularly the high molecular weight NF220, are robustly phosphorylated. Separated in a gel and immunoreacted with antibodies specific for squid NF, different phosphorylation patterns between axoplasm and GFL are drastically different (Fig. 6C). In fact, phosphorylation of most proteins in the giant axon is conspicuously compartmentalized; it is topographically regulated with most activity restricted to the axoplasm of the giant axon but very little in soma, GFL (Grant et al., 1999; Grant & Pant, 2004; Fig. 6B and C).

Two major proteins in the axon are highly phosphorylated and HMW (high molecular weight complex) probably consisting of cross-linked NF220 and with itself (Fig. 6C). Dephosphorylation of this fraction by different concentrations of alkaline phosphatase yields an end point at 190 kDa, presumed to be the nonphosphorylated NF220 to be present in the cell bodies.

7. NF-ASSOCIATED PROTEIN KINASES

It is not surprising that squid NFs, with multiple phosphorylation acceptor sites, are targeted by and associated with several different protein kinases. The numerous endogenously phosphorylated KSP repeats in the tail domain of NF220 implicate proline-directed kinases such as Cdks, GSK3,

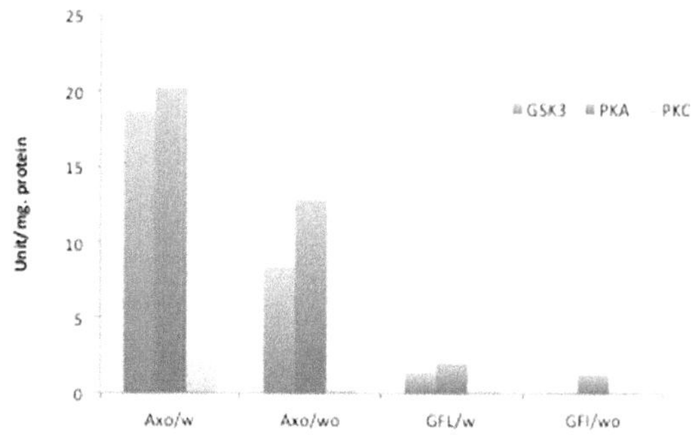

Figure 7 Sample of squid kinase activity in GFL and axoplasm. Activities of three kinases GSK3, PKA, and PKC, with (w) and without (wo) phosphatase inhibitors. Major differences between GFL and axoplasm (Axo) are seen in GSK3 and PKA kinases.

Erk1,2, and JNK among others. It should be noted that CK1 was identified as a principle kinase associated with axonal NFs (Floyd et al., 1991). In Fig. 7, an example of differences in activities of three protein kinases, GSK3, PKA, and PKC in lysates of GFL and axoplasm, is shown. The activities are greatest in axoplasm with (w) and without (wo) phosphatase inhibitors. This difference, in part, may explain the endogenous phosphorylation patterns of axoplasm compared to GFL.

8. SQUID NFs ARE INCLUDED IN COMPARTMENT-SPECIFIC MULTIMERIC PROTEIN COMPLEXES: P13suc1 AFFINITY CHROMATOGRAPHY

As previously emphasized, topographic regulation of NF protein phosphorylation is attributed, in part, to kinase activity differences in axonal versus soma compartments. Localization of kinases, regulators, substrates, and phosphatases has been explored (Floyd et al., 1991; Gallant et al., 1986; Grant et al., 1999; Llinas, Gruner, Sugimori, McGuinness, & Greengard, 1991; Morfini et al., 2004) and shown that these are organized into compartment-specific multimeric complexes (Grant et al., 1999; Takahashi et al., 1995; Veeranna, Shetty, Takahashi, Grant, & Pant, 2000), and it is likely that phosphorylation is locally regulated in such complexes. One approach to preparing such complexes is the technique of affinity chromatography using P13suc1-coated beads which specifically bind to cdc2-like kinases (such as cdk5), usually associated with such complexes (Brizuela, Draetta, & Beach, 1987; Draetta, Brizuela, Potashkin, & Beach, 1987; Grant et al., 1999; Takahashi et al., 1995).

Homogenates of axoplasm and GFL (25 axons, GFLs per 1 ml extraction buffer C (see above), centrifuged 10,000 rpm 20 min, 4 °C, then resuspended in buffer and spun to collect supernatants. The protein complexed with P13suc1 affinity beads were analyzed as follows:

Beads are prewashed in extraction buffer before collected in a 1:1 (v/v) ratio of packed beads to buffer. An aliquot (1–3 ml) of supernatant mixed with beads in ratio of 5:1 (v/v) is incubated with shaking at 4° C for 3 h. The beads are collected by centrifuging at 10,000 rpm for 2 min, then washed four times in wash buffer (PBS, 1% Triton X-100, and 10% glycerol). Packed beads are resuspended in extraction buffer and used for SDS-PAGE, Western blots, and kinase assays.

A comparison of Coomassie-stained gels and Western blots of axoplasm and GFL lysates exhibits similar repertoires of proteins except for the NF220 and HMW NF (Fig. 8A); similar cytoskeletal and kinase protein patterns are evident. When comparing the P13 extracts, however, profound differences are seen in Western blots of cytoskeletal proteins which are more evident in the P13 axoplasm extracts (Fig. 8B) The endogenous and exogenous phosphorylating activities of P13 AXO are 6 to more than 20 times greater than the GFL P13 extracts (Grant et al., 1999; Takahashi et al., 1995). These observations suggest that the compartmental-specific patterns of phosphorylation in giant axons are in part due to differences in the structure of multimeric complexes of cytoskeletal substrates, kinases, regulators, and phosphatases, although evidence for differences in the latter is not that clear. Subsequent studies have shown that tyrosine phosphatase may be more prevalent in the GFL than in axoplasm and may contribute to the lower levels of overall phosphorylation (Grant & Pant, 2004).

The abundance of NF proteins in axoplasm macromolecular complexes with their many putative phosphorylation acceptor sites no doubt accounts for the robust phosphorylation expression. Structure of the complex, where substrates with multiple phosphorylation sites and kinases intimately interact, facilitates the sequential order of multisite phosphorylation by different kinases (Roach, 1991). Although NF subunits are synthesized in cell bodies, a transient head domain phosphorylation by PKA of the NF60/70 and NF190 in the cell body may inhibit tail domain phosphorylations until the N-terminal site is dephosphorylated as NF proteins are transported past the axon hillock into the axon (Zheng, Li, Veeranna & Pant, 2003). PKA is indeed one of the principal kinases localized in cells of the soma.

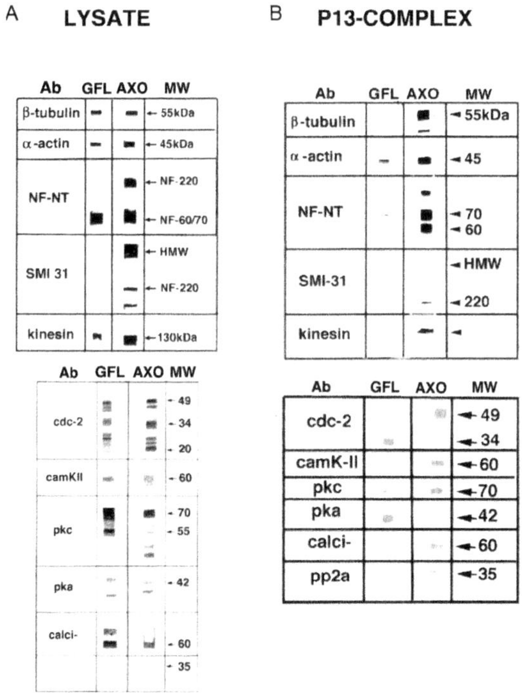

Figure 8 Comparison of protein expression in P13 multimeric complexes of GFL and axoplasm. (A) Western blot analysis of lysates extracted from axoplasm and GFL. Upper panels show expression of cytoskeletal proteins with most expression of NFs in axoplasm. Lower panels: similarity of protein kinase expression. cdc2, cyclic-like kinase; camKII, cam kinase; calci., calcineurin a phosphatase, pp2a, protein phosphatase 2. In all Western blots, the identified bands at the molecular weights shown consistently displayed the most robust expression at various antibody dilutions. (B).Panels on right show expression of cytoskeletal and kinase proteins in P13suc1 complexes from GFL and axoplasm lysates whose expression is shown on the left. Symbols same as in lysate panels. *Modified from figs 5 and 8, respectively, Grant et al. (1999).*

REFERENCES

Abercrombie, R. F., Masukawa, L. M., Sjodin, R. A., & Livengood, D. (1981). Uptake and release of 45Ca by Myxicola axoplasm. *The Journal of General Physiology*, *78*, 413–429.

Adams, D. J., & Gillespie, J. I. (1988). The actions of L-glutamate at the postsynaptic membrane of the squid giant synapse. *The Journal of Experimental Biology*, *140*, 535–548.

Allen, R. D., Metuzals, J., Tasaki, I., Brady, S. T., & Gilbert, S. P. (1982). Fast axonal transport in squid giant axon. *Science*, *218*, 1127–1129.

Altamirano, A. A., Breitwieser, G. E., & Russell, J. M. (1999). Activation of Na+, K+, Cl- cotransport in squid giant axon by extracellular ions: Evidence for ordered binding. *Biochimica et Biophysica Acta*, *1416*, 195–207.

Arnold, J. M. (1990). Evolution and intelligence of the cephalopads. In D. L. Gilbert, W. J. William, & R. J. Arnold (Eds.), *Squid as experimental animals*. New York: Plenum Press.

Augustine, G. J., Charlton, M. P., & Smith, S. J. (1985). Calcium entry into voltage-clamped presynaptic terminals of squid. *The Journal of Physiology, 367*, 143–162.

Bearer, E. L., DeGiorgis, J. A., Bodner, R. A., Kao, A. W., & Reese, T. S. (1996). Evidence for myosin motors on organelles in squid axoplasm. *Proceedings of the National Academy of Sciences of the United States of America, 93*(12), 6064–6068.

Brady, S. T., Lasek, R. J., & Allen, R. D. (1982). Fast axonal transport in extruded axoplasm from squid giant axon. *Science, 218*, 1129–1131.

Brady, S. T., Lasek, R. J., & Allen, R. D. (1985). Video microscopy of fast axonal transport in extruded axoplasm: A new model for study of molecular mechanisms. *Cell Motility, 5*, 81–101.

Brinley, F. J., Jr., & Tiffert, T. (1978). Kinetics of calcium accumulation by mitochondria, studied in situ, in squid giant axons. *FEBS Letters, 91*, 25–29.

Brizuela, L., Draetta, G., & Beach, D. (1987). p13suc1 acts in the fission yeast cell division cycle as a component of the p34cdc2 protein kinase. *EMBO Journal, 6*(11), 3507–3514.

Brown, A., & Lasek, R. J. (1993). Neurofilaments move apart freely when released from the circumferential constraint of the axonal plasma membrane. *Cell Motility and the Cytoskeleton, 26*(4), 313–324.

Clay, J. R. (1998). Excitability of the squid giant axon revisited. *Journal of Neurophysiology, 80*, 903–913.

Clay, J. R., & Kuzirian, A. M. (2000). Localization of voltage-gated K(+) channels in squid giant axons. *Journal of Neurobiology, 45*, 172–184.

Cohen, R. S., Pant, H. C., House, S., & Gainer, H. (1987). Biochemical and immunocytochemical characterization and distribution of phosphorylated and nonphosphorylated subunits of neurofilaments in squid giant axon and stellate ganglion. *Journal of Neuroscience, 7*, 2056–2074.

Draetta, G., Brizuela, L., Potashkin, J., & Beach, D. (1987). Identification of p34 and p13, human homologs of the cell cycle regulators of fission yeast encoded by cdc2+ and suc1+. *Cell, 50*(2), 319–325.

Floyd, C. C., Grant, P., Gallant, P. E., & Pant, H. C. (1991). Principal neurofilament-associated protein kinase in squid axoplasm is related to casein kinase I. *The Journal of Biological Chemistry, 266*(8), 4987–4994.

Galbraith, J. A., Reese, T. S., Schlief, M. L., & Gallant, P. E. (1999). Slow transport of unpolymerized tubulin and polymerized neurofilament in the squid giant axon. *Proceedings of the National Academy of Sciences of the United States of America, 96*, 11589–11594.

Gallant, P. E., Pant, H. C., Pruss, R. M., & Gainer, H. (1986). Calcium-activated proteolysis of neurofilament proteins in the squid giant neuron. *Journal of Neurochemistry, 46*, 1573–1581.

Gilly, W. F., & Armstrong, C. M. (1982). Divalent cations and the activation kinetics of potassium channels in squid giant axons. *The Journal of General Physiology, 79*, 965–996.

Gotow, T., & Tanaka, J. (1994). Phosphorylation of neurofilament H subunit as related to arrangement of neurofilaments. *Journal of Neuroscience Research, 37*, 691–713.

Gotow, T., Tanaka, T., Nakamura, Y., & Takeda, M. (1994). Dephosphorylation of the largest neurofilament subunit protein influences the structure of crossbridges in reassembled neurofilaments. *Journal of Cell Science, 107*(Pt. 7), 1949–1957.

Grant, P., Diggins, M., & Pant, H. C. (1999). Topographic regulation of cytoskeletal protein phosphorylation by multimeric complexes in the squid giant fiber system. *Journal of Neurobiology, 40*, 89–102.

Grant, P., & Pant, H. C. (2000). Neurofilament protein synthesis and phosphorylation. *Journal of Neurocytology, 29*, 843–872.

Grant, P., & Pant, H. C. (2004). Topographic regulation of phosphorylation in giant neurons of the squid, Loligo pealei: Role of phosphatases. *Journal of Neurobiology, 58*, 514–528.

Grant, P., Tseng, D., Gould, R. M., Gainer, H., & Pant, H. C. (1995). Expression of neurofilament proteins during development of the nervous system in the squid Loligo pealei. *The Journal of Comparative Neurology, 356*, 311–326.

Hodgkin, A. L., & Huxley, A. F. (1952a). The components of membrane conductance in the giant axon of Loligo. *The Journal of Physiology, 116*, 473–496.

Hodgkin, A. L., & Huxley, A. F. (1952b). Currents carried by sodium and potassium ions through the membrane of the giant axon of Loligo. *The Journal of Physiology, 116*, 449–472.

Hodgkin, A. L., & Huxley, A. F. (1952c). The dual effect of membrane potential on sodium conductance in the giant axon of Loligo. *The Journal of Physiology, 116*, 497–506.

Jaffe, H., Sharma, P., Grant, P., & Pant, H. (2001). Characterization of the phosphorylation sites of the squid (Loligo pealei) high-molecular-weight neurofilament protein from giant axon axoplasm. *Journal of Neurochemistry, 76*, 1022–1031.

Laemmli, U. K., Beguin, F., & Gujer-Kellenberger, G. (1970). A factor preventing the major head protein of bacteriophage T4 from random aggregation. *Journal of Molecular Biology, 47*(1), 69–85.

Llinas, R. R. (1982). Calcium in synaptic transmission. *Scientific American, 247*, 56–65.

Llinas, R., Gruner, J. A., Sugimori, M., McGuinness, T. L., & Greengard, P. (1991). Regulation by synapsin I and Ca(2+)-calmodulin-dependent protein kinase II of the transmitter release in squid giant synapse. *The Journal of Physiology, 436*, 257–282.

Martin, R. (1965). On the structure and embryonic development of the giant fibre system of the squid Loligo vulgaris. *Zeitschrift für Zellforschung und Mikroskopische Anatomie, 67*, 77–85.

Martin, R. (1996). The structure of the neurofilament cytoskeleton in the squid giant axon and synapse. *Journal of Neurocytology, 25*, 547–554.

Martin, R., Door, R., Ziegler, A., Warchol, W., Hahn, J., & Breitig, D. (1999). Neurofilament phosphorylation and axon diameter in the squid giant fibre system. *Neuroscience, 88*, 327–336.

Martin, R., & Miledi, R. (1975). A presynaptic complex in the giant synapse of the squid. *Journal of Neurocytology, 4*, 121–129.

Metuzals, J., Fishman, H. M., & Robb, I. A. (1995). The neurofilamentous network-smooth endoplasmic reticulum complex in transected squid giant axon. *The Biological Bulletin, 189*, 216–218.

Metuzals, J., Hodge, A. J., Lasek, R. J., & Kaiserman-Abramof, I. R. (1983). Neurofilamentous network and filamentous matrix preserved and isolated by different techniques from squid giant axon. *Cell and Tissue Research, 228*, 415–432.

Morfini, G., Szebenyi, G., Brown, H., Pant, H. C., Pigino, G., DeBoer, S., et al. (2004). A novel CDK5-dependent pathway for regulating GSK3 activity and kinesin-driven motility in neurons. *The EMBO Journal, 23*, 2235–2245.

Otis, T. S., & Gilly, W. F. (1990). Jet-propelled escape in the squid Loligo opalescens: Concerted control by giant and non-giant motor axon pathways. *Proceedings of the National Academy of Sciences of the United States of America, 87*, 2911–2915.

Pant, H. C., Gallant, P. E., & Gainer, H. (1986). Characterization of a cyclic nucleotide- and calcium-independent neurofilament protein kinase activity in axoplasm from the squid giant axon. *The Journal of Biological Chemistry, 261*, 2968–2977.

Pant, H. C., Veeranna, & Grant, P. (2000). Regulation of axonal neurofilament phosphorylation. *Current Topics in Cellular Regulation, 36*, 133–150.

Pratt, M. M. (1986). Stable complexes of axoplasmic vesicles and microtubules: Protein composition and ATPase activity. *The Journal of Cell Biology, 103*, 957–968.

Roach, P. J. (1991). Multisite and hierarchal protein phosphorylation. *Journal of Biological Chemistry, 266*(22), 14139–14142 (Review).
Rosenthal, J. J., & Gilly, W. F. (2003). Identified ion channels in the squid nervous system. *Neurosignals, 12*, 126–141.
Shea, T. B., & Chan, W. K. (2008). Regulation of neurofilament dynamics by phosphorylation. *The European Journal of Neuroscience, 27*, 1893–1901.
Shea, T. B., & Lee, S. (2011). Neurofilament phosphorylation regulates axonal transport by an indirect mechanism: A merging of opposing hypotheses. *Cytoskeleton (Hoboken, N.J.), 68*, 589–595.
Sternberger, L. A., & Sternberger, N. H. (1983). Monoclonal antibodies distinguish phosphorylated and nonphosphorylated forms of neurofilaments in situ. *Proceedings of the National Academy of Sciences of the United States of America, 80*(19), 6126–6130.
Szaro, B. G., Pant, H. C., Way, J., & Battey, J. (1991). Squid low molecular weight neurofilament proteins are a novel class of neurofilament protein. A nuclear lamin-like core and multiple distinct proteins formed by alternative RNA processing. *The Journal of Biological Chemistry, 266*, 15035–15041.
Takahashi, M., Amin, N., Grant, P., & Pant, H. C. (1995). P13suc1 associates with a cdc2-like kinase in a multimeric cytoskeletal complex in squid axoplasm. *Journal of Neuroscience, 15*, 6222–6229.
Tytell, M., Pant, H. C., Gainer, H., & Hill, W. D. (1990). Characterization of the distinctive neurofilament subunits of the soma and axon initial segments in the squid stellate ganglion. *Journal of Neuroscience Research, 25*, 153–161.
Tytell, M., Zackroff, R. V., & Hill, W. D. (1988). Axonal neurofilaments differ in composition and morphology from those in the soma of the squid stellate ganglion. *Cell Motility and the Cytoskeleton, 9*, 349–360.
Veeranna, G. J., Shetty, K. T., Takahashi, M., Grant, P., & Pant, H. C. (2000). Cdk5 and MAPK are associated with complexes of cytoskeletal proteins in rat brain. *Brain Research. Molecular Brain Research, 76*, 229–236.
Way, J., Hellmich, M. R., Jaffe, H., Szaro, B., Pant, H. C., Gainer, H., et al. (1992). A high-molecular-weight squid neurofilament protein contains a lamin-like rod domain and a tail domain with Lys-Ser-Pro repeats. *Proceedings of the National Academy of Sciences of the United States of America, 89*, 6963–6967.
Westerfield, M., & Joyner, R. W. (1982). Postsynaptic factors controlling the shape of potentials at the squid giant synapse. *Neuroscience, 7*, 1367–1375.
Williamson, R., & Chrachri, A. (2004). Cephalopod neural networks. *Neurosignals, 13*, 87–98.
Young, J. Z. (1938). The functioning of the giant nerve fibres of the squid. *The Journal of Experimental Biology, 15*, 170–185.
Young, J. Z. (1939). Fused neurons and synaptic contacts in the giant nerve fibres of cephalopods. *Philosophical Transactions of the Royal Society of London B, 229*, 465–503.
Young, J. Z. (1973). The giant fibre synapse of Loligo. *Brain Research, 57*, 457–460.
Zheng, Y. L., Li, B. S., Veeranna, & Pant, H. C. (2003). Phosphorylation of the head domain of neurofilament protein (NF-M): A factor regulating topographic phosphorylation of NF-M tail domain KSP sites in neurons. *Journal of Biological Chemistry, 278*(26), 24026–24032.

CHAPTER TWENTY-THREE

Using *Xenopus* Embryos to Study Transcriptional and Posttranscriptional Gene Regulatory Mechanisms of Intermediate Filaments

Chen Wang, Ben G. Szaro[1]

Department of Biological Sciences, University at Albany, State University of New York, Albany, New York, USA

[1]Corresponding author: e-mail address: bgs86@albany.edu

Contents

1. Introduction 636
   1.1 Intermediate Filament Genes as Markers for Cellular Differentiation 636
   1.2 Analysis of Neuronal Intermediate Filament Gene *cis*-Regulatory Elements 638
   1.3 *Xenopus* as a Model System for Studies of Gene Function and Regulation of Gene Expression Using an Intact Vertebrate Organism 638
2. Preparation of Expression Plasmids 641
   2.1 Description of gIGFP Reference and DsRed2 Test Plasmids 641
   2.2 Two-Step Cloning Strategy for Generating New Expression Plasmids 642
   2.3 Preparation of Plasmid DNA for Microinjection into Early *Xenopus* Embryos 642
3. Microinjection of *Xenopus* Embryos 644
   3.1 Useful *Xenopus* References 644
   3.2 Instrumentation 645
   3.3 Obtaining and Injecting Embryos 647
4. Assays for Gene Expression 649
   4.1 Reporter Protein Expression 649
   4.2 Reporter RNA Expression 650
5. Assays for Effects of *cis*-Regulatory Elements on RNA Processing, Trafficking, and Translation 651
   5.1 RNA Processing Efficiency 651
   5.2 Efficiency of Export of RNA from the Nucleus 652
   5.3 Analysis of mRNA Translation by Polysomal Profiling 653
6. Conclusions 654
   6.1 Pearls and Pitfalls 654

Acknowledgments 654

References 654

Methods in Enzymology, Volume 568
ISSN 0076-6879
http://dx.doi.org/10.1016/bs.mie.2015.07.012

Abstract

Intermediate filament genes exhibit highly regulated, tissue-specific patterns of expression during development and in response to injury. Identifying the responsible *cis*-regulatory gene elements thus holds great promise for revealing insights into fundamental gene regulatory mechanisms controlling tissue differentiation and repair. Because much of this regulation occurs in response to signals from surrounding cells, characterizing them requires a model system in which their activity can be tested within the context of an intact organism conveniently. We describe methods for doing so by injecting plasmid DNAs into fertilized *Xenopus* embryos. A prokaryotic element for site-specific recombination and two dual *HS4* insulator elements flanking the reporter gene promote penetrant, promoter-typic expression that persists through early swimming tadpole stages, permitting the observation of fluorescent reporter protein expression in live embryos. In addition to describing cloning strategies for generating these plasmids, we present methods for coinjecting test and reference plasmids to identify the best embryos for analysis, for analyzing reporter protein and RNA expression, and for characterizing the trafficking of expressed reporter RNAs from the nucleus to polysomes. Thus, this system can be used to study the activities of *cis*-regulatory elements of intermediate filament genes at multiple levels of transcriptional and posttranscriptional control within an intact vertebrate embryo, from early stages of embryogenesis through later stages of organogenesis and tissue differentiation.

1. INTRODUCTION

1.1 Intermediate Filament Genes as Markers for Cellular Differentiation

A hallmark characteristic of intermediate filament genes is their cell type and developmental stage-specific expression (Bennett et al., 1978; Franke, Schmid, Osborn, & Weber, 1978; Schmid et al., 1979). This characteristic has led to the extensive use of intermediate filament mRNA and protein expression both as identifiers of the origins of metastasized tumors (Bannasch, Zerban, Schmid, & Franke, 1980) and as markers for the state of differentiation of specific cell types (Cochard & Paulin, 1984; Godsave, Anderton, & Wylie, 1986; Tapscott, Bennett, Toyama, Kleinbart, & Holtzer, 1981). In developing mammalian neurons, for example, undifferentiated neuroepithelial cells initially express nestin (Lendahl, Zimmerman, & McKay, 1990), followed by vimentin (Bignami, Raju, & Dahl, 1982; Cochard & Paulin, 1984). As neuroblasts of the central nervous system (CNS) differentiate into neurons, expression of these intermediate filament genes is followed first by expression of α-internexin (Kaplan, Chin, Fliegner, & Liem, 1990), and of peripherin in phylogenetically old

CNS neurons (Escurat, Djabali, Gumpel, Gros, & Portier, 1990; Parysek & Goldman, 1988; Troy, Brown, Greene, & Shelanski, 1990), together with low levels of the light (*nefl*) and medium (*nefm*) neurofilament triplets (Carden, Trojanowski, Schlaepfer, & Lee, 1987; Nixon & Shea, 1992). Later, as axons emerge, expression of the heavy neurofilament (*nefh*) follows, and expression of all three neurofilament triplet genes subsequently increases as axons make contact with their postsynaptic targets, become myelinated, and expand in caliber (Schlaepfer & Bruce, 1990).

This progression is phylogenetically conserved among vertebrates, as evidenced by a similar progression occurring with only slight variations during the development of *Xenopus* neurons (Charnas, Szaro, & Gainer, 1992; Gervasi, Stewart, & Szaro, 2000; Godsave et al., 1986; Hemmati-Brivanlou, Mann, & Harland, 1992; Lin & Szaro, 1994; Sharpe, 1988; Szaro & Gainer, 1988; Szaro, Lee, & Gainer, 1989; Undamatla & Szaro, 2001). Moreover, many of the features of this progression are coupled with axon outgrowth, because they are largely recapitulated in successfully regenerating optic axons of both fish and frog, in which they are modulated by interactions between the regrowing axons and cells encountered along the visual pathway (Asch et al., 1998; Gervasi, Thyagarajan, & Szaro, 2003; Glasgow, Druger, Fuchs, Lane, & Schechter, 1994; Niloff, Dunn, & Levine, 1998; Zhao & Szaro, 1994, 1995).

Whereas the variations in intermediate filament gene expression seen among neuronal cell subtypes (e.g., spinal motor neurons vs. interneurons) are directly related to differences in gene transcription, changes in expression that are modulated by interactions between axons and the surrounding cells are significantly influenced through posttranscriptional gene regulatory mechanisms (Ananthakrishnan, Gervasi, & Szaro, 2008; Ananthakrishnan & Szaro, 2009; Schwartz, Shneidman, Bruce, & Schlaepfer, 1994). Aberrant posttranscriptional regulation of neuronal intermediate filament gene expression also underlies many of the pathological changes associated with neurodegenerative diseases such as amyotrophic lateral sclerosis (Gallo et al., 2005; Lin, Zhai, Cañete-Soler, & Schlaepfer, 2004; Lin et al., 2003; Szaro & Strong, 2010; Thyagarajan, Strong, & Szaro, 2007). Thus, characterizing the *cis*-regulatory gene elements responsible, along with the *trans*-acting factors that bind them, has great potential for providing key insights into the molecular mechanisms underlying both the differentiation of neurons and the development and maintenance of healthy axons. Because these events are influenced by the complex cellular environments occupied by developing neurons, such characterization is best done within an intact developing nervous system.

1.2 Analysis of Neuronal Intermediate Filament Gene *cis*-Regulatory Elements

The first efforts to identify neuronal intermediate filament gene *cis*-regulatory elements were performed in cell lines. Studies in PC12 cells, for example, identified elements required for the induction of *peripherin* expression by nerve growth factor (Chang & Thompson, 1996; Leonard, Gorham, Cole, Greene, & Ziff, 1988), and studies comparing expression between neuronal and nonneuronal cell lines identified elements that modulate levels of neuronal intermediate filament gene transcription. However, these studies provided few insights into the elements responsible for neuron-specific expression (Budhram-Mahadeo et al., 1995; Pospelov, Pospelova, & Julien, 1994). The construction of transgenic mouse lines later demonstrated the power of analyzing neuronal intermediate filament gene *cis*-regulatory elements in an intact organism by revealing that elements required for neuronal expression lay within intragenic as well as upstream regulatory sequences (Beaudet, Cote, Houle, & Julien, 1993; Beaudet et al., 1992; Belecky-Adams, Wight, Kopchick, & Parysek, 1993; Uveges, Shan, Kramer, Wight, & Parysek, 2002; Zimmerman et al., 1994). Despite the power of this approach, the difficulty and expense of generating transgenic mouse lines have ultimately compromised its general utility, raising the need for alternative model systems.

1.3 *Xenopus* as a Model System for Studies of Gene Function and Regulation of Gene Expression Using an Intact Vertebrate Organism

Since its first use in medical laboratories for testing pregnancy in the 1930s (Elkan, 1938), the South African claw-toed frog, *Xenopus laevis*, has become a standard laboratory organism for studies in cell, molecular, and developmental biology. In more recent years, a sister species, *Xenopus tropicalis*, has added genetic resources to the repertoire of tools used by *Xenopus* investigators, chiefly due to its smaller genome and shorter generation time (Amaya, Offield, & Grainger, 1998).

The ease of rearing *Xenopus* under laboratory conditions and of obtaining hundreds of fertile embryos on demand, together with its rapid, well-characterized external development (Nieuwkoop & Faber, 1994), has promoted its adoption as a model organism for studying early vertebrate development. Initial studies used *Xenopus* embryos for detailed analyses of the time course of early developmental events (Hughes, 1957), for testing the developmental capacity of isolated cellular nuclei (Fischberg,

Gurdon, & Elsdale, 1959; Gurdon, Elsdale, & Fischberg, 1958), and for comparing the developmental capacities of embryonic cells reared in isolation against those grown in contact with other cells (Corner, 1964; Hunt, 1969). With the advent of convenient techniques for injecting a variety of macromolecules into early stage embryos (Jacobson & Hirose, 1978), *Xenopus* also became an attractive system for studying gene function during vertebrate development. Gene products can be readily inactivated by injecting antibodies (Warner, Guthrie, & Gilula, 1984), synthetic RNAs encoding dominant negatives (Christian, Edelstein, & Moon, 1990; Wright, Cho, Hardwicke, Collins, & De Robertis, 1989), and antisense morpholino oligonucleotides (Nutt, Bronchain, Hartley, & Amaya, 2001). One can also over- or mis-express proteins, individually or in combination, by injecting *in vitro* transcribed RNAs (Harvey & Melton, 1988; Kintner, 1988; Krieg & Melton, 1984), as well as exogenous DNAs (Rusconi & Schaffner, 1981; Vize, Hemmati-Brivanlou, Harland, & Melton, 1991).

As these methods of analyzing gene function in *Xenopus* embryos were being developed, the prospect that injected DNAs could also be used to study the *in vivo* activity of *cis*-regulatory gene elements was also explored. Years of using cell lines for such studies had indicated that model systems in which expression could be tested in the intact organism were needed to reveal the full nature of these elements. The ease of injecting macromolecules into newly fertilized *Xenopus* embryos and of analyzing subsequent expression in this externally developing embryo made *Xenopus* especially attractive. Unfortunately, early studies demonstrated that in only a very few instances was it possible to obtain reporter expression from injected plasmids bearing *cis*-regulatory sequences at the correct developmental stages (e.g., gastrula-specific protein 17 (Krieg & Melton, 1985)), in the right tissues (e.g., cardiac actin (Mohun, Taylor, Garrett, & Gurdon, 1989) and epidermal keratin (Jonas, Snape, & Sargent, 1989)), or by providing appropriate environmental stimuli (e.g., raising the temperature to activate expression of the *hsp70* promoter (Krone & Heikkila, 1989)). For the vast majority of elements tested, reporter expression was more typically "mosaic," a catch-all term indicating that expression levels varied extensively in places where it was expected and often inappropriately appeared elsewhere (Vize et al., 1991). The weaker the promoter elements were, the more difficult it was to discern their *in vivo* activity. Even in instances in which the temporal and spatial specificity of expression from the injected plasmid was relatively good, the injected DNA, along with its expression, was generally lost by tailbud stages, long before most cells fully differentiated and the genes characterizing this final state were expressed (Andres, Muellener, & Ryffel, 1984; Marini, Etkin, & Benbow, 1988). Thus, these limitations severely curtailed the utility of this simple, convenient method for analyzing *cis*-regulatory gene elements, and they subsequently spurred the development of methods for obtaining more reliable expression into later stages by making fully transgenic animals.

These latter methods are mostly based on achieving stable integration of exogenous DNAs into the genome (e.g., restriction enzyme-mediated integration of DNA (Amaya & Kroll, 1999; Kroll & Amaya, 1996; Kroll & Gerhart, 1994; Sparrow, Latinkic, & Mohun, 2000), transposon-mediated transgenesis (Hamlet et al., 2006; Sinzelle et al., 2006; Yergeau et al., 2009), and I-*Sce*I meganuclease-mediated transgenesis (Ishibashi, Love, & Amaya, 2012; Ogino, McConnell, & Grainger, 2006; Pan, Chen, Loeber, Henningfeld, & Pieler, 2006)). Although such transgenesis methods have proved successful for characterizing the activity of *cis*-regulatory gene elements, they are technically more involved and thus, for most investigators, less convenient than using plasmid injection. Moreover, in the F_0-injected embryos, which permit analysis within a few days rather than the months it takes to raise transgenic lines, these methods still yield enough mosaic expression that they become most useful for analyzing expression in the F_1 generation. Thus, methods that would permit more effective analysis of reporter expression in F_0 animals have been sought after.

Several methods have since emerged for use with F_0 animals. Injection of bacterial artificial chromosomes (BACs) into fertilized *Xenopus* eggs has proved useful for screening promoters before proceeding to make transgenic lines (Fish, Nakayama, & Grainger, 2012), but BACs are harder to work with than plasmids. Thus, efforts have been made to improve the fidelity of expression from injected plasmids. One such method coinjects into fertilized eggs a plasmid containing a prokaryotic *attB* element together with RNA encoding an integrase from ϕC31 bacteriophage, which facilitate integration into the genome (Allen & Weeks, 2005). Flanking the reporter gene and its regulatory elements with dual *β-globin Hypersensitive Site 4* (*HS4*) insulator elements from chicken further helps to minimize the influence on expression from endogenous regulatory elements surrounding the integration site (Allen & Weeks, 2005; Chung, Whiteley, & Felsenfeld, 1993; Love et al., 2011; Pikaart, Recillas-Targa, & Felsenfeld, 1998; West, Gaszner, & Felsenfeld, 2002). In developing similar plasmids to study *cis*-regulatory gene elements of the *Xenopus nefm* gene, we serendipitously discovered that orienting both the flanking dual *HS4* insulator elements inward, toward the transgene, yielded more robust, promoter-typic reporter expression, which persisted well into feeding tadpole stages, even without coinjecting the integrase RNA (Wang & Szaro, 2015). In that study, to control for variations in the injection procedure, a second reference plasmid containing these elements, but yielding a standardized green fluorescent protein (GFP) expression pattern, was coinjected with a similarly constructed test plasmid containing *DsRed2* and *cis*-regulatory elements of interest. Embryos exhibiting the most faithful expression of the GFP reference plasmid provided the best indicator of the activity of the coinjected DsRed2 test plasmid's *cis*-regulatory elements. Also, a two-step cloning strategy was developed to overcome the tendency of these plasmids to undergo recombination during their initial construction, enabling one to generate the plasmids reliably using standard cloning methods and conventional strains of *E. coli* (e.g., subcloning efficiency DH5α). Here, we present protocols for generating and using such plasmids to analyze both transcriptional and post-transcriptional activities of *cis*-regulatory gene elements in *Xenopus* embryos. The method permits analyzing expression into early feeding tadpole stages (e.g., at least through st. 43–46), which is much later than what has been possible to be achieved with other plasmid-injection approaches.

2. PREPARATION OF EXPRESSION PLASMIDS

2.1 Description of glGFP Reference and DsRed2 Test Plasmids

For our studies of *cis*-regulatory elements from the *nefm* gene, we have used a reference plasmid (pSPORT1[*attB/Ins1/*(1.5 kb) *Xenopus nefm2 promoter/glGFP/β-globin* 3′-UTR/*Ins2*]) that contains DNA encoding GreenLantern green fluorescent protein (*glGFP,* 0.7 kb, originally extracted from pGreenlantern-1, GibcoBRL) under the control of a 1.5 kb promoter from the *Xenopus nefm2* gene (Roosa, Gervasi, & Szaro, 2000). The 3′-UTR (2.2 kb) originates from rabbit *β-globin*, and the *Ins1* and *Ins2* elements are dual insulators, each consisting of two single *HS4* insulator core (250 bp) elements joined in tandem (Chung, Bell, & Felsenfeld, 1997; Chung et al., 1993). The dual insulator elements differ from one another in their orientations, with each pointing inward toward the transgene elements. The *attB* element helps promote persistence of the injected DNA. A description of how these plasmids were generated, together with the expected expression patterns obtained when they are injected into 2-cell stage embryos can be found in Wang and Szaro (2015). In brief, the reference plasmid generates predominantly neural expression, with some additional weak expression in tail somites (Fig. 1). Test plasmids are analogous to the reference plasmid, except that *glGFP* is replaced by *DsRed2* (derived from pDsRed2-C1; Clontech, Mountain View, CA), and various regulatory elements are used in place of either the promoter, 3′-UTR, or both, depending on whether the elements are expected to be involved in transcriptional or posttranscriptional control. Introns and various 5′-UTRs can also be engineered into the plasmid for testing, using the same two-step cloning strategy described below.

2.2 Two-Step Cloning Strategy for Generating New Expression Plasmids

Because of the high propensity for plasmids bearing dual flanking, inverted *HS4* insulators to undergo recombination during cloning (Sekkali et al., 2008; Tran & Vleminckx, 2014), we developed a two-step cloning strategy (Fig. 2) to generate new plasmids, reducing this propensity significantly (Wang & Szaro, 2015). In brief, an intermediate cloning vector (pBluescriptII[*attB/Ins1*]) is used for placing a promoter of interest downstream of the *attB* and *Ins1* elements by directional cloning into appropriate restriction sites within the multiple cloning site of this intermediate vector (e.g., *Spe*I and *Not*I). Once in the plasmid, the promoter is then extracted by high-fidelity PCR, using primers that add *Nhe*I sites flanking the *attB* and new promoter elements. After digesting this PCR product with *Nhe*I, the piece is then ligated into a second host vector prepared with *Nhe*I (pSPORT1[*Nhe*I/*glGFP*/*β-globin* 3′-UTR/*Ins2*]) to create the final plasmid used for injection. To confirm the fidelity of the plasmids used for injection they are typically sequenced. The vectors (pBluescriptII[*attB/Ins1*] and pSPORT1[*Nhe*I/*glGFP*/*β-globin* 3′-UTR/*Ins2*]), as well as the reference plasmid (pSPORT1[*attB*/*Ins1*/(1.5 kb) *Xenopus nefm2 promoter*/*glGFP*/*β-globin* 3′-UTR/*Ins2*]), are available from the authors, upon request.

2.3 Preparation of Plasmid DNA for Microinjection into Early *Xenopus* Embryos

Plasmid DNA should be prepared using kits that yield high-quality DNA (e.g., Plasmid Maxi Kit, Qiagen, Valencia, CA; PureYield Miniprep, Promega, Madison, WI) from *E. coli* strains that yield both high-quality DNA and low levels of endotoxins (e.g., DH5α). Typically, equal amounts of test and reference plasmids (75 pg each) are mixed together for injection,

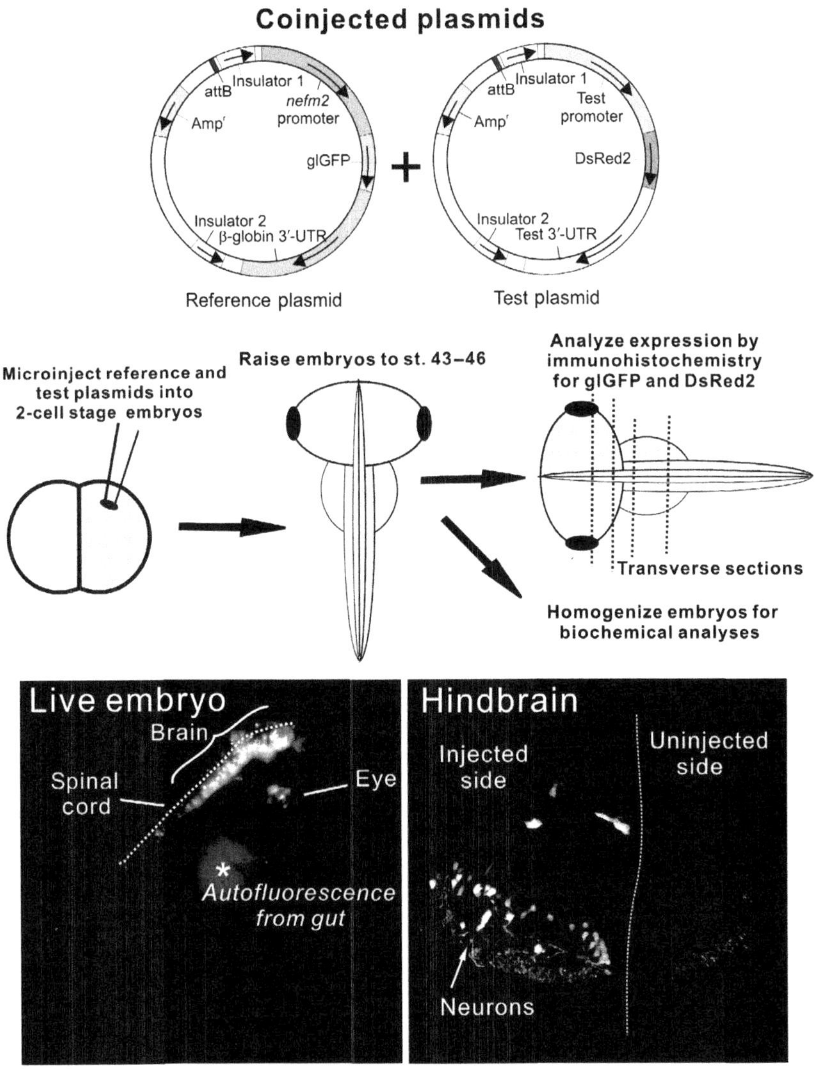

Figure 1 (Top) A test plasmid expressing DsRed2 and *cis*-regulatory elements of interest is coinjected with a reference plasmid expressing glGFP and containing an *nefm*2 promoter and the 3′-UTR from *β-globin*. Both plasmids contain an *attB* element and flanking dual *HS4* insulator elements. (Middle) The two plasmids are coinjected at the 2-cell stage and resultant embryos are reared to early swimming tadpole stages (st. 43–46) for observation and biochemical analyses. (Bottom) Examples of glGFP expression derived from a reference plasmid containing 1.5 kb of upstream regulatory sequence from the *Xenopus nefm*2 gene, illustrated in a live tadpole and in a section of hindbrain processed for glGFP immunofluorescence. (See the color plate.)

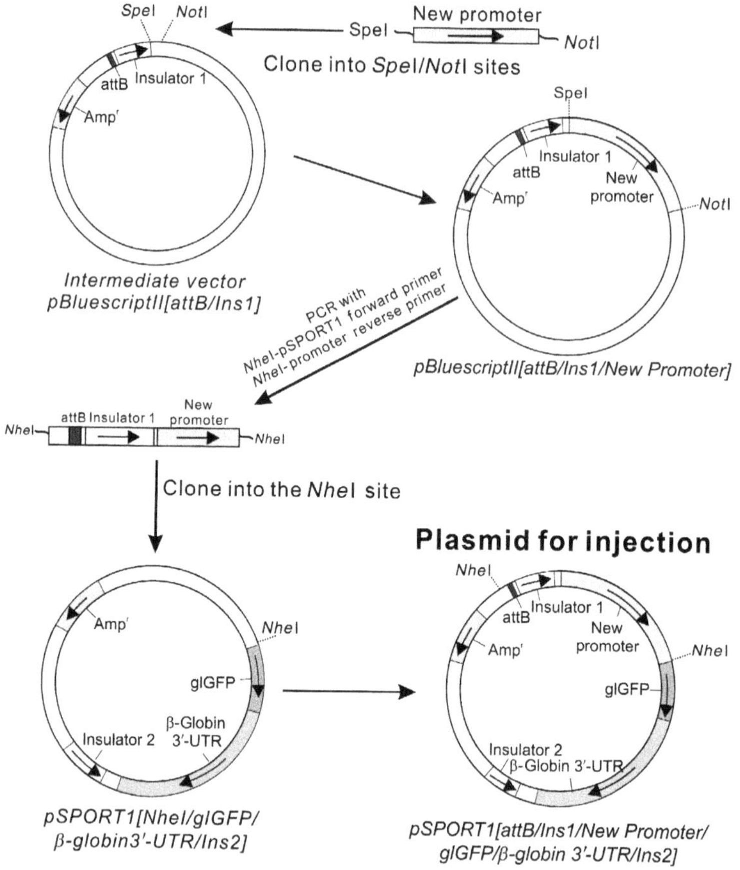

Figure 2 A two-step cloning procedure to reduce DNA recombination during construction of plasmids containing the flanking *HS4* dual insulators oriented inward toward the transgene. For more details, see Section 2.2 and Wang and Szaro (2015). *Figure adapted from Wang and Szaro (2015).*

yielding a combined final DNA concentration of 150 pg in 10 nl per injected embryo.

3. MICROINJECTION OF *XENOPUS* EMBRYOS

3.1 Useful *Xenopus* References

Because *Xenopus* is widely used as a model system for development, many valuable resources are available for the investigator. Currently, the best portal for finding these is the *Xenbase* Web site (www.xenbase.org) (Bowes et al., 2008; James-Zorn et al., 2013). It contains links to the latest versions of the *X. laevis* and *X. tropicalis* genome databases, BLAST engines for analyzing and searching for the *Xenopus* orthologs of genes from other species, data on gene expression patterns, numerous links to the Web sites of *Xenopus* researchers, lists of *Xenopus* suppliers, and an ever expanding compendium of useful protocols for working with the organism. Helpful, more detailed information for those less familiar with using *Xenopus* can be found in the references listed on *Xenbase*, as well as in the following publications:

1. Useful information on raising and keeping *Xenopus* in the lab can be found in Green (2009).
2. The normal developmental table of Nieuwkoop and Faber (1994) is indispensable for staging embryos. The time needed to reach a particular developmental stage given in this book is accurate for *X. laevis* at 22 °C; the time for other temperatures can be estimated by dividing the time given in Nieuwkoop and Faber by a correction factor ($R = 0.095T$ (°C) − 1.104, Lin & Szaro, 1994). Tables comparing developmental rates between *X. laevis* and *X. tropicalis* are also available on the *Xenbase* Web site (Khokha et al., 2002).
3. Books containing protocols useful for studying early developmental stages of *Xenopus* are edited by Kay and Peng (1991) and Sive, Grainger, and Harland (2000).
4. Useful methods exist for whole-mount immunostaining (Klymkowsky & Hanken, 1991) and for whole-mount *in situ* hybridization (Harland, 1991; O'Keefe, Melton, Ferreiro, & Kintner, 1991; Shain & Zuber, 1996).

3.2 Instrumentation

1. Injection micropipettes are pulled from glass capillary tubes (Kwik-Fil Borosilicate Glass Capillaries, TW-150-4; World Precision Instruments, Sarasota, FL) using a commercial electrode puller (Narishige, PP-830). The tips are beveled on a grinding wheel (Narishige, EG-4) at an angle of 35° and a final outer diameter of 5–6 μm (Fig. 3C). Smaller tips tend to clog and require higher injection pressures, which harm the embryos, whereas larger ones may damage the embryo and promote too much backflow.
2. The injection apparatus (Fig. 3A) consists of (1) a good dissecting microscope (e.g., Leica/Wild M3B) with an ocular micrometer in the eyepiece for measuring drop size and a fiber optic light source (e.g., Schott Fostec), (2) a joystick micromanipulator (e.g., Narishige,

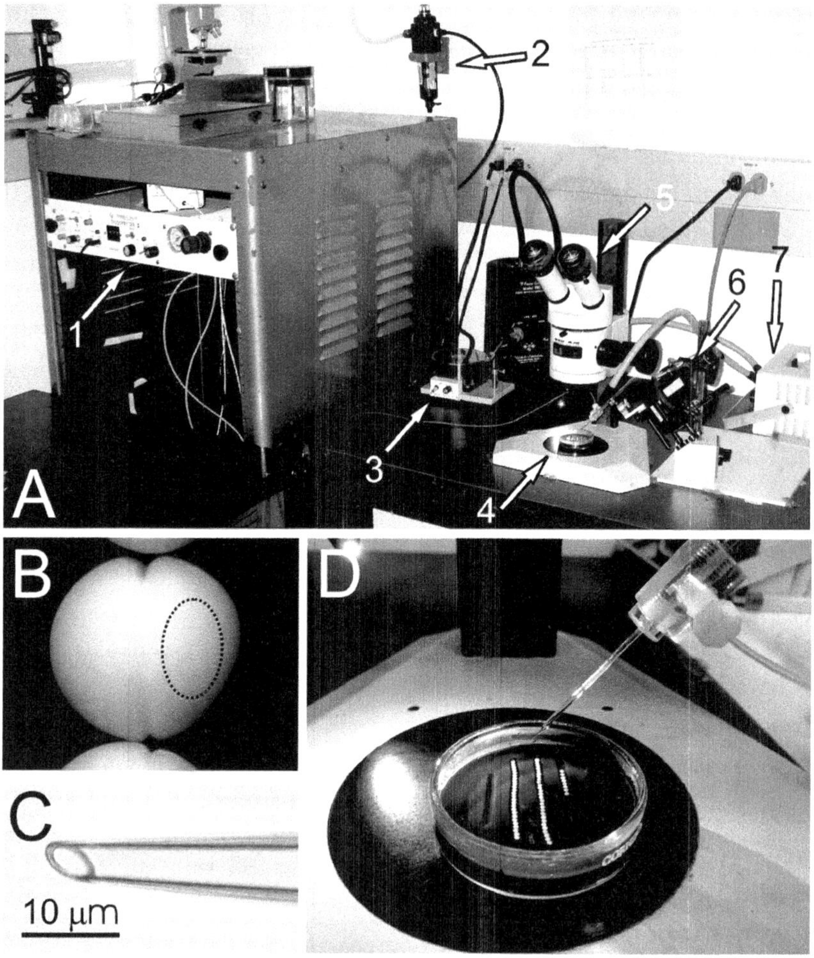

Figure 3 (A) An overview of the apparatus for injecting *Xenopus* embryos. (1) A rack-mounted Picospritzer II (General Valve Corporation, Fairfield, NJ), which controls the duration and pressure of the air pulse; (2) filter on the house air supply (Speedaire; Dayton Electric Manufacturing Co., Niles, IL), which provides pressure and is connected to the rear of the Picospritzer; (3) apparatus for beveling injection pipettes (EG-4; Narishige International USA, East Meadow, NY); (4) platform (IP; Narishige) and dish for holding embryos; (5) stereo microscope (Wild M3B; Leica Microsystems, Buffalo Grove, IL); (6) joystick micromanipulator (MN-151 and GJ-1; Narishige); (7) fiber optic light source (Schott North America, Inc., Southbridge, MA). (B) A 2-cell stage albino embryo; the ellipse indicates the approximate site of injection. (C) Beveled micropipette. (D) Embryos for injection are placed into grooves made on the surface of a wax-filled Petri dish filled with 5% Ficoll in HBS.

MN-151) mounted onto a metal baseplate (e.g., Narishige, IP) with a magnetic stand (e.g., Narishige, GJ-1) on a stable surface, and (3) a device for delivering controlled air pulses to the injection micropipette (e.g., Picospritzer II, General Valve Corporation, Fairfield, NJ). This apparatus may be connected either to a tank of dry nitrogen or to the house air supply via a filter, which prevents contaminating the air line with oil and moisture. Tips within the appropriate size range should yield the correct injection volume at a relatively low pressure of ~15–20 psi.

3. A 60-mm Petri dish containing melted beeswax blackened with a few grams of bone black works well for holding the embryos. The small end of a borosilicate Pasteur pipette sealed by fire polishing is heated in an alcohol lamp for making small indentations and grooves in the surface of the beeswax that are suitable for holding the embryos in place.
4. A fluorescence dissecting microscope with the appropriate filters is needed to visualize the fluorescent reporter proteins in live embryos.

3.3 Obtaining and Injecting Embryos

Many laboratories use *in vitro* fertilization of oocytes to obtain embryos (Heasman, Holwill, & Wylie, 1991), but doing so requires sacrificing the males. We prefer instead to obtain embryos through natural spawnings (Gurdon, 1967) induced by injecting the male and female with human chorionic gonadotropin (HCG; CG-10, Sigma-Aldrich). This conserves the males and yields eggs at a steady pace throughout the morning of the injections. Sexually mature, gravid females are selected for their red cloaca, and males for their black nuptial pads, which are located on the inner surface of their forepaws and forearms. More detailed instructions for how to inject HCG into the dorsal lymph sac of *Xenopus* can be found on the *Xenbase* Web site (http://wiki.xenbase.org/xenwiki/index.php/inducing_egg_laying_via_hCG_injection_(Vize_lab)) and in Cross and Powers (2008).

1. The night before, males and females are injected into the dorsal lymph sac with ~500–1000 and ~800–1200 U HCG, respectively, using a 1-ml tuberculin syringe and a 27-gauge needle.
2. After injection, each pair is placed into a small aquarium (e.g., polycarbonate hamster cages) containing dechlorinated tap water at a level just sufficient to cover both animals when they are clasped. A heavy metal lid is placed on top to prevent them from escaping, and a towel or lab coat is draped over the top so that they will not be disturbed by activities in the room. The combination of light and rising temperature (~22–24 °C) helps to initiate spawning in the early morning, approximately 12 h after

the adults are injected. Once begun, spawnings generally continue for several hours, allowing fertilized eggs to be continually collected.

3. To obtain embryos for injection, fertilized eggs are removed beginning at 15–20 min after they are laid by drawing them from the bottom of the tank into a 10-ml plastic serological pipette from which the tapered end has been broken off. Several dozen of newly fertilized eggs are collected into a 4½″ diameter glass culture dish (Carolina Biological) filled with dechlorinated tap water, which is then poured off and replaced with buffered dejellying solution (0.39 g dithiothreitol; 3.1 g Tris base (Sigma); 500 ml dechlorinated tap water). Eggs are gently swirled for 2 min, until the jelly comes loose—any longer makes them too soft to inject. The dejellying solution is then poured off and the eggs are rinsed 8–10 times in dechlorinated tap water. At least one dish of uninjected embryos from each mating pair is set aside to monitor the vitality of the spawning (>90% normal development is typical for a healthy spawning).
4. Prior to injecting the embryos, the micropipette must be filled with DNA solution and calibrated. Depending on the number of eggs to be injected, 1.5–3 μl of solution are pipetted into 20 μl of clean mineral oil deposited onto a glass microscope slide covered with Parafilm. The top end of the micropipette is inserted into a piece of 1.5 mm i.d. Tygon tubing connected to a #18 gauge needle mounted onto a 5-ml syringe. The injection solution is then drawn into the micropipette, using the syringe for suction. The injection micropipette is then removed from the tubing and mounted into the halter of the micropipette holder. (Alternatively, the micropipette can be back-filled using a gel-loading pipette tip.) For calibration, the tip is inserted into the mineral oil droplet while observing it under the dissecting microscope and a drop is ejected with a test pulse (~15–20 psi; ~10 ms). The volume is determined by measuring its diameter with an ocular micrometer. The pulse duration is then adjusted until the desired injection volume of 10 nl is attained. Injection volumes exceeding 25 nl produce abnormal embryos, whereas volumes less than 5 nl disperse poorly. Calibration must be performed each time a new micropipette is used.
5. The injection dish is filled with 5% Ficoll (Type 400, Sigma-Aldrich) dissolved in HEPES-buffered Steinberg's solution (HBS: 58.22 m*M* NaCl, 0.67 m*M* KCl, 0.34 m*M* $Ca(NO_3)_2$, 0.83 m*M* $MgSO_4$, 10 m*M* HEPES, pH 7.6; HBS is autoclaved or filtered through a 0.22-μm filter) supplemented with 500 U/l penicillin-G and 375 U/l streptomycin

(Sigma-Aldrich), and the sorted 2-cell stage embryos are placed in the wells or grooves for injection (Fig. 3D). The micropipette is lowered above the embryo and several pulses of solution are ejected into the Ficoll solution to ensure that the tip is unclogged. Immediately after the last pulse, the micropipette is inserted into the embryo. The needle is oriented at ~45° angle and targeted to the animal cap (see Fig. 3B). Once the tip has penetrated the cortex of the embryo, solution is pulsed once into the embryo and the tip is withdrawn. Upon withdrawing the tip, a second pulse is then delivered to eject any material that has entered the micropipette from the injected embryo.

6. Injected embryos are transferred to an untreated 60-mm polystyrene Petri dish containing 5% Ficoll/HBS (<30 embryos to a dish) at 20–22 °C. At Nieuwkoop and Faber stage (st.) 9, embryos are transferred first to 100% HBS, then successively for 1–5 min each through 70% and 50% HBS, and finally into 20% of 2 × Ca^{++} HBS for rearing at 20–22 °C to the desired stage. Waiting any longer will prevent embryos from gastrulating normally.

4. ASSAYS FOR GENE EXPRESSION

Because genes are under both transcriptional and posttranscriptional control, the activity of *cis*-regulatory elements should be characterized by assaying for both reporter RNA and fluorescent protein itself. The methods for doing so are all based on standard, widely used molecular biological protocols for detecting RNA and protein, and thus, we concentrate here on the details and variations that facilitate their use with *Xenopus*.

4.1 Reporter Protein Expression

Both glGFP and DsRed2 are readily visible in live embryos viewed under a fluorescence stereo microscope. Until the yolk clears, *Xenopus* embryos are translucent, and fluorescence can be hard to observe in deeper structures. After st. 41 (~3 days) and until metamorphosis, they become more transparent, making it relatively easy to observe fluorescence throughout the living embryo. Once reflexes develop, it is necessary to anesthetize them lightly with tricaine methanesulfonate (MS-222, Sigma-Aldrich) for imaging. Because the embryos are easily overdosed at these early stages, only as little MS-222 as is necessary to quiet movement is used, and they are removed as soon as they stop moving. Effective doses typically range from 1:20,000 at the earliest tailbud stages to 1:5000 at later tadpole stages.

1. For a more complete assessment of expression, observations of live embryos are supplemented with immunohistochemistry because of the time delay between the onset of expression and the emergence of fluorescence. Whole mount immunostaining is quite effective through st. 43, allowing one to avoid both the labor and difficulty of sectioning younger, more fragile embryos. Standardized protocols (Dent, Polson, & Klymkowsky, 1989) are used with 1% dimethyl sulfoxide added to all antibody solutions to improve penetration (Gervasi et al., 2000; Szaro, Grant, Lee, & Gainer, 1991).
2. For initial fixation, anesthetized embryos are immersed in 4% paraformaldehyde/2% sucrose in 0.1 *M* sodium phosphate buffer, pH 7.4 (PB), for 2 h. After washing out the fixative with PB, embryos are then postfixed in Dent's fixative (80% methanol/20% DMSO), immediately prior to immunostaining, to facilitate penetration of the antibodies.
3. Useful primary antibodies for detecting glGFP and DsRed2 are polyclonal antibodies made in goat (GFP antibody, Rockland) and rabbit (Living Colors DsRed Polyclonal Antibody, Clontech), respectively, both diluted at 1:500. Species-appropriate Alexa Fluor 488 or 546 antibodies (Life Technologies, used at 1:1000) can be used as secondary antibodies. Processed embryos are effectively cleared for observation by treating them with benzyl alcohol:benzyl benzoate (1:2) immediately prior to observation.
4. Embryos older than st. 35/36 can also be cryosectioned, mounted on slides, and immunostained using standard procedures. For sectioning, the time for initial fixation by immersion should be extended to overnight.
5. Expression of protein can also be monitored by Western blot using standard protocols, with each embryo yielding ~100 μg of total protein.

4.2 Reporter RNA Expression

RNA expression can be quantified using qRT-PCR and standard protocols (e.g., Power SYBR PCR Green Master Mix, Applied Biosystems).

1. For a single biological replicate, total RNA is extracted from 10 embryos using a Polytron PT3000 (Kinematica AG) and 2-ml microcentrifuge tubes for homogenization, and an RNeasy Plus Mini Kit (Qiagen) to extract the RNA. Typical yields range from 30 to 40 μg. Prior to the reverse transcriptase reaction, the RNA samples are treated extensively with DNase I (e.g., RQ1 RNase-free DNase I, Promega) to remove residual injected plasmid, and the RNA is recovered by phenol/

chloroform extraction and ethanol precipitation (Davis, Kuehl, & Battey, 1994).

2. The cDNA for each individual qRT-PCR reaction is prepared from 1 μg of the extracted RNA using oligo dT_{12-18} (2 pmol, Invitrogen) to prime the reverse transcriptase reaction (SuperScript III, Invitrogen). The primers used in the subsequent pPCR reaction are *glGFP* forward, 5′-CCG ACC ATT ATC AAC AGA ACA CTC CAA-3′; *glGFP* reverse, 5′-TGG GTG GAC AGG TAA TGG TTG T-3′; *DsRed2* forward, 5′-GTC ATC ACC GAG TTC ATG CGC TTC-3′; *DsRed2* reverse, 5′-TTG TAG ATG AAG CAG CCG TCC TGC-3′. A no-RT control should be included with each sample to ensure that any plasmid DNA remaining from the injections was effectively removed by the DNase I treatment.

5. ASSAYS FOR EFFECTS OF *CIS*-REGULATORY ELEMENTS ON RNA PROCESSING, TRAFFICKING, AND TRANSLATION

This system can also be used to assay the activities *in vivo* of elements involved in regulating gene expression at the posttranscriptional level. Sufficient embryos can be generated for analyzing, for example, effects on the efficiency of the splicing of particular introns (Section 5.1), as well as nucleocytoplasmic export (Section 5.2), and the loading of RNA onto polysomes for translation (Section 5.3). For these studies, introns of intermediate filament genes can be cloned directly into the *DsRed2* sequences test plasmid, and the *β-globin* 3′-UTR can be replaced with that of an intermediate filament gene and further modified to test individual *cis*-regulatory elements.

5.1 RNA Processing Efficiency

To ensure efficient recovery of long, intron-bearing RNAs, it is critical to isolate total RNA by homogenization in guanidine isothiocyanate (GITC) buffer followed by ultracentrifugation through CsCl (Ananthakrishnan et al., 2008).

1. Ten embryos are homogenized in 1–2 ml of 4 *M* GITC (4 *M* guanidine isothiocyanate, 30 m*M* sodium acetate, pH 6.0, 140 m*M* 2-mercaptoethanol) using a Polytron and 2-ml microcentrifuge tubes. This solution should be prepared using RNase-free water and reagents, vacuum filtered through a 0.22-μm cellulose nitrate filter, and stored at room temperature in a bottle wrapped in aluminum foil.

2. The homogenate is diluted to a volume of 3 ml and overlaid atop a pad of 4 ml CsCl buffer (5.7 *M* CsCl, 30 m*M* sodium acetate, pH 6.0) in polyallomer ultracentrifuge tubes (e.g., Beckman #344059 tubes, SW 41 Ti rotor). Additional GITC buffer is added to fill the tubes to the top, and the tubes are spun for 18–20 h ($110{,}000 \times g$, 20 °C). A successful run is marked by the presence of lipid atop the GITC, a band of protein at the CsCl/GITC interface, one or more bands containing DNA within the CsCl, and an RNA pellet at the bottom.
3. Solution overlying the RNA is decanted and the pellet resuspended in 200 μl RNase-free water. The RNA is further treated with DNase I, then cleaned by phenol/chloroform extraction followed by ethanol precipitation as in Section 4.2, before it is resuspended in 100 μl RNase-free water.
4. Afterward, the yield is determined spectrophotometrically (A_{260}/A_{280}), and 1 μg is reverse transcribed for qRT-PCR as in Section 4.2, except that an additional 2 pmol of gene-specific reverse primer derived from the 3′-UTR is added to improve the efficiency of the reaction, and the temperature is increased to 55 °C.
5. The resultant cDNA is then analyzed by qRT-PCR using the same primers to the fluorescent reporters as in Section 4.2 to quantify the total amount of each respective RNA present, as well as by an additional reaction with primers that span the 3′-end of the splice site of the intron, ~200 nt apart, to quantify the amount of nascent (unspliced) reporter RNA. The forward primer is selected from within the intron itself, whereas the reverse primer is selected from the exon, downstream of the splice site. Processing efficiency is then determined by taking the ratio of the nascent (unspliced) to the total product and subtracting it from 100%.

5.2 Efficiency of Export of RNA from the Nucleus

To quantify the efficiency of RNA export from the nucleus, the nuclear and cytosolic fractions are first separated by low-speed centrifugation prior to RNA extraction (Ananthakrishnan et al., 2008; Meyuhas, Bibrerman, Pierandrei-Amaldi, & Amaldi, 1996). Homogenization is done using ground glass homogenizers (e.g., 2 ml Pyrex Ten Broek, #7727-2) instead of a Polytron to keep nuclei intact.

1. Embryos are homogenized in ~10 volumes (0.5 ml for 10–20 embryos) of ice-cold Polysomal Buffer A (PBA: 25 m*M* NaCl, 5 m*M* $MgCl_2$,

25 m*M* Tris adjusted to pH 7.5 at 4 °C), and the resulting homogenate is briefly spun at low speed in a microcentrifuge (285 × *g*, 2 min, 4 °C).

2. RNA is extracted from both the pellet and the supernatant (RNeasy Plus Mini kit, Qiagen) and the fluorescent reporter RNA is quantified by qRT-PCR as in Section 4.2.

5.3 Analysis of mRNA Translation by Polysomal Profiling

To determine effects of *cis*-regulatory elements on mRNA translation, the RNA is collected from embryos and analyzed by polysome profiling (Meyuhas et al., 1996).

1. Embryos (10–30) are homogenized in 500 μl PBA (titrate the pH at 4 °C)/10 m*M* vanadyl ribonucleoside complex (New England Biolabs, to inhibit ribonucleases) and spun at low speed in the microcentrifuge as in Section 5.2. The resulting supernatant is then layered on to a 10 ml linear sucrose gradient (e.g., 5–56% w/w) in PBA on ice, and then spun in the ultracentrifuge (169,000 × *g*, 4°C) for 2 h in a swinging bucket rotor (e.g., Beckman SW41 Ti). Fractions (0.5 ml) are collected in 2-ml microcentrifuge tubes. Note that to preserve the polysomes on the mRNA, all solutions must be kept cold (0–4 °C) during this step.
2. To each fraction, 50 μl of 3 *M* sodium acetate (pH 6.0), 20 μl of 0.5 *M* Na_2EDTA (pH 8.0), 2 μl linear acrylamide (Ambion), and 1.5 ml of 95% ethanol are added and mixed well to precipitate the complexes. Tubes are placed at −80 °C overnight or until needed, then spun at 4 °C in a microcentrifuge for 1 h to precipitate complexes thoroughly. The pellet is then rinsed with ethanol, respun to remove the ethanol, and briefly dried under vacuum.
3. The pellet is resuspended at room temperature in 200 μl of 0.5% SDS buffer (0.5% SDS, 0.1 *M* NaCl, 1 m*M* EDTA, 10 m*M* Tris at pH 7.4), then extracted with phenol/chloroform and precipitated with ethanol (Davis et al., 1994) to remove the SDS.
4. The pellet is resuspended in 49 μl DNase I reaction buffer and 1 U (1 μl) of RNase-free DNase I (e.g., RQ1 RNase-free DNase I, Promega) and incubated 30 min at 37 °C to digest any residual, contaminating injected plasmid. The reaction is terminated by phenol/chloroform extraction, followed by the addition of 2 μl linear acrylamide (Ambion) and ethanol precipitation.
5. After precipitation, the pellet is resuspended in 20 μl RNase-free water. From this, 2 μl is used to quantify total RNA (A_{260}/A_{280}, Nanodrop),

and 9 μl is used for qRT-PCR. 1 μl of the 20 μl of cDNA generated is used to quantify the fluorescent reporter RNA as in Section 4.2.

6. CONCLUSIONS

A model system for analyzing the activity of *cis*-regulatory gene elements within the context of an intact vertebrate embryo is presented. The method can be used to quantify protein and mRNA expression levels, as well as for analyzing effects on RNA processing, trafficking, and translation.

6.1 Pearls and Pitfalls

Maintaining a breeding colony that is well fed and free of disease is essential for obtaining healthy spawnings, which are key for success. Early monitoring of the injected embryos and isolating those that were successfully injected and healthy from those that were damaged or unlabeled is also critical. Coinjecting the glGFP reference plasmid permits rapid screening for successfully injected embryos under the fluorescence dissecting microscope. Although the modified plasmids described here dramatically reduce mosaic expression, because some mosaicism nonetheless still occurs, those embryos exhibiting the most representative expression patterns of the reference plasmid should be chosen for subsequent analysis (Wang & Szaro, 2015). Most downstream analytic techniques can be done with as few as 10 embryos, but it is advisable to inject at least 70 embryos with each construct to ensure that enough embryos exhibiting appropriate expression of the reference plasmid survive to yield samples for several biological replicates that can be pooled for statistical analysis. Finally, for analyzing RNAs, it is important to use solutions, centrifuge tubes, and pipette tips that are RNase free.

ACKNOWLEDGMENTS

This work was supported by a grant from the National Science Foundation (IOS 1257449, B. G. S.) and a predoctoral fellowship from the American Association of University Women (2012–2013 International Fellowship, C. W.).

REFERENCES

Allen, B. G., & Weeks, D. L. (2005). Transgenic *Xenopus laevis* embryos can be generated using ϕC31 integrase. *Nature Methods*, *2*, 975–979.

Amaya, E., & Kroll, K. L. (1999). A method for generating transgenic frog embryos. *Methods in Molecular Biology*, *97*, 393–414.

Amaya, E., Offield, M. F., & Grainger, R. M. (1998). Frog genetics: *Xenopus tropicalis* jumps into the future. *Trends in Genetics, 14*, 253–255.

Ananthakrishnan, L., Gervasi, C., & Szaro, B. G. (2008). Dynamic regulation of middle neurofilament (NF-M) RNA pools during optic nerve regeneration. *Neuroscience, 153*, 144–153.

Ananthakrishnan, L., & Szaro, B. G. (2009). Transcriptional and translational dynamics of light neurofilament subunit RNAs during *Xenopus laevis* optic nerve regeneration. *Brain Research, 1250*, 27–40.

Andres, A.-C., Muellener, D. B., & Ryffel, G. U. (1984). Persistence, methylation and expression of vitellogenin gene derivatives after injection into fertilized eggs of *Xenopus laevis*. *Nucleic Acids Research, 12*, 2283–2302.

Asch, W. S., Leake, D., Canger, A. K., Passini, M. A., Argenton, F., & Schechter, N. (1998). Cloning of zebrafish neurofilament cDNAs for plasticin and gefiltin: Increased mRNA expression in ganglion cells after optic nerve injury. *Journal of Neurochemistry, 71*, 20–32.

Bannasch, P., Zerban, H., Schmid, E., & Franke, W. W. (1980). Liver tumors distinguished by immunofluorescence microscopy with antibodies to proteins of intermediate-sized filaments. *Proceedings of the National Academy of Sciences of the United States of America, 77*, 4948–4952.

Beaudet, L., Charron, G., Houle, D., Tretjakoff, I., Peterson, A., & Julien, J.-P. (1992). Intragenic regulatory elements contribute to transcriptional control of the neurofilament light gene. *Gene, 116*, 205–214.

Beaudet, L., Cote, F., Houle, D., & Julien, J.-P. (1993). Different post transcriptional controls for the human neurofilament light and heavy genes in transgenic mice. *Molecular Brain Research, 18*, 23–31.

Belecky-Adams, T., Wight, D. C., Kopchick, J. J., & Parysek, L. M. (1993). Intragenic sequences are required for cell type-specific and injury-induced expression of the rat peripherin gene. *Journal of Neuroscience, 13*, 5056–5065.

Bennett, G. S., Fellini, S. A., Croop, J. M., Otto, J. J., Bryan, J., & Holtzer, H. (1978). Differences among 100-Å filament subunits from different cell types. *Proceedings of the National Academy of Sciences of the United States of America, 75*, 4364–4368.

Bignami, A., Raju, T., & Dahl, D. (1982). Localization of vimentin, the nonspecific intermediate filament protein, in embryonal glia and in early differentiating neurons. *Developmental Biology, 91*, 286–295.

Bowes, J. B., Snyder, K. A., Segerdell, E., Gibb, R., Jarabek, C., Noumen, E., et al. (2008). Xenbase: A *Xenopus* biology and genomics resource. *Nucleic Acids Research, 36*, D761–D767.

Budhram-Mahadeo, V., Morris, P. J., Lakin, N. D., Theil, T., Ching, G. Y., Lillycrop, K. A., et al. (1995). Activation of the alpha-internexin promoter by the Brn-3a transcription factor is dependent on the N-terminal region of the protein. *Journal of Biological Chemistry, 270*, 2853–2858.

Carden, M. J., Trojanowski, J. Q., Schlaepfer, W. W., & Lee, V. M. Y. (1987). Two-stage expression of neurofilament polypeptides during rat neurogenesis with early establishment of adult phosphorylation patterns. *Journal of Neuroscience, 7*, 3489–3504.

Chang, L., & Thompson, M. A. (1996). Activity of the distal positive element of the peripherin gene is dependent on proteins binding to an Ets-like recognition site and a novel inverted repeat site. *Journal of Biological Chemistry, 271*, 6467–6475.

Charnas, L. R., Szaro, B. G., & Gainer, H. (1992). Identification and developmental expression of a novel low molecular weight neuronal intermediate filament protein in *Xenopus laevis*. *Journal of Neuroscience, 12*, 3010–3024.

Christian, J. L., Edelstein, N. G., & Moon, R. T. (1990). Overexpression of wild-type and dominant negative mutant vimentin subunits in developing *Xenopus* embryos. *The New Biologist, 2*, 700–711.

Chung, J. H., Bell, A. C., & Felsenfeld, G. (1997). Characterization of the chicken β-globin insulator. *Proceedings of the National Academy of Sciences of the United States of America, 94,* 575–580.

Chung, J. H., Whiteley, M., & Felsenfeld, G. (1993). A 5′ element of the chicken β-globin domain serves as an insulator in human erythroid cells and protects against position effect in *Drosophila. Cell, 74,* 505–514.

Cochard, P., & Paulin, D. (1984). Initial expression of neurofilaments and vimentin in the central and peripheral nervous system of the mouse embryo in vivo. *Journal of Neuroscience, 4,* 2080–2094.

Corner, M. (1964). Localization of capacities for functional development in the neural plate of *Xenopus laevis. Journal of Comparative Neurology, 123,* 243–255.

Cross, M. K., & Powers, M. (2008). Obtaining eggs from *Xenopus laevis* females. *Journal of Visualized Experiments, 18,* 890.

Davis, L. G., Kuehl, W. M., & Battey, J. F. (1994). *Basic methods in molecular biology.* Norwalk: Appleton and Lange.

Dent, J. A., Polson, A. G., & Klymkowsky, M. W. (1989). A whole-mount immunocytochemical analysis of the expression of the intermediate filament protein vimentin in *Xenopus. Development, 105,* 61–74.

Elkan, E. R. (1938). The *Xenopus* pregnancy test. *British Medical Journal, 2.* 1253–1274.2.

Escurat, M., Djabali, K., Gumpel, M., Gros, F., & Portier, M.-M. (1990). Differential expression of two neuronal intermediate-filament proteins, peripherin and the low-molecular-mass neurofilament protein (NF-L), during the development of the rat. *Journal of Neuroscience, 10,* 764–784.

Fischberg, M., Gurdon, J. B., & Elsdale, T. R. (1959). Nuclear transfer in amphibia and the problem of the potentialities of the nuclei of differentiating tissues. *Experimental Cell Research, 6 Supplement,* 161–178.

Fish, M. B., Nakayama, T., & Grainger, R. M. (2012). Simple, fast, tissue-specific bacterial artificial chromosome transgenesis in *Xenopus. Genesis, 50,* 307–315.

Franke, W. W., Schmid, E., Osborn, M., & Weber, K. (1978). Different intermediate-sized filaments distinguished by immunofluorescence microscopy. *Proceedings of the National Academy of Sciences of the United States of America, 75,* 5034–5038.

Gallo, J.-M., Jin, P., Thornton, C. A., Lin, H., Robertson, J., D'Souza, I., et al. (2005). The role of RNA and RNA processing in neurodegeneration. *Journal of Neuroscience, 25,* 10372–10375.

Gervasi, C., Stewart, C.-B., & Szaro, B. G. (2000). *Xenopus laevis* peripherin (XIF3) is expressed in radial glia and proliferating neural epithelial cells as well as in neurons. *Journal of Comparative Neurology, 423,* 512–531.

Gervasi, C., Thyagarajan, A., & Szaro, B. G. (2003). Increased expression of multiple neurofilament mRNAs during regeneration of vertebrate central nervous system axons. *Journal of Comparative Neurology, 461,* 262–275.

Glasgow, E., Druger, R. K., Fuchs, C., Lane, W. S., & Schechter, N. (1994). Molecular cloning of gefiltin (ON1): Serial expression of two new neurofilament mRNAs during optic nerve regeneration. *EMBO Journal, 13,* 297–305.

Godsave, S., Anderton, B. H., & Wylie, C. (1986). The appearance and distribution of intermediate filament proteins during differentiation of the central nervous system, skin and notochord in *Xenopus laevis. Journal of Embryology and Experimental Morphology, 97,* 201–223.

Green, S. L. (2009). *The laboratory* Xenopus *sp..* Boca Raton: CRC Press.

Gurdon, J. B. (1967). African clawed frogs. In F. H. Wilt & N. K. Wessels (Eds.), *Methods in developmental biology* (pp. 75–84). New York: T.Y. Crowell.

Gurdon, J. B., Elsdale, T. R., & Fischberg, M. (1958). Sexually mature individuals of *Xenopus laevis* from the transplantation of single somatic nuclei. *Nature, 182,* 64–65.

Hamlet, M. R., Yergeau, D. A., Kuliyev, E., Takeda, M., Taira, M., Kawakami, K., et al. (2006). Tol2 transposon-mediated transgenesis in *Xenopus tropicalis*. *Genesis*, *44*, 438–445.

Harland, R. M. (1991). *In situ* hybridization: An improved whole mount method for *Xenopus* embryos. In L. Wilson, B. K. Kay, & H. B. Peng (Eds.), *Methods in cell biology: Vol. 36. Xenopus laevis: Practical uses in cell and molecular biology* (pp. 685–696). San Diego: Academic Press.

Harvey, R. P., & Melton, D. A. (1988). Microinjection of synthetic Xhox-1A homeobox mRNA disrupts somite formation in developing *Xenopus* embryos. *Cell*, *53*, 687–697.

Heasman, J., Holwill, S., & Wylie, C. C. (1991). Fertilization of cultured *Xenopus* oocytes and use in studies of maternally inherited molecules. In L. Wilson, B. K. Kay, & H. B. Peng (Eds.), *Methods in cell biology: Vol. 36. Xenopus laevis: Practical uses in cell and molecular biology* (pp. 214–231). San Diego: Academic Press.

Hemmati-Brivanlou, A., Mann, R. W., & Harland, R. M. (1992). A protein expressed in the growth cones of embryonic vertebrate neurons defines a new class of intermediate filament protein. *Neuron*, *9*, 417–428.

Hughes, A. (1957). The development of the primary nervous system in *Xenopus laevis* (Daudin). *Journal of Anatomy*, *91*(Part 3), 323–338.

Hunt, P. M. (1969). Effects of rotating neural tissue and underlying mesoderm in *Xenopus laevis* embryos. *Acta Embryologiae Experimentalis (Palermo)*, *2*, 211–229.

Ishibashi, S., Love, N. R., & Amaya, E. (2012). A simple method of transgenesis using I-sce I meganuclease in *Xenopus*. *Methods in Molecular Biology*, *917*, 205–218.

Jacobson, M., & Hirose, G. (1978). Origin of the retina from both sides of the embryonic brain: A contribution to the problem of crossing at the optic chiasma. *Science*, *202*, 637–639.

James-Zorn, C., Ponferrada, V. G., Jarabek, C. J., Burns, K. A., Segerdell, E. J., Lee, J., et al. (2013). Xenbase expansion and updates of the *Xenopus* model organism database. *Nucleic Acids Research*, *41*, D865–D870.

Jonas, E. A., Snape, A. M., & Sargent, T. D. (1989). Transcriptional regulation of a *Xenopus* embryonic epidermal keratin gene. *Development*, *106*, 399–405.

Kaplan, M. P., Chin, S. S. M., Fliegner, K. H., & Liem, R. K. H. (1990). Alpha-internexin, a novel neuronal intermediate filament protein, precedes the low molecular weight neurofilament protein (NF-L) in the developing rat brain. *Journal of Neuroscience*, *10*, 2735–2748.

Kay, B. K., & Peng, H. B. (1991). *Methods in cell biology: Vol. 36; Xenopus laevis: Practical uses in cell and molecular biology*. San Diego: Academic Press.

Khokha, M. K., Chung, C., Bustamante, E. L., Gaw, L. W., Trott, K. A., Yeh, J., et al. (2002). Techniques and probes for the study of *Xenopus tropicalis* development. *Developmental Dynamics*, *225*, 499–510.

Kintner, C. (1988). Effects of altered expression of the neural cell adhesion molecule, N-CAM, on early neural development in *Xenopus* embryos. *Neuron*, *1*, 545–555.

Klymkowsky, M. W., & Hanken, J. (1991). Whole-mount staining of *Xenopus* and other vertebrates. In L. Wilson, B. K. Kay, & H. B. Peng (Eds.), *Methods in cell biology: Vol. 36. Xenopus laevis: Practical uses in cell and molecular biology* (pp. 420–443). San Diego: Academic Press.

Krieg, P. A., & Melton, D. A. (1984). Functional messenger RNAs are produced by SP6 *in vitro* transcription of cloned cDNAs. *Nucleic Acids Research*, *12*, 7057–7070.

Krieg, P. A., & Melton, D. A. (1985). Developmental regulation of a gastrula-specific gene injected into fertilized *Xenopus* eggs. *EMBO Journal*, *4*, 3463–3471.

Kroll, K. L., & Amaya, E. (1996). Transgenic *Xenopus* embryos from sperm nuclear transplantations reveal FGF signaling requirements during gastrulation. *Development*, *122*, 3173–3183.

Kroll, K. L., & Gerhart, J. C. (1994). Transgenic *X. laevis* embryos from eggs transplanted with nuclei of transfected cultured cells. *Science, 266*, 650–654.

Krone, P. H., & Heikkila, J. J. (1989). Expression of microinjected hsp 70/CAT and hsp 30/CAT chimeric genes in developing *Xenopus laevis* embryos. *Development, 106*, 271–281.

Lendahl, U., Zimmerman, L. B., & McKay, R. D. G. (1990). CNS stem cells express a new class of intermediate filament protein. *Cell, 60*, 585–595.

Leonard, D. G. B., Gorham, J. D., Cole, P., Greene, L. A., & Ziff, E. B. (1988). A nerve growth factor-regulated messenger RNA encodes a new intermediate filament protein. *Journal of Cell Biology, 106*, 181–193.

Lin, W., & Szaro, B. G. (1994). Maturation of neurites in mixed cultures of spinal cord neurons and muscle cells from *Xenopus laevis* embryos followed with antibodies to neurofilament proteins. *Journal of Neurobiology, 25*, 1235–1248.

Lin, H., Zhai, J., Cañete-Soler, R., & Schlaepfer, W. W. (2004). 3′ Untranslated region in a light neurofilament (NF-L) mRNA triggers aggregation of NF-L and mutant superoxide dismutase 1 proteins in neuronal cells. *Journal of Neuroscience, 24*, 2716–2726.

Lin, H., Zhai, J., Nie, Z., Wu, J., Meinkoth, J. L., Schlaepfer, W. W., et al. (2003). Neurofilament RNA causes neurodegeneration with accumulation of ubiquitinated aggregates in cultured motor neurons. *Journal of Neuropathology and Experimental Neurology, 62*, 936–950.

Love, N. R., Thuret, R., Chen, Y., Ishibashi, S., Sabherwal, N., Paredes, R., et al. (2011). Transgenesis: A cross-species, modular transgenesis resource. *Development, 138*, 5451–5458.

Marini, N. J., Etkin, L. D., & Benbow, R. M. (1988). Persistence and replication of plasmid DNA microinjected into early embryos of *Xenopus laevis*. *Developmental Biology, 127*, 421–434.

Meyuhas, O., Bibrerman, Y., Pierandrei-Amaldi, P., & Amaldi, F. (1996). Analysis of polysomal RNA. In P. A. Krieg (Ed.), *A laboratory guide to RNA: Isolation, analysis, and synthesis* (pp. 65–81). New York: Wiley-Liss.

Mohun, T. J., Taylor, M. V., Garrett, N., & Gurdon, J. B. (1989). The CArG promoter sequence is necessary for muscle-specific transcription of the cardiac actin gene in *Xenopus* embryos. *EMBO Journal, 8*, 1153–1161.

Nieuwkoop, P. D., & Faber, J. (1994). *Normal table of* Xenopus laevis *(Daudin)*. New York: Garland Publishing.

Niloff, M. S., Dunn, R. J., & Levine, R. L. (1998). The levels of retinal mRNA for gefiltin, a neuronal intermediate filament protein, are regulated by the tectum during optic nerve regeneration in the goldfish. *Molecular Brain Research, 61*, 78–89.

Nixon, R. A., & Shea, T. B. (1992). Dynamics of neuronal intermediate filaments: A developmental perspective. *Cell Motility and the Cytoskeleton, 22*, 81–91.

Nutt, S. L., Bronchain, O. J., Hartley, K. O., & Amaya, E. (2001). Comparison of morpholino based translational inhibition during the development of *Xenopus laevis* and *Xenopus tropicalis*. *Genesis, 30*, 110–113.

Ogino, H., McConnell, W. B., & Grainger, R. M. (2006). Highly efficient transgenesis in *Xenopus tropicalis* using *I-SceI* meganuclease. *Mechanisms of Development, 123*, 103–113.

O'Keefe, H. P., Melton, D. A., Ferreiro, B., & Kintner, C. (1991). *In situ* hybridization. In L. Wilson, B. K. Kay, & H. B. Peng (Eds.), *Methods in cell biology: Vol. 36. Xenopus laevis: Practical uses in cell and molecular biology* (pp. 443–695). San Diego: Academic Press.

Pan, F. C., Chen, Y., Loeber, J., Henningfeld, K., & Pieler, T. (2006). *I-SceI* meganuclease-mediated transgenesis in *Xenopus*. *Developmental Dynamics, 235*, 247–252.

Parysek, L. M., & Goldman, R. D. (1988). Distribution of a novel 57 kDa intermediate filament (IF) protein in the nervous system. *Journal of Neuroscience, 8*, 555–563.

Pikaart, M. J., Recillas-Targa, F., & Felsenfeld, G. (1998). Loss of transcriptional activity of a transgene is accompanied by DNA methylation and histone deacetylation and is prevented by insulators. *Genes and Development, 12*, 2852–2862.
Pospelov, V. A., Pospelova, T. V., & Julien, J.-P. (1994). AP-1 and Krox-24 transcription factors activate the neurofilament light gene promoter in P19 embryonal carcinoma cells. *Cell Growth and Differentiation, 5*, 187–196.
Roosa, J. R., Gervasi, C., & Szaro, B. G. (2000). Structure, biological activity of the upstream regulatory sequence, and conserved domains of a middle molecular mass neurofilament gene of *Xenopus laevis*. *Molecular Brain Research, 82*, 35–51.
Rusconi, S., & Schaffner, W. (1981). Transformation of frog embryos with a rabbit β-globin gene. *Proceedings of the National Academy of Sciences of the United States of America, 78*, 5051–5055.
Schlaepfer, W. W., & Bruce, J. (1990). Simultaneous up-regulation of neurofilament proteins during the postnatal development of the rat nervous system. *Journal of Neuroscience Research, 25*, 39–49.
Schmid, E., Tapscott, S., Bennett, G. S., Croop, J., Fellini, S. A., Holtzer, H., et al. (1979). Differential location of different types of intermediate-sized filaments in various tissues of the chicken embryo. *Differentiation, 15*, 27–40.
Schwartz, M. L., Shneidman, P. S., Bruce, J., & Schlaepfer, W. W. (1994). Stabilization of neurofilament transcripts during postnatal development. *Molecular Brain Research, 27*, 215–220.
Sekkali, B., Tran, H. T., Crabbe, E., De Beule, C., Van Roy, F., & Vleminckx, K. (2008). Chicken β-globin insulator overcomes variegation of transgenes in *Xenopus* embryos. *The FASEB Journal, 22*, 2534–2540.
Shain, D. H., & Zuber, M. X. (1996). Sodium dodecyl sulfate (SDS)-based whole-mount in situ hybridization of *Xenopus laevis* embryos. *Journal of Biochemical and Biophysical Methods, 31*, 185–188.
Sharpe, C. R. (1988). Developmental expression of a neurofilament-M and two vimentin-like genes in *Xenopus laevis*. *Development, 103*, 269–277.
Sinzelle, L., Vallin, J., Coen, L., Chesneau, A., Du Pasquier, D., Pollet, N., et al. (2006). Generation of transgenic *Xenopus laevis* using the Sleeping Beauty transposon system. *Transgenic Research, 15*, 751–760.
Sive, H. L., Grainger, R. M., & Harland, R. M. (2000). *Early development of* Xenopus laevis*: A laboratory manual.* Cold Spring Harbor: Cold Spring Harbor Laboratory Press.
Sparrow, D. G., Latinkic, B., & Mohun, T. J. (2000). A simplified method of generating transgenic *Xenopus*. *Nucleic Acids Research, 28*, E12.
Szaro, B. G., & Gainer, H. (1988). Immunocytochemical identification of non-neuronal intermediate filament proteins in the developing *Xenopus laevis* nervous system. *Developmental Brain Research, 43*, 207–224.
Szaro, B. G., Grant, P., Lee, V. M. Y., & Gainer, H. (1991). Inhibition of axonal development after injection of neurofilament antibodies into a *Xenopus laevis* embryo. *Journal of Comparative Neurology, 308*, 576–585.
Szaro, B. G., Lee, V. M. Y., & Gainer, H. (1989). Spatial and temporal expression of phosphorylated and non-phosphorylated forms of neurofilament proteins in the developing nervous system of *Xenopus laevis*. *Brain Research. Developmental Brain Research, 48*, 87–103.
Szaro, B. G., & Strong, M. J. (2010). Post-transcriptional control of neurofilaments: New roles in development, regeneration and neurodegenerative disease. *Trends in Neurosciences, 33*, 27–37.
Tapscott, S. J., Bennett, G. S., Toyama, Y., Kleinbart, F., & Holtzer, H. (1981). Intermediate filament proteins in the developing chick spinal cord. *Developmental Biology, 86*, 40–54.

Thyagarajan, A., Strong, M. J., & Szaro, B. G. (2007). Post-transcriptional control of neurofilaments in development and disease. *Experimental Cell Research, 313*, 2088–2097.
Tran, H. T., & Vleminckx, K. (2014). Design and use of transgenic reporter strains for detecting activity of signaling pathways in *Xenopus*. *Methods, 66*, 422–432.
Troy, C. M., Brown, K., Greene, L. A., & Shelanski, M. L. (1990). Ontogeny of the neuronal intermediate filament protein, peripherin, in the mouse embryo. *Neuroscience, 36*, 217–237.
Undamatla, J., & Szaro, B. G. (2001). Differential expression and localization of neuronal intermediate filament proteins within newly developing neurites in dissociated cultures of *Xenopus laevis* embryonic spinal cord. *Cell Motility and the Cytoskeleton, 49*, 16–32.
Uveges, T. E., Shan, Y., Kramer, B. E., Wight, D. C., & Parysek, L. M. (2002). Intron 1 is required for cell type-specific, but not injury responsive, peripherin gene expression. *Journal of Neuroscience, 22*, 7959–7967.
Vize, P. D., Hemmati-Brivanlou, A., Harland, R. M., & Melton, D. A. (1991). Assays for gene function in developing *Xenopus* embryos. In L. Wilson, B. K. Kay, & H. B. Peng (Eds.), *Methods in cell biology: Vol. 36. Xenopus laevis: Practical uses in cell and molecular biology* (pp. 368–388). San Diego: Academic Press.
Wang, C., & Szaro, B. G. (2015). A method for using direct injection of plasmid DNA to study *cis*-regulatory element activity in F_0 *Xenopus* embryos and tadpoles. *Developmental Biology, 398*, 11–23.
Warner, A. E., Guthrie, S. C., & Gilula, N. B. (1984). Antibodies to gap junctional protein selectively disrupt junctional communication in the early amphibian embryo. *Nature, 311*, 127–131.
West, A. G., Gaszner, M., & Felsenfeld, G. (2002). Insulators: Many functions, many mechanisms. *Genes and Development, 16*, 271–288.
Wright, C. V. E., Cho, K. W. Y., Hardwicke, J., Collins, R. H., & De Robertis, E. M. (1989). Interferences with function of a homeobox gene in *Xenopus* embryos produces malformations of the anterior spinal cord. *Cell, 59*, 81–93.
Yergeau, D. A., Johnson Hamlet, M. R., Kuliyev, E., Zhu, H., Doherty, J. R., Archer, T. D., et al. (2009). Transgenesis in *Xenopus* using the Sleeping Beauty transposon system. *Developmental Dynamics, 238*, 1727–1743.
Zhao, Y., & Szaro, B. G. (1994). The return of phosphorylated and nonphosphorylated epitopes of neurofilament proteins to the regenerating optic nerve of *Xenopus laevis*. *Journal of Comparative Neurology, 343*, 158–172.
Zhao, Y., & Szaro, B. G. (1995). The optic tract and tectal ablation influence the composition of neurofilaments in regenerating optic axons of *Xenopus laevis*. *Journal of Neuroscience, 15*, 4629–4640.
Zimmerman, L., Parr, B., Lendahl, U., Cunningham, M., McKay, R., Gavin, B., et al. (1994). Independent regulatory elements in the nestin gene direct transgene expression to neural stem cells or muscle precursors. *Neuron, 12*, 11–24.

CHAPTER TWENTY-FOUR

Intermediate Filaments in *Caenorhabditis elegans*

Noam Zuela, Yosef Gruenbaum[1]
Department of Genetics, The Alexander Silberman Institute of Life Sciences, The Hebrew University of Jerusalem, Jerusalem, Israel
[1]Corresponding author: e-mail address: gru@vms.huji.ac.il

Contents

Abstract

More than 70 different genes in humans and 12 different genes in *Caenorhabditis elegans* encode the superfamily of intermediate filament (IF) proteins. In *C. elegans*, similar to humans, these proteins are expressed in a cell- and tissue-specific manner, can assemble into heteropolymers and into 5–10 nm wide filaments that account for the principal structural elements at the nuclear periphery, nucleoplasm, and cytoplasm. At least 5 of the 11 cytoplasmic IFs, as well as the nuclear IF, lamin, are essential. In this chapter, we will include a short review of our current knowledge of both cytoplasmic and nuclear IFs in *C. elegans* and will describe techniques used for their analyses.

ABBREVIATIONS

AEBSF 4-(2-aminoethyl) benzenesulfonyl fluoride hydrochloride
DAPI 4′,6-diamidino-2-phenylindole
DTT dithiothreitol
FRAP fluorescence recovery after photobleaching

Methods in Enzymology, Volume 568
ISSN 0076-6879
http://dx.doi.org/10.1016/bs.mie.2015.09.020

GAFP glial fibrillary acidic protein
GFP green fluorescent proteins
IFs intermediate filaments
NPCs nuclear pore complexes
SDS-PAGE sodium dodecyl sulfate polyacrylamide gel electrophoresis

1. INTRODUCTION

Caenorhabditis elegans cells contain the three filament systems: microtubules, microfilaments, and intermediate filaments (IFs), similar to mammalian cells. The genome of *C. elegans* contains 11 genes encoding for cytoplasmic IFs (Karabinos, Schmidt, Harborth, Schnabel, & Weber, 2001), and one gene encoding for a nuclear IF (Riemer, Dodemont, & Weber, 1993). Based on sequence comparisons, the IF proteins can be divided into six groups: Group 1 includes IFA-1,2,3,4; Group 2 includes IFB-1,2; Group 3 includes IFC-1,2; Group 4 includes IFD-1,2; Group 5 includes IFP-1; and Group 6 includes Ce-lamin (Carberry, Wiesenfahrt, Windoffer, Bossinger, & Leube, 2009; Karabinos et al., 2001; Riemer et al., 1993). Like all other IFs, these *C. elegans* proteins have a short N-terminal head domain, a central rod domain that is highly conserved in length and includes four α-helical regions made of coiled-coils (termed 1A, 1B, 2A, and 2B) separated by three linker regions, and a C-terminal tail domain. Like all other invertebrate IFs, Coil 1B of all the *C. elegans* IF proteins contains six heptad repeats (42 amino acids) that are present in all lamins and are missing from all vertebrate cytoplasmic IF proteins, thus making invertebrate IFs more closely related to nuclear lamins than the vertebrate IFs. In addition, six *C. elegans* IF proteins contain a globular domain in their tail domain, similar to the globular domain in the lamin tail domains.

2. ESSENTIAL ROLES OF CYTOPLASMIC IFs IN *C. ELEGANS*

Mammals have over 70 IF genes. With the exception of lamins, they are all expressed in a cell- and tissue-specific manner. For example, keratins are present in epithelial cells and vimentin is present in mesenchymal, endothelial, and hematopoietic cells. Muscle cells express desmin, neuronal cells express the neurofilament triplet proteins, neuroglia cells express glial fibrillary acidic protein (GAFP), and so on. Some cells have more than one filamentous IF network. The IF proteins have specific roles such as maintaining cell stiffness, regulating mechanosensing, forming lamellipodia, regulating cell migration, and allowing cell adhesion (Gruenbaum & Aebi, 2014). Similarly, the 11 cytoplasmic IFs in *C. elegans* have tissue-specific patterns of expression. For example, IFA-1 is expressed in marginal cells of the pharynx, excretory cells, the vulva, uterus and rectum, innervated invaginations of the cuticle, some neurons, and the pharyngeal–intestinal valve. IFB-1 is expressed in the intestine, hypodermis, and in the pharyngeal junctions of early larvae. The complete description of IF genes in *C. elegans*, their protein distribution, molecular mass, and closest homologous human IF, appears in Table 1.

Knockdowns of the 11 cytoplasmic IF genes have revealed that 5 of them are essential for survival. Knockdown of *ifb-1* causes late embryonic lethality with phenotypes of epidermal fragility, abnormal epidermal morphogenesis, and muscle detachment. Reductions in *ifb-1* levels cause morphogenetic defects and defective outgrowth of the excretory cell (Karabinos, Schulze, Schunemann, Parry, & Weber, 2003; Woo, Goncharov, Jin, & Chisholm, 2004). Downregulation of *ifa-3* causes lethality at the late embryo/larval L1 stage with morphogenesis phenotypes that are indistinguishable from those of embryos lacking *ifb-1*. Downregulation of *ifa-2* causes lethality only at the larval L4 stage. The damage localizes to hypodermal hemidesmosomes and there is also muscle detachment (Hapiak et al., 2003). Thus, IFB-1 and IFA-3 are likely the major IFs in embryonic epidermal attachment structures, while IFA-2 is a major IF in the larval epidermal attachment structures. Knockdown of *ifc-1* or *ifd-2* at mild heat shock (25 °C) conditions leads to embryonic lethality and early larval arrest with defects in epidermal morphogenesis and muscle attachment (Karabinos, Schünemann, & Weber, 2004).

3. ESSENTIAL ROLES OF LAMIN IN *C. ELEGANS*

Lamins are nuclear IF proteins that are conserved in all multicellular animals. They form a protein scaffold at the nuclear periphery, termed the nuclear lamina. A small fraction of lamins is also present in the nucleoplasm. Mammals have three lamin genes termed *LMNA*, *LMNB1*, and *LMNB2* that encode four major isoforms: lamins A, lamin C (A-type lamins), lamin B1, and lamin B2 (B-type lamins) (Stuurman, Heins, & Aebi, 1998). The B-type lamins are permanently farnesylated at their C terminus, while the mature A-type lamins are not. Each metazoan cell

Table 1 Intermediate filaments in *C. elegans*

Gene	Protein	Molecular Mass (kDa)	Expression Pattern	Lethality	Function	*H. sapiens* Homology
Cytoplasmic intermediate filaments:						
ifa-1	IFA-1 isoform a	66.5	Marginal cells of the pharynx, excretory cells, uterus, amphid sensory neurons, tail neurons and other unidentified neurons, pharyngeal intestinal valve, vulva, rectum	Early larvea	Structural component of the cytoskeleton, larval development, and normal intestinal morphology	Isoform C of Prelamin-A/C (88.3%)
	IFA-1 isoform	66.4				
	IFA-1 isoform c	65.5				
	IFA-1 isoform	68.5				
ifa-2	IFA-2 (MUA-6)	67.1	Hypodermis, uterus, vulva, rectum, amphids	Early larvea	Hypodermal integrity, attachment of muscles to the body wall, normal positioning of excretory canals and muscles	Lamin B1 (87.3%)
ifa-3	IFA-3	66.5	Hypodermis (embryos and larvea), ventral nerve cord (larvea), rectum	Late embryonic stage/early larvea	Viability after L1/L2, attachment of the cuticle to the hypodermis, correct positioning of excretory canals and body muscles, locomotion	Lamin B1 (89.3%)

ifa-4	IFA-4	66.3	Pharyngeal intestinal valve, excretory cells, intestine of dauer larvea, rectum, some neurons of the tail	Nonessential	Unknown	Lamin B1 (88.5%)
ifb-1	IFB-1 isoform a	63.7	Hypodermis, amphids (weak), pharyngeal marginal cells, excretory cells (weak), pharyngeal intestinal valve, vulva, uterus (weak), rectum	Late embryonic stage	Embryonic development, epidermal morphogenesis, attachment of muscle to the body wall	Isoform C of Prelamin-A/C (86.2%)
	IFB-1 isoform b	67.1	Hypodermis, amphids, pharynx, excretory cells (weak), pharyngeal intestinal valve, vulva, uterus, rectum, intestine of dauer larvea, tail neurons		Embryonic development, epidermal morphogenesis, attachment of cuticle to hypodermis	
IFB-1 is always coexpressed with a member of the IFA family and forms heteropolymers with either one of its members						
ifb-2	IFB-2 isoform a	61.6	Intestine	Embryonic stage	Unknown	Lamin B2 (80.6%)
	IFB-2 isoform c	59.8				
	IFB-2 isoform	65.5				
ifc-1	IFC-1	55.9	Intestine, hypodermis, pharyngeal junction of early larvea	Early larvea (10%)	Epidermal morphogenesis, muscle attachments	65 kDa protein (71.6%)

Continued

Table 1 Intermediate filaments in *C. elegans*—cont'd

Gene	Protein	Molecular Mass (kDa)	Expression Pattern	Lethality	Function	*H. sapiens* Homology
ifc-2	IFC-2 isoform a	127.6	Cytoplasm of intestinal cells, desmosomes of intestinal and pharyngeal cells	Nonessential	Normal movement, growth, body size and shape, vulval integrity, cuticle strength	Uncharacterized protein (28.9%)
	IFC-2 isoform	141.5				
	IFC-2 isoform c	78.2				
	IFC-2 isoform	70.1				
ifd-1	IFD-1 isoform a	66.7	Intestine	Nonessential	Structural component of the cytoskeleton	Isoform C of Prelamin-A/C (81.2%)
	IFD-1 isoform	70.1				
ifd-2	IFD-2	51.6	Intestine	Nonessential	Structural component of the cytoskeleton	Uncharacterized protein (79.9%)
ifp-1	IFP-1 (IFE-1)	89	Intestine	Nonessential	Structural component of the cytoskeleton	Lamin B2 (48.7%)
Nuclear intermediate filaments:						
lmn-1	LMN-1	64.1	Nuclear envelope of all cells (not mature sperm cells), nuclear interior (some adult and embryonic cells)	Embryonic stage	Chromatin organization, cell cycle progression, chromosome segregation, correct spacing of NPCs, nuclear shape, signal transduction	Lamin B1 (93.1%)

expresses at least one B-type lamin. The vertebrate A-type lamins probably evolved only during the process of vertebrate evolution. They make separate filamentous networks that are required, among other things, for regulating nuclear mechanoresistance. The A- and B-type lamins often make different protein complexes (Gruenbaum & Aebi, 2014). The vertebrate A-type lamins are not essential, while B-type lamins are essential in human cells (Harborth, Elbashir, Bechert, Tuschl, & Weber, 2001). In contrast, B-type lamins are not essential during mouse embryogenesis or in certain *Drosophila* tissues (Kim, Zheng, & Zheng, 2013). The A-type lamins are among the most mutated human genes, and their mutations lead to numerous heritable diseases, termed laminopathies, that affect muscle, fat, neuron, bone, and skin tissues and range from muscular dystrophies to lipodystrophies neuropathies and early aging diseases (Worman & Bonne, 2007). Most laminopathies are autosomal dominant.

C. elegans cells express only one lamin gene, termed *lmn-1*, which encodes a single protein isoform (Ce-lamin). This lamin is a B-type lamin that is permanently farnesylated and is expressed in all cells during development (Liu et al., 2000). Compared to vertebrate B-type lamins, the N-terminal domain of Ce-lamin is 14 residue longer, its coil 2 is 14 residues shorter and the C-terminal domain is 25 residues shorter (Riemer et al., 1993). Ce-lamin also has many hallmarks of A-type lamins, including a role in maintaining nuclear structure, a presence in the nuclear interior of a fraction of the Ce-lamins and an interaction with proteins that in mammals interact specifically with A-type lamins (Lyakhovetsky & Gruenbaum, 2014). *lmn-1* is an essential gene, since its downregulation leads to embryonic lethality. The effect of *lmn-1* downregulation is already observed as early as the first zygotic divisions with phenotypes that include misshaped nuclei, nuclear membrane lobulation and invagination, clustering of nuclear pore complexes (NPCs), and several chromatin aberrations common in cancer cells such as interchromosome nuclear connections (bridges), abnormal chromatin condensation, redistribution of heterochromatin, and abnormal segregation of chromosomes (Cohen et al., 2002; Liu et al., 2000). Some embryos manage to escape the lethal effects of *lmn-1*(RNAi) probably due to their being laid outside the most effective time window of RNAi's effect. These animals are mostly sterile, displaying a dramatic reduction in the number of germ cells and an increase in oocyte abnormalities, including multiple nuclei and large vacuoles (Liu et al., 2000). Downregulation of *lmn-1* also affects the localization of other components of the nuclear envelope including emerin, LEM-2, UNC-84, and BAF-1 but not matefin/

SUN-1 (Barkan et al., 2012; Gruenbaum, Lee, Liu, Cohen, & Wilson, 2002; Lee et al., 2002; Liu et al., 2003; Margalit, Liu, Fridkin, Wilson, & Gruenbaum, 2005). Downregulation of *lmn-1* at the adult stage leads to a reduced lifespan (Haithcock et al., 2005). Interestingly, *C. elegans* expressing Ce-lamin mutated in conserved residues, which in humans cause laminopathic diseases, share many phenotypes with the human diseases, thus making *C. elegans* a useful model organism to study these diseases (Bank & Gruenbaum, 2011).

4. ASSEMBLY OF *C. ELEGANS* LAMINS

Depending on the assembly conditions, bacterially expressed and purified Ce-lamin can form *in vitro* either 10-nm wide filaments or paracrystalline arrays (Foeger et al., 2006; Karabinos, Schunemann, Meyer, Aebi, & Weber, 2003). An *in vitro* assembly model of Ce-lamin, based on electron microscopy negative staining and cryo-electron tomography studies, suggested a hierarchal order of assembly wherein lamins first form dimers which then polymerize to form a polar head-to-tail linear polymer (Heitlinger et al., 1991). Next, lateral assembly of two head-to-tail polymers interacts in an antiparallel manner to form a four molecule wide nonpolar protofilament (Ben-Harush et al., 2009) which further assemble either into IF-like, 10 nm filaments (Fig. 1), composed of three or four protofilaments or into paracrystalline arrays (Fig. 1) (Ben-Harush et al., 2009; Wiesel et al., 2008). To better understand how Ce-lamins assemble *in vivo* in somatic cells, Ce-lamin was ectopically expressed *ex vivo* in *Xenopus laevis* oocytes. Cryo-electron tomography analysis revealed that Ce-lamins assemble into flexible protofilaments that interact with each other and exhibit a diameter of 5–6 nm (Grossmann et al., 2012). The latter experiments showed that protofilaments are the basic assembly units *in vivo* and that the 10 nm IF-like lamin filament structures of Ce-lamins represent only one form of assembly out of several assembly possibilities, which may adapt to different functional requirements and mechanical stresses (Grossmann et al., 2012). Determining the structure of lamin *in vivo* in somatic cells is a major task for future studies.

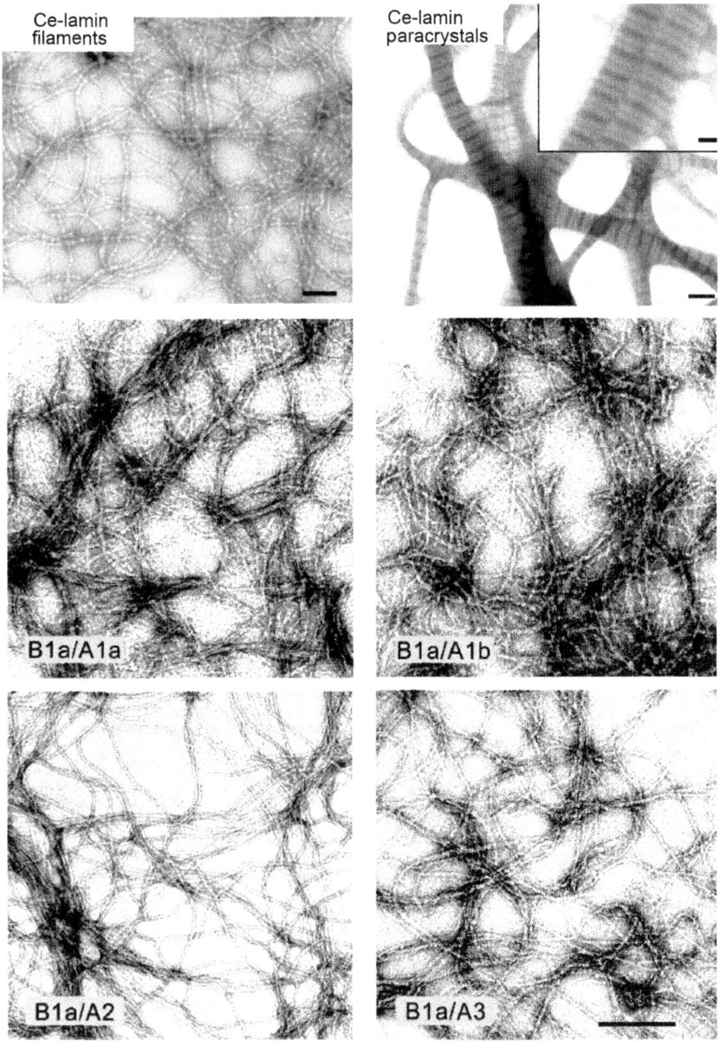

Figure 1 Top panels: Electron micrographs of negatively stained Ce-lamin assembled *in vitro* under conditions of filament assembly (left) or paracrystals (right). The scale bar represents 100 nm. Bottom panels: Electron micrographs of negatively stained IFs assembled *in vitro* from equal amounts of the recombinant proteins B1a/A1a, B1a/A1b, B1a/A2, and B1a/A3. The scale bar represents 200 nm. *Top panels: Figure taken from Wiesel et al. (2008). Bottom panels: Figure taken from Karabinos, Schulze, et al. (2003).*

5. ASSEMBLY OF *C. ELEGANS* IFs

Karabinos, Schulze, et al. (2003) used bacterially expressed IFA-1, IFA-2, IFA-3, IFA-4, and IF-B1 to study the *in vitro* assembly properties of *C. elegans* IF proteins. Ten nanometer filaments formed when IFB-2 was mixed with equal amounts of IFA-1, IFA-2, IFA-3, or IFA-4, demonstrating assembly of filaments composed of heteropolymers (Fig. 1). The heteropolymer formation is supported by the coexpression in the same cells of IFB-1 with different IFA proteins and by the overlapping phenotypes seen between IFB-1 and IFA protein mutations (see Section 2). Heteropolymeric assembly is quite common in mammalian IFs and was best studied in the keratin system. *C. elegans* IFs differs from mammalian IFs, since there is only one protein of one IF type, which assembles with four closely related proteins of the second IF type (Goldman, 2001).

Very little is known about the roles of posttranslational modifications in the process of cytoplasmic IF assembly. Work by Kaminsky et al. (2009) provided new insights into the role of SUMOylation in regulating the assembly of IFB-1 and its heteropolymers *in vivo*. Depletion of SUMO or mutating the SUMO acceptor site on IFB-1 causes a reduction of this cytoplasmic soluble pool, leading to a decrease in the exchange rate between the cytoplasmic pool and the existing filaments. This in turn leads to aberrant filament organization and to the formation of ectopic filaments in the lateral epidermis. The latter observations show that posttranslational modifications of *C. elegans* IFs play an important role in maintaining the IF network.

6. METHODS

6.1 Ce-Lamin Assembly *In Vitro*

1. Protein expression

   BL21DE3 Codon Plus bacteria are transformed with a plasmid encoding Ce-lamin fused to 6 × His. To express the lamin protein, a bacterial culture grown overnight in LB, is diluted 1:10 in 500 ml 2YT medium

   2YT medium (1 l):

   Bacto tryptone 16 g

   Bacto yeast extract 10 g

   NaCl 5 g

   Adjust pH to 7.0 with NaOH

The resulting culture is grown to an OD_{600} of 0.6–0.8, whereupon protein induction is initiated by the addition of IPTG to a final concentration of 0.5 m*M*. The bacterial culture is then incubated for 3 h at 37 °C. Bacteria are then harvested by 20 min centrifugation at 6000 × *g*.

2. Protein purification

2.1 The bacterial pellet is resuspended in 5 ml of lysis buffer for every 500 ml of the starting bacterial culture.

Lysis buffer:

20 m*M* Tris–HCl pH 7.4
200 m*M* NaCl
5 m*M* Benzamidin
0.1 m*M* AEBSF
3.4 μg/ml Aprotinin

2.2 Bacterial cell lysis is obtained using a microfluidics device (model number-LV1, Newton, MA) according to the manufacturer's instructions.

2.3 Centrifuge the crude cell extract at 10,000 × *g* for 10 min at 4 °C. Discard the supernatant.

2.4 Wash the inclusion bodies pellet as follows: resuspend the pellet in 5 ml of Triton buffer and then centrifuge at 4 °C for 10 min at 10,000 × *g*. Discard the supernatant. The pellet with the inclusion bodies contains most of the lamin molecules.

Triton buffer:

20 m*M* Tris–HCl pH 7.4
200 m*M* NaCl
5 m*M* Benzamidin
0.1 m*M* AEBSF
3.4 μg/ml Aprotenin
1% Triton X-100

2.5 Resuspended the pellet in lysis buffer (second wash).

2.6 Centrifuge at 4 °C for 10 min at 10,000 × *g*. Discard the supernatant.

2.7 Dissolve the inclusion bodies by suspending the pellet in 5 ml of freshly made urea buffer.

Urea buffer:

8 *M* urea
50 m*M* Tris–HCl pH 8
50 m*M* NaCl

2.8 Centrifuge for 10 min at 10,000 × *g* at 4 °C. Retain the supernatant. Remove a 20-μl aliquot into SDS-loading buffer to test the level of Ce-lamin induction.

2.9 Add to the urea-soluble extract (supernatant from step 2.8) 0.5 ml of packed nickel beads prewashed twice with 10 ml of column loading buffer:

Colum loading buffer:

8 *M* urea
50 m*M* Tris–HCl pH 8
20 m*M* imidazole pH 7.4
150 m*M* NaCl
1 m*M* Dithiothreitol (DTT)

2.10 Gently rotate the suspension for 1 h at 4 °C.

2.11 Spin down the beads at 500 × *g*. Transfer the supernatant (unbound material) to a fresh tube.

2.12 Resuspend the nickel beads in 5 ml of column loading buffer and transfer to a Econo-Column (BioRad Inc.).

2.13 Drain the column (do not allow it to completely dry) and then wash the beads with 20 ml of column loading buffer.

2.14 Ce-lamin-6His is eluted from the column by one 0.25 ml portion (fraction 1) followed by four 0.5 ml portions (fractions 2–5) of Elution buffer. The protein typically elutes to fractions 2 and 3 (Fig. 2A).

Elution buffer:

8 *M* urea
50 m*M* Tris–HCl pH 8
200 m*M* imidazole pH 7.4
150 m*M* NaCl
1 m*M* DTT

2.15 The specific IF fractions are pooled and resolved by 8% sodium dodecyl sulfate polyacrylamide gel electrophoresis (SDS-PAGE) to confirm their purity.

3. Assembly of Ce-lamin filaments

For filament assembly, the purified protein at a final concentration of 0.1 mg/ml is dialyzed for 4 h at a ratio of 1:500 (V:V) against a buffer containing 2 m*M* Tris–HCl pH 9, 1 m*M* DTT followed by a second 1 h dialysis against a buffer containing 15 m*M* Tris–HCl pH 7.4, 1 m*M* DTT. Extending the second dialysis to 12 h results

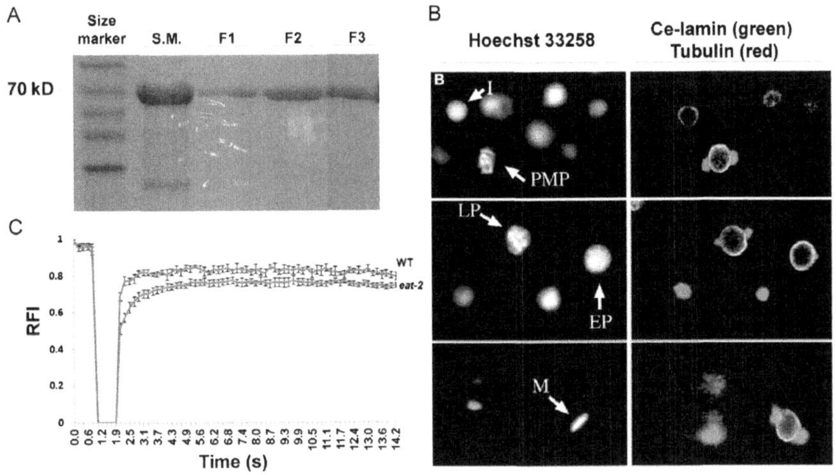

Figure 2 (A) Bacterially expressed 8XHis-tagged Ce-lamin was dissolved in 8 *M* urea and purified on a nickel column. The different fractions were loaded on 8% SDS-PAGE gel and stained with Coomassie Brilliant Blue. S.M., starting material; F1–3, first three elution fractions of the nickel column. (B) Indirect immunofluorescence of *C. elegans* embryos stained with DAPI for DNA and by specific antibodies for Ce-lamin (green) and tubulin (red). Mitosis stages (white arrows): I, interphase; PMP; prometaphase; LP, late prophase; EP, early prophase; M, metaphase. (C) Measurements of relative fluorescence intensity (RFI) during FRAP of BAF-1::GFP in wild type (blue) or eat-2 (red) larvae at the first larval stage (L1). Error bars represent SEM. *Panel (B) is taken from figure 4B in Lee et al. (2000)* Mol. Biol. Cell *11, 3089–3099. Panel (C) is taken from fig. 1A in Bar et al. (2014).* Mol. Biol. Cell *25, 1127–1136.* (See the color plate.)

in the formation of more elaborate filamentous structures (Foeger et al., 2006).

4. Assembly of a Ce-lamin paracrystalline array

   For paracrystalline assembly, the purified protein at final concentration of 0.1 mg/ml is dialyzed for 1 h at a ratio of 1:500 (V:V) against a buffer containing 15 m*M* Tris–HCl pH 7.4, 1 m*M* DTT followed by a second 1 h dialysis against a buffer containing 20 m*M* $CaCl_2$, 25 m*M* Tris–HCl pH 9, 1 m*M* DTT.

5. Preparation of samples for electron microscopy analysis

   An aliquot (7 μl) of the dialyzed Ce-lamin solution is absorbed onto glow-discharged carbon-coated copper grids for 1 min and negatively stained with 2% uranyl acetate. The grids (Formvar carbon film on 300 Mesh Copper, EMS. Cat. FCF300Cu) are visualized using a Phillips EM 410 transmission electron microscope.

6.2 Assembly of *C. elegans* Cytoplasmic IF A1, A2, A3, and B1 *In Vitro*

1. Protein expression

The different *C. elegans* cytoplasmic IF encoding plasmids are used to transform bacteria. BL21DE3 Codon Plus, pCRT7/CT-TPOP bacteria (Invitrogene), Pkk388-1 bacteria (CLON-TECH), or pET23 bacteria (Novagene) can be used for protein expression.

2. Protein purification

 2.1 Dissolve IF-containing inclusion bodies by resuspending the pellet in freshly made urea buffer:

 8 *M* urea

 10 m*M* $NaHPO_4$

 1 m*M* 2-β-mercaptoethanol pH 6.6

 2.2 Separate the various cytoplasmic IF species by cation-exchange chromatography on a Mono-S (GE Healthcare Bio-Science Corp., NJ) employing a linear salt gradient of 0–400 m*M* NaCl.

 2.3 The specific IF fractions are pooled and resolved by 8% SDS-PAGE gel to confirm their purity.

3. Assembly of cytoplasmic IF filaments

Purified recombinant IFB-1, together with IFA-1, IFA-2, IFA-3, or IFA-4 are dialyzed for 20 h at room temperature against a buffer containing 10 m*M* Tris–HCl pH 7.2 and 1 m*M* 2-mercaptoethanol. IFB-1/IFA-1, IFB-1/IFA-2, and IFB-1/IFA-3 are dialyzed at a final concentration of 0.4, 0.6, and 0.8 mg/ml, respectively.

4. Preparation of samples for electron microscopy analysis

Aliquots (7 μl) of the dialyzed cytoplasmic IF solutions are absorbed onto glow-discharged carbon-coated copper grids for 1 min and negatively stained with 2% uranyl acetate. The carbon-coated grids (Formvar carbon film on 300 Mesh Copper, EMS. Cat. FCF300Cu) are visualized using a Phillips EM 410 transmission electron microscope (Karabinos, Schulze, et al., 2003).

6.3 Indirect Immunofluorescence Assays Used to Analyze IFs in *C. elegans*

1. Preparation of polylysine covered slides

Place 100 μl of the polylysine solution (Sigma 0.1% w/v Cat#P8920) on clean slides (Super Frost/Plus, Menzelglass 041300).

After 3–5 min, remove excess solution using a 0–200 μl Gilson and allow slides to completely dry.

2. Staining protocol
   2.1 Put a 9 μl drop of M9 buffer on the polylysine coated area of the slide and place in this drop 6–10 *C. elegans* worms at the desired developmental stage.
   2.2 Place a 40 mm long coverslip on top of the drop and gently tap with forceps to spread the water and to allow eggs to spill out of the worm. Use long forceps to hold the slide and put it in the liquid nitrogen until the temperature equilibrates. Remove the slide from the liquid nitrogen and quickly remove the coverslip by flicking it off. Freeze-cracking of the worms allows better antibody penetration.
   2.3 Immediately place the slides in cold methanol (−20 °C) for 10 min.
   2.4 Place the slides in 4% paraformaldehyde (Electron Microscope Sciences, Hatfield PA, Cat#15710) diluted in PBST:
   For 1 l
   NaCl 8 g
   KCL 0.2 g
   $NaHPO_4$ 1.44 g
   $KHPO_4$ 0.24 g
   Adjust pH to 7.0 with HCl
   Add Triton X-100 to 0.25%
   Incubate slides for 10 min at room temperature.
   2.5 Wash slides 3 ×, 5 min each, with PBST in a slide container while gently shaking.
   2.6 Block nonspecific binding by placing the slides for 15 min in PBST solution containing 0.1% low-fat milk at room temperature.
   2.7 Put the slide on parafilm in a small humidity chamber. Put 50 μl of the primary antibody, diluted to the recommended dilution in blocking solution, on the part of the slide containing the worms. Incubate overnight at 4 °C on a horizontal shaker. For lower volume of antibodies, cover the slide with a coverslip.
   2.8 Wash slides twice, each for 1 h, with PBST in a slide container while gently shaking.
   2.9 Put 50 μl of the corresponding secondary antibody diluted to the recommended dilution in PBST on the part of the slide containing the worms. Incubate for 3 h in a humid chamber on a horizontal shaker at room temperature.

2.10 Wash slides twice for 1 h with PBST in a slide container while gently shaking.

2.11 For DNA staining, place slides in the humidity chamber and put 300 μl of 4′,6-diamidino-2-phenylindole (DAPI) diluted 1:1000 in PBST on the worm containing part of the slide, incubate for 10 min on a shaker, then wash for 10 min in PBST.

2.12 Place a drop of 2% *N*-propyl gallate in glycerol or a commercial mounting media on the stained worms, cover with a 22 × 22 mm #1 coverslip and seal with paraffin or nail polish.

2.13 Staining is visualized using a fluorescence microscope.

An example of indirect immunofluorescence staining of *C. elegans* early embryos appears in Fig. 2B.

6.4 Fluorescence Recovery After Photobleaching

1. Sample preparation

   Place worms at the middle of the fourth larval stage on 4% agarose pads. Thirty minutes before the initiation of the experiment anesthetized the worms using 0.1% tricaine and 0.01% tetramisole, or paralyze the worms with a 2 m*M* Levamisol solution.

2. Fluorescence recovery after photobleaching (FRAP) procedure

   Image worms using a confocal microscope in a single *Z* plane with 1% laser power transmission. Select a defined region of interest and photobleach it with 100% laser power transmission. Take a series of images at brief intervals before, during and after the photobleaching.

3. Analysis

   Measure fluorescence levels in the area of interest (ROI), the background area (BG) and the total nuclear area (Tot) as a function of time before and after the bleaching process. Normalize values according to the following formula: $((ROI_t - BG_t)/(Tot_t - BG_t)) \times ((ROI_{t0} - BG_{t0})/(Tot_{t0} - BG_{t0}))$. A typical experiment contains at least 10 worms of each strain/treatment tested.

An example of FRAP analysis of BAF-1::GFP appears in Fig. 2C.

7. SUMMARY

The 11 *C. elegans* genes that encode cytoplasmic IFs show tissue-specific patterns of expression and at least five of them are essential for the animal's survival. Many of the *C. elegans* IF proteins make heteropolymers, similar to mammalian cytoplasmic IFs. The single nuclear IF gene is evolutionary conserved and essential. This evolutionary conservation together with the relative simplicity of the system and the availability of the diverse genetics tools make *C. elegans* a powerful model to study the organization and functions of IFs, as well as studying IF-associated proteins.

While the lamin gene has been studied in detail, relatively little is known about the organization and functions of the cytoplasmic IFs. Applying the above mentioned techniques with the addition of whole organism approaches can shed light on the formation of filaments and their functions *in vivo* and can assist in explaining why disruptions of the human homologues lead to heritable diseases.

ACKNOWLEDGMENTS

We gratefully acknowledge funding from the Muscular Dystrophy Association (MDA), the Binational Israel-USA Science Foundation (BSF 2007215) and the Niedersachsen-Israeli Research Cooperation program.

REFERENCES

Bank, E. M., & Gruenbaum, Y. (2011). *Caenorhabditis elegans* as a model system for studying the nuclear lamina and laminopathic diseases. *Nucleus, 2*, 350–357.

Barkan, R., Zahand, A. J., Sharabi, K., Lamm, A. T., Feinstein, N., et al. (2012). Ce-emerin and LEM-2: Essential roles in *Caenorhabditis elegans* development, muscle function, and mitosis. *Molecular Biology of the Cell, 23*, 543–552.

Ben-Harush, K., Wiesel, N., Frenkiel-Krispin, D., Moeller, D., Soreq, E., et al. (2009). The supramolecular organization of the *C. elegans* nuclear lamin filament. *Journal of Molecular Biology, 386*, 1392–1402.

Carberry, K., Wiesenfahrt, T., Windoffer, R., Bossinger, O., & Leube, R. E. (2009). Intermediate filaments in *Caenorhabditis elegans*. *Cell Motility and the Cytoskeleton, 66*, 852–864.

Cohen, M., Tzur, Y. B., Neufeld, E., Feinstein, N., Delannoy, M. R., et al. (2002). Transmission electron microscope studies of the nuclear envelope in *Caenorhabditis elegans* embryos. *Journal of Structural Biology, 140*, 232–240.

Foeger, N., Wiesel, N., Lotsch, D., Mucke, N., Kreplak, L., et al. (2006). Solubility properties and specific assembly pathways of the B-type lamin from *Caenorhabditis elegans*. *Journal of Structural Biology, 155*, 340–350.

Goldman, R. D. (2001). Worms reveal essential functions for intermediate filaments. *Proceedings of the National Academy of Sciences of the United States of America, 98*, 7659–7661.

Grossmann, E., Dahan, I., Stick, R., Goldberg, M. W., Gruenbaum, Y., & Medalia, O. (2012). Filaments assembly of ectopically expressed *Caenorhabditis elegans* lamin within Xenopus oocytes. *Journal of Structural Biology, 177*, 113–118.

Gruenbaum, Y., & Aebi, U. (2014). Intermediate filaments: A dynamic network that controls cell mechanics. *F1000Prime Reports, 6*, 54.

Gruenbaum, Y., Lee, K. K., Liu, J., Cohen, M., & Wilson, K. L. (2002). The expression, lamin-dependent localization and RNAi depletion phenotype for emerin in *C. elegans*. *Journal of Cell Science, 115*, 923–929.

Haithcock, E., Dayani, Y., Neufeld, E., Zahand, A. J., Feinstein, N., et al. (2005). Age-related changes of nuclear architecture in *Caenorhabditis elegans*. *Proceedings of the National Academy of Sciences of the United States of America*, *102*, 16690–16695.
Hapiak, V., Hresko, M. C., Schriefer, L. A., Saiyasisongkhram, K., Bercher, M., & Plenefisch, J. (2003). mua-6, a gene required for tissue integrity in *Caenorhabditis elegans* encodes a cytoplasmic intermediate filament. *Developmental Biology*, *263*, 330–342.
Harborth, J., Elbashir, S. M., Bechert, K., Tuschl, T., & Weber, K. (2001). Identification of essential genes in cultured mammalian cells using small interfering RNAs. *Journal of Cell Science*, *114*, 4557–4565.
Heitlinger, E., Peter, M., Haner, M., Lustig, A., Aebi, U., & Nigg, E. A. (1991). Expression of chicken lamin B2 in Escherichia coli: Characterization of its structure, assembly, and molecular interactions. *The Journal of Cell Biology*, *113*, 485–495.
Kaminsky, R., Denison, C., Bening-Abu-Shach, U., Chisholm, A. D., Gygi, S. P., & Broday, L. (2009). SUMO regulates the assembly and function of a cytoplasmic intermediate filament protein in *C. elegans*. *Developmental Cell*, *17*, 724–735.
Karabinos, A., Schmidt, H., Harborth, J., Schnabel, R., & Weber, K. (2001). Essential roles for four cytoplasmic intermediate filament proteins in *Caenorhabditis elegans* development. *Proceedings of the National Academy of Sciences of the United States of America*, *98*, 7863–7868.
Karabinos, A., Schulze, E., Schunemann, J., Parry, D. A., & Weber, K. (2003). In vivo and in vitro evidence that the four essential intermediate filament (IF) proteins A1, A2, A3 and B1 of the nematode *Caenorhabditis elegans* form an obligate heteropolymeric IF system. *Journal of Molecular Biology*, *333*, 307–319.
Karabinos, A., Schunemann, J., Meyer, M., Aebi, U., & Weber, K. (2003). The single nuclear lamin of *Caenorhabditis elegans* forms in vitro stable intermediate filaments and paracrystals with a reduced axial periodicity. *Journal of Molecular Biology*, *325*, 241–247.
Karabinos, A., Schünemann, J., & Weber, K. (2004). Most genes encoding cytoplasmic intermediate filament (IF) proteins of the nematode *Caenorhabditis elegans* are required in late embryogenesis. *European Journal of Cell Biology*, *83*, 457–468.
Kim, Y., Zheng, X., & Zheng, Y. (2013). Proliferation and differentiation of mouse embryonic stem cells lacking all lamins. *Cell Research*, *23*, 1420–1423.
Lee, K. K., Starr, D., Liu, J., Cohen, M., Han, M., et al. (2002). Lamin-dependent localization of UNC-84, a protein required for nuclear migration in *Caenorhabditis elegans*. *Molecular Biology of the Cell*, *13*, 892–901.
Liu, J., Lee, K. K., Segura-Totten, M., Neufeld, E., Wilson, K. L., & Gruenbaum, Y. (2003). MAN1 and emerin have overlapping function(s) essential for chromosome segregation and cell division in *Caenorhabditis elegans*. *Proceedings of the National Academy of Sciences of the United States of America*, *100*, 4598–4603.
Liu, J., Rolef-Ben Shahar, T., Riemer, D., Spann, P., Treinin, M., et al. (2000). Essential roles for *Caenorhabditis elegans* lamin gene in nuclear organization, cell cycle progression, and spatial organization of nuclear pore complexes. *Molecular Biology of the Cell*, *11*, 3937–3947.
Lyakhovetsky, R., & Gruenbaum, Y. (2014). Studying lamins in invertebrate models. *Advances in Experimental Medicine and Biology*, *773*, 245–262.
Margalit, A., Liu, J., Fridkin, A., Wilson, K. L., & Gruenbaum, Y. (2005). *A lamin-dependent pathway that regulates nuclear organization, cell cycle progression and germ cell development*. London: John Wiley & Sons, Ltd.
Riemer, D., Dodemont, H., & Weber, K. (1993). A nuclear lamin of the nematode *Caenorhabditis elegans* with unusual structural features; cDNA cloning and gene organization. *European Journal of Cell Biology*, *62*, 214–223.
Stuurman, N., Heins, S., & Aebi, U. (1998). Nuclear lamins: Their structure, assembly, and interactions. *Journal of Structural Biology*, *122*, 42–66.

Wiesel, N., Mattout, A., Melcer, S., Melamed-Book, N., Herrmann, H., et al. (2008). Laminopathic mutations interfere with the assembly, localization and dynamics of nuclear lamins. *Proceedings of the National Academy of Sciences of the United States of America*, *105*, 180–185.

Woo, W. M., Goncharov, A., Jin, Y., & Chisholm, A. D. (2004). Intermediate filaments are required for *C. elegans* epidermal elongation. *Developmental Biology*, *267*, 216–229.

Worman, H. J., & Bonne, G. (2007). "Laminopathies": A wide spectrum of human diseases. *Experimental Cell Research*, *313*, 2121–2133.

CHAPTER TWENTY-FIVE

Mechanical Probing of the Intermediate Filament-Rich *Caenorhabditis Elegans* Intestine

Oliver Jahnel*, Bernd Hoffmann†, Rudolf Merkel†, Olaf Bossinger*,[1], Rudolf E. Leube*,[2]

*Institute of Molecular and Cellular Anatomy, RWTH Aachen University, Aachen, Germany
†Institute of Complex Systems, ICS-7: Biomechanics, Jülich, Germany
[2]Corresponding author: e-mail address: rleube@ukaachen.de

Contents

1. Introduction 682
2. Imaging of Intermediate Filaments in *C. elegans* Intestines by Epifluorescence Microscopy 686
3. Outline of Intestinal Rings in *C. elegans* by a Fluorescent Apical Junction Reporter 688
   3.1 Confocal Laser Scanning Microscopy of Worms 690
4. Dissection of Intestines and Vitality Testing 691
   4.1 Viability Testing of the Dissected Intestine 691
5. Experimental Setup for Micropipette Measurements 695
   5.1 Forging of Pipettes 696
   5.2 Calibrating Pipette Setup 697
   5.3 Analysis of Intestinal Mechanics 699
6. Outlook 702
Acknowledgments 703
References 703

Abstract

It is commonly accepted that intermediate filaments have an important mechanical function. This function relies not only on intrinsic material properties but is also determined by dynamic interactions with other cytoskeletal filament systems, distinct cell adhesion sites, and cellular organelles which are fine-tuned by multiple signaling pathways. While aspects of these properties and processes can be studied *in vitro*, their full complexity can only be understood in a viable tissue context. Yet, suitable and easily accessible model systems for monitoring tissue mechanics at high precision are rare. We show that the dissected intestine of the genetic model organism *Caenorhabditis elegans* fulfills this requirement. The 20 intestinal cells, which are arranged in an invariant

[1] Current address: Molecular Cell Biology, Institute of Anatomy I, University of Cologne, 50937 Cologne, Germany.

Methods in Enzymology, Volume 568
ISSN 0076-6879
http://dx.doi.org/10.1016/bs.mie.2015.08.030

fashion, are characterized by a dense subapical mesh of intermediate filaments that are attached to the *C. elegans* apical junction. We present procedures to visualize details of the characteristic intermediate filament–junctional complex arrangement in living animals. We then report on methods to prepare intestines with a fully intact intermediate filament cytoskeleton and detail procedures to assess their viability. A dual micropipette assay is described to measure mechanical properties of the dissected intestine while monitoring the spatial arrangement of the intermediate filament system. Advantages of this approach are (i) the high reproducibility of measurements because of the uniform architecture of the intestine and (ii) the high degree of accessibility allowing not only mechanical manipulation of an intact tissue but also control of culture medium composition and addition of drugs as well as visualization of cell structures. With this method, examination of worms carrying mutations in the intermediate filament system, its interacting partners and its regulators will become feasible.

ABBREVIATIONS

C. elegans *Caenorhabditis elegans*
CeAJ *C. elegans* apical junction
CFP cyan fluorescent protein
int intestinal ring
PBS phosphate-buffered saline

1. INTRODUCTION

Cells experience a wide range of mechanical stresses in their native tissue environment. This is especially true for epithelial cells, which are responsible for efficient and continuous barrier formation while being subjected to forces from the outside and inside of the body. Major players in the integration and dissipation of these forces are the intermediate filaments together with their plasma membrane attachment sites by acting as mechanical shock absorbers, especially at large deformations (Beil et al., 2003; Fudge et al., 2008). The *in vitro* observations, that single-intermediate filaments can be stretched up to 3.6-fold before breaking (Kreplak, Bär, Leterrier, Herrmann, & Aebi, 2005) and that intermediate filaments respond to high strains by hardening (e.g., Janmey, Euteneuer, Traub, & Schliwa, 1991; Lin et al., 2010), are ideal properties to support this function. Observations on genetically modified cells provide further evidence. Thus, deletion of vimentin intermediate filaments leads to reduced stiffness and impaired mechanical stability of fibroblasts resulting in reduced migration and contraction (Brown, Hallam, Colucci-Guyon, & Shaw, 2001; Eckes et al., 1998; Wang & Stamenović, 2000). Expression of different desmin intermediate filament mutants in rat fibroblasts leads to altered cell stiffness as determined by atomic force microscopy (Plodinec et al., 2011). Similarly, epithelial cells expressing keratin mutants (Lulevich, Yang, Rivkah Isseroff, & Liu, 2010) and keratin-free keratinocytes are less stiff but become more motile (Ramms et al., 2013; Seltmann, Fritsch, Kas, & Magin, 2013; Seltmann, Roth, et al., 2013). The occurrence of blister-forming diseases in patients carrying point mutations in keratin genes is probably the most compelling evidence for a mechanical function of epithelial intermediate filaments (recent review in Homberg & Magin, 2014). However, the underlying mechanisms are still not fully understood. Besides a direct mechanical contribution of keratin filaments themselves, perturbed signaling and adhesion are also considered to be important factors in the development of mechanical deficiencies (Kröger et al., 2013; Liovic et al., 2008; Russell, Ross, & Lane, 2010; Seltmann, Cheng, Wiche, Eriksson, & Magin, 2015).

In the recent past considerable efforts have been undertaken to measure mechanical properties of the cellular intermediate filament system. Most emphasis has been on the investigation of single cells using microchannels (Rolli, Seufferlein, Kemkemer, & Spatz, 2010), microfluidic optical stretchers (Seltmann, Fritsch, et al., 2013), and atomic force microscopy (Lulevich et al., 2010; Ramms et al., 2013; Walter, Busch, Seufferlein, & Spatz, 2011) to probe whole-cell mechanical properties from the outside, while particle-tracking microrheology (Sivaramakrishnan, DeGiulio, Lorand, Goldman, & Ridge, 2008) and magnetic tweezers (Ramms et al., 2013) have been used for testing cytoplasmic viscoelasticity. These studies were complemented by analyses of the keratin cytoskeleton in fixed and partially extracted cells (e.g., Beil et al., 2003; Paust, Paschke, Beil, & Marti, 2013; Walter et al., 2011). Besides encountering a very high degree of cell-to-cell variability, these analyses neglected the tight coupling of epithelial cells to each other and the extracellular matrix. For epithelial cell sheets, analyses have been restricted to reconstituted cell monolayers using dispase assays to examine tissue cohesion (Kröger et al., 2013) and to the study of keratin network dynamics in cells grown on flexible substrates (Beriault et al., 2012; Felder et al., 2008; Fois et al., 2013; Fudge et al., 2008; Hecht et al., 2012).

Thus, a major current need is the investigation of the mechanical properties of the epithelial intermediate filament cytoskeleton in its native tissue environment. Such an enterprise should taken into account the multidimensional intermediate filament network organization, which is determined by its cell type-specific 3D arrangement, turnover dynamics, association with other filament systems, attachment to defined adhesion sites, and the multiple dynamic interactions with other cellular components. The complex nature of vertebrate tissues and the substantial efforts needed to isolate, cultivate, and genetically modify them are major obstacles. Therefore, using the genetic model organism *Caenorhabditis elegans* may offer a simplified approach. Its highly invariant body plan, the simplicity and speed to grow and propagate clonal worm lines and the easiness of genetic manipulation together with the abundance of intermediate filaments in the epithelial tissues of *C. elegans* are all in favor of using this very well-characterized model organism. A proof of principle was recently provided by the elegant work of Michel Labouesse and his group which uncovered an epidermal mechanotransduction pathway in the epidermis resulting in intermediate filament phosphorylation (Zhang et al., 2011). Remarkably, they were able to show that application of mechanical pressure on *C. elegans* embryos can trigger this pathway. Our own interest is in the epithelial cytoskeleton of the *C. elegans* intestine (Carberry et al., 2012; Carberry, Wiesenfahrt, Windoffer, Bossinger, & Leube, 2009). The approximately 800 μm long tube-like intestine (midgut) extends from the pharynx (foregut) to the hindgut and consists of 20 cells, which are involved in nutrient uptake and secretion. They are arranged as a single-cell layer forming nine intestinal rings (ints) surrounding the ellipsoid lumen (recent review in McGhee, 2013).

The *C. elegans* genome contains 11 cytoplasmic intermediate filament genes coding for at least 14 polypeptides because of differential splicing (Carberry et al., 2009; Dodemont, Riemer, Ledger, & Weber, 1994; Karabinos, Schmidt, Harborth, Schnabel, & Weber, 2001) and a single gene for a nuclear lamin (Gruenbaum, Lee, Liu, Cohen, & Wilson, 2002; Liu et al., 2000). Of the 11 genes encoding cytoplasmic intermediate filament polypeptides, six are predominantly if not exclusively transcribed in the intestine. They encode for IFB-2, IFC-1, IFC-2, IFD-1, IFD-2, and IFP-1 (Carberry et al., 2009). All are characterized by a central α-helical rod domain with subdomains L1, L12, and L2 encompassing the characteristic heptad repeat needed to form the stable coiled-coil dimers (Carberry et al., 2009; Dodemont et al., 1994). The polymerization properties of the intestinal intermediate filaments have not been investigated to date. A unique feature of the cytoplasmic *C. elegans* intermediate filament polypeptides is the presence of a 42 amino acid insertion in coil 1b that is typical for protostomia and the nuclear lamins. The B-type intermediate filament polypeptides also contain an Ig-like domain in their carboxytermini. Evolutionary analyses show that the intestinal intermediate filament polypeptides of *C. elegans* are unique and differ from those expressed in other nematodes, invertebrates, and vertebrates (Table 1).

Table 1 OrthoMCL Clusters of *C. elegans* Intestinal Intermediate Filaments, *C. elegans* Intermediate Filaments in Other Tissues and *C. elegans* Lamin (in Cooperation with P. Schiffer, Biocenter University of Cologne)

			IFB-2 (intestine)	IFC-1 (intestine)	IFC-2 (intestine)	IFD-1 (intestine)	IFD-2 (intestine)	IFP-1 (intestine)	IFA-1/MUA-6/ IFA-3/IFA-4/IFB-1 (other tissues)	LMN-1 (lamin)
Nematoda (clades)	V	*C. elegans*	1	1	1	1	1	1	5	1
	V	*C. briggsae*	1	1	1	1	1	1	5	1
	V	*C. remanei*	1	1	1	1	1	1	5	1
	V	*C. angaria*	1	-	-	1	-	1	4	2
	V	*P. pacificus*	-	-	-	-	-	-	5	1
	IV	*B. xylophylus*	-	-	-	-	-	-	4	1
	IV	*M. hapla*	-	-	-	-	-	-	1	1
	III	*D. immitis*	-	-	-	-	-	-	4	1
	III	*L. loa*	-	-	-	-	-	-	4	1
	III	*A. suum*	-	-	-	-	-	-	3	1
	III	*B. malayi*	-	-	-	-	-	-	3	1
	II	***E. brevis***	**-**	**-**	**-**	**-**	**-**	**-**	**1**	**2**
	I	***T. spiralis***	**-**	**-**	**-**	**-**	**-**	**-**	**2**	**1**
	I	***R. culicivorax***	**-**	**-**	**-**	**-**	**-**	**-**	**1**	**1**
Other species	Nematomorpha	*Gordius spec.*	-	-	-	-	-	-	3	3
	Tardigrada	*H. dujardini*	-	-	-	-	-	-	-	3
	Hexapoda	*D. melanogaster*	-	-	-	-	-	-	-	2
	Hexapoda	*A. pisum*	-	-	-	-	-	-	-	2
	Hexapoda	*T. castaneum*	-	-	-	-	-	-	-	1
	Chelicerata	*T. urticae*	-	-	-	-	-	-	-	1
	Lophotrochozoa	*C. capitata*	-	-	-	-	-	-	-	1
	Crustacea	*D. pulex*	-	-	-	-	-	-	-	1
	Myriapoda	*S. maritima*	-	-	-	-	-	-	-	1
	Echinodermata	*S. purpuratus*	-	-	-	-	-	-	-	1
	Chordata	*B. floridae*	-	-	-	-	-	-	-	1
	Vertebrata	*H. sapiens*	-	-	-	-	-	-	-	3

The presence of the *C. elegans* intestinal intermediate filaments seems to be linked to the endotube, a prominent and mechanically resilient structure that is characteristic for *C. elegans*. The endotube is a clearly discernable dense region in electron micrographs, which is located just below the organelle-free terminal web that anchors the microvillar actin bundles (Bossinger, Fukushige, Claeys, Borgonie, & McGhee, 2004; McGhee, 2013; Munn & Greenwood, 1984). The intestinal intermediate filaments are highly enriched in the subapical region and have been localized to the endotube by immunoelectron microscopy (Bossinger et al., 2004). The endotube is anchored to the *C. elegans* apical junction (CeAJ). This apical cell–cell adhesion site is ultrastructurally homogenous but encompasses several molecular subdomains (recent review in Pásti & Labouesse, 2014).

In this communication, we will first describe, how the intermediate filaments and cell junctions are organized as a contiguous sheath surrounding the intestinal lumen and will then go on to describe a simple method to isolate functionally intact intestines, whose mechanical properties can be quantitatively examined using a dual pipette assay. Possible applications of this system will be briefly outlined at the end.

2. IMAGING OF INTERMEDIATE FILAMENTS IN *C. elegans* INTESTINES BY EPIFLUORESCENCE MICROSCOPY

The following protocol is routinely used to examine the intermediate filament distribution in worms producing fluorescently labeled reporters. Figure 1 shows a live L4 larva, in which IFB-2::CFP fusion proteins localize specifically to the subapical periluminal domain of the intestine (Carberry et al., 2009; Hüsken et al., 2008). The continuous tube-like sleeve surrounding the intestinal lumen from the first to ninth int is readily seen.

Materials

- Viable L4 larvae of strain BJ49 *kcIs6 [ifb2::cfp]IV*
- Levamisole (Sigma-Aldrich, cat no. 31742)
- 2.5% (w/v) agarose (Biozym, cat no. 840004) in distilled H_2O
- Microscope slides (e.g., Menzel-Gläser, cat no. cut edges AA00000102E)
- Vaseline
- Platinum wire (VWR, cat no. 631-7101) to prepare a worm pick
- Phosphate-buffered saline (PBS; Sigma-Aldrich, cat no. D8537)

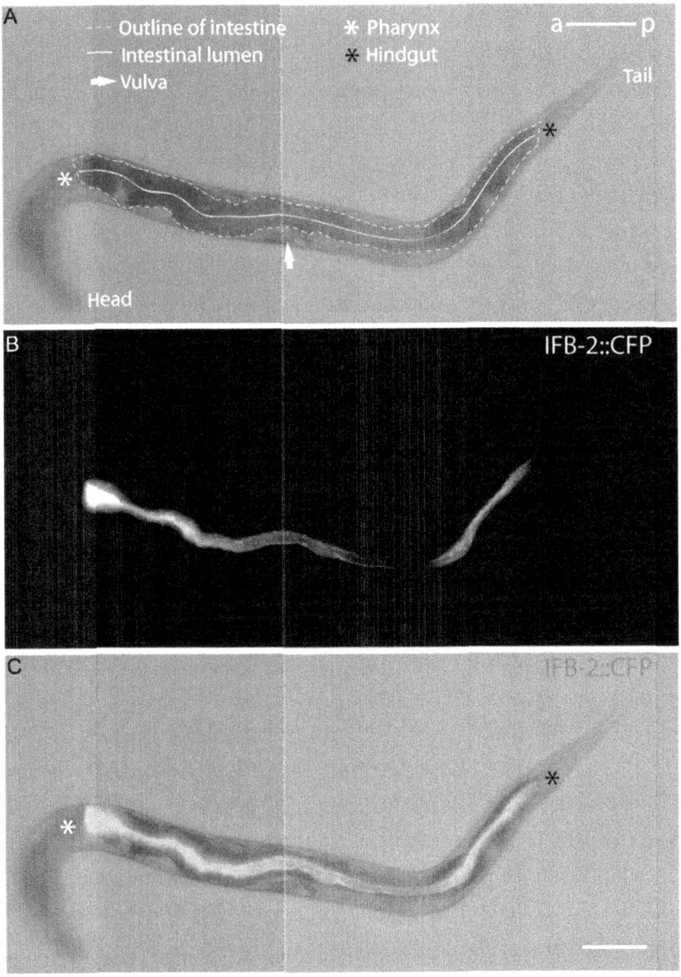

Figure 1 Fluorescence distribution of an IFB-2::CFP reporter in an L4 larva of strain BJ49. The fluorescence is concentrated in a dense network in the apical cytoplasm of intestinal cells entirely surrounding the intestinal lumen. (A) Phase-contrast micrograph, (B) corresponding fluorescence micrograph and merged overlay with fluorescence in cyan (light gray in the print version) in (C). The fluorescence signal is restricted to the intestine beginning after the pharyngeal terminal bulb (white asterisk) and ending before the hindgut (black asterisk). The dashed line shows the contours of the intestinal cells, the continuous line shows the position of the intestinal lumen. Arrow, vulva; a, anterior; p, posterior. Scale bar: 50 μm.

Equipment

- Fluorescence microscope (e.g., Zeiss ApoTome 2, Carl Zeiss)
- 20 × Objective (e.g., Plan-Apochromat 20 ×/0.8 M27, Carl Zeiss)

Methods

1. To prepare agarose pads, melt the 2.5% agarose solution and place a drop between two microscope slides to flatten. After hardening, slides are separated and the pad adhering to one of the slides is ready for use.
2. Transfer L4 larvae with the worm pick into a 20-μl drop of PBS containing 1 m*M* Levamisole that had been placed on the agarose pad. Put a coverslip on top and ensure that all edges of the coverslip are completely sealed with vaseline to prevent evaporation.
3. Record phase-contrast images and IFB-2::CFP fluorescence using the CFP (cyan fluorescent protein) filter set.

3. OUTLINE OF INTESTINAL RINGS IN *C. elegans* BY A FLUORESCENT APICAL JUNCTION REPORTER

For understanding the mechanical coupling of the cytoplasmic intermediate filament cytoskeleton in the intestine, a detailed characterization of the spatial arrangement of the CeAJ is crucial. Using fluorescently labeled components of the CeAJ intermediate filament attachment sites can be easily outlined. The example shown in Fig. 2A and B presents the fluorescence of an mCherry-labeled reporter for the CeAJ component DLG-1 (*C. elegans* homologue of the *Drosophila melanogaster* tumor supressor gene discs large) revealing a ladder-type pattern in the intestine (Bossinger, Klebes, Segbert, Theres, & Knust, 2001; Firestein & Rongo, 2001; McMahon, Legouis, Vonesch, & Labouesse, 2001). Since its presence is not restricted to the intestine (Fig. 2A), fluorescence signals outside of the pharynx, intestine, and hindgut were removed to prepare Fig. 2A′. In this cropped image, the fluorescence signal in the ints is much better visible. While the first int consists of four cells, all others are made up of two cells each. They are referred to as int1 to int9 from anterior to posterior (Sulston, Schierenberg, White, & Thomson, 1983). Careful inspection shows that (i) the rings are twisted from anterior to posterior with respect to each other in a clockwise direction and (ii) not all half rings are in register with the largest stagger in the middle of the worm. These properties are reason for the characteristic ladder pattern (see also Hermann, Leung, & Priess, 2000; Leung, Hermann, & Priess, 1999). The high-magnification images in Fig. 2B and B′ highlight the resulting arrangement of the DLG-1::mCherry-labeled CeAJ in ints6–8. The simplified 3D

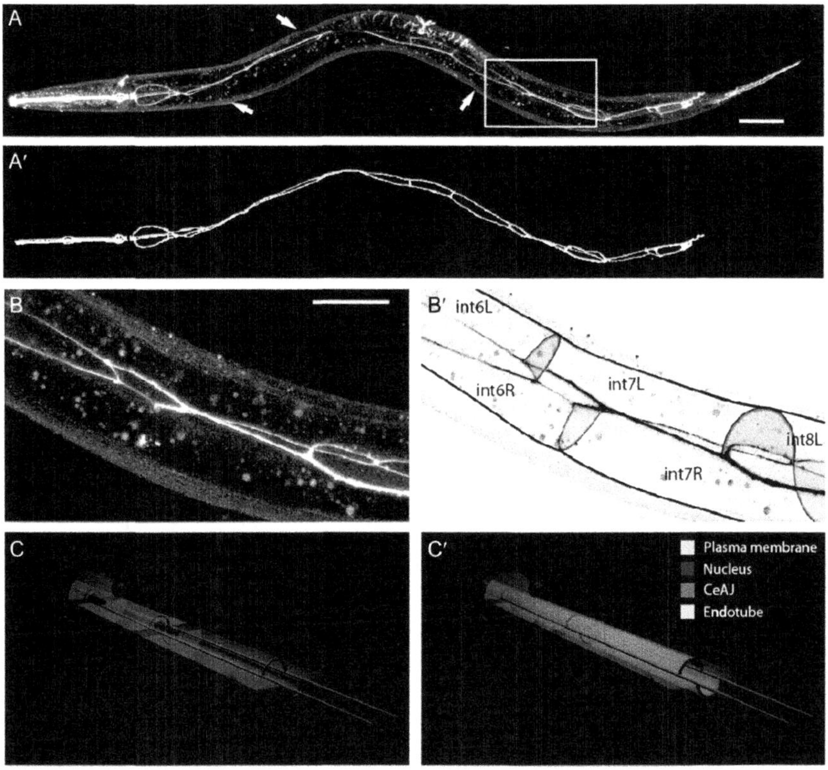

Figure 2 Detection and modeling of the intermediate filament-associated apical cell contacts in the *C. elegans* intestine. (A) Composite micrograph depicting the fluorescence of the CeAJ marker DLG-1::mCherry in the intestine of an L4 *C. elegans* larva as a single-projection view of stitched images. Note that slight movements of the worm resulted in misalignments at borders of image stacks (arrows). Since DLG-1::mCherry is not restricted to the intestine but is also detectable in other tissues including pharynx, rectum, seam cells, hypodermis, spermatheca, and vulva, only focal planes 6–16 of the entire stack of 20 focal planes are shown to improve visualization of the intestinal fluorescence. Panel (A′) shows a cropped version of the micrograph in A to further alleviate examination of the intestinal fluorescence pattern (see also corresponding animation in Movie 1 (http://dx.doi.org/10.1016/bs.mie.2015.08.030)). (B) High magnification taken from A (boxed area) of a region encompassing int-6 to int-8. Details of the typical ladder pattern can be seen, which are caused by the staggered and clockwise twisted arrangement of the intestinal cells. Panel (B′) shows an inverse version of B and also includes the position of the basolateral cell borders, which were drawn manually with the help of simultaneously recorded phase-contrast images (not shown). Panels (C) and (C′) Simplified 3D reconstructions of part of the intestine depicting the CeAJ without (C) and with the attached intermediate filament-rich endotube (C′). The corresponding animation in Movie 2 (http://dx.doi.org/10.1016/bs.mie.2015.08.030) highlights further details. Scale bars: 50 μm in A (same magnification in A′) and 25 μm in B (same magnification in B′). (See the color plate.)

reconstructions in Fig. 2C and C′ further demonstrate the CeAJ localization with respect to the cell body and depict, how it serves as an attachment site for the circumferential intermediate filament-rich endotube.

3.1 Confocal Laser Scanning Microscopy of Worms

Viable reporter worms are immobilized on glass slides for high-resolution confocal laser scanning fluorescence microscopy. To record the fluorescence in entire worms, multiple image stacks have to be prepared, assembled, and stitched. Small movements of the animals may pose problems (arrows in Fig. 2A). Imaging parameters have to be adjusted to the intensity and distribution of the reporter construct used. Multiple options are available for 3D reconstruction and visualization.

Materials

- *C. elegans* strain BJ246 *Is[dlg-1p::dlg-1::mcherry]* (kindly provided by Dr. Andrea Hutterer)

Equipment

- Confocal laser scanning fluorescence microscope (e.g., LSM710 Duo, Carl Zeiss) equipped with a helium–neon laser
- 63 × Objective (e.g., Objective Plan-Apochromat 63 ×/1.40 Oil DIC M27, Carl Zeiss)
- Fiji 1.4 software (National Institutes of Health)
- Amira 5.4 software (Visage Imaging)
- Blender 2.73 (Blender Foundation)

Methods

1. Transfer viable L4 larvae as described in Section 2 into a drop of PBS containing 1 m*M* Levamisole on a microscope slide.
2. Mount prepared sample on the microscope stage.
3. Detect fluorescence of DLG-1::mCherry with the 543 nm line of the helium–neon laser and set laser intensity to 50%. Monitor the emitted light between 592 and 722 nm with 1.72 airy units. Acquire images in multiple focal planes with an image size of 1024 × 1024 pixel (pixel size 0.13 μm) and record with a grayscale range of 16-bit using a dwell time of 6.3 μs for each pixel.
4. Stitch images using the ZEN software (Carl Zeiss) or any other available software (e.g., Fiji) to obtain a single-image file of the entire worm.
5. Fiji software can be used for image projection and Amira software for 3D reconstruction of the fluorescent signal. For preparing a 3D model of the intestine, use the open-source 3D computer graphics software Blender.

4. DISSECTION OF INTESTINES AND VITALITY TESTING

Tissue mechanics are dependent on the functioning of all cellular components. It is therefore critical to prepare viable intestinal tissue and to assess its functionality. With comparatively little experience, viable intestines can be prepared with this simple procedure. We prefer to use L4 larvae because they can be easily identified in the dissection stereo microscope. In addition, the well-defined L4 stage is rather short (~10 h at 22 °C) and thereby improves comparability between different worms. It is important to handle the dissected intestine with utmost care to avoid local damage, which would be detrimental to mechanical measurements.

Materials

- Viable L4 larvae
- Coverslip
- Leibovitz's L-15 medium liquid with GlutaMAX™-I (Life Technologies, cat no. 31415-029) supplemented with 10% fetal calf serum (FCS; Life Technologies, cat no. 10270-106) and osmolarity adjusted to 350 ± 5 mOsm with 60% sucrose (Sigma-Aldrich, cat no. S9379) using an osmometer
- Worm pick
- Scalpel

Equipment

- Stereo microscope (e.g., BMS 141 Bino Zoom; Breukhoven B.V.)
- Osmometer (Osmomat 030, Gonotec)

Methods

1. Transfer L4 larva with the worm pick into a drop of medium on a coverslip.
2. Cut the larva at the pharynx just anterior of the posterior bulb using a sharp scalpel.
3. A large part of the intestine will be pushed out from the body cavity because of the high internal pressure. In most instances, the gonads will be squeezed out at the same time. Both remain attached to the body of the worm.
4. Worms can be stored until use (see below).

4.1 Viability Testing of the Dissected Intestine

It is known that the stressed intestine elicits a fluorescent wave from the anterior to the posterior preceding cell death (Coburn et al., 2013; Coburn & Gems, 2013). This phenomenon is explained by massive release of fluorescence from autofluorescent intestinal granules. These granules are known to be lysosome-related organelles, which contain a glycosylated form of anthranilic acid, a blue fluorophore responsible for the blue fluorescence emitted by the intestinal granules. The fluorescent wave is induced by the calcium-dependent calpain–cathepsin necrotic cell death cascade and is transmitted to adjacent posterior intestinal cells via INX-16 channels. The presence of characteristic autofluorescent granules can therefore be taken as an indication of cell viability. Figure 3 presents fluorescence recordings of dissected intestines immediately after preparation and after 2 h of storage in cell culture medium. In both situations, multiple distinct autofluorescent granules are seen in the isolated intestines at the beginning of the sequence identical to the *in situ* situation (Coburn et al., 2013; Coburn & Gems, 2013).

Upon repeated irradiation for imaging, however, granular fluorescence disappears and is substituted by a diffuse cytoplasmic fluorescence. These changes are propagated from cell to cell in an anterior-to-posterior direction as is typical for the fluorescent wave (Fig. 3). We interpret this as a stress reaction to the repeated radiation. Extended storage increases radiation sensitivity (compare Fig. 3A with B). Note, however, that appearance of the fluorescent wave is not directly linked to cell death. Instead, another kind of granular fluorescence appears within approximately 15 min. It is not known if these novel granular structures are newly emerging intestinal granules formed through processes such as the endocytosis of substances from the cell culture medium (Bossinger, Wiegner, & Schierenberg, 1996), other types of vesicles or nonvesicular aggregates. After a lag period, fluorescence is completely lost coinciding with uptake of trypan blue as a definite sign of cell death (Fig. 4). The lag period may last for more than 30 min in the freshly prepared intestine but may be as short as 5 min after storage (Fig. 3A and B). Taken together, we recommend to only use intestines with typical autofluorescent granules for mechanical measurements and to keep irradiation at a minimum.

Materials

- Dissected intestines (see previous section)
- Microscope slides
- Leibovitz's L-15 medium
- 0.4% Trypan blue solution (Sigma-Aldrich, cat no. 93595-50ML)

Equipment

- Fluorescence microscope (e.g., Zeiss ApoTome 2, Carl Zeiss)
- 20 × Objective (e.g., Plan-Apochromat 20 ×/0.8 M27, Carl Zeiss)

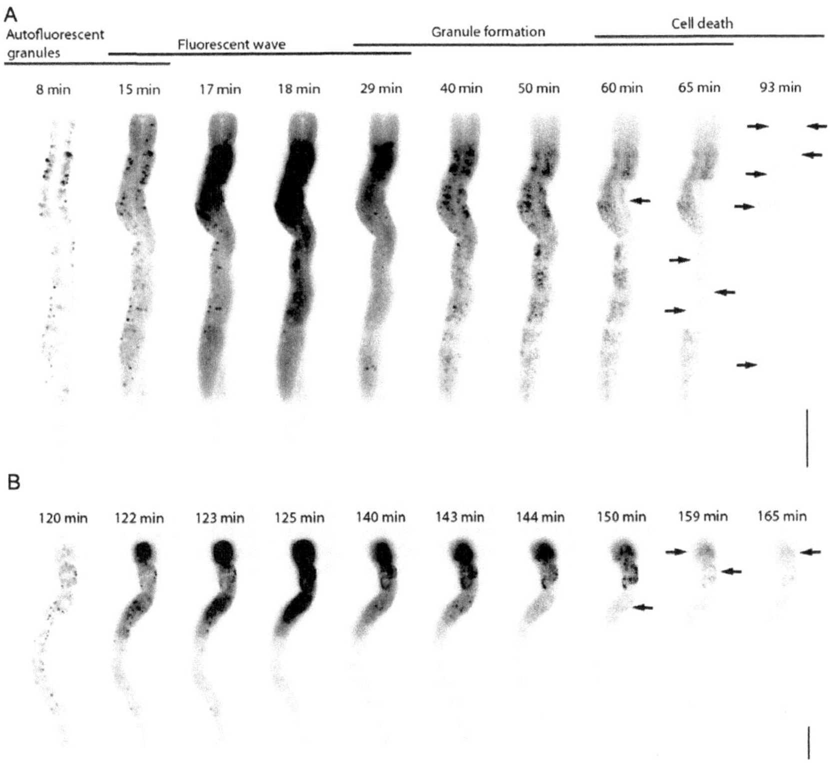

Figure 3 Induction of a fluorescent wave in dissected intestines. The images show fluorescence recordings (grayscale inverted) of an intestine starting either 8 or 120 min after dissection (A and B, respectively). The corresponding complete time series are provided in Movies 3 and 4 (http://dx.doi.org/10.1016/bs.mie.2015.08.030). Note the characteristic autofluorescent intestinal granules in the cytoplasm at the beginning of both time series. The autofluorescence of the granules disappears giving rise to increased and diffuse fluorescence. This event is propagated from cell to cell in anterior-to-posterior direction and is referred to as the fluorescent wave. It is followed by appearance of another kind of granular fluorescence (granule formation) and subsequent complete loss of fluorescence upon cell death (arrows). Note that even after 120 min of storage in cell culture medium autofluorescent granules are still clearly visible at the beginning of the recording indicating viability. Induction of the fluorescent wave and cell death, however, occur much more rapidly. Scale bars: 50 μm.

Methods

1. Transfer the dissected worms and adhering internal organs to another drop of medium onto a microscope slide.
2. Record cytoplasmic granular fluorescence using the DAPI filter set.
3. Monitor fluorescence at 30 s intervals.

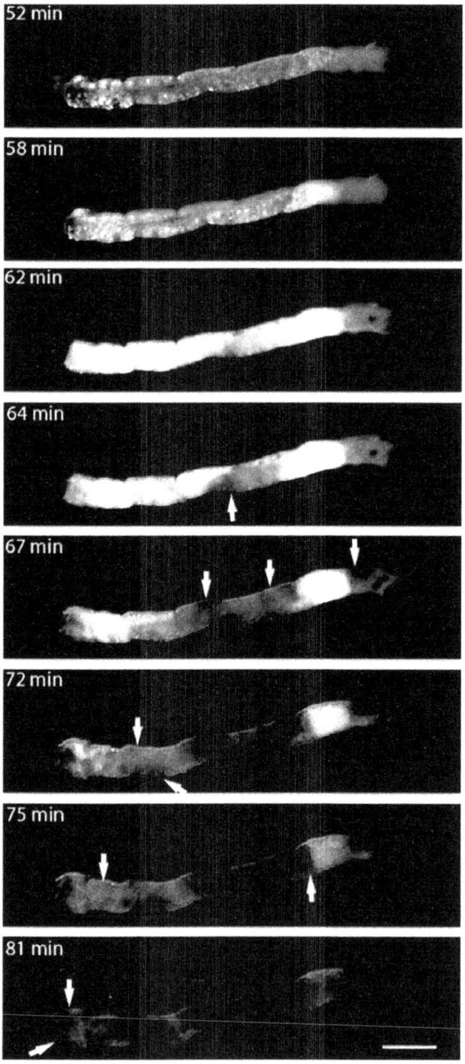

Figure 4 Monitoring cell death by trypan blue staining (dark blue) in dissected intestines. The images are overlays of the phase contrast and fluorescence micrographs in a dissected intestine. The intestine was stored for 52 min in culture medium before recording. The images are taken from a time-lapse recording that is presented in Movie 5 (http://dx.doi.org/10.1016/bs.mie.2015.08.030). Note that a fluorescent wave (white) is initiated at 58 min and that loss of fluorescence coincides with uptake of trypan blue into dying cells (arrows). Scale bar: 50 μm. (See the color plate.)

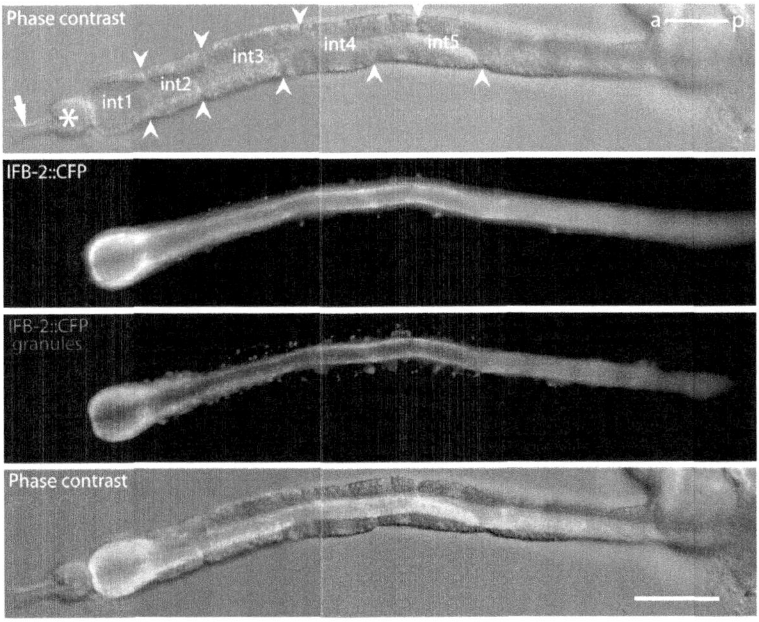

Figure 5 IFB-2::CFP fluorescence in a viable dissected intestine of reporter strain BJ49. Top: phase contrast. Arrow, pharyngeal isthmus; asterisk, terminal pharyngeal bulb; arrowheads, cell borders of intestinal cells between the intestinal rings int1, int2, int3, int4, and int5. Upper middle: corresponding fluorescence recording of IFB-2:: CFP. Lower middle: overlay of IFB-2::CFP fluorescence (cyan) and autofluorescent intestinal granules (false red color), which indicate full vitality of the dissected intestine. Bottom: overlay of phase contrast and IFB-2::CFP fluorescence. Note that the distribution of IFB-2::CFP in the dissected intestine is comparable to its distribution in a living L4 larva (Fig. 1). Scale bar: 50 μm. (See the color plate.)

4. To check for cell death, add equal volume of 0.4% trypan blue.

The above method can be used to assess vitality in dissected intestines that produce fluorescent reporters such as those shown in Figs. 1 and 2. Figure 5 shows normal-appearing intestinal granules in the dissected intestine of an IFB-2::CFP reporter strain. The fluorescent fusion protein localizes specifically to the periluminal domain as seen in the intact worm (Fig. 1).

5. EXPERIMENTAL SETUP FOR MICROPIPETTE MEASUREMENTS

Micropipettes are well-established tools for mechanical probing (Evans, 1989; Mitchison & Swann, 1954; Needham, 1993). Careful preparation of micropipettes is essential for high-precision measurements. They have to be tailored to fully attach to the isolated intestine without any leakage and without damaging the brittle cells. The setup for the dual pipette assay needs to be calibrated for each new pipette and needs to be readjusted in between measurements (Ligezowska et al., 2011).

5.1 Forging of Pipettes

Materials

- Borosilicate glass capillaries, 1 mm outer and 0.5 mm inner diameter (Hilgenberg GmbH)

Equipment

- A microforge (own fabrication) comprising (i) a heating filament made of platinum with a drop of low-melting glass solder attached, (ii) a well-regulated current source, (iii) a mechanical micromanipulator for micro-needle maneuvering, and (iv) a stereo microscope (Zhelev, Needham, & Hochmuth, 1994)
- Micropipette puller P-97 Flaming Brown (Sutter Instrument Corporation)

Methods

1. With the help of the micropipette puller, borosilicate glass capillaries are pulled into microneedles with an inner diameter of approximately 2–10 μm using the following settings: heat 551, pull 0, velocity 140, time 200, and pressure 500. Please note that parameters have to be adjusted to your specific instrumentation and lab conditions. The manufactured microneedles are very thin. Handle with care.
2. To obtain a flat and axisymmetric tip of 2–10 μm inner diameter, the microneedle is opened at the end using the microforge. Heat the drop of glass solder slightly above melting temperature. Position the microneedle near the molten drop. Immerse microneedle horizontally into the fluid drop until the tip reaches the desired inner diameter. Let the solder glass cool down. Upon retraction, the microneedle breaks at the surface of the now solid drop of glass. Adjusting temperature and judging pipette diameter during insertion require a bit of experience.
3. The tip of the resulting micropipette is smoothed by bringing it close to the heating wire of the microforge without contacting it directly.
4. Forged micropipettes can be stored in an upright position and dust-free environment until use.

5.2 Calibrating Pipette Setup

Materials

- Silica microspheres ($d = 1.53$ μm, Polysciences, Inc., cat no. 24327-15)
- Leibovitz's L-15 medium

Equipment

- Micropipette flushing needle MicroFil 28 g, 97 mm (cat no. MicroFil MF28G-5, World Precision)
- Bright field light microscope (e.g., Axiovert 200, Carl Zeiss)
- CCD camera (e.g., Sensicam QE, PCO)
- Three-axis water hydraulic fine micromanipulator (e.g., MHW-3 and MX-35A, Narishige Co.)
- A hydraulic pressure control system (own fabrication)
- Culture chamber (own fabrication; 3 mm high glass chamber with a total volume of 2 ml and openings on both sides for pipette access)
- Syringes
- Syringe filter units (Merck Millipore Ltd., cat no. SLGP033RS)
- Tubing and luers
- Pipette holders

Methods

The experimental setup described below is shown schematically in Fig. 6.

1. Install the hydraulic pressure control system consisting of an adjustable water reservoir attached to a 1.6-m long vertical track driven by a geared DC motor (for details, see Dieluweit et al., 2010) and connect the water reservoir to a pipette holder through a plastic tube. Attach the pipette holder to the three-axis hydraulic micromanipulator.
2. Connect the second pipette holder to a syringe and attach it to the second three-axis hydraulic micromanipulator.
3. Fill both systems with water and ensure that absolutely no air bubbles remain in the system.
4. Fill the micropipettes with medium using a syringe with a syringe filter using a micropipette flushing needle and mount the micropipettes onto the pipette holders. Press some fluid out of each pipette to remove remaining air.
5. Carefully insert the micropipettes laterally into the culture chamber and position them above the sample using the micromanipulators.
6. For calibrating the stretching pipette, adjust the height of the water reservoir and identify the position where no net flow in or out the stretching micropipette is observed. This is accomplished by monitoring the movement of beads that are dispersed in the medium. Even tiny air

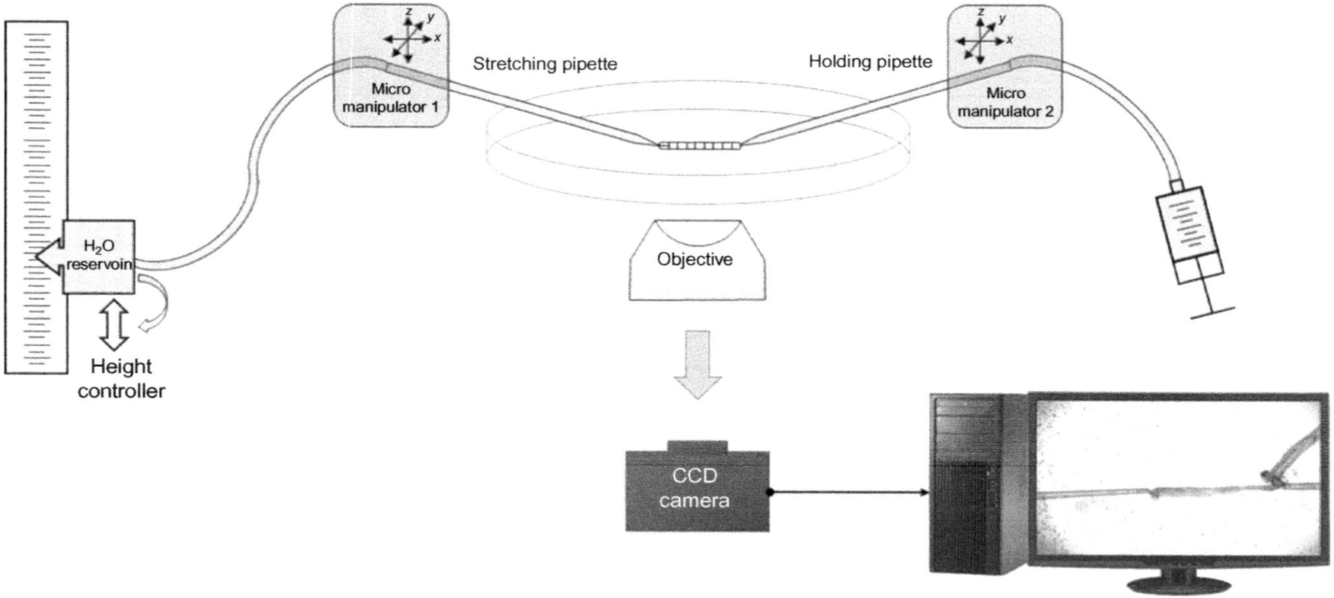

Figure 6 Scheme of experimental setup to measure viscoelasticity of dissected intestines using micropipettes. The dissected intestine is placed in a culture chamber with cell culture medium. With the help of micromanipulators, the posterior end close to the remains of the worm is attached to a holding pipette and the anterior end to a stretching pipette. The latter is connected to a height-adjustable water reservoir for precisely regulating pressure. The intestine is viewed from below with an inverted microscope. Images are recorded with a CCD camera.

bubbles in the system result in unstable zero pressure. Set this position as zero pressure position. Knowing camera pixel size, inner diameter can be calculated from phase-contrast images before experimental start.

5.3 Analysis of Intestinal Mechanics

The following paragraph describes a typical experiment to test the viscoelastic properties of the dissected intestine. Figure 7 illustrates a regimen of repeated stretching experiments at the same force and incrementally increasing forces.

Materials

- Dissected *C. elegans* intestine (section 4)
- Silica microspheres ($d = 1.53$ μm, cat no. 24327-15, Polysciences, Inc.)
- Leibovitz's L-15 medium

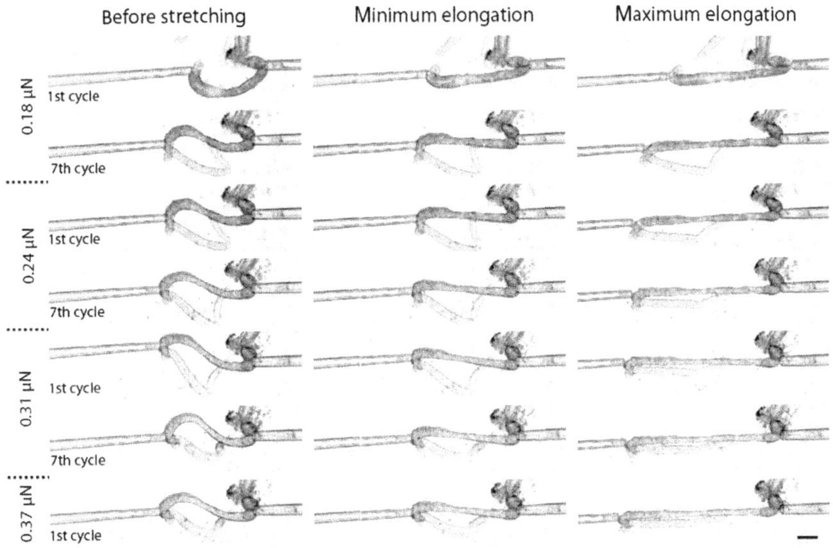

Figure 7 Recording of repeated and incrementally increasing stretching of a dissected intestine. The phase-contrast images are taken from an experiment performed on a single intestine. The stretching pipette (inner diameter 8.9 μm) is to the left, the holding pipette to the right. Parts of the attached worm can be seen next to the holding pipette, and part of the pharynx can be seen located at the left in front of the first intestinal ring. In addition, the gonads, which were extruded together with the intestine, are also visible in the background. To calibrate the suction pressure and to assess that the pipettes are tightly attached to the intestine, beads are used (black dots). Stretching resulted in elongation, which was increasing with repeated stretching and with elevating suction pressure. Note that intestinal length is reduced between cycles but does not recover completely. The entire image series is provided in Movie 6 (http://dx.doi.org/10.1016/bs.mie.2015.08.030). Scale bar: 50 μm.

Equipment

- Culture chamber (details in 5.2)
- Fluorescence microscope (Axiovert 200, Carl Zeiss)

Methods

1. Transfer dissected intestines with adhering internal organs and the remaining worm into another drop of medium in the culture chamber. Disperse a few indicator beads into the medium.
2. Mount the culture chamber onto the microscope stage and calibrate the stretching pipette (5.2). Afterward, the calibrated stretching micropipette should not be moved vertically any more.
3. Grab the dissected intestine with the holding pipette on the posterior end by aspiration and attach the stretching pipette to the anterior end. Take care of proper sealing.
4. Set the desired aspiration force of the stretching pipette by adjusting the height of the water reservoir. By calculating the difference between the adjusted position and position 0, the aspiration pressure can be calculated using Pascal's law $\mathit{\Delta} P = \rho g \mathit{\Delta} h$, where ΔP is the aspiration pressure, ρ is the fluid density, g is the earth's gravitational acceleration, and Δh is the height difference between position 0 and the adjusted position. Aspiration force F_{Asp} can be calculated according to $F_{\text{Asp}} = \pi R^2 \mathit{\Delta} P$ with R being the inner radius of the stretching pipette and ΔP the aspiration pressure.
5. Adjust the position of the holding pipette without stretching the intestine until both pipettes are aligned along the same axis within the focal plane of the microscope.
6. Start the recording of the microscopic images and stretch the intestine by moving the stretching pipette slowly away from the holding pipette.
7. Stretch the intestine until it detaches from the stretching pipette and document the whole-stretching process for additional time after detachment depending on scientific question.
8. Record epifluorescence and phase-contrast images during stretching process.

Dissected intestines of *C. elegans* strains, in which intermediate filaments are fluorescently labeled, are highly suitable to study mechanical stress-dependent behavior during micropipette-dependent stretching. Figure 8 presents a combination of fluorescence and bright field recordings of a dissected L4 intestine of strain BJ49, which expresses an IFB-2::CFP reporter. Upon stretching, the intestinal endotube shows a similar viscoelastic behavior as the stretched tissue.

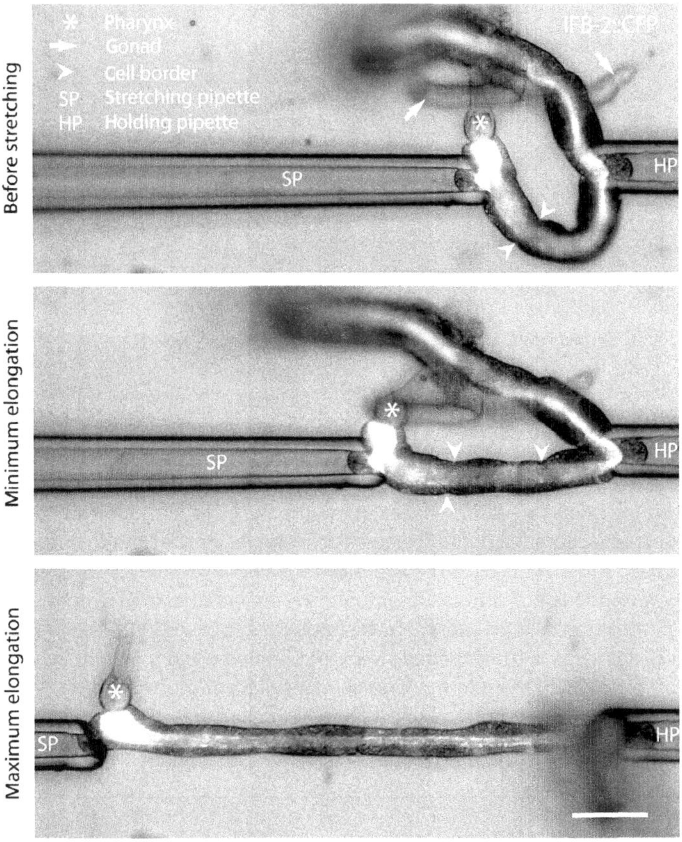

Figure 8 Coordinated stretching of the intermediate filament-rich endotube and cells of a dissected intestine. The intestine was prepared from an IFB-2::CFP reporter worm of strain BJ49. The combined fluorescence and phase-contrast images show that controlled micropipette stretching elongates the intermediate filament system in conjunction with the intestine without any visible damage. Asterisk, pharyngeal bulb; arrowheads, cell borders; SP, stretching pipette; HP, holding pipette. Scale bar: 50 μm.

To estimate the mechanical contribution of the intermediate filament-rich endotube to tissue integrity, the pharynx of a dissected intestine was cut off in order to deliberately generate a weak point (asterisk in Fig. 9). Figure 9 shows the IFB-2::CFP fluorescence during repeated stretches of the dissected intestine at high-aspiration force. The dissected intestine experienced a particularly high degree of deformation during stretching at the attachment positions of the pipettes. Cells weakened at these positions. Even when their cytoplasm was sucked into the pipettes, the intermediate filament system remained fully intact (arrow, arrowhead).

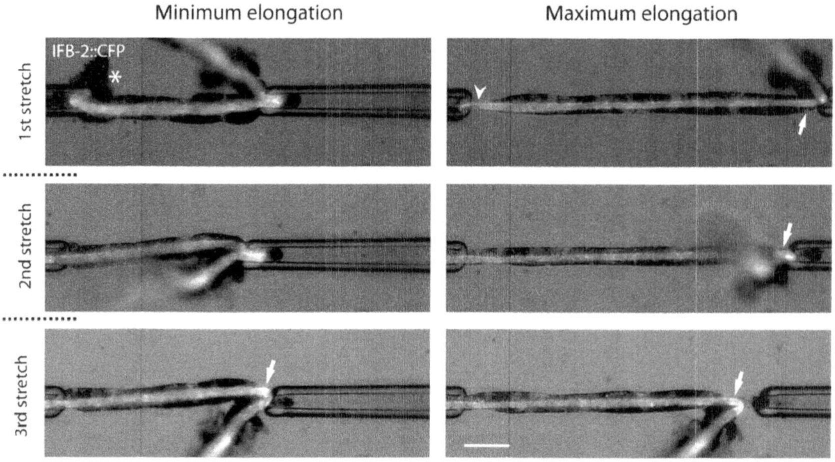

Figure 9 Maintenance of the intermediate filament-rich endotube after cell disruption. Cutting off the pharynx induces a weak point into the intestinal tissue architecture leading to substantial damage of the first intestinal ring (asterisk). Note that during the first stretch cycle cellular remains of the first intestinal ring are ripped off by the left pipette, but the endotube is not disrupted (arrowhead). At the attachment point of the right pipette parts of the cells including the endotube are aspirated into the pipette. During stretching, damages to the attached cells are observed up to the point, where they are peeled off uncovering the intermediate filament-rich endotube (arrows). Note that the endotube shows no visible signs of damage. Scale bar: 50 μm.

6. OUTLOOK

The examples presented highlight the advantages of the dissected *C. elegans* intestine as a model system for the examination of tissue mechanics and reveal its unique properties. They further highlight the extreme mechanical resilience of the *C. elegans* intermediate filament system, which withstands forces up to 1 μN and can be stretched to more than 150% without rupture even when the rest of the cell is pulled away. Our data further demonstrate that the intestine does not fully resume its original length after stretching. Taken together, we conclude that both an elastic and a plastic component determine intestinal mechanics. These observations considerably extend the report of Beriault et al. (2012), who imaged the keratin cytoskeleton in cultured cells upon extensive stretching and the work by Fois et al. (2013), Hecht et al. (2012), and Felder et al. (2008), who examined biochemical and morphological alterations of the keratin cytoskeleton during and after stretching. Our observations furthermore demonstrate that very stable and resilient cell–cell junctions control the organization of the

C. elegans intermediate filament system, since dissociation at cell–cell borders did not occur. Our setup offers an ideal system to study, how and to which degree the intermediate filament system and its associated structures contributes to the mechanical properties of the intestine. It may help to characterize and identify auxiliary and other systems that contribute to the unique mechanical resilience of the *C. elegans* intestine. The use of genetically modified worms will greatly help in this endeavor. Mutants are freely available (e.g., *Caenorhabditis* Genetics Center), or gene expression can be downregulated through RNA interference (RNAi) by feeding (using, e.g., the clones available in the Ahringer RNAi-library (Geneservice)). Alternatively, drugs can be used to selectively disrupt cytoskeletal components or signaling pathways. The experimental setup allows monitoring the localization of components of the CeAJ and the apical cytoskeleton using fluorescent reporter strains during mechanical manipulation. The described experimental system also shows promise for examining the mechanosensory response of the intestinal epithelium. Combining stretching experiments with local measurements of enzyme activity, ion concentration and pH will further expand our understanding of the complex organization and regulation of the intermediate filament cytoskeleton and to describe a biological function of the different cell adhesion complexes (Bossinger et al., 2001; Pásti & Labouesse, 2014; Pilipiuk, Lefebvre, Wiesenfahrt, Legouis, & Bossinger, 2009; Waaijers, Ramalho, Koorman, Kruse, & Boxem, 2015).

ACKNOWLEDGMENTS

This work was supported by a scholarship from the Jürgen Manchot-Stiftung to O.J. and by grants from the German Research Council to R.M. (ME 1458/6-3), O.B. (BO 1061/11-3), and R.L. (LE566 14-3).

REFERENCES

Beil, M., Micoulet, A., von Wichert, G., Paschke, S., Walther, P., Omary, M. B., et al. (2003). Sphingosylphosphorylcholine regulates keratin network architecture and visco-elastic properties of human cancer cells. *Nature Cell Biology*, *5*, 803–811.

Beriault, D. R., Haddad, O., McCuaig, J. V., Robinson, Z. J., Russell, D., Lane, E. B., et al. (2012). The mechanical behavior of mutant K14-R125P keratin bundles and networks in NEB-1 keratinocytes. *PLoS One*, *7*, e31320.

Bossinger, O., Fukushige, T., Claeys, M., Borgonie, G., & McGhee, J. D. (2004). The apical disposition of the Caenorhabditis elegans intestinal terminal web is maintained by LET-413. *Developmental Biology*, *268*, 448–456.

Bossinger, O., Klebes, A., Segbert, C., Theres, C., & Knust, E. (2001). Zonula adherens formation in Caenorhabditis elegans requires dlg-1, the homologue of the Drosophila gene discs large. *Developmental Biology*, *230*, 29–42.

Bossinger, O., Wiegner, O., & Schierenberg, E. (1996). Embryonic gut differentiation in nematodes: Endocytosis of macromolecules and its experimental inhibition. *Roux's Archives of Developmental Biology*, *205*, 494–497.

Brown, M. J., Hallam, J. A., Colucci-Guyon, E., & Shaw, S. (2001). Rigidity of circulating lymphocytes is primarily conferred by vimentin intermediate filaments. *The Journal of Immunology*, *166*, 6640–6646.

Carberry, K., Wiesenfahrt, T., Geisler, F., Stöcker, S., Gerhardus, H., Überbach, D., et al. (2012). The novel intestinal filament organizer IFO-1 contributes to epithelial integrity in concert with ERM-1 and DLG-1. *Development*, *139*, 1851–1862.

Carberry, K., Wiesenfahrt, T., Windoffer, R., Bossinger, O., & Leube, R. E. (2009). Intermediate filaments in Caenorhabditis elegans. *Cell Motility and the Cytoskeleton*, *66*, 852–864.

Coburn, C., Allman, E., Mahanti, P., Benedetto, A., Cabreiro, F., Pincus, Z., et al. (2013). Anthranilate fluorescence marks a calcium-propagated necrotic wave that promotes organismal death in C. elegans. *PLoS Biology*, *11*, e1001613.

Coburn, C., & Gems, D. (2013). The mysterious case of the C. elegans gut granule: Death fluorescence, anthranilic acid and the kynurenine pathway. *Frontiers in Genetics*, *4*, 151.

Dieluweit, S., Csiszár, A., Rubner, W., Fleischhauer, J., Houben, S., & Merkel, R. (2010). Mechanical properties of bare and protein-coated giant unilamellar phospholipid vesicles. A comparative study of micropipet aspiration and atomic force microscopy. *Langmuir*, *26*, 11041–11049.

Dodemont, H., Riemer, D., Ledger, N., & Weber, K. (1994). Eight genes and alternative RNA processing pathways generate an unexpectedly large diversity of cytoplasmic intermediate filament proteins in the nematode Caenorhabditis elegans. *The EMBO Journal*, *13*, 2625.

Eckes, B., Dogic, D., Colucci-Guyon, E., Wang, N., Maniotis, A., Ingber, D., et al. (1998). Impaired mechanical stability, migration and contractile capacity in vimentin-deficient fibroblasts. *Journal of Cell Science*, *111*, 1897–1907.

Evans, E. A. (1989). Structure and deformation properties of red blood cells: Concepts and quantitative methods. *Methods in Enzymology*, *173*, 3–35.

Felder, E., Siebenbrunner, M., Busch, T., Fois, G., Miklavc, P., Walther, P., et al. (2008). Mechanical strain of alveolar type II cells in culture: Changes in the transcellular cytokeratin network and adaptations. *American Journal of Physiology. Lung Cellular and Molecular Physiology*, *295*, L849–L857.

Firestein, B. L., & Rongo, C. (2001). DLG-1 is a MAGUK similar to SAP97 and is required for adherens junction formation. *Molecular Biology of the Cell*, *12*(11), 3465–3475.

Fois, G., Weimer, M., Busch, T., Felder, E. T., Oswald, F., von Wichert, G., et al. (2013). Effects of keratin phosphorylation on the mechanical properties of keratin filaments in living cells. *The FASEB Journal*, *27*, 1322–1329.

Fudge, D., Russell, D., Beriault, D., Moore, W., Lane, E. B., & Vogl, A. W. (2008). The intermediate filament network in cultured human keratinocytes is remarkably extensible and resilient. *PLoS One*, *3*, e2327.

Gruenbaum, Y., Lee, K. K., Liu, J., Cohen, M., & Wilson, K. L. (2002). The expression, lamin-dependent localization and RNAi depletion phenotype for emerin in C. elegans. *Journal of Cell Science*, *115*, 923–929.

Hecht, E., Knittel, P., Felder, E., Dietl, P., Mizaikoff, B., & Kranz, C. (2012). Combining atomic force-fluorescence microscopy with a stretching device for analyzing mechanotransduction processes in living cells. *The Analyst*, *137*, 5208.

Hermann, G. J., Leung, B., & Priess, J. R. (2000). Left-right asymmetry in C. elegans intestine organogenesis involves a LIN-12/Notch signaling pathway. *Development*, *127*, 3429–3440.

Homberg, M., & Magin, T. M. (2014). Beyond expectations: Novel insights into epidermal keratin function and regulation. In K. W. Jeon (Ed.), *International review of cell and molecular biology: Vol. 311* (pp. 265–306): San-Diego: Academic Press. http://dx.doi.org/10.1016/b978-0-12-800179-0.00007-6.

Hüsken, K., Wiesenfahrt, T., Abraham, C., Windoffer, R., Bossinger, O., & Leube, R. E. (2008). Maintenance of the intestinal tube in Caenorhabditis elegans: The role of the intermediate filament protein IFC-2. *Differentiation, 76*, 881–896.

Janmey, P. A., Euteneuer, U., Traub, P., & Schliwa, M. (1991). Viscoelastic properties of vimentin compared with other filamentous biopolymer networks. *The Journal of Cell Biology, 113*, 155–160.

Karabinos, A., Schmidt, H., Harborth, J., Schnabel, R., & Weber, K. (2001). Essential roles for four cytoplasmic intermediate filament proteins in Caenorhabditis elegans development. *Proceedings of the National Academy of Sciences of the United States of America, 98*, 7863–7868.

Kreplak, L., Bär, H., Leterrier, J. F., Herrmann, H., & Aebi, U. (2005). Exploring the mechanical behavior of single intermediate filaments. *Journal of Molecular Biology, 354*, 569–577.

Kröger, C., Loschke, F., Schwarz, N., Windoffer, R., Leube, R. E., & Magin, T. M. (2013). Keratins control intercellular adhesion involving PKC-α-mediated desmoplakin phosphorylation. *The Journal of Cell Biology, 201*, 681–692.

Leung, B., Hermann, G. J., & Priess, J. R. (1999). Organogenesis of the Caenorhabditis elegans intestine. *Developmental Biology, 216*, 114–134.

Ligezowska, A., Boye, K., Eble, J. A., Hoffmann, B., Klösgen, B., & Merkel, R. (2011). Mechanically enforced bond dissociation reports synergistic influence of Mn2+ and Mg2+ on the interaction between integrin α7β1 and invasin. *Journal of Molecular Recognition, 24*, 715–723.

Lin, Y.-C., Yao, N. Y., Broedersz, C. P., Herrmann, H., MacKintosh, F. C., & Weitz, D. A. (2010). Origins of elasticity in intermediate filament networks. *Physical Review Letters, 104*, 058101. http://dx.doi.org/10.1103/PhysRevLett.104.058101.

Liovic, M., Lee, B., Tomic-Canic, M., D'Alessandro, M., Bolshakov, V. N., & Lane, E. B. (2008). Dual-specificity phosphatases in the hypo-osmotic stress response of keratin-defective epithelial cell lines. *Experimental Cell Research, 314*, 2066–2075.

Liu, J., Ben-Shahar, T. R., Riemer, D., Treinin, M., Spann, P., Weber, K., et al. (2000). Essential roles for Caenorhabditis elegans lamin gene in nuclear organization, cell cycle progression, and spatial organization of nuclear pore complexes. *Molecular Biology of the Cell, 11*, 3937–3947.

Lulevich, V., Yang, H., Rivkah Isseroff, R., & Liu, G. (2010). Single cell mechanics of keratinocyte cells. *Ultramicroscopy, 110*, 1435–1442.

McGhee, J. D. (2013). The Caenorhabditis elegans intestine. *Wiley Interdisciplinary Reviews: Developmental Biology, 2*, 347–367.

McMahon, L., Legouis, R., Vonesch, J.-L., & Labouesse, M. (2001). Assembly of C. elegans apical junctions involves positioning and compaction by LET-413 and protein aggregation by the MAGUK protein DLG-1. *Journal of Cell Science, 114*, 2265–2277.

Mitchison, J. M., & Swann, M. M. (1954). The mechanical properties of the cell surface I. The cell elastimeter. *Journal of Experimental Biology, 31*, 443–460.

Munn, E. A., & Greenwood, C. A. (1984). The occurrence of submicrovillar endotube (modified termina web) and associated cytoskeletal structures in the intestinal epithelia of nematodes. *Philosophical Transactions of the Royal Society of London. Series B, Biological Sciences, 306*, 1–18.

Needham, D. (1993). Measurement of interbilayer adhesion energies. In N. Duzgunees (Ed.), *Methods in enzymology: Vol. 220* (pp. 111–129): Oxford: Academic Press. http://dx.doi.org/10.1016/0076-6879(93)20078-h.

Pásti, G., & Labouesse, M. (2014). Epithelial junctions, cytoskeleton, and polarity. *WormBook: The Online Review of C. elegans Biology, 4*, 1–35.

Paust, T., Paschke, S., Beil, M., & Marti, O. (2013). Microrheology of keratin networks in cancer cells. *Physical Biology*, *10*, 065008.

Pilipiuk, J., Lefebvre, C., Wiesenfahrt, T., Legouis, R., & Bossinger, O. (2009). Increased IP3/Ca2+ signaling compensates depletion of LET-413/DLG-1 in C. elegans epithelial junction assembly. *Developmental Biology*, *327*, 34–47.

Plodinec, M., Loparic, M., Suetterlin, R., Herrmann, H., Aebi, U., & Schoenenberger, C.-A. (2011). The nanomechanical properties of rat fibroblasts are modulated by interfering with the vimentin intermediate filament system. *Journal of Structural Biology*, *174*, 476–484.

Ramms, L., Fabris, G., Windoffer, R., Schwarz, N., Springer, R., Zhou, C., et al. (2013). Keratins as the main component for the mechanical integrity of keratinocytes. *Proceedings of the National Academy of Sciences of the United States of America*, *110*, 18513–18518.

Rolli, C. G., Seufferlein, T., Kemkemer, R., & Spatz, J. P. (2010). Impact of tumor cell cytoskeleton organization on invasiveness and migration: A microchannel-based approach. *PLoS One*, *5*, e8726.

Russell, D., Ross, H., & Lane, E. B. (2010). ERK involvement in resistance to apoptosis in keratinocytes with mutant keratin. *Journal of Investigative Dermatology*, *130*, 671–681.

Seltmann, K., Cheng, F., Wiche, G., Eriksson, J. E., & Magin, T. M. (2015). Keratins stabilize hemidesmosomes through regulation of β4-integrin turnover. *Journal of Investigative Dermatology*, *135*, 1609–1620. http://dx.doi.org/10.1038/jid.2015.46.

Seltmann, K., Fritsch, A. W., Kas, J. A., & Magin, T. M. (2013). Keratins significantly contribute to cell stiffness and impact invasive behavior. *Proceedings of the National Academy of Sciences of the United States of America*, *110*, 18507–18512.

Seltmann, K., Roth, W., Kröger, C., Loschke, F., Lederer, M., Hüttelmaier, S., et al. (2013). Keratins mediate localization of hemidesmosomes and repress cell motility. *Journal of Investigative Dermatology*, *133*, 181–190.

Sivaramakrishnan, S., DeGiulio, J. V., Lorand, L., Goldman, R. D., & Ridge, K. M. (2008). Micromechanical properties of keratin intermediate filament networks. *Proceedings of the National Academy of Sciences of the United States of America*, *105*, 889–894.

Sulston, J. E., Schierenberg, E., White, J. G., & Thomson, J. N. (1983). The embryonic cell lineage of the nematode Caenorhabditis elegans. *Developmental Biology*, *100*, 64–119.

Waaijers, S., Ramalho, J. J., Koorman, T., Kruse, E., & Boxem, M. (2015). The C. elegans Crumbs family contains a CRB3 homolog and is not essential for viability. *Biology Open*, *4*, 276–284.

Walter, N., Busch, T., Seufferlein, T., & Spatz, J. P. (2011). Elastic moduli of living epithelial pancreatic cancer cells and their skeletonized keratin intermediate filament network. *Biointerphases*, *6*, 79–85.

Wang, N., & Stamenović, D. (2000). Contribution of intermediate filaments to cell stiffness, stiffening, and growth. *American Journal of Physiology. Cell Physiology*, *279*, C188–C194.

Zhang, H., Landmann, F., Zahreddine, H., Rodriguez, D., Koch, M., & Labouesse, M. (2011). A tension-induced mechanotransduction pathway promotes epithelial morphogenesis. *Nature*, *471*, 99–103.

Zhelev, D. V., Needham, D., & Hochmuth, R. M. (1994). A novel micropipet method for measuring the bending modulus of vesicle membranes. *Biophysical Journal*, *67*, 720–727.

CHAPTER TWENTY-SIX

Using *Drosophila* for Studies of Intermediate Filaments

Jens Bohnekamp*, Diane E. Cryderman†, Dylan A. Thiemann†, Thomas M. Magin*[,1], Lori L. Wallrath†[,1]

*Institute of Biology and Translational Center for Regenerative Medicine, University of Leipzig, Leipzig, Germany
†Department of Biochemistry, University of Iowa, Iowa City, Iowa, USA
[1]Corresponding authors: e-mail address: thomas.magin@uni-leipzig.de; lori-wallrath@uiowa.edu

Contents

1. Introduction 708
2. Methods 711
   2.1 Generation of Stocks Expressing Intermediate Filaments 711
   2.2 Assay for Overt Phenotypes and Effects on Viability 714
   2.3 Measure Levels of Intermediate Filament Expression 717
   2.4 Cytological Analysis of Intermediate Filament Formation 718
   2.5 Live Cell Imaging of Dissected Tissues 722
3. Conclusions 723
References 723

Abstract

Drosophila melanogaster is a useful organism for determining protein function and modeling human disease. *Drosophila* offers a rapid generation time and an abundance of genomic resources and genetic tools. Conservation in protein structure, signaling pathways, and developmental processes make studies performed in *Drosophila* relevant to other species, including humans. *Drosophila* models have been generated for neurodegenerative diseases, muscular dystrophy, cancer, and many other disorders. Recently, intermediate filament protein diseases have been modeled in *Drosophila*. These models have revealed novel mechanisms of pathology, illuminated potential new routes of therapy, and make whole organism compound screens feasible. The goal of this chapter is to outline steps to study intermediate filament function and model intermediate filament-associated diseases in *Drosophila*. The steps are general and can be applied to study the function of almost any protein. The protocols outlined here are for both the novice and experienced *Drosophila* researcher, allowing the rich developmental and cell biology that *Drosophila* offers to be applied to studies of intermediate filaments.

Methods in Enzymology, Volume 568
ISSN 0076-6879
http://dx.doi.org/10.1016/bs.mie.2015.08.028

ABBREVIATIONS

DGRC *Drosophila* Genomics Resource Center
DSHB Developmental Studies Hybridoma Bank
EBS epidermolysis bullosa simplex
HSE heat shock element
IF intermediate filament
IRP inverted repeats
K14 keratin 14
K5 keratin 5
UAS upstream activating sequence
wt wild type

1. INTRODUCTION

For over 100 years, *Drosophila melanogaster* has been a premier model organism, providing valuable insights on molecular processes that occur during development (Kohler, 1994). Recent advances in genetic and genomic tools have made it easier than ever before to identify the function of a particular protein using *Drosophila*. Mutant phenotypes associated with loss of a wild-type protein and/or the expression of a mutant version of a protein typically reflect defects in specific signaling pathways and developmental processes. Thus, relatively rapid experimentation can be performed to identify the function of a protein.

Two general strategies have been used for modeling human disease in *Drosophila* (Bouleau & Tricoire, 2015; Gonzalez, 2013; Martinez-Morentin et al., 2015; Okray et al., 2015; Somers, Nguyen, Lumb, Batterham, & Perry, 2015; Yu & Bonini, 2011). The first strategy required the identification of a *Drosophila* orthologue to the human disease gene using BLAST (http://blast.ncbi.nlm.nih.gov/Blast.cgi) and similar bioinformatics programs. If the cross-species gene relationship is unclear, the human homologue can be tested for rescue of the *Drosophila* mutant, provided there is a phenotype associated with the mutant (Callaerts et al., 1999). If an orthologue is identified, both null alleles and designer mutant alleles can be made via the CRISPR/Cas9 system (Bassett & Liu, 2014a, 2014b; Beumer & Carroll, 2014; Sebo, Lee, Peng, & Guo, 2014; Xu et al., 2015). Alternatively, the human disease mutations can be modeled into a transgene and expressed in an otherwise wild-type background or in a genetic background carrying a null allele for the gene of interest. A second strategy is used when an orthologue of the human disease gene does not exist in *Drosophila* (Bilen & Bonini,

2007; Bohnekamp et al., 2015; Warrick et al., 2005). Transgenes encoding wild-type and mutant versions of the human disease gene are expressed in *Drosophila*, either globally or in specific tissues. Then, the organism is analyzed for mutant phenotypes. Due to the tremendous history of *Drosophila* research, the resulting phenotypes often provide insights on the biological pathways that are dysregulated. Both of these strategies have been used to generate *Drosophila* models of intermediate filament-associated diseases (Bohnekamp et al., 2015; Dialynas et al., 2012, 2015; Dialynas, Speese, Budnik, Geyer, & Wallrath, 2010; Schulze et al., 2005).

Many eukaryotes possess both nuclear and cytoplasmic intermediate filaments (Fig. 1A). Nuclear intermediate filaments, called lamins, are thought to be the most ancient intermediate filament (Peter & Stick, 2012, 2015) and contain a nuclear localization sequence that directs localization to the nucleus. Genes encoding cytoplasmic intermediate filaments are highly diversified by gene duplication mechanisms, such that the human genome contains 70 different intermediate filament genes that are expressed in tissue-specific and developmentally regulated patterns, with *keratin* genes forming the largest class (Homberg & Magin, 2014).

As intermediate filaments, lamins and keratins share a similar protein domain organization, composed of a head, alpha-helical rod, and tail domain (Fig. 1B). Both proteins dimerize and polymerize into filaments that form networks in their respective cellular compartments (Fig. 1A) (Fuchs & Weber, 1994). Lamins form a nuclear network that provides structural support for the nucleus and organizes the genome by making contacts with chromatin (Wilson & Berk, 2010). Keratins form a cytoplasmic network that provides mechanical strength to cells and participates in the regulation of growth, inflammation, and organelle function, by local regulation of protein activities and interaction with cell adhesion complexes (Coulombe, Kerns, & Fuchs, 2009; Homberg & Magin, 2014; Loschke, Seltmann, Bouameur, & Magin, 2015; Omary, Ku, Strnad, & Hanada, 2009).

Mutations in ~20 of the 54 mammalian intermediate filament genes give rise to tissue-specific disorders, compromising the integrity of skin, muscle (heart and skeletal), connective, and neuronal tissues (http://www.interfil.org/) (Szeverenyi et al., 2008). The large number of cytoplasmic intermediate filament proteins, the limited knowledge of their associated proteins, and the complexity of disease phenotypes represent considerable obstacles for understanding mechanisms of pathogenesis. The use of *Drosophila* to study the function of intermediate filaments provides a fresh approach to overcome these obstacles. The fact that genomic and biochemical studies

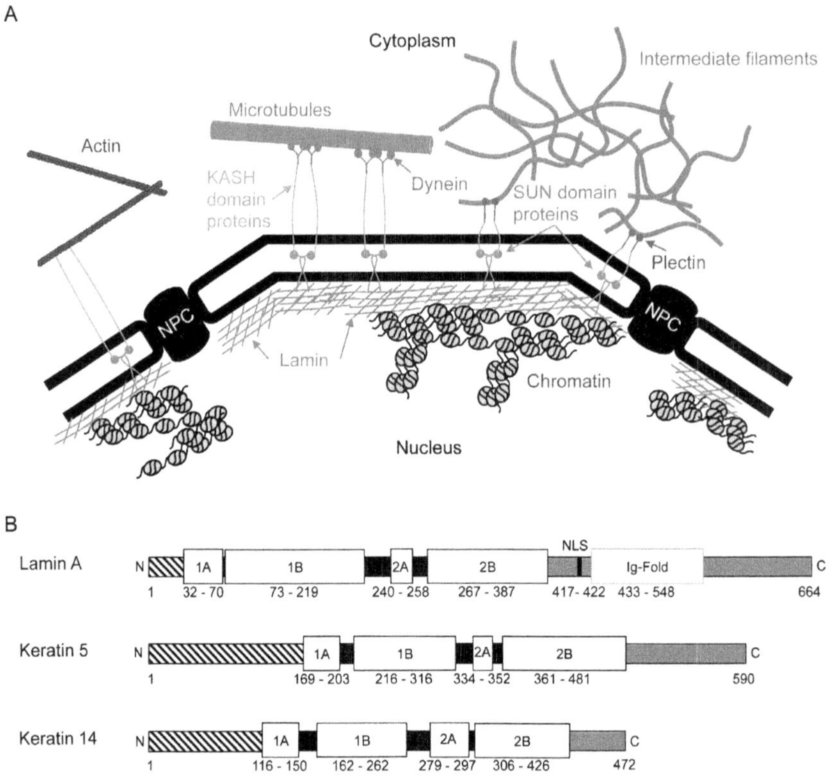

Figure 1 Cellular localization and protein domain structure of intermediate filaments. (A) Diagram of a cell showing intermediate filaments and their connections to other proteins. Lamins, the nuclear intermediate filaments, organize the chromatin for proper gene expression. In addition, lamins interact with other nuclear envelope proteins such as SUN domain proteins, components of the LINC complex (Kim, Birendra, & Roux, 2015), which bridge the nucleoskeleton and the cytoskeleton. Cytoplasmic intermediate filaments, such as keratins, form networks in the cytoplasm that provide structural integrity to tissues. (B) The domain structure of human lamin A and keratin 5 and 14. These proteins possess a domain structure common to intermediate filaments that is composed of a small head domain (diagonal stripes), a coil-coil rod domain (black with white boxes) and a tail domain (green). Lamins possess a nuclear localization sequence (NLS) and an Ig-fold domain, which differentiates them from keratins (Peter & Stick, 2015). The length of each domain is indicated according to the Human Intermediate Filament Database (http:/www.interfil.org). (See the color plate.)

have revealed the presence of cytoplasmic intermediate filaments in all bilaterians excluding arthropods offers the unique opportunity to exploit expression of mammalian intermediate filaments in *Drosophila* to study their regulation and function in an unbiased manner.

The benefit of studying intermediate filaments in *Drosophila* is apparent. Mutations in the two *Drosophila* genes encoding lamins, *lamin Dmo*, and *Lamin C*, revealed their essential nature and produced phenotypes that shed light on their functions (Lyakhovetsky & Gruenbaum, 2014). In addition, mutations in the human *LMNA* gene, encoding A-type lamins, have been modeled in the *Drosophila Lamin C* gene and expressed in muscle (Dialynas et al., 2012, 2015, 2010). Consequently, the flies developed muscular dystrophy. Analysis of the defective muscles revealed that the Nrf2/Keap-1 antioxidant response pathway was activated by mutant lamins. Immunohistochemical staining of human muscle biopsy tissues expressing the analogous mutation showed activation of the Nrf2/Keap-1 pathway, validating the *Drosophila* findings and suggesting new targets for therapy (Dialynas et al., 2012, 2015). In addition, transgenes encoding mutations in the human *keratin 14 (K14)* gene, which cause the blistering skin disorder epidermolysis bullosa simplex (EBS), were globally expressed in combination with *a keratin 5 (K5)* transgene, encoding a dimerization partner. Expression resulted in semi-lethality with rare adult survivors possessing a blistered wing phenotype (Bohnekamp et al., 2015). Analysis of *Drosophila* tissues revealed a collapse of the keratin network into cytoplasmic protein aggregates and tissue fragility, similar to that observed for the human disease condition (Coulombe et al., 2009). Thus, *Drosophila* is a proven model for studies of intermediate filaments. The following sections include methods for generating and analyzing *Drosophila* stocks that express intermediate filaments.

2. METHODS

2.1 Generation of Stocks Expressing Intermediate Filaments

The life cycle of *D. melanogaster* and methods for culturing have been thoroughly described elsewhere (Ashburner, Golic, & Hawley, 2005). In short, embryos are laid in the media at the bottom of a culture vessel, develop into larvae that crawl up the sides of the culture vessel, pupate on the walls of the vessel, and emerge from the pupal cases as pharate adults. The entire life cycle takes ~11 days when cultured at the standard growth conditions of 25 °C and 65% humidity.

Here, we describe methods to generate transgenic stocks expressing intermediate filaments. These stocks are valuable to understand mechanisms of network assembly and to identify molecular defects associated with mutant versions of intermediate filaments, particularly those associated with human disease.

Stock generation

1. A cDNA encoding the intermediate filament protein is cloned into a *D. melanogaster* P-element expression vector. There are many expression vectors available through the Drosophila Genomics Resource Center (DGRC, https://dgrc.cgb.indiana.edu/Home). A commonly used vector is pUAST that contains five binding sites for the yeast Gal4 transcription factor (Fig. 2A, top) (Duffy, 2002). The resulting transgenic flies, called "responders," are crossed to "expresser" flies, which express the yeast Gal4 transcription factor in a temporal and/or tissue-specific pattern (Fig. 2B, top). The progeny will express the transgene in the same pattern as the Gal4 transcription factor (Fig. 2B, top). A collection of hundreds of characterized Gal4 driver stocks that express Gal4 in one or multiple tissues are available at the Bloomington Stock Center (http://flystocks.bio.indiana.edu/). Another frequently used vector is pCaSpeR (Fig. 2A, bottom; Pirrotta, 1988). This vector possesses a heat shock promoter, which allows for rapid induction of expression following heat shock treatment (Fig. 2B, bottom). Typically, one 45 min treatment at 37 °C is sufficient to express the exogenous protein. The frequency and duration of heat shock is often empirically derived to achieve the desired level of expression. It is common for heat shock treatments to be delivered once or twice per day to maintain expression throughout development. If it is desirable to tag the protein of interest at the amino or carboxy terminus, the DGRC has transformation vectors possessing a variety of tags such as eGFP, eCFP, Venus, 3X HA, 6X Myc, and 3X FLAG. Tagged versions of proteins can be expressed from the *hsp70* promoter (inducible expression) or the *actin 5C* promoter (constitutive expression). Most transformation vectors possess a *white*$^+$ gene as a selectable marker (Fig. 2A) and a *white*$^-$ stock is used as a host. The *white*$^+$ gene encodes a transporter protein that allows pigment to be deposited in the eye, allowing for rapid selection of transformants based on eye color.
2. The recombinant transformation vector is used for germline transformation (Rubin & Spradling, 1982). Detailed protocols for injecting embryo to germline transformants are available (Hackmann, Joedicke, Panneels, & Sinning, 2015; Spradling, 1986). Alternatively, the transformation vector can be submitted to a company such as BestGene (http://thebestgene.com/), Tefor Infrastructure (http://fly-facility.com/), The Transgenic Drosophila Fly Core at Brigham and Women's Hospital

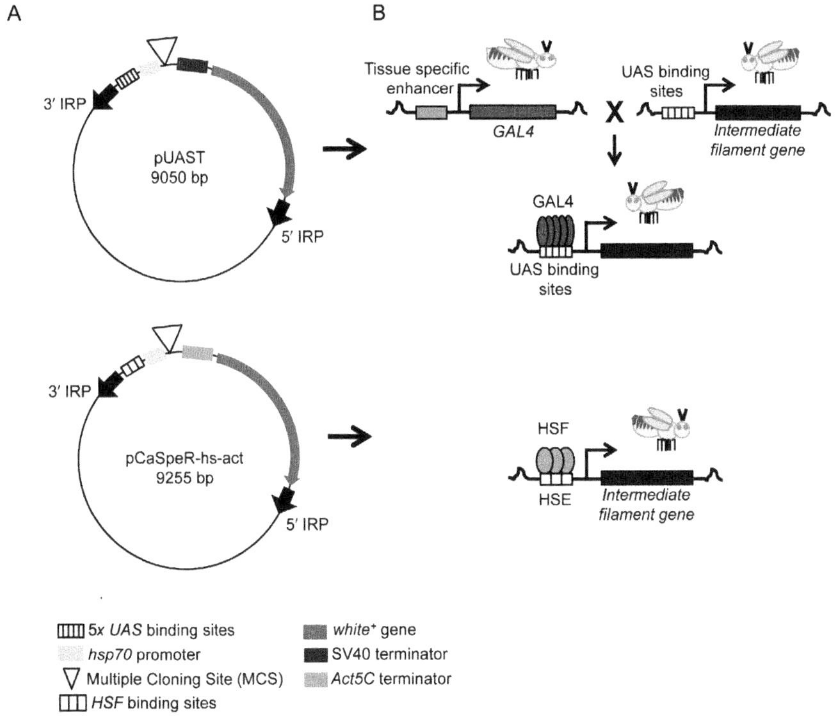

Figure 2 Diagram of transformation vectors and corresponding transgenes. (A) *Drosophila* transformation vectors commonly used for ectopically expressing a protein of interest. The pUAST vector contains a 5x upstream activating sequence (UAS) that binds Gal4 and a minimal *hsp70* promoter that lacks heat shock elements (HSEs) (top). The pCas-hs-act possesses HSEs that bind heat shock factor (HSF), resulting in transcriptional activation under stress conditions. Growth at different temperatures (18–29 °C) allows for "rheostat" control of gene expression, with more expression at higher temperature (bottom). Both vectors contain a *white*$^{+}$ gene as a selectable eye color marker following injection into a *white*$^{-}$ host strain. In addition, both vectors possess inverted repeats (IRPs) that allow for transposition into the genome in the presence of transposase, which is generated from a coinjected plasmid lacking IRPs. (B) Diagram of transgenes used to ectopically express an intermediate filament or other gene of interest. Transformants generated using the pUAST vector, called "responders," are crossed to flies from stocks called "drivers." Each driver stock expresses the Gal4 transcription factor from a tissue-specific enhancer (top). The resulting progeny express the desired protein in the given tissue. Hundreds of driver stocks are available through the Bloomington Drosophila Stock Center.

(http://www.partners.org/researchcores/researchcores_xtra%20folder/transgenic/drosophila_MGH.asp), and UniHua (http://www.unihuaii.com/English/), which will perform embryo injections. Briefly, the transformation vector and a helper plasmid that encodes transposase are coinjected into the posterior ends of developing embryos (the site of germ cell formation). Since P-element transposition is limited to the germline, adults that arise from the injected embryos do not express the reporter gene in somatic tissues. Therefore, the resulting adults are backcrossed to the host stock and the progeny are screened for expression of the reporter gene. Using the vectors shown above (Fig. 2A), a white-eyed host stock (*w*−) is used for injection; transformants are selected based on yellow to dark red eye phenotypes. The color variation correlates with the level of expression of the *white*$^{+}$ gene, which is influenced by the chromatin state at the genomic insertion site. Investigators can elect to set up the crosses and screen for transformants or this can be provided through commercial sources listed above.

Female *Drosophila* store sperm from prior matings; therefore, all crosses must be performed with "virgin" females (Ashburner et al., 2005). While the sexual history of the females is not directly ascertained, a young female (within 8–10 h after emerging from the pupal case) is likely to be a virgin. These young females have specific phenotypes that allow them to be easily scored and collected for deliberate matings (Ashburner et al., 2005).

3. P-element insertion is fairly random throughout the genome and typically produces only a single insertion of the transgene (Liao, Rehm, & Rubin, 2000). Once transformants have been identified, they are used in a series of crosses to generate homozygous transgenes (Fig. 3). If the P-element insertion did not disrupt an essential gene, a homozygous stock is generated and remains stable indefinitely. If the P-element insert is homozygous lethal, then the transgene can be maintained by placing the P-element bearing chromosome over a balancer chromosome (e.g., *TM3*, Fig. 3), which contains multiple inverted regions that suppress recombination (Ashburner et al., 2005; Greenspan, 2004).

2.2 Assay for Overt Phenotypes and Effects on Viability

Once a stable transgenic stock has been generated, it can be used for expression analysis. In the case of the pUAST vector, the transgenic stock is crossed to Gal4 "driver" stocks to achieve different temporal and spatial patterns of intermediate filament gene expression. In the case of the pCaSperR vector, the transgenic stock is given heat shock treatments to induce intermediate filament formation. Following expression, the investigator can examine the flies for overt phenotypes. Alterations in body patterning and eye and wing morphology are easy to score and can be examined using the light

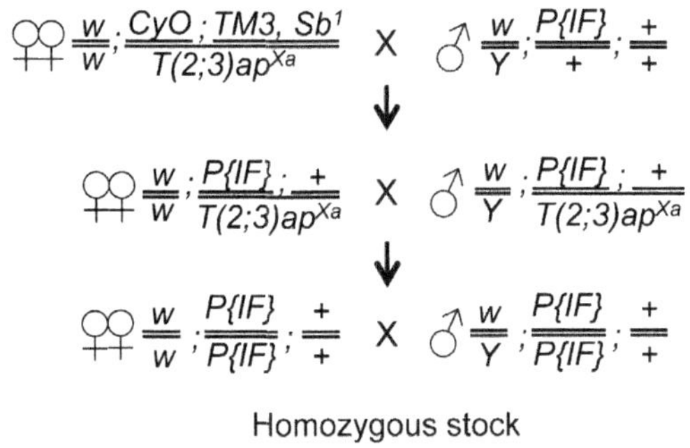

Figure 3 Genetic crossing scheme to generate flies homozygous for the intermediate filament (IF) transgene. Virgin females possessing a compound second and third chromosome marked with *ap*^{Xa} (jagged wing phenotype), *T(2;3)ap*^{Xa}, over a second chromosome balancer chromosome marked with *CyO* (curly wing phenotype), and a third chromosome balancer named *TM3* that is marked with *Sb*^{1} (thick bristle phenotype) are crossed to males possessing an IF transgene. Resulting siblings of the designated genotype are crossed to each other. Progeny that are homozygous for the IF transgene are crossed to each other to make a stable stock.

microscope. Physical abnormalities are compared to those in the literature, often providing clues to the developmental pathways disrupted.

To test for effects of intermediate filament gene expression on viability, the transgene is placed over a marked balancer chromosome, preferably one that can be scored at the larval stage, and the resulting stock is crossed to the desired Gal4 driver stock. The numbers of progeny with and without the balancer chromosome are scored. There are numerous balancers that carry transgenes that encode fluorescent and/or visible markers available at the Bloomington Stock Center (http://flystocks.bio.indiana.edu/Browse/balancers/balancer_main.htm) that can be used for this purpose. If there are no effects on viability, balancer and nonbalancer chromosome possessing flies should be of equal numbers. If there are effects on viability, balancer possessing flies should be greater in number than those without a balancer.

Testing for effects on viability

1. Set up the cross with the flies of interest at 25 °C. Allow flies to mate freely for 2 days, then allow adults to lay embryos for 3 h on apple juice agar plates (4.5 g agar, 2.5 g sucrose in 200 ml apple juice).
2. After 24 h, select the first instar larvae possessing either the balancer or the nonbalancer chromosome, place in separate 350 mm Petri dishes containing Whatman filter paper moistened with water and mashed cornmeal/sucrose media, and allow to develop at 25 °C.

3. The number of individuals at each developmental stage is counted daily. Larval stages can be identified by the morphology of mouth hooks and/or anterior spiracles (Roberts, 1986).

If expression of the intermediate filament causes embryonic lethality, the number of first instar larvae with a balancer chromosome will be higher than those lacking a balancer chromosome. For example, expression of wild-type heterodimerization partners K5 and K14 had no effect on viability relative to controls (Fig. 4). In contrast, expression of wild-type K5 and mutant K14 R125C caused reduced viability at the pupal and adult stages (Fig. 4), with rare adult survivors possessing a blistered wing phenotype (Bohnekamp et al., 2015). K14 R125C is the most frequent amino acid substitution that causes the blistering skin disorder EBS (Szeverenyi et al., 2008).

If the lethal phase occurs post embryonically, then the number of larvae, pupae, or adults possessing the chromosome with the intermediate filament transgene (i.e., nonbalancer chromosome) will be less than expected. Control experiments should be simultaneously set up in which wild-type flies are crossed to the stock possessing the balancer chromosome and lacking the intermediate filament transgene. The progeny from this cross is used to calculate the expected ratio of balancer to nonbalancer larvae.

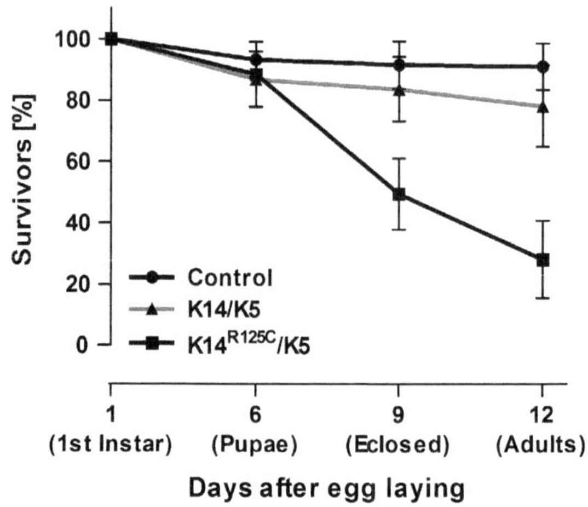

Figure 4 Analysis of lethal phase. Survival curves of flies expressing wild-type keratins K14/K5 or K14^{R125C}/K5 by the ubiquitous *Act5C-Gal4* driver. The *CyO-GFP* balancer chromosome-bearing flies were used as controls. Lethality was analyzed at the time points indicated as days after egg laying. Mean standard deviation for $n \geq 6$ is indicated.

2.3 Measure Levels of Intermediate Filament Expression

RNA isolation

1. Collect six larvae or adults, five muscle fillets, or 50–100 pairs of salivary glands (depending on the tissue used for IF expression analysis) for RNA isolation. This will yield approximately 50–100 μg of total RNA.
2. Grind tissue in 500 μl TRIzol (Life Technologies #15596-026) using a disposable pestle (Sigma #Z359947) and centrifuge for 10 min in a microfuge at 4 °C. Place clear supernatant in a fresh microfuge tube.
3. Perform phase separation, and RNA precipitation according to the manufacturer's protocol (Life Technologies). The use of Phaselock tubes (5Prime #2302830) is highly recommended during the phase separation step to avoid contaminating genomic DNA.
4. Resuspend the resulting pellet in 10–25 μl RNAse-free H_2O and determine the concentration of nucleic acid using a Nanodrop (Thermo Scientific) or spectrophotometer. The desired optical density at 260/280 should be greater than or equal to 1.8.

DNase I treatment of RNA and generation of cDNA

1. Treat 1–2 μg total RNA with DNase I (Life Technologies #18068-015) according to manufacturer's protocol at room temperature for 15 min.
2. Treat with 1/10 volume of 25 m*M* EDTA and incubate at 65 °C for 10 min to inactivate the DNase I.
3. Use 4–8 μl of DNase I-treated total RNA to make cDNA using First Strand cDNA Synthesis kit (Life Technologies #11904-018) per manufacturer's protocol.

Quantitative reverse transcriptase PCR (RT-PCR)

1. Use 0.5–1 μl of cDNA (from above) per RT-PCR reaction with gene specific primers and control primers such as those corresponding to the *RpL32* or *Gapdh1* gene. Primer sequences for these genes and most others are listed in the Fly Primer Bank on the Drosophila RNAi Screening Center website (http://www.flyrnai.org).
2. Perform RT-PCR reactions in triplicate using Power SYBER Green PCR Master Mix (Applied Biosciences #4367659) in a total volume of 25 μl per well in a 96-well microtiter plate.
3. Perform a two-step or three-step PCR reaction using conditions appropriate for the specific primers, up to 40 cycles.

Note: A control RT-PCR reaction lacking reverse transcriptase should be performed for each independent RNA sample to ensure for the absence of contaminating genomic DNA.

Protein extraction and western analysis

1. Grind with a disposable pestle (Sigma #Z359947) four to six whole larvae, larval muscle fillets, adults, or 100 pairs of salivary glands, depending on the tissue of interest, in 60–100 μl Laemmli buffer (Laemmli, 1970).
2. Boil samples at 95 °C for 10–15 min. Spin 3 min in a microcentrifuge to pellet tissue. Load 10–15 μl of the supernatant onto an SDS–PAGE gel and perform electrophoresis until Bromophenol Blue dye migrates to the bottom of the gel.
3. Transfer the proteins to a nitrocellulose membrane (Amersham #10600002) via electrophoretic transfer.
4. Block the membrane with 3–4% dry milk in PBST (137 m*M* NaCl, 2.7 m*M* KCl, 10 m*M* Na_2HPO_4, 2 m*M* KH_2PO_4, 0.1% Tween-20).
5. Incubate with the primary antibody of interest overnight at 4 °C with 1% dry milk in PBST on a rotary shaker (Labline).
6. Wash the membrane three times with PBST in a dish for 5 min each on a rotary shaker.
7. Incubate the membrane with the appropriate secondary antibody labeled with horseradish peroxidase (HRP, Thermo Scientific) for 1–2 h at room temperature in 1% dry milk in PBST on rotary shaker.
8. Wash as step 6.
9. Use the HRP detection kit (Thermo Scientific #34080) according to manufacturer's directions and expose the membrane to X-ray film or analyze using a Epi Chemi II Darkroom (UVP).

Notes: For sample quantitation, the membrane can be stripped with a stripping agent (Invitrogen # 21059) and incubated with an antibody to a control protein that should not be altered by the expression of exogenous intermediate filaments. Antibodies to GAPDH and many other *Drosophila* proteins are available from the University of Iowa Developmental Studies Hybridoma Bank (DSHB, http://dshb.biology.uiowa.edu), Santa Cruze Biotechnology, Inc. (http://www.scbt.com), and Cell Signaling Technology (http://www.cellsignal.com).

Some primary antibodies produce high background under certain blocking conditions, but not others (Johansen et al., 2009). If background bands are apparent, other blocking protocols should be attempted such as the use of bovine serum albumin in place of dry milk.

2.4 Cytological Analysis of Intermediate Filament Formation

To study intermediate filament organization in *Drosophila*, it is convenient to use large cell types that can be easily dissected and provide high resolution for

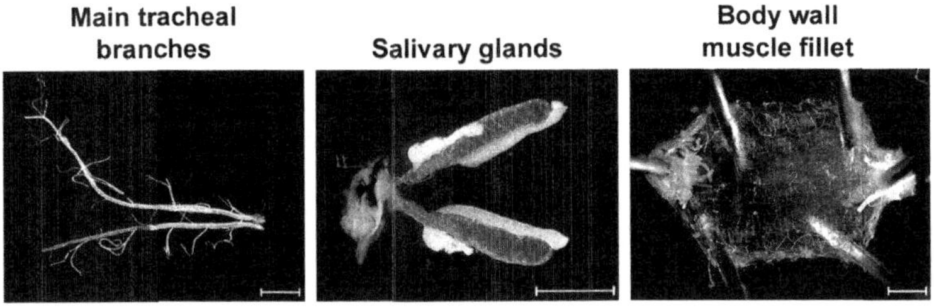

Figure 5 Dissected tissues from third instar larvae. The main tracheal branches (left) and a pair of salivary glands (with attached mouth hooks and some fat body tissue) (middle) that were dissected from a third instar larva. A third instar larva was pinned down, cut open and the organs removed to reveal the body wall muscle (right). The epidermis resides under the muscles. All images are displayed with the anterior on the left and posterior on the right. Scale bars = 0.5 mm. (See the color plate.)

confocal microscopy. However, it is also possible to use smaller cells types, such as diploid cells that make up the imaginal discs (sacs of cells that give rise to specific adult structures, such as the eye and wing). Here, we describe the dissection of main tracheal branches, salivary glands, larval body wall muscles, and epidermis, which all contain large cells (Fig. 5). In addition, we show examples of their use to study intermediate filament formation (Fig. 6).

Isolation of main tracheal branches

1. Select third instar larvae from culture vessel and rinse with PBS (137 m*M* NaCl, 2.7 m*M* KCl, 10 m*M* Na_2HPO_4, 1.8 m*M* KH_2PO_4, pH 7.4).
2. Place rinsed larvae in a Petri dish containing PBS. Using a dissecting microscope for visualization, stabilize the larvae by holding one pair of forceps (DUMONT® type 5, LH79.1) at the head region just below the anterior spiracles.
3. Using a second pair of forceps, pull off the head including the anterior spiracles, so that organs spill out. Then, rip open epidermis/cuticle in the close proximity to the posterior spiracles until the posterior part is no longer connected to the anterior part by epidermis/cuticle. Take care not to harm the main tracheal branches.
4. Remove the pair of forceps that are holding the spiracles with the remaining epidermis/cuticle and carefully pull the two main tracheal branches out, while holding the anterior part with the second pair of forceps. If desired, the remaining epidermis/cuticle can be removed by pulling on the spiracles with one pair of forceps while simultaneously holding the remaining epidermis/cuticle with the second pair of forceps (Fig. 5, left).

As an example of immunohistochemical staining of tracheal cells, Fig. 6A shows the obligate heterodimers K5 and K14 that form a cytoplasmic

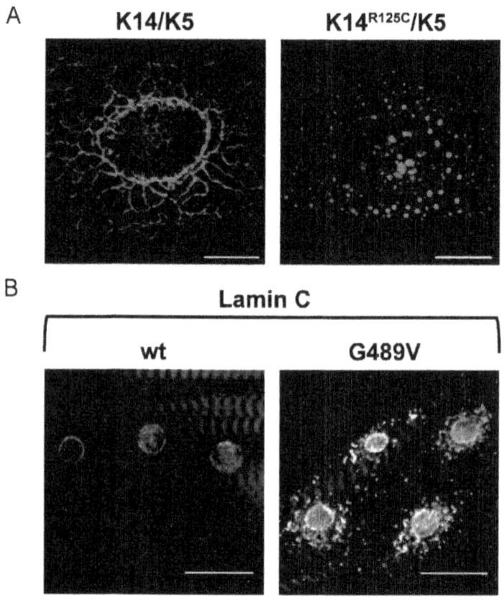

Figure 6 Immunohistochemistry of intermediate filaments in tracheal and muscle cells (A) Localization of keratins in tracheal cells. Tracheal cells expressing keratins were stained with antibodies to K14 (Magin lab; red) and DAPI (blue). Expression of the wild-type K14/K5 resulted in keratin network formation. In contrast, expression of $K14^{R125C}/K5$ resulted in cytoplasmic keratin aggregation. Scale bar = 10 μm. (B) Localization of lamins in larval body wall muscle. Larval body wall muscle stained with antibodies to Lamin C (DSHB # LC28.26; green), DAPI (blue), and phalloidin, which stains actin (red). Muscle-expressing wild-type Lamin C shows localization to the nuclear envelope (left). In contrast, muscles expressing mutant Lamin C (G498V) shows cytoplasmic Lamin C aggregation (right). Scale bar = 40 μm. (See the color plate.)

network in tracheal cells, similar to that observed in mammalian cells (Fig. 6A, left; Anton-Lamprecht & Schnyder, 1982). In contrast, wild-type K5 and K14 R125C form cytoplasmic aggregates (Fig. 6A, right). Similar aggregates are observed in mammalian systems (Anton-Lamprecht & Schnyder, 1982). Thus, cytological analyses can be used to test mutants for the ability to form filaments *in vivo*.

Isolation of salivary glands

1. Select third instar larvae from the walls of the culture vessel and rinse with PBS.
2. Transfer rinsed larvae to a Petri dish containing PBS. With one pair of forceps, grab the tip of the mouth hooks while holding the body with a second pair of forceps toward the posterior end of the animal. Carefully pull the mouth hooks until the salivary glands are exposed from the anterior end.

3. Separate the salivary glands from the remaining tissues (e.g., brain, eye-antennal discs) and dissect away the fat body and other associated tissues (Fig. 5, middle).

Isolation of larval body wall muscle

1. Place a drop of cold muscle dissection buffer (128 m*M* NaCl; 5 m*M* Hepes, pH 7.4; 2 m*M* KCl; 35 m*M* sucrose) into a magnetic dissection chamber (Ramachandran & Budnik, 2010).
2. Using the forceps, select a wandering third instar larva and place it in the drop of muscle dissection buffer.
3. Pin down the anterior and the posterior ends of the larva, stretching the organism lengthwise.
4. Using spring scissors (Fisher/Fine Science Tools vannas #15000-00) make a horizontal incision on the dorsal side of the larva, near the anterior end. Place one blade of the scissors into the incision and make a vertical cut along the dorsal midline.
5. Open the incision and pin down the right and left flaps of tissue away from body.
6. Add several drops of muscle dissection buffer to the inner body cavity, causing the organs to float. Remove the tracheal system using forceps to remove the remainder of the organs by gently dislodging them from the body cavity. During this process, you might need to stretch and repin the body wall both horizontally and vertically (Fig. 5, right).

As an example of immunohistochemical staining of larval body wall muscle, Fig. 6B shows wild-type Lamin C localization to the nuclear envelope (Fig. 6B, left). In contrast, the G489V amino acid substitution mutant froms cytoplasmic aggregates (Fig. 6B, right). The G489V substitution was modeled after a *LMNA* mutation that causes muscular dystrophy in humans (http://www.umd.be/LMNA/W_LMNA/clinics.shtml) (Dialynas et al., 2012). Thus, immunohistochemical analyses can be used to determine the cellular compartmentalization and aggregation properties of mutant lamins.

Isolation of epidermis

1. Dissect the body wall muscles according to the procedures described above.
2. Carefully remove the muscle fibers while the muscle fillet is still attached to the Petri dish. The remaining material is the epidermis.

Notes: It is nearly impossible to completely remove all of the muscle fibers from the underlying epidermis; however, this typically does not present a problem for most applications. If the epidermal sample needs to be fixed for immunofluorescence, it is easier to remove the muscle after fixation.

Immunohistochemistry of dissected tissue

1. Fix dissected tissue with 4% formaldehyde in 1 × PBS for at least 15 min. Wash the animal twice in PBS for 5 min. Fixed larvae can be stored for up to 2 days at 4 °C in a 1.5-ml microfuge tube containing 1 ml 1 × PBS.
2. Wash sample three times with 1 ml 1 × PBS for 5 min in a dissecting dish (Fisher Scientific # 21–379) on rotary shaker.
3. Wash sample three times with 1 ml permeabilization buffer (PBS 1x; 0.5% TX-100; 5 m*M* $MgCl_2$) for 5 min on rotary shaker.
4. Incubate the tissue with primary antibody using a minimum volume of 50 μl overnight at 4 °C (or 1 h at room temperature). Dilute the antibody in permeabilization buffer that contains 0.5% boiled and filtered fish skin gelatin. Make the 10% fish skin gelatin by warming 45% Fish Skin Gelatin (Sigma, G7765) to 65 °C for 30 min, diluting to 10% in sterilized dH_2O, heating the solution until boiling, and then removing from heat.Filter the hot solution through a 0.45-μm filter using a 10–20 ml syringe and store the final solution at 4 °C.
5. Repeat wash step 2.
6. Incubate the tissue with a secondary antibody, diluted in a minimum volume of 50 μl of permeabilization buffer with 0.5% fish skin gelatin.
7. Repeat wash step 3.
8. Repeat wash step 2.
9. Place the tissue in a 20-μl drop of Vectashield™ (Vector Laboratories) on a microscope slide. Carefully place a coverslip over the tissue, making sure to avoid bubbles between the slide and the coverslip. This step should be performed under the dissecting scope.
10. Seal the coverslip with nail polish and allow to thoroughly dry prior to viewing under a microscope.

2.5 Live Cell Imaging of Dissected Tissues

Drosophila is highly amenable to live cell imaging using fluorescent confocal microscopy. Here, we describe how live cell imaging is performed using an inverted confocal microscope (LSM 780, Carl Zeiss, Jena, Germany).

1. Third instar larvae expressing fluorescent protein-tagged intermediate filament proteins are dissected and organs of interest are transferred to a glass bottom dish (Thermo Scientific, 150680) filled with PBS.

2. Using forceps, position a glass coverslip over the dissected organs to prevent floating; this also brings the organs in tight contact with the glass bottom. The fluorescence tagged intermediate filament proteins are easily observed within the dissected tissues.

Note: For prolonged observation, *Drosophila* cell culture media (Biowest LLC, L0207) should be used to maintain physiological status.

3. CONCLUSIONS

Drosophila is a proven model for studies of intermediate filaments (Bohnekamp et al., 2015; Dialynas et al., 2012, 2015). The *in vivo* environment allows for analysis of network assembly and disassembly, including live imaging. The fact that mutations in intermediate filaments cause overt mutant phenotypes provides insights into disease mechanisms and biological pathways altered during pathogenesis. In addition, mutant phenotypes allow for genetic suppressors to be identified. These suppressors offer novel avenues of therapy. *Drosophila* is a proven model for whole organism compound screens that allow for designer genetic backgrounds (Das & Cagan, 2013; Rudrapatna, Bangi, & Cagan, 2014). Compounds that ameliorate phenotypes induced by mutant keratins are candidates for therapy as described for other nonintermediate filament proteins (Gasque, Conway, Huang, Rao, & Vosshall, 2013; Poidevin, Zhang, & Jin, 2015). Thus, at the level of basic biology and deciphering pathomechanisms, *Drosophila* has a lot to offer for studies of intermediate filaments.

REFERENCES

Anton-Lamprecht, I., & Schnyder, U. W. (1982). Epidermolysis bullosa herpetiformis Dowling-Meara. Report of a case and pathomorphogenesis. *Dermatologica, 164*(4), 221–235.

Ashburner, M., Golic, K. G., & Hawley, R. S. (2005). *Drosophila: A laboratory handbook* (2nd ed.). Cold Spring Harbor, NY: Cold Spring Harbor Laboratory Press.

Bassett, A., & Liu, J. L. (2014a). CRISPR/Cas9 mediated genome engineering in Drosophila. *Methods, 69*(2), 128–136. http://dx.doi.org/10.1016/j.ymeth.2014.02.019.

Bassett, A. R., & Liu, J. L. (2014b). CRISPR/Cas9 and genome editing in Drosophila. *Journal of Genetics and Genomics, 41*(1), 7–19. http://dx.doi.org/10.1016/j.jgg.2013.12.004.

Beumer, K. J., & Carroll, D. (2014). Targeted genome engineering techniques in Drosophila. *Methods, 68*(1), 29–37. http://dx.doi.org/10.1016/j.ymeth.2013.12.002.

Bilen, J., & Bonini, N. M. (2007). Genome-wide screen for modifiers of ataxin-3 neurodegeneration in Drosophila. *PLoS Genetics, 3*(10), 1950–1964. http://dx.doi.org/10.1371/journal.pgen.0030177.

Bohnekamp, J., Cryderman, D. E., Paululat, A., Baccam, G. C., Wallrath, L. L., & Magin, T. M. (2015). A Drosophila model of epidermolysis bullosa simplex. *The Journal of Investigative Dermatology*. http://dx.doi.org/10.1038/jid.2015.129.

Bouleau, S., & Tricoire, H. (2015). Drosophila models of Alzheimer's disease: Advances, limits, and perspectives. *Journal of Alzheimer's Disease*, *45*(4), 1015–1038. http://dx.doi.org/10.3233/JAD-142802.

Callaerts, P., Munoz-Marmol, A. M., Glardon, S., Castillo, E., Sun, H., Li, W. H., et al. (1999). Isolation and expression of a Pax-6 gene in the regenerating and intact Planarian Dugesia(G)tigrina. *Proceedings of the National Academy of Sciences of the United States of America*, *96*(2), 558–563.

Coulombe, P. A., Kerns, M. L., & Fuchs, E. (2009). Epidermolysis bullosa simplex: A paradigm for disorders of tissue fragility. *The Journal of Clinical Investigation*, *119*(7), 1784–1793. http://dx.doi.org/10.1172/JCI38177.

Das, T. K., & Cagan, R. L. (2013). A Drosophila approach to thyroid cancer therapeutics. *Drug Discovery Today: Technologies*, *10*(1), e65–e71. http://dx.doi.org/10.1016/j.ddtec.2012.09.004.

Dialynas, G., Flannery, K. M., Zirbel, L. N., Nagy, P. L., Mathews, K. D., Moore, S. A., et al. (2012). LMNA variants cause cytoplasmic distribution of nuclear pore proteins in Drosophila and human muscle. *Human Molecular Genetics*, *21*(7), 1544–1556. http://dx.doi.org/10.1093/hmg/ddr592.

Dialynas, G., Shrestha, O. K., Ponce, J. M., Zwerger, M., Thiemann, D. A., Young, G. H., et al. (2015). Myopathic lamin mutations cause reductive stress and activate the Nrf2/Keap-1 pathway. *PLoS Genetics*, *11*(5), e1005231. http://dx.doi.org/10.1371/journal.pgen.1005231.

Dialynas, G., Speese, S., Budnik, V., Geyer, P. K., & Wallrath, L. L. (2010). The role of Drosophila Lamin C in muscle function and gene expression. *Development*, *137*(18), 3067–3077. http://dx.doi.org/10.1242/dev.048231.

Duffy, J. B. (2002). GAL4 system in Drosophila: A fly geneticist's Swiss army knife. *Genesis*, *34*(1-2), 1–15. http://dx.doi.org/10.1002/gene.10150.

Fuchs, E., & Weber, K. (1994). Intermediate filaments: Structure, dynamics, function, and disease. *Annual Review of Biochemistry*, *63*, 345–382. http://dx.doi.org/10.1146/annurev.bi.63.070194.002021.

Gasque, G., Conway, S., Huang, J., Rao, Y., & Vosshall, L. B. (2013). Small molecule drug screening in Drosophila identifies the 5HT2A receptor as a feeding modulation target. *Scientific Reports*, *3*, srep02120. http://dx.doi.org/10.1038/srep02120.

Gonzalez, C. (2013). Drosophila melanogaster: A model and a tool to investigate malignancy and identify new therapeutics. *Nature Reviews Cancer*, *13*(3), 172–183. http://dx.doi.org/10.1038/nrc3461.

Greenspan, R. J. (2004). *Fly pushing: The theory and practice of Drosophila genetics* (2nd ed.). Cold Spring Harbor, NY: Cold Spring Harbor Laboratory Press.

Hackmann, Y., Joedicke, L., Panneels, V., & Sinning, I. (2015). Expression of membrane proteins in the eyes of transgenic Drosophila melanogaster. *Methods in Enzymology*, *556*, 219–239. http://dx.doi.org/10.1016/bs.mie.2014.12.012.

Homberg, M., & Magin, T. M. (2014). Beyond expectations: Novel insights into epidermal keratin function and regulation. *International Review of Cell and Molecular Biology*, *311*, 265–306. http://dx.doi.org/10.1016/B978-0-12-800179-0.00007-6.

Johansen, K. M., Cai, W., Deng, H., Bao, X., Zhang, W., Girton, J., et al. (2009). Polytene chromosome squash methods for studying transcription and epigenetic chromatin modification in Drosophila using antibodies. *Methods*, *48*(4), 387–397. http://dx.doi.org/10.1016/j.ymeth.2009.02.019.

Kim, D. I., Birendra, K. C., & Roux, K. J. (2015). Making the LINC: SUN and KASH protein interactions. *Biological Chemistry*, *396*(4), 295–310. http://dx.doi.org/10.1515/hsz-2014-0267.

Kohler, R. E. (1994). *Lords of the fly: Drosophila genetics and the experimental life*. Chicago and London: The University of Chicago Press.

Laemmli, U. K. (1970). Cleavage of structural proteins during the assembly of the head of bacteriophage T4. *Nature*, *227*(5259), 680–685.

Liao, G. C., Rehm, E. J., & Rubin, G. M. (2000). Insertion site preferences of the P transposable element in Drosophila melanogaster. *Proceedings of the National Academy of Sciences of the United States of America*, *97*(7), 3347–3351. http://dx.doi.org/10.1073/pnas.050017397.

Loschke, F., Seltmann, K., Bouameur, J. E., & Magin, T. M. (2015). Regulation of keratin network organization. *Current Opinion in Cell Biology*, *32*, 56–64. http://dx.doi.org/10.1016/j.ceb.2014.12.006.

Lyakhovetsky, R., & Gruenbaum, Y. (2014). Studying lamins in invertebrate models. *Advances in Experimental Medicine and Biology*, *773*, 245–262. http://dx.doi.org/10.1007/978-1-4899-8032-8_11.

Martinez-Morentin, L., Martinez, L., Piloto, S., Yang, H., Schon, E. A., Garesse, R., et al. (2015). Cardiac deficiency of single cytochrome oxidase assembly factor scox induces p53-dependent apoptosis in a Drosophila cardiomyopathy model. *Human Molecular Genetics*. http://dx.doi.org/10.1093/hmg/ddv106.

Okray, Z., de Esch, C. E., Van Esch, H., Devriendt, K., Claeys, A., Yan, J., et al. (2015). A novel fragile X syndrome mutation reveals a conserved role for the carboxy-terminus in FMRP localization and function. *EMBO Molecular Medicine*, 7(4), 423–437. http://dx.doi.org/10.15252/emmm.201404576.

Omary, M. B., Ku, N. O., Strnad, P., & Hanada, S. (2009). Toward unraveling the complexity of simple epithelial keratins in human disease. *The Journal of Clinical Investigation*, *119*(7), 1794–1805. http://dx.doi.org/10.1172/JCI37762.

Peter, A., & Stick, R. (2012). Evolution of the lamin protein family: What introns can tell. *Nucleus*, *3*(1), 44–59.

Peter, A., & Stick, R. (2015). Evolutionary aspects in intermediate filament proteins. *Current Opinion in Cell Biology*, *32*, 48–55. http://dx.doi.org/10.1016/j.ceb.2014.12.009.

Pirrotta, V. (1988). Vectors for P-mediated transformation in Drosophila. *Biotechnology*, *10*, 437–456.

Poidevin, M., Zhang, F., & Jin, P. (2015). Small-molecule screening using Drosophila models of human neurological disorders. *Methods in Molecular Biology*, *1263*, 127–138. http://dx.doi.org/10.1007/978-1-4939-2269-7_10.

Ramachandran, P., & Budnik, V. (2010). Dissection of Drosophila larval body-wall muscles. *Cold Spring Harbor Protocols*, *2010*(8). http://dx.doi.org/10.1101/pdb.prot5469. pdb prot5469.

Roberts, D. B. (1986). Basic Drosophila care and techniques. In D. B. Roberts (Ed.), *Drosophila: A practical approach* (pp. 1–38). Oxford England and Washington DC: IRL Press. Chapter 1.

Rubin, G. M., & Spradling, A. C. (1982). Genetic transformation of Drosophila with transposable element vectors. *Science*, *218*(4570), 348–353.

Rudrapatna, V. A., Bangi, E., & Cagan, R. L. (2014). A Jnk-Rho-Actin remodeling positive feedback network directs Src-driven invasion. *Oncogene*, *33*(21), 2801–2806. http://dx.doi.org/10.1038/onc.2013.232.

Schulze, S. R., Curio-Penny, B., Li, Y., Imani, R. A., Rydberg, L., Geyer, P. K., et al. (2005). Molecular genetic analysis of the nested Drosophila melanogaster lamin C gene. *Genetics*, *171*(1), 185–196. http://dx.doi.org/10.1534/genetics.105.043208.

Sebo, Z. L., Lee, H. B., Peng, Y., & Guo, Y. (2014). A simplified and efficient germline-specific CRISPR/Cas9 system for Drosophila genomic engineering. *Fly (Austin)*, *8*(1), 52–57. http://dx.doi.org/10.4161/fly.26828.

Somers, J., Nguyen, J., Lumb, C., Batterham, P., & Perry, T. (2015). In vivo functional analysis of the Drosophila melanogaster nicotinic acetylcholine receptor Dalpha6 using the insecticide spinosad. *Insect Biochemistry and Molecular Biology*. http://dx.doi.org/10.1016/j.ibmb.2015.01.018.

Spradling, A. C. (1986). P element-mediated transformation. *Drosophila: A practical approach* (pp. 175–197). Oxford England and Washington DC: IRL Press. Chapter 8.

Szeverenyi, I., Cassidy, A. J., Chung, C. W., Lee, B. T., Common, J. E., Ogg, S. C., et al. (2008). The human intermediate filament database: Comprehensive information on a gene family involved in many human diseases. *Human Mutation*, *29*(3), 351–360. http://dx.doi.org/10.1002/humu.20652.

Warrick, J. M., Morabito, L. M., Bilen, J., Gordesky-Gold, B., Faust, L. Z., Paulson, H. L., et al. (2005). Ataxin-3 suppresses polyglutamine neurodegeneration in Drosophila by a ubiquitin-associated mechanism. *Molecular Cell*, *18*(1), 37–48. http://dx.doi.org/10.1016/j.molcel.2005.02.030.

Wilson, K. L., & Berk, J. M. (2010). The nuclear envelope at a glance. *Journal of Cell Science*, *123*(Pt 12), 1973–1978. http://dx.doi.org/10.1242/jcs.019042.

Xu, J., Ren, X., Sun, J., Wang, X., Qiao, H. H., Xu, B. W., et al. (2015). A toolkit of CRISPR-based genome editing systems in Drosophila. *Journal of Genetics and Genomics*, *42*(4), 141–149. http://dx.doi.org/10.1016/j.jgg.2015.02.007.

Yu, Z., & Bonini, N. M. (2011). Modeling human trinucleotide repeat diseases in Drosophila. *International Review of Neurobiology*, *99*, 191–212. http://dx.doi.org/10.1016/B978-0-12-387003-2.00008-2.

AUTHOR INDEX

Note: Page numbers followed by "*f*" indicate figures, and "*t*" indicate tables.

A

Aach, T., 61–62, 68–72, 70*f*, 234–235, 420–421
Abe, M., 177
Abercrombie, R.F., 617
Abraham, C., 686–688
Abumuhor, I.A., 480
Acton, T.B., 5–6
Adam, S.A., 89*t*, 193–194, 197–198, 202–205, 392*t*, 409–411, 582–583
Adams, A.D., 490–491
Adams, D.H., 542
Adams, D.J., 618–619
Adams, P.D., 16–17, 19
Adams, S.R., 60
Adamson, E.D., 75
Adelman, W.J., 616–617, 619–620, 620*f*
Adessi, C., 281–282
Adjei, A.A., 568–569
Aebi, U., 4–7, 8*t*, 10*t*, 11–12, 19–25, 38*t*, 41, 43–44*t*, 51, 52*f*, 86–87, 114, 140–141, 179, 198–200, 233, 262–264, 267, 305–309, 339, 354–355, 394, 411, 447–448, 450–451, 462–463, 480, 490, 558–559, 582–583, 662–668, 682–683
Afonine, P.V., 16, 19
Agbulut, O., 540–542, 546
Ahmad, F., 561*t*
Ahmadi, B., 264
Aimoto, S., 89*t*
Ajiro, K., 89*t*
Akamatsu, W., 512–513
Akgul, B., 231*t*, 235–237
Akita, Y., 89*t*
Akiyama, M., 513–514
Akoyev, V., 589–590
Alam, C., 114, 141, 223, 305, 309, 352–354, 365–370, 378, 582–583
Alam, C.M., 355–358, 365–370, 366*t*
Alam, H., 357–358, 362–364
Alam, P., 355–358, 365–370, 366*t*
Alba, A., 496
Alber, T., 6–7
Albers, K., 237–238, 330
Albrecht, H., 513
Alcalai, R., 561*t*
Alessi, D.R., 568–569
Aletta, J.M., 489–490
Alizadeh, A., 583–584
Allen, B.G., 640–641
Allen, E.H., 167*t*, 170
Allen, R.D., 616, 618–619
Allman, E., 691–692
Alonso, A., 143
Alonso-Basanta, M., 521*t*
Alonso-Martin, S., 543
Alousi, S., 584–585
Altamirano, A.A., 616
Altenbach, C., 21–22
Altmannsberger, M., 221
Altshuler, P.J., 357–361, 366*t*
Alvarado, D.M., 318
Alvarez, R., 496
Amaldi, F., 652–654
Amano, M., 89*t*, 100–102, 105–106
Amara, S.G., 573
Amaya, E., 638–640
Ambrogio, C., 88, 391, 405
Ameen, N.A., 366*t*
Amidi, F., 376*t*
Amin, N., 616–617, 628–629
Amit, M., 436
Amoh, Y., 513
Ampudia, R.M., 496
Amsterdam, A., 593
Anand, P., 569
Ananthakrishnan, L., 637, 651–653
Ananthakrishnan, R., 44*t*
Anders, J., 319, 322–323
Andersen, P., 6–7, 262, 271*t*
Andersson, A.-C., 547
Andersson, D., 529–530
Andersson-Lendahl, M., 452, 594

Anderton, B.H., 472, 636–637
Andley, U.P., 588*t*
Ando, S., 87–88, 89*t*, 102–103, 104*f*
Andres, A.-C., 639–640
Andres, D.A., 560, 561*t*
Andresen, B.S., 238–240
Andrews, P.D., 244–246, 247*f*
Angelastro, J.M., 484–485
Angeletti, R., 489–490
Antila, M., 563
Antohi, S., 198–200
Antonescu, C.R., 141–142
Antoniou, C., 80
Anton-Lamprecht, I., 238, 719–721
Antonsson, P., 11–12
Antony, M.L., 191*t*, 193–194
Anver, M., 560–562, 561*t*
Aoyagi, K., 362–364
Apiwattanakul, M., 496
Appel, R.D., 260, 271*t*
Arakawa, K., 563–565
Arbustini, E., 10*t*
Archer, T.D., 640
Argenton, F., 637
Arigo, A.P., 582–585
Arimura, N., 89*t*
Arimura, T., 559–560, 561*t*, 563, 567–568, 567*f*
Arin, M.J., 237–238
Armstrong, C.M., 616
Arner, A., 452, 594
Arnett, E.E., 582–584, 603
Arnold, J.M., 616–617, 619–620, 620*f*
Arnost, K., 240
Arora, J.K., 588*t*
Arora, P., 190, 202–205
Arthur, P.G., 569
Arutyunov, A., 310, 329–330, 339
Aryal, P., 518–520
Asadullah, K., 573
Asch, W.S., 487, 637
Asghar, M.N., 365–370
Ashburner, M., 711–712, 714
Aspenstrom, P., 43–44*t*, 50–51
Asselin, E., 357–358, 362–364
Athanasiadis, N.C., 429
Atherton, A.J., 442
Athlan, E.S., 490
Atkinson, S.D., 167*t*, 170
Attanasio, A., 494
Augusteyn, R.C., 586
Augustine, G.J., 616
Austen, M.D., 309
Austin, C.P., 573
Austin, E.V., 165–170
Avruch, J., 391, 405
Azhar, S., 358–361, 359*t*
Aziz, A., 7, 8*t*, 21, 25–26, 262–263, 480
Azumi, N., 141–142

B

Baas, F., 559
Bacakova, L., 411–412
Baccam, G.C., 708–709, 711, 716, 723
Baden, H.P., 338
Bader, B.L., 143, 366*t*
Badowski, C., 221–247
Bahadur, R.P., 264
Bahinski, A., 439
Baines, A.J., 569
Bairoch, A., 260, 271*t*
Bajar, B.T., 80
Baker, H., 490–491, 493–494
Baker, M.A., 391, 405–406, 409–411, 420–421
Baker, P.M., 563
Balasubramanian, L., 208–209
Baliban, R.C., 116*t*
Ballard, C.C., 16
Ballard, J., 568–569
Balmain, A., 310, 318–319, 326
Balogh, J., 451–452
Bangi, E., 723
Bank, E.M., 667–668
Banker, G., 487–488
Banks, E.R., 141–142
Banks, P.M., 141–142
Banna, C.D., 540–542
Bannasch, P., 636–637
Bannbers, E., 231*t*, 235–237
Bao, X., 718
Bar, H., 7, 23–24, 38*t*, 41, 411, 450–451
Bär, H., 682–683
Bär, J., 487–488
Barber, J.D., 261, 271*t*
Barberis, L., 88, 391, 405

Barbiroli, A., 10*t*
Barcellos-Hoff, M.H., 588*t*
Bardag-Gorce, F., 373, 376*t*
Bardsley, W.F., 244
Bargagna-Mohan, P., 188–214, 397–398
Baribault, H., 61–62, 143, 352–353, 365–372, 366*t*
Barkan, R., 667–668
Barlan, K., 396, 417–418
Barnes, E.A., 478–480
Barnes, P.J., 569
Barnes, R.H., 560, 561*t*
Barnett, M., 589–590
Barry, D.M., 463–464, 466, 471–473
Barry, G.S., 463–464, 466, 471–473
Barry, K.J., 471–473
Bartlett, C.A., 487
Barton, G.J., 5–6, 261, 271*t*
Bartoo, M.L., 416
Baselga, E., 237
Bassett, A.R., 708–709
Basseville, M., 490–491
Basu, A., 412
Batazzi, A.S., 310
Bates, S., 569
Batterham, P., 708–709
Battey, J., 617, 621–622
Battey, J.F., 650, 653
Battifora, H., 141–142
Bauer, J.W., 231*t*
Bauer, M., 311, 337
Bauer, R.A., 197
Baumeister, W., 6–7
Bavelloni, A., 89*t*
Beach, D., 101
Beaudet, L., 478–479, 638
Beaulieu, J.-M., 490, 493–495
Beazley, L.D., 487
Bécane, H.M., 559
Bechert, K., 595*t*, 663–667
Becker, B., 538–540, 543, 546
Becourt, C., 496
Bedina-Zavec, A., 358, 359*t*
Beebe, D.C., 589–590
Beer, H.D., 310
Beermann, M.L., 487
Beese, C., 157
Beggs, A.H., 538, 543
Behnisch, T., 487–488
Beil, M., 38*t*, 43–44*t*, 52–53, 61, 305–309, 354–355, 412–413, 682–683
Bekaii-Saab, T., 568–569
Belecky-Adams, T., 638
Bell, A.C., 641–642
Bellin, R.M., 447–451, 538–543, 546
Benarous, R., 492
Benbow, R.M., 639–640
Bender, L., 50, 188–190, 191*t*, 193–197, 201–205, 397–398
Benedetti, E.L., 585
Benedetto, A., 691–692
Ben-Harush, K., 668
Bening-Abu-Shach, U., 670
Beningo, K.A., 412
Benmohammed, S., 227–229
Bennett, B.L., 569, 571*t*
Bennett, G.S., 636–637
Benoliel, A.M., 44*t*
Ben-Shahar, T.R., 684
Benson, D., 487–488
Ben-Yaakov, K., 190
Bera, S., 487–488
Bercher, M., 663
Bereiter-Hahn, J., 244–245
Beretti, F., 89*t*
Berger, I., 452
Bergeron, C., 495
Bergmans, H.E., 12–13
Bergo, M.O., 558–559
Beriault, D., 682–683
Beriault, D.R., 62–64, 244–245, 683, 702–703
Beric-Maskarel, K., 495
Berika, M.Y., 244
Berk, J.M., 709
Berman, H.M., 266, 271*t*
Bernat, B.A., 568–569
Bernd, A., 244–245
Bernot, K.M., 305, 306–307*f*, 339
Berry, E.C., 434–435
Bertacchini, J., 89*t*
Bertaud, J., 41
Berthault, N., 89*t*
Berthold, C.-H., 512
Bertrand, A.T., 560–562, 561*t*
Bertschy-Meier, D., 88, 391, 405

Betsholtz, C., 512
Betts, J.C., 569
Beumer, K.J., 708–709
Beuvin, M., 560, 561*t*
Bhat, N., 560–562, 561*t*
Bhat, T.N., 266, 271*t*
Bhate, A.V., 357–358, 362–364
Bhatnagar, A., 591
Bhatt, H., 560–562, 561*t*
Bhattacharya, S.S., 391, 420–421, 584–585
Bhosle, R.C., 543
Bi, D., 290
Bian, C., 10*t*, 20, 27–28, 262–263
Bian, W., 510–511, 584–585
Bibliowicz, J., 585, 592–593
Bibrerman, Y., 652–654
Biernat, J., 87–88
Bigio, E.H., 488
Bignami, A., 636–637
Bilak, M.M., 539–540, 546
Bilak, S.R., 539–540, 546
Bilen, J., 708–709
Binder, H.J., 357–358, 365–370, 366*t*
Birendra, K.C., 710*f*
Birkenkamp-Demtroder, K., 362–364
Bischoff, J.R., 101
Bischoff, R., 583
Biscocho, N., 429
Bjornstad, K.A., 588*t*
Blais, K., 510–511
Blalock, T.D., 208–209
Blankenship, T., 583–584
Blau, H.M., 437–438
Bleaken, B.M., 193–194, 585, 591
Blobel, G., 558–559
Bloch, R.J., 428–429, 540, 542–543, 546
Bloemendal, H., 584–585
Blokzijl, F., 357–358
Blouin, R., 61–62
Blumenberg, M., 309
Blumenstein, R., 573
Blundell, T.L., 17–18
Bo, T., 190
Bober, B.G., 463–464, 466, 471–473
Bocina, I., 584–585
Bock, G., 377–378
Boeda, B., 42
Boespflug-Tanguy, O., 165
Boggon, T.J., 18
Bogoyevitch, M.A., 569
Bohl, B.P., 80
Bohnekamp, J., 708–723
Boitard, C., 496
Bokoch, G.M., 80
Bolhuis, P.A., 559
Bolis, L., 240
Bollati, M., 10*t*
Bolognesi, M., 10*t*
Bolshakov, S., 188–190, 193–196
Bolshakov, V.N., 229–234, 231*t*, 240, 682–683
Bonakdar, N., 43–44*t*, 51
Bond, M., 543
Bongrand, P., 44*t*
Bonifas, J.M., 165, 166*f*, 221, 310
Bonini, N.M., 708–709
Bonne, G., 166*f*, 167*t*, 170, 559–560, 563, 567–568, 567*f*, 570–574, 572*t*, 576–577, 663–667
Boor, P., 114, 141, 223, 305, 309, 352–354, 378, 582–583
Borges, A.M., 357–358
Borgonie, G., 686
Borisuth, N.S., 586
Borisy, G.G., 60, 393, 595*t*
Bork, P., 116*t*
Borman, S.K., 202–205
Born, D., 262, 271*t*
Bórquez, D.A., 511*f*, 514–515
Borrill, Z., 569
Bosco, P., 463–464
Boselli, F., 452
Bosl, M.R., 366*t*, 370
Bossinger, O., 662, 682–703
Bouameur, J.E., 357–358, 709
Bouchard, J.-P., 494
Boudreau, R.L., 365
Bouhassira, E.E., 586–587, 589*f*
Bouleau, S., 708–709
Boulton, M.E., 591
Bourgon, L., 61–62
Bousquet, O., 38*t*, 41
Bouvet, M., 513
Bova, G.S., 513–514
Bowden, P.E., 164–165, 266–267, 352–353
Bowen, L., 589–590

Bowes, J.B., 644–645
Boxem, M., 702–703
Boyd, A.E., 60
Boye, K., 695–696
Bozanic, D., 584–585
Bradford, Y., 592–593
Bradley, A., 428–429, 436–437, 451–452
Bradley, E.W., 573
Brady, S.T., 462–464, 473, 616, 618–619
Braendgaard, H., 488
Brand, M., 593–594
Brandenberger, R., 11–12
Brandon, E.P., 518–520
Brandt, B.L., 439
Bransfield, K., 575
Brault, J.J., 88
Brautigan, D.L., 50
Bravo, A., 366*t*, 370–371
Bravo, N.S., 309
Braynen, W., 565
Breckler, J., 540
Breitig, D., 617
Breitwieser, G.E., 616
Brennan, L., 586–587, 589*f*
Brenner, D.A., 365–370
Brenner, M., 165
Bretaña, N.A., 116*t*
Brettel, M., 4–6, 23–24, 391, 394, 480, 594
Breuer, B., 143
Bridges, A.J., 568–569, 571*t*
Bringans, S.D., 287*f*, 288, 290
Brinley, F.J., 618–619
Brion, J.P., 472
Brissette, J., 338
Brocheriou, V., 538–540
Broday, L., 670
Brodsky, G.L., 559
Brody, B.A., 490–491
Broedersz, C.P., 38*t*, 41, 411, 682–683
Brohl, D., 366*t*
Bromley, E.H., 582–583
Bronchain, O.J., 638–639
Broome, R.L., 366*t*
Brose, K., 445
Brosius, M., 538–540, 543
Brown, A., 60, 420–421, 481, 616–617, 619–620, 620*f*
Brown, D.C., 141–142
Brown, H., 628
Brown, K., 493–494, 636–637
Brown, M., 573–574
Brown, M.J., 43–44*t*, 50, 682–683
Brown, R.A., 238–241, 239*f*, 242*f*
Brownless, J., 472
Bruce, J., 636–637
Bruneau, B.G., 434
Brunette, D.M., 244–245
Brunt, E.M., 151, 177, 366*t*, 373–377
Bruton, J., 451–452
Bryan, J., 636–637
Bryson, K., 5–6, 261, 271*t*
Bryson, W.G., 287*f*, 288, 290
Bu, H., 92*t*
Buchan, D.W., 5–6
Bucki, R., 411–412
Buckova, H., 227
Budamagunta, M.S., 7, 8*t*, 21, 23–26, 262–263, 391, 480
Budhram-Mahadeo, V., 638
Budnik, V., 708–709, 711, 721
Buehler, M.J., 38*t*, 41
Buhle, L., 558–559
Bunkoczi, G., 16, 19
Burchill, S.A., 569
Burden, J., 395–396
Burer, A., 22–23, 264, 391
Burgon, P.G., 561*t*
Burgstaller, G., 243–244
Burikhanov, R., 188–190
Burkhard, P., 7, 10*t*, 19, 114, 198–200, 262–263, 267, 480, 582–583
Bürki, K., 75
Burn, D., 488
Burnham, N.A., 48*t*
Burns, K.A., 644–645
Burt, D., 585–586
Busch, T., 62–64, 683, 702–703
Busch-Nentwich, E.M., 452
Bussow, H., 237–238
Bustamante, E.L., 645
Butcher, E.C., 365–370, 366*t*
Butin-Israeli, V., 582–583
Butler, J.P., 44*t*
Byfield, F.J., 411–412
Byrne, C., 309
Byun, J., 213–214

C

Cabreiro, F., 691–692
Cadrin, M., 357–358, 362–364
Cafiso, D.S., 21–22
Cagan, R.L., 723
Cai, L., 520
Cai, W., 718
Cairns, N.J., 488
Callaerts, P., 708–709
Calvisi, D.F., 513–514
Cameron, J., 451–452
Camesano, T.A., 48*t*
Campbell, J., 463–464, 472–473
Campbell, K.P., 543
Campbell, L.E., 366*t*, 370
Camu, W., 494
Canada, F.J., 188–190, 193–194, 197–198
Canamucio, M., 225*t*
Candi, E., 322–323
Cañete-Soler, R., 637
Canger, A.K., 487, 637
Cao, K., 560
Cao, R., 290
Cao, T., 237–238
Capecchi, M.R., 464
Capetanaki, Y., 428–430, 436–438, 445–446, 450–453, 583, 585
Capo, C., 44*t*
Caput, D., 558–559
Carberry, K., 662, 683–684, 686–688
Carden, M.J., 462, 636–637
Carlomagno, Y., 494
Carlsson, L., 429, 512, 540, 543
Carmo-Fonseca, M., 238
Carnevale, K., 490
Carpenter, C., 471–473
Carpenter, S., 50
Carrascal, J., 496
Carrasco, M.J., 188–190, 193–194, 197–198
Carrasco, S., 80
Carrillo, J., 496
Carroll, D., 708–709
Carson, A.A., 38*t*, 41
Carte, J.M., 583
Carter, J.M., 584–585
Casanova, M.L., 358, 362–365, 366*t*, 370–371
Case, D.A., 266–267, 271*t*
Casselman, J., 585–586
Casselman, J.T., 584–585
Cassidy, A.J., 4–5, 165, 222, 259, 271*t*, 310, 355*t*, 709–710, 716
Castellani, C., 378–379
Casteras-Simon, M., 451
Castillo, E., 708–709
Castrillon, J., 546–547
Cattin, M.E., 560–562, 561*t*
Caulin, C., 366*t*
Cavalcante, L.A., 540–542
Cavanagh, H.D., 198–200, 202–205
Cazales, M., 105
Cebers, A., 41
Cebrián-Silla, A., 529–530
Cecena, G., 143, 359*t*
Celis, J.E., 512
Cenni, V., 89*t*
Cerda, J., 594
Chadan, S., 490
Chae, J., 586
Chakabarti, P., 264
Chalfie, M., 60
Chamberlain, C., 80
Chamberlain, J.L., 496
Chambers, S.P., 593
Chamcheu, J.C., 231*t*, 235–237
Chan, L.L., 286–288
Chan, W.K., 617
Chan, W.T., 452
Chan, Y., 586
Chan, Y.-M., 330
Chandar, S., 429, 560–562, 561*t*
Chandra, D., 591
Chandrasekaran, S., 521*t*
Chang, E., 588*t*
Chang, K.-W., 511–514, 511*f*
Chang, L., 195–196, 198–200, 396, 417–418, 490–491, 638
Chang, P.Y., 588*t*
Chang, W., 558–559
Chang, W.C., 116*t*
Chang, Y.-F., 511*f*, 513–514
Chapman, C., 521*t*
Chapman, H.D., 496
Chapman, S.J., 583
Charlton, M.P., 616

Charnas, L.R., 637
Charrier, E.E., 36–54
Charron, G., 638
Charron, P., 10*t*
Chaudhary, N., 558–559
Chaukar, D.A., 357–358, 362–364
Chaunavel, A., 488–489
Chaurasia, S.S., 205–207
Che, X., 190
Cheatham, T.E., 266–267, 271*t*
Chen, A.C., 582–584, 603
Chen, C.-C., 511*f*, 513–514
Chen, C.S., 411–412
Chen, H.-L., 510–511, 516–517
Chen, J., 165–170, 167*t*, 237–238
Chen, J.F., 43–44*t*, 50
Chen, L., 164–182, 366*t*, 584–585
Chen, L.-H., 511–512, 516–517
Chen, P., 290
Chen, S.C., 588*t*
Chen, S.P., 593
Chen, V.B., 16, 19
Chen, X.-x, 488–489
Chen, Y., 366*t*, 370–374, 462, 481, 486–487, 640–641
Chen, Y.-W., 511–512, 516–517
Chen, Y.-Y., 488
Chen, Z., 520
Cheng, C.K., 315*t*
Cheng, F., 682–683
Cheng, J., 237–238, 309
Cheng, L., 510–513
Cheraud, Y., 538–542
Chernoivanenko, I.S., 202–205, 391, 396–397, 417–418
Chernyatina, A.A., 4–28, 262–263, 267, 480
Chesneau, A., 640
Cheung, J., 518–520
Cheung, J.K., 198–200
Cheung, K.J., 241–243
Chevessier, F., 451–452
Chi, F., 190
Chi, Y.I., 5–6, 27–28
Chien, C.-L., 480, 484–485, 487–488
Chin, L.K., 411–412
Chin, S.S.M., 478–480, 484–487, 490, 494, 636–637
Ching, G.Y., 478–481, 484–485, 488–489, 638
Chiquet, M., 11–12
Chisholm, A.D., 663, 670
Chiu, C.P., 437–438
Chiu, F.C., 478–480
Cho, K.W.Y., 638–639
Choate, D.M., 490–491
Choh, V., 583–584
Choi, D.-Y., 490–491
Choi, J.C., 559–560, 563, 564*f*, 566, 567*f*, 569–574, 572*t*, 576–577
Choi, J.W., 362–364
Chomiki, N., 24
Chou, C.C., 38*t*
Chou, C.F., 121, 133
Chou, Y.-H., 95, 101–102, 260–261, 392*t*, 393–394, 396, 417–418, 420–421, 511*f*, 512, 514–516, 518, 521*t*, 522–523, 527
Choudhary, C., 115–117, 130–131
Choudhary, S.K., 428–429, 437–438
Chrachri, A., 617–618
Christensen, B.N., 591
Christensen, R., 362–364
Christian, J.L., 638–639
Chruscinski, A., 431
Chu, Y.W., 141–142, 241–243
Chua, K.L., 452
Chuang, A.-Y., 513–514
Chung, B.M., 7, 10*t*, 42, 114, 262–263, 310, 329–330, 339, 582–583
Chung, B.-M., 61–62
Chung, C., 645
Chung, C.W., 4–5, 165, 222, 259, 271*t*, 310, 355*t*, 709–710, 716
Chung, J.H., 640–642
Chylack, L.T., 588*t*
Cianciola, N., 543
Ciniselli, C.M., 157
Ciombor, K.K., 568–569
Cipolat, S., 444
Cízková, D., 513
Claeys, A., 708–709
Claeys, M., 686
Clark, A.R., 584–585
Clark, J.I., 583–586
Clarke, C., 442
Clay, J.R., 616

Claycomb, W.C., 439
Clayton, C., 569
Cleland, M.M., 140–141, 188–190, 191*t*, 193–195, 197–198, 201–205, 582–583
Clemen, C.S., 451–452
Clerens, S., 281–282, 284*f*, 287*f*, 288, 290, 291*f*, 293*f*, 294, 295*f*, 297–299, 299*f*
Cleveland, D.W., 462–464, 473
Clevers, H., 357–358
Cmejla, R., 355–357
Cmejlova, J., 355–357
Coats, S.E., 229–234, 231*t*, 238–241, 243–244
Coburn, C., 691–692
Cochard, P., 462, 636–637
Coen, L., 640
Cogli, L., 496
Cohen, C., 267
Cohen, G.N., 18
Cohen, M., 667–668, 684
Cohen, P., 568–569
Cohen, R.S., 616–617, 622
Cohen, S., 88
Cohlberg, J.A., 480
Cohn, J., 558–559
Colakoglu, G., 420–421
Cole, C., 5–6, 261, 271*t*
Cole, G.J., 510–511
Cole, P., 489–491, 638
Coletti, D., 543
Collard, J.F., 463
Collin, C., 309
Collins, R.H., 638–639
Collins, S., 190
Colodner, K.J., 166*f*
Colucci-Guyon, E., 42–50, 43–44*t*, 166*f*, 198–200, 413–414, 429, 451–452, 472, 682–683
Colwell, H., 568–569
Common, J.E.A, 165, 221–247, 259, 271*t*, 310, 355*t*, 709–710, 716
Conlin, T., 592–593
Conner, A.S., 445
Conors, R.J., 198–200
Conrad, M., 594
Contag, C.H., 167*t*, 170
Contreras-Vallejos, E., 511*f*, 514–515
Conway, S., 723
Cooke, A., 496
Corbo, M., 494
Cordeiro, M.F., 195–196, 198–200
Cornellison, C.D., 282, 284*f*, 294
Corner, M., 638–639
Corrado, L., 494
Cosgrove, L., 357–358
Costa, C., 88, 391, 405
Côté, F., 463, 478–479, 638
Cote, R.J., 148
Cottrell, J.R., 589–590
Couillard-Despres, S., 462–464
Coulombe, P.A., 7, 10*t*, 38*t*, 42, 61–62, 86–87, 114–117, 121, 124, 140–141, 164–170, 166*f*, 167*t*, 221, 237–238, 246, 262–263, 305–340, 352–355, 415, 582–583, 709, 711
Coulthard, L.R., 569
Couprie, J., 20, 27–28
Cova, E., 494
Covert, M.W., 80
Cowan, N.J., 478–479
Cowie, D.B., 18
Cowtan, K., 19, 267, 271*t*
Cowtan, K.D., 16
Crabbe, E., 642
Cragg, G.M., 188–190
Craveur, P., 116*t*
Crawford, T.O., 463–464, 473
Crews, C.M., 188–190
Crewther, W.G., 262–264
Crick, F.H.C., 258, 267
Crinière, E., 488–489
Critchley, D.R., 540, 543, 546
Crocker, J.C., 417–418
Croizat, B., 489–490
Croop, J.M., 636–637
Cross, M.K., 647–649
Crum, J., 463–464, 466, 472–473
Cruz, K., 41
Cryderman, D.E., 708–723
Csiszár, A., 697
Cubitt, A.B., 60
Cuccurullo, A., 88, 391, 405
Cudney, B., 15
Cuenda, A., 568–569
Cui, C., 490
Cunningham, C.C., 44*t*

Cunningham, M., 512–513, 638
Curio-Penny, B., 708–709
Curran, K.L., 586
Curtis, M.A., 542
Cvekl, A., 586–587, 589*f*
Czonstke, M., 43–44*t*, 51

D

D'Alessandro, M., 172–173, 682–683
Dabiri, B.E., 44*t*, 53
Dafni, U., 442
Dagenais, S., 188–190
Dagvadorj, A., 543
Dahan, I., 668
Dahl, D., 636–637
Dahlstrand, J., 510–511
Dahm, R., 86–87, 582–583, 585–586, 603
Daigle, N., 310, 362–364
Dale, J.M., 463–464, 466, 471–473
D'Alessandro, M., 229–234, 231*t*, 238–241, 243–244
Damron, D.S., 540, 543, 546
Dang, T., 221–247
Dange, P.P., 357–358
Daniel, C., 513
Daniel, E.N., 355–358, 365–370, 366*t*
Daniel, K.W., 190
Danoviz, M.E., 529–530
Danuser, G., 392*t*, 395–396, 399, 403, 403*f*
Darden, T., 266–267, 271*t*
Darmon, Y.M., 143, 359*t*
Das, K., 478–480
Das, T.K., 723
Dauer, W.T., 559–560
Daumas-Duport, C., 540–542
Davey, D.D., 141–142
Davey, N.E., 262, 271*t*
David-Ferreira, J.F., 238
Davidoff, M.S., 513
Davidson, K.M., 366*t*, 370
Davidson, L., 233
Davidson, M.W., 392*t*, 395–396, 399–400, 403, 403*f*
Davidson, P.M., 114
Davies, A., 417–418
Davies, A.M., 240
Davies, B.S., 560, 561*t*
Davies, P., 87–88
Davis, H.A., 365
Davis, I.W., 16, 19
Davis, J., 429
Davis, L.G., 650, 653
Davis, N., 543–546, 551*f*
Davos, C.H., 429, 443, 451–452
Day, K., 529–530
Dayani, Y., 667–668
De Beule, C., 642
De Boni, U., 50
de Brevern, A.G., 116*t*
de Bruijn, E., 452
de Esch, C.E., 708–709
de Jong, W., 584–585
de Koster, C.G., 560
de Medeiros, F.W., 205–207
de Néchaud, B., 489–492
de Pablo, Y., 529–530
De Pons, J., 259, 271*t*
De Robertis, E.M., 638–639
de Souza Martins, S.C., 540–542, 546
De Thonel, A., 517–518, 520
de Waegh, S.M., 462–464, 473
de Winter, F., 518–520, 519*f*, 528–529
Deb-Choudhury, S., 280–300
Debes, C., 259–260, 352–353
DeBoer, S., 628
Dechat, T., 86–87, 164–165, 193–194, 391
Decker, S.J., 568–569, 571*t*
Decostre, V., 559–560, 563, 567–568, 567*f*
Dedes, J., 376*t*
Deerinck, T.J., 463–464, 472–473
DeFelice, L.J., 240
Degani, Y., 289–290
Dege, C., 496
Degenstein, L., 237–238, 309
DeGiulio, J.V., 44*t*, 51–52, 357–358, 683
Dehais, C., 488–489
Delannoy, M.R., 667–668
DeLano, W.L., 8*t*
Delarasse, C., 166*f*
Delcarpio, J.B., 439
Delgado-Olguin, P., 434
Della Gaspera, B., 166*f*
Delorenzi, M., 262, 271*t*
Delsert, C., 493–494
Dempsey, G.T., 399
Deng, C.X., 464

Deng, H., 718
Denison, C., 670
Denk, H., 140–160, 353, 372–378, 376*t*
Dennis, G., 565
Dent, J.A., 650
Denton, C.L., 571*t*
Deokule, S.P., 188–190, 191*t*, 194–205, 199*f*, 201*f*
DePianto, D., 165–170, 167*t*, 237–238, 310
DePianto, D.J., 310
Der Perng, M., 582–585
Desmarais, D., 490–491
Desmond, P.F., 543
Desmoulière, A., 542
D'Eustachio, P., 478–479
Devi, P.U., 197
Devilliers, G., 585
Devriendt, K., 708–709
DFedtke, B., 391, 394
Dhaliwal, D., 586
Dhaliwal, R., 586
Dhe-Paganon, S., 5–6, 27–28
Dhouailly, D., 221
Di Barletta, M.R., 559
Di Iorio, E., 227–229
Di Nunzio, F., 227–229
Di, C.-G., 528–529
Dialynas, G., 708–709, 711, 721, 723
Diana, J., 496
Dickas, V., 225*t*
Dickson, D.W., 87–88, 488
Dickson, M.A., 229
Didonna, A., 392*t*, 395*f*, 398
Diehl, L., 365–370
Dieluweit, S., 697
Dietl, P., 683, 702–703
Diggins, M., 616–617, 626*f*, 627–629, 630*f*
Diguet, N., 540–542, 546
Dikdan, G., 588*t*
Dimova, N., 189*f*, 191*t*, 193–197
Ding, F., 80
Ding, L., 392*t*, 395–396, 399, 403, 403*f*
Dingli, F., 89*t*
Dinkel, H., 262, 271*t*
Dinkova-Kostova, A.T., 165–170, 167*t*, 237–238
Diokmetzidou, A., 428–454
Discher, D.E., 411–412
Distler, J.H., 324–325
Divo, M., 141, 352–353
Dixon, D., 446
Djabali, K., 490–491, 493–494, 636–637
Dlugosz, A.A., 310
Do, H., 147–148
Dobb, M.G., 264
Dobrovic, A., 147–148
Dodd, A., 593
Dodemont, H., 662, 667–668, 684
Dodson, E.J., 16
Doetschman, T.C., 436, 464
Dogan, B., 391, 405, 420–421
Dogic, D., 42–50, 44*t*, 198–200, 413–414, 682–683
Doherty, J.R., 640
Dokholyan, N.V., 80
Dokukin, M., 48*t*
Domenjoud, L., 143
Dominguez, B., 518–520, 519*f*, 528–529
Dominov, J.A., 512–513, 521*t*
Donner, L.R., 488–489
Donno, C., 496
Dooley, C.M., 452
Doolittle, R.F., 262
Door, R., 617
Dorn, G.W., 451–452
Doroudchi, M.M., 492, 494–495
Dos Santos, G., 391, 405–406, 409–411, 420–421
Dotto, G.P., 338
Douglas, S.H., 575
Dow, L.E., 513–514
Dowling, L.M., 263–264
Dozier, C., 105
Draper, B.W., 593
Drenth, J., 17–18
Drenth, J.P., 378–379
Drinker, M., 12
Drozdetskiy, A., 5–6
Droz-Georget Lathion, S., 227–229
Druger, R.K., 487, 637
D'Souza, I., 637
Du Pasquier, D., 640
Duan, S., 362–364
Ducker, W.A., 44*t*, 50–51
Duclos, S., 365–370
Ducray, F., 488–489

Dudley, D.T., 568–569, 571*t*
Duffy, J.B., 712
Duncan, G., 588*t*, 589–590
Dundas, S.R., 238–241, 239*f*, 242*f*
Dunia, I., 585
Dunlop, D.S., 60, 481
Dunlop, S.A., 487
Dunn, N., 592–593
Dunn, R.J., 637
Durand, K., 488–489
Duran-Moreno, M., 529–530
Durham, H.D., 50, 167*t*, 170, 492, 494–496
Duszynski, J., 444
Dutertre, S., 105
Duvic, M., 309
Duyckaerts, C., 488
Duzgunees, N., 695–696
Dyer, J.M., 282, 287*f*, 288, 290, 291*f*, 293*f*, 294, 295*f*, 297–299, 299*f*
Dziedzic, D.C., 588*t*

E

Eady, R.A., 225*t*, 583
Eaton, G.R., 21–22
Eaton, S.S., 21–22
Ebert, S., 44*t*
Eble, J.A., 695–696
Echols, N., 16, 19
Eckert, B.S., 50
Eckes, B., 42–50, 44*t*, 198–200, 413–414, 682–683
Eckhart, L., 260–261
Edelstein, N.G., 638–639
Eden, A., 436
Edström, L., 451–452, 513
Egeland, D.B., 439
Eguchi, G., 586
Eguchi, Y., 586
Ehrlicher, A.J., 42–50, 44*t*, 396–397, 415–420
Ehrmann, J., 513
Eichmann, D.A., 365
Eichmuller, S., 311, 337
Eichner, R., 309, 354–355
Eigenbrod, S., 488–489
Eils, R., 43–44*t*, 50, 193–194
Eisen, A.Z., 309
Eistetter, H., 436
Ekker, S.C., 593
Elbashir, S.M., 595*t*, 663–667
Elder, G.A., 463–464
El-Heliebi, A., 366*t*, 370–374
Eliasson, C., 512
Elkan, E.R., 638
Ellis, M., 584–585
Ellisman, M.H., 488–489
Elsdale, T.R., 638–639
Emsley, P., 16, 19, 267, 271*t*
Engel, J., 11–12
Engel, K.B., 148–149
Engler, A., 411–412
English, L.S., 127, 238
Engvall, E., 225*t*
Enikolopov, G., 513, 529–530
Enomoto, A., 105–106
Enomoto, M., 105
Epler, R.G., 225*t*
Epstein, E.H., 165, 166*f*, 221, 310
Epstein, J.I., 513–514
Ergin, B., 157
Erglis, K., 41
Eriksson, J.E., 42, 50, 86–87, 91, 114, 193–194, 357–358, 363*f*, 365–371, 366*t*, 391, 510–530, 682–683
Escaich, S., 166*f*
Escalante-Alcalde, D., 560–562, 561*t*
Escurat, M., 490–491, 493–494, 636–637
Eshhar, Z., 238
Eshkind, L.G., 143
Esquivel, C.O., 358–361, 359*t*
Esson, D.W., 208–209
Esteves, R.A., 237–238, 310
Esue, O., 38*t*, 41
Etienne-Manneville, S., 42, 241–243
Etkin, L.D., 639–640
Euskirchen, G., 60
Euteneuer, U., 38*t*, 41, 411, 583, 682–683
Evans, D.J., 310
Evans, E.A., 695–696
Evans, M.J., 464
Evans, P.R., 16
Evans, R.J., 568–569
Evans, R.M., 42–50, 87, 450–451
Evert, M., 513–514
Ewald, A.J., 241–243
Eyer, J., 38*t*, 462, 582–583
Eyk, J.E.V., 61–62

F

Faber, J., 638–639, 645
Fabich, H., 42–50, 44*t*, 396–397, 415–418
Fabris, G., 53, 682–683
Fabry, B., 44*t*
Faggianelli, N., 452
Faigle, W., 89*t*
Faiz, M., 529–530, 542
Fajer, P., 27
Falasco, L., 494
Fallet, C., 540–542
Falsey, R.R., 193–194
Fan, Y., 492
Fashena, D., 592–593
Fassler, R., 311, 337
Fatkin, D., 559
Faundez, V., 417–418
Fauquet, M., 490–491
Faure, R., 366*t*
Fauser, C., 11–12
Faust, L.Z., 708–709
Favalli, V., 10*t*
Feany, M.B., 166*f*
Federici, T., 540–542
Fedtke, B., 4–6, 23–24
Feinstein, N., 667–668
Fekete, B., 542, 545
Felder, E., 683, 702–703
Felder, E.T., 62–64, 683, 702–703
Feldman, C.L., 574
Fellini, S.A., 636–637
Felsenfeld, G., 640–642
Feng, L., 359*t*
Feng, N., 352–353, 359*t*, 366*t*, 370–371
Feng, X., 121, 324–325, 333
Feng, Z., 266, 271*t*
Fernandez, A., 515–516
Fernandez, R., 89*t*
Fernandez-Acenero, M.J., 366*t*
Ferrari, S., 227–229
Ferraris, S.E., 517–518, 519*f*, 520, 521*t*, 523, 525, 527
Ferreira, R., 452
Ferreiro, B., 645
Fetsch, J.F., 141–142
Fickert, P., 372–373, 376*t*
Field, S.L., 575
Fiering, S., 446
Figueroa, Y., 366*t*
Filion, M., 490–491
Filip, S., 513
Filliatreau, G., 490
Fine, A., 445
Fini, M.E., 202–205
Fink, L.M., 87
Fire, A., 446
Firestein, B.L., 688–690
Fischberg, M., 638–639
Fischer, A.H., 312
Fischer, C., 543
Fischer, H., 260–261
Fischer, H.P., 221
Fischer, S., 462–463
Fish, M.B., 640–641
Fisher, D.Z., 558–559
Fisher, M., 586
Fisher, S.K., 540–542
Fishman, H.M., 617
Fiszman, M., 492
Fitzgerald, P., 21, 24–25
FitzGerald, P.G., 21–26, 22*f*, 391, 583–586
Fitzpatrick, T.B., 309
Flanagan, L.A., 411–412
Flannery, K.M., 708–709, 711, 721, 723
Flavell, D., 478–479
Fleischhauer, J., 697
Fleiszig, S.M., 310
Fleming, T.P., 588*t*
Fliegner, K.H., 478–481, 486–487, 636–637
Flitney, F.W., 61–64
Floering, L.M., 190
Flores, M.A., 167*t*, 170
Flores, R., 484–485, 488
Florin, A., 365–370
Floudas, C.A., 116*t*
Floyd, C.C., 616–617, 627–628
Flügge, U.I., 296
Foa, C., 44*t*
Foeger, N., 668, 672
Fois, G., 62–64, 683, 702–703
Foisner, R., 164–165
Foitzik, K., 311, 337
Foley, A., 434
Folland, E.D., 574
Folmer, J., 463–464
Folpe, A.I., 141–142

Fong, L.G., 558–560
Fontana, R.J., 165
Foote, L.J., 225*t*
Fort, C., 490–491
Fortelius, L.E., 357–358, 362–370, 366*t*
Fortier, A.M., 357–358, 362–364
Fouquet, B., 594
Fox, C.H., 146–148
Fox, N.C., 488
Fraley, S., 376*t*
Franco, C.A., 542, 546
Frank, M.H., 513–514, 521*t*
Franke, W.W., 4–5, 22–24, 60, 75, 121, 142–143, 146, 147*f*, 225*t*, 238, 260–261, 309, 352–353, 359*t*, 366*t*, 370–371, 376*t*, 636–637
Franz, T., 366*t*, 370
Frappier, T., 484–485
Fraser, R.D.B., 261–264, 265*f*, 268–270
Frazer, K., 592–593
Fredberg, J.J., 44*t*
Fredriksson, S., 545
Freedberg, I.M., 309
French, B.A., 373, 376*t*
French, S.W., 146, 372–377, 376*t*
Frenkiel-Krispin, D., 668
Frenneaux, M., 559
Freudenmann, M., 143
Frezza, C., 444
Fridkin, A., 667–668
Friedrich, M.G., 589–590
Friedrich, V.L., 463–464
Frisén, J., 513
Fritsch, A.W., 44*t*, 53, 310, 357–358, 682–683
Froment, C., 105
Froncisz, W., 24
Fu, A.K.Y., 518–520
Fu, D.J., 167*t*, 170
Fu, J.D., 434–435
Fu, P., 376*t*
Fu, W.-Y., 518–520
Fu, Y., 594
Fuchs, C., 487, 637
Fuchs, E., 86–87, 164–165, 166*f*, 221, 237–238, 262–263, 305–310, 330, 332–335, 354–355, 462–463, 489–490, 583, 709, 711
Fuchsbichler, A., 145–146, 145*t*, 353, 372–378, 376*t*
Fudge, D.S., 583–584, 603, 682–683
Fujii, T., 487–488
Fujiki, H., 50
Fujimoto, J., 463–464, 466, 473
Fujinaka, H., 355–357
Fujita, E., 310
Fujita, M., 89*t*
Fukuda, H., 89*t*
Fukui, H.N., 588*t*
Fukushige, T., 686
Fukushima, K., 193–194
Fukusho, E., 89*t*
Funaki, M., 411–412
Fung, K.Y., 357–358
Fyhrquist, F., 574–575

G

Gabbiani, G., 542–543, 545–546
Gabriele, S., 44*t*, 53
Gabrielson, E., 241–243
Gadeaud, E., 488–489
Gage, F.H., 518–520
Gainer, H., 616–617, 618*f*, 621–624, 621*f*, 623*f*, 628, 637, 650
Galbraith, J.A., 618–619
Galie, P.A., 411–412
Gallant, P.E., 616–619, 627–628
Gallo, J.-M., 637
Gammon, L., 231*t*, 235–237
Ganesh, T., 50, 188–190, 191*t*, 193–197, 201–205, 397–398
Gangadaran, P., 357–358, 362–364
Gao, C.Y., 590
Gao, J., 538–540
Gao, N., 513–514
Gao, W., 565
Gao, X., 584–585
Gao, Y.S., 417–418
Garabatos, N., 496
Garcia, M., 237
Garcia, M.L., 462–473
Garcia, V.B., 463–464, 466, 473
García-Belda, P., 529–530
García-Pelagio, K.P., 543
García-Verdugo, J.M., 529–530
Gardiner, E.M., 80

Gareau, J.R., 115–117, 129
Garesse, R., 708–709
Garman, E., 15
Garrett, N., 639–640
Garrod, D.R., 244, 357–358, 363*f*
Gartner, C., 88
Garzon, B., 188–190, 193–194, 197–198
Gasque, G., 723
Gasteiger, E., 260, 271*t*
Gaszner, M., 640–641
Gatter, K.C., 141–142
Gattiker, A., 260, 271*t*
Gaudreau, P., 309
Gavin, A.C., 115–117
Gavin, B., 512–513, 638
Gaw, L.W., 645
Gay, R.E., 324–325
Gay, S., 324–325
Gayarre, J., 188–190
Gearhart, J.D., 463
Gearing, M., 488
Gehart, H., 357–358
Gehringer, M.M., 127
Geiger, B., 60, 225*t*, 238
Geisler, F., 683–684
Geisler, N., 7, 8*t*, 19–20, 188–190, 262–263, 394, 406–409, 462–463
Gelfand, V.I., 193–194, 197–198, 202–205, 391–421
Gems, D., 691–692
Gentil, B.J., 167*t*, 170, 496
Georges, P.C., 38*t*, 411–412
Gerace, L., 558–559
Gerdes, A.M., 451–452
Gerhardus, H., 683–684
Gerhart, J.C., 640
Gervasi, C., 637, 641–642, 650–653
Geske, R., 436–437
Gester, T., 231*t*, 235–237
Geyer, P.K., 708–709, 711
Gharbi, S., 188–190
Ghori, N., 366*t*, 370–371
Gia, H.P., 496
Giancotti, F.G., 225*t*
Gibb, J.M., 472
Gibb, R., 644–645
Gibbs, A.J., 260–261
Gibbs, D., 266–267
Giese, A., 488–489
Gijbels, M., 560, 561*t*
Gilbert, D.L., 616–617, 619–620, 620*f*
Gilbert, L.A., 362–364
Gilbert, S., 310, 357–358, 362–364, 366*t*
Gilbert, S.P., 616, 618–619
Gilles, R., 240
Gillespie, J.I., 618–619
Gilliland, G., 266, 271*t*
Gilly, W.F., 616–618
Gil-Perotín, S., 529–530
Gilquin, B., 20, 27–28
Gilula, N.B., 638–639
Giorgi, C., 444
Giraudel, A., 281–282
Girkin, J.M., 582–603
Girton, J., 718
Gish, R., 165, 359*t*
Gizang-Ginsberg, E., 490–491
Gladilin, E., 43–44*t*, 50, 193–194
Glardon, S., 708–709
Glasgow, E., 487, 637
Glass, D.J., 88
Glogauer, M., 44*t*
Gnad, F., 116*t*
Godbout, M.J., 309
Godi, M., 494
Godsave, S., 636–637
Gogg-Kamerer, M., 140–160
Gohlke, H., 266–267, 271*t*
Gokhin, D.S., 582–584, 603
Goldberg, M.W., 588*t*, 589–590, 668
Goldfarb, L., 543
Goldie, K., 462–463
Goldie, K.N., 4–6, 23–24, 391, 394
Goldman, A., 89*t*
Goldman, A.E., 558–559, 582–583
Goldman, J.E., 165
Goldman, R.D., 44*t*, 50–52, 60–64, 86–87, 95, 101–102, 140–141, 202–205, 241–243, 357–358, 363*f*, 391–421, 490–491, 493–494, 511*f*, 514–516, 518, 521*t*, 522–523, 527, 558–559, 582–583, 636–637, 670, 683
Golic, K.G., 711–712, 714
Golik, B., 471–473
Gomi, H., 472
Goncharov, A., 663

Gong, Z., 452
Gonzalez, C., 708–709
Gonzalez, M.E., 165–170
Gonzalez, P., 43–44*t*, 50, 193–194
Gonzalez-Gonzalez, E., 167*t*, 170
Goodell, M.A., 445
Goodman, S.L., 60, 238
Goodwill, W., 540, 542, 546
Gordesky-Gold, B., 708–709
Gorham, J.D., 489–491, 493–494, 638
Gorlia, T., 488–489
Goto, H., 86–106, 358–361, 585
Goto, T., 89*t*
Gotow, T., 41, 463–464, 472–473, 617
Gou, J.P., 41
Goudeau, B., 451
Gould, R.M., 622–624, 623*f*
Gould, T.W., 518–520, 519*f*, 528–529
Gourov, A., 463–464
Gowing, G., 494
Graham, L., 337–338
Graham, M., 452
Graham, R.M., 134, 373, 376*t*, 377–378
Grainger, R.M., 638, 640–641, 645
Granato, M., 593
Granger, B.L., 538, 540, 542, 546, 591
Grant, P., 616–629, 650
Grasbon-Frodl, E., 488–489
Gräter, F., 259–260, 352–353
Graves, D.J., 447–448
Graw, J., 584–585, 588*t*
Gray, J., 586
Green, H., 121, 233, 309
Green, S.L., 645
Greenberg, H.B., 352–353, 359*t*, 366*t*, 370–371
Greene, L.A., 484–485, 489–491, 493–494, 636–638
Greenfield, E.A., 91, 105
Greengard, P., 628
Greenspan, R.J., 714
Greenwood, C.A., 686
Gregersen, N., 238–240
Gregor, M., 243–244
Greiling, T.M., 586
Grenier, A., 365–370
Grevesse, T., 44*t*, 53
Gribbon, C., 588*t*
Griffin, J.W., 463–464, 473
Griffin, M., 411–412
Griggs, N.W., 117–120, 130–131, 132*f*
Grin, B., 86–87, 188–190, 191*t*, 193–195, 197–198, 201–205, 391, 392*t*, 394–398, 395*f*, 412–413, 417–418
Groelz, D., 157
Grogan, S.P., 43–44*t*, 50
Groppi, V.E., 164–182
Gros, F., 489–494, 636–637
Gros-Louis, F., 494
Gross, A., 165, 359*t*, 366*t*, 371
Gross, J.M., 585, 592–593
Gross, L.A., 60
Grosse-Kunstleve, R.W., 16–17
Grossmann, E., 668
Grosveld, F., 478–479
Grosvenor, A.J., 282, 284*f*, 294
Grotendorst, G.R., 208–209
Grube, D., 148–149
Gruber, M., 17, 262
Gruenbaum, Y., 114, 662–677, 684, 711
Grund, C., 60, 75, 143, 238, 366*t*, 370, 372
Gruner, J.A., 628
Grunwald, A., 409–411
Gu, C.H., 38*t*, 41
Gu, Y.-y., 488–489
Guck, J., 44*t*
Gudjonsson, S.A., 379
Gudkova, A., 451–452
Guérette, D., 510–511, 538
Gui, H., 164–182
Guicheney, P., 538–540
Guillaudeau, A., 488–489
Guldiken, N., 352–379
Gullberg, M., 545, 547
Gumpel, M., 490–491, 493–494, 636–637
Gunaherath, G.M., 193–194
Gunatilaka, A.A., 193–194
Gunawardena, J., 116*t*
Gundersen, G.G., 60, 395–396, 558–559
Gündisch, S., 157
Gunduz, U., 396–397
Gunhaga, L., 589–591
Guo, M., 42–50, 44*t*, 391–421
Guo, T., 290–291
Guo, Y., 708–709
Gupta, T., 238–241, 239*f*, 242*f*

Gurdon, J.B., 638–640, 647–649
Gurney, M.E., 494
Guschel, M., 244–245
Gústafsdóttir, S.M., 545
Gustafson, D.L., 571*t*
Gustke, N., 87–88
Guthrie, S.C., 638–639
Guyon, J.R., 538–543
Guz, N., 48*t*
Guzenko, D., 4–28, 262–263
Guzman, C., 38*t*, 41, 411
Guzman, G., 190
Gygi, S.P., 88, 670

H

Ha, M.H., 89*t*, 95, 96*f*, 101–103
Haas, P., 141–142
Haase, I., 322
Haase, K., 47*f*
Haber, R., 543
Habtezion, A., 114, 353, 355–358, 362–370, 366*t*, 582–583
Hackmann, Y., 712
Haddad, O., 62–64, 244–245, 683, 702–703
Haegeman, G., 190, 191*t*, 193–194
Haffter, P., 593
Hagaman, J.R., 464
Hagemann, L.J., 464
Hagemann, T.L., 165–170
Hagio, M., 513–514
Hahm, E.R., 190, 191*t*, 193–194
Hahn, J., 617
Hahn, K.M., 80
Hahn, S.A., 362–364
Hahn, W.C., 229
Haigler, H.T., 21–22
Haines, R.L., 305, 310
Hainzl, S., 231*t*, 237
Haithcock, E., 667–668
Haley, J., 478–480
Hall, W.W., 193–194
Hallam, J.A., 43–44*t*, 50, 682–683
Hamer, K., 357–358
Hamill, K.J., 243
Hamlet, M.R., 640
Hammers, H.J., 188–190, 191*t*, 194–195
Hammerschmidt, M., 593
Hammouda, E.H., 559
Hamza, A., 188–190, 189*f*, 191*t*, 193–197, 205–207, 206*f*, 397–398
Han, M., 667–668
Han, M.C., 310
Han, S.-K., 488
Hanada, S., 165, 352–354, 366*t*, 373–379, 709
Handjiski, B., 311, 337
Häner, M., 4–6, 23–24, 391, 394, 480, 668
Hanisch, A., 97, 102*f*, 105–106
Hanken, J., 645
Hannoush, R.N., 126
Hao, Y., 520
Hapiak, V., 663
Haque, M.S., 588*t*
Harada, M., 134, 146, 177, 366*t*, 370–378, 376*t*
Haraguchi, T., 595*t*
Harborth, J., 595*t*, 662–667, 684
Harbury, P.B., 6–7
Hardiville, S., 133
Hardwicke, J., 638–639
Harland, D.P., 280–300
Harland, R.M., 637–640, 645
Harmala-Brasken, A.S., 50
Harman, A.M., 487
Harper, S., 290
Harrap, B.S., 264
Harris, J.R., 309, 315*t*
Harrison, S.W., 446
Hart, G.W., 133
Hartley, K.O., 638–639
Hartmann, A., 542
Hartmann, D., 353, 372, 374–377
Hartwig, J., 38*t*, 41
Harvey, R.P., 638–639
Harvey, S.A., 452
Hashim, G., 484–485
Hashimoto, R., 89*t*
Hashimoto, T., 225*t*
Hasib, L., 513
Hategan, A., 411–412
Hatzfeld, M., 262–263, 305–309, 330, 354–355
Haudenschild, D.R., 43–44*t*, 50
Havaki, S., 442
Hawley, R.S., 711–712, 714
Hay, M., 50

Hayashi, Y., 434
Hayashi-Takanaka, Y., 105
Hayat, M.A., 317–318
Haybaeck, J., 143, 355, 366*t*, 370–374, 376*t*, 546–547
Hayman, G.T., 259, 271*t*
Hays, A.P., 494
Hazan, R., 142–143, 146, 147*f*
He, B.P., 495
He, C.Z., 494
He, T., 370, 511*f*, 514–516, 518, 519*f*, 520, 521*t*, 522–523, 525, 527
Head, R., 357–358
Heagerty, A.H., 221, 225*t*, 310
Heasman, J., 647–649
Heath, J.P., 446, 585
Hebbar, N., 188–190
Hebert, A., 165, 166*f*, 221, 310
Hecht, E., 683, 702–703
Hedberg, B., 451–452
Hegde, K.R., 589–590
Heid, H., 145*t*, 376*t*
Heid, H.W., 309
Heikkila, J.J., 639–640
Heilmann, C.J., 560
Heim, R., 60
Heinemann, U., 16
Heins, S., 462–463, 663–667
Heitlinger, E., 668
Helbling-Leclerc, A., 560, 561*t*, 563
Helenius, T.O., 352–379
Helfand, B.T., 86–87, 193–194, 201–205, 391, 392*t*, 394–397, 412–413, 493–494
Heliö, T., 563
Hellman, J., 514–515
Hellmich, M.R., 617, 621–622, 621*f*
Helmes, M., 445
Hemmati-Brivanlou, A., 637–640
Henderson, P., 391, 405, 420–421
Hendrix, M.J., 141–142, 241–243
Henkelman, M., 518–520, 519*f*, 528–529
Henningfeld, K., 640
Hennings, H., 323
Hennrich, M.L., 115–117
Henrick, K., 19
Herawi, M., 513–514
Herberich, G., 61–62, 68–72, 70*f*, 234–235
Herbstritt, C.J., 188–190, 191*t*, 194–195
Hergenrother, P.J., 193–194
Hergt, M., 7, 8*t*, 19–20, 262–263, 400
Hermann, G.J., 688–690
Herrmann, H., 4–7, 8*t*, 10*t*, 19–25, 38*t*, 41, 43–44*t*, 51, 52*f*, 86–87, 114, 117–120, 129–130, 140–141, 179, 188–190, 191*t*, 193–195, 197–205, 233, 260–264, 267, 305–309, 315*t*, 339, 354–355, 366*t*, 391, 394, 400, 406–409, 411–413, 420–421, 447–448, 450–451, 462–463, 480, 490, 521*t*, 582–585, 594, 668, 669*f*, 682–683
Herron, A.J., 559–560, 563, 567–568, 567*f*
Herzenberg, L.A., 446
Hess, J.F., 4–28, 262–263, 480, 583–584, 603
Hesse, M., 75, 86–87, 143, 233, 305, 366*t*, 370, 376*t*
Heyninck, K., 190, 191*t*, 193–194
Heys, K.R., 589–590
Hibi, K., 397–398
Hickerson, R.P., 167*t*, 170, 231*t*, 237, 310, 318–319, 326
Hijikata, H., 513–514, 520
Hijikata, T., 543, 546
Hilgers, R.D., 542
Hill, A.J., 227
Hill, W.D., 617, 618*f*, 626*f*
Hirahara, F., 225*t*
Hirako, Y., 539–542
Hirano, A., 488
Hirano, H., 89*t*
Hirokawa, N., 60
Hirono, Y., 88, 89*t*
Hirose, G., 638–639
Hiroshima, Y., 513
Hirsch, N., 586
Hisanaga, S., 95, 101–102
Hisano, Y., 593–594
Hitomi, K., 134
Ho, C.-L., 60, 395–396, 481, 484–485, 487–488, 490
Ho, M., 512–513
Hobbs, R.P., 305, 310, 582–583
Hochgreb, T., 452
Hochmuth, R.M., 696
Hochstein, B., 38*t*, 41, 411
Hockfield, S., 510–511

Hodge, A.J., 617
Hodges, H., 514–515
Hodgkin, A.L., 616
Hodgkins, I.J., 286–288
Hoekstra, W.P., 12–13
Hoffman, B.D., 417–418
Hoffman, P.N., 463
Hoffmann, B., 682–703
Hoffmann, E.K., 240
Hofmann, I., 4–5, 22–23, 260–261
Hofner, M., 428–429, 436
Hogervorst, F., 225*t*
Holbrook, K.A., 309, 323
Holekamp, N.M., 589–590
Holland, R., 225*t*
Hollrigl, A., 428–429, 436
Holmes, D., 244
Holmström, M., 563
Holstein, A.F., 513
Holton, J.L., 488
Holtzer, H., 583, 636–637
Holwell, T.A., 42–50
Holwill, S., 647–649
Homberg, M., 114, 582–583, 682–683, 709
Homma, S., 167*t*, 170, 559–560, 563, 564*f*, 566, 567*f*, 569–574, 572*t*, 576–577
Hong, S.W., 362–364
Hoogland, C., 260, 271*t*
Hookway, C., 193–194, 197–198, 202–205, 391–421
Hope, C.M., 80
Hopkins, N., 593
Hopkinson, S.B., 65–66, 243
Horimoto, K., 105–106
Horn, H.M., 227, 238–240
Hornbeck, P.V., 116*t*, 117, 118*t*
Horne, P., 492
Horwitz, B., 238
Hosack, D.A., 565
Hosokawa, T., 517–518, 520
Hosono, M., 134
Hossenlopp, P., 20, 27–28
Hou, D., 362–364
Houben, S., 697
Houle, D., 638
Hourdé, C., 543
House, S., 616–617, 622
Houseweart, M.K., 463–464
Howe, D.G., 592–593
Howe, W.E., 75
Hresko, M.C., 663
Hsiao, P.-J., 511–512, 516–517
Hsieh, S., 464, 473
Hu, L., 359*t*
Hu, Y.Z., 588*t*
Huang, B.-M., 511–514, 511*f*
Huang, F.L., 588*t*
Huang, G.N., 434
Huang, H., 582–583
Huang, J., 198–200, 513–514, 521*t*, 723
Huang, K.L., 188–190, 193–196
Huang, K.Y., 116*t*
Huang, L.-d., 488–489
Huang, P., 593–594
Huang, S., 464
Huang, S.H., 190
Huang, Y., 434
Huang, Y.J., 5–7, 8*t*, 25, 262–263, 480
Hubbell, W.L., 21–22, 24
Huc, C., 492, 496
Huch, M., 357–358
Hudson, E.A., 589–590
Hughes, A., 638–639
Hughey, J.J., 80
Hugo, C., 513
Huh, C.G., 515–516
Huiatt, T.W., 447–449, 538–540, 543, 546
Huisman, E.M., 41
Hull, C.C., 589–590
Hull, P.R., 231*t*
Hunt, P.M., 638–639
Hurme, T., 512–515, 518
Hüsken, K., 686–688
Hutcheson, A.M., 582–585, 603
Huttelmaier, S., 243–244, 682–683
Hutton, E., 237–238, 309
Hutton, M.E., 165, 166*f*, 221, 237–238, 310, 583
Huxley, A.F., 616
Hwang, W.Y., 594
Hyde, J.S., 24
Hyder, C.L., 91, 114, 358–361, 513–514, 521*t*
Hylock, R.H., 365

I

Ibaraki, N., 588*t*
Idbaih, A., 488–489
Idei, T., 38*t*, 41
Idler, W.W., 264
Ieda, M., 434–435
Igaki, H., 362–364
Iismaa, S.E., 134
Ikeda, M., 488
Ikegami, M., 225*t*
Ikegami, Y., 105
Ikegawa, M., 395–396
Ikeshita, S., 193–194
Ilagan, E., 310, 329–330, 339
Illmensee, K., 75
Imajoh-Ohmi, S., 89*t*, 95, 96*f*, 101–103
Imamura, M., 538–540, 543, 546
Imamura, S., 309
Imani, R.A., 708–709
Imanishi, S.Y., 358–361, 517–518, 520
Inada, H., 89*t*, 310
Inagaki, M., 86–106, 126–127, 190, 191*t*, 193–195, 197–200, 202–205, 203*f*, 213–214, 358–361
Inagaki, N., 89*t*, 102–103, 104*f*
Ingber, D., 42–50, 44*t*, 198–200, 413–414, 682–683
Ingber, D.E., 42–50, 44*t*, 582–583
Ingle, A.D., 357–358
Iniguez-Lluhi, J.A., 117–120, 129–130
Ino, Y., 229
Inoko, A., 89*t*, 105–106, 202–205, 585
Inoue, R., 586
Inoue, T., 586
Intong, L.R., 224
Iozzo, R.V., 365–370, 366*t*
Ip, F.C.F., 518–520
Ip, W., 538–540, 543, 546
Ireland, M.E., 584–586
Isas, J.M., 21–22
Ischenko, I., 490–491
Ishibashi, S., 640–641
Ishida-Yamamoto, A., 221, 310, 583
Ishiguro, K., 89*t*
Ishikawa, H., 583
Ishikawa, R., 543, 546
Ishikawa, T., 366*t*, 370
Ishiwata, T., 513–514
Ishizu, H., 488
Iskratsch, T., 193–194, 202–205
Isobe, T., 89*t*
Isokawa, K., 543, 546
Isoniemi, K.O., 358–361
Israeli, E., 392*t*, 395*f*, 398
Ito, M., 103
Itohara, S., 472
Itskovitz-Eldor, J., 436
Ivanek, R., 355–357
Ivaninskii, S., 7, 188–190, 406–409
Ivanyi, I., 260, 271*t*
Ivaska, J., 42, 86–87
Iwanaga, K., 445
Iwata, S., 167*t*, 170, 559–560, 563, 564*f*, 566, 567*f*, 569–574, 572*t*, 576–577
Iyer, S.A., 208–209
Iyer, S.V., 165, 357–358
Izawa, I., 88, 89*t*, 91–95, 96*f*, 101–102, 103*f*, 126–127, 197–198
Izhawa, I., 310
Izmiryan, A., 538–542, 546
Izquierdo, C., 496

J

Jacka, E.M., 165–170
Jackson, B.W., 75
Jacob, J.T., 310
Jacobs, R., 512–513
Jacobson, K.A., 312
Jacobson, M., 638–639
Jacomy, H., 463, 495
Jaenicke, R., 584–585
Jaffe, H., 511*f*, 514–515, 617, 621–622, 621*f*
Jahn, D., 560
Jahnel, O., 682–703
Jain, M., 445
Jaitovich, A., 117–120
Jakobsdottir, M., 379
Jalouli, M., 510–511
James, J., 244–246, 247*f*, 582–584
James, J.L., 238–241, 239*f*, 242*f*
James-Zorn, C., 644–645
Janig, E., 145–146, 353, 374–378
Janin, J., 264
Janmey, P.A., 38*t*, 41–50, 44*t*, 53, 391–421, 583, 682–683
Jansen, J.F., 141–142

Janssens, S., 490–491
Jarabek, C.J., 644–645
Jaros, E., 488
Jarrin, M., 582–603
Jarvius, J., 545, 547
Jarvius, M., 547
Jayasena, C.S., 452
Jayasundera, V.B., 372
Jeffery, W.R., 586
Jeney, S., 38*t*, 41, 411
Jensen, K., 481
Jensen, M.H., 42–50, 44*t*, 396–397, 415–420, 582–583
Jensen, P.K., 238–240
Jensen, T.G., 238–240
Jensen, U.B., 238–240
Jeon, K.W., 682–683
Jeong, H.J., 105
Jerabkova, B., 227
Jester, J.V., 198–200, 202–205
Ji, S., 521*t*
Ji, X., 366*t*
Jia, L., 521*t*
Jiang, W.Q., 512
Jiang, Y., 362–364
Jin, P., 637, 723
Jin, Y., 663
Jin, Z., 512–513
Jing, N., 584–585
Jing, R., 540, 542–546
Joanne, P., 543
Joedicke, L., 712
Jogia, D., 429, 560, 561*t*
Johansen, K.M., 718
Johansson, C.B., 513
Johnson Hamlet, M.R., 640
Johnson, A.B., 165
Johnson, D.A., 165–170
Johnson, F.B., 146–148
Johnson, G.V., 134
Johnson, L.N., 17–18
Johnson, O., 16
Johnson, P.F., 490–491
Johnson, R.G., 591
Jonas, E.A., 639–640
Jones, D.L., 569
Jones, D.R., 574
Jones, D.T., 5–6, 261, 271*t*
Jones, J.C., 65–66, 243
Jones, L.N., 264
Jones, M.R., 462–473
Jones, P.S., 594
Jones, Y., 488–489
Jong, A., 190
Jong, T.D., 190
Jonkman, M.F., 243
Jonkmann, M.F., 229–234, 231*t*, 238–241, 243–244
Jorcano, J.L., 143
Josephs, K.A., 488
Joyner, R.W., 616
Ju, J.H., 362–364
Julien, J.-P., 190, 462–464, 478–479, 481, 488, 490, 493–495, 638
Jungel, A., 324–325
Juni, A., 595*t*
Jurkko, R., 563
Just, S., 452

K

Kaartinen, M., 563
Kabsch, W., 16
Kachinsky, A.M., 512–513, 521*t*
Kahn, M., 354–355
Kaibuchi, K., 89*t*
Kaileh, M., 193–194
Kaiserman-Abramof, I.R., 617
Kajita, M., 395–396
Kalaparthi, V., 48*t*
Kalimo, H., 511*f*, 514–517
Källberg, M., 261, 271*t*
Kallioniemi, O., 141–142
Kamimura, J., 338
Kaminsky, R., 670
Kammerer, R.A., 11–12
Kan, D., 488
Kandow, C.E., 412
Kane, D.A., 593
Kane, L.A., 115–117
Kang, C., 463–464
Kang, H., 518
Kang, J.S., 362–364
Kania, J., 10*t*, 20, 27–28, 262–263
Kantorow, M., 586–587, 589*f*
Kanumuri, V., 60, 481
Kao, G.D., 521*t*

Kap, M., 157
Kapinos, L.E., 7, 10*t*
Kaplan, D.R., 490–491
Kaplan, E.L., 576
Kaplan, M.P., 478–481, 484–487, 636–637
Karabinos, A., 585–586, 662–663, 668, 669*f*, 670, 674, 684
Karantza, V., 141
Karginov, A.V., 80
Karpov, V., 490–491
Karpova, A., 487–488
Karsenti, D., 436
Karten, H.J., 594
Kas, J.A., 38*t*, 41, 44*t*, 53, 310, 357–358, 682–683
Kasahara, K., 86–106, 190, 191*t*, 193–195, 197–200, 202–205, 203*f*, 213–214
Kasahara, M., 193–194
Kasas, S., 38*t*, 41, 411
Kashiwagi, Y., 89*t*
Kashofer, K., 157, 374–377, 376*t*
Kaspar, R.L., 167*t*, 170, 237, 310, 318–319, 326
Kasper, M., 584–585
Katagata, Y., 362–364
Katsimpoulas, M., 429
Katus, H.A., 450–451
Katz, M., 436
Kaufman, M.H., 464
Kaufmann, E., 462–463
Kaufmann, P., 494
Kaur, H., 205–207
Kawachi, Y., 225*t*
Kawaguchi, A., 512–513
Kawahara, K., 513–514
Kawajiri, A., 89*t*, 97, 100–102, 105–106
Kawakami, K., 640
Kawamura, T., 362–364
Kawana, S., 513–514
Kawasaki, H., 89*t*
Kawasaki, K., 462
Kay, B.K., 638–640, 645, 647–649
Kay, J.N., 593
Keeffe, E.B., 358–361, 359*t*
Kegler, D., 490–491, 493–494
Keire, P., 529–530
Kelley, L.A., 261, 271*t*
Kemkemer, R., 62–64, 683
Kemler, R., 436
Kemppainen, K., 358–361
Kendrew, J.C., 258
Kennedy, A.R., 225*t*
Kennedy, B.K., 164–165
Kennel, S.J., 225*t*
Kenner, L., 376*t*
Kerns, M.L., 165–170, 167*t*, 237–238, 310, 709, 711
Kerr, J.P., 540, 542–543, 546
Keshav, K., 565
Keski-Oja, J., 61–62
Kettleborough, R.N., 452
Keyser, K.T., 594
Khan, P.A., 538
Khanamiryan, L., 539–542
Khaw, P.T., 195–196, 198–200
Khokha, M.K., 645
Khosla, C., 134, 373, 376*t*, 377–378
Khoury, G.A., 116*t*
Khuan Abby Ho, Y., 188–190, 189*f*, 191*t*, 193–197, 397–398
Khuon, S., 61–64, 392*t*, 393–394, 420–421, 521*t*
Kidd, G.J., 464, 473
Kidd, M.E., 86–87, 193–194, 391, 420–421
Kidson, S.H., 586
Kieckhaefer, J.E., 560
Kilkenny, A.E., 322–323, 338
Kim, D.I., 710*f*
Kim, H., 190, 202–205
Kim, J.U., 428–429, 436
Kim, K.B., 188–190, 189*f*, 191*t*, 193–197, 205–207, 206*f*
Kim, M.S., 7, 10*t*, 262–263
Kim, N.E., 582–584, 603
Kim, P.S., 6–7
Kim, R., 411–412
Kim, S., 310, 362–364
Kim, S.H., 191*t*, 193–194
Kim, Y., 663–667
Kim, Y.E., 188–190, 189*f*, 191*t*, 193–197, 397–398
Kimelberg, H.K., 240
Kimes, B.W., 439
Kimmel, C.B., 592–593
Kimura, H., 105
Kintner, C., 638–639, 645

Kippenberger, S., 244–245
Kirfel, J., 237–238
Kirmse, R., 38*t*, 412–413
Kirschner, M.W., 558–559
Kishaba, Y., 512–513
Kitamura, M., 134
Kitamura, T., 462
Kitenbergs, G., 41
Kittas, C., 442
Kitzmann, M., 515–516
Kivistö, S., 563
Kiyama, H., 89*t*
Kiyono, T., 105, 310
Klausegger, A., 231*t*
Klebes, A., 688–690, 702–703
Kleeberger, W., 513–514
Kleinbart, F., 636–637
Klein-Hitpass, L., 165–170, 167*t*, 237
Klier, F.G., 75
Kline, E., 50, 188–190, 191*t*, 193–197, 201–205, 397–398
Klingstedt, T., 374–377, 376*t*
Klocke, R., 513
Klocker, H., 92*t*
Klosen, P., 490–491
Klösgen, B., 695–696
Klukas, K.A., 591
Klymkowsky, M.W., 42–50, 450–451, 645, 650
Knittel, P., 683, 702–703
Knott, G., 227–229
Knust, E., 688–690, 702–703
Kobatake, Y., 89*t*
Kobayashi, T., 466
Kobilka, B.K., 431
Kobori, K., 105–106, 585
Koch, M., 683–684
Koch, T.M., 43–44*t*, 51
Kochin, V., 89*t*, 91, 114, 517–518, 520
Koehn, H., 290, 291*f*, 293*f*, 297
Kohler, R.E., 708
Kojima, S., 241–243, 393, 395–397, 415–417, 595*t*
Kokorin, A.I., 25
Kolár, Z., 513
Kolbas, O., 290
Kolbe, T., 143, 376*t*
Koller, B.H., 464
Kollmannsberger, P., 44*t*
Kollmar, M., 585–586
Kölsch, A., 61, 420–421
Komai, S., 89*t*
Komel, R., 358, 359*t*
Komine, M., 309
Kon, O.L., 290–291
Kon, T., 585, 600
Kong, L., 92*t*
Konig, K., 365–370
Konings, G., 560, 561*t*
Konishi, H., 89*t*
Kono, N., 563–565
Kononen, J., 141–142
Konoshita, J.H., 588*t*
Koorman, T., 702–703
Kopan, R., 309
Kopchick, J.J., 638
Kopeckova, L., 227
Korc, M., 513–514
Korde, N., 38*t*, 53
Kornhauser, J.M., 116*t*, 117, 118*t*
Kosako, H., 89*t*, 100–102, 105–106
Koshikawa, N., 225*t*
Kosik, K.S., 87–88
Kostareva, A., 23–24, 450–452
Kostavasili, I., 451–452
Kostavassili, I., 429
Koster, S., 582–583
Kota, P., 80
Kothe, S., 366*t*
Kotula, E., 89*t*
Kouloumenta, A., 428–429, 452–453
Kowalczyk, A.P., 417–418
Kowalsman, N., 190
Kozlov, S.V., 560, 561*t*
Kramer, B.E., 638
Krams, S.M., 358–361, 359*t*
Kranz, C., 683, 702–703
Krausz, E., 586, 588*t*
Kreplak, L., 4–5, 7, 22–23, 38*t*, 41, 179, 354–355, 411, 447–448, 450, 668, 672, 682–683
Krepler, R., 225*t*
Krieg, P.A., 638–640, 652–654
Krieger, A., 488–489
Krimm, I., 20, 27–28
Krishnan, S., 357–358, 365–370, 366*t*

Krissinel, E., 19
Kriz, J., 190, 493–494
Kröger, C., 243–244, 358–370, 366*t*, 682–683
Kroll, K.L., 640
Krone, P.H., 639–640
Kros, J.M., 488–489
Krsinic, G.K., 281
Krsinic, G.L., 282
Krug, M., 16
Kruse, E., 702–703
Kryzer, T.J., 496
Ksiezak-Reding, H., 87–88
Ku, N.O., 61–62, 89*t*, 114, 117–121, 124–129, 128*f*, 133–134, 140–141, 164–165, 166*f*, 177, 238, 260–261, 305, 310, 329–330, 352–355, 355*t*, 357–361, 359*t*, 365–372, 366*t*, 378, 462–463, 472–473, 480, 709
Kubben, N., 560, 561*t*
Kubicka, S., 309
Kucerova, J., 240
Kuck, J.F., 588*t*
Kuck, K.D., 588*t*
Kucukoglu, O., 359*t*, 366*t*, 370–374
Kuczmarski, E.R., 140–141, 391–421
Kuehl, W.M., 650, 653
Kufferath, I., 355, 546–547
Kuga, T., 89*t*
Kuhlmann, M., 513
Kuijpers, H.J., 225*t*
Kuisk, I.R., 428–429, 437–438, 445–446
Kulesh, D.A., 143, 359*t*
Kulik, A.J., 38*t*, 41, 411
Kuliyev, E., 640
Kulkarni, A.B., 468
Kumar, A., 60, 462–464, 472–473, 481, 486–487, 490
Kumar, N., 190
Kumar, V., 310
Kumari, V., 191*t*, 193–194
Kunii, E., 513–514, 520
Kunkel, L.M., 538, 540–543
Kuo, C.H., 202–205
Kuo, T.-L., 511–512, 516–517
Kupfer, H., 238
Kuppusamy, P., 195–196
Kural, C., 392*t*, 396
Kurowska-Stolarska, M., 324–325
Kuscuoglu, D., 352–379
Kusenda, B.J., 365
Kusubata, M., 88, 89*t*, 95, 101–102
Kuszak, J., 585
Kuzin, A.P., 7, 8*t*, 25, 262–263, 480
Kuzirian, A.M., 616
Kvarnstrom, S.M., 355–358, 365–370, 366*t*
Kwan, R., 117–120, 130–131, 132*f*, 167*t*, 171, 178–179, 182, 357–361, 366*t*
Kyte, J., 262

L

Labouesse, M., 683–684, 686, 688–690, 702–703
Lacène, E., 560, 561*t*, 563
Lacina, L., 221–247
Lackinger, E., 142–143, 146, 147*f*, 376*t*
Lackner, C., 151, 374
Laemmli, U.K., 566, 718
Lahat, G., 188–190, 193–196
Lähdeniemi, I.A.K., 352–379
Lahtela-Kakkonen, M., 190, 191*t*
Lai, R.Y., 569
Lakin, N.D., 638
Lakonishok, M., 396, 417–418
Laliberte, C.L., 518–520, 519*f*, 528–529
Lam, R., 10*t*, 20, 27–28, 262–263
Lamb, N.J., 515–516
Lamm, A.T., 667–668
Lammerding, J., 114
Lamzin, V.S., 18
Landegren, U., 355, 546–547
Landis, S.C., 573
Landmann, F., 683–684
Landon, F., 490–492
Landsbury, A., 86–87, 582–586, 603
Landt, O., 378–379
Landwehr, R., 11–12
Lane, E.B., 60, 62–64, 164–165, 221–247, 266–267, 305–310, 339, 352–353, 358, 359*t*, 583, 682–683, 702–703
Lane, H.C., 565
Lane, M.D., 305, 306–307*f*
Lane, W.S., 487, 637
Lang, A.B., 225*t*
Lang, R., 231*t*

Langbein, L., 141, 164–165, 260–261, 309, 317, 352–353
Langen, R., 21–22
Langer, R., 157
Langheinrich, U., 593
Langhofer, M., 65–66
Langowski, J., 22–23, 264, 391, 412–413
Lankford, T.K., 225*t*
Lanson, N.A., 439
LaPash Daniels, C.M., 165–170
Lapierre, L.R., 510–511
Lapointe, L., 490–491
Lara, M.F., 167*t*, 170
Larcher, F., 237, 366*t*
Larcher, J.C., 540–542, 546
Larivière, R.C., 494–495
Larson, A.R., 513–514, 521*t*
Larson, M.H., 362–364
Lasek, R.J., 486–487, 616–620, 620*f*
Latham, V., 116*t*, 117, 118*t*
Latinkic, B., 640
Lattanzi, G., 89*t*
Lau, A.W.C., 417–418
Lau, B., 290, 291*f*, 293*f*, 297
Lau, D.L., 191*t*, 193–194
Laufer, C., 452
Laulederkind, S.J., 259, 271*t*
Laurent, S., 494
Lauricella, A.M., 15
Lautscham, L., 43–44*t*, 51
Lawe, D., 490–491
Lawniczak, J., 584–585
Lawrence, J.B., 558–559
Lawrenson, J.G., 589–590
Layton, W.J., 188–190
Lazar, A.M., 193–194, 202–205
Lazarides, E., 538, 540, 542, 546, 591
Lazaro, G., 513–514, 521*t*
Lazaro, J.B., 515–516
Lazzeroni, L.C., 165, 378–379
Le Bihan, M.C., 560–562, 561*t*
Leachman, S.A., 231*t*, 237
Leahy, D.J., 7, 10*t*, 262–263
Leake, D., 487, 637
Lebedev, A.A., 19
Lebiedzinska, M., 444
Lecomte, M.J., 490–491
Leconte, L., 490–491
Lederer, M., 243–244, 682–683
Ledger, N., 684
Leduc, C., 241–243
Lee, B., 240, 682–683
Lee, B.T., 4–5, 165, 222, 259, 271*t*, 310, 355*t*, 709–710, 716
Lee, C.H., 7, 10*t*, 262–263, 306*f*, 333, 335–337
Lee, C.-W., 513–514, 521*t*
Lee, D., 338
Lee, D.K., 362–364
Lee, D.M., 189*f*, 190, 191*t*, 193–197, 205–207, 206*f*
Lee, E., 294, 295*f*, 297–299, 299*f*, 490–491
Lee, H., 511–512
Lee, H.B., 708–709
Lee, H.K., 565
Lee, J., 190, 191*t*, 193–194, 644–645
Lee, J.-A., 510–511
Lee, J.E., 406
Lee, J.K., 362–364
Lee, K.-F., 518–520, 519*f*, 528–529
Lee, K.K., 667–668, 684
Lee, K.M., 362–364
Lee, M.K., 463
Lee, S., 617
Lee, T.Y., 116*t*
Lee, V.M.-Y., 462–464, 473, 494, 636–637, 650
Lee, W., 190, 202–205
Lee, W.-C., 488
Lefebvre, C., 702–703
Legouis, R., 688–690, 702–703
Lehnart, S.E., 570–574, 572*t*
Lehner, M., 143, 372–373
Lehto, V.P., 61–62
Lehuen, A., 496
Lei, L., 190, 191*t*, 193–195, 197–200, 202–205, 203*f*, 213–214
Leidenius, R., 574–575
Leifeld, L., 366*t*
Leigh, I.M., 221, 225*t*, 229–234, 231*t*, 238–241, 243–244, 309–310, 583
Leight, J.L., 38*t*
Leimena, C., 429, 560–562, 561*t*
Leitgeb, S., 143, 366*t*, 370, 372

Leitner, A., 38*t*
Lemonnier, M., 492
Lendahl, U., 510–515, 518, 636–638
Lengler, J., 588*t*
Lennarz, W.J., 305, 306–307*f*
Leonard, D.G.B., 489–491, 638
Leonard, J.M., 117–120, 130–131, 132*f*
Leonard, M., 586
Leong, F.T., 575
Lepinoux-Chambaud, C., 582–583
LeShoure, R., 134
Leslie Pedrioli, D.M., 167*t*, 170
Leslie, A.G., 16
Leslie, B.J., 193–194
Lessard, J.C., 237–238, 310, 318–319, 326
Letai, A., 165, 166*f*, 221, 310
Leterrier, J.F., 38*t*, 41, 53, 540–542, 682–683
Lettner, T., 231*t*, 237
Letunic, I., 116*t*
Leube, R.E., 60–80, 234–235, 305–309, 354–355, 358–361, 365–372, 366*t*, 420–421, 662, 682–703
Leuchowius, K.-J., 547
Leung, B., 688–690
Leung, C.L., 494
Levada, K., 359*t*, 366*t*, 371–372
Levavasseur, F., 463, 488
Levine, R.L., 637
Levy, E., 478–479
Levy, R., 48*t*
Lewis, G.P., 213–214, 540–542
Lewis, S.A., 478–479
Ley, C.A., 490–491
Lezcano, C., 513–514, 521*t*
Li, A., 488
Li, C., 463–464, 473
Li, C.J., 362–364
Li, D.H., 366*t*
Li, H., 428–429, 437–438, 445–446
Li, J., 373, 376*t*
Li, L., 324–325, 513, 540, 542, 546
Li, M., 452, 513–514, 594
Li, S., 86–87, 294, 311, 337, 397–398
Li, T., 512–513
Li, W.H., 708–709
Li, X., 488–489
Li, Y., 708–709
Li, Z., 429, 451–452, 512, 539–543, 546, 588*t*
Liang, S., 290
Liang, S.M., 406–409
Liao, C.X., 447–448
Liao, G.C., 714
Liao, H., 167*t*, 170
Liao, J., 89*t*, 114, 121, 124, 126–127, 140–141, 238, 260–261, 305, 329–330, 354–355, 357–358, 359*t*, 362–364, 366*t*, 462–463, 472–473
Liao, R., 445
Libowitz, A.A., 193–194, 585, 591
Lichtenberg-Kraag, B., 87–88
Lichti, U., 319, 322–323
Lidov, H.G., 538–540, 543
Lieber, J.G., 451
Liem, R.K.H., 60, 395–396, 462, 478–497, 636–637
Lienau, T.C., 378
Ligezowska, A., 695–696
Likhtenshtein, G.I., 25
Lill, J.R., 126
Lillycrop, K.A., 638
Lim, J.K., 358–361, 359*t*
Lim, L., 89*t*, 96*f*, 98–102, 105–106
Lim, S.S., 60, 393
Lim, T.M., 452
Lim, W.A., 362–364
Lima, C.D., 115–117, 129
Lin, D., 589–590
Lin, F., 558–559
Lin, H., 490–491, 637
Lin, L.R., 588*t*
Lin, S.P., 588*t*, 593–594
Lin, W., 464, 473, 488, 518–520, 637, 645
Lin, Y.-C., 38*t*, 41, 80, 290, 411, 682–683
Linberg, K.A., 213–214
Lincoln, B., 44*t*
Lindenbaum, M., 463
Lindqvist, J., 510–530
Ling, G., 311, 337
Liovic, M., 62–64, 229–234, 231*t*, 240, 266–267, 358, 359*t*, 682–683
Lipton, A.M., 488
Liu, C.S., 589–590
Liu, G., 682–683
Liu, H.-N., 492

Liu, J., 188–190, 193–196, 667–668, 684, 708–709
Liu, L., 164–182, 434, 512–513
Liu, T.-C., 487–488
Liu, W., 259, 271*t*
Liu, W.Y., 593
Liu, X.-M., 528–529
Liu, Y., 27, 50, 86–87, 188–190, 191*t*, 193–197, 201–205, 397–398, 582–583, 585–586, 603
Liu, Z., 290
Livengood, D., 617
Llinas, R.R., 616, 628
Lloyd, C., 237–238, 309
Lo, S.C., 286–288
Lobsiger, C.S., 463–464, 472–473
Lockett, T., 357–358
Loeber, J., 640
Loew, D., 89*t*
Loh, S.H., 452
Lohkamp, B., 267, 271*t*
Loitsch, S., 244–245
Loll, P.J., 12
Long, F., 19
Longenecker, G., 468, 515–516
Longley, M.A., 237–238
Looi, K., 167*t*, 171, 178–179, 182, 358
Loomis, P.A., 393–394
Loparic, M., 43–44*t*, 51, 52*f*, 682–683
Lorand, L., 44*t*, 51–52, 134, 357–358, 683
Loranger, A., 310, 357–358, 362–370, 366*t*
Lorenz, J.N., 451–452
Lorie, E.P., 231*t*, 235–237
Lorincz, A.L., 325–326
Loschke, F., 243–244, 357–358, 682–683, 709
Lotsch, D., 668, 672
Lotz, M.K., 43–44*t*, 50
Love, D.R., 593
Love, F., 518
Love, N.R., 640–641
Love, R., 417–418
Lovering, R.M., 366*t*, 370, 543
Lovicu, F.J., 590
Lowe, J.S., 488
Lowery, J., 391–421
Lowthert, L.A., 89*t*, 121, 124, 329–330, 354–355
Lu, C.T., 116*t*
Lu, H., 75, 165–170, 167*t*, 237, 261, 271*t*
Lu, J.T., 165–170, 559
Lu, K.-S., 487–488
Lu, P., 513–514
Lu, S.C., 366*t*
Lu, X.-y., 488–489
Lu, Z., 429
Lubensky, T.C., 412, 417–418
Lubsen, N.H., 584–585
Luczak, J., 43–44*t*, 51
Lüdke, A., 513
Luecke, H., 21–22
Luft, J.R., 15
Lukka, R., 512–515, 518
Lulevich, V., 682–683
Lumb, C., 708–709
Luna, G., 213–214, 540–542
Lunardi, J., 281–282
Lund, L.M., 540, 542–543, 546
Lundegaard, C., 6–7, 262, 271*t*
Lunny, D.P., 310, 366*t*, 370
Luo, H., 260–261, 520
Luo, R., 266–267, 271*t*
Luo, X., 434
Luo, Z.J., 391, 405
Lupas, A.N., 6–7, 17, 262, 271*t*
Lupinetti, J., 540, 542–543, 546
Lussier, M., 309
Lustig, A., 7, 11–12, 188–190, 406–409, 668
Lustig, R.A., 521*t*
Lyakhovetsky, R., 667–668, 711
Lynch, C.D., 193–194, 202–205
Lysz, T.W., 588*t*

M

Ma, A.S.P., 325–326
Ma, L.L., 38*t*, 41, 86–87, 246, 310, 334–335, 586
Ma, Z.Y., 588*t*
Maaloum, M., 48*t*
Maar, T., 240
Mabuchi, T., 362–364
Macaluso, F.P., 478–480
Macarthur, K.M., 521*t*
MacCallum, S.F., 366*t*, 370
Maccario, H., 233
Macgregor, G.R., 446

Machaidze, G., 7
Macioce, P., 484–485
Mackenzie, I.C., 225*t*
Mackenzie, I.R.A., 488
MacKenzie, R., 417–418
Mackintosh, F.C., 38*t*, 41, 411, 417, 682–683
Maclaren, J.A., 281
MacRae, C., 559
MacRae, T.P., 262–264
Maeder, M.L., 594
Maeno, K., 513–514, 520
Magdeldin, S., 355–357
Magee, G.J., 227
Maghnouj, A., 362–364
Magin, T.M., 44*t*, 53, 61–62, 75, 86–87, 114, 143, 164–170, 167*t*, 233, 237–238, 305–310, 352–355, 357–372, 366*t*, 376*t*, 378, 582–583, 682–683, 708–723
Mahadevan, D., 193–194
Mahajan, V., 353, 372, 374–377, 376*t*
Mahammad, S., 42–50, 44*t*, 140–141, 188–190, 191*t*, 193–195, 197–198, 201–205, 391, 392*t*, 394–398, 395*f*, 412–413, 415–418
Mahanti, P., 691–692
Mahmood, H., 44*t*
Maier, H.J., 429, 443
Maillet, J.-C., 521*t*
Mainka, A., 43–44*t*, 51
Maisel, H., 584–586
Maki, M., 134
Makihara, H., 105–106
Maksym, G.N., 44*t*
Mallet, J., 494–495
Malmgren, K., 542, 545
Mandelkow, E.M., 87–88
Mandell, J., 487–488
Manenti, S., 585
Maniotis, A., 42–50, 44*t*, 198–200, 413–414, 682–683
Mann, D.L., 451–452
Mann, M., 115–117, 116*t*, 130–131
Mann, R.W., 637
Manne, U., 148
Manser, E., 89*t*, 96*f*, 98–102, 105–106
Mansergh, F.C., 591
Mansilla, F., 362–364
Mansour, S.L., 464
Mao, F.F., 528–529
Mao, X., 412
Marcantonio, J.M., 588*t*, 589–590
Marceau, N., 61–62, 310, 357–358, 362–364, 366*t*, 376*t*
Marcus, A.I., 190, 191*t*, 193–194
Marek, J., 227
Marekov, L.N., 22–23, 260–264, 391, 512
Marg, B., 411–412
Margalit, A., 667–668
Margolick, J.B., 121
Marie, Y., 488–489
Marini, N.J., 639–640
Marinos, E., 442
Markl, J., 594
Marron, M.T., 193–194
Marschall, H.U., 372–373
Marsh, V., 568–569
Marszalek, J.R., 463
Marti, O., 38*t*, 683
Martin, R., 617–618
Martinet, D., 227–229
Martinez, A.E., 188–190, 193–194, 197–198
Martinez, L., 708–709
Martinez-Morentin, L., 708–709
Martinez-Palacio, J., 366*t*
Martys, J.L., 60, 395–396
Masaki, S., 582–585, 603
Mashukova, A., 362–365
Mason, C.A., 480
Mason, D.Y., 141–142
Massart, C., 560, 561*t*, 563
Masson, G., 379
Masukawa, L.M., 617
Masur, S.K., 198–200
Matej, R., 542
Mathew, J., 366*t*
Mathews, K.D., 708–709, 711, 721, 723
Mathias, R.T., 583–584, 603
Matsubara, D., 512–513
Matsuda, Y., 513–514
Matsui, S., 89*t*, 96*f*, 98–102, 105–106
Matsumoto, N., 89*t*
Matsumura, K., 225*t*
Matsuoka, Y., 87–88, 89*t*, 95, 101–103
Matsushita, K., 89*t*

Matsuura, K., 445
Matsuyama, M., 87, 105–106, 585
Matsuzawa, K., 89*t*
Mattina, S., 463–464, 473
Mattout, A., 668, 669*f*
Matveeva, E.A., 202–205, 396–397
Maul, G., 558–559
Maurer, M., 311, 337
Mavilio, F., 227–229
Mavroidis, M., 428–429, 451–453
Mayer, R.J., 569
Mazzalupo, S., 325
McAbney, J.P., 588*t*
McAvoy, J.W., 589–590
McCarthy, B., 563
McConnell, W.B., 640
McCormack, P.L., 568–569
McCormick, M.E., 38*t*
McCoy, A.J., 16–17
McCuaig, J.V., 62–64, 244–245, 583–584, 603, 683, 702–703
McCulloch, A.D., 429
McCulloch, C.A., 190, 202–205
McDaniel, K.M., 141–142, 241–243
McDermott, M.F., 569
McGhee, J.D., 683–684, 686
McGowan, K.M., 309, 315*t*, 330
McGrath, J.A., 583
McGraw, T.S., 487
McGuffin, L., 261, 271*t*
McGuinness, T.L., 628
McIntyre, G.A., 260–261
McKay, I.A., 311, 337
McKay, R.D.G., 510–513, 636–638
McKeegan, K., 543
McKeith, I.G., 488
McKeon, F.D., 558–559
McKern, N.M., 264
McKie, N., 588*t*
McLachlan, A.D., 262–264
McLean, J.R., 492, 496
McLean, W.H., 140–141, 164–165, 309–310, 415
McMahon, A.C., 429, 560, 561*t*
McMahon, A.P., 436–437
McMahon, L., 688–690
McMillan, J.R., 309
McNamara, M., 588*t*
McNeil, J.A., 558–559
McNicholas, S., 19
McPherson, A., 15
McReynolds, M.A., 490
Mearns, B.M., 134
Medalia, O., 668
Meder, B., 452
Meder, L., 365–370
Medina, L., 490–491
Meding, S., 157
Mege, J.L., 44*t*
Mehta, V.V., 589–590
Meier, M., 7, 8*t*, 19–20, 262–263, 400
Meier, P., 576
Meijer, D., 478–479
Meinkoth, J.L., 637
Melamed-Book, N., 668, 669*f*
Melcer, S., 668, 669*f*
Melino, G., 322–323
Mellodew, K., 514–515
Mellone, S., 494
Melton, D.A., 638–640, 645
Meltzer, S.J., 589–590
Melzak, K.A., 48*t*
Mendez, M.G., 42–50, 44*t*, 86–87, 193–194, 201–205, 241–243, 391, 392*t*, 393–397, 411–418, 493–494
Meng, W., 290–291
Menko, A.S., 193–194, 585–586, 591
Mericskay, M., 429, 451–452
Merkel, R., 682–703
Merlini, L., 559
Merlino, G., 366*t*, 370–371
Merz, K.M., 266–267, 271*t*
Messing, A., 165
Mestroni, L., 559
Mesturoux, L., 488–489
Meta, M., 558–559
Metsärinne, K., 574–575
Metuzals, J., 616–619
Meyer, M., 668
Meyers, C.A., 315*t*
Meyuhas, O., 652–654
Mezulis, S., 261, 271*t*
Michael, R., 584–585
Michael, S., 262, 271*t*
Michaelevski, I., 190
Michel, B.A., 324–325

Michel, M., 309
Michele, D.E., 543
Michibayashi, C., 586
Michie, S.A., 166*f*, 177, 352–353, 365–372, 366*t*
Michurina, T.V., 513–514
Mickle, J.P., 487
Micoulet, A., 43–44*t*, 52–53, 682–683
Middendorff, R., 513
Miettinen, M., 141–142
Migheli, A., 494
Mignot, C., 166*f*
Mikhailov, A., 50, 60, 395–396, 511*f*, 514–517
Mikhaylova, M., 487–488
Miklavc, P., 683, 702–703
Miledi, R., 617–618
Miletic, D., 492
Millan, J.L., 586
Millecamps, S., 494–495
Miller, C.C.J., 472
Miller, J.B., 512–513, 521*t*
Miller, R., 174
Miller, R.K., 392*t*
Milligan, B., 281
Millward, G.R., 264
Milner, D.J., 428–429, 436–438, 451–452, 583
Milstone, L.M., 231*t*
Milting, H., 513
Min, Y.S., 362–364
Minguez, P., 116*t*
Minin, A.A., 202–205, 396–397
Minneci, F., 5–6
Minor, W., 16
Miocic, S., 559
Mishra, L.C., 188–190
Misiorek, J.O., 352–379
Mitake, S., 487–488
Mitchison, J.M., 695–696
Mittal, B., 60
Miwa, S., 513
Miyai, K., 365–370, 366*t*
Miyamae, Y., 89*t*
Miyasaka, H., 60
Miyata, T., 512–513
Miyatake, Y., 193–194
Miyazaki, K., 492
Miyazaki, Y., 463–464, 473
Mizaikoff, B., 683, 702–703
Mizuno, D., 417
Mizuno, Y., 538–543
Mizushima, H., 225*t*
Mizushima, K., 540
Mizutani, H., 96*f*, 101–102, 103*f*
Moch, M., 60–80, 234–235
Modderman, P.W., 225*t*
Moeller, D., 668
Moffitt, J.R., 399
Mogensen, M.M., 62–64
Mohammed, H.H., 366*t*, 370
Mohan, R., 188–214
Mohseni, P., 518–520, 519*f*, 528–529
Mohun, T.J., 639–640
Moir, R.D., 60–64, 202–205, 392*t*, 395–396
Mokhtari, K., 488–489
Mokrý, J., 513, 584–585
Moll, I., 309
Moll, R., 141, 221, 225*t*, 309, 352–353
Molloy, J.E., 416
Molnar, A., 374
Monaghan, P., 442
Monckton, D.G., 588*t*
Monsma, P.C., 420–421
Montanez, E., 311, 337
Monteiro, M.J., 463
Montelione, G.T., 5–6
Montgomery, C., 436–437
Mooberry, S.L., 170–171
Moon, R.T., 638–639
Moon, T.J., 44*t*
Moons, D.S., 372
Moore, C.B., 310
Moore, H.M., 148–149
Moore, J.E., 167*t*, 170
Moore, J.R., 42–50, 44*t*, 396–397, 415–420
Moore, S.A., 708–709, 711, 721, 723
Moore, T., 391, 420–421, 584–585
Moore, W., 682–683
Moossa, A.R., 513
Morabito, L.M., 708–709
Morales-Nebreda, L., 391, 405–406, 409–411, 420–421
Moran, J., 240
Morcos, P.A., 593

Morfini, G., 628
Mori, H., 563–565
Mori, M., 366*t*, 370
Moritoh, C., 89*t*, 102–103
Moriyama, K., 225*t*
Morley, S.M., 172–173, 229, 231*t*, 238–241, 239*f*, 242*f*, 243–244, 310
Mornon, J.P., 20, 27–28
Morreale de Escobar, G., 366*t*, 370–371
Morris, E.J., 582–583
Morris, J.R., 486–487
Morris, P.J., 638
Morrow, J.P., 559–560, 563, 564*f*, 566, 567*f*, 569–574, 572*t*, 576–577
Mor-Vaknin, N., 188–190, 189*f*, 191*t*, 193–197, 397–398
Mounkes, L.C., 560, 561*t*
Mouquet, F., 445
Moustafa, T., 372–373
Moustakas, A., 231*t*, 235–237
Mouyobo, C.E., 231*t*, 235–237
Moya, K.L., 490
Moyle, P.M., 126
Moynihan, P.F., 574
Mrksich, M., 582–583
Mrock, L., 586
Muchir, A., 165–170, 167*t*, 558–577
Mücke, N., 7, 22–24, 38*t*, 41, 264, 391, 411–413, 450–451, 668, 672
Muellener, D.B., 639–640
Mueller, U., 16
Muir, T.W., 126
Mukai, S., 193–194
Müller, D., 513
Muller, J., 244–245
Muller, M., 391, 420–421, 584–585
Muller, S.A., 4–6, 23–24, 391, 394, 462–463
Muller-Rover, S., 311, 337
Mulligan, R.C., 445
Mullins, M.C., 593
Muma, N.A., 493–494
Mun, J.J., 310
Munick, M.D., 594
Munir, M.I., 428–429, 437–438
Munn, E.A., 686
Muñoz, V., 266–267, 271*t*
Munoz-Marmol, A.M., 708–709
Muntasser, S., 495
Muntoni, F., 559
Murata, S., 366*t*
Murauer, E.M., 237
Murayama, A., 512–513
Muriel, J., 543
Murphy, A.J., 518–520, 519*f*, 528–529
Murray, B., 116*t*, 117, 118*t*
Murray, B.W., 569, 571*t*
Murray, C.I., 61–62
Murray, M.E., 411–413
Murrell, D.F., 224
Murthy, S.N., 140–141, 201–205, 391, 392*t*, 394–398, 395*f*, 412–413
Mushynski, W.E., 167*t*, 170, 462–463, 478–479, 490, 494–495
Musil, L.S., 586
Muto, A., 593

N

Naami, A., 542
Nagai, T., 445
Nagaishi, M., 488–489
Nagalingam, A., 195–196
Nagashima, K., 560–562, 561*t*
Nagashima, Y., 225*t*
Nagata, K., 88, 89*t*, 91–95, 96*f*, 97–102, 105–106
Nagel, R.M., 15
Nagle, R.B., 141–142, 241–243
Nagy, P.L., 165–170, 559, 708–709, 711, 721, 723
Nahon, P., 372
Naini, A., 494
Naito, A.T., 445
Naito, Y., 595*t*
Naito, Z., 513–514
Nakade, S., 593–594
Nakamichi, I., 366*t*, 370–372
Nakamura, A., 543, 546
Nakamura, F., 190, 202–205
Nakamura, K., 586
Nakamura, N., 510–511
Nakamura, T., 588*t*
Nakamura, Y., 89*t*, 617
Nakano, T., 103
Nakayama, T., 640–641
Nakazato, Y., 488–489
Nakazawa, N., 513–514

Nam, K., 362–364
Nam, Y.J., 434
Namikawa, K., 89*t*
Nan, L., 373
Nasevicius, A., 593
Nathrath, W.B., 225*t*
Navajas, D., 44*t*
Navsaria, H., 172–173, 221, 229, 231*t*, 238–240, 243–244, 310
Needham, D., 695–696
Neelakantan, A., 208–209
Nekrasova, O.E., 391, 417–418
Nelson, C., 601*f*
Nelson, W.G., 309
Nemecek, S., 584–585
Neuber, C., 560–562, 561*t*
Neufeld, E., 667–668
Nevo, J., 42, 86–87
Newman, C., 411–412
Newman, D.J., 188–190
Ng, Y.Z., 221–247
Ngai, K.L., 95, 101–102
Nguyen, J., 708–709
Nguyen, M.D., 492, 494–495
Ni, G., 573–574
Nicolet, S., 6–7, 8*t*, 20–21, 24–25, 262–263, 267, 480
Nie, Z., 637
Niel, F., 560, 561*t*, 563
Nielsen, M., 6–7, 262, 271*t*
Nielsen, M.E., 513–514
Nielsen, M.L., 115–117
Nieman, M., 451–452
Nieminen, M.I., 370
Nieuwkoop, P.D., 638–639, 645
Nigam, R., 259, 271*t*
Nigg, E.A., 100–102, 105–106, 668
Nijveen, H., 260–261
Niki, T., 512–513
Nikitina, N.V., 586
Nikolova, V., 429, 560, 561*t*
Nikolova-Krstevski, V., 560–562, 561*t*
Niloff, M.S., 637
Nilsson, K.P., 374–377, 376*t*
Nimmo, E.R., 391, 405, 420–421
Nishi, Y., 87, 89*t*, 103, 104*f*
Nishida, Y., 130–131
Nishikawa, Y., 193–194
Nishimura, Y., 585, 600
Nishizawa, K., 87–88, 89*t*
Nishizawa, M., 310
Nishizawa, Y., 225*t*, 539–542
Nitschke, R., 542
Nixon, R.A., 481, 636–637
Nobusawa, S., 488–489
Noe, E., 166*f*
Noensie, E., 332–333
Noetzel, E., 542
Nolan, G.P., 446
Nomoto, T., 585, 600
Nomura, F., 89*t*
Nordeen, S.K., 451
Nordgren, N., 43–44*t*, 50–51
Noumen, E., 644–645
Nowak, R.B., 582–584, 603
Nozaki, M., 366*t*, 370
Nozaki, N., 89*t*, 105
Nugent, T.C., 5–6
Nukina, N., 87–88
Nuraldeen, R., 353, 372, 374–377
Nutt, S.L., 638–639
Nyström, J.H., 352–379

O

O'Leary, E.C., 569, 571*t*
Oberle, E., 141–142
Oblinger, M.M., 493–494
O'Brien, K.F., 538, 543
Ochi, H., 586
O'Connor, M.D., 589
Odaka, C., 366*t*
Odermatt, B.F., 225*t*
Oelschlager, D.K., 148
Oeste, C.L., 188–190, 193–194, 197–198
Offield, M.F., 638
Ogawara, M., 89*t*, 95, 96*f*, 101–103, 104*f*
Ogg, S.C., 4–5, 165, 222, 259, 271*t*, 310, 355*t*, 709–710, 716
Ogino, H., 586, 640
Oguri, T., 89*t*
Oh, S., 362–364
Ohashi, H., 105
Ohga, R., 593–594
Ohki-Hamazaki, H., 595*t*
Ohmuro-Matsuyama, Y., 105
Ohnishi, M., 466

Ohno, S., 89*t*, 193–194
Ohoka, A., 395–396
Ohsawa, F., 105
Ohshima, T., 515–516
Oikonomopoulos, A., 445
Oka, T., 585, 600
Okabe, S., 60
Okada, M., 462
Okamoto, H., 588*t*, 593–594
Okamoto, K., 540–542
O'Keefe, H.P., 645
Okray, Z., 708–709
Okumura, E., 95, 101–102
Okumura, M., 539–542
Okumura-Noji, K., 487–488
Okuzaki, D., 193–194
Oliva, J., 376*t*
Olivé, M., 543
Olmsted, J., 50
Olsson, C., 545
Omary, M.B., 43–44*t*, 52–53, 61–62, 89*t*, 91, 114–135, 140–141, 164–182, 197–198, 238, 259–261, 305, 310, 329–330, 352–361, 355*t*, 359*t*, 365–378, 366*t*, 415, 462–463, 472–473, 514–515, 530, 582–583, 682–683, 709
O'Neill, A., 366*t*, 370, 543
Ooukayoun, J.P., 309
Opal, P., 86–87, 392*t*, 393–394, 420–421
Oprescu, A.-M., 496
Orenstein, J.M., 337–338
Orger, M.B., 593
Oriolo, A.S., 358, 362–365, 366*t*
Orndahl, C., 542, 545
Orr, A.W., 543–546, 551*f*
Ortiz, M., 411–412
Orwin, D.F.G., 281–282, 283*f*
Osborn, M., 221, 636–637
Oshima, R.G., 75, 143, 166*f*, 177, 359*t*, 365–372, 366*t*
Osinska, H., 451–452
Osmanagic-Myers, S., 164–165, 243–244
Osterreicher, J., 513
Ostlund, C., 20, 27–28, 166*f*
Osumi, N., 510–511
Oswald, F., 62–64, 683, 702–703
Ota, S., 593–594
Otali, D., 148
Otis, T.S., 617–618
Otsuka, F., 225*t*
Otto, J.J., 636–637
Otwinowski, Z., 16
Ouellet, M., 366*t*
Owaribe, K., 225*t*
Oyama, T., 445
Ozaki, Y., 193–194
Ozasa, H., 513–514, 520
Ozawa, T., 488

P

Pachter, J.S., 462, 478–479
Pack, C.G., 89*t*
Padilla, G.P., 7, 8*t*, 19–20, 262–263, 400
Page-McCaw, P.S., 593
Paladini, R.D., 309, 332–333
Pallari, H.-M., 42, 86–87, 91, 114, 193–194, 391, 511*f*, 514–520, 519*f*, 521*t*, 522–523, 525, 527–529
Paller, A.S., 165, 166*f*, 221, 310
Palmer, I., 406–409
Palmer, J.E., 569
Pan, F.C., 640
Pan, X., 115–117, 305, 582–583
Pan, Y., 540, 542–546
Panagopoulou, P., 451–452
Pandey, C., 290–291
Pandit, T., 589–590
Pang, L., 568–569, 571*t*
Pang, N.N., 43–44*t*, 50
Panjikar, S., 18
Panneels, V., 712
Pant, H.C., 91, 126–127, 511*f*, 514–517, 616–629
Papadopoulos, A., 560, 561*t*
Papathanasiou, S., 428–430, 443
Paradis, G., 445
Paranthan, R.R., 189*f*, 190, 191*t*, 193–197, 205–207, 206*f*
Parca, L., 116*t*
Paredes, R., 640–641
Paris, S., 488–489
Parisi, A.F., 574
Park, D., 528–529
Park, H., 89*t*, 126–129
Parker, K.K., 44*t*, 53
Parkhurst, A., 586

Parlakian, A., 543
Parr, B., 512–513, 638
Parry, D.A.D., 86–87, 256–273, 352–353, 391, 480, 512, 663, 669*f*, 670, 674
Parthasarathy, V., 18
Parysek, L.M., 490–491, 493–494, 636–638
Pasantes-Morales, H., 240
Paschke, S., 43–44*t*, 52–53, 682–683
Pashley, R.M., 44*t*, 50–51
Pasquali, C., 88, 391, 405
Passini, M.A., 637
Pásti, G., 686, 702–703
Pastor, X., 496
Patchornik, A., 289–290
Patel, P., 50, 188–190, 191*t*, 193–197, 201–205, 397–398
Patel, T.R., 7, 8*t*, 19–20, 262–263, 400
Paton, I.R., 585–586
Patterson, G.H., 395–396
Patton, E.E., 593
Paul, M., 527–528, 538–552
Paulin, D., 429, 451–452, 462, 512, 538–540, 542–543, 546, 636–637
Paulsen, B.S., 540–542, 546
Paulson, H.L., 708–709
Paulsson, J., 547
Paululat, A., 708–709, 711, 716, 723
Paus, R., 311, 337
Paust, T., 38*t*, 683
Pavlidis, P., 559–560, 563, 565, 567–568, 567*f*
Pawelzyk, P., 38*t*, 41
Peckl-Schmid, D., 237
Peden, A.A., 417–418
Pei, Y., 594
Peking, P., 237
Pekny, M., 540, 542, 546
Pekny, T., 542
Pelham, R.J., 411–412
Pellegrini, G., 227–229
Pelling, A.E., 47*f*
Pellissier, J.F., 538–540
Peltekian, E., 540–542
Peltola, O., 365–371, 366*t*
Pena, S.D., 50
Peng, H.B., 638–640, 645, 647–649
Peng, J., 261, 271*t*
Peng, Y., 708–709
Penner, J., 365–370, 366*t*
Percy, M.E., 495
Perera, N.M., 233
Perez-Sala, D., 188–190, 193–194, 197–198
Perkins, B.D., 585
Perlson, E., 190
Perng, M.D., 583–586
Perou, C.M., 141–142
Perozo, E., 27
Perrot, R., 392*t*, 395*f*, 398, 495
Perry, J., 512–513
Perry, M.M., 585–586
Perry, R.H., 488
Perry, T., 708–709
Pestonjamasp, K.N., 80
Peter, A., 538, 585–586, 709, 710*f*
Peter, M., 668
Peterhoff, C.M., 462, 490
Peters, B., 75, 237–238
Peters, J., 6–7
Petersen, B., 6–7, 262, 271*t*
Petersen, T.N., 6–7, 262, 271*t*
Peterson, A., 462, 638
Petrak, J., 355–357
Petroll, W.M., 198–200, 202–205
Pettersson, T., 43–44*t*, 50–51
Pezzulo, T., 494
Pfister, O., 445
Piatigorsky, J., 585–586
Pieler, T., 640
Pieper, F., 585
Pierandrei-Amaldi, P., 652–654
Pietras, K., 545
Pigino, G., 628
Pihl-Lundin, I., 231*t*, 235–237
Pikaart, M.J., 640–641
Pilipiuk, J., 702–703
Piloto, S., 708–709
Pina-Paz, S., 310, 318–319, 326
Pincon-Raymond, M., 429, 451–452
Pincus, Z., 691–692
Pino, G., 590
Pintar, J.E., 481
Pinton, P., 444
Pipas, J.M., 588*t*
Pirrotta, V., 712
Piskorowski, R.A., 546–547
Pitre, A., 543–546, 551*f*

Pittenger, J.T., 21
Pittock, S.J., 496
Piwko-Czuchra, A., 311, 337
Pizard, A., 561*t*
Pizzamiglio, S., 157
Pizzolato, G., 542–543, 545–546
Planas, R., 496
Plancha, C.E., 238
Planko, L., 165–170, 167*t*, 237
Plenefisch, J., 663
Plessmann, U., 462–463
Plodinec, M., 43–44*t*, 51, 52*f*, 682–683
Plotnikov, A., 567–568
Plowman, J.E., 280–300, 354–355
Pogoda, K., 411–412
Pohl, E., 582–585
Poidevin, M., 723
Pollet, N., 640
Pollheimer, M.J., 366*t*, 370
Polson, A.G., 650
Pommepuy, I., 488–489
Ponce, J.M., 708–709, 711, 723
Ponchel, F., 575
Ponferrada, V.G., 644–645
Poosch, M., 584–585
Porcu, M., 559
Porter, R.M., 305–309, 339
Portet, S., 412–413
Portier, M.-M., 489–491, 493–494, 636–637
Pospelov, V.A., 638
Pospelova, T.V., 638
Potterton, L., 19
Poujeol, C., 490–491
Poul, M.A., 515–516
Pournin, S., 429, 451–452
Powell, H.R., 16
Powers, M., 647–649
Praetzel-Wunder, S., 260–261
Prahlad, V., 60, 202–205, 260–261, 392*t*, 395–396, 512
Prasher, D.C., 60
Prats, C., 429
Pratt, M.M., 618–619
Preisegger, K.H., 376*t*
Prescott, A.R., 62–64, 582–585, 588*t*, 603
Prescott, Q., 391, 420–421, 584–585
Price, D., 121, 305, 354–355, 493–494
Price, J., 365–370, 366*t*
Price, M.G., 540, 542
Priess, J.R., 688–690
Procaccia, S., 567–568
Procaccino, D., 518
Prochazkova, M., 511*f*, 514–515
Procter, J., 5–6
Progida, C., 496
Prudner, B.C., 540, 543, 546
Pruss, R.M., 478–479, 489–490, 616, 628
Psarras, S., 428–429
Puca, A.A., 538, 543
Puertas, M.C., 496
Pugh, J., 493–494
Purcell, C., 489–490
Purkis, P.E., 225*t*
Puz, S., 429, 436
Pylvänäinen, J.W., 513–514, 521*t*

Q

Qi, L.S., 362–364
Qi, X., 434
Qian, L., 434–435
Qiao, H.H., 708–709
Qin, Z., 41
Qu, B., 585–586
Quaranta, M., 105
Quinlan, R.A., 86–87, 121, 582–603
Quitschke, W., 594
Qvarnström, S.M., 513–514, 521*t*

R

Rabilloud, T., 281–282
Rainer, I., 377–378
Raju, P., 494
Raju, T., 636–637
Rakic, J.M., 589–590
Ralston, E., 429
Ramachandran, P., 721
Ramaekers, F.C., 225*t*
Ramalho, J.J., 702–703
Raman, S.V., 563
Ramana, K.V., 591
Ramburan, A., 141–142
Ramdial, P.K., 141–142
Ramirez, A., 366*t*, 370–371
Ramírez, M., 529–530
Ramli, U.S., 290, 291*f*, 293*f*, 297
Rammensee, S., 38*t*, 53

Ramms, L., 53, 682–683
Ramspacher, C., 452
Rantanen, J., 512–515, 518
Rao, M.V., 60, 462–464, 466, 472–473, 481, 486–487, 490
Rao, Y., 723
Rappoport, J.Z., 392*t*, 395–396, 399, 403, 403*f*
Rathje, L.S.Z., 43–44*t*, 50–51
Ravindranath, P.P., 188–190, 191*t*
Read, R.J., 16–17
Real, F.X., 378–379
Rebehmed, J., 116*t*
Recchia, A., 227–229
Recillas-Targa, F., 640–641
Reddan, J.R., 586, 588*t*
Reddan, P.R., 588*t*
Reddy, P.P., 487–488
Reddy, T.R., 488–489
Reddy, V.N., 588*t*
Reed, P.W., 366*t*, 370
Reese, T.S., 618–619
Regot, S., 80
Rehen, S.K., 540–542, 546
Rehm, E.J., 714
Rehm, M., 359*t*, 366*t*, 371
Reichelt, J., 237–238, 322
Reichenzeller, M., 86–87, 233
Reidler, D., 582–583
Reilly, S.A., 167*t*, 170, 570–574, 572*t*, 576–577
Reipert, S., 243–244
Reischauer, B., 157
Ren, S., 560, 561*t*
Ren, X., 708–709
Renou, L., 560, 561*t*
Renz, M., 415, 418–420
Resau, J.H., 589–590
Resneck, W.G., 366*t*, 370
Restle, D., 42–50, 44*t*, 411–414
Resurreccion, E.Z., 121, 365–370, 366*t*
Reuter, I., 514–515
Reuter, U., 310, 358–361, 365–370, 366*t*
Reynaert, H., 542
Reynoso, J., 513
Reyon, D., 594
Reza, H.M., 586
Rheinwald, J.G., 233
Rhim, J.S., 588*t*
Rhodes, G., 14–17
Rhodes, J.D., 588*t*
Ribeiro-da-Silva, A., 494
Riccio, M., 89*t*
Richards, J., 584–585
Richardson, J.B., 529–530
Richardson, R.M., 202–205
Richter, M., 310
Rickelt, S., 429, 443
Ridderstrale, K., 547
Ridge, K.M., 44*t*, 51–52, 86–87, 117–120, 193–194, 241–243, 357–358, 391–421, 683
Riegman, P.H., 157
Riemer, D., 662, 667–668, 684
Riethmacher, D., 513
Ringertz, N.R., 512
Riopel, C.L., 121
Risling, M., 513
Rives, M.L., 546–547
Rivkah Isseroff, R., 682–683
Robb, I.A., 617
Robert, A., 193–194, 197–198, 202–205, 391–421
Roberts, D.B., 716
Roberts, M.W.G., 513–514, 521*t*
Robertson, D., 442
Robertson, J., 490, 492, 494–496, 569, 637
Robidoux, J., 190
Robinson, K.A., 494
Robinson, Z.J., 62–64, 244–245, 683, 702–703
Robson, R.M., 447–451, 538–543, 545–546
Rochat, A., 227–229
Rockney, D.E., 165–170
Rodger, J., 487
Rodier, F., 264
Rodriguez, D., 165, 683–684
Roeber, S., 488–489
Roederer, M., 446
Rogaeva, E., 492
Rogel, M.R., 117–120, 241–243, 391, 405–406, 409–411, 420–421
Rogers, K.R., 23–24
Rogers, M.A., 260–261
Rohena, C.C., 170–171

Rolef-Ben Shahar, T., 667–668
Roller, P.P., 146–148
Rolli, C.G., 62–64, 683
Romer, L.H., 241–243
Rommerscheidt-Fuss, U., 365–370
Ronfard, V., 229
Rongo, C., 688–690
Ronnlund, D., 43–44*t*, 50–51
Roop, D.R., 237–238, 305, 315*t*, 322–323, 338, 478–479
Roosa, J.R., 641–642
Rose, J., 312
Rose, M., 542
Rosenthal, J.J., 616
Ross, H., 244–246, 682–683
Rossor, M.N., 488
Rossow, M.J., 193–194, 197–198, 202–205, 392*t*
Roth, S., 452
Roth, W., 243–244, 310, 682–683
Rothman, A.L., 165, 166*f*, 221, 310
Rott, L.S., 121
Rottman, J.N., 560, 561*t*, 573–574
Rotty, J.D., 42, 114, 310, 318–319, 326, 327*f*, 328, 335–337, 582–583
Roux, K.J., 710*f*
Roux, P.P., 115–117
Rowat, A.C., 38*t*, 41, 411
Roy, A., 582–585
Roy, P.S., 540, 543, 546
Royal, A., 309, 490–491
Ruan, J., 10*t*, 20, 27–28, 262–263
Rubin, G.M., 712, 714
Rubner, W., 697
Ruco, L.P., 337
Rudrapatna, V.A., 723
Ruf-Zamojski, F., 452
Rugg, E.L., 172–173, 221, 227, 229, 231*t*, 238–240, 243–244, 310
Rui, L., 117–120, 130–131, 132*f*
Ruiz, A., 188–190, 191*t*, 194–195
Ruoslahti, E., 225*t*
Rusconi, S., 638–639
Rushton, W.A.H., 462
Rusiñol, A.E., 558–559
Russell, D., 62–64, 240, 244–246, 247*f*, 682–683, 702–703
Russell, J.M., 616
Russell, M.A., 540, 542–543, 546
Russell, P., 586, 588*t*
Ruttner, J.R., 225*t*
Rydberg, L., 708–709
Rydenhag, B., 542, 545
Ryffel, G.U., 639–640

S

Sabbe, L., 193–194
Sabherwal, N., 640–641
Safo, P., 490–491
Saga, S., 88, 89*t*
Sager, P.R., 50
Sah, R.L., 582–584, 603
Sahlgren, C.M., 50, 511*f*, 512–518, 519*f*, 520, 521*t*, 522–523, 525, 527
Said, M.I., 15
Saito, T., 89*t*
Saiyasisongkhram, K., 663
Sakaguchi, T., 462
Sakamoto, K., 589–590
Sakamoto, Y., 42
Sakata, S.T., 569, 571*t*
Sakoda, S., 89*t*
Sakuma, T., 593–594
Sakurai, M., 89*t*, 100–102, 105–106
Salas, P.J., 358, 366*t*
Salazar, G., 417–418
Sale, K., 27
Salikhov, K.M., 21–22
Salomon, G.D., 479
Saltiel, A.R., 568–569, 571*t*
Samadi, R., 594
Sambuughin, N., 543
Sander, J.D., 594
Sanders, E.J., 591
Sandilands, A., 366*t*, 370, 582–585, 603
Sanelli, T., 492
Sanger, J.M., 60
Sanger, J.W., 60
Sano, M., 362–364
Sanoudou, D., 429
Santha, M., 490–491
Santoriello, C., 593
Saraga-Babic, M., 584–585
Saravanamuthu, S.S., 590
Sarge, K.D., 129
Sargent, T.D., 639–640

Sarria, A.J., 451
Sasaki, D.T., 569, 571*t*
Sasaki, T., 60, 89*t*, 462, 481, 486–487, 490, 559
Sasaoka, T., 540
Satelli, A., 86–87, 397–398
Sato, C., 87
Sato, T., 357–358
Sauer, F., 44*t*
Sauerbruch, T., 366*t*
Saunders, D.N., 362–364
Saunter, C.D., 584–585, 590, 601*f*
Savard, P.E., 538
Sawamoto, K., 512–513
Sawant, S.S., 357–358, 362–364
Sax, C.M., 586
Saxel, O., 437–438
Saxena, N.K., 195–196
Schafer, G., 92*t*
Schaffner, W., 638–639
Schechter, N., 487, 594, 637
Schiavo, G., 496
Schier, A.F., 593–594
Schierenberg, E., 688–690, 692–695
Schiffer, D., 494
Schiller, D.L., 142–143, 146, 147*f*, 225*t*
Schinkinger, S., 44*t*
Schips, T.G., 429, 443
Schirmer, E.C., 165
Schlaepfer, W.W., 462, 636–637
Schlief, M.L., 618–619
Schliwa, M., 38*t*, 41, 411, 583, 682–683
Schlossarek, S., 560–562, 561*t*
Schmid, E., 60, 75, 238, 636–637
Schmidt, C.F., 417
Schmidt, H., 662, 684
Schmidt, R., 322–323
Schmidt, W., 436
Schmitt-Graeff, A., 542
Schnabel, R., 662, 684
Schnabl, B., 365–370
Schneider, J.L., 357–358
Schnoelzer, M., 376*t*
Schnölzer, M., 560
Schnyder, U.W., 238, 719–721
Schoenenberger, C.-A., 43–44*t*, 51, 52*f*, 682–683
Schon, E.A., 708–709
Schonthaler, H.B., 585–586
Schopferer, M., 38*t*, 41, 411
Schott, C., 157
Schramm, S., 560
Schreiber, K.H., 164–165
Schriefer, L.A., 663
Schröder, R., 143, 366*t*, 370, 372
Schroter, C., 87–88
Schuldiner, M., 436
Schulthess, T., 11–12
Schulz, H.U., 378–379
Schulze, E., 663, 669*f*, 670, 674
Schulze, S.R., 708–709
Schumacher, J., 7, 10*t*, 262–263
Schünemann, J., 663, 668, 669*f*, 670, 674
Schutz, J., 451–452
Schütz, W., 560
Schwartz, K., 166*f*
Schwartz, M.L., 637
Schwarz, N., 53, 60–80, 366*t*, 371–372, 682–683
Schweitzer, S.C., 42–50, 450–451
Schweizer, J., 164–165, 317, 352–353
Scobioala, S., 513
Scorrano, L., 444
Scott, W.G., 267, 271*t*
Sebo, Z.L., 708–709
Seeberger, T., 583–584
Seftor, E.A., 141–142, 241–243
Seftor, R.E., 141–142, 241–243
Segbert, C., 688–690, 702–703
Seger, R., 190, 202–205, 567–568
Segerdell, E.J., 644–645
Segura-Totten, M., 667–668
Sejersen, T., 450–452, 512–513, 543, 594
Sekimata, M., 89*t*, 102–103
Sekkali, B., 642
Selkoe, D.J., 87–88
Seltmann, K., 44*t*, 53, 243–244, 310, 357–358, 682–683, 709
Selvin, P.R., 392*t*, 396
Sempe, P., 496
Senden, T.J., 44*t*, 50–51
Sengupta, K., 409–411
Seo, J., 267
Sera, F., 570–573
Sereti, K.I., 445
Sernett, S.W., 538–543, 546

Serpinskaya, A.S., 392*t*, 396, 417–418
Serrano, L., 266–267, 271*t*
Seufferlein, T., 62–64, 683
Sevinc, E., 542
Sewram, V., 141–142
Sewry, C., 559
Sexton, C.J., 172–173, 229, 231*t*, 238–241, 239*f*, 242*f*, 243–244
Shabbir, S.H., 582–583
Shah, S.B., 429, 463–464, 466, 472–473
Shain, D.H., 645
Shaked, M., 190
Shan, J., 167*t*, 170, 570–574, 572*t*
Shan, Y., 638
Shaner, N.C., 395–396
Shapiro, L., 18
Sharabi, K., 667–668
Sharada, A.C., 197
Sharma, D., 195–196
Sharma, P., 617, 621
Sharma, S., 38*t*, 41, 411
Sharpe, C.R., 637
Shatunov, A., 451
Shaw, C., 190, 191*t*, 193–195, 197–200, 202–205, 203*f*, 213–214
Shaw, G., 174, 462, 487–488, 492, 494–495
Shaw, R., 174
Shaw, S., 43–44*t*, 50, 682–683
Shea, T.B., 91, 126–127, 487, 617, 636–637
Shearer, A., 529–530
Shearer, R.F., 362–364
Sheehy, N., 193–194
Sheetz, M.P., 193–194, 202–205
Shefer, G., 529–530
Shelanski, M.L., 479, 489–490, 493–494, 636–637
Sheldrick, G.M., 19
Shemanko, C.S., 172–173, 229, 231*t*, 238–240, 243–244
Shen, H., 463, 471–473
Shen, J., 290
Shen, M.Y., 582–583
Shenoy, R., 569
Sherman, B.T., 565
Shetty, K.T., 628
Shevinsky, L.H., 75
Shi, S.-R., 148–149
Shibata, M., 88, 89*t*
Shifrin, Y., 190, 202–205
Shikata, K., 89*t*
Shim, S., 362–364
Shima, H., 89*t*
Shimi, T., 89*t*, 409–411
Shimizu, H., 487–488
Shimoyama, M., 259, 271*t*
Shintou, T., 585, 600
Shipman, R.L., 24–25
Shirahata, A., 397–398
Shirahatti, N., 193–194
Shiurba, R., 510–511
Shneidman, P.S., 637
Shoelson, S.E., 5–6, 27–28
Shrestha, O.K., 708–709, 711, 723
Shui, Y.B., 589–590
Shumaker, D.K., 86–87, 193–194, 201–205, 391–421
Sidky, M., 478–479
Siebenbrunner, M., 683, 702–703
Siegel, M., 134, 373, 376*t*, 377–378
Siegfried, C., 589–590
Signer, E., 585–586
Sihag, R.K., 91, 126–127
Sillje, H., 97, 102*f*, 105–106
Silvander, J.S.G., 352–379
Silverberg, S.G., 141–142
Sim, J.H., 362–364
Simon, R., 374–377, 376*t*
Simon, S., 582–585
Simonneau, M., 490–491
Simpson, K.W., 391, 405, 420–421
Sinagra, G., 559
Sinensky, M.S., 558–559
Sing, Y., 141–142
Singer, P.A., 359*t*
Singh, B.B., 188–190
Singh, D., 569
Singh, S.K., 357–358, 365–370, 366*t*
Singh, S.V., 190, 195–196
Singla, A., 372
Sinning, I., 712
Sinzelle, L., 640
Siriwardena, D., 195–196, 198–200
Sitikov, A., 241–243, 357–358
Sivaramakrishnan, S., 44*t*, 51–52, 357–358, 683
Sive, H.L., 645

Sjöberg, G., 23–24, 450–452, 512–513, 543
Sjodin, R.A., 617
Sjöqvist, M., 529–530
Skalli, O., 527–528, 538–552
Skov Jensen, S., 560–562, 561*t*
Skrzypek, E., 116*t*, 117, 118*t*
Slingsby, C., 584–585
Slochower, D.R., 41
Slotta-Huspenina, J., 157
Smear, M.C., 593
Smith, A.J., 133
Smith, F.J., 167*t*, 170, 227, 231*t*, 366*t*, 370
Smith, J.L., 15
Smith, S., 446, 585
Smith, S.D., 205–207
Smith, S.J., 616
Smith, T.A., 266–268, 269*f*
Smyth, L., 569
Snape, A.M., 639–640
Snider, N.T., 89*t*, 91, 114–135, 167*t*, 171, 178–179, 182, 197–198, 305, 354–355, 357–358, 372, 377–378, 514–515, 530, 582–583
Snyder, K.A., 644–645
So, P., 165, 378–379
Soares, D.C., 391, 405, 420–421
Socolow, P., 478–480
Soderberg, O., 547
Soellner, P., 121
Soetikno, R.M., 366*t*
Sohl, G., 237–238, 366*t*
Sokolov, I., 48*t*
Solimando, L., 409–411
Solomon, F.E., 197
Somers, J., 708–709
Somerville, M.J., 495
Son, Y.-J., 518
Sonawala, A., 488–489
Song, K., 434
Song, L., 27
Song, S., 86–87, 582–583, 585–586, 603
Soni, P.N., 241–243
Sonnenberg, A., 225*t*
Sorensen, C.B., 238–240
Soreq, E., 668
Soriano, M.E., 429, 443
Soriano, P., 436–437
Sotiropoulou, C., 442
Soumaka, E., 429
Spann, P., 667–668, 684
Sparks, E.A., 563
Sparrow, D.G., 640
Sparrow, J.C., 416
Spatz, J.P., 62–64, 683
Speaker, T.J., 167*t*, 170
Speakman, P.T., 264
Speed, T., 262, 271*t*
Speese, S., 708–709, 711
Speicher, D.W., 289–290
Speicher, K., 290
Spencer, C.I., 434
Spencer, P.H., 480
Spencer-Dene, B., 496
Spengler, R.M., 365
Spicer, A., 583–584
Spielmann, H.P., 560, 561*t*
Spinelli, L., 233
Splittgerber, U., 188–190
Spradling, A.C., 712
Springer, R., 53, 682–683
Spurej, G., 376*t*
Spurny, M., 366*t*, 370
Srinawasan, J., 484–485
Srinivasan, C., 188–190, 189*f*, 191*t*, 193–205, 199*f*, 201*f*
Srinivasan, K.K., 197
Srivastava, D., 434–435
Srivastava, S.K., 591
Stacey, S.N., 379
Stakeberg, J., 512
Stallworth, B.S., 439
Stambrook, P.J., 490
Stamenović, D., 42–50, 43–44*t*, 682–683
Stanley, J.R., 315*t*
Stark, G.R., 289–290
Starr, D., 667–668
Stary, M., 428–429, 436
Steed, E., 452
Steel, J.B., 225*t*
Steiner, B., 87–88
Steiner, R.A., 19
Steinert, P.M., 22–23, 86–87, 260–264, 266–267, 305, 315*t*, 322–323, 338, 391, 392*t*, 393–394, 420–421, 478–479, 512, 558–559
Steklov, N., 43–44*t*, 50

Stepp, M.A., 193–194, 225*t*, 585, 591
Sternberg, M.J.E., 261, 271*t*
Sternberger, L.A., 87–88
Sternberger, N.H., 87–88
Sterneck, E., 490–491
Steven, A.C., 262–263, 305
Stevens, C., 391, 405, 420–421
Stevens, H.P., 309
Stevenson, W., 463–464, 466, 471–473
Stewart, C.-B., 637, 650
Stewart, C.L., 560, 561*t*
Stewart, M., 262–264
Stick, R., 538, 585–586, 668, 709, 710*f*
Stock, D.W., 586
Stock, J., 6–7, 262, 271*t*
Stockard, C.R., 148
Stöcker, S., 683–684
Stockigt, F., 451–452
Stoger, U., 372–373
Stojan, J., 266–267, 358, 359*t*
Stoler, A., 309
Stone, M.R., 366*t*, 370
Stoppacciaro, A., 337
Storoni, L.C., 16–17
Straight, A.F., 174
Stramer, B.M., 202–205
Straube-West, K., 393–394
Strbak, V., 240
Streit, W.J., 487
Strelkov, S.V., 4–28, 114, 188–190, 198–200, 262–263, 267, 394, 406–409, 450, 480, 582–583
Strickland, D., 80
Strnad, P., 61–64, 114, 117–120, 133–134, 140–141, 146, 165, 177, 223, 259–260, 305, 309, 352–379, 582–583, 709
Stromer, M.H., 539–540, 546
Strong, J., 366*t*, 370
Strong, M.J., 492, 494–495, 637
Strucksberg, K.H., 451–452
Studier, F.W., 12–13, 18
Stumptner, C., 140–160, 353, 372–378, 376*t*
Stuurman, N., 663–667
Styers, M.L., 417–418
Su, H.-T., 511–512, 516–517
Su, M.G., 116*t*
Su, P.-H., 511–514, 511*f*
Suetterlin, R., 43–44*t*, 51, 52*f*, 682–683
Suganuma, H., 395–396
Sugimori, M., 628
Sugimura, K., 395–396
Sugimura, Y., 134
Suhr, R., 514–515
Sulem, P., 379
Sullivan, T., 560–562, 561*t*
Sulston, J.E., 688–690
Sultana, A., 406
Sultana, S., 540–542, 546
Summer, R., 445
Sun, D., 60
Sun, H., 708–709
Sun, J., 126, 164–182, 708–709
Sun, J.S., 89*t*
Sun, N., 543, 546
Sun, T.T., 121, 221, 309
Sun, Y.H., 593
Sung, H.-K., 518–520, 519*f*, 528–529
Suzuki, A., 488–489
Suzuki, E., 264
Suzuki, S., 225*t*
Suzuki, T., 487–488
Sviripa, V.M., 188–190
Swann, M.M., 695–696
Sweeney, L., 569
Swensson, O., 309–310
Sylvestersen, K.B., 115–117
Szaro, B.G., 487, 493–494, 617, 621–622, 621*f*, 636–654
Szebenyi, G., 628
Szeverenyi, I., 4–5, 165, 222, 259, 271*t*, 310, 355*t*, 709–710, 716
Sztul, E., 417–418

T

Tabernero, L., 244
Tachibana, K., 95, 101–102
Tachibana, T., 89*t*
Tada, K., 586
Tada, T., 487–488
Taira, M., 640
Takahashi, F., 595*t*
Takahashi, H., 362–364
Takahashi, K., 38*t*, 41, 309
Takahashi, M., 89*t*, 97, 616–617, 628–629
Takahashi, T., 87–88, 89*t*, 102–103, 310

Takai, Y., 89*t*, 102–103
Takakuwa, O., 513–514, 520
Takamura, H., 225*t*
Takashita, N., 512–513
Takeda, M., 213–214, 617, 640
Takemoto, D.J., 589–590
Takemura, M., 472
Taketo, M.M., 366*t*, 370
Takeuchi, Y., 89*t*, 95, 96*f*, 101–102
Takizawa, K., 366*t*
Talalay, P., 165–170, 167*t*, 237–238
Tal-Singer, R., 569
Tam, C., 310
Tamai, Y., 365–370, 366*t*
Tamiya, S., 588*t*
Tamura, K., 89*t*
Tamura, S., 466
Tan, J.C., 429, 560–562, 561*t*
Tan, T.S., 221–247
Tan, W., 434
Tanabe, K., 89*t*, 100–102, 105–106
Tanaka, H., 86–106, 585
Tanaka, J., 89*t*, 617
Tanaka, T., 225*t*, 617
Tang, H.Y., 289–290
Tao, G.-F., 260–261
Tao, G.Z., 114, 121, 125–127, 134, 140–141, 167*t*, 171, 177–179, 182, 238, 353–358, 355*t*, 365–372, 366*t*, 378, 462–463, 472–473
Taoka, M., 89*t*
Taphoorn, M., 488–489
Tapscott, S.J., 636–637
Tardieu, A., 584–585
Tardin, C., 417
Tasaki, I., 616, 618–619
Tatsu, Y., 105
Tatsuka, M., 89*t*, 97, 100–102, 105–106
Taura, C., 88, 89*t*
Tawk, M., 540–542
Taylor, C.R., 148–149
Taylor, G.L., 16
Taylor, J., 601*f*
Taylor, M.V., 639–640
Teduka, K., 513–514
Teplyakov, A., 16–17
Terada, S., 488
Terada, Y., 100–102, 105–106
Terao, E., 490–491
Terkeltaub, R.A., 134, 373, 376*t*, 377–378
Terrault, N.A., 359*t*
Terse, A., 511*f*, 514–515
Tesser, P., 594
Tezcan, O., 396–397
Thaiparambil, J.T., 50, 188–190, 191*t*, 193–197, 201–205, 397–398
Theaker, J.M., 141–142
Theil, T., 638
Theres, C., 688–690, 702–703
Thibault, P., 115–117
Thiemann, D.A., 708–723
Tholozan, F.M., 584–585, 588*t*, 589–590
Thomas, A., 281–282, 284*f*, 294
Thomas, C., 487, 496
Thomas, G., 48*t*
Thomas, K.R., 464
Thompson, A., 190, 191*t*, 193–195, 197–200, 202–205, 203*f*, 213–214
Thompson, K., 188–190, 191*t*, 194–205, 199*f*, 201*f*
Thompson, M.A., 489–492, 638
Thompson, T.G., 538–540, 543
Thomson, J.N., 688–690
Thon, N., 488–489
Thorleifsson, G., 379
Thornell, L.E., 429, 512, 543
Thorner, P.S., 512–513
Thornton, C.A., 637
Thorsen, K., 362–364
Thueringer, A., 143, 355, 374–377, 376*t*, 546–547
Thuret, R., 640–641
Thyagarajan, A., 637
Tian, L., 518
Tian, W., 513
Tibarewal, P., 233
Tibshirani, R., 141–142
Tidman, M.J., 238–240
Tiffert, T., 618–619
Tikkanen, I., 574–575
Timms, J., 188–190
Titeux, M., 538–542
Titterington, L., 315*t*
Tittle, R.K., 585, 592–593
Tiwari, R., 357–358, 362–364
Tjostheim, S., 492

Toca-Herrera, J.L., 48*t*
Toda, M., 202–205
Togashi, H., 89*t*
Toivola, D.M., 50, 114, 117–121, 125–127, 133–134, 140–141, 146, 164–165, 177, 223, 238, 260–261, 305, 309, 352–379, 462–463, 472–473, 582–583
Tokui, T., 87, 95, 101–103
Toledo-Aral, J.-J., 490–491
Toman, O., 355–357
Tomarev, S.I., 585–586, 588*t*
Tome, Y., 513
Tomic-Canic, M., 229–234, 231*t*, 240, 309, 682–683
Tomita, M., 563–565
Tomonaga, T., 89*t*
Tomono, Y., 89*t*, 102–103, 105
Tomooka, Y., 510–511
Tong, X., 310, 317
Toomes, C., 575
Topping, D.L., 357–358
Török, C., 513
Torok, N., 309
Torvaldson, E., 89*t*, 518, 519*f*, 520, 521*t*, 523, 525, 527
Tow, D.E., 574
Toyama, Y., 636–637
Trabut, V., 105
Tran, D., 429, 445–446, 451–452
Tran, H.T., 642
Tran, K., 157, 586
Trapp, B.D., 464, 473
Traub, P., 38*t*, 41, 411, 448–449, 583, 682–683
Trauner, M., 376*t*
Trautwein, C., 372
Travis, K., 44*t*
Treiber, M., 378–379
Treinin, M., 667–668, 684
Trejdosiewicz, L.K., 60, 238
Trejo, H.E., 241–243
Tremblay, M.J., 366*t*
Tretjakoff, I., 638
Trevor, K., 359*t*
Tricoire, H., 708–709
Trifilieff, P., 546–547
Trinh le, A., 452
Tripathi, B.J., 586
Tripathi, R.C., 586
Trojanowski, J.Q., 462, 494, 636–637
Troken, J.R., 241–243, 391, 405–406, 409–411, 420–421
Trott, K.A., 645
Troy, C.M., 493–494, 636–637
Troyanovsky, S.M., 225*t*
Trucchi, B., 189*f*, 191*t*, 193–197
Trueb, B., 225*t*
Truscott, R.J., 589–590
Tsai, L., 188–190, 191*t*, 193–195, 197–198, 201–205
Tsai, S.Q., 594
Tschaharganeh, D.F., 513–514
Tseng, D., 622–624, 623*f*
Tseng, Y., 38*t*, 41
Tsien, R.Y., 60
Tsikitis, M., 428–430
Tsim, K.W., 518–520
Tsonis, P.A., 586
Tsoupri, E., 428–429
Tsuchiya, K., 488
Tsujimura, K., 87, 89*t*, 95, 96*f*, 101–103
Tsujimura, Y., 539–542
Tsunematsu, Y., 588*t*
Tsutsumi, O., 89*t*, 102–103
Tsutsumi, R., 586
Tsuzuki, T., 466
Tu, P.-H., 463–464, 473, 494
Tu, Y., 60, 560, 561*t*
Tucker, P.A., 18
Turksen, K., 237–238, 309
Tuschl, T., 594, 595*t*, 663–667
Tytell, M., 617, 618*f*, 626*f*
Tyurin-Kuzmin, P.A., 391, 417–418
Tzivion, G., 391, 405
Tzur, Y.B., 667–668

U

Überbach, D., 683–684
Uchida, A., 89*t*, 420–421
Uchikado, H., 488
Uchiumi, A., 362–364
Uchiyama, Y., 463–464, 472–473
Udupa, N., 197
Uehara, F., 513
Uemura, T., 513–514, 520
Ui-Tei, K., 595*t*

Uldschmid, A., 97, 102*f*, 105–106
Undamatla, J., 493–494, 637
Urata, Y., 586
Urban, T.J., 165
Urich, D., 391, 405–406, 409–411, 420–421
Urizar, E., 546–547
Uryu, K., 488
Usachov, V., 165, 259–260, 352–353, 366*t*, 370–374
Usukura, J., 539–542
Utreras, E., 511*f*, 514–515
Uveges, T.E., 638
Uwanogho, D.A., 514–515
Uyama, N., 542

V

Vagin, A.A., 16–17, 19
Vahlquist, A., 266–267
Vaittinen, S., 511*f*, 512–518
Vale, R.D., 60, 202–205, 395–396
Valenzuela, D.M., 88
Valickova, J., 227
Vallee, R., 50
Vallin, J., 640
van Boxtel, R., 357–358
van de Kemp, A., 585
van de Rijn, M., 141–142
van den Berg, T.J., 584–585
van den Bosch de Aguilar, P., 490–491
van den Hoogenhof, M.M., 560, 561*t*
van der Kooi, A.J., 559
van der Veen, C., 311, 337
Van der Veken, P., 190, 191*t*
van Die, I.M., 12–13
Van Dyke, M., 6–7, 262, 271*t*
van Eeden, F., 452
Van Esch, H., 708–709
Van Eyk, J.E., 115–117
van Haelst, U.J., 225*t*
van Marle, J., 584–585
van Meegen, M., 559
van Oosten, A., 411–412
Van Roey, K., 262, 271*t*
Van Rossen, E., 542
Van Roy, F., 642
Van Stralen, S., 583–584, 603
van Weeghel, M., 560, 561*t*
Vanden Berghe, W., 190, 191*t*, 193–194
Vandromme, M., 515–516
Varela, A.E., 429
Varela, E., 451–452
Varma, S.D., 589–590
Varnous, S., 559–560, 561*t*, 563
Vassar, R., 237–238, 310, 583
Vassear, M., 143
Vasseur, M., 359*t*
Vatsellas, G., 428–430
Vedantham, V., 434
Veeranna, 462, 481, 486–487, 515–516, 617
Veeranna, G.J., 628
Vegners, R., 38*t*, 41
Veigel, C., 416
Verardo, M.R., 213–214
Verderio, P., 157
Verdin, E., 130–131
Vernon, J.A., 281–282, 287*f*, 288, 290
Verstegen, M.M., 357–358
Vesely, K., 227
Vidal, M., 237–238
Viebahn, C., 584–585
Viero, S., 512–513
Viertler, C., 140–160
Vijayaraj, P., 237–238, 358–361, 365–370, 366*t*
Vikstrom, K.L., 60, 392*t*, 393
Villa, M.C., 496
Villalón, E., 462–473
Villanueva, C., 366*t*
Villegas, J., 376*t*
Vincent, M., 538
Virtanen, I., 61–62
Virtanen, M., 231*t*, 235–237
Vishwasrao, H.D., 546–547
Vize, P.D., 638–640
Vleminckx, K., 642
Vogel, B., 452
Vogl, A.W., 682–683
Voloudakis-Baltatzis, I., 442
von Wichert, G., 43–44*t*, 52–53, 62–64, 682–683, 702–703
Voncken, J.W., 560, 561*t*
Vonesch, J.-L., 688–690
Vooturi, S., 188–190, 191*t*, 194–205, 199*f*, 201*f*
Vorgias, C.E., 448–449

Voss, J.C., 4–28, 262–263, 391, 480
Vosshall, L.B., 723
Vrensen, G.F., 584–585, 589–590
Vrielink, A., 417–418
Vyoral, D., 355–357

W

Waaijers, S., 702–703
Wada, F., 134
Wada, H., 445
Wada, Y., 89*t*
Wagenmaker, E.R., 372
Wagner, E., 80
Wagner, L.M., 589
Wagner, O.I., 38*t*, 53
Wahlby, C., 547
Wahlstrom, A.M., 558–559
Wakamatsu, Y., 510–511
Wakely, P.E., 141–142
Wald, F.A., 358, 362–365, 366*t*
Walde, S., 451
Walker, J.L., 193–194, 585, 591
Walko, G., 243–244
Wallace, P., 584–586
Wallrath, L.L., 708–723
Walls, R.J., 282
Wally, V., 231*t*, 237
Walter, N., 683
Walther, P., 38*t*, 43–44*t*, 52–53, 682–683, 702–703
Wan, W., 148
Wang, C., 636–654
Wang, D.S., 174
Wang, F., 305–340
Wang, G., 362–364
Wang, H., 261, 264, 271*t*, 582–584, 603
Wang, J.H., 518–520
Wang, J.P., 10*t*, 20, 27–28, 262–263
Wang, J.Y., 588*t*
Wang, L., 60, 166*f*, 190, 420–421, 561*t*, 591
Wang, N., 42–50, 43–44*t*, 198–200, 413–414, 682–683
Wang, S.Q., 188–190, 193–196, 261, 271*t*, 399, 431
Wang, X., 190, 237–238, 290, 362–364, 451–452, 582–584, 708–709
Wang, Y., 593
Wang, Y.H., 41
Wang, Y.L., 411–412, 588*t*
Wang, Y.P., 593
Wang, Z., 261, 271*t*, 317
Wanninger, F., 143, 366*t*, 370, 372
Warchol, W., 617
Ward, C.M., 473
Ward, J.M., 515–516
Ward, W.W., 60
Warden, C.H., 24
Ware, C.F., 366*t*
Warner, A.E., 638–639
Warnock, L., 569
Warrick, J.M., 708–709
Wasielewski, P., 588*t*
Wass, M.N., 261, 271*t*
Watanabe, K., 585, 600
Watkins, S.C., 540
Watts, S.A., 148
Wawersik, M.J., 86–87, 325, 332–333
Waxman, S.G., 462
Way, J., 617, 621–622, 621*f*
Weatheritt, R.J., 262, 271*t*
Weber, C., 462–463
Weber, K., 86–87, 164–165, 221, 262–263, 305–309, 330, 354–355, 377–378, 462–463, 489–490, 595*t*, 636–637, 662–668, 669*f*, 670, 674, 684, 709
Weber, S., 452
Webster, C., 437–438
Wedig, T., 7, 8*t*, 19–20, 22–23, 38*t*, 41, 188–190, 191*t*, 193–195, 197–198, 201–205, 262–264, 305–309, 339, 391, 394, 400, 411, 420–421, 584–585
Weeks, D.L., 640–641
Weeks, S.D., 12
Weerasinghe, S.V., 117–120, 129–130, 167*t*, 171, 178–179, 182, 357–361, 366*t*
Wegener, A., 582–585, 603
Wehman, A.M., 593
Wei, Q., 582–583
Weibrecht, I., 547
Weiglein, A.H., 372–373
Weimer, M., 62–64, 683, 702–703
Weinberg, R.A., 229
Weinert, B.T., 130–131
Weisleder, N., 429, 451–452
Weiss, M.S., 16, 18
Weiss, R.A., 309

Weissig, H., 266, 271*t*
Weissman, J.S., 362–364
Weitz, D.A., 38*t*, 41, 391–421, 582–583, 682–683
Weitzer, G., 428–429, 436, 451–452, 583
Welsh, M., 584–585
Wen, L., 362–364
Wen, Q., 38*t*, 41, 48*t*, 53, 412
Wendschlag, N., 188–190, 189*f*, 191*t*, 193–197, 397–398
Weng, C.-C., 511–512, 516–517
Weng, Y.C., 492
Werb, Z., 241–243
Were, F., 366*t*, 370–371
Werlang-Perurena, A., 488–489
Werner, E.D., 5–6, 27–28
Wertz, I.E., 126
Wessel, D., 296
Wessels, N.K., 647–649
West, A.G., 640–641
Westbrook, J., 266, 271*t*
Westerblad, H., 451–452
Westerfield, M., 616
West-Mays, J.A., 590
Westphal, H., 588*t*
Wetzels, R.H., 225*t*
Weyer, L., 495
White, C.L., 488
White, D.C.S., 416
White, D.E., 569
White, J.G., 688–690
Whiteley, M., 640–642
Whiting, J., 146–148
Wiche, G., 243–244, 682–683
Widengren, J., 43–44*t*, 50–51
Widestrand, A., 529–530
Wieckowski, M.R., 444
Wiegner, O., 692–695
Wiese, P.J., 463–464, 466, 471–473
Wiesel, N., 668, 669*f*, 672
Wiesenfahrt, T., 662, 683–684, 686–688, 702–703
Wight, D.C., 638
Wilhelmsson, U., 529–530, 540, 542, 545–546
Wilkins, M.H.F., 257
Willard, L., 589–590
Willenbacher, N., 38*t*, 41
Williams, B.J., 543–545
Williams, P.J., 464
Williamson, R., 617–618
Williamson, T.L., 463–464
Willoughby, V., 488–489
Wills-Karp, M., 310, 329–330, 339
Wilson, K.L., 638–640, 645, 647–649, 667–668, 684, 709
Wilson, L.T., 373
Wilson, N.J., 227
Wilson, S.E., 205–207
Wilson-Heiner, M., 365–370, 366*t*
Wilt, F.H., 647–649
Windoffer, R., 53, 60–80, 234–235, 305–309, 354–355, 358–361, 365–372, 366*t*, 420–421, 662, 682–684, 686–688
Wingfield, P.T., 406–409
Winn, M.D., 16–17
Winter, L., 451–452
Winterhalter, K.H., 225*t*
Wirtz, D., 38*t*, 41, 246, 310, 334–335
Wistbacka, N., 510–530
Wistow, G.J., 584–585
Wizeman, J., 188–190, 191*t*, 194–205, 199*f*, 201*f*
Wohlenberg, C., 310
Wolf, C.M., 561*t*
Wolff, A., 492
Wolff, C., 157
Wolff, K., 309
Wolff, M.R., 559
Wolfley, J.R., 15
Wöll, S., 61
Wolpert, L., 256
WolV, K., 309
Won, G.J., 583–584
Wong, J., 493–494
Wong, N.K., 495
Wong, P.C., 310, 317, 328, 339, 462–463
Wong, Z.-R., 511–514, 511*f*
Woo, W.M., 663
Wood, T.L., 481
Woodcock-Mitchell, J., 309
Woods, I.G., 593–594
Woods, J.L., 281–282, 283*f*
Wooley, C.F., 563
Worman, H.J., 165–170, 166*f*, 167*t*, 558–577, 663–667

Wormstone, I.M., 588*t*, 589–590
Wortzel, I., 202–205
Wottawah, F., 44*t*
Wride, M.A., 583–584, 591
Wright, C.J., 568–569
Wright, C.V.E., 638–639
Wright, T.L., 165, 358–361, 359*t*
Wu, C.H., 190
Wu, D., 452, 582–583
Wu, J.J., 584–585, 589–590, 637
Wu, J.Y., 229
Wu, K.C., 260–261, 512
Wu, K.K., 510–511, 516–517
Wu, W., 167*t*, 170, 559–560, 563, 564*f*, 566, 567*f*, 569–574, 572*t*, 576–577, 582–603
Wu, Y., 373
Würflinger, T., 61, 420–421
Wydner, K.L., 558–559
Wylie, C.C., 636–637, 647–649

X

Xanthou, G., 429
Xiang, A.P., 528–529
Xiao, A., 593–594
Xiao, S., 492, 496
Xiao, X.H., 560–562, 561*t*
Xie, H., 294
Xie, Q., 164–182
Xie, X.S., 399
Xu, B.W., 355–357, 708–709
Xu, C., 10*t*, 20, 27–28, 262–263
Xu, H., 513–514
Xu, J., 708–709
Xu, W., 569, 571*t*
Xu, Z., 463–464, 473
Xue, W., 513–514
Xue, Z.G., 538–540, 542, 546

Y

Yablonka-Reuveni, Z., 529–530
Yadav, M.C., 586
Yaffe, D., 437–438
Yager, C., 471–473
Yamada, S., 38*t*, 41, 86–87, 246, 310, 334–335
Yamada, T., 588*t*
Yamada, Y., 563–565
Yamagata, K., 105
Yamaguchi, T., 89*t*, 91, 97, 102*f*, 105–106, 126–127
Yamakawa, H., 539–542
Yamamoto, M., 165–170, 167*t*, 513
Yamamoto, T., 513–514
Yamamoto, Y., 586
Yamanishi, K., 309
Yamashita, M., 362–364
Yan, B., 309
Yan, J., 708–709
Yan, Y., 481, 510–513
Yanagawa, Y., 466
Yanagida, M., 89*t*, 100–102, 105–106
Yang, C., 586–587, 589*f*
Yang, H.-Y., 511–512, 558–559, 682–683, 708–709
Yang, J., 510–511, 518–520, 519*f*, 528–529, 565, 584–585
Yang, M., 513
Yang, S.H., 558–559
Yang, W., 362–364
Yang, X., 233
Yang, Y., 586–587, 589*f*
Yano, S., 513
Yano, T., 88, 89*t*
Yanuka, O., 436
Yao, N.Y., 38*t*, 41, 310, 329–330, 339, 411, 682–683
Yao, Z., 362–364
Yasuda, K., 586
Yasui, Y., 89*t*, 96*f*, 97–102, 105–106
Yates, C.M., 261, 271*t*
Yazdanbaksh, K., 478–479
Ye, S., 488–489
Yeh, J., 645
Yeh, T.C., 568–569
Yen, S.H., 87–88, 479
Yeo, L.S., 560–562, 561*t*
Yergeau, D.A., 640
Yeung, T., 411–412
Yi, E.H., 362–364
Yin, X., 463–464, 473
Yodh, A.G., 412
Yokoo, H., 488–489
Yokota, O., 488
Yokota, Y., 188–190
Yokoyama, T., 96*f*, 97, 101–102, 102–103*f*, 105–106

Yonemura, S., 105–106, 585
Yoon, K.H., 61–64
Yoon, M., 60–64, 202–205, 392*t*, 395–396
Yoshida, Y., 355–357
Yoshimura, T., 134
Young, C., 115–117
Young, D., 463–464, 472–473
Young, G.H., 708–709, 711, 723
Young, J.Z., 616–618, 618*f*
Young, L., 582–603
Young, S.G., 558–560
Yu, Q.C., 237–238, 309
Yu, Z., 708–709
Yuan, A., 60, 462–464, 472–473, 481, 486–487, 490
Yuan, J., 447–448
Yuan, Q.X., 376*t*
Yuasa, K., 543, 546
Yuh, C.-H., 510–511, 516–517
Yung, W.-H., 518–520
Yurchenco, P., 311, 337
Yuspa, S.H., 319, 322–323, 338

Z

Zaccone, P., 496
Zackroff, R.V., 617, 626*f*
Zahand, A.J., 667–668
Zahreddine, H., 683–684
Zatloukal, B., 355, 546–547
Zatloukal, K., 140–160, 353, 355, 366*t*, 370, 372–378, 376*t*, 546–547
Zayner, J., 80
Zehorai, E., 567–568
Zelenka, P.S., 588*t*, 590
Zeller, R., 312
Zerban, H., 636–637
Zhai, J., 637
Zhan, C.G., 189*f*, 190, 191*t*, 193–197, 205–207, 206*f*
Zhan, Q., 513–514, 521*t*
Zhan, X.H., 188–190, 191*t*, 194–195
Zhang, B., 116*t*, 117, 118*t*, 593–594
Zhang, F., 723
Zhang, H., 121, 290–291, 513–514, 560, 683–684
Zhang, J., 582–583, 588*t*
Zhang, L., 193–194, 528–529, 585, 591
Zhang, Q., 86–87, 582–586, 603
Zhang, S.J., 294, 451–452
Zhang, T., 6–7, 582–583
Zhang, W., 188–190, 718
Zhang, X., 193–194, 202–205, 521*t*
Zhang, Y., 92*t*, 129, 355–357, 540, 542–543, 546
Zhao, B., 496
Zhao, J., 478–497
Zhao, L., 542
Zhao, Y., 487, 568–569, 637
Zhao, Z., 355–357, 513–514
Zhelev, D.V., 696
Zheng, K., 512–513
Zheng, X., 663–667
Zheng, Y., 663–667
Zhong, B.H., 121, 125, 134, 165, 177, 353–355, 355*t*, 365–372, 366*t*, 378–379, 520
Zhou, C., 53, 682–683
Zhou, J., 290
Zhou, M., 593–594
Zhou, Q., 121, 125, 127, 134, 165, 177, 238, 352–355, 355*t*, 359*t*, 366*t*, 370–371, 378–379
Zhou, T., 290
Zhou, X., 141, 164–165, 352–353, 359*t*
Zhou, Y.Y., 431
Zhu, C.-q., 488–489
Zhu, H., 640
Zhu, Q.S., 188–190, 193–196, 462–464, 488
Zhu, W.Z., 431
Zhu, X., 512–513
Zhu, Y., 290–291, 362–364
Zhu, Z., 593–594
Zhuang, X., 399
Ziegler, A., 617
Zieman, A., 305–340
Ziff, E.B., 489–494, 638
Zigrino, P., 165–170, 167*t*, 237
Ziman, B., 431
Zimbelmann, R., 7, 8*t*, 19–20, 188–190, 262–263, 394, 406–409
Zimek, A., 305
Zimmerman, L.B., 510–513, 586, 636–638
Zinman, L., 492
Zinovieva, R.D., 585–586
Ziol, M., 372

Zirbel, L.N., 708–709, 711, 721, 723
Zoltoski, R.K., 582–585, 603
Zon, L.I., 593
Zuber, M.X., 645
Zuela, N., 662–677
Zuo, Y., 518
Zupancic, T., 358, 359*t*
Zwerger, M., 708–709, 711, 723

SUBJECT INDEX

Note: Page numbers followed by "*f*" indicate figures, and "*t*" indicate tables.

A

Adult cardiomyocyte culture
- calcium reintroduction, 433
- cardiomyocyte dissociation, 433
- heart perfusion, 433
- pelleted cardiomyocytes, 433–434
- perfusion apparatus, setup, 432–433
- protocol, 431
- setup, 431

AGADIR, 266–267, 271*t*
Aggregate intracellular forces, 418–420
A-kinase-anchoring protein (AKAP), 543. *See also* Synemin
AMBER, 266–267, 271*t*
Amyotrophic lateral sclerosis (ALS), 494–495
Antigen retrieval (AR) methods, 148, 149*t*, 151
Astrocyte, 542
Astrocytoma, 542–545
Atomic force microscope (AFM)
- to determine cell stiffness, 48*t*
- indentation method, 52*f*
- modes ofmeasurement, 47*f*

attB, 641–642
Aurora-B
- cleavage furrow-specific vimentin phosphorylation, 103*f*
- vimentin phosphorylation sites, 94*f*

Avian erythrocytes, 540
Axonal radial growth, 463
Axoplasm. *See also* Squid neurofilaments (NF)
- fractionation and preparation, 624–625
- isolation, 619–620, 620*f*
- phosphorylation assay, 625–627
- soma (GFL) *vs.*, 627

B

Baby hamster kidney (BHK) cells, 358–361
Bacterial artificial chromosomes (BACs) injection, 640–641
Bioinformatics, IF
- α-helical segments, 6–7
- structural features of, 5–6
- tripartite organization, 5

Biolayer interferometry (BLI)
- analysis of interaction, 407*f*
- determination, 406
- instruments, 406
- protein purification, 406–409

BLAST engines, 644–645
Blastocyst
- injection, 468
- Krt8-YFP detection, 78*f*

2B2 mimetic peptide, 394, 395*f*
Bulk flow analysis, 70–72
1-butyl-3-methylimidazolium chloride, 294

C

Caenorhabditis elegans
- Ce-lamin assembly *in vitro*, 670–673
- fluorescence recovery after photobleaching, 676
- genome of, 662
- intermediate filaments
  - apical cell, 688–690, 689*f*
  - assembly, 670, 674
  - confocal laser scanning microscopy, 690
  - epifluorescence microscopy, 686–688, 687*f*
  - epithelial cells, 682–683
  - essential roles, 662–663, 664*t*
  - fluorescence microscopy assays, 674–676
  - intestines dissection, 691–695, 693–695*f*
  - mechanical properties, 683–684
  - micropipette measurements, 695–701
  - observations, 702–703
  - OrthoMCL clusters, 685*t*
  - polymerization properties, 684

Caenorhabditis elegans (*Continued*)
proteins groups, 662
vimentin deletion, 682–683
vitality testing, 691–695
in vitro observations, 682–683
lamins
assembly, 668–669
B-type, 663–668
Ce-lamin, 667–668
lmn-1, 667–668
mammals genes, 663–667
vertebrate A-type, 663–667
Calcium-depleted serum, 323
Calcium switch media, 323
CaMKII
activity of, 102–103
signaling, 104*f*
Cardiac fibrosis, MAP kinase activity assessment, 575–576
C2012 cells, 437–439
cDNA encoding, *Drosophila melanogaster*, 712, 717
Ce-lamin paracrystalline array, assembly, 673
Cell culture, SEKs, 362*f*
baby hamster kidney cells, 358–361
CRISPR Type II system, 362–364
downregulation, 362–365, 363*f*
epithelial–mesenchymal transition, 358
overexpression, 358–361, 359*t*
short hairpin RNAs, 362–364
short interfering RNAs, 362–365
Cell culture, skin keratins, 319–330
Cell line systems
cardiomyocyte, 439
skeletal muscle, 437–439
Cell migration
collective cell movement, 243–244
single-cell movement, 243
study, 241–243
withaferin A, 201–202, 201*f*
Cell spreading
cytoplasmic IF mutations, 240–241, 242*f*
withaferin A, 202–205
Cellular techniques
nestin expression, in cell lines
transient overexpression, 522–525
transient siRNA transfections, 521–522
nestin, signaling property analysis tools
biochemical methods, 527–528
microscopy-based techniques, 525–527
Charcot-Marie-Tooth type 2B, 496
CIS-regulatory elements, *Xenopus* embryos
mRNA translation analysis, 653–654
RNA export from nucleus, 652–653
RNA processing efficiency, 651–652
COILS, 262, 271*t*
Collagen contraction, 198–201
Compartment-specific phosphorylation. *See* Squid neurofilaments (NF)
Conduction velocity, 472–473
Confocal laser scanning microscopy
proximal ligation assays, 550
worms, 690
COOT, 267, 271*t*
Cosedimentation assay, 336
Crystallization, 14
conditions for, 15–16
experiments stages, 14–15
IF protein fragments, 15*f*
Cyclin-dependent kinase 5 (Cdk5). *See also* Nestin
ACh agonist-mediated activation, 518–520
activity of, 515–516
in differentiating myoblasts, 518
increased myogenic differentiation, 518, 519*f*
interaction, 517
myogenic differentiation, 518
nestin-mediated scaffolding of, 518, 520
soluble pool, 518
Cytoplasmic intermediate filament
assembly, 674
essential roles
in *C. elegans*, 662–663
knockdowns of, 663
Cytoplasmic intermediate filament mutations
experimental model systems, 227–238
3D cultures, disease modeling in, 235–237
fluorescence time-lapse imaging, 234–235
mouse models uses, 237–238

simple culture systems, disease modeling in, 229–234
keratin identification
deoxyribonucleic acid collection, 226–227
diagnosis and candidate gene, 224–226
molecular identification of, 226–227
skin biopsies, to keratinopathies, 223–224
keratin, in stress assays
aggregates and misfolded protein, 238
cell motility and migration, 241–244, 245*f*
cell spreading, 240–241, 242*f*
heat stress, 238–239
mechanical stress, 244–246, 247*f*
osmotic stress, 240
link between human diseases and, 221–222
in vivo consequences, 222–223

D

Danio rerio model. *See* Zebrafish
Data mining methodologies, IF chains
genome sequences, 259
heptad substructure recognition, 262–263
model building, 267–268
molecular dynamics, 266–267
secondary structure prediction, 261
sequence analysis, 259–261
solvent exposure, 262
Desmin, 89*t*, 92*t*
AFM indentation method for, 52*f*
animal model, mutation studies
knockout and knockin, 451–452
transgenic mice, 451–452
zebrafish morphants and knock-in lines, 452
cell line systems
cardiomyocyte cell lines, 439
skeletal muscle, 437–439
deletion mutants, from bacteria, 449–450
embryoid body differentiation, 436–437
expression, in whole mouse embryos, 445–446
flow cytometry analysis, 445
identification, associated proteins
GST pull-down assay, 453
immunoprecipiation, 453–454
yeast 2-hybrid system, 452–453
IF, mechanical properties, 42–51
immunoelectron microscopy
double immunogold labeling, 443
postembedding immunogold labeling, 442
preembedding immunogold labeling, 443
tissue preparation, 442
immunofluorescence staining
adherent cells, 441–442
OCT embedded tissue, 440
isolation, from cells and tissues, 448–449
neonatal cardiac fibroblasts, generation of cardiomyocytes
direct transition, 435–436
isolation, 434–435
primary culture systems
adult cardiomyocyte culture, 431–434
neonatal cardiomyocyte culture, 430–431
production and purification of, GST-fused protein, 449–450
role of, 429–430
subcellular fractionation and immunoblotting, 444
in vitro assembly, 450
in vivo assembly, 450–451
Diabetes, 496
3,5-Diethoxycarbonyl-1,4-dihydrocollidine (DDC), 372–373
3D imaging of Keratin intermediate filaments. *See* Keratin intermediate filaments, 3D imaging
Dimethyl sulfoxide (DMSO), 195
Disease model
in 3D cultures, 235–237
in α-internexin
neurodegenerative disorder, 488–489
NIFID, 488
in peripherin
ALS, 494–495
and diabetes, 496
neurodegenerative diseases, 496
in simple culture systems
advantages and disadvantages, 229, 230*t*
pathomimetic cell lines generation, 229–234
published human cell line models, 231*t*

Double-label immunofluorescence, 143–146, 144*f*, 145*t*
Drosophila melanogaster
intermediate filaments, 709, 711
cellular localization, 710*f*
cytological analysis, 718–722
cytoplasmic, 709–710
embryonic lethality, 716
epidermis, 721
germline transformation, 712
immunohistochemistry, 720*f*
larval body wall muscle, 721
lethal phase, 716, 716*f*
live cell imaging, 722–723
pCaSpeR vector, 712, 714–715
P-element insertion, 714
protein domain structure, 710*f*
protein extraction, 718
pUAST vector, 712, 714–715
RNA isolation, 717
RT-PCR, 717
salivary glands isolation, 720
stock generation, 711–714
tracheal branches isolation, 719
mutant phenotypes, 708
strategy, 708–709
Drug screening
cell system selection, 172–173
cell transduction and validation set up, 172*f*
high-throughput apparatus, 173*f*
libraries and vendors for, 179–182, 180*t*
materials, 173
mechanism of action, 178–179
methods, 174–176
validation, 176–178
DTT, 286
Dulbecco's modified eagle medium (DMEM), 65

E

Echocardiography, MAP kinase activity assessment, 573–574
Electron microscopic immunocytochemistry. *See* Immunoelectron microscopy
Electron microscopy
Ce-lamin, 673
C. elegans, 674
Electron paramagnetic resonance (EPR). *See* Site-directed spin labeling (SDSL-EPR) technique
ELISA, phospho-specific antibody characterization, 96*f*, 98–99
Embryoid body (EB)
diagnosis, 224–226
differentiation, 436
generation, of cardiomyocytes through, 438*f*
protocol for, 436
reproducible differentiation, 437
subtypes, 225*t*
trypsinization, 437
Emery–Dreifuss muscular dystrophy, 559
Epidermis
Drosophila melanogaster, 721
palmar-plantar, 309
Epidermolysis bullosa simplex (EBS), 221
Epifluorescence microscopy, *Caenorhabditis elegans*, 686–688, 687*f*
Epithelial cells
C. elegans, intermediate filaments, 682–683
Epithelial–mesenchymal transition (EMT), 358
European Committee for Standardization, 148–149
ExPASsy, 260, 271*t*
Ex vivo lens models
chick lens-dissociated epithelial model, 591–592
organ/tissue culture conditions, 589–590
rat ex vivo lens fiber cell explants, 591
rat lens epithelium explants, 590–591
Eye lens, 582–583. *See also* Intermediate filaments (IF)
beaded filaments, 585–586
body, 584–585
diagram of, 584*f*
ex vivo lens models, 589–592
vimentin, 584–585
in vitro lens models, 586–589
in vivo lens models, 592–594

F

Far-western assay, 336
FFPE tissue. *See* Formalin-fixed paraffin-embedded (FFPE) tissue

Fibrosis, ocular injury model of
 alkali burn corneal injury, 209–210
 tissue harvesting, 210–211
Fiji software, 68
Filament assembly and transport, 472–473
Filensin, 583
Fluorescence recovery after photobleaching (FRAP), 676
Force spectrum microscopy (FSM), 418–420
Formalin-fixed paraffin-embedded (FFPE) tissue
 keratins IHC, 151–156, 154*t*
Frozen tissue, Keratin in, 142–146

G

Genome-wide association studies, 379
Germline transformation, *Drosophila melanogaster*, 712
GFAP. *See* Glial fibrillary acidic protein (GFAP)
 mutations,
Giant axonal neuropathy (GAN), 398
Glial fibrillary acidic protein (GFAP), 193–194
β-globin hypersensitive site, 640–641
GraphPad Software, 69
Green fluorescent protein (GFP), 60
 transfection with, vimentin IFs, 395–396
Guanidine isothiocyanate (GITC), 651–652

H

Hanks-HEPES imaging medium, 65
H9c2, 439
Heat stress, 238–239
High-glycine-tyrosine proteins (HGTPs), 282–283
High salt extract (HSE), 121–124, 354–355
High-speed sedimentation assay, 334
HL-1, 439
HPV55, 95
Human Intermediate Filament Database (HIFD), 222, 227
Human T-cell leukemia virus type 1 (HTLV-1), 488–489
Human vimentin fragments, 8*t*
HV55, 95

I

IEF prefractionation, 296–299
Immunoblotting
 neurofilaments protein phosphorylation analysis, 469–471
 phospho-specific antibody characterization, 96*f*, 99–100
Immunoelectron microscopy
 double immunogold labeling, 443
 postembedding immunogold labeling, 442
 preembedding immunogold labeling, 443
 tissue preparation, 442
Immunofluorescence
 adherent cells, 441–442
 histochemistry, for EB diagnosis, 224–226
 OCT embedded tissue, 440
Immunohistochemistry (IHC)
 cell spreading, 205
 of dissected tissue, 722
 Drosophila melanogaster, 720*f*
 Keratins, 141–143
  in FFPE tissue, 151–156, 154*t*
  on paraffin-embedded tissue, 146–156, 150*t*, 152*f*, 159–160
  PAXgene-fixed paraffin-embedded tissues, 157–159
  tissue fixatives on, 157*f*
Immunoprecipitation (IP), 516
 keratins, 329–330
 PTM, 124–126
 synemin, 546
Immuoblotting
 MAP kinase signal assessment, 565–566
Indirect immunofluorescence
 polylysine covered slides preparation, 674
 staining protocol, 675
Intermediate filament (IF) chains
 α- and β-contents, 261
 assembly, 263–264
 coiled-coil rod domain model, 267
 heptad/hendecad substructure, 262–263
 imaging techniques, 268–270
 interchain ionic interactions, 262–263
 model structures, 267–268
 mutational effects, 266–267
 protein assembly, 263–264
 sequence comparisons, 260–261

Intermediate filament (IF) chains (*Continued*)
sequences and preliminary characterization, 259–260
structural/functional motifs, 262
tertiary structure, 266
Intermediate filaments (IFs)
assembly process, 4–5
bioinformatics
amino acid sequence, 5
α-helical segments, 6–7
structural features of, 5–6
tripartite organization, 5
Caenorhabditis elegans
apical cell, 688–690, 689*f*
confocal laser scanning microscopy, 690
epifluorescence microscopy, 686–688, 687*f*
epithelial cells, 682–683
intestines dissection, 691–695, 693–695*f*
mechanical properties, 683–684
micropipette measurements, 695–701
observations, 702–703
OrthoMCL clusters, 685*t*
polymerization properties, 684
vimentin deletion, 682–683
vitality testing, 691–695
in vitro observations, 682–683
C. elegans
assembly, 670
cytoplasmic, 674
fluorescence microscopy assays, 674–676
proteins groups, 662
Drosophila melanogaster, 709, 711
cytological analysis, 718–722
lethal phase, 716, 716*f*
live cell imaging, 722–723
pCaSpeR vector, 712, 714–715
protein extraction, 718
pUAST vector, 712, 714–715
RNA isolation, 717
RT-PCR, 717
stock generation, 711–714
drugs identification
libraries and vendors, 179–182, 180*t*
mechanism of action, 178–179
screening methods, 171–176
validation, 176–178
eye lens, 582–583
beaded filaments, 585–586
body, 584–585
diagram of, 584*f*
ex vivo lens models, 589–592
vimentin, 584–585
in vitro lens models, 586–589
in vivo lens models, 592–594
gene expression
cellular differentiation, as marker, 636–637
cis-regulatory elements analysis, 638
identified, 583
keratin, 146, 151
mechanical properties
atomic force microscopy, 48*t*
keratins, 51–53
neurofilaments, 53
vimentin and desmin, 42–51
mutations, 165
unbiased drug screening, 170–171
phosphoregulation on structure, 86*f*
phosphorylation of, 140–141
posttranslational modifications
assembly and disassembly dynamics of, 114, 115*t*
biochemical analysis, 120–126
chemical and pharmacological approaches, 117–120
cross-talk mechanisms, 117
high salt extraction, 121–124
immunoprecipitation of, 124–126
limitations, 115–117
monitoring methods, 126–134
structural components, 114
tools, 115–117, 116*t*
protein, 510–511
actins and tubulins, 164–165
feature of, 164–165
prototype, domains and consequences, 166*f*
proteins family, 140–141
reorganization, 87
SDSL-EPR technique
data interpretation and impact, 24–26

on human vimentin, 22*f*
limitations and outlook, 26–27
sample preparation and measurements, 22–24
site-specific phosphorylation, 88, 89*t*
vimentin
bioimaging, in living zebrafish, 599–603
RNA interference, 594, 595*t*
siRNA, 594, 596
transient manipulation, in zebrafish, 597–599
vimentin expression, 86–87
viscoelasticity
characterize *in vitro*, 38*t*
mechanical properties, 36–37
molecular mechanisms, 41
network mechanics, 41
persistence length, 36–37
schematic diagram of, 37*f*
X-ray crystallography
crystallization, 14–16
data collection, 16
design, of fragments, 7–11
experimental phasing, 17–18
expression, 11–13
limitations, 19–20
phasing by molecular replacement, 16–17
purification, 13–14
structure and impact, 20–21
structure refinement and validation, 19
α-Internexin
cDNA clones of, 497*t*
in disease
neurodegenerative disorder, 488–489
NIFID, 488
functions, 487–488
purification, 479
structure and assembly, 479–481
tissue and developmental expression
commercial antibodies, 484–485, 485*t*
embryonic mouse brain, immunostaining, 481, 482*f*, 483–484
IHC staining, 482–483
mRNA levels study, 481
preparation and paraffin infiltration, 483
in situ hybridization, 486–487
type IV neuronal intermediate filament protein, 478–479
Intracellular activity, vimentin IFs role, 417–418
In vitro lens models
epithelial cell lines, 586–587
generation, of mature lens cells, 587–589
mammalian lens epithelial cell lines, 588*t*
stem cell technology, 587
three-step differentiation, 589*f*
In vivo lens models, 592–594
In vivo mouse models
nestin function, 528–530

J

JNK inhibitors, 569
Jpred3, 261, 271*t*

K

KeraDyn software, 70–72
KeraMove software, 67–69
Keratin
antigen retrieval techniques, 148
bile duct epithelia, 143
conformation-dependent epitopes, 146
double-label IIF, 143–146, 144*f*, 145*t*
epitope-specific staining of, 147*f*
European Committee for Standardization, 148–149
formalin pigments, 147–148
in frozen tissue, 142–146
functions of, 140–141
intermediate filament, 146, 151
Mallory-Denk bodies, 146
on mouse tissue cryo-sections, 143–146
profiling, 141
types, 140–141
Keratin(s). *See also* Simple epithelial keratins (SEKs)
acidic extraction, 286–288
alkaline extraction, of whole fibers, 281–282
cell culture studies, 319–330
chemical digestion of resistant membranes, 289–290
cytoplasmic network, 709
differential extraction, of whole fibers

Keratin(s) (*Continued*)
chaotrope concentration, 282–285
reductant concentration, 286
solution pH effect, 286
filaments
assembly, 305–309
functions and associated skin disorders, 310
genes
attributes, differential regulation, and disease association of, 307*f*
encoding, 305
expression, in interfollicular, 309
family, 306*f*
palmar-plantarepidermis, 309
IF, mechanical properties, 51–53
immunoprecipitation, 329–330
ionic liquid-assisted extraction
digestion with $BMIM^{+}Cl^{-}$, 296
IEF prefractionation, 296–299
KAP isolation, 294–295
keratins isolation, 296
isolation of a-layer, 288
Keratin 8 (Krt8-YFP), 75, 89*t*, 92*t*
Keratin 18, 89*t*, 92*t*
mouse skin samples
cell culture studies, 338
collection, 337
harvest, for RNA and proteins, 318–319
isolation, 311–318
isolation and analysis of, proteins and RNA from, 338–339
morphological studies, 337–338
in vitro studies with, 339
peptide extraction, 290–294
proteolytic digestion, of whole fiber, 288
sodium deoxycholate and 018 empore™, 290–294
in tracheal cells, 720*f*
types, 352–353
in vitro methods
cosedimentation assay, 336
expression, 331
far-western assay, 336
formation, of type I–type II, 333
high-speed sedimentation assay, 334
low-speed sedimentation assay, 335
materials, 330–331
purification, 332
reconstitution, 333
TEM, 334
Keratin-associated proteins (KAPs)
extractability, 282–286
isolation, 294–295
Keratin intermediate filaments, 3D imaging
assembly and disassembly, cycle of, 62*f*
in cultured cells
acquisition of image, 67
bulk flow analysis, 70–72
cells preparation, 66
keratin network normalization, 67–68
keratin turnover measurement, 72–74
measuring keratin movement, 68–69
medium preparation, 65
microscopes, 64
surface coating, 65–66
morphology, 63*f*
in murine preimplantation embryos
embryo collection and cultivation, 75–76
image processing and analysis, 78–79
preimplantation embryos, 76–77
optimized for bleaching, 73–74
Keratin mutations
experimental model systems, 227–238
3D cultures, disease modeling in, 235–237
fluorescence time-lapse imaging, 234–235
mouse models uses, 237–238
simple culture systems, disease modeling in, 229–234
identification
deoxyribonucleic acid collection, 226–227
diagnosis and candidate gene, 224–226
molecular identification of, 226–227
skin biopsies, to keratinopathies, 223–224
link between human diseases and, 221–222
in stress assays
aggregates and misfolded protein, 238
cell motility and migration, 241–244, 245*f*
cell spreading, 240–241, 242*f*

heat stress, 238–239
mechanical stress, 244–246, 247*f*
osmotic stress, 240
in vivo consequences, 222–223, 223*f*
Keratinocytes
calcium switch protocol to induce, differentiation, 322–323
gene transfer protocols for, 324
growth media for, 322
immunofluorescence staining, 325–326
immunoprecipitation, 329–330
isolation and analysis, of RNA and proteins from, 326–329
isolation, from newborn mouse skin, 320–321
materials, 319–320
scratch wounding assay for, 323–324
skin explants ex vivo, 325
Keratinopathy, 221
skin biopsies to identify skin biopsies to identify, 223–224
transgenic mouse models, 237–238
Keratins immunohistochemistry (IHC), 141–143
in FFPE tissue, 151–156, 154*t*
on paraffin-embedded tissue
formalin stabilizes, 147–148
issues, 159–160
optimization, 146–156
preanalytical parameters, 148–149, 150*t*, 151, 152*f*
PAXgene-fixed paraffin-embedded tissues, 157–159
tissue fixatives on, 157*f*
KLH, 97

L

Lamin(s), 709
assembly, 668–669
B-type, 663–668
Ce-lamin, 667–668
in larval body wall muscle, 720*f*
lmn-1, 667–668
mammals genes, 663–667
vertebrate A-type, 663–667
Lamin A/C, 89*t*, 92*t*
Lamin A/C *(LMNA)* gene mutations, cardiomyopathy-causing
knockout and knockin mouse models, 561*t*
Lmna$^{\Delta 32/+}$, 560–562
Lmna$^{H222P/H222P}$ mice, 560, 563
Lmna$^{N195K/N195K}$ mice, 560
Lamin B2, 89*t*, 92*t*
Laminopathy, 559
Larval body wall muscle
Drosophila melanogaster, 721
lamins in, 720*f*
Lethal phase, *Drosophila melanogaster*, 716, 716*f*
Live cell imaging, *Drosophila melanogaster*, 722–723
Liver diseases
MDBs—keratin aggregates in, 372
biochemical analysis, 377–378
D12450B diet, 373
3,5-diethoxycarbonyl-1,4-dihydrocollidine, 372–373
H&E stain, 374–377
imaging and quantification of, 374–377
network reorganization, 375*f*
SEKs, 353
LMNA gene mutations. *See* Lamin A/C *(LMNA)* gene mutations, cardiomyopathy-causing
Lmna$^{H222P/H222P}$ mice
MAP kinase activity assessment
cardiac fibrosis, 575–576
echocardiography, 573–574
inhibition of activities, 574
natriuretic peptide A secretion, 574–575
survival prolongation, 576
MAP kinase inhibitor treatment studies
JNK inhibitors, 569
MEK1/2 inhibitors, 568–569
p38α inhibitors, 569
treatment protocols, 570
MAP kinase signaling, altered
cardiac transcriptomic analysis, 563–565
immuoblotting assessment, 565–566
nuclear translocation analysis, 567–568
Loligo stellate ganglion, 618*f*
Low-speed sedimentation assay, 335
Lysine acetylation, 130–133
Lysine–serine–proline (KSP), 462–463

M

Mallory-Denk bodies, 134
Mallory–Denk bodies (MDBs), 372
biochemical analysis, 377–378
D12450B diet, 373
3,5-diethoxycarbonyl-1,4-dihydrocollidine, 372–373
H&E stain, 374–377
imaging and quantification of, 374–377
network reorganization, 375*f*
Mallory-Denk bodies (MDBs), keratins, 146
MAP kinase inhibitor, in *Lmna*$^{H222P/H222P}$ mice
activity assessment
cardiac fibrosis, 575–576
echocardiography, 573–574
inhibition of activities, 574
natriuretic peptide A secretion, 574–575
survival prolongation, 576
altered signal analysis
cardiac transcriptomic analysis, 563–565
immuoblotting assessment, 565–566
nuclear translocation analysis, 567–568
treatment studies
JNK inhibitors, 569
MEK1/2 inhibitors, 568–569
p38α inhibitors, 569
treatment protocols, 570
MARCOIL, 262, 271*t*
Mechanical stress, 244–246, 247*f*
MEK1/2 inhibitors, 568–569
Microinjection
vimentin and mimetic peptides, 393
Xenopus embryos
instrumentation, 645–647
obtaining and injecting embryos, 647–649
useful *Xenopus* references, 644–645
zebrafish embryos
embryos preparation, 598–599
preparation of, 597*f*
vimentin morpholinos preparation, 598
vimentin mRNA preparation, 598
zebrafish, 599
Micropipette, *Caenorhabditis elegans*, 695–696
calibrating setup, 697–699
cell disruption, 702*f*
dissected intestines, 698*f*, 701*f*
forging, 696
intestinal mechanics analysis, 699–701, 699*f*
Mimetic peptides, 393–394
Mitogen-activated protein kinase inhibitors. *See* MAP kinase inhibitor, in *Lmna*$^{H222P/H222P}$ mice
Morpholinos, 593
Motion analysis, keratin, 68–69
Mouse embryonic fibroblasts (mEFs), 415–417
Mouse skin samples
harvest, for RNA and proteins
materials, 318
preparation and analysis of, 318–319
isolation
harvesting adult and newborn mice, 311–312
materials, 311
for routine morphological study, 312–317
for in situ hybridization, 317
for transmission electron microscopy, 317–318
MPV55, 95
Murine preimplantation embryos, keratin 3D imaging
embryo collection and cultivation, 75–76
image processing and analysis, 78–79
preimplantation embryos, 76–77
MV55, 95
Myoepithelial cells, synemin, 542
Myofibroblast transformation, 198–201
Myopathies, 543

N

Natriuretic peptide A secretion, 574–575
nefm gene, 641–642
Neonatal cardiac fibroblasts (NCFs)
direct transition
transdifferentiation efficiency evaluation, 436
transduction into iCM, 435–436

virus generation, suggestions, 435
isolation, 434–435
Nerve development, 473
Nestin
Cdk5
ACh agonist-mediated activation, 518–520
acts upstream, 517
in differentiating myoblasts, 518
increased myogenic differentiation, 518, 519*f*
interaction, 517
myogenic differentiation, 518
nestin-mediated scaffolding of, 518, 520
soluble pool, 518
commercial cell lines with, 521*t*
distribution, 512
expression, dynamic regulation of
in liver cancer tumorigenesis, 513–514
in mesenchymal stem cells, 514
promoter, 512–513
tissue stress and injury, 513
in tumors, 513–514
function
cellular techniques, 520–528
in vivo mouse models, 528–530
isoforms, 511–512
during neuronal cell death
Cdk5, 515–516
direct involvement, 516
immunoprecipitation study, 516
Nes-S isoform, 516–517
in zebrafish, 516–517
phosphorylation, 514–515
posttranslational modifications, strict control of, 514–515
protein structure of, 511*f*
transient overexpression of, 522–523
unique features, 510–511
vimentin, 512
Nes-640, transient overexpression of
in C2012 cells using JetPEI, 523–524
in SW13 cells by electroporation, 525
NetSurfP, 262, 271*t*
Neurodegenerative diseases
α-internexin, 488–489
peripherin, 496
Neurofilament heavy chain (NFH), 479–480
Neurofilament medium chain (NFM), 479–481
Neurofilaments (NFs)
composition, 462
heavy (NF-H), 463
IF, mechanical properties, 53
light (NF-L), 463
medium (NF-M), 463–464
protein phosphorylation by genetic manipulation
applications, 471–473
blastocyst injection, 468
breeding selection, 468
ES cells via homologous recombination, 466
immunohistochemistry, analysis by, 469–471
mice/genotyping cell line generation, 468–469
site-directed mutagenesis, 465–466
targeting construct, 466
subunit proteins, 462–463
Neurofilament triplet proteins (NFTPs), 481
Neuromuscular junctions (NMJs), 518–520, 519*f*
Neuronal intermediate filament inclusion disease (NIFID), 488
Neuronal intermediate filament protein
α-internexin, 478–479
peripherin, 489–490
NFH, 89*t*
NFL, 89*t*
NFM, 92*t*
NheI, 642
NIFID. *See* Neuronal intermediate filament inclusion disease (NIFID)
2-Nitro-5-thiocyano-benzoic acid (NTCB), 289–290
NLRP3, 409–411
NMR spectroscopy, 266
Nuclear envelope, 558–559
Nuclear translocation analysis, 567–568

O

Ocular injury model, of fibrosis
alkali burn corneal injury, 209–210

Ocular injury model, of fibrosis (*Continued*)
tissue harvesting, 210–211
O-linked glycosylation, 133–134
Optical-Tweezer active microrheology, 415–417
Osmotic stress, 240

P

p38α inhibitors, 569
Paraffin
dehydration and embedding, 314
fresh frozen tissue samples preparation, 316
H&E staining, 314
immunohistochemistry, 316
preparation, of embedded tissues, 312–317
PAXgene-fixed paraffin-embedded (PFPE) tissues
keratins IHC, 157–159
pCaSpeR vector
Drosophila melanogaster, 712, 714–715
P012 cells, 638
Peripherin, 89*t*
cDNA clones of, 497*t*
in disease
ALS, 494–495
and diabetes, 496
neurodegenerative diseases, 496
functions, 493–494
structure and assembly, 490
tissue and developmental expression
in adults, 490–491
commercial antibodies, 485*t*
dorsal root ganglia, 491, 492*f*
environmental factors, 490–491
functional significance, 492
gene, 492, 493*f*
type III neuronal IF protein, 489–490
PFPE tissues. *See* PAXgene-fixed paraffin-embedded (PFPE) tissues
Phakinin, 583
Phosphorylation
PTM, 126–127
K8 serine and tyrosine, biochemical analysis, 128*f*
materials, 127–129
phospho-site-specific antibodies, 126–127, 127*t*
vimentin phosphorylation
in mitosis, 101–102
in signal transduction, 102–104
PhosphoSitePlus, 117
Phospho-specific antibody. *See* Site- and phosphorylation state-specific antibodies
Photoactivatable/convertible protein
dynamics using, 400–403
filament severing and reannealing, 402
filament transport, 401–402
fusions with, 399–400
optimizing imaging, 401
parameters, 401
subunit exchange, 403
Photomultiplier tube (PMT), 67
Phyre2, 261, 271*t*
P-NFH, 622
Polyacrylamide (PAA) gels, 412
Polysomal profiling, mRNA translation analysis, 653–654
Polyvinylidine difluoride (PVDF), 471
Posttranslational modifications (PTM). *See also* Intermediate filaments (IF)
assembly and disassembly dynamics of, 114, 115*t*
biochemical analysis
high salt extraction, 121–124
immunoprecipitation, 124–126
materials and reagents, 120–121
chemical and pharmacological approaches, 117–120
cross-talk mechanisms, 117
high salt extraction, 121–124
immunoprecipitation of, 124–126
limitations, 115–117
lysine acetylation, 130–133
O-linked glycosylation, 133–134
phosphorylation
K8 serine and tyrosine, biochemical analysis, 128*f*
materials, 127–129
phospho-site-specific antibodies, 126–127, 127*t*
structural components, 114
sumoylation

biochemical and immunofluorescence analysis, 131*f*
lamins, 129
of mammalian, 129
materials, 129–130
tools, 115–117, 116*t*
transamidation, 134
pPEP-TEV expression vector, 11*f*
Primary cell culture systems
rabbit corneal fibroblasts, 207–208
rabbit Tenon's capsule fibroblasts, 208–209
Promoter elements, 639–640
Pronase E, 288
Protein expression
Ce-lamin, 670
C. elegans, 674
Protein kinase C (PKC), 102–103
Protein purification
Ce-lamin, 671
C. elegans, 674
recombinant proteins, 13*f*
Proximal ligation assays (PLA)
amplification step, 546–547
confocal scanning laser microscope, 550
control reactions, 551
primary antibody, 547–548
probes, 549
region of interest tool, 550
steps, 544*f*, 546–547
U87 human glioblastoma cells, 551*f*
PSIPRED, 261, 271*t*
P13SU01 affinity chromatography
abundance of, 629
affinity chromatography, 628
Coomassie-stained gels *vs.* Western blots, 629
homogenates, 629
vs. protein expression, 630*f*
pUAST vector, 712, 714–715
PYMOL, 268, 271*t*

R

Rabbit corneal fibroblasts, 207–208
Rabbit Tenon's capsule fibroblasts (RbTCFs)
primary cell cultures, from ocular tissues, 208–209
α-SMA expression in, 199*f*
RaptorX, 261, 271*t*
Reporter protein expression, *Xenopus* embryos, 649–650
Reporter RNA expression, *Xenopus* embryos, 650–651
Reverse transcriptase PCR (RT-PCR), 650–651, 717
Rhokinase
phosphorylation during mitosis, 102*f*
vimentin phosphorylation sites, 94*f*
RNA
extraction, 318
precipitation, 318
RNA interference (RNAi), 594, 595*t*

S

Salivary glands, 720
Scratch wounding assay, 323–324
SDS-PAGE, 448*f*, 625
Selective plane illumination microscopy (SPIM), 599–600, 601*f*
Selumetinib, 576
Ser55, 101–102, 102*f*
Ser71
cleavage furrow-specific vimentin phosphorylation, 103*f*
immunoblotting, 100
phosphorylation during mitosis, 102*f*
Ser72
anti-vimentin-pSer72, 94*f*
cleavage furrow-specific vimentin phosphorylation, 103*f*
immunoblotting, 100
phosphorylation during mitosis, 102*f*
Short hairpin RNA (shRNA), 396–397
Short interfering RNAs (siRNAs), SEK, 362–365
Simple epithelial keratins (SEKs), 352–353. *See also* Keratins
antibodies, 356*t*
cell culture, 362*f*
baby hamster kidney cells, 358–361
CRISPR Type II system, 362–364
downregulation, 362–365, 363*f*
epithelial–mesenchymal transition, 358
overexpression, 358–361, 359*t*
short hairpin RNAs, 362–364

Simple epithelial keratins (SEKs) (*Continued*)
short interfering RNAs, 362–365
DHPLC, 378
function, 353
genetic mouse models, 365–372
genome-wide association studies, 379
high salt extract, 354–355
inherited variations, 378–379
isolation, 354–357
liver diseases, 353
Mallory–Denk bodies, 372
antibodies, 376*t*
biochemical analysis, 377–378
D12450B diet, 373
3,5-diethoxycarbonyl-1,4-dihydrocollidine, 372–373
H&E stain, 374–377
imaging and quantification of, 374–377
network reorganization, 375*f*
molecular weights, 355*t*
in transgenic animals, 353
SiRNA
preparation and storage of, 596
transfection of cells with, 596
Site- and phosphorylation state-specific antibodies
applications, 105
characterization
ELISA, 96*f*, 98–99
immunoblotting, 96*f*, 99–100
immunostaining, 100
production
phosphopeptide to carrier protein conjugation, 97
synthetic phosphopeptide, 91–97
vimentin phosphorylation
in mitosis, 101–102
in signal transduction, 102–104
Site-directed mutagenesis, 465–466
Site-directed spin labeling (SDSL-EPR) technique
data interpretation and impact
coiled-coil regions, identification, 25, 27*f*
line shape, 25
periodicity, 24–25
vimentin dimer region, structure of, 25, 26*f*
on human vimentin, 22*f*
limitations and outlook, 26–27
nitroxide spin label, 21–22
sample preparation and measurements, 22–24
Small-interfering RNA (siRNA), 396–397
Small ubiquitin-like modifier (SUMO), 12
Sodium deoxycholate (SDC), 290
Soluble bead binding assay, 409–411
Soluble vimentin (sVim)
covalent modification, 188–190
Western blot analysis of, 199*f*
Squid neurofilaments (NF)
antibodies, 622, 623*f*
axoplasm isolation, 619–620
compartment-specific multimeric complexes, 628–629
developmental regulation
fractionation, of axoplasm and preparation, 624–625
immunohistochemical expression, 624*f*
phosphorylation, 622–624
phosphorylation assay, 625–627
SDS-PAGE, 625
soma *vs.* axoplasm, 627
Western blots, 625
dynamic axonal lattice, 617
genes, 621
giant axon, 616–618
giant fiber system, 617–619
NF220, 617
posttranslational modifications, 617
protein kinases, 627–628
P13SU01 affinity chromatography
abundance of, 629
affinity chromatography, 628
Coomassie-stained gels *vs.* Western blots, 629
homogenates, 629
vs. protein expression, 630*f*
stellate ganglion, 616–618
Stress assays
aggregates and misfolded protein, 238
cell motility and migration, 241–244, 245*f*
cell spreading, 240–241, 242*f*
heat stress, 238–239
mechanical stress, 244–246, 247*f*
osmotic stress, 240

Striated muscle cells, synemin, 540, 542
Sumoylation
  biochemical and immunofluorescence analysis, 131*f*
  lamins, 129
  of mammalian, 129
  materials, 129–130
Synemin
  with Akt activity, 543–545
  antibodies, 545
  in astrocytoma cells, 542–545
  in avian erythrocytes, 540
  cell distribution, 540–543, 541*t*
  cell lines, 546
  genetic ablation, 543
  immunoprecipitation protocols, 546
  interactions, 543–545
  isoforms, 538–539, 539*f*
  myoepithelial cells, 542
  properties, 539–540
  proximal ligation assays
    amplification step, 546–547
    confocal scanning laser microscope, 550
    control reactions, 551
    primary antibody, 547–548
    probes, 549
    region of interest tool, 550
    steps, 544*f*, 546–547
    U87 human glioblastoma cells, 551*f*
  purified, 546
  roles, 539–540
  structure, 538–539, 539*f*
  tissues rich in, 546

T

Thr316, 514–515
Tissue dissection, *Drosophila melanogaster*
  immunohistochemistry, 722
  live cell imaging, 722–723
Total internal reflection fluorescence (TIRF) microscopy, 400
Tracheal branches, *Drosophila melanogaster*, 709, 711
Transdifferentiation, 434
Transmission electron microscopy (TEM)
  keratin micrograph, 290 294
  mouse skin samples, 317–318
  wool cuticle layers, 290–294
Trichocyte keratins, 280, 281. *See also* Keratins
Trifluoroacetic acid (TFA), 291
Tris(2-carboxyethyl)phosphine (TCEP), 286–288
TURBOFRODO, 267, 271*t*

U

Unit-length-filament (ULF), 403, 405, 480

V

Vimentin, 89*t*, 92*t*, 512
  AFM indentation method for, 52*f*
  basic structure, 391
  bioimaging, in living zebrafish, 599–603
  disruption
    2B2 mimetic peptide, 394, 395*f*
    dominant-negative forms, 395–396
    giant axonal neuropathy, 398
    microinjection, 393
    mimetic peptides, 393–394
    silencing expression of, cell, 396–397
    transfection, with green fluorescent protein label, 395–396
    withaferin A, 397–398
  diverse cellular functions of, 391, 392*t*
  drugs and proteins disruption, 50–51
  eye lens, IF, 584–585
  human vimentin fragments, crystal structures, 8*t*
  IF in *C. elegans*, 682–683
  interactions, 405–411
    biolayer interferometry, 406–409
    soluble bead binding assay between, 409–411
  knockout models, 42–50
  mechanical properties, 411–414
    collagen gel contraction studies, 413–414
    three-dimensional substrate studies, 411–413
  morpholinos preparation, 598
  mRNA preparation, 598
  mutated desmin to disruption, 51
  phosphorylation
    Aurora-B and Rhokinase, sites for, 94*f*
    in mitosis, 101–102

Vimentin (*Continued*)
in signal transduction, 102–104
phosphorylation status of, 197–198
photoactivatable/convertible protein
dynamics using, 400–403
filament severing and reannealing, 402
filament transport, 401–402
fusions with, 399–400
optimizing imaging, 401
parameters, 401
subunit exchange, 403
quantification of
filament transport measurement, 403–404
subunit exchange measurement, 405
regulates, 198–200
RNA interference, 594, 595*t*
roles, in cell mechanics, 415–420
affect intracellular activity, 417–418
aggregate intracellular forces, 418–420
cytoplasmic mechanics, active microrheology, 415–417
siRNA, 594, 596
structure, 26*f*
transient manipulation, in zebrafish, 597–599
wild-type, 23
Viscoelasticity
characterize *in vitro*, 38*t*
mechanical properties, 36–37
molecular mechanisms, 41
network mechanics, 41
persistence length, 36–37
schematic diagram of, 37*f*

W

Western blot
cell spreading, 205
collagen contraction, 201
Coomassie-stained gels *vs.*, 629
myofibroblast transformation, 201
squid neurofilaments, 625
withaferin A, 212–213
Withaferin A (WFA)
binding process, 188–190
binding targets, 193–194
chemical structure, 189*f*
distribution, in rabbit eyes, 196–197
formulation, for animal studies, 195–196
GFAP, 193–194
immunohistochemistry analysis, 211–212
NF-κB inhibitory activity, 190
ocular injury model, of fibrosis
alkali burn corneal injury, 209–210
tissue harvesting, 210–211
phosphorylation status, 193–194
primary cell culture systems
rabbit corneal fibroblasts, 207–208
rabbit tenon's capsule fibroblasts, 208–209
properties and uses, in cell culture, 195
soluble vimentin, 188–190
systemic delivery, 197
targeting IFs *in vitro*
cell migration, 201–202, 201*f*
cell spreading, 202–205
collagen contractile activity, 198–201
dynamic properties, 197–198
myofibroblast transformation, 198–201
targeting IFs *in vivo*, antifibrotic activity, 205–207
targets and activities, 191*t*
vimentin, 193–194
vimentin IFs, 397–398
western blot analysis, 212–213
Withania somnifera, 188–190

X

Xenopus embryos
assays for gene expression
reporter protein expression, 649–650
reporter RNA expression, 650–651
BACs injection, 640–641
CIS-regulatory elements
mRNA translation analysis, 653–654
RNA export from nucleus, 652–653
RNA processing efficiency, 651–652
exogenous DNA integration, 640
gene expression studies, 639–640
microinjection
instrumentation, 645–647
obtaining and injecting embryos, 647–649
useful *Xenopus* references, 644–645

plasmid
cloning strategy, 642
DNA preparation, 642–644
glGFP reference and DsRed2 test plasmids, 641–642
X-ray crystallography
crystallization, 14–16
data collection, 16
design, of fragments, 7–11
experimental phasing, 17–18
expression, 11–13
limitations, 19–20
phasing by molecular replacement, 16–17
purification, 13–14
structure
and impact, 20–21
refinement and validation, 19

Z

Zebrafish
bioimaging in living
imaging preparation, 602
live cell imaging, 602–603
material required, 600–601
sample preparation, 602
SPIM, 599–600
CRISPR/Cas system, 594
development, 592–593
genome, 593
by morpholino microinjection, 599
embryos preparation, 598–599
preparation of, 597*f*
vimentin morpholinos preparation, 598
vimentin mRNA preparation, 598
transient genetic manipulations, 593–594
transient reverse genetic techniques, 593

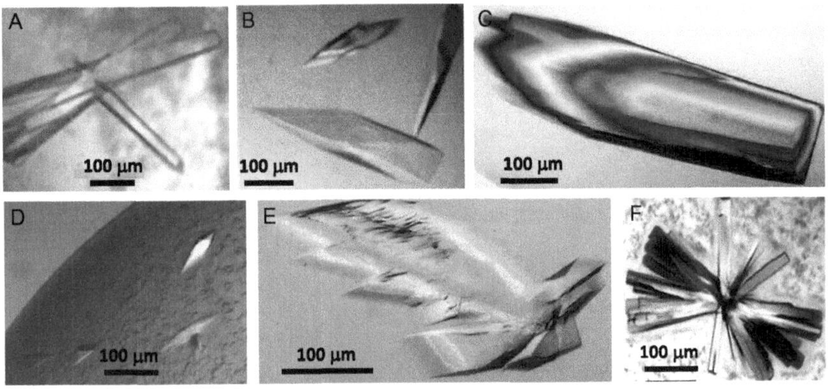

Anastasia A. Chernyatina *et al.*, Figure 3 Examples of optimized crystals of IF protein fragments. (A) Vimentin fragment including coil1A, L1 and part of coil1B (PDB code 3SSU). (B) The same fragment with a stabilizing mutation Y117L (PDB code 3S4R). (C)–(E) Three related fragments each corresponding to a major part of vimentin coil1B (PDB codes 3SWK, 4YV3, and 4YPC, respectively). (F) First half of vimentin coil2 with the L265C mutation (PDB code 3TRT).

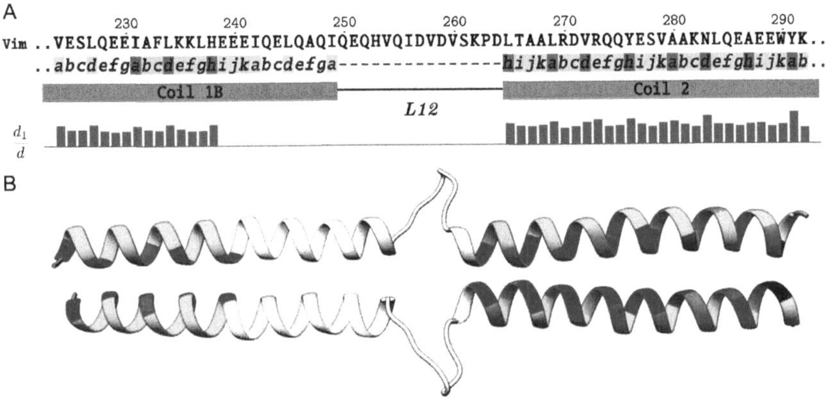

Anastasia A. Chernyatina *et al.*, Figure 5 Structure of the vimentin dimer region (residues 224–292) encompassing the C-terminal part of coil1B, linker 12, and the beginning of coil2. (A) The amino acid sequence, with the heptad (yellow) and hendecad (violet) repeats of the coiled coil indicated above. The d_1/d values calculated from the EPR experiments are shown as a bar graph. (B) Ribbon model of the same region, based on crystallographic structures of the coiled-coil segments and modeling of the L12 linker (Chernyatina et al., 2015). The model is colored according to the d_1/d values from light blue (lowest) to dark blue (highest); white = no data.

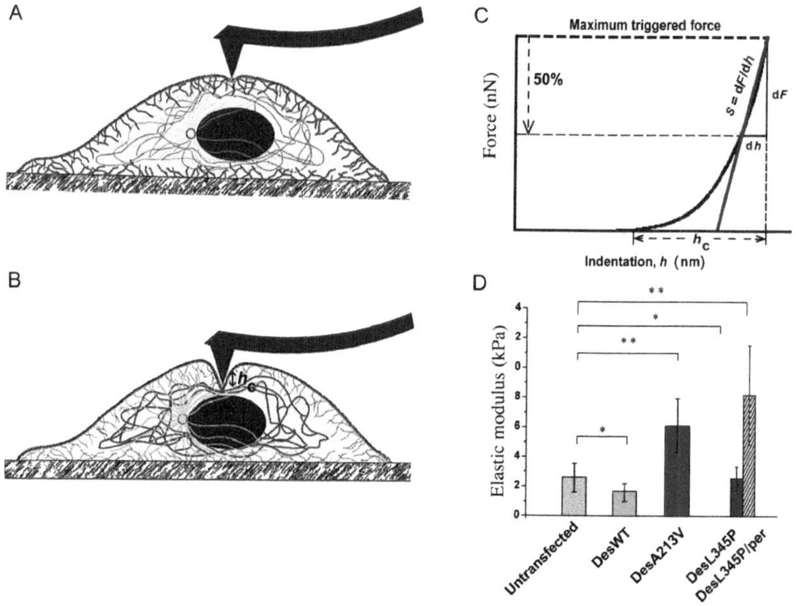

Elisabeth E. Charrier and Paul A. Janmey, Figure 3 AFM indentation method for analyzing desmin and vimentin IF nanomechanics in cells. (A) When indenting a cell, the AFM tip first encounters the actin cytoskeleton (blue) below the plasma membrane and then (B) the intermediate filament network (red). (C) The retracting AFM force curve specifies the cell's response to the force *F* applied to indent the cell to a depth h_C. The force curve can be divided into two main segments. The lower segment corresponds to the response of the actin cytoskeleton beneath the plasma membrane, whereas the upper segment of the curve predominantly represents the response of the deeper intermediate filament network. A linear fit to the upper 50% of the force curve (red) is used to determine the elastic modulus. (D) Elastic modulus (E_s) of untransfected cells (rat-2 fibroblasts), cells transfected with WT desmin–GFP (DesWT) and cells expressing two types of desmin point mutants, DesA213V, which forms filaments and DesL345P, which does not. Solid bars denote the average stiffness of the whole cell or the region away from the nucleus, and the cross-hatched bar denotes the perinuclear area. The statistical analysis shows mean values of E_s and standard deviation (*$p < 0.05$, **$p < 0.0001$) (Plodinec et al., 2011).

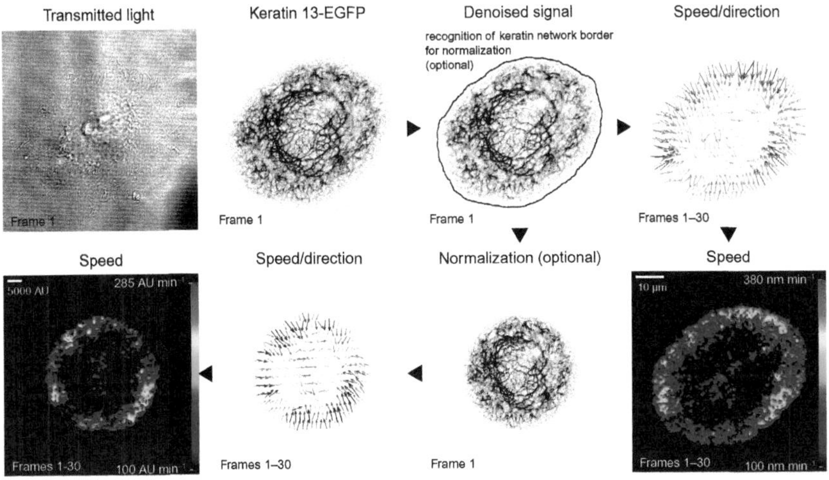

Nicole Schwarz *et al.*, Figure 3 Calculation of keratin speed from time-lapse recordings. The pictures show steps of recording and analyzing keratin motion in a single AK13-1 cell. AK13-1 cells producing keratin 13-EGFP were plated on a laminin 332-rich matrix for 52 h prior to imaging. Phase contrast (transmitted light) and keratin 13-EGFP fluorescence (inverse presentation) were recorded for 15 min every 30 s in the bottom plane. The background was reduced by Anscombe curvelet transform-based denoising (cf. Moch et al., 2013). Optionally, the overall keratin network shape was automatically delineated (black outline) for normalization into a standard circular shape with a defined diameter. A comparison of the nonnormalized and the normalized recordings is shown in Movie 3 (http://dx.doi.org/10.1016/bs.mie.2015.07.034). The results from motion analysis were then used to prepare vector maps to depict the direction of movement (corresponding to direction of arrows, which is mostly toward the cell interior). The vectors furthermore show the speed of keratin movement, which corresponds to length and thickness of the vectors. In addition, the speed is shown in detail in heat maps by a color scale as indicated in the images (the speed values correspond to mean speed per pixel).

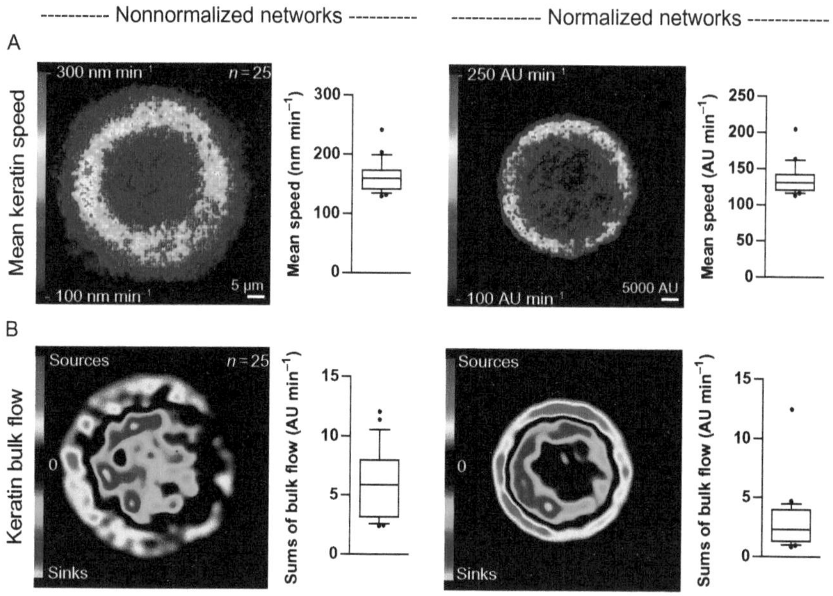

Nicole Schwarz *et al.*, Figure 4 Quantitative measurement of keratin speed (A) and bulk flow (B) in nonnormalized and normalized fluorescence recordings of AK13-1 cells expressing keratin 13-EGFP. Fluorescence was recorded in the bottom plane of the cells by confocal microscopy at 30 s intervals for 15 min. The data show the compiled results of standardized measurements of 25 single cell recordings. They are presented as heat maps with subcellular resolution and as whisker box plots (10–90% percentiles). In (A), the heat maps reveal that keratins move faster in the cell periphery than in the cell center underneath the nucleus. The diagrams show that keratin filaments are moving with a median speed of 160 nm min^{-1} before normalization and with 131 AU min^{-1} after normalization of the network shape. In (B), the heat maps show that keratin filaments are primarily assembled in the cell periphery and disassembled in the perinuclear area. Between these two zones keratin is transported and no net keratin assembly or disassembly is detected. The keratin bulk flow can also be described in diagrams as shown in AU. Note that the AU results in diagrams are only of value when different populations/conditions are compared with each other or when data distribution is of interest (e.g., Gaussian distribution).

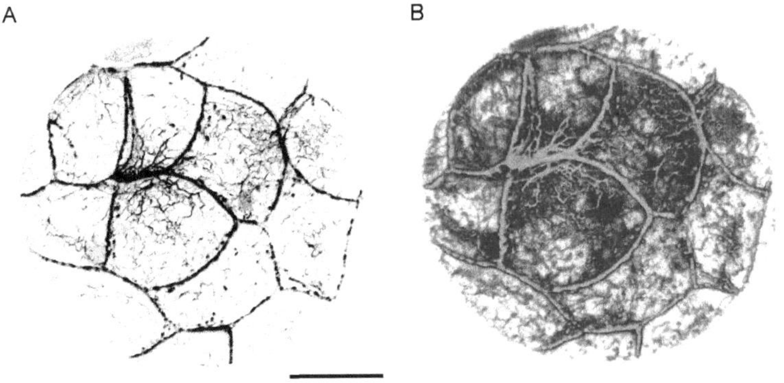

Nicole Schwarz *et al.*, Figure 8 Detection of Krt8-YFP in a late blastocyst by confocal fluorescence microscopy. The maximum intensity projection (A) and 3D reconstruction (B) of the recorded fluorescence (25 focal planes, 1 μm steps) show an extensive network throughout the trophectoderm layer. Relative fluorescence intensity in (B) is color coded with light green being the strongest and deep blue being the weakest signal. An animation of the reconstruction is presented in Movie 5. Scale bar: 20 μm.

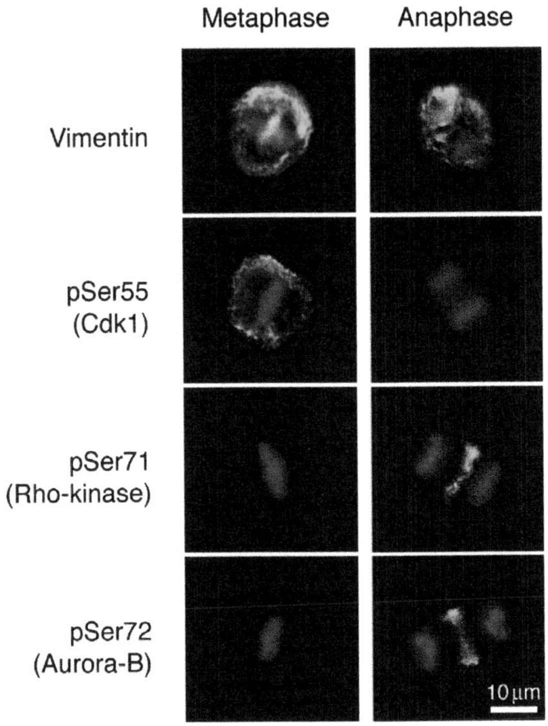

Hidemasa Goto *et al.*, Figure 4 Site-specific vimentin phosphorylation during mitosis. U251 glioma cells were stained with anti-vimentin, anti-vimentin-pSer55, anti-vimentin-pSer71, or anti-vimentin-pSer72 (green). Chromosomes were also stained with propidium iodide (red). *Reproduced from ©Yamaguchi et al. (2005) (originally published in* The Journal of Cell Biology, *http://dx.doi.org/10.1083/jcb.200504091).*

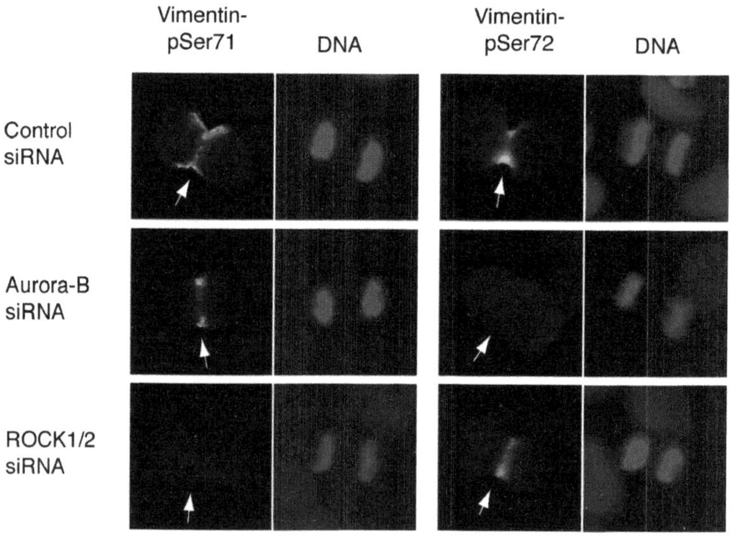

Hidemasa Goto *et al.*, Figure 5 Cleavage furrow-specific vimentin phosphorylation at Ser71 or Ser72 is impaired in HeLa cells treated with small interfering RNA(s) (siRNA) specific to Rho-kinase or Aurora-B, respectively. Arrows indicate the position of cleavage furrow. *Reproduced from ©Yokoyama et al. (2005) (originally published in* Genes to Cells, *http://dx.doi.org/10.1111/j.1365-2443.2005.00824.x).*

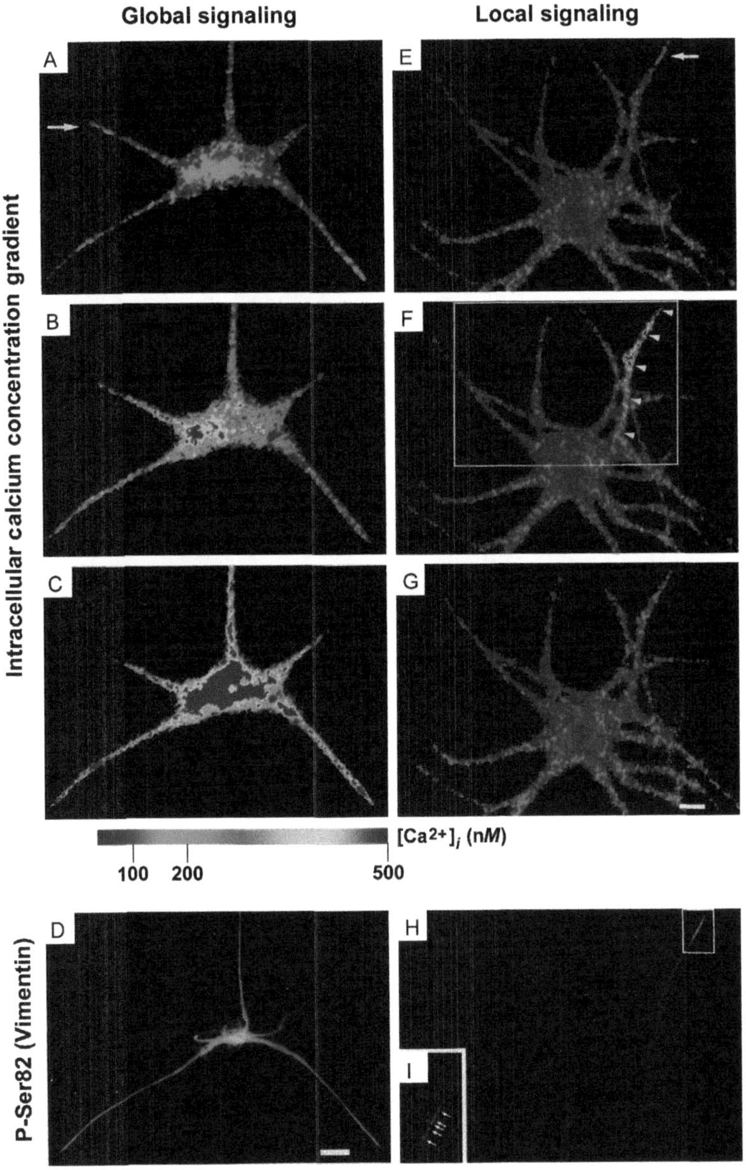

Hidemasa Goto *et al.*, Figure 6 Local and global signaling of CaMKII defined by the area of Ca^{2+} signals. Intracellular distribution of Ca^{2+} signaling in an astrocyte was monitored by fura-2-based Ca^{2+} microscopy (A–C and E–F). $[Ca^{2+}]_i$ in an astrocyte before (A and E), or at 0.5 min (B and F), 1.5 min (C), or 4 min (G) after the local application of 10 μ*M* $PGF_{2\alpha}$ for 0.25 min. At 5 min after the $[Ca^{2+}]_i$ measurement, each cell was fixed and then stained with anti-vimentin-pSer82 (D, H, and I; green). A magnified image in a rectangle (H) are shown in (I). Arrows (A and E) indicate sites of $PGF_{2\alpha}$ application. Arrowheads indicate the process showing Ca^{2+} signaling (F), or the area in which vimentin-Ser82 phosphorylation is elevated (I). Bars, 20 μm. *Reproduced from ©Inagaki et al. (1997) (originally published in* The Journal of Biological Chemistry, *272: 25195–25199).*

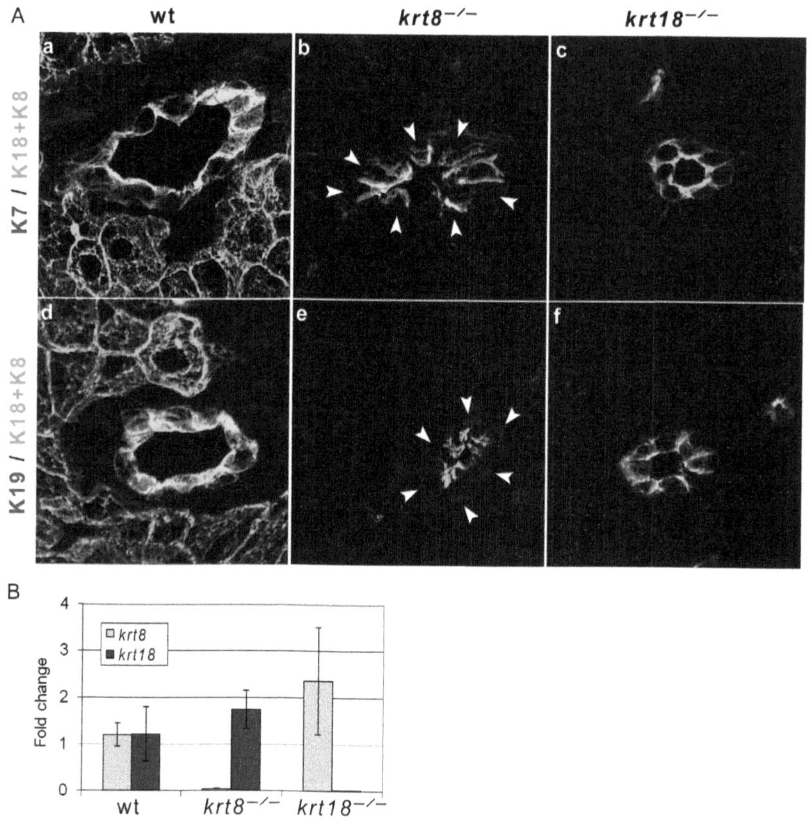

Cornelia Stumptner ***et al.*****, Figure 1** Expression (A and B) and localization (A) of keratins in *krt8*$^{-/-}$, *krt18*$^{-/-}$ and wild-type (wt) mice. (A) Double-label IIF microscopy of mouse hepatocytes and bile ducts with Keratin 7 (K7 red)/Keratin 8 or Keratin 18 (K8 or K18 green) (a–c) and with Keratin 19 (K19 red)/Keratin 8 or Keratin 18 (K8 or K18 green) antibodies (d–f). In wt mice (a and d) hepatocytes show a regular Keratin 8/18 network which is lost in hepatocytes of keratin knockout mice (b, c, e, f) in which one keratin partner is missing. In bile duct epithelia of wt mice (a and d) Keratin 7 and Keratin 19 (both red) and Keratin 8 and Keratin 18 (both green) form a cytoplasmic filament network. Note that in *krt8*$^{-/-}$ mice only Keratin 7 (b) but not Keratin 19 (e) is detectable and is restricted to the apical and lateral portion of the bile duct cells (arrowheads indicate the basal cell portion of bile ducts in b, e). (B) Quantitative real time RT-PCR analysis of *krt8* and *krt18* expression in livers of *krt8*$^{-/-}$, *krt18*$^{-/-}$ and wt mice performed using TaqMan® probe-based gene expression analysis. Keratin expression values are normalized to the house-keeping gene TATAbox binding protein and represent mean values of three to five mice each.

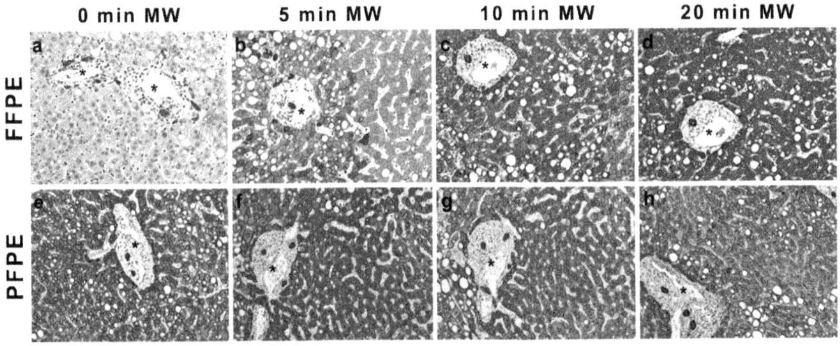

Cornelia Stumptner *et al.*, Figure 4 Effect of different tissue fixatives on keratin IHC. Tissue samples from a human liver were fixed in parallel with formalin (a-d) and PAXgene (e-h) for 24 h before further tissue processing and paraffin embedding. Five micrometer thick sections were used for IHC staining using the antibodies to Keratin 8/18 according to the optimized protocol as shown in Fig. 3B. Different times of microwave (MW) treatment of sections (0, 5, 10, and 20 min; pH 9.0) were tested for antigen retrieval. Asterisks indicate portal tracts (a-h).

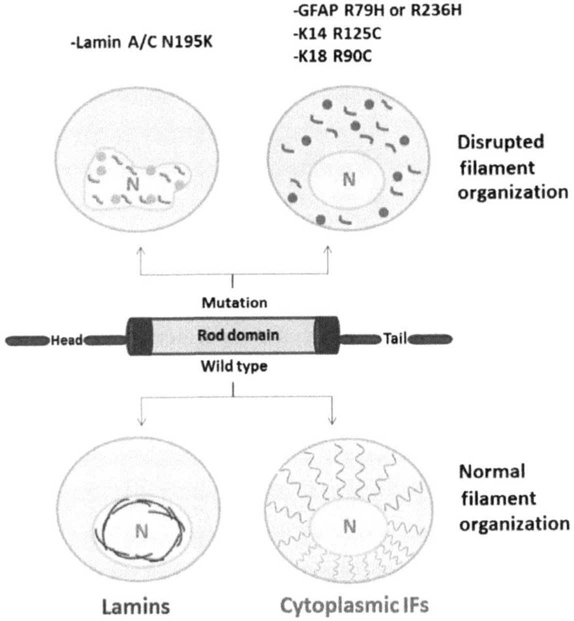

Jingyuan Sun *et al.*, Figure 1 Prototype IF protein domains and consequences of IF mutation on filament organization. The schematic shows the three IF protein domains: a central α-helical coiled-coil relatively conserved "rod" domain (310–350 amino acids) that is flanked by N- and C-terminal non-α-helical "head" and "tail" domains (of variable length depending on the IF protein) which, in turn, provide the exceptional structural diversity among IFs. The mutations responsible for the most severe IF-pathy phenotypes are typically located at ultraconserved helix-initiation and helix-termination motifs at the beginning and end of the rod domain that cause disruption of the filamentous organization into short filaments or dots. Examples of IF mutations that cause the type of filament disruption that is schematically shown include lamin A/C N195K mutation (causes Emery–Dreifuss muscular dystrophy) (Ostlund, Bonne, Schwartz, & Worman, 2001), GFAP R79H or R236H mutation (both cause Alexander disease) (Mignot et al., 2007; Wang, Colodner, & Feany, 2011), K14 R125C mutation (causes EBS) (Bonifas et al., 1991; Coulombe et al., 1991), and K18 R90C (predisposes to liver injury) (Ku, Michie, Oshima, & Omary, 1995). N, nucleus.

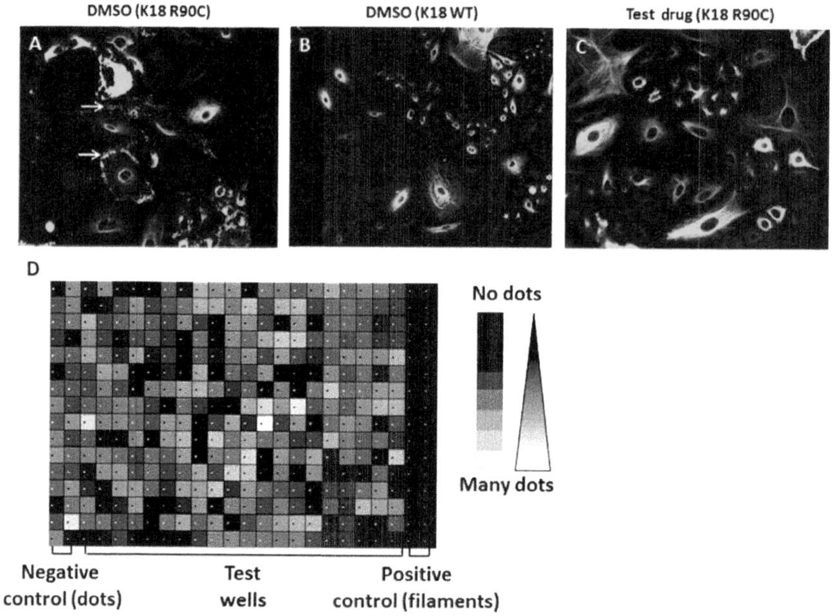

Jingyuan Sun *et al.*, Figure 4 Examples of drug screening phenotypes and a typical heat map as visualized by ImageExpress. (A) Negative control cells which were transduced with GFP-K18 R90C lentivirus and treated with DMSO for 48 h (arrows highlight the dots). (B) Positive control cells which were transduced with GFP-K18 WT lentivirus and treated with DMSO for 48 h. (C) GFP-K18 R90C lentivirus-transduced cells treated with a compound that results in normalization of the disrupted keratins. (D) The first two columns (far left) of the 384-well plat represent GFP-mutant IF-transduced cells treated with DMSO for 48 h as a negative control (primarily red, or non-green color), while the two columns on the far right represent GFP-WT IF-transduced cells treated with DMSO as a positive control. The middle columns represent GFP-mutant IF-transduced cells with different drug library compounds after 48 h. Green indicates few dots/well, while red indicates mainly dots/well. Several wells in the middle columns are bright green, indicating the added compound may have corrected the IF mutant phenotype or is too toxic in those wells (i.e., dead cells will also have few dots and represent false positives).

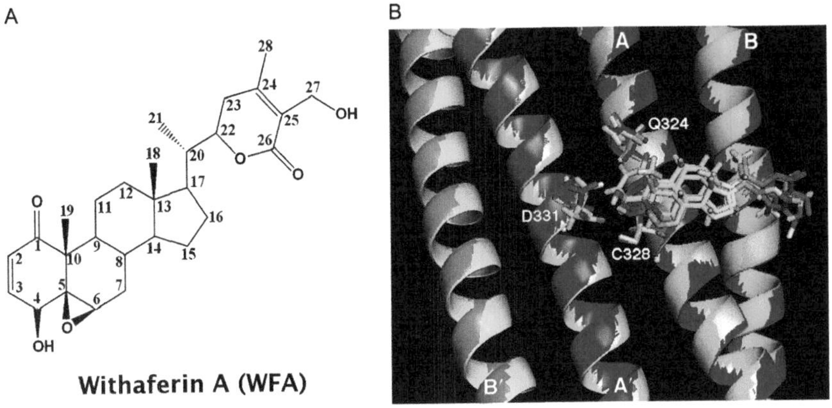

Royce Mohan and Paola Bargagna-Mohan, Figure 1 (A) Chemical structure of WFA. (B) Molecular model of the WFA-binding site in the 2B rod domain of tetrameric type III IFs. Shown is a multicolor composite overlap of WFA docked individually in the binding pockets of vimentin, GFAP, and desmin protein structures, respectively, that identifies a similar binding mode for WFA in these IFs. The key amino acid residues cysteine (C328), aspartic acid (D331), and glutamine (Q324) are numbered with respect to vimentin, for simplicity (Bargagna-Mohan et al., 2010, 2012). *Reproduced with permission as a modified version from Bargagna-Mohan et al. (2007).*

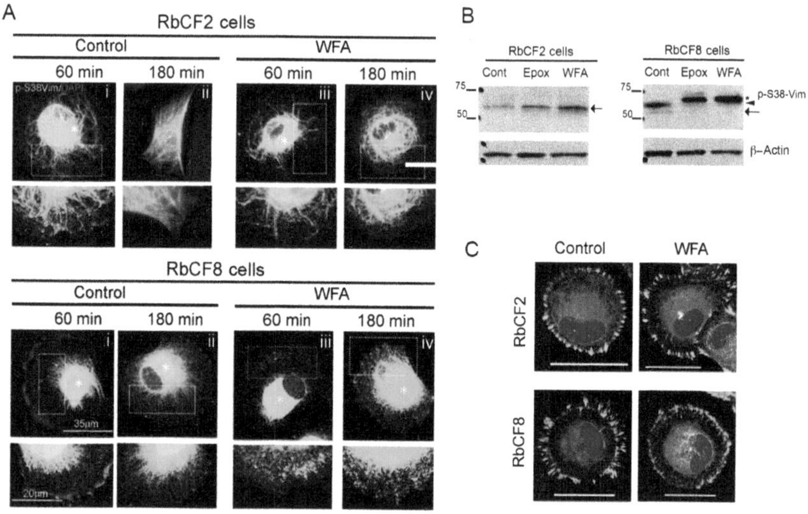

Royce Mohan and Paola Bargagna-Mohan, Figure 4 (A) Immunohistochemistry of pSer38Vim staining (green) in RbCF2 and RbCF8 cells. Scale bar = 35 μm. Inset panels represent 60 × magnification of selected areas. Scale bar = 20 μm. (B) Western blot analysis of pSer38Vim in soluble extracts from RbCF2 and RbCF8 cells treated in presence or absence of WFA for 30 min. Asterisk marks the 67-kDa high molecular weight pSer38Vim hyperphosphorylated species, and the arrow and arrowheads represent the 57- and 61-kDa pSer38Vim bands, respectively. (C) Immunohistochemistry of RbCF2 and RbCF8 cells plated for 30 min followed by treatment in presence or absence of 1 μ*M* WFA for 1 h. Cells were fixed and stained with paxillin (green), vimentin (red), and DAPI (blue). Scale bar = 35 μm. Antibodies (rabbit anti-pSer38Vim antibody (A and B) and mouse anti-vimentin monoclonal V9 (C) from Santa Cruz Biotechnology; mouse anti-β-actin monoclonal from Sigma (B); rabbit anti-paxillin monoclonal (C) from Abcam) were employed. *Modified version from original published images in Bargagna-Mohan et al. (2015).*

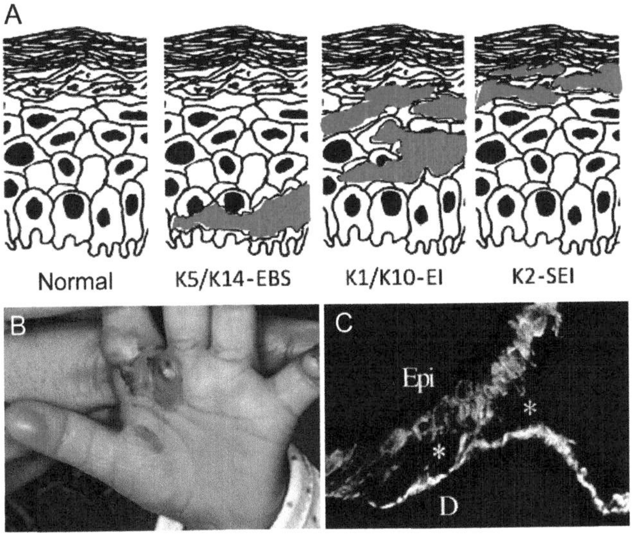

Tong San Tan *et al.*, Figure 1 The *in vivo* consequences of keratin mutations. (A) Diagrammatic examples of planes of epidermal fracture caused by different keratin mutations (EBS: epidermolysis bullosa simplex, EI: epidermolytic ichthyosis, SEI: superficial epidermolytic ichthyosis). Cell fragility (red) is closely correlated with expression range of the mutated keratin and is diagnostically indicative. (B) Clinical presentation of EBS, Dowling-Meara type (EBS-DM). (C) Histological frozen section of EBS-DM biopsy showing immunofluorescence staining of blister area using monoclonal antibody RCK107 to K14 (white). Cleavage plane leaves K14 material on both upper and lower aspects of blister. Epi: Epidermis; D: Dermis; * Intracellular cleavage of basal keratinocytes. *Panel (B) and (C): from Mark Koh, Kandang Kerbau Women's and Children's Hospital, Singapore.*

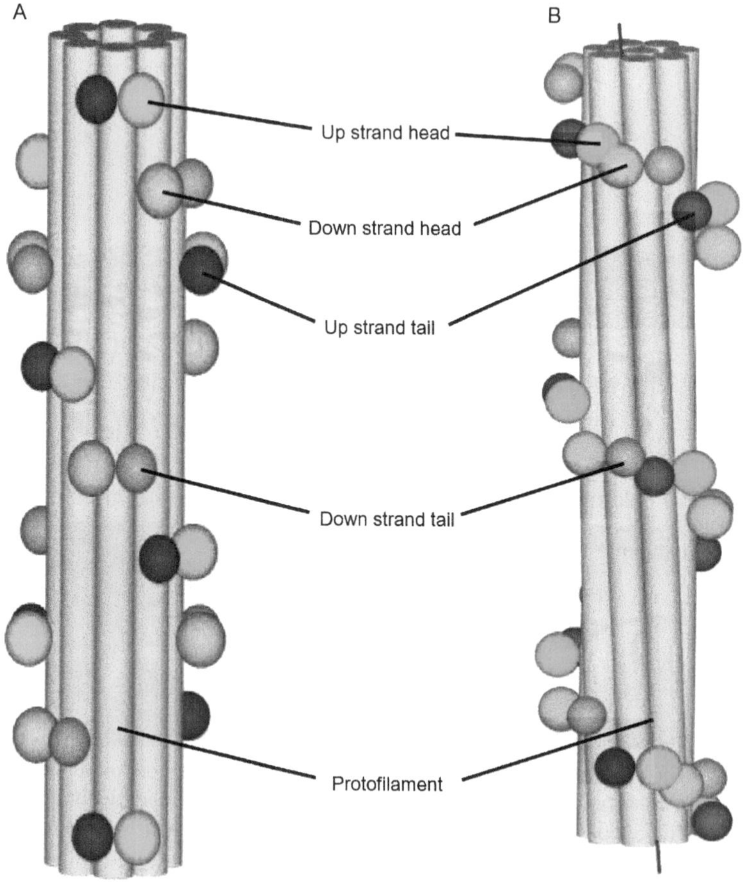

David A.D. Parry, Figure 1 Diagrammatic representation of the structures of (A) "reduced" hair keratin IF and (B) "oxidized" hair keratin IF. The α-helical coiled-coil rod domains of the protofilaments, which are composed of antiparallel molecular strands, are colored yellow. The head domains are represented by green and blue spheres for the up and down strands, respectively, and the tail domains are represented by red and orange spheres in the up and down strands, respectively. The dispositions of the head and tail domains differ significantly between the two structures. In the "reduced" structure, the head and tail domains lie on a two-start left-hand helix that would appear at low resolution to give a diagonal banding pattern of spacing 22 nm. In contrast, in the "oxidized" structure, the head and tail domains are in much closer spatial proximity and appear to be distributed on a helix of pitch length 23.5 nm. Also, there is some radial compaction of the "oxidized" structure relative to the "reduced" one. Details of the radial, axial, and azimuthal coordinates of the protofibrils are given in the text. *Reprinted from Fraser and Parry (2005) with permission from Elsevier.*

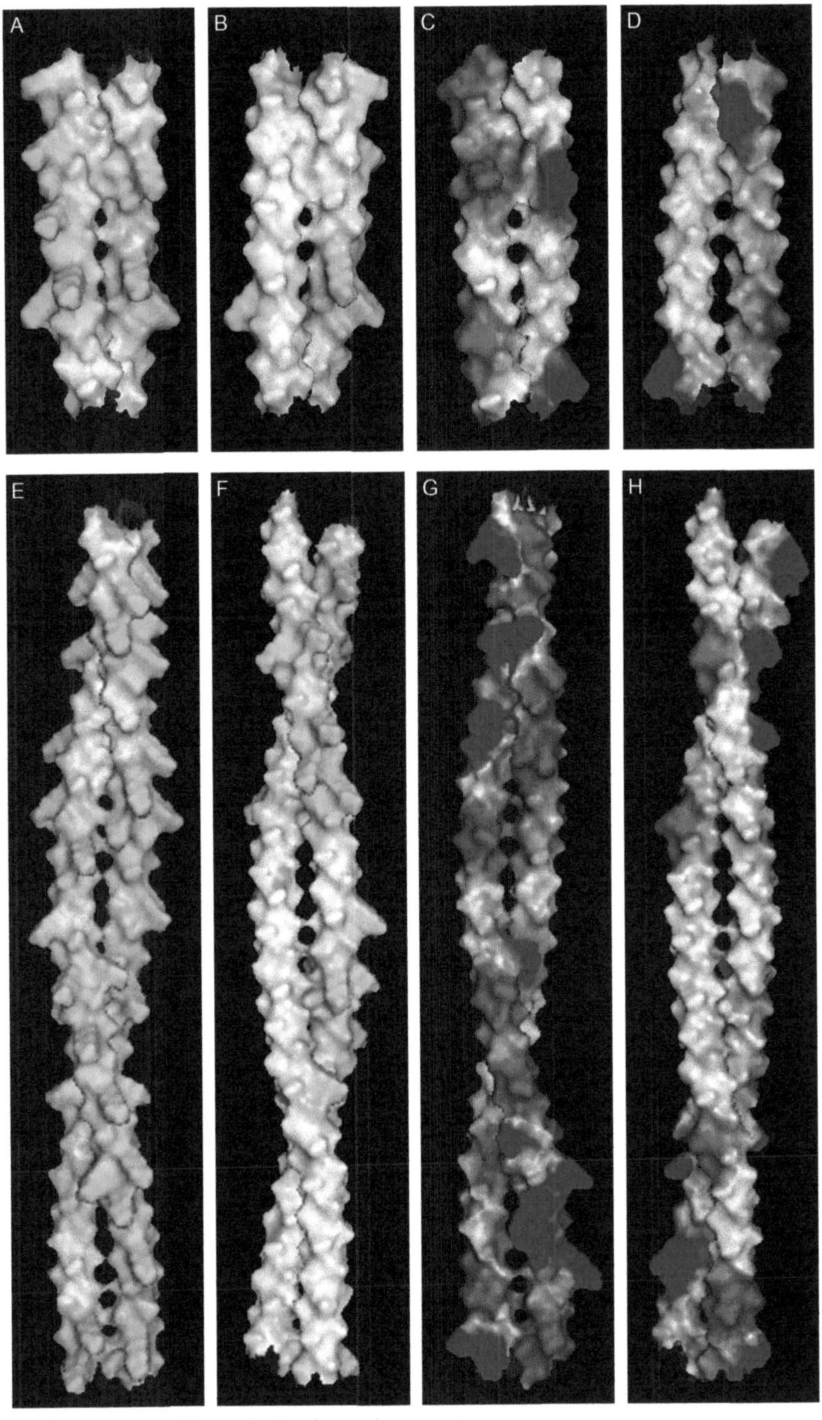

David A.D. Parry, Figure 2 See legend on next page.

David A.D. Parry, Figure 2 One face of segment 1A (face A) in K35/K85 trichocyte keratin (A) is more apolar than the opposite one (face B) (B), and also the differences in the charged residues that occurred between the Type I and Type II chains were largely confined to face A (C) rather than face B (D). In addition, in (E), it was shown that one face of segment 1B in K10/K1 epidermal keratin (face C) displayed the bulk of the apolar residues when compared to that present in its opposite face D (F). Face C also contained the bulk of the charged residue differences that occurred between the two chains (G) compared to face D (H). These results indicate the probable internal face of these segments and those sequence features that play an especially important part in assembly. Details of the methods used (PYMOL) and the color scheme employed are given in Smith and Parry (2008). Apolar residues are colored green, and difference profiles for acidic and for basic residues are colored red and blue, respectively. All other residues are represented by alanines (white). *Reprinted from Smith and Parry (2008) with permission from Elsevier.*

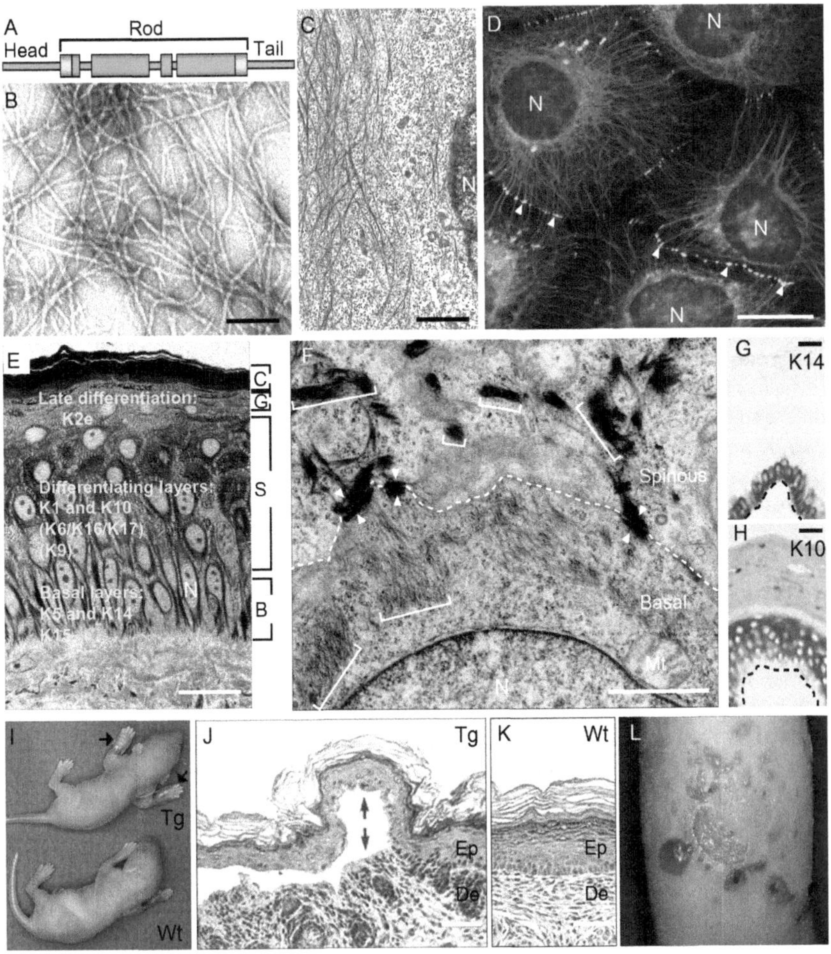

Fengrong Wang *et al.*, Figure 2 Attributes, differential regulation, and disease association of keratins. (A) Tripartite domain structure shared by all keratin and other intermediate filament (IF) proteins. A central α-helical "rod" domain acts as a key determinant of self-assembly and is flanked by nonhelical "head" and "tail" domains at the N-terminus and C-terminus, respectively. The ends of the rod domain contain 15–20 amino acid regions, here shown is yellow that are highly conserved among all IFs. (B) Visualization of filaments, reconstituted *in vitro* from purified K5 and K14, by negative staining and electron microscopy. Bar, 125 nm. (C) Ultrastructure of the cytoplasm of epidermal cells in primary culture as shown by transmission electron microscopy. Keratin filaments are abundant and tend to be organized in large bundles of loosely packed filaments in the cytoplasm. Bar, 5 μm. (D) Triple-labeling for keratin (red) and desmoplakin (green), a desmosome component, and DNA (blue) by indirect immunofluorescence of epidermal cells in culture. Keratin filaments are organized in a network that spans the entire cytoplasm and are attached to desmosomes at points of cell–cell contacts (arrowheads). Bar, 30 μm. N, nucleus. (E) Histological cross section of

(Continued)

Fengrong Wang *et al.*, Figure 2—Cont'd resin-embedded human trunk epidermis, revealing the basal (B), spinous (S), granular (G), and cornified (C) compartments. The differentiation-dependent distribution of keratin proteins in the epidermis is indicated. Bar, 50 µm. N, nucleus. (F) Ultrastructure of the boundary between the basal and suprabasal cells in mouse trunk epidermis as seen by routine transmission electron microscopy. The sample, from which this micrograph was taken, is oriented in the same manner as (E). Organization of keratin filaments as loose bundles (brackets in basal cell) correlates with the expression of K5–K14 in basal cells, whereas the formation of much thicker and electron-dense filament bundles (brackets in spinous cell) reflects the onset of K1–K10 expression in early differentiating keratinocytes. Arrowheads point to desmosomes. Bar, 1 µm. N, nucleus. (G and H) Differential distribution of keratin epitopes on human skin tissue cross sections (similar to E) as visualized by an antibody-based detection method. K14 occurs in the basal layer, where the epidermal progenitor cells reside (G). K10 primarily occurs in the differentiating suprabasal layers of epidermis (H). Dashed line, basal lamina. Bar, 100 µm. (I) Newborn mouse littermates. The top mouse is transgenic (Tg) and expresses a mutated form of K14 in its epidermis. Unlike the control pup below (Wt), this transgenic newborn shows extensive blistering of its front paws (arrows). (J and K) Hematoxylin and eosin (H&E)-stained histological cross section through paraffin-embedded newborn mouse skin similar to those shown in (I). Compared with the intact skin of a control littermate (K, Wt), the epidermis of the K14 mutant expressing transgenic pup (J, Tg) shows intraepidermal cleavage within the basal layer, where the mutant keratin is expressed (opposing arrows). Bar, 100 µm. (L) Leg skin in a patient with the Dowling-Meara form of epidermolysis bullosa simplex. Several skin blisters are grouped in a herpetiform pattern. *Reproduced from Coulombe & Bernot, 2004.*

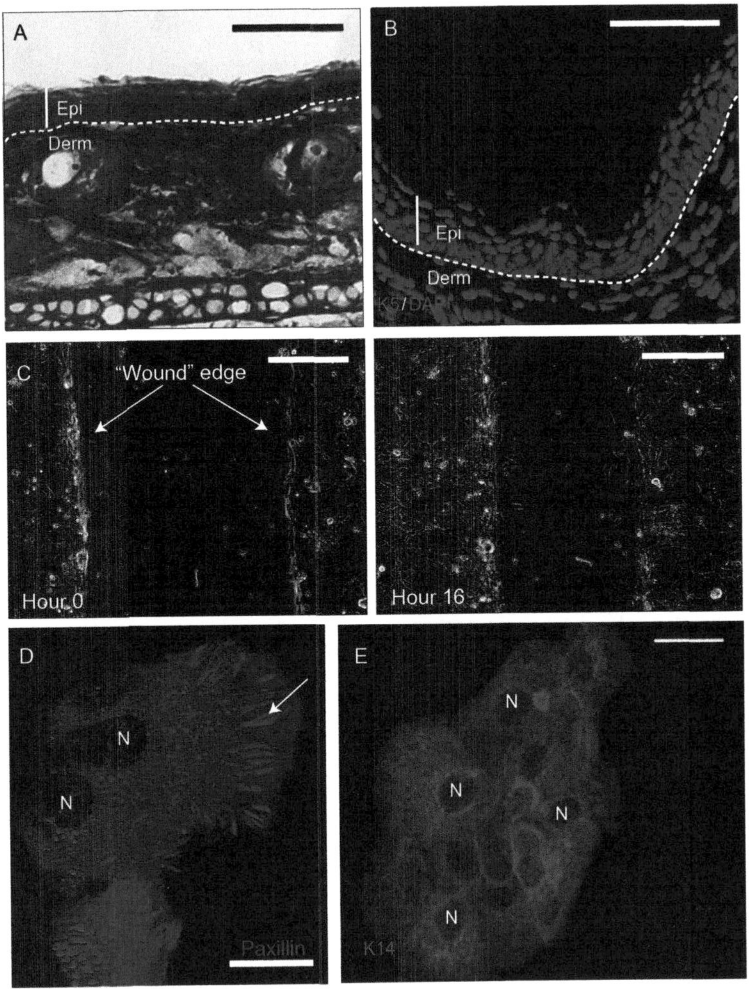

Fengrong Wang *et al.*, Figure 3 Various analyses of skin keratins utilizing mouse tissue and cultured primary keratinocytes. (A) Hematoxylin and eosin stain of fresh-frozen adult mouse ear tissue (4 months old). Dotted line marks the boundary between the epidermis (Epi) and dermis (Derm). Bar, 50 μm. (B) Fresh-frozen front paw tissue of a 2-month-old mouse processed for immunofluorescence of basal keratin K5. Note the restriction of K5 to the basal (progenitor) layer of keratinocytes. Dotted line marks the boundary between the epidermis (Epi) and dermis (Derm). Bar, 50 μm. (C) Live cell images of "wounding assay" using WT mouse keratinocytes in primary culture. Freshly isolated keratinocytes were plated in chamber slides with culture inserts. The "wound" was introduced by removing culture inserts when cells were 100% confluent. Phase contrast imaging was performed with a Zeiss Axio Observer Z1 microscope equipped with Zeiss EC Plan-Neofluar 10×/0.3 Ph1 objective for 16 h. Bar, 200 μm. (D) Transient

(Continued)

Fengrong Wang *et al.*, Figure 3—Cont'd expression of mCherry fluorescence protein (mCherry)-tagged paxillin in mouse primary keratinocytes. Freshly isolated keratinocytes were transfected with a plasmid encoding mCherry-tagged paxillin using the nucleofection method before plating in chamber slides with culture inserts. After removing the culture inserts, keratinocytes were allowed to migrate for at least 8 h before imaging using a Zeiss Axio Observer Z1 fluorescence microscope equipped with Zeiss EC Plan-Neofluar 40 × objective. Keratinocytes shown in this panel were located at the leading edge in "wounding assay." Arrow points to a paxillin-positive focal adhesion. Bar, 20 μm. N, nucleus. (E) Immunofluorescence staining of keratin K14 in mouse skin keratinocytes in primary culture. Keratinocytes were isolated from newborn mouse pups and cultured in mKer media for 2 days. They were then fixed with 4% PFA and permeabilized with 0.5% Triton/PBS. Bar, 50 μm. N, nucleus.

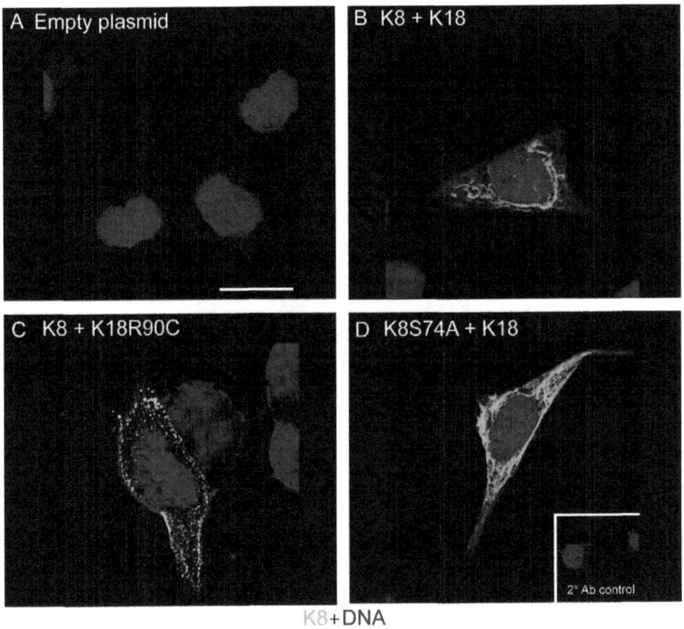

Pavel Strnad *et al.*, Figure 1 Overexpression of WT SEKs and SEK variants in cells. Immunofluorescence staining for K8 (Troma I, 1:1500, Developmental Studies Hybridoma Bank, University of Iowa, IA) was performed in MEFvimKO cells transfected with empty plasmid (A), WT K8/K18 (B) and K8/18 variants together with their WT partner keratin (C) and (D) as indicated in the figure. The transfection of cells was carried out via electroporation, and the cells were fixed in acetone at −20 °C for 10 min. DNA was stained with Draq5 (1:5000), and the cells were visualized with a Leica TCS SP5 confocal microscope. The insert in panel (D) shows signals from cells stained with secondary antibody and Draq5. Scale bar = 20 μm.

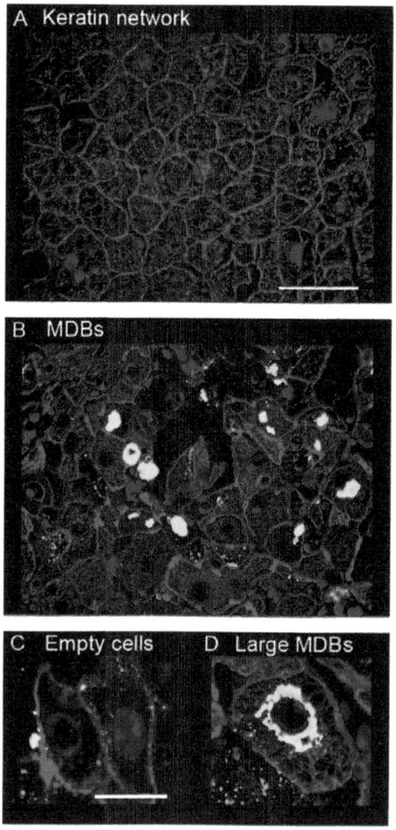

Pavel Strnad *et al.*, Figure 3 Keratin network reorganization in a mouse model of Mallory–Denk body (MDB) formation. Double immunofluorescence staining for K8/K18 (antibody 8592-red) and p62 (antibody GP62-C, Progen; green; see Table 5 for details) depicts the normal appearance of the SEK network in untreated C57BL/6 mice (A), as well as the SEK reorganization in animals fed with 0.1% 3,5-diethoxycarbonyl-1,4-dihydrocollidine (DDC) for 10 weeks (B)–(D). Note that some cells display only residual nonfilamentous K8/K18 signal in the cell periphery (empty cells; C); while others contain large MDBs that appear as yellow inclusions due to the double-labeling with K8/K18 and p62 (D). ProLong Gold mounting medium containing DAPI (Invitrogen) was used to visualize the nuclei. Scale bars: A,B = 50 μm; C,D = 20 μm.

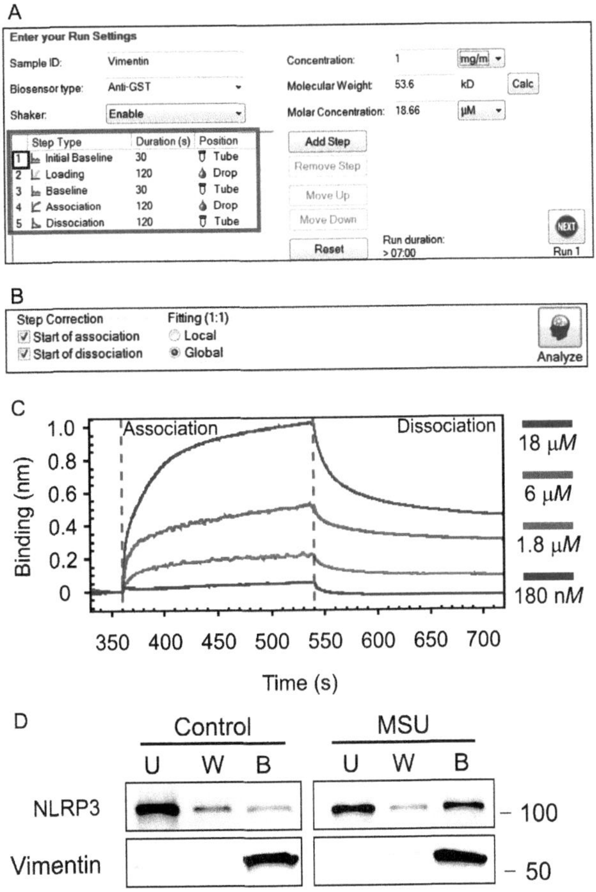

Karen M. Ridge *et al.*, Figure 4 Biolayer interferometric analysis of interaction between vimentin and NLRP3 (NACHT, LRR, and PYD domains-containing protein 3). (A) Screenshot of the BLItz Pro software in which the settings for an experiment are shown. The only value that needs to be provided is the Molar Concentration, which can be altered for each run. If the concentration and molecular weight of the protein being used as the analyte is provided, the Molar Concentration value will be calculated. The red square shows where the Step Type, Duration, and Position can be specified for the experiment. (B) Settings for binding analysis: Start of Association and Start of Dissociation checked, and Global Fitting selected. (C) Determination of association and dissociation rates between a bait protein (NLRP3) attached to a biosensor and various concentrations of vimentin diluted in BLI kinetics buffer. (D) Examination of binding between NLRP3 and vimentin. Cell extracts were prepared from vimentin-null bone marrow-derived macrophages (BMDMs). Some cells were treated with monosodium urate (MSU) to activate the BMDMs, while others were treated with saline (control). S-tag, His-tag vimentin was added to these extracts and binding of NLRP3 was examined (U, unbound; W, washed; B, bound). Binding was evaluated by chemiluminescence.

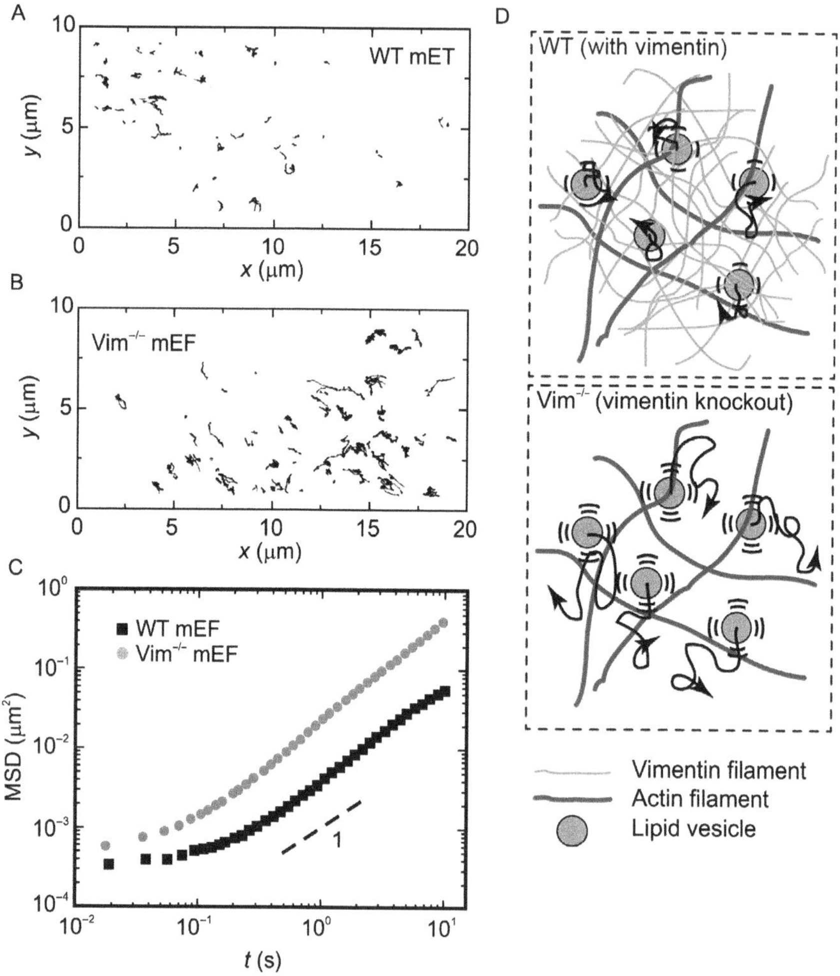

Karen M. Ridge *et al.*, Figure 8 Intracellular movement of endogenous vesicles and protein complexes inside wild-type (WT) and *Vimentin*$^{-/-}$ (*Vim*$^{-/-}$) mouse embryonic fibroblasts (mEFs). (A, B) Ten-second trajectories of endogenous vesicles and protein complexes in the cytoplasm of (A) WT mEFs and (B) *Vim*$^{-/-}$ mEFs. These refractive objects are visualized by bright-field microscopy. (C) Calculation of the mean squared displacement of vesicles and protein complexes shows that these organelles move faster in the *Vim*$^{-/-}$ mEFs than in the WT mEFs. (D) Illustration of random organelle movement in networks with and without vimentin. In the WT cells, the vimentin network constrains the diffusive-like movement of organelles; in the *Vim*$^{-/-}$ cells, organelles move more freely.

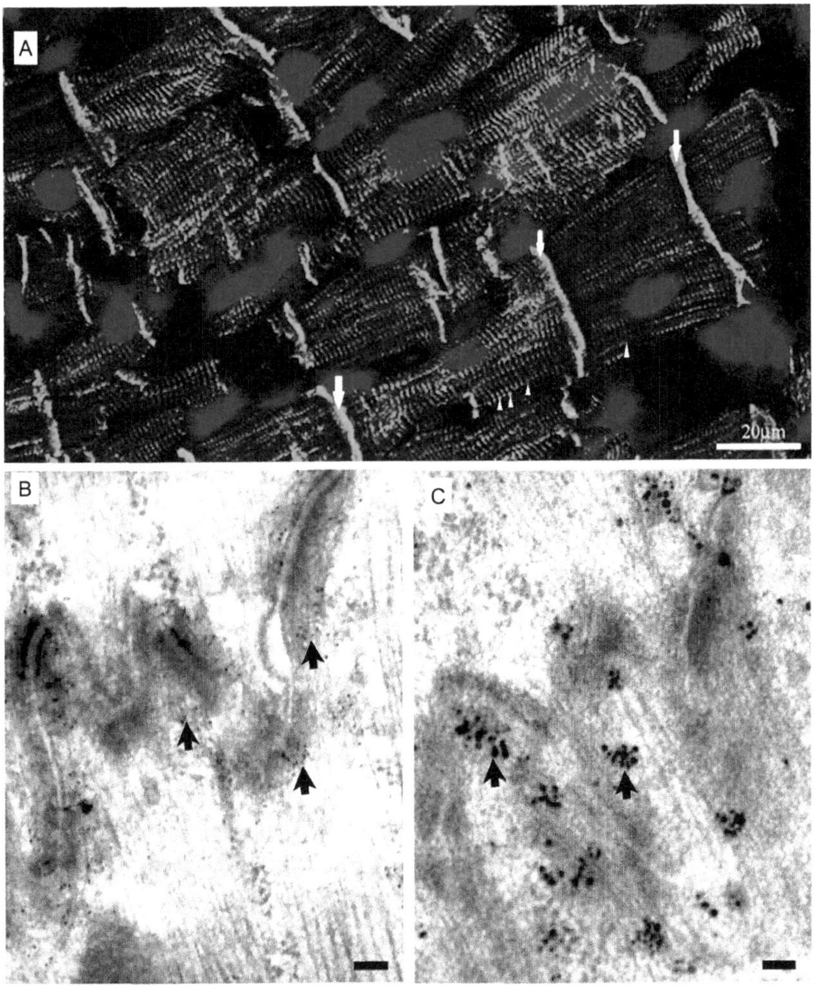

Antigoni Diokmetzidou *et al.*, Figure 4 Localization of desmin in the myocardium by immunofluorescence (A) and immunoelectron microscopy (B and C). (A) Frozen sections of human myocardium are stained with anti-desmin. White arrows point to Intercalated Disks and arrowheads to Z-disks. (B and C) Ultrathin sections of mouse myocardium stained with anti-desmin and decorated by 15 nm colloidal gold particles (A) or labeling with silver amplification (B), both pointed by white arrows. Only a part of an Intercalated Disk is shown in both cases. Scale bar: 2 μm.

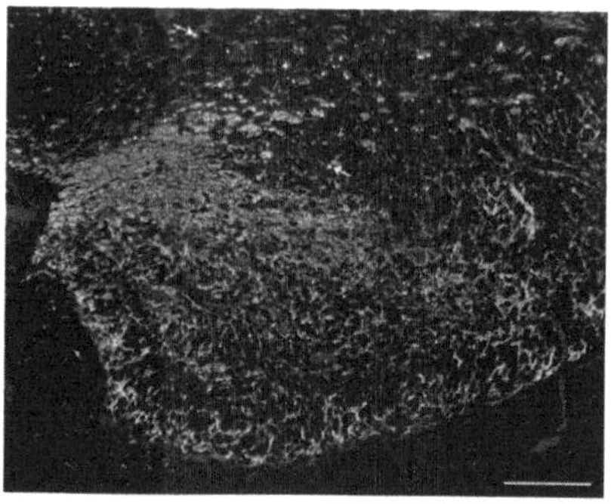

Jian Zhao and Ronald K.H. Liem, Figure 1 Double immunofluorescence of sections of embryonic brain (E16) of an animal with a N98SNFL knock-in mutation with antibodies specific for NFL (red) and α-internexin (green). NFL positive neurons show aggregates that also contain α-internexin (arrows). However, a number of α-internexin positive processes can be observed that are NFL-negative and are therefore not aggregated. Bars = 100 μm.

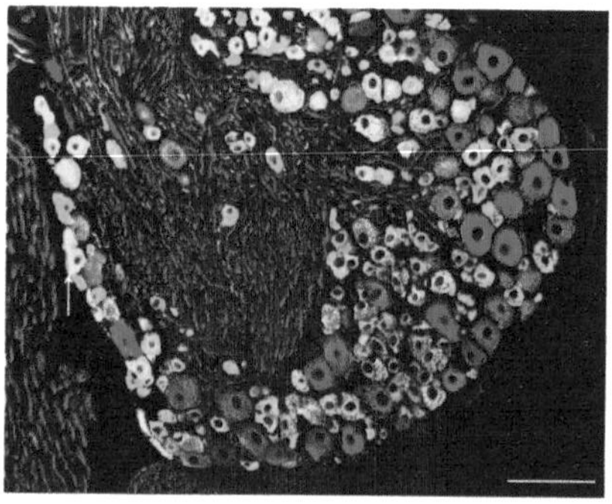

Jian Zhao and Ronald K.H. Liem, Figure 2 Double immunofluorescence of sections of dorsal root ganglia from adult mice using antibodies specific for NFL (red) and peripherin (green). Most neurons show strong staining of only one of NFL and peripherin and little staining of the other. Arrow depicts a neuron showing coexpression of both NFL and peripherin. Axons show preferential staining of NFL. Bars = 100 μm.

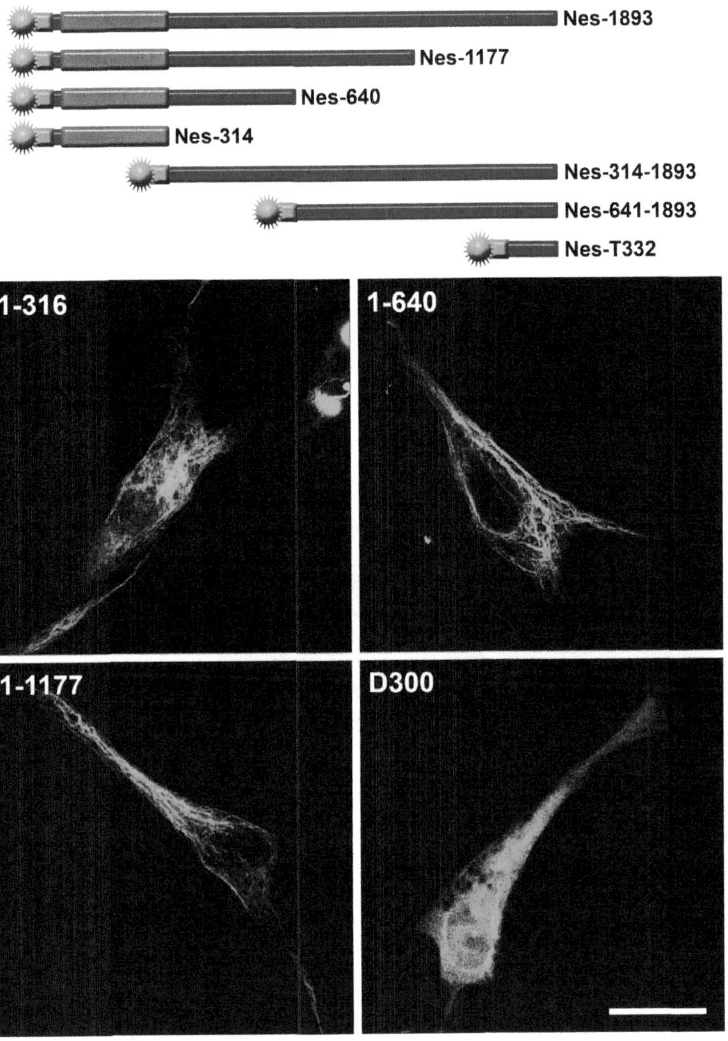

Julia Lindqvist *et al.*, Figure 4 Nestin deletion mutant constructs. Differentially truncated, N-terminally GFP-tagged rat nestin constructs can be employed to effectively investigate the physiological roles of nestin. The position of the GFP-tag is highlighted in green and the rod domain in orange. The short nestin head (N-terminus) and long tail (C-terminus) are blue. The mutants include the following constructs: full-length nestin (Nes-1893), nestin tail truncations (Nes-1177, Nes-640), and nestin lacking the C-terminal tail completely (Nes-314). N-terminally truncated nestin constructs (lacking the rod domain) include Nes-314-1893, Nes-641-1893, Nes-T332, the latter construct containing only the last 332 amino acids of the C-terminus. The morphology obtained with these constructs is exemplified by images from BHK-21 cells that were transfected with four different representative nestin constructs. In cells transfected with the construct containing the head and rod domain of nestin, Nes-314, filaments can be seen to form throughout the cells. Likewise, the constructs Nes-640 and Nes-1177 also give rise to filamentous structures. Mutants lacking the rod-domain, like Nes-T332(D300), fail to form filamentous structures (similar morphology was obtained with all truncations lacking the rod domain). The images were acquired with Leica TCS SP5 confocal microscope and are presented as maximum projections of the GFP signal. Scale bar is 20 μm.

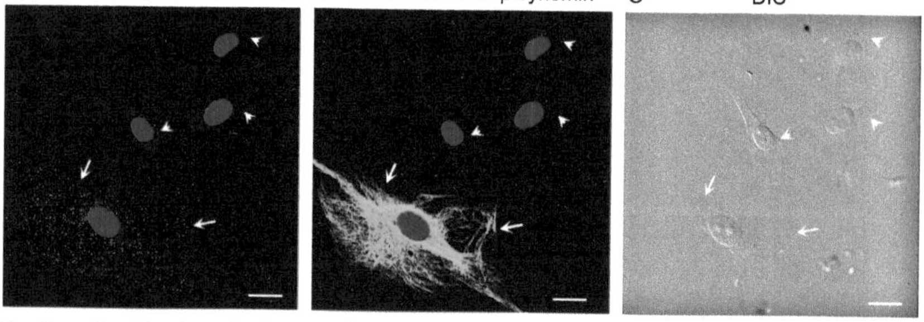

Madhumita Paul and Omar Skalli, Figure 3 PLA demonstrating the interaction between β-synemin and PP2A A subunit in U87 human glioblastoma cells (obtained from the ATCC), which was previously reported by our laboratory (Pitre et al., 2012). U87 cells were transfected with β-synemin tagged with the V5 epitope and PLA was performed with mouse anti-V5 and rabbit anti-PP2A A subunit (A). Following the PLA reaction, cells were stained with anti-mouse IgG conjugated to Alexa Fluor 488 to identify the cells transfected with the synemin construct as they contain a filamentous synemin network (light gray filaments) (B). Nuclear DNA was stained with DAPI and appears dark gray on the micrographs (A and B). PLA dots in the cell transfected with synemin (arrow) demonstrate interaction between synemin and PP2A A subunit. The absence of PLA dots in the cells not expressing the synemin construct (arrowhead) indicates the specificity of the reaction. (C) DIC, differential interference contrast. Bars = 10 μm.

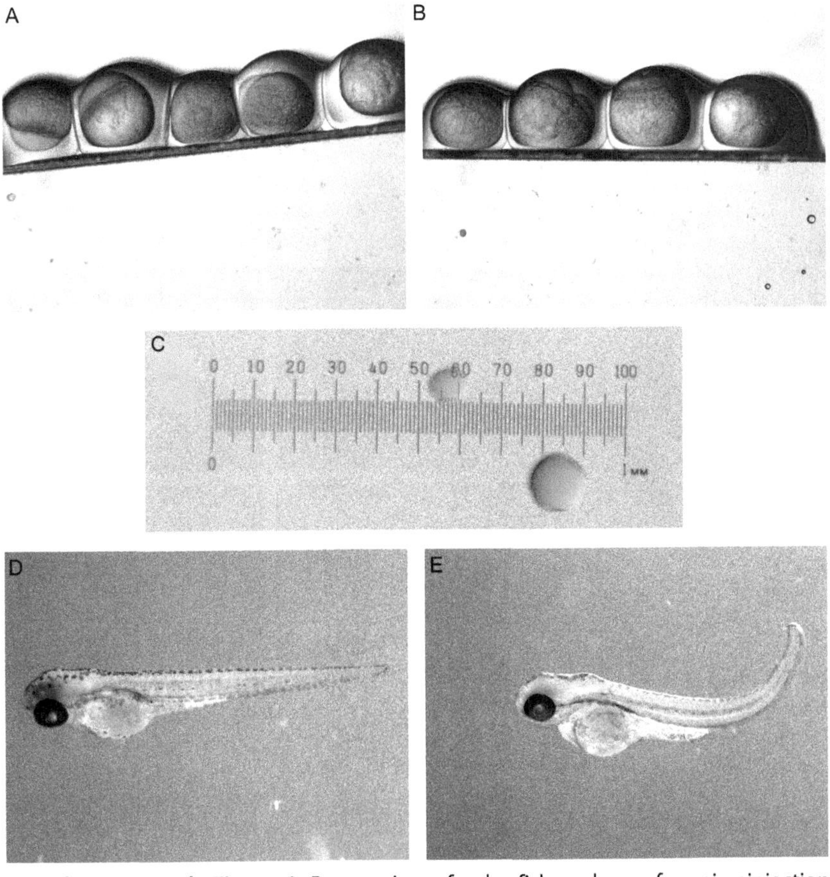

Miguel Jarrin *et al.*, Figure 3 Preparation of zebrafish embryos for microinjection. (A) One-cell stage embryos were selected and transferred into a Petri dish and then aligned against a microscope slide prior to microinjection. (B) Two-cell stage embryos prepared for microinjection. (C) Calibration of the microinjection volume. The volume to be microinjected is measured using mineral oil and a micrometer. (D) Day 3 postfertilization control embryo with no morphological defects. (E) Day 3 microinjected embryos showing morphological defects. Effectiveness of the microinjection is confirmed by phenotypic analyses followed by protein expression analysis by immunoblotting.

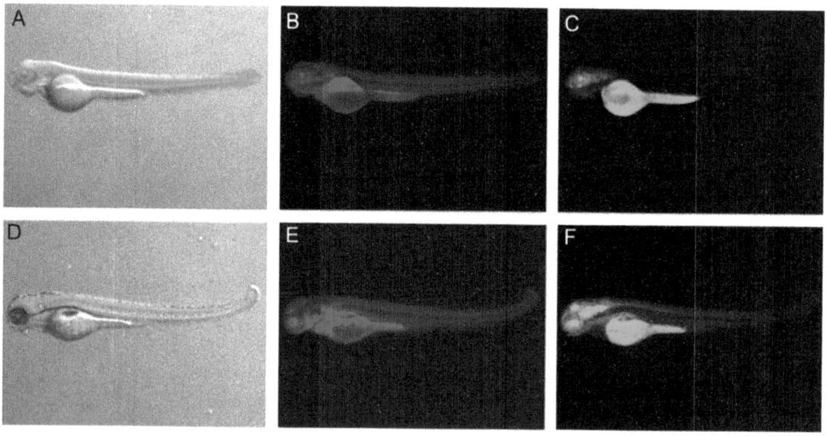

Miguel Jarrin *et al.*, Figure 4 Staining of the embryos *in vivo* with vital dyes. The use of different vital dyes is an invaluable technique for bioimaging applicable to any zebrafish embryo. In this figure, we have summarized the representative staining of the day 3 embryos. (A)–(C) Control day 3 embryo. (D)–(F) Day 3 microinjected embryo. Embryos were staining for 1 h with coumarin 6 (C) and (F), and BODIPY TR (B) and (E) to detect cell membranes.

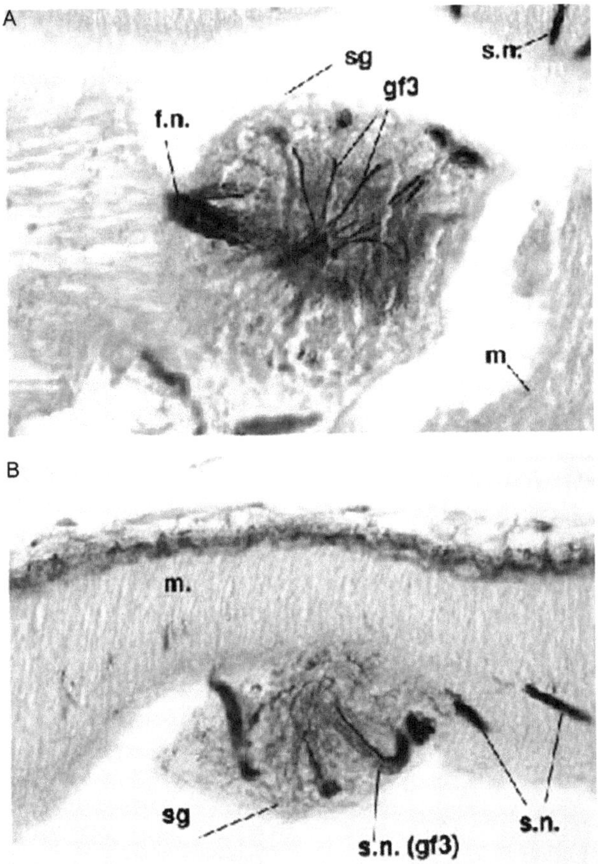

Philip Grant and Harish C. Pant, Figure 5 Immunohistochemical expression of NF proteins in stellate ganglion of squid hatchling. (A) Horizontal section of stage 30 squid hatchling stained with SMI 31, antibody which detects the NF220 phosphorylated NF. f.n., fin nerve; sg, stellate ganglion cell bodies; gf3 are giant axons in stellar nerves; s.n., stellar nerve in mantle muscle, m. (B) Horizontal section of stage 30 hatchling showing expression of NF–NT antibody which stains rod region shared by all NF proteins (NF60, NF70, and NF220), m, mantle muscle; sg, cell bodies in stellate ganglion; s.n., stellar nerve containing giant axon (gf3).

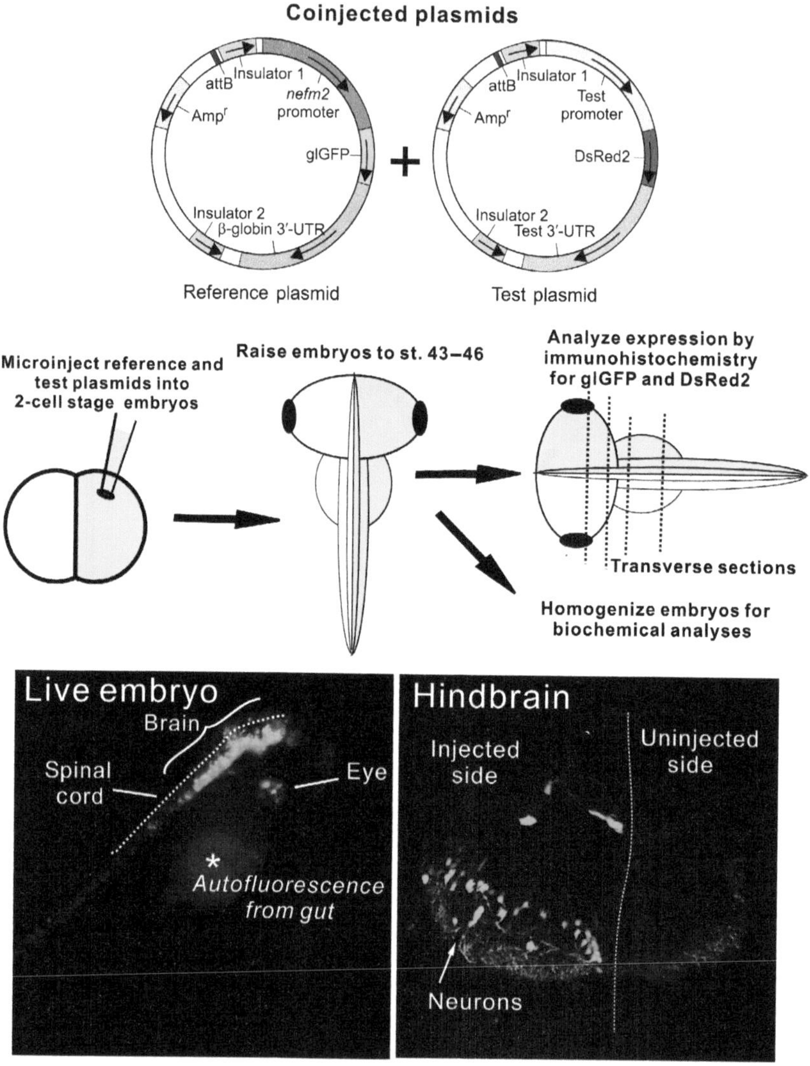

Chen Wang and Ben G. Szaro, Figure 1 (Top) A test plasmid expressing DsRed2 and *cis*-regulatory elements of interest is coinjected with a reference plasmid expressing glGFP and containing an *nefm*2 promoter and the 3′-UTR from *β-globin*. Both plasmids contain an *attB* element and flanking dual *HS4* insulator elements. (Middle) The two plasmids are coinjected at the 2-cell stage and resultant embryos are reared to early swimming tadpole stages (st. 43–46) for observation and biochemical analyses. (Bottom) Examples of glGFP expression derived from a reference plasmid containing 1.5 kb of upstream regulatory sequence from the *Xenopus nefm*2 gene, illustrated in a live tadpole and in a section of hindbrain processed for glGFP immunofluorescence.

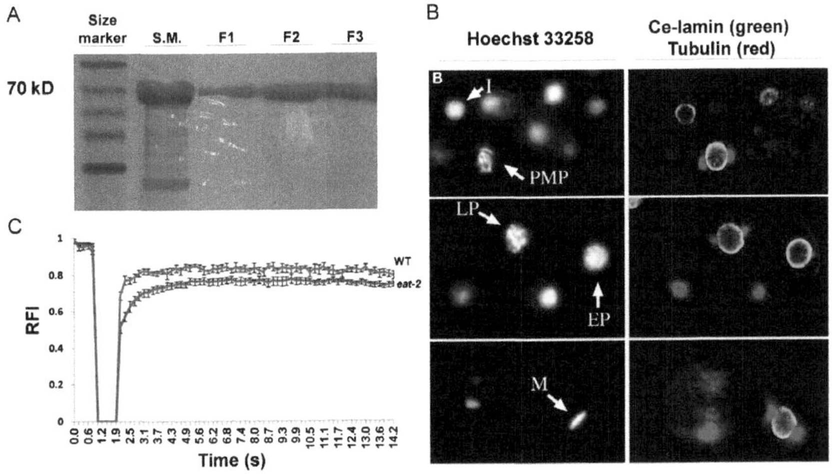

Noam Zuela and Yosef Gruenbaum, Figure 2 (A) Bacterially expressed 8XHis-tagged Ce-lamin was dissolved in 8 *M* urea and purified on a nickel column. The different fractions were loaded on 8% SDS-PAGE gel and stained with Coomassie Brilliant Blue. S.M., starting material; F1–3, first three elution fractions of the nickel column. (B) Indirect immunofluorescence of *C. elegans* embryos stained with DAPI for DNA and by specific antibodies for Ce-lamin (green) and tubulin (red). Mitosis stages (white arrows): I, interphase; PMP; prometaphase; LP, late prophase; EP, early prophase; M, metaphase. (C) Measurements of relative fluorescence intensity (RFI) during FRAP of BAF-1::GFP in wild type (blue) or eat-2 (red) larvae at the first larval stage (L1). Error bars represent SEM. *Panel (B) is taken from figure 4B in Lee et al. (2000)* Mol. Biol. Cell *11, 3089–3099. Panel (C) is taken from fig. 1A in Bar et al. (2014).* Mol. Biol. Cell *25, 1127–1136.*

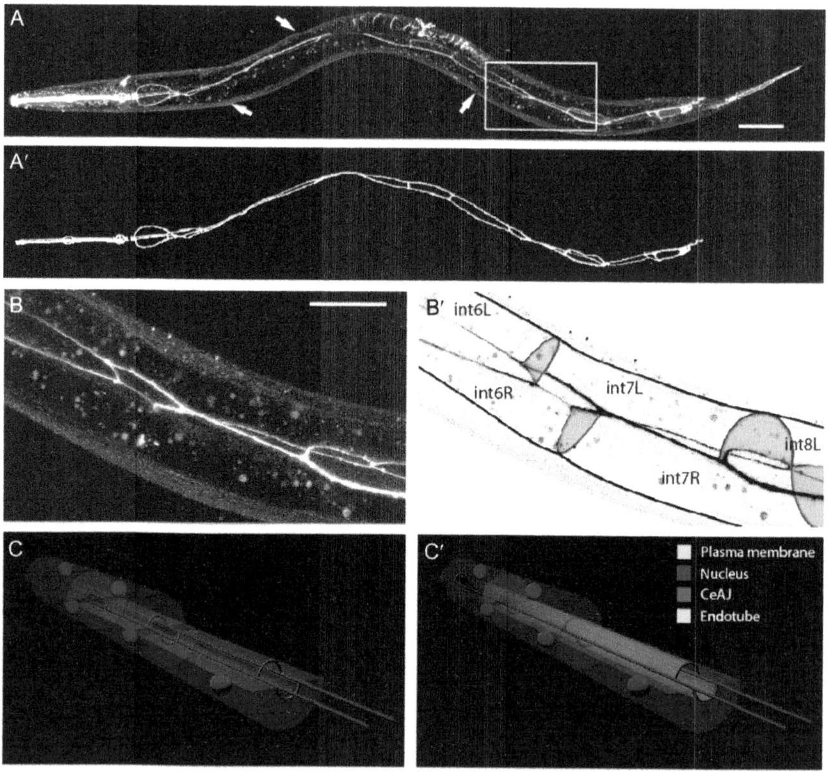

Oliver Jahnel *et al.*, Figure 2 Detection and modeling of the intermediate filament-associated apical cell contacts in the *C. elegans* intestine. (A) Composite micrograph depicting the fluorescence of the CeAJ marker DLG-1::mCherry in the intestine of an L4 *C. elegans* larva as a single-projection view of stitched images. Note that slight movements of the worm resulted in misalignments at borders of image stacks (arrows). Since DLG-1::mCherry is not restricted to the intestine but is also detectable in other tissues including pharynx, rectum, seam cells, hypodermis, spermatheca, and vulva, only focal planes 6–16 of the entire stack of 20 focal planes are shown to improve visualization of the intestinal fluorescence. Panel (A′) shows a cropped version of the micrograph in A to further alleviate examination of the intestinal fluorescence pattern (see also corresponding animation in Movie 1 (http://dx.doi.org/10.1016/bs.mie.2015.08.030)). (B) High magnification taken from A (boxed area) of a region encompassing int-6 to int-8. Details of the typical ladder pattern can be seen, which are caused by the staggered and clockwise twisted arrangement of the intestinal cells. Panel (B′) shows an inverse version of B and also includes the position of the basolateral cell borders, which were drawn manually with the help of simultaneously recorded phase-contrast images (not shown). Panels (C) and (C′) Simplified 3D reconstructions of part of the intestine depicting the CeAJ without (C) and with the attached intermediate filament-rich endotube (C′). The corresponding animation in Movie 2 (http://dx.doi.org/10.1016/bs.mie.2015.08.030) highlights further details. Scale bars: 50 μm in A (same magnification in A′) and 25 μm in B (same magnification in B′).

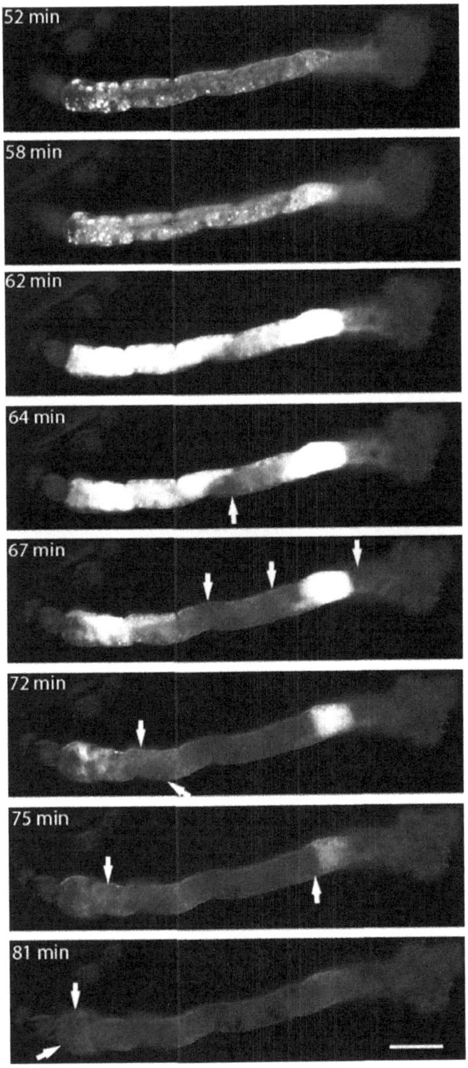

Oliver Jahnel *et al.*, Figure 4 Monitoring cell death by trypan blue staining (dark blue) in dissected intestines. The images are overlays of the phase contrast and fluorescence micrographs in a dissected intestine. The intestine was stored for 52 min in culture medium before recording. The images are taken from a time-lapse recording that is presented in Movie 5 (http://dx.doi.org/10.1016/bs.mie.2015.08.030). Note that a fluorescent wave (white) is initiated at 58 min and that loss of fluorescence coincides with uptake of trypan blue into dying cells (arrows). Scale bar: 50 μm.

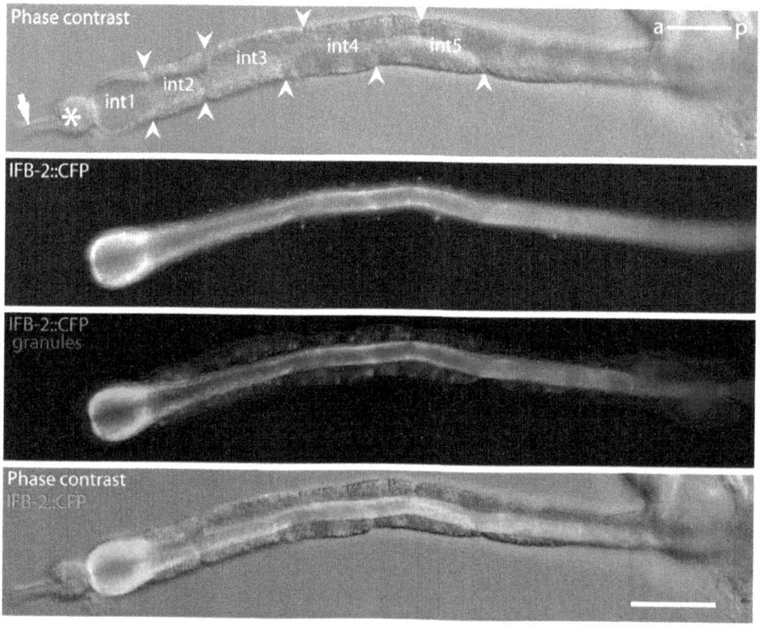

Oliver Jahnel *et al.*, Figure 5 IFB-2::CFP fluorescence in a viable dissected intestine of reporter strain BJ49. Top: phase contrast. Arrow, pharyngeal isthmus; asterisk, terminal pharyngeal bulb; arrowheads, cell borders of intestinal cells between the intestinal rings int1, int2, int3, int4, and int5. Upper middle: corresponding fluorescence recording of IFB-2::CFP. Lower middle: overlay of IFB-2::CFP fluorescence (cyan) and autofluorescent intestinal granules (false red color), which indicate full vitality of the dissected intestine. Bottom: overlay of phase contrast and IFB-2::CFP fluorescence. Note that the distribution of IFB-2::CFP in the dissected intestine is comparable to its distribution in a living L4 larva (Fig. 1). Scale bar: 50 μm.

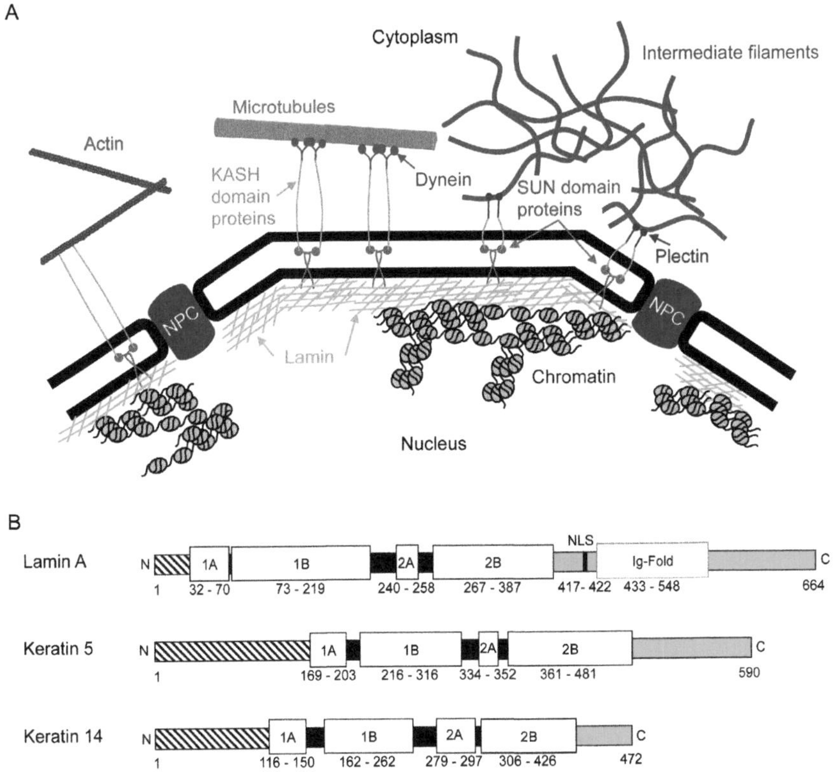

Jens Bohnekamp *et al.*, Figure 1 Cellular localization and protein domain structure of intermediate filaments. (A) Diagram of a cell showing intermediate filaments and their connections to other proteins. Lamins, the nuclear intermediate filaments, organize the chromatin for proper gene expression. In addition, lamins interact with other nuclear envelope proteins such as SUN domain proteins, components of the LINC complex (Kim, Birendra, & Roux, 2015), which bridge the nucleoskeleton and the cytoskeleton. Cytoplasmic intermediate filaments, such as keratins, form networks in the cytoplasm that provide structural integrity to tissues. (B) The domain structure of human lamin A and keratin 5 and 14. These proteins possess a domain structure common to intermediate filaments that is composed of a small head domain (diagonal stripes), a coil–coil rod domain (black with white boxes) and a tail domain (green). Lamins possess a nuclear localization sequence (NLS) and an Ig-fold domain, which differentiates them from keratins (Peter & Stick, 2015). The length of each domain is indicated according to the Human Intermediate Filament Database (http://www.interfil.org).

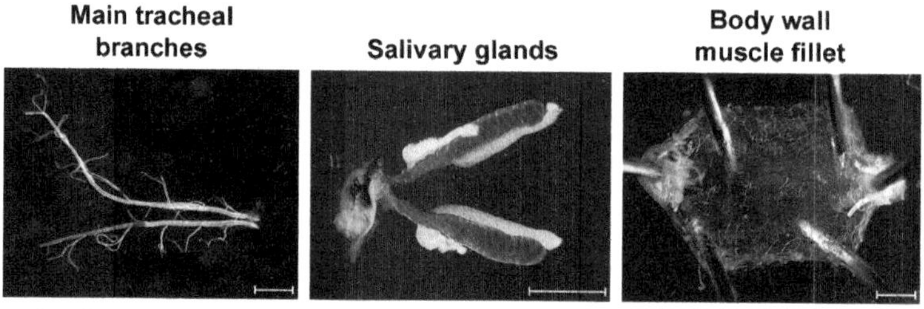

Jens Bohnekamp *et al.*, Figure 5 Dissected tissues from third instar larvae. The main tracheal branches (left) and a pair of salivary glands (with attached mouth hooks and some fat body tissue) (middle) that were dissected from a third instar larva. A third instar larva was pinned down, cut open and the organs removed to reveal the body wall muscle (right). The epidermis resides under the muscles. All images are displayed with the anterior on the left and posterior on the right. Scale bars = 0.5 mm.

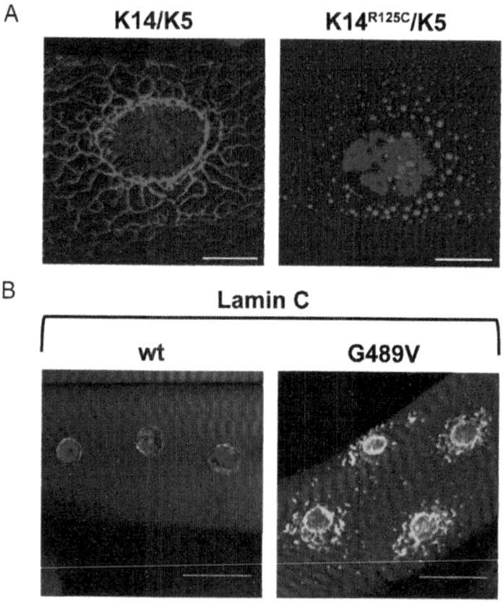

Jens Bohnekamp *et al.*, Figure 6 Immunohistochemistry of intermediate filaments in tracheal and muscle cells (A) Localization of keratins in tracheal cells. Tracheal cells expressing keratins were stained with antibodies to K14 (Magin lab; red) and DAPI (blue). Expression of the wild-type K14/K5 resulted in keratin network formation. In contrast, expression of $K14^{R125C}$/K5 resulted in cytoplasmic keratin aggregation. Scale bar = 10 μm. (B) Localization of lamins in larval body wall muscle. Larval body wall muscle stained with antibodies to Lamin C (DSHB # LC28.26; green), DAPI (blue), and phalloidin, which stains actin (red). Muscle-expressing wild-type Lamin C shows localization to the nuclear envelope (left). In contrast, muscles expressing mutant Lamin C (G498V) shows cytoplasmic Lamin C aggregation (right). Scale bar = 40 μm.

Edwards Brothers Malloy
Ann Arbor MI. USA
January 20, 2016